T0200560

Spanish Dictionary of Telecommunications
Diccionario Inglés de Telecomunicaciones

Routledge Spanish Dictionary of Telecommunications
Diccionario Inglés de Telecomunicaciones

Spanish–English/English–Spanish
Español–Inglés/Inglés–Español

Spanish terminology/Terminología española
Emilio G. Muñiz Castro
Centro Iberoamericano de Terminología (IBEROTERM)

London and New York

First published 1998
by Routledge
11 New Fetter Lane, London EC4P 4EE

Simultaneously published in the USA and Canada by Routledge
29 West 35th Street, New York, NY 10001

Typeset in Monotype Times, Helvetica Neue and Bauer Bodoni by Routledge

British Library Cataloguing-in-Publication Data
A catalogue record for this book is available from the British Library

Library of Congress Cataloging-in-Publication Data
Applied for

ISBN 0-415-15266-6

Spanish Dictionary of Telecommunications
Diccionario Inglés de Telecomunicaciones

Commissioning Editor/Dirección Técnica
Martin Barr

Project Manager/Dirección del Proyecto
Gemma Belmonte Talero

Editorial Manager/Dirección Editorial
Janice McNeillie

Project Editor/Edición del Proyecto
Lisa Carden

Editorial/Redacción
Justine Bird Anne Convery Richard Cook Alison Crann
Michael Janes Kathryn Phillips-Miles Anna Reid Mary Rigby
Robert Timms Mar Villanueva Donald Watt

Programme Manager/Dirección del Programa
Sinda López

Systems/Sistemas Informáticos
Omar Raman Simon Thompson

Acknowledgements/Agradecimientos

The English term list is based on our database of terminology first published in the *Routledge French Dictionary of Telecommunications*, 1997. We gratefully acknowledge the original contribution of Stewart Wittering.

La lista de términos ingleses está extraída de nuestra base de datos terminológicos que fue publicada por primera vez en el *Diccionario Francés de Telecomunicaciones Routledge*, en 1997. Agradecemos la contribución original de Stewart Wittering.

Emilio G. Muñiz Castro MA gained his first degree, and subsequently a master's from *L'Université Catholique*, Paris. He is founder and director of the IBEROTERM terminology centre and the BIBTERM terminology bank. He was vice president of the International Federation of Translators from 1980 to 1984, and is the author of a number of specialist dictionaries.

Emilio G. Muñiz Castro MA obtuvo una licenciatura y una maestría de *L'Université Catholique*, París. Es el fundador y director del centro de terminología IBEROTERM y el banco de terminología BIBTERM. Fue vicepresidente de la Federación Internacional de Traductores entre 1980 y 1984, y es autor de un número de diccionarios especializados.

Contents/Índice de contenidos

Features of the dictionary/ Estructura del diccionario

Genders are indicated at Spanish nouns — *Se indican los géneros para los sustantivos españoles*

oscilador *m* GEN oscillator; ~ **en anillo** ELECTRÓN ring oscillator; ~ **de circuitos T en paralelo** ELECTRÓN parallel-tee oscillator; ~ **de cristal de regulación por tensión** (*VCXO*) ELECTRÓN voltage-controlled crystal oscillator (*VCXO*); ~ **de cuarzo** ELECTRÓN quartz oscillator; ~ **de diapasón** TEC RAD tuning fork oscillator drive; ~ **de doble T** COMPON, ELECTRÓN twin-T oscillator; ~ **estándar** COMPON, ELECTRÓN standard oscillator; ~ **excitado por choque** ELECTRÓN shock-excited oscillator, TEC RAD ringing oscillator, shock-excited oscillator; ~ **de frecuencia de batido** ELECTRÓN, TRANS beat-frequency oscillator (*BFO*); ~ **de funcionamiento libre** ELECTRÓN free-running oscillator; ~ **local** ELECTRÓN, TEC RAD local oscillator (*LO*); ~ **maestro estabilizado** (*SMO*) PRUEBA stabilized master oscillator (*SMO*); ~ **de mando** TEC RAD switch driver; ~ **de onda de retorno** TELEFON backward wave oscillator (*BWO*); ~ **de onda de seno o coseno** ELECTRÓN sine-cosine oscillator; ~ **paramétrico** ELECTRÓN parametric oscillator; ~ **de relajación** ELECTRÓN relaxation oscillator; ~ **de reloj** ELECTRÓN clock oscillator; ~ **transitrón** COMPON, ELECTRÓN transitron oscillator; ~ **de variación de fase** ELECTROTEC phase-shift oscillator; ~ **con voltaje controlado** ELECTRÓN voltage-controlled oscillator (*VCO*)

Subject area labels given in alphabetical order show appropriate translations — *Los indicadores de materia ordenados alfabéticamente muestran la traducción apropiada*

Compound terms are listed alphabetically at the first element — *Los compuestos siguen el orden alfabético a partir de su primer elemento*

oscilar *vi* GEN oscillate

oscilatorio *adj* GEN oscillatory

oscilógrafo: ~ **catódico** *m* ELECTRÓN oscillograph tube

osciloscopio *m* ELECTRÓN oscilloscope

oscurecer *vt* TV darken

OSI *abr* (*interconexión de sistemas abiertas*) DATOS, INFO, TELEFON, TRANS OSI (*open systems interconnection*)

Cross-references from abbreviations to their full forms are shown for both the Spanish and English translation — *Las remisiones entre las abreviaturas y sus formas plenas aparecen tanto para los términos españoles como para los ingleses*

OTF *abr* (*frecuencia óptima de tráfico*) TEC RAD OTF (*optimum traffic frequency*)

otorgado *adj* TV awarded

oval *adj* GEN oval

ovalidad *f* GEN ovality

OWF *abr* (*frecuencia óptima de trabajo*) TEC RAD OWF (*optimum working frequency*)

oxidación *f* COMPON scaling; ~ **interna en fase de vapor** ELECTRÓN inside vapour (*AmE* vapor) phase oxidation (*IVPO*)

American English variants are given — *Se indican las variantes del inglés americano*

Las remisiones entre las abreviaturas y sus formas plenas aparecen tanto para los términos españoles como para los ingleses

Las variantes del inglés británico y americano se dan de forma completa y aparecen indicadas pertinentemente

Los compuestos siguen el orden alfabético a partir de su primer elemento

Los homógrafos son diferenciados mediante un número superíndice

Los contextos ofrecen información suplementaria

Se indican los géneros para los sustantivos españoles

LA *abbr* (*local application*) TELEP LA (*aplicación local*)

label *n* GEN *marking* etiqueta *f*; ~ **frame** GEN unidad *f* de identificación; ~ **holder** GEN portaetiquetas *m*; ~ **printer** PER APP impresora *f* de etiquetas

labelling *BrE* (*AmE* **labeling**) *n* GEN etiquetado *m*

laboratory *n* GEN laboratorio *m*; ~ **technician** GEN técnico *m* de laboratorio; ~ **test** TEST prueba *f* de laboratorio; ~ **test conditions** *n pl* TEST condiciones *f pl* de laboratorio

lace *vt* PARTS *cables* coser con tiretas

laced: ~ **cable fan** *n* PARTS abanico *m* de cable cosido

lacquer *n* PARTS laca *f*

ladder: ~**-type filter** *n* RADIO filtro *m* en escalera

lag *n* COMP, ELECT retraso *m*

lagging: ~ **phase shift** *n* POWER desplazamiento *m* de fase retardada

Lambert: ~**'s cosine law** *n* ELECT, TRANS ley *f* de coseno de Lambert

Lambertian: ~ **emission** *n* ELECT, TRANS emisión *f* lambertiana; ~ **radiator** *n* ELECT, TRANS *fibre optics* transmisor *m* de Lambert, radiador *m* lambertiano; ~ **reflector** *n* ELECT, TRANS reflector *m* lambertiano; ~ **source** *n* ELECT, TRANS generador *m* de Lambert, generador *m* lambertiano

laminate[1] *n* PARTS material *m* laminar; ~ **sheet** ELECT, PARTS panel *m* laminado

laminate[2] *vt* PARTS laminar

laminated: ~ **plastic** *n* PARTS plástico *m* laminado

laminating *n* PARTS laminación *f*

lamination *n* PARTS laminación *f*

lamp: ~ **base** *n* POWER casquillo *m* de lámpara; ~ **cap** *n* POWER casquillo *m* de lámpara; ~ **display panel** *n* PER APP, TELEP panel *m* de lámparas; ~ **extractor** *n* GEN arrancador *m* de lámpara; ~ **holder** *n* POWER portalámparas *m*; ~ **lens** *n* POWER lente *f* de lámpara; ~ **panel** *n* PER APP, TELEP panel *m* de lámparas; ~ **signal** *n* PER APP, TELEP *indicating device* señal *f* luminosa; ~**-sleeve pliers** *n pl* PARTS alicates *m pl* para retirar lámparas; ~ **strip** *n* POWER regleta *f* de lámparas; ~ **test** *n* POWER, TEST prueba *f* de lámparas

LAN *abbr* (*local area network*) COMP, DATA, TELEP red *f* de área local; ~ **manager** *n* COMP, DATA, TELEP gestor *m* de red de área local

Cross-references from abbreviations to their full form are shown for both the Spanish and English translation

British English and American English variants are given in full and labelled accordingly

Compound terms are listed alphabetically at the first element

Superscript numbers denote homographs

Contexts give supplementary information

Genders are indicated at Spanish nouns

Using the dictionary

Range of coverage

This specialist dictionary contains a broad base of terminology from all areas of telecommunications and related fundamental scientific fields. As such, we have sought to include the essential vocabulary of areas such as Parts & Materials, Telephony & Telegraphy and Electronics.

Placement & ordering of entries

All terms, including hyphenated compounds, are entered at their first element, except where that element is an article, preposition, conjunction, pronoun or other delexicalized word. In such cases, the term is entered at the next valid element.

When this element is itself a headword with a technical sense of its own, compound forms follow the simple form, which is replaced by a swung dash. If the first element is not translated itself, a colon precedes the compounds:

fusion: ~ **splice** *n* TRANS *fibre optics* empalme *m* por fusión

Within nests, articles, pronouns and prepositions are ignored in determining the sequence of open compounds and phrases. For example:

abonado *m* GEN subscriber, TV underwriter; ~ **a bordo** TEC RAD on-board subscriber; ~ **comercial** TEC RAD business subscriber; ~ **dado de baja** TELEFON ceased subscriber; ~ **a línea telefónica internacional** TELEFON international subscriber dialling (*AmE* dialing); ~ **a PBX** TELEFON PBX subscriber; ~ **con prioridad** TELEFON priority subscriber; ~ **de prueba** PRUEBA test customer; ~ **que llama** TELEFON caller; ~ **telefónico** TELEFON telephone subscriber; ~ **de vagabundeo** TEC RAD roaming subscriber

Abbreviations and acronyms written in upper case appear after entry terms of the same form written in lower case.

Terms containing figures and symbols are alphabetized according to their expansion when written out in full.

Parts of speech

Part of speech labels are given at all entries except illustrative phrases. At translations, genders are given to all Spanish nouns. In cases where nested terms have the same part of speech as the headword, parts of speech are not repeated. For a complete list of these labels and their expansions, please see page xix.

When terms with exactly the same form fall into different part of speech categories, the different senses are distinguished by a superscript number immediately following the headword. The sequence for these categories is: abbreviation, adjective, adverb, noun, verb and phrase. For example:

delete[1]: ~ **abstract-operation** *n* DATA operación *f* abstracta de borrado; ~ **channel** *n* COMP canal *m* de borrado
delete[2] *vt* GEN borrar, anular

Illustrative phrases follow a ◆ at the end of the entry to which they refer. For example:

cancelar *vt* GEN cancel;◆ ~ **el efecto de** DATOS, TELEFON override

Ordering of translations

Every term is accompanied by one or more subject area labels indicating the area in which it is used. For a complete list of these areas and their expansions, please see pages xix–xx.

When the same term is used in more than one area, multiple labels are shown as appropriate, and listed in alphabetical order. If a term has the same translation in more than one area, this translation is given after the sequence of labels. For example:

simplex *adj* RADIO, TELEP, TRANS simplex

When a term has different translations according to the area in which it is used, however, the relevant translation is given after each subject area label or set of labels. For example:

shift *n* COMP *on keyboard* decalaje *m*, ELECT, PER APP, TELEP, TRANS desplazamiento *m*

Supplementary information

Additional contextualizing information about the usage of a term is often given, and can take the form of:

1. the typical subject or object of a verb. For example:

book *vt* APPL *call*, TELEP retener

2. nouns typically used with an adjective. For example:

bilinear *adj* PER APP *scanner* bilineal

3. supplementary information to a subject area label. For example:

análisis: ~ **de modo de fallo** (*FMA*) PRUEBA *seguridad de funcionamiento* fault-mode analysis (*FMA*)

4. a paraphrase or broad equivalent. For example:

tar *n* DATA *file compression* brea *f*

When different translations apply in the same subject area, contextual information is also used to show which translation is appropriate in each case.

Where no real equivalent of a headword exists in the target language, an explanatory translation is given and is distinguished by a change in typeface. For example:

defruiter *n* RADIO supresor de señales de retorno asíncronas

Cross-references

Both British English and American English terms are covered in the dictionary, and Latin American spelling variants are also included in the Spanish coverage. These variants are differentiated by geographical labels, which are listed in full, along with their expansions, on page xix.

Spelling variants are cross-referred to the preferred spelling, where full information is given. For example:

computador *AmL ver computadora AmL*

computadora *AmL f* (*cf ordenador Esp*) INFO computer

In the case of geographical spelling variants, the American English form is cross-referred to the British English form.

The American variant is then shown again as a cross-reference at the British English headword:

colour *BrE* (*AmE* **color**) *n* GEN color *m*

Geographical lexical variants are cross-referred, with translations given at each entry:

earthing *BrE n* (*cf grounding AmE*) ELECT, POWER conexión *f* a tierra

grounding *AmE n* (*cf earthing BrE*) ELECT, POWER conexión *f* a tierra

Both spelling and lexical variants appear as translations. Lexical variants are shown in full, for example:

conexión: **~ a tierra** ELECTRÓN bonding, earthing *BrE*, grounding *AmE*, ELECTROTEC bonding, earthing *BrE*, earthing arrangement *BrE*, grounding arrangement *AmE*, grounding *AmE*

Where a spelling variant exists in a translation, the British English translation appears in full without a geographical label. The American English variant is shown and labelled appropriately. For example:

programa: **~ de cable** TELEFON cable programme (*AmE* program)

If more than one variant appears in a translation, both the British English and the American English forms are shown in full:

señalización: **~ del punto central** SEÑAL centre-point signalling (*AmE* center-point signaling); **~ del punto medio** SEÑAL midpoint signalling (*AmE* signaling)

Abbreviations are cross-referred to their full forms and vice versa, with the appropriate translations given at each entry. For example:

CF *abr* (*función de control*) CONTROL, DATOS, INFO, TELEFON CF (*control function*)

función: **~ de control** (*CF*) CONTROL, DATOS, INFO, TELEFON control function (*CF*)

If a source language abbreviation does not have a target language equivalent, the translation of the expanded form is given:

RES *abbr* (*reserved field*) ELECT campo *m* reservado

When an abbreviation has different full forms according to subject areas, the relevant expanded forms appear after each label or set of labels. For example:

ATM *abbr* APPL, COMP (*asynchronous transfer mode*) ATM (*modo de transferencia asíncrona*), DATA, TELEP, TEST (*abstract data method*) ATM (*metódo de datos resumidos*)

Consejos para la utilización de este diccionario

Selección del vocabulario

Este diccionario especializado tiene un acervo terminológico que abarca todos los dominios de las telecomunicaciones. Hemos procurado incluir el vocabulario esencial de materias tales como Componentes y Materiales, Telefonía y Telegrafía, y Electrónica.

Orden y colocación de las entradas

Todos los términos, incluyendo los compuestos formados mediante un guión, se ordenan a partir de su primer elemento. Cuando el primer elemento es un artículo, una preposición, una conjunción, un pronombre u otra palabra delexicalizada, el término se ordena según el próximo elemento válido.

Cuando varios compuestos participan de un lexema inicial común de carácter técnico, el primer elemento se reemplaza por una virgulilla. Si el primer elemento no tiene una traducción, los términos compuestos formados a partir de dicho elemento aparecen precedidos por dos puntos. Por ejemplo:

fusion: ~ **splice** *n* TRANS *fibre optics* empalme *m* por fusión

Los artículos, los pronombres y las preposiciones no se toman en cuenta a la hora de determinar el orden de los términos compuestos o las frases. Por ejemplo:

abonado *m* GEN subscriber, TV underwriter; ~ **a bordo** TEC RAD on-board subscriber; ~ **comercial** TEC RAD business subscriber; ~ **dado de baja** TELEFON ceased subscriber; ~ **a línea telefónica internacional** TELEFON international subscriber dialling (*AmE* dialing); ~ **a PBX** TELEFON PBX subscriber; ~ **con prioridad** TELEFON priority subscriber; ~ **de prueba** PRUEBA test customer; ~ **que llama** TELEFON caller; ~ **telefónico** TELEFON telephone subscriber; ~ **de vagabundeo** TEC RAD roaming subscriber

Las abreviaturas y los acrónimos escritos con mayúsculas aparecen tras términos idénticos escritos en minúsculas.

Los términos que incluyen cifras y símbolos se ordenan alfabéticamente, como si estuvieran escritos en letra.

Categorías gramaticales

Tras cada término, menos en las frases ilustrativas, se indica la categoría gramatical mediante abreviatura. El género aparece para cada traducción española que posea la categoría de nombre. En los casos en que un término dé origen a otros, no se repetirá la abreviatura si la categoría gramatical no cambia. Puede consultarse una lista completa de las categorías gramaticales y sus abreviaturas en la página xix.

En el caso de términos cuyo elemento inicial es idéntico pero pertenecen a diferentes categorías gramaticales, las acepciones se distinguen por la inclusión de un supraíndice tras la entrada. El orden seguido es el de abreviatura, adjetivo, adverbio, nombre, verbo y frase. Por ejemplo:

delete[1]: ~ **abstract-operation** *n* DATA operación *f* abstracta de borrado; ~ **channel** *n* COMP canal *m* de borrado
delete[2] *vt* GEN borrar, anular

Las frases ilustrativas van precedidas por un ◆ y se encuentran al final de las entradas a las que se refieren. Por ejemplo:

cancelar *vt* GEN cancel;◆ ~ **el efecto de** DATOS, TELEFON override

Orden de las traducciones

Cada término va acompañado de uno o más indicadores de materia que especifican el ámbito de uso. En las páginas xix–xx se encuentra una lista completa de tales indicadores.

Cuando un mismo término se utiliza en varios ámbitos, se consignan alfabéticamente los indicadores pertinentes. Cuando la traducción de un término es la misma en varios ámbitos, dicha traducción aparece tras la secuencia de indicadores. Por ejemplo:

simplex *adj* RADIO, TELEP, TRANS simplex

Cuando un término permite diferentes traducciones según el ámbito en que se utiliza, se ofrece la traducción apropiada tras cada indicador o conjunto de indicadores de materia. Por ejemplo:

shift *n* COMP *on keyboard* decalaje *m*, ELECT, PER APP, TELEP, TRANS desplazamiento *m*

Información complementaria

En muchos casos se ofrece información adicional para precisar el uso de un término. Dicha información puede consistir en:

1. el sujeto u objeto típicos de un verbo. Por ejemplo:

book *vt* APPL *call*, TELEP retener

2. nombres usados habitualmente con un adjetivo. Por ejemplo:

bilinear *adj* PER APP *scanner* bilineal

3. información complementaria del indicador de materia. Por ejemplo:

análisis: ~ **de modo de fallo** (*FMA*) PRUEBA *seguridad de funcionamiento* fault-mode analysis (*FMA*)

4. una paráfrasis o equivalente general. Por ejemplo:

tar *n* DATA *file compression* brea *f*

En los casos en que existen traducciones diferentes en un mismo ámbito, se utiliza la información contextual para mostrar cuál es la traducción más apropiada en cada caso.

Cuando no existe el equivalente real de un término en la otra lengua, se da una traducción explicativa, la cual aparece con tipografía diferente. Por ejemplo:

defruiter *n* RADIO supresor de señales de retorno asíncronas

Remisiones

Este diccionario incluye términos del inglés británico así como del inglés americano. La cobertura española de este diccionario también incluye variantes ortográficas de América Latina. Se indica siempre su procedencia geográfica mediante indicador. Para ver una lista completa de estos indicadores y de sus correspondencias, por favor consúltese la página xix.

Las variantes ortográficas remiten a la forma más frecuente, donde se ofrece una información completa. Por ejemplo:

computador *AmL ver computadora AmL*

computadora *AmL f* (*cf ordenador Esp*) INFO computer

Las variantes ortográficas del inglés americano remiten a las entradas respectivas del inglés británico. La forma del inglés americano aparece como remisión en la entrada correspondiente del inglés británico. Por ejemplo:

colour *BrE* (*AmE* **color**) *n* GEN color *m*

Las variantes geográficas de carácter léxico se remiten ofreciéndose las traducciones completas en cada entrada. Por ejemplo:

earthing *BrE n* (*cf grounding AmE*) ELECT, POWER conexión *f* a tierra

grounding *AmE n* (*cf earthing BrE*) ELECT, POWER conexión *f* a tierra

Las variantes geográficas, tanto ortográficas como léxicas, aparecen cuando son traducciones. Las variantes léxicas se ofrecen íntegramente:

conexión: **~ a tierra** ELECTRÓN bonding, earthing *BrE*, grounding *AmE*, ELECTROTEC bonding, earthing *BrE*, earthing arrangement *BrE*, grounding arrangement *AmE*, grounding *AmE*

Cuando una variante ortográfica existe en la traducción, la traducción correspondiente al inglés británico se ofrece íntegramente sin indicador geográfico. La variante del inglés americano aparece con su indicador geográfico apropiado. Por ejemplo:

programa: **~ de cable** TELEFON cable programme (*AmE* program)

Cuando existe más de una variante en una traducción, las formas, tanto del inglés británico como del inglés americano, aparecen íntegramente:

señalización: **~ del punto central** SEÑAL centrepoint signalling (*AmE* center-point signaling); **~ del punto medio** SEÑAL midpoint signalling (*AmE* signaling)

Las abreviaturas remiten a las formas íntegras y viceversa, y las traducciones apropiadas aparecen en cada entrada. Por ejemplo:

CF *abr* (*función de control*) CONTROL, DATOS, INFO, TELEFON CF (*control function*)

función: **~ de control** (*CF*) CONTROL, DATOS, INFO, TELEFON control function (*CF*)

Cuando una abreviatura de la lengua de origen no tiene un equivalente en la lengua de destino, la traducción que se ofrece es la de la forma íntegra. Por ejemplo:

RES *abbr* (*reserved field*) ELECT campo *m* reservado

Cuando una abreviatura tiene varias formas íntegras según el ámbito en que se usa, la forma íntegra se ofrece tras cada indicador o conjunto de indicadores de materia:

ATM *abbr* APPL, COMP (*asynchronous transfer mode*) ATM (*modo de transferencia asíncrona*), DATA, TELEP, TEST (*abstract data method*) ATM (*metódo de datos resumidos*)

Abbreviations used in this dictionary/ Abreviaturas utilizadas en este diccionario

Parts of speech/Categorías gramaticales

abbr	abbreviation	abreviatura
abr	abreviatura	abbreviation
adj	adjective	adjetivo
adv	adverb	adverbio
f	feminine	femenino
f pl	feminine plural	femenino plural
fra	frase	phrase
m	masculine	masculino
m pl	masculine plural	masculino plural
n	noun	sustantivo
n pl	noun plural	sustantivo plural
phr	phrase	frase
vi	intransitive verb	verbo intransitivo
v refl	verbo reflexivo	reflexive verb
vt	transitive verb	verbo transitivo

Geographic codes/Indicadores geográficos

AmE	American English	Inglés americano
AmL	América Latina	Latin America
BrE	British English	Inglés británico
Esp	España	Spain

Subject area labels/Indicadores de materia

APPL	Applications & Services	Aplicaciones y Servicios
CLIM	Climate Control	Climatología
COMP	Computing	Informática
CONTR	Control Technology	Tecnología de control
DATA	Data	Tratamiento de datos
ELECT	Electronics	Electrónica
GEN	General Telecommunications	Telecomunicaciones Generales
PARTS	Parts & Materials	Componentes y Materiales

PER APP	Terminal & Peripheral Apparatus	Periféricos y Terminales
POWER	Power Systems	Electromecánica y Electrotecnia
RADIO	Radio-based Technologies	Tecnologías radiodirigidas
SIG	Signalling	Señalización
SWG	Switching	Conmutación y Conexión
TELEP	Telephony & Telegraphy	Telefonía y Telegrafía
TV	Television & Video	Televisión y Vídeo
TEST	Testing Procedures	Métodos de prueba
TRANS	Transmission	Transmisiones

APLIC	Aplicaciones y Servicios	Applications & Services
CLIMAT	Climatología	Climate Control
CONMUT	Conmutación y Conexión	Switching
COMPON	Componentes y Materiales	Parts & Materials
CONTROL	Tecnología de control	Control Technology
DATOS	Tratamiento de datos	Data
ELECTRÓN	Electrónica	Electronics
ELECTROTEC	Electromecánica y Electrotecnia	Power Systems
GEN	Telecomunicaciones Generales	General Telecommunications
INFO	Informática	Computing
PERIF	Periféricos y Terminales	Terminal & Peripheral Apparatus
PRUEBA	Métodos de prueba	Testing Procedures
SEÑAL	Señalización	Signalling
TEC RAD	Tecnologías radiodirigidas	Radio-based Technologies
TELEFON	Telefonía y Telegrafía	Telephony & Telegraphy
TRANS	Transmisiones	Transmission
TV	Televisión y Vídeo	Television & Video

Registered trademarks®

Every effort has been made to label terms which we believe constitute trademarks. The legal status of these, however, remains unchanged by the presence or absence of any such label.

Marcas registradas®

Hemos hecho el máximo esfuerzo para señalar los términos que estimamos protegidos por un registro de marca. Sin embargo, la ausencia o presencia de esta mención no surte efecto sobre su estado legal.

ESPAÑOL–INGLÉS
SPANISH–ENGLISH

A

AAL *abr* (*envoltura de adaptación de un ATM*) DATOS AAL (*ATM adaptation layer*)

AARQ *abr* (*petición A-asociada*) DATOS, TELEFON AARQ (*A-associate request*)

abanico: **~ de cable cosido** *m* COMPON laced cable fan

aberración: **~ esférica** *f* ELECTRÓN, TRANS spherical aberration; **~ óptica** *f* TRANS optical aberration

abierto *adj* GEN unrestricted, open

ablandador *m* COMPON softener

abonado *m* GEN subscriber, TV underwriter; **~ a bordo** TEC RAD on-board subscriber; **~ comercial** TEC RAD business subscriber; **~ dado de baja** TELEFON ceased subscriber; **~ a línea telefónica internacional** TELEFON international subscriber dialling (*AmE* dialing); **~ a PBX** TELEFON PBX subscriber; **~ con prioridad** TELEFON priority subscriber; **~ de prueba** PRUEBA test customer; **~ que llama** TELEFON caller; **~ telefónico** TELEFON telephone subscriber; **~ de vagabundeo** TEC RAD roaming subscriber

abonarse *v refl* GEN subscribe

abono *m* GEN subscription; **~ al teléfono** TELEFON telephone subscription

aboquillar *vt* ELECTRÓN crimp

abrazadera *f* ELECTRÓN clamp; **~ de anclaje de base** ELECTRÓN base support bracket; **~ de cable** ELECTRÓN, TRANS *tendido de cables* cable clamp; **~de toma de masa** ELECTROTEC earth clamp *BrE*, ground clamp *AmE*

abrillantamiento *m* INFO brightening

abrir *vt* DATOS unbind, ELECTRÓN, ELECTROTEC open, *grifos, llaves, interruptores* turn on, open

absorbedor: **~ de ondas** *m* COMPON, ELECTRÓN, ELECTROTEC surge absorber

absorción *f* INFO takeover; **~ de carga** ELECTRÓN energy absorption; **~ de microondas** TEC RAD microwave absorption; **~ de resonancia** ELECTRÓN resonance absorption; **~ sonora** GEN sound absorption

abstracto *adj* DATOS, INFO abstract

ABU *abr* (*Unión de Radiodifusión de Asia y el Pacífico*) TEC RAD ABU (*Asia-Pacific Broadcasting Union*)

acabado *m* INFO finish; **~ texturado** COMPON texture finish

acaecimiento: **~ de recepción** *m* DATOS receive event

acaparador *m* INFO grabber

acarreo: **~ circular** *m* INFO end-around carry; **~ de lectura anticipada** *m* ELECTRÓN look-ahead carry; **~ rápido** *m* INFO high-speed carry

accesibilidad: **~completa** *f* CONMUT full availability; **~ limitada** *f* CONMUT limited availability

acceso *m* GEN access; **~ aéreo** DATOS overhead access (*OHA*); **~ aleatorio** INFO random access; **~ aleatorio a la memoria** INFO random memory access; **~ a archivos** DATOS, INFO file access; **~ básico** TELEFON, TRANS basic access; **~ de código múltiple** DATOS multiple-code access; **~ condicional** TV conditional access; **~ de conversación** TELEFON talk spurt; **~ denegado condicionalmente** CONMUT conditionally toll-denied access; **~ derivado primario** TELEFON primary rate access; **~ digital integrado** (*IDA*) TRANS integrated digital access (*IDA*); **~ directo** DATOS, INFO, PERIF direct access; **~ directo a la memoria** (*DMA*) TELEFON direct memory access (*DMA*); **~ directo a operador** TELEFON attendant override; **~ a disco compacto fotográfico** INFO photo-CD access; **~ de entrada** DATOS inlet access; **~ especializado del enlace de datos** DATOS data-link specialized access; **~ fácil** INFO easy access; **~ fiable forzado mts** DATOS mts-forced reliable access; **~ fiable mts** DATOS mts-reliable access; **~ y gestión de transferencia de archivos** (*FTAM*) DATOS, INFO *proceso de datos a distancia*, TELEFON file-transfer access and management (*FTAM*); **~ integrado** TELEFON integrated access; **~ léxico** GEN lexical access; **~ a una línea ocupada** TELEFON busyoverride; **~ múltiple** TEC RAD multiple access; **~ múltiple de detección de la portadora** DATOS carrier-sense multiple access (*CSMA*); **~ múltiple de dirección de impulso** (*PAMA*) TEC RAD pulse-address multiple access (*PAMA*); **~ múltiple de división de código** (*CDMA*) TEC RAD code-division multiple access (*CDMA*); **~ múltiple con división de frecuencia ortogonal** (*ODFMA*) TEC RAD orthogonal frequency-division multiple access (*ODFMA*); **~ múltiple por división de**

frecuencias (*AMDF*) CONMUT, SEÑAL, TELEFON, TRANS frequency-division multiple access (*FDMA*); ~ **por palabra clave** INFO keyword access; ~ **rápido** INFO fast access; ~ **a la RDSI** TELEFON ISDN access; ~ **a la red** DATOS, TEC RAD, TRANS network access; ~ **a la red interurbana AUTOVON** TELEFON AUTOVON trunk access; ~ **remoto** DATOS, TELEFON remote access; ~ **remoto a base de datos** (*RDA*) DATOS, TELEFON remote database access (*RDA*); ~ **remoto a documentos** DATOS, TELEFON remote document access; ~ **restringido** TV limited access; ~ **de salida** CONMUT outlet access; ~ **semipermanente** DATOS semipermanent access; ~ **con tarjeta inteligente** APLIC smartcard access; ~ **universal S** TELEFON S-universal access; ~ **de usuario** GEN user access; ~ **en velocidad primaria a la RDSI** TELEFON ISDN primary-rate access

accesorio *m* GEN fitting; ~ **eléctrico** COMPON electrical fitting; ~ **MCS** TELEFON MCS attachment

acción: ~ **automática general** *f* DATOS general auto-action; ~ **continua** *f* CONTROL continuous action; ~ **diferida** *f* GEN delayed action; ~ **de dos posiciones** *f* CONTROL two-step action; ~ **de entidad funcional** *f* DATOS functional entity action (*FEA*); ~ **escalonada** *f* CONTROL step action; ~ **local** *f* ELECTROTEC local action; ~ **por modulación** *f* CONTROL modulating action; ~ **de muestreo y retención** *f* CONTROL sample and hold action; ~ **a niveles múltiples** *f* CONTROL multistep action; ~ **retardada** *f* GEN delayed action

accionador *m* TEC RAD actuator; ~ **de antena** TEC RAD antenna drive motor; ~ **DMA virtual** DATOS virtual DMA driver

accionamiento *m* SEÑAL operation; ~ **automático** DATOS auto-action; ~ **a distancia** GEN remote control; **de** ~ **térmico** *adj* GEN thermally-operated

ACE *abr* (*elemento de conexión de acceso*) DATOS, TELEFON ACE (*access connection element*)

aceite: ~ **antioxidante** *m* GEN rust-protecting oil; ~ **mineral** *m* COMPON mineral oil

aceitera *f* GEN oil can, oil nipple

aceleración: ~ **radial** *f* INFO radial acceleration

acelerador *m* COMPON, ELECTRÓN accelerator; ~ **intermedio** INFO booster

aceptación: ~ **de direcciones extensas** *f* DATOS long-address acceptance; ~ **inicial** *f* PRUEBA initial acceptance; ~ **in situ** *f* TEC RAD on-site acceptance; ~ **mundial** *f* GEN worldwide acceptance; ~ **de pedido** *f* GEN order taking; ~ **de selección rápida** *f* TELEFON fast-select acceptance; ~ **de sobrecarga** *f* CONMUT, TELEFON overflow accept

aceptor *m* DATOS, ELECTRÓN acceptor

acero: ~ **para herramientas** *m* COMPON tool steel; ~ **para muelles** *m* COMPON spring steel; ~ **de nitruración** *m* COMPON nitriding steel

ACID *abr* (*atomicidad, consistencia, aislamiento y durabilidad*) DATOS ACID (*atomicity, consistency, isolation and durability*)

acierto *m* TEC RAD hit

acimut *m* TEC RAD acimuth

acometida: ~ **de antena** *f* TEC RAD antenna lead-in

acontecimiento: ~ **de telefonía entrelazada hacia atrás** *m* (*BITE*) TELEFON backward-interworking telephony event (*BITE*)

acoplado *adj* ELECTRÓN coupled, ganged, *base, diodo, transistor* implanted

acoplador *m* ELECTRÓN coupler; ~ **acústico** DATOS, TELEFON acoustic coupler; ~ **central de alarma** CONTROL central alarm coupler; ~ **cerrado** TELEFON closed coupler; ~ **de corriente alterna** ELECTROTEC alternating current coupler; ~ **direccional** CONMUT, DATOS, TRANS directional coupler; ~ **en estrella** TRANS star coupler; ~ **de filtro** ELECTRÓN filter coupler; ~ **de guíaondas** TEC RAD, TRANS waveguide coupler; ~ **monomodal** TRANS single-mode coupler; ~ **de múltiplex por división en longitud de onda** TELEFON, TRANS wavelength division multiplex coupler; ~ **múltiplex principal** TELEFON main multiplex coupler; ~ **de procesamiento de señal vocal** TELEFON voice-signal processing coupler; ~ **roscado** COMPON, ELECTRÓN threaded coupler; ~ **de selección de modo** TRANS mode-selective coupler; ~ **de señalización multiprotocolaria** SEÑAL multiprotocol signalling (*AmE* signaling) coupler; ~ **en T** COMPON, ELECTRÓN T coupler; ~ **WDM** TRANS WDM coupler; ~ **en Y** ELECTRÓN, TRANS Y coupler

acoplamiento *m* GEN coupling, grouping; ~ **acústico** TELEFON acoustic coupling; ~ **en derivación** ELECTROTEC bridge connection; ~ **de eje** COMPON shaft coupling; ~ **de eslabón** TEC RAD link coupling; ~ **interetápico** ELECTRÓN interstage coupling; ~ **maestro** GEN jumper king; ~ **magnético** ELECTRÓN magnet coupling; ~ **de modos** TRANS mode coupling; ~ **mutuo** ELECTRÓN mutual coupling; ~ **de onda** ELECTRÓN wave coupling; ~ **parásito** ELECTRÓN parasitic coupling; ~ **RC** ELECTRÓN RC coupling; ~ **por resistencia** ELECTRÓN resistance coupling; ~ **por resistencia-capacitancia**

ELECTRÓN resistance-capacitance coupling; ~ **roscado** COMPON screw coupling; ~ **en serie** ELECTRÓN, ELECTROTEC series connection; ~ **por transformador** ELECTROTEC transformer coupling

acoplar *vt* INFO couple

ACPM *abr* (*máquina de asociación de protocolo de control*) DATOS ACPM (*association control protocol machine*)

ACSE *abr* (*elemento para el control de asociación*) DATA, TELEP ACSE (*association control service element*)

ACTE *abr* (*Comité Asesor para Equipo de Telecomunicaciones*) TELEFON ACTE (*Advisory Committee for Telecommunications Equipment*)

actitud: ~ **de satélite** *f* TEC RAD, TV satellite attitude

activación *f* GEN activation; ~ **por la voz** CONTROL, PERIF, TELEFON voice activation

activador: ~ **de bus** *m* ELECTRÓN *de elementos lógicos* bus driver

actividad *f* GEN activity; ~ **colectiva** GEN ensemble activity; ~ **de núcleo** GEN core activity

actividades: ~ **de apoyo** *f pl* INFO support activities

activo *adj* GEN active

actuación: ~ **de apoyo del mantenimiento** *f* TELEFON maintenance support performance; ~ **de operador** *f* TELEFON attendant action

actualización *f* GEN catching-up, updating; ~ **de localización** TEC RAD location updating; ~ **rápida** DATOS fast update; ~ **retroactiva** COMPON, ELECTRÓN retrofit

actualizar *vt* GEN update

acuerdo: ~ **de boceto** *m* GEN outline agreement; ~ **de comunicaciones** *m* GEN communications agreement; ~ **de creación de red** *m* TV network establishment agreement; ~ **de explotación de red** *m* TV network operating agreement; ~ **marco** *m* GEN framework agreement; ~ **de sincronización en paralelo** *m* ELECTRÓN parallel clocking arrangement; ~ **sub-Clos** *m* TELEFON, TRANS sub-Clos arrangement

acumulación: ~ **de trabajo** *f* INFO backlog

acumulador *m* ELECTROTEC accumulator, storage battery; ~ **de plomo** ELECTROTEC lead storage battery, lead accumulator

acuse: ~ **de recibo** *m* TELEFON acknowledgement

acústica: ~ **arquitectónica** *f* TELEFON room acoustics

adaptación *f* ELECTRÓN, INFO matching; ~ **especial** GEN customizing, tailoring; ~ **de impedancias** TEC RAD impedance matching; ~ **de ruta de orden inferior** TELEFON lower-order path adaptation (*LPA*); ~ **de sección** DATOS, TELEFON section adaptation; ~ **de trayecto de orden superior** TELEFON, TRANS higher-order path adaptation

adaptado: ~ **especialmente** *adj* GEN customized, tailored

adaptador *m* GEN adaptor; ~ **de anillo** DATOS ring adaptor; ~ **de aparato** INFO device adaptor; ~ **de borne** DATOS, PERIF, TELEFON terminal adaptor (*TA*); ~ **de bus** CONMUT bus adaptation unit; ~ **de cartucho único** INFO single-cartridge adaptor; ~ **de electricidad de CA** INFO AC power adaptor; ~ **gráfico de colores** (*CGA*) INFO colour (*AmE* color) graphics adaptor (*CGA*); ~ **de gráficos destacados** (*EGA*) INFO enhanced graphics adaptor (*EGA*); ~ **guía-coaxil** TRANS waveguide-to-coax adaptor; ~ **hembra-hembra** ELECTRÓN, INFO female-female adaptor; ~ **integrado** INFO integrated adaptor; ~ **de interfaz** INFO, TELEFON interface adaptor; ~ **intermedio** ELECTRÓN, TELEFON medium adaptor (*MA*); ~ **múltiple montable** ELECTRÓN gang mountable adaptor; ~ **para RDSI** TELEFON ISDN adaptor; ~ **simétrico/asimétrico** ELECTRÓN balun adaptor; ~ **terminal B de la red de transmisión digital de RDSI-BA** TELEFON B-ISDN terminal adaptor (*TA-LB*); ~ **de tipo pasante de alimentación transversal** ELECTRÓN bulkhead feed-through adaptor

adaptar: ~ **especialmente** *vt* GEN customize, tailor

adecuarse: ~ **a las especificaciones** *fra* GEN meet the specifications

adelgazar *vt* INFO, PRUEBA reduce

adherido *adj* ELECTRÓN stuck

adherirse *v refl* ELECTRÓN stick

adhesión *f* ELECTRÓN adhesion; ~ **de armadura** ELECTRÓN armature stick

adhesivo: ~ **anaeróbico** *m* ELECTRÓN anaerobic adhesive

adiatérmico *adj* ELECTROTEC nonconducting

aditamento *m* TELEFON attachment

adjudicación *f* TELEFON *de clases de servicio* assignment; ~ **de espectro** TEC RAD spectrum allocation; ~ **de frecuencias** TEC RAD frequency allotment; ~ **de nombres** INFO *de etiquetas* tagging; ~ **de trayectoria** CONMUT path allocation

adjuntar *vt* GEN enclose

administración: ~ **de base de datos** *f* DATOS, INFO database management; ~ **de estación** *f* TRANS station management; ~ **de garante** *f*

TELEFON guarantor administration; ~ **local** *f* TV local authority; ~ **de red** *f* GEN network management; ~ **reguladora** *f* GEN regulatory administration; ~ **de telecomunicaciones** *f* TELEFON telecommunications administration; ~ **temporal** *f* TELEFON terminal administration; ~ **de tránsito** *f* TELEFON, TRANS transit administration

administrador: ~ **de red** *m* INFO network administrator

admitancia *f* ELECTRÓN, ELECTROTEC admittance; ~ **en cortocircuito** ELECTRÓN, ELECTROTEC short-circuit admittance; ~ **efectiva de entrada** ELECTRÓN, ELECTROTEC effective input admittance; ~ **efectiva de salida** ELECTRÓN, ELECTROTEC effective output admittance; ~ **eléctrica** ELECTRÓN, ELECTROTEC electrical admittance; ~ **de electrodo** ELECTRÓN, ELECTROTEC electrode admittance; ~ **de entrada** ELECTRÓN, ELECTROTEC input admittance; ~ **de salida** ELECTRÓN, ELECTROTEC output admittance

adornar *vt* DATOS *página Web* embellish

adquisición: ~ **de blanco** *f* TEC RAD target acquisition; ~ **de conocimiento** *f* APLIC, INFO knowledge acquisition; ~ **de datos** *f* DATOS data acquisition; ~ **en modo de búsqueda** *f* (*SMA*) DATOS search mode acquisition (*SMA*); ~ **y sincronización de trama** *f* TRANS frame acquisition and synchronization; ~ **de vídeo** *f* INFO video acquisition

adulterante *m* COMPON dopant

adyacente: **no** ~ *adj* TEC RAD nonadjacent

AEI *abr* (*invocación de la unidad de aplicación*) DATOS AEI (*application-entity invocation*)

AENOR (*Asociación española de normalización y certificación*) GEN AENOR (*Spanish Association of Normalization and Certification*)

aerosol *m* TEC RAD aerosol

afirmación: ~ **de valor de atributo** *f* (*AVA*) DATOS attribute-value assertion (*AVA*)

afirmado *adj* TELEFON asserted

agarradera *f* INFO grab

agarrotado *adj* DATOS, ELECTRÓN, TEC RAD jammed

agarrotar *vt* DATOS, ELECTRÓN, TEC RAD jam

agarrotarse *v refl* DATOS, ELECTRÓN, TEC RAD jam

AGC *abr* (*control automática de ganancia*) ELECTRÓN, TELEFON, TRANS AIS (*automatic gain control*); ~ **con corrección anticipante** *m* TRANS feedforward AGC

agencia: ~ **privada autorizada** *f* (*RPOA*) TELEFON recognized private operating agency (*RPOA*)

Agencia: ~ **Espacial Europea** *f* (*ESA*) TEC RAD European Space Agency (*ESA*)

agenda *f* INFO diary, notebook; ~ **electrónica** INFO notebook computer, notepad computer

agente *m* DATOS, TELEFON agent; ~ **anticorrosivo** GEN rust-preventing agent; ~ **endurecedor** COMPON *plástico* setting agent; ~ **telemático** DATOS, TELEFON telematic agent (*TLMA*); ~ **de transferencia de mensajes** DATOS, TELEFON message transfer agent (*MTA*); ~ **usuario** TELEFON user agent (*UA*); ~ **de usuario de EDI** TELEFON EDI user agent (*EDI-UA*); ~ **de usuario de mensajes por voz** TELEFON voice-messaging user agent; ~ **de usuario de sistema de mensajería por voz** (*VMGS-UA*) DATOS, TELEFON voice-messaging system user agent (*VMGS-UA*); ~ **de usuario de sistema de mensajería interpersonal** DATOS interpersonal messaging system user agent; ~ **de usuarios de sistema de mensajería por voz** APLIC voice messaging system user agent (*VMGS-UA*)

aglomerante: ~ **de nailon** *m* ELECTRÓN nylon binder

agrupación *f* TRANS bunching; ~ **de canales** TRANS channel pool; ~ **redundante de discos independientes** INFO redundant array of independent disks

agrupaciones: ~ **de enlace de adaptadores balun** *f pl* ELECTRÓN linking clusters of balun adaptors

agrupado *adj* ELECTRÓN linked, TELEFON blocked, TRANS bunched

aguja *f* COMPON, ELECTRÓN, PERIF pin, DATOS, TELEFON pointer, INFO pointer, stylus; ~ **de cadeneo** GEN arrow

agujereado *adj* COMPON, ELECTRÓN holed

agujero *m* ELECTRÓN hole; ~ **de cable** TRANS cable hole; ~ **de componente** TRANS finger hole; ~ **de fijación** GEN mounting hole; ~ **de pasante de conexión** ELECTRÓN through-connection hole; ~ **de paso** ELECTRÓN via hole; ~ **de posición** ELECTRÓN tooling hole; ~ **de posicionamiento** ELECTRÓN *para circuitos impresos* locating hole; ~ **sin nudo** COMPON, ELECTRÓN *para circuitos impresos* landless hole

ahorro: ~ **de energía** *m* GEN energy saving; ~ **de tiempo** *m* GEN time saving

ahusamiento *m* ELECTRÓN, TEC RAD taper

AI *abr* (*inteligencia artificial*) APLIC, CONMUT, INFO AI (*artificial intelligence*)

AIS *abr* (*señal de indicación de alarma*) PRUEBA, TELEFON AIS (*alarm indication signal*)

aislador: ~ **pasamuros** *m* ELECTROTEC lead-in insulator

aislamiento *m* COMPON, ELECTRÓN, ELECTROTEC insulation; ~ **acústico** TELEFON sound insulation; ~ **por cinturón** TRANS *de cable* belt insulation; ~ **de conductor** COMPON *de cable* core wrap, TRANS *de cable* conductor insulation; ~ **mineral** ELECTRÓN mineral insulation; ~ **multicapa** (*MLI*) ELECTRÓN multilayer isolation (*MLI*); ~ **de polarización cruzada** TEC RAD cross-polarization isolation, TRANS cross-polar isolation; ~ **polarizante** TEC RAD, TRANS polarizing isolation; ~ **de poliolefina** ELECTRÓN polyolefin insulation; ~ **sólido extruido** GEN *cables* extruded solid insulation; ◆ **sin** ~ ELECTRÓN uninsulated

aislante[1] *adj* COMPON, ELECTRÓN, ELECTROTEC insulating

aislante[2] *m* COMPON, ELECTRÓN, ELECTROTEC insulant, insulator, isolator, TEC RAD, TRANS isolator

aislar *vt* GEN insulate

AJ *abr* (*antiinterferencias*) TEC RAD AJ (*anti-jamming*)

ajeno *m* INFO, TRANS aliasing

ajustable *adj* GEN settable

ajustador: ~ **de cable** *m* TRANS cable trimmer; ~ **de ganancia activado por el ruido** *m* (*NOGAD*) ELECTRÓN noise-operated gain adjusting device (*NOGAD*)

ajuste *m* INFO, TELEFON *sobre una máquina* setting, adjustment, *de moldes* setting-up; ~ **de alarma** PRUEBA alarm setting; ~ **aproximado** GEN coarse setting; ~ **a cero** PRUEBA setting to zero; ~ **fino** ELECTRÓN, ELECTROTEC, INFO trimming; ~ **fino de sintonización** GEN fine tuning; ~ **forzado** GEN drive fit; ~ **de frecuencias** TEC RAD frequency tracking; ~ **de imagen** DATOS framing; ~ **incorrecto** GEN maladjustment; ~ **inicial pleno** (*IFLU*) PERIF initial full line-up (*IFLU*); ~ **no normalizado** (*NSS*) DATOS, TELEFON nonstandard setup (*NSS*); ~ **de núcleo** INFO core flush; ~ **del periodo de la señal de llamada** PERIF ringing-period adjustment; ~ **preciso** GEN fine adjustment; ~ **de secuencia** CONMUT sequence adjustment; ~ **del varactor** ELECTRÓN varactor tuning

ajustes: ~ **de impresión** *m pl* INFO print settings

alabeo *m* COMPON twist, *de formaciones* winding, ELECTRÓN twist, edge warping, ELECTROTEC winding

alambre *m* COMPON, ELECTRÓN, TRANS *cables* cable; ~ **acanalado** TRANS grooved cable; ~ **de cobre** GEN copper wire; ~ **de conexión** ELECTRÓN connecting wire; ~ **para conexiones** ELECTRÓN hook-up wire; ~ **de drenaje** GEN drain wire; ~ **estañado** GEN tinned wire; ~ **neutro** GEN neutral wire

alargamiento *m* GEN strain; ~ **hasta rotura** COMPON rupture strain

alarma *f* DATOS, PRUEBA, TELEFON alarm; ~ **de atenuación de palabra singular** DATOS unique word loss alarm; ~ **de averías** TELEFON failure alarm; ~ **cíclica** PRUEBA cyclic alarm; ~ **contra robos** PRUEBA burglar alarm; ~ **de corte de suministro eléctrico** ELECTROTEC mains-failure alarm; ~ **de desconexión** ELECTROTEC power-off alarm; ~ **a distancia** TEC RAD, TELEFON remote alarm; ~ **de distribución de potencia** ELECTROTEC power-distribution alarm; ~ **de fusible** PRUEBA *electricidad* fuse alarm; ~ **de fusible de estante** TELEFON shelf-fuse alarm; ~ **de fusible de fila de bastidores** COMPON, ELECTROTEC suite-fuse alarm; ~ **del grupo de portadoras** TRANS carrier-group alarm; ~ **importante** PRUEBA major alarm; ~ **independiente** PERIF *contestador* independent alarm; ~ **indicadora de unidad** TELEFON unit-indicating alarm; ~ **menor** PRUEBA minor alarm; ~ **de retorno** PRUEBA backward alarm; ~ **de servicio** TELEFON service alarm; ~ **de sobretensión** ELECTROTEC overvoltage alarm; ~ **de subtensión** ELECTROTEC undervoltage alarm; ~ **urgente** TELEFON urgent alarm; ~ **visual** PRUEBA visual alarm

alas: ~ **de encaminamiento** *f pl* ELECTRÓN routing wings

álbum: ~ **de discos compactos fotográficos** *m* INFO photo-CD portfolio

alcance *m* PRUEBA scope, TEC RAD spread, TELEFON reach; ~ **de antena** TEC RAD antenna range; ~ **límite de radar** TEC RAD radar range; **de** ~ **medio** *adj* GEN mid-range; ~ **óptimo de sintonización mecánica** ELECTRÓN optimum mechanical tuning range; ~ **de rueda de modulación** INFO modulation wheel range; ~ **de satélite** TEC RAD, TV satellite range; ~ **de sintonización mecánica** ELECTRÓN mechanical tuning range; ~ **de visibilidad óptica** TEC RAD line-of-sight range

aleación *f* COMPON alloy; ~ **de aluminio** COMPON aluminium (*AmE* aluminum) alloy; ~ **de cobre** COMPON copper alloy; ~ **eutéctica** COMPON eutectic alloy; ~ **de níquel** COMPON nickel alloy

aleatoriedad *f* GEN randomization

aleatorio *adj* GEN random

aleatorizar *vt* GEN *tono* randomize

alerta *f* PRUEBA, TEC RAD alerting; ~ **barco-tierra** TEC RAD ship-to-shore alerting; ~ **entre barcos** TEC RAD ship-to-ship alerting; ~ **pasiva**

TEC RAD passive alerting; ~ **tierra-barco** TEC RAD shore-to-ship alerting

aleta: ~ **adelantada** *f* ELECTRÓN flying lead; ~ **de refrigeración** *f* COMPON cooling flange

alfabético *adj* GEN alphabetical

alfabetización *f* GEN alphabetization

alfabetizar *vt* GEN alphabetize

alfabeto *m* DATOS alphabet; ~ **fonético** TEC RAD phonetic alphabet; ~ **internacional nº 5** (*IA5*) DATOS, TELEFON international alphabet no. 5 (*IA5*); ~ **de referencia internacional** (*IRA*) TELEFON international reference alphabet (*IRA*); ~ **telegráfico** TELEFON telegraph alphabet; ~ **telegráfico internacional nº 2** (*ITA2*) DATOS, TELEFON international telegraph alphabet no. 2 (*ITA2*)

alfafotográfico *adj* TV alphaphotographic

alfageométrico *adj* TV alphageometric

alfamosaico *adj* TV alphamosaic

algorítmico *adj* INFO algorithmic

algoritmo: ~ **de árbol espaciador** *m* (*STA*) ELECTRÓN spanning-tree algorithm (*STA*); ~ **de grupo de expertos en cinematografía** *m* DATOS motion picture experts group algorithm; ~ **MPEG** *m* DATOS MPEG algorithm; ~ **de portadora única** *m* INFO, TEC RAD, TRANS single-carrier algorithm; ~ **de Rivest, Shamir, Adleman** *m* (*RSA*) TELEFON Rivest, Shamir, Adleman algorithm (*RSA*)

alias *m* GEN alias

alicates: ~ **ajustables** *m pl* GEN slip-joint pliers; ~ **de corte** *m pl* GEN cutting pliers; ~ **de corte diagonal** *m pl* GEN diagonal cutting pliers; ~ **de corte frontal** *m pl* GEN end cutter; ~ **de corte lateral** *m pl* GEN side cutters, side-cutting pliers; ~ **paralelos** *m pl* GEN parallel pliers; ~ **pelacables** *m pl* ELECTRÓN stripping pliers; ~ **de precintar** *m pl* GEN sealing pliers; ~ **de punta angular** *m pl* GEN curved-nose pliers; ~ **de punta larga** *m pl* GEN long-nose pliers, snipe-nose pliers; ~ **para retirar lámparas** *m pl* GEN lamp-sleeve pliers; ~ **para retirar lentes** *m pl* GEN lens pliers

alimentación *f* ELECTRÓN *de circuitos híbridos* firing, GEN feed; ~ **Cutler** TEC RAD Cutler feed; ~ **directa** TELEFON direct feed; ~ **dividida** ELECTROTEC, TELEFON split feed; ~ **eléctrica ininterrumpible** (*UPS*) ELECTROTEC uninterruptible power supply (*UPS*); ~ **eléctrica ininterrumpida** ELECTROTEC uninterrupted power supply; ~ **escalonada** TEC RAD swan feed; ~ **frontal** GEN front feed; ~ **del guíaondas del haz** TEC RAD beam waveguide feed; ~ **hoja a hoja** PERIF *de impresoras* sheet feed; ~ **hoja por hoja** DATOS, INFO *de impresora, digitalizador* single-page feeding, single-sheet feeding; ~ **individual** TELEFON individual current feed; ~ **de línea** DATOS, INFO, TELEFON line feed; ~ **de micrófono** ELECTROTEC, TELEFON microphone current feed; ~ **de micrófono de carbón** ELECTROTEC, TELEFON carbon microphone current feed; ~ **de papel** PERIF paper feed; ~ **por pernos** PERIF pin feed; ~ **de tensión** ELECTROTEC voltage feed; ~ **en el vértice** TEC RAD *de antena de reflector* vertex feed

alimentador: ~ **de cuello de cisne** *m* TEC RAD gooseneck feed; ~ **de hojas** *m* PERIF sheet feeder, paper-feed unit; ~ **de papel** *m* PERIF paper-feed unit, sheet feeder; ~ **primario de antena** *m* TEC RAD antenna feed; ~ **de tarjetas** *m* INFO card hopper

alimentar *vt* GEN feed

alineación: ~ **básica de imagen** *f* TRANS basic frame alignment (*BFA*); ~ **de caracteres** *f* TELEFON character alignment; ~ **de frecuencias** *f* TEC RAD frequency alignment; ~ **inicial** *f* TELEFON initial alignment; ~ **de llamadas** *f* TELEFON call alignment; ~ **óptica** *f* TEC RAD boresight; ~ **de precisión** *f* DATOS *de digitalizador* precision framing; ~ **de trama** *f* TRANS frame alignment; ~ **de unidad de señal** *f* SEÑAL signal unit alignment

alineamiento *m* GEN alignment; ~ **de fase** ELECTRÓN phase alignment; ~ **de múltiples bastidores** TRANS multiframe alignment (*MFA*)

alinear *vt* GEN align

alisado[1] *adj* ELECTROTEC, INFO, TEC RAD smooth

alisado[2] *m* ELECTROTEC, INFO, TEC RAD smoothing

aliviar *vt* INFO, TELEFON relieve

alma *f* GEN core; ~ **de cable** GEN cable core

almacén *m* GEN magazine; ~ **de conmutación progresiva** CONMUT progressive switching magazine; ~ **de datos** DATOS data store; ~ **individual** DATOS individual store; ~ **de terminación de línea** TRANS line-terminating magazine

almacenado *adj* DATOS, INFO, TEC RAD stored

almacenamiento *m* DATOS, INFO, TEC RAD storing; ~ **de acceso lento** INFO slow-access storage; ~ **asociativo** INFO associative storage; ~ **borrable** INFO erasable store; ~ **de cifras** TELEFON digit storing; ~ **de contenido direccionable** DATOS content-addressable storage; ~ **en directorio** TELEFON directory store; ~ **en disco** INFO disk store; ~ **en disco magnético** INFO disk storage; ~ **estable** DATOS, INFO nonvolatile storage; ~ **de gran**

capacidad DATOS bulk storage, mass storage; ~ **indeleble** INFO, TV nonerasable storage; ~ **intermedio de información de vídeo** DATOS, INFO, TV video data buffering; ~ **intermedio de multitrama** DATOS multiframe buffering; ~ **de la lectura del contador** TELEFON meter account save; ~ **en memoria intermedia** DATOS, ELECTRÓN buffering; ~ **de mensaje de correo electrónico** DATOS *Internet* e-mail message store; ~ **de núcleos** INFO core storage; ~ **rápido** DATOS fast storage; ~ **y recuperación de imagen** DATOS image storage and retrieval; ~ **de tambor** CONMUT drum store; ~ **de traslación** CONMUT translation store

almacenar *vt* DATOS, INFO, PERIF, TEC RAD store

almohadilla: ~ **pasante** *f* ELECTROTEC feed-through pad; ~ **de pérdida de inserción** *f* TRANS insertion loss pad

almohadillas: ~ **R o T** *f pl* TELEFON *en extensiones de teléfonos* R or T pads

alojamiento: ~ **de componente** *m* ELECTRÓN component housing; ~ **de herramienta con bloqueo de engarce** *m* ELECTRÓN crimp ratchet tool frame

alquiler: ~ **de líneas** *m* PERIF line rental

alt. *abr* (*alternativa*) DATOS alt. (*alternative*)

alta: **de ~ fidelidad** *adj* GEN high-fidelity; **de ~ frecuencia** *adj* ELECTROTEC high-frequency; **de ~ potencia** *adj* ELECTROTEC high-power; **de ~ prioridad** *adj* GEN high-priority; **de ~ resistencia** *adj* ELECTROTEC high-resistance; **de ~ resolución** *adj* (*hi-res*) INFO, TV high-resolution (*hi-res*); **de ~ seguridad** *adj* GEN high-security

altavoz *m* INFO, TEC RAD, TELEFON loudspeaker; ~ **de cabecera** TELEFON pillow loudspeaker; ~ **de cono** COMPON cone loudspeaker; ~ **incorporado** INFO, PERIF built-in speaker, built-in loudspeaker; ~ **para el registro medio** ELECTRÓN mid-range speaker

alterable *adj* INFO alterable

alteración: ~ **de color** *f* GEN discolouration (*AmE* discoloration); ~ **cromática de extremos** *f* ELECTROTEC fringing; ~ **de tamaño** *f* CONMUT size alteration

alternadamente *adv* GEN alternately

alternado *adj* TEC RAD alternated

alternador: ~ **CA** *m* ELECTROTEC AC generator; ~ **de corriente alterna** *m* ELECTROTEC alternating current generator; ~ **de campo giratorio** *m* ELECTROTEC, TEC RAD, TELEFON rotating-field alternator

alternativa *f* (*alt.*) DATOS alternative (*alt.*)

altímetro *m* PRUEBA, TELEFON altimeter; ~ **de sonda** TEC RAD height indicator

alto *adj* GEN high; **de ~ rendimiento** GEN high-performance; **de ~ riesgo** GEN high-risk; **de ~ voltaje** ELECTROTEC high-voltage

altura *f* GEN height; ~ **aparente** TEC RAD apparent height; ~ **del apogeo** TEC RAD altitude of the apogee; ~ **de arco** PRUEBA height of arch; ~ **del componente** ELECTRÓN *de circuitos impresos* component height; ~ **del conducto portacables** TEC RAD duct height; ~ **efectiva** TEC RAD effective height; ~ **efectiva de antena** TEC RAD effective antenna height; ~ **efectiva sobre la media del terreno** TEC RAD effective height above average terrain; ~ **de soldadura** COMPON soldering height; ~ **de tono normal** INFO standard pitch; ~ **virtual** TEC RAD virtual height

ALU *abr* (*unidad aritmética y lógica*) INFO ALU (*arithmetic and logic unit*)

alumbrado: ~ **de emergencia** *m* ELECTROTEC emergency lighting

aluminio *m* GEN aluminium (*AmE* aluminum)

amarre *m* GEN *de cables* lashing

ambigüedad *f* GEN ambiguity

ámbito: ~ **de frecuencia** *m* TEC RAD, TRANS frequency domain; **de ~ nacional** *adj* GEN nationwide; **en el ~ profesional** *adj* PERIF professional-level

AMDF *abr* (*acceso múltiple por división de frecuencias*) CONMUT, SEÑAL, TELEFON, TRANS FDMA (*frequency-division multiple access*)

amenaza: ~ **pasiva** *f* DATOS passive threat

AMI *abr* (*inversión alternativa de señales*) SEÑAL, TELEFON, TRANS AMI (*alternate mark inversion*)

aminoplástico *m* COMPON aminoplastic

amortiguación *f* TEC RAD muting; ~ **de colores** INFO colour (*AmE* color) muting; ~ **de eco de recepción** TELEFON receive echo loss; ~ **de flexión** TRANS bending loss; ~ **de sonido** TELEFON sound attenuation

amortiguador: ~ **de chispas** *m* ELECTROTEC spark quench; ~ **de fibra** *m* TRANS fibre (*AmE* fiber) buffer; ~ **de goma** *m* GEN rubber shock-absorber; ~ **de horizontal** *m* COMPON, CONMUT selecting-bar damper; ~ **de inercia** *m* ELECTROTEC inertia damper; ~ **de sonido** *m* GEN sound absorber, sound attenuator

amortiguamiento *m* GEN decay; ~ **crítico** TELEFON *acústica* critical damping

amortiguar *vt* ELECTROTEC quench, attenuate, absorb, TRANS damp

amortización *f* GEN amortization

amperaje *m* CONMUT, DATOS size, ELECTROTEC

de fusibles, INFO, SEÑAL amperage; ~ **de servicio** ELECTRÓN ampere rating
amperímetro *m* PRUEBA ammeter; ~ **de mordazas** PRUEBA clip-on ammeter
amperio *m* ELECTROTEC ampere; **~-vuelta** ELECTROTEC ampere-turn
ampersand *m* INFO ampersand
ampliabilidad *f* INFO expandability, extendibility, TELEFON extendibility
ampliable *adj* INFO expandable, extendible, TELEFON extendible
ampliación *f* INFO zoom-in
ampliado *adj* INFO extended, expanded, TELEFON extended
ampliador: ~ **de memoria** *m* INFO memory extender
ampliar *vt* GEN extend, amplify
amplificación *f* GEN amplification; ~ **de FI** TEC RAD, TRANS IF amplification; ~ **de fibra de erbio impurificada** TRANS erbium-doped fibre (*AmE* fiber) amplification; ~ **máxima** TELEFON singing point; ~ **no lineal** TRANS nonlinear amplification; ~ **de onda** ELECTRÓN wave amplification; ~ **de potencia** DATOS, INFO, TRANS power amplification
amplificado *adj* ELECTRÓN, ELECTROTEC, TV *circuito* amplified
amplificador *m* GEN amplifier; ~ **con bajo nivel de ruidos** ELECTRÓN, TEC RAD, TRANS low-noise amplifier; ~ **de banda ancha** ELECTRÓN wideband amplifier, TRANS broadband amplifier; ~ **en bucle** PRUEBA loop amplifier; ~ **de campo cruzado** TRANS crossed-field amplifier; ~ **compensador** ELECTRÓN buffer amplifier; ~ **en contrafase** ELECTRÓN, ELECTROTEC push-pull amplifier; ~ **con corrección anticipante** ELECTRÓN feedforward amplifier; ~ **Darlington** ELECTRÓN Darlington amplifier; ~ **DCME** TELEFON DCME gain; ~ **de desbloqueo periódico** ELECTRÓN gated amplifier; ~ **de fibra de erbio impurificada** TRANS erbium-doped fibre (*AmE* fiber) amplifier; ~ **de filtro** ELECTRÓN filter amplifier; ~ **final** ELECTRÓN final amplifier; ~ **de FI secundario** TEC RAD, TRANS secondary IF amplifier; ~ **de guíaondas progresivo** ELECTRÓN, TEC RAD, TRANS travelling-waveguide (*AmE* traveling-waveguide) amplifier; ~ **de interrupción** ELECTRÓN chopper amplifier; ~ **de inversión** ELECTROTEC inverting amplifier; ~ **de láser diódico semiconductor** (*SLDA*) ELECTRÓN semiconductor laser diode amplifier (*SLDA*); ~ **de línea** TV line amplifier; ~ **magnético** ELECTROTEC transductor amplifier; ~ **de medición** ELECTRÓN measuring amplifier; ~ **para medición del ruido** TELEFON noise-measuring amplifier; ~ **de ondas progresivas** ELECTRÓN, TEC RAD, TRANS travelling-wave (*AmE* traveling-wave) amplifier; ~ **operacional** ELECTRÓN, TRANS operational amplifier; ~ **parafásico** ELECTRÓN paraphase amplifier; ~ **paramétrico** TRANS parametric amplifier (*paramp*); ~ **de pequeña señal** ELECTRÓN, SEÑAL small-signal amplifier; ~ **de potencia** (*PA*) DATOS, INFO, TRANS power amplifier (*PA*); ~ **previo de sintetizador** INFO synthesizer driver; ~ **de realimentación negativa** ELECTRÓN negative-feedback amplifier; ~ **reflejo** ELECTRÓN reflex amplifier; ~ **regulado por tensión** ELECTROTEC voltage-controlled amplifier; ~ **de RF** TELEFON RF amplifier; ~ **de semiconductores** COMPON, ELECTRÓN semiconductor amplifier; ~ **sin transformador de salida** ELECTROTEC single-ended amplifier; ~ **sumador gradual** ELECTROTEC summing-scaling amplifier; ~ **de teclado** INFO keyboard amplifier; ~ **telegráfico** TELEFON telegraph magnifier; ~ **de tubo de ondas progresivas** ELECTRÓN, TRANS travelling-wave (*AmE* traveling-wave) tube amplifier; ~ **de voltaje** ELECTROTEC voltage amplifier; ~ **de voz** ELECTRÓN voice amplifier
amplificar *vt* GEN amplify, extend
amplio: ~ **surtido** *m* GEN wide range
amplitud *f* ELECTRÓN, INFO amplitude, swing, TEC RAD, TELEFON amplitude; ~ **aérea** ELECTROTEC, TRANS overhead span; ~ **básica** TEC RAD basic amplitude; ~ **de contenido** DATOS content length; ~ **de controlador de intervalo** INFO breath-controller range; ~ **de cresta a cresta** GEN peak-to-peak amplitude; ~ **de curva primitiva** INFO pitchbend range; ~ **de desplazamiento** ELECTRÓN displacement amplitude; ~ **de espectro** TEC RAD spectrum amplitude; ~ **del haz** TEC RAD, TRANS beam width; ~ **de impulso** TELEFON pulse magnitude, TRANS pulse amplitude; ~ **máxima** TELEFON peak amplitude; ~ **de ojo** TEC RAD eye width; ~ **óptima de escala** PRUEBA optimum scale range; ~ **relativa** TELEFON *de eco elemental* relative amplitude; ~ **de repetidor** TRANS repeater span; ~ **de señal** SEÑAL signal amplitude
análisis *m* GEN analysis; ~ **de árbol de fallos** DATOS, PRUEBA fault-tree analysis; ~ **de averías** PRUEBA failure analysis, fault analysis; ~ **de camino crítico** GEN *gestión de proyectos* critical-path analysis; ~ **de campo próximo** TEC RAD, TRANS near-field analysis; ~ **de**

dominio de frecuencia DATOS frequency-domain analysis; ~ **de errores** DATOS error analysis; ~ **de esfuerzos** PRUEBA stress analysis; ~ **espectral** PRUEBA spectral analysis; ~ **de facturación** APLIC charging analysis; ~ **de flujo de datos** DATOS data-flow analysis; ~ **granulométrico** ELECTRÓN mesh analysis; ~ **de imagen** TV image analysis; ~ **de longitud numérica** TELEFON analysis for number length; ~ **de mantenimiento** PRUEBA maintenance analysis; ~ **de modo de fallo** (*FMA*) PRUEBA *seguridad de funcionamiento* fault-mode analysis (*FMA*); ~ **multidimensional de datos** DATOS multidimensional data analysis; ~ **de Pareto** PRUEBA Pareto analysis; ~ **preliminar** TELEFON preanalysis; ~ **prosódico** DATOS prosodic analysis; ~ **de protocolos** DATOS protocol analysis; ~ **de riesgos** PRUEBA risk analysis, hazard analysis; ~ **de secuencia** GEN sequence analysis; ~ **de señales** SEÑAL signal analysis; ~ **de sistemas** INFO, PRUEBA systems analysis; ~ **de tarifa** TELEFON tariff analysis; ~ **de tráfico** DATOS, PRUEBA, TELEFON traffic analysis; ~ **de voz** TELEFON speech analysis

analista *m* GEN analyst; ~ **de procesamiento de datos electrónicos** GEN EDP analyst; ~**-programador** GEN analyst-programmer (*AmE* analyst-programer); ~ **de sistemas** GEN systems analyst

analizador *m* INFO scanner, ELECTRÓN, SEÑAL analyser (*AmE* analyzer); ~ **de categorías** TELEFON category analyser (*AmE* analyzer); ~ **de distorsión de señal** (*SDA*) PRUEBA signal distortion analyser (*AmE* analyzer) (*SDA*); ~ **de espectro** PRUEBA, TEC RAD spectrum analyser (*AmE* analyzer); ~ **de imagen** TV image analyser (*AmE* analyzer); ~ **de línea** PRUEBA line analyser (*AmE* analyzer); ~ **de lógica** DATOS logic analyser; ~ **lógico** DATOS logic analyser; ~ **de módem** PRUEBA modem tester; ~ **de onda** PRUEBA wave analyser (*AmE* analyzer); ~ **de redes** PRUEBA network analyser (*AmE* analyzer); ~ **de señales** INFO, SEÑAL signal analyser (*AmE* analyzer); ~ **de tráfico** DATOS, PRUEBA, TELEFON traffic analyser (*AmE* analyzer); ~ **de transistores** PRUEBA transistor analyser (*AmE* analyzer)

analizar *vt* GEN analyse (*AmE* analyze)

analógico *adj* GEN analog; ~ **a digital** (*A-D*) INFO analog-to-digital (*A-D*); ~**-digital** INFO analog-digital

anamorfosis *m* GEN anamorphosis

ancho: ~ **de antena** *m* TEC RAD antenna width; ~ **de banda** *m* DATOS, ELECTRÓN, TEC RAD, TRANS, TV bandwidth; ~ **de banda de coherencia** *m* TEC RAD, TRANS coherence bandwidth; ~ **de banda espectral** *m* TEC RAD, TRANS spectral bandwidth; ~ **de banda limitado** *m* DATOS limited bandwidth; ~ **de banda de mitad de potencia** *m* TEC RAD, TRANS half-power bandwidth; ~ **de banda necesario** *m* TEC RAD, TRANS necessary bandwidth; ~ **de banda nominal** *m* INFO, TV nominal bandwidth; ~ **de banda de punta** *m* ELECTRÓN nose bandwidth; ~ **de banda de ruido general efectivo** *m* TEC RAD, TRANS width of effective overall noise band; ~ **de banda sobre demanda** *m* INFO bandwidth on demand; ~ **de banda de trabajo** *m* TEC RAD, TRANS operating bandwidth; ~ **de banda x dB** *m* TEC RAD x dB bandwidth; ~ **de canal nominal** *m* TV nominal channel width; ~ **de conducto** *m* TEC RAD duct thickness; ~ **del haz de mitad de potencia** *m* TEC RAD, TRANS half-power beamwidth; ~ **de línea** *m* ELECTRÓN *circuitos impresos* line width; ~ **de pulso** *m* DATOS, ELECTRÓN, TEC RAD, TELEFON, TRANS pulse width (*PW*)

anchura *f* INFO, TEC RAD width; ~ **de conductor** ELECTRÓN conductor width; ~ **espectral** TRANS spectral width; ~ **de trazo** INFO stroke width

anclaje *m* INFO anchor

anclajes: ~ **para cable** *m pl* COMPON cable ties

andén *m* ELECTRÓN, TRANS platform

anfitrión *m* DATOS host

ángel *m* TEC RAD radar angel

ángulo: ~ **acimutal** *m* TEC RAD angle of azimuth; ~ **de alimentación** *m* TRANS *óptica de fibras* launch angle; ~ **apuntador** *m* TEC RAD pointing angle; ~ **de Brewster** *m* TRANS *óptica de fibras* Brewster's angle; ~ **crítico** *m* TRANS critical angle; ~ **de depresión** *m* TEC RAD angle of depression; ~ **de deriva** *m* TEC RAD squint angle; ~ **de elevación** *m* TEC RAD, TRANS angle of elevation, elevation angle; ~ **de evitación** *m* TEC RAD avoidance angle; ~ **exocéntrico** *m* TEC RAD exocentric angle; ~ **de fase** *m* ELECTROTEC, INFO phase angle; ~ **geocéntrico** *m* TEC RAD geocentric angle; ~ **de giro** *m* TRANS *óptica de fibras* angle of rotation; ~ **de Hall** *m* ELECTRÓN *semiconductores* Hall angle; ~ **de incidencia** *m* TRANS angle of incidence; ~ **de llegada** *m* TRANS angle of arrival; ~ **de picado** *m* TEC RAD angle of tilt; ~ **de radiación** *m* ELECTRÓN, TRANS radiation angle; ~ **de recepción** *m* TRANS acceptance angle; ~ **respecto de eje** *m* TEC RAD angle from the axis; ~ **de salida** *m*

ELECTRÓN output angle, TRANS angle of departure, output angle; ~ **de situación** *m* TRANS angle of site; ~ **sólido del haz** *m* TEC RAD beam area; ~ **sólido del haz del enlace del alimentador** *m* TEC RAD feeder-link beam area; ~ **topocéntrico** *m* TEC RAD topocentric angle; ~ **de tránsito** *m* ELECTRÓN transit angle; ~ **de trayectoria de vuelo** *m* TEC RAD flight-path angle; ~ **de vértice** *m* TEC RAD vertex angle; ~ **de visión** *m* GEN viewing angle

anidar *vt* INFO, PRUEBA nest

anillo *m* COMPON, DATOS, ELECTRÓN ring; ~ **colector** ELECTROTEC slip ring; ~ **de conexión** CONMUT jumper ring; **de ~ doble** *adj* ELECTRÓN dual-ring; ~ **de fijación** COMPON fixing ring; ~ **de guardia** TELEFON guard ring; ~ **guíahilos** COMPON *cableado* jumper-wire eyelet; ~ **híbrido** TEC RAD hybrid ring; ~ **en O** COMPON O ring; ~ **obturador** COMPON gasket ring; ~ **protector de varias escalas** ELECTRÓN multirange grommet; ~ **de retardo** COMPON *relés* delay slug, slug; ~ **de roscar** COMPON threading ring; ~ **sellador** COMPON sealing ring; ~ **separador** COMPON spacing ring

animación: ~ **facial** *f* INFO facial animation; ~ **funcional** *f* INFO behavioural animation; ~ **con sonido** *f* DATOS *de digitalizador* sound animation; ~ **por teclado de registro de imágenes** *f* INFO keyframe animation

anisócrono *adj* GEN anisochronous

anisotrópico *adj* GEN anisotropic

anodizar *vt* ELECTRÓN anodize

ánodo *m* ELECTRÓN anode

anotación *f* GEN notation; ~ **vocal** INFO, TELEFON voice annotation

anotado *adj* TELEFON recorded

anotador *m* INFO notepad

ANS *abr* (*tono de contestación*) DATOS ANS (*answer tone*)

ANSI *abr* (*Instituto Nacional Americano de Normas*) GEN ANSI (*American National Standards Institute*)

antena *f* INFO, TV aerial (*AE*), TEC RAD aerial (*AE*), antenna, command antenna; ~ **en abanico** TEC RAD fan-beam antenna; ~ **acimutal** TEC RAD azimuth antenna; ~ **acodada** TEC RAD bend aerial; ~ **acústica** TEC RAD acoustic antenna; ~ **adaptativa** TEC RAD adaptive antenna; ~ **de alimentación focal** TEC RAD focal-fed antenna; ~ **anexa** TEC RAD co-located antenna; ~ **de ángulo cóncavo** TEC RAD corner antenna; ~ **de ángulo de conmutación de lóbulo** TEC RAD squint antenna; ~ **en anillo** TEC RAD ring antenna; ~ **de apertura** TEC RAD aperture antenna; ~ **de apertura alimentada por guíaondas** TEC RAD waveguide-fed aperture antenna; ~ **en array de radiación transversal** TEC RAD broadside array antenna; ~ **de artesa** TEC RAD hogtrough antenna; ~ **bicónica** TEC RAD biconical antenna; ~ **de bocina** TEC RAD flare antenna, horn antenna; ~ **de campo E** TEC RAD E-field antenna; ~ **Cassegrain** TEC RAD Cassegrain antenna; ~ **en cigarro** TEC RAD cigar aerial, cigar antenna; ~ **cilíndrica de reflector** TEC RAD cylindrical reflector antenna; ~ **de cilindro parabólico** TEC RAD parabolic cylinder antenna; ~ **colectiva** TEC RAD community aerial, TV collective aerial, community aerial, community antenna; ~ **compensada** TEC RAD cosecant-squared antenna; ~ **conforme** TEC RAD conformal antenna; ~ **corregida por exploración** TEC RAD scanning-corrected antenna; ~ **en cortina** TEC RAD antenna curtain; ~ **de cuadro** TEC RAD loop aerial, loop antenna; ~ **cuasiomnidireccional** TEC RAD quasi-omnidirectional antenna; ~ **de diedro** PRUEBA dihedral antenna; ~ **dieléctrica de varillas** TEC RAD polyrod; ~ **dipolo** TEC RAD dipole aerial, dipole antenna; ~ **dipolo doblado** TEC RAD folded-dipole antenna; ~ **directiva** DATOS directive antenna, TEC RAD directional aerial; ~ **dirigida** TEC RAD directional antenna; ~ **de disco** TEC RAD disk antenna; ~ **de doble reflector** TEC RAD double-reflector antenna; ~ **de elementos en fase** TEC RAD phased-array antenna; ~ **de embudo** TEC RAD hoghorn; ~ **empotrada** TEC RAD flush antenna; ~ **esférica** TEC RAD spherical antenna; ~ **en el espacio libre** TEC RAD free-space antenna; ~ **espiral** TEC RAD spiral antenna; ~ **estándar económica** TEC RAD economic standard antenna; ~ **de estándar mínimo** TEC RAD minimum-standard antenna; ~ **exploradora** INFO, PERIF scanner, TEC RAD scanner, scanning antenna; ~ **exterior** TEC RAD outdoor aerial, TV rooftop aerial; ~ **de ferrita** TEC RAD ferrite antenna; ~ **giratoria** TEC RAD rotatable aerial, rotatable antenna; ~ **de guíaondas** TEC RAD waveguide antenna; ~ **de guíaondas ranurado** TEC RAD slotted-waveguide antenna; ~ **del haz** TEC RAD beam aerial, beam antenna; ~ **del haz perfilado** TEC RAD shaped-beam antenna; ~ **helicoidal** TEC RAD corkscrew antenna, helix aerial, helix antenna; ~ **de hilo** TEC RAD wire aerial, wire antenna; ~ **horizontal** TEC RAD horizontal antenna; ~ **individual** TV individual aerial; ~ **integral** TEC RAD integral antenna; ~ **de látigo** TEC RAD whip antenna; ~ **de látigo**

de cuarto de onda TEC RAD quarter-wave whip antenna; ~ **de látigo en torniquete** TEC RAD turnstile whip antenna; ~ **de lente** TEC RAD lens antenna; ~ **logarítmica** TEC RAD log-periodic antenna; ~ **magnética** TEC RAD ferrite antenna; ~ **en mariposa** TEC RAD antenna butterfly; ~ **matricial** TEC RAD array antenna; ~ **monopolo** TEC RAD monopole antenna, unipole antenna; ~ **multifrecuencia** TEC RAD multifrequency antenna; ~ **multihaz** TEC RAD multiple-beam antenna; ~ **multilobular** TEC RAD multilobe aerial; ~ **omnidireccional** TEC RAD omnidirectional aerial, omnidirectional antenna; ~ **de onda progresiva** TEC RAD, TRANS travelling-wave (*AmE* traveling-wave) aerial, travelling-wave (*AmE* traveling-wave) antenna; ~ **con parábola achatada** TEC RAD cheese antenna; ~ **parabólica** TEC RAD dish aerial, dish antenna, parabolic aerial, parabolic antenna; ~ **parabólica desplazada del haz perfilado** TEC RAD shaped-beam offset parabolic antenna; ~ **periscópica** TEC RAD periscope antenna; ~ **de plato** TEC RAD dish aerial, dish antenna; ~ **de puntos múltiples** TEC RAD multispot antenna; ~ **de radar** TEC RAD radar antenna; ~ **radárica** TEC RAD radar aerial; ~ **ranurada** TEC RAD slot antenna; ~ **de ranuras cruzadas** TEC RAD crossed-slot antenna; ~ **de rastreo** TEC RAD tracking antenna; ~ **de recepción** TEC RAD receiving antenna; ~ **receptora** TEC RAD receiver antenna; ~ **receptora de satélite** TEC RAD satellite receiver antenna; ~ **rectificadora** TEC RAD rectifying antenna; ~ **rectilínea** TEC RAD rectilinear antenna; ~ **de red planar de transductores** TEC RAD planar-array antenna; ~ **reflectante** TEC RAD reflecting antenna; ~ **de reflector** TEC RAD reflector antenna; ~ **reflectora de alimentación frontal** TEC RAD front-fed reflector antenna; ~ **reflectora cuadriculada** TEC RAD gridded reflector antenna; ~ **de reflector Gregoriana** TEC RAD Gregorian reflector antenna, Gregorian reflector aerial; ~ **de reflector múltiple** TEC RAD multiple-reflector antenna; ~ **de reflector parabólico** TEC RAD parabolic reflector antenna; ~ **replegada** TEC RAD bow-tie antenna; ~ **rómbica** TEC RAD rhombic antenna; ~ **de rotación inversa electrónica** TEC RAD electronically-despun antenna; ~ **de satélite artificial** TEC RAD, TV satellite aerial, satellite antenna; ~ **de TDRSS** TEC RAD antenna TDRSS; ~ **de televisión** TV television aerial; ~ **de televisión sólo para recepción** TEC RAD, TV television receive-only antenna; ~ **toroidal** TEC RAD toroidal antenna; ~ **de transmisión de banda estrecha** TEC RAD narrow-beam transmit antenna; ~ **de transmisión de eurohaz** TEC RAD eurobeam transmit antenna; ~ **de transmisión por vía satélite** TEC RAD satellite transmit antenna; ~ **transmisora** TEC RAD transmitting antenna; ~ **de trébol** TEC RAD cloverleaf antenna; ~ **TVRO** TEC RAD, TV TVRO antenna; ~ **en V** TEC RAD V-shaped antenna; ~ **de varilla** TEC RAD rod antenna; ~ **de varillas telescópicas** TEC RAD fishbone antenna; ~ **en V inclinada** TEC RAD sloping-V antenna; ~ **Yagi** TEC RAD Yagi aerial, Yagi antenna

antenas: ~ **parabólicas adosadas** *f pl* TEC RAD back-to-back parabolic antennas

anteproyecto *m* GEN draft project

anticorrosivo *m* GEN rust preventive

antiestático *adj* ELECTRÓN anti-static

antiinterferencias *f pl* (*AJ*) TEC RAD anti-jamming (*AJ*)

antioscilación *f* TEC RAD anti-sloshing

antirreflectante *adj* (*AR*) GEN anti-reflection (*AR*)

antisolapamiento *m* INFO, TRANS anti-aliasing

anulación *f* ELECTRÓN negation, GEN deletion, INFO negation, TEC RAD suppression

anulado *adj* ELECTRÓN negated, GEN deleted, INFO negated, TEC RAD suppressed

anular *vt* ELECTRÓN negate, GEN delete, INFO negate, TEC RAD suppress

anunciar *vt* INFO usher in

anuncio: ~ **comercial** *m* TV commercial; ~ **publicitario** *m* TV commercial; ~ **supervisor** *m* TELEFON supervisory announcement

AOQ *abr* (*calidad media de salida*) PRUEBA AOQ (*average outgoing quality*)

AOQL *abr* (*promedio límite de calidad de subidas*) PRUEBA AOQL (*average outgoing quality limit*)

apagachispas *m* ELECTROTEC spark quencher

apagón *m* TEC RAD *servicio eléctrico* blackout

aparamétrico *adj* GEN nonparametric

aparato *m* GEN mechanism; ~ **de abonado** TELEFON subscriber's station; ~ **de aguja** TELEFON pointer instrument; ~ **alimentador** INFO, TEC RAD, TRANS feeder; ~ **auxiliar** TELEFON auxiliary set; ~ **de comunicación** ELECTRÓN talk set; ~ **de control de eco** TEC RAD, TELEFON echo-control device; ~ **facsímil** PERIF, TELEFON facsimile apparatus, fax machine; ~ **de fax** PERIF, TELEFON fax machine; ~ **de fax de grupo 1** APLIC, PERIF, TELEFON group 1 facsimile apparatus, group 1 fax machine; ~ **de intercomunicación telefónica** PERIF, TELEFON intercommunication telephone set; ~ **de operador** TELEFON attendant's set; ~ **para probar líneas eléctricas** PRUEBA line tester; ~ **en prueba** PRUEBA device under test; ~ **receptor** INFO, TELEFON receiving device; ~ **de sobremesa** PERIF tabletop device; ~ **de telecomunicaciones para sordos** (*TDD*) GEN telecommunications device for the deaf (*TDD*); ~ **de telefax del grupo 4** APLIC, PERIF group 4 facsimile apparatus; ~ **telefónico** PERIF, TELEFON handset; ~ **telefónico especial** PERIF, TELEFON special telephone set; ~ **telefónico estándar** TELEFON standard telephone set; ~ **telefónico de mesa** PERIF table instrument; ~ **telefónico de pared** PERIF, TELEFON wall-mounted telephone set; ~ **telefónico de recepción amplificada** TELEFON receive-amplified handset telephone; ~ **telefónico de sobremesa** TELEFON table set; ~ **telefónico de teclado** PERIF, TELEFON keypad telephone set; ~ **terminal** DATOS, INFO, PERIF, TELEFON terminating equipment

aparcado *adj* CONMUT, TELEFON *llamada* parked

aparcar *vt* CONMUT, TELEFON park

apareamiento *m* ELECTRÓN twinning

aparecer: **no** ~ *vi* GEN fail to appear

APCI *abr* (*información de control del protocolo de aplicación*) DATOS APCI (*application protocol control information*)

APD *abr* (*fotodiodo de avalancha*) COMPON, DATOS, TRANS APD (*avalanche photodiode*)

APDU *abr* (*unidad de datos del protocolo de aplicación*) DATOS, TELEFON APDU (*application protocol data unit*); ~ **de ABORTO-A** *m* DATOS A-ABORT APDU; ~ **de liberación-petición-A** *m* (*RLRQ*) DATOS A-release-request APDU (*RLRQ*); ~ **de liberación-respuesta-A** *m* (*RLRE*) DATOS A-release-response APDU (*RLRE*)

apéndice *m* COMPON, INFO tab

aperiódico *adj* GEN aperiodic

apertura *f* GEN aperture; ~ **numérica de lanzamiento** ELECTRÓN, TRANS launch numerical aperture (*LNA*); ~ **de ojo** TRANS eye opening

API *abr* (*interfaz del programa de aplicación*) DATOS API (*application program interface*)

ápice *m* GEN apex

apilable *adj* COMPON, INFO stackable

apilamiento *m* COMPON, INFO stacking

apilar *vt* COMPON, INFO stack

aplazado *adj* INFO deferred

aplicación *f* GEN application; ~ **de comunicación** TELEFON communication application; ~ **de elementos de programación** APLIC, DATOS, INFO software application; ~ **de hipertexto** INFO hypertext application; ~ **local** (*LA*) TELEFON local application (*LA*)

aplicado *adj* ELECTRÓN applied

aplicador: ~ **de obturador** *m* COMPON sealant applicator

aplicar *vt* GEN apply

apoápside *m* TEC RAD apoapsis

apoderado: ~ **de usuarios de directorio** *m* (*DUA*) APLIC, DATOS directory user agent (*DUA*)

apogeo *m* TEC RAD apogee

apoyar: ~ **sobre** *vt* COMPON bear on, lie against

apoyo *m* GEN backup, support; ~ **de cable** TRANS cable support; ~ **empotrado** PERIF wall bracket

apremio *m* INFO prompt

aprendizaje: ~ **analógico** *m* INFO analogical learning; ~ **auxiliada por computadora** *AmL f* (*cf enseñanza auxiliada por ordenador Esp*), (*CAL*) INFO computer-assisted learning (*CAL*); ~ **auxiliada por ordenador** *Esp f* (*cf enseñanza auxiliada por computadora AmL*), (*CAL*) INFO computer-assisted learning (*CAL*); ~ **basada en la computadora** *AmL f* (*cf enseñanza basada en el ordenador Esp*) INFO computer-based learning; ~ **basada en el ordenador** *Esp f* (*cf enseñanza basada en la computadora AmL*) INFO computer-based learning

apriete *m* GEN tightening

aprobación: ~ **de aptitud** *f* PRUEBA *calidad* capability approval

aprobado: **no** ~ *adj* PERIF *aparato* nonapproved

aproximación: ~ **asintomática** *f* GEN asymptomatic approximation; ~ **de Chebichef** *f* TRANS Chebyshev approximation; ~ **por radar de precisión** *f* TEC RAD precision radar approach

APSK *abr* (*tecleado de cambio en la relación de fase de amplitud*) SEÑAL APSK (*amplitude phase-shift keying*)

aptitud: ~ **para funcionar** *f* APLIC, TELEFON serviceability

apto: ~ **para funcionar** *adj* GEN serviceable

apuntador *m* TV teleprompter *AmE*, autocue *BrE*

AR *abr* (*antirreflectante*) GEN AR (*anti-reflection*)

ARABSAT *abr* (*Organización Arabe de Comunicaciones por Satélite*) TEC RAD ARABSAT (*Arab Satellite Communications Organization*)

arandela *f* COMPON washer; ~ **abierta** COMPON slotted washer; ~ **aislante** COMPON, ELECTRÓN grommet; ~ **aislante exterior** ELECTRÓN outboard grommet; ~ **aislante interior** ELECTRÓN inboard grommet; ~ **Belleville** COMPON Belleville spring; ~ **elástica** COMPON spring washer; ~ **espaciadora** COMPON spacing washer; ~ **de mica** ELECTRÓN mica washer; ~ **plana** GEN plane washer

árbol *m* DATOS, ELECTRÓN, INFO tree; ~ **B** DATOS B tree; ~ **base** DATOS root tree; ~ **de comportamiento** DATOS behaviour (*AmE* behavior) tree; ~ **de decisión** DATOS decision tree; ~ **denominador** DATOS naming tree; ~ **de diodos** ELECTRÓN diode tree; ~ **de fallos** PRUEBA *seguridad de funcionamiento* fault tree; ~ **de información de directorio** DATOS, TELEFON directory information tree (*DIT*); ~ **local** DATOS local tree; ~ **de nombre de registro jerárquico** DATOS registration-hierarchical name tree; ~ **de relés** COMPON relay tree

arborizado *adj* DATOS, ELECTRÓN, INFO treed

archivar *vt* DATOS, INFO archive, TELEFON *télex* put on file

archivo: ~ **binario** *m* TELEFON binary file; ~ **de documento** *m* DATOS *Internet* document file; ~ **de espera** *m* TELEFON queue file; ~ **lógico** *m* DATOS logical file; ~ **PCX multipágina** *m* (*DCX*) DATOS multipage PCX file (*DCX*); ~ **de sonido** *m* DATOS sound file; ~ **de texto** *m* DATOS text file; ~ **de trabajo** *m* INFO work file

archivos *m pl* GEN archives; **en** ~ *adj* ELECTRÓN in rolls

arco: ~ **eléctrico** *m* ELECTROTEC electric arc; ~ **de servicio** *m* TEC RAD service arc; ~ **visible** *m* TEC RAD visible arc

área: ~ **de abono** *f* DATOS, INFO, TEC RAD, TELEFON, TV subscription area; ~ **de asociación** *f* DATOS *SDL* association area, INFO correction area; ~ **de captación** *f* TEC RAD capture area; ~ **de captación de centralita grupal** *f* CONMUT group switching center catchment area *AmE*; ~ **de cobertura libre de interferencias** *f* TEC RAD interference-free coverage area; ~ **de cobertura nominal** *f* TEC RAD nominal coverage area; ~ **de coherencia** *f* TRANS coherence area; ~ **de comunicación urbana** *f* TELEFON local calling area (*LCA*); ~ **de eco** *f* TEC RAD radar cross-section; ~ **efectiva de antena** *f* TEC RAD antenna effective area; ~ **de entrada** *f* INFO lead-in area; ~ **extendida** *f* TEC RAD wide area; ~ **funcional de gestión de sistemas** *f* INFO systems management functional area; ~ **de impresión** *f* INFO printable area; ~ **imprimible** *f* INFO printable area; ~ **de indicador de encaminamiento** *f* TELEFON routing-number area; ~ **de localización** *f* TEC RAD location area; ~ **de numeración** *f* TELEFON numbering area; ~ **de pista de impresión** *f* ELECTRÓN printing track area; ~ **de plan de numeración** *f* (*NPA*) TELEFON numbering plan area (*NPA*); ~ **de propagación** *f* GEN area of propagation; ~ **de recepción** *f* TEC RAD reception area; ~ **de servicio digital centralizado** *f* DATOS centralized DSA; ~ **de servicio del enlace del alimentador** *f* TEC RAD feeder-link service area; ~ **de trabajo** *f* GEN area of work, work area; ~ **urbana** *f* TV urban area; ~ **de uso pleno** *f* ELECTRÓN plenum-use area; ~ **virtual** *f* INFO virtual area

argumento *m* DATOS argument; ~ **de parámetro simple** TELEFON simple parameter argument; ~ **de perigeo** TEC RAD argument of perigee

armado *adj* GEN armoured (*AmE* armored), reinforced

armadura *f* GEN reinforcement; ~ **de cable** TRANS cable armour (*AmE* armor)

armario: ~ **de cables** *m* TRANS cable cabinet; ~ **de conexiones** *m* ELECTRÓN wiring closet; ~ **de controles** *m* CONTROL control cabinet; ~ **del convertidor** *m* ELECTROTEC converter cabinet; ~ **de distribución** *m* TRANS distribution cabinet; ~ **de distribución de cables** *m* ELECTROTEC cable distribution cabinet; ~ **de herramientas** *m* GEN tool cabinet; ~ **interconectado** *m* TELEFON cross-connect cabinet; ~ **satélite** *m* ELECTRÓN satellite closet; ~ **de subdistribución** *m* TRANS subdistribution cupboard

armarios: ~ **de acumuladores** *m pl* TRANS battery cabinets

armazón *m* COMPON, INFO framework; ~ **de estante** TELEFON shelf chassis; ~ **de selectores** CONMUT, TELEFON selector rack; ~ **de vertical** COMPON vertical unit base

armonía *f* INFO unison

armónico *m* ELECTRÓN harmonic; ~ **superior** INFO overtone

arnés *m* ELECTRÓN harness

arpegiador *m* INFO arpeggiator

arpegio *m* INFO arpeggio

arquitectura *f* INFO architecture; ~ **en árbol y en rama** DATOS, INFO tree and branch architecture; ~ **del contenido del documento** DATOS document content architecture (*DCA*); ~ **de documento administrativo** (*ODA*) DATOS, INFO office document architecture (*ODA*); ~ **estratiforme** INFO layered architecture; ~ **en estrella** TRANS, TV star architecture; ~ **de gestión de la red** TRANS network management architecture; ~ **industrial estándar** (*ISA*) INFO *procesamiento remoto de datos* industry standard architecture (*ISA*); ~ **informática de los sistemas de operaciones** INFO operations systems computing architecture; ~ **de memoria** INFO memory architecture; ~ **en paralelo** TRANS parallel architecture; ~ **de procesador** DATOS, INFO processor architecture; ~ **de protocolo telemático** (*TPA*) TELEFON telematic protocol architecture (*TPA*); ~ **de la red de transporte Ericsson** (*ETNA*) TRANS Ericsson transport network architecture (*ETNA*); ~ **de sistema distribuido independiente** DATOS stand-alone DSA; ~ **del sistema de división de función** CONMUT function division system architecture; ~ **sistólica** APLIC, TRANS systolic architecture

arrancador: ~ **de lámpara** *m* GEN lamp extractor

arrancar *vt* INFO boot, start up

arranque *m* INFO boot-up; ~ **del aparato** TELEFON device start; ~ **en bucle** TELEFON loop start; ~ **a distancia** ELECTRÓN remote boot; ~ **de estrella-triángulo** ELECTROTEC star-delta starting; ~ **de señal de llamada** TELEFON ring start

arrastrar: ~ **y eliminar** *vt* INFO drag and drop

arrastre: ~ **de imagen** *m* GEN dragging; ~ **en paralelo** *m* INFO parallel carry

array *m* GEN array; ~ **de diodos** ELECTRÓN diode array

arreglo: ~ **de avería** *m* ELECTRÓN damage recovery

arrendamiento *m* GEN leasing

arriendo *m* GEN lease

arrollador: ~ **de papel** *m* PERIF paper winder

arrollamiento: ~ **espiral helicoidal** *m* ELECTRÓN helix spiral wrap; ~ **de inducido** *m* ELECTROTEC armature winding; ~ **de retención** *m* ELECTROTEC holding winding; ~ **de unión** *m* COMPON, ELECTROTEC tie wrap

arrollamientos: ~ **antagonistas** *m pl* ELECTROTEC counter-acting winding

arseniuro: ~ **de galio** *m* (*GaAs*) COMPON gallium arsenide (*GaAs*)

artefacto *m* GEN appliance

artesa *f* ELECTRÓN trough

articulación *f* TELEFON articulation; ~ **de antena múltiple** TEC RAD array hinge; ~ **silábica** TELEFON syllable articulation

artículo: ~ **reparado** *m* PRUEBA repaired item

artículos: ~ **no conformes** *m pl* PRUEBA non-conforming items

artificial *adj* DATOS, ELECTRÓN, INFO artificial

asa *f* ELECTROTEC transport handle, bow, INFO grip

ASCII *abr* (*Código Estándar Americano para Intercambio de Información*) DATOS, INFO, TELEFON ASCII (*American Standard Code for Information Interchange*)

ASE *abr* (*elemento del servicio de aplicación*) DATOS, TELEFON ASE (*application service element*); ~ **del usuario** *m* DATOS user-ASE

aseguramiento: ~ **de calidad** *m* PRUEBA quality assurance

asegurar *vt* GEN ensure

asesor: ~ **de clientela** *m* GEN customer advisor

asesoramiento: ~ **de servicio sucesivo** *m* TELEFON follow-on service advice

asfalto *m* GEN asphalt

asiento: ~ **de cable** *m* TRANS cable lug; ~ **de rodamiento** *m* COMPON bearing seat

asignación *f* GEN allocation, assignment; ~ **de almacenamiento** INFO storage allocation; ~ **de canal** TELEFON channel dedication; ~ **de clave** INFO key assignment; ~ **de controlador de intervalo** INFO breath-controller assignment; ~**por demanda** TEC RAD demandassignment; ~ **dinámica** INFO dynamic allocation; ~ **dinámica dememoria** INFO dynamic storage allocation; ~ **dinámica de recursos** INFO dynamic resourceallocation; ~ **de frecuencias** TEC RAD frequency assignment; ~ **de órbita** TEC RAD assignment of orbit; ~ **de parámetro** INFO parameter assignment; ~ **de pedal de sostén** INFO sustain-pedal assignment; ~ **de rueda de modulación** INFO modulation-wheel assignment; ~ **de salida** INFO output assignment; ~ **de tareas paralelas** ELECTRÓN parallel tasking; ~ **de valor** INFO value assignment

asignado: **no** ~ *adj* INFO unassigned

asignador: ~ **de ruta-B** *m* DATOS B-router

asignar *vt* GEN allocate

asimetría *f* GEN *de distribución* skewness

asimétrico *adj* ELECTROTEC unbalanced, unsymmetrical, INFO asymmetric, TELEFON unsymmetrical

asimilación: ~ **progresiva** *f* GEN gradual phasing-in

asíncrono *adj* DATOS, INFO, TRANS asynchronous

asistencia: ~ **a diseño** *f* INFO design aid

asistente *m* TELEFON attendant

asistido *adj* INFO assisted

asistir *vt* GEN assist

ASK *abr* (*tecleado de cambio de amplitud*) SEÑAL ASK (*amplitude-shift keying*)

ASN *abr* (*número muestral promedio*) PRUEBA *calidad* ASN (*average sample number*)

ASO *abr* (*objeto del servicio de aplicación*) DATOS, TELEFON ASO (*application service object*)

asociación *f* GEN association; ~ **de**

aplicaciones DATOS application association; ~ **N** DATOS N association

Asociación: ~ **Española de Normalización y Certificación** *f* (*AENOR*) GEN Spanish Institute for Normalization and Certification (*AENOR*); ~ **Europea de Fabricantes de Ordenadores** *f* (*ECMA*) INFO European Computer Manufacturers Association (*ECMA*)

aspereza *f* GEN roughness

aspirador: ~ **de estaño** *m* ELECTRÓN surplus solder collector

astable *adj* ELECTRÓN *elementos lógicos* astable

astático *adj* GEN *física* astatic

astriónica *f* ELECTRÓN astrionics

atajo *m* INFO shortcut

ataque *m* INFO attack

atar *vt* GEN lash, lash down

atarjea *f* ELECTRÓN troughing

atasco *m* ELECTRÓN bottleneck

atender: ~ **a** *vt* TELEFON *necesidades del consumidor* cater for

atenuación *f* INFO, TEC RAD, TRANS attenuation, fading; ~ **de banda de paso** INFO, TEC RAD, TELEFON, TRANS pass-band attenuation; ~ **de bloqueo** TELEFON suppression loss; ~ **compuesta** TELEFON composite loss, TRANS composite loss, effective loss; ~ **debida a espacio abierto** TEC RAD, TRANS loss relative to free space; ~ **de eco** TELEFON echo loss; ~ **del eco del impulso** TELEFON pulse echo attenuation; ~ **del eco en el receptor** TELEFON listener echo loss; ~ **de equilibrado de eco** TELEFON echo balance return loss; ~ **equivalente de nitidez** TELEFON equivalent articulation loss; ~ **en el espacio libre** TEC RAD, TRANS free-space attenuation; ~ **por interacción** TRANS interaction loss; ~ **de intermodulación** TRANS *electrónica* intermodulation attenuation; ~ **local** TEC RAD site attenuation; ~ **logarítmica normal** TEC RAD *móvil* log-normal fading; ~ **de nivel de salida** INFO output-level attenuation; ~ **de palabra singular** DATOS unique-word loss; ~ **posamplificador** ELECTRÓN post-amplifier loss; ~ **progresiva de filtro** ELECTRÓN filter roll-off; ~ **por recodo** TRANS loss around a corner; ~ **de reflejos** TEC RAD clutter attenuation; ~ **por reflexión** TELEFON, TRANS return loss; ~ **de regularidad** TELEFON regularity loss; ~ **de retorno** TRANS backward attenuation; ~ **de retorno transversal** (*TRL*) TELEFON transverse return loss (*TRL*); ~ **de señal** SEÑAL signal attenuation; ~ **de sistema** GEN system loss; ~ **de a-t-b de transmisión** TELEFON, TRANS transmission loss of a-t-b; ~ **de tensión** ELECTROTEC voltage attenuation; ~ **transductiva** TELEFON transducer loss; ~ **de transmisión directa** TELEFON, TRANS through-transmission loss; ~ **de trayectoria** TEC RAD, TELEFON path attenuation

atenuador *m* (*STC*) CONTROL sensitivity time control (*STC*), INFO attenuator, fader, TRANS attenuator; ~ **absorbente** TRANS absorptive attenuator; ~ **de aleta** TEC RAD flap attenuator; ~ **calibrado** TRANS calibrated attenuator; ~ **de guíaondas** TEC RAD, TRANS waveguide attenuator; ~ **de guillotina** TEC RAD guillotine attenuator; ~ **de lámina** TEC RAD vane attenuator; ~ **lasérico** ELECTRÓN laser attenuator; ~ **de nivel** INFO level attenuator; ~ **de nivel de salida** INFO output-level attenuator; ~ **de pistón** TEC RAD piston attenuator; ~ **reactivo** TEC RAD reactive attenuator; ~ **resistivo** ELECTRÓN resistive attenuator; ~ **rotativo** TEC RAD rotary attenuator

atenuar *vt* INFO, TEC RAD, TRANS attenuate, fade

ATI *abr* (*total medio inspeccionado*) PRUEBA *calidad* ATI (*average total inspected*)

atmósfera: ~ **radioeléctrica normal** *f* TEC RAD standard radio atmosphere; ~ **de referencia para refracción** *f* TEC RAD reference atmosphere for refraction

atomicidad *f* DATOS atomicity; ~, **consistencia, aislamiento y durabilidad** (*ACID*) DATOS atomicity, consistency, isolation and durability (*ACID*)

átomo *m* INFO atom

atonal *adj* INFO atonal

atrapaondas *m* TEC RAD wave trap

atribución: ~ **de baliza respondedora** *f* TEC RAD transponder allotment; ~ **de espacio** *f* GEN space allocation; ~ **de ruido** *f* TRANS noise allocation

atributo *m* GEN attribute; ~ **de campo definido** (*DDA*) TELEFON defined-domain attribute (*DDA*); ~ **definido por campo** TELEFON domain-defined attribute; ~ **general** DATOS general attribute; ~ **nodal** INFO node attribute; ~ **unívoco** DATOS single-valued attribute; ~ **de valor predeterminado** DATOS set-valued attribute; ~ **de vídeo** TV video attribute

atributos: ~ **de archivo** *m pl* TELEFON file attributes; ~ **de fichero BFT** *m pl* TELEFON BFT file attributes; ~ **de sector** *m pl* INFO sector attributes

ATU *abr* (*unidad de sintonización de antena*) TEC RAD ATU (*antenna tuning unit*)

audibilidad *f* GEN audibility

audible *adj* GEN audible
audio *adj* GEN audio
Audio: **~-CD** *m* INFO CD-Audio
audioconferencia *f* TEC RAD audioconference
audiovisual *adj* TV audiovisual
auditoría: **~ de calidad** *f* PRUEBA quality audit
aullar *vi* TELEFON, TRANS howl
aullido *m* TELEFON, TRANS howling; **~, canto, silbido** TELEFON howling
aumentar *vt* GEN boost, enhance
aumento: **~ de capacidad** *m* DATOS capacity increase; **con ~ de potencia** *adj* ELECTRÓN powered-up; **~ de presión** *m* GEN pressure increase; **~ de tráfico** *m* TELEFON traffic growth
aural *adj* INFO aural
auricular *m* PERIF, TELEFON earphone, earpiece, receiver cap
auriculares *m pl* PERIF, TELEFON headphones, receiver headset
autentificación *f* DATOS, SEÑAL authentication; **~ convincente** DATOS strong authentication; **~ por entidades de par** DATOS peer-entity authentication; **~ de identidad de par** DATOS peer-identity authentication; **~ del origen de datos** DATOS data-origin authentication; **~ del origen de la información** DATOS data-origin authentication; **~ simple** DATOS simple authentication
autoadaptable *adj* INFO self-adapting
autoajustable *adj* INFO self-adjusting
autoalerta *f* DATOS auto-alert
autocontrol *adj* GEN self-checking
autocorrección *f* DATOS, INFO, SEÑAL, TELEFON auto-correction
autocorrelación *f* GEN auto-correlation
autoelevación *f* INFO bootstrap
automáticamente *adv* GEN automatically
automático *adj* GEN automatic
automatización *f* CONTROL automation; **~ finita** DATOS finite automation; **~ de oficina** GEN office automation
automatizar *vt* GEN automate
autonomizar *vt* CONTROL automize
autónomo *adj* ELECTRÓN, INFO self-contained, autonomous, TELEFON autonomous
autooscilante *adj* ELECTRÓN self-oscillating
autor *m* GEN author
autoretención: **de ~** *adj* GEN self-locking
autoridad: **~ de certificación** *f* DATOS certification authority; **~ denominadora** *f* DATOS naming authority; **~ de direccionamiento de la red** *f* DATOS network-addressing authority; **~ patrocinadora** *f* GEN sponsoring authority; **~ de registro internacional** *f* DATOS international registration authority
autoridades: **~ de la radiodifusión** *f pl* TEC RAD broadcasting authority
autorización *f* GEN authorization; **~ de conexión** TV operating authorization
autorizado *adj* GEN authorized, commissioned
autorizar *vt* GEN authorize, commission
autorreposición *f* INFO self-resetting
autosincrónico *adj* ELECTRÓN, ELECTROTEC selsyn
auxiliar[1] *adj* INFO ancillary, backup
auxiliar[2]: **~ personal digital** *m* INFO personal digital assistant; **~ para programación gráfica** *m* INFO graphic-programming aid
AVA *abr* (*afirmación de valor de atributo*) DATOS AVA (*attribute-value assertion*)
avalancha *f* ELECTRÓN avalanche
avance *m* COMPON feeding, DATOS preview, ELECTROTEC feeding, INFO feeding, preview; **~ de fase** TRANS phase advance; **~ de fase transmitida** TRANS transmitted-phase advance; **~ de hoja** TELEFON form feed
AVD *abr* (*circuito de voz y datos alternados*) TRANS AVD (*alternate voice-data circuit*)
avellanado *m* GEN countersinking
avellanar *vt* COMPON ream
avería *f* GEN damage, fault; **~ de máquina** INFO equipment failure
averiado *adj* COMPON broken, damaged, out-of-order
averiar *vt* GEN break, damage
avisador: **~ alfanumérico** *m* TEC RAD alphanumeric pager; **~ de bloqueo** *m* TELEFON blocking alarm; **~ de interceptación** *m* TELEFON intercept announcer
aviso *m* TELEFON message, notice, advice; **~ de asignaciones** DATOS assignment message; **~ de averías** ELECTROTEC fault reporting; **~ a los aviadores** TEC RAD notice to airmen; **~ de carga incremental** TELEFON incremental-charge advice; **~ de conexión** TELEFON hook flash; **~ de facturación** APLIC charge advice; **~ de servicio de retorno** TELEFON return-service advice; **~ de servicio tasado** TELEFON paid-service advice
AX: **~ doble** *m* COMPON, ELECTRÓN twin AX
axial *adj* DATOS, ELECTRÓN, INFO, TRANS axial
axioma *m* DATOS *LDS* axiom
ayuda: **~ a diagnóstico** *f* PRUEBA diagnostic aid; **~ especial** *f* INFO balloon help; **~ mnemotécnica** *f* TELEFON mnemonic; **~ supervisora** *f* CONTROL, PRUEBA supervisory aid
ayudante *m* GEN assistant

B

baja[1]: **de ~ frecuencia** *adj* TEC RAD low-frequency; **de ~ resistencia** *adj* ELECTROTEC low-resistance; **de ~ tensión** *adj* ELECTRÓN low-voltage

baja[2] *f* TEC RAD deregistration

bajada *f* TELEFON, TEC RAD down lead, lead-in; **~ de antena** TEC RAD, TELEFON antenna lead-in, lead-in

bajante *m* TEC RAD *de antena* down lead

bajar *vt* INFO move down

bajo[1] *adj* GEN low

bajo[2] *m* INFO bass

balance: **~ de potencia** *m* ELECTROTEC, TRANS *óptica de fibras* power budget

balanceado *adj* ELECTRÓN, ELECTROTEC, TEC RAD, TRANS balanced, equalized

balancear *vt* ELECTRÓN, ELECTROTEC, TEC RAD, TRANS balance, equalize

balanceo: **~ de antena** *m* TEC RAD antenna swing-off

baliza: **~ por impulsos** *f* TEC RAD pulsed beacon; **~ de radar** *f* TEC RAD radar beacon

balun *m* ELECTRÓN, ELECTROTEC balun

banco: **~ de contactos alineados** *m* TELEFON straight bank; **~ de contactos salteados** *m* TELEFON slipped bank; **~ de datos** *m* DATOS data bank; **~ de imágenes** *m* TV image bank; **~ de memoria** *m* INFO computer bank, memory bank; **~ de pruebas** *m* PRUEBA proving stand

banda[1]: **~ ancha** *f* DATOS, INFO, TELEFON, TRANS broad band; **~ base** *f* TEC RAD, TRANS, TV base band; **~ C** *f* TEC RAD C band; **~ ciudadana** *f* TEC RAD citizens' band (*CB*); **~ de conducción** *f* ELECTRÓN *semiconductores* conduction band; **~ crítica** *f* TELEFON critical band; **~ de energía** *f* COMPON *semiconductores* energy band; **~ estrecha** *f* TRANS narrow band; **~ de etiquetado de designación transparente** *f* ELECTRÓN transparent-designation labelling (*AmE* labeling) strip; **~ de expansión** *f* TEC RAD expansion band; **~ de frecuencia asignada** *f* TEC RAD assigned-frequency band; **~ de frecuencias** *f* TEC RAD, TRANS frequency band; **~ de frecuencias de sintonización** *f* ELECTRÓN frequency tuning range; **~ de frecuencias vocales** *f* TRANS speech band; **~ de frecuencia vocal** *f* TRANS voice-frequency band; **~ horaria de velocidad** *f* TELEFON rate timeband; **~ de incertidumbre de frecuencias** *f* TRANS frequency-uncertainty band; **~ lateral** *f* (*SB*) TEC RAD, TELEFON, TRANS sideband (*SB*); **~ lateral doble** *f* (*DSB*) TEC RAD, TELEFON, TRANS double sideband (*DSB*); **~ lateral independiente** *f* (*ISB*) TEC RAD, TRANS independent sideband (*ISB*); **~ lateral inferior** *f* TEC RAD, TRANS lower sideband; **~ lateral con portadora** *f* TEC RAD, TELEFON, TRANS sideband with carrier; **~ lateral residual** *f* (*RSB*), (*VSB*) TEC RAD, TRANS residual sideband (*RSB*), vestigial sideband (*VSB*), TV residual sideband (*RSB*); **~ lateral superior** *f* (*USB*) TEC RAD, TRANS upper sideband (*USB*); **~ lateral única** *f* TEC RAD, TRANS single sideband; **~ piloto** *f* DATOS format tape; **~ de protección** *f* TEC RAD guard band; **~ protectora** *f* GEN carrying strap; **~ radioemisora** *f* TEC RAD broadcasting band; **~ de rotulación de designación transparente** *f* COMPON transparent-designation labelling (*AmE* labeling) strip; **~ de ruido general efectivo** *f* TRANS effective overall noise band; **~ satélite** *f* TEC RAD satellite band; **~ térmica** *f* COMPON heat strip; **~ de tono incorporado** *f* TEC RAD tone in band; **~ de valencia** *f* COMPON valence band; **~ de videofrecuencia** *f* TRANS, TV videofrequency band; **~ de voz** *f* DATOS voice band

banda[2]: **en ~** *adj* SEÑAL, TRANS in-band

bandas *f pl* INFO bands

bandeja: **~ de absorción de impacto** *f* COMPON shock-absorbing tray; **~ de cable** *m* TRANS cable shelf; **~ de herramientas** *f* DATOS, INFO tool palette; **~ mezcladora** *f* ELECTRÓN mixing tray; **~ de papel** *f* PERIF paper tray

bandera: **~ de datos nuevos** *f* DATOS new data flag (*NDF*); **~ reglamentaria de segmentación** *f* DATOS segmentation permitted flag (*SPF*)

banderola: **~ de estado** *f* INFO status flag

BAPTA *abr* (*conjunto de marcación y transferencia de energía*) ELECTROTEC BAPTA (*bearing and power transfer assembly*)

baquelita *f* COMPON bakelite

Bark *m* INFO Bark

barniz: **~ de inmersión** *m* GEN *pinturas* dipping varnish

barnizar: ~ **con pistola** *vt* COMPON spray-finish

barra *f* COMPON, INFO, TELEFON bar; ~ **de clavijas** ELECTROTEC plug bar; ~ **colectora** ELECTROTEC, INFO, TELEFON bus; ~ **colectora de CC** ELECTROTEC DC bus; ~ **colectora de cobre** ELECTROTEC copper busbar; ~ **colectora para fuerza** ELECTROTEC power busbar; ~ **colectora principal** ELECTROTEC main busbar; ~ **de contactos** ELECTROTEC contact bar; ~ **de cortocircuito** ELECTRÓN shorting bar; ~ **de distribución** ELECTROTEC busbar, distribution busbar, INFO, TELEFON busbar; ~ **de distribución de CC** ELECTROTEC DC busbar; ~ **doble** GEN *símbolos* double solidus; ~ **espaciadora** COMPON spacing bar, INFO, PERIF space bar; ~ **hendida** INFO split bar; ~ **de herramientas** DATOS, INFO *Internet* toolbox; ~ **de impresión** COMPON, TELEFON print bar, writing bar, writing edge; ~ **de interconexión de celdas** ELECTROTEC cell-interconnection bar; ~ **de menú** INFO menu bar; ~ **de metalización** ELECTRÓN plating bar; ~ **de montaje** COMPON mounting bar; ~ **neutra** ELECTROTEC neutral bar; ~ **oblicua** GEN solidus; ~ **oblicua invertida** INFO inverse solidus; ~ **positiva** ELECTROTEC positive bar; ~ **de refuerzo** COMPON stiffening bar; ~ **de relés** COMPON relay bar, relay mounting bar; ~ **de selección de nivel** CONMUT level-selecting bar; ~ **selectora** COMPON, CONMUT selecting bar; ~ **de soporte** COMPON supporting bar; ~ **de suelo** COMPON floor bar, floor rail; ~ **taladradora** COMPON boring bar; ~ **tensora** COMPON stay bar; ~ **de tipos** TELEFON type bar; ~ **de título** INFO title bar

barrado *adj* TEC RAD, TELEFON barred

barrena *f* TEC RAD roll-off

barrer *vt* ELECTRÓN, TEC RAD sweep

barrera: ~ **de potencial** *f* COMPON potential barrier

barrido: ~ **de frecuencia** *m* GEN glide; ~ **intermedio** *m* TEC RAD interscan; ~ **de izquierda a derecha** *m* TV horizontal sweep; ~ **lineal** *m* TV line sweep; ~ **de puntos entrelazados** *m* TV dot-interlace sweep; ~ **vertical** *m* INFO vertical scan, TV field sweep

báscula: ~ **biestable con una entrada** *f* ELECTRÓN single-control bistable trigger circuit; ~ **monoestable** *f* COMPON, ELECTRÓN single-shot flip-flop

basculador: ~ **de desviación** *m* CONMUT branch flip-flop; ~ **gatillado** *m* COMPON, ELECTRÓN triggered flip-flop; ~ **orientado** *m* COMPON, ELECTRÓN steered flip-flop; ~ **transistorizado** *m* COMPON, ELECTRÓN transistor flip-flop

bascular *vt* CONMUT, ELECTRÓN, ELECTROTEC, INFO *interruptor* toggle

base *f* COMPON, ELECTRÓN *de tubo electrónico* base, INFO root; ~ **de conocimiento** APLIC, INFO knowledge base; ~ **de datos** DATOS, INFO database; ~ **de datos de abonados y análisis** TELEFON subscriber and analysis database; ~ **de datos de gestión** DATOS management database; ~ **de datos gráficos** INFO graphics database; ~ **de datos relacional** DATOS relational database (*RDB*); ~ **de datos semántica** DATOS semantic database; ~ **de datos de superficie y terreno** TEC RAD terrain and surface database; ~ **de datos temporal** DATOS temporal database; ~ **factual** ELECTRÓN factual basis; ~ **de hechos** INFO facts base; ~ **de impulsos** TELEFON pulse base; ~ **de información para dirección de empresas** DATOS management information base (*MIB*); ~ **de información para la gestión de la seguridad** DATOS security management information base (*SMIB*); ~ **de montaje** COMPON mounting base; ~ **de plantilla de cable** TRANS formboard base; ~ **de tiempo de línea** INFO line output; ~ **de tiempos gatillada** ELECTRÓN triggered time-base; ~ **de transistor** COMPON, ELECTRÓN transistor base; ~ **de tubo electrónico** COMPON, ELECTRÓN electronic tube base; ~ **de válvula** ELECTRÓN valve base; ◆ **sobre una ~ secundaria** TEC RAD on a secondary basis

bastidor: ~ **de acumulador** *m* ELECTROTEC battery rack; ~ **alámbrico** *m* TELEFON wired rack; ~ **de ampliación** *m* TELEFON *técnicas de montaje* extension rack; ~ **para aparatos diversos** *m* (*MAR*) TELEFON miscellaneous apparatus rack (*MAR*); ~ **auxiliar** *m* ELECTRÓN auxiliary equipment rack; ~ **auxiliar múltiple** *m* TELEFON submultiframe (*SMF*); ~ **de cables** *m* TRANS cable rack; ~ **de conmutadores** *m* CONMUT, TELEFON switch rack; ~ **de contadores** *m* TELEFON meter rack; ~ **de continuación** *m* COMPON continuation rack; ~ **de convertidor** *m* ELECTROTEC converter rack; ~ **de distribución** *m* ELECTROTEC distribution rack; ~ **de dos caras** *m* TELEFON double-sided rack; ~ **de equipo básico** *m* COMPON basic rack, *sistema de montaje* basic-equipment rack; ~ **de equipo de fuerza** *m* ELECTROTEC power rack; ~ **estrecho** *m* TELEFON slim rack; ~ **inalámbrico** *m* COMPON, TELEFON unwired rack; ~ **independiente** *m* TELEFON stand-alone rack; ~ **de placas** *m* COMPON *técnicas de montaje* board frame; ~ **de selectores** *m* CONMUT switch rack; ~ **unilateral** *m* TELEFON single-sided rack

batería *f* ELECTROTEC battery; ~ **anódica** ELECTROTEC anode battery; ~ **de carga equilibrada** ELECTROTEC floating battery; ~ **de carga plena** ELECTROTEC fully-charged battery; ~ **de célula seca** ELECTROTEC dry-cell battery; ~ **central** TELEFON central battery; ~ **de central** ELECTROTEC *para suministro de energía* exchange battery; ~ **de comprobación** INFO check battery; ~ **de conversación** TEC RAD, TELEFON talk battery; ~ **de filtros** INFO filterbank; ~ **de filtros para el análisis** INFO analysis filterbank; ~ **de gel** ELECTROTEC gel battery; ~ **de larga duración** ELECTROTEC long-life battery; ~ **local** TELEFON local battery; ~ **negativa** ELECTROTEC negative battery; ~ **de placa** ELECTROTEC plate battery; ~ **positiva** ELECTROTEC positive battery; ~ **de pruebas abstractas estandarizadas de referencia** DATOS, PRUEBA reference standardized abstract test-suite; ~ **de pruebas abstractas seleccionadas** (*SATS*) DATOS, PRUEBA selected abstract test-suite (*SATS*); ~ **de pruebas ejecutables seleccionadas** (*SETS*) DATOS, PRUEBA selected executable test-suite (*SETS*); ~ **de refuerzo** ELECTROTEC booster battery; ~ **de reserva** ELECTROTEC backup battery; ~ **solar** ELECTROTEC solar battery; ~ **tampón** ELECTROTEC buffer battery

batidos: ~ **por minuto** *m pl* INFO beats per minute

bayoneta *f* COMPON, ELECTRÓN, ELECTROTEC bayonet

BB *abr* (*boletín electrónico*) DATOS, INFO, TELEFON BB (*bulletin board*)

BBS *abr* (*sistema de boletíns electrónicos*) DATOS, INFO, TELEFON BBS (*bulletin board system*)

BCS *abr* (*sistema de cableado BULL*) DATOS BCS (*BULL cabling system*)

BFT *abr* (*transferencia de archivos binarios*) TELEFON BFT (*binary file transfer*)

biblioteca *f* GEN library; ~ **de central** GEN exchange library; ~ **de datos** DATOS library of data; ~ **implícita** DATOS default library; ~ **de instalación** GEN installation library; ~ **de programas** INFO program library; ~ **de señales** CONMUT signal library

bidireccional *adj* INFO, TEC RAD, TELEFON, TRANS, TV bidirectional

bidireccionalidad *f* INFO, TEC RAD, TELEFON, TRANS, TV two-way directionality

bien: ~ **recibido** *adj* DATOS successfully received

biestabilidad *f* GEN bistability

bifurcación *f* INFO jump, TRANS branch-off

bilateral *adj* GEN bilateral

bilineal *adj* PERIF *escáner* bilinear

bimetálico *adj* COMPON bimetal

binarios: ~ **cifrados** *m pl* (*SB1*) DATOS scrambled binary ones (*SB1*); ~ **descifrados** *m pl* (*USB1*) DATOS unscrambled binary ones (*USB1*)

BIOS *abr* (*sistema básico de entrada/salida*) PRUEBA BIOS (*basic input/output system*)

BIP *abr* (*paridad intercalada de bitio*) DATOS BIP (*bit interleaved parity*)

BIP-X *abr* (*paridad X intercalada de bitio*) DATOS BIP-X (*bit interleaved parity X*)

birrefringencia *f* DATOS, TRANS birefringence

BIS *abr* (*servicio independiente de portador*) TELEFON BIS (*bearer independent service*)

bisagra *f* ELECTRÓN butt

biselado *m* COMPON bevelling (*AmE* beveling)

biselar *vt* COMPON bevel

bit: ~ **indicador** *m* DATOS flag bit; ~ **menos significativo** *m* (*BMS*) DATOS least significant bit (*LSB*); ~ **de paridad** *m* DATOS, TEC RAD parity bit; ~ **de rastreo** *m* CONMUT tracer bit; ~ **de rebose** *m* DATOS, TELEFON overhead bit; ~ **de signo** *m* INFO sign bit

BIT *abr* (*prueba básica de interconexión*) DATOS BIT (*basic interconnection test*)

BITE *abr* (*acontecimiento de telefonía entrelazada hacia atrás*) TELEFON BITE (*backward-interworking telephony event*)

bitímbrico *adj* INFO bi-timbral

bitio: ~ **de banderola** *m* DATOS tag bit; ~ **de comando/respuesta** *m* (*C/R*) TELEFON command/response bit (*C/R*); ~ **delimitador** *m* TELEFON framing bit; ~ **erróneo** *m* INFO erroneous bit; ~ **final** *m* TELEFON final bit; ~ **de indicación de prioridad** *m* DATOS priority-indication bit; ~ **indicador inverso** *m* INFO backward-indicator bit; ~ **de información aplicado** *m* TELEFON applied data bit; ~ **de señalización** *m* SEÑAL, TRANS signalling (*AmE* signaling) bit; ~ **de servicio** *m* DATOS housekeeping bit; ~ **de sincronización** *m* TELEFON synchronization bit

bitios: ~ **por color** *m pl* PERIF *de escáner* bits per colour (*AmE* color); ~ **por pulgada** *m pl* (*bpp*) INFO bits per inch (*bpi*); ~ **por segundo** *m pl* (*bps*) DATOS bits per second (*bps*); ~ **de servicio** *m pl* INFO service bits; ~ **transmitidos anteriormente** *m pl* TRANS earlier-transmitted bits

Bitnet *m* DATOS Bitnet

bivocal *m* TELEFON speech plus duplex

blanco *m* ELECTRÓN, TEC RAD target; ~ **nominal**

TELEFON nominal white; ~ **pasivo** TEC RAD skin target

blindado: **no ~** *adj* ELECTRÓN nonarmoured (*AmE* nonarmored)

blindaje *m* COMPON shielding, TRANS *de cables* armour (*AmE* armor), armouring (*AmE* armoring) ; ~ **de acero** COMPON steel armour (*AmE* armor); ~ **de aluminio-poliéster de Beldfoil** ELECTRÓN Beldfoil aluminium-polyester (*AmE* aluminum-polyester) shield; ~ **de fleje de acero** COMPON steel tape armour (*AmE* armor); ~ **de trenza de cobre estañado** COMPON, ELECTRÓN tinned copper-braid shield

blocaje: ~ **de zona de Fresnel** *m* TRANS Fresnel zone blockage

bloque *m* GEN block, INFO *de información* chunk; ~ **de circuitos troncales** CONMUT junctor block; ~ **de código** DATOS code block; ~ **de conducto múltiple** COMPON, TRANS multiple-duct block; ~ **conector de cubo** COMPON socket-connector block; ~ **de conexión** GEN connection unit; ~ **de conocimiento** INFO knowledge chunk; ~ **de contactos de espigas** COMPON, ELECTRÓN pin contact unit; ~ **de contactos de horquilla** ELECTRÓN fork contact unit; ~ **de contactos de manguito** ELECTROTEC sleeve contact unit; ~ **de contactos de patillas planas** COMPON, ELECTRÓN blade contact unit; ~ **de coordenadas** TELEFON block of parameters; ~ **de datos** DATOS data block; ~ **de datos retenidos** DATOS retained-data block; ~ **descriptor de orden** (*CDB*) DATOS command-descriptor block (*CDB*); ~ **de distribución** ELECTRÓN *dispositivos de conexión* distribution block, ELECTROTEC *suministro de energía* distribution terminal; ~ **emisor** ELECTRÓN source block; ~ **de energía** TEC RAD power cube; ~ **de enlace ascendente** TEC RAD uplink block; ~ **erróneo** (*EB*) DATOS erroneous block (*EB*); ~ **de frenado** COMPON brake block; ~ **funcional** TELEFON *en SDL* functional block; ~ **de funciones** CONMUT function block; ~ **de impresión** PERIF printing block; ~ **de memoria** INFO storage block; ~ **de memoria de datos** CONMUT *código* data store block; ~ **de memoria de programa** CONMUT, INFO program-store block; ~ **de mensajes CRC** (*CMB*) TELEFON CRC message block (*CMB*); ~ **paramétrico** TELEFON parameter block; ~ **de placa** PERIF plate block; ~ **de prueba de ruta** TELEFON route-test block; ~ **de puenteado** COMPON, ELECTROTEC strapping block; ~ **de resonador** GEN resonator block; ~ **de un solo conducto** COMPON, TRANS single-duct block; ~ **superior** DATOS frame overhead; ~ **de terminales** DATOS, ELECTRÓN tag block; ~ **de terminales de tornillo** ELECTROTEC screw terminal block; ~ **de texto** DATOS word block, TELEFON text block

bloqueado *adj* TELEFON busied out

bloqueo *m* GEN blocking, *de oscilador* lock; ~ **a distancia** CONTROL remote blocking; ~ **de elementos de programación** INFO software lockout; ~ **de enlace de señalización** SEÑAL signalling-link (*AmE* signaling-link) blocking; ~ **de línea** TELEFON line lockout; ~ **de llamada interurbana** TELEFON toll-call barring; ~ **de micrófono** PERIF *de contestador* microphone cutoff; ~ **de pantalla** INFO screen locking; ~ **de retorno** TELEFON return block; ~ **de ruta** TELEFON route blocking

bloques: ~ **adicionables optativos** *m pl* DATOS optionally-droppable blocks; ~ **no separables** *m pl* DATOS nondroppable blocks

BLU *abr* (*banda lateral única*) TEC RAD, TELEFON, TRANS SSB (*single sideband*)

BMS *abr* (*bit menos significativo*) DATOS LSB (*least significant bit*)

BNF *abr* (*impreso Backus-Naur*) DATOS BNF (*Backus-Naur form*)

bobina *f* ELECTROTEC *de alambre* coil; ~ **de autoinducción** ELECTROTEC choke; ~ **de barra selectora** COMPON, CONMUT selecting-bar coil; ~ **de bloqueo** ELECTROTEC choking coil; ~ **calefactora** ELECTROTEC heat coil; ~ **de carga** TRANS loading coil; ~ **de carga de antena** TEC RAD antenna-loading coil; ~ **correctora** ELECTROTEC peaking coil; ~ **de drenaje** ELECTROTEC drainage coil; ~ **de electroimán** ELECTRÓN magnet coil; ~ **de enfoque** ELECTRÓN focusing coil; ~ **exploradora** TEC RAD search coil; ~ **híbrida** ELECTROTEC hybrid coil; ~ **de inducción** ELECTRÓN induction coil; ~ **de inductancia** ELECTROTEC inductance coil; ~ **de núcleo de hierro** ELECTRÓN *electromagnetismo* iron-core coil; ~ **de relé** COMPON, ELECTROTEC relay coil; ~ **de repetición** TELEFON repeat coil; ~ **repetidora** TELEFON repeating coil; ~ **de retención** ELECTROTEC holding coil; ~ **de yugo** COMPON, ELECTRÓN yoke coil; ~ **de yugo de desviación** COMPON, ELECTRÓN deflection yoke

bobinado: ~ **auxiliar** *m* ELECTROTEC auxiliary winding; ~ **de relé** *m* COMPON relay winding

bobinas: ~ **de campo** *f pl* TEC RAD field coils

boca *f* PERIF *de conducto* mouthpiece

boceto *m* GEN outline

bocina *f* TEC RAD horn; ~ **con corrección de**

fase TEC RAD phase-corrected horn; ~ **radiante** TEC RAD radiating horn
bodega: ~ **de almacenamiento de cables** *f* TRANS cable-storage hold; ~ **para cable** *f* TRANS cable tank
boleador *m* INFO trackball
boletín *m* INFO newsletter, TELEFON bulletin; ~ **electrónico** (*BB*) DATOS, INFO, TELEFON bulletin board (*BB*)
bolsillo: **de** ~ *adj* INFO pocket, TEC RAD pocket, pocket-sized
bomba *f* ELECTROTEC pump
bombear *vt* ELECTROTEC pump
booleano *adj* DATOS, INFO Boolean
boquilla *f* TEC RAD nozzle; ~ **disparadora** ELECTRÓN trigger nozzle; ~ **de embudo** TELEFON funnel mouthpiece
borde[1]: ~ **exterior** *m* INFO outer rim; ~ **de ficha** *m* ELECTRÓN *circuitos impresos* card edge; ~ **frontal de impulso** *m* ELECTRÓN pulse leading edge; ~ **del haz** *m* TEC RAD beam edge; ~ **posterior** *m* GEN rear edge; ~ **de trazo** *m* INFO stroke edge
borde[2]: **en el** ~ **anterior** *fra* GEN at the leading edge
bordo: **a** ~ *adj* TEC RAD in-flight
borne: ~ **de conexión** *m* ELECTRÓN connector terminal; ~ **neutro** *m* ELECTROTEC neutral terminal; ~ **positivo** *m* ELECTROTEC positive terminal; ~ **de salida** *f* ELECTROTEC output terminal
borrado *m* INFO, PERIF, TEC RAD erasing, erasure, deleting, TELEFON blanking; ~ **de memoria** INFO memory cleaning, memory erasure; ~ **de mensajes** PERIF *contestador* erasure of messages; ~ **selectivo** DATOS selective erasing
borrador: ~ **a escala** *m* GEN *dibujo técnico* draft to scale; ~ **de esquema** *m* GEN *dibujo técnico* draft sketch; ~ **de norma internacional** *m* GEN draft international standard
borrar *vt* INFO, PERIF, TEC RAD erase, wipe out, TELEFON blank out
borrosidad *f* ELECTRÓN smearing, TV *digitalizador* fuzziness
borroso *adj* INFO blurred
botón *m* GEN button; ~ **de ampliación** INFO expansion button; ~ **de anotador** INFO annotation button; ~ **de bloqueo** TELEFON blocking button; ~ **de códigos** TELEFON recall button; ~ **de comunicación secreta** DATOS, PERIF, TELEFON privacy button, secrecy button; ~ **por defecto** INFO default button; ~ **de demostración** INFO demo button; ~ **ejecutar** INFO execute button; ~ **de enlace** INFO tie button; ~ **de llamada** PERIF, TELEFON push button; ~ **de mando** GEN control knob; ~ **de navegación** INFO *en Internet* navigation button; ~ **de puesta a tierra** CONMUT *PBX* earthing button *BrE*, grounding button *AmE*; ~ **pulsador** INFO, PERIF *de contestador* push button; ~ **de reposición** GEN restoring button; ~ **salir** INFO exit button; ~ **de tierra** CONMUT earthing button *BrE*, grounding button *AmE*
bovedilla *f* ELECTRÓN overhang
boya: ~ **reflectora de radar** *f* TEC RAD radar reflector buoy
bpp *abr* (*bitios por pulgada*) INFO bpi (*bits per inch*)
bps *abr* (*bitios por segundo*) DATOS bps (*bits per second*)
brazo: ~ **de contacto deslizante** *m* CONMUT wiper arm; ~ **de reactancia** *m* TEC RAD nondissipative stub
brea *f* DATOS tar
brida *f* COMPON fishplate; ~ **ancha** ELECTRÓN wide flange; ~ **estrecha** ELECTRÓN narrow flange; ~ **plana** TEC RAD plain connector
brillo *m* TV glare
broca: ~ **helicoidal** *f* COMPON twist drill
bronce *m* COMPON, ELECTRÓN brass; ~ **de aluminio** COMPON aluminium (*AmE* aluminum) bronze; ~ **niquelado** COMPON nickel brass
bruñidor *m* GEN polishing tool
BSI *abr* (*Instituto Británico de Normas*) GEN BSI (*British Standards Institution*)
BT *abr* (*clase de transferencia de documentos en masa*) TELEFON BT (*document bulk transfer class*)
BTM *abr* (*clase de transferencia y manipulación de documentos en masa*) TELEFON BTM (*document bulk transfer and manipulation class*)
B3ZS *abr* (*código bipolar con sustitución de tres ceros*) DATOS, SEÑAL, TELEFON B3ZS (*bipolar code with three-zero substitution*)
bucle *m* GEN loop; ~ **de acceso rápido** DATOS rapid-access loop; ~ **de alimentación** ELECTROTEC feeding loop; ~ **de atenuación de tensiones** ELECTRÓN stress-relief loop; ~ **básico** INFO basic loop; ~ **de bloqueo de demora** ELECTRÓN delay-lock loop; ~ **cerrado de realimentación** ELECTRÓN feedback closed loop; ~ **de cinta** ELECTRÓN belt loop; ~ **de comprobación** TELEFON check loop; ~ **de control** CONTROL control loop; ~ **de control de errores** TELEFON error-control loop; ~ **de Costas** TELEFON Costas loop; ~ **enganchado** ELECTRÓN *oscilador* locked loop; ~ **de evento principal** INFO main event loop; ~ **fibroóptico** TRANS fibre-optic (*AmE* fiber-optic) loop; ~ **de**

histéresis ELECTRÓN hysteresis cycle; ~ **de indicación multipunto** DATOS multipoint indication loop; ~ **local** TEC RAD, TELEFON local loop; ~ **de medición** PRUEBA measuring loop; ~ **de minicalibre** TELEFON minigauge (*AmE* minigage) loop; ~ **de patrón único** TELEFON unigauge (*AmE* unigage) loop; ~ **rotatorio** DATOS revolver loop; ~ **sincronizado por fase óptica** ELECTRÓN phase-locked loop; ◆ **en** ~ INFO looped

búfer: ~ **de edición para rellamada** *m* INFO recall-edit buffer; ~ **de recepción** *m* INFO receive buffer

buque: ~ **cablero** *m* TRANS cable boat; ~ **para tendido de cables submarinos** *m* TRANS cable-laying ship

bus: ~ **bifilar** *m* TV two-wire bus; ~ **de control de energía** *m* ELECTROTEC power control bus; ~ **espaciador** *m* DATOS space bus; ~ **de estación multiprocesadora** *m* TELEFON multiprocessor station bus; ~ **local** *m* INFO local bus; ~ **de memoria de programa** *m* CONMUT, INFO program store bus; ~ **pasivo** *m* GEN passive bus; ~ **principal** *m* ELECTROTEC main bus; ~ **de prueba de potencia** *m* ELECTROTEC power-test bus

busca: ~ **de personas por tono más voz** *f* TEC RAD tone-plus-voice paging

buscador *m* DATOS browser; ~ **de emisor de códigos** TELEFON code-sender finder; ~ **de llamada** CONMUT line finder; ~ **registrador** CONMUT register finder

buscapersonas: ~ **de agilidad de frecuencias** *m* TEC RAD frequency agility pager; ~ **con mensajes** *m* GEN message pager; ~ **metropolitano** *m* TEC RAD city paging; ~ **con pantalla** *m* TEC RAD display paging; ~ **de valor añadido** *m* TEC RAD value-added paging; ~ **con visualización de mensaje** *m* TEC RAD message display pager

buscar/reemplazar *vt* DATOS find/replace

búsqueda *f* ELECTROTEC hunting, INFO browsing, search, TEC RAD search; ~ **de archivo** DATOS file search; ~ **por contexto** INFO context search; ~ **dicotómica** INFO binary search, dichotomizing search; ~ **encadenada** INFO chaining search; ~ **de línea** TELEFON line hunting; ~ **de nota** INFO note search; ~ **por PBX** TELEFON PBX hunting; ~ **de personas por altavoz** TEC RAD loudspeaker paging, voice paging; ~ **y rescate** (*SRR*) TEC RAD search and replace (*SRR*); ~ **por RR** TELEFON RR hunting; ~ **textual** DATOS text search

buzón *m* DATOS mailbox; ~ **para llamadas de emergencia** TELEFON emergency call box; ~ **telefónico** APLIC, PERIF, TELEFON voice mailbox; ~ **para textos** APLIC text mailbox

byte *m* DATOS, ELECTRÓN, INFO, TELEFON byte

C

C³ *abr* (*cubo C*) INFO C³ (*C-cube*)
CA *abr* (*corriente alterna*) GEN AC (*alternating current*)
caballón *m* TEC RAD border
cabecera: **~ de árbol** *f* DATOS, INFO tree header; **~ de intercambio** *f* DATOS interchange header; **~ de mensaje** *f* DATOS message header
cabellera: **~ de cometa** *f* INFO comet tail
cabeza *f* INFO *unidad de disco* head; **~ de cable** TRANS cable-distribution head, cable sleeve; **~ de cola** CONMUT head of the queue; **~ de lectura de FM** INFO FM stack; **~ de limpieza** TELEFON cleaning head; **con ~ de plástico** *adj* ELECTRÓN plastic-tipped; **~ de red** GEN network head; **~ de registro** GEN recording head; **~ reproductora** INFO playback head; **~ de RF** TEC RAD RF head; **~ de sensor solar** TEC RAD sun sensor head
cabezal: **~ de tubos** *m* (*HDR*) DATOS, ELECTRÓN, INFO *caldera acuotubular* header (*HDR*)
cabina: **~ telefónica pública** *f* TELEFON public call box, public telephone booth
cable *m* GEN cable, *fase* lead; **~ de aceite fluido** GEN oil-filled cable; **~ de acoplamiento** COMPON jumper wire; **~ aéreo** ELECTROTEC overhead cable, TRANS aerial cable, overhead cable; **~ aislado** COMPON, ELECTRÓN, ELECTROTEC insulated cable; **~ con aislamiento de papel impregnado** TRANS paper-insulated mass-impregnated cable; **~ de alimentación** ELECTROTEC power cord; **~ alimentador** DATOS, ELECTRÓN, ELECTROTEC, INFO power cable, TRANS feeding cable; **~ apantallado** COMPON, TELEFON screened cable; **~ de batería** ELECTROTEC battery cable; **~ bifilar** COMPON, ELECTRÓN, TRANS two-conductor cable; **~ bipolar** COMPON, ELECTRÓN twin cable; **~ blindado** COMPON shielded cable, shielded cord, ELECTROTEC shielded cable; **~ de bus** COMPON *técnicas de equipamiento* bus cable; **~ de calefacción** COMPON, ELECTROTEC heating cable; **~ de campo radial** GEN radial-field cable; **~ para canalizaciones** GEN duct cable; **~ de circuitos de cuatro parejas** TRANS quad-pair cable; **~ coaxial** TRANS coax; **~ coaxial delgado** COMPON, TRANS thin coaxial cable; **~ coaxial grueso** COMPON, TRANS thick coaxial cable; **~ de cobre** TRANS copper cable; **~ conductor con dos blindajes aislados** COMPON triax cable; **~ de conductor dúplex** COMPON, ELECTRÓN twin conductor cable; **~ de conductores metálicos** TRANS metallic-conductor cable; **~ de conductor metálico** TRANS metal conductor cable; **~ de conectador de línea** INFO line connector cord; **~ de conexión** COMPON connecting cable, connection cable, ELECTRÓN, TRANS patch cord; **~ de conexión de filtro** ELECTRÓN crossover patch cable; **~ de construcción para grandes amperajes** (*HDBC*) ELECTRÓN heavy-duty building cable (*HDBC*); **~ de construcción con guía de luz** COMPON lightguide building cable; **~ cuádruple de pares** TRANS quad cable; **~ con cubierta de plomo** ELECTROTEC lead-sheathed cable; **~ de desconexión** ELECTRÓN breakout cable; **~ deshilachado** ELECTRÓN frayed wiring; **~ de distribución** ELECTROTEC *suministro de energía* distribution cable; **~ de distribución local** TRANS local distribution cable; **~ de distribución nominal** ELECTRÓN plenum-rated cable; **~ eléctrico** INFO electric cord; **~ enchufable** ELECTRÓN plug-in cable; **~ encintado** TRANS *cables* belted cable; **~ enterrado** TRANS buried cable; **~ de fibra óptica ranurado** TRANS slotted-core optical fibre (*AmE* fiber) cable; **~ krarupizado** COMPON, TRANS krarup cable; **~ de mando** CONTROL, TRANS control cable; **~ de material no absorbente** TRANS paper-insulated non-draining cable; **~ de mezclador** COMPON jumbler cable; **~ monoconductor** COMPON, ELECTRÓN single-conductor cable; **~ multifibro** TRANS multifibre (*AmE* multifiber) cable; **~ múltiple** CONMUT multiple cable; **~ de múltiples conductores** TRANS multiconductor cable; **~ múltiple vertical** CONMUT vertical multiple cable; **~ de multiplicación de salidas** ELECTRÓN fan-out cable; **~ no cargado** TELEFON nonloaded cable; **~ de pares** GEN paired cable; **~ de pares combinados** TRANS multiple twin cable; **~ de pérdidas de acoplamiento de mejora** TRANS grading-coupling loss cable; **~ portador** TRANS messenger cable, messenger wire; **~ presurizado** GEN pressurized cable; **~ princi-**

pal fibroóptico ELECTRÓN fibre-optic (*AmE* fiber-optic) backbone; ~ **de protección** TRANS protection cable; ~ **con protección contra la intemperie** COMPON weather-sealed cable; ~ **R** TELEFON R-wire; ~ **ranurado** TRANS slotted-core cable; ~ **de red** TRANS network cable; ~ **relleno** GEN filled cable; ~ **relleno de gelatina** COMPON jelly-filled cable; ~ **de retenida** COMPON stay wire, TEC RAD guy wire; ~ **revestido compacto** TRANS tight-jacketed cable; ~ **de señalización** SEÑAL signalling (*AmE* signaling) cable; ~ **de una sola fibra** COMPON single-fibre (*AmE* single-fiber) cable; ~ **de soldador** COMPON splicer cable; ~ **submarino** TRANS submarine cable; ~ **subterráneo** COMPON, ELECTRÓN underground cable; ~ **superconductor** TRANS superconductor cable; ~ **de suspensión** COMPON suspension strand; ~ **de suspensión central** TRANS central suspension cable; ~ **sustentador** TEC RAD support cable, TRANS bearer cable; ~ **de telecomunicaciones** COMPON telecommunications cable; ~ **terminal** ELECTROTEC stub cable; ~ **termosensor** COMPON heat-trace cable; ~ **terrestre** TRANS land cable; ~ **tetrafilar** ELECTRÓN four-wire cable; ~ **de tiras** TRANS ribbon cable; ~ **trenzado** COMPON, ELECTRÓN, TRANS braided cable; ~ **triaxial** COMPON triaxial cable; ~ **tubular** GEN pipe cable; ~ **unipolar** COMPON single-core cable; ◆ **sin** ~ GEN cordless

cableado *m* COMPON cabling, stranding, *conexión* wiring, ELECTRÓN stranding, ELECTROTEC *conexión* wiring, TRANS cable set; ~ **aéreo** ELECTROTEC, TRANS overhead cabling; ~ **de bastidor** TELEFON rack cabling; ~ **de cobre** ELECTRÓN copper cabling; ~ **de espiga a espiga** COMPON, ELECTRÓN pin-to-pin wiring; ~ **impreso** ELECTRÓN *por la técnica del circuito impreso* printed wiring; ~ **preformado** ELECTRÓN wiring harness; ~ **de punto a punto** COMPON point-to-point wiring; ~ **uniforme** COMPON, TEC RAD, TELEFON uniform wiring

cableaje *m* COMPON, TRANS cabling

cablear *vt* GEN wire

cables: ~ **preensamblados** *m pl* TRANS preassembled cables; ~ **de reparación** *m pl* TRANS repairing cables

CAD *abr* (*diseño asistido por ordenador*) INFO CAD (*computer-aided design*)

CAD/CAM *abr* (*diseño y fabricación asistidos por ordenador*) INFO CAD/CAM (*computer-aided design and computer-aided manufacturing*)

cadena *f* CONMUT chain, DATOS network, string, INFO, TELEFON network; ~ **de aviso de servicio** DATOS *transmisión de caracteres* service string advice; ~ **elemental** INFO elementary stream; ~ **de filtro múltiple** TRANS multiple-filter chain; ~ **fuente** INFO source stream; ~ **internacional** TELEFON international chain; ~ **de llamadas** TELEFON call string; ~ **de llamada de tres direcciones** TELEFON three-way call chaining; ~ **de mando de vuelo** TEC RAD attitude control system; ~ **nula** DATOS null string; ~ **de referencia** APLIC, PERIF key stream; ~ **regional en A** TELEFON A-leg; ~ **de relés** COMPON relay chain; ~ **de selección** TELEFON selection allotter; ~ **tetrafilar** TELEFON four-wire chain; ~ **de texto** TELEFON text string

cadencia *f* TEC RAD cadence, TELEFON repetition frequency; ~ **de impulsos** ELECTRÓN, SEÑAL, TELEFON, TRANS pulse recurrence frequency

CAE *abr* (*ingeniería asistida por ordenador*) INFO CAE (*computer-aided engineering*)

caída: ~ **de presión** *f* GEN pressure drop; ~ **de tensión** *f* ELECTRÓN, ELECTROTEC voltage drop; ~ **de tensión en un tubo** *f* ELECTRÓN tube voltage drop

caja *f* GEN cash-box; ~ **acústica de montaje mural** TELEFON wall cabinet; ~ **alta** GEN upper case; ~ **de auricular** PERIF, TELEFON receiver case; ~ **blanda** GEN soft case; ~ **de bobinas de carga** TRANS loading-coil case, loading-coil pot; ~ **de bornes** ELECTROTEC terminal box; ~ **de cable de un solo par** COMPON single-pair cable box; ~ **de cojinete** COMPON bearing housing; ~ **de conexión** TELEFON terminal box; ~ **de conmutación** CONMUT switch unit; ~ **de derivación** COMPON, ELECTRÓN, ELECTROTEC, TRANS junction box; ~ **de desconexión** ELECTRÓN breakout box; ~ **de discos** INFO disk box; ~ **de distribución** ELECTROTEC distribution box, TELEFON connecting box; ~ **de ecos** TEC RAD echo box; ~ **de empalme** GEN cable box; ~ **de empalme terminal** COMPON termination joint box; ~ **para fichas** TELEFON card holder; ~ **de fusibles** ELECTROTEC fuse box; ~ **de la parte superior del aparato** TV *equipo conector de discriminación* set-top box; ~ **de pinturas** DATOS palette; ~ **de protección** COMPON protective case; ~ **de resistencias de décadas** ELECTROTEC decade resistance box; ~ **de selección** INFO selection box; ~ **terminal mural** ELECTRÓN, ELECTROTEC, INFO, TELEFON wall terminal box

cajetilla: ~ **de devolución de moneda** *f* PERIF *de teléfono público* return flap

cajetín: ~ **de devolución de moneda** *m* PERIF *de teléfono público* refund chute, coin-return cup

CAL *abr* (*aprendizaje auxiliada por ordenador*) INFO CAL (*computer-assisted learning*)

calafatear *vt* GEN *hierro* caulk

calafateo *m* GEN *de hierro* caulking; ~ **de RFI** TEC RAD RFI caulking

calcinación *f* ELECTRÓN, TELEFON *cocción, tostión* burning

calcular *vt* GEN calculate

cálculo *m* GEN calculation; ~ **de la potencia de enlace** TEC RAD, TRANS link power budget

calefacción *f* GEN heating

calentador *m* ELECTRÓN heater

calentar *vt* DATOS, INFO, TEC RAD, TELEFON, TV *caldera* serve

calibración *f* PRUEBA gauging (*AmE* gaging)

calibraciones *f pl* INFO settings

calibrado *m* PRUEBA calibration

calibrador *m* PRUEBA gauge (*AmE* gage) ; ~ **de alambres** COMPON wire gauge (*AmE* gage); ~ **de cursor** PRUEBA slide calliper (*AmE* caliper); ~ **de profundidad** PRUEBA *dispositivos de medición* depth gauge (*AmE* gage)

calibrar *vt* PRUEBA calibrate

calibre: ~ **de alambres** *m* COMPON wire gauge (*AmE* gage) ; ~ **de cursor** *m* PRUEBA sliding calliper (*AmE* caliper); ~ **de espesor** *m* PRUEBA *metrología* feeler gauge (*AmE* gage); ~ **exterior** *m* PRUEBA snap gauge (*AmE* gage); ~ **hembra de roscas** *m* PRUEBA thread gauge (*AmE* gage) ; ~ **de mordaza** *m* PRUEBA snap gauge (*AmE* gage) ; ~ **para rosca** *m* PRUEBA thread gauge (*AmE* gage)

calidad: ~ **asistida por ordenador** *f* (*CAQ*) INFO computer-aided quality (*CAQ*); ~ **de la imagen** *f* INFO picture quality, TV image quality, picture quality; ~ **inaceptable** *f* PRUEBA unacceptable quality; ~ **inferior de servicio** *f* PRUEBA lower quality of service; ~ **de letra** *f* INFO letter quality; ~ **media de salida** *f* (*AOQ*) PRUEBA average outgoing quality (*AOQ*); ~ **de servico** *f* (*QOS*) GEN quality of service (*QOS*); ~ **del sonido** *f* INFO, TEC RAD, TV sound quality; ~ **telefónica** *f* ELECTRÓN, TELEFON *circuito* voice grade; ~ **de transmisión** *f* PRUEBA, TELEFON, TRANS transmission performance, transmission quality; ~ **de transmisión de la voz** *f* ELECTRÓN, TELEFON voice transmission quality

calificación: ~ **de obligaciones** *f* GEN credit rating

calificador *m* DATOS qualifier; ~ **de la referencia del receptor** DATOS recipient-reference qualifier

calimetría *f* PRUEBA qualimetry

calle *f* TEC RAD lane

calor *m* GEN heat

CAM *abr* (*fabricación asistida por ordenador*) INFO CAM (*computer-aided manufacturing*)

cámara *f* GEN camera; ~ **anecoica** PRUEBA *acústica* anechoic chamber, anechoic room, free-field room; ~ **de cables** TRANS cable chamber; ~ **de gránulos de carbón** COMPON *micrófonos* carbon-granule chamber; ~ **sin reverberaciones** PRUEBA echoless chamber; ~ **subterránea** TRANS underground chamber

cambiable *adj* GEN convertible, removable

cambiador *m* ELECTROTEC changer; ~ **de fase** ELECTRÓN, SEÑAL phase changer; ~ **de frecuencia** ELECTROTEC frequency changer

cambiar *vt* GEN change; ◆ ~ **de color** GEN discolour (*AmE* discolor); ~ **de nombre de** INFO *fichero* rename

cambio: ~ **a cifras** *m* PERIF, TELEFON figure shift; ~ **de código** *m* DATOS, TELEFON escape code; ~ **de código en los datos** *m* (*EID*) DATOS escape in data (*EID*); ~ **de conexiones** *m* COMPON wiring change; ~ **de control** *m* CONTROL control change; ~ **de disco** *m* INFO disk change; ~ **de frecuencia** *m* TEC RAD frequency changing; ~ **de lado** *m* CONMUT side shifting; ~ **a letras** *m* TELEFON letter shift; ~ **de línea** *m* TELEFON *télex* line shift; ~ **de nombre** *m* INFO renaming; ~ **del reloj** *m* PERIF clock change; ~ **del sentido de la voz** *m* TEC RAD speech switchover; ~ **de soportes** *m* DATOS media conversion; ~ **volumétrico** *m* INFO volume change

camino: ~ **de acceso** *m* INFO, TRANS, TV access path; ~ **crítico** *m* GEN *planificaciones* critical path; ~ **de derivación** *m* DATOS derivation path; ~ **digital de referencia hipotética** *m* (*HRDP*) TEC RAD, TELEFON, TRANS hypothetical reference digital path (*HRDP*); ~ **elegido** *m* DATOS *Internet* chosen path; ~ **de navegación por hipertexto** *m* INFO hypertext navigation path; ~ **tangencial de onda** *m* TEC RAD tangential wave path; ~ **de transmisión** *m* TRANS transmission highway

camisa *f* COMPON *máquinas* sleeve, ELECTRÓN, ELECTROTEC jacket, *máquinas* sleeve, TRANS *de ladrillos refractarios* jacket; ~ **de fibra** COMPON, TRANS fibre (*AmE* fiber) jacket; ~ **de protección** ELECTRÓN protective sheath; ~ **refrigerante** GEN cooling jacket

campana *f* ELECTRÓN *fragua, laboratorio* hood

campaña: ~ **agresiva por correo** *f* GEN mail shot; ~ **de publicidad** *f* GEN publicity campaign

campo *m* DATOS domain; ~ **de acción MCS** TELEFON MCS domain; ~ **de actividades** GEN field of activities; ~ **de antenas** TEC RAD antenna farm; ~ **auxiliar superior** TEC RAD, TRANS upper subfield; ~ **de búsqueda** DATOS, INFO search field; ~ **de clase de sobrecarga** TELEFON overload class field; ~ **coercitivo** ELECTRÓN *electromagnetismo* coercive field; ~ **de conexión** ELECTROTEC connection field; ~ **de conocimiento** GEN field of knowledge; ~ **de control** CONTROL control range; ~ **de control UIH** TELEFON UIH control field; ~ **cruzado** TELEFON crossed field; ~ **de datos de usuario** DATOS user data-field; ~ **distante** GEN distant field; ~ **de dominante** TEC RAD keynote field; ~ **eléctrico** ELECTRÓN electric field; ~ **de entrada de texto** INFO text-entry field; ~ **evanescente** ELECTRÓN, TRANS evanescent field; ~ **de exploración** TELEFON scanning field; ~ **finito** SEÑAL finite field; ~ **de fondo** TV bottom field; ~ **de fugas** ELECTRÓN leakage field; ~ **giratorio** ELECTROTEC, TEC RAD, TELEFON rotating field; ~ **de identificador de protocolos** DATOS protocol identifier field; ~ **de iluminancia** ELECTROTEC illuminance range; ~ **indicador de longitud** DATOS length-indicator field; ~ **inductor** ELECTRÓN induction field; ~ **de información facsímil** (*FIF*) DATOS, INFO facsimile information field (*FIF*); ~ **de información de la señalización** SEÑAL signalling (*AmE* signaling) information field; ~ **lejano** GEN far field; ~ **magnetizante** ELECTRÓN magnetizing field; ~ **de medición** PRUEBA measuring range; ~ **modal** TRANS *óptica de fibras* mode-field; ~ **de modo de operación** (*campo-M*) PERIF mode-of-operation field (*M-field*); ~ **del número de control** CONMUT check-number field; ~ **oscuro** TRANS dark field; ~ **perturbador** TRANS noise field; ~ **próximo** TEC RAD, TRANS near field; ~ **repercutido** TELEFON reverberated field; ~ **de representación visual** INFO display field; ~ **reservado** ELECTRÓN reserved field (*RES*); ~ **de retención** ELECTRÓN holding range; ~ **reverberante** TELEFON reverberant field; ~ **S1** CONMUT S1-field; ~ **S2** CONMUT S2-field; ~ **de selección de vínculo de señalización** SEÑAL signalling-link (*AmE* signaling-link) selection field; ~ **terminal** TV top field; ~ **de terminales** GEN terminal field; ~ **de tipo n** PERIF *pbx* n-field; ~ **de tolerancia del alma** TRANS core tolerance field; ~ **de tolerancia del chapado** TRANS cladding tolerance field; ~ **uniforme** GEN uniform field; ~ **de utilidad de la red** TELEFON network utility field; ~ **visual** (*FOV*) TEC RAD field of view (*FOV*)

campo-M *abr* (*campo de modo de operación*) PERIF M-field (*mode-of-operation field*)

canal *m* GEN *hidráulica* chute, TEC RAD, TV channel; ~ **de abonado** TELEFON *interfaz DTE/DCE multiplexado* subscriber channel; ~ **de acceso** TELEFON, TRANS, TV *telefonía* access channel; ~ **de acceso en alternativa** TELEFON alternate-access channel; ~ **de acceso restringido** TV limited-access channel; ~ **adyacente** TEC RAD, TELEFON adjacent channel; ~ **de agrupamiento de señalización** SEÑAL signalling (*AmE* signaling) grouping channel (*SGC*); ~ **aislado** TRANS blanked-off channel; ~ **de ajuste inicial** SEÑAL, TRANS setup channel; ~ **de alta calidad** TV premium channel; ~ **amorfo** TRANS amorphous channel; ~ **de asignación** TRANS assignment channel; ~ **B** DATOS B-channel, TELEFON channel B, TRANS *canal de información* B-channel; ~ **de banda ancha** TRANS wideband channel; ~ **básico** INFO basic channel; ~ **de cable de bus** COMPON *técnicas de equipamiento* bus cable chute; ~ **para cable de fuerza** TELEFON power cable chute; ~ **de cables** TRANS *tendido de cables* cable chute; ~ **de cable de señales** TELEFON signal cable chute; ~ **de calidad funcional normal** TRANS standard-grade performance channel; ~ **de calidad superior** TV premium channel; ~ **colateral ortogonal** TEC RAD orthogonal co-channel; ~ **de comentario** TRANS commentary channel; ~ **de comunicación de datos** DATOS, TRANS data-communication channel; ~ **de comunicaciones** TV communication channel; ~ **de conexión de armario** CONMUT *instalación de equipos* cabinet connection chute; ~ **de control** CONTROL control channel; ~ **de control de cifrado** TRANS encryption control channel; ~ **de control embebido** DATOS embedded control channel; ~ **de control por inversión** TELEFON reversal control channel; ~ **de control de la red** TRANS network control channel; ~ **de control y retardo** TRANS control-and-delay channel; ~ **de corriente portadora** TELEFON, TRANS bearer channel; ~ **D** DATOS D-channel, TELEFON channel D, D-channel; ~ **de datos** DATOS data channel; ~ **de datos de voz alternado** TRANS alternate voice-data channel; ~ **derivado** TEC RAD frequency-derived channel; ~ **desvanescente de Rician** ELECTRÓN Rician-fading channel; ~ **de distribución** DATOS *digitalizador* distribution channel; ~ **dividido** CONMUT split channel; ~

emisor TRANS transmitting channel; ~ **de entrada/salida** INFO I/O channel; ~ **de estado** TELEFON status channel; ~ **de fila** TELEFON row chute; ~ **de frecuencias** TEC RAD frequency channel; ~ **gaussiano** DATOS Gaussian channel; ~ **de grupo de expertos en cinematografía** TV motion picture experts group channel; ~ **guíaondas con difusión de titanio** TRANS titanium waveguide channel; ~ **de imagen** TV picture channel; ~ **interferente** TRANS interfering channel; ~ **lateral** TELEFON side chute; ~ **lateral secundario** TEC RAD second adjacent channel; ~ **libre marcado** CONMUT marked idle channel; ~ **de llamadas** TEC RAD calling channel; ~ **local** TV local channel; ~ **lógico** TELEFON logical channel; ~ **de mantenimiento** TRANS maintenance channel; ~ **de mantenimiento de operadores de red** TRANS network-operators' maintenance channel (*NOMC*); ~ **MCS** TELEFON MCS channel; ~ **de memoria de datos** CONMUT *código* data store channel; ~ **mixto análogo-digital** TELEFON mixed analog-digital channel; ~ **MPEG** TV MPEG channel; ~ **multihaz** TEC RAD multiray channel; ~ **multiplexor** TRANS multiplexer channel; ~ **no interpolado** TRANS noninterpolated channel; ~ **operativo embebido** DATOS embedded operations channel (*EOC*); ~ **orbital** TEC RAD orbital slot; ~ **de polarización cruzada** TRANS cross-polar channel; ~ **principal** CONMUT main chute; ~ **privado MCS** TELEFON MCS private channel; ~ **de procesamiento de transacciones** DATOS transaction processing channel; ~ **de programas de radiodifusión** TEC RAD broadcast programme (*AmE* program) channel; ~ **de radiodifusión** TEC RAD broadcast channel; ~ **de Rayleigh** DATOS Rayleigh channel; ~ **de recepción** INFO receive channel, receiving channel, reception channel, TRANS receive channel; ~ **de registro** TELEFON recording channel; ~ **de retransmisión de prueba** TV preview channel; ~ **de retroceso** TELEFON, TRANS backward channel; ~ **de RF** TEC RAD RF channel; ~ **de salida** INFO output channel; ~ **por satélite** TEC RAD, TV satellite channel; ~ **de señalización** SEÑAL, TELEFON signalling (*AmE* signaling) channel; ~ **de señalización no dedicado** TRANS nondedicated signalling (*AmE* signaling) channel; ~ **de señal de satélite** SEÑAL satellite signalling (*AmE* signaling) channel; ~ **de servicio** TELEFON speaker circuit; ~ **de servicio técnico** TRANS engineering service channel; ~ **de sobrecarga** TRANS overload channel; ~ **de sonido** TRANS audiochannel, TV sound channel; ~ **telegráfico** TELEFON telegraph channel; ~ **de teletipo** TRANS teletype-order channel; ~ **de televisión** TV television channel; ~ **de televisión por vía satélite** TEC RAD, TV satellite television channel; ~ **temático** TV theme channel; ~ **de trabajo** TEC RAD operating channel; ~ **de tráfico** DATOS traffic channel, TEC RAD traffic channel, traffic path; ~ **de transmisión** TEC RAD, TRANS, TV transmission channel; ~ **transmultiplexor** ELECTRÓN, TELEFON, TRANS transmultiplexer channel; ~ **transversal** COMPON *técnicas de ensamblado* cross chute; ~ **único por portadora** (*SCPC*) TEC RAD single channel per carrier (*SCPC*); ~ **de velocidad mitad** TRANS half-rate channel; ~ **virtual** SEÑAL, TRANS virtual channel; ~ **virtual de señalización** SEÑAL, TRANS signalling (*AmE* signaling) virtual channel; ~ **virtual de señalización de emisión** TRANS broadcast signaling virtual channel *AmE*; ~ **virtual de señalización de radiodifusión** TRANS broadcast signalling (*AmE* signaling) virtual channel (*BSVC*); ~ **de voz** TELEFON voice channel, TRANS speech channel, voice channel

canalización *f* TEC RAD ducting, TRANS channelling (*AmE* channeling); ~ **para cables** TRANS cable trough; ~ **circular** ELECTROTEC ring main

cancelación *f* GEN cancellation; ~ **de eco** ELECTRÓN, TEC RAD, TRANS echo cancellation; ~ **de localización** TEC RAD location cancellation; ~ **selectiva** TEC RAD selective cancellation

cancelador: ~ **de ecos** *m* ELECTRÓN, TEC RAD, TRANS echo canceller (*AmE* canceler)

cancelar *vt* GEN cancel; ◆ ~ **el efecto de** DATOS, TELEFON override

canción: ~ **de propaganda** *f* INFO demo song

canilla *f* CLIMAT, COMPON spout

cañón: ~ **electrónico de Pierce** *m* ELECTRÓN Pierce gun; ~ **electrónico sumergido** *m* ELECTRÓN immersed gun; ~ **de inyección magnetrónica** *m* ELECTRÓN magnetron injection gun; ~ **de lectura** *m* ELECTRÓN reading gun

cantidad: ~ **de datos** *f* DATOS data amount; ~ **media de inspección** *f* PRUEBA *calidad* average amount of inspection

caos *m* ELECTRÓN chaos

capa *f* ELECTRÓN coating, GEN layer, TRANS coating; ~ **agotada del emisor** COMPON *semiconductores* emitter depletion layer; ~ **de aplicación** APLIC, DATOS, TRANS application layer; ~ **de bobinado** ELECTROTEC winding layer; ~ **de canalización** TEC RAD ducting layer; ~ **de detención** ELECTRÓN barrier layer; ~ **E esporádica** TEC RAD sporadic E layer; ~ **del electrón** GEN electron shell; ~ **de elementos de cableado** COMPON layer of cabling elements; ~ **endurecida** GEN hard case; ~ **de enlace** TRANS link layer; ~ **de enlace de datos** DATOS, TRANS data-link layer; ~ **epitaxial** COMPON epitaxial layer; ~ **F** INFO, TEC RAD F-layer; ~ **intrínseca** ELECTRÓN *semiconductores* intrinsic layer; ~ **ionosférica** TEC RAD ionospheric layer; ~ **del modelo de interconexión abierta de sistemas** INFO OSI model layer; ~ **de modelo OSI** DATOS OSI model layer; ~ **N** DATOS, ELECTRÓN N layer; ~ **protectora** ELECTRÓN resist foil; ~ **protectora contra soldadura** COMPON solder resist layer; ~ **de red conmutada** CONMUT, TRANS, TV OSI network layer; ~ **de red OSI** DATOS OSI network layer; ~ **de revestimiento** COMPON, ELECTRÓN sheathing layer; ~ **de separación de colores** INFO colour (*AmE* color) separation overlay; ~ **de sesión** DATOS, TRANS session layer; ~ **superpuesta** INFO overlay; ~ **de transición** COMPON depletion layer; ~ **de transmisión** DATOS transport layer, TRANS transmission layer, transport layer; ~ **de transporte OSI** DATOS OSI transport layer; ~ **de unidades** COMPON *cables* layer of units

capacidad *f* TEST capability, capacity; ~ **de absorción de humedad** GEN moisture-absorbing capacity; ~ **de ampliación de la memoria** INFO memory expansion capacity; ~ **del cable** TRANS cable capacity; ~ **del canal** TRANS channel capacity; ~ **de corrección de errores por ráfagas** DATOS error burst-correcting capability; ~ **de crecimiento** GEN growth capability; ~ **de digitalización** PERIF *reconocimiento de caracteres* scanning capability; ~ **de distorsión de la cuantificación** TELEFON, TRANS quantizing distortion power; ~ **de eclipse** TEC RAD eclipse capability; ~ **efectiva** TELEFON effective capacity; ~ **de las funciones locales** CONMUT local function capabilities; ~ **de impresión** PERIF print capacity; ~ **de indicación de la posición del móvil** TEC RAD roaming capability; ~ **de inductancia** ELECTRÓN inductance capacity; ~ **interruptora** ELECTROTEC interrupting capacity; ~ **de manipulación del tráfico** DATOS, TELEFON traffic-handling capacity; ~ **máxima de memoria** PERIF *contestador* maximum memory capacity; ~ **de memoria** DATOS storage capacity; ~ **de nota simultánea** INFO *reproducción de sonido* simultaneous note capacity; ~ **de petición de mensajes** DATOS, TELEFON facility request message (*FAR*); ~ **del portador** TELEFON bearer capability, TRANS bearer capacity; ~ **de procesamiento** DATOS, INFO processing power; ~ **de proceso** PRUEBA process capability; ~ **de retención** INFO hold facility; ~ **de servicio** APLIC, TELEFON serveability; ~ **sintonizadora** ELECTRÓN, INFO, TEC RAD tuning capability; ~ **terminal** TEC RAD top capacitive loading; ~ **de tráfico** TRANS traffic capacity; ~ **de transmisión** DATOS transmission capability

capacitancia *f* ELECTRÓN capacitance; ~ **de la capa agotada del emisor** COMPON *semiconductores* emitter depletion layer capacitance; ~ **de difusión** COMPON *semiconductores* diffusion capacitance; ~ **efectiva de entrada** ELECTROTEC effective input capacitance; ~ **efectiva de salida** ELECTROTEC effective output capacitance; ~ **eléctrica** ELECTRÓN electrical capacitance; ~ **electródica** ELECTRÓN electrode capacitance; ~ **del encapsulado** ELECTRÓN *semiconductores* case capacitance; ~ **inductiva** ELECTRÓN inductance capacitance; ~ **interelectródica** ELECTRÓN interelectrode capacitance; ~ **parásita** ELECTROTEC stray capacitance; ~ **de salida** ELECTRÓN output capacitance; ~ **para señal débil** ELECTRÓN small-signal capacitance; ~ **en una unión** COMPON, ELECTRÓN *semiconductores* junction capacitance

capacitivo *adj* ELECTRÓN capacitive

capacitor *m* ELECTROTEC capacitor, INFO condenser; ~ **de aceite** ELECTRÓN oil-filled capacitor; ~ **acelerador** COMPON, ELECTROTEC speed-up capacitor; ~ **apagachispas** COMPON, ELECTROTEC spark-quenching capacitor; ~ **bipolar** COMPON, ELECTROTEC two-gang capacitor; ~ **de capa delgada** COMPON, ELECTRÓN thin-film capacitor; ~ **en derivación** COMPON, ELECTROTEC shunt capacitor; ~ **filtrador** COMPON, ELECTROTEC smoothing capacitor; ~ **de mica plateada** COMPON, ELECTRÓN silver mica capacitor; ~ **MP** ELECTROTEC MP capacitor; ~ **mylar** ELECTRÓN mylar capacitor; ~ **de papel metalizado** ELECTROTEC metallized-paper (*AmE* metalized-paper) capacitor; ~ **pasante** ELECTROTEC feedthrough capacitor; ~ **patrón** ELECTROTEC standard capacitor; ~ **de película metálica** ELECTRÓN metal-film capacitor; ~ **de película metalizada** ELECTRÓN metallized-film (*AmE* metalized-film) capacitor; ~ **de tipo NPO** ELECTROTEC NPO-type capacitor; ~ **tubular** COMPON, ELECTROTEC tubular capacitor; ~

variable de aire COMPON, ELECTROTEC variable air-capacitor

cápsula *f* GEN capsule; **~ microfónica** PERIF transmitter inset; **~ con patillas** ELECTRÓN leaded chip-carrier; **~ sin patillas** (*LCC*) ELECTRÓN leadless chip-carrier (*LCC*)

captación *f* INFO capture, TELEFON *del sonido* picking up; **~ de componentes parásitos** TEC RAD stray pick-up; **~ de vídeo** INFO video capture

captador: **~ de iones** *m* ELECTRÓN ion trap

captura: **~ de datos con teclado** *f* INFO keyboarding

CAQ *abr* (*calidad asistida por ordenador*) INFO CAQ (*computer-aided quality*)

carácter *m* APLIC, DATOS, INFO, TELEFON character; **~ de alimentación de línea** DATOS line-feed character; **~ de borrado** INFO deletion character; **~ cambiador de código** DATOS shift-in character; **~ de cambio de línea** (*NL*) DATOS new-line character (*NL*); **~ de cancelación de bloque** INFO block-cancel character; **~ de código** TELEFON code character; **~ de continuación** TELEFON continuation character; **~ de control** INFO, TELEFON control character; **~ de control de errores** TEC RAD error-check character; **~ de control de impresión** DATOS print-control character; **~ de corrección** TELEFON correction character; **~ de desplazamiento hacia fuera** DATOS shift-out character; **~ determinante de formato** DATOS, TELEFON format effector; **~ de ejecución** TELEFON execution character; **~ de empaquetadura** DATOS stuffing character; **~ de escape** DATOS escape character; **~ de espacio** DATOS space character (*SP*); **~ especial** DATOS special character; **~ final** DATOS final character; **~ de formato** DATOS, INFO layout character; **~ gráfico** DATOS graphic character; **~ de mando** DATOS command character; **~ nulo** (*NUL*) DATOS null character (*NUL*); **~ de reconocimiento negativo** TELEFON negative acknowledge character; **~ de relleno** DATOS filler character; **~ de retorno del carro** GEN carriage return character; **~ de retroceso** INFO backspace character; **~ de señal** SEÑAL character of signal; **~ de separación** DATOS gap character; **~ de signo** INFO sign character; **~ de signo binario** INFO sign-binary character; **~ de sincronización** DATOS synchronization character; **~ de sustitución** DATOS substitute character, INFO joker, wildcard character; **~ de tabulación horizontal** DATOS, TEC RAD horizontal tabulation character (*HT*); **~ de tabulación vertical** DATOS, TEC RAD vertical tabulation character (*VT*)

caracteres: **~ dinámicamente redefinibles** *m pl* (*DRCS*) INFO dynamically-redefinable characters (*DRCS*); **~ por pulgada** *m pl* (*cpi*) INFO characters per inch (*cpi*)

característica *f* GEN characteristic; **~ de amplitud de fase** TRANS phase-amplitude characteristic; **~ de apoyo del servicio** TELEFON service support performance; **~ espectral** ELECTRÓN, TRANS spectral characteristic; **~ espectral de la eficiencia cuántica** ELECTRÓN, TRANS quantum-efficiency spectral characteristic; **~ del fallo de potencia** ELECTROTEC, INFO, TELEFON service reliability performance; **~ de fiabilidad del servicio** PRUEBA service reliability performance; **~ de frecuencia** TEC RAD frequency characteristic; **~ funcional** GEN functional characteristic, functional feature; **~ de llamada virtual** DATOS, TELEFON virtual-call characteristic; **~ de operabilidad del servicio** PRUEBA service operability performance; **~ de recepción** TEC RAD receive characteristic; **~ de rendimiento** PRUEBA performance characteristic; **~ de respuesta con onda rectangular** ELECTRÓN, SEÑAL, TRANS square-wave response characteristic; **~ de sensibilidad espectral relativa** ELECTRÓN relative spectral sensitivity characteristic; **~ de señal** SEÑAL signal characteristic; **~ de sobrecontrol** TELEFON override feature; **~ de transferencia de señal luminosa** ELECTRÓN light-signal transfer characteristic; **~ de transmisión** TRANS transmission characteristic

características: **~ anódo-catódo** *f pl* ELECTRÓN *semiconductores* anode-cathode characteristics; **~ de disponibilidad del servicio** *f pl* TELEFON service availability performance; **~ de dominio frecuencial** *f pl* TEC RAD frequency-domain characteristics; **~ de la red** *f pl* PRUEBA network performance (*NP*); **~ de salida** *f pl* ELECTRÓN output characteristics

carburo: **~ metálico curvado** *m* ELECTRÓN bend carbide

carcinotrón *m* ELECTRÓN carcinotron

carga *f* APLIC charge, *horno* charging, DATOS uploading, ELECTRÓN, ELECTROTEC charge, INFO load, TEC RAD, TRANS loading; **~ automática** INFO auto-load; **~ de base** ELECTRÓN *semiconductores* base charge, TELEFON base load; **~ de batería** ELECTROTEC battery charging; **~ por bobinas** TRANS coil loading; **~ de canal** TEC RAD, TRANS channel loading; **~ de cartucho** INFO cartridge loading;

~ **compartida** TELEFON loadsharing; ~ **compensadora** ELECTROTEC *suministro de energía* equalizing charge; ~ **de conexión** TV connection charge; ~ **continua** GEN *cables* continuous loading; ~ **eléctrica** ELECTRÓN electrical charge; ~ **de electrón** ELECTRÓN electron charge; ~ **equilibrada** TEC RAD matched load; ~ **espacial** ELECTRÓN space charge; ~ **flotante** ELECTROTEC *baterías* floating charge; ~ **frontal** GEN front loading; ~ **inicial** TELEFON initial charge; ~ **de intensidad de tráfico** TELEFON traffic intensity load; ~ **de juego de resortes** COMPON spring-set load; ~ **lenta** ELECTROTEC trickle charge, trickle charger; ~ **líquida** TEC RAD water load; ~ **nominal** ELECTRÓN nominal load; ~ **de operación** ELECTROTEC normal charging; ~ **óptima** TEC RAD optimum load; ~ **rápida** ELECTROTEC *de baterías* rapid charge; ~ **reactiva** ELECTRÓN reactive load; ~ **remota** PRUEBA remote loading; ~ **resistiva** ELECTRÓN, ELECTROTEC resistive load; ~ **de salida** ELECTROTEC output load; ~ **de servicio de elementos múltiples** APLIC multielement service charge; ~ **de sistema** GEN system load; ~ **sobre el suelo** GEN *construcción* floor load; ~ **de un teléfono móvil** TELEFON mobile-station charge; ~ **unidad** TELEFON unit charge; ~ **útil** GEN *comunicación por satélite* payload; ◆ **sin** ~ ELECTROTEC, INFO unloaded

cargable: ~ **por teleproceso** *adj* DATOS *Internet* downloadable, ELECTRÓN downline-loadable, downloadable, INFO downloadable

cargado *adj* GEN loaded

cargador *m* ELECTROTEC *baterías* charging set, *hornos* charger; ~ **automático inicial** INFO bootstrap loader; ~ **de batería** APLIC, ELECTROTEC charging unit, battery unit; ~ **de compartimento único** ELECTROTEC single-compartment charger; ~ **de documento** DATOS *digitalizador* document loader; ~ **del modulador** ELECTRÓN modulator charger; ~ **de pila** ELECTROTEC battery charger

cargar *vt* DATOS upload, ELECTRÓN, TELEFON charge; ◆ ~ **una llamada a** APLIC charge a call to; ~ **para prueba** PRUEBA test-load

cargo[1]: ~ **al abonado llamado** *m* APLIC called-line charging; ~ **por alquiler** *m* GEN rental charge; ~ **por cobro** *m* TELEFON collection charge; ~ **de conexión inicial** *m* APLIC initial connection charge; ~ **por preaviso** *m* TELEFON preavis charge

cargo[2]: **sin** ~ *fra* TELEFON no charge

carrera *f* COMPON, INFO *de émbolo* stroke

carrete *m* COMPON spool, DATOS reel, ELECTROTEC bobbin, coil former; ~ **de cable** TRANS cable reel; ~ **para enrollar cables** TRANS cable drum; ~ **de guía** TELEFON guide spool

carrito: ~ **de estantes** *m* COMPON shelved trolley

carro: ~ **de barra de tipos** *m* TELEFON type-bar carriage; ~ **portabaterías** *m* ELECTROTEC battery cart

carta: ~ **de intenciones** *f* GEN *comercio* letter of intent; ~ **de resolución** *f* TV resolution chart

cartografía: ~ **medioambiental** *f* INFO environment mapping

cartuchera *f* INFO cartridge box

cartucho *m* INFO cartridge; ~ **de carga** INFO loading cartridge; ~ **de cuarto de pulgada** (*QIC*) TV quarter-inch cartridge (*QIC*); ~ **de prestaciones** INFO performance cartridge; ~ **de programa** INFO program cartridge

casa: ~ **inteligente** *f* TV hi-tech home; ~ **matriz** *f* GEN home office, parent company

casar *vt* ELECTRÓN mate

cascada *f* GEN cascade

casco: ~ **con auriculares** *m* TELEFON headset; ~ **de dos auriculares independientes** *m* TELEFON split headphones; ~ **de protección** *m* GEN safety helmet

caseta *f* GEN house

casete *f* GEN cassette

casilla: ~ **de abonado** *f* TV subscriber box; ~ **de control** *f* INFO check box

caso: ~ **imprevisto** *m* PRUEBA unforeseen event; ~ **más desfavorable** *m* PRUEBA worst case; ~ **no identificado** *m* PRUEBA unidentified event; ~ **de prueba abstracta parametrizada** *m* DATOS, PRUEBA parameterized abstract test case; ~ **de prueba ejecutable parametrizada** *m* DATOS, PRUEBA parameterized executable test case

casquillo *m* COMPON, TRANS ferrule; ~ **de cojinete** COMPON bearing shell; ~ **de lámpara** ELECTROTEC lamp base, lamp cap; ~ **de rosca** COMPON screw base; ~ **roscado** ELECTRÓN threaded bushing

CAT *abr* (*prueba asistida por ordenador*) INFO CAT (*computer-aided testing*)

catálogo *m* GEN catalogue (*AmE* catalog); ~ **de disco compacto fotográfico** INFO photo CD catalogue (*AmE* catalog); ~ **de fallos** TELEFON error catalogue (*AmE* catalog), failure catalogue (*AmE* catalog)

categoría *f* GEN status

categorías: ~ **para la conmutación de paquetes** *f pl* TELEFON packet layer

catenular *vt* INFO catenate

cátodo *m* COMPON, ELECTRÓN cathode; ~ **de**

caldeo directo COMPON, ELECTRÓN directly-heated cathode; ~ **equipotencial** ELECTRÓN indirectly-heated cathode; ~ **líquido** ELECTRÓN pool cathode; ~ **de tungsteno toriado** ELECTRÓN thoriated-tungsten cathode

CAU *abr* (*unidad de acceso controlado*) TELEFON CAU (*controlled-access unit*)

caucho *m* COMPON rubber; ~ **blando** COMPON soft rubber

causa: ~ **de la alteración** *f* PRUEBA *fiabilidad* cause of trouble; ~ **del fallo** *f* PRUEBA *seguridad de funcionamiento* failure cause

cavidad *f* ELECTRÓN cavity, TELEFON hollowness, TRANS cavity; ~ **resonante** ELECTRÓN, TRANS resonant cavity; ~ **TB** TEC RAD, TRANS transmitter blocker cell

CBT *abr* (*formación basada en ordenadores*) INFO CBT (*computer-based training*)

CC *abr* (*corriente continua*) ELECTROTEC DC (*direct current*)

CCB *abr* (*placa de conexión de cable*) TRANS *normas de construcción* CCB (*cable connection board*)

CCITT *abr* (*Comité Consultivo Internacional de Telefonía y Telegrafía*) TELEFON CCITT (*International Telegraph and Telephone Consultative Committee*)

CCR *abr* (*servicio de red de compromisos, acuerdos y recuperación*) DATOS CCR (*commitment, concurrency, and recovery*)

CD *abr* (*disco compacto*) INFO CD (*compact disc*); ~ **multicapa** *m* DATOS multilayer CD

CDB *abr* (*bloque descriptor de orden*) DATOS CDB (*command-descriptor block*)

CDE *abr* (*codificado convolucional y diferencialmente*) DATOS CDE (*convolutionally and differentially encoded*)

cdf *abr* (*función de distribución acumulativa*) DATOS cdf (*cumulative distribution function*)

CDF *abr* (*conmutación de desplazamiento de fase*) ELECTROTEC, TEC RAD, TRANS PSK (*phase-shift keying*)

CD-I *abr* (*disco compacto interactivo*) INFO CD-I (*compact disc interactive*)

CDMA *abr* (*acceso múltiple de división de código*) TEC RAD CDMA (*code-division multiple access*)

CD-SOTR *abr* (*sistema de operación de CD-ROM en tiempo real*) INFO CD-RTOS (*CD-ROM real-time operating system*)

CE *abr* (*Comunidad Europea*) GEN, TEC RAD EC (*European Community*)

cebo: ~ **eléctrico de cantidad** *m* COMPON, ELECTROTEC battery fuse

celda: ~ **de memoria** *f* DATOS storage cell; ~ **en paraguas** *f* TEC RAD umbrella cell; ~ **sectorial** *f* TEC RAD sector cell

célula *f* ELECTROTEC, TEC RAD cell; ~ **de cajero automático** DATOS, INFO ATM cell; ~ **electrolítica** ELECTRÓN electrolytic cell, pot; ~ **fotoconductiva** ELECTRÓN photoconductive cell; ~ **fotosensible** DATOS *de digitalizador*, ELECTRÓN photosensitive cell; ~ **fotovoltaica** COMPON photovoltaic cell; ~ **de lluvia** TEC RAD rain cell; ~ **TB** TEC RAD, TRANS TB cell; ~ **TR** TEC RAD TR cell

celular *adj* GEN cellular

CEN *abr* (*Comité Europeo para la Normalización*) GEN CEN (*European Committee for Standardization*)

CENELEC *abr* (*Comité Europeo para la Normalización Electrotécnica*) GEN CENELEC (*European Committee for Electrotechnical Standardization*)

cenit *m* TEC RAD zenith

central *f* TELEFON exchange; ~ **de acceso internacional** CONMUT, TELEFON international gateway exchange; ~ **automática** TELEFON unmanned exchange; ~ **automática rural** CONMUT rural automatic exchange; ~ **automática unitaria** CONMUT unit automatic exchange (*UAX*); ~ **autónoma** CONMUT stand-alone exchange; ~ **de circuitos conmutados** CONMUT circuit-switched exchange; ~ **de comunicaciones** CONMUT central exchange trunk; ~ **de conmutación para mensajes internacionales** CONMUT, DATOS international data switching exchange (*IDSE*); ~ **de conmutación de paquetes** CONMUT, TELEFON packet-switching exchange (*PSE*); ~ **de conmutación de datos** *m* CONMUT, DATOS data-switching exchange; ~ **de destino** CONMUT terminating exchange, TELEFON called exchange, destination exchange, destination office, terminating exchange; ~ **digital integrada** CONMUT integrated digital exchange; ~ **distribuida geográficamente** CONMUT geographically-distributed exchange; ~ **eléctrica** ELECTROTEC electric power station; ~ **electrónica de relé de láminas** COMPON, CONMUT, ELECTROTEC reed-relay electronic exchange; ~ **holográfica** CONMUT holographic exchange; ~ **de identificación** (*XID*) TELEFON exchange of identification (*XID*); ~ **internacional** CONMUT international exchange; ~ **internacional de acceso a la conmutación de paquetes** CONMUT international packet-switching gateway exchange; ~ **interurbana principal** CONMUT main trunk switching centre (*AmE* center); ~ **local** CONMUT, TELEFON local exchange (*LE*); ~ **nº 7** CONMUT no. seven exchange; ~ **no atendida** TELEFON unattended exchange; ~ **de pasillo** CONMUT gateway exchange; ~ **periférica** (*CP*) TELEP peripheral exchange (*PE*); ~ **primaria** CONMUT home exchange, host exchange, TEC RAD host exchange; ~ **principal** CONMUT main exchange; ~ **principal de línea troncal internacional** CONMUT main international-trunk switching centre (*AmE* center); ~ **de la RDSI** CONMUT ISDN exchange; ~ **rotativa** CONMUT rotary exchange; ~ **rural** TELEFON rural exchange; ~ **satélite** CONMUT satellite exchange; ~ **de servicio** CONMUT serving exchange; ~ **de servicios digitales integrados** CONMUT integrated digital services exchange; ~ **de subzona** CONMUT subzone centre (*AmE* center); ~ **tándem de enlace** TELEFON juction tandem exchange; ~ **de telecomunicaciones** TELEFON telecommunications exchange; ~ **telefónica** CONMUT, TELEFON central exchange, central exchange switch, exchange switchboard, switching exchange; ~ **telefónica comunitaria** CONMUT community dial office *AmE*; ~ **telefónica dependiente** CONMUT, TELEFON dependent exchange; ~ **telefónica discriminante de satélite** CONMUT discriminating satellite exchange; ~ **telefónica intermedia de enlace** CONMUT junction tandem exchange; ~ **telefónica marítima por satélite** CONMUT, TEC RAD maritime satellite switching centre (*AmE* center); ~ **telefónica pública** CONMUT, TELEFON public telephone exchange *BrE*, central office switch *AmE*; ~ **telefónica urbana** TELEFON local central office; ~ **télex** CONMUT, TELEFON telex exchange; ~ **terminal** TELEFON, TRANS terminal exchange; ~ **terminal internacional** TELEFON international terminal exchange; ~ **tipo** TELEFON type exchange; ~ **de tránsito** CONMUT subzone centre (*AmE* center), transit exchange, DATOS, TELEFON transit exchange; ~ **de tránsito intercontinental** TELEFON intercontinental transit exchange; ~ **de tránsito internacional** (*INTTR*) CONMUT, TELEFON international transit exchange (*INTTR*)

centralita: ~ **automática multiservicio** *f* CONMUT multiservice automatic exchange; ~ **combinada local/tránsito** *f* CONMUT combined local/transit exchange; ~ **continental** *f* CONMUT continental exchange; ~ **local de marcado directo** *f* TEC RAD key system; ~ **pública** *f* TELEFON public exchange; ~ **subordinada** *f* TELEFON subordinate exchange; ~ **telecomandada** *f* TELEFON remote-controlled exchange

centralizar *vt* GEN centralize

centralógrafo *m* TELEFON centralograph

centro *m* GEN centre (*AmE* center), hub; ~ **del alma** TRANS core centre (*AmE* center); ~ **de apoyo multisatélite** (*MSSS*) TEC RAD multi-satellite support centre (*AmE* center) (*MSSS*); ~ **de cálculo** DATOS data-processing centre (*AmE* center), INFO computing centre (*AmE* center); ~ **de cálculo de costes** GEN cost centre (*AmE* center); ~ **de calibrado** PRUEBA *metrología* calibration centre (*AmE* center); ~ **candidato de conmutación de mensajes** TEC RAD candidate MSC; ~ **de chapado** TRANS cladding centre (*AmE* center); ~ **de comunicaciones** TRANS communications centre (*AmE* center); ~ **de conexión** TV connection centre (*AmE* center); ~ **de conmutación** CONMUT, TELEFON, TV switching centre (*AmE* center); ~ **de conmutación de circuitos** CONMUT circuit-switching centre (*AmE* center); ~ **de conmutación de grupo** CONMUT group switching centre (*AmE* center) (*GSC*); ~ **de conmutación internacional** CONMUT, TELEFON international switching centre (*AmE* center); ~ **de conmutación marítima** CONMUT, TELEFON maritime switching centre (*AmE* center); ~ **de conmutación de mensajes** CONMUT message-switching centre (*AmE* center); ~ **de conmutación móvil** (*MSC*) CONMUT, TELEFON mobile switching centre (*AmE* center) (*MSC*); ~ **de conmutación de posición inicial** TEC RAD home switching centre (*AmE* center); ~ **de conmutación de satélite** (*SSC*) CONMUT, TEC RAD satellite switching centre (*AmE* center) (*SSC*); ~ **de conmutación de servicios móviles** CONMUT, TELEFON mobile services switching centre (*AmE* center); ~ **de conmutación de télex** CONMUT, TELEFON telex switching centre (*AmE* center); ~ **de conmutación de tránsito** CONMUT, DATOS, TELEFON transit switching centre (*AmE* center); ~ **de contratación de programas** TV program booking centre (*AmE* center) ; ~ **de control de operaciones** PRUEBA operations control centre (*AmE* center); ~ **de control de red** PRUEBA network control centre (*AmE* center); ~ **de control de satélites** TEC RAD satellite control centre (*AmE* center) (*SCC*); ~ **de coste** GEN cost centre (*AmE* center); ~ **director** CONMUT controlling exchange; ~ **de distribución** ELECTROTEC, TV distribution centre (*AmE* center); ~ **de empalmes montable en pared**

(*WSC*) ELECTRÓN wall-mountable splice centre (*AmE* center) (*WSC*); **~ de explotaciones** GEN operations centre (*AmE* center); **~ de explotación y mantenimiento de red** PRUEBA network operation and maintenance centre (*AmE* center); **~ de facturación** APLIC billing centre (*AmE* center), charging centre (*AmE* center); **~ de gestión** PRUEBA management centre (*AmE* center); **~ de gestión de red** PRUEBA network management centre (*AmE* center) (*NMC*); **~ de grabación** CONMUT, TELEFON recording centre (*AmE* center); **~ de grabación de datos** DATOS data-collection centre (*AmE* center); **~ del haz** TEC RAD beam centre (*AmE* center); **~ de interconexión montable en pared** (*WIC*) ELECTRÓN wall-mountable interconnect centre (*AmE* center) (*WIC*); **~ internacional de mantenimiento de la transmisión** (*ITMC*) PRUEBA international transmission-maintenance centre (*AmE* center) (*ITMC*); **~ de investigación** GEN research centre (*AmE* center); **~ de investigación de comunicaciones** GEN communications research centre (*AmE* center); **~ de mantenimiento de conmutación internacional** (*ISMC*) PRUEBA international switching maintenance centre (*AmE* center) (*ISMC*); **~ de mantenimiento orientado al cliente** PRUEBA customer-access maintenance centre (*AmE* center) (*CAMC*); **~ marítimo** TELEFON maritime centre (*AmE* center); **~ marítimo de conmutación por satélite** (*MSSC*) TEC RAD maritime satellite switching centre (*AmE* center) (*MSSC*); **~ del núcleo** TRANS core centre (*AmE* center); **~ de operación** TELEFON operating centre (*AmE* center); **~ de operaciones** GEN operations centre (*AmE* center); **~ de operaciones, administración y mantenimiento** PRUEBA, TELEFON operation, administration and maintenance centre (*AmE* center); **~ de operaciones y mantenimiento** PRUEBA, TELEFON operation and maintenance centre (*AmE* center); **~ de operación de vehículo espacial de comunicaciones** TEC RAD communications spacecraft operation centre (*AmE* center); **~ operativo** TELEFON operational centre (*AmE* center); **~ operativo remoto** TEC RAD remote operating centre (*AmE* center); **~ principal de conmutación internacional** CONMUT main international switching centre (*AmE* center); **~ principal de distribución internacional** CONMUT main international switching centre (*AmE* center); **~ de procesamiento de datos** DATOS data-processing center *AmE*; **~ de programas sonoros internacional** (*ISPC*) TV international sound-programme centre (*AmE* international sound-program center) (*ISPC*); **~ de programas de televisión internacional** TV international television programme centre (*AmE* international television program center); **~ público de telefax** TELEFON public facsimile station; **~ de recepción de fallos** PRUEBA fault reception centre (*AmE* center) (*FRC*); **~ de reclamaciones** TELEFON fault centre (*AmE* center); **~ regional** TELEFON regional centre (*AmE* center); **~ de registro de facturación** TELEFON charging-recording centre (*AmE* center); **~ de reserva de programas** (*PBC*) TV programme booking centre (*AmE* program booking center) (*PBC*); **~ de servicio de abonados** TELEFON subscriber's service centre (*AmE* center); **~ de servicio de videotexto** APLIC, INFO, TELEFON, TV videotexto service centre (*AmE* center); **~ de servidor** GEN server centre (*AmE* center); **~ de superficie de referencia** TELEFON, TRANS reference-surface centre (*AmE* center); **~ de televisión internacional** TV international television centre (*AmE* center); **~ terminal internacional** TELEFON international terminal centre (*AmE* center); **~ terminal nacional** TELEFON national terminal centre (*AmE* center); **~ de tránsito internacional** TRANS *telecomunicaciones* gateway; **~ de tratamiento de datos** DATOS data-processing centre (*AmE* center)

Centro: **~ Nacional para Aplicaciones de Superinformática** *m* (*NCSA*) DATOS *Internet* National Center for Supercomputing Applications (*NCSA*)

cepillo *m* INFO plane

cero *m* INFO zero; **~ eléctrico** PRUEBA *aparato de medición* electrical zero; **~ mecánico** PRUEBA mechanical zero

cerradura: **~ de llave** *f* INFO key lock; **~ de seguridad** *f* PRUEBA safety lock

cerrar *vt* GEN close, turn off

cerrojo *m* ELECTRÓN latch lock

certificación *f* GEN certification

certificado *m* GEN certificate; **~ de aprobación** GEN certificate of approval; **~ de aptitud** PRUEBA *calidad* certificate of compliance; **~ de calibrado** PRUEBA *metrología* calibration certificate; **~ de calidad SQS** PRUEBA SQS quality certificate; **~ de conformidad** PRUEBA *calidad* certificate of conformance, certificate of conformity

cese: **~ repentino** *m* PRUEBA cutback

cesta *f* DATOS basket

cesura *f* DATOS caesura

CD-ROM *abr* (*disco compacto de memoria sólo de lectura*) GEN CD-ROM (*compact disk read-only memory*)
CF *abr* (*función de control*) CONTROL, DATOS, INFO, TELEFON CF (*control function*)
CFM *abr* (*modulación de compresión-expansión y de frecuencia*) TRANS CFM (*companding and frequency modulation*)
CFR *abr* (*confirmación por recibir*) DATOS, TELEFON CFR (*confirmation to receive*)
CGA *abr* (*adaptador gráfico de colores*) INFO CGA (*colour graphics adaptor*)
chapa *f* COMPON veneer; **~ frontal** ELECTRÓN faceplate; **~ de marcado** GEN marking plate; **~ metálica** COMPON sheet metal; **~ de salida simple** COMPON, ELECTRÓN single-gang faceplate
chapado: **~ deprimido** *m* TRANS depressed cladding; **~ de fibra óptica** *m* TRANS fibre-optic (*AmE* fiber-optic) cladding
charla *f* DATOS *Internet*, TEC RAD, TELEFON talk
chasis *m* GEN chassis
chasquido: **~ de manipulación** *m* TEC RAD, TELEFON *telegrafía* key click
chillido *m* TEC RAD squealing
chincheta *f* INFO *gráficos* thumbnail
chip *m* DATOS, ELECTRÓN chip
chirrido *m* TEC RAD, TRANS chirping
chispa *f* ELECTROTEC spark
chispeo *m* ELECTROTEC sparking
choque *m* ELECTRÓN, ELECTROTEC *electrodeposición* striking; **~ de desconexión** TELEFON clearing collision
chorro: **~ de conversación** *m* TEC RAD talk spurt; **~ de voz** *m* TELEFON voice spurt
CIAE *abr* (*circuito integrado de aplicación específica*) COMPON, ELECTRÓN ASIC (*application-specific integrated circuit*)
ciber *adj* DATOS *Internet* cyber
cibercultura *f* DATOS *Internet* cyberculture
ciberespacio *m* DATOS *Internet* cyberspace
cibernauta *m* DATOS *Internet* cybernaut
cibernética *f* DATOS cybernetics
ciclado: **~ de temperatura** *m* CLIMAT, PRUEBA temperature cycling
cíclico *adj* GEN cyclical
ciclo *m* INFO, TELEFON cycle; **~ de carga** ELECTROTEC charge cycle; **~ cronometrado** INFO clock cycle; **~ de descarga** ELECTROTEC discharge cycle; **~ de histéresis** ELECTRÓN hysteresis cycle; **~ de prueba** PRUEBA testing cycle; **~ de secuencia obligada** SEÑAL compelled cycle; **~ de trabajo** ELECTROTEC, TEC RAD compelled cycle
ciego *adj* INFO blind
cierre *m* COMPON *semiconductores* latch-up; **~ de presión** COMPON snap locking; **~ de rosca** COMPON screwlock
cifra *f* GEN *documentación* figure; **~ binaria** TRANS binary figure; **~ de disponibilidad** CONMUT availability figure; **~ de idioma** TELEFON language digit; **~ de ruido puntual** TEC RAD spot noise figure; **~ de ruido del receptor** TEC RAD receiver-noise figure; **~ de unidad** GEN units digit
cifrado *m* DATOS, SEÑAL encipherment; **~ enlace por enlace** DATOS link-by-link encipherment; **~ de enlaces** GEN *criptología* link encryption; **~ de extremo a extremo** DATOS end-to-end encipherment; **~ de la voz** INFO, TELEFON speech ciphering
cifrar *vt* GEN encrypt
cifras: **~ de determinación de tarifa** *f pl* TELEFON fee digits
CIG-DTC *abr* (*identificación DTC de abonado que llama*) TELEFON CIG-DTC (*calling-subscriber identification-DTC*); **~ de orden de instalación no normalizada** *m* (*NSC-CIG-DTC*) DATOS nonstandard facility command CIG-DTC (*NSC-CIG-DTC*)
cilindro: **~ de impresión** *m* PERIF print barrel
CIM *abr* (*fabricación informatizada*) INFO CIM (*computer-integrated manufacturing*)
cincha *f* INFO cinching
cinta *f* GEN tape, ribbon; **~ aislante** ELECTROTEC electrical tape; **~ de audio digital de cabeza rotatoria** (*R-DAT*) DATOS, INFO rotary-head digital audio tape (*R-DAT*); **~ auxiliar** INFO backing tape; **~ de conductores flexibles** ELECTROTEC flexstrip; **~ de control de carro** PERIF *impresoras* carriage-control tape; **~ de designación** GEN *marcaje* designating tape; **~ perforada** TELEFON perforated tape, punched tape; **~ de serpentina** INFO streamer; **~ de vídeo** INFO, TV videotape
CIRC *abr* (*código Reed-Solomon intercruzado*) INFO CIRC (*cross-interleaved Reed-Solomon code*)
circuitería: **~ de computadora** *AmL f* (*cf circuitería de ordenador Esp*) INFO computer circuitry; **~ de ordenador** *Esp f* (*cf circuitería de computadora AmL*) INFO computer circuitry
circuito *m* GEN circuit; **~ de abonado** TELEFON subscriber circuit; **~ de acceso** TELEFON access circuit; **~ activador** ELECTRÓN trigger circuit; **~ de alarma** PRUEBA, TELEFON alarm circuit; **~ de alarma de averías** TELEFON failure alarm circuit; **~ de alarma de fallos** TELEFON fault alarm circuit; **~ amplificador** ELECTROTEC amplifier circuit; **~ amplificado**

TRANS amplified circuit; ~ **arrendado** TELEFON leased circuit; ~ **arrendado a tiempo parcial** TRANS part-time leased circuit; ~ **automático internacional** TELEFON international automatic circuit; ~ **auxiliar** ELECTROTEC auxiliary circuit; ~ **de banda ancha** TRANS wideband circuit; ~ **bidireccional** ELECTRÓN two-way circuit; ~ **bipolar** ELECTRÓN bipolar circuit; ~ **de bucle** ELECTROTEC loop circuit; ~ **de bucle cerrado** ELECTRÓN closed-loop circuit; ~ **de bucle de realimentación** ELECTRÓN, INFO, TEC RAD feedback loop; ~ **en celosía** ELECTROTEC lattice network; ~ **cerrado** TRANS loopback; ~ **cerrado de vídeo** TV video loop; ~ **de codificación** SEÑAL coding circuit; ~ **combinatorio** ELECTRÓN combinatorial circuit; ~ **de comparación** ELECTRÓN comparison circuit; ~ **de compensación** ELECTRÓN compensation circuit, INFO buffer circuit; ~ **de conexión dentro de la oficina** CONMUT intraoffice junctor circuit *AmE*, interchange junctor circuit *BrE*; ~ **de conferencias** TELEFON conference circuit; ~ **configurador de impulsos** ELECTRÓN pulse-shaping circuit; ~ **de conmutación de antena** TEC RAD antenna switching circuit; ~ **conmutado** CONMUT switched circuit; ~ **contador** ELECTRÓN counting circuit; ~ **continental** TELEFON continental circuit; ~ **de control** CONTROL control circuit; ~ **de cordón de control de puente** TELEFON bridge control cord circuit; ~ **de cordón flexible** PRUEBA, TELEFON cord circuit; ~ **de corrección de impulsos** TRANS pulse-correction circuit; ~ **de cuatro hilos conductores** TELEFON four-wire circuit; ~ **Darlington** ELECTRÓN Darlington circuit; ~ **de datos** DATOS, TRANS data circuit; ~ **de decisión** TELEFON decision circuit; ~ **desfasador** ELECTROTEC phase-shifting circuit; ~ **detenedor** TELEFON stopper circuit; ~ **diferencial** ELECTROTEC differential circuit; ~ **de diodos** ELECTRÓN diode circuit; ~ **con dispositivo de carga acoplada** DATOS charge-coupled device circuit; ~ **de ecualizador** TRANS equalizer circuit; ~ **eléctrico** ELECTRÓN electrical circuit, ELECTROTEC electric circuit; ~ **de enclavamiento** ELECTRÓN latch circuit; ~ **de enlace** ELECTRÓN interface circuit, TELEFON junction circuit; ~ **de enlace entre centralitas telefónicas automáticas** TRANS *programas sonoros* inter-PABX tie circuit; ~ **equivalente de Thévenin** ELECTROTEC Thévenin equivalent circuit; ~ **estabilizador de tensión** ELECTRÓN voltage-stabilizing circuit; ~ **de excitación** ELECTROTEC energizing circuit; ~ **excitador** ELECTROTEC driving circuit; ~ **de extensión** TELEFON extension circuit; ~ **fantasma** TELEFON, TRANS phantom circuit; ~ **de fijación de amplitud** ELECTRÓN clamping circuit; ~ **generador de paridad** CONMUT parity-generating circuit; ~ **híbrido** ELECTRÓN hybrid circuit; ~ **identificador** ELECTRÓN identifier circuit; ~ **IGE** ELECTRÓN, TELEFON, TRANS LSI circuit; ~ **impreso** COMPON, ELECTRÓN printed circuit; ~ **impreso de capas múltiples** ELECTRÓN multilayer printed circuit; ~ **de integración a escala muy grande** ELECTRÓN, TELEFON, TRANS very large-scale integration circuit; ~ **de integración a gran escala** ELECTRÓN, TELEFON, TRANS large-scale integration circuit; ~ **de integración de escala ultragrande** ELECTRÓN ultralarge scale integration circuit; ~ **integrado** COMPON, ELECTRÓN integrated circuit; ~ **integrado de aplicación específica** (*CIAE*) COMPON, ELECTRÓN application-specific integrated circuit (*ASIC*); ~ **integrado de capas gruesas** COMPON, ELECTRÓN thick-layer integrated circuit; ~ **integrado de contorno pequeño** (*SOIC*) ELECTRÓN small-outline integrated circuit (*SOIC*); ~ **integrado con efecto de campo** DATOS field-effect integrated circuit; ~ **integrado Hall** ELECTRÓN Hall integrated circuit; ~ **integrado híbrido** ELECTRÓN hybrid integrated circuit; ~ **integrado monolítico milimétrico** ELECTRÓN millimetric monolithic integrated circuit; ~ **integrado de película gruesa** COMPON, ELECTRÓN thick-film integrated circuit; ~ **integrador** ELECTRÓN integrating circuit; ~ **de intercambio** TRANS interchange circuit; ~ **de intercepción troncal** TELEFON intercept trunk; ~ **de interconexión** TELEFON interconnection circuit; ~ **de interconexión de línea** (*LIC*) CONMUT line interface circuit (*LIC*); ~ **intercontinental** TELEFON intercontinental circuit; ~ **de interconexión de línea de abonado** (*SLIC*) CONMUT subscriber line interface circuit (*SLIC*); ~ **internacional** TELEFON international circuit; ~ **internacional de TV de destino múltiple** TV international multiple-destination television circuit; ~ **internacional de programas de sonido de destino múltiple** TV international multiple-destination sound-programme (*AmE* sound-program) circuit; ~ **internacional dedicado** TELEFON international leased circuit; ~ **de inversor** ELECTRÓN inverter circuit; ~ **inversor** ELECTRÓN NOT element; ~ **de kiloflujo** APLIC kilostream circuit; ~ **largo** TELEFON long circuit; ~ **de línea de abonado** CONMUT, TELEFON subscriber line circuit (*SLC*); ~ **de llamada virtual** CONMUT virtual call switched virtual circuit; ~ **lógico bipolar** ELECTRÓN bipolar logic; ~ **lógico de emisor acoplado** (*CLEA*) DATOS, ELECTRÓN emitter-coupled logic (*ECL*); ~ **lógico de transistor a transistor** (*CLTT*) ELECTRÓN transistor-transistor logic (*TTL*); ~ **de mantenimiento de nivel** ELECTRÓN *electricidad* holding circuit; ~ **marítimo de satélite** TEC RAD maritime satellite circuit; ~ **marítimo terrestre** TELEFON maritime terrestrial circuit; ~ **matricial** ELECTRÓN matrix circuit; ~ **de medición** PRUEBA measuring circuit; ~ **de megaflujo** TRANS megastream circuit; ~ **de memoria** ELECTRÓN memory circuit; ~ **metálico** TELEFON metallic circuit; ~ **monofilar** COMPON, ELECTRÓN single-wire circuit; ~ **MSI** TELEFON MSI circuit; ~ **NO** ELECTRÓN NOT circuit; ~ **no inductivo** ELECTROTEC noninductive circuit; ~ **no lineal** TRANS nonlinear circuit; ~ **NO-O** ELECTRÓN *disyuntivo negativa* NOR circuit; ~ **no recíproco** ELECTRÓN nonreciprocal circuit; ~ **NO-SI-ENTONCES** ELECTRÓN NOT-IF-THEN element; ~ **NO-Y** ELECTRÓN NAND gate; ~ **O** ELECTRÓN OR circuit; ~ **de ocupación** TELEFON seizure circuit; ~ **O-EXCLUSIVO** ELECTRÓN EXCLUSIVE-OR element; ~ **O-INCLUSIVO** ELECTRÓN INCLUSIVE-OR circuit; ~ **de ondas portadoras** TRANS carrier circuit; ~ **de órdenes de voz** TELEFON, TRANS voice-order wire; ~ **O de señales N** *f* INFO OR of N signals; ~ **opticoelectrónico integrado** ELECTRÓN integrated optoelectronic circuit; ~ **óptico integrado** TRANS integrated optical circuit (*IOC*); ~ **de órdenes** TRANS order wire; ~ **oscilante** ELECTRÓN, ELECTROTEC oscillation circuit; ~ **en paralelo** ELECTRÓN parallel circuit; ~ **pasivo** ELECTRÓN passive circuit; ~ **permisivo** DATOS permissive circuit; ~ **de pila patrón** ELECTRÓN standard cell circuit; ~ **portador** DATOS bearer circuit; ~ **primario** TRANS backbone circuit; ~ **de programa de sonido** TV sound-programme (*AmE* sound-program) circuit; ~ **de programa de sonido internacional** TV international sound-programme (*AmE* sound-program) circuit; ~ **de protección** PRUEBA protection circuit; ~ **de prueba de señal** PRUEBA, SEÑAL signal test circuit; ~ **en puente** ELECTRÓN, ELECTROTEC bridge circuit; ~ **de puerta** ELECTRÓN *elementos lógicos* gate circuit; ~ **puerta de mayoría** ELECTRÓN majority gate; ~ **puramente resistivo** ELECTROTEC purely-resistive circuit; ~ **RCD** ELECTRÓN RCD circuit; ~ **real** TELEFON side circuit; ~ **de reconocimiento** ELECTRÓN recognition circuit; ~ **de recuperación de la temporización** ELECTRÓN, TRANS timing-recovery circuit; ~ **de recuperación del reloj** ELECTRÓN clock-recovery circuit; ~ **de referencia hipotético** (*HRC*) TEC RAD, TELEFON, TRANS hypothetical reference circuit (*HRC*); ~ **de referencia hipotética para telefonía** TELEFON hypothetical reference circuit for telephony; ~ **de referencia hipotética para transmisión de programas de sonido** TEC RAD hypothetical reference circuit for sound-programme (*AmE* sound-program) transmissions; ~ **de referencia hipotética terrestre** TEC RAD terrestrial hypothetical reference circuit; ~ **de refrigeración** CLIMAT refrigeration circuit; ~ **de relación** ELECTRÓN ratio circuit; ~ **de reserva** TELEFON reserve circuit; ~ **resistivo** ELECTRÓN resistive circuit; ~ **de resonancia paralelo** ELECTRÓN, ELECTROTEC parallel resonant circuit; ~ **resonante** ELECTRÓN, ELECTROTEC resonant circuit; ~ **resonante en serie** ELECTRÓN, ELECTROTEC series resonant circuit; ~ **de retardo** ELECTRÓN delay circuit; ~ **de salida** ELECTRÓN output circuit; ~ **secreto** DATOS, TELEFON privacy circuit; ~ **de seguridad** TELEFON guard circuit; ~ **semiduplex** TELEFON half-duplex circuit; ~ **de señal** CONMUT, SEÑAL, TELEFON line circuit, signal circuit, TRANS signal path; ~ **en serie** ELECTRÓN series circuit; ~ **de servicio ómnibus** TELEFON omnibus service circuit; ~ **de servicio técnico** TRANS engineering service circuit; ~ **de servicio de terminal múltiple** TELEFON multiterminal service circuit; ~ **simétrico** ELECTRÓN, ELECTROTEC push-pull circuit; ~ **simple de multipunto** TELEFON simple multipoint circuit; ~ **sin relación** ELECTRÓN ratioless circuit; ~ **de sintonía escalonada** ELECTRÓN stagger-tuned circuit; ~ **sintonizado** ELECTROTEC tuned circuit; ~ **sintonizador** ELECTRÓN tuner circuit, tuning circuit; ~ **de sonido de programa** TRANS programme (*AmE* program) sound circuit; ~ **sumador** DATOS adder; ~ **superfantasma** TRANS double-phantom circuit, superphantom circuit; ~ **superpuesto** TELEFON superposed circuit; ~ **de supervisión de central propia** CONMUT own-exchange supervisory circuit; ~ **tanque** ELECTRÓN tank circuit; ~ **de telecomunicación** TEC RAD, TELEFON telecommunication circuit; ~ **telefónico**

internacional TELEFON, TRANS international telephone circuit; ~ **telefónico privado escalonado** TRANS speech-grade private wire; ~ **telegráfico** TELEFON telegraph circuit; ~ **telegráfico balanceado superfantasma** TELEFON double-phantom balanced telegraph circuit; ~ **telegráfico de frecuencia vocal** TELEFON, TRANS voice-frequency telegraph circuit; ~ **telegráfico equilibrado superfantasma** TELEFON superphantom balanced telegraph circuit; ~ **telegráfico fantasma** TELEFON phantom telegraph circuit; ~ **de televisión** TV television circuit; ~ **de televisión internacional** TV international television circuit; ~ **temporizador** GEN timing circuit; ~ **térmico equivalente** COMPON *semiconductores* equivalent thermal network; ~ **terminal** ELECTROTEC, TELEFON terminal circuit; ~ **terminal de central** (*ETC*) CONMUT exchange terminal circuit (*ETC*); ~ **de tipo tetrafilar** TELEFON four-wire type circuit; ~ **de traducción** TELEFON translation circuit; ~ **de tráfico** TELEFON traffic circuit; ~ **trampa** TEC RAD wave trap, trap circuit; ~ **de transferencia de mando** PRUEBA override circuit; ~ **de tránsito intercontinental** TELEFON intercontinental transit circuit; ~ **trifásico** ELECTROTEC three-phase circuit; ~ **troncal de PBX** TELEFON PBX trunk circuit; ~ **troncal de salida** TELEFON outgoing trunk circuit; ~ **SI-ENTONCES** ELECTRÓN *elementos lógicos* IF-THEN element; ~ **troncal** TEC RAD trunking; ~ **ULSI** ELECTRÓN ULSI circuit; ~ **umbral** ELECTRÓN threshold gate; ~ **de uso privado de datos** TRANS data private wire; ~ **de varias etapas** ELECTRÓN multistage circuit; ~ **virtual** DATOS, TRANS virtual circuit; ~ **virtual canal D** DATOS, TELEFON D-channel virtual circuit; ~ **VLSI** COMPON, ELECTRÓN VLSI circuit; ~ **de voz** TELEFON, TRANS speech circuit; ~ **de voz y datos alternados** (*AVD*) TRANS alternate voice-data circuit (*AVD*); ~ **Y** ELECTRÓN *elementos lógicos* AND circuit

circuitos: ~ **de noria** *m pl* CONMUT *registro de desplazamiento* bucket brigade device (*BBD*); ~ **prefundidos** *m pl* ELECTRÓN gate array circuit

circulación: ~ **de los materiales** *f* CONMUT, DATOS, INFO, TELEFON, TRANS *operaciones, talleres* routing

circulador *m* TRANS circulator; ~ **híbrido** TEC RAD hybrid circulator

circularidad: **no** ~ *f* TELEFON *de revestimiento, de núcleo*, TRANS *de revestimiento, núcleo* noncircularity; **no** ~ **del núcleo** *f* TELEFON, TRANS noncircularity of core; **no** ~ **del revestimiento** *f* TELEFON, TRANS noncircularity of cladding; **no** ~ **de la superficie de referencia** *f* TELEFON noncircularity of reference surface

círculo: ~ **de calidad** *m* PRUEBA quality circle; ~ **primitivo** *m* GEN pitch circle

circunferencia: ~ **interna** *f* GEN root circle

circunvolucionar *vt* TEC RAD *órbita* circularize

CISE *abr* (*equipo complejo de conjunto de instrucciones*) DATOS CISE (*complex instruction set equipment*)

CL *abr* (*sin conexiones*) GEN CL (*connectionless*)

claridad *f* GEN clarity

clase *f* GEN brand, category; ~ **de comando** TELEFON command class; ~ **de conmutación de tarifas** TELEFON tariff-switching class; ~ **de documento** GEN *documentación* class of document; ~ **de emisión** TEC RAD class of emission; ~ **de error del protocolo** DATOS protocol error class; ~ **funcional** (*FC*) TELEFON functional class (*FC*); ~ **funcional A** (*FCA*) TELEFON functional class A (*FCA*); ~ **funcional B** (*FCB*) TELEFON functional class B (*FCB*); ~ **de HDLC** TELEFON class of HDLC; ~ **no disponible de servicio u opción** DATOS service-or-option not-available class; ~ **no operativa de servicio u opción** DATOS service-or-option not-implemented class; ~ **de objeto** DATOS object class; ~ **de objeto gestionado** DATOS managed-object class (*MOC*); ~ **de operaciones** DATOS operation class; ~ **de producción** DATOS throughput class, thruput class *AmE*; ~ **de protocolo** DATOS protocol class; ~ **de recurso no disponible** DATOS resource-unavailable class; ~ **de servicio incondicional** DATOS unconditional class of service; ~ **sin notificación** DATOS *de servicio* no notification class; ~ **de transferencia de documentos en masa** (*BT*) TELEFON document bulk transfer class (*BT*); ~ **de transferencia y manipulación de documentos en masa** (*BTM*) TELEFON document bulk transfer and manipulation class (*BTM*)

clasificación *f* GEN classification; ~ **de bloques** INFO block sort; ~ **de los fallos** PRUEBA *fiabilidad* classification of failures; ~ **prioritaria** TELEFON priority classification

clasificador *m* ELECTROTEC sorter, INFO arranger, sorter; ~ **automático** INFO auto-arranger

clasificar *vt* GEN classify

clave *f* INFO code element; ~ **alfabética** INFO alphabet key; ~ **de autorización** TELEFON authorization key; ~ **de búsqueda** INFO search key; ~ **de código** TEC RAD code key; ~ **de**

control DATOS check bit; ~ **de control del módem** DATOS modem control key; ~ **de grupo** ELECTRÓN group key; ~ **de interflujo** TELEFON interflow key; ~ **de interrogación** INFO polling key; ~ **pública** TEC RAD public key; ~ **real** INFO *COBOL* actual key; ~ **de referencia del receptor** DATOS, INFO recipient's reference password

clavija *f* COMPON, ELECTRÓN free connector; ~ **bipolar para corriente alterna** INFO alternative current socket; ~ **de código** TEC RAD *móviles* code plug; ~ **de conexión** COMPON, COMPON, PERIF jack; ~ **de contacto** ELECTRÓN *dispositivos de conexión* contact pin; ~ **DIN** INFO DIN jack, DIN plug; ~ **de distribución** ELECTROTEC *suministro de energía* distribution plug; ~ **de espigas** COMPON free-blade connector, ELECTRÓN free pin connector, *dispositivos de conexión* free blade connector; ~ **de horquilla** COMPON, ELECTRÓN *dispositivos de conexión* free fork connector; ~ **de manguito** COMPON, ELECTRÓN *dispositivos de conexión* free sleeve connector; ~ **para medición** PRUEBA measuring plug; ~ **metálica de doble eje** ELECTRÓN *f* metal twinax plug; ~ **modular modificada** (*MMP*) ELECTRÓN modified modular plug (*MMP*); ~ **de ocupación** CONMUT break jack; ~ **de puenteado** COMPON, ELECTROTEC strapping plug; ~ **de punta cónica** ELECTRÓN *dispositivos de conexión* banana jack, banana plug, PRUEBA banana plug; ~ **de tres conductores** COMPON three-conductor plug

CLEA *abr* (*circuito lógico de emisor acoplado*) DATOS, ELECTRÓN ECL (*emitter-coupled logic*)

cliché: ~ **de producción** *m* ELECTRÓN production master

cliente: ~ **comercial** *m* GEN business customer; ~ **de un laboratorio de pruebas** *m* DATOS client of a test laboratory

cliente/servidor *m* INFO client/server

CLIP *abr* (*presentación de la identificación de la línea que llama*) TELEFON CLIP (*calling-line identification presentation*)

cliquear *vt* INFO click

clistrón: ~ **de variación equilibrada** *m* ELECTRÓN floating drift-tube klystron

clon *m* GEN clone

cloropreno *m* GEN chloroprene

cloruro: ~ **de polivinilo** *m* ELECTRÓN polyvinyl chloride

CLR *abr* (*indicador de sonoridad de un circuito*) TELEFON CLR (*circuit-loudness rating*)

CLTT *abr* (*circuito lógico de transistor a transistor*) ELECTRÓN TTL (*transistor-transistor logic*)

CLUT *abr* (*tabla de consulta de colores*) INFO CLUT (*colour look-up table*)

CMB *abr* (*bloque de mensajes CRC*) TELEFON CMB (*CRC message bloc*)

co: ~**-polar** *adj* TEC RAD co-polar; ~**-polarizado** *adj* TEC RAD co-polarized

coaxial[1]: ~ **delgado** *m* COMPON, TRANS thin coax; ~ **grueso** *m* COMPON, TRANS thick coax

coaxial[2] *m* ELECTRÓN coaxial

cobertura *f* TEC RAD, TV coverage; ~ **dentro de un edificio** TEC RAD in-building coverage; ~ **general** TEC RAD blanket coverage; ~ **global** TEC RAD global coverage; ~ **del haz del punto de exploración** TEC RAD spot beam coverage; ~ **hemisférica** TEC RAD hemispherical coverage; ~ **de sílice** COMPON silica coating; ~ **superpuesta** TEC RAD overcoverage; ~ **vertical** TEC RAD vertical coverage; ~ **volumétrico** TEC RAD volumetric coverage

cobrar: ~ **en destino** *vt* APLIC, TELEFON reverse the charges *BrE*, call collect *AmE*

cobre: ~ **electrolítico** *m* GEN electrolytic copper; ~ **estañado** *m* COMPON, ELECTRÓN tinned copper

cobro: ~ **revertido** *m* APLIC reversed charging, TELEFON reversed charging, transferred charging

cocanal *m* TEC RAD co-channel

cociente: ~ **de cancelación** *m* TEC RAD cancellation ratio

codec: ~ **de subbanda** *m* INFO sub-band codec

códice: ~ **legal mu** *m* TRANS mu law codec

codificación *f* DATOS encoding, INFO coding, coding process, encoding, TEC RAD cyphering, TELEFON coding process, encoding, TRANS coding, encoding; ~ **por adaptación** SEÑAL adaptive coding; ~ **de anchura** TEC RAD *modulación por amplitud de impulso* width coding; ~ **aritmética** DATOS, INFO arithmetic coding; ~ **básica** INFO basic encoding rules (*BER*); ~ **binaria** SEÑAL binary coding; ~ **bit a bit** DATOS bit-by-bit encoding; ~ **de bloque** TRANS block coding; ~ **convolucional** INFO convolutional coding; ~ **convolucional de media unidad de velocidad** DATOS rate one half convolutional coding; ~ **de corrección de error** DATOS error correction coding; ~ **dentro del cuadro** TV intraframe coding; ~ **dentro de la trama** DATOS in-frame coding; ~ **para la detección de errores** DATOS error-detection coding; ~ **en enrejado** DATOS trellis coding; ~ **de entropía** SEÑAL entropy coding; ~ **de fase** DATOS phase encoding; ~ **fotográfica** INFO,

TELEFON photographic coding; ~ **de frecuencia** TEC RAD frequency encoding; ~ **de Golay** DATOS Golay coding; ~ **de imagen** DATOS image coding; ~ **de impulsos** DATOS, TRANS pulse coding; ~ **intertrama** DATOS interframe coding; ~ **por intervalos** TRANS gap coding; ~ **de longitud de tirada** (*RLE*) INFO run-length encoding (*RLE*); ~ **por mínimo desplazamiento de fase por filtrado gaussiano** SEÑAL, TELEFON Gaussian-filtered minimum shift keying (*GMSK*); ~ **non-intra** INFO non-intra coding; ~ **pictórica** INFO, TELEFON picture coding; ~ **predictiva** DATOS predictive coding; ~ **progresiva** INFO, TELEFON progressive coding; ~ **secuencial compatible progresiva** TELEFON progressive-compatible-sequential coding; ~ **secuencial de progresión simple** TELEFON single-progression sequential coding; ~ **de señal compuesta** TRANS composite signal coding; ~ **de subbanda** SEÑAL sub-band coding; ~ **de subbanda ajustable de precisión** (*PASC*) INFO precision adaptive sub-band coding (*PASC*); ~ **por submuestreo múltiple** (*MUSE*) DATOS multiple subsampling encoding (*MUSE*); ~ **uniforme** TELEFON, TRANS uniform encoding; ~ **vocal** INFO, TEC RAD, TELEFON vocoding; ~ **de la voz** INFO, TEC RAD, TELEFON speech coding, speech encoding

codificado *adj* DATOS, INFO, TRANS, TV encoded; ~ **convolucional y diferencialmente** (*CDE*) DATOS convolutionally and differentially encoded (*CDE*); **no** ~ TELEFON uncoded

codificador *m* INFO, TRANS coder, encoder; ~ **activado por la voz** INFO, TEC RAD, TELEFON voice-operated coder; ~ **angular** DATOS angle encoder; ~ **de canales** TEC RAD *móviles* channel coder; ~ **CELP** SEÑAL CELP coder; ~ **convolucional** DATOS convolutional encoder; ~ **de interfaz** TRANS interface coder; ~ **de modos** TRANS mode scrambler; ~ **de múltiples frecuencias** TRANS multiple-frequency encoder; ~ **de señales ADPCM** TELEFON ADPCM encoder; ~ **vocal** INFO, TEC RAD, TELEFON voice coder; ~ **de la voz** INFO, TEC RAD, TELEFON speech encoder; ~ **de la voz para comunicación secreta** TEC RAD, TELEFON speech scrambler

codificador/descodificador *m* (*codec*) INFO, TRANS coder/decoder (*codec*); ~ **de vídeo** INFO video codec

codificar *vt* DATOS, INFO, TRANS, TV code, encode

código: ~ **abreviado** *m* TELEFON *télex* short code; ~ **de acceso de operadora** *m* TELEFON attendant-access code; ~ **de acceso a la selección automática interurbana** *m* TELEFON subscriber trunk dialling (*AmE* dialing) access code; ~ **alfabético** *m* INFO alphabet code; ~ **AMI** *m* TRANS AMI code; ~ **antiguo** *m* TELEFON old code; ~ **de autorización** *m* TELEFON authorization code; ~ **de avería** *m* TELEFON fault code; ~ **de barras** *m* APLIC, DATOS bar code; ~ **baudot** *m* INFO baudot code; ~ **bifásico** *m* APLIC *RDSI* biphase code; ~ **binario PCM** *m* DATOS, ELECTRÓN, TELEFON PCM binary code; ~ **binario rectilíneo** *m* DATOS, ELECTRÓN straight binary code; ~ **binario reflejado** *m* DATOS, ELECTRÓN reflected binary code; ~ **bipolar** *m* SEÑAL bipolar code; ~ **bipolar con sustitución de tres ceros** *m* (*B3ZS*) DATOS, SEÑAL, TELEFON bipolar code with three-zero substitution (*B3ZS*); ~ **de bloqueo** *m* DATOS, TELEFON interlock code; ~ **de bloques** *m* INFO block code; ~ **cablegráfico bivalente** *m* TELEFON, TRANS two-condition cable code; ~ **de carácter** *m* INFO character code; ~ **de central** *m* GEN exchange code; ~ **cíclico** *m* DATOS cyclic code; ~ **cíclico de bloque** *m* DATOS cyclic block code; ~ **de cinco unidades** *m* TELEFON five-unit code; ~ **de color** *m* GEN colour (*AmE* color) code; ~ **complementario** *m* TRANS complementary code; ~ **continental** *m* TELEFON continent code; ~ **convolucional** *m* INFO convolution code, convolutional code; ~ **convolucional disruptivo** *m* INFO punctured convolutional code; ~ **convolutivo** *m* INFO convolutive code; ~ **de corrección de errores** *m* DATOS, SEÑAL, TELEFON error-correcting code; ~ **corrector de errores** *m* DATOS, INFO, SEÑAL, TELEFON error-correction code (*ECC*); ~ **de corriente doble** *m* TELEFON, TRANS double-current code; ~ **4B3T** *m* TRANS code 4B3T; ~ **de datos** *m* DATOS data code; ~ **de detección de errores** *m* (*EDC*) DATOS, INFO, TELEFON error-detecting code; ~ **de diagnóstico** *m* TELEFON diagnostic code; ~ **de disparidad emparejada** *m* TELEFON paired-disparity code; ~ **DNTX** *m* SEÑAL DNTX code; ~ **dos de cinco** *m* DATOS two-out-of-five code; ~ **de empresa** *m* PERIF *PBX* company code; ~ **encadenado** *m* DATOS concatenated coding (*CC*); ~ **de encaminamiento** *m* TELEFON routing code; ~ **de enlace de señalización** *m* SEÑAL signalling-link (*AmE* signaling-link) code; ~ **de error** *m* TELEFON error code; ~ **de error de bitios** *m* INFO bit error code; ~ **especificado** *m* DATOS given code; ~ **esquemático** *m* INFO skeletal code; ~ **de estado y control** *m* CONTROL, DATOS status and control code; ~ **de ética** *m*

GEN code of ethics; ~ **frame** *m* DATOS frame code; ~ **de Golay de media unidad de velocidad** *m* DATOS rate 1/2 Golay code; ~ **de Gray** *m* DATOS Gray code; ~ **de Hamming** *m* DATOS Hamming code; ~ **Hollerith** *m* INFO Hollerith code; ~ **de identificación** *m* DATOS, TELEFON identification code; ~ **de identificación de la empresa** *m* PERIF *PBX* company identity code; ~ **de identificación de portador** *m* DATOS, TELEFON bearer identification code; ~ **de identificación del recibidor** *m* DATOS, TELEFON recipient identification code; ~ **de identificación de la red de datos** *m* DATOS, TELEFON data-network identification code (*DNIC*); ~ **de identificación de la red télex** *m* DATOS telex network identification code; ~ **identificador** *m* DATOS, TELEFON *teleimpresora* answer-back code; ~ **de igual longitud** *m* DATOS, TELEFON equal-length code; ~ **interpretativo** *m* INFO interpretive code; ~ **de inversión alternada de marcas** *m* SEÑAL, TELEFON, TRANS alternate mark inversion code; ~ **en línea** *m* DATOS, TELEFON, TRANS line code; ~ **de mandato** *m* DATOS, TELEFON command code; ~ **de máquina** *m* INFO machine code; ~ **de marcación** *m* TELEFON dialling (*AmE* dialing) code; ~ **Morse** *m* TELEFON Morse code; ~ **Morse internacional** *m* TELEFON international Morse code; ~ **multifrecuencia de secuencia obligada** *m* SEÑAL compelled multi-frequency code; ~ **de múltiples bastidores** *m* TRANS multiframe alignment code; ~ **de nivel N** *m* TELEFON N-level code; ~ **no binario** *m* DATOS nonbinary code; ~ **no transmitir** *m* (*código SDTNX*) SEÑAL *m* do-not-transmit code (*SDTNX code*); ~ **NRZ** *m* DATOS, TRANS NRZ code; ~ **de operación** *m* INFO operation code; ~ **de país** *m* DATOS, TELEFON country code; ~ **de país de los datos** *m* DATOS data country code; ~ **postal** *m* INFO post code; ~ **PRB** *m* DATOS PRB code; ~ **de prioridad de procesamiento** *m* DATOS, INFO processing priority code; ~ **de prioridad de transmisión** *m* DATOS, TRANS transmission priority code; ~ **de protección de errores** *m* DATOS error-protection code; ~ **de punto de destino** *m* DATOS, TELEFON destination point code; ~ **de punto de origen** *m* TELEFON originated point code; ~ **de punto de señalización** *m* SEÑAL signalling-point (*AmE* signaling-point) code; ~ **de puntos fijos** *m* DATOS, TELEFON point code; ~ **de redundancia cíclica** *m* DATOS, TELEFON cyclic redundancy code; ~ **redundante** *m* SEÑAL redundant code; ~ **Reed-Solomon** *m* INFO Reed-Solomon code; ~ **Reed-Solomon intercruzado** *m* (*CIRC*) INFO cross-interleaved Reed-Solomon code (*CIRC*); ~ **de respuesta** *m* TELEFON answer code; ~ **de retorno a la conexión normal** *m* SEÑAL changeback code; ~ **RS** *m* INFO RS code; ~ **SDTNX** *m* DATOS SDTNX code; ~ **secreto** *m* TEC RAD secret code; ~ **de selección** *m* TELEFON selection code; ~ **selectivo de no transmitir** *m* DATOS selective do not transmit code; ~ **de señal** *m* SEÑAL signal code; ~ **de señalamiento de control** *m* SEÑAL control signalling (*AmE* signaling) code; ~ **de señalización de línea** *m* (*LSC*) DATOS, SEÑAL line signalling (*AmE* signaling) code (*LSC*); ~ **de servicio** *m* CONMUT, TELEFON service code; ~ **de sincronización** *m* DATOS, INFO synchronization code; ~ **de teclado** *m* PERIF keyset; ~ **telegráfico** *m* TELEFON cable code, telegraph code; ~ **telegráfico bivalente** *m* TELEFON, TRANS two-condition telegraph code; ~ **telegráfico trivalente** *m* TELEFON, TRANS three-condition telegraph code; ~ **télex de destino** *m* (*TDC*) DATOS, TELEFON telex destination code (*TDC*); ~ **temporal longitudinal** *m* (*LTC*) INFO longitudinal time code (*LTC*); ~ **de tiempo de intervalo vertical** *m* (*VITC*) DATOS, INFO vertical interval time code (*VITC*); ~ **de tiempos de la SMPTE** *m* INFO, TV SMPTE time code; ~ **trivalente para cable** *m* TELEFON, TRANS three-condition cable code; ~ **vocal** *m* TRANS speech code; ~ **de zona** *m* TELEFON area code

Código: ~ **Estándar Americano para Intercambio de Información** *n* (*ASCII*) DATOS, INFO, TELEFON American Standard Code for Information Interchange *m* (*ASCII*)

codo: ~ **H** *m* TEC RAD H-bend, H-corner

coeficiente *m* GEN coefficient; ~ **de atenuación** TEC RAD, TRANS attenuation coefficient; ~ **de atenuación de imagen** TRANS image-attenuation coefficient; ~ **de calidad** PRUEBA, TEC RAD figure of merit; ~ **de cambio de fase** GEN *acústica* phase-change coefficient; ~ **de cambio de fase de imágenes** TRANS image phase-change coefficient; ~ **de dispersión** TEC RAD, TRANS *óptica de fibras* scattering coefficient; ~ **de dispersión debida al material** TRANS material dispersion coefficient; ~ **de error binario residual** TEC RAD residual binary error rate; ~ **de expansión lineal** GEN *física* coefficient of linear expansion; ~ **de extinción** ELECTRÓN, TRANS extinction ratio (*EX*); ~ **de fase** GEN phase coefficient; ~ **de Hall** ELECTRÓN *semiconductores* Hall coefficient; ~

de impulsos TELEFON pulse duty factor; ~ **de onda estacionaria** (*ROE*) TEC RAD, TRANS standing-wave ratio (*SWR*); ~ **de propagación** ELECTRÓN, TEC RAD, TRANS propagation coefficient; ~ **de reflexión** TEC RAD, TRANS reflection coefficient; ~ **de reflexión especular** TRANS specular reflection coefficient; ~ **de reflexión de irregularidad** TELEFON irregularity reflection coefficient; ~ **de relleno** DATOS filling coefficient; ~ **de ruido** PRUEBA, TEC RAD, TRANS noise figure; ~ **de seguridad** GEN confidence coefficient; ~ **de temperatura** CLIMAT temperature coefficient; ~ **de tiempo de subida** COMPON rise-time coefficient; ~ **de transferencia de imagen** DATOS, TRANS image transfer coefficient; ~ **de transmisión** TRANS *óptica de fibras* transmission coefficient

coercividad *f* ELECTRÓN coercivity

COFDM *abr* (*multiplex de división de frecuencia ortogonal codificada*) TRANS COFDM (*coded orthogonal frequency division multiplex*)

cognitivo *adj* GEN cognitive

coherencia *f* GEN coherence, consistency; ~ **espacial** TRANS spatial coherence; ~ **de fase** TEC RAD, TRANS phase coherence; ~ **de la frecuencia** TEC RAD frequency coherence; ~ **de modulación** TELEFON modulation coherence; ~ **de onda** ELECTRÓN wave coherence; ~ **temporal** TRANS temporal coherence

cohesión *f* GEN cohesion

cohete: ~ **de lanzamiento** *m* TEC RAD launch rocket

coincidencia *f* TELEFON coincidence

coincidente *adj* TELEFON coincident

cojinete: ~ **de bolas** *m* COMPON ball bearing; ~ **de largo recorrido** *m* TEC RAD long-path bearing; ~ **de piedra dura** *m* COMPON, ELECTROTEC jewel bearing

cola *f* GEN queue; ~ **de espera** TELEFON waiting queue; ~ **de espera limitada** TELEFON limited-waiting queue; ~ **de fibra de empalme** COMPON splicing pigtail; ~ **en servidor múltiple** DATOS, INFO multiple-server queue; ~ **de un solo servidor** DATOS, INFO single-server queue

colchón *m* GEN cushion

colección: ~ **de términos** *f* GEN *lingüística* collection of terms

colectivo *adj* GEN corporate

colector *m* ELECTRÓN, ELECTROTEC collector, commutator; ~ **de datos** DATOS data sink; ~ **de transistor** ELECTRÓN transistor collector

colgado *adj* INFO, TELEFON on-hook

colgar *vi* TELEFON replace the handset

colimación *f* TRANS collimation

colimador *m* TRANS collimator

colisión *f* ELECTRÓN collision, TEC RAD clashing; ~ **electrónica** ELECTRÓN electron collision

collarín *m* ELECTRÓN *de tubo de rayos catódicos* neck; ~ **de borne** ELECTRÓN clamp terminal

colocación *f* GEN placement, placing

colofonia *f* ELECTRÓN *soldadura* colophony

colón *m* INFO colon

color *m* GEN colour (*AmE* color); ~ **primario sustractivo** INFO subtractive primary colour (*AmE* color); **de ~ sólido** *adj* GEN colourfast (*AmE* colorfast)

coloración *f* GEN coloration, colorization

colores: ~ **contrastantes** *m pl* TV contrasting colours (*AmE* colors)

ColorSence® *m* INFO ColorSence®

ColorSync® *m* INFO ColorSync®

columna *f* GEN *documentación* column; ~ **de espigas** ELECTRÓN pin column

columnas: ~ **múltiples** *f pl* DATOS *Internet* multiple columns

coma *f* INFO comma; ~ **flotante** INFO floating point

COMA *abr* (*ML COM de la ramificación A*) TELEFON COMA (*ML COM of A branch*)

COMB *abr* (*ML COM de ramificación B*) TELEFON COMB (*ML COM of B branch*)

combarse *v refl* COMPON, ELECTROTEC wind

combinación: ~ **binaria** *f* INFO, TELEFON bit combination; ~ **de carácter especial** *f* DATOS special character combination; ~ **heurística** *f* DATOS heuristic mix; ~ **de pista** *f* INFO track mix; ~ **de prestaciones** *f* INFO performance combination

combinador *m* ELECTRÓN combiner

combinar *vt* ELECTRÓN combine

combinatorio *adj* GEN combinational

comentario *m* GEN comment

comercialización: ~ **por nichos** *f* GEN niche marketing

comienzo *m* GEN start; ~ **de llamada** DATOS beginning of call demand; ~ **de mensaje** DATOS beginning of message (*BOM*), PERIF message beginning; ~ **de la pulsación** INFO beginning of pulse

comisión: ~ **básica** *f* TELEFON basic charge; ~ **de mejoramiento** *f* INFO upgrade board

Comisión: ~ **Federal de Comunicación** *f* (*FCC*) DATOS Federal Communication Commission (*FCC*)

comisionista *m* INFO broker

Comité: ~ **Asesor para Equipo de Telecomunicaciones** *m* (*ACTE*) TELEFON

Advisory Committee for Telecommunications Equipment (*ACTE*); ~ **Consultivo Internacional de Telefonía y Telegrafía** *m* (*CCITT*) GEN International Telegraph and Telephone Consultative Committee (*CCITT*); ~ **12 del Instituto Nacional Americano de Normas** *m* DATOS American National Standards Institute committee 12; ~ **Europeo para la Normalización** *m* (*CEN*) GEN European Committee for Standardization (*CEN*); ~ **Europeo para la Normalización Electrotécnica** *m* (*CENELEC*) GEN European Committee for Electrotechnical Standardization (*CENELEC*); ~ **Nacional del Sistema de Televisión** *m* INFO National Television System Committee

cómo: **¿~ me recibe?** *fra* TEC RAD how do you read me?

comodidad: ~ **telefónica** *f* TEC RAD, TELEFON phoning comfort

compacidad *f* GEN compactness

compactar *vt* INFO compact

compacto *adj* GEN compact, solid

compañía: ~ **de explotación de la red** *f* TV operating company; ~ **operadora** *f* GEN operating company; ~ **telefónica regional de sistema Bell** *f* TELEFON regional Bell operating company

compansión: ~ **silábica** *f* SEÑAL, TRANS syllabic companding

comparación *f* GEN comparison; ~ **lógica** INFO logical comparison

comparador *m* ELECTRÓN comparator; ~ **de señales** TEC RAD signal comparator; ~ **de tensión criogénica** ELECTRÓN cryogenic current comparator

compartición *f* TEC RAD sharing; ~ **de dispositivo periférico** *m* DATOS peripheral device sharing

compartimiento *m* COMPON compartment, GEN, INFO *acto* sharing, division; ~ **para placa de circuitos** COMPON *técnicas de montaje* board compartment; ~ **de recursos** INFO resource sharing

compás: ~ **descendente** *m* INFO downbeat

compatibilidad *f* GEN *propiedad*, INFO compatibility; ~ **ascendente** INFO upward compatibility; ~ **de los datos** INFO data compatibility; ~ **dimensional** GEN dimensional compatibility

compatible *adj* GEN, INFO compatible; ~ **con aparato auditivo** PERIF *contestador* hearing-aid compatible

compendio *m* GEN *formación* compendium

compensación *f* GEN compensation; ~ **de arrastre** TELEFON drift compensation; ~ **de distorsión característica** DATOS, TELEFON characteristic distortion compensation; ~ **de fase** ELECTRÓN, ELECTROTEC phase compensation, phase equalization; ~ **de luminancia** TELEFON luminance compensation; ~ **de movimiento** TV motion compensation; ~ **reveladora** TRANS compromise equalization; ~ **en la trayectoria eléctrica** TRANS electrical path equalization

compensador: ~ **de distorsión trapezoidal** *m* TV line keystone waveform; ~ **estrecho** *m* TRANS *óptica de fibras* tight buffer; ~ **de fase** *m* ELECTRÓN, ELECTROTEC phase equalizer; ~ **de sintonía** *m* COMPON, ELECTRÓN trimmer capacitor; ~ **de transmisión** *m* TELEFON, TRANS transmission buffer

compensar *vt* GEN compensate

competencia *f* GEN competition; ~ **regulada** GEN regulated competition

competitividad *f* GEN competitiveness

competitivo *adj* GEN competitive

compilación *f* GEN compilation

compilador *m* INFO compiler

compilar *vt* DATOS *directorio* compile

complejidad *f* GEN complexity

complejo: ~ **de conmutación** *m* CONMUT, TELEFON switching complex; ~ **de lanzamiento** *m* TEC RAD launch complex; ~ **de red de conmutación** *m* CONMUT switching network complex

complemento *m* GEN complement

completamente: ~ **cargado** *adj* GEN fully-loaded

completar *vt* GEN complete

completo *adj* GEN complete, full

componente *m* GEN component; ~ **de acceso a la red** DATOS, TELEFON network access component; ~ **de acceso a los servicios de la red** TELEFON network service access component; ~ **C** DATOS C component; ~ **de CC** TELEFON DC component; ~ **de CC inútil** ELECTRÓN nonuseful DC component; ~ **de conmutación** COMPON, CONMUT, ELECTROTEC, TELEFON switching component; ~ **de control** CONTROL control component; ~ **cromático** TELEFON colour (*AmE* color) component; ~ **de crominancia** TV chrominance component; ~ **de diafonía ininteligible** TELEFON unintelligible crosstalk component; ~ **eliminador de interferencias** PRUEBA noise suppression component; ~ **espectral** COMPON, TRANS spectral component; ~ **estándar** COMPON, ELECTRÓN standard component; ~ **impreso** ELECTRÓN printed component; ~ **de**

intermodulación TRANS intermodulation component; ~ **de invocación del servicio** TELEFON service invocation component; ~ **de luminancia** INFO luminance component; ~ **M** DATOS M component; ~ **de montaje exterior** COMPON surface-mounted component; ~ **O** DATOS O component; ~ **de ondas portadoras** ELECTRÓN carrier component; ~ **ondulatorio** ELECTROTEC ripple component; ~ **pasivo** COMPON, ELECTRÓN passive component; ~ **de restricciones** DATOS constraints part; ~ **de señal** SEÑAL signal component; ~ **de servicio** INFO service component; ~ **soldado** ELECTRÓN flying-lead component; ~ **de utilización de la red** DATOS, TELEFON network utilization component; ~ **de utilización de los servicios de la red** DATOS, TELEFON network service utilization component; ~ **Z** TEC RAD Z component

componentes: ~ **análogos multiplexados** *m pl* DATOS, TV multiplexed analog components; ~ **de diafonía inteligible** *m pl* TELEFON intelligible crosstalk components; ~ **fuera de banda inofensivos** *m pl* TELEFON harmless out-of-band components; ~ **fuera de banda perjudiciales** *m pl* TELEFON harmful out-of-band components

comportamiento *m* DATOS behaviour (*AmE* behavior); ~ **del abonado** TELEFON subscriber's dialling (*AmE* dialing) habits; ~ **de la capacidad de tráfico** TELEFON trafficability performance; ~ **por defecto** DATOS default behaviour (*AmE* behavior); ~ **progresivo** INFO, TELEFON progressive behaviour (*AmE* behavior); ~ **secuencial** INFO, TELEFON sequence behaviour (*AmE* behavior); ~ **del tráfico** TELEFON traffic performance

composición *f* GEN composition; ~ **de tarifas** APLIC, TELEFON tariff structure; ~ **tipográfica** DATOS typesetting

compositor: ~ **de ritmo** *m* INFO rhythm composer

compra: ~ **concentrada** *f* GEN one-stop shopping; ~ **a distancia** *f* APLIC, TV teleshopping

comprehensivo *adj* GEN comprehensive

compresión *f* GEN compression; ~ **de ancho de banda** DATOS bandwidth compression; ~ **de blanco** INFO, TELEFON, TRANS white crushing; ~ **de datos** DATOS data compression; ~ **dentro del cuadro** TV intraframe compression; ~ **entre imágenes** TV interframe compression; ~ **espacial** INFO spatial compression; ~-**expansión** DATOS, INFO, TELEFON, TRANS companding; ~-**expansión de bloque** INFO block companding; ~ **facsímil** TELEFON facsimile compression; ~ **de fichero** DATOS file compression; ~ **de imagen** DATOS, TV image compression; ~ **de impulso** ELECTRÓN pulse compression; ~ **de señales** SEÑAL signal compression; ~ **temporal** DATOS, INFO temporal compression; ~ **de velocidad binaria** DATOS bit-rate compression; ~ **de voz** TELEFON speech compression; ~ **de voz digital** TELEFON digital speech compression

compresor *m* DATOS compresseur *m*, TRANS compressor; ~-**expansor** DATOS, SEÑAL compander, compressor-expander

comprimir *vt* DATOS compress, INFO squeeze, *fichero* zip, TELEFON, TRANS compress; ◆ ~-**expandir** DATOS, INFO, TELEFON, TRANS compand, compress-expand

comprobación *f* GEN audit; ~ **automática** PRUEBA built-in check; ~ **de caracteres** TELEFON character check; ~ **cíclica** ELECTRÓN repetitive testing; ~ **de componentes** PRUEBA component-level testing; ~ **de conformidad** DATOS, PRUEBA conformance testing; ~ **durante la instalación** PRUEBA in-process testing; ~ **de estado** TEC RAD status check; ~ **de funcionamiento** PRUEBA functional testing; ~ **de grupo** DATOS group testing; ~ **integrada** DATOS embedded testing; ~ **de seguridad** DATOS *de transmisiones* security audit; ~ **simulada** INFO desk checking; ~ **de sobrecarga** INFO overflow check; ~ **de suma** INFO summation check

comprobador *m* PRUEBA tester; ~ **rutinario** PRUEBA routiner

compromiso *m* DATOS commitment

compuesto *m* GEN compound; ~ **de disipador térmico** ELECTROTEC heat sink compound; ~ **hidrófugo** COMPON water-blocking compound

COMPUSEC *abr* (*seguridad informática*) DATOS, INFO COMPUSEC (*computer security*)

computación *AmL f* (*cf informática Esp*) INFO computing; ~ **colaboradora** INFO collaborative computing

computador *AmL ver computadora AmL*

computadora *AmL f* (*cf ordenador Esp*) INFO computer; ~ **agenda** INFO palmtop computer; ~ **central** INFO host computer, mainframe computer; ~ **central de videotexto** INFO, TELEFON, TV videotext host; ~ **de control** INFO controlling computer; ~ **con cuadro de pinzas** INFO clipboard computer; ~ **especializada** INFO special purpose computer; ~ **especializada integrada en un equipo** INFO embedded computer; ~ **frontal** CONMUT, INFO front-end computer; ~

incorporada INFO built-in computer; ~ **de mano** INFO hand-held computer; ~ **neural** INFO neural computer; ~ **personal** (*CI*) INFO personal computer (*PC*); ~ **con programación reducida** DATOS, INFO reduced-instruction-set computer (*RISC*); ~ **con programa de control de conmutador matricial** CONMUT matrix switch control software machine; ~ **con programa de control del complejo de conmutación** CONMUT switching complex control software machine; ~ **con programa de control del enlace por PCM** TELEFON PCM link control software machine; ~ **con programa para la gestión de supletorios** TELEFON auxiliaries management software machine, auxiliary management software machine; ~ **con programa de mensajes de trabajo y funcionamiento** CONTROL operation and maintenance messaging software machine; ~ **virtual** INFO virtual machine

computadorizado *AmL adj* (*cf informatizado Esp*) INFO computerized

computar *vt* INFO compute

cómputo *m* INFO computation; ~ **múltiple** TELEFON multimetering; ~ **simple** TELEFON single-fee metering; ~ **total** APLIC, INFO, TEC RAD, TELEFON total charge

común *adj* GEN common, ordinary

comunicación *f* GEN communication; ~ **amiga** DATOS handshaking; ~ **bidireccional** TEC RAD, TELEFON, TV two-way communication; ~ **bifurcada** TELEFON forked working; ~ **binaria sincrónica** DATOS binary synchronous communication (*BSC*); ~ **colectiva** APLIC, TELEFON conference call; ~ **de datos** TRANS data communication; ~ **en los dos sentidos** TELEFON both-way communication; ~ **de entrada** TEC RAD, TRANS inbound communication; ~ **de extremo a extremo** TELEFON end-to-end communication; ~ **de facturación a un tercero** TELEFON third-party billing call; ~ **falsa** TELEFON wrong connection; ~ **interurbana no tasada** TELEFON freefone call *BrE*, toll-free call *AmE*; ~ **monomediática** TV monomedia communication; ~ **multiuso** TELEFON multiuse communication; ~ **no verbal** DATOS nonverbal communication; ~ **personal** TELEFON personal call; ~ **de punto a área** TEC RAD, TRANS point-to-area communication; ~ **de punto a multipunto** TEC RAD, TRANS point-to-multipoint communication; ~ **punto a punto** TEC RAD, TRANS point-to-point communication; ~ **por ráfagas meteóricas** TEC RAD meteor-burst communication; ~ **saliente** TEC RAD, TRANS outbound communication; ~ **sin tasa de duración** TELEFON untimed call; ~ **submarina** TEC RAD, TELEFON underwater communication; ~ **con tasa de duración** TELEFON timed call; ~ **télex de servicio** TELEFON, PERIF service telex call; ~ **tripartita** APLIC, TELEFON three-party conference; ~ **unidireccional** CONMUT, TEC RAD, TELEFON, TV one-way communication; ~ **verbal automática del servicio de cargo** ELECTRÓN automatic verbal announcement of charges service; ~ **virtual** DATOS, TELEFON virtual communication

comunicaciones: ~ **por cable** *f pl* APLIC cable communications; ~ **comerciales** *f pl* GEN business communications; ~ **diferidas** *f pl* ELECTRÓN deferrals; ~ **espaciales** *f pl* GEN space communications; ~ **hombre-máquina** *f pl* (*MMC*) INFO man-machine communications (*MMC*); ~ **internacionales** *f pl* TELEFON international communications; ~ **móviles** *f pl* TEC RAD mobile communications; ~ **personales** *f pl* TRANS personal communications; ~ **por satélite** *f pl* TEC RAD, TELEFON, TV satellite communications

comunicante *m* TELEFON communicator

comunicar *vi* GEN communicate

comunicarse *v refl* GEN communicate; ◆ ~ **a distancia** GEN telecommunicate

comunidad: ~ **de radiofaros respondedores** *f* TEC RAD community of transponders

Comunidad: ~ **Europea** *f* (*CE*) GEN, TEC RAD European Community (*EC*)

concatenación *f* DATOS concatenation

conceder *vt* APLIC grant; ◆ ~ **licencia a** APLIC license

concedido *adj* GEN granted

concentración *f* GEN concentration; ~ **en una etapa de conmutación** TELEFON concentration in a switching stage; ~ **iónica** ELECTRÓN ion concentration; ~ **de vapor** *m* CLIMAT vapour (*AmE* vapor) concentration

concentrador *m* ELECTRÓN, TRANS concentrator; ~ **de acceso** TRANS access concentrator; ~ **anexo a la central** TELEFON co-located exchange concentrator; ~ **de cableado en bucle** (*LWC*) ELECTRÓN loop wiring concentrator (*LWC*); ~ **de calor** INFO heat concentrator; ~ **de canales** TRANS channel concentrator; ~ **de central** CONMUT exchange concentrator; ~ **de centralita remoto** TELEFON remote exchange concentrator; ~ **colocado** CONMUT collocated concentrator; ~ **de conexiones** ELECTRÓN wiring concentrator; ~ **de datos** DATOS data concentrator; ~ **digital local** TELEFON local digital concentrator; ~ **digital remoto** TELEFON remote digital concentrator; ~ **de líneas** TELEFON, TRANS line concentrator, line concentrator, stand-alone concentrator; ~ **de líneas remoto** TRANS remote line concentrator; ~ **remoto** TRANS remote concentrator

concentrar *vt* DATOS bundle

concentricidad *f* GEN concentricity

concéntrico *adj* GEN concentric

concepto *m* GEN concept; ~ **de camino principal** TEC RAD major-path concept

conceptual *adj* GEN conceptual

concluir *vt* GEN conclude

concordancia: ~ **con un modelo** *f* APLIC, INFO pattern matching

condensado *adj* DATOS compressed

condensador *m* ELECTRÓN, ELECTROTEC capacitor; ~ **de acoplamiento** ELECTRÓN, ELECTROTEC coupling capacitor; ~ **bipolar** COMPON, ELECTROTEC two-gang condenser; ~ **de desacoplo** ELECTRÓN, ELECTROTEC decoupling capacitor; ~ **electrolítico** COMPON electrolytic capacitor; ~ **electroquímico** COMPON electrochemical capacitor; ~ **de filtro** COMPON, ELECTROTEC filter capacitor; ~ **de mica** COMPON, ELECTRÓN mica capacitor; ~ **de neutralización** ELECTRÓN neutralizing capacitor; ~ **de óxido** COMPON, ELECTRÓN oxide capacitor; ~ **de papel** ELECTROTEC paper capacitor; ~ **de película delgada** COMPON, ELECTRÓN thin-layer capacitor; ~ **de tantalio** COMPON, ELECTRÓN tantalum capacitor; ~ **en una unión** COMPON *semiconductores*, ELECTRÓN junction capacitor

condición *f* GEN condition; ~ **A** TELEFON condition A; ~ **para activación de puerta** ELECTRÓN *elementos lógicos* gate condition; ~ **de entrada** DATOS entry status; ~ **de falla** TELEFON fault condition; ~ **inicial** ELECTRÓN initial condition; ~ **de régimen permanente** ELECTRÓN, TRANS steady-state condition; ~ **de reposo entre caracteres** TELEFON intercharacter rest condition; ~ **tono ausente** ELECTRÓN, INFO, TEC RAD, TELEFON *f* tone-off condition; ~ **tono presente** ELECTRÓN, INFO, TEC RAD, TELEFON *f* tone-on condition; ~ **Z** TELEFON condition Z

condicionalmente: ~ **obligatorio** *adj* TELEFON conditionally mandatory (*CM*)

condicionamiento: ~ **modificado** *m* DATOS modified constraint; ~ **para señal** *m* SEÑAL signal conditioning

condiciones: ~ **del espacio libre** *f pl* TEC RAD free-space conditions; ~ **de espera** *f pl* GEN stand-by conditions; ~ **de establecimiento del estado estacionario** *f pl* ELECTRÓN steady-state launching conditions; ~ **de funcionamiento** *f pl* ELECTRÓN operational conditions, ELECTROTEC service conditions; ~ **de laboratorio** *f pl* PRUEBA laboratory test conditions; ~ **de recepción** *f pl* INFO receive conditions, reception conditions; ~ **de referencia** *f pl* PRUEBA reference conditions; ~ **secundarias** *f pl* TELEFON *de modulación* auxiliary conditions; ~ **de señalización** *f pl* SEÑAL signalling (*AmE* signaling) conditions; ~ **de servicio** *f pl* ELECTRÓN operating conditions, TELEFON operating conditions, operational conditions; ~ **de transmisión** *f pl* INFO, TRANS transmission conditions

conducción *f* GEN conduction; ~ **electrónica** ELECTRÓN electron conduction; ~ **elevada** TEC RAD elevated duct; ~ **extrínseca** COMPON, ELECTRÓN *semiconductores* extrinsic conduction; ~ **intrínseca** COMPON, ELECTRÓN *semiconductores* intrinsic conduction; ~ **de tipo B** ELECTRÓN *semiconductores* B-type conduction; ~ **de tipo N** COMPON N-type conduction

conductancia *f* ELECTRÓN, ELECTROTEC conductance; ~ **eléctrica** ELECTRÓN electrical conductance; ~ **de electrodo** ELECTRÓN electrode conductance; ~ **mutua** ELECTRÓN mutual conductance

conductibilidad *f* GEN conductibility

conducto *m* COMPON duct, ELECTRÓN duct, funnel, TRANS duct; ~ **ascendente de cables** TRANS cable runway, cable riser; ~ **cableado** TRANS cabling duct; ~ **de cables** TRANS cable raceway; ~ **de chapa metálica** COMPON sheet metal duct; ~ **elevado** TEC RAD elevated duct; ~ **interior** ELECTRÓN inner duct; ~ **multitubular** GEN duct route; ~ **portacables** COMPON, ELECTRÓN, TRANS conduit; ~ **preestirado encogido en frío** ELECTRÓN prestretched cold-shrink tubing

conductor[1] *adj* ELECTRÓN, ELECTROTEC conducting, conductive

conductor[2] *m* GEN conductor; ~ **aislado** COMPON, ELECTRÓN insulated conductor; ~ **central** TRANS central conductor; ~ **de cobre** ELECTRÓN copper conductor; ~ **compactado** TRANS *cables* compacted conductor; ~ **común de acoplamiento mutuo universal** (*GPIB*) DATOS general-purpose interface bus (*GPIB*); ~ **doble** GEN *conductores eléctricos* double conductor; ~ **de doble torcido** COMPON twisted pair; ~ **de doble torcido blindado**

ELECTRÓN, TRANS shielded twisted pair (*STP*); ~ **dúplex** GEN twin conductor; ~ **estampado** ELECTRÓN printed conductor; ~ **estañado** COMPON, ELECTRÓN tinned conductor; ~ **exterior** TELEFON, TRANS outer conductor; ~ **de fase** ELECTROTEC phase conductor; ~ **flexible** ELECTRÓN flex; ~ **flexible de casquillo** ELECTRÓN flex-ferrule; ~ **flexible de la red de CA** INFO AC mains cord; ~ **de guíaondas** COMPON waveguide wire; ~ **de hilos múltiples** COMPON, ELECTRÓN multistrand wire, multiwire conductor; ~ **hueco** COMPON, TRANS *cables* hollow conductor; ~ **interior** TELEFON, TRANS inner conductor; ~ **de metal desnudo** GEN plain conductor; ~ **negativo** ELECTRÓN negative wire; ~ **neutral** ELECTRÓN neutral conductor; ~ **nivelado** ELECTRÓN flush conductor; ~ **de oropel** COMPON, ELECTRÓN tinsel conductor; ~ **perfilado** COMPON shape conductor, shaped conductor; ~ **plano reversible** ELECTRÓN reversible blade; ~ **de puesta a tierra** ELECTROTEC grounding conductor *AmE*, earthing conductor *BrE*; ~ **de retorno** ELECTROTEC return conductor; ~ **con revestimiento metálico** ELECTRÓN metal-clad conductor; ~ **de sección sectorial** COMPON sector-shaped conductor; ~ **segmentado** ELECTRÓN milliken conductor; ~ **de la señal de llamada** TELEFON ring wire; ~ **de tierra** ELECTROTEC earthing conductor *BrE*, grounding conductor *AmE*; ~ **único** COMPON, ELECTRÓN single conductor

conductores: ~ **radiales** *m pl* ELECTROTEC radial leads

conductos *m pl* ELECTRÓN raceways

conectabilidad *f* TV connectability

conectable *adj* INFO luggable

conectado *adj* GEN connected, on, plugged in; **no** ~ GEN nonconnected, off, unplugged; ~ **con** GEN *Internet*, TELEFON connected to; ~ **permanentemente** GEN permanently connected; ~ **a voltaje** ELECTROTEC connected to voltage

conectar *vt* GEN connect, TELEFON put through; ◆ ~ **en un bucle** ELECTROTEC connect in a loop; ~ **en cascada** ELECTRÓN connect in cascade; ~ **a una impedancia terminal** GEN *cable* terminate; ~ **negativo** ELECTROTEC connect negative; ~ **en paralelo** ELECTRÓN, ELECTROTEC connect in parallel; ~ **en serie** ELECTRÓN connect in series

conectarse *v refl* GEN connect; ◆ ~ **al sistema** INFO log on, log in

conector *m* GEN connector; ~ **acoplable** COMPON *óptica de fibras* intermateable connector; ~ **blindado** ELECTRÓN shielded connector; ~ **de borde impreso** ELECTRÓN printed-edge connector; ~ **de bus** CONMUT bus connection unit; ~ **de circuito** TELEFON circuit junctor; ~ **de contactos múltiples** COMPON, ELECTRÓN multipin connector; ~ **de datos universal** (*UDC*) ELECTRÓN universal data connector (*UDC*); ~ **DIN** INFO DIN connector; ~ **enchufable** COMPON, ELECTRÓN plug-in connector; ~ **de entrada** TEC RAD in-connector; ~ **especial** COMPON special connector; ~ **de espigas** ELECTRÓN free-blade connector; ~ **de espigas redondas** COMPON, ELECTRÓN free-pin connector; ~ **de estación en bucle** ELECTRÓN loop-station connector; ~ **híbrido** TRANS *óptica de fibras* hybrid connector; ~ **de interfaz de medios** COMPON media-interface connector; ~ **macho** GEN plug; ~ **multifibro** TRANS multifibre (*AmE* multifiber) connector; ~ **multipunto** DATOS multipoint connector; ~ **precableado** ELECTRÓN prewired connector; ~ **de regleta** COMPON, ELECTRÓN strip connector; ~ **de salida** GEN out connector; ~ **terminal** ELECTRÓN terminal connector

conectores: ~ **anexos** *m pl* ELECTRÓN associated connectors; ~ **bicónicos** *m pl* ELECTRÓN biconic connectors; ~ **complementarios** *m pl* GEN mating connectors; ~ **de la matriz** *m pl* ELECTRÓN array connectors; ~ **de modo múltiple** *m pl* TRANS multiple-mode connectors

conexión *f* GEN connection; ~ **del abonado** TELEFON subscriber's lead-in; ~ **adicional** TELEFON Connect-Additional; ~ **AFC** TEC RAD AFC pull-in; ~ **alámbrica directa** ELECTRÓN, ELECTROTEC wire through connection; ~ **arrollada** GEN wrapped connection; ~ **arrollada para datos** ELECTRÓN data wrap plug; ~ **bidireccional** INFO two-way link; ~ **en bucle de trombón** TELEFON trombone loop connection; ~ **de cables** TRANS cable connection; ~ **de cajero automático** INFO ATM connection; ~ **de camino de menor orden** TRANS lower-order path connection (*LPC*); ~ **de canal virtual** DATOS, TV virtual-channel connection (*VCC*); ~ **con computadora** *AmL* (*cf conexión con ordenador Esp*) TRANS computer link; ~ **conmutada** DATOS *Internet* dial-up connection; ~ **continental** TELEFON continental connection; ~ **por corrientes portadoras** TRANS carrier-current connection; ~ **en curso** TELEFON connection in progress; ~ **de los datos** DATOS data connection; ~

desordenada DATOS *Internet* wild connection; ~ **digital de extremo a extremo** ELECTRÓN end-to-end digital connection; ~ **directa de componentes fuera de banda perjudiciales** TELEFON harmful out-of-band components direct through-connection; ~ **dividida** TELEFON split connection; ~ **de enlace de datos** DATOS data-link connection (*DLC*); ~ **enrollada** COMPON wire-wrap connection; ~ **por enrollamiento** COMPON wire-wrapping; ~ **entre capas** COMPON, ELECTRÓN *circuitos impresos* interlayer connection; ~ **escalonada** ELECTROTEC step-by-step connection; ~ **en espera** TELEFON camp-on; ~ **en espiral** ELECTRÓN, TRANS *óptica de fibras* pigtail; ~ **en estrella** ELECTROTEC star connection, wye connection, Y-connection, TRANS star connection; ~ **fácil** GEN easy connection; ~ **en frecuencia portadora** TRANS carrier-frequency connection; ~ **en grupo** TELEFON group link; ~ **de guíaondas** TEC RAD, TRANS waveguide connection; ~ **híbrida** TEC RAD hybrid junction; ~ **de indicador** CONMUT pointer connection; ~ **inductiva** ELECTROTEC impedance bond; ~ **inicial** TELEFON Connect-Initial; ~ **intercontinental** TELEFON intercontinental connection; ~ **interfacial** ELECTRÓN interfacial connection, INFO interface connection; ~ **internacional** TELEFON international connection; ~ **internacional de TV de destino múltiple** TV international multiple-destination television connection; ~ **de larga distancia** TELEFON long-distance connection; ~ **de la matriz** ELECTRÓN array connection; ~ **MCS** TELEFON MCS connection; ~ **mediante clavija y jack** ELECTROTEC plug-and-jack connection; ~ **multienlace** TEC RAD multilink connection; ~ **multipunto** DATOS multipoint connection; ~ **N** TELEFON N-connection; ~ **con ordenador** *Esp* (*cf conexión con computadora AmL*) TRANS computer link; ~ **orientada** TELEFON directed connection; ~ **en paralelo** ELECTROTEC parallel connection; ~ **PPP de Internet** DATOS PPP Internet connection; ~ **de presentación** DATOS presentation connection; ~ **para procesamiento de datos** DATOS data-processing connection; ~ **con programa de sonido e imagen** TV sound and television programme (*AmE* program) connection; ~ **de programa sonoro internacional** TV international sound-programme (*AmE* program) connection; ~ **de puente** COMPON strapping; ~ **en puente** TRANS bridge splice; ~ **de punto a multipunto** APLIC point-to-multipoint connection; ~ **de ranura múltiple** TELEFON multislot connection; ~ **a la red** DATOS network connection, ELECTROTEC mains connection, mains supply, TELEFON network connection; ~ **de referencia hipotética** TELEFON hypothetical reference connection; ~ **roscada** ELECTROTEC screwed connection; ~ **secreta** DATOS, TELEFON privacy connection; ~ **semiduplex** TELEFON half duplex connection; ~ **semipermanente** TELEFON semipermanent connection; ~ **en serie** ELECTRÓN connection in series; ~ **con el servicio nocturno** CONMUT night-service connection; ~ **de sesión** DATOS, TRANS session connection; ~ **simplex por canales conjugados** TELEFON *dúplex* two-way simplex connection; ~ **sin marcación** APLIC, TELEFON nondialled (*AmE* nondialed) connection; ~ **de sistemas de empresa** (*ESCON*) INFO *IBM* enterprise-systems connection (*ESCON*); ~ **soldada** COMPON soldered connection; ~ **en T** TRANS *cables* branch joint; ~ **telefónica internacional** TELEFON international telephone connection; ~ **de televisión internacional** TV international television connection; ~ **a tierra** ELECTRÓN bonding, earthing *BrE*, grounding *AmE*, ELECTROTEC bonding, earthing *BrE*, earthing arrangement *BrE*, grounding arrangement *AmE*, grounding *AmE*; ~ **por torsión** ELECTRÓN twisting; ~ **por torsión cruzada** TRANS *cables* cross-twisting; ~ **de transporte** TELEFON transport connection; ~ **transversal** ELECTRÓN, ELECTROTEC, TELEFON through connection; ~ **transversal de hilo remachado** ELECTRÓN, ELECTROTEC *circuitos impresos* clinched-wire through connection; ~ **a través de centralita** CONMUT connection through an exchange; ~ **de trayecto de orden superior** DATOS, TRANS higher-order path connection (*HPC*); ~ **con trayectoria virtual** DATOS virtual-path connection (*VPC*); ~ **trombón** TELEFON trombone connection; ~ **virtual** TELEFON virtual connection; ~ **volante** ELECTROTEC strapping; ◆ **sin ~ a tierra** ELECTROTEC ungrounded *AmE*, unearthed *BrE*

conexionado: ~ **impreso** *m* COMPON printed wiring; ~ **preformado** *m* ELECTRÓN harness

conexiones: ~ **con hilo desnudo** *f pl* TELEFON bare wiring

conferencia *f* GEN conference; ~ **controlada por operador** TELEFON attendant-controlled conference; ~ **de estación** TEC RAD station conference; ~ **de mando multipunto** DATOS multipoint command reference; ~ **por operador** TELEFON attendant conference; ~ **por vía satélite** TV satellite conference

Conferencia: ~ **Europea de Administración Postal y Telecomunicaciones** *f* (*ECPT*) GEN European Conference of Postal and Telecommunications Administrations (*ECPT*)
confiabilidad *f* GEN reliability
confianza: ~ **comercial** *f* GEN commercial confidence
confidencial *adj* GEN confidential
configuración *f* GEN configuration; ~ **en anillo** TRANS ring configuration; ~ **de antena** TEC RAD antenna configuration; ~ **de bloque** DATOS framing pattern, frame pattern; ~ **de brazo oscilante** ELECTRÓN swing-arm feature; ~ **de conferencia** TV conferencing configuration; ~ **de difracción** TRANS diffraction pattern; ~ **de directividad horizontal** TEC RAD horizontal directivity pattern; ~ **del electrodo** ELECTRÓN electrode configuration; ~ **en estrella** TRANS star configuration; ~ **y gestión de datos** DATOS configuration and data management; ~ **matriz** ELECTRÓN matrix configuration; ~ **de memoria** INFO memory configuration; ~ **en miniestrella** TV mini-star configuration; ~ **operativa** DATOS operating environment; ~ **de pantallas múltiples** TV multiscreen configuration; ~ **de perforaciones** ELECTRÓN *circuito impreso* hole pattern; ~ **de pixel** INFO pixel map; ~ **del recorrido inverso de un bucle** ELECTRÓN loopback configuration; ~ **de la red** TELEFON network configuration; ~ **de señal** ELECTRÓN, SEÑAL signal shaping; ~ **de símbolos** DATOS symbol pattern; ~ **de sistema** INFO system configuration; ~ **del sistema bus** INFO bus configuration; ~ **total** CONMUT total configuration; ~ **de la unidad binaria** DATOS bit mapping
configurar *vt* GEN configure
confirmación *f* DATOS, TELEFON confirmation; ~ **de los datos enviados por datagrama** DATOS, SEÑAL, TELEFON datagram delivery data confirmation; ~ **de desconexión** TELEFON clearing confirmation; ~ **de entrega** TELEFON delivery confirmation; ~ **de mensaje** DATOS message confirmation; ~ **primaria** DATOS, TELEFON primitive confirm; ~ **de recepción** (*COR*) DATOS, TELEFON confirmation of receipt (*COR*); ~ **por recibir** (*CFR*) DATOS, TELEFON confirmation to receive (*CFR*)
confirmado: **no** ~ *adj* DATOS nonconfirmed
confirmar *vt* GEN confirm; ◆ ~ **recepción** DATOS, SEÑAL, TELEFON acknowledge
conflicto *m* GEN dispute
confluencia *f* GEN confluence
conformador: ~ **de impulsos** *m* TELEFON pulse shaper
conformidad *f* GEN conformance, conformity; ~ **de los dispositivos** TELEFON conformance of devices; ~ **subordinada** DATOS dependent conformance
confusión *f* GEN confusion
congestión *f* CONMUT congestion; ~ **de espectro** TEC RAD spectrum congestion; ~ **de recepción** TELEFON reception congestion; ~ **de red** TELEFON, TRANS network congestion; ~ **de ruta** TELEFON route congestion
conglomerado: ~ **de celdillas** *m* TEC RAD cluster of cells; ~ **de redes** *m* TELEFON network cluster
cónico *adj* COMPON tapered, TELEFON conical
conificar: ~ **el extremo de** *vt* COMPON, ELECTRÓN, INFO, TRANS *tubos* tag
conjunción: ~ **negativa** *f* ELECTRÓN nonconjunction
conjunto: ~ **de alternativas** *m* DATOS set of alternatives; ~ **de caracteres codificados** *m* TELEFON coded character set; ~ **de contexto definido** *m* DATOS defined-context set; ~ **de dispositivos mecánicos** *m* GEN gadgetry; ~ **electrónico de rotación** *m* (*DCE*) ELECTRÓN despun control electronics (*DCE*); ~ **para ensamblar** *m* GEN preassembly kit; ~ **de filtros** *m* DATOS *Internet* filter set; ~ **de fusible** *m* ELECTROTEC fuse unit; ~ **impreso** *m* ELECTRÓN printed-board assembly; ~ **de marcación y transferencia de energía** *m* (*BAPTA*) ELECTROTEC bearing and power transfer assembly (*BAPTA*); ~ **de programas** *m* CONMUT, INFO program package; ~ **de puertas** *m* GEN gate array; ~ **rectificador de semiconductores** *m* COMPON semiconductor rectifier stack; ~ **de resorte bimetálico** *m* COMPON bimetal spring set; ~ **terminal** *m* TELEFON local end; ~ **de unidades lógicas** *m* DATOS logic array
conjuntor *m* GEN junctor; ~ **de contestación** TELEFON answer jack; ~ **general** COMPON, ELECTRÓN branching jack; ~ **rotatorio** GEN rotary switch; ~ **terminal** COMPON, CONMUT, TELEFON terminating junctor
conjuntos: ~ **de caracteres codificados ISO de 7 y 8 bitios** *m pl* TELEFON ISO 7-bit and 8-bit coded character sets
conmutabilidad *f* GEN exchangeability
conmutable *adj* GEN exchangeable
conmutación *f* GEN commutation, switching, toggling, changeover; ~ **activada por la voz** CONMUT, CONTROL voice-operated switching; ~ **bifilar** CONMUT two-wire switching; ~ **de canal** CONMUT channel switching; ~ **de circuitos**

CONMUT, TELEFON circuit switching; ~ **de cuatro hilos** CONMUT four-wire switching; ~ **de datos** CONMUT, DATOS, ELECTRÓN data switching; ~ **de dial** INFO dial switching; ~ **de división espacial** CONMUT space-division switching; ~ **de división de frecuencia** TELEFON frequency division switching, TRANS frequency-division duplexing; ~ **de elementos** INFO cell switching; ~ **de emergencia** TELEFON emergency changeover; ~ **entre células** TEC RAD intercell switching; ~ **espacial** CONMUT space switching, spatial switching; ~ **de grupo** CONMUT group switching; ~ **de haz** TEC RAD beam switching; ~ **de línea** TELEFON line switching; ~ **de llamada en curso** CONMUT call-in-progress switching; ~ **de longitud de onda** CONMUT wavelength switching; ~ **de mensajes** CONMUT message switching; ~ **a modo de espera** GEN changeover to standby; ~ **en multiplex espacial** CONMUT, TELEFON spatial multiplex switching; ~ **por paquetes** CONMUT, DATOS, TELEFON, TRANS packet switching (*PS*); ~ **de precisión** INFO fine switching; ~ **de protección** CONMUT, TELEFON protection switching; ~ **rápida de paquetes** CONMUT, DATOS, TRANS fast packet switching; ~ **por redundancia** PRUEBA redundancy switching; ~ **de retardo de fase** (*CDF*) SEÑAL, TEC RAD phase-shift keying (*PSK*); ~ **de satélite** CONMUT satellite switching, TEC RAD satellite switching; ~ **de tarifas** TELEFON tariff switching; ~ **de enlace de datos** ELECTRÓN data-link switching; ~ **de variación de fase cuaternaria** SEÑAL, TEC RAD quaternary phase-shift keying

conmutador[1]: ~ **de acoplamiento de carácter** *m* CONMUT character coupling switch; ~ **activado por la voz** *m* COMPON, CONMUT voice-operated switch; ~ **de adaptación de tensión** *m* ELECTROTEC voltage adaptor switch; ~ **de anillo** *m* TEC RAD ring switch; ~ **ATM** *m* CONMUT ATM switch; ~ **de banda ancha** *m* CONMUT broadband switch; ~ **de banda estrecha** *m* CONMUT narrow-band switch; ~ **basculante de programas** *m* CONMUT, INFO program flip-flop switch; ~ **bifilar** *m* CONMUT two-wire switch; ~ **bifilar doble** *m* CONMUT four-wire switch; ~ **de bit** *m* CONMUT bit switch; ~ **de botón** *m* GEN plunger key; ~ **de canales** *m* TEC RAD channel switch; ~ **de caracteres** *m* CONMUT character switch; ~ **de central telefónica** *m* TELEFON central exchange switch; ~ **de circuito** *m* CONMUT circuit switch; ~ **de circuitos** *m* CONMUT circuit-switching unit; ~ **de circuito virtual** *m* CONMUT virtual-circuit switch (*VCS*); ~ **compensable** *m* CONMUT, ELECTROTEC trickable switch; ~ **de corte** *m* GEN cutoff switch; ~ **de datos** *m* CONMUT, DATOS, ELECTRÓN data switch; ~ **deslizante** *m* COMPON slide switch; ~ **de división de longitud de onda** *m* CONMUT wavelength division switch; ~ **e/t** *m* TEC RAD e/t switch; ~ **de elementos** *m* ELECTROTEC *baterías* cell switch; ~ **de enclavamiento** *m* TEC RAD interlock switch; ~ **en estrella** *m* CONMUT star switch; ~ **de fax** *m* DATOS fax switch; ~ **funcional** *m* GEN function switch; ~ **de fusibles** *m* ELECTROTEC fuse switch; ~ **de grupo** *m* CONMUT envelope switch; ~ **de guíaondas** *m* CONMUT waveguide switch; ~ **híbrido** *m* CONMUT hybrid switch; ~ **de horquilla** *m* PERIF hook switch; ~ **por inercia** *m* ELECTROTEC impact switch; ~ **inversor** *m* ELECTROTEC reversing switch; ~ **de línea** *m* CONMUT line switch; ~ **local** *m* CONMUT, TELEFON local switch; ~ **de luces** *m* INFO light switch; ~ **manual de monocordio** *m* CONMUT, TELEFON single-cord switchboard; ~ **matricial** *m* CONMUT matrix switch; ~ **de memoria del programa** *m* CONMUT, INFO program store switch; ~ **de mensajes** *m* CONMUT message switch; ~ **metropolitano** *m* CONMUT metropolitan switch; ~ **mezclador de vídeo** *m* CONMUT video switch; ~ **de modulación** *m* TEC RAD modulation switch; ~ **móvil aislado con fibra** *m* CONMUT moving fibre (*AmE* fiber) switch; ~ **de octetos** *m* CONMUT byte switch; ~ **operado por onda portadora** *m* TEC RAD carrier-operated switch; ~ **óptico de acoplador direccional** *m* CONMUT directional coupler optical switch; ~ **óptico integrado** *m* CONMUT integrated optical switch; ~ **de paquete** *m* CONMUT, DATOS, TELEFON, TRANS packet switch; ~ **de paquetes de datos de voz** *m* CONMUT voice-data packet switch; ~ **de ponderación del ruido** *m* ELECTRÓN noise-weighing switch; ~ **principal** *m* ELECTROTEC main switch; ~ **de programa** *m* CONMUT, INFO program switch; ~ **de puesta a tierra** *m* ELECTROTEC earthing switch *BrE*, grounding switch *AmE*; ~ **de punto de cruce** *m* CONMUT cross-point switch; ~ **de punto de transmisión de señales** *m* CONMUT, SEÑAL, TELEFON, TRANS signal transfer point switch; ~ **rápido de circuitos** *m* CONMUT fast circuit switch; ~ **rápido de paquetes** *m* CONMUT, DATOS, TRANS fast packet switch (*FPS*); ~ **de la RDSI** *m* CONMUT ISDN switch; ~ **de resorte** *m* CONMUT, ELECTROTEC spring-loaded switch; ~ **rural** *m* CONMUT rural switch; ~ **S** *m* CONMUT,

TELEFON S switch; ~ **de saltos** *m* TEC RAD hopping switch; ~ **de satélite** *m* TEC RAD satellite switch; ~ **de secuencia** *m* CONMUT, ELECTROTEC sequence switch; ~ **de servicio nocturno** *m* CONMUT night-service key; ~ **STP** *m* CONMUT STP switch; ~ **de STP** *m* SEÑAL STP switch; ~ **telegráfico manual** *m* CONMUT, TELEFON telegraph switchboard; ~ **de tiempo** *m* ELECTROTEC meter changeover clock; ~ **de tipo telefónico** *m* CONMUT keyswitch; ~ **TR** *m* CONMUT, TEC RAD, TRANS TR switch; ~ **universal** *m* CONMUT universal switch; ~ **de voz-datos integrados** *m* CONMUT integrated voice-data switch

conmutador[2]: **sin** ~ *fra* CONMUT, ELECTRÓN, ELECTROTEC, TELEFON switchless

conocimiento: ~ **del producto** *m* GEN product awareness

CONP *abr* (*protocolo de capa de red orientado a la conexión*) TRANS CONP (*connection-oriented network-layer protocol*)

CONS *abr* (*servicio de red orientado a la comunicación*) APLIC CONS (*connection-oriented mode network service*)

consecutivo *adj* GEN consecutive

consejo: ~ **de telecomunicaciones** *m* TELEFON telecommunications board

conservación: ~ **de mensajes** *f* PERIF, TELEFON saving of messages; ~ **ordinaria** *f* GEN routine maintenance

consola *f* GEN console; ~ **de carguero** TELEFON trader's turret; ~ **de comunicación** DATOS communication console; ~ **para el montaje** COMPON mounting bracket; ~ **de sistema** TELEFON system console; ~ **con una clave por línea** APLIC, PERIF key-per-line console *BrE*, key-per-trunk console *AmE*; ~ **de visualización** INFO, TV display console

consolidación *f* TRANS consolidation

consonante *f* GEN *lingüística* consonant

constancia: ~ **del flujo luminoso** *f* TV lumen maintenance

constante *f* GEN constant; ~ **de atenuación** TEC RAD attenuation constant, TRANS attenuation constant, attenuation factor; ~ **de atenuación de imagen** TRANS image attenuation constant; ~ **de comparación** GEN criterion; ~ **de difusión** COMPON *semiconductores* diffusion constant; ~ **de disociación electrolítica** GEN electrolytic dissociation constant; ~ **de error** CONTROL error constant; ~ **de fase** GEN phase constant; ~ **figurada** INFO figurative constant; ~ **de un instrumento de medición** PRUEBA *metrología* instrument constant; ~ **de propagación** ELECTRÓN, TEC RAD, TRANS propagation constant; ~ **de transferencia de imagen** DATOS, TRANS image transfer constant

construcción *f* GEN assembly; ~ **de antena** TEC RAD antenna building

constructor *m* GEN maker

consulta *f* GEN consultation; ~ **alternativa** TELEFON alternation on inquiry, inquiry reciprocation; ~ **a distancia** GEN remote consultation; ~ **de tablas** GEN table look-up

consultor: ~ **de telecomunicaciones** *m* GEN telecommunications consultant

consultoría *f* GEN consultancy

consumo *m* GEN consumption, usage; ~ **de potencia** ELECTROTEC power drain

contabilización: ~ **selectiva** *f* TELEFON selective accounting

contacto *m* COMPON, ELECTRÓN, ELECTROTEC contact; ~ **de acción rápida** COMPON snap-action contact; ~ **de acompañamiento** COMPON, ELECTROTEC trailing contact; ~ **auxiliar** ELECTROTEC auxiliary contact; ~ **de banco** TELEFON bank contact; ~ **bipolar** COMPON double-break contact; ~ **de borde** ELECTRÓN edge board contact; ~ **de conmutación sin interrupción** ELECTRÓN make-before-break contact; ~ **de cruce** CONMUT crosspoint; ~ **de cruce de banda ancha** CONMUT broadband crosspoint; ~ **de cruce bifilar doble** CONMUT four-wire crosspoint; ~ **de cruce CMOS** TELEFON CMOS crosspoint; ~ **de cuchilla** ELECTROTEC blade contact; ~ **deslizante** COMPON sliding contact; ~ **de doble cierre** COMPON *contactos eléctricos* double-make contact; ~ **de emisión de impulsos** TELEFON pulsing contact; ~ **del equipo del abonado** TELEFON subscriber equipment keeper; ~ **de espigas** COMPON, ELECTRÓN pin contact; ~ **de espigas planas** COMPON blade contact; ~ **de frotamiento** COMPON wiping contact; ~ **de horquilla** ELECTRÓN fork contact; ~ **impreso** ELECTRÓN printed contact; ~ **impreso de borde de placa** ELECTRÓN printed edge-board contact; ~ **de impulsión** COMPON passing contact; ~ **de láminas** COMPON *interruptores, relés* reed contact; ~ **de manguito** COMPON sleeve contact; ~ **de mercurio** ELECTRÓN mercury contact; ~ **óhmico** COMPON ohmic contact; ~ **de presión** COMPON pressure contact; ~ **puntual** COMPON point contact; ~ **de relé** COMPON relay contact; ~ **de resorte** COMPON spring contact; ~ **de retención** ELECTROTEC holding contact; ~ **de ruptor** CONMUT, ELECTROTEC contact point; ~ **semiconductor**

COMPON, ELECTRÓN semiconductor contact; ~ **simple** COMPON single contact

contactor *m* CONMUT, ELECTRÓN contactor, ELECTROTEC circuit closer

contactos: ~ **dobles** *m pl* COMPON twin contacts

contador *m* GEN counter; ~ **de abonado** PERIF subscriber's meter, TELEFON subscriber metering, subscriber's meter; ~ **del abonado** TELEFON subscriber's call meter; ~ **binario** ELECTRÓN binary counter; ~ **de bloqueo** TELEFON blocking counter; ~ **de bloques con acuse de recibo** TELEFON block-acknowledged counter; ~ **de bloques completos** TELEFON block-completed counter; ~ **de congestiones de ruta** CONMUT, TELEFON route-congestion counter; ~ **de cuadro I** TELEFON I-frame counter; ~ **de datos en bruto** DATOS raw-data counter; ~ **de décadas** ELECTRÓN *elementos lógicos* decade counter; ~ **de decenas** ELECTRÓN tens counter; ~ **domiciliario del abonado** PERIF, TELEFON subscriber's check meter; ~ **de estadística** PRUEBA statistics counter; ~ **de fallos** TELEFON error counter, failure counter; ~ **de frecuencia** PRUEBA frequency counter; ~ **de Geiger** PRUEBA Geiger counter; ~ **de Geiger-Müller** PRUEBA Geiger-Müller counter; ~ **de horas de funcionamiento** APLIC, PRUEBA hour meter; ~ **de impulsos** ELECTRÓN, SEÑAL pulse counter, TELEFON pulse meter; ~ **de impulsos de cómputo** TELEFON tariff pulse meter; ~ **de inducción** ELECTROTEC induction meter; ~ **de instrucciones** INFO program counter; ~ **de Johnson** ELECTRÓN Johnson counter; ~ **de mensajes recibidos** PERIF *contestador* messages received counter; ~ **módulo-n** ELECTRÓN modulo-n counter; ~ **para muestras gaseosas** PRUEBA gas sample counter tube; ~ **de muestras líquidas** TEC RAD liquid-sample counter; ~ **del nivel de ocupación** TELEFON traffic level counter; ~ **de ocupaciones** TELEFON seizure counter; ~ **de perturbaciones** TELEFON disturbance counter; ~ **privado de abonado** PERIF, TELEFON subscriber's private meter; ~ **de rectificador** ELECTROTEC, PRUEBA rectifier meter; ~ **de reexpedición** CONMUT redirecting counter; ~ **de retransmisión** TELEFON redrive counter; ~ **de ruta congestionada** CONMUT congested-route counter; ~ **de sobrecarga** TELEFON overflow meter; ~ **tarificador de llamadas** APLIC call-charge counter; ~ **de tráfico** ELECTRÓN, PRUEBA, TELEFON traffic meter; ~ **de transporte ondulante** PRUEBA ripple-through counter; ~ **de unidades** ELECTRÓN units counter

contar[1]: ~ **hacia atrás** *vi* GEN count down

contar[2]: ~ **el tiempo de** *vt* GEN time

contenedor *m* GEN container (*CT*); ~ **de celdas** ELECTROTEC *baterías* cell container; ~ **de elemento de batería** ELECTROTEC battery cell container; ~**-n** APLIC, DATOS, TELEFON container-n (*C-n*)

contenido: ~ **binario equivalente** *m* TELEFON equivalent binary content; ~ **de decisión** *m* INFO decision content; ~ **de humedad** *m* GEN moisture content

contestación *f* TELEFON answer

contestador: ~ **automático** *m* TELEFON answering machine, answer machine, answerphone, Ansaphone®; ~ **automático de acceso remoto** *m* TELEFON remote-access answering machine; ~ **de código** *m* TELEFON code-answering device, code-answering unit; ~ **de comprobación de continuidad** *m* TELEFON continuity-check responder; ~ **estático** *m* PERIF, TELEFON announcement-only facility; ~ **con fax** *m* PERIF, TELEFON fax answerphone; ~ **con sintetizador de voz** *m* TELEFON answering device with voice synthesizer

contexto *m* DATOS, INFO context; ~ **de aplicaciones** GEN applications context; ~ **por defecto** DATOS default context; ~ **de la presentación** DATOS presentation context; ~ **de tarificación de llamadas** TELEFON call-charge context

contextual *adj* DATOS, INFO *lingüística* contextual

contiguo *adj* GEN next

continuación: ~ **de mensaje** *f* DATOS continuation of message (*COM*)

continuo *adj* INFO continuous

contorneamiento *m* ELECTROTEC creepage, flashover

contorneo *m* ELECTROTEC bypassing, INFO contouring

contorno *m* INFO contouring; ~ **de grado A** TEC RAD A-grade contour; ~ **de grado B** TEC RAD B-grade contour; ~ **de sonoridad** TELEFON isophon, loudness contour

contracción *f* COMPON shrinkage, shrinking

contracorriente *adj* ELECTRÓN, ELECTROTEC, TELEFON upstream

contrafase *f* ELECTRÓN, ELECTROTEC anti-phase, TEC RAD push-pull

contramedida *f* GEN counter-measure

contrapeso *m* GEN counterweight

contrapresión *f* GEN *física* back pressure, counterpressure

contrarrestar *vt* GEN counteract
contraseña *f* DATOS, INFO, PERIF password
contraste *m* GEN contrast; **~ del índice de refracción** TRANS refractive-index contrast; **~ de luminancia** TV luminance contrast; **~ máximo** TV contrast range; **~ radiográfico** TEC RAD radiographic contrast
contratante: **~ principal** *m* GEN main contractor
contratista *m* GEN contractor
contratuerca *f* ELECTRÓN locknut
contravención *f* GEN infringement
contribución: **~ del nivel de interferencia** *f* TEC RAD interference level contribution
control *m* GEN monitoring; **~ de acceso** DATOS, PRUEBA access control; **~ de acceso medio** CONTROL medium access control (*MAC*); **~ de acceso de señal de llamada** ELECTRÓN ring-access control; **~ de actitud** TEC RAD attitude control; **~ de amplitud cuantificada** TELEFON amplitude-quantized control; **~ asimétrico** CONTROL, TRANS single-ended control; **~ de atenuación** TRANS attenuation control; **~ de los auriculares** TELEFON headphone monitoring; **~ automática de ganancia** (*AGC*) TELEFON automatic gain control (*AGC*); **~ de batería** ELECTROTEC battery monitor; **~ biterminal** TRANS double-ended control; **~ por bloques** INFO block check; **~ de calidad de proceso** PRUEBA process quality control; **~ de canal** TRANS channel control; **~ de carga de línea** TELEFON line load control; **~ del circuito** GEN circuit control; **~ de contraste** TV contrast control; **~ de conversación** CONTROL, TELEFON speech control; **~ de cursor** INFO slider control; **~ de datos nulos** INFO null-data check; **~ de dinámica** CONTROL volume-range control; **~ a distancia** CONTROL, INFO, TEC RAD, TELEFON telecontrol; **~ de eco** TEC RAD, TELEFON echo control; **~ de encaminamiento** DATOS routing control; **~ de encriptación** DATOS scrambling control; **~ del enlace de datos** DATOS data-link control; **~ de error de cabecera** DATOS header error control (*HEC*); **~ de error del enlace de señalización** SEÑAL signalling-link (*AmE* signaling-link) error monitoring; **~ espectral a distancia** (*RSM*) TEC RAD remote spectrum-monitoring (*RSM*); **~ de espera** TELEFON queue control; **~ estadístico de la calidad** PRUEBA statistical quality control; **~ de expresión** INFO expression control; **~ de fase** ELECTRÓN phase control; **~ de fase generalizado** PRUEBA generalized phase control; **~ de flujo de tráfico de señalización** DATOS, SEÑAL signalling-traffic (*AmE* signaling-traffic) flow control; **~ de formato** INFO format control; **~ de frecuencias** ELECTROTEC, TRANS frequency control; **~ de ganancia** ELECTRÓN gain control; **~ genérico del flujo** CONTROL, TRANS generic flow control (*GFC*); **~ de iluminación** INFO, TV brilliance control; **~ de inclinación orbital** TEC RAD orbital inclination control; **~ de intensidad** TV intensity control; **~ de intensidad de tráfico** DATOS, TELEFON traffic flow control; **~ de interrupción** TELEFON interruption control; **~ de intervalo** INFO breath control; **~ en lazo cerrado** CONTROL, INFO feedback control; **~ local** TELEFON local record; **~ lógico de enlace** CONTROL logical link control (*LLC*); **~ de memoria** DATOS storage control; **~ del microprocesador** CONMUT microprocessor control; **~ del módem** DATOS modem control; **~ por muestreo** PRUEBA spot check; **~ multivariable** PRUEBA multivariable control; **~ de nivel** INFO level control; **~ de nivel acústico** APLIC sound-level control; **~ de nivel automático** INFO automatic level control; **~ de nivel de ruido** TELEFON noise level control; **~ de nivel de sonido** PERIF sound-level control; **~ óptimo** CONTROL optimal control; **~ de la orientación** CONTROL orientation control; **~ panorámico** INFO pan control; **~ de parámetro de uso** (*UPC*) APLIC usage parameter control (*UPC*); **~ por polarización** INFO bias control; **~ de posición** TEC RAD station keeping; **~ de potencia del emisor** TEC RAD transmitter power control; **~ de potencia del trayecto de subida** TEC RAD up-path power control; **~ de proceso** CONTROL, PRUEBA process control; **~ de programación** INFO desk checking; **~ programado** CONTROL programmed control; **~ remoto** PERIF *de contestador automático* remote controller; **~ remoto de antena** TEC RAD antenna remote control; **~ rotativo** COMPON rotary control; **~ secuencial** CONMUT, DATOS, INFO, TELEFON sequencing; **~ de señal de identificación** CONMUT, TEC RAD pilot-tone system; **~ de silenciamiento** TEC RAD squelch control; **~ de sistema** INFO system control; **~ del sistema de comunicaciones** (*CSC*) GEN communications system control (*CSC*); **~ de sobrecarga** TELEFON overload control; **~ táctil** INFO touch control; **~ de la tasa de error por unidad de señal** SEÑAL, TELEFON signal unit error-rate monitoring; **~ del tiempo** INFO tempo control; **~ del tiempo de conducción** ELECTRÓN

semiconductores conduction time control; ~ **de tolerancias** GEN limit check; ~ **de tono** CONTROL, INFO tone control; ~ **de tráfico** DATOS, TELEFON traffic control; ~ **de trémolo** INFO tremolo control; ~ **de unidad** DATOS, ELECTRÓN, INFO device control; ~ **de volumen** CONTROL, INFO volume control; ~ **por voz** PERIF speech control

controlabilidad *f* GEN controllability

controlado: ~ **por DTE** *adj* TELEFON DTE-controlled; **no** ~ **por DTE** *adj* TELEFON DTE-uncontrolled; ~ **localmente** *adj* CONTROL, TRANS locally controlled; ~ **manualmente** *adj* CONTROL manually-controlled

controlador *m* INFO controller; ~ **alámbrico de frecuencia** CONTROL wired sequence controller; ~ **alámbrico de secuencia** INFO wired sequence controller; ~ **del bus** INFO bus arbitrator; ~ **de comunicaciones de datos** DATOS data communications controller; ~ **de dispositivo** DATOS, ELECTRÓN, INFO device controller, device driver; ~ **entrada/salida** INFO input/output controller; ~ **de la estación base** TEC RAD base station controller; ~ **de grupo** DATOS group controller; ~ **de impresora** DATOS printer controller; ~ **de intervalo** INFO breath controller; ~ **de nodo terminal** (*TNC*) TRANS terminal node controller (*TNC*); ~ **periférico** DATOS, ELECTRÓN, INFO peripheral controller; ~ **de polarización** ELECTRÓN polarization controller; ~ **de secuencias programado** INFO programmed sequence controller; ~ **de teclado** INFO keyboard controller; ~ **del tiempo** INFO tempo controller; ~ **de tráfico** DATOS, TELEFON traffic controller; ~ **de la unidad de disco** DATOS, INFO disk-drive controller

conurbación *f* GEN conurbation

convección: ~ **libre** *f* CLIMAT free convection

convergencia *f* DATOS, TRANS convergence; ~ **de transmisión** TRANS transmission convergence

conversación *f* GEN conversation; ~ **con aviso de llamada** TELEFON messenger call; ~ **imposible** TEC RAD conversation impossible; ~ **tridireccional** PERIF, TEC RAD, TELEFON three-way conversation

conversión *f* GEN conversion; ~ **de código** ELECTRÓN, TELEFON, TRANS code conversion; ~ **a código bipolar** TRANS conversion to bipolar code; ~ **de datos** DATOS data conversion; ~ **de fichero** DATOS file conversion; ~ **de frecuencia** TEC RAD, TRANS frequency conversion, frequency translation, TV frequency conversion; ~ **de frecuencia descendente** TRANS frequency down-conversion; ~ **de hipertexto** INFO hypertext conversion; ~ **de imagen** TV image conversion; ~ **de señales** SEÑAL, TELEFON signal conversion; ~ **telegráfica** TELEFON telegraph conversion

conversor: ~ **de energía** *m* ELECTROTEC energy converter

convertidor *m* ELECTROTEC converter; ~ **A-D** INFO, SEÑAL, TELEFON A-D converter (*ADC*); ~ **de alimentación** DATOS, INFO power converter; ~**-ascendente** DATOS, TEC RAD up converter; ~ **de código** INFO, TRANS code converter; ~ **del contador de carga** APLIC charge-metering converter; ~ **de conversión doble** TRANS double-conversion converter; ~ **D-A** GEN D-A converter; ~ **de datos** INFO data converter; ~ **de exploración** TEC RAD scan converter; ~ **de frecuencia** ELECTRÓN, ELECTROTEC, TRANS, TV frequency converter; ~ **interfacial** DATOS interface converter; ~ **de interfaz de señalización** SEÑAL signal interface converter; ~ **de paralelo a serie** ELECTRÓN, TELEFON parallel-to-serial converter; ~ **paralelo-serie** ELECTRÓN, TELEFON parallel-to-serial converter; ~ **de protocolo** DATOS protocol converter; ~ **de puente** ELECTROTEC bridge converter; ~ **de refuerzo** ELECTROTEC booster converter; ~ **de RF a IF** TEC RAD down converter; ~ **de señales** SEÑAL, TELEFON signal converter; ~ **de señalización** SEÑAL signalling (*AmE* signaling) converter; ~ **del sistema de televisión** TV television system converter; ~ **telegráfico** TELEFON telegraph converter; ~ **de tensión** ELECTROTEC voltage converter; ~ **de tono a MIDI** INFO pitch-to-MIDI converter

convertir *vt* DATOS, INFO convert; ◆ ~ **a código bipolar** TRANS convert to bipolar code; ~ **información en paralelo a información en serie** PERIF intercouple; ~ **de paralelo a serie** INFO destaticize

cooperar *vi* GEN interwork

coordinación *f* GEN coordination

coordinador: ~ **de obligaciones** *m* DATOS commitment coordinator

coordinar *vt* GEN coordinate

copia *f* GEN print-out; ~ **local** TRANS local copy; ~ **en pantalla** INFO, TV soft copy, screen copy; ~ **de seguridad de datos** DATOS data backup; ~ **de seguridad de un fichero** INFO file backup

copiar *vt* GEN *documentación* copy

coprocesador *m* INFO coprocessor

coproducción *f* TV coproduction

COR *abr* (*confirmación de recepción*) DATOS, TELEFON COR (*confirmation of receipt*)

corchete *m* GEN square bracket; ~ **de anclaje**

COMPON strain relief bracket; ~ **de montaje de anclaje** COMPON strain-relief mounting bracket
cordón: ~ **de alimentación** *m* ELECTRÓN power cord, INFO line cord, power cord; ~ **de aparato telefónico** *m* APLIC, PERIF handset cord; ~ **conector** *m* TRANS cord set; ~ **de conexión a la red** *m* ELECTRÓN, ELECTROTEC line cord; ~ **de extensión** *m* ELECTROTEC extension cord; ~ **de instrumento** *m* PERIF instrument cord
cornucopia *f* TEC RAD cornucopia
corona *f* ELECTROTEC corona
corpus *m* DATOS, INFO corpus
correa: ~ **de transmisión** *f* COMPON driving belt
corrección: ~ **automática** *f* INFO auto-correction; ~ **de compensación de línea** *f* TV line bend correction; ~ **del contorno** *f* INFO contour correction; ~ **de declive en tiempo de campo** *f* TV field tilt correction; ~ **de desplazamiento del aislamiento** *f* ELECTRÓN insulation displacement termination; ~ **de la distorsión trapecial del barrido vertical** *f* TV field keystone correction; ~ **de errores** *f* GEN error correction; ~ **de fallos** *f* PRUEBA *seguridad de funcionamiento* fault correction; ~ **de impulsos** *f* SEÑAL, TRANS pulse correction; ~ **previa** *f* ELECTRÓN pre-equalization; ~ **de pruebas** *f* PERIF proofreading; ~ **de sombra** *f* TV shading correction
corrector: ~ **de forma de onda** *m* ELECTRÓN waveform corrector; ~ **de impulso** *m* TELEFON pulse corrector; ~ **ortográfico** *m* INFO spell-checker
corredera: ~ **articulada** *f* INFO built slide
corredor *m* TEC RAD corridor
corregir *vt* GEN correct; ◆ ~ **un fallo** TELEFON clear a fault; ~ **pruebas** PERIF proofread
correlación *f* GEN correlation; ~ **espacio-temporal** CONMUT, DATOS space-time correlation; ~ **de reflexión** INFO reflection mapping
correlador *m* GEN correlator
correo *m* INFO mail
correspondencia: ~ **de servicio telegráfico** *f* TELEFON telegraph service correspondence; ~ **telefónica** *f* (*VM*) APLIC, TELEFON voice mail (*VM*)
corriente *f* GEN current; ~ **de aislamiento entre filamento calefactor-cátodo** ELECTRÓN heater-cathode insulation current; ~ **de alimentación** ELECTROTEC feeding current; ~ **alterna** (*CA*) GEN alternating current (*AC*); ~ **alterna de descarga** ELECTRÓN *de protector* alternating discharge current; ~ **alterna monofásica** ELECTROTEC single-phase alternating current; ~ **de ánodo** ELECTRÓN anode current; ~ **de antena** TEC RAD antenna current; ~ **de arco** ELECTRÓN arc current; ~ **por avería** ELECTRÓN fault current; ~ **de base** ELECTRÓN *semiconductores* base current; ~ **de carga** ELECTROTEC *baterías* charging current; ~ **de carga flotante** ELECTROTEC *baterías* float-charging current; ~ **catódica mínima** ELECTRÓN minimum cathode current; ~ **CBO** DATOS CBO stream; ~ **de choque nominal** ELECTROTEC nominal impulse discharge current; ~ **en circuito cerrado** ELECTROTEC closed-circuit current; ~ **circular** TELEFON ring current; ~ **del colector** ELECTRÓN *semiconductores* collector current; ~ **continua** (*CC*) ELECTROTEC direct current (*DC*); ~ **continua orientada al bitio** DATOS continuous bit oriented stream; ~ **continua permanente en estado de conducción** ELECTRÓN *tiristores* continuous direct on-state current; ~ **de corte** ELECTRÓN *transistores* cutoff current; ~ **de corte de puerta** ELECTRÓN *tiristores* gate turn-off current; ~ **de cortocircuito** ELECTROTEC short-circuit current; ~ **de cresta de sobrecarga accidental** ELECTRÓN, ELECTROTEC surge forward current; ~ **desactivada** ELECTRÓN off-state current; ~ **de descarga de impulsos** ELECTRÓN *de protector* impulse discharge current; ~ **en diente de sierra** ELECTRÓN sawtooth current; ~ **directa de cresta repetitiva** ELECTRÓN, ELECTROTEC repetitive-peak forward current; ~ **directa de cresta repetitiva en estado de conducción** ELECTRÓN, ELECTROTEC repetitive-peak onstage current; ~ **directa de cresta repetitiva en estado de no conducción** ELECTRÓN, ELECTROTEC repetitive-peak offstage current; ~ **directa media** ELECTRÓN, ELECTROTEC mean forward current; ~ **directa de sobrecarga** ELECTRÓN, ELECTROTEC overload forward current; ~ **doble** TELEFON *télex* double current; ~ **eléctrica** ELECTROTEC electric current; ~ **electrónica** ELECTRÓN, ELECTROTEC electron current; ~ **del emisor** ELECTRÓN *semiconductores* emitter current; ~ **de encendido** ELECTRÓN heater starting current; ~ **de encendido de puerta** ELECTRÓN *tiristores* gate turn-on current; ~ **de enganche** ELECTROTEC *tiristores* latching current; ~ **de entrada** ELECTRÓN inrush current; ~ **por la envoltura** PRUEBA sheath current; ~ **estacionaria** ELECTRÓN standing current; ~ **en estado de conducción** COMPON, ELECTRÓN on-state current; ~ **de excitación** ELECTROTEC driving current; ~ **de filamento** ELECTRÓN

filament current, heater current; ~ **fotoeléctrica** ELECTRÓN, ELECTROTEC, TRANS photocurrent; ~ **fuerte** ELECTROTEC heavy current; ~ **de gas** ELECTRÓN gas current; ~ **de gran amperaje** ELECTROTEC power current; ~ **humectante** ELECTRÓN, ELECTROTEC wetting current; ~ **de inactividad máxima** ELECTRÓN maximum non-operate current; ~ **inducida** ELECTROTEC induced current; ~ **inicial de filamento** ELECTRÓN filament starting current, filament surge current; ~ **inversa de base** ELECTRÓN *semiconductores* base reverse current; ~ **inversa de bloqueo** ELECTRÓN reverse-blocking current; ~ **inversa de electrodo** ELECTRÓN inverse electrode current; ~ **inversa resistiva** ELECTRÓN resistive reverse current; ~ **iónica** ELECTRÓN ionic current; ~ **de irradiación-saturación** ELECTRÓN, ELECTROTEC irradiation-saturation current; ~ **de irrupción** ELECTROTEC *tiristores* breakover current; ~ **de liberación** ELECTROTEC release current; ~ **de limitación** ELECTROTEC limiting current; ~ **limitada por la carga espacial** ELECTRÓN space-charge limited current; ~ **de línea** INFO line current; ~ **de llamada** TELEFON ringing current; ~ **de llamada codificada** TELEFON coded ringing current; ~ **luminiscente de protector** ELECTRÓN glow current of protector; ~ **luminosa** ELECTRÓN, ELECTROTEC, TRANS light current; ~ **magnetizante** ELECTRÓN magnetizing current; ~ **de no disparo de puerta** ELECTRÓN *tiristores* gate non-trigger current; ~ **nominal de descarga alterna** ELECTROTEC nominal alternating discharge current; ~ **de ondas lumínicas** ELECTRÓN lightwave stream; ~ **operatoria mínima** ELECTRÓN minimum operate current; ~ **oscura electródica** ELECTRÓN electrode dark current; ~ **parásita** ELECTRÓN sneak current, ELECTROTEC eddy current; ~ **de pérdida a tierra** ELECTRÓN leakage current; ~ **perturbadora** TELEFON noise current; ~ **de placa** ELECTRÓN plate current; ~ **de polarización** ELECTRÓN bias current, ELECTROTEC polarization current; ~ **de polarización de rejilla** ELECTRÓN grid current; ~ **principal** ELECTROTEC main current; ~ **de proceso** INFO run stream; ~ **pulsatoria** ELECTROTEC pulsating current; ~ **reactiva** ELECTROTEC reactive current; ~ **de recuperación en sentido inverso** ELECTROTEC reverse-recovery current; ~ **reflejada** TRANS reflected current; ~ **de reflexión total** ELECTRÓN total reflector current; ~ **residual** ELECTRÓN dark current; ~ **residual del emisor** ELECTRÓN *semiconductores* emitter cutoff current; ~ **residual de puerta** ELECTRÓN *tiristores* gate cutoff current; ~ **de retención** ELECTRÓN hold current, *semiconductores* holding current; ~ **de retorno** TRANS return current; ~ **de revestimiento catódico** ELECTRÓN cathode-covering current; ~ **de salida** ELECTRÓN output current; ~ **de saturación por tensión** ELECTRÓN, ELECTROTEC voltage saturation current; ~ **de señal** ELECTRÓN signal current; ~ **de servicio** ELECTROTEC operating current; ~ **simple** ELECTROTEC, TRANS single current; ~ **de sonda** ELECTRÓN probe current; ~ **submarina** ELECTROTEC undercurrent; ~ **del substrato** ELECTRÓN substrate current; ~ **de transferencia de cebado** ELECTRÓN starter transfer current; ~ **umbral** ELECTRÓN, ELECTROTEC, TRANS threshold current; ~ **en vacío** ELECTROTEC no-load current; ~ **vagabunda** ELECTROTEC stray current

corrientes: ~ **fuertes** *f pl* ELECTROTEC heavy current

corrimiento *m* PRUEBA creep; ~ **de resina** ELECTRÓN resin smear

corroído *adj* GEN corroded

corrosión *f* GEN corrosion

cortador: ~ **de cubierta** *m* COMPON *de cable* sheath cutter

cortadora *f* GEN cutter

cortar *vt* INFO *con el punzón sacabocados* punch out; ◆ ~ **el energía de** ELECTRÓN power down; ~ **y pegar** DATOS, INFO cut and paste

corte *m* GEN cutoff, *rueda dentada* cutting; ~ **por filtración** INFO filter cutoff; ~ **de la señal de llamada** TELEFON ring tripping; ~ **en el suministro de corriente** ELECTROTEC power cut; ~ **de suministro eléctrica** ELECTROTEC mains failure; ~ **de transmisión** PRUEBA, TRANS transmission breakdown

cortocircuitación: ~ **leve** *f* ELECTROTEC soft shorting

cortocircuitar *vi* ELECTRÓN, ELECTROTEC, INFO short

cortocircuito *m* (*S/C*) ELECTRÓN, ELECTROTEC, INFO short circuit (*S/C*), short; ~ **de equipo** ELECTROTEC hard shorting

cosecante *f* TEC RAD cosecant

coser: ~ **con tiretas** *vt* COMPON *correas* lace

cosido: ~ **de cable** *m* TRANS cable forming

coste *Esp* (*AmL* **costo**) *m* GEN cost; ~ **de la calidad del productor** PRUEBA producer's quality cost; ~ **de calidad del usuario** PRUEBA user's quality cost; ~ **completo de utilización** APLIC *calidad* life cycle cost; ~ **de**

explotación GEN operating expense; ~ **original** GEN *economía* first cost; ~ **de prevención** PRUEBA prevention cost; ~ **de producción** GEN cost of production, production cost; ~ **de tasación** PRUEBA *calidad* appraisal cost; ~ **total** GEN total cost; ~ **total de posesión** GEN *calidad* cost of ownership

costes *Esp* (*AmL* **costos**) *m pl* GEN costs; ~ **administrativos** GEN clerical costs; ~ **de funcionamiento** GEN running costs

costo *AmL ver coste Esp*

costos *AmL ver costes Esp*

costura: ~ **de cables** *f* TRANS cable lacing

cota: ~ **superior** *f* TRANS upper boundary

cotejar *vt* DATOS, INFO collate

cotización *f* GEN *de los precios* quotation

CP *abr* TELEFON (*central periférica*) CP (*peripheral exchange*), (*central primaria*) CP (*primary exchange*), (*central pública*) CP (*public exchange*)

CPE *abr* (*equipo en locales del abonado*) PERIF CPE (*customer premises equipment*)

cpi *abr* (*caracteres por pulgada*) INFO cpi (*characters per inch*)

CPL *abr* (*lenguaje combinado de programación*) PERIF CPL (*combined programming language*)

C/R *abr* (*bitio de comando/respuesta*) TELEFON C/R (*command/response bit*)

craquelado *m* TELEFON crackle

CRC *abr* (*verificación de redundancia cíclica*) DATOS, TELEFON CRC (*cyclic redundancy check*)

CRE *abr* (*equivalente de referencia corregido*) TELEFON CRE (*corrected reference equivalent*)

creación *f* GEN creation

crear *vt* GEN create

creciente *adj* GEN increasing

crecimiento: ~ **de mercado** *m* GEN market growth

credenciales *f pl* DATOS credentials

cremallera *f* ELECTROTEC, INFO, TELEFON rack

crepitación *f* TELEFON crackling noise, scratching noise

cresta *f* ELECTRÓN peak; ~ **de tensión** ELECTROTEC voltage peak; ~ **de tensión inversa** (*PIV*) ELECTROTEC peak inverse voltage (*PIV*); ~ **de tráfico** TELEFON peak traffic, traffic peak

criogenia *f* GEN *física* cryogenics

criotrón *m* ELECTRÓN cryotron

criptograma *m* DATOS, TEC RAD, TELEFON, TV cryptogram

criptón *m* COMPON *química* krypton

cristal *m* GEN crystal, *materiales* glass; ~ **de control** ELECTRÓN *electrónica* control crystal; ~ **líquido direccionado al plasma** ELECTRÓN plasma-addressed liquid crystal; ~ **de ventana** INFO windowpane

cristales: ~ **líquidos** *m pl* COMPON, ELECTRÓN, INFO liquid crystals; ~ **líquidos nemáticos** *m pl* ELECTRÓN nematic liquid crystals

criterio: ~ **de decisión** *m* PRUEBA *calidad* decision function; ~ **del fallo** *m* PRUEBA *seguridad de funcionamiento* failure criterion; ~ **de rechazo** *m* PRUEBA rejective condition

criticidad: ~ **del fallo** *f* PRUEBA *seguridad de funcionamiento* failure criticality

cromaticidad *f* TV chromaticity

cromático *adj* TV chromatic

cromeado *m* GEN chrominance

cronización *f* GEN timing

cronograma *m* GEN timing diagram; ~ **principal** TEC RAD master time-plan

cronómetro *m* GEN timer

croquis *m* ELECTRÓN outline drawing; ~ **a escala** GEN *dibujo técnico* dimensional draft

cruce: ~ **de carreteras en T** *m* TEC RAD T-junction

cruz *f* INFO *cursor* cross hair

CSC *abr* (*control del sistema de comunicaciones*) GEN CSC (*communications system control*)

CSE: ~ **remoto** *m* TELEFON remote CSE

CSG *abr* (*geometría constructiva tridimensional*) INFO CSG (*constructive solid geometry*)

CSI-DIS *abr* (*identificación DID del abonado llamado*) TELEFON CSI-DIS (*called-subscriber identification DIS*); ~ **de instalación no normalizada** *m* (*NSF-CSI-DIS*) DATOS non-standard facility CSI-DIS (*NSF-CSI-DIS*)

CSU *abr* (*unidad de almacenamiento y conversión*) DATOS CSU (*conversion and storage unit*)

CTCSS *abr* (*sistema continuo de silenciador de tono controlado*) TEC RAD CTCSS (*continuous tone-controlled squelch system*)

CTR *abr* (*respuesta a continuar para corregir*) DATOS CTR (*response to continue to correct*)

CTSM *abr* (*manual de especificación de comprobación de conformidad*) DATOS CTSM (*conformance testing specification manual*)

cuadrante: ~ **alfa** *m* INFO alpha dial; ~ **de visualización** *m* DATOS display board

cuadratura: **en** ~ *adj* ELECTROTEC in quadrature

cuadrete *m* GEN quad; ~ **de cable** TRANS cable quad

cuadrícula: ~ **de sombra** *f* ELECTRÓN shadow grid

cuadripolo *m* ELECTRÓN, ELECTROTEC, TRANS

four-terminal network, two-terminal pair, two-terminal-pair network

cuadro *m* COMPON, CONMUT, INFO board; ~ **aditivo** INFO add-in board; ~ **de análisis** GEN *documentación* analysis table; ~ **de audífonos** PERIF phones jack; ~ **de características** GEN *documentación* data sheet; ~ **del circuito** ELECTRÓN, INFO circuit board; ~ **de conexiones** TELEFON, TV connection box; ~ **conmutador automático-manual** CONMUT, PERIF auto-manual switchboard; ~ **DCME** TELEFON DCME frame; ~ **descriptivo del fallo** TELEFON fault pattern, fault syndrome; ~ **de diálogo de archivo estándar** DATOS, INFO standard file dialog box; ~ **de diálogo del zoom** INFO zoom box; ~ **de fusibles** ELECTROTEC fuse holder, fuse panel; ~ **interno** TV intraframe; ~ **interurbano** CONMUT toll switch; ~ **múltiple** CONMUT multiple frame, multiple switchboard; ~ **de paletas** ELECTROTEC paddle board; ~ **de pinzas** INFO clipboard; ~ **de rechazo de trama** (*FRMR*) TELEFON frame reject frame (*FRMR*); ~ **de terminales** TELEFON terminal board; ~ **de vídeo** INFO video board

cuádruplex *m* ELECTRÓN quadplex

cualidades: ~ **técnicas de la telefonía** *f pl* TELEFON telephony performance

cualificación: ~ **de información para autorización** *f* DATOS authorization-information qualifier

cualificador: ~ **de autorización de la información** *m* SEÑAL authorization-information qualifier; ~ **de código de identificación** *m* DATOS identification-code qualifier; ~ **de la unidad de aplicación** *m* DATOS application-entity qualifier

cuantificación *f* TRANS quantization, quantizing; ~ **de bloques** TRANS block quantization; ~ **de datos** TELEFON data quantization; ~ **de riesgos** PRUEBA risk quantification; ~ **de una señal** SEÑAL signal quantization; ~ **uniforme** TELEFON, TRANS uniform quantizing

cuantificador *m* TRANS quantizer

cuantificar *vt* TRANS quantize

cuantización *f* INFO, TELEFON, TRANS quantizing; ~ **solar** SEÑAL solar quantization; ~ **de vector de estado finito** SEÑAL finite-state vector quantization; ~ **vectorial** DATOS, SEÑAL vector quantization

cuarteto *m* DATOS quartet

cuartil *m* GEN quartile; ~ **inferior** GEN *estadística* lower quartile; ~ **superior** GEN upper quartile

cuarto *m* INFO room; ~ **de vuelta** ELECTRÓN quarter turn

cuba: ~ **de amalgamación** *f* INFO pan; ~ **electrolítica** *f* ELECTRÓN *química* electrolytic cell

cubierta *f* GEN covering; ~ **aislante** COMPON, ELECTRÓN, ELECTROTEC sleeving; ~ **de atenuación acústica** COMPON, TELEFON sound-attenuating cover; ~ **de batería** ELECTROTEC battery case; ~ **de cable** TRANS cable cap, cable jacket, cable sheath; ~ **del cable** TRANS cable covering; ~ **de dos piezas** COMPON, TELEFON two-piece cover; ~ **estrecha** TRANS *óptica de fibras* tight jacket; ~ **de extremo de fila** TELEFON row-end cover; ~ **lateral** TELEFON side cover; ~ **de plomo** GEN *cables* lead sheath, lead sheathing; ~ **portadora de protección** GEN protective carrying case; ~ **protectora** COMPON protective cover; ~ **protectora de plástico** INFO protective plastic coating; ~ **protectora de tubos de rayos catódicos** COMPON cover plate; ~ **de repetidor** SEÑAL repeater deck; ~ **de una sola pieza** TELEFON one-piece cover

cubierto: ~ **de hoja plisada** *adj* TRANS pleated-foil covered

cubo *m* GEN hub; ~ **C** (C^3) INFO C-cube (C^3); ~ **portabobina** TELEFON reel hub

cubrir: ~ **las necesidades de** *vt* GEN meet the needs of

cuchilla: ~ **pelacables** *f* COMPON stripping knife

cuenta: ~ **atrás** *f* GEN countdown; ~ **de epoxia** *f* ELECTRÓN epoxy bead; ~ **de gestión** *f* GEN management account; ~ **principal** *f* GEN major account

cuentagotas *m* INFO eyedropper

cuentatiempo *m* TELEFON running time meter

cuerda: ~ **de bajo** *f* INFO bass chord; ~ **de bajo automático** *f* INFO auto bass chord; ~ **de montaje** *f* ELECTRÓN mounting cord

cuerpo *m* GEN body; ~ **cilíndrico** COMPON shell; ~ **fotoemisor** TV light emitter; ~ **de prueba** DATOS, PRUEBA test body; ~ **de relé** DATOS relay entity

culata: ~ **del relé** *f* COMPON, ELECTRÓN, TELEFON relay yoke

culote: ~ **de lámpara** *m* COMPON valve base

cúmetro *m* PRUEBA Q-meter

cumplir *vt* GEN *requisitos* meet; ◆ ~ **con el plazo fijado** GEN meet the deadline

cuna *f* PERIF, TEC RAD cradle

cuota *f* GEN fee; ~ **de abono** GEN subscription fee; ~ **de conexión** TELEFON connection charge, nonrecurrent charge; ~ **de conexión**

a la red TELEFON network connection fee; ~ **única** TELEFON nonrecurrent fee

curso: **en** ~ *adj* GEN ongoing

cursor *m* INFO cursor, slide contact, slider, TEC RAD cursor; ~ **destellante** INFO flashing cursor

curva: ~ **de amplitud** *f* INFO amplitude curve; ~ **de Bézier** *f* INFO Bézier curve; ~ **de cable** *f* TRANS cable bend; ~ **característica operativa** *f* GEN operating characteristic curve; ~ **de corriente de descarga de la tensión** *f* TELEFON *de protector* voltage discharge-current curve; ~ **de duración de tensión de ruptura de impulsos** *f* TELEFON *de protector* impulse spark-over voltage-time curve; ~ **de eco** *f* TELEFON echo curve; ~ **de enmascaramiento** *f* TELEFON masking curve; ~ **envolvente** *f* INFO envelope curve; ~ **de error** *f* DATOS error pattern; ~ **exponencial descendente** *f* INFO falling exponential curve; ~ **de frecuencias** *f* TRANS frequency curve; ~ **de funcionamiento** *f* ELECTRÓN operating curve; ~ **gamma** *f* APLIC, PERIF *digitalizador* gamma curve; ~ **gaussiana** *f* PRUEBA Gaussian curve; ~ **de histéresis** *f* ELECTRÓN *electrónica* hysteresis curve; ~ **lineal descendente** *f* INFO falling linear curve; ~ **M** *f* TEC RAD M-curve; ~ **OC** *f* GEN OC curve; ~ **primitiva** *f* INFO pitchbend; ~ **de reparto de resistencia** *f* ELECTRÓN taper curve; ~ **de réplica de frecuencia** *f* ELECTRÓN, TEC RAD frequency response curve

curvadora *f* COMPON bender

curvar *vt* GEN bend

curvatura: ~ **de cinta** *f* INFO wraparound

D

D *abr* (*valor asignable por defecto*) DATOS D (*defaultable*)

D-A *abr* (*digital a analógico*) GEN D-A (*digital-to-analog*)

DAC *abr* (*dispositivo acoplado por carga*) ELECTRÓN CCD (*charge-coupled device*)

daño *m* ELECTRÓN prejudice; **~ por excavación** APLIC excavation damage; **~ por excavadora** APLIC excavator damage; **~ en la faceta** TRANS *óptica de fibras* facet damage

dar: **~ baja prioridad a** *vt* GEN give low priority to; **~ elevada prioridad a** *vt* GEN give high priority to; **~ prioridad a** *vt* GEN give priority to

datagrama *m* DATOS *Internet*, SEÑAL, TELEFON datagram

datos *m pl* GEN data; **~ del abonado que llama** TELEFON calling-party data; **~ ASCII** INFO ASCII data; **~ bajo voz** (*DUV*) TEC RAD data under voice (*DUV*); **~ básicos** GEN basic data; **~ de bobinado** ELECTROTEC winding data; **~ CD-I** INFO CD-I data; **~ de control** INFO control data; **~ de descripción de tarea** TELEFON task-description data; **~ digitales** DATOS, INFO digital data; **~ editados** DATOS bound data; **~ eléctricos** TRANS *óptica de fibras* electrical data; **~ de encaminamiento** TELEFON route-choice data, routing data; **~ de estado** DATOS, PRUEBA status data; **~ de explotación** TELEFON operational data; **~ de función** DATOS function data; **~ de gestión** DATOS management data; **~ de gestión de la red** INFO network-management data; **~ gráficos** TV image data; **~ de instalación** GEN *documentación* installation data; **~ Manchester** INFO Manchester data; **~ de mantenimiento N para usuario** DATOS N-user maintenance data; **~ en masa** INFO bulk data; **~ no analizados** DATOS raw data; **~ personales** GEN personal data; **~ recibidos** (*RD*) DATOS receive data (*RD*), received data; **~ de salida** DATOS output data; **~ secuenciales** DATOS, INFO sequence data; **~ semipermanentes** TELEFON semipermanent data; **~ de sincronización** INFO synchronization data; **~ sobre acción atómica** DATOS atomic-action data; **~ sobre la calidad de servicio** DATOS, INFO performance data; **~ sobre el sistema** INFO system data; **~ técnicos** GEN engineering data; **~ de terreno** TEC RAD terrain data; **~ transmitidos** DATOS transmitted data; **~ de usuario** DATOS, INFO user data; **~ de usuario N** TELEFON N-user data; **~ sobre voz** *f* TEC RAD data above voice

dB *abr* (*decibelio*) GEN dB (*decibel*)

dBm *abr* (*decibelios por encima de un milivatio*) TELEFON dBm (*decibels above one milliwatt*)

DCE *abr* (*conjunto electrónico de rotación*) ELECTRÓN DCE (*despun control electronics*)

DCME *abr* (*equipo de multiplicación del circuito digital*) TELEFON DCN (*digital circuit multiplication equipment*)

DCN *abr* (*desconectar*) GEN DCN (*disconnect*)

DCR *abr* (*grabadora de datos de casete*) PERIF DCR (*data cassette recorder*)

DCS *abr* (*señal de mando digital*) DATOS DCS (*digital command signal*); **~ de identificación de abonado emisor** *m* (*TCI-DCS*) DATOS, TELEFON transmitting-subscriber identification-DCS (*TCI-DCS*)

DCTE *abr* (*terminal de circuito de datos*) TRANS DCTE (*data circuit terminating equipment*)

DCU *abr* (*unidad de distribución y control*) CONTROL DCU (*distribution and control unit*)

DCX *abr* (*archivo PCX multipágina*) DATOS DCX (*multipage PCX file*)

DD *abr* (*doble densidad*) INFO DD (*double density*)

DDA *abr* (*atributo de campo definido*) TELEFON DDA (*defined-domain attribute*)

debilitación: **~ en recepción** *f* TELEFON receive loss

debilitamiento: **~ de señal** *m* SEÑAL signal weakening

decadencia: **~ orbital** *f* TEC RAD orbital decay

decibelio *m* (*dB*) GEN decibel (*dB*)

decibelios: **~ por encima de un milivatio** *m pl* (*dBm*) TELEFON decibels above one milliwatt (*dBm*)

decimal *adj* GEN decimal

decisegundo: **~ sin error** *m* TELEFON error-free decisecond

decisión: **~ heurística** *f* DATOS heuristic decision; **~ de paso** *f* DATOS pass verdict; **~ de programa** *f* DATOS soft decision; **~ relacionada con la circuitería** *f* ELECTRÓN hard

decision; ~ **de transacción** *f* DATOS transaction commitment
decisivo: **no** ~ *adj* DATOS inconclusive
declaración *f* DATOS declaration, statement, INFO, TELEFON statement; ~ **de conformidad de instalación** (*ICS*) DATOS implementation-conformance statement (*ICS*); ~ **de conformidad de objeto gestionado** (*MOCS*) DATOS managed-object conformance statement (*MOCS*); ~ **detallada** APLIC itemized statement; ~ **final** TELEFON end statement
decodificación: ~ **de decisión de programa** *f* DATOS soft-decision decoding; ~ **Viterbi** *f* TELEFON Viterbi decoding
decodificador: ~ **ADPCM** *m* TELEFON ADPCM decoder
decodificar *vt* GEN decode
decreciente *adj* GEN *orden* descending
decuantización *f* INFO dequantization
dedo *m* DATOS *Internet* finger; ~ **de guía** TELEFON guide finger; ~ **selector** COMPON, CONMUT selecting finger
deducido *adj* GEN deducted
defasaje: ~ **de nivel** *m* TELEFON level shifting
defase *f* ELECTROTEC phase displacement
defecto *m* GEN bug, flaw; ~ **de aislamiento** ELECTROTEC insulation fault; ~ **de cálculo** PRUEBA *dependabilidad* design failure; ~ **corregible** PRUEBA correctable fault; ~ **de enrollamiento** COMPON wrapping fault; ~ **de fabricación** PRUEBA manufacturing defect; ~ **menor** PRUEBA minor defect
defectuoso *adj* GEN defective
defensa: ~ **central** *f* TELEFON central defence (*AmE* defense)
definición *f* DATOS, ELECTRÓN, INFO definition, resolution; ~ **en distancia** TEC RAD range discrimination; ~ **de exploración** TELEFON scanning pitch; ~ **genérica** DATOS generic definition; ~ **de imagen** DATOS *de digitalizador*, TV sharpness, picture definition; ~ **de recepción** TELEFON reception definition; ~ **de transmisión** TELEFON, TRANS transmission definition
deflector *m* TEC RAD deflector
deflexión: ~ **de aguja** *f* TELEFON *de dispositivo indicador* needle deflection
deformación *f* COMPON strain, ELECTRÓN, INFO *piezas fundidas* warping; ~ **facial** INFO facial warping; ~ **de forma libre** (*FFD*) INFO free-form deformation (*FFD*)
deformar *vt* COMPON, ELECTRÓN, INFO warp
degradación *f* GEN degradation; ~ **catastrófica** TELEFON *óptica de fibras* catastrophic degradation; ~ **de ganancia** TEC RAD gain degradation; ~ **con garbo** TEC RAD, TRANS graceful degradation; ~ **de señal** SEÑAL signal degradation (*SD*)
degradar *vt* GEN downgrade
degresivo *adj* GEN degressive
dejar: ~ **fuera** *vt* GEN leave out; ~ **fuera de servicio** *vt* GEN put out of order, render inoperative
DEL *abr* (*diodo electroluminiscente*) COMPON, ELECTRÓN, TRANS LED (*light-emitting diode*); ~ **superluminiscente** *m* COMPON, ELECTRÓN, TRANS superluminescent LED (*SLD*)
delaminación *f* ELECTRÓN *de circuito impreso* delamination
deletrear *vt* GEN spell
deletreo *m* GEN spelling
delimitador *m* GEN delimiter; ~ **aritmético** INFO arithmetic delimiter
delineamiento: ~ **de rayos** *m* INFO ray tracing
delta *m* ELECTRÓN *célula magnética* delta
demanda: ~ **de desconexión** *f* TELEFON clearing request; ~ **de mercado** *f* GEN market demand; ~ **obligatoria de fin de conversación** *f* SEÑAL compulsory clearing request; ~ **de testigo de mando multipunto** *f* DATOS multipoint demand token claim; ~ **de tráfico** *f* TEC RAD traffic demand
demodulación: ~ **de frecuencia** *f* TRANS frequency demodulation
demodulador: ~ **de frecuencia** *m* TRANS frequency demodulator
demora *f* GEN delay; ~ **en la cola de espera** TELEFON queueing delay; ~ **logística** PRUEBA *seguridad de funcionamiento*, TELEFON logistic delay; ~ **de procesamiento** DATOS, INFO processing delay; ~ **en la respuesta** TELEFON answering delay; ~ **rotacional** INFO rotational delay
demostración *f* INFO demonstration, demo
denegar *vt* INFO deny
denominación *f* GEN denomination
densidad: ~ **acústica** *f* TELEFON sound density; ~ **aparente** *f* GEN *provisión y distribución* bulk density; ~ **de corriente umbral** *f* ELECTRÓN, ELECTROTEC, TRANS threshold-current density; ~ **de energía** *f* GEN energy density; ~ **de energía/ruido** *f* TRANS energy-to-noise density; ~ **de errores** *f* DATOS error density; ~ **espectral** *f* DATOS, TRANS spectral density; ~ **de exploración** *f* TELEFON scanning density; ~ **de flujo** *f* ELECTRÓN flux density; ~ **de flujo de fuerza utilizable como referencia** *f* TEC RAD reference-usable power-flux density; ~ **de flujo de potencia media** *f* TEC RAD medium-power flux density; ~ **de flujo de potencia utilizable** *f*

TEC RAD usable-power flux density; ~ **de frecuencia** *f* GEN frequency density; ~ **de grabación** *f* DATOS packing density, recording density, ELECTRÓN packing density; ~ **de llamadas** *f* TELEFON calling rate; ~ **mínima de flujo energético utilizable** *f* TEC RAD minimum usable power-flux density; ~ **de pista** *f* INFO track density; ~ **de potencia** *f* ELECTROTEC power density; ~ **de probabilidad** *f* GEN probability density; ~ **relativa** *f* GEN relative density; ~ **de tráfico** *f* TELEFON traffic density

dependencia: ~ **de modo de funcionamiento** *f* DATOS mode dependency

deposición *f* GEN deposition; ~ **axial de vapor** ELECTRÓN vapour (*AmE* vapor) axial deposition (*VAD*); ~ **lateral de plasma** ELECTRÓN lateral plasma deposition; ~ **de vapor exterior** COMPON outer vapour (*AmE* vapor) deposition; ~ **de vapor químico** ELECTRÓN chemical vapour (*AmE* vapor) deposition; ~ **de vapor químico órgano-metálico** ELECTRÓN, TRANS metal-organic chemical vapour (*AmE* vapor) deposition

deprimir *vt* INFO depress

depuración: ~ **de programa** *f* DATOS program debugging

depurar *vt* DATOS debug

derechos: ~ **de exclusividad** *m pl* TV exclusive rights; ~ **de radiodifusión** *m pl* TV broadcasting rights; ~ **de suscripción** *m pl* GEN rights

deriva *f* GEN drift; ~ **de frecuencia** ELECTRÓN, TEC RAD frequency drift; ~ **de frecuencia normalizada** TEC RAD, TRANS normalized frequency drift; ~ **giroscópica** TEC RAD gyroscopic drift; ~ **de órbita** TEC RAD orbit drift

derivación *f* COMPON shunt, ELECTROTEC shunting, INFO bypass; ~ **de bifurcación** ELECTRÓN branching bypass; ~ **e inserción** TRANS drop and insert; ~ **de línea** ELECTRÓN line tapping; ~ **de red fina** COMPON, ELECTRÓN thin-net tap

derivado *m* GEN derivative

derivar *vt* GEN derive

derrame *m* GEN break

derrumbe *m* COMPON collapse

DES *abr* (*norma de cifrado de datos*) DATOS DES (*data encryption standard*)

desacentuación *f* ELECTRÓN de-emphasis

desacoplamiento *m* ELECTRÓN, ELECTROTEC decoupling

desactivación *f* GEN, INFO inoperability, disablement

desactivado *adj* GEN, INFO disabled, inoperable

desactivar *vt* GEN, INFO deactivate, disable

desacuerdos: ~ **en el número de bitios** *m pl* DATOS number of bit disagreements

desadaptación: ~ **de tipo** *f* INFO *imprenta* type mismatch

desagrupamiento: ~ **de carga espacial** *m* ELECTRÓN space-charge debunching

desagrupar *vt* ELECTRÓN debunch, INFO ungroup

desajuste: ~ **de frecuencia normalizado** *m* TEC RAD normalized frequency departure

desalineación: ~ **mecánica de punto** *f* ELECTRÓN mechanical spot misalignment; ~ **de trama** *f* DATOS *digitalizador* deframing

desarmar *vt* GEN dismantle

desarrollar *vt* GEN develop

desarrollo *m* GEN development

desbarbar *vt* GEN cut back

desbloquear *vt* ELECTRÓN, TEC RAD *canal de radio*, TELEFON deblock, unblock

desbloqueo *m* ELECTRÓN, TEC RAD *de canal de radio*, TELEFON deblocking, unblocking; ~ **de vínculo de señalización** SEÑAL signalling-link (*AmE* signaling-link) unblocking

desbordamiento *m* GEN overflow; ~ **porcentual** TELEFON percentage overflow

descafilar *vt* INFO rub

descansar: ~ **sobre** *vt* COMPON bear against

descarga *f* ELECTRÓN, ELECTROTEC, INFO, TELEFON discharge; ~ **atmosférica** CLIMAT lightning discharge; ~ **atmosférica indirecta** CLIMAT indirect strike; ~ **disruptiva** GEN *electricidad* disruptive breakdown; ~ **de gas** ELECTROTEC gas discharge; ~ **por ionización acumulativa** ELECTRÓN avalanche breakdown; ~ **luminiscente** ELECTRÓN glow discharge; ~ **por medio de tecla** INFO key release; ~ **de memoria principal** TEC RAD *programas* roll-out; ~ **en penacho** ELECTROTEC brush discharge; ~ **sin mantenimiento automático** ELECTRÓN non-self-maintained discharge

descargado *adj* ELECTRÓN, ELECTROTEC, INFO, TELEFON unloaded

descargar *vt* ELECTRÓN, ELECTROTEC, INFO, TELEFON unload

descendente *adj* INFO downward

descentrado: ~ **de eje** *adj* TEC RAD off-axis

descentralizado *adj* GEN decentralized

descentralizar *vt* GEN decentralize

descifrado *m* GEN deciphering, decoding

desciframiento *m* DATOS unscrambling, GEN deciphering, decoding, INFO unscrambling

descifrar *vt* DATOS unscramble, GEN decipher, decode, INFO unscramble

descodificación *f* GEN deciphering, decoding; ~ **dudosa** DATOS hard-decision decoding; ~ **de**

realimentación ELECTRÓN feedback decoding; ~ **de umbral** DATOS threshold decoding; ◆ ~ **uno de cinco** ELECTRÓN one-out-of-five decode
descodificador *m* GEN decoder; ~ **para colocar encima del televisor** TV set-top decoder; ~ **de decisiones difíciles** SEÑAL hard-decision decoder; ~ **FEC de velocidad mitad** SEÑAL half-rate FEC decoder; ~ **interfacial** TRANS interface decoder; ~ **de línea** TRANS line decoder; ~ **de máxima probabilidad** *f* TEC RAD maximum-likelihood decoder
descolgado *adj* TELEFON off-hook
descolgamiento *m* INFO, PERIF, TEC RAD, TELEFON off-hooking
descomposición *f* COMPON decomposition
descompresión *f* DATOS, INFO decompression; ~ **de fichero** DATOS, INFO file decompression
descomprimir *vt* DATOS, INFO unzip, decompress
desconectado *adj* GEN unplugged, disconnected; ~ **de tierra** ELECTROTEC unearthed *BrE*, ungrounded *AmE*
desconectar *vt* (*DCN*) GEN unplug, disconnect (*DCN*)
desconectarse: ~ **del sistema** *vi* INFO log off, log out
desconexión *f* INFO clearing, disconnection, TEC RAD withdrawal, TELEFON clearing, disconnection; ~ **de abonado llamado** CONMUT, TELEFON called-subscriber release; ~ **de abonado que llama** CONMUT, TELEFON calling-subscriber release; ~ **automática** ELECTROTEC auto power off, INFO auto power off, auto-stop; ~ **en cadena** INFO chain clear; ~ **de circuito secreto** DATOS, INFO, TELEFON privacy release; ~ **doble** INFO, TELEFON both-party release; ~ **espuria** (*SSO*) PRUEBA spurious switch-off (*SSO*); ~ **de montaje** ELECTRÓN mounting cut-out; ~ **prematura** INFO, TELEFON premature release; ~ **de primer abonado** INFO, TELEFON first-subscriber release; ~ **de prioridad** TELEFON priority disconnection; ~ **simple** INFO, TELEFON first-party release
descongelación *f* GEN de-icing, defrosting
descongelar *vt* GEN de-ice, defrost
descorrelación *f* GEN decorrelation
descripción *f* GEN description; ~ **de aparato** TELEFON device description; ~ **de bloque funcional** TELEFON functional-block description; ~ **de célula de mantenimiento** DATOS maintenance-cell description (*MCD*); ~ **de circuito** GEN *documentación* circuit description; ~ **de función** GEN *documentación* function description; ~ **funcional** (*FD*) GEN *en SDL* functional description (*FD*); ~ **de funcionamiento** TELEFON operational description; ~ **general** TELEFON overview; ~ **de proceso** TELEFON process description; ~ **de señalización** SEÑAL signalling (*AmE* signaling) description; ~ **de símbolo** GEN symbol description
descriptación *f* GEN decryption
descronometrado *adj* ELECTRÓN unclocked
descubrir *vt* GEN uncover
desembragar *vt* GEN disengage
desenchufar *vt* GEN unplug
desencriptar *vt* GEN descramble
desenganchar *vt* PERIF, TELEFON unhook
desenrollamiento *m* GEN unwrapping
desenrollar *vt* GEN unwrap
desensamblar *vt* GEN disassemble, take apart
desensibilización *f* ELECTRÓN desensitization
desequilibrado *adj* ELECTROTEC, INFO unbalanced
desequilibrio *m* ELECTROTEC, INFO imbalance; ~ **de capacidad** TELEFON capacity imbalance; ~ **de distribución de tráfico** CONMUT, TELEFON traffic-distribution imbalance; ~ **de impedancia** ELECTRÓN, TELEFON impedance imbalance; ~ **de intensidad de tráfico** CONMUT, TELEFON traffic-load imbalance; ~ **de tráfico** CONMUT, TELEFON traffic imbalance
desfasador *m* ELECTRÓN, ELECTROTEC, TEC RAD, TRANS phase shifter; ~ **de guíaondas** TEC RAD waveguide-phase changer, waveguide-phase shifter; ~ **rotativo** TEC RAD rotary-phase changer, rotary-phase shifter
desfasaje *m* ELECTRÓN, ELECTROTEC, TEC RAD, TRANS phase shift, phase change
desfase *m* ELECTRÓN phase displacement
desfiltrado *m* DATOS *de digitalizador* descreening
desgasificación *f* GEN *de lámparas eléctricas* degassing
desgasificar *vt* GEN *lámparas eléctricas* degas
desguace *m* COMPON scrapping
deshielo *m* GEN de-icing
designación: ~ **de cable** *f* TRANS cable designation; ~ **de función** *f* GEN function designation; ~ **de plantilla de cable** *f* TRANS formboard designation; ~ **de referencia** *f* INFO reference designation
designado *m* GEN assignee
designador: ~ **de código internacional** *m* (*ICD*) DATOS international code designator (*ICD*)
designar *vt* GEN designate
desigualdad *f* GEN inequality; ~ **de ganancia** ELECTRÓN gain inequality; ~ **de retardo de**

luminancia-crominancia TV chrominance-luminance delay inequality
desincronización: ~ **de sonido de película** *f* INFO, TV picture-sound desynchronization
desintonización *f* TEC RAD, TV detuning
desintonizar *vt* TEC RAD, TV detune
desionización *f* ELECTRÓN de-ionization
deslizamiento: ~ **controlado** *m* TEC RAD, TELEFON controlled slip; ~ **de imagen** *m* TV frame slip; ~ **de línea** *m* TV line slip; ~ **sin control** *m* TEC RAD, TELEFON uncontrolled slip; ~ **de tráfico binario de transmisión** *m* TEC RAD, TRANS transmission-bit slip
desmagnetización *f* ELECTRÓN demagnetization, demagnetizing, INFO erasing
desmagnetizar *vt* ELECTRÓN demagnetize, INFO erase
desmodulación *f* ELECTRÓN demodulation; ~ **de fase** ELECTRÓN phase demodulation; ~ **incoherente** DATOS noncoherent demodulation
desmodulador *m* ELECTRÓN, INFO, TRANS demodulator; ~ **de extensión umbral** ELECTRÓN, TEC RAD threshold-extension demodulator; ~ **telegráfico** TELEFON telegraph demodulator
desmodular *vt* ELECTRÓN, INFO, TRANS demodulate
desmoldeadora *f* COMPON *fundición* stripping machine
desmontable *adj* GEN detachable
desmontaje: ~ **de paquetes de datos** *m* DATOS, TELEFON packet disassembly
desmontar *vt* GEN demount, dismount
desmultiplexado *m* INFO, TEC RAD, TRANS demultiplexing
desmultiplexor *m* INFO, TEC RAD, TRANS demultiplexer; ~ **por división en longitud de onda** TELEFON, TRANS *óptica de fibras* wavelength division demultiplexer; ~ **de onda guiada** (*GWD*) TRANS guided-wave demultiplexer (*GWD*)
desnivel: ~ **de frecuencia no preciso** *m* TEC RAD nonprecision frequency offset; ~ **de sombra** *m* INFO shadow offset
desnudo *adj* ELECTROTEC uninsulated
desoldar *vt* COMPON, ELECTRÓN unsolder
desorientación *f* TEC RAD depointing
despachador *m* GEN dispatcher
despachar *vt* GEN dispatch
despaquetificador *m* DATOS, TELEFON depacketizer
despegue *m* GEN separation
desperfecto *m* GEN impairment
desplazado *adj* GEN off-set, displaced
desplazamiento *m* GEN displacement, shift; ~ **angular** TELEFON, TRANS *óptica de fibras* angular displacement; ~ **en cuadrifase** SEÑAL, TEC RAD quadriphase shift keying; ~ **estrecho** TEC RAD narrow shift; ~ **de fase retardada** ELECTROTEC lagging-phase shift; ~ **de frecuencia** TEC RAD frequency hopping, TRANS frequency shift; ~ **en frecuencia** TRANS frequency displacement; ~ **de frecuencia de onda portadora** TRANS carrier-frequency offset; ~ **lateral** TEC RAD lateral displacement; ~ **lógico** INFO end-around shift, logic shift; ~ **longitudinal** GEN longitudinal displacement; ~ **mecánico de punto** ELECTRÓN mechanical spot displacement; ~ **de octava** INFO octave shift; ~ **en paralelo** TRANS parallel displacement; ~ **respecto a la frecuencia central** TEC RAD displacement from centre (*AmE* center) frequency; ~ **vertical** GEN vertical displacement
desplegar *vt* TEC RAD *antena* deploy
despliegue *m* TEC RAD *de antena* deployment; ~ **inicial** GEN initial deployment
despojos *m pl* INFO waste
despolarización *f* TEC RAD depolarization
despromediación *f* GEN de-averaging
despupinizar *vt* ELECTROTEC *cables* de-load
desregulación *f* GEN *economía* deregulation
destino *m* GEN destination
destornillador *m* GEN screwdriver; ~ **helicoidal de trinquete** COMPON spiral-ratchet screwdriver; ~ **para tornillos ranurados** COMPON slotted screwdriver; ~ **Torx** COMPON Torx screwdriver
desvanecimiento *m* INFO, TEC RAD, TV fade; ~ **de enlace ascendente** TEC RAD uplink fade; ~ **por lluvia** TEC RAD rain fading; ~ **por multitrayectoria** TEC RAD multipath fading; ~ **de Rayleigh** TELEFON Rayleigh fading; ~ **de sonido** TEC RAD, TV sound fade-out; ♦ **sin** ~ GEN nonfading
desventaja *f* GEN disadvantage
desviación *f* GEN deviation; ~ **de aguja** TELEFON *de indicador* pointer deflection; ~ **de atenuación** TRANS attenuation deviation; ~ **característica** GEN standard deviation; ~ **de diámetro de núcleo** TELEFON, TRANS core diameter deviation; ~ **de diámetro de superficie de referencia** TELEFON, TRANS reference-surface diameter deviation; ~ **de diámetro de superficie de revestimiento** TELEFON cladding-surface diameter deviation; ~ **de fase** GEN phase deviation; ~ **de frecuencia** TEC RAD, TRANS frequency departure, frequency deviation; ~ **de frecuencia máxima** TEC RAD,

TRANS peak-frequency deviation; ~ **lateral** TEC RAD lateral deviation; ~ **media** PRUEBA mean deviation; ~ **observada** PRUEBA observed deviation; ~ **de sistema** CONTROL system deviation; ~ **de tráfico** TELEFON traffic rerouting

desviar *vt* TELEFON divert, reroute

desvío: ~ **en caso de ocupado** *m* TELEFON diversion on busy; ~ **cuando no hay contestación** *m* TELEFON diversion on no reply; ~ **de frecuencia** *m* GEN frequency swing; ~ **de llamada** *m* TELEFON diversion of call; ~ **de llamada incondicional** *m* TELEFON unconditional call-forwarding; ~ **temporal de llamada** *m* PERIF *contestador automático* temporary call-forwarding

detalles: ~ **de destino** *m pl* DATOS *Internet* destination details

detección *f* GEN detection, INFO pinpointing; ~ **de averías** DATOS fault location, PRUEBA, TELEFON fault detection; ~ **de borde** DATOS edge detection; ~ **de calidad de señal de datos** DATOS, SEÑAL data-signal quality detection; ~ **de colisión** ELECTRÓN collision detection; ~ **cuadrática** ELECTRÓN square-law detection; ~ **a distancia** GEN remote sensing; ~ **de errores** DATOS error detection; ~ **homodina** TRANS homodyne detection; ~ **de impulso** SEÑAL pulse detection; ~ **por lápiz fotosensible** INFO light-pen detection; ~ **por manipulación** CONTROL, DATOS manipulation detection; ~ **de marcador** DATOS detection of marker; ~ **de palabra singular** DATOS unique-word detection; ~ **de palabra singular ausente** DATOS unique-word missed detection; ~ **de portadora de datos** DATOS data-carrier detection (*DCD*); ~ **de señales** SEÑAL signal detection

detectado: **no** ~ *adj* INFO, PRUEBA undetected

detectar *vt* ELECTRÓN, GEN detect, INFO pinpoint, PRUEBA sense

detectividad *f* ELECTRÓN, PRUEBA, TRANS detectivity; ~ **normalizada** ELECTRÓN, PRUEBA, TRANS *óptica de fibras* normalized detectivity

detector *m* GEN *desmodulación* detector; ~ **de autocorrelación** TEC RAD autocorrelation detector; ~ **de calor** TV heat detector; ~ **de campo oscuro** TRANS *óptica de fibras* dark-field detector; ~ **de cero** TEC RAD null detector; ~ **de cresta** GEN peak detector; ~ **cuadrático** ELECTRÓN square-law detector; ~ **de datos de banda de frecuencias vocales** DATOS voice-band data detector; ~ **de dispositivo de carga acoplada** PERIF *escáner* CCD detector; ~ **de errores** DATOS error detector; ~ **de fallo en el canal de datos** TELEFON data-channel failure detector; ~ **de fallo en la portadora de datos** TRANS data-carrier failure detector; ~ **de fase** ELECTRÓN phase detector; ~ **de filtrado y muestreo** ELECTRÓN filter-and-sample detector; ~ **de grietas** PRUEBA crack detector; ~ **luminoso** ELECTRÓN, TV light detector; ~ **matriz** PERIF *digitalizador* matrix detector; ~ **de medición** PRUEBA measuring detector; ~ **de palabra** TEC RAD, TELEFON speech detector; ~ **de pérdida de alineación de trama** TELEFON loss-of-frame-alignment detector; ~ **de presencia de intrusos** APLIC intruder-presence detector; ~ **de radar** TEC RAD radar-warning receiver; ~ **de ruidos** TELEFON noise detector; ~ **de señales** SEÑAL signal detector; ~ **de señalización** SEÑAL signalling (*AmE* signaling) detector; ~ **de señal de línea recibida de canal de datos** DATOS, TRANS data-channel received-line signal detector; ~ **de temperatura por resistencia** CLIMAT resistive-temperature detector

detener *vt* INFO halt

detergente *m* GEN detergent

deteriorado *adj* TELEFON mutilated, TRANS damaged, impaired

deterioro *m* TELEFON mutilation, TRANS impairment; ~ **de la calidad** PRUEBA deterioration of quality

determinación: ~ **de órbita** *f* TEC RAD orbit determination; ~ **de ruta** *f* TELEFON route determination; ~ **de sentido** *f* INFO cross-check; ~ **de tarifa** *f* TELEFON tariff determination

determinante: ~ **lateral** *m* CONMUT side-determining

determinar: ~ **la dirección** *fra* TEC RAD take a bearing

deuterio *m* GEN deuterium

deuterón *m* GEN deuteron

devanado: ~ **bifilar** *m* ELECTROTEC bifilar winding; ~ **de fase** *m* COMPON phase winding

devolver *vt* GEN *dinero* pay back; ◆ ~ **a alguien la llamada** TELEFON phone sb back; ~ **la llamada** TELEFON phone back

día: ~ **de calendario juliano modificado** *m* TEC RAD modified Julian day

diafonía: ~ **entre las dos direcciones** *f* TRANS go-to-return crosstalk; ~ **ininteligible** *f* TELEFON unintelligible crosstalk; ~ **múltiple** *f* TELEFON babble

diafragma *m* COMPON diaphragm

diagnosis *f* PRUEBA diagnosis; ~ **de averías** PRUEBA fault diagnosis

diagnosticar *vt* PRUEBA diagnose

diagnóstico *m* GEN diagnosis; **~ detallado** GEN detailed diagnosis, fine diagnosis; **~ de error** DATOS error diagnosis

diagnósticos *m pl* INFO diagnostics

diagrama *m* GEN diagram; **~ de árbol en bloques** DATOS *SDL* block-tree diagram; **~ de Bode** CONTROL, ELECTROTEC Bode diagram; **~ de campo próximo** TEC RAD near-field pattern; **~ de central** CONMUT, ELECTROTEC exchange diagram; **~ de circuito** CONMUT, ELECTRÓN circuit diagram; **~ de circuito y cableado** CONMUT, ELECTRÓN circuit-and-wiring diagram; **~ de circulación de control** DATOS *SDL* control-flow diagram; **~ de circulación de señales** SEÑAL signal-flow diagram; **~ de cobertura** TEC RAD coverage diagram; **~ de conexión** GEN connection diagram; **~ de conexiones** COMPON wiring diagram; **~ de conjuntos** GEN block diagram; **~ de consecuencia de fallo** GEN failure-consequence diagram; **~ de difracción de campo próximo** TEC RAD, TRANS near-field diffraction pattern; **~ de difracción de Fraunhofer** TRANS Fraunhofer diffraction pattern; **~ de difracción de Fresnel** TRANS Fresnel diffraction pattern; **~ direccional de antena** TEC RAD antenna radiation pattern; **~ de directividad de antena** TEC RAD antenna directivity diagram; **~ de dispersión** INFO scatter diagram; **~ de distribución de cables** TRANS cable distribution diagram; **~ de estructuración** INFO setup diagram; **~ de flujo** GEN flow chart, flow diagram; **~ de flujo de datos** INFO data flow chart, data flow diagram; **~ de ganancia de antena** TEC RAD antenna gain pattern; **~ de inspección** PRUEBA *calidad* inspection diagram; **~ de instalación** GEN installation diagram; **~ lógico** GEN logic diagram; **~ de marcha** INFO running diagram; **~ de número terminal** GEN terminal number diagram; **~ de Nyquist** CONTROL Nyquist diagram; **~ en ocho** TEC RAD figure-of-eight diagram; **~ de ojo** ELECTRÓN, TRANS eye diagram, eye pattern, eyeshape pattern; **~ de ojos** TRANS eye pattern; **~ de operaciones** DATOS operations chart; **~ pictórico** GEN pictorial diagram; **~ de predicción de frecuencia** TEC RAD frequency-prediction chart; **~ de puenteado** ELECTROTEC strapping diagram; **~ de puntos** GEN *estadística* dot diagram; **~ de radiación** TEC RAD, TRANS radiation pattern; **~ de radiación de campo próximo** TEC RAD, TRANS near-field radiation pattern; **~ de radiación de equilibrio** TRANS equilibrium radiation pattern; **~ de red** TELEFON network diagram; **~ de reflectancia** TV reflectance diagram; **~ reticulado** GEN gridiron pattern; **~ de señalización** SEÑAL signalling (*AmE* signaling) diagram; **~ sintáctico** TELEFON syntax diagram; **~ de sistema** INFO system diagram; **~ de tablero** INFO plugging chart; **~ de tendido de hilos** COMPON wire-running diagram; **~ terminal** GEN terminal diagram; **~ de tiempos** GEN timing chart; **~ de transición de estados** PRUEBA state-transition diagram; **~ en zigzag** GEN zigzag pattern

diagramar *vt* INFO lay out

diálogo *m* DATOS, TV dialog; **~ de apertura** ELECTRÓN log-in dialog; **~ exterior de salida** TELEFON output outside dialog; **~ hombre-máquina** INFO man-machine dialog

diámetro: **~ de campo de modos** *m* TRANS mode field diameter; **~ de capa** *m* COMPON *cables* layer diameter; **~ efectivo teórico** *m* COMPON theoretical pitch diameter; **~ exterior** *m* GEN outer diameter; **~ del haz** *m* TRANS *óptica de fibras* beam diameter; **~ medio de alma** *m* TRANS average core diameter; **~ medio de chapeado** *m* TRANS average cladding diameter; **~ medio de superficie de referencia** *m* (*DRav*) TELEFON average reference-surface diameter (*DRav*); **~ normal** *m* TRANS stock diameter; **~ de núcleo** *m* TELEFON, TRANS core diameter; **~ primitivo** *m* GEN pitch diameter; **~ de revestimiento** *m* TRANS *óptica de fibras* cladding diameter; **~ de superficie de referencia** *m* TELEFON, TRANS reference-surface diameter

diapasón *m* INFO tuning fork

diapositiva: **~ en color** *f* GEN colour (*AmE* color) slide

diario *m* GEN *documentación* diary, journal; **~ de recuperaciones** DATOS recovery log

dibit: **~ doble descifrado** *m* (*S1*) DATOS unscrambled double dibit (*S1*)

dibujo: **~ del acabado de una superficie** *m* COMPON surface-finish drawing; **~ de cableado** *m* COMPON wiring drawing; **~ de capas** *m* COMPON, ELECTRÓN *circuitos impresos* layer drawing; **~ de conexión** *m* GEN terminal drawing; **~ de conjunto de central** *m* GEN exchange-assembly drawing; **~ detallado** *m* GEN *documentación* detail drawing; **~ a escala** *m* GEN scale drawing, basic-size drawing, dimensional drawing; **~ funcional** *m* GEN designation drawing; **~ de implantación** *m* GEN *documentación* layout drawing; **~ de instalación** *m* GEN installation drawing; **~ de instalación de central** *m* GEN *documentación*

exchange-installation drawing; ~ **lineal** *m* DATOS *digitalizador* line drawing; ~ **de numeración** *m* GEN numbering drawing; ~ **de número terminal** *m* GEN terminal-number drawing; ~ **de perforación** *m* ELECTRÓN *circuitos impresos* hole drawing
dieléctrico *adj* ELECTRÓN dielectric
diferencia *f* GEN difference; ~ **de capacidad** TELEFON difference of capacity; ~ **de fase** ELECTRÓN, ELECTROTEC phase difference; ~ **de frecuencia** TEC RAD frequency difference; ~ **de frecuencia normalizada** TEC RAD normalized frequency difference; ~ **en la hora del reloj** GEN clock-time difference; ~ **de índice de refracción ESI** TELEFON ESI refractive-index difference; ~ **de potencial** ELECTROTEC potential difference; ~ **de temperatura** CLIMAT temperature difference; ~ **de trama temporal** INFO, TEC RAD, TRANS timescale difference
diferenciación *f* TELEFON *de impulsos* differentiation
diferencial *adj* GEN differential
diferenciar *vt* GEN differentiate
diferido *adj* GEN deferred
difracción *f* TRANS diffraction; ~ **múltiple** TRANS multiple diffraction; ~ **de onda** ELECTRÓN wave diffraction
difundido: **muy** ~ *adj* GEN widespread
difundir *vt* GEN diffuse
difusión *f* GEN diffusion; ~ **de programas de entretenimiento** TV entertainment broadcasting
digitación *f* INFO fingering
digital: ~ **a analógico** *adj* (*D-A*) GEN digital-to-analog (*D-A*); ~ **no interpolado** *adj* TRANS digital non-interpolated
digitalización *f* GEN digitalization, digitization; ~ **dirigida** INFO directed scan; ~ **de frecuencia** TEC RAD, TRANS frequency scan; ~ **por haz lasérico** INFO laser-beam scanning; ~ **de señal** SEÑAL, TRANS signal digitalization; ~ **de tabla** GEN table scanning; ~ **de voz** INFO, TELEFON voice digitization
digitalizado *adj* GEN digitized, digitalized
digitalizador: ~ **del haz dirigido** *m* INFO directed-beam scanner; ~ **de imagen** *m* INFO image scanner; ~ **de imágenes** *m* INFO, TV video digitizer; ~ **manual** *m* PERIF hand-held scanner; ~ **de RF** *m* TEC RAD RF scanner
digitalizar *vt* GEN digitalize, digitize
dígito *m* DATOS, TELEFON, TRANS digit; ~ **binario** TRANS binary digit; ~ **de centena** INFO hundreds digit; ~ **de decenas** INFO tens digit; ~ **hexadecimal** DATOS, INFO hexadecimal digit; ~ **justificable** TELEFON, TRANS justifiable digit; ~ **de justificación** TELEFON, TRANS justifying digit; ~ **de paridad** INFO parity digit; ~ **de relleno** DATOS stuffing digit; ~ **de servicio** TELEFON, TRANS service digit; ~ **de servicio de justificación** TELEFON, TRANS justification service digit; ~ **de signo** INFO sign digit
dígitos: ~ **de información de ruta** *m pl* TELEFON route information digits; ~ **de recorrido** *m pl* CONMUT routing digits
dilatador: ~ **de impulsos** *m* TELEFON, TRANS pulse stretcher
diluyente *m* COMPON thinner
dimensión: ~ **fractal** *f* INFO fractal dimension; ~ **observada** *f* PRUEBA observed dimension; ~ **de separación** *f* TELEFON spacing dimension
dimensionamiento: ~ **adecuado** *m* ELECTRÓN right-sizing
dimensiones: ~ **básicas** *f pl* GEN basic dimensions
dinámico *adj* GEN dynamic
dinamitero *m* INFO blaster
diodo *m* COMPON, ELECTRÓN, TELEFON, TRANS diode; ~ **de aproximación secundario** COMPON, ELECTRÓN second approximation diode; ~ **de avalancha** ELECTRÓN avalanche diode; ~ **bidireccional** ELECTRÓN *semiconductores* diac; ~ **de Burrus** COMPON, TELEFON, TRANS Burrus diode; ~ **de capacidad variable** COMPON, ELECTRÓN *semiconductores* capacitance diode; ~ **de comunicación** CONMUT, ELECTRÓN switchover diode; ~ **conmutador** COMPON, CONMUT, ELECTRÓN switching diode; ~ **de contacto de punta** COMPON point-contact diode; ~ **de cristal** ELECTROTEC crystal diode; ~ **de doble base** COMPON *semiconductores* double-base diode; ~ **electroluminiscente** GEN light-emitting diode (*LED*); ~ **electroluminiscente con emisión de borde** TRANS *óptica de fibras* edge-emitting LED; ~ **de emisión de borde** ELECTRÓN, TRANS edge-emitting diode; ~ **emisor de luz** GEN light-emitting diode (*LED*); ~ **emisor de luz superluminiscente** COMPON, ELECTRÓN, TRANS *óptica de fibras* superluminescent light-emitting diode; ~ **de fijación** ELECTRÓN clamping diode; ~ **de fijación por corriente continua** TEC RAD DC clamp diode; ~ **fijador de nivel** ELECTRÓN clamp diode; ~ **fotoemisor de emisión marginal** ELECTRÓN, TRANS edge-emitting light-emitting diode; ~ **generador de ruido** ELECTRÓN noise-generator diode; ~ **láser** (*LD*) ELECTRÓN, TRANS laser diode (*LD*); ~ **láser DFB** TRANS *óptica de fibras* DFB laser diode; ~ **láser FP** TRANS *óptica de fibras* FP laser diode; ~ **láser de inyección**

TRANS injection laser diode (*ILD*); ~ **planar** COMPON planar diode; ~ **de plasma** ELECTRÓN plasma diode; ~ **de punta de oro** COMPON, ELECTRÓN *semiconductores* gold-bonded diode; ~ **rectificador de avalancha controlada** ELECTRÓN *semiconductores* controlled-avalanche rectifier diode; ~ **rectificador de semiconductor** COMPON, ELECTRÓN semiconductor rectifier diode; ~ **rectificador de toma de tierra** ELECTROTEC earthing probe *BrE*, grounding probe *AmE*; ~ **de referencia de voltaje** COMPON, ELECTRÓN voltage reference diode; ~ **regulador de voltaje** COMPON voltage regulator diode; ~ **restaurador de CC** TEC RAD DC restorer diode; ~ **de ruido ideal** ELECTRÓN ideal-noise diode; ~ **de Schottky** COMPON Schottky diode; ~ **semiconductor** COMPON, ELECTRÓN semiconductor diode; ~ **de señal** ELECTRÓN signal diode; ~ **de silicio** COMPON silicon diode; ~ **superradiante** COMPON, ELECTRÓN, TRANS superradiant diode (*SRD*); ~ **túnel** COMPON, ELECTRÓN tunnel diode; ~ **de una sola unión** COMPON *semiconductores* unijunction diode; ~ **de unión** COMPON junction diode; ~ **unitunel** COMPON, ELECTRÓN unitunnel diode; ~ **Zener** COMPON Zener diode

díplex *m* COMPON, TEC RAD, TELEFON, TRANS diplex

diplexor *m* COMPON, TEC RAD, TELEFON, TRANS *antena* diplexer

dipolo *m* TEC RAD dipole; ~ **ranurado** TEC RAD slot-fed dipole; ~ **de semionda** TEC RAD half-wave dipole

dirección *f* GEN direction, address, TEC RAD steering, TELEFON address; ~ **abreviada** TELEFON shortened address; ~ **absoluta** INFO absolute address; ~ **de blanco** TEC RAD target direction; ~ **de bloque lógico** (*LBA*) INFO logical block address (*LBA*); ~ **de bloqueo** ELECTRÓN *semiconductores* blocking direction; ~ **del cable** TELEFON cable address; ~ **calculada** INFO generated address; ~ **de CD-ROM** INFO CD-ROM address; ~ **compartida** DATOS party address; ~ **de conversación** TELEFON speech direction; ~ **de correo electrónica** DATOS e-mail address; ~ **de destino** TELEFON destination address; ~ **de encaminamiento** CONMUT, DATOS routing address; ~ **de encaminamiento invertido** CONMUT, DATOS reverse-routing address; ~ **de facturación** TELEFON billing address; ~ **de flujo** GEN flow direction; ~ **genérica** DATOS generic address; ~ **indexada** INFO indexed address; ~ **indirecta** INFO, TELEFON indirect address; ~ **de instalación** TELEFON installation address; ~ **de instrucción** INFO instruction address; ~ **N** DATOS N address; ~ **normal de flujo** GEN normal flow direction; ~ **N respondiendo** DATOS responding N address; ~ **de número llamado** DATOS called N address; ~ **de presentación** DATOS presentation address; ~ **de punto de acceso al servicio de transporte** DATOS transport-service access-point address; ~ **de punto de acceso del servicio N** DATOS, TELEFON N-service access-point address; ~ **de punto de acceso al servicio de red** DATOS network-service access-point address; ~ **de punto de acceso al servicio de transporte** TELEFON, TRANS transport-service access-point address; ~ **de punto de incorporación a subred** DATOS subnetwork point-of-attachment address; ~ **de recepción** PRUEBA, TRANS receive direction; ~ **de red** DATOS network address; ~ **redireccionada** TELEFON redirection address; ~ **relativa** INFO relative address; ~ **reubicable** INFO, TELEFON relocatable address; ~ **rotativa** TELEFON rotary direction; ~ **simbólica** INFO symbolic address; ~ **de sistema** TELEFON system address; ~ **de subred** DATOS subnetwork address; ~ **telegráfica** TELEFON telegraphic address; ~ **de terminal de datos** DATOS data-terminal-equipment address; ~ **URL** DATOS *Internet* bookmark; ~ **de usuario** TELEFON user address; ~ **virtual** INFO virtual address; ~ **de voz** TELEFON direction of speech

direccionabilidad: ~ **del haz** *f* TEC RAD beam steerability

direccional: ~ **efectivo** *m* INFO effective address

direccionamiento *m* DATOS, INFO addressing; ~ **de base** INFO base address; ~ **calculado** INFO hash coding; ~ **implícito** INFO implied addressing; ~ **indirecto** INFO indirect addressing; ~ **inmediato** INFO immediate addressing; ~ **progresivo automático** INFO one-ahead addressing; ~ **de red** TRANS network addressing; ~ **de registro** CONMUT register addressing; ~ **simbólico** INFO symbolic addressing; ~ **virtual** INFO virtual addressing

directividad *f* TEC RAD directivity

directo: ~ **al disco** *adv* INFO direct to disk

director *m* GEN manager; ~ **comercial de mercadotecnia** GEN marketing manager; ~ **de conferencia colectiva** APLIC conference call chairperson; ~ **de conferencia múltiple** APLIC conference call chairperson; ~ **técnico** INFO technical manager

directorio *m* GEN directory; ~ **en cadena** INFO chain directory; ~ **de operación** INFO opera-

tion directory; ~ **de papel** TELEFON paper directory; ~ **raíz** INFO root directory
directriz *f* GEN guideline
discado: ~ **automático** *m* TELEFON auto-dialling (*AmE* auto-dialing); ~ **de corto código** *m* TELEFON short-code dialling (*AmE* dialing); ~ **de marcación con pulsador** *m* PERIF, TELEFON push-button dial; ~ **de precaptación de línea** *m* TELEFON preseizure dialling (*AmE* dialing); ~ **por señales de frecuencia vocal** *m* PERIF, TELEFON voice-frequency dialling (*AmE* dialing)
disco: ~ **de cambio de fase** *m* DATOS phase-change disk; ~ **de cifras** *m* PERIF digit plate; ~ **compacto fotográfico** *m* (*PCD*) DATOS, INFO photo CD, photo compact disc (*PCD*); ~ **compacto de memoria sólo de lectura** *m* (*CD-ROM*) GEN compact disk read-only memory (*CD-ROM*); ~ **compacto multicapa** *m* DATOS multilayer compact disc; ~ **compacto óptico multicapa** *m* DATOS multilayer optical compact disc; ~ **compacto reescribible** *m* DATOS, INFO rewritable compact disc; ~ **compartido** *m* INFO shared disk; ~ **de conexión** *m* DATOS *Internet*, INFO connection disk; ~ **dactilar** *m* PERIF, TELEFON dial disk *AmE*, finger wheel *BrE*, dialler (*AmE* dialer); ~ **duro** *m* INFO hard disk; ~ **flexible** *m* INFO floppy disk; ~ **flexible reversible** *m* INFO reversible flexible disk; ~ **marcador** *m* PERIF, TELEFON dial disk *AmE*, finger wheel *BrE*; ~ **MO** *m* DATOS, ELECTRÓN MO disk; ~ **óptico reescribible** *m* DATOS, INFO rewritable optical disk
disconformidad *f* PRUEBA nonconformity
discordancia *f* GEN mismatch
discordancias: ~ **de bitios** *f pl* DATOS bit disagreements
discrecional *adj* INFO discretionary
discriminación *f* PRUEBA discrimination; ~ **de antena** TEC RAD antenna discrimination; ~ **de mensaje** INFO message discrimination; ~ **de mensaje de señalización** SEÑAL signalling-message (*AmE* signaling-message) discrimination; ~ **de polarización cruzada** TEC RAD, TRANS cross-polar discrimination
discriminador *m* CONMUT, DATOS, ELECTRÓN discriminator; ~ **de duración de impulsos** TEC RAD pulse-length discriminator; ~ **de progresión de eventos** DATOS event-forwarding discriminator; ~ **de protocolo** DATOS protocol discriminator; ~ **telegráfico** PRUEBA two-tone detector
diseñador: ~ **industrial** *m* GEN industrial designer
diseño *m* GEN design; ~ **asistido por ordenador** (*CAD*) INFO computer-aided design (*CAD*); ~ **de circuito** GEN circuit design; ~ **de experimentos** GEN *estadística* experiment design; ~ **y fabricación asistidos por ordenador** (*CAD/CAM*) INFO computer-aided design and computer-aided manufacturing (*CAD/CAM*); ~ **de fichero** DATOS file layout; ~ **industrial** APLIC industrial design; ~ **lógico** GEN logic design; ~ **mecánico** GEN mechanical design; ~ **de unidades** CONMUT, PERIF unit design
disgregación *f* GEN detachment
disipación: ~ **de calor** *f* ELECTRÓN *física* power dissipation; ~ **de electrodo** *f* ELECTRÓN electrode dissipation; ~ **de potencia** *f* ELECTRÓN *física* heat dissipation
disipador: ~ **de calor** *m* ELECTRÓN heat dissipator, heat sink; ~ **térmico** *m* ELECTRÓN heat sink, heat dissipator
disipar *vt* ELECTROTEC dissipate
disipativo *adj* INFO, TELEFON lossy
disminución[1]: ~ **de capacidad** *f* DATOS capacity decrease
disminución[2]: **con ~ de potencia** *fra* ELECTRÓN powered-down
disociación: ~ **electrolítica** *f* GEN *química* electrolytic dissociation; ~ **electroquímica** *f* GEN electrochemical dissociation
disparador *m* ELECTRÓN ripcord; ~ **de Schmitt** ELECTRÓN Schmitt trigger
disparar *vt* ELECTRÓN, INFO, PRUEBA trigger; ◆ ~ **la llamada** TELEFON trip the ringing
disparidad *f* TELEFON disparity
disparo *m* ELECTRÓN, INFO, PRUEBA trigger, *action* triggering
dispersiómetro *m* PRUEBA scatterometer
dispersión *f* ELECTRÓN, TEC RAD dispersion, TRANS backscattering, dispersion; ~ **anómala** TRANS *óptica de fibras* anomalous dispersion; ~ **de energía** TEC RAD energy dispersal; ~ **por fibra** TRANS fibre (*AmE* fiber) scattering; ~ **de frecuencias** TEC RAD, TRANS frequency offset; ~ **de guíaondas** TEC RAD, TRANS waveguide dispersion; ~ **de impulsos** ELECTRÓN, TELEFON, TRANS pulse dispersion, pulse spreading; ~ **ionosférica** TEC RAD ionospheric scatter; ~ **de material** TEC RAD material scattering, TRANS material dispersion; ~ **de onda** ELECTRÓN wave dispersion; ~ **de perfil** TRANS profile dispersion; ~ **de precipitación** TEC RAD precipitation scatter; ~ **de Raman** TRANS *óptica de fibras* Raman scattering; ~ **de Raman estimulada** (*SRS*) TEC RAD stimulated Raman scattering (*SRS*); ~ **de Rayleigh** DATOS, TELEFON, TRANS *física* Rayleigh

scattering; ~ **de tiempo por trayectos múltiples** TEC RAD multipath time dispersion; ~ **troposférica** TEC RAD tropospheric scatter; ~ **de velocidad de grupo** (*GVD*) TRANS group-velocity dispersion (*GVD*)

display: ~ **matricial** *m* ELECTRÓN matrix display

disponer *vt* INFO arrange; ◆ ~ **en capas** TRANS layer

disponibilidad *f* GEN availability; ~ **asintótica** PRUEBA *seguridad de funcionamiento*, TELEFON asymptotic availability; ~ **de circuito dedicado** TELEFON availability of leased circuit; ~ **de circuitos** TEC RAD, TELEFON circuit availability; ~**/indisponibilidad** TELEFON availability/unavailability; ~**/indisponibilidad asintótica** TELEFON asymptotic availability/unavailability; ~**/indisponibilidad media** TELEFON mean availability/unavailability; ~ **de equipo** TELEFON equipment availability; ~ **media** TELEFON mean availability; ~ **de programas** TV programme (*AmE* program) availability

disponible *adj* GEN available; ~ **en espera** DATOS ready stand by; ~ **en línea** DATOS ready on-line; ~ **en stock** INFO in-stock

disposición *f* GEN arrangement; ~ **de cables** COMPON disposition of wires; ~ **en capas** TRANS laying; ~ **de digitalización en zigzag** INFO zigzag scanning order; ~ **de sincronización de ondulación** ELECTRÓN ripple-clocking arrangement; ~ **en zigzag** GEN staggering

dispositivo *m* GEN device, feature; ~ **de acceso a la red** DATOS network gateway; ~ **acoplado por carga** (*DAC*) ELECTRÓN charge-coupled device (*CCD*); ~ **de adaptación** ELECTRÓN matching device; ~ **de alimentación** INFO feeding device; ~ **para aplicación de recepción** DATOS receiving-application entity; ~ **apuntador** INFO pointing device; ~ **de bifurcación en T** COMPON *óptica de las fibras fototransmisoras* T-branching device; ~ **de bifurcación en Y** TRANS Y-branching device; ~ **de búsqueda** DATOS, INFO search engine; ~ **de conducción de tráfico** DATOS, TELEFON traffic-carrying device; ~ **de conexión** ELECTRÓN connecting device; ~ **de conmutación** COMPON, CONMUT, ELECTROTEC, TELEFON switching device; ~ **de conmutación óptico mecánico** CONMUT mechanical optical switch; ~ **de contacto** CONMUT, ELECTRÓN contact unit; ~ **de control** CONTROL control device; ~ **de control de la conexión de señalización** (*SCCP*) SEÑAL, TELEFON signalling (*AmE* signaling) connection control part (*SCCP*); ~ **de control de errores** DATOS error-control device (*ECD*); ~ **de control de señalización** SEÑAL signalling (*AmE* signaling) control part; ~ **corrector de distorsiones geométricas** INFO timing-correction device; ~ **de corriente de llamada** TELEFON ring-current set, ringing-current set; ~ **de corrientes portadoras** TRANS carrier-current device; ~ **de disparo** ELECTRÓN, ELECTROTEC trip; ~ **eliminador de interferencias** PRUEBA noise-suppressor device; ~ **de encaminamiento** DATOS routing device; ~ **de entrada/salida** INFO, TELEFON I/O device; ~ **de entrada vocal** INFO voice-input device; ~ **de fijación** COMPON jig; ~ **de fijación para cableado** COMPON wiring fixture, wiring jig; ~ **de fijación para el transporte** GEN transport locking device; ~ **de iluminación** TEC RAD illuminating device; ~ **de inferencia** APLIC inference engine; ~ **de inyección de carga** ELECTROTEC charge-injection device; ~ **lógico** ELECTRÓN logic device; ~ **de maniobra** CONTROL operating device; ~ **de mediación** INFO mediation device (*MD*); ~ **de multiplicación de salidas** ELECTRÓN fan-out device; ~ **nominal telefónico** PERIF telephony-rated device; ~ **OAS** COMPON, ELECTRÓN, TRANS SAW device; ~ **periférico** DATOS, INFO, PERIF peripheral device; ~ **de protección contra golpes** TELEFON anti-shock device; ~ **de protección de red** (*NPD*) TELEFON network-protection device (*NPD*); ~ **de prueba** PRUEBA testing device; ~ **de puesta en cola** TELEFON queueing device; ~ **radiante** TEC RAD radiating device; ~ **rectificador** ELECTROTEC rectifying device; ~ **de recurso compartido** DATOS, INFO shared-resource device; ~ **de registro de imágenes** INFO frame grabber; ~ **de relleno** DATOS stuffing device; ~ **de salida** INFO output device; ~ **semiconductor** COMPON, ELECTRÓN semiconductor device; ~ **de señales de llamada y de tonos** SEÑAL, TELEFON tone-and-ringing set; ~ **silenciador** TEC RAD muting device; ~ **de soporte lógico** DATOS, INFO software tool; ~ **telemedidor** DATOS telemeasuring device; ~ **terminal de fibra óptica** TRANS fibre-optic (*AmE* fiber-optic) terminal device; ~ **terminal de recepción de fibra óptica** TRANS receive fibre-optic (*AmE* fiber-optic) terminal device; ~ **de transferencia de carga** ELECTRÓN charge-transfer device; ~ **virtual** INFO virtual device; ~ **de visualización** DATOS, INFO display device

disquete *m* INFO diskette, disk

disquetera *f* INFO disk drive

disrupción: **~ primaria** *f* ELECTRÓN *semiconductores* first breakdown

distancia[1]: **a ~** *adv* GEN remotely

distancia[2] *f* GEN distance; **~ aislante** ELECTROTEC insulation gap; **~ de arrastre** ELECTRÓN drive distance; **~ a los bordes** ELECTRÓN edge distance; **~ de contorneo** ELECTROTEC creeping distance; **~ entre estaciones de base** TEC RAD *comunicación móvil* base-station separation; **~ entre las hileras** TELEFON row spacing; **~ entre pistas** INFO track pitch; **~ explosiva** ELECTROTEC spark gap; **~ de Hamming** INFO Hamming distance; **~ del horizonte radioeléctrico** TEC RAD radio-horizon distance; **~ media de reutilización** TEC RAD mean re-use distance; **~ oblicua** TEC RAD slant range; **~ de reutilización en el mismo canal** TEC RAD *móvil terrestre* co-channel re-use distance; **~ de seguridad** ELECTROTEC flashover clearance; **~ de señal** SEÑAL signal distance; **~ de visión** GEN viewing distance

distanciamiento: **~ de antena radiogoniométrica** *m* TEC RAD range-finding antenna spacing

distinguir *vt* GEN distinguish

distorsión *f* DATOS, INFO, TEC RAD, TELEFON, TV distortion; **~ de amplitud** TRANS amplitude distortion; **~ de amplitud de fase** ELECTRÓN, TRANS phase-amplitude distortion; **~ armónica de enésimo orden** TELEFON nth-order harmonic distortion; **~ de armónico** SEÑAL waveform distortion; **~ asimétrica** ELECTRÓN, TELEFON asymmetrical distortion; **~ de atenuación** TELEFON, TRANS attenuation distortion; **~ cíclica** TELEFON cyclic distortion; **~ en cojín** TV pincushion distortion; **~ de contraste en los bordes** INFO edge flare; **~ de cuantificación** TELEFON, TRANS quantizing distortion; **~ de desplazamiento de fase** TRANS phase-shift distortion; **~ de desviación** ELECTRÓN, TELEFON bias distortion; **~ en la emisión** TELEFON, TRANS transmitter distortion; **~ de fase** ELECTRÓN, TRANS phase distortion; **~ de fase de frecuencia** ELECTRÓN, TRANS phase-frequency distortion; **~ de forma de onda en tiempo de campo** DATOS field-time waveform distortion; **~ geométrica** TV geometric distortion; **~ de imagen** TV tilt; **~ de impulsos** TELEFON pulse distortion; **~ individual** TELEFON individual distortion; **~ intermodal** TRANS intermodal distortion; **~ de intermodulación** TRANS intermodulation distortion; **~ intramodal** TRANS intramodal distortion; **~ de línea** ELECTRÓN S-distortion, TELEFON fortuitous distortion; **~ de luminancia en tiempo de línea** INFO line-time luminance distortion; **~ lumínica en tiempo de campo** INFO field-time luminance distortion; **~ de medias tintas** TV half-tone distortion; **~ modal** TRANS modal distortion; **~ de modos** TRANS mode distortion; **~ multimodal** TRANS multimode distortion; **~ de punto explorador** ELECTRÓN spot distortion; **~ de retardo de grupo** *m* TRANS group-delay distortion; **~ de segundo armónico** ELECTRÓN second-harmonic distortion; **~ de señal** GEN signal distortion; **~ telegráfica** TELEFON telegraph distortion; **~ de tercer armónico** (*3HD*) ELECTRÓN third-harmonic distortion (*3HD*); **~ trapezoidal** ELECTRÓN trapezoidal distortion; **~ de voz** TELEFON speech distortion

distorsionado *m* DATOS, INFO, TEC RAD, TELEFON, TV scrambling

distorsionador *m* DATOS, INFO, TEC RAD, TELEFON, TV scrambler

distorsionar *vt* DATOS, INFO, TEC RAD, TELEFON, TV scramble

distribución *f* GEN distribution; **~ de amplitud de ruido** TRANS noise-amplitude distribution (*NAD*); **~ automática de llamadas** TELEFON automatic call distribution (*ACD*); **~ binomial** GEN *estadística* binomial distribution; **~ cableada** TV cabled distribution; **~ de cables** TRANS cable distribution; **~ de cadencia** TRANS clock-rate distribution; **~ de carteles** INFO posteriation; **~ de clave de encriptación** TEC RAD encryption-key distribution; **~ de datos** DATOS, TELEFON data distribution; **~ espectral** TRANS *óptica de fibras* spectral distribution; **~ en estrella** TRANS star distribution; **~ de fallos** TELEFON failure distribution; **~ de frecuencias** GEN frequency distribution; **~ de fuerza** ELECTROTEC power distribution; **~ gaussiana** PRUEBA Gaussian distribution; **~ del índice de refracción** TRANS *óptica de fibras* refractive-index distribution; **~ indirecta** TEC RAD indirect distribution; **~ logarítmica normal** TEC RAD log-normal shadowing; **~ de mensajes** TELEFON message distribution; **~ del mensaje de señalización** SEÑAL signalling-message (*AmE* signaling-message) distribution; **~ en modalidad de equilibrio** TRANS equilibrium-mode distribution; **~ en modo de desequilibrio** TRANS nonequilibrium-mode distribution; **~ de modos** TRANS mode distribution; **~ mundial por ráfaga de referencia** DATOS global reference burst distribution; **~ normal** GEN normal distribution; **~ de probabilidad** GEN probability distribution; ~ **de probabilidad de amplitud** SEÑAL amplitude-probability distribution; ~ **de programas** TV programme (*AmE* program) distribution; ~ **de propaganda** TV leafleting; ~ **ramificada** TRANS tree distribution; ~ **de señales** SEÑAL signal distribution; ~ **de señales de reloj** TRANS clock distribution; ~ **de tiempo de retención** CONMUT holding-time distribution; ~ **de tráfico** CONMUT, TELEFON traffic distribution

distribuidor *m* GEN supplier; ~ **aleatorio** TELEFON jumping allotter; ~ **central de impulsos** CONMUT central pulse distributor; ~ **de exploración** ELECTRÓN scanner distributor; ~ **de líneas urbanas** TELEFON local-line concentrator; ~ **de recepción** TEC RAD receiver distributor; ~ **de tramas** TRANS frame distributor; ~ **de transmisor múltiple** TRANS multiple-transmitter distributor

distribuir *vt* GEN distribute, deliver

disuadir *vt* INFO deter

disyuntor *m* ELECTROTEC protector

divergencia: ~ **del haz** *f* TEC RAD, TRANS beam divergence; ~ **de la salida** *f* ELECTRÓN output divergence

diversidad *f* GEN diversity; ~ **por dos haces** TEC RAD two-ray diversity; ~ **de encaminamiento** TRANS routing diversity; ~ **de frecuencias** TEC RAD, TRANS frequency diversity; ~ **multihaz** TEC RAD multiray diversity; ~ **de recepción** TEC RAD receiver diversity

diversificación *f* GEN diversification

dividir *vt* GEN divide; ◆ ~ **en sílabas** INFO divide into syllables

divisibilidad *f* GEN divisibility

divisible *adj* GEN divisible

división *f* GEN division; ~ **de banda por código circular** DATOS rolling-code band splitting; ~ **de celdas** TEC RAD cell splitting; ~ **de código** TEC RAD, TELEFON code division; ~ **de escala** PRUEBA scale division; ~ **en el espacio** CONMUT, TELEFON space division; ~ **de frecuencia** TRANS frequency division; ~ **de muestra** TRANS sample division; ~ **en sílabas** INFO *procesamiento de voz* division into syllables; ~ **de trabajos** CONMUT job division

divisor *m* GEN divider; ~ **de archivo** TV file divider; ~ **de frecuencias** TRANS frequency divider; ~ **del haz** INFO, TRANS beam splitter; ~ **de potencia** ELECTROTEC, TEC RAD power divider; ~ **de tensión** ELECTROTEC voltage divider

DLSAP *abr* (*punto de acceso al servicio de enlace de datos*) DATOS DLSAP (*data-link service access point*)

DMA *abr* (*acceso directo a la memoria*) INFO DMA (*direct memory access*)

doblado: ~ **automático** *m* INFO auto-bend

doblador: ~ **de muelles de contacto** *m* COMPON spring adjuster, spring bender, PRUEBA contact-spring adjuster, contact-spring bender, spring adjuster, spring bender; ~ **de tensión** *m* ELECTROTEC voltage doubler; ~ **de voltaje** *m* ELECTRÓN doubler

doblaje *m* GEN dub, dubbing

doblar *vt* GEN dub

doble *adj* INFO dual; ~ **click** *m* DATOS, INFO double click; ~ **control** *m* INFO double check; ~ **densidad** *f* (*DD*) INFO double density (*DD*)

documentación *f* GEN documentation; ~ **de instalación** GEN installation documentation

documento: ~ **de cableado** *m* COMPON wiring document; ~ **de embarque** *m* GEN shipping document; ~ **de especificación de comprobación de conformidad** *m* DATOS conformance-testing specification document; ~ **de estructura** *m* INFO structure document; ~ **fuente** *m* INFO source document; ~ **de hipertexto** *m* INFO hypertext document; ~ **de identidad** *m* GEN *seguridad* identity card; ~ **de origen** *m* DATOS, INFO original document; ~ **SGML** *m* INFO SGML document; ~ **suplementario** *m* GEN supplementary document; ~ **de teletexto** *m* TELEFON teletext document; ~ **vectorial** *m* DATOS *Internet* vector document; ~ **de videotexto** *m* INFO videotext document; ~ **de Web** *m* DATOS, INFO *Internet* Web document

domicilio: ~ **de abonado que llama** *m* TELEFON calling-party address; ~ **de número que llama** *m* DATOS calling-N address

dominio: ~ **denominador** *m* DATOS naming domain; ~ **de dirección de red** *m* DATOS network-addressing domain; ~ **de gestión** *m* DATOS, TELEFON management domain; ~ **mundial de dirección de red** *m* DATOS global network addressing domain; ~ **público** *m* DATOS, INFO *Internet* public domain

domótica *f* APLIC domotics

donador *m* COMPON donor

dosificar *vt* GEN dose

dosis *f* GEN dosage

dotación: ~ **de personal** *f* GEN staffing

DRav *abr* (*diámetro medio de superficie de referencia*) TELEFON DRav (*average reference-surface diameter*)

DRCS *abr* (*caracteres dinámicamente redefinibles*) INFO DRCS (*dynamically-redefinable characters*)

DSA: ~ **cooperativa** *f* DATOS cooperating DSA

DSB *abr* (*banda lateral doble*) TEC RAD, TRANS DSB (*double sideband*)

DTC *abr* (*retardo de tránsito acumulado*) DATOS CTD (*cumulative transit delay*)

DTE *abr* (*equipo terminal de datos*) DATOS, PERIF, TELEFON DTE (*data terminal equipment*)

DTMF *abr* (*multifrecuencia de tono dual*) DATOS DTMF (*digital tone multifrequency*)

DTS *abr* (*reloj fechador descodificador*) INFO DTS (*decoding time stamp*)

DUA *abr* (*apoderado de usuarios de directorio*) APLIC, DATOS DUA (*directory user agent*)

dupleto *m* DATOS, INFO two-bit byte

dúplex[1] *adj* GEN duplex, full-duplex

dúplex[2]: **~ completo** *m* TELEFON full-duplex

duplexador: **~ de radar** *m* TEC RAD radar duplexer

duplexor *m* CONMUT, TEC RAD, TRANS duplexer

duplicación *f* GEN duplication

duplicador: **~ de frecuencia** *m* TRANS frequency doubler

duplicar *vt* GEN duplicate

durabilidad *f* COMPON *resistencia*, DATOS, PRUEBA durability

duración *f* GEN duration; **~ de amortiguamiento de impulso** ELECTRÓN, TELEFON pulse decay time; **~ de ciclo de producción** INFO run duration; **~ de una conversación** TELEFON conversation time; **~ facturable de comunicación por télex** TELEFON chargeable duration of telex call; **~ facturada** APLIC, TELEFON charged duration; **~ de fuga** TELEFON leak duration; **~ de impulso** DATOS, ELECTRÓN, TEC RAD, TELEFON, TRANS pulse duration, pulse length; **~ de llamada** TELEFON length of call; **~ de mantenimiento de órbita** TEC RAD station-keeping lifetime; **~ máxima de aviso** PERIF *contestador* maximum announcement duration; **~ máxima de mensaje** PERIF *contestador* maximum message duration; **~ promedio de las llamadas** TELEFON average call duration; **~ de restablecimiento** COMPON, DATOS, ELECTRÓN recovery time; **~ de retorno** TRANS restoring time; **~ de señal** SEÑAL signal duration; **~ de señal de llamada** TELEFON ringing duration; **~ teórica** TELEFON theoretical duration; **~ de vida** PRUEBA lifetime; **~ de vida operativa** PRUEBA, TELEFON operational lifetime

dureza *f* COMPON toughness; **~ de radiación** COMPON radiation hardness

durmiente *adj* INFO dormant

duro *adj* COMPON tough

DUV *abr* (*datos bajo voz*) TEC RAD DUV (*data under voice*)

E

EB *abr* (*bloque erróneo*) DATOS EB (*erroneous block*)
ebonita *f* COMPON ebonite
ECMA *abr* (*Asociación Europea de Fabricantes de Ordenadores*) INFO ECMA (*European Computer Manufacturers Association*)
eco *m* GEN echo; **~ de amplitud corregida** TELEFON amplitude-corrected echo; **~ de blanco** TEC RAD target echo; **~ captado por hablante** TELEFON talker echo; **~ de circunvalación terrestre** TEC RAD round-the-world echo; **~ con corrección de fase** TELEFON phase-corrected echo; **~ elemental** TELEFON elementary echo; **~ falso** TEC RAD false echo; **~ final de recepción** TELEFON receive-end echo; **~ lateral** TEC RAD side echo; **~ de nubes** TEC RAD cloud return; **~ parásito** TEC RAD angel, unwanted echo; **~ posterior** TEC RAD back echo; **~ de radar** TEC RAD radar echo; **~ en el receptor** TELEFON listener echo; **~ de retorno** TELEFON backward echo; **~ de segunda vuelta** TEC RAD second-time-around echo; **~ de terreno** TEC RAD land return; **~ de tiras antirradar** TEC RAD chaff echo
ECO *abr* (*pedido de modificación técnica*) COMPON ECO (*engineering change order*)
ecómetro: **~ de impulsos** *m* TELEFON pulse echo meter
econo: **~ de precipitación** *m* TEC RAD precipitation return
economía: **~ de escala** *f* GEN economy of scale; **~ de riesgo** *f* PRUEBA risk economy
ecos: **~ por exploración** *m pl* TEC RAD hits per scan; **~ de lluvia** *m pl* TEC RAD rain clutter; **~ de mar** *m pl* TEC RAD sea clutter
ECPT *abr* (*Conferencia Europea de Administración Postal y Telecomunicaciones*) GEN ECPT (*European Conference of Postal and Telecommunications Administrations*)
ecuación: **~ de radar** *f* TEC RAD radar equation; **~ de regresión** *f* GEN regression equation; **~ de Sellmeir** *f* TRANS *óptica de fibras* Sellmeir's equation
ecualizador *m* (*EQL*) CONTROL, ELECTRÓN, ELECTROTEC, TELEFON, TRANS equalizer (*EQL*); **~ de amplitud** INFO, TRANS amplitude equalizer; **~ con realimentador de decisión** ELECTRÓN decision-feedback equalizer; **~ de tensión de celdas** ELECTROTEC cell-voltage equalizer
EDC *abr* (*código de detección de errores*) DATOS, INFO, TELEFON error-detecting code
EDI *abr* (*indicador de ecos por efecto Doppler*) TEC RAD EDI (*Echo Doppler indicator*)
EDI-AU *abr* (*unidad de acceso EDI*) TELEFON EDI-AU (*EDI access unit*)
edición *f* GEN *industria* publishing, *libro, nueva edición* edition, *acto* editing; **~ en cadena** INFO chain edit; **~ de datos** DATOS data editing; **~ de diagramas** GEN diagram issue; **~ de información** INFO data editing; **~ de textos** APLIC, INFO text editing
edificio: **~ inteligente** *m* APLIC intelligent building
editar *vt* GEN *publicar* publish, *corregir* edit
editor *m* DATOS, INFO editor; **~ de enlace** INFO linkage editor; **~ de textos** DATOS, INFO text editor; **~ de textos PCD** INFO PCD writer
EDQ *abr* (*electrodinámica cuántica*) ELECTRÓN QED (*quantum electrodynamics*)
educación: **~ a distancia** *f* TV distance learning, teletuition
educacional *adj* DATOS, GEN, INFO, TELEFON educational
EETDN *abr* (*negociación de retardo de tránsito de extremo a extremo*) DATOS, TELEFON EETDN (*end-to-end transit delay negotiation*)
EF *abr* (*instalación de entrada*) ELECTRÓN EF (*entrance facility*)
efectividad *f* GEN effectiveness
efecto *m* GEN effect; **~ de antena** TEC RAD antenna effect; **~ de apantallamiento** GEN screen effect, screening effect; **~ de arco iris** TV strobe effect; **~ de arrastre de frecuencias** ELECTRÓN frequency-pulling effect; **~ de avalancha** ELECTRÓN *semiconductores* avalanche effect; **~ de barrido** INFO glide effect; **~ blindaje** TEC RAD, TV shielding effect; **~ de bloqueo** GEN blocking effect; **~ de cabellera de cometa** INFO comet tail effect; **~ de campo** DATOS, ELECTRÓN *transistores* field effect; **~ de captación** ELECTRÓN capture effect; **~ corona** ELECTROTEC corona effect; **~ cuántico múltiple** ELECTRÓN multiple quantum well; **~ de curva primitiva** INFO pitchbend effect; **~ de desfase** INFO phase-shifting effect; **~ deslizante** INFO

slide effect; ~ **Doppler** GEN Doppler effect; ~ **de eco** INFO, TELEFON echo effect; ~ **eco** TV multiple image; ~ **electroacústico** ELECTRÓN acousto-electric effect; ~ **de ensombrecimiento de onda** TEC RAD wave-shadowing effect; ~ **de estricción** COMPON pinch effect; ~ **estroboscópico** ELECTRÓN strobing, INFO strobe effect; ~ **Faraday** TRANS Faraday effect; ~ **fotoconductor** ELECTRÓN photoconductive effect; ~ **fotoemisivo** ELECTRÓN, TRANS photoemissive effect; ~ **fotovoltaico** ELECTRÓN photovoltaic effect; ~ **de funcionamiento** INFO performance effect; ~ **Hall** ELECTRÓN Hall effect; ~ **de iluminación concentrada** INFO spotlight effect; ~ **Joule** ELECTRÓN, ELECTROTEC Joule effect; ~ **Kelvin** ELECTROTEC Kelvin effect; ~ **Kerr** TRANS *óptica de fibras* Kerr effect; ~ **local** TELEFON side tone; ~ **magnetostrictivo** ELECTRÓN magnetostrictive effect; ~ **microfónico** ELECTRÓN microphonic effect, microphonism; ~ **Miller** ELECTRÓN Miller effect; ~ **mosaico** INFO mosaic effect; ~ **negativo** INFO negative effect; ~ **de niebla** INFO fog effect; ~ **de noche** TEC RAD night effect; ~ **de obstáculo** TELEFON obstacle effect; ~ **ondulatorio** ELECTROTEC fringe effect; ~ **pelicular** ELECTROTEC skin effect; ~ **persistente** INFO lingering effect; ~ **perturbador** GEN perturbing effect; ~ **Pockels** TRANS *óptica de fibras* Pockels effect; ~ **poligonal** TEC RAD mesh effect; ~ **propagado por conducción** ELECTRÓN conducted effect; ~ **de quantum Hall** ELECTRÓN quantum Hall effect; ~ **radial** INFO spoking; ~ **recíproco** GEN interplay; ~ **de retardo** INFO delay effect; ~ **Rocky Point** ELECTRÓN rocky-point effect; ~ **S** ELECTRÓN S-effect; ~ **Schottky** ELECTRÓN Schottky effect; ~ **Seebeck** ELECTRÓN Seebeck effect; ~ **de sinergía** GEN synergy effect; ~ **sistemático** PRUEBA systematic effect; ~ **sonoro** INFO, TELEFON sound effect; ~ **Thomson** GEN *física* Thomson effect; ~ **de tipo ajuste** INFO phasing-type effect; ~ **túnel** COMPON tunnel action, tunnel effect, ELECTRÓN tunnelling (*AmE* tunneling); ~ **de Wiegand** DATOS Wiegand effect

efector *m* INFO effector

efectos: **contra ~ locales** *adj* TELEFON anti-sidetone

eficacia: ~ **de acoplamiento** *f* TRANS coupling efficiency; ~ **de coste** *f* GEN cost effectiveness; ~ **en función del coste** *f* GEN cost efficiency

eficiencia: ~ **por bitio** *f* DATOS bit efficiency; ~ **de canal** *f* TRANS channel efficiency; ~ **de un código** *f* SEÑAL code efficiency; ~ **cuántica** *f* SEÑAL quantum efficiency; ~ **de emisión termoiónica** *f* ELECTRÓN thermionic-emission efficiency; ~ **de emisor** *f* COMPON *semiconductores* emitter efficiency; ~ **espectral** *f* TRANS spectral efficiency; ~ **de espectro** *f* TEC RAD, TRANS spectrum efficiency; ~ **de la fuente de energía** *f* ELECTROTEC, TRANS source-power efficiency; ~ **radiante** *f* TRANS radiant efficiency; ~ **de respuesta de respondedor** *f* TEC RAD transponder-reply efficiency; ~ **en el tubo** *f* ELECTRÓN, PRUEBA tube efficiency

EFIR *abr* (*filtro de respuesta de impulso finito igualmente ondulado*) ELECTRÓN EFIR (*equiripple finite-impulse response filter*)

EGA *abr* (*adaptador de gráficos destacados*) INFO EGA (*enhanced graphics adaptor*)

EID *abr* (*cambio de código en los datos*) DATOS EID (*escape in data*)

EIRP *abr* (*potencia equivalente radiada isotrópicamente*) TEC RAD EIRP (*equivalent isotropically radiated power*)

eje *m* COMPON axle, INFO axis, TEC RAD line of centre (*AmE* center); ~ **de balanceo** TEC RAD roll axis; ~ **de crominancia** TV chrominance axis; ~ **de elevador** ELECTRÓN riser shaft; ~ **de fibra** TRANS fibre (*AmE* fiber) axis; ~ **de haz** TEC RAD beam axis; ~ **longitudinal** TEC RAD centre (*AmE* center) line; ~ **de luminancia** INFO luminance axis; ~ **óptico** TRANS optic axis; ~ **semimayor** TEC RAD semimajor axis; ~ **de visación** TV line-of-sight

ejecución *f* INFO *de programa* running; ~ **de conformidad** DATOS conforming implementation; ~ **de entrada de discriminador** DATOS discriminator input object; ~ **de módem** DATOS modem turnaround; ~ **de trabajo** INFO job execution

ejecutante *m* DATOS performer

ejecutar *vt* INFO perform

ejecutivo: ~ **de ventas** *m* GEN sales executive

elasticidad *f* GEN resilience, elasticity; ~ **a la interferencia** TRANS resilience to interference; ~ **de sistema** INFO system resilience

elastómetro *m* PRUEBA elastometer

ELCD *abr* (*pantalla electroluminiscente de película delgada*) COMPON, ELECTRÓN TFEL (*thin-film electroluminescent display*)

electreto *m* COMPON, ELECTROTEC electret

electricidad *f* GEN electricity; ~ **de CA** INFO AC power; ~ **de CC** INFO DC power; ~ **para consumo doméstico** ELECTROTEC household electricity; ~ **estática** ELECTROTEC static, static electricity

electricista *m* GEN electrician

electrificación *f* PRUEBA electrification
electroacústica *f* TELEFON electroacoustics
electroacústico *adj* TELEFON electroacoustic
electrodeposición *f* ELECTRÓN plating-up
electrodinámica: ~ **cuántica** *f* (*EDQ*) ELECTRÓN quantum electrodynamics (*QED*)
electrodinámico *adj* ELECTRÓN, TELEFON *electromagnetismo* electrodynamic
electrodo *m* ELECTRÓN, ELECTROTEC electrode; ~ **de cebado** ELECTRÓN, ELECTROTEC starting electrode; ~ **cebador** ELECTRÓN, ELECTROTEC primer electrode; ~ **de disparo** COMPON, ELECTRÓN, ELECTROTEC trigger electrode; ~ **de enfoque** ELECTRÓN, ELECTROTEC focusing electrode; ~ **poligonal** ELECTROTEC, ELECTRÓN mesh electrode; ~ **de señal** ELECTRÓN, ELECTROTEC signal electrode
electroimán *m* COMPON, ELECTROTEC electromagnet; ~ **de concentración** COMPON, ELECTRÓN focusing magnet; ~ **de relé** COMPON, ELECTRÓN relay magnet
electrólito *m* ELECTRÓN electrolyte
electroluminiscencia *f* ELECTRÓN, TRANS electroluminescence
electromagnético *adj* GEN electromagnetic
electromecánica *f* GEN electromechanics
electromecánico *adj* GEN electromechanical
electrón *m* GEN electron; ~ **de conducción** ELECTRÓN *semiconductores* conduction electron; ~ **interno** ELECTRÓN *física* inner-shell electron; ~ **periférico** ELECTRÓN outer electron; ~ **positivo** GEN positive electron
electrónica *f* GEN electronics; ~ **cuántica** ELECTRÓN quantum electronics; ~ **de integración** ELECTRÓN integrated electronics; ~ **molecular** ELECTRÓN molecular electronics; ~ **de potencia** ELECTROTEC power electronics
electrónico *adj* GEN electronic
electronvoltio *m* ELECTROTEC, GEN electron volt
electroóptico *adj* CONMUT, ELECTRÓN, TRANS electro-optic, electro-optical
electroplastia *f* ELECTRÓN *química* electroforming
electroquímica *f* GEN electrochemistry
electrostática *f* GEN electrostatics
electrostático *adj* COMPON, ELECTRÓN, ELECTROTEC, PERIF, PRUEBA, TELEFON electrostatic
electrotecnia *f* GEN electrical engineering
electrotécnico *m* GEN electrical engineer
elegir *vt* INFO select
elemento *m* GEN element; ~ **de acceso a la red** TRANS network gateway element; ~ **de almacenamiento** DATOS storage element; ~ **de almacenamiento de datos** DATOS datastore element; ~ **básico** GEN basic element; ~ **de batería** ELECTROTEC battery cell; ~ **de bifurcación en T** ELECTRÓN *óptica de las fibras fototransmisoras* T-branching device; ~ **de bifurcación en Y** ELECTRÓN Y-branching device; ~ **de cableado** TRANS cabling element; ~ **central de resistencia mecánica** TRANS *cable* central strength member; ~ **de comparación** CONTROL comparing element; ~ **compensador** CONTROL compensating element; ~ **de comunicación en tránsito** DATOS, TRANS transit-connection element (*TCE*); ~ **de conexión de acceso** (*ACE*) DATOS, TELEFON access-connection element (*ACE*); ~ **de conexiones** DATOS connection element (*CE*); ~ **de conmutación** CONMUT, ELECTRÓN switching element; ~ **de construcción** GEN *tecnología de diseño* building element, construction element; ~ **para el control de asociación** (*ACSE*) DATOS, TELEFON association control service element (*ACSE*); ~ **de control final** CONTROL final controlling element; ~ **de datos de carácter codificado** TELEFON coded-character data element; ~ **detectable** INFO detectable element; ~ **detector** PRUEBA sensing element; ~ **de diagnóstico** TELEFON diagnostic element; ~ **de efecto Hall** ELECTRÓN *semiconductores* Hall-effect device; ~ **de equivalencia** ELECTRÓN IF-AND-ONLY-IF element; ~ **excitado** TEC RAD driven element; ~ **de filtro** DATOS filter item; ~ **de firma** (*SIG*) SEÑAL, TELEFON signature element (*SIG*); ~ **de fuerza contraelectromotriz** ELECTROTEC counter cell, counter-electromotive force cell; ~ **de identidad** ELECTRÓN *elementos lógicos* identity element, identity gate; ~ **identificador** TELEFON identity element; ~ **de imagen** ELECTRÓN, INFO, TV picture element; ~ **infográfico** INFO display element; ~ **de lengüeta** COMPON reed element; ~ **lógico** ELECTRÓN logic element, INFO functor; ~ **lógico programable y borrable** (*EPLD*) ELECTRÓN erasable programmable logic device (*EPLD*); ~ **de marcación** TELEFON indicating element; ~ **mayoritario** ELECTRÓN majority element; ~ **de modulación** TELEFON, TRANS modulation element; ~ **O** ELECTRÓN OR element; ~ **objeto** TEC RAD target cell; ~ **O-INCLUSIVO** ELECTRÓN *elementos lógicos* INCLUSIVE-OR element; ~ **óptico holográfico** DATOS holographic optical element; ~ **orbital** TEC RAD orbital element; ~ **parásito** TEC RAD parasitic element; ~ **pictórico** (*PE*) INFO, TELEFON pictorial ele-

ment (*PE*); ~ **pictórico de ~ de control** TELEFON control element PE; ~ **procesador de la línea de abonado** TELEFON subscriber line PE; ~ **radiante** TEC RAD radiating element; ~ **de red** TRANS network element; ~ **de red del servicio de aplicación de gestión** DATOS network management application service element (*NM-ASE*); ~ **de regulación** ELECTROTEC *baterías* end cell; ~ **de respuesta firmado** TELEFON signed response element; ~ **de señal** SEÑAL, TELEFON signal element; ~ **señalizador de restitución de modulación** SEÑAL modulation restitution signal element; ~ **de señal telegráfica** SEÑAL telegraph-signal element; ~ **de servicio** DATOS, TELEFON, TRANS service element; ~ **de servicio de acceso y gestión de transferencia de ficheros** DATOS *de archivos* file-transfer access and management service element (*FTAMSE*); ~ **de servicio de administración de mensajes** (*MASE*) DATOS message-administration service element (*MASE*); ~ **de servicio de aplicación** (*ASE*) DATOS, TELEFON application service element (*ASE*); ~ **de servicio de aplicación de gestión de sistemas** INFO systems-management application service element; ~ **de servicio de aplicación de procesamiento de transacciones** (*TPASE*) DATOS transaction-processing application service element (*TPASE*); ~ **de servicio de entrega de mensajes** (*MDSE*) DATOS message-delivery service element (*MDSE*); ~ **de servicio de manejo remoto** DATOS, TELEFON remote-operation service element; ~ **de servicio de presentación de mensaje** (*MSSE*) DATOS message-submission service element (*MSSE*); ~ **de servicio de recuperación de mensajes** (*MRSE*) DATOS message-retrieval service element (*MRSE*); ~ **de servicio de transferencia** DATOS reliable-transfer service element; ~ **de servicio de transferencia de mensajes** (*MTSE*) DATOS message-transfer service element (*MTSE*); ~ **de sustitución** ELECTROTEC fuse link; ~ **de umbral** ELECTRÓN threshold element; ~ **unitario** GEN unit element; ~ **universal** INFO universal element; ~ **de usuario** (*UE*) DATOS, TELEFON user element (*UE*)

elevación *f* GEN elevation

elevada: **de ~ impedancia** *adj* ELECTRÓN high-impedance

eliminación: ~ **de ecos** *f* ELECTRÓN, TEC RAD, TRANS echo cancellation; ~ **de espiras** *f* GEN deconvolution; ~ **de interferencias** *f* ELECTRÓN interference suppression, TEC RAD, TRANS interference suppression, noise cancellation; ~ **de línea oculta** *f* INFO hidden-line removal; ~ **progresiva** *f* TEC RAD phasing-out; ~ **de registro de localización** *f* TEC RAD location deregistration; ~ **de silencio** *f* PRUEBA, TELEFON, TRANS silence elimination; ~ **de zincado** *f* COMPON zinc stripping

eliminador: ~ **de batería** *m* ELECTROTEC battery eliminator; ~ **de ecos** *m* ELECTRÓN, TEC RAD, TRANS echo canceller (*AmE* canceler); ~ **de interferencias** *m* ELECTROTEC interference eliminator, TEC RAD interference canceller (*AmE* canceler); ~ **de matriz** *m* ELECTRÓN matrix remover; ~ **de módem** *m* DATOS modem eliminator; ~ **de modos de revestimiento** *m* TRANS cladding-mode stripper; ~ **de ruidos** *m* ELECTROTEC noise eliminator; ~ **de sobrevoltaje** *m* ELECTROTEC surge suppressor

elipse *f* GEN ellipse

elíptico *adj* GEN elliptic, elliptical

emanaciones: ~ **comprometedoras** *f pl* TEC RAD *seguridad* compromising emanations

embalse *m* ELECTRÓN reservoir

embarque *m* TEC RAD shipping

émbolo *m* COMPON piston

EMF *abr* (*fuerza electromotriz*) ELECTRÓN, ELECTROTEC EMF (*electromotive force*)

emisible *adj* DATOS, INFO scriptable

emisión *f* GEN broadcast (*BC*); ~ **en banda lateral residual** TEC RAD, TRANS vestigial-sideband emission; ~ **de banda lateral única** TEC RAD, TRANS single-sideband emission; ~ **en BLU** TEC RAD SSB emission; ~ **de campo** ELECTRÓN field emission; ~ **de coseno** TRANS cosine emission; ~ **deseada** ELECTRÓN, SEÑAL, TEC RAD wanted emission; ~ **de documento** GEN document issue; ~ **espontánea** ELECTRÓN, TRANS spontaneous emission; ~ **espuria** TEC RAD spurious emission; ~ **espuria de satélite** TEC RAD spurious satellite emission; ~ **estimulada** ELECTRÓN, TRANS stimulated emission; ~ **falsa** TRANS spurious emission; ~ **de frecuencia normal** TEC RAD standard-frequency emission; ~ **de impulsos de CA** TELEFON AC pulsing, alternating current pulsing; ~ **de indicativo** DATOS, TELEFON answer-back; ~ **lambertiana** TRANS *óptica de fibras* Lambertian emission; ~ **luminosa** PRUEBA light output; ~ **no deseada** TEC RAD unwanted emission; ~ **normal de señal horaria** TEC RAD standard time-signal emission; ~ **parásita** ELECTRÓN spurious emission, stray emission; ~ **parásita conducida** ELECTRÓN conducted spurious emission; ~ **perturbadora de confusión** TEC

RAD confusion jamming; ~ **de portadora reducida** TEC RAD reduced-carrier emission; ~ **por satélite** TEC RAD satellite broadcast; ~ **simultánea** TEC RAD, TV simulcast, simultaneous transmission; ~ **SSB** TRANS SSB emission; ~ **termoiónica** ELECTRÓN thermionic emission; ~ **de TV** TV TV broadcasting; ~ **por vía satélite** TV satellite broadcast

emisiones: ~ **espurias de banda lateral** *f pl* TEC RAD, TELEFON, TRANS sideband splatter

emisividad *f* ELECTRÓN, TEC RAD, TRANS emissivity

emisor *m* GEN emitter; ~ **automático de indicativo** DATOS answer-back unit; ~ **de chispa** TEC RAD spark transmitter; ~ **de código** TELEFON code sender; ~ **de código de tarifa** TELEFON tariff-message sender; ~ **de código de tono** TELEFON tone-code sender; ~ **de impulsos de cómputo** TELEFON meter-pulse sender; ~ **de indicativo** TELEFON answer-back unit; ~ **de intercambio** DATOS interchange sender; ~ **interferente intencional** TEC RAD jamming transmitter; ~ **de llamadas** TELEFON keysender; **~-receptor multifrecuencia** SEÑAL, TELEFON MF sender-receiver, multifrequency sender-receiver; ~ **de señales** SEÑAL, TRANS signal sender; ~ **de socorro** TEC RAD emergency transmitter; ~ **de transistor** ELECTRÓN transistor emitter; ~ **de transmisión** DATOS, TRANS transmission sender

emisora: ~ **de TV** *f* TV TV broadcaster

emitancia *f* ELECTRÓN, TEC RAD, TRANS emittance; ~ **luminosa** (*M*) TRANS luminous emittance (*M*); ~ **radiante** ELECTRÓN, TRANS radiant emittance; ~ **radiante espectral** ELECTRÓN, TRANS spectral radiant emittance

emitir *vt* GEN, TEC RAD emit

empalizada *f* ELECTRÓN railings

empalmador *m* COMPON splicer; ~ **de cables** TRANS cable jointer

empalme *m* GEN joint; ~ **de cables** TRANS cable joining, cable joint, cable splice; ~ **de fibra óptica** TRANS fibre-optic (*AmE* fiber-optic) splice; ~ **por fusión** TRANS fusion splice; ~ **mecánico** TRANS mechanical splice; ~ **recto** COMPON, ELECTRÓN straight joint; ~ **terminal** COMPON termination joint

empaquetado *m* GEN wrapping

empaquetador: **~-desempaquetador** *m* TRANS packetizer-depacketizer

empaquetar *vt* GEN pack

emparejamiento *m* TV twinning

emparrillado: ~ **de puesta a tierra** *m* TEC RAD, TRANS grounding mat *AmE*, earthing mat *BrE*

emplazamiento *m* GEN siting; ~ **exterior** TV outside location; ~ **orbital** TEC RAD orbital location

empotrar *vt* DATOS, ELECTRÓN, INFO embed

empresa: ~ **de asistencia técnica** *f* GEN service company; ~ **de explotación de red** *f* TELEFON network-operating company; ~ **operadora de red** *f* TELEFON network-operating company, network operator; ~ **de programación** *f* DATOS, INFO software company

emrp *abr* (*potencia efectiva radiada en monopolio*) TEC RAD emrp (*effective monopole-radiated power*)

emulación *f* ELECTRÓN, INFO emulation; ~ **de terminal** DATOS *Internet* terminal emulation

emulador *m* ELECTRÓN, INFO emulator; ~ **de protocolos** DATOS protocol emulator; ~ **de terminal** DATOS terminal emulator

emular *vt* ELECTRÓN, INFO emulate

EN *abr* (*Euronorma*) GEN EN (*Euronorm*)

encabezador: ~ **de secuencia** *m* INFO sequence header

encabezamiento *m* DATOS heading; ~ **de carta** GEN letterhead; ~ **de documento** GEN document heading; ~ **de formulario** GEN form heading; ~ **de grupo funcional** DATOS functional-group header

encajar *vt* GEN engage

encaminador *m* DATOS, ELECTRÓN, TELEFON, TRANS router; ~ **en puente** DATOS bridge router

encaminamiento *m* CONMUT, DATOS, TELEFON, TRANS routing; ~ **adaptable** TELEFON adaptive routing; ~ **alternativo** SEÑAL, TELEFON second-attempt routing; ~ **dependiente del estado y del tiempo** TRANS state- and time-dependent routing; ~ **de mensaje** TRANS message route; ~ **de mensaje de señalización** SEÑAL signalling-message (*AmE* signaling-message) routing; ~ **multiprotocolario** INFO, TELEFON multiprotocol routing; ~ **multiprotocolo** DATOS, TRANS multiprotocol routing; ~ **normal** TELEFON normal routing; ~ **normal de señalización** SEÑAL normal routing of signalling (*AmE* signaling); ~ **en la red** DATOS network routing; ~ **de referencia** CONMUT reference routing; ~ **de señalización** SEÑAL signalling (*AmE* signaling) routing; ~ **de tráfico** TELEFON traffic handling, traffic routing

encapsulación *f* GEN encapsulation

encapsulado[1] *adj* GEN encapsulated

encapsulado[2] *m* TRANS packaging, flatpack; ~ **plano de cuadrete** (*QFP*) TRANS quad flatpack (*QFP*); ~ **con una sola línea de**

conexiones (*MBS*) GEN single in-line package (*SIP*)
encapsulamiento *m* GEN encapsulating
encargo: **por ~** *adv* TV on demand
encauzamiento: **~ de tráfico** *m* TELEFON traffic handling
encendido: **~ por cebador** *m* ELECTRÓN primer ignition
enchavetado *adj* ELECTRÓN, INFO keyed
enchufado *adj* GEN plugged-in
enchufe *m* ELECTROTEC socket outlet, GEN plug; **~ de anclaje** TV cable socket; **~ de CA** ELECTROTEC, INFO AC socket; **~ hembra** ELECTROTEC, INFO, PRUEBA female plug; **~ macho** ELECTROTEC, INFO, PRUEBA male plug; **~ mural** ELECTROTEC, ELECTROTEC, INFO, TELEFON wall socket; **~ mural de alimentación de CA** TV AC-mains wall socket
encintadora *f* ELECTROTEC lapper
encintar *vt* COMPON *cable* wrap
encogimiento *m* COMPON shrinkage
encomendar *vt* TELEFON commission
encontrar *vt* DATOS, INFO find
encontrar/sustituir *vt* DATOS, INFO find/replace
encorvar *vt* GEN curve
encriptación *f* DATOS, SEÑAL, TEC RAD, TELEFON, TRANS, TV encryption; **~ de clave pública** TEC RAD public-key encryption
encriptador *m* DATOS, SEÑAL, TEC RAD, TELEFON, TRANS, TV encryptor
encuadramiento: **~ de textura** *m* INFO texture mapping
encuadre *m* INFO mapping; **~ de modelo** INFO pattern mapping
endurecedor *m* COMPON *plásticos* hardening agent
endurecer *vt* CONTROL, PRUEBA *condiciones* tighten
endurecimiento *m* CONTROL, PRUEBA *de condiciones* tightening
energía *f* GEN energy; **~ de bastidor** ELECTROTEC, TELEFON rack power; **~ eléctrica** GEN power; **~ de fuga de punta** ELECTRÓN spike-leakage energy; **~ de ionización** ELECTRÓN ionization energy; **~ radiante** ELECTRÓN, TRANS radiant energy; **~ de reserva** ELECTROTEC stand-by power; **~ retrodispersada** ELECTROTEC, TEC RAD, TELEFON, TRANS back scatter
énfasis *m* INFO emphasis
enfocado *adj* GEN focused
enfoque *m* ELECTRÓN *de tubo de rayos catódicos* focusing
enfriamiento: **~ en paralelo** *m* CLIMAT parallel cooling
enfriar *vt* GEN cool
enganche: **~ de blanco** *m* TEC RAD target lock-on
engranado *m* INFO meshing
engranaje *m* INFO *equipo* gear
enlace *m* GEN liaison, link; **~ de acceso** TELEFON access link; **~ de alimentador** TEC RAD, TRANS feeder link; **~ de alimentador de ~ descendente** TEC RAD down-link feeder link; **~ ascendente** TEC RAD, TELEFON uplink; **~ de blanco** INFO target link; **~ conmutado** CONMUT, DATOS *comunicación de datos* circuit-switched connection; **~ de contribución** TEC RAD contribution link; **~ creado por autor** INFO author-created link; **~ creado por usuario** INFO user-created link; **~ cronológico** INFO chronological link; **~ de datos** SEÑAL data link; **~ de datos de señalización** SEÑAL signalling-data (*AmE* signaling-data) link; **~ descendente** TEC RAD, TELEFON down-link; **~ de distribución** TEC RAD distribution link; **~ de diversidad angular** DATOS angle-diversity link; **~ de doble búsqueda a larga distancia** ELECTRÓN long-distance dual-homing link; **~ de ensamblado de grupo secundario quince** TELEFON fifteen-supergroup assembly link; **~ entre estudio y transmisor** (*STL*) TEC RAD studio-transmitter link (*STL*); **~ entre posiciones** TELEFON interposition trunk; **~ entre procesadores** INFO interprocessor link; **~ entre satélites** (*ISL*) TEC RAD intersatellite link (*ISL*), satellite-to-satellite link, TV satellite-to-satellite link; **~ F/O** ELECTROTEC, TRANS F/O link; **~ fibroóptico** ELECTROTEC fibre-optic (*AmE* fiber-optic) link; **~ de grupo cuaternario** TELEFON supermastergroup link; **~ de grupo especializado internacional** TELEFON international leased group link; **~ de grupos** INFO cluster link; **~ de grupos secundarios** TELEFON supergroup link; **~ en grupo terciario** TELEFON mastergroup link; **~ con hipertexto** INFO hypertext link; **~ internacional** TELEFON international link; **~ internacional de programas sonoros de destino múltiple** TELEFON international multiple-destination sound-programme (*AmE* sound-program) link; **~ internacional de TV de destino múltiple** TV international multiple-destination television link; **~ de larga distancia** TELEFON intertoll trunk; **~ en línea** TELEFON line link; **~ más desfavorable** TRANS worst link; **~ matricial** TELEFON matrix link; **~ multisatélite** TEC RAD, TV multisatellite link; **~ oficina central** CONMUT central office trunk

AmE, central exchange trunk *BrE*; ~ **organizacional** INFO organizational link; ~ **de origen** INFO source link; ~ **por palabra clave** INFO keyword link; ~ **de paquetes de datos** DATOS, TRANS data-packet link; ~ **por PCM** TELEFON PCM link; ~ **principal de distribución** TRANS primary distribution link; ~ **de programa sonoro internacional** TV international sound-programme (*AmE* sound-program) link; ~ **receptor** INFO destination link; ~ **de referencia** INFO referential link; ~ **regenerativo** TRANS regenerative link; ~ **de repetidor en puente de fibra óptica** ELECTRÓN fibre-optic (*AmE* fiber-optic) bridge-repeater link; ~ **por satélite** TEC RAD, TV satellite link; ~ **de satélite de reflexión múltiple** TEC RAD multihop satellite link; ~ **semipermanente** TELEFON semipermanent link; ~ **de señalización** SEÑAL, TELEFON signalling (*AmE* signaling) link; ~ **de señal de reserva** SEÑAL reserve signalling (*AmE* signaling) link; ~ **de sitio cruzado** TRANS cross-site link; ~ **de televisión internacional** TV international television link; ~ **de transmisión** TELEFON, TRANS transmission link; ~ **transversal** TELEFON transverse junction

enlaces: ~ **entre soportes de transmisión** *m pl* TRANS interfacility links

enlazador *m* INFO linker

enlazamiento *m* INFO linkage

enmascaramiento *m* DATOS masquerade, TELEFON, TRANS, TV masking; ~ **antirradar** TEC RAD radar camouflage; ~ **de ruido** TRANS noise masking

enmasillar *vt* COMPON putty

enrejado: ~ **de alambre** *m* COMPON wire netting

enrollamiento: ~ **de llamada** *m* ELECTRÓN ring wrap

enroscamiento: ~ **de borde** *m* INFO edge curl

enrutamiento: ~ **alternativo** *m* GEN alternative routing; ~ **alternativo de señales** *m* SEÑAL alternative routing of signalling (*AmE* signaling); ~ **de tráfico enlace por enlace** *m* TELEFON link-by-link traffic routing

ensamblado: ~**-desensemblado de paquete facsímil** *m* DATOS facsimile packet assembly-disassembly

ensamblador *m* INFO assembler; ~ **cruzado** INFO cross assembler; ~**-desensamblador de paquetes** (*PAD*) DATOS, TELEFON packet assembler-disassembler (*PAD*)

ensamblaje *m* GEN assembly

ensanchamiento: ~ **de espectro de salto de frecuencia** *m* TEC RAD frequency-hopping spread spectrum; ~ **de impulso** *m* ELECTRÓN, TELEFON, TRANS pulse widening, pulse broadening; ~ **de pulsos** *m* ELECTRÓN, TELEFON, TRANS pulse widening, pulse broadening

ensayo: ~ **de adaptación ambiental** *m* CLIMAT environmental test; ~ **alfa** *m* PRUEBA alpha trial; ~ **ambiental** *m* GEN environmental test; ~ **por aplicación de esfuerzos escalonados** *m* PRUEBA step-stress test; ~ **de campo** *m* PRUEBA, TEC RAD field trial; ~ **para comparación** *m* INFO benchmark testing; ~ **de comprobación de canal** *m* PRUEBA channel-check test; ~ **destructivo** *m* PRUEBA destructive testing; ~ **de fatiga** *m* PRUEBA fatigue test; ~ **de homologación** *m* GEN qualification test; ~ **manométrico** *m* GEN pressure test; ~ **de muestreo** *m* PRUEBA sampling test; ~ **no destructivo** *m* TRANS *óptica de fibras* proof test; ~ **preliminar** *m* PRUEBA preliminary test; ~ **subjetivo** *m* PRUEBA subjective test; ~ **de tensión** *m* PRUEBA voltage test; ~ **de tipo** *m* PRUEBA type test

enseñanza: ~ **asistida por computadora** *AmL f* (*cf enseñanza asistida por ordenador Esp*) INFO computer-assisted instruction; ~ **asistida por ordenador** *Esp f* (*cf enseñanza asistida por computador AmL*) INFO computer-assisted instruction

entalladura *f* TEC RAD notch

entalpia *f* GEN *física* enthalpy

entidad *f* GEN entity; ~ **de aplicación** GEN application entity; ~ **de aplicación emisora** DATOS sending-application entity; ~ **de aplicación iniciadora de asociación** DATOS association-initiating application entity; ~ **de aplicación respondedora de asociación** DATOS association-responding application entity; ~ **funcional** DATOS functional entity (*FE*); ~ **para la gestión de capas** (*LME*) DATOS layer-management entity (*LME*); ~ **de mantenimiento** DATOS maintenance entity; ~ **de mantenimiento de instalación de cliente** APLIC customer-installation maintenance entity (*CIME*); ~ **N** DATOS N-entity; ~ **par** DATOS, TEC RAD peer entity; ~ **de red** DATOS network entity; ~ **de transporte** DATOS transport entity

entorno *m* DATOS, INFO environment; ~ **administrativo** GEN office environment; ~ **cableado en estrella** ELECTRÓN star-wired environment; ~ **de configuración de interfaz** (*ICE*) TELEFON interface-configuration environment (*ICE*); ~ **industrial** GEN industrial environment; ~ **de manipulación de tráfico de mensajes** (*MHE*) APLIC, DATOS, TELEFON message-handling environment (*MHE*); ~ **de mensajería por intercambio de datos electrónicos** APLIC, DATOS, TELEFON EDI-messaging environment (*EDIME*); ~ **de mensajería interpersonal** (*IPME*) APLIC, DATOS, TELEFON interpersonal-messaging environment (*IPME*); ~ **de mensajería vocal** APLIC, DATOS, TELEFON voice-messaging environment (*VMGE*); ~ **de mensajería por voz** (*VMGE*) APLIC, DATOS, TELEFON voice-messaging environment (*VMGE*); ~ **OSI** DATOS OSI environment; ~ **RDSI** DATOS ISDN environment

entr. *abr* (*de entretenimiento*) DATOS *Internet* rec. (*recreational*)

entrada *f* INFO input; ~ **de alimentación anódica total** ELECTRÓN total plate-power input; ~ **en blanco** DATOS blank entry; ~ **de cables** TRANS cable inlet, cable entry; ~ **en comunicación** DATOS, TEC RAD handshake; ~ **de contenido reenviado** DATOS returned-content entry; ~ **de directorio** DATOS directory entry; ~ **disparadora** ELECTRÓN trigger input; ~ **expresa para guía de luz** (*LXE*) ELECTRÓN lightguide express entry (*LXE*); ~ **hija** DATOS child entry; ~ **de informe entregado** DATOS delivered-report entry; ~ **de línea** INFO line input; ~ **por lotes a distancia** (*RBE*) DATOS remote-batch entry (*RBE*); ~ **de mensaje entregado** DATOS delivered-message entry; ~ **de micrófono** ELECTRÓN microphone input; ~ **mono** INFO, PERIF mono input; ~ **multibitios** DATOS multibit input; ~ **orientada** ELECTRÓN steered input; ~ **de origen** DATOS parent entry; ~ **de parámetro de formato** INFO format-parameter input; ~ **principal** DATOS main entry; ~ **de selección** TELEFON selection input; ~ **de teletrabajo** DATOS remote-job entry (*RJE*); ~ **de tierra de señalización** SEÑAL, TRANS signal earth input *BrE*, signal ground input *AmE*; ~ **de unidades** ELECTRÓN units input

entradas: ~ **brutas** *f pl* GEN gross revenue; ~ **y salidas** *f pl* DATOS *Internet* ins and outs

entrada/salida *f* (*E/S*) ELECTRÓN, INFO, TELEFON input/output (*I/O*); ~ **virtual** (*VIO*) INFO virtual input/output (*VIO*)

entrante[1] *adj* (*I/C*) GEN incoming (*I/C*)

entrante[2]: ~ **de pared** *m* GEN recess

entrega *f* GEN delivery; ~ **aplazada** TELEFON delayed delivery

entregar *vt* GEN deliver

entrehierro: ~ **de cabeza** *m* INFO head gap

entrelazado[1] *adj* TEC RAD, TELEFON, TRANS, TV interlaced

entrelazado[2] *m* GEN interworking

entrelazamiento *m* TEC RAD, TELEFON, TRANS, TV interlacing; ~ **de impulsos** TELEFON, TRANS pulse interlacing

entrelazar *vt* TEC RAD, TELEFON, TRANS, TV interlace

entretenimiento: **de** ~ *adj* (*entr.*) DATOS *Internet* recreational (*rec.*)

entropía *f* GEN entropy; ~ **media por carácter** DATOS mean entropy per character; ~ **negativa** DATOS negentropy

entusiasta: ~ **de computadora** *AmL m* (*cf entusiasta de ordenador Esp*) INFO computer enthusiast; ~ **de ordenador** *Esp m* (*cf entusiasta de computadora AmL*) INFO computer enthusiast

enumerar *vt* DATOS, INFO list

enunciado: ~ **de conformidad de sistema** *m* (*SCS*) DATOS, PRUEBA system-conformance statement (*SCS*)

envainado *m* ELECTROTEC canning

envasado *m* GEN packaging

envejecimiento *m* GEN ageing

enviar *vt* GEN forward; ◆ ~ **por fax** GEN fax, telecopy; ~ **por telegrama** TELEFON send by wire

envío: ~ **automático** *m* DATOS auto-forward; ~ **de clave de frecuencia vocal** *m* TELEFON voice-frequency key-sending; ~ **EDI** *m* DATOS EDI forwarding; ~ **de llamada incondicional** *m* TELEFON unconditional call-forwarding; ~ **de llamada de instalación** *m* TELEFON *ISDN* installation call-forwarding; ~ **de tiempo suprimido** *m* DATOS deleted-time sending; ~ **de transmisión** *m* TELEFON, TRANS transmission sending; ~ **VM** *m* DATOS VM forwarding

envoltura *f* COMPON housing, CONMUT, DATOS, INFO envelope, TEC RAD shroud; ~ **de adaptación de un ATM** (*AAL*) GEN ATM adaptation layer (*AAL*); ~ **de una sola fibra** COMPON single-fibre (*AmE* single-fiber) jacket; ~ **termoencogible** COMPON, TRANS thermo-shrinkable sheathing; ~ **de tubo** COMPON tube wrap

envolvente: ~ **de señal** *m* SEÑAL signal envelope; ~ **de timbre** *m* INFO timbre envelope; ~ **de volumen** *m* INFO volume envelope

envuelta *f* COMPON, ELECTRÓN jacketing; ~ **con ramificación de salida** ELECTRÓN fan-out jacketing

EOM *abr* (*fin de mensaje*) DATOS, TELEFON EOM (*end of message*)

EOP *abr* (*fin de procedimiento*) DATOS, TELEFON EOP (*end of procedure*)

EOQC *abr* (*Organización Europea de Control de*

Calidad) GEN EOQC (*European Organization for Quality Control*)

EOR-EOM *abr* (*fin de retransmisión EOM*) DATOS, TELEFON EOR-EOM (*end of retransmission EOM*)

EOR-EOP *abr* (*fin de retransmisión EOP*) DATOS, TELEFON EOR-EOP (*end of retransmission EOP*)

EOR-MPS *abr* (*fin de retransmisión MPS*) DATOS, TELEFON EOR-MPS (*end of retransmission MPS*)

EOR-NULL *abr* (*fin de retransmisión NULL*) DATOS, TELEFON EOR-NULL (*end of retransmission NULL*)

EOR-PRI-EOM *abr* (*fin de retransmisión PRI-EOM*) DATOS, TELEFON EOR-PRI-EOM (*end of retransmission PRI-EOM*)

EOR-PRI-EOP *abr* (*fin de retransmisión PRI-EOP*) DATOS, TELEFON EOR-PRI-EOP (*end of retransmission PRI-EOP*)

EOR-PRI-MPS *abr* (*fin de retransmisión PRI-MPS*) DATOS, TELEFON EOR-PRI-MPS (*end of retransmission PRI-MPS*)

EOS *abr* (*fin de selección*) DATOS, TELEFON EOS (*end of selection*)

EOT *abr* (*fin de transmisión*) DATOS EOT (*end of transmission*)

EP: **~ de carga en ejecución** *m* TELEFON charging-in-progress PE; **~ de categoría terminal** *m* TELEFON terminal-category PE; **~ de cuadro de distribución** *m* CONMUT switchboard PE; **~ emisor y receptor de señalización combinada** *m* SEÑAL combined signalling (*AmE* signaling) sender and receiver PE; **~ emisor de señalización** *m* SEÑAL signalling (*AmE* signaling) sender PE; **~ de equipo terminal** *m* PERIF terminal-equipment PE; **~ de módulo de conmutación** *m* CONMUT switching-module PE; **~ receptor de señalización** *m* SEÑAL signalling (*AmE* signaling) receiver PE; **~ de símbolo de incertidumbre** *m* TELEFON uncertainty-symbol PE; **~ de trayecto de conexión** *m* CONMUT switching-path PE; **~ de trayectoria de conmutación reservada** *m* TELEFON reserved switching-path PE

epílogo: **~ de procedimiento** *m* TELEFON procedure epilogue

epitaxia *f* COMPON epitaxy; **~ de fase de vapor órgano-metálico** ELECTRÓN metal-organic vapour-phase (*AmE* vapor-phase) epitaxy

EPLD *abr* (*elemento lógico programable y borrable*) ELECTRÓN EPLD (*erasable programmable logic device*)

epoxi *adj* COMPON, ELECTRÓN *plásticos*, TRANS *cables de fibra* epoxy

epóxido *m* COMPON *plásticos* epoxide

EPROM *abr* (*memoria sólo de lectura eléctricamente programable*) ELECTRÓN EPROM (*electrically-programmable read-only memory*)

EQL *abr* (*ecualizador*) GEN EQL (*equalizer*)

equidistante *adj* GEN equidistant

equilibrado[1] *adj* ELECTROTEC, INFO balanced; **~ a tierra** ELECTROTEC balanced-to-earth

equilibrado[2] *m* DATOS, ELECTROTEC, INFO, TEC RAD, TRANS balancing

equilibrio *m* GEN balance, counterbalance; **~ de impedancia** ELECTRÓN, TELEFON impedance balance; **~ de línea** TRANS line balance; **~ longitudinal** ELECTROTEC longitudinal balance

equipado: **~ con** *adj* GEN fitted with

equipar *vt* GEN equip

equipo *m* GEN equipment; **~ de abonado** COMPON subscriber equipment, PERIF customer equipment (*CEQ*), TELEFON subscriber equipment; **~ auxiliar** INFO auxiliary equipment; **~ de ayuda** DATOS, INFO helpware; **~ bivocal** TELEFON speech-plus-duplex equipment; **~ de canales** TRANS channel equipment; **~ de carga** APLIC charging equipment; **~ de circuito** TELEFON circuit equipment; **~ de cola de espera** TELEFON queueing equipment; **~ de coma flotante** INFO floating-point hardware; **~ complejo de conjunto de instrucciones** (*CISE*) DATOS complex instruction set equipment (*CISE*); **~ de comprobación** PRUEBA, TELEFON testing equipment; **~ de comunicación de datos** PERIF data-communications equipment; **~ de conexión** GEN connection equipment; **~ de conmutación** CONMUT, TELEFON switching equipment; **~ de conmutación y distribución** COMPON, CONMUT, ELECTROTEC, TELEFON switchgear; **~ de conmutación por la voz** CONMUT voice-switching equipment; **~ de control** CONTROL controlling equipment, DATOS monitoring equipment; **~ de control de clima** CLIMAT *ingeniería ambiental* climate-control equipment; **~ de control de fila de bastidores** TELEFON suite-supervision equipment; **~ convertidor de señales** SEÑAL, TELEFON signal-conversion equipment; **~ de cuadro conmutador** CONMUT, TELEFON switchboard equipment; **~ de detección** TEC RAD detection equipment; **~ eléctrico** DATOS electrical equipment; **~ eliminador de interferencia** TEC RAD interference-suppression equipment; **~ emisor** INFO, TRANS transmitting equipment; **~ de empalmar**

INFO splicing facility; ~ **de encriptación** GEN encrypting equipment; ~ **de escucha** TELEFON tapping equipment; ~ **de extensión de canales** TRANS channel-extension equipment; ~ **frontal** TEC RAD front-end equipment; ~ **fuera de servicio** GEN equipment disabled; ~ **de gestión de sistemas** (*SME*) INFO systems-management equipment (*SME*); ~ **de grabación de datos** DATOS data-collection equipment; ~ **de instalación** GEN *ingeniería de planta* installation kit; ~ **de instalaciones no normalizadas** TELEFON nonstandard facilities equipment; ~ **para instalar en automóviles** TEC RAD *mobicom* car-installation kit; ~ **de intercepción** TELEFON interception equipment; ~ **de interconexión** TELEFON interconnection equipment; ~ **de interfono** PERIF interphone equipment; ~ **de limpieza** GEN cleaning kit; ~ **de línea** TRANS line equipment; ~ **de línea de abonado** DATOS, INFO, TEC RAD, TELEFON, TV subscriber's line equipment; ~ **de línea larga** TRANS long-line equipment; ~ **en locales de abonado** (*CPE*) PERIF customer-premises equipment (*CPE*); ~ **de maniobra** CONTROL control equipment; ~ **de mantenimiento** GEN maintenance equipment; ~ **de mantenimiento de instalación de cliente** PRUEBA client-installation maintenance equipment; ~ **de medición** PRUEBA measuring equipment; ~ **de medición de ruido** PRUEBA noise-measuring equipment; ~ **de medida de llamadas** APLIC call-metering equipment; ~ **de megafonía** TEC RAD public-address equipment; ~ **modular de línea** CONMUT line-module equipment; ~ **multiplex PCM** TELEFON, TRANS PCM-multiplex equipment; ~ **de no direccionamiento de red** TELEFON network-nonaddressing equipment; ~ **no duplicado** TELEFON nonduplicated equipment; ~ **periférico** GEN peripheral equipment; ~ **periférico remoto** (*RPE*) INFO remote peripheral equipment (*RPE*); ~ **de posición de operadora** TELEFON position set; ~ **de potencia** ELECTROTEC power equipment; ~ **de presurización de cables** GEN cable-pressurization equipment; ~ **de procesamiento de cheques** APLIC cheque-processing (*AmE* check-processing) equipment; ~ **de procesamiento de imagen** DATOS, TV image-processing equipment; ~ **de protección** ELECTROTEC protection equipment; ~ **de protección de módem** DATOS, PERIF modem-protection equipment; ~ **protegido contra el polvo** ELECTRÓN dustproof equipment; ~ **de prueba** DATOS, INFO testware; ~ **de prueba incorporado** PRUEBA, TELEFON built-in test equipment; ~ **para prueba de línea** PRUEBA line-test set; ~ **de radar** TEC RAD radar head, radar set; ~ **receptor** INFO destination equipment; ~ **de reserva** TELEFON stand-by equipment; ~ **de respuesta** TELEFON answering equipment; ~ **de señales de tráfico** SEÑAL traffic-signalling (*AmE* traffic-signaling) equipment; ~ **de señalización de líneas** TELEFON line signalling (*AmE* signaling) equipment; ~ **de tarificación central** APLIC central-charging equipment; ~ **tarificador de llamadas** APLIC call-charge equipment; ~ **de telecomunicaciones** GEN telecommunications equipment; ~ **telefónico con teclado** PERIF, TELEFON key telephone set; ~ **de terminación de línea** TELEFON line-terminating equipment; ~ **terminal** (*TE*) DATOS, INFO, PERIF, TELEFON terminal equipment (*TE*); ~ **terminal de bastidores** TELEFON suite terminal equipment; ~ **terminal B de la RDSI** TELEFON B-ISDN terminal equipment; ~ **terminal de circuito de datos** DATOS, PERIF data-circuit terminal equipment; ~ **terminal de datos** (*DTE*) DATOS, PERIF, TELEFON data terminal equipment (*DTE*); ~ **terminal de datos ocupado** TELEFON DTE busy; ~ **terminal digital de abonado** TELEFON subscriber digital termination; ~ **terminal móvil de procesamiento de datos** TEC RAD mobile data-processing terminal equipment; ~ **terminal para el proceso de datos** DATOS, PERIF data-processing terminal equipment; ~ **terminal de red** TELEFON, TRANS network terminal equipment (*NTE*); ~ **terminal de señalización** (*punto de entrada*) TELEFON signalling (*AmE* signaling) terminal equipment (*STE*); ~ **de translación de grupo primario** TELEFON primary group translating equipment; ~ **de transmisión** TRANS transmission equipment; ~ **de transposición de grupo secundario** TELEFON supergroup translating equipment; ~ **tropicalizado** CLIMAT, COMPON tropicalized equipment; ~ **univocal** TELEFON speech-plus-simplex equipment; ~ **de vendedores** GEN sales force; ~ **de verificación** PRUEBA check-out equipment; ~ **de vídeo** TV video equipment; ~ **de videocomunicación** TV videocommunication equipment

equipos: ~ **compartidos** *m pl* DATOS, INFO *Internet* shareware

equivalencia *f* GEN equivalence; ~ **lógica** ELECTRÓN *elementos lógicos* IF-AND-ONLY-

IF operation; **no ~** ELECTRÓN *de elementos lógicos* nonequivalence

equivalente: **~ mínimo admisible** *m* TRANS minimum net loss; **~ de referencia** *m* TELEFON reference equivalent; **~ de referencia corregido** *m* (*CRE*) TELEFON corrected-reference equivalent (*CRE*)

ergómetro *m* PRUEBA power meter

ergonomía *f* INFO ergonomics

ergonómico *adj* INFO ergonomic

Erlangio *m* ELECTRÓN, TELEFON Erlang

erosión *f* COMPON erosion; **~ de contactos** ELECTROTEC contact erosion

erp *abr* (*potencia efectiva radiada*) TEC RAD erp (*effective radiated power*)

ERR *abr* (*respuesta para finalizar la transmisión*) DATOS ERR (*response for end of transmission*)

errata *f* GEN misprint

error *m* GEN error; **~ de alineación** TRANS alignment fault; **~ de amplitud** INFO amplitude error; **~ de apreciación** PRUEBA judgment error; **~ de aproximación** GEN approximation error; **~ de balanceo** TEC RAD swing error; **~ de bitio** TRANS bit error; **~ de cableado** COMPON wiring error; **~ de cálculo** GEN calculation error, miscalculation; **~ de calibrado** PRUEBA calibration error; **~ de calibrado de ganancia** ELECTRÓN gain calibration error; **~ de carga** INFO loading error; **~ de comunicación** GEN communication error; **~ de concentricidad en el revestimiento del alma** TELEFON, TRANS core-cladding concentricity error; **~ de concentricidad superficial de referencia al núcleo** TRANS core-reference surface concentricity error; **~ corregido de resistencia equivalente** TELEFON corrected equivalent resistance error; **~ cuadrantal** TEC RAD quadrantal error; **~ de cuantificación** TRANS quantization error; **~ debido a la inercia** PRUEBA inertia error; **~ de distanciamiento de antena** TEC RAD antenna spacing error; **~ doble** TELEFON double error; **~ de estimación** GEN estimation error; **~ de excentricidad** TEC RAD pointer centring (*AmE* centering) error; **~ de facturación** APLIC charging error; **~ de fuera de sincronismo** TRANS out-of-sync error; **~ de funcionamiento** INFO operation error; **~ humano** GEN human error, mistake; **~ instrumental** PRUEBA loop alignment error, *metrología* instrumental error; **~ de interpolación** PRUEBA *metrología* interpolation error; **~ de intervalo de tiempo máximo** DATOS maximum time interval error (*MTIE*); **~ intrínseco** PRUEBA *metrología* intrinsic error; **~ irrecuperable** INFO irrecoverable error; **~ en juego de ensayo ejecutable** DATOS, PRUEBA executable test-case error; **~ de lectura** INFO misreading, reading error; **~ de linealidad** COMPON linearity error; **~ local** TEC RAD site error; **~ de marcación** TELEFON dialling (*AmE* dialing) error; **~ máximo de intervalo relativo** DATOS maximum relative time interval error (*MRTIE*); **~ de mecanografía** INFO typing mistake; **~ de medición** PRUEBA measurement error; **~ medio** PRUEBA mean error; **~ de método** PRUEBA error of method; **~ de muestreo** TRANS sampling error; **~ de noche** TEC RAD night error; **~ de observación** PRUEBA observation error; **~ octantal** TEC RAD octantal error; **~ de ortografía** GEN spelling mistake; **~ de paralaje** PRUEBA parallax error; **~ parásito** PRUEBA parasitic error; **~ de polarización** TEC RAD polarization error; **~ de predicción** INFO prediction error; **~ de procedimiento** DATOS procedural error; **~ de programación** DATOS, INFO programming error; **~ de propagación** TEC RAD propagation error; **~ de protocolo** TELEFON protocol error; **~ reducido** PRUEBA reduced error; **~ relativo** PRUEBA relative error; **~ de reloj** TRANS clock error; **~ remoto** DATOS remote error; **~ residual** DATOS, INFO, TELEFON residual error; **~ de resistencia equivalente** TELEFON equivalent resistance error; **~ de seguimiento** INFO mistracking, TEC RAD tracking error; **~ semicircular** TEC RAD semicircular error; **~ de separación** TEC RAD *de antenas* spacing error; **~ simple** PRUEBA single error; **~ de sincronización** TRANS synchronization error; **~ sistemático** PRUEBA systematic error; **~ de tecleado** APLIC, DATOS, PERIF keying error; **~ de temperatura** CLIMAT, PRUEBA temperature error; **~ típico** GEN standard error; **~ tipo de polarización** TEC RAD, TRANS standard wave error; **~ de traducción** DATOS translation error; **~ de transmisión** PRUEBA, TRANS transmission error; **~ de transmisor** TEC RAD, TRANS transmitter error

E/S *abr* (*entrada/salida*) ELECTRÓN, INFO, TELEFON I/O (*input/output*)

ESA *abr* (*Agencia Espacial Europea*) TEC RAD ESA (*European Space Agency*)

ESC *abr* (*escape*) DATOS, INFO, TELEFON ESC (*escape*)

escala *f* GEN *de dibujo* scale; **~ de cargas** APLIC scale of charges; **~ por cinta cortada** TELEFON, TRANS torn-tape relay; **~ de ganancia** ELECTRÓN gain scale; **~ de grises** DATOS, TELEFON grey (*AmE* gray) scale; **~ de matices** TELEFON tone wedge; **~ móvil** INFO sliding scale; ~ **de Pantone** ELECTRÓN Pantone scale; ~ **primitiva** TELEFON pitch scale; ~ **de reducción de cinco puntos** PRUEBA five-point impairment scale; ~ **de sonoridad** TELEFON loudness scale; ~ **de temple constante** INFO equal-tempered scale; ~ **temporal coordinada** TEC RAD coordinated time scale; ~ **de tiempo atómica** TEC RAD atomic time scale

escalado *adj* GEN scalar

escalera: ~ **de ruedas** *f* COMPON *suministros y distribución* rolling-track ladder

escalón: ~ **de la cuantificación** *m* TRANS quantization step

escalonadamente *adv* GEN stepwise

escalonamiento: ~ **de frecuencias** *m* TEC RAD frequency staggering

escáner *m* INFO, PERIF scanner; ~ **de imágenes en color** PERIF colour (*AmE* color) image scanner

escape *m* (*ESC*) DATOS, INFO, TELEFON escape (*ESC*)

escasez *f* GEN shortage; ~ **de capacidad** GEN capacity shortage

escena: **en** ~ *adj* (*ONS*) TELEFON on-scene (*ONS*)

esclavo *m* ELECTRÓN slave

escobilla *f* GEN *de carbón* brush; ~ **de contacto** CONMUT wiper

ESCON *abr* (*conexión de sistemas de empresa*) INFO *IBM* ESCON (*enterprise-systems connection*)

escribir: ~ **con guión** *vt* INFO hyphenate

escritura: ~ **continua** *f* INFO word wrapping; ~ **por pasos** *f* INFO step writing; ~ **por teclado** *f* TV stroke writing

escucha *f* GEN eavesdropping, listening, listening-in; ~ **clandestina** TELEFON tapping; ~ **en grupo** TELEFON group listening; ◆ **estar a la** ~ INFO, TEC RAD be on stand-by

escuchar *vi* INFO listen; ◆ ~ **clandestinamente** ELECTRÓN, TELEFON tap

escudo: ~ **antimagnético** *m* DATOS anti-magnetic shield; ~ **solar** *m* TEC RAD sunshield

esfuerzo *m* COMPON stress; ~ **ambiental** CLIMAT environmental stress; ~ **de fluencia** GEN *resistencia* flow stress; ~ **funcional** PRUEBA *seguridad de funcionamiento* functional stress; ~ **de tracción** COMPON tensile stress

esmaltado *adj* GEN enamelled (*AmE* enameled)

esmalte *m* COMPON enamel; ~ **de secado en horno** COMPON, TV stoving lacquer

esméctico *adj* ELECTRÓN smectic

espaciado *m* TEC RAD spacing; ~ **de canales** TEC RAD channel spacing; ~ **interlineal** DATOS interline spacing

espaciamiento *m* INFO, TV spacing; ~ **de frecuencia** TEC RAD frequency spacing

espacio *m* GEN space; ~ **para bobinado** ELECTROTEC winding space; ~ **de cebado** ELECTRÓN trigger gap; ~ **cromático** TV colour (*AmE* color) space; ~ **en disco** INFO disk space; ~ **de imagen** INFO image space; ~ **interlineal** INFO *tratamiento de textos* line space; ~ **de memoria** INFO memory space; ~ **de memoria de imagen** INFO image-storage space; ~ **Morse** TELEFON Morse space; ~ **muestral** PRUEBA sample space; ~ **objeto** INFO object space; ~ **panorámico** DATOS panoramic space; ~ **principal de descarga** ELECTRÓN main gap; ~ **de protección** TEC RAD guard space; ~ **próximo a la Tierra** TEC RAD near-Earth space; ~ **remoto** TEC RAD deep space; ~ **para texto** INFO text space; ~ **para título** GEN title space

espacistor *m* COMPON, ELECTRÓN spacistor

especialista: ~ **en informática** *m* GEN computer scientist

especificación *f* PRUEBA specification; ~ **ambiental** GEN environmental specification; ~ **de bloque funcional** TELEFON functional-block specification; ~ **de control** INFO control statement; ~ **y diseño de función** CONMUT function specification and design; ~ **de estructura** GEN structure specification; ~ **de fabricación** GEN product specification; ~ **de fecha y hora** PERIF *contestador telefónico automático* date-time stamping; ~ **formal** DATOS formal specification; ~ **de funcionamiento** GEN functional specification; ~ **inicial de intercambio de gráficos** (*IGES*) INFO initial graphics exchange specification (*IGES*); ~ **de inspección** PRUEBA inspection specification; ~ **de materiales** GEN material specification; ~ **de objetivo** PRUEBA target specification; ~ **de plantilla de cable** TRANS formboard specification; ~ **de trayecto de señales** SEÑAL marshal document

especificaciones: ~ **básicas** *f pl* GEN basic specifications; ~ **de conformidad** *f pl* PRUEBA conformity specifications; ~ **de destino** *f pl* GEN *documentación* assignment specifications; ~ **de interfaz de arrastre de la red** *f pl* (*NDIS*) ELECTRÓN network-drive interface specifications (*NDIS*); ~ **de tarea** *f pl* GEN *documentación* task specifications

especificado *adj* PRUEBA specified

especificar *vt* PRUEBA specify

específico *adj* GEN specific

espectador *m* INFO, TV viewer

espectro *m* INFO, TEC RAD, TRANS spectrum; ~ **admisible fuera de banda** TEC RAD permissible out-of-band spectrum; ~ **de armónicos** INFO spectrum of harmonics; ~ **de Brillouin** DATOS Brillouin spectrum; ~ **deseado** TEC RAD desired spectrum; ~ **de difusión híbrido** TEC RAD hybrid spread spectrum; ~ **extendido** (*SS*) TEC RAD spread spectrum (*SS*); ~ **de flujo** ELECTRÓN flux pattern; ~ **de frecuencias** TEC RAD frequency spectrum; ~ **de impulso** TELEFON, TRANS pulse spectrum; ~ **de líneas** ELECTRÓN, TRANS line spectrum; ~ **de radiodifusión** TEC RAD, TRANS broadcasting spectrum; ~ **sonoro** TEC RAD, TELEFON, TRANS sound spectrum; ~ **de tren de impulsos** TELEFON pulse-train spectrum

espectrograma *m* TEC RAD, TELEFON, TRANS spectrogram; ~ **de sonidos** TEC RAD, TELEFON, TRANS sound spectrogram

espectrometría *f* TEC RAD, TRANS spectrometry; ~ **de absorción** TEC RAD absorption spectrometry

espectrómetro *m* TEC RAD, TRANS spectrometer

espectroscopía: ~ **infrarroja transformada de Fourier** *n* (*FTIR*) PRUEBA Fourier transform infrared spectography (*FTIR*)

espejo: ~ **dicroico** *m* TRANS dichroic mirror; ~ **esférico** *m* TRANS spherical mirror; ~ **de segunda superficie** *m* TEC RAD second-surface mirror

espera[1]: ~ **de DCE** *f* TELEFON DCE waiting; ~ **de DTE** *f* TELEFON DTE waiting

espera[2]: **a la** ~ *fra* TELEFON on hold

esperanza: ~ **de vida** *f* APLIC life expectancy, lifetime expectancy

espesor *m* COMPON thickness; ~ **de material** COMPON material thickness; ~ **de pared** COMPON wall thickness; ~ **de revestimiento electrolítico de agujero** ELECTRÓN *circuito impreso* hole-plating thickness; ~ **total de placa** COMPON, ELECTRÓN total board thickness

espiga *f* COMPON spigot; ~ **de conector** COMPON, ELECTRÓN plug pin; ~ **de conexión** ELECTRÓN connecting pin, connecting tag; ~ **de enrollado** COMPON wrapping pin; ~ **de plantilla de cable** TRANS formboard peg; ~ **de soldar** COMPON soldering lug, soldering tag

espionaje: ~ **industrial** *m* GEN industrial espionage; ~ **telefónico** *m* TELEFON wiretapping

espiral: **en** ~ *adj* GEN spiral

espuma: ~ **amortiguadora de impacto** *f* ELECTRÓN shock-absorbing foam; ~ **de poliestireno expandido** *f* COMPON expanded-polystyrene foam

esquema *m* INFO outline; ~ **de agrupación** GEN *documentación* grouping diagram; ~ **de bloques de fiabilidad** PRUEBA reliability block diagram; ~ **de cableado** GEN *documentación* cable drawing; ~ **de colores** INFO colour (*AmE* color) scheme; ~ **de disposición** ELECTRÓN *circuitos impresos* layout sketch; ~ **de distribución de cables** GEN *documentación* cable-distribution table; ~ **de montaje de cables** TRANS cable-mounting drawing; ~ **de distribución de potencia** ELECTROTEC power-distribution diagram; ~ **de precios** APLIC pricing scheme; ~ **de recorrido de cables** GEN *documentación* cable-way drawing; ~ **sinóptico** PRUEBA survey diagram; ~ **de terminal de cables** TRANS cable-terminal drawing; ~ **de vías troncales** TELEFON trunking diagram

estabilidad *f* GEN stability; ~ **absoluta** TELEFON absolute stability; ~ **de circuito** TRANS circuit stability; ~ **dimensional** COMPON dimensional stability; ~ **de fase** ELECTRÓN phase stability; ~ **de frecuencia** TEC RAD frequency stability; ~ **vertical** TV vertical hold

estabilización: ~ **de precios** *f* APLIC price fixing

estabilizador: ~ **de tensión** *m* ELECTRÓN, ELECTROTEC voltage stabilizer

estabilizar *vt* GEN stabilize

estable *adj* GEN stable

establecer *vt* GEN establish; ◆ ~ **una comunicación** TELEFON put through a call; ~ **una comunicación directa con** TELEFON establish a direct link with

establecimiento: **en el** ~ *adv* GEN on the premises

estación: ~ **de acoplamiento** *f* ELECTRÓN docking station; ~ **de acoplamiento simple** *f* (*SAS*) ELECTRÓN single-attachment station (*SAS*); ~ **alimentada a distancia** *f* TELEFON dependent station; ~ **alimentadora** *f* TRANS power-feeding station; ~ **autoalimentada** *f* ELECTROTEC power-feeding station; ~ **de barco** *f* TEC RAD ship station; ~ **base** *f* TEC RAD, TELEFON base station; ~ **de cinta** *f* INFO tape station; ~ **de comunicaciones** *f* TEC RAD communication station; ~ **de consulta** *f* DATOS inquiry station; ~ **de control** *f* CONTROL, INFO, TELEFON control station; ~ **de control de circuito** *f* TELEFON circuit control station; ~ **de control de operadoras auxiliares** *f* TELEFON auxiliaries control station; ~ **de control de sistema** *f* TELEFON system control station; ~ **costera** *f* TEC RAD, TELEFON coast station, coastal station, shore station; ~ **costera de un centro marítimo** *f* TELEFON maritime centre (*AmE* center) shore station; ~

de cubo *f* TEC RAD hub station; ~ **de detección** *f* INFO, PRUEBA sensing station; ~ **a distancia** *f* DATOS remote station; ~ **de embarcación de salvamento** *f* TEC RAD survival-craft station; ~ **de emisión de señal horaria normal** *f* TEC RAD standard time-signal emission station; ~ **espacial** *f* TEC RAD, TV space station; ~ **espacial de radiodifusión por satélite** *f* TEC RAD broadcasting-satellite space station; ~ **de frecuencia normal y señal horaria** *f* TEC RAD standard-frequency and time-signal station; ~ **fuera de red** *f* TELEFON, TV off-net station; ~ **fuera de servicio** *f* TRANS unserviceable station; ~ **de indicación de posición de móvil** *f* TEC RAD roaming station; ~ **para llamadas de socorro** *f* PERIF emergency-call stations; ~ **matriz de control** *f* TELEFON matrix control station; ~ **mixta** *f* INFO combined station; ~ **mixta de transmisión/recepción** *f* TEC RAD combined transmitting/receiving station; ~ **móvil** *f* TEC RAD, TELEFON mobile station (*MS*); ~ **móvil radioeléctrica** *f* TEC RAD mobile radio station; ~ **móvil terrestre** *f* TEC RAD, TELEFON land mobile station (*LMS*); ~ **no atendida** *f* TRANS unattended station; ~ **de pasillo** *f* TEC RAD gateway station; ~ **principal** *f* TEC RAD host station; ~ **principal de control** *f* TELEFON main control station; ~ **puntual** *f* TRANS point station; ~ **que llama** *f* TELEFON calling station; ~ **de radiodifusión** *f* TEC RAD broadcast station; ~ **de radio enlace** *f* TEC RAD radio-relay station; ~ **de radiofaro indicadora de posición de emergencia** *f* TEC RAD emergency position-indicating radio-beacon station; ~ **de recepción** *f* TV receiver station; ~ **de recorrido descendente** *f* TRANS down-range station; ~ **con registros de reserva** *f* TEC RAD grandfathered station; ~ **repetidora** *f* TEC RAD relay station, TELEFON, TRANS repeater station; ~ **repetidora autoalimentada** *f* TRANS power-feeding repeater station; ~ **repetidora autónoma** *f* TRANS directly-powered repeater station; ~ **repetidora para enlace móvil-base** *f* TEC RAD mobile-to-base relay station; ~ **repetidora principal** *f* TELEFON main repeater station; ~ **repetidora subordinada** *f* TELEFON dependent repeater station; ~ **repetidora de suministro de potencia** *f* ELECTROTEC power-feeding repeater station; ~ **de seguimiento remoto de centrales** *f* PRUEBA central remote monitoring station; ~ **de sincronización y base de tiempo** *f* TELEFON synchronization and time-base station; ~ **de subcontrol de circuito** *f* TELEFON circuit subcontrol station; ~ **subdirectriz** *f* TELEFON subcontrol station; ~ **supervisora general** *f* TELEFON general supervisory station; ~ **supletoria** *f* CONMUT *PBX* extension set; ~ **telefónica para emergencias** *f* TELEFON emergency telephone station; ~ **terminal de línea** *f* TRANS line-terminal station; ~ **terrena costera** *f* TEC RAD coastal earth station; ~ **terrestre** *f* TEC RAD land station, terrestrial station, TELEFON land station; ~ **totalmente restringida** *f* CONMUT fully-restricted station; ~ **de trabajo** *f* DATOS, INFO, PERIF work station; ~ **de trabajo de imágenes fotográficas** *f* INFO photo-imaging workstation

estacionario *adj* GEN stationary

estadística *f* GEN statistics; ~ **de funcionamiento** TELEFON operating statistics; ~ **de tráfico** DATOS, TEC RAD, TELEFON, TRANS traffic statistics

estadísticas: ~ **de facturación** *f pl* APLIC charging statistics

estadístico *m* GEN statistic; ~ **de orden** GEN order statistic

estado *m* GEN state; ~ **de acumulador** ELECTROTEC battery condition; ~ **de alerta** TELEFON state of alert; ~ **de alineación de trama** TRANS frame-alignment state; ~ **de avería** TELEFON fault state; ~ **de bloqueo inverso** ELECTRÓN reverse-blocking state; ~ **de canal** TRANS channel state; ~ **de comprobación inicial** DATOS initial-testing state; ~ **de conversación** TEC RAD, TELEFON talking state; ~ **de espera** INFO waiting state; ~ **estacionario** GEN steady state; ~ **final** DATOS final state; ~ **finito** DATOS, SEÑAL finite state; ~ **de funcionamiento** PRUEBA operating state; ~ **inestable** ELECTRÓN unstable state; ~ **inicial** DATOS, ELECTRÓN initial state; ~ **límite** PRUEBA *seguridad de funcionamiento* limiting state; ~ **lógico** ELECTRÓN logic state; ~ **mesomorfo** ELECTRÓN mesomorph state; ~ **no operativo** PRUEBA nonoperating state, outage, TELEFON down-state, nonoperating state, outage; ~ **no peligroso** PRUEBA nonhazardous state; ~ **no requerido** PRUEBA *seguridad de funcionamiento* free state; ~ **de ocupación** TELEFON seizure condition; ~ **de prueba** PRUEBA testing state; ~ **de reposo** ELECTRÓN rest state, ELECTROTEC quiescent state, rest state, TELEFON, TRANS rest state; ~ **de reserva** PRUEBA stand-by state; ~ **de revisión** GEN revision state; ~ **de saturación** ELECTRÓN saturation state; ~ **secundario** INFO substatus; ~ **secundario de indicación multipunto** DATOS multipoint indication secondary status; ~ **de señal** GEN signal state;

~ **de señalización** SEÑAL signalling (*AmE* signaling) state; **de ~ sólido** *adj* COMPON solid-state; ~ **tampón** ELECTRÓN buffer state; ~ **de temperatura limitada** CLIMAT, ELECTRÓN temperature-limited state; ~ **de transición** TELEFON state of transition; ~ **transitorio** TELEFON transitional state; ~ **transversal** TRANS *óptica de fibras* cross state; ~ **útil** PRUEBA, TELEFON up-state; ◆ **en ~ de conducción** ELECTRÓN *transistores* conducting, on

estafeta *f* GEN courier

estampación *f* COMPON *de plásticos* embossing; ~ **con estarcido** COMPON, INFO screen printing

estampador *m* COMPON stamper

estanco *adj* GEN sealed

estándar: ~ **de hecho** *m* DATOS de facto standard

estante *m* TELEFON shelf

estañado *m* COMPON tinning

estaño *m* COMPON tin; ~ **para soldar** COMPON soldering tin; ~ **de soldar con núcleo de resina** ELECTRÓN resin-core solder

estator *m* COMPON, ELECTRÓN stator

estatus: ~ **de puesta a uno** *m* INFO set status

estenografía *f* INFO shorthand writing

estéreo[1] *adj* INFO, PERIF, TELEFON, TV stereo

estéreo[2]: ~ **MS** *m* TV MS stereo

estereofonía *f* INFO, TV stereophony

estereofónico *adj* INFO, PERIF, TELEFON, TV stereophonic

estereoscópico *adj* INFO, TV stereoscopic

estilete *m* APLIC stylus, INFO pen

estilo *m* GEN style; ~ **claro** INFO plain style; ~ **de fuente tipográfica** INFO font style; ~ **de letra de imprenta** PERIF typeface

estimación *f* GEN estimate, estimation; ~ **incorrecta** GEN, error of judgment, misjudgment; ~ **de movimiento** TV motion estimation; ~ **de riesgos** PRUEBA risk assessment

estimado *adj* PRUEBA estimated

estímulo: ~ **de colores** *m* INFO colour (*AmE* color) stimulus

estipulación *f* GEN stipulation

estirar *vt* GEN stretch

estocástico *adj* GEN stochastic

estrangulador *m* ELECTRÓN restrictor

estrategia *f* GEN strategy; ~ **de encaminamiento de tráfico** TELEFON traffic-routing strategy

estratificación *f* TRANS *de cableaje* bedding

estratificador: ~ **de inversión** *m* TRANS inversion layer

estrella: ~ **D** *f* PRUEBA D-star; ~ **doble pasiva** *f* TRANS passive double star

estribo *m* ELECTROTEC U-link

estropear *vt* TELEFON mutilate

estructura *f* GEN structure; ~ **de antena** TEC RAD antenna structure; ~ **en árbol** DATOS, INFO, TRANS tree structure; ~ **en árbol y en rama** DATOS, ELECTRÓN tree and branch structure; ~ **en cadena** INFO chain structure; ~ **de capa de aplicación** TRANS application-layer structure; ~ **combinada de pastillas de sílice ranuradas** ELECTRÓN composite structure of grooved silicon chips; ~ **de control** CONTROL control structure; ~ **de documento** GEN document structure; ~ **en estrella** TRANS, TV star structure; ~ **de información de gestión** (*SMI*) DATOS structure of management information (*SMI*); ~ **mecánica** GEN mechanical structure; ~ **mecánica de centrales** TELEFON exchange mechanical structure; ~ **mecánica de fila** TELEFON row mechanical structure; ~ **de multitrama** DATOS multiframe structure; ~ **de onda lenta** ELECTRÓN slow-wave structure; ~ **paralela** PRUEBA parallel structure; ~ **de red** TELEFON network structure; ~ **de red primaria** TRANS backbone structure; ~ **de soporte de antena** TEC RAD antenna-supporting structure; ~ **submicrónica** GEN submicron structure; ~ **de tabla** GEN table structure; ~ **de tráfico** TELEFON traffic pattern; ~ **de trama** TRANS frame structure

estructuración *f* GEN structuration, structuring

estuche: ~ **de protección** *m* GEN carrying case

estudio *m* GEN study; ~ **de caso práctico** GEN case study; ~ **de emplazamiento** TEC RAD site investigation; ~ **de factibilidad** TEST feasibility study; ~ **local** TV local studio; ~ **de mercado** GEN market survey; ~ **objeto** GEN object study; ~ **posterior** (*FS*) DATOS further study (*FS*); ~ **de rumbo** TELEFON route analysis; ~ **de tráfico** TELEFON traffic study

etapa *f* GEN stage; ~ **de ampliación** TELEFON extension step; ~ **de amplificación** ELECTRÓN amplification stage; ~ **de concentración** TRANS concentration stage; ~ **de conmutación** CONMUT, TELEFON switching stage, selector stage; ~ **de distribución** TRANS distribution stage; ~ **espacial** CONMUT, TELEFON space stage; ~ **de excitación** ELECTRÓN driving stage; ~ **excitadora** ELECTRÓN driver stage; ~ **de expansión** TRANS expansion stage; ~ **mezcladora** TEC RAD mixer stage; ~ **remota de conmutación** CONMUT remote switching stage; ~ **de repetidor canalizada** TRANS channelized stage of repeater; ~ **S** CONMUT, TELEFON S stage; ~ **de salida** ELECTRÓN output stage, TEC RAD final stage, output stage, TELEFON final stage; ~ **de selección** CONMUT, TELEFON selection stage; ~ **T** TELEFON T stage; ~ **terminal** CONMUT terminating stage; ~ **de tránsito** TELEFON transit stage

ETC *abr* (*circuito terminal de central*) CONMUT ETC (*exchange terminal circuit*)

ETF *abr* (*modulación ocho a catorce*) INFO ETF (*eight to fourteen modulation*)

Ethernet *m* DATOS Ethernet; ~ **delgado** ELECTRÓN thin Ethernet

etilcelulosa *f* COMPON ethyl cellulose

etiqueta *f* GEN label; ~ **en blanco** ELECTRÓN blank label; ~ **de cabecera** INFO HDR label; ~ **de cable** TRANS cable tag; ~ **de encaminamiento** TELEFON routing label; ~ **de seguridad** DATOS *de transmisiones* security label; ~ **de título** INFO header label; ◆ **sin ~** INFO unlabelled (*AmE* unlabeled)

etiquetado *m* GEN labelling (*AmE* labeling)

etiquetador *m* INFO tagger

etiquetar *vt* INFO tag

ETM *abr* (*modulación ocho a diez*) INFO ETM (*eight to ten modulation*)

ETNA *abr* (*Arquitectura de la Red de Transporte Ericsson*) TRANS ETNA (*Ericsson Transport Network Architecture*)

EURB *abr* (*Unión Europea de Radiodifusión*) TEC RAD EURB (*European Union for Radio Broadcasting*)

Euronorma *f* (*EN*) GEN Euronorm (*EN*)

evaluación *f* GEN assessment, evaluation; ~ **de distancia** GEN assessment of distance; ~ **de fiabilidad** PRUEBA reliability measure; ~ **de proveedores** PRUEBA supplier evaluation, vendor appraisal; ~ **de riesgos** PRUEBA risk evaluation, risk assessment

evaluar *vt* GEN assess, evaluate

evaporación *f* GEN evaporation

evento: ~ **cualificado** *m* DATOS qualified event; ~ **de justificación de puntero** *m* DATOS pointer justification event (*PJE*); ~ **no cualificado** *m* DATOS unqualified event; ~ **de prueba sintácticamente inválida** *m* DATOS, PRUEBA syntactically-invalid test event; ~ **de referencia de la capa de paquetes** *m* DATOS, TELEFON packet-layer reference event; ~ **semánticamente inválido** *m* DATOS, PRUEBA semantically invalid event

EWOS *abr* (*Seminario Europeo para Sistemas Abiertos*) GEN EWOS (*European Workshop for Open Systems*)

ex: ~**-directorio** *adj* (*XD*) GEN ex-directory *BrE* (*XD*), unlisted *AmE*

exactitud *f* PRUEBA trueness; ~ **de reproducción** ELECTRÓN, TEC RAD fidelity

examen *m* GEN examination

examinar *vt* GEN examine

excavadora: ~ **de zanjas para cables** *f* GEN cable plough (*AmE* plow)

excederse *v refl* INFO overrun

excéntrico *adj* INFO offbeat

excitación *f* ELECTRÓN, TELEFON excitation; ~ **por choque** ELECTRÓN, TEC RAD shock excitation; ~ **por impulsos** ELECTRÓN impulse excitation; ~ **de oscilador estabilizado por línea** ELECTRÓN line-stabilized oscillator drive

excitador: ~ **de relé** *m* COMPON, ELECTROTEC relay driver

excitar *vt* ELECTRÓN excite

excluido *adj* GEN excluded

excluir *vt* GEN exclude

exclusión *f* ELECTRÓN *elementos lógicos* exclusion

excrecencia *f* ELECTRÓN outgrowth

existencias: ~ **de repuestos** *f pl* COMPON spare-parts inventory

exitancia: ~ **radiante** *f* ELECTRÓN, TRANS *en punto de superficie* radiant exitance

expandir *vt* DATOS, INFO unpack, unzip, expand

expansión *f* GEN expansion; ~ **de ancho de banda** TEC RAD, TRANS bandwidth expansion; ~ **en árbol** DATOS, INFO tree expansion; ~ **de espectro** TEC RAD spectrum expansion, spectrum spreading; ~ **de señal** SEÑAL signal expansion

expectativa *f* GEN *estadística* expectation

expediente: ~ **de sistema** *m* DATOS *Internet* system dossier

experiencia: ~ **de trabajo** *f* TELEFON operational experience

experimento: ~ **factorial** *m* PRUEBA factorial experiment

experto: ~ **de campo** *m* INFO domain expert

expirar *vi* GEN expire

explícito *adj* GEN explicit

exploración *f* GEN scanning; ~ **ampliada** TEC RAD expanded sweep; ~ **de campo** INFO field scan; ~ **de campo próximo** TEC RAD, TRANS near-field scanning; ~ **en distancia** TEC RAD range search; ~ **de haz** TEC RAD beam scanning; ~ **de línea** INFO, TEC RAD line scan, line scanning; ~ **de nivel** TELEFON level hunt; ~ **progresiva** INFO progressive scanning; ~ **por puntos sucesivos** TV dot interlace scanning; ~ **sectorial** TEC RAD sector scanning; ~ **vectorial**

INFO vector scanning; ~ **en velocidad** TEC RAD speed search, velocity search
explorador: ~ **de frecuencias** *m* TRANS frequency scanner; ~ **de imagen** *m* PERIF, TV image scanner; ~ **de transparencias** *m* DATOS, INFO slide scanner
exploradores: ~ **en oposición** *m pl* TEC RAD back-to-back scanners
explotación: ~ **en cocanal** *f* TEC RAD co-channel operation; ~ **en dos frecuencias** *f* SEÑAL, TEC RAD two-frequency operation; ~ **por portadoras distintas** *f* TEC RAD spaced carrier operation
exponencial *adj* GEN exponential
exponente *m* GEN exponent
exponer *vt* TV expose
exportación *f* INFO export
exportar *vt* DATOS export
exposición *f* GEN *comercio* exhibition
expresión *f* GEN expression; ~ **aritmética** INFO arithmetical expression; ~ **técnica** GEN technical expression; ~ **de valor indeterminado** DATOS any-value expression
expulsar *vt* DATOS, INFO *disco* eject
extender *vt* INFO, TELEFON expand
extensibilidad *f* INFO, TELEFON expandability, extensibility
extensible *adj* INFO, TELEFON expandable
extensión *f* (*X*) GEN extension (*X*); ~ **abierta** TELEFON unbarred extension; ~ **de código** TELEFON code extension; ~ **de código gráfico** TELEFON graphic-code extension; ~ **de difusión** COMPON diffusion length; ~ **de dirección de red** (*NAE*) DATOS network-address extension (*NAE*); ~ **de error** TELEFON error spread; ~ **de fractura** COMPON, PRUEBA length of fracture; ~ **fuera del edificio** (*OPX*) TELEFON off-premises extension (*OPX*); ~ **prioritaria** TELEFON priority extension; ~ **con rodillo** ELECTRÓN pin-out; ~ **de señal** SEÑAL signal extension; ~ **de servicio hotelero** CONMUT, TELEFON hotel-service extension; ~ **supletoria** TELEFON extension set
extensor *m* INFO expander; ~ **de bucle** TELEFON loop extender; ~ **de FM** INFO FM expander; ~ **de guíaondas** TEC RAD line lengthener; ~ **de línea** TEC RAD line stretcher
externo *adj* GEN external
extra *adj* GEN, INFO, TELEFON extra
extracción: ~ **de contorno** *f* INFO contour extraction; ~ **de señal** *f* SEÑAL, TRANS signal extraction
extractor: ~ **de placas** *f* PRUEBA *herramientas* board-extraction tool, board extractor
extraer *vt* INFO extract
extraño *adj* ELECTROTEC, INFO, TELEFON extraneous
extrapolación *f* INFO extrapolation
extrapolar *vt* INFO extrapolate
extremidad: ~ **no reflejante** *f* TEC RAD matched termination
extremo: ~ **de cable** *m* TRANS cable butt, cable head, cable termination; ~ **de hilo** *m* COMPON wire end
extruir *vt* GEN *cables* extrude
extrusión *f* GEN *de cables* extrusion
eyector *m* ELECTRÓN knockout
EyePhone *m* INFO EyePhone

F

F *abr* (*faradio*) ELECTRÓN, ELECTROTEC F (*farad*)
fabricación *f* GEN manufacturing; **~ asistida por computadora** *AmL* (*CAM*), (*cf fabricación asistida por ordenador Esp*) INFO computer-aided manufacturing (*CAM*); **~ asistida por ordenador** *Esp* (*CAM*), (*cf fabricación asistida por computadora AmL*) INFO computer-aided manufacturing (*CAM*); **~ de chapas** COMPON *hierro* plating; **~ informatizada** (*CIM*) INFO computer-integrated manufacturing (*CIM*)
fabricante *m* GEN manufacturer, producer; **~ de computadoras** *AmL* (*cf fabricante de ordenadores Esp*) INFO computer manufacturer; **~ de ordenadores** *Esp* (*cf fabricante de computadoras AmL*) INFO computer manufacturer
faceta *f* TRANS *óptica de fibras* facet
facilidad: **~ de manejo** *f* GEN user-friendliness, consumer-friendliness; **~ de prueba** *f* PRUEBA testability; **~ de uso** *f* GEN user-friendliness, consumer-friendliness
facsímil *m* GEN facsimile
factible *adj* PRUEBA feasible
factor *m* GEN factor; **~ de absorción** TEC RAD absorption factor; **~ de aceleración posdeflexión** ELECTRÓN post-deflection acceleration factor; **~ de acoplamiento** TRANS *óptica de fibras* coupling factor; **~ de actividad vocal** TELEFON speech-activity factor; **~ de agudeza** TELEFON peakedness factor; **~ de amplificación** ELECTROTEC amplification factor, mu factor; **~ de amplificación de tensión** ELECTROTEC voltage-amplification factor; **~ de antena** TEC RAD aerial factor, antenna factor; **~ de arrastre** ELECTRÓN pulling figure; **~ de atenuación** ELECTROTEC loss factor; **atenuación de reflexión de eco de impulso** TELEFON pulse-echo return loss; **~ de cilindro** TELEFON drum factor; **~ clave** TV key factor; **~ de contenido de gas** ELECTRÓN gas-content factor; **~ de cooperación** TELEFON factor of cooperation; **~ de corrección** PRUEBA correction factor; **~ de degradación** TEC RAD degradation figure; **~ de desmultiplicación de impulsos** GEN scaling factor; **~ de direccionalidad de antena** TEC RAD aerial-directivity factor, antenna-directivity factor; **~ de disipación** ELECTROTEC dissipation factor; **~ de disponibilidad operativa** PRUEBA operational-availability factor; **~ de distorsión** ELECTRÓN distortion factor; **~ efectivo de radio terrestre** TEC RAD effective Earth-radius factor; **~ de empuje** ELECTRÓN pushing figure; **~ de equipo** PRUEBA practice factor; **~ de equipo de operadores** PRUEBA crew factor; **~ de forma** PRUEBA form factor; **~ influyente** PRUEBA influencing factor; **~ de interferencia** TRANS interferer; **~ de luminancia** INFO luminance factor; **~-m** ELECTRÓN m-factor; **~ de modulación** TRANS modulation factor; **~ de multiplicación de error** TELEFON error-multiplication factor; **~ de multiplicación de gas** ELECTRÓN gas-multiplication factor; **~ de potencia** ELECTROTEC power factor; **~ Q** ELECTROTEC Q factor; **~ de recepción** TELEFON take-up factor; **~ de recubrimiento** TEC RAD coverage factor; **~ de reflexión** ELECTRÓN, TEC RAD, TRANS reflection factor; **~ de reflexión de amplitud** TRANS *óptica de fibras* amplitude-reflection factor; **~ de reflexión transversal** TRANS transverse-reflection factor; **~ de ruido** PRUEBA, TEC RAD, TRANS noise factor; **~ de ruido puntual** TEC RAD spot-noise factor; **~ de sensibilidad** TEC RAD sensitivity factor; **~ de sombra** TEC RAD shadow factor; **~ de vacío** ELECTRÓN vacuum factor
factoraje *m* INFO, PRUEBA *matemáticas* factoring
factorial *m* INFO, PRUEBA *matemáticas* factorial
facturable *adj* APLIC chargeable
facturación *f* APLIC billing, charging, invoicing; **~ centralizada** APLIC centralized billing; **~ por conceptos** APLIC itemized billing; **~ detallada** APLIC detailed billing; **~ individual** APLIC individual charging; **~ a tercero** APLIC third-party billing
faja *f* INFO band, TEC RAD, TV channel
fallo *m* GEN failure, fault; **~ de calentamiento** ELECTROTEC *de soldadura* heating fault; **~ catastrófico** PRUEBA catastrophic failure; **~ en la continuidad** PRUEBA continuity fault; **~ corregido** PRUEBA corrected fault; **~ crítico** PRUEBA critical failure; **~ por debilidad** PRUEBA weakness failure; **~ por desgaste** PRUEBA wear-out failure; **~ de diseño** PRUEBA design fault; **~ de encendido** ELECTROTEC misfire; **~**

esporádico PRUEBA sporadic failure; ~ **de fabricación** PRUEBA manufacturing failure, manufacturing fault; ~ **humano** PRUEBA human error; ~ **individual** PRUEBA individual fault; ~ **inicial** PRUEBA early failure; ~ **intermitente** PRUEBA intermittent fault, intermittent failure; ~ **de línea** TELEFON line fault; ~ **por manipulación indebida** PRUEBA mishandling failure; ~ **mayor** PRUEBA major failure; ~ **no reparable** PRUEBA noncorrectable fault; ~ **no significativo** PRUEBA nonrelevant failure; ~ **en onda piloto de grupo** TRANS group-pilot failure; ~ **de potencia** ELECTROTEC, INFO, TELEFON group-pilot failure; ~ **progresivo** PRUEBA drift failure; ~ **de punto único** PRUEBA single-point failure; ~ **que deteriora la función** PRUEBA function-degrading failure, function-degrading fault; ~ **que impide la función** PRUEBA function-preventing failure, function-preventing fault; ~ **que no afecta a la función** PRUEBA function-permitting failure, function-permitting fault; ~ **que perturba la función** PRUEBA function-disturbing failure, function-disturbing fault; ~ **de red** (*ND*) SEÑAL, TELEFON, TRANS network default (*ND*), network failure; ~ **de red en señal de bucle local** SEÑAL network fault in local loop signal; ~ **repentino** PRUEBA sudden failure; ~ **de retención de imagen** TV image-retention fault; ~ **de señalización** SEÑAL signalling (*AmE* signaling) fault; ~ **sistemático** PRUEBA systematic failure; ~ **temporal** PRUEBA temporary fault, temporary malfunction; ~ **total repentino** PRUEBA sudden total failure; ~ **transitorio** PRUEBA volatile fault; ~ **por uso indebido** PRUEBA misuse failure

falsa: ~ **imagen** *f* INFO, TV streaking; ~ **imitación** *f* ELECTRÓN imitative deception

falso: ~ **arranque** *m* PRUEBA false start

falta: ~ **de enlace de señalización** *f* SEÑAL signalling-link (*AmE* signaling-link) failure; ~ **de registro** *f* TEC RAD nonregistration; ~ **de señal** *f* SEÑAL signal failure (*SF*)

familia: ~ **de caracteres** *f* DATOS, INFO character font; ~ **de caracteres básica** *f* DATOS, INFO basic character font

fantastrón *m* ELECTRÓN phantastron

fantomización *f* TRANS phantoming

FAQ *abr* (*preguntas formuladas con frecuencia*) INFO FAQ (*frequently-asked questions*)

faradio *m* (*F*) ELECTRÓN, ELECTROTEC farad (*F*)

faro *m* TEC RAD beacon

fasaje *m* TV phasing

fase *f* GEN phase; ~ **de amplificación** ELECTRÓN, ELECTROTEC amplifier stage; ~ **ascendente** TEC RAD ascent phase; ~ **de buscador de llamada** TRANS line-finder stage; ~ **de compilación** INFO compiling phase; ~ **de control de red** ELECTRÓN, TELEFON network-control phase; ~ **de datos** DATOS, TRANS data phase; ~ **de ejecución** INFO execute phase; ~ **de espera** PRUEBA stand-by phase; ~ **de generación de frecuencia** TRANS frequency-generating stage; ~ **nemática** ELECTRÓN nematic phase; ~ **de transferencia de datos** DATOS, TRANS data-transfer phase; ~ **de transferencia de datos transparentes** DATOS, TRANS transparent-data-transfer phase; ~ **de transmisión de datos** DATOS, TRANS data-transmission phase; ~ **de transmisión de enlace ascendente** TEC RAD, TRANS uplink transmission-phase

fases: ~ **sucesivas de llamada** *f pl* TELEFON successive call phases

FAT *abr* (*tabla de adjudicación de archivo*) DATOS FAT (*file allocation table*)

fatiga *f* COMPON fatigue

favor *m* ELECTRÓN favour (*AmE* favor)

fax *m* GEN fax; ~ **de grupo 1** PERIF, TELEFON group 1 fax; ~ **módem** INFO, PERIF, TELEFON fax modem

faxear *vt* GEN fax

faz: ~ **de fibra** *f* TRANS fibre (*AmE* fiber) end face

FBI *abr* (*intervalo de supresión de campo*) TV FBI (*field blanking interval*)

FC *abr* (*clase funcional*) TELEFON FC (*functional class*)

FCA *abr* (*clase funcional A*) TELEFON FCA (*functional class A*)

FCB *abr* (*clase funcional B*) TELEFON FCB (*functional class B*)

FCC *abr* (*Comisión Federal de Comunicación*) DATOS FCC (*Federal Communications Commission*)

FCS *abr* (*secuencia de control de trama*) DATOS, TELEFON FCS (*frame check sequence*)

FD *abr* (*descripción funcional*) GEN FD (*functional description*)

FDT *abr* (*técnica de descripción formal*) DATOS FDT (*formal description technique*)

fecha *f* GEN date; ~ **de calendario juliano modificada** (*MJD*) TEC RAD modified Julian date (*MJD*); ~ **de conexión** GEN setup date; ~ **de fabricación** GEN date of manufacture; ~ **y hora de transmisión** DATOS date and time of transmission; ~ **de instalación** GEN installation date; ~ **de intercambio** DATOS interchange date; ~ **ordinal** GEN ordinal date; ~ **según el calendario juliano** TEC RAD Julian date

ferroaleación *f* COMPON ferroalloy

ferrocobalto *m* COMPON ferrocobalt
ferromanganeso *m* COMPON ferromanganese
ferroníquel *m* COMPON ferronickel
ferrosilicio *m* COMPON ferrosilicon
FFD *abr* (*deformación de forma libre*) INFO FFD (*freeform deformation*)
FFIA *abr* (*Formato de Fichero de Intercambio-Audio*) TEC RAD FFIA (*Audio-Interchange File Format*)
FI *abr* (*identificador de formato*) ELECTROTEC, TELEFON FI (*format identifier*)
fibra *f* GEN fibre (*AmE* fiber); ~ **con dispersión nivelada** TRANS dispersion-flattened fibre (*AmE* fiber); ~ **dopada con lantánido** COMPON, TRANS lanthanide-doped fibre (*AmE* fiber); ~ **de fusible progresivo** TRANS fuse-tapered fibre (*AmE* fiber); ~ **de guía débil** TRANS weakly-guiding fibre (*AmE* fiber); ~ **hasta el domicilio** TRANS fibre (*AmE* fiber) to the home (*FTTH*); ~ **hasta el edificio** TRANS fibre (*AmE* fiber) to the building (*FTTB*); ~ **hasta la oficina** TRANS fibre (*AmE* fiber) to the office (*FTTO*); ~ **de índice escalonado** COMPON, TRANS step index fibre (*AmE* fiber); ~ **de índice escalonado equivalente** TRANS equivalent step index fibre (*AmE* fiber); ~ **de índice graduado** COMPON, ELECTRÓN, TRANS grading-index fibre (*AmE* fiber); ~ **de índice gradual** COMPON, ELECTRÓN, TRANS graded-index fibre (*AmE* fiber); ~ **de índice de ley potencial** TRANS power-law index fibre (*AmE* fiber); ~ **de índice nivelador** COMPON, ELECTRÓN, TRANS grading-index fibre (*AmE* fiber); ~ **de lanzamiento** TRANS launching fibre (*AmE* fiber); ~ **monomodo** COMPON, TRANS single-mode fibre (*AmE* fiber); ~ **multimodo** COMPON, TRANS multimode fibre (*AmE* fiber); ~ **óptica de índice escalonado** COMPON, TRANS step index optical fibre (*AmE* fiber); ~ **óptica multimodo de núcleo grande** TRANS large-core multiple-mode optic fibre (*AmE* fiber); ~ **PCS** COMPON, TRANS PCS-fibre (*AmE* PCS-fiber); ~ **de plástico** COMPON, TRANS plastic fibre (*AmE* fiber), all-plastic fibre (*AmE* fiber); ~ **para pozo de registro de canalización** TRANS fibre (*AmE* fiber) to the curb (*FTTC*); ~ **premoldeada** COMPON, TRANS preformed fibre (*AmE* fiber); ~ **de sílice** COMPON, TRANS silica fibre (*AmE* fiber), all-silica fibre (*AmE* fiber); ~ **soplada** COMPON, TRANS blown fibre (*AmE* fiber); ~ **al suelo** TRANS fibre (*AmE* fiber) to the floor (*FTTF*); ~ **de vidrio** COMPON, TRANS fibreglass (*AmE* fiberglass), glass fibre (*AmE* fiber)
ficha: ~ **de cabecera** *f* INFO header card; ~ **de control** *f* CONTROL control card; ~ **para extensión** *f* PERIF extension socket
fichero *m* DATOS, INFO file; ~ **de ayuda** INFO help file; ~ **binario** DATOS binary file; ~ **cerrado** INFO locked file; ~ **de control de recuperación de trabajos** INFO job-recovery control file; ~ **de direcciones de archivos** TELEFON file of file addresses; ~ **de enlace** INFO link file; ~ **de imagen** DATOS image file; ~ **inactivo** DATOS storage file; ~ **de intercambio de datos** DATOS, INFO data-exchange file; ~ **de recuperación de trabajos** INFO job-recovery file; ~ **de Web** DATOS, INFO Web file
fidelidad *f* ELECTRÓN, TEC RAD fidelity
Fidonet *m* DATOS *Internet* Fidonet
FIF *abr* (*campo de información facsímil*) DATOS, INFO FIF (*facsimile information field*)
figura: ~ **binaria** *f* INFO binary figure
fijación *f* DATOS binding; ~ **de cable** COMPON cable holder; ~ **de indicador** COMPON, CONMUT selecting-finger base, selecting-finger support; ~ **de precios** APLIC price-fixing; ~ **de tarifa** APLIC tariff setting
fijador: ~ **de cable** *m* COMPON cable grip; ~ **de frecuencia** *m* TEC RAD frequency lock; ~ **de hilos** *m* COMPON wire fastener; ~ **de mayúsculas** *m* INFO, PERIF shift lock
fijar *vt* DATOS secure, INFO fix, freeze, secure, TELEFON fix
fijo *adj* DATOS secured, INFO fixed, frozen, secured, TELEFON fixed
fila *f* INFO, TELEFON row; ~ **de bastidores** TELEFON rack row; ~ **de cuadros conmutadores** CONMUT, TELEFON switchboard row, switchboard suite; ~ **de espigas** ELECTRÓN pin row; ~ **posterior** TELEFON rear row; ~ **de salida** ELECTRÓN output row
filamento *m* COMPON, ELECTRÓN filament
filtración *f* GEN filtering; ~ **por adaptiva** TRANS adaptive filtering; ~ **fuera de canal** TRANS out-of-channel filtering; ~ **de varias velocidades** TRANS multirate filtering
filtrado *m* GEN filtering; ~ **de canal** TV channel filtering; ~ **de coseno elevado** GEN raised-cosine filtering; ~ **lógico** INFO logic filtering; ~ **multidimensional** ELECTRÓN multidimensional filtering
filtrar *vt* GEN filter
filtro *m* GEN filter; ~ **en anillo** TEC RAD ring filter; ~ **antiinterferencia** TEC RAD wave trap; ~ **antiparasitario** ELECTROTEC, INFO line filter; ~ **antirradiación** DATOS anti-radiation filter; ~ **antirreflectante** DATOS anti-glare filter; ~ **de bandas sin atenuación de imagen** ELECTRÓN image parameter filter; ~ **de blanqueo**

ELECTRÓN whitening filter; **~ de bloqueo de banda** TRANS band-elimination filter, band-rejection filter; **~ con bobina de entrada** ELECTRÓN L-section filter; **~ de compensación** ELECTRÓN weighting filter; **~ para contraste** TV contrast filter; **~ controlado por tensión** ELECTROTEC voltage-controlled filter; **~ corrector** DATOS shaping filter; **~ de cristal** TEC RAD crystal filter; **~ de cuarzo** ELECTRÓN quartz filter; **~ de cuatro polos** ELECTRÓN four-pole filter; **~ dicroico** ELECTRÓN, TRANS dichroic filter; **~ digital multidimensional** ELECTRÓN multidimensional digital filter; **~ direccional** ELECTRÓN directional filter; **~ EFIR** (*filtro de respuesta de impulso finito igualmente ondulado*) ELECTRÓN, TRANS EFIR filter (*equiripple finite impulse response filter*); **~ eliminador de banda** TRANS bandstop filter; **~ eliminador de interferencias** ELECTRÓN, TRANS interference filter; **~ de entalladura** ELECTRÓN, TRANS notch filter; **~ en escalera** TEC RAD ladder-type filter; **~ espectral** TRANS spectral filter; **~ de fase mínima** TRANS minimum-phase filter; **~ de fase sincronizada** ELECTRÓN phase-locked filter; **~ de FI** TEC RAD, TRANS IF filter; **~ FIR** (*filtro de respuesta de impulso finito*) ELECTRÓN, TRANS FIR filter (*finite impulse response filter*); **~ de formación de señal** ELECTRÓN, SEÑAL signal-shaping filter; **~ de frecuencias** ELECTRÓN, TEC RAD frequency filter; **~ de función** TRANS mode filter; **~ IIR** (*filtro de respuesta de impulso infinito*) ELECTRÓN, TRANS IIR filter (*infinite impulse response filter*); **~ incorporado** ELECTRÓN built-in filter; **~ interdigital** TRANS interdigital filter; **~ intermedio de rojo** ELECTRÓN red pass filter; **~ de llave** INFO key filter; **~ m-derivado** TEC RAD m-derived filter; **~ de modo de reflexión** TEC RAD reflection-mode filter; **~ de modo resonante** TEC RAD resonant-mode filter; **~ multibanda** TRANS multiband filter; **~ de nivelación** ELECTROTEC smoothing filter; **~ no recurrente** DATOS, TRANS nonrecursive filter; **~ pasivo** ELECTRÓN passive filter; **~ paso banda** TRANS bandpass filter (*BPF*); **~ de respuesta de impulso finito** (*filtro FIR*) ELECTRÓN, TRANS finite impulse response filter (*FIR filter*); **~ de respuesta de impulso finito igualmente ondulado** (*filtro EFIR*) ELECTRÓN, TRANS equiripple finite impulse response filter (*EFIR filter*); **~ de respuesta de impulso infinito** (*filtro IIR*) ELECTRÓN, TRANS infinite impulse response filter (*IIR filter*); **~ de semicelosía** ELECTRÓN, TRANS half-lattice filter; **~ transversal** TRANS transversal filter; **~ de visión** INFO viewing filter

fin: **~ de comunicación** *m* TELEFON end of communication; **~ de diálogo** *m* TELEFON end of dialog; **~ de energía de vida** *m* ELECTROTEC end-of-life power; **~ de impulso** *m* INFO end of pulse; **~ de línea** *m* PERIF *impresoras* end of line; **~ de mensaje** *m* (*EOM*) TELEFON end of message (*EOM*); **~ de movimiento** *m* TELEFON end of transaction; **~ de plazo** *m* GEN deadline; **~ de procedimiento** *m* (*EOP*) TELEFON end of procedure (*EOP*); **~ de programa de interrupción de procedimiento** *m* (*PRI-EOP*) TELEFON procedure interrupt EOP (*PRI-EOP*); **~ de retransmisión EOM** *m* (*EOR-EOM*) TELEFON end of retransmission EOM (*EOR-EOM*); **~ de retransmisión EOP** *m* (*EOR-EOP*) TELEFON end of retransmission EOP (*EOR-EOP*); **~ de retransmisión MPS** *m* (*EOR-MPS*) TELEFON end of retransmission MPS (*EOR-MPS*); **~ de retransmisión NULL** *m* (*EOR-NULL*) TELEFON end of retransmission NULL (*EOR-NULL*); **~ de retransmisión PRI-EOM** *m* (*EOR-PRI-EOM*) TELEFON end of retransmission PRI-EOM (*EOR-PRI-EOM*); **~ de retransmisión PRI-EOP** *m* (*EOR-PRI-EOP*) TELEFON end of retransmission PRI-EOP (*EOR-PRI-EOP*); **~ de retransmisión PRI-MPS** *m* (*EOR-PRI-MPS*) TELEFON end of retransmission PRI-MPS (*EOR-PRI-MPS*); **~ de salida** *m* TELEFON end of output; **~ de selección** *m* (*EOS*) TELEFON end of selection (*EOS*); **~ de transmisión** *m* (*EOT*) TRANS end of transmission (*EOT*)

final: **~ de cabeza de red** *m* TV network head-end; **~ de conexión de canal virtual** *m* DATOS, TRANS, TV virtual-channel connection endpoint (*VCCE*); **~ de conexión de trayectoria virtual** *m* DATOS, TRANS, TV virtual-path connection endpoint (*VPCE*); **~ de mensaje de interrupción de procedimiento** *m* (*PRI-EOM*) TELEFON procedure interrupt EOM (*PRI-EOM*); **~ de programa de interrupción de procedimiento** *m* TRANS procedure interrupt EOP (*PRI-EOP*)

finalidad: **~ de campo** *f* PRUEBA field purpose

física: **~ de estado sólido** *f* GEN solid-state physics

flanco: **~ de filtro** *m* ELECTRÓN filter flank; **~ de impulso** *m* TELEFON pulse edge; **~ posterior** *m* ELECTRÓN trailing edge; **~ posterior de impulso** *m* ELECTRÓN pulse trailing edge

FLCD *abr* (*pantalla de cristal líquido*

ferroeléctrico) INFO FLCD (*ferro-electric liquid crystal display*)

flecha: ~ **vertical** *f* INFO vertical arrow

fleje: ~ **de acero** *m* COMPON steel tape

flexibilidad *f* GEN flexibility; ~ **de reacción al error** INFO error resilience; ~ **de red primaria** ELECTRÓN backbone flexibility

flexible *adj* GEN flexible

flota: ~ **de apoyo** *f* TEC RAD subfleet

flotilla: ~ **de radiocomunicación móvil** *f* TEC RAD mobile radio fleet; ~ **de vehículos** *f* TEC RAD vehicle fleet

fluctuación *f* GEN fluctuation; ~ **de fase de reloj** ELECTRÓN clock jitter; ~ **de fase de tiempo de tránsito** ELECTRÓN transit-time jitter; ~ **de punto explorador** TV spot wobble; ~ **de temporización** TRANS timing jitter; ~ **de tensión** ELECTROTEC voltage fluctuation; ~ **de tiempo de tránsito** ELECTRÓN transit-time spread

fluctuar *vi* GEN fluctuate

flujo: ~ **de bitio** *m* DATOS bit stream; ~ **de datos** *m* DATOS data stream; ~ **descodificado** *m* DATOS decoded stream; ~ **electrónico** *m* ELECTRÓN, ELECTROTEC electron flow; ~ **energético** *m* ELECTRÓN, TRANS radiant flux; ~ **luminoso acodado** *m* ELECTRÓN knee-luminous flux; ~ **de potencia interferente** *m* TRANS interfering power flux; ~ **de trabajos** *m* INFO job stream

flujómetro *m* CONTROL, PRUEBA flow meter, fluxmeter

fluorescencia *f* TEC RAD, TRANS *óptica de fibras* fluorescence

FM *abr* INFO (*gerencia de informática*) FM (*facilities management*), TELEFON (*frecuencia media*) MF (*medium frequency*)

FMA *abr* (*análisis de modo de fallo*) PRUEBA FMA (*fault-mode analysis*)

FN *abr* (*notificación reenviada*) TELEFON FN (*forwarded notification*)

folleto: ~ **de instrucciones** *m* GEN instruction booklet

fondo *m* GEN background; ~ **común de interpolación** TRANS interpolation pool; ~ **invertido** TV inverted background; ~ **de pantalla** INFO screen background, display background

fonema *m* INFO speech sound

fonio *m* TELEFON phon

forma: ~ **de cableado** *f* COMPON wiring form; ~ **libre** *f* INFO free form; ~ **de montaje** *f* COMPON mode of mounting; ~ **normalizada** *f* INFO normalized form; ~ **de onda** *f* GEN waveform, wave shape; ~ **de onda activadora** *f* TEC RAD triggering waveform; ~ **de onda compuesta** *f* ELECTRÓN composite waveform; ~ **de onda de modulación por impulsos codificados** *f* INFO PCM waveform; ~ **de onda no sinusoidal** *f* INFO non-sine-wave waveform

formación: ~ **basada en computadoras** *AmL f* (*CBT*), (*cf formación basada en ordenadores Esp*) INFO computer-based training (*CBT*); ~ **basada en ordenadores** *Esp f* (*CBT*), (*cf formación basada en computadoras AmL*) INFO computer-based training (*CBT*); ~ **de cable** *f* COMPON cable forming; ~ **de reborde** *f* INFO lipping; ~ **de umbrales** *f* ELECTRÓN, ELECTROTEC thresholding

formante *m* TELEFON *acústica* formant

formateado *m* DATOS, INFO formatting

formato *m* GEN format; ~ **de cuadro** TV frame format; ~ **de documento** DATOS document format; ~ **de intercambio de documentos** DATOS document interchange format; ~ **de intercambio de documentos administrativos** DATOS office-document interchange format; ~ **de intercambio de ficheros JPEG** (*JFIF*) INFO JPEG-file interchange format (*JFIF*); ~ **de intercambio gráfico** INFO graphics interchange format; ~ **de interfaz de gráficos** (*GIF*) DATOS *Internet* graphics interchange format (*GIF*); ~ **intermedio común dividido en cuatro** (*QCIF*) INFO quarter common intermediate format (*QCIF*); ~ **internacional X.121** DATOS international X.121 format; ~ **internacional de numeración de datos** DATOS international data number format; ~ **libre** DATOS free format; ~ **no numerado** TELEFON unnumbered format; ~ **de papel** PERIF paper format; ~ **de paquete de datos** DATOS packet format; ~ **de publicación de referencia** (*RPF*) DATOS reference publication format (*RPF*); ~ **de punto de imagen** INFO pel aspect ratio; ~ **de texto muy elaborado** (*RTF*) DATOS, INFO rich text format (*RTF*); ~ **vertical** DATOS vertical format

Formato: ~ **de Fichero de Intercambio-Audio** *m* DATOS Audio-Interchange File Format; ~ **de Fichero para Intercambio de Datos Acústicos** *m* INFO *Internet* Audio-Interchange File Format

fórmula *f* INFO formula

formulario: ~ **continuo** *m* PERIF continuous form; ~ **impreso** *m* PERIF printed form

foro *m* DATOS *Internet* forum

forrado: ~ **de PVC** *adj* COMPON, ELECTRÓN PVC-jacketed

forro *m* COMPON sheath; ~ **en espiral** COMPON spiral wrap

forzar: ~ **la entrada** *vi* TELEFON break in
fosforescencia *f* ELECTRÓN, TV phosphorescence
foso *m* COMPON, INFO pit
fotoacoplador *m* ELECTRÓN photocoupler
fotoaislante *m* ELECTRÓN photoisolator
fotocátodo *m* ELECTRÓN photocathode
fotocomposición *f* TV photocomposition
fotoconductividad *f* ELECTRÓN, TRANS photoconductivity
fotodetector *m* TRANS photodetector; ~ **de diodos** TRANS diode photodetector
fotodiodo *m* COMPON, ELECTRÓN, TRANS photodiode; ~ **de avalancha** (*APD*) COMPON, ELECTRÓN, TRANS avalanche photodiode (*APD*)
fotoeléctrico *adj* ELECTRÓN photoelectric
fotoemisión *f* ELECTRÓN, TRANS photoemission
fotolitografía *f* INFO photolithography
fotoluminescencia *f* TRANS *óptica de fibras* photoluminescence
fotomáscara *f* ELECTRÓN photomask
fotómetro *m* TV exposure meter
fotomultiplicador *m* ELECTRÓN photomultiplier
fotón *m* TRANS photon
fotónica *f* TRANS photonics
fotorrealista *adj* INFO photorealistic
fototransistor *m* COMPON, TRANS phototransistor
fototubo: ~ **lleno de gas** *m* ELECTRÓN gas-filled phototube
fotovideotexto *m* DATOS photovideotext
FOV *abr* (*campo visual*) TEC RAD FOV (*field of view*)
FPU *abr* (*unidad de coma flotante*) INFO FPU (*floating-point unit*)
fracción: ~ **de empaquetado** *f* DATOS packing fraction; ~ **de muestreo** *f* TRANS sampling fraction
fractura *f* COMPON fracture
fragilidad *f* COMPON brittleness
fragmentación: ~ **de partículas** *f* INFO particle shredding
franjas: ~ **de interferencia** *f pl* TRANS *óptica de fibras* interference fringe pattern
frase *f* DATOS, INFO phrase
fraseo *m* DATOS, INFO phrasing
fraude: ~ **informático** *m* INFO computer fraud
frecuencia *f* GEN frequency; ~ **acústica** GEN voice frequency (*VF*); ~ **asignada** TEC RAD assigned frequency; ~ **con barrido** TEC RAD swept frequency; ~ **de barrido horizontal** TV horizontal-sweep frequency; ~ **de basculación** CONMUT switching rate; ~ **de base** ELECTRÓN, INFO base frequency, basic frequency; ~ **de batido** ELECTRÓN, TRANS beat frequency; ~ **de captación** ELECTRÓN capture frequency; ~ **característica** TEC RAD characteristic frequency; ~ **de cierre** ELECTRÓN cutoff frequency; ~ **compartida** TEC RAD shared frequency; ~ **contrastada** TEC RAD standard frequency; ~ **de corte** TEC RAD quench frequency; ~ **de corte resistiva** ELECTRÓN resistive cutoff frequency; ~ **crítica** ELECTRÓN, TRANS critical frequency, threshold frequency; ~ **crítica de banda** TRANS critical band rate; ~ **designada** TRANS designated frequency; ~ **de destellos** ELECTROTEC flashing rate; ~ **de exploración** TV dot frequency; ~ **de imágenes** INFO picture frequency, TEC RAD image frequency, TV field frequency, picture frequency, frame frequency, frame rate; ~ **infraacústica** GEN subaudio frequency; ~ **infratelefónica** TELEFON subtelephone frequency; ~ **intercalada** TEC RAD interspersed frequency; ~ **lateral** TEC RAD side frequency; ~ **lenta** ELECTRÓN idler frequency; ~ **de línea** INFO, TV line frequency; ~ **de línea de exploración** TV scanning-line frequency; ~ **de línea horizontal** TV horizontal-line frequency; ~ **de marcha continua** ELECTRÓN free-running frequency; ~ **más alta** TEC RAD higher frequency; ~ **máxima utilizable** (*MUF*) TEC RAD maximum usable frequency (*MUF*); ~ **mínima utilizable** TEC RAD lowest usable frequency (*LUF*); ~ **de modulación** TELEFON, TRANS modulating frequency, modulation frequency; ~ **de muestreo** TRANS sampling frequency, sampling rate; **frecuencia** ~ GEN multiple frequency; ~ **muy alta** (*VHF*) TRANS very high frequency (*VHF*); ~ **nominal** TRANS centre (*AmE* center) frequency; ~ **normalizada** TEC RAD, TRANS normalized frequency; ~ **de ondas portadoras** TRANS carrier frequency; ~ **óptima de trabajo** (*OWF*) TEC RAD optimum working frequency (*OWF*); ~ **óptima de tráfico** (*OTF*) TEC RAD optimum traffic frequency (*OTF*); ~ **de oscilación** ELECTRÓN oscillation frequency, oscillator frequency; ~ **de oscilación libre** ELECTRÓN free-running oscillator frequency; ~ **de palabras** DATOS word rate; ~ **piloto** CONTROL control frequency; ~ **de portadora de sonido** TV sound-carrier frequency; ~ **de red** ELECTROTEC mains frequency; ~ **de referencia** TEC RAD, TRANS reference frequency; ~ **de repetición** TELEFON repetition rate; ~ **de repetición de impulsos** ELECTRÓN, SEÑAL, TELEFON, TRANS pulse-repetition frequency (*PRF*), pulse-repetition rate (*PRR*); ~

resonante ELECTRÓN, TEC RAD, TELEFON resonant frequency, resonance frequency; ~ **de subportadora** TEC RAD, TELEFON, TRANS subcarrier frequency; ~ **de trabajo** TEC RAD operating frequency; ~ **ultraacústica** GEN super-audio frequency; ~ **ultraalta** (*UHF*) TV ultrahigh frequency (*UHF*); ~ **ultratelefónica** TELEFON, TRANS super-telephone frequency

frecuencímetro *m* PRUEBA frequency meter

frente: ~ **de componentes** *m* COMPON, ELECTRÓN *circuitos impresos* component front; ~ **inclinado** *m* COMPON, ELECTRÓN tilted front; ~ **de onda** *m* GEN wavefront

frisa: ~ **de guíaondas** *f* COMPON, TEC RAD, TRANS waveguide shim

FRMR *abr* (*cuadro de rechazo de trama*) TELEFON FRMR (*frame reject frame*)

FSO *abr* (*función del sistema operativo*) PRUEBA, TELEFON OSF (*operating system function*)

FT *abr* (*transformación de Fourier*) PRUEBA FT (*Fourier transformation*)

FTAM *abr* (*acceso y gestión de transferencia de archivos*) DATOS, INFO FTAM (*file transfer access and management*)

FTIR *abr* (*espectroscopía infrarroja transformada de Fourier*) PRUEBA FTIR (*Fourier transform infrared spectrography*)

FTP *abr* (*protocolo de transferencia de archivos*) DATOS *Internet* FTP (*file transfer protocol*)

fuente *f* GEN source; ~ **de alimentación** ELECTROTEC power supply; ~ **de alimentación de batería** ELECTROTEC battery power supply; ~ **de alimentación de CA** ELECTROTEC alternating-current power supply; ~ **de alimentación de emergencia** ELECTROTEC emergency power supply; ~ **de corriente alterna** ELECTROTEC alternating-current power supply; ~ **DCE** TELEFON DCE source; ~ **de energía** ELECTROTEC power source; ~ **de energía de reserva** ELECTROTEC backup power source; ~ **interferente** TEC RAD interfering source; ~ **de luz** INFO, TV light source; ~ **de luz direccional** INFO directional light source; ~ **de luz distante** INFO distant light source; ~ **remota** DATOS remote resource; ~ **de ruidos** ELECTROTEC noise source; ~ **de señales de sincronismo de multiplexor** TRANS multiplexer timing source; ~ **de sonido ululado incorporado** ELECTRÓN built-in warble-tone source; ~ **sonora** INFO, TELEFON sound source; ~ **de tipos** DATOS type font

fuera[1]: ~ **de alcance** *adj* TEC RAD out of range, out-of-reach; ~ **de banda** *adj* (*OOB*) ELECTRÓN, TEC RAD out of band (*OOB*); ~ **de haz** *adj* TEC RAD off-beam; ~ **de servicio** *adj* APLIC out of service, out of order, out of action

fuera[2]: ~ **de línea** *adv* DATOS, INFO, TELEFON offline

fuerza: ~ **de arranque** *f* ELECTRÓN pull-off strength; ~ **ciclomotora** *f* (*cmf*) TEC RAD cyclomotive force (*cmf*); ~ **electromotriz** *f* (*EMF*) ELECTRÓN, ELECTROTEC electromotive force (*EMF*); ~ **de expansión** *f* GEN *física* expansion force; ~ **de pelado** *f* COMPON stripping force; ~ **de pulsión de tecla** *f* INFO key-touch force; ~ **de tracción** *f* COMPON, PRUEBA pulling force, tensile force

fuga *f* ELECTRÓN, ELECTROTEC leakage, INFO leak

función *f* GEN function; ~ **de alarma** PRUEBA alarm function; ~ **de aplicación** DATOS application function; ~ **de autentificación** DATOS authentication function; ~ **de canal virtual relacionada con la conexión** (*VCCRF*) DATOS, TRANS virtual-channel connection-related function (*VCCRF*); ~ **de captación** TELEFON pick-up facility; ~ **colorimétrica** INFO colorimetric function; ~ **de comunicación** DATOS communication capability; ~ **de comunicación de mensajes** DATOS message-communication function (*MCF*); ~ **conexión-desconexión** PERIF on-off function; ~ **de conmutación de acceso al servicio** DATOS, CONMUT, TELEFON service-access switch function; ~ **de conteo** ELECTRÓN counting function; ~ **de contestador automático** PERIF, TELEFON answer-machine function; ~ **de control** (*CF*) CONTROL, DATOS, INFO, TELEFON control function (*CF*); ~ **de control de asociación múltiple** (*MACF*) DATOS multiple-association control function (*MACF*); ~ **de control de asociación simple** DATOS single-association control function; ~ **de control de presentación** TELEFON presentation control function; ~ **de convergencia dependiente de red secundaria** DATOS subnetwork-dependent convergence function (*SNDCF*); ~ **de coordinación específica de servicio** DATOS service-specific coordination function; ~ **de corrección** PRUEBA correction function; ~ **de creación de códigos de comprobación** DATOS hash function; ~ **DCME** TELEFON DCME function; ~ **de desplazamiento** INFO shift function; ~ **de densidad de probabilidad** (*PDF*) DATOS probability density function (*PDF*); ~ **despertador** PERIF *de contestador* snooze function; ~ **de Dirac** TELEFON *impulsos* Dirac function; ~ **de directorio N** DATOS Ndirectory function; ~ **de disponibilidad** PRUEBA availability function; ~ **de distribución** DATOS, PRUEBA *estadística* distribution function; ~ **de distribución acumulativa** (*cdf*) DATOS cumulative distribution function (*cdf*); ~ **de distribución de tiempo de vida** PRUEBA life-length distribution function; ~ **de elemento de red** TRANS network element function (*NERF*); ~ **de encaminamiento y de relevo** CONMUT relaying and routing function; ~ **de enlace de señalización** SEÑAL signalling-link (*AmE* signaling-link) function; ~ **interoperación de facsímil** TELEFON facsimile interworking function; ~ **escalón unitario** TELEFON unit step; ~ **de aplicación de gestión** DATOS management application function (*MAF*); ~ **de gestión de red** TRANS network management function; ~ **de gestión de sistemas** INFO systems management function; ~ **de interoperación** TELEFON interworking function; ~ **de interoperación de paquetes de télex** (*TPIWF*) DATOS, TELEFON telex-packet interworking function (*TPIWF*); ~ **inversa** GEN inverse function; ~ **de conmutación** CONMUT, TELEFON switching function; ~ **de mantenibilidad** PRUEBA maintainability function; ~ **de mantenimiento** PRUEBA maintenance function; ~ **de mediación** INFO mediation function (*MF*); ~ **N** DATOS N function; ~ **de nivel superior** TRANS higher-level function (*HLF*); ~ **de orden** DATOS command function; ~ **de paridad** TELEFON parity function; ~ **por pasos** TELEFON step function; ~ **potencial** ELECTROTEC potential function; ~ **primitiva de indicación** DATOS indication primitive; ~ **de transferencia de modulación** ELECTRÓN modulation transfer function; ~ **en rampa** TELEFON ramp function; ~ **de registrador** TELEFON, TRANS register function; ~ **relacionada con la conexión** DATOS, TRANS connection-related function (*CRF*); ~ **de renovación** PRUEBA renewal function; ~ **de reserva** ELECTROTEC backup function; ~ **de secuencia** CONMUT sequence function; ~ **de seguimiento** TEC RAD tracing function; ~ **de servicio** DATOS, INFO housekeeping function; ~ **de sintonización** TEC RAD tuning function; ~ **de sistema operativo** (*FSO*) PRUEBA, TELEFON operating-system function (*OSF*); ~ **de trabajo** GEN work function; ~ **de tránsito relacionada con la conexión** (*TCRF*) DATOS, TRANS transit connection-related function (*TCRF*); ~ **de trayectoria virtual relacionada con la conexión** (*VPCRF*) DATOS, TRANS virtual-path connection-related function (*VPCRF*); ~ **de umbral** ELECTRÓN threshold function

funcionalidad *f* GEN functionality; ~ **fiable** PRUEBA trusted functionality; ~ **de subred** DATOS subnetwork functionality

funcionamiento *m* GEN performance, operation; ~ **con ambas corrientes** TELEFON double-current working; ~ **en banda limitada** TRANS bandwidth-limited operation; ~ **en bucle cerrado** CONTROL closed-loop control; ~ **cíclico** PRUEBA cycling; ~ **en circuito cerrado** TRANS closed-circuit working; ~ **en contrafase** ELECTRÓN, ELECTROTEC push-pull operation; ~ **de lóbulo lateral** TEC RAD side-lobe performance; ~ **de módem** PERIF modem performance; ~ **en modo de agotamiento** COMPON *semiconductores* depletion-mode operation; ~ **en modo de enriquecimiento** COMPON *semiconductores* enhancement-mode operation; ~ **de polarización cruzada** TRANS cross-polar performance; ~ **sin carga** GEN no-load operation

funcionar *vi* GEN operate

funciones: ~ **de gestión de enlace de señalización** *f pl* SEÑAL signalling-link (*AmE* signaling-link) management functions; ~ **de gestión de red de señalización** *f pl* SEÑAL signalling-network (*AmE* signaling-network) management functions; ~ **de gestión de ruta de señalización** *f pl* SEÑAL signalling-route (*AmE* signaling-route) management functions; ~ **de gestión de tráfico de señalización** *f pl* SEÑAL signalling-traffic (*AmE* signaling-traffic) management functions; ~ **de manipulación de mensaje de señalización** *f pl* SEÑAL signalling-message (*AmE* signaling-message) handling functions; ~ **de red de señalización** *f pl* SEÑAL signalling-network (*AmE* signaling-network) functions

fundente *m* COMPON flux

fundición *f* ELECTRÓN, ELECTROTEC blowing

fundido: ~ **encadenado** *m* TV crossfade

fundirse *v refl* ELECTRÓN, ELECTROTEC blow

fusible *m* COMPON, ELECTRÓN, ELECTROTEC fuse; ~ **de alambre** COMPON, ELECTROTEC wire fuse; ~ **de alimentación** COMPON, ELECTROTEC feeding fuse; ~ **de aparato** COMPON, ELECTROTEC equipment fuse; ~ **de batería** COMPON, ELECTROTEC battery fuse; ~ **de cartucho** COMPON, ELECTROTEC cartridge fuse; ~ **de distribución** COMPON distribution fuse; ~ **de estante** COMPON, ELECTROTEC shelf fuse; ~ **de fila de bastidores** COMPON, ELECTROTEC suite fuse; ~ **de filtro** COMPON, ELECTROTEC filter fuse; ~ **fundido** COMPON, ELECTROTEC blown fuse; ~ **de fusión cerrada** ELECTROTEC enclosed fuse-link; ~ **de hilo delgado** COMPON, ELECTROTEC fine-wire fuse; ~ **de línea de fuerza** COMPON, ELECTROTEC power fuse; ~ **miniatura** COMPON, ELECTROTEC miniature fuse; ~ **de potencia** COMPON, ELECTROTEC power fuse; ~ **principal** COMPON, ELECTROTEC main fuse; ~ **de red** COMPON, ELECTROTEC mains fuse; ~ **de repetidor** ELECTROTEC repeater fuse; ~ **en tubo de vidrio** COMPON, ELECTRÓN, ELECTROTEC glass-tube fuse; ~ **tubular** COMPON, ELECTRÓN, ELECTROTEC tube fuse

fusión *f* ELECTRÓN *circuitos impresos* fusing, INFO merging, TRANS *óptica de fibras* fusing; ~ **térmica** ELECTRÓN *circuitos impresos* flow melting

fusionar *vt* ELECTRÓN fuse, INFO merge, TRANS fuse

G

GaAs *abr* (*arseniuro de galio*) COMPON GaAs (*gallium arsenide*)

gabinete: **~ de subdistribución** *m* TRANS subdistribution cabinet; **~ de telecomunicaciones** *m* ELECTRÓN telecommunications closet (*TC*)

galería: **~ de cables** *f* TRANS cable vault, cable cellar

galvanómetro *m* PRUEBA galvanometer

gama *f* INFO gamut, PRUEBA scope, range, TV gamut; **~ de audibilidad** TELEFON audibility range; **~ azimutal** TEC RAD range of bearings; **~ de frecuencias** TEC RAD, TRANS frequency range; **~ de modos** ELECTRÓN moding; **~ de señales de sincronización** TEC RAD *oscilador* lock-in range

gamma *f* ELECTRÓN, TV gamma

ganancia *f* GEN gain; **~ de antena** TEC RAD aerial gain, antenna gain; **~ de antena de transmisión por vía satélite** TEC RAD satellite-transmit aerial gain, satellite-transmit antenna gain; **~ de borde de cobertura** TEC RAD edge-of-coverage gain; **~ de centro de haz** TEC RAD beam centre (*AmE* center) gain; **~ compuesta** TRANS composite gain; **~ directiva** TEC RAD directive gain; **~ efectiva** TRANS effective gain; **~ en el eje** TEC RAD on-axis gain; **~ impeditiva** TEC RAD obstacle gain; **~ de inserción** TRANS insertion gain; **~ de interpolación** TRANS interpolation gain (*IG*); **~ logarítmica** CONTROL logarithmic gain; **~ óptima** ELECTRÓN optimum gain; **~ de potencia** ELECTRÓN power gain; **~ de potencia de antena** TEC RAD aerial power gain, antenna power gain; **~ de potencia incremental** ELECTRÓN incremental power gain; **~ de recepción** TEC RAD receive gain; **~ por reflexión** TRANS reflection gain; **~ de saturación** ELECTRÓN saturation gain; **~ para señal débil** ELECTRÓN small-signal gain; **~ de tensión** ELECTROTEC voltage gain; **~ de transcodificación** ELECTRÓN transcoding gain; **~ de transductor** TELEFON transducer gain; **~ unidad** ELECTRÓN unity gain; **~ unitaria** ELECTRÓN unit gain

gancho *m* TELEFON hook; **~ conmutador** PERIF, TEC RAD switch hook

garantía: **sin ~** *fra* INFO unsecured

garra *f* COMPON *para argolla, anilla, asa de cajas* dog

Gb *abr* (*gigabitio*) DATOS, INFO Gb (*gigabyte*)

gc *abr* (*gigaciclo*) PRUEBA gc (*gigacycle*)

GDI *abr* (*interfaz de estación de presentación visual*) INFO GDI (*graphic device interface*)

gel *m* COMPON gel

gemelos *m pl* INFO twins

generación *f* GEN generation; **~ de caracteres** INFO character generation; **~ de dirección** INFO address generation; **~ de impulsos de cómputo** TELEFON meter-pulse generation; **~ de onda piloto de grupo** TRANS group-pilot generation; **~ de onda de referencia** TEC RAD, TRANS reference-pilot generation; **~ de ondas** ELECTRÓN wave generation; **~ de palabra** INFO, TEC RAD, TELEFON speech generation; **~ de pares** COMPON pair generation; **~ de señal de llamada** TELEFON tone ringing; **~ de sistema** INFO system generation; **~ de tabla** GEN table generation; **~ de tonos** ELECTRÓN, INFO tone generation

generador *m* GEN generator; **~ de aplicación** DATOS application generator; **~ de CA** ELECTROTEC AC generator; **~ de caracteres** INFO character generator; **~ de CC** ELECTROTEC DC generator; **~ de código** TELEFON code generator; **~ de compiladores** INFO compiler generator; **~ de curvas** INFO curve generator; **~ en diente de sierra** ELECTRÓN sawtooth generator; **~ de efectos especiales** INFO special-effects generator; **~ de frecuencia** ELECTRÓN, ELECTROTEC, TELEFON frequency generator; **~ de frecuencia inestable** ELECTROTEC sliding-frequency generator; **~ de frecuencia de línea** TV line-frequency generator; **~ de función** INFO function generator; **~ Hall** ELECTRÓN *semiconductores* Hall generator; **~ de imágenes de microondas con sensor especial** (*SSMI*) ELECTRÓN special-sensing microwave imager (*SSMI*); **~ de imagen de prueba** TV pattern generator; **~ de impulsos** ELECTRÓN, INFO, TELEFON, TRANS pulser, pulse generator; **~ de impulsos de apertura** TELEFON break-pulsing generator; **~ de impulsos de cómputo** TELEFON meter-pulse generator; **~ de impulsos rectangulares** ELECTRÓN square-pulse generator; ~ **de Lambert** ELECTRÓN, TRANS *óptica de fibras* Lambertian source; ~ **lambertiano** ELECTRÓN, TRANS *óptica de fibras* Lambertian source; ~ **de medición de tiempo** TRANS timing generator; ~ **de MF** TELEFON MF generator; ~ **de modulación** TRANS modulation generator; ~ **multifrecuencia** SEÑAL, TELEFON multifrequency generator; ~ **de onda cuadrada** ELECTRÓN, SEÑAL square-wave generator; ~ **de onda envolvente** INFO envelope generator; ~ **de ondas** ELECTRÓN wave generator; ~ **de portadoras** TRANS carrier supply; ~ **de programas** INFO program generator; ~ **de rayos X** ELECTRÓN X-ray source; ~ **de ruido** INFO, TRANS noise generator, noise-generating device; ~ **de señales** PRUEBA, SEÑAL signal generator; ~ **de señal de interrogación** (*ISG*) TEC RAD interrogation-signal generator (*ISG*); ~ **de señal de llamada** TELEFON ringing generator; ~ **de tarifa** TELEFON tariff generator; ~ **de temporización de re~** (*RTG*) DATOS regenerator-timing generator (*RTG*); ~ **de tensión de llamada** TELEFON ringing-voltage generator; ~ **de tensión de referencia** ELECTROTEC reference-voltage generator; ~ **de tono** ELECTRÓN, INFO tone generator; ~ **de tono de llamada** TELEFON ring-tone generator; ~ **de tono de tictac** TELEFON tick-tone generator; ~ **de tramas** TRANS frame generator

geoestacionario *adj* TEC RAD geostationary

geometría: ~ **constructiva tridimensional** *f* (*CSG*) INFO constructive solid geometry (*CSG*); ~ **de fibras** *f* TRANS fibre (*AmE* fiber) geometry

geosincrónico *adj* TEC RAD geosynchronous

gerencia: ~ **de informática** *f* (*FM*) INFO facilities management (*FM*)

gestión *f* GEN management; ~ **asistida por computadora** *AmL* (*cf gestión asistida por ordenador Esp*) INFO computer-aided management, computer-assisted management; ~ **asistida por ordenador** *Esp* (*cf gestión asistida por computadora AmL*) INFO computer-aided management, computer-assisted management; ~ **de calidad total** (*TQM*) PRUEBA total-quality management (*TQM*); ~ **de crisis** PRUEBA crisis management; ~ **de datos** DATOS, INFO data management (*DM*); ~ **de emplazamiento de posición** TEC RAD location management; ~ **de frecuencia** TEC RAD frequency management; ~ **informatizada** INFO computer-aided management, computer-assisted management; ~ **de instalaciones** APLIC facilities management (*FM*); ~ **de memoria** DATOS memory management; ~ **en niveles** DATOS layer management; ~ **de objeto** INFO object management; ~ **periférica** PERIF peripheral management; ~ **de red** TELEFON network management; ~ **de red internacional** TELEFON international network management; ~ **remota de documentos** TELEFON remote document management; ~ **de riesgos** GEN risk management; ~ **de sistema** INFO system management; ~ **de sistema de comunicaciones** APLIC communications-system management; ~ **de sistemas** APLIC systems management; ~ **y supervisión de red** PRUEBA network supervision and management; ~ **por teclado** TELEFON key management

gestor: ~ **de archivo de red** *m* (*NFM*) INFO network file manager (*NFM*); ~ **de base de datos** *m* DATOS, INFO database manager; ~ **de canal privado MCS** *m* TELEFON MCS private channel manager; ~ **de compresión de imagen** *m* DATOS, TV image-compression manager; ~ **de comunicaciones de usuario** *m* (*UCM*) INFO user-communications manager (*UCM*); ~ **de impresión** *m* INFO printer driver; ~ **de línea de fax** *m* TELEFON facsimile-line manager; ~ **de memoria** *m* INFO memory manager; ~ **de red de área local** *m* DATOS LAN manager; ~ **de servidor terminal** *m* (*TSM*) DATOS terminal-server manager (*TSM*); ~ **de tipos** *m* INFO font manager

GFSK *abr* (*tecleo de inversión de frecuencia con filtro gaussiano*) SEÑAL GFSK (*Gaussian-filtered frequency-shift keying*)

GHz *abr* (*gigahertzio*) GEN GHz (*gigahertz*)

GI *abr* (*identificador de grupo*) TELEFON GI (*group identifier*)

GIF *abr* (*formato de interfaz de gráficos*) DATOS *Internet* GIF (*graphics interchange format*)

gigabit *m* DATOS, INFO gigabit

gigabitio *m* (*Gb*) DATOS, INFO gigabyte (*Gb*)

gigaciclo *m* (*gc*) PRUEBA gigacycle (*gc*)

gigahertzio *m* (*GHz*) GEN gigahertz (*GHz*)

girador *m* TEC RAD gyrator

girar[1] *vt* INFO rotate, spin

girar[2] *vi* INFO rotate, spin

giratorio *adj* GEN revolving

giro: ~ **cablegráfico** *m* TELEFON cable transfer

girofrecuencia *f* TEC RAD gyrofrequency; ~ **transversal** TEC RAD transverse gyrofrequency

girotrón *m* ELECTRÓN gyrotron

GL *abr* (*longitud de grupo*) TELEFON GL (*group length*)

GMT *abr* (*hora media de Greenwich*) GEN GMT (*Greenwich Mean Time*)
GO *abr* (*óptica geométrica*) TRANS GO (*geometric optics*)
gobernabilidad *f* TEC RAD steerability
gobernable *adj* TEC RAD steerable
goniómetro: ~ **de rayos catódicos** *m* TEC RAD cathode-ray direction finder
gopher *m* DATOS, INFO gopher
GPIB *abr* (*conductor común de acoplamiento mutuo universal*) DATOS GPIB (*general-purpose interface bus*)
GPS *abr* (*sistema de localización global*) TEC RAD GPS (*global positioning system*)
grabación *f* GEN recording; ~ **de datos** DATOS data recording; ~ **por modulación de fase** DATOS phase-modulation recording; ~ **por retorno a cero** INFO return-to-zero recording; ~ **con retorno a un estado de referencia** INFO return-to-reference recording; ~ **de voz** INFO, TELEFON voice print
grabado *m* INFO writing
grabador: ~ **de cinta** *m* TELEFON tape recorder; ~ **de mesa** *m* TEC RAD *móvil* table cassette; ~ **de un solo canal** *m* PERIF single-channel recorder; ~ **de videocasete** *m* TV videocassette recorder (*VCR*)
grabadora: ~ **de cinta de vídeo** *f* TV videotape recorder (*VTR*); ~ **de datos de casete** *f* (*DCR*) PERIF data-cassette recorder (*DCR*)
grabar *vt* GEN record, etch
gradaciones: ~ **intermedias** *f pl* TELEFON half-toning
gradiente: ~ **de gravedad** *m* TEC RAD gravity gradient; ~ **normal de vertical de refractividad** *m* TEC RAD standard refractivity vertical gradient; ~ **de potencial** *m* ELECTROTEC potential gradient
grado: ~ **aceptado de distorsión** *m* TELEFON conventional degree of distortion; ~ **de bloqueo de servicio** *m* TELEFON blocking grade of service; ~ **de coherencia** *m* TRANS degree of coherence, coherence degree; ~ **de distorsión arranque-parada** *m* TELEFON degree of start-stop distortion; ~ **de distorsión arranque-parada bruta** *m* TELEFON degree of gross start-stop distortion; ~ **de distorsión arranque-parada sincrónica** *m* TELEFON degree of synchronous start-stop distortion; ~ **de distorsión inherente** *m* TELEFON degree of inherent distortion; ~ **de distorsión isócrona** *m* TELEFON degree of isochronous distortion; ~ **de distorsión paralela anisócrona prematura** *m* TELEFON degree of early anisochronous parallel distortion; ~ **de distorsión paralela anisócrona tardía** *m* TELEFON degree of late anisochronous parallel distortion; ~ **de distorsión de prueba normalizada** *m* TELEFON degree of standardized-test distortion; ~ **de distorsión en el servicio** *m* TELEFON degree of distortion in service; ~ **de error binario** *m* INFO binary error rate; ~ **de libertad** *m* INFO degree of freedom; ~ **de luminosidad** *m* DATOS *digitalizador* nuance of lightness; ~ **de modulación** *m* TRANS degree of modulation; ~ **de oscuridad** *m* DATOS *de digitalizador* nuance of darkness; ~ **de separación entre filas** *m* DATOS row pitch; ~ **de servicio** *m* PRUEBA grade of service
graduación *f* PRUEBA graduation; ~ **de escala** PRUEBA scale mark; ~ **por teclado** INFO keyboard scaling
graduar *vt* PRUEBA graduate
gráfica *f* GEN chart
gráfico *m* GEN graph; ~ **de barras** INFO *estadísticas* bar graph; ~ **de dispersión** INFO scatter graph; ~ **de fase** ELECTRÓN phase plot; ~ **de recorridos** PRUEBA range chart; ~ **de sectores** INFO *estadísticas* pie chart; ~ **de valores acumulados** PRUEBA *calidad* cumulative-sum chart; ~ **Y-O** DATOS AND-OR graph
gráficos: ~ **de computadora** *AmL m pl* (*cf gráficos de ordenador Esp*) INFO, TV computer graphics; ~ **de imágenes** *m pl* TV image graphics; ~ **de ordenador** *Esp m pl* (*cf gráficos de computadora AmL*) INFO, TV computer graphics
gramática: ~ **concreta** *f* DATOS concrete grammar; ~ **gráfica concreta** *f* DATOS concrete graphical grammar
gran: ~ **computadora** *AmL f* (*cf gran ordenador Esp*) INFO mainframe, mainframe computer; ~ **ordenador** *Esp m* (*cf gran computadora AmL*) INFO mainframe, mainframe computer
granularidad *f* TV granularity
granulometría *f* TELEFON grading
gránulo *m* TV granule; ~ **de carbón** COMPON *micrófonos* carbon granule
grapa: ~ **guardacable** *f* COMPON pull-relief clamp; ~ **de sujeción** *f* COMPON *tendido de cables* anchoring bracket; ~ **de sujeción de envoltura** *f* COMPON sheath-retention clamp
grieta *f* COMPON, PRUEBA crack; ~ **por tensiones** COMPON stress crack
grupo *m* GEN group, cluster; ~ **adaptador de señalización** SEÑAL signalling (*AmE* signaling) adaptor group; ~ **alterno** INFO alternate group; ~ **alterno de líneas troncales finales** CONMUT, TELEFON alternate final-trunk group; ~ **de atributos** INFO attribute group; ~ **de atributos**

mutuamente excluyentes INFO mutually-exclusive attribute group; ~ **básico** TELEFON, TRANS basic group; ~ **bidireccional** TRANS both-way group; ~ **de cables** GEN cable group; ~ **cerrado de abonados** TELEFON closed subscriber group; ~ **cerrado de usuarios** TELEFON closed user group (*CUG*); ~ **cerrado de usuarios con acceso de salida** TELEFON closed user group with outgoing access; ~ **cerrado de usuarios de videotexto** INFO, TV videotext closed user group; ~ **de circuitos** CONMUT, ELECTRÓN, TELEFON circuit group; ~ **de circuitos aleatorios equivalente** TELEFON equivalent random-circuit group; ~ **de circuitos preferente** CONMUT, ELECTRÓN, TELEFON first-choice circuit group; ~ **de circuitos totalmente abastecidos** TELEFON fully-provided circuit group; ~ **de circuitos de transferencia** TELEFON interposition trunk group; ~ **conectable** COMPON slide-in unit; ~ **de conexión directa** TRANS through-group; ~ **de contacto de ruptura** CONMUT break-contact group; ~ **cuaternario** TELEFON supermastergroup; ~ **de datos** DATOS data group; ~ **por defecto** DATOS default group; ~ **detectable** INFO detectable group; ~ **doble** GEN twin group; ~ **electrógeno de reserva** ELECTROTEC stand-by power plant; ~ **de emisión doble** DATOS multicast group; ~ **enésimo de elección** TELEFON nth-choice group; ~ **de enlace de llamada de anotadora y de salida** TELEFON recording-completing trunk group; ~ **de enlace de señalización** SEÑAL signalling-link (*AmE* signaling-link) group; ~ **de enlaces de matrices** TELEFON group of matrix links; ~ **de escobillas de contacto** CONMUT wiper set; ~ **final de circuitos** TELEFON final circuit group; ~ **de gestión de objeto** (*OMG*) GEN object-management group (*OMG*); ~ **invocado** DATOS invoked group; ~ **de líneas** CONMUT, TRANS line group; ~ **de líneas interurbanas de intercambio** TELEFON interchange trunk group; ~ **de líneas troncales de emergencia** TELEFON emergency trunk group; ~ **de líneas troncales entre tándems** TELEFON intertandem trunk group; ~ **de líneas troncales finales de ruta alterna** CONMUT, TELEFON alternate-route final trunk group; ~ **de líneas troncales de una misma central** TELEFON intraoffice trunk group *AmE*, interchange trunk group *BrE*; ~ **de líneas troncales parlantes** TELEFON announcement trunk group; ~ **de líneas troncales primarias de larga distancia** TELEFON intertoll primary trunk group; ~ **de líneas troncales de servicios auxiliares** TELEFON auxiliary-services trunk group; ~ **de muelles de contacto de dos pasos** COMPON two-stage contact spring set, two-step contact spring set; ~ **de muelles de contacto horizontal** COMPON selecting-bar spring set; ~ **de muelles de funcionamiento secuencial** COMPON sequence spring set; ~ **PBX** TELEFON PBX group; ~ **preferente** CONMUT, ELECTRÓN, TELEFON first-choice group; ~ **de relés** COMPON, CONTROL, TELEFON relay set; ~ **de reserva** GEN stand-by set; ~ **secundario** TELEFON, TRANS supergroup; ~ **secundario básico** TELEFON, TRANS basic supergroup; ~ **secundario de conexión directa** TRANS through supergroup; ~ **terciario** TELEFON, TRANS mastergroup; ~ **terciario de conexión directa** TRANS through mastergroup; ~ **de tonos y auxiliares** TELEFON tones and auxiliaries group; ~ **troncal de larga distancia** TELEFON intertoll trunk group; ~ **de unidad administrativa** CONTROL administrative unit group; ~ **de visualización** INFO display group

GSM *abr* (*sistema mundial para comunicaciones móviles*) TEC RAD GSM (*global system for mobile communications*)

GTS *abr* (*sistema mundial de telecomunicaciones*) GEN GTS (*global telecommunications system*)

guardapolvos *m* COMPON dust cover

guardar *vt* INFO, TELEFON save

guarnición *f* COMPON brake lining; ~ **de caucho** COMPON rubber gasket; ~ **de conducto** COMPON, ELECTRÓN raceway cover

GUI *abr* (*interfaz de usuario gráfica*) INFO GUI (*graphical user interface*)

guía *f* GEN guide; ~ **administrativa** DATOS office directory; ~ **de conversación** TELEFON speaking guide; ~ **de deslizamiento** COMPON, INFO, TEC RAD, TELEFON, TRANS running bar; ~ **de luz** ELECTRÓN, TRANS lightguide; ~ **orbital** TEC RAD orbital guidance; ~ **de referencia** INFO reference guide; ~ **rítmica** INFO rhythmic guide; ~ **de rodillo** COMPON platen guide; ~ **tridimensional** TRANS three-dimensional guide; ~ **de usuario** GEN user guide, user's guide

guíahilos *m* COMPON wire guide, jumper-wire guide

guíaondas *m* ELECTRÓN, TEC RAD, TRANS waveguide; ~ **en espiral** TRANS spiral waveguide; ~ **de haz** TEC RAD beam waveguide; ~ **no recíproca** *f* TEC RAD, TRANS nonreciprocal waveguide; ~ **ondulado interiormente** *f* TEC RAD, TRANS ridge waveguide; ~ **de película delgada** TRANS thin-film waveguide; ~ **progresiva** *f* ELECTRÓN, TEC RAD, TRANS travelling (*AmE* traveling) waveguide; ~ **ranurado** TEC RAD, TRANS slotted waveguide; ~ **tabicado** TEC RAD, TRANS septate waveguide; ~ **unifilar** TEC RAD single-wire transmission line

guión *m* INFO dash, hyphen

GVD *abr* (*dispersión de velocidad de grupo*) TRANS GVD (*group velocity dispersion*)

GWD *abr* (*desmultiplexor de onda guiada*) TRANS GWD (*guided-wave demultiplexer*)

H

H *abr* (*henrio*) ELECTRÓN H (*henry*)
habla *f* TEC RAD, TELEFON speech
hacer: ~ **una llamada a cobro revertido** *vt* APLIC, TELEFON call collect *AmE*, reverse the charges *BrE*
halo *m* TV halo, halation
halógenos: **sin** ~ *fra* COMPON halogen-free
hardware *m* INFO hardware
hartleyo *m* ELECTRÓN Hartley
haz *m* ELECTRÓN, ELECTROTEC, TEC RAD, TRANS, TV beam; ~ **en abanico** ELECTROTEC fan-shaped beam; ~ **de circuitos de ruta única** TELEFON only-route circuit group; ~ **direccional** TEC RAD directional beam; ~ **dirigido** TV spot beam; ~ **de electrones** ELECTRÓN electron beam; ~ **de enlace descendente** TEC RAD down-link beam; ~ **de fibras** TRANS fibre (*AmE* fiber) bundle; ~ **en forma de paleta** TEC RAD paddle-shaped beam; ~ **gaussiano** TRANS Gaussian beam; ~ **global** TEC RAD global beam; ~ **de guíahilos** COMPON wire-guide beam; ~ **de láser** ELECTRÓN laser beam; ~ **localizado** TEC RAD spot beam; ~ **luminoso** ELECTRÓN light beam; ~ **molecular** TV molecular beam; ~ **en pincel** TEC RAD pencil beam
HBT *abr* (*transistor bipolar de heterounión*) ELECTRÓN, COMPON HBT (*heterojunction bipolar transistor*)
HDBC *abr* (*cable de construcción para grandes amperajes*) ELECTRÓN HDBC (*heavy duty building cable*)
HDL *abr* (*lenguaje de descripción del soporte físico*) INFO HDL (*hardware description language*)
HDR *abr* (*cabezal de tubos*) DATOS, INFO HDR (*header*)
HDTM *abr* (*módulo de transmisión en semiduplex*) TRANS HDTM (*half-duplex transmission module*)
hélice *f* GEN helix
helicoidal *adj* GEN helical
henrio *m* (*H*) ELECTRÓN henry (*H*)
heptodo *m* ELECTRÓN heptode
hermano *m* INFO sibling
hermeticidad *f* COMPON imperviousness
hermético *adj* COMPON hermetic
hermetización: ~ **de desconexión** *f* COMPON breakout jacketing
herrajes *m pl* COMPON fittings
herramienta *f* COMPON tool; ~ **para cortar y pelar** COMPON cutting and stripping tool; ~ **de corte** COMPON *alambres* cutting tool, cleaving tool; ~ **de desenrollamiento** COMPON unwrapping tool; ~ **dobladora** COMPON bending tool; ~ **para engarzar** COMPON crimp tool; ~ **de enrollado** COMPON wrapping tool; ~ **de limpieza** COMPON scouring tool; ~ **de pelado** COMPON stripping tool; ~ **para retirar la cubierta** COMPON sheath-stripping tool; ~ **separadora de trazos** COMPON stroke-cleaving tool; ~ **torneadora** COMPON turning tool
hertzio *m* (*Hz*) GEN hertz (*Hz*)
heterócrono *adj* TRANS heterochronous
heterodino *adj* TEC RAD *dispositivo* heterodyne
heterogeneidad *f* GEN heterogeneity
heterogéneo *adj* GEN heterogeneous
heterounión *f* ELECTRÓN, COMPON, TRANS heterojunction; ~ **doble** TRANS double heterojunction; ~ **simple** TRANS single heterojunction
heurística *f* DATOS heuristics
hexodo *m* ELECTRÓN hexode
hi-res *abr* (*de alta resolución*) INFO, TV hi-res (*high-resolution*)
híbrido *m* CONMUT, ELECTRÓN, INFO, TEC RAD, TRANS hybrid
hidrófono *m* PERIF hydrophone
hidrometeoro *m* TEC RAD hydrometeor
hierro: ~ **de base** *m* COMPON base iron; ~ **de fondo** *m* COMPON base iron; ~ **tensor de soporte** *m* COMPON guy attachment; ~ **vertical** *m* COMPON vertical iron; ~ **vertical exterior** *m* COMPON outer vertical iron
hija *f* INFO daughter
hijo *m* INFO child
hilera: ~ **de estirar** *f* COMPON die
hilo *m* COMPON, ELECTRÓN, ELECTROTEC, TELEFON wire; ~ **activo** CONMUT release wire; ~ **aramid** COMPON aramid yarn; ~ **blindado** COMPON shielded wire; ~ **para bobinado** COMPON winding wire; ~ **de cómputo** TELEFON meter wire; ~ **conectado a un enchufe** TELEFON T wire; ~ **de conversación** COMPON speech wire; ~ **desnudo** COMPON,

ELECTRÓN, ELECTROTEC bare wire; ~ **esmaltado** COMPON enamelled (*AmE* enameled) wire; ~ **de estaño para soldar** COMPON soldering wire; ~ **flexible de conexión** COMPON pigtail; ~ **fusible** ELECTROTEC fuse wire; ~ **de lectura** TELEFON reading wire; ~ **de llamada** TELEFON call wire; ~ **de ocupación** TELEFON seizure wire; ~ **pelado** COMPON stripped wire; ~ **positivo** TELEFON positive wire; ~ **de punta** COMPON *cable* tip; ~ **recocido** COMPON *termotratamiento* annealed wire; ~ **de retorno** ELECTROTEC return wire; ~ **sensor** COMPON sensor wire; ~ **único** COMPON, ELECTRÓN single wire

hipercurso *m* INFO hypercourse

hiperdatos *m pl* DATOS hyperdata

hiperdocumento *m* INFO hyperdocument

hiperfrecuencia *f* TEC RAD superhigh frequency (*SHF*)

hipergrupo *m* TRANS hypergroup; ~ **de conexión directa** TRANS through hypergroup

hiperlibro *m* INFO hyperbook

hipermedia *m* DATOS, INFO *Internet* hypermedia

hiperobjeto *m* INFO hyperobject

hiperpista *f* INFO hypertrail

hipertexto *m* DATOS, INFO hypertext

hipervisor *m* TELEFON hypervisor

hipótesis *f* GEN hypothesis

hipsómetro *m* TRANS level measuring set

histéresis *f* ELECTRÓN hysteresis

histograma *m* PRUEBA histogram; ~ **de barras** PRUEBA bar histogram

HLR *abr* (*registro de posición inicial*) TEC RAD HLR (*home location register*)

hogar: ~ **con receptor de televisión** *m* TV television home

hoja: ~ **de árbol** *f* DATOS, INFO tree leaf; ~ **de cálculo** *f* INFO spreadsheet; ~ **de codificación** *f* GEN *documentación* coding sheet; ~ **delgada de metal** *f* COMPON, ELECTROTEC foil; ~ **de especificaciones** *f* GEN *documentación* data sheet; ~ **de normas** *f* GEN standard sheet; ~ **preimpregnada** *f* ELECTRÓN prepreg

holografía *f* CONMUT, DATOS, ELECTRÓN holography

homócrono *adj* TRANS homochronous

homogeneidad *f* GEN homogeneity

homogéneo *adj* GEN homogeneous

homologación *f* GEN qualification approval, type approval

homounión *f* COMPON, ELECTRÓN, TRANS homojunction

hora: ~ **de arranque de tubo** *f* ELECTRÓN tube starting time; ~ **cargada media** *f* TELEFON mean busy hour; ~ **de llamada** *f* TV ringing time; ~ **de máxima audiencia** *f* TELEFON peak hour; ~ **media de Greenwich** *f* (*GMT*) GEN Greenwich Mean Time (*GMT*); ~ **ocupada promedio en CCS** *f* TELEFON average busy hour in CCS; ~ **universal** *f* GEN universal time; ~ **universal coordinada** *f* (*HUC*) GEN universal time coordinated (*UTC*); ~ **de verificación** *f* PRUEBA check-out time

horario: ~ **de programas** *m* TV programme (*AmE* program) schedule; ~ **de radiodifusión** *m* TEC RAD broadcasting schedule

horizontal: ~ **de nivel** *f* CONMUT level-selecting horizontal bar

horizonte: ~ **de radar** *m* TEC RAD radar horizon

horma: ~ **de cable** *f* COMPON cable formboard

horquilla *f* TELEFON switch hook

HRC *abr* (*circuito de referencia hipotético*) TEC RAD, TELEFON, TRANS HRC (*hypothetical reference circuit*)

HRDP *abr* (*camino digital de referencia hipotética*) TEC RAD, TELEFON, TRANS HRDP (*hypothetical reference digital path*)

HTML *abr* (*lenguaje de referencia de hipertexto*) DATOS, INFO HTML (*hypertext markup language*)

HTTP *abr* (*protocolo de transferencia de hipertexto*) DATOS, INFO *Internet* HTTP (*hypertext transfer protocol*)

HUC *abr* (*hora universal coordinada*) GEN UTC (*universal time coordinated*)

huella *f* COMPON, ELECTRÓN indentation

humedad *f* GEN moisture

humedecimiento *m* ELECTRÓN *de contactos* wetting; ~ **en mercurio** ELECTRÓN mercury-wetting

humidifugacia *f* GEN moisture resistance

HV *abr* (*alta tensión*) ELECTRÓN, ELECTROTEC HV (*high voltage*)

Hz *abr* (*hertzio*) GEN Hz (*hertz*)

I

I *abr* (*instalado*) GEN I (*implemented*)
I & D *abr* (*investigación y desarrollo*) PRUEBA R & D (*research and development*)
IA5 *abr* (*alfabeto internacional nº 5*) DATOS, TELEFON IA5 (*international alphabet no. 5*)
IAT *abr* (*tiempo atómico internacional*) TEC RAD IAT (*international atomic time*)
I/C *abr* (*entrante*) GEN I/C (*incoming*)
ICC *abr* (*portadora incrementalmente coherente*) TV ICC (*incrementally coherent carrier*)
ICCM *abr* (*interfuncionamiento por diagrama de control de llamadas*) TELEFON ICCM (*interworking by call-control mapping*)
ICD *abr* (*designador de código internacional*) DATOS ICD (*international code designator*)
ICE *abr* (*entorno de configuración del interfaz*) TELEFON ICE (*interface configuration environment*)
icono *m* INFO icon; **~ de enlace** INFO link icon; **~ pictórico** INFO pictorial icon
iconoscopio *m* ELECTRÓN iconoscope; **~ de imagen** ELECTRÓN image iconoscope
ICRD *abr* (*redirección y deflexión de llamadas de red interna*) TRANS ICRD (*internetwork call redirection and deflection*)
ICS *abr* (*declaración de conformidad de la instalación*) DATOS ICS (*implementation conformance statement*)
ID *abr* (*identificación*) GEN ID (*identification*)
IDA *abr* (*acceso digital integrado*) TRANS IDA (*integrated digital access*)
identidad: **~ de abonado al servicio móvil** *f* TEC RAD mobile subscriber identity; **~ de estación de barco** *f* TEC RAD, TELEFON ship-station identity; **~ de estación costera** *f* TEC RAD, TELEFON coast-station identity, coastal-station identity; **~ de estación llamada** *f* TEC RAD, TELEFON called-station identity; **~ de grupo llamante** *f* DATOS, TELEFON group call identity; **~ de línea llamada** *f* TELEFON called-line identity; **~ de línea que llama** *f* TELEFON calling-line identity; **~ de red** *f* TELEFON network identity; **~ de red de tránsito RPOA** *f* TELEFON RPOA transit-network identity; **~ de selección** *f* TELEFON selection identity; **~ temporal de estación móvil** *f* TEC RAD, TELEFON temporary mobile-station identity
identificación *f* (*ID*) GEN identification (*ID*); **~ de abonado llamado** TELEFON called-subscriber identification (*CSI*); **~ de abonado que llama** TELEFON calling-subscriber identification (*CIG*); **~ de área de localización** TEC RAD, TELEFON location-area identification; **~ de blanco** TEC RAD target identification; **~ de calle** TEC RAD lane identification; **~ de central** TELEFON exchange identification; **~ por defecto** DATOS default identification; **~ DIS de abonado llamado** TELEFON called-subscriber identification-DIS (*CSI-DIS*); **~ DTC de abonado que llama** TELEFON calling-subscriber identification-DTC (*CIG-DTC*); **~ de emisor** DATOS, TELEFON sender identification; **~ de emisor de intercambio** DATOS, TELEFON interchange-sender identification; **~ de función** TELEFON function identification; **~ de línea llamada** TELEFON called-line identification (*CDLI*); **~ de línea que llama** TELEFON calling-line identification; **~ de llamadas maliciosas** TELEFON malicious-call identification; **~ de multiplexión** (*MID*) TRANS multiplexing identification (*MID*); **~ de naves propias** (*IFF*) TEC RAD, TELEFON identification friend or foe (*IFF*); **~ de número que llama** TELEFON calling-number identification; **~ de parámetros** TELEFON parameter identification; **~ de plan de numeración** DATOS, TELEFON numbering-plan identification; **~ de plano numérico** DATOS, TELEFON number-plan identification; **~ de receptor de intercambio** DATOS, TELEFON interchange-receiver identification; **~ de red de tránsito** TRANS transit-network identification; **~ de riesgos** PRUEBA risk identification; **~ de servicio** APLIC service identification; **~ de usuario de MCS** TELEFON MCS-user identification; **~ de usuario de red** (*NUI*) DATOS, TELEFON network-user identification (*NUI*); **~ de voz** APLIC, CONTROL voice identification
identificador *m* GEN identifier; **~ de árbol** DATOS, INFO tree identifier; **~ de autorización y formato** DATOS, SEÑAL authority and format identifier (*AFI*); **~ de canal virtual** DATOS, TRANS virtual-channel identifier (*VCI*); **~ de comando** TELEFON command identifier; **~ de conformidad de adjudicación** DATOS,

TELEFON submission identifier; ~ **de conjunto de datos** DATOS data-set identifier (*DSI*); ~ **por defecto** DATOS default identifier; ~ **de destino** DATOS destination identifier; ~ **de dominio inicial** DATOS initial-domain identifier (*IDI*); ~ **de enlace de datos** DATOS data-link connection identifier (*DLCI*); ~ **de formato** (*FI*) ELECTROTEC, TELEFON format identifier (*FI*); ~ **de fuente** TELEFON source identifier; ~ **de grupo** (*GI*) TELEFON group identifier (*GI*); ~ **de grupo de parámetros** (*PGI*) INFO parameter-group identifier (*PGI*); ~ **de invocación de proceso de aplicación** DATOS application-process invocation identifier; ~ **de invocación de unidad de aplicación** DATOS application-entity invocation identifier; ~ **de objeto** DATOS object identifier; ~ **de plan de numeración** DATOS numbering-plan identifier; ~ **de protocolos** (*PRT-ID*) DATOS protocol identifier (*PRT-ID*); ~ **de punto de acceso a servicio** DATOS service-access point identifier (*SAPI*); ~ **de punto final de conexión N** DATOS, TELEFON N-connection end-point identifier; ~ **de punto final de conexión de transporte** DATOS, TELEFON transport-connection end-point identifier; ~ **de punto final terminal** DATOS terminal end-point identifier (*TEI*); ~ **de registro** DATOS registration identifier; ~ **de respuesta** TELEFON response identifier (*RI*); ~ **de sector de transacción** DATOS, TELEFON transaction-branch identifier; ~ **sintáctico** DATOS, TELEFON syntax identifier; ~ **de transacción** DATOS, TELEFON transaction identifier; ~ **de trayectoria virtual** DATOS, TRANS virtual-path identifier (*VPI*); ~ **de usuario de MCS** TELEFON MCS-user ID

I/F *abr* (*interfaz*) GEN I/F (*interface*)

IFF *abr* (*identificación de naves propias*) TEC RAD, TELEFON IFF (*identification friend or foe*)

IFLU *abr* (*medición plena inicial*) PRUEBA IFLU (*initial full line-up*)

IGE *abr* (*integración a gran escala*) COMPON, ELECTRÓN, TELEFON, TRANS LSI (*large-scale integration*)

IGES *abr* (*especificación inicial de intercambio de gráficos*) INFO IGES (*initial graphics exchange specification*)

ignición *f* ELECTRÓN ignition

ignitrón *m* ELECTRÓN ignitron

igualación *f* GEN equalization; ~ **de retardo de grupo** TEC RAD, TRANS group-delay equalization; ~ **de tiempo de propagación** TRANS propagation-time equalization

igualador: ~ **de retardos** *m* INFO delay equalizer

ILS *abr* (*sistema de aterrizaje por instrumentos*) TEC RAD ILS (*instrument landing system*)

iluminación *f* ELECTROTEC, TV lighting; ~ **a contraluz** ELECTRÓN, INFO backlighting

iluminado *adj* INFO, ELECTRÓN, TELEFON illuminated, lit

iluminar *vt* INFO, ELECTRÓN, TELEFON illuminate; ◆ ~ **a contraluz** ELECTRÓN, INFO backlight

imagen *f* GEN picture, image, field; ~**-b** INFO, TV b-picture; ~ **binaria** DATOS binary image; ~ **borrosa en movimiento** TV motion blur; ~ **de campo-b** INFO, TV b-field picture; ~ **de campo-i** INFO, TV i-field picture; ~ **codificada bidireccionalmente predictiva** INFO, TV bidirectionally-predictive coded picture; ~ **codificada predictiva en ambos sentidos** INFO, TV bidirectionally-predictive coded picture; ~ **de columna-b** INFO, TV b-frame picture; ~ **de copia impresa** INFO, PERIF hard-copy image; ~ **cromática** INFO, TV colour (*AmE* color) image; ~ **dentro de una imagen** (*PIP*) INFO picture in picture (*PIP*); ~ **desvanecida** INFO dissolved image; ~ **de detalle** INFO inset picture; ~ **a dos niveles** INFO bi-level image; ~ **electrónica de mapa** TEC RAD, TV video map; ~ **explorada** PERIF scanned image; ~ **fantasma** TV ghost; ~ **fija** INFO, TV still frame; ~**-i** INFO, TV i-picture; ~ **instantánea** INFO snapshot; ~ **intracodificada** INFO, TV intracoded picture; ~ **de memoria** INFO core image; ~ **de memoria ejecutable** TELEFON executable memory image; ~ **multinivel** INFO multilevel image; ~ **multinivel limitada** INFO limited multilevel image; ~**-p** INFO, TV p-picture; ~ **de pantalla** INFO screen image; ~ **predictiva codificada** INFO, TV predictive-coded picture; ~ **reconstruida** INFO reconstructed picture; ~ **de referencia** INFO, TV reference picture; ~ **de referencia futura** INFO, TV future-reference picture; ~ **de referencia pasada** INFO, TV past-reference picture; ~ **sintética** DATOS, TV synthetic image; ~ **tridimensional** TV three-dimensional image; ~ **virtual** DATOS, TV virtual image; ~ **de visualización transitoria** INFO, TV soft-copy image

imaginería: ~ **informática** *f* INFO computer imagery

imán *m* ELECTRÓN magnet; ~ **de avance** TELEFON stepping magnet; ~ **de reposición** CONMUT resetting magnet

impar *adj* GEN odd

impedancia *f* ELECTRÓN, ELECTROTEC, INFO, TEC RAD impedance; ~ **de adaptación** ELECTRÓN matching impedance; ~ **de cortocircuito** ELECTROTEC short-circuit impedance; ~ **efec-**

tiva de entrada ELECTROTEC effective-input impedance; ~ **efectiva de salida** ELECTROTEC effective-output impedance; ~ **de electrodo** ELECTRÓN electrode impedance; ~ **de entrada de antena** TEC RAD aerial-input impedance, antenna-input impedance; ~ **de fuente** ELECTROTEC source impedance; ~ **de imagen** ELECTRÓN image impedance; ~ **inversa** ELECTRÓN inverse impedance; ~ **mutua** ELECTRÓN mutual impedance; ~ **de onda** ELECTRÓN, TEC RAD wave impedance; ~ **de onda característica** ELECTRÓN, TEC RAD characteristic wave impedance; ~ **de salida** ELECTRÓN, ELECTROTEC output impedance; ~ **en serie** ELECTRÓN, ELECTROTEC series impedance; ~ **terminal** ELECTRÓN terminal impedance, terminating impedance

imperfección: ~ **pictórica** *f* TV picture blemish

impermeabilidad *f* GEN impermeability

impermeabilización *f* COMPON waterproofing

impermeable *adj* GEN impermeable, impervious

implantación: ~ **de iones** *f* ELECTRÓN ion implantation

implantador *m* ELECTRÓN implanter

implementación *f* GEN implementation

implementar *vt* GEN implement

implícito *adj* DATOS implicit

impolarizado *adj* ELECTROTEC nonpolarized

importación *f* DATOS import

importado *adj* DATOS imported

importar *vt* DATOS import

imprecisión *f* PRUEBA inaccuracy

impregnación *f* COMPON impregnation

impregnador *m* COMPON *química* impregnating agent

impregnar *vt* COMPON impregnate

impresión *f* APLIC, INFO, PERIF printing; ~ **de contenido de memoria** INFO memory printout; ~ **de error** INFO error printout, fault printout; ~ **láser** PERIF laser printing; ~ **línea por línea** PERIF line printing; ~ **múltiple** ELECTRÓN multiple pattern; ~ **negativa** ELECTRÓN negative pattern; ~ **sonora de día y fecha** PERIF *de contestador automático* vocal date-time stamping

impreso *m* DATOS, INFO, TEC RAD form; ~ **Backus-Naur** (*BNF*) DATOS Backus-Naur form (*BNF*)

impresor: ~ **de disco compacto fotográfico** *m* PERIF photo-CD writer

impresora *f* GEN printer; ~ **de cambio de fase** PERIF phase-change printer; ~ **de chorro de tinta** PERIF inkjet printer; ~ **de cinta** PERIF tape printer; ~ **por deposición iónica** PERIF ion-deposition printer; ~ **electrográfica** PERIF electrographic printer; ~ **de etiquetas** PERIF label printer; ~ **con gran capacidad de almacenamiento** PERIF LCS printer; ~ **de impacto** PERIF impact printer; ~ **láser** PERIF laser printer; ~ **de LED** PERIF LED printer; ~ **de línea** PERIF line printer; ~ **matricial** PERIF matrix printer; ~ **de páginas** PERIF page printer; ~ **de puntos** PERIF dot-matrix printer; ~ **sin impacto** PERIF nonimpact printer

imprimir *vt* APLIC, INFO, PERIF print

impulsado *adj* ELECTROTEC powered

impulsión *f* ELECTRÓN, TELEFON, TRANS impulsing

impulso *m* GEN pulse; ~ **activador** ELECTRÓN, INFO trigger pulse; ~ **aperiódico** GEN nonrecurring pulse; ~ **de apertura** TELEFON break pulse; ~ **básico de cómputo** TELEFON basic metering pulse; ~ **bipolar** TELEFON bipolar pulse; ~ **de carga** APLIC charging impulse; ~ **de cierre** TELEFON make pulse; ~ **de cómputo** TELEFON meter pulse; ~ **de control** CONTROL control pulse; ~ **en coseno** TRANS cosine pulse; ~ **en coseno cuadrado** TRANS cosine-squared pulse; ~ **en diente de sierra** ELECTRÓN sawtooth pulse; ~ **disparador de línea** ELECTRÓN triggering lead pulse; ~ **de error** DATOS error pulse; ~ **estroboscópico automático** TEC RAD walking strobe pulse; ~ **de fijación** TEC RAD *radar* strobe pulse; ~ **gaussiano** TELEFON, TRANS Gaussian pulse; ~ **de igualación** TV equalizing pulse; ~ **inhibidor** ELECTRÓN inhibiting pulse; ~ **interferente** ELECTRÓN interference pulse; ~ **de interrogación** TEC RAD interrogating pulse; ~ **luminoso** TV light pulse; ~ **de marcación por disco telefónico** TELEFON dial pulse; ~ **de muestreo** PRUEBA sample pulse; ~ **negativo** ELECTRÓN negative pulse; ~ **de onda sinusoidal** ELECTRÓN, INFO, TELEFON sine-wave pulse; ~ **parásito** TEC RAD spurious count, TELEFON pulse spike; ~ **de puerta** ELECTRÓN *elementos lógicos* gate pulse; ~ **de puesta a uno** CONTROL set pulse; ~ **rectangular** TELEFON rectangular pulse; ~ **de referencia** TEC RAD framing pulse; ~ **reflejado** TEC RAD reflected pulse; ~ **de reposición** ELECTRÓN resetting pulse, restoring pulse; ~ **de salida** INFO output pulse; ~ **de semilínea** TV half-line pulse; ~ **en seno cuadrado** TRANS sine-squared pulse; ~ **de sincronización** ELECTRÓN clock pulse; ~ **de sincronización vertical** TV field-sync pulse; ~ **triangular** ELECTRÓN triangular pulse; ~ **unipolar**

TELEFON unipolar pulse; ~ **de violación** TRANS violation pulse

impulsor *m* GEN driver; ~ **de cartucho de cinta** PERIF tape-cartridge drive

impulsos: ~ **por segundo** *m pl* TEC RAD, TELEFON pulses per second (*pps*)

impureza *f* ELECTRÓN impurity

IMS *abr* (*interconexión multiseñal*) DATOS MSI (*multisignal interconnect*)

IN *abr* (*red inteligente*) APLIC, TRANS IN (*intelligent network*)

inactivo *adj* GEN inoperative, inactive

inalámbrico *adj* INFO, TEC RAD, TELEFON wireless

inaplicable *adj* (*N/A*) DATOS, TELEFON not applicable (*N/A*)

inauguración *f* TEC RAD inauguration

inaugurar *vt* TEC RAD inaugurate

incertidumbre: ~ **de medición** *f* PRUEBA uncertainty of measurement

incidencia *f* PRUEBA incidence

incisión *f* GEN incision

inclinación *f* TEC RAD inclination; ~ **de carácter** APLIC *reconocimiento de caracteres* character skew; ~ **negativa** TELEFON negative tilt; ~ **de onda** TEC RAD wave tilt; ~ **de techo de impulso** TELEFON pulse droop

inclusión *f* ELECTRÓN *circuitos impresos* inclusion

incoherencia *f* TRANS incoherence

incoherente *adj* TRANS incoherent

incompatible *adj* DATOS incompatible

inconsistencia *f* INFO inconsistency

incorrecto *adj* TELEFON *modulación, restitución, señal* incorrect

incrementar *vt* GEN increase

incremento *m* GEN increment

indefinido *adj* GEN undefined

independencia: ~ **de datos** *f* INFO data independence; ~ **de secuencia de bitios** *f* INFO bit-sequence independence

indexación *f* GEN indexing

indicación *f* GEN indication, *marcaje* designation; ~ **de alarma posterior selectiva** PRUEBA selective back-alarm indication; ~ **de alarma remota** PRUEBA remote-alarm indication (*RAI*); ~ **de ausencia de envío de datagrama** DATOS, SEÑAL, TELEFON datagram non-delivery indication; ~ **de avería** PRUEBA fault indication; ~ **de bloqueo** TELEFON blocking indication; ~ **de borrado de DCE** TELEFON DCE clear indication; ~ **de cero con aguja** DATOS null-pointer indication (*NPI*); ~ **de concatenación** ELECTRÓN concatenation indication (*CI*); ~ **de contador** TELEFON meter indication, meter-value indication; ~ **de desconectar N** *m* DATOS N-disconnect indication; ~ **de escape** TELEFON escape indication; ~ **de estatus** PRUEBA status indication; ~ **de fallo** *m* PRUEBA failure indication; ~ **de llamada no atendida** PERIF, TELEFON unanswered-call indication; ~ **de mensaje en espera** TELEFON message-waiting indication; ~ **multipunto de comunicación cero** DATOS multipoint indication zero communication; ~ **de número que llama** TELEFON calling-number indication; ~ **de petición de bloque paramétrico** TELEFON parameter-block request indication; ~ **de posición de móvil** TEC RAD roaming; ~ **primaria** DATOS primitive indication; ~ **recordatoria** TELEFON reminder indication; ~ **de respuesta** TELEFON answer indication; ~ **de retardo de tránsito** DATOS transit-delay indication; ~ **de rotura** PRUEBA indication of fracture; ~ **de sonoridad de circuito** SEÑAL circuit-loudness rating; ~ **visual** PRUEBA visual indication

indicador *m* GEN indicator; ~ **de abandono de red** PRUEBA network-discard indicator; ~ **de abonado que llama** TELEFON caller display; ~ **acústico** INFO beeper, TELEFON ringing set; ~ **de aguja** CONMUT switch indicator; ~ **de altitud y posición** TEC RAD height-position indicator; ~ **de apantallamiento** TV screening indicator (*SI*); ~ **automático visual** TV autocue *BrE*, teleprompter *AmE*; ~ **de averías** PRUEBA fault display; ~ **de base de clavija** TELEFON plug-sleeve indicator; ~ **de batido** INFO beat indicator; ~ **de campo** TELEFON field indicator; ~ **de capacidad** DATOS facility indicator; ~ **de carga/no carga** TELEFON charge/no charge indicator; ~ **de categoría de abonado que llama** TELEFON calling-party category indicator; ~ **de cero** PRUEBA null meter; ~ **de clase de usuario** TELEFON user-class indicator; ~ **de código de país** TELEFON country-code indicator; ~ **de congestión** TEC RAD congestion indicator; ~ **contestando la categoría de abonado que llama** TELEFON calling-party category response indicator; ~ **de control** INFO check indicator; ~ **de control de protocolo** DATOS protocol-control indicator; ~ **de crédito** TELEFON credit indicator; ~ **de datos de extremo a extremo** DATOS end-to-end information indicator; ~ **DCC/CNIC** SEÑAL DCC/DNIC indicator; ~ **de deposición de bloques** DATOS block-dropping indicator; ~ **de desbordamiento** INFO overflow indicator; ~ **de destino** TELEFON destination indicator; ~

de desviación de frecuencia PRUEBA frequency-deviation meter; ~ **de ecos por efecto Doppler** (*EDI*) TEC RAD Echo Doppler indicator (*EDI*); ~ **de enrutamiento alternativo** TELEFON alternative-routing indicator; ~ **de estado de circuito** CONMUT circuit-state indicator; ~ **de estado de operador** TELEFON attendant-status indicator; ~ **de falta de papel** PERIF paper-out switch; ~ **de fin de conversación** TELEFON clearing indicator; ~ **de flujo** TELEFON flow indicator; ~ **de identidad de línea llamada** TELEFON called-line identity indicator; ~ **de información solicitada** DATOS solicited-information indicator; ~ **lateral** CONMUT side indicator; ~ **de liberación** DATOS, SEÑAL release indicator; ~ **de línea que llama** TELEFON calling-line indicator (*CLI*); ~ **de llamada** PERIF call display, call indicator; ~ **de llamada a grupo de usuarios cerrado** TELEFON closed user-group call indicator; ~ **de llamada nacional** TELEFON national call indicator; ~ **de llamada redireccionada** TELEFON redirection-address indicator; ~ **de longitud** DATOS length indicator (*LI*); ~ **de longitud de campo** TELEFON field-length indicator; ~ **luminoso** PERIF light indicator; ~ **de mensaje recibido** PERIF message-received indicator; ~ **de método de extremo a extremo** DATOS end-to-end method indicator; ~ **de método SCCP** SEÑAL, TELEFON SCCP-method indicator; ~ **de modificaciones** DATOS modification indicator; ~ **de nivel de líquidos** TELEFON liquid-level indicator; ~ **de nivel de aceite** TELEFON oil-level gauge (*AmE* gage); ~ **de ocupación** TELEFON seizure indicator; ~ **de ondas estacionarias** PRUEBA standing-wave meter; ~ **de origen** TELEFON origin indicator; ~ **de petición** TELEFON request indicator; ~ **de petición de identificación de línea llamada** TELEFON called-line identification-request indicator; ~ **de pila** INFO stack pointer; ~ **de plan de numeración** DATOS, TELEFON numbering-plan indicator; ~ **de posición de móvil** TEC RAD roamer; ~ **de potencia** ELECTRÓN, ELECTROTEC power indicator; ~ **de presentación** DATOS presentation indicator; ~ **de razón de reexpedición original** TELEFON original redirection reason indicator; ~ **de recomendación** DATOS recommendation indicator; ~ **de recorrido** ELECTRÓN tracker; ~ **de reexpedición** CONMUT redirecting indicator; ~ **de restricción en la presentación de eventos** DATOS event-presentation restriction indicator; ~ **restringido de presentación de dirección** DATOS address-presentation restricted indicator; ~ **de ruido** TELEFON noise indicator; ~ **de salida** INFO output indicator; ~ **de secuencia de fases** ELECTROTEC phase-sequence indicator; ~ **de selección RPOA** TELEFON RPOA selection indicator; ~ **de señal** SEÑAL, TRANS signal indicator; ~ **de señal de llamada** TELEFON ringing-pilot indicator; ~ **de servicio** TELEFON service indicator; ~ **de siete segmentos** CONTROL seven-segment display; ~ **solicitando la categoría de abonado que llama** TELEFON calling-party category-request indicator; ~ **de sonoridad de circuito** (*CLR*) TELEFON circuit-loudness rating (*CLR*); ~ **de supresor de eco** TELEFON echo-suppressor indicator; ~ **de timbre** TELEFON ring indicator; ~ **de tipo de circuito** TELEFON nature-of-circuit indicator; ~ **de usuario a usuario** DATOS user-to-user indicator; ~ **de velocidad** TEC RAD speed indicator, velocity indicator; ~ **de velocidad acimutal** TEC RAD azimuth-speed indicator; ~ **visual** INFO, PRUEBA visual indicator, TELEFON star indicator, visual indicator

indicar *vt* GEN indicate

indicativo: ~ **de encaminamiento** *m* TELEFON routing number; ~ **de identificación** *m* TEC RAD background signature

índice *m* GEN index; ~ **acumulado de fuga** TRANS cumulative-leakage index; ~ **de borrado** INFO erasing rate; ~ **de calidad de transmisión** PRUEBA, TELEFON, TRANS transmission-performance rating; ~ **de caracteres gráficos básicos de teletexto** TELEFON teletext basic-graphic character index; ~ **de caracteres gráficos de teletexto** TELEFON teletext graphic character index; ~ **de caracteres de teletexto** TELEFON teletext character index; ~ **de carga anual** TELEFON annual-charge ratio; ~ **de correspondencia de formas** CONMUT pattern-correspondence index; ~ **de crecimiento** GEN growth rate; ~ **de desvanecimiento** TEC RAD fading rate; ~ **de error de bitios en períodos largos** TELEFON long-term bit error rate; ~ **de error de caracteres** DATOS character-error ratio; ~ **escalonado equivalente** TELEFON equivalent-step index (*ESI*); ~ **de frecuencia** TRANS frequency ratio; ~ **de función de control básico de teletexto** TELEFON teletext basic-control function index; ~ **de función de control de teletexto** TELEFON teletext control-function index; ~ **gradual** ELECTRÓN, TRANS graded index; ~ **de grupo** TRANS group index; ~ **de justificación nominal** TELEFON nominal-justification rate;

~ **de llamadas falsas** TELEFON false-calling rate; ~ **local** TELEFON local index; ~ **máximo de justificación** TELEFON maximum-justification rate; ~ **de modulación** TEC RAD modulation index; ~ **de nitidez** TEC RAD, TELEFON, TRANS articulation index; ~ **de nitidez en logátomos** TEC RAD, TELEFON, TRANS logatom articulation index; ~ **de refracción** ELECTRÓN, TEC RAD, TRANS, TV refractive index, refraction index; ~ **de refracción modificado** TEC RAD modified refractive index; ~ **de sobrecorriente** ELECTROTEC overcurrent factor

indisponibilidad *f* PRUEBA outage, unavailability; ~ **media** PRUEBA mean unavailability; ~ **de satélite** TEC RAD, TV satellite outage

inducción *f* ELECTRÓN, ELECTROTEC induction; ~ **mutua** ELECTRÓN mutual induction

inducir *vt* ELECTRÓN, ELECTROTEC induce

inductancia *f* ELECTRÓN, ELECTROTEC inductance; ~ **magnetizante** ELECTRÓN magnetizing inductance; ~ **mutua** ELECTRÓN mutual inductance; ~ **en serie** ELECTRÓN, ELECTROTEC series inductance

inductivo *adj* ELECTRÓN, ELECTROTEC inductive

inductor *m* COMPON, ELECTRÓN, ELECTROTEC inductor

inercia *f* ELECTROTEC, PRUEBA inertia

inestabilidad *f* DATOS volatility, ELECTRÓN, TEC RAD instability; ~ **de frecuencia** TEC RAD, TELEFON frequency instability, frequency uncertainty; ~ **de imagen** INFO, TEC RAD, TRANS, TV jitter; ~ **de realimentación** GEN feedback instability

inestructurado *adj* INFO unstructured, nonstructured

inexactitud *f* PRUEBA inaccuracy

inferencia: ~ **de grupo por defecto** *f* DATOS default group inference

inferencias: ~ **lógicas por segundo** *f pl* (*LI/S*), (*LIPS*) PRUEBA logical inferences per second (*LI/S, LIPS*)

infinitamente: ~ **variable** *adj* CONTROL infinitely variable

inflamable *adj* GEN ignitable

influencia *f* GEN influence

información *f* GEN information; ~ **de alineación de trama** TRANS framing information; ~ **de autentificación** DATOS, SEÑAL authentication information; ~ **de autorización** DATOS, SEÑAL authorization information; ~ **de carga** APLIC charging information; ~ **de codificación** INFO coding information; ~ **colateral** INFO side information; ~ **de control** INFO control information; ~ **de control de protocolo** DATOS, TELEFON protocol-control information; ~ **de control de protocolo de aplicación** (*APCI*) DATOS, TELEFON application-protocol control information (*APCI*); ~ **de control de protocolo N** DATOS, TELEFON N-protocol control information; ~ **de control de protocolos AAL** DATOS, TELEFON AAL-protocol control information (*AAL-PCI*); ~ **de directorio** APLIC, TELEFON directory enquiries *BrE*, directory assistance *AmE*; ~ **estadística sobre tráfico de cuadros** ELECTRÓN statistical frame traffic information; ~ **extra de instrumentación de protocolos para comprobación** (*PIXIT*) DATOS, PRUEBA protocol implementation extra information for testing (*PIXIT*); ~ **de facturación** APLIC charging data, billing information; ~ **de localización** TEC RAD location information; ~ **en masa** INFO bulk information; ~ **no numerada** DATOS unnumbered information (*UI*); ~ **parásita** INFO trash; ~ **preparatoria** DATOS, TELEFON housekeeping information; ~ **Q** PRUEBA *sobre calidad* Q-information; ~ **de reserva** DATOS backup information; ~ **de ritmo** DATOS timing information; ~ **de señalización** SEÑAL signalling (*AmE* signaling) information; ~ **significativa** DATOS meaningful information; ~ **de sincronización** INFO synchronization information; ~ **sobre abonados** APLIC, TELEFON directory information *AmE*, directory enquiries *BrE*; ~ **sobre control de protocolo N** DATOS, TELEFON N-protocol control information; ~ **sobre direccionamiento de protocolo N** DATOS N-protocol addressing information; ~ **sobre dirección de protocolo de red** DATOS network-protocol address information; ~ **sobre la entrada** DATOS entry information; ~ **sobre el servicio** DATOS service information; ~ **de temporización** TRANS timing information; ~ **de tono** TELEFON tone-coded information; ~ **de usuario a usuario** DATOS user-to-user information (*UUI*); ~ **de vídeo** DATOS, TV video data

informática *Esp f* (*cf computación AmL*) INFO computing; ~ **en colaboración** INFO collaborative computing; ~ **doméstica** INFO home computing; ~ **integrada** DATOS integrated data processing; ~ **con tolerancia de fallos** INFO fault-tolerant computing; ~ **de usuario final** INFO end-user computing

informatización *f* INFO computerization

informatizado *Esp adj* (*cf computadorizado AmL*) INFO computerized

informe *m* GEN report; ~ **de averías** PRUEBA

fault report; ~ **de errores** PRUEBA error report; ~ **de prueba de conformidad de protocolo** (*PCTR*) DATOS, PRUEBA protocol-conformance test report (*PCTR*); ~ **de prueba de conformidad de sistema** (*SCTR*) DATOS, PRUEBA system-conformance test report (*SCTR*)

infométrica *f* DATOS infometrics

infraestructura *f* GEN infrastructure; ~ **industrial única** TV single industrial infrastructure; ~ **de red** TRANS network infrastructure

infrarrojo[1] *adj* (*IR*) GEN infrared (*IR*)

infrarrojo[2] *m* (*IR*) GEN infrared (*IR*); ~ **de onda corta** TEC RAD shortwave infrared

infrasonido *m* INFO infrasound

ingeniería *f* GEN engineering; ~ **asistida por computadora** *AmL* (*CAE*), (*cf ingeniería asistida por ordenador Esp*) INFO computer-aided engineering (*CAE*); ~ **asistida por ordenador** *Esp* (*CAE*), (*cf ingeniería asistida por computadora AmL*) INFO computer-aided engineering (*CAE*); ~ **de conocimiento** INFO knowledge engineering; ~ **de corriente fotoeléctrica** ELECTROTEC light-current engineering; ~ **de corriente fuerte** ELECTROTEC heavy-current engineering; ~ **de fiabilidad** PRUEBA reliability engineering; ~ **lingüística** INFO linguistic engineering; ~ **de potencia** ELECTROTEC electrical engineering, power engineering; ~ **de red** TELEFON network engineering; ~ **de sistemas** GEN systems engineering; ~ **de soporte lógico** INFO software engineering; ~ **de soporte lógico auxiliada por computadora** *AmL* (*cf ingeniería de soporte lógico auxiliada por ordenador Esp*) INFO computer-aided software engineering (*CASE*); ~ **de soporte lógico auxiliada por ordenador** *Esp* (*cf ingeniería de soporte lógico auxiliada por computadora AmL*) INFO computer-aided software engineering (*CASE*); ~ **de telecomunicaciones** GEN communications engineering, telecommunications engineering

ingeniero *m* GEN engineer; ~ **de apoyo de sistemas** GEN systems-support engineer; ~ **de conocimiento** INFO knowledge engineer; ~ **de máquinas** GEN hardware engineer; ~ **de programación** DATOS, INFO software engineer; ~ **de sistemas** GEN systems engineer; ~ **de sistemas CAD/CAM** INFO CAD/CAM systems engineer; ~ **de sonido** GEN sound engineer; ~ **de telecomunicaciones** GEN telecommunications engineer

ingresos: ~ **de explotación** *m pl* APLIC operating revenue

inhibición *f* ELECTRÓN *en circuitos lógicos* inhibition

inhibir *vt* ELECTRÓN *en circuitos lógicos* inhibit

iniciación *f* INFO bootstrapping; ~ **de registro** INFO log-in, log-on

iniciador: ~ **de asociaciones** *m* DATOS association initiator; ~ **N** *m* DATOS N initiator

inicialización *f* GEN initialization

inicializar *vt* GEN initialize

iniciar *vt* INFO initialize, initiate; ~ **la sesión** INFO log in, log on

inicio: ~ **de comando de documento** *m* DATOS command document start; ~ **de la conexión** *m* DATOS connection overhead

inmunidad: ~ **a las interferencias** *f* TEC RAD immunity to interference, interference immunity; ~ **al ruido** *f* TRANS immunity to noise

inmutable *adj* GEN immutable

inoperatividad: ~ **de procesador** *f* TELEFON processor outage

inoxidable *adj* COMPON rustproof

inscrito *adj* DATOS, INFO listed

inserción *f* GEN insertion; ~ **en órbita** TEC RAD orbit insertion; ~ **de receptor** PERIF, SEÑAL, TELEFON receiver insertion; ~ **de señal** SEÑAL, TRANS signal insertion

inserto *m* COMPON, INFO, TRANS insert, TV inbetween

insonorización *f* GEN sound insulation, soundproofing

insonorizar *vt* GEN soundproof

inspección *f* PRUEBA inspection; ~ **por atributos** PRUEBA inspection by attributes; ~ **durante la instalación** PRUEBA in-process inspection; ~ **de entrega** PRUEBA delivery inspection; ~ **final** PRUEBA final inspection; ~ **por muestreo** PRUEBA sampling inspection; ~ **normal** PRUEBA normal inspection; ~ **preventiva** PRUEBA preventive inspection; ~ **en primera presentación** PRUEBA original inspection; ~ **de proceso** PRUEBA process inspection; ~ **de recepción** PRUEBA receiving inspection; ~ **reducida** PRUEBA reduced inspection; ~ **de selección** PRUEBA screening inspection; ~ **por variables** PRUEBA inspection by variables; ~ **visual** PRUEBA visual inspection; ~ **volante** PRUEBA patrol inspection

inspeccionar *vt* PRUEBA inspect

instalación *f* GEN installation; ~ **de abonado** TELEFON subscriber installation, subscriber's installation; ~ **automática** ELECTRÓN autoinstallation; ~ **de cable desnudo** TELEFON skinny-wire set; ~ **de emergencia** APLIC emergency installation; ~ **de entrada** (*EF*) ELECTRÓN entrance facility (*EF*); ~ **de equipo** INFO equipment installation; ~ **de espera**

permitida TELEFON waiting-allowed facility; ~ **de fax de abonado** TELEFON subscriber's fax position; ~ **de modificación D-bitio** INFO D-bit modification facility; ~ **de montaje-desmontaje de paquete de datos** DATOS, TELEFON packet assembly-disassembly facility; ~ **de montaje mural** PERIF wall-mounted installation; ~ **móvil** TEC RAD mobile installation; ~ **multipunto centralizada** TELEFON centralized multipoint facility; ~ **N** TELEFON N-facility; ~ **no normalizada** (*NSF*) TELEFON nonstandard facility (*NSF*); ~ **en prueba** (*IUT*) PRUEBA implementation under test (*IUT*); ~ **con puesta a tierra** ELECTROTEC earthing arrangement *BrE*, grounding arrangement *AmE*; ~ **de referencia** TELEFON reference installation; ~ **de registro de DTE inactiva** TELEFON DTE-inactive registration facility; ~ **de señales de tráfico** SEÑAL traffic signalling (*AmE* signaling) installation; ~ **subterránea** APLIC underground installation; ~ **de terminal para transmisión de datos** PERIF terminal-installation for data transmission; ~ **de terminal teletexto multiestación** PERIF multistation teletext terminal installation; ~ **terminal de abonado** PERIF subscriber terminal installation

instalado *adj* (*I*) GEN implemented (*I*)

instalador *m* GEN fitter, installer; **~-distribuidor** GEN installer-distributor

instalar *vt* GEN install

instantáneo *adj* GEN instantaneous

instante *m* GEN instant

instantes: ~ **ideales** *m pl* TELEFON ideal instants

Instituto: ~ **Europeo de Investigación y Estudios Estratégicos** *m* GEN European Institute of Research and Strategic Studies; ~ **Europeo de Normas de Telecomunicaciones** *m* GEN European Telecommunications Standards Institute (*ETSI*)

instrucción *f* GEN instruction; ~ **de caracteres** INFO character instruction; ~ **de decisión** INFO decision instruction; ~ **de dirección n-más-una** INFO n-plus-one address instruction; ~ **de dirección una-más-una** INFO one-plus-one address instruction; ~ **de discriminación** INFO discrimination instruction; ~ **efectiva** INFO effective instruction; ~ **de ejecución** INFO execute statement; ~ **en hilera** INFO in-line instruction; ~ **lógica** INFO logic instruction; ~ **de máquina** INFO computer instruction; ~ **de pausa** INFO pause instruction; ~ **de una sola dirección** INFO one-address instruction; ~ **TTCN** DATOS, INFO TTCN statement; ~ **vacía** INFO no-operation instruction

instrucciones: ~ **de alineamiento** *f pl* TRANS line-up instructions; ~ **de bobinado** *f pl* ELECTROTEC winding instructions; ~ **de empleo** *f pl* GEN instructions for use; ~ **de mantenimiento** *f pl* GEN maintenance instructions; ~ **de operación** *f pl* GEN operating directions; ~ **paso a paso** *f pl* INFO step-by-step instructions; ~ **para el uso** *f pl* GEN operating instructions

instrumental: ~ **para alineación óptica** *m* TEC RAD boresight facility

instrumento: ~ **de análisis** *m* PRUEBA analysis meter; ~ **de bobina móvil** *m* ELECTROTEC moving-coil instrument; ~ **electromagnético** *m* ELECTROTEC moving-iron instrument; ~ **de imán permanente con bobina móvil** *m* ELECTROTEC moving-coil permanent-magnet instrument; ~ **indicador de cresta** *m* PRUEBA peak-reading instrument; ~ **de inserción** *m* GEN insertion tool; ~ **de integración-medición** *m* PRUEBA integrating-measuring instrument; ~ **de medición** *m* PRUEBA measuring instrument; ~ **de medición eléctrica** *m* PRUEBA electric-measuring instrument; ~ **de medición indicador** *m* PRUEBA indicating-measuring instrument; ~ **de precisión** *m* PRUEBA precision instrument; ~ **de registro** *m* PRUEBA recording instrument; ~ **de verificación** *m* PRUEBA verifying instrument

insuficiencia: ~ **de memoria** *f* INFO memory shortage

integración *f* GEN integration, INFO embedding; ~ **de datos de voz** DATOS voice-data integration; ~ **a escala muy grande** (*IMGE*) COMPON, ELECTRÓN, TELEFON, TRANS very large-scale integration (*VLSI*); ~ **a gran escala** (*IGE*) COMPON, ELECTRÓN, TELEFON, TRANS large-scale integration (*LSI*); ~ **a mediana escala** COMPON, ELECTRÓN, TELEFON, TRANS medium-scale integration; ~ **a pequeña escala** (*IPE*) COMPON, ELECTRÓN, TELEFON, TRANS small-scale integration (*SSI*); ~ **en red** TELEFON networking; ~ **de sistemas** DATOS systems integration

integrador *m* DATOS, ELECTRÓN, TEC RAD integrator; ~ **adicionador** ELECTRÓN summing integrator; ~ **de sistemas** DATOS systems integrator

integral *adj* GEN integral

integrar *vt* GEN integrate, INFO embed

integridad *f* DATOS, TELEFON, TRANS integrity; ~ **de datos** DATOS data integrity; ~ **de secuencia numérica** TELEFON digit-sequence integrity

inteligencia: ~ **artificial** *f* (*AI*) APLIC, CONMUT, INFO artificial intelligence (*AI*)
inteligente *adj* INFO smart
inteligibilidad: ~ **de voz** *f* INFO, TEC RAD, TELEFON speech intelligibility
intensidad *f* GEN intensity; ~ **de campo** ELECTROTEC, PRUEBA, TEC RAD field intensity, field strength; ~ **de campo utilizable como referencia** TEC RAD reference-usable field strength; ~ **de envolvente** TEC RAD duct intensity; ~ **de fallo** PRUEBA failure intensity; ~ **de llamadas** TELEFON call intensity; ~ **de lluvia** TEC RAD rainfall rate; ~ **media de fallo** PRUEBA mean-failure intensity; ~ **media de tráfico** TELEFON average-traffic intensity; ~ **mínima de campo utilizable** TEC RAD minimum-usable field strength; ~ **radiante** ELECTRÓN, TRANS radiant intensity; ~ **de señal** SEÑAL signal strength; ~ **sonora** GEN sound intensity; ~ **de tráfico** CONMUT, TELEFON traffic intensity, traffic load; ~ **de tráfico de abonado** TELEFON subscriber-traffic rate; ~ **de tráfico aleatorio equivalente** TELEFON equivalent random traffic intensity
intensificación: ~ **de tráfico** *f* CONMUT, TELEFON traffic loading
intento: ~ **de acceso fraudulento** *m* TEC RAD fraudulent-access attempt; ~ **efectivo de llamada** *m* TEC RAD, TELEFON effective-call attempt; ~ **fructífero de llamada** *m* TEC RAD, TELEFON successful-call attempt; ~ **frustrado** *m* TEC RAD, TELEFON unsuccessful attempt; ~ **de llamada perdido** *m* TEC RAD, TELEFON lost-call attempt; ~ **de llamada terminado** *m* TEC RAD, TELEFON completed-call attempt; ~ **de ocupación de una línea** *m* TELEFON bid for a line
intentos: ~ **de ocupación por circuito y por hora** *m pl* TELEFON bids per circuit per hour
interacción *f* GEN interaction; ~ **bilateral alternada** DATOS two-way alternate interaction; ~ **de gestión** DATOS management interaction; ~ **de materia iónica** ELECTRÓN ion-matter interaction; ~ **en monólogo** DATOS monologue interaction
interactividad *f* GEN interactivity
interactivo *adj* GEN interactive
intercalación *f* GEN interleaving; ~ **de bitios** DATOS bit interleaving; ~ **de bloques** TEC RAD block interleaving; ~ **de impulsos** TELEFON, TRANS pulse interleaving
intercalado *adj* GEN interleaved
intercalador *m* GEN interleaver
intercalar *vt* GEN interleave
intercambiabilidad *f* GEN interchangeability
intercambiable *adj* GEN interchangeable; **no** ~ GEN noninterchangeable
intercambiador: ~ **de iones** *m* ELECTRÓN ion exchanger
intercambio *m* GEN interchange; ~ **de autentificación** DATOS, SEÑAL authentication exchange; ~ **de datos** DATOS data interchange (*TDI*); ~ **de datos comerciales** DATOS trade-data interchange (*TDI*); ~ **de datos de comercio de las Naciones Unidas** (*UNTDI*) DATOS United Nations trade-data interchange (*UNTDI*); ~ **entre centrales** TELEFON interexchange; ~ **de estado de terminal de referencia** (*RTSI*) TEC RAD reference terminal status interchange (*RTSI*); ~ **de información** DATOS data interchange; ~ **iónico** ELECTRÓN ion exchange; ~ **de paquetes de Internet** (*IPX*) CONMUT, DATOS Internet packet exchange (*IPX*); ~ **de protocolo de red interna** DATOS internetwork protocol exchange
interceptación: ~ **especial** *f* TELEFON special intercept; ~ **de llamadas** *f* TELEFON interception of calls
interceptar *vt* ELECTRÓN, TELEFON tap
intercomunicación *f* TELEFON intercommunication (*intercom*)
interconectar *vt* GEN interconnect
interconexión *f* GEN interconnection; ~ **de componente periférico** INFO peripheral-component interconnect (*PCI*); ~ **cruzada** ELECTRÓN cross-strapping; ~ **digital** CONMUT digital cross-connect; ~ **multiseñal** (*IMS*) DATOS multisignal interconnect (*MSI*); ~ **de redes** DATOS network interconnection; ~ **con red telefónica local** ELECTRÓN local-network interconnect; ~ **troncal digital** ELECTRÓN digital trunk network; ~ **universal S** TELEFON S-universal interface
interfaz *f* (*I/F*) GEN interface (*I/F*); ~ **de adaptador de aparato** INFO device-adaptor interface; ~ **de casete** INFO cassette interface; ~ **de cinta** INFO tape interface; ~ **de circuito** INFO circuit interface; ~ **de circuito general** INFO general-circuit interface; ~ **de circuito de interconexión** TRANS tie-circuit interface; ~ **de computadora** *AmL* (*cf interfaz de ordenador Esp*) DATOS, INFO computer interface; ~ **de computadora inteligente** *AmL* (*cf interfaz de ordenador inteligente Esp*) DATOS, INFO intelligent-computer interface; ~ **computadora-PBX** *AmL* (*cf interfaz ordenador-PBX Esp*) INFO computer-PBX interface (*CPI*); ~ **contradireccional** TRANS contradirectional interface; ~ **de control de medios** (*MCI*) INFO media-control interface (*MCI*); ~ **de**

datos distribuido por cobre DATOS, TRANS copper-distributed data interface; ~ **de datos distribuidos por fibra** DATOS, TRANS fibre-distributed (*AmE* fiber-distributed) data interface; ~ **de disco rígido** INFO hard-disk interface; ~ **de estación de presentación visual** (*GDI*) INFO graphic-device interface (*GDI*); ~ **física para sincronismo de multiplexor** TRANS multiplexer-timing physical interface (*MTPI*); ~ **de frecuencia básica** DATOS basic-rate interface; ~ **gráfico** INFO graphical interface; ~ **hombre-máquina** INFO man-machine interface; ~ **icónica** INFO icon interface; ~ **de mantenimiento** TELEFON maintenance interface; ~ **matriz-enlace** TELEFON matrix-link interface; ~ **de nodo de red** (*NNI*) TRANS network-node interface (*NNI*); ~ **no de red** TELEFON non-network interface; ~ **normalizada** GEN standard interface; ~ **de ordenador** *Esp* (*cf interfaz de computadora AmL*) DATOS, INFO computer interface; ~ **de ordenador inteligente** *Esp* (*cf interfaz de computadora inteligente AmL*) DATOS, INFO intelligent-computer interface; ~ **ordenador-PBX** *Esp* (*cf interfaz computadora-PBX AmL*) INFO computer-PBX interface (*CPI*); ~ **en paralelo** DATOS parallel interface; ~ **de pequeño sistema informático** (*SCSI*) INFO small computer system interface (*SCSI*); ~ **de programa de aplicación** (*API*) DATOS, INFO application-program interface (*API*); ~ **de programación** DATOS, INFO programming interface; ~ **de red** DATOS, INFO, TELEFON network interface; ~ **de señalización** SEÑAL signal interface; ~ **de tarifa básica** TELEFON basic-rate interface; ~ **terrestre** TEC RAD, TRANS terrestrial interface; ~ **de transmisión** INFO, TRANS transmission interface; ~ **en U** TELEFON U interface; ~ **UCN-CNE** TELEFON UCN-CNE interface; ~ **universal** DATOS, INFO general-purpose interface; ~ **de usuario** APLIC user interface; ~ **de usuario gráfica** (*GUI*) INFO graphical user interface (*GUI*); ~ **usuario-red** DATOS, TRANS user-network interface (*UNI*); ~ **vocal de usuario** APLIC vocal user interface

interferencia *f* GEN interference; ~ **admisible** TEC RAD, TRANS permissible interference; ~ **de banda lateral** TEC RAD, TELEFON, TRANS sideband interference; ~ **de canal propio** TEC RAD, TRANS co-channel interference; ~ **constructiva** TRANS *óptica de fibras* constructive interference; ~ **cuasiimpulsora** TEC RAD quasi-impulsive interference; ~ **destructiva** TRANS *óptica de fibras* destructive interference; ~ **de diversión** TEC RAD deception jamming; ~ **de eclipse** TEC RAD eclipse interference; ~ **eléctrica** INFO electrical interference; ~ **de enlace de línea portadora** TRANS carrier-link interference; ~ **entre canales** TRANS crosstalk; ~ **de frecuencia única** SEÑAL, TEC RAD, TELEFON single-frequency interference; ~ **interzonal** TEC RAD interzone interference; ~ **perjudicial** TEC RAD harmful interference; ~ **radiada** ELECTRÓN radiated interference; ~ **selectiva** ELECTRÓN, TEC RAD selective jamming; ~ **solar** TEC RAD solar interference; ~ **de trayectos múltiples** TEC RAD multipath interference

interferometría *f* PRUEBA interferometry; ~ **con línea de base muy larga** (*VLBI*) PRUEBA very long baseline interferometry (*VLBI*); ~ **plana** TEC RAD, TRANS slab interferometry; ~ **transversal** TRANS transverse interferometry

interferómetro *m* PRUEBA interferometer; ~ **luminoso** DATOS light interferometer

interflujo *m* TELEFON interflow

interfono *m* PERIF, TELEFON interphone, intercommunication system

interfuncionabilidad *f* GEN interworkability

interfuncionamiento: ~ **por diagrama de control de llamadas** *m* (*ICCM*) TELEFON interworking by call-control mapping (*ICCM*); ~ **entre redes** *m* TELEFON interworking between networks; ~ **mediante acceso de puerta** *m* (*IPA*) DATOS interworking by port access (*IPA*); ~ **de plan de numeración** *m* DATOS, TELEFON numbering-plan interworking; ~ **de señalización** *m* SEÑAL signalling (*AmE* signaling) interworking; ~ **de servicio** *m* TELEFON service interworking; ~ **de videotexto** *m* INFO, TV videotext interworking

intermedio *adj* GEN intermediate

intermitencia *f* GEN intermittence

intermodulación *f* TRANS intermodulation; ~ **ionosférica** TEC RAD ionospheric cross-modulation; ~ **pasiva** (*PIM*) TRANS passive intermodulation (*PIM*)

Internet *m* DATOS, INFO Internet

interno *adj* GEN internal

interoficinal *adj* TELEFON interoffice

interoperabilidad *f* GEN interoperability

interpolación *f* PERIF, PRUEBA, TRANS interpolation; ~ **de señales vocales** TRANS interpolation of speech signals; ~ **de voz digital** TRANS digital speech interpolation

interpolado *adj* PERIF, PRUEBA, TRANS interpolated

interpolador *m* PERIF, PRUEBA, TRANS interpolator

interpolar *vt* PERIF, PRUEBA, TRANS interpolate
interponer *vt* TRANS interpose
interposición *f* DATOS, TELEFON interposition
interpretabilidad *f* GEN interpretability
interpretable *adj* GEN interpretable
interpretación *f* GEN interpretation; ~ **semántica** DATOS semantic interpretation
interpretar *vt* GEN interpret
intérprete *m* INFO interpreter
interrogación *f* GEN interrogation, polling; ~ **a distancia** PERIF *de contestador automático* remote interrogation; ~ **de estaciones de barco** TEC RAD ship polling; ~ **de invitación a transmitir** DATOS invitation-to-transmit polling; ~ **múltiple** INFO multipolling; ~ **remota por telecomando** PERIF *de contestador automático* remote-interrogation by telecommand
interrogador *m* TEC RAD interrogator
interrogar *vt* GEN interrogate
interrogatorio *m* INFO quiz
interrumpir *vt* GEN interrupt
interrupción *f* GEN interruption; ~ **de llamada en curso** TELEFON interruption of call in progress; ~ **de operación** TELEFON operation interruption; ~ **de potencia** ELECTROTEC power outage; ~ **de procedimiento negativa** DATOS procedure interrupt negative (*PIN*); ~ **de procedimiento positiva** DATOS procedure interrupt positive; ~ **de proceso negativa** TELEFON procedure interrupt negative (*PIN*); ~ **de proceso positiva** TELEFON procedure interrupt positive; ~ **de red** ELECTRÓN network disruption; ~ **de señal** TELEFON breakdown signal; ~ **de servicio** PRUEBA, TELEFON service interruption; ~ **en el servicio técnico** PRUEBA technical breakdown; ~ **temporal** PRUEBA, TELEFON temporary interruption; ~ **de transmisión** TRANS interruption of transmission
interruptor *m* GEN interrupter, switch; ~ **automático** ELECTRÓN, ELECTROTEC circuit breaker; ~ **auxiliar** CONMUT, INFO auxiliary switch; ~ **basculante** ELECTROTEC tumbler switch; ~ **de botón de presión** CONMUT push-button switch; ~ **de cuchilla** ELECTROTEC knife switch; ~ **a distancia** CONMUT remote switching; ~ **de encendido/apagado** ELECTROTEC power on/off switch; ~ **de fin de carrera** ELECTROTEC limit switch; ~ **de gancho** PERIF, TELEFON cradle switch; ~ **de membrana** ELECTRÓN membrane switch, membrane keyswitch; ~ **de potencia** ELECTROTEC power switch; ~ **principal** ELECTROTEC main circuit breaker; ~ **de protector de motor** ELECTROTEC motor-protector switch; ~ **de red** ELECTROTEC mains switch, on-off switch; ~ **de rueda moleteada** CONMUT, ELECTRÓN thumbwheel switch; ~ **de señal de llamada** TELEFON ringing interruptor
intersincronización *f* TV genlocking
interurbano *adj* TELEFON intercity
intervalo *m* GEN interval, gap; ~ **de alineación de trama** TELEFON, TRANS frame-alignment time slot; ~ **audible** TELEFON range of audibility; ~ **de borrado de línea** TV line-blanking interval; ~ **de borrado vertical** (*VBI*) TV vertical-blanking interval (*VBI*); ~ **de confianza** GEN confidence interval; ~ **de cuantificación** TELEFON quantizing interval; ~ **desprotegido** TELEFON unguarded interval; ~ **de encendido** ELECTRÓN starter gap; ~ **de energía** COMPON *semiconductores* energy gap; ~ **entre bloques** DATOS interblock gap; ~ **de exploración** TEC RAD scanning interval; ~ **de facturación** APLIC charging interval; ~ **de horario** TEC RAD, TV timetable slot; ~ **de impulsos** TELEFON, TRANS pulse interval, pulse spacing; ~ **de impulsos de cómputo** TELEFON meter-pulse interval; ~ **de interacción** ELECTROTEC interaction gap; ~ **de mantenimiento** PRUEBA maintenance interval; ~ **mínimo** TELEFON minimum interval; ~ **mínimo aceptable** TELEFON minimum acceptable interval; ~ **de muestreo** TRANS sampling interval; ~ **de reintento** TEC RAD retry interval; ~ **de supresión de campo** (*FBI*) TV field-blanking interval (*FBI*); ~ **de supresión de imagen** TV horizontal-blanking interval; ~ **de tarifa** TELEFON tariff time slot; ~ **de tiempo de canal** TELEFON, TRANS channel time slot; ~ **de tiempo de dígito justificable** TELEFON, TRANS justifiable digit time slot; ~ **de tiempo de señalización** SEÑAL signalling (*AmE* signaling) time slot; ~ **de tiempo de un solo dígito** TELEFON digit time slot; ~ **de tiempo de tarifa modulada** DATOS modulated-tariff time slot; ~ **de tiempo de verificación de símbolo** TRANS check-symbol time slot; ~ **de tolerancia** PRUEBA tolerance range; ~ **unitario** PRUEBA unit interval
interzonal *adj* ELECTRÓN, TEC RAD interzonal
intracodificación *f* INFO, TV intracoding
intrasladable *adj* INFO unrelocatable
introducción *f* DATOS entry; ~ **de datos** INFO data entry; ~ **progresiva** TEC RAD phasing-in
introducir *vt* DATOS enter
intrusión *f* ELECTRÓN, TELEFON intrusion
INTTR *abr* (*central de tránsito internacional*) CONMUT, TELEFON INTTR (*international transit exchange*)

inválido *adj* GEN invalid
inventario *m* GEN inventory
inversión *f* GEN inversion; ~ **alterna de bitios** TRANS alternate bit inversion; ~ **alterna de dígitos** TRANS alternate digit inversion; ~ **alternativa de señales** (*AMI*) SEÑAL, TELEFON, TRANS alternate mark inversion (*AMI*); ~ **de fase** ELECTROTEC phase inversion; ~ **de frecuencia** TEC RAD frequency inversion; ~ **de líneas** TELEFON line reversal; ~ **de marca de código** TRANS code-mark inversion (*CMI*); ~ **de polaridad** ELECTROTEC pole changing, pole reversal; ~ **de polaridad de línea** TRANS line-polarity inversion, line-polarity reversal; ~ **de temperatura** TEC RAD temperature inversion
inversor: ~ **de fase** *m* ELECTROTEC phase inverter; ~ **de polaridad** *m* ELECTROTEC pole changer, pole inverter
invertir *vt* GEN invert
investigación *f* GEN examination, research; ~ **aplicada** PRUEBA applied research; ~ **básica** PRUEBA basic research; ~ **y desarrollo** (*I & D*) PRUEBA research and development (*R & D*); ~ **heurística** DATOS heuristic search; ~ **de tráfico** TELEFON traffic research
investigador *m* GEN researcher
investigar *vt* GEN investigate
invitación: ~ **a transmitir** *f* DATOS invitation to transmit
invocación: ~ **de entidad N** *f* DATOS N-entity invocation; ~ **de proceso de aplicación** *f* DATOS application-process invocation; ~ **de TPSU** *f* DATOS TPSU invocation; ~ **de unidad de aplicación** *f* (*AEI*) DATOS application-entity invocation (*AEI*); ~ **de usuario de servicio de procesamiento de transacciones** *f* (*TPSUI*) DATOS transaction-processing service user invocation (*TPSUI*)
invocar *vt* TEC RAD, TELEFON invoke
inyección: ~ **lograda** *f* TEC RAD *de satélite* successful injection
inyector: ~ **de aceite** *m* COMPON oil gun
ion *m* ELECTRÓN, TEC RAD ion
ionización *f* ELECTRÓN, TEC RAD ionization; ~ **esporádica** TEC RAD sporadic ionization
ionograma *m* TEC RAD ionogram
ionoscopio *m* ELECTRÓN ionoscope
ionosfera *f* TEC RAD ionosphere; ~ **superior** TEC RAD upper ionosphere
ionosonda *f* TEC RAD ionosonde
IPA *abr* (*interfuncionamiento mediante acceso de puerta*) DATOS IPA (*interworking by port access*)
IPE *abr* (*integración a pequeña escala*) COMPON, ELECTRÓN, TELEFON, TRANS SSI (*small-scale integration*)
IPME *abr* (*entorno de mensajería interpersonal*) TELEFON IPME (*interpersonal messaging environment*)
IPMS *abr* (*sistema de mensajería interpersonal*) TELEFON IPMS (*interpersonal messaging system*)
IPN *abr* (*notificación interpersonal*) TELEFON IPN (*interpersonal notification*)
IPX *abr* (*intercambio de paquetes de Internet*) CONMUT, DATOS IPX (*Internet packet exchange*)
IR *abr* (*infrarrojo*) GEN IR (*infrared*)
IRA *abr* (*alfabeto de referencia internacional*) TELEFON IRA (*international reference alphabet*)
irradiación *f* ELECTRÓN, TEC RAD, TRANS radiation, irradiation
irradiancia *f* ELECT, POWER, TRANS irradiance; ~ **espectral** GEN spectral irradiance
irregularidad: ~ **de servicio** *f* PRUEBA service irregularity
irrestricto *adj* GEN unrestricted
irreversible *adj* GEN irreversible, nonreversible
ISA *abr* (*arquitectura industrial estándar*) DATOS, INFO ISA (*industry standard architecture*)
ISB *abr* (*banda lateral independiente*) TEC RAD ISB (*independent sideband*)
ISG *abr* (*generador de señal de interrogación*) TEC RAD ISG (*interrogation signal generator*)
ISI *abr* (*perturbación entre símbolos*) INFO, TRANS, TV ISI (*intersymbol interference*)
ISL *abr* (*enlace entre satélites*) TEC RAD, TV inter-satellite link (*ISL*), satellite-to-satellite link
isla *f* COMPON, ELECTRÓN *circuitos impresos* land
ISLS *abr* (*supresión del lóbulo lateral del interrogador*) TEC RAD ISLS (*interrogator sidelobe suppression*)
ISMC *abr* (*centro de mantenimiento de conmutación internacional*) CONMUT ISMC (*international switching maintenance centre*)
ISO *abr* (*Organización Internacional de Normalización*) GEN ISO (*International Standards Organization*)
isócrono *adj* TRANS isochronous
isotrópico *adj* ELECTRÓN, TEC RAD isotropic
ISPC *abr* (*centro de programas sonoros internacional*) TV ISPC (*international sound-programme centre*)
IST *abr* (*prueba de satélite integrada*) PRUEBA IST (*integrated satellite test*)
ISU *abr* (*unidad de señal inicial*) SEÑAL ISU (*initial signal unit*)
ITA2 *abr* (*alfabeto telegráfico internacional nº 2*) DATOS, TELEFON ITA2 (*international telegraph alphabet no. 2*)

itinerante *adj* TEC RAD itinerant
itinerario: ~ **básico** *m* INFO base line
ITMC *abr* (*centro internacional de mantenimiento de la transmisión*) PRUEBA, TRANS ITMC (*international transmission-maintenance centre*)
ITS *abr* (*sistemas inteligentes de enseñanza asistida*) INFO ITS (*intelligent tutoring systems*)
ITV *abr* (*televisión educativa*) TV ITV (*instructional television*)
IUT *abr* (*instalación en prueba*) PRUEBAS IUT (*implementation under test*)
IVHS *abr* (*sistemas inteligentes para vehículos en carretera*) TEC RAD IVHS (*intelligent vehicle highway systems*)

J

jack: ~ **de casco** *m* COMPON, PERIF headset jack; ~ **de escucha** *m* COMPON, TELEFON listening jack; ~ **modular modificado** *m* (*MMJ*) COMPON, ELECTRÓN modified modular jack (*MMJ*); ~ **en paralelo** *m* COMPON, ELECTRÓN parallel jack; ~ **de salida** *m* COMPON, INFO output jack; ~ **de salida blindado de chapa frontal triple** *m* COMPON, ELECTRÓN shielded triple-faceplate outlet jack; ~ **de servicio** *m* COMPON, TELEFON service jack; ~ **de voz normal** *m* COMPON, TELEFON standard-voice jack

jefe: ~ **de instalación** *m* GEN installation site manager; ~ **de obras** *m* GEN construction manager; ~ **de pruebas** *m* PRUEBA testing manager; ~ **de tráfico** *m* DATOS, TELEFON traffic superintendent

jerarquía *f* GEN hierarchy; ~ **confusa** INFO tangled hierarchy

jerárquico *adj* GEN hierarchical

JFET *abr* (*transistor con efecto de campo de enlace*) COMPON, ELECTRÓN JFET (*junction field effect transistor*)

JFIF *abr* (*formato de intercambio de ficheros JPEG*) INFO JFIF (*JPEG file interchange format*)

joystick *m* INFO, PERIF joystick

JSM *abr* (*módulo de almacenamiento del puente de conexión*) ELECTRÓN JSM (*jumper storage module*)

juego: ~ **automático** *m* INFO auto-play; ~ **de caracteres** *m* INFO character set, font; ~ **de caracteres de sistema** *m* INFO system font; ~ **de disparo** *m* INFO trigger play; ~ **de ensayo ejecutable** *m* PRUEBA executable test case; ~ **de ensayo genérico** *m* PRUEBA generic test case; ~ **de herramientas** *m* COMPON toolkit; ~ **de instrucciones** *m* INFO instruction set; ~ **de matrices** *m* COMPON, ELECTRÓN die set; ~ **de piezas de repuesto** *m* COMPON spare-parts kit; ~ **de relés de distribución** *m* CONMUT distribution relay set; ~ **de relés de línea de enlace** *m* CONMUT junction-line relay set; ~ **de relés de línea urbana** *m* CONMUT exchange-line junction set; ~ **de relés de posición** *m* CONMUT position relay set; ~ **de resortes de contacto de ruptura** *m* CONMUT break-contact spring set; ~ **de selectores** *m* TELEFON rank of selectors

junta *f* GEN joint; ~ **de dilatación** COMPON expansion joint; ~ **de estanqueidad** COMPON gasket; ~ **giratoria** COMPON rotating joint; ~ **de guíaondas** COMPON waveguide gasket; ~ **hermética de cierre** COMPON closure gasket; ~ **hermética de plomo** COMPON lead seal; ~ **óptica** TRANS optical splice; ~ **roscada** COMPON screw joint; ~ **soldada** COMPON soldered joint

juntera *f* COMPON jointer

juntor: ~ **de línea de abonado** *m* TELEFON subscriber-line junctor

justificación *f* INFO, TELEFON, TRANS justification; ~ **negativa** TELEFON, TRANS negative justification; ~ **positiva** TELEFON, TRANS positive justification; ~ **positiva-negativa** TELEFON, TRANS positive-negative justification

justificar *vt* INFO, TELEFON, TRANS justify

K

kaledia *f* INFO kaledia
kc *abr* (*kilociclo*) PRUEBA kc (*kilocycle*)
kHz *abr* (*kilohertzio*) PRUEBA kHz (*kilohertz*)
kilobaudio *m* TRANS kilobaud
kilociclo *m* (*kc*) PRUEBA kilocycle (*kc*)
kilohertzio *m* (*kHz*) PRUEBA kilohertz (*kHz*)
kiloinstrucciones: **~ por segundo** *f pl* (*KIPS*) INFO kilo-instructions per second (*KIPS*)
kilooperaciones: **~ por segundo** *f pl* (*KOPS*) INFO kilo-operations per second (*KOPS*)
KIPS *abr* (*kiloinstrucciones por segundo*) INFO KIPS (*kilo-instructions per second*)
klistron *m* ELECTRÓN, TEC RAD klystron; **~ de cavidad múltiple** ELECTRÓN multicavity klystron; **~ de reflexión** ELECTRÓN reflex klystron
KMS *abr* (*sistema de gestión del conocimiento*) INFO KMS (*knowledge management system*)
KOPS *abr* (*kilooperaciones por segundo*) INFO KOPS (*kilo-operations per second*)
krarupización *f* TRANS krarupization
krarupizar *vt* TRANS krarupize, load continuously

L

LA *abr* (*aplicación local*) TELEFON LA (*local application*)

laboratorio *m* GEN laboratory

laca *f* COMPON lacquer, shellac; **~ blanca** COMPON white lacquer; **~ de inmersión** COMPON dipping lacquer; **~ protectora** COMPON resist lacquer; **~ protectora contra soldadura** COMPON solder-resist lacquer; **~ de retoque** COMPON touch-up lacquer; **~ de soldadura** COMPON soldering lacquer

laguna *f* COMPON *semiconductores* vacancy

lámina: **~ de conexión** *f* ELECTRÓN, ELECTROTEC bonding sheet; **~ de contactos** *f* TELEFON bank of contacts; **~ de desgaste** *f* COMPON wear blade; **~ de entrehierro** *f* COMPON residual plate; **~ de fusible** *f* ELECTROTEC fuse strip

laminación *f* COMPON lamination, laminating

laminar *vt* COMPON laminate

lámpara: **~ de alarma de averías** *f* PRUEBA failure-alarm lamp; **~ de descarga luminiscente** *f* ELECTRÓN glow-discharge lamp; **~ de efluvios** *f* ELECTRÓN glow lamp; **~ de estado** *f* PRUEBA status lamp; **~ de fin de conversación** *f* TELEFON clearing lamp; **~ fluorescente** *f* ELECTRÓN fluorescent lamp; **~ indicadora** *f* PRUEBA indicator lamp; **~ indicadora de avería** *f* PRUEBA fault-alarm lamp; **~ de línea libre** *f* TELEFON free-line signal (*FLS*); **~ de llamada** *f* PERIF calling lamp, TELEFON line lamp; **~ de mano** *f* PRUEBA inspection lamp; **~ de ocupación** *f* TELEFON seizure lamp; **~ portátil** *f* PRUEBA inspection lamp; **~ de posición** *f* TELEFON locating lamp; **~ de recepción** *f* TELEFON receive lamp; **~ de recepción de datos** *f* TELEFON receive-data lamp; **~ de respuesta** *f* TELEFON answer lamp; **~ de señalización** *f* INFO display light

lanzadera *f* INFO shuttle; **~ espacial** TEC RAD, TV space shuttle

lapeado *m* COMPON lapping

lápiz: **~ fotosensible** *m* INFO, PERIF light pen; **~ de lectura** *m* PERIF *código de barras* wand reader

laquear: **~ con pistola** *vt* COMPON spray-lacquer

laringófono *m* TEC RAD, TELEFON laryngophone, throat microphone

láser *m* ELECTRÓN, PERIF, PRUEBA, TRANS laser; **~ de diodo** COMPON, ELECTRÓN, TRANS diode laser; **~ de doble guía sintonizable** ELECTRÓN tunable twin-guide laser; **~ de frecuencia única** ELECTRÓN, TRANS single-frequency laser; **~ monomodal** ELECTRÓN, TRANS single-mode laser; **~ multimodo** ELECTRÓN, TRANS multimode laser; **~ de rayos X** ELECTRÓN X-ray laser; **~ semiconductor** COMPON, ELECTRÓN, TRANS semiconductor laser; **~ sincronizado por inyección** ELECTRÓN injection-locked laser; **~ TTG** ELECTRÓN TTG laser

lata *f* ELECTRÓN can

latencia *f* INFO latency

lazo: **~ de sincronización de fase** *m* ELECTRÓN phase-locked loop (*PLL*)

LBA *abr* (*dirección de bloque lógico*) INFO LBA (*logical block address*)

LCC *abr* (*cápsula sin patillas*) ELECTRÓN LCC (*leadless chip-carrier*)

LD *abr* (*diodo láser*) ELECTRÓN, TRANS LD (*laser diode*)

lector *m* GEN reader; **~ de caracteres** APLIC character reader; **~ de código de barras** PERIF bar-code scanner; **~ de discos compactos** INFO CD reader; **~ de discos flexibles** INFO floppy-disk reader; **~ de identificadores** PERIF badge reader; **~ de microformas** PERIF microform reader; **~ de tarjetas** PERIF card reader; **~ de tarjetas inteligentes** PERIF smartcard reader

lectora: **~ de curvas** *f* INFO curve follower; **~ de lista de control** *f* DATOS tally reader

lector/impresor: **~ de microformas** *m* PERIF microform reader/printer

lectura *f* GEN read, reading; **~ de caracteres** APLIC character reading; **~ de contacto mecánico** TV mechanical-contact reading; **~ destructiva** INFO destructive read; **~ por diodo electroluminiscente** INFO LED read-out; **~ a distancia** GEN remote reading; **~ remota** GEN remote reading; **~ remota de medidor** TV remote meter reading; **~ de salida** INFO read-out; **~ terminal** ELECTROTEC, PRUEBA terminated reading; **~ de trama temporal** INFO, TRANS timescale reading; **~ visual** INFO visual read-out

leer *vt* GEN read

lenguaje *m* GEN language; **~ de base de datos**

DATOS, INFO database language; ~ **CCITT de alto nivel** TELEFON CCITT high-level language; ~ **de consulta** DATOS query language; ~ **de control** ELECTRÓN, INFO, TELEFON control language; ~ **de control de red** (*NCL*) ELECTRÓN, TELEFON network control language (*NCL*); ~ **CPL/1** INFO CPL/1 language; ~ **de descripción** DATOS description language; ~ **de descripción de datos** DATOS, INFO data-description language; ~ **de descripción de documento** DATOS document-description language; ~ **de descripción y especificación** (*SDL*) DATOS, INFO specification and description language (*SDL*); ~ **de descripción de página** (*PDL*) DATOS, INFO page-description language (*PDL*); ~ **de descripción de soporte físico** (*HDL*) INFO hardware-description language (*HDL*); ~ **ensamblador** INFO assembly language; ~ **especializado** INFO special-purpose language; ~ **de especificación** GEN specification language; ~ **estándar generalizado de referencia** (*SGML*) INFO standard generalized markup language (*SGML*); ~ **estratificado** INFO stratified language; ~ **fuente** INFO source language; ~ **hombre-máquina** (*MML*) INFO man-machine language (*MML*); ~ **de interrogación de texto íntegro estructurado** DATOS structured full-text query language; ~ **de interrogación estructurado** (*SQL*) DATOS structured query language (*SQL*); ~ **de manipulación de datos** DATOS data-manipulation language; ~ **de máquina** INFO machine language; ~ **objeto** INFO object language, target language; ~ **de órdenes** INFO command language; ~ **de órdenes de imprimir** (*PCL*) DATOS print-command language (*PCL*); ~ **PL/M** INFO PL/M language; ~ **de programa de control** INFO control-program language; ~ **de referencia de hipertexto** (*HTML*) DATOS, INFO hypertext markup language (*HTML*); ~ **técnico** GEN technical language

lengüeta *f* COMPON reed

lente: ~ **de caja oval** *f* TEC RAD egg-box lens; ~ **escalonada** *f* TEC RAD zoned lens; ~ **de lámpara** *f* ELECTROTEC lamp lens; ~ **de placas paralelas** *f* TEC RAD parallel-plate lens; ~ **plana E** *f* TEC RAD E-plane lens; ~ **de retardo** *f* TEC RAD delay lens

leva: ~ **de impulsos** *f* PERIF pulse cam disc (*AmE* disk), pulse disc (*AmE* disk)

ley *f* GEN law; ~ **A** TRANS A-law; ~ **de codificación** TELEFON, TRANS coding law, encoding law; ~ **de coseno de Lambert** ELECTRÓN, TRANS *óptica de fibras* Lambert's cosine law; ~ **de emisión de coseno** TRANS *óptica de fibras* cosine-emission law; ~ **de Faraday** ELECTROTEC Faraday's law; ~ **de Gauss** PRUEBA *estadística* Gauss's law; ~ **de Joule** ELECTRÓN, ELECTROTEC Joule's law; ~ **de Laplace** GEN *matemática, física* Laplace's law; ~ **de Snell** TRANS Snell's law

leyes: ~ **de Kirchhoff** *f pl* ELECTRÓN, ELECTROTEC Kirchhoff's laws

lezna *f* COMPON broach

LFN *abr* (*nombre de fichero lógico*) INFO LFN (*logical file name*)

LIB *abr* (*placa interfaz de línea*) CONMUT LIB (*line interface board*)

liberación: ~ **de abonado llamado** *f* CONMUT, TELEFON called-subscriber release

libre *adj* TELEFON idle

LIC *abr* (*circuito de interconexión de línea*) CONMUT LIC (*line interface circuit*)

licencia: ~ **de explotación** *f* GEN operating licence (*AmE* license); ~ **de producción** *f* PRUEBA production permit

líder: ~ **de grupo funcional** *m* DATOS functional group leader

LIFO *abr* (*último en entrar, primero en salir*) INFO LIFO (*last in, first out*)

ligadura *f* COMPON, INFO, TELEFON tie; ~ **simple** COMPON single lashing

lima: ~ **de aguja** *f* COMPON needle file; ~ **de bonete** *f* COMPON barrette file

limitación *f* GEN limitation, constraint; ~ **de cresta** TELEFON, TRANS peak limiting

limitado *adj* GEN restricted

limitador *m* ELECTRÓN, ELECTROTEC limiter; ~ **de conversaciones** TELEFON conversation-time limiter; ~ **de Schmitt** ELECTRÓN Schmitt limiter; ~ **de tensión** ELECTROTEC voltage limiter

limitar *vt* GEN limit

límite *m* GEN limit, boundary; ~ **de amplitud de respuesta** TRANS amplitude-response limit; ~ **de aviso** CONTROL warning limit; ~ **de carácter** APLIC *reconocimiento de caracteres* character boundary; ~ **de célula** TEC RAD cell boundary; ~ **de clase** PRUEBA *estadística* class limit; ~ **de cobertura estadística** PRUEBA statistical-coverage limit; ~ **de confianza** GEN confidence limit; ~ **de elasticidad** COMPON tensile-yield limit; ~ **entre celdas** TEC RAD cell border; ~ **de fluencia** PRUEBA *resistencia* creep modulus; ~ **inferior** PRUEBA lower limit; ~ **de objeto gestionado** DATOS managed-object boundary; ~ **de potencia** ELECTRÓN power limit; ~ **de regulación** PRUEBA setting range; ~ **de rendimiento** COMPON yield limit; ~

de ruido TEC RAD noise limit; ~ **superior** PRUEBA upper limit; ~ **superior de audibilidad** TEC RAD, TELEFON threshold of discomfort; ~ **de tolerancia** PRUEBA tolerance limit

límites: ~ **de control** *m pl* PRUEBA control limits; ~ **de escala** *m pl* PRUEBA scale range; ~ **de exposición** *m pl* GEN exposure limits; ~ **de zona** *m pl* TEC RAD zone boundaries

línea *f* TEC RAD, TELEFON line, trunk; ~ **de abonado** TELEFON subscriber loop, subscriber's line, subscriber line; ~ **de abonado averiada** TELEFON subscriber-line out of order; ~ **de abonado ocupada** TELEFON subscriber-line busy; ~ **aérea** ELECTROTEC, TRANS overhead line; ~ **alquilada** (*LS*) DATOS, TELEFON, TRANS leased line (*LS*); ~ **con alternación de fase** TV phase-alternate line; ~ **artificial** TRANS line simulator; ~ **de base aumentada** TEC RAD augmented baseline; ~ **bidireccional** TRANS both-way line; ~ **de cambio de fecha** TEC RAD data line; ~ **de central** CONMUT, TELEFON exchange line; ~ **central de pruebas de transmisión** (*TTT*) PRUEBA, TELEFON, TRANS transmission-test trunk (*TTT*); ~ **central de trazo** INFO stroke centre (*AmE* center) line; ~ **de comandos** INFO command line; ~ **compartida** TELEFON party line; ~ **compartida por dos abonados** TELEFON two-party line; ~ **de comportamiento** DATOS behaviour (*AmE* behavior) line; ~ **conformadora de impulsos** TEC RAD pulse-forming line; ~ **conmutada** CONMUT, TELEFON dial-up line; ~ **de construcción emparedada** TRANS sandwich line; ~ **directa** TELEFON, TRANS through line; ~ **doble** APLIC double-circuit line; ~ **de enlace** CONMUT, TELEFON, TRANS tie line; ~ **entre cuadros conmutadores** TRANS interswitchboard line; ~ **especializada** TELEFON specialized line; ~ **espectral** TRANS spectrum line; ~ **de exploración** INFO, TV scan line, scanning line; ~ **de extensión** CONMUT extension line; ~ **externa privada** CONMUT, TELEFON *PBX* exclusive external line; ~ **de flujo** GEN flow line; ~ **H** TEC RAD H line; ~ **de identificación** DATOS identification line; ~ **internacional** TELEFON international line; ~ **interurbana de intercambio** TELEFON interchange trunk; ~ **de larga distancia** TELEFON, TRANS long-distance line; ~ **libre** TELEFON spare line, unoccupied line; ~ **local** TELEFON local line; ~ **de mantenimiento** TELEFON line of maintenance; ~ **marcada como prioritaria** TELEFON priority-marked line; ~ **de órdenes de ingeniería** TELEFON engineering order wire (*EOW*); ~ **de petición de orden** INFO prompt line; ~ **plana** TEC RAD stripline; ~ **preprogramada** TELEFON preprogrammed line; ~ **de previo pago** TELEFON coin-box line; ~ **privada** TELEFON tie trunk; ~ **pública** TELEFON public line; ~ **radiante de sector anular** TEC RAD annular-sector radiating line; ~ **ranurada** TEC RAD slotted line; ~ **ranurada de medida** TEC RAD slotted measuring line; ~ **de referencia de espaciado de caracteres** INFO character-spacing reference line; ~ **de retardo** DATOS delay line; ~ **de retardo acústico** DATOS acoustic-delay line; ~ **de servicio** ELECTRÓN operating line, TELEFON service line; ~ **de servicio compartido** TELEFON shared-service line; ~ **subterránea** TRANS underground line; ~ **suelta** INFO orphan; ~ **superconductora** TRANS superconductor line; ~ **telefónica** TELEFON telephone line; ~ **telefónica principal** TELEFON main telephone line; ~ **terrestre** TELEFON, TRANS land line; ~ **de tierra** TEC RAD baseline; ~ **de transmisión** TEC RAD, TRANS transmission line; ~ **de transmisión con transposiciones** TELEFON, TRANS transposed transmission line; ~ **de transporte de energía** ELECTROTEC power line, power-transmission line; ~ **con transposiciones** TELEFON, TRANS transposed line; ~ **trazada con gramil** COMPON scribed line; ~ **trifilar** CONMUT, TELEFON three-wire line; ~ **de triple acceso** CONMUT, TELEFON triple-access line; ~ **troncal de arranque en bucle** TELEFON loop-start trunk; ~ **troncal de desconexión** TELEFON release trunk; ~ **troncal de enlaces de desconexión** (*RLT*) TELEFON release-link trunk (*RLT*); ~ **troncal entre tándems** TELEFON intertandem trunk; ~ **troncal de grupo intermarcador** TELEFON intermarker group trunk; ~ **troncal interlocal** TELEFON interlocal trunk; ~ **troncal interna de oficina** TELEFON intraoffice trunk *AmE*, intrachange trunk *BrE*; ~ **troncal parlante** TELEFON announcement trunk; ~ **troncal de pruebas de transmisión** (*TTT*) PRUEBA, TELEFON, TRANS transmission-test trunk (*TTT*); ~ **de unión** TELEFON junction line; ~ **vacía** TELEFON blank line; ~ **vectorial** TEC RAD vector line; ~ **de visión** TEC RAD line of sight; ◆ **en** ~ INFO, TELEFON on-line

lineal *adj* GEN linear; **no** ~ GEN nonlinear

linealidad *f* GEN linearity; **no** ~ GEN nonlinearity; **no** ~ **de línea** TEC RAD line nonlinearity

linealizador *m* TRANS linearizer

líneas: ~ **mixtas** *f pl* TELEFON mixed lines; ~ **múltiples a la misma dirección** *f pl* TELEFON multiple lines at the same address; ~ **por pantalla** *f pl* TV lines per screen

LIPS *abr* (*inferencias lógicas por segundo*) PRUEBA LIPS (*logical interferences per second*)

líquido *m* GEN liquid; ~ **para frenos** COMPON brake fluid

LI/S *abr* (*inferencias lógicas por segundo*) PRUEBA LI/S (*logical inferences per second*)

lista: ~ **de cables** *f* TRANS cable list; ~ **de comprobación de clientes** *f* DATOS client checklist; ~ **de conexiones enrolladas** *f* COMPON wire-wrap connection list; ~ **de configuración de canales** *f* TRANS channel-mapping list; ~ **de control** *f* PRUEBA checklist; ~ **de desplazamiento ascendente** *f* DATOS, INFO push-up list; ~ **de desplazamiento descendente** *f* DATOS, INFO push-down list; ~ **de direcciones** *f* DATOS mailing list; ~ **de distribución** *f* DATOS, INFO, TELEFON distribution list (*DL*); ~ **de documentos** *f* GEN document list; ~ **de grupos de cables** *f* TRANS *documentación* cable group list; ~ **de palabras clave** *f* INFO keyword list; ~ **de ruta** *f* TELEFON route list

listado[1]: **no** ~ *adj* APLIC, TELEFON ex-directory *BrE*, unlisted *AmE*

listado[2] *m* DATOS, INFO listing

listar: **sin** ~ *fra* TELEFON unlisted

listo: ~ **para activar** *adj* GEN ready to activate; ~ **para datos** *adj* TRANS ready for data; ~ **para enviar** *adj* DATOS ready for sending

literal *m* INFO literal

litio *m* COMPON lithium

LIU *abr* (*unidad de interconexión de guía de luz*) ELECTRÓN LIU (*lightguide interconnection unit*)

liviano *adj* ELECTRÓN lightweight

llamada *f* GEN call, TELEFON call; ~ **abreviada** TELEFON abbreviated dialling (*AmE* dialing); ~ **de agenda** APLIC, TELEFON diary call; ~ **de alarma** APLIC, TELEFON alarm call; ~ **con aviso previo** TELEFON preavis call; ~ **a cobro revertido** TELEFON transferred-charge call; ~ **en cola de espera** TELEFON waiting call; ~ **de conferencia múltiple** APLIC, TELEFON conference calling; ~ **en curso** INFO *de subrutina* calling into execution; ~ **desde la red hacia fuera** TRANS on-net to off-net call; ~ **de dirección múltiple** TELEFON multiaddress calling; ~ **directa** TELEFON direct call; ~ **doble** TELEFON double call; ~ **de emergencia** APLIC, TELEFON emergency call; ~ **de emisión doble** DATOS multicast call; ~ **entrante/~ saliente** TELEFON dial in/dial out; ~ **falsa** TELEFON false call; ~ **fantasma** TRANS ghost call; ~ **frustrada** TELEFON unsuccessful call; ~ **general** TELEFON general call; ~ **global** TEC RAD global call; ~ **gratuita** TELEFON freephone call *BrE*, toll-free call *AmE*; ~ **de grupo** TELEFON group call; ~ **inefectiva** TELEFON ineffective call; ~ **intermitente** TELEFON interrupted ringing; ~ **interna** TELEFON inside call; ~ **interurbana** TELEFON toll call; ~ **de larga distancia** TELEFON long-distance call; ~ **de línea troncal de tarifa urbana** TELEFON local charge-rate trunk call; ~ **local** TELEFON local call; ~ **lograda** TELEFON successful call; ~ **maliciosa** TELEFON malicious call; ~ **metropolitana** TELEFON city call; ~ **nacional** TELEFON inland call; ~ **no atendida** PERIF, TELEFON unanswered call; ~ **a número de grupo** TELEFON group number call; ~ **de operador a operador** TELEFON attendant-to-attendant calling; ~ **PBX** TELEFON PBX call; ~ **perdida** CONMUT, TELEFON, TRANS lost call; ~ **de persona a persona** TELEFON person-to-person call; ~ **con prioridad** TELEFON priority call; ~ **de procedimiento** TEC RAD procedure call; ~ **de prueba de extremo a extremo** PRUEBA, TELEFON end-to-end test call; ~ **recordatoria** APLIC, TELEFON reminder call; ~ **redireccionada** TELEFON redirected call; ~ **de repetición** TELEFON repeat call; ~ **retenida** APLIC, TELEFON call hold; ~ **de salida** TELEFON outgoing call; ~ **selectiva** TELEFON selective calling, selective ringing; ~ **de servicio** TELEFON service call; ~ **simplificada** TELEFON abbreviated dialling (*AmE* dialing); ~ **simultánea** TELEFON simultaneous call; ~ **de tarifa urbana** TELEFON local charge-rate call; ~ **con tarjeta de crédito** TELEFON credit-card call; ~ **por teletexto** TELEFON teletext call; ~ **de télex** TELEFON telex call; ~ **terminada** TELEFON completed call; ~ **por tono** TELEFON tone call; ~ **tridireccional** TEC RAD, TELEFON three-way call; ~ **urgente** TELEFON urgent call; ~ **a usuario de servicio CMISE** DATOS invoking CMISE service user; ~ **virtual** DATOS, TELEFON virtual call; ◆ **hacer una ~ a cobro revertido** *vt* APLIC, TELEFON call collect *AmE*, reverse the charges *BrE*

llamadas: **por** ~ *adv* TEC RAD, TELEFON on a per-call basis

llamador *m* GEN caller, requester

llamar[1] *vt* GEN call, TELEFON call, phone, telephone; ◆ ~ **a alguien** TELEFON call someone; ~ **al teléfono** TELEFON phone, telephone; ~ **por teléfono** TELEFON phone, telephone; ~ **sin cargo** TELEFON call toll-free;

~ **por el teléfono interno** TELEFON call over the intercom

llamar[2] *vi* GEN call, TELEFON call, phone, telephone

llave: ~ **de apertura** *f* INFO unlock key; ~ **fija** *f* COMPON spanner *BrE*, wrench *AmE*; ~ **de tuercas** *f* COMPON spanner *BrE*, wrench *AmE*

llenar *vt* GEN fill

lleno *adj* GEN full

LME *abr* (*entidad para la gestión de capas*) DATOS LME (*layer management entity*)

LMSS *abr* (*servicio móvil terrestre de satélite*) TEC RAD, TELEFON, TRANS LMSS (*land-mobile satellite service*)

lm/W *abr* (*lumen por watt*) PRUEBA lm/W (*lumen per watt*)

lóbulo *m* ELECTRÓN, TEC RAD *antenas* lobe; ~ **lateral** TEC RAD side lobe; ~ **lateral de ángulo grande** TEC RAD wide-angle side lobe; ~ **lateral de antena** TEC RAD aerial side lobe, antenna side lobe; ~ **posterior** TEC RAD back lobe; ~ **principal** TEC RAD main lobe

localización *f* GEN localization, location, locating; ~ **automática** INFO auto-locate; ~ **de avería** PRUEBA fault localization, fault isolation, fault tracing; ~ **de cables** TRANS cable localization; ~ **general** PRUEBA general localization; ~ **de llamadas maliciosas** TELEFON malicious-call tracing; ~ **no supervisada** ELECTRÓN nonsupervised location; ~ **de onda portadora** TRANS carrier detect; ~ **de posición inicial** INFO home spot; ~ **vía satélite** TEC RAD satellite location

localizador *m* GEN locater, locator, TEC RAD pager; ~ **de cables** TRANS cable locater; ~ **de coordenadas** INFO locater, locator; ~ **de emisoras** ELECTRÓN intercept receiver; ~ **de recursos uniforme** (*URL*) DATOS *Internet* uniform resource locator (*URL*); ~ **de renglón** TELEFON line finder

localizar *vt* GEN locate

log *abr* (*logaritmo*) TEC RAD log (*logarithm*)

logaritmo *m* (*log*) TEC RAD logarithm (*log*)

logátomo *m* TEC RAD, TELEFON *acústica* logatom

lógica *f* GEN logic; ~ **acoplada de emisor positivo** INFO positive emitter-coupled logic; ~ **de control de memoria** INFO memory-control logic; ~ **débil** INFO soft logic; ~ **de decisión mayoritaria** ELECTRÓN majority-decision logic; ~ **formal** INFO formal logic; ~ **de generación de caracteres** INFO character-generation logic; ~ **de núcleos** INFO core logic; ~ **plurívoca** ELECTRÓN many-valued logic; ~ **temporal** DATOS temporal logic; ~ **de trama vertical** TV vertical-grating logic

logística *f* GEN logistics; ~ **de apoyo al mantenimiento** PRUEBA maintenance-support logistics; ~ **asistida por computadora** *AmL* (*cf logística asistida por ordenador Esp*) INFO computer-aided logistics; ~ **asistida por ordenador** *Esp* (*cf logística asistida por computadora AmL*) INFO computer-aided logistics; ~ **de sistema central de multiproveedores** GEN multivendor host support

longevidad *f* COMPON, DATOS, PRUEBA longevity

longitud *f* GEN length, TEC RAD *posición* longitude; ~ **de bloque** DATOS frame length; ~ **de cable** TRANS cable length; ~ **de coherencia** TRANS coherence length; ~ **de corte** COMPON *cables* cutting length; ~ **de equilibrio** TRANS equilibrium length; ~ **de fila** TELEFON row length; ~ **de grupo** (*GL*) TELEFON group length (*GL*); ~ **de kevlar** ELECTRÓN length of kevlar; ~ **de línea de exploración total** TELEFON total scanning line length; ~ **de línea de exploración utilizable** TELEFON usable scanning line length; ~ **de lóbulo** ELECTRÓN lobe length; ~ **máxima** TV maximum length; ~ **de onda** TEC RAD, TRANS wavelength; ~ **de onda de corte** TEC RAD, TRANS cutoff wavelength; ~ **de onda eléctrica** TEC RAD electrical wavelength; ~ **de onda de entrada** ELECTRÓN, TRANS threshold wavelength; ~ **de onda en la guía** TEC RAD, TRANS guide wavelength; ~ **de onda de intensidad máxima** TEC RAD, TRANS peak-intensity wavelength; ~ **de palabra** DATOS word length; ~ **de parámetro** INFO, TELEFON parameter length; ~ **de paso** INFO step length; ~ **de paso de hélice** COMPON *cables* length of lay; ~ **de registro** DATOS register length; ~ **de ruptura** COMPON, PRUEBA breaking length; ~ **de trayecto eléctrico** TRANS electrical-path length; ~ **de vano** ELECTROTEC length of span

longitudinalmente *adv* GEN longitudinally

lo: ~ **que se ve es** ~ **que se obtiene** *fra* (*WYSIWYG*) INFO what you see is what you get (*WYSIWYG*)

lote *m* INFO, PRUEBA batch, lot

LRS *abr* (*sistema de referencia en bucle*) TELEFON LRS (*loop reporting system*)

LS *abr* (*línea alquilada*) DATOS, TELEFON, TRANS LS (*leased line*)

LSC *abr* (*código de señalización de línea*) DATOS, SEÑAL LSC (*line signalling code*)

LSU *abr* (*unidad de señal solitaria*) SEÑAL LSU (*lone signal unit*)

LTC *abr* (*código temporal longitudinal*) INFO LTC (*longitudinal time code*)

LU *abr* (*unidad lógica*) DATOS, INFO, TELEFON LU (*logic unit*)

lugar: ~ **de instalación** *m* GEN installation site; ~ **de trabajo** *m* GEN place of work, workplace

lumen *m* PRUEBA lumen; ~ **por watt** (*lm/W*) PRUEBA lumen per watt (*lm/W*)

luminancia *f* ELECTRÓN, INFO, TV luminance

luminescencia *f* ELECTRÓN, TRANS luminescence

luminiscente *adj* ELECTRÓN, TRANS luminescent

luminosidad *f* INFO, TV brightness, brilliance

luminoso *adj* GEN luminous

lupa: ~ **de aumento** *f* PERIF *pantalla* magnifying glass

luz *f* GEN light; ~ **anódica** ELECTRÓN positive glow; ~ **de aviso** CONTROL warning lamp; ~ **intermitente** PRUEBA, TV flashing light; ~ **parpadeante** TV flickering light; ~ **visible** ELECTRÓN visible light

LWC *abr* (*concentrador de cableado en bucle*) ELECTRÓN LWC (*loop wiring concentrator*)

LXE *abr* (*entrada expresa para guía de luz*) ELECTRÓN LXE (*lightguide express entry*)

M

M *abr* (*emitancia luminosa*) TRANS M (*luminous emittance*)
mA *abr* (*miliamperio*) ELECTRÓN mA (*milliamp*)
MACF *abr* (*función de control de asociación múltiple*) DATOS MACF (*multiple association control function*)
macro *m* DATOS, INFO macro
macrobloque *m* INFO macroblock
macrocélula *f* TEC RAD macrocell
macroflexión *f* TRANS macrobending
macrogenerador *m* INFO macrogenerator
macroinstrucción *f* DATOS, INFO macroinstruction
macrosegmentación *f* TEC RAD macrosegmentation
madera: ~ **contrachapada** *f* COMPON plywood
MADT *abr* (*transistor de microaleación difusa*) ELECTRÓN MADT (*micro-alloy diffused transistor*)
maestro *m* ELECTRÓN master
magnético *adj* GEN magnetic
magnetismo *m* GEN magnetism; ~ **remanente** ELECTRÓN residual magnetism
magnetización *f* ELECTRÓN magnetization
magnetoconductividad *f* ELECTRÓN magnetoconductivity
magnetoóptico *adj* (*MO*) ELECTRÓN, INFO magneto-optic (*MO*)
magnetoresistencia *f* ELECTRÓN magnetoresistance
magnetoscopio *m* ELECTRÓN magnetoscope
magnetostricción *f* ELECTRÓN magnetostriction
magnetrón *m* ELECTRÓN, TEC RAD magnetron; ~ **de impulsos** ELECTRÓN, TEC RAD pulsed magnetron; ~ **sintonizado por tensión** ELECTRÓN, TEC RAD voltage-tuned magnetron
magnitud: ~ **influyente** *f* PRUEBA influence quantity
malla *f* ELECTRÓN *cristales* lattice, *de tubo de almacenamiento* mesh; ~ **de campo** ELECTRÓN field mesh; ~ **unidireccional** INFO one-way link
mancha: ~ **iónica** *f* TV ion burn
mancomunidad *f* TRANS *de canales* pool
mandar: ~ **a distancia** *vt* CONTROL telecommand
mando: ~ **de desandar un bucle** *m* TV loopback command; ~ **local** *m* CONTROL, TRANS local control; ~ **manual** *m* INFO hand control; ~ **de trinquete** *m* ELECTRÓN ratchet control
mandril *m* COMPON driftpin
mandrilar *vt* COMPON bore
manejabilidad *f* GEN manageability
manejable *adj* GEN manageable, handy
manejo: ~ **de datos** *m* DATOS data handling
manga: ~ **de empalme de cables** *f* TRANS cable distribution head
mango: ~ **del aparato telefónico** *m* TELEFON handset handle
manguito: ~ **de apriete** *m* COMPON clamping sleeve; ~ **de empalme** *m* COMPON jointing sleeve; ~ **de empalme de cable** *m* COMPON, TRANS cable sleeve; ~ **de empalme rotativo** *m* COMPON, ELECTRÓN rotary ferrule; ~ **indicador** *m* COMPON, TELEFON plug-sleeve indicator; ~ **de plomo** *m* COMPON *cables* lead sleeve; ~ **roscado** *m* COMPON threaded bushing; ~ **termoencogible** *m* COMPON heat-shrinkable sleeve; ~ **del tubo de impregnación** *m* COMPON potting boot
manipulación *f* GEN handling; ~ **de amplitud** SEÑAL amplitude keying; ~ **por desplazamiento de frecuencia** (*MDF*) SEÑAL frequency shift keying (*FSK*); ~ **por desviación de fase binaria** SEÑAL binary phase-shift keying (*BPSK*); ~ **por desviación de fase correlativa** SEÑAL correlative phase-shift keying; ~ **de mensajes** APLIC, TELEFON message handling; ~ **por secuencia de fase de cuaternaria de haz perfilado** SEÑAL phase-shaped quaternary phase-shift keying; ~ **todo-nada** (*OOK*) SEÑAL on-off keying (*OOK*)
manipulador *m* CONMUT handler, TELEFON keyer; ~ **cerrado** TELEFON key down; ~ **dactilográfico** APLIC, PERIF keyboard sender; ~ **desconectado** INFO key off; ~ **del documento** TELEFON document loader; ~ **de memoria de datos** CONMUT data store handler; ~ **Morse** SEÑAL Morse key; ~ **de paquetes de datos** (*PH*) DATOS packet handler (*PH*)
manivela *f* PERIF crank
manómetro *m* PRUEBA manometer
manos: ~ **libres** *adj* GEN hands-free
mantener *vt* GEN maintain
mantenibilidad *f* GEN maintainability

mantenimiento *m* GEN maintenance; **~ de archivos** INFO file maintenance; **~ centralizado** CONTROL centralized maintenance; **~ controlado** CONTROL controlled maintenance; **~ correctivo** CONTROL, PRUEBA corrective maintenance; **~ correctivo controlado** CONTROL, PRUEBA controlled corrective maintenance; **~ deficiente** PRUEBA poor maintenance; **~ a distancia** CONTROL, INFO telemaintenance; **~ de equipos de radio** ELECTROTEC, TEC RAD radio-equipment maintenance; **~ en el exterior** PRUEBA off-site maintenance; **~ in situ** PRUEBA on-site maintenance; **~ libre de averías** PRUEBA trouble-free maintenance; **~ local** PRUEBA local maintenance; **~ no programado** PRUEBA unscheduled maintenance; **~ de órbita** TEC RAD station keeping; **~ pasivo** PRUEBA passive maintenance; **~ preventivo** PRUEBA preventive maintenance; **~ de primera línea** PRUEBA first-line maintenance; **~ programado** PRUEBA scheduled maintenance; **~ que deteriora la función** PRUEBA function-degrading maintenance; **~ que impide la función** PRUEBA function-preventing maintenance; **~ que no afecta a la función** PRUEBA function-permitting maintenance; **~ de la red** APLIC network maintenance, network service (*NS*); **~ relativo a la función** PRUEBA function-affecting maintenance

manual: **~ de especificación de comprobación de conformidad** *m* (*CTSM*) PRUEBA conformance-testing specification manual (*CTSM*); **~ de funcionamiento** *m* GEN operation manual; **~ de instalación** *m* GEN installation manual; **~ de instrucciones** *m* GEN instruction manual; **~ de operación y mantenimiento** *m* GEN operation and maintenance manual; **~ de referencia** *m* GEN reference manual; **~ técnico** *m* INFO technical manual; **~ del usuario** *m* GEN user manual, user's manual

mapa *m* INFO map; **~ de asignaciones** DATOS assignment map; **~ de enlaces** INFO link map; **~ de Karnaugh** GEN Karnaugh map; **~ de red** CONMUT, TELEFON network map

maqueta *f* GEN mock-up

máquina: **~ de asociación de protocolo de control** *f* (*ACPM*) DATOS association control protocol machine (*ACPM*); **~ de avisos** *f* PERIF, TELEFON announcement machine; **~ para conexión arrollada** *f* COMPON wrapping machine; **~ de distribución de mensaje de software** *f* INFO software message distribution machine; **~ de facsímil** *f* PERIF, TELEFON facsimile machine; **~ de fax del grupo 1** *f* PERIF, TELEFON group 1 fax machine; **~ de fax del grupo 2** *f* PERIF, TELEFON group 2 fax machine; **~ de fax del grupo 3** *f* PERIF, TELEFON group 3 fax machine; **~ de fax del grupo 4** *f* PERIF, TELEFON group 4 fax machine; **~ de llamada** *f* TELEFON ringing machine; **~ medidora automática** *f* (*ATM*) DATOS automatic teller machine (*ATM*); **~ para pelar hilos** *f* COMPON insulation stripper; **~ de protocolo de canal** *f* DATOS channel protocol machine (*CPM*); **~ de protocolo de procesamiento de transacciones** *f* (*TPPM*) DATOS transaction processing protocol machine (*TPPM*); **~ de protocolos** *f* DATOS protocol machine; **~ de recepción** *f* TRANS receive machine; **~ de software** *f* INFO software machine; **~ de software de gestión de protocolos** *f* TELEFON protocol handling software machine

MAR *abr* (*bastidor para aparatos diversos*) TELEFON MAR (*miscellaneous apparatus rack*)

marca: **~ de búsqueda** *f* INFO search mark; **~ de comienzo de cinta magnética** *f* INFO beginning-of-tape marker; **~ establecida** *f* INFO set mark; **~ estroboscópica** *f* ELECTRÓN strobe; **~ final** *f* INFO end mark; **~ de grupo** *f* DATOS group mark

marcación *f* GEN marking, TELEFON dialling (*AmE* dialing); **~ abreviada** TELEFON speed calling, speed dialling (*AmE* dialing); **~ de ausencia de operador** TELEFON off-duty state marking, unattended mode marking; **~ en bucle** TELEFON loop dialling (*AmE* dialing); **~ de canal** TEC RAD channel marking; **~ de categoría de abonado** TELEFON class marking; **~ de error** PRUEBA error marking, failure marking; **~ de línea vacante** TELEFON vacancy marking; **~ manual** TELEFON dial-up; **~ por nombre** TELEFON single-button calling; **~ de prioridad** TELEFON priority marking; **~ de ruta libre** TELEFON route-idle marking; **~ de un solo dígito** TELEFON single-digit dialling (*AmE* dialing); **~ por teclas** PERIF, TELEFON push-button dialling (*AmE* dialing); **~ de toma previa** TELEFON preseizure dialling (*AmE* dialing)

marcado: **~ de averías** *m* PRUEBA fault marking; **~ de buril** *m* COMPON scribed mark; **~ de desconexión en bucle** *m* TELEFON loop-disconnect dialling (*AmE* dialing); **~ directo internacional** *m* TELEFON international direct dialling (*AmE* dialing) (*IDD*); **~ directo internacional a distancia** *m* TELEFON international direct distance dialling (*AmE* dialing) (*IDDD*); **~ inmediato** *m* TELEFON immediate dialling

(*AmE* dialing); ~ **en modo combinado** *m* TELEFON mixed-mode dialling (*AmE* dialing)

marcador *m* GEN marker; ~ **de alcance** TEC RAD range marker; ~ **de buscador del registrador** CONMUT register-finder marker; ~ **de categorías** TELEFON category marker; ~ **de cola** INFO trail marker; ~ **de enlace** INFO link marker; ~ **estroboscópico en escalón** TEC RAD step strobe marker; ~ **de página** DATOS, INFO *Internet* page marker; ~ **predefinido** DATOS predefined marker; ~ **de ruta** TELEFON route marker

marcaje: ~ **por cómputo simple** *m* TELEFON unit-charged marking

marcar *vt* INFO tag, TELEFON dial; ◆ ~ **manualmente** TELEFON dial up; ~ **el tiempo** DATOS time-tag

marcha *f* GEN pace

marco *m* COMPON, INFO, TELEFON, TRANS frame; ~ **de control** DATOS control frame; ~ **de montaje** COMPON mounting frame; ~ **MPEG codificado** INFO coded MPEG frame; ~ **de página** DATOS, INFO page frame; ~ **de relé** TELEFON relay frame; ~ **repartidor** TRANS distribution frame; ~ **repartidor de dos caras** TRANS double-sided distribution frame

margarita: ~ **de impresión** *f* PERIF printwheel, daisywheel

margen *m* GEN margin; ~ **de ajuste** GEN range of adjustment; ~ **de alargamiento** TRANS *óptica de fibras* expansion window; ~ **de descodificación** TEC RAD decoding margin; ~ **de enlace** TRANS link margin; ~ **de error** PRUEBA margin of error; ~ **de protección** TEC RAD protection margin; ~ **de referencia** PRUEBA reference range; ~ **de regulación** GEN range of regulation; ~ **total de protección equivalente** TEC RAD overall equivalent protection margin

marginal *adj* GEN marginal

masa *f* GEN mass; ~ **del electrón** ELECTRÓN electron mass

máscara *f* TELEFON, TRANS mask; ~ **del impulso** TELEFON, TRANS pulse mask; ~ **de retardo de grupo** TRANS group-delay mask; ◆ **sin** ~ INFO unmasked

MASE *abr* (*elemento de servicio de administración de mensajes*) DATOS MASE (*message administration service element*)

masilla *f* COMPON putty

máser *m* ELECTRÓN maser

mástil *m* TEC RAD mast; ~ **de antena** TEC RAD aerial mast, antenna mast

MATD *abr* (*retardo máximo aceptable de tránsito*) DATOS MATD (*maximum acceptable transit delay*)

material *m* GEN material; ~ **a base de revestimiento metálico** COMPON metal-clad base material; ~ **básico** INFO basic material; ~ **débilmente magnético** COMPON, ELECTRÓN soft magnetic material; ~ **de fuerte remanencia magnética** COMPON, ELECTRÓN hard magnetic material; ~ **laminar** COMPON laminate; ~ **para mechas** COMPON wicking; ~ **de relleno** COMPON filler

matiz *m* INFO, TV hue

matriz *f* GEN matrix; ~ **de acceso** TELEFON access matrix; ~ **de cavidad única** COMPON single-cavity mould (*AmE* mold); ~ **de Chireix** TEC RAD Chireix array; ~ **de cinta** ELECTRÓN ribbon matrix; ~ **de conexión** CONMUT connection matrix; ~ **de conmutación** CONMUT switching matrix; ~ **de conmutación central** CONMUT central switching matrix; ~ **de conmutación óptica integrada** CONMUT integrated optical switching matrix; ~ **de cuantificación** INFO quantization matrix; ~ **de diseño** ELECTRÓN *circuitos impresos* artwork master; ~ **de ferrita** INFO ferrite matrix; ~ **de núcleo de ferrita** INFO ferrite core matrix; ~ **de núcleos** INFO core matrix; ~ **pasiva** INFO passive matrix; ~ **de referencia** INFO key matrix; ~ **de rejilla de espigas** ELECTRÓN pin-grid array; ~ **retrorreflectora** TEC RAD retroreflector array; ~ **de teclado** INFO keyboard matrix; ~ **de tráfico** TELEFON traffic matrix; ~ **de vídeo gráfico** (*VGA*) INFO video graphics array (*VGA*)

MATV *abr* (*televisión de antena colectiva*) TV MATV (*master antenna television*)

MAU *abr* (*unidad de acceso multiestación*) DATOS MAU (*multistation access unit*)

máxima: ~ **apertura numérica teórica** *f* ELECTRÓN, TRANS maximum theoretical numerical aperture

maximización *f* GEN maximization

maximizar *vt* GEN maximize

Mb *abr* (*megabyte*) DATOS Mb (*megabyte*), Mbyte (*megabyte*)

Mbps *abr* (*millones de bitios por segundo*) DATOS, INFO, TRANS Mbps (*millions of bits per second*)

MBS *abr* (*unidad de señal de sincronización multibloque*) SEÑAL MBS (*multiblock synchronization signal unit*)

MCI *abr* (*interfaz de control de medios*) INFO MCI (*media control interface*)

MCS *abr* (*sistema de comunicaciones multipunto*) TELEFON MCS (*multipoint communications *system*); ~ **anulación de mando multipunto** *m* DATOS multipoint command negating MCS

MCSAP *abr* (*punto de acceso del servicio MCS*) TELEFON MCSAP (*MCS service access point*); ~ **de control** *m* TELEFON control MCSAP

MCW *abr* (*ondas continuas moduladas*) TEC RAD MCW (*modulated continuous waves*)

MDF *abr* SEÑAL (*manipulación por desplazamiento de frecuencia*) FSK (*frequency shift keying*), TELEFON, TRANS (*multiplexado por división de frecuencias*) FDM (*frequency-division multiplexing*)

MDS *abr* (*señal detectable mínima*) TRANS MDS (*minimum detectable signal*)

MDSE *abr* (*elemento de servicio de entrega de mensajes*) DATOS MDSE (*message delivery service element*)

mecánica *f* GEN mechanics; ~ **cuántica** ELECTRÓN quantum mechanics; ~ **ondulatoria** GEN wave mechanics

mecanismo *m* GEN mechanism, INFO drive; ~ **de autentificación** DATOS authentication mechanism; ~ **de captación de la voz** INFO speech capture device; ~ **de conmutación** CONMUT, TELEFON switching mechanism; ~ **de control** CONTROL controlling device; ~ **de desconexión** ELECTRÓN release mechanism, trigger mechanism; ~ **del fallo** PRUEBA failure mechanism; ~ **impulsador de cinta** INFO tape drive; ~ **de protección contra tirones** COMPON strain relief mechanism; ~ **selector** CONMUT, TELEFON selector mechanism; ~ **de señal sonora** TEC RAD bleeper

mechón *m* COMPON swab

media: ~ **aritmética** *f* GEN arithmetic mean, arithmetic average; ~ **aritmética ponderada** *f* GEN arithmetic weighted mean; ~ **de bits por muestra** *f* DATOS average bits per sample; ~ **clavija** *f* COMPON half-size plug; ~ **cuadrática** *f* GEN root mean square (*rms*); ~ **geométrica** *f* PRUEBA geometric mean; ~ **de horas-hombre de mantenimiento** *f* PRUEBA mean maintenance man-hours; ~ **de opinión** *f* INFO mean opinion score (*MOS*)

mediana *f* GEN median

medición *f* GEN measurement, PRUEBA gauge (*AmE* gage); ~ **de aislamiento** ELECTROTEC insulation measurement; ~ **de atenuación** TRANS attenuation measurement; ~ **de carga unidad** TELEFON unit charge metering; ~ **ecométrica** TELEFON echometric measurement; ~ **de la frecuencia** PRUEBA frequency measurement; ~ **de índice de errores** DATOS error-rate measurement; ~ **inicial** APLIC initial metering; ~ **de la intensidad de campo** ELECTROTEC field strength measurement; ~ **plena inicial** (*IFLU*) PRUEBA initial full line-up (*IFLU*); ~ **de ruido** TELEFON noise measurement; ~ **de tráfico** ELECTRÓN, PRUEBA, TELEFON traffic metering, traffic measurement

medida: ~ **de comparación** *f* PRUEBA comparison measurement; ~ **correctiva** *f* GEN corrective measure; ~ **de desviación** *f* PRUEBA deviation measurement; ~ **de disponibilidad** *f* PRUEBA availability measure; ~ **de inteligibilidad** *f* TEC RAD articulation score; ~ **de llamadas** *f* TELEFON call metering; ~ **de sonoridad objetiva** *f* TELEFON objective loudness rating; ~ **de transmisión** *f* PRUEBA, TRANS transmission measurement

medidas *f pl* GEN measures; ~ **de seguridad** PRUEBA security arrangements; ~ **sobre el terreno** PRUEBA field measurements; ~ **de urgencia** PRUEBA emergency action

medidor *m* GEN meter; ~ **de bujía-pie** PRUEBA footcandle meter; ~ **de campo** PRUEBA field intensity meter; ~ **de deformación** COMPON strain gauge (*AmE* gage); ~ **de fase** ELECTROTEC phase meter; ~ **de frecuencia lámina** PRUEBA reed frequency meter; ~ **de frecuencia de lámina vibrante** PRUEBA vibrating-reed frequency meter; ~ **de iluminancia** PRUEBA illuminance meter; ~ **de intensidad de campo** PRUEBA field strength meter; ~ **de intensidad de la señal** SEÑAL signal strength meter; ~ **de intervalo de tiempo** PRUEBA time-interval meter; ~ **de llamadas** TELEFON call meter; ~ **de luminancia** PRUEBA luminance meter; ~ **de posición sintonizadora** ELECTRÓN tuning position meter; ~ **de potencia** PRUEBA power meter; ~ **de ruido** TELEFON noise meter, noise-measuring device; ~ **en unidades S** PRUEBA S-meter; ~ **de vataje del par termoeléctrico** PRUEBA thermocouple wattmeter; ~ **de visibilidad** PRUEBA visibility meter

medio *m* GEN medium; ~ **de acceso al videotexto** TELEFON, TV videotext gateway; ~ **láser** ELECTRÓN, TRANS laser medium; ~ **refringente** GEN refractor; ~ **de transmisión** TRANS, TV transmission medium

medios *m pl* GEN media

medir *vt* GEN measure

megabitio *m* DATOS megabit

megabyte *m* (*Mb*) DATOS megabyte (*Mb*, *Mbyte*)

megaflop *f* INFO megaflop

megalips *m pl* INFO megalips

megaóhmetro *m* PRUEBA Megger®, megohmmeter

mejoramiento *m* GEN improvement, upgrading, upgrade

mejorar *vt* GEN improve, INFO upgrade

membrana *f* COMPON membrane; ~ **basilar** TEC RAD basilar membrane; ~ **del kevlar** TEC RAD kevlar skin

memorándum: ~ **de acuerdo** *m* (*MOU*) GEN memorandum of understanding (*MOU*)

memoria *f* GEN memory, storage; ~ **de acceso aleatorio** (*RAM*) DATOS, INFO random access memory (*RAM*); ~ **de acceso rápido** DATOS, INFO cache storage, quick-access memory; ~ **de almacenamiento de imágenes codificadas** INFO coded-image storage space; ~ **anexa** ELECTROTEC *circuitos impresos* bump; ~ **de archivo virtual** DATOS virtual file store; ~ **de asignación de canal** TRANS channel allocation store; ~ **de banco** INFO bank memory; ~ **caché** DATOS, INFO cache store; ~ **de categorías** TELEFON category store, class-mark store; ~ **central** DATOS main storage; ~ **de cinta** INFO tape memory; ~ **de conservación de registros** CONMUT register-preserve memory; ~ **de contenido direccionable** INFO content-addressable memory; ~ **de contenido direccionado** INFO content-addressed memory; ~ **de control** INFO control memory, control storage; ~ **criogénica** INFO cryogenic storage; ~ **de diez números** PERIF ten-number memory; ~ **dinámica** DATOS, INFO buffer memory; ~ **direccionable de datos** DATOS, INFO data-addressable memory; ~ **de disco** INFO disk memory; ~ **estable** DATOS, INFO nonvolatile memory, nonvolatile storage; ~ **expandida** DATOS, INFO expanded memory; ~ **de extensión** INFO extension store; ~ **de facturación** APLIC charging store; ~ **de funcionamiento** INFO performance memory; ~ **holográfica** DATOS holographic storage; ~ **de imagen** CONTROL picture store; ~ **indeleble** INFO nonerasable memory; ~ **indirecta** PERIF *contestador* indirect memory; ~ **inestable** DATOS, INFO volatile memory, volatile storage; ~ **de intercambio** DATOS core of exchange; ~ **intermedia** DATOS, INFO buffer storage, cache memory; ~ **intermedia de edición** INFO function edit buffer; ~ **intermedia de monedero** TELEFON coin buffer; ~ **intermedia de retransmisión** INFO, TRANS retransmission buffer; ~ **de lectura inversa** DATOS, INFO push-down storage; ~ **de masa** INFO bulk store; ~ **matricial** INFO matrix storage; ~ **de mensaje** DATOS, TELEFON message store; ~ **de mensaje EDI** DATOS EDI message store; ~ **de mensajería por voz** (*VM-MS*) DATOS, TELEFON voice messaging store (*VM-MS*); ~ **de mensaje de sistema de mensajería por voz** (*VMGS-MS*) DATOS, TELEFON voice messaging system message store (*VMGS-MS*); ~ **múltiple** TEC RAD multiple memory; ~ **de núcleos de ferrita** INFO ferrite core storage; ~ **de núcleos magnéticos** INFO ferrite storage; ~ **de programa** INFO program store; ~ **de reaprovisionamiento** INFO reorder buffer; ~ **del receptor** INFO destination memory; ~ **de reserva** INFO backup store; ~ **de semiconductores** DATOS semiconductor storage; ~ **de señales vocales** TELEFON speech memory; ~ **sólo de lectura** (*ROM*) INFO read-only memory (*ROM*); ~ **sólo de lectura eléctricamente programable** (*EPROM*) ELECTRÓN electrically-programmable read-only memory (*EPROM*); ~ **sólo de lectura programable borrable por medios eléctricos** ELECTRÓN electrically-erasable programmable read-only memory; ~ **de texto** INFO text memory; ~ **de trabajo** INFO utility memory; ~ **de la vía de transmisión** CONMUT path memory; ~ **virtual** DATOS, INFO virtual memory, virtual storage; ~ **volátil** DATOS, INFO volatile memory, volatile storage

memorización: ~ **de datos de voz** *f* INFO storing of voice data; ~ **de dígitos** *f* TELEFON storing of digits; ~ **de mensajes** *f* DATOS message storing; ~ **provisional de los datos** *f* DATOS data buffering

memorizado *adj* DATOS, INFO stored

memorizar *vt* DATOS, INFO store

mensaje *m* GEN message; ~ **aceptado de solicitud de cancelación de sistema** DATOS, TELEFON facility cancellation request accepted message; ~ **aceptado de solicitud de registro de sistema** DATOS, TELEFON facility registration request accepted message; ~ **almacenable** PERIF, TELEFON storable message; ~ **de borrado y error** DATOS blank-and-burst message; ~ **del canal de servicio** PRUEBA service channel message; ~ **de confusión** DATOS confusion message; ~ **de control** INFO control message; ~ **de correo electrónico** DATOS e-mail message; ~ **de datos binarios** TELEFON binary data message; ~ **de datos en cuadrícula** DATOS grid data message; ~ **de desconexión** PRUEBA clearing message; ~ **de error** DATOS error message; ~ **de estado** TEC RAD status message; ~ **de estado del circuito** TELEFON circuit state message; ~ **estándar de datos de la OTAN** TEC RAD standard NATO data message; ~ **de facsímil** TELEFON facsimile message; ~ **de fallo** PRUEBA fault message; ~ **de**

fallo eliminado PRUEBA fault-cleared message; ~ **de gestión de bloque** DATOS frame management message; ~ **de gestión consolidada de enlace entre capas** DATOS consolidated link-layer management message; ~ **guía oral** PERIF *de contestador automático* voice prompt; ~ **hablado** PERIF *de contestador automático*, TELEFON spoken message; ~ **de información de usuario a usuario** DATOS user-to-user information message; ~ **inicial de dirección** CONMUT, SEÑAL, TELEFON initial address message; ~ **por intercambio de datos electrónicos** DATOS, TELEFON EDI message (*EDIM*); ~ **de interrogación de un grupo de circuitos** CONMUT circuit-group query message (*CQM*); ~ **de liberación** TEC RAD release message (*REL*); ~ **de nueva asignación** DATOS new assignment message; ~ **radiodifundido** TEC RAD broadcast message; ~ **rechazado de solicitud de cancelación de sistema** TELEFON facility cancellation request rejected message; ~ **rechazado de solicitud de registro de sistema** TELEFON facility registration request rejected message; ~ **de rechazo de modificación de llamadas** TELEFON call-modification reject message; ~ **de respuesta de un grupo de circuitos** CONMUT circuit-group query response message (*CQR*); ~ **de segmento único** (*SSM*) DATOS single-segment message (*SSM*); ~ **de señal** SEÑAL signal message; ~ **de señalización** SEÑAL signalling (*AmE* signaling) message; ~ **de sistema rechazado** DATOS facility-rejected message (*FRJ*); ~ **de solicitud de cancelación de sistema** TELEFON facility cancellation request message; ~ **de solicitud de registro de sistema** TELEFON facility registration request message; ~ **supervisor** TELEFON supervisory message; ~ **por télex** TELEFON telex message; ~ **con texto** TEC RAD text message; ~ **de una unidad** TELEFON one-unit message; ~ **de unidades múltiples** TELEFON multiunit message; ~ **verbal** TELEFON verbal message; ~ **vocal** DATOS, TELEFON vocal message; ~ **por voz** DATOS, TELEFON voice message

mensajería: ~ **independiente del proveedor** *f* (*VIM*) TEC RAD vendor-independent messaging (*VIM*); ~ **por intercambio de datos electrónicos** *f* DATOS, TELEFON EDI messaging (*EDIMG*); ~ **interpersonal** *f* TELEFON interpersonal messaging

ménsula *f* COMPON bracket

menú *m* INFO menu; ~ **desplegable** INFO drop-down menu; ~ **de enlaces** INFO links menu; ~ **de separación** INFO tear-off menu

mercado: ~ **de nicho** *m* GEN niche market

mesa: ~ **de control** *f* CONTROL, TV control desk; ~ **de observación** *f* CONTROL observation desk; ~ **de supervisión de tráfico** *f* TELEFON traffic supervision desk

mesócrono *adj* TELEFON, TRANS mesochronous

metaestable *adj* ELECTRÓN metastable

metal: ~ **fino** *m* COMPON precious metal; ~ **inestable** *m* COMPON base metal; ~ **no ferroso** *m* COMPON nonferrous metal; ~ **precioso** *m* COMPON precious metal

metalenguaje *m* INFO metalanguage

metalización *f* COMPON metallization (*AmE* metalization); ~ **por contacto** COMPON contact metallization (*AmE* metalization); ~ **selectiva** COMPON pattern plating; ~ **total del panel** COMPON panel plating; ~ **al vacío** COMPON vacuum metallization (*AmE* metalization)

metalizado[1] *adj* COMPON metallized (*AmE* metalized)

metalizado[2]: ~ **al vacío** *m* COMPON vacuum metallization (*AmE* metalization)

metarrazonamiento *m* INFO metareasoning

metódico *adj* GEN methodic

método *m* GEN method; ~ **de acceso** CONMUT, TELEFON access method; ~ **de acceso virtual** (*VAM*) CONMUT virtual access method (*VAM*); ~ **de alineación óptica** TEC RAD boresight method; ~ **de alteración** TRANS cutback technique; ~ **alternativo de pruebas** PRUEBA alternative test method; ~ **atributivo** PRUEBA attribute method; ~ **básico de corrección de errores** PRUEBA basic error correction method; ~ **de búsqueda de averías** PRUEBA fault-tracing procedure; ~ **del camino crítico** GEN critical-path method; ~ **de la central** TELEFON exchange method; ~ **de cero** PRUEBA null method; ~ **de claves** DATOS, INFO hashing; ~ **de comparación** PRUEBA comparison method; ~ **complementario de medición** PRUEBA complementary method of measurement; ~ **de comprobación RSE** DATOS RSE test method; ~ **de comprobación transversal** PRUEBA transverse test method; ~ **coordinado de comprobación interna de una capa** PRUEBA coordinated single-layer embedded test method; ~ **de los dos puntos de control** TEC RAD two-control-point method; ~ **del eco de impulsos** TRANS pulse echo method; ~ **estadístico** PRUEBA statistical method; ~ **estructural** PRUEBA structural method; ~ **de facturación** APLIC, TELEFON charging method; ~ **heurístico** DATOS heuristic method; ~ **incorporado de prueba a distancia de una capa** PRUEBA remote single-layer embedded test method; ~ **indirecto de medición** PRUEBA indirect method of measurement; ~ **local refractado** TRANS refracted near-end method; ~ **de medición** PRUEBA measurement method; ~ **de medición de coincidencias** PRUEBA coincidence method of measurement; ~ **de medición por comparación** PRUEBA comparison method of measurement; ~ **de medición por desviación** PRUEBA deflection method of measurement; ~ **de medición por resonancia** PRUEBA resonance method of measurement; ~ **de medición por transposición** PRUEBA transposition method of measurement; ~ **de medida por contacto** PRUEBA contact method of measurement; ~ **de modulación AM nulo** TEC RAD null AM modulation method; ~ **de ocultación** INFO concealment method; ~ **preventivo de control de errores de transmisión cíclicos** TRANS preventive cyclic retransmission error control method; ~ **de programación** DATOS, INFO programming method; ~ **de prueba coordinado** PRUEBA coordinated test method; ~ **de prueba coordinado de una capa** PRUEBA coordinated single-layer test method; ~ **de prueba a distancia** PRUEBA remote test method; ~ **de prueba a distancia de una sola capa** PRUEBA remote single-layer test method; ~ **de prueba incorporado de capa única local** PRUEBA local single-layer embedded test method; ~ **de prueba local** PRUEBA local test method; ~ **de prueba del recorrido inverso de un bucle** PRUEBA loop-back test method; ~ **de prueba de referencia** PRUEBA reference test method; ~ **de rayos refractados** TRANS refracted-ray method; ~ **de recuento de defectos** PRUEBA defect-counting method; ~ **de reflexión de Fresnel** TRANS Fresnel reflection method

metro: ~ **universal** *m* PRUEBA universal meter

metrología *f* PRUEBA metrology

mezcla: ~ **de modos** *f* TRANS mode mixing

mezclador *m* GEN mixer, INFO, TEC RAD, TELEFON, TV scrambler; ~ **de modos** TRANS mode mixer

mezclar *vt* GEN mix, INFO, TEC RAD, TELEFON, TV scramble

MF *abr* GEN (*modulación de fase*) PM (*phase modulation*), (*multifrecuencia*) MF (*multi-frequency*), TEC RAD, TELEFON, TRANS (*modulación de frecuencia*) FM (*frequency modulation*)

MFBE *abr* (*modulación de frecuencia de banda estrecha*) TEC RAD, TRANS NBFM (*narrow-band frequency modulation*)

MHE *abr* (*entorno de manipulación del tráfico de mensajes*) APLIC, DATOS, TELEFON MHE (*message handling environment*)

microcasete *f* PERIF *para contestador automático* microcassette

microchip *m* ELECTRÓN microchip

microcinta *f* ELECTRÓN microribbon

microcircuito *m* ELECTRÓN microcircuit; ~ **integrado** ELECTRÓN integrated microcircuit; ~ **de procesamiento de señal digital** *f* DATOS digital signal-processing chip

microcomputadora *AmL f* (*cf microordenador Esp*) INFO microcomputer

microconjunto *m* ELECTRÓN micro-assembly

microelectrónica *f* ELECTRÓN microelectronics

microficha *f* GEN microfiche

microflexión *f* TRANS microbending

micrófono *m* PERIF, TELEFON handset, microphone, mike; ~ **con amplificador** PERIF, TELEFON amplified handset; ~ **antirruidos** TELEFON noise-cancelling (*AmE* noise-canceling) microphone; ~ **de carbón** PERIF carbon microphone; ~ **de cinta** PERIF, TELEFON ribbon microphone; ~ **de condensador** PERIF, TELEFON condenser microphone; ~ **direccional** PERIF, TELEFON directional microphone; ~ **de electreto** PERIF, TELEFON electret microphone; ~ **electrostático** PERIF, TELEFON capacitor microphone; ~ **de mesa** PERIF, TELEFON desk microphone; ~ **de solapa** PERIF, TELEFON lapel microphone, breastplate microphone; ~ **sonda** PERIF, TELEFON probe microphone

microforma *f* DATOS microform

microimagen *f* DATOS micro-image

microinterruptor *m* CONMUT, ELECTRÓN microswitch

microonda *f* TEC RAD microwave

microordenador *Esp m* (*cf microcomputadora AmL*) INFO microcomputer

micropelícula *f* GEN microfilm

microplaca: ~ **elevada** *f* INFO lifted chip

microprocesador *m* INFO microprocessor

microprogramación *f* INFO microprogramming

microranura *f* TRANS microslot

microscopía: ~ **electrónica** *f* ELECTRÓN electron microscopy

microscopio: ~ **electrónico** *m* ELECTRÓN electron microscope; ~ **electrónico de exploración** *m* (*SEM*) ELECTRÓN scanning electron microscope (*SEM*); ~ **óptico** *m* TV light microscope

microsegmentación *f* TEC RAD microsegmentation

microsegundo *m* GEN microsecond

microseparador *m* ELECTRÓN microstripper

microsintonía *f* ELECTRÓN microtune, microtuning
microteléfono *m* PERIF, TELEFON handset telephone; ~ **alámbrico** PERIF, TELEFON wired handset; ~ **inalámbrico** PERIF, TELEFON cordless handset; ~ **de prueba** PRUEBA, TELEFON test handset; ~ **de radio celular** TEC RAD cellular radio handset
microtonal *adj* INFO microtonal
MID *abr* (*identificación de multiplexión*) TRANS MID (*multiplexing identification*)
migración *f* GEN migration; ~ **eléctrica** ELECTROTEC electric migration; ~ **de los iones** ELECTRÓN ion migration
migrar *vi* GEN migrate
miliamperio *m* (*mA*) ELECTRÓN milliamp (*mA*)
milisegundo *m* GEN millisecond
millones: ~ **de bitios por segundo** *m pl* (*Mbps*) DATOS, INFO, TRANS millions of bits per second (*Mbps*)
miniaturización *f* GEN miniaturization
miniaturizado *adj* GEN miniaturized
miniaturizar *vt* GEN miniaturize
miniclavija *m* PERIF mini jack
minidisco *m* INFO mini disk; ~ **flexible** INFO mini floppy
minienrollado *m* ELECTRÓN mini-wrapping
minimización *f* INFO minimization
minimizar *vt* INFO minimize
minúscula *f* GEN lower case
minuto: ~ **degradado** *m* DATOS, TEC RAD degraded minute
mira: ~ **de nivelación** *f* GEN hub
MIS *abr* (*sistema de información para gestión*) DATOS MIS (*management information system*)
MJD *abr* (*fecha del calendario juliano modificada*) TEC RAD MJD (*modified Julian date*)
MLI *abr* (*aislamiento multicapa*) ELECTRÓN MLI (*multilayer isolation*)
MLP *abr* (*procedimiento multienlace*) DATOS MLP (*multilink procedure*)
MMC *abr* (*comunicaciones hombre-máquina*) INFO MMC (*man-machine communications*)
MMJ *abr* (*jack modular modificado*) COMPON, ELECTRÓN MMJ (*modified modular jack*)
MML *abr* (*lenguaje hombre-máquina*) INFO MML (*man-machine language*)
MMP *abr* (*clavija modular modificada*) ELECTRÓN MMP (*modified modular plug*)
MO *abr* (*magnetoóptico*) ELECTRÓN, INFO MO (*magneto-optic*)
MOCS *abr* (*declaración de conformidad del objeto gestionado*) DATOS MOCS (*managed-object conformance statement*)
modalidad: ~ **electromagnética** *f* TRANS electromagnetic mode; ~ **transparente** *f* DATOS transparent mode; ~ **virtual** *f* INFO virtual mode
modelación *f* GEN modelling (*AmE* modeling); ~ **del haz** ELECTRÓN, TEC RAD beam forming
modelar *vt* INFO, TEC RAD shape
modelización: ~ **de datos** *f* DATOS data modelling (*AmE* modeling)
modelo *m* GEN model; ~ **de color CMY** INFO CMY colour (*AmE* color) model; ~ **de color RGB** INFO, TV RGB colour (*AmE* color) model; ~ **compartimentado** PRUEBA compartmental model; ~ **de conexión** CONMUT switching pattern; ~ **de directividad vertical** TEC RAD vertical directivity pattern; ~ **eléctrico** ELECTRÓN electrical model; ~ **estándar de datos de la OTAN** TEC RAD standard NATO data model; ~ **general** TELEFON general model; ~ **hipotético** GEN *estadística* assumed model; ~ **de iluminación** INFO lighting model; ~ **de lluvia** TEC RAD rain model; ~ **OSI** DATOS OSI model; ~ **psicoacústico** INFO psychoacoustic model; ~ **reducido** GEN reduced model; ~ **de referencia** DATOS reference model; ~ **de referencia de protocolos** DATOS protocol reference model; ~ **de reutilización** TEC RAD reuse pattern; ~ **de ritmo** INFO rhythm pattern; ~ **de servicio de emisión doble** DATOS multicast service model; ~ **de trama de hilos** COMPON wire frame model
módem *m* (*modulador/desmodulador*) GEN modem (*modulator/demodulator*); ~ **de canal** TRANS channel modem; ~ **conmutado** DATOS dial-up modem; ~ **de contestación** DATOS answering modem; ~ **de modo múltiple** TRANS multiple-mode modem; ~ **nulo** DATOS null modem; ~ **óptico inteligente** DATOS intelligent optical modem; ~ **que llama** DATOS calling modem; ~**-radio** TEC RAD radio modem
modernización *f* GEN modernization
modificable *adj* INFO modifiable
modificación *f* GEN modification; ~ **de anuncio remoto** PERIF *contestador automático* remote announcement modification; ~ **del diseño** GEN design modification; ~ **de función** GEN function change, function modification; ~ **de llamada entrante** TELEFON incoming call modification; ~ **de puenteado** ELECTROTEC restrapping
modificador *m* INFO modifier; ~ **de datos** INFO data modifier; ~ **de instrucciones** INFO instruction modifier
modo *m* GEN mode; ~ **de acceso a memoria** INFO memory access mode; ~ **acoplado** TRANS coupled mode; ~ **de adquisición rutinario**

DATOS routine acquisition mode; ~ **de ampliación** TELEFON extension mode; ~ **asociado** TELEFON associated mode; ~ **automático** DATOS auto-mode; ~ **combinado** GEN mixed mode; ~ **de conexión** DATOS connection mode; ~ **de continuación** TELEFON continuation mode; ~ **de control** TEC RAD control mode, monitoring mode; ~ **de control de altitud determinada por radar** TEC RAD radar altitude control mode; ~ **de control compartido** CONTROL, DATOS shared-control mode; ~ **de control polarizado** DATOS polarized-control mode; ~ **con corrección de errores** TELEFON error correction mode; ~ **cuasiasociado** TELEFON quasi-associated mode; ~ **desconectado** TELEFON disconnected mode; ~ **desconectado normal** DATOS normal disconnected mode; ~ **de desconexión en bucle** PERIF *contestador* loop-disconnect mode; ~ **E** TEC RAD E mode; ~ **echoplex** TELEFON ecoplex mode; ~ **de efecto túnel** TRANS tunnelling (*AmE* tunneling) mode; ~ **equilibrado asincrónico de puesta a uno** (*SABM*) TELEFON set asynchronous balanced mode (*SABM*); ~ **equilibrado asincrónico de puesta a uno extendido** (*SABME*) TELEFON set asynchronous balanced mode extended (*SABME*); ~ **de espera** PRUEBA, TELEFON stand-by mode; ~ **de estado inicial** INFO initial condition mode; ~ **de facturación** APLIC, TELEFON charging mode; ~ **de fallo** PRUEBA *seguridad de funcionamiento* failure mode, fault mode, TELEFON failure mode; ~ **funcional** TELEFON functional mode; ~ **de funcionamiento** GEN mode of operation; ~ **gráfico** INFO graphics mode; ~ **guiado** TRANS *óptica de fibras* guided mode; ~ **de guíaondas** TEC RAD, TRANS waveguide mode; ~ **H** TEC RAD H mode; ~ **híbrido** ELECTRÓN hybrid mode; ~ **lateral** TRANS side mode; ~ **ligado** TRANS bound mode; ~ **linealmente polarizado** TRANS linearly-polarized mode (*LP mode*); ~ **local** TELEFON local mode; ~ **longitudinal simple** TRANS single longitudinal mode; ~ **magnético transversal** (*TM*) TEC RAD, TRANS transverse magnetic mode (*TM*); ~ **de mensaje** DATOS message mode; ~ **de menú** TELEFON menu mode; ~ **de montaje** TELEFON mounting mode; ~ **multicíclico** DATOS multicyclic mode; ~ **multilongitudinal** TELEFON, TRANS multilongitudinal mode (*MLM*); ~ **de múltiples destinos** DATOS multidestination mode; ~ **negativo de señalización** SEÑAL nonmode of signalling (*AmE* signaling); ~ **no asociado** TELEFON nonassociated mode; ~ **no ligado** TRANS unbound mode; ~ **de no repetición de explotación en dos frecuencias** TELEFON two-frequency operation non-repeater mode; ~ **normal** DATOS normal mode; ~ **de núcleo** TRANS *óptica de fibras fotoconductoras* core mode; ~ **OM** PERIF OM mode; ~ **de operación** TELEFON operational mode; ~ **de operación cuasiasociado** TELEFON quasi-associated mode of operation; ~ **de operación de fondo** TEC RAD background mode of operation; ~ **de operación no asociado** TELEFON nonassociated mode of operation; ~ **de operación totalmente disociada** TELEFON fully-dissociated mode of operation; ~ **de oscilación** TEC RAD mode of oscillation; ~ **de polarización lineal** TRANS linearly polarized mode; ~ **de propagación** TRANS *óptica de fibras* propagation mode; ~ **de propagación guiado** TEC RAD, TRANS trapped mode; ~ **de radiación** ELECTRÓN, TRANS radiation mode; ~ **rápido** DATOS *de digitalización* rapid mode; ~ **de recepción** TEC RAD receiving mode; ~ **repetidor automático de explotación en dos frecuencias** TELEFON two-frequency operation automatic repeater mode; ~ **de reserva** PRUEBA backup mode; ~ **de respuesta normal** DATOS normal-response mode; ~ **de retrobucle de satélite** TEC RAD satellite loopback mode; ~ **de revestimiento** TRANS *óptica de fibras* cladding mode; ~ **de señalización cuasiasociado** SEÑAL, TELEFON quasi-associated mode of signalling (*AmE* signaling); ~ **de señalización no asociado** SEÑAL nonassociated mode of signalling (*AmE* signaling); ~ **sin inversión de letra** INFO, PERIF *teclado* unshifted mode; ~ **de sobrecarga DCME** TELEFON DCME overload mode; ~ **TDMA** SEÑAL, TEC RAD, TELEFON, TRANS TDMA mode; ~ **TE** TRANS TE mode; ~ **TEM** TRANS TEM mode; ~ **TM** TEC RAD, TRANS TM mode; ~ **de trabajo no duplicado** TELEFON nonduplicated working mode; ~ **transverso** TRANS *óptica de fibras* transverse mode; ~ **troposférico** TEC RAD tropospheric mode; ~ **de vibración** ELECTRÓN vibration mode; ~ **de vigilancia** TEC RAD watching mode; ~ **X.410-1984** DATOS X.410-1984 mode

modulación *f* GEN modulation; ~ **de la altura** INFO pitch modulation; ~ **de amplitud** ELECTRÓN, INFO, SEÑAL, TRANS, TV amplitude modulation (*AM*); ~ **de amplitud positiva** ELECTRÓN, TV positive amplitude modulation; ~ **de amplitud de rejilla** ELECTRÓN grid amplitude modulation; ~ **de ángulo** TRANS angle modulation; ~ **de ángulo de gran**

desviación TRANS wide-deviation angle modulation; ~ **automática** TRANS automodulation; ~ **binaria** DATOS binary modulation; ~ **codificada en enrejado** (*TCM*) DATOS, TRANS trellis-coded modulation (*TCM*); ~ **de compresión-expansión y de frecuencia** (*CFM*) TRANS companding and frequency modulation (*CFM*); ~ **de datos** INFO data modulation; ~ **delta CDF-2** SEÑAL delta modulation PSK-2; ~ **delta compandida** SEÑAL, TRANS companded delta modulation (*CDM*); ~ **de espectro expandido** TEC RAD spread-spectrum modulation; ~ **espuria** TRANS spurious modulation; ~ **de fase** (*MF*) GEN phase modulation (*PM*); ~ **fragmentada** TELEFON curbed modulation; ~ **de frecuencia** (*MF*) TEC RAD, TRANS, TV frequency modulation (*FM*); ~ **de frecuencia de banda estrecha** (*MFBE*) TEC RAD, TRANS narrowband frequency modulation (*NBFM*); ~ **de frecuencia de impulsos** TELEFON pulse frequency modulation (*PFM*); ~ **de frecuencia regularizada** TEC RAD, TRANS tamed frequency modulation; ~ **de frecuencia de subportadora** TELEFON subcarrier frequency modulation; ~ **del impulso** INFO, TELEFON, TRANS pulse modulation; ~ **de impulsos** TELEFON modulation of pulses; ~ **por impulsos** TRANS modulation by pulses; ~ **de impulsos en amplitud** TELEFON, TRANS pulse amplitude modulation; ~ **por impulsos codificados** (*PCM*) DATOS, INFO, TELEFON, TRANS pulse-code modulation (*PCM*); ~ **de impulsos cuantificados** TRANS quantized pulse modulation; ~ **de impulsos en duración** (*PDM*), (*PWM*) DATOS, TEC RAD, TELEFON, TRANS pulse duration modulation (*PDM*), pulse width modulation (*PWM*); ~ **de impulsos en intervalo** TELEFON pulse-interval modulation; ~ **de impulsos en posición** (*PPM*) DATOS, TELEFON, TRANS pulse position modulation (*PPM*); ~ **de impulsos en el tiempo** (*PTM*) TELEFON, TRANS pulse time modulation (*PTM*); ~ **por inversión de fase** TELEFON phase-inversion modulation; ~ **múltiple** TRANS multiple modulation; ~ **por mutación de frecuencia** SEÑAL frequency-exchange signalling (*AmE* signaling); ~ **negativa de amplitud** ELECTRÓN, TV negative amplitude modulation; ~ **negativa de frecuencia** ELECTRÓN negative frequency modulation; ~ **no coherente por barrido de frecuencia** TRANS noncoherent swept-tone modulation; ~ **ocho a catorce** (*ETF*) INFO eight to fourteen modulation (*ETF*); ~ **ocho a diez** (*ETM*) INFO eight to ten modulation (*ETM*); ~ **por la palabra** CONTROL, PERIF, TELEFON voice control; ~ **parásita** TEC RAD spurious modulation; ~ **de portadors de impulsos** TELEFON pulsed carrier modulation; ~ **de portadora suprimida** ELECTROTEC quiescent carrier modulation; ~ **por pseudorruido** DATOS, TRANS pseudonoise modulation; ~ **con referencia fija** TELEFON modulation with a fixed reference; ~ **simple** TEC RAD simple modulation; ~ **telegráfica** TELEFON telegraph modulation; ~ **del timbre** INFO timbre modulation; ~ **transicional** TEC RAD multistate modulation; ~ **triangular** SEÑAL, TRANS delta modulation; ~ **a varios niveles** TEC RAD multilevel modulation; ~ **de velocidad** ELECTRÓN velocity modulation; ~ **vocal de banda angosta** TRANS narrow-band voice modulation (*NBVM*)

modulador *m* ELECTRÓN, TRANS modulator; **~/desmodulador** (*módem*) INFO, TRANS modulator/demodulator; ~ **doble en anillo** ELECTRÓN ring modulator; ~ **de fase** TRANS phase modulator; ~ **Hall** ELECTRÓN *semiconductores* Hall modulator; ~ **de luz espacial** ELECTRÓN spatial light modulator (*SLM*); ~ **telegráfico** TELEFON telegraph modulator

modular[1] *adj* GEN modular

modular[2] *vt* GEN modulate

modularidad *f* GEN modularity

modularización *f* GEN modularization

módulo *m* GEN module; ~ **de almacenamiento del puente de conexión** (*JSM*) ELECTRÓN jumper storage module (*JSM*); ~ **de ampliación** TELEFON extension module; ~ **de compresibilidad** COMPON *resistencia* bulk modulus; ~ **de concentración** ELECTRÓN concentration module; ~ **del conmutador electrónico** PRUEBA sampler module; ~ **de construcción** GEN *tecnología del diseño* building module; ~ **conversor** ELECTRÓN converter module; ~ **de distribución** ELECTROTEC *suministro de energía* distribution module; ~ **de elasticidad** COMPON modulus of elasticity; ~ **enchufable** COMPON, ELECTRÓN, INFO plug-in module; ~ **de energía del bastidor** ELECTROTEC, TELEFON rack power module; ~ **de fijación de lóbulo** ELECTRÓN lobe attachment module; ~ **de función** CONMUT function module; ~ **de identidad del abonado** (*SIM*) TEC RAD subscriber identity module (*SIM*); ~ **de interfaz** GEN interface module (*IM*); ~ **de interfaz de líneas** CONMUT line interface module; ~ **de interfaz terrestre**

TRANS terrestrial interface module; ~ **de líneas** CONMUT line module; ~ **objeto** INFO object module; ~ **periférico** PERIF peripheral module; ~ **de refracción** TEC RAD refractive modulus; ~ **de la sección para flexión del plástico** COMPON bending strength; ~ **de sonido** INFO sound module; ~ **de transmisión en semiduplex** (*HDTM*) TRANS half-duplex transmission module (*HDTM*); ~ **de tratamiento de enlaces** DATOS, INFO link-handling module; ~ **de Young** COMPON Young's modulus

molde: ~ **seccionado** *m* COMPON split mould (*AmE* mold)

moldeado *adj* GEN moulded (*AmE* molded)

momentáneo *adj* GEN momentary

momento *m* GEN moment, point in time; ~ **de arranque** GEN time of start; ~ **de conexión** COMPON turn-on time; ~ **cinético** TEC RAD angular momentum; ~ **de FI** ELECTRÓN, TEC RAD IF stage

monitor *m* GEN monitor; ~ **de digitalización múltiple** TV multiscan monitor; ~ **de fase** ELECTROTEC phase monitor; ~ **de forma de onda** CONTROL waveform monitor; ~ **de frecuencia** TEC RAD frequency monitor; ~ **de fuga a tierra** ELECTROTEC earth leakage guard *BrE*, ground leakage guard *AmE*; ~ **de nivel** TRANS level monitor; ~ **de proporción de errores** TELEFON error-rate monitor; ~ **de ráfagas** PRUEBA burst monitor; ~ **de tensión** CONTROL, ELECTROTEC voltage monitor; ~ **de transmisión en modo continuo** PRUEBA continuous-mode transmission monitor

monitorización: ~ **a distancia** *f* PRUEBA remote monitoring

monocelular *adj* TEC RAD monocellular

monocristal *adj* ELECTRÓN single-crystal

monocromador *m* TRANS monochromator

monocromático *adj* TRANS monochromatic

monocromo *adj* GEN monochrome

monoestable *adj* ELECTRÓN monostable

monofásico *adj* ELECTROTEC single-phase

monofonía *f* TEC RAD monophony

monofónico *adj* GEN monophonic

monolítico *adj* ELECTRÓN monolithic

monomodo *adj* ELECTRÓN monomode

monoscopio *m* ELECTRÓN monoscope

montado: ~ **en agujero** *adj* ELECTRÓN hole-mounted; ~ **en bastidor** *adj* ELECTRÓN rack-mounted; ~ **en un vehículo** *adj* TEC RAD vehicle-mounted

montador: ~ **electricista** *m* GEN electrical fitter; ~ **de equipo** *m* GEN equipment assembler; ~ **técnico** *m* GEN technical fitter

montaje *m* GEN assembly, mounting; ~ **en agujero** ELECTRÓN *circuitos impresos* hole mounting; ~ **de bastidor de siete ranuras** COMPON, ELECTRÓN seven-slot rack-mounting assembly; ~ **de cables** TRANS cable assembly; ~ **de cables ópticos** TRANS optical cable assembly; ~ **en contrafase** ELECTRÓN, ELECTROTEC push-pull arrangement; ~ **de culata** COMPON, ELECTRÓN yoke assembly; ~ **de dos B más D** TRANS two-B-plus-D arrangement; ~ **embutido** GEN flush mounting, recessed mounting; ~ **mural** ELECTRÓN, PERIF, TELEFON wall mounting; ~ **de panel de circuito integrado** ELECTRÓN integrated circuit panel assembly; ~ **de sensor solar** (*SSA*) TEC RAD sun sensor assembly (*SSA*); ~ **sobre platina** COMPON, ELECTROTEC strip-mounted set; ~ **en el suelo** GEN floor mounting; ~ **de taladrado** COMPON, ELECTRÓN boring fixture

montante *m* TRANS bearer; ~ **del bastidor** TELEFON rack stanchion, rack upright; ~ **inicial** TELEFON first upright

montar *vt* GEN assemble, mount, set up

montura: ~ **de piedra dura** *f* COMPON, ELECTROTEC jewel mount; ~ **de termistor** *f* COMPON, PRUEBA thermistor mount

morfeado *m* INFO morphing

mosaico *m* ELECTRÓN, PRUEBA mosaic; ~ **alfa** PRUEBA alpha mosaic

mostrar *vt* GEN show

motivo: ~ **de tramado** *m* INFO *gráficos* dithering

motor: ~ **de avance por pasos** *m* ELECTROTEC stepper motor; ~ **diesel** *m* GEN diesel engine; ~ **fónico** *m* TEC RAD phonic motor; ~ **de impulsión** *m* GEN driving motor

motorizado *adj* ELECTRÓN powered

MOU *abr* (*memorándum de acuerdo*) GEN MOU (*memorandum of understanding*)

mover *vt* DATOS, INFO move

movilidad: ~ **iónica** *f* ELECTRÓN *química* ionic mobility

MPA *abr* (*red multihaz de elementos en fase*) TEC RAD MPA (*multibeam phased array*)

MPBTL *abr* (*pérdida mínima admisible de transmisión básica*) DATOS MPBTL (*minimum permissible basic transmission loss*)

MPDU *abr* (*unidad de datos de protocolo de mensaje*) DATOS MPDU (*message protocol data unit*)

MPIL *abr* (*nivel máximo permisible de interferencia*) DATOS MPIL (*maximum permissible interference level*)

MPS *abr* (*señal multipágina*) DATOS, TELEFON MPS (*multipage signal*); ~ **de interrupción de**

procedimiento *m* (*PRI-MPS*) DATOS, TELEFON procedure interrupt MPS (*PRI-MPS*)
MPSK *abr* (*tecleo de cambio de fase multinivel*) SEÑAL MPSK (*multilevel phase-shift keying*)
MRSE *abr* (*elemento del servicio de recuperación de mensajes*) DATOS MRSE (*message retrieval service element*)
MS *abr* (*estación móvil*) TEC RAD, TELEFON MS (*mobile station*)
MSC *abr* (*centro de conmutación móvil*) SWG MSC (*mobile switching centre*); **~ de pasillo** *m* CONMUT gateway MSC; **~ de sujeción** *m* TEC RAD anchor MSC
MSRN *abr* (*número de localización automática de estación móvil*) TELEFON MSRN (*mobile station roaming number*)
MSSC *abr* (*centro marítimo de conmutación por satélite*) TEC RAD MSSC (*maritime satellite switching centre*)
MSSE *abr* (*elemento de servicio de presentación de mensaje*) DATOS MSSE (*message submission service element*)
MSSS *abr* (*centro de apoyo multisatélite*) TEC RAD MSSS (*multisatellite support centre*)
MSSW *abr* (*onda de superficie magnetostática*) ELECTRÓN MSSW (*magnetostatic surface wave*)
MTBF *abr* (*tiempo medio entre fallos*) PRUEBA MTBF (*mean time between failures*)
MTBI *abr* (*tiempo medio entre interrupciones*) INFO MTBI (*mean time between interruptions*)
MTBO *abr* (*tiempo medio entre revisiones*) PRUEBA MTBO (*mean time between overhauls*)
MTFF *abr* (*tiempo medio hasta el primer fallo*) INFO MTFF (*mean time to first failure*)
MTP *abr* (*parte de la transferencia de mensaje*) DATOS, TELEFON MTP (*message transfer part*)
MTSE *abr* (*elemento del servicio de transferencia de mensajes*) DATOS MTSE (*message transfer service element*)
MTSO *abr* (*oficina de conmutación de teléfono móvil*) TEC RAD MTSO (*mobile telephone switching office*)
MTSR *abr* (*tiempo medio hasta el restablecimiento del servicio*) PRUEBA MTSR (*mean time to service restoration*)
MTTR *abr* (*tiempo medio hasta el restablecimiento*) PRUEBA MTTR (*mean time to recovery*)
muaré *m* TV moiré
mudo *adj* INFO mute
muelle: **~ alzador** *m* CONMUT lifting spring; **~ de hilo** *m* CONMUT wire spring; **~ impulsor** *m* GEN operating spring
muerto *adj* ELECTROTEC dead
muesca *f* INFO indent, indentation; **~ de polaridad** ELECTRÓN polarizing slot; **~ posicionadora** GEN key groove, keyway; **~ de posicionamiento** ELECTRÓN *circuitos impresos* locating notch; **~ de puesta a tierra** ELECTRÓN earthing indent *BrE*, grounding indent *AmE*; **~ para toma de tierra** ELECTRÓN earthing indent *BrE*, grounding indent *AmE*
muestra *f* GEN sample; **~ aleatoria simple** PRUEBA simple random sample; **~ estratificada** PRUEBA stratified sample; **~ geométrica** PRUEBA *calidad* geometric sample; **~ orientada** PRUEBA *calidad* directional sample; **~ parcial** PRUEBA part sample; **~ de producto** PRUEBA specimen; **~ puntual** PRUEBA spot sample; **~ reconstruida** TELEFON reconstructed sample; **~ representativa** GEN representative sample; **~ sesgada** PRUEBA *calidad* biased sample; **~ de voz** TRANS speech sample
muestras: **~ por segundo** *f pl* TRANS samples per second
muestreador *m* GEN sampler
muestrear *vt* GEN sample
muestreo *m* GEN sampling; **~ aleatorio** ELECTRÓN random sampling; **~ en cadena** PRUEBA *calidad* chain sampling; **~ continuo multinivel** PRUEBA multilevel continuous sampling; **~ continuo simple** PRUEBA single-level continuous sampling; **~ estratificado** PRUEBA stratified sampling; **~ a granel** PRUEBA *calidad* bulk sampling; **~ integral** TRANS integral sampling; **~ multietapa** PRUEBA multistage sampling, nested sampling; **~ múltiple** TRANS multiple sampling, multisampling; **~ de Nyquist** INFO Nyquist sampling; **~ óptimo** TRANS optimal sampling; **~ y retención** CONTROL sample and hold; **~ simple** PRUEBA single sampling; **~ sistemático** PRUEBA systematic sampling; **~ por universos** GEN cluster sampling
MUF *abr* (*frecuencia máxima utilizable*) TEC RAD MUF (*maximum usable frequency*); **~ básica** *f* TEC RAD basic MUF; **~ operativa** *m* TEC RAD operational MUF
muldex *m* TELEFON muldex
multiacoplador: **~ de escucha** *m* TEC RAD receiver multicoupler
multiárea *adj* TEC RAD multiarea
multibloque *adj* TELEFON multiblock
multicanal *adj* TRANS multichannel
multidireccionamiento *m* DATOS multiaddressing
multiemplazamiento *adj* GEN multi-site
multifacetado *adj* GEN multifaceted

multifrecuencia *f* (*MF*) SEÑAL, TELEFON, TRANS *modo de señalización* multifrequency (*MF*); ~ **de tono dual** (*DTMF*) DATOS digital tone multifrequency (*DTMF*)
multifunción *adj* GEN multifunction
multihaz *m* TEC RAD multiple beam
multilateral *adj* GEN multilateral
multilineal *adj* GEN multiline
multilocal *adj* TEC RAD multisite
multimedia *adj* INFO, TV multimedia
multímetro *m* PRUEBA multimeter
multipista *adj* DATOS multitrack
múltiple[1] *adj* GEN multiple
múltiple[2]: ~ **de hilo desnudo** *m* TELEFON bare-wire multiple; ~ **de selectores** *m* CONMUT selector multiple, TELEFON selector multiple, switch multiple; ~ **de verticales** *m* CONMUT vertical unit multiple
múltiplex *m* CONMUT, DATOS, INFO, TEC RAD, TRANS multiplex; ~ **por división de frecuencia** TELEFON, TRANS frequency-division multiplex; ~ **de división de frecuencia ortogonal codificada** (*COFDM*) TRANS coded orthogonal frequency division multiplex (*COFDM*); ~ **por división en longitud de onda** TELEFON, TRANS wavelength division multiplex; ~ **de estación de intercontrol** TELEFON intercontrol station multiplex
multiplexación: ~ **continua** *f* TEC RAD continuous multiplexing; ~ **de división direccional** *f* TRANS directional division multiplexing
multiplexado: ~ **por división de frecuencias** *m* (*MDF*) TELEFON, TRANS frequency-division multiplexing (*FDM*); ~ **de frecuencia** *m* TELEFON, TRANS frequency multiplexing
multiplexador: ~ **de datos** *m* INFO data multiplexer; ~ **de memoria de datos** *m* CONMUT *códigos* data store multiplexer; ~ **remoto** *m* TRANS remote multiplexer
multiplexión *f* CONMUT, DATOS, INFO, TEC RAD, TRANS multiplexing; ~ **asíncrona por división en el tiempo** TELEFON, TRANS asynchronous time-division multiplexing; ~ **por división de longitud de onda** (*WDM*) TELEFON, TRANS wavelength division multiplexing (*WDM*); ~ **del paquete de datos** TEC RAD packet multiplexing
multiplexor *m* (*MUX*) CONMUT, DATOS, INFO, TEC RAD, TRANS multiplexer (*MUX*); ~ **de acceso de estación de control** TELEFON control-station access multiplex; ~ **de conmutación** CONMUT switching multiplexer; ~ **de datos secundario** DATOS, TRANS second data multiplexer; ~ **por división en longitud de onda** TELEFON, TRANS wavelength division multiplexer; ~ **del emisor** TRANS transmitter-multiplexer; ~ **estadístico de conmutación** CONMUT switching statistical multiplexer; ~ **de primera orden** DATOS, TRANS first-order multiplexer; ~ **de primer dato** DATOS, TRANS first-data multiplexer; ~ **sumar-restar** TRANS add-drop multiplexer
multiplicación: ~ **debida al gas** *f* ELECTRÓN gas multiplication; ~ **de error** *f* TELEFON error multiplication
multiplicado *adj* TELEFON multiplied
multiplicador *m* GEN multiplicator, multiplier; ~ **binario** DATOS binary multiplier; ~ **de frecuencia** TRANS frequency multiplier; ~ **Hall** ELECTRÓN *semiconductores* Hall multiplier
multipolar *adj* ELECTROTEC multipolar, multipole
multiprocesador *m* DATOS, INFO multiprocessor
multiprogramación *f* INFO multiprogramming
multiproveedor *m* GEN multivendor
multipunto *adj* GEN multipoint
multiregional *adj* GEN multiregional
multirelé *adj* TEC RAD multirelay
multisesión *f* INFO multisession
multisincrónico *adj* TV multisynchronous
multitarea[1] *adj* DATOS, INFO multitask
multitarea[2] *f* DATOS, INFO multitasking
multitímbrico *adj* INFO multitimbral
multitrama *f* DATOS, TELEFON, TRANS multiframe
multiusuario *adj* GEN multiuser
multivibrador *m* (*MV*) ELECTRÓN multivibrator (*MV*); ~ **monoestable** ELECTRÓN one-shot multivibrator, single-shot element
multivisión *f* INFO slide show
MUSE *abr* (*codificación por submuestreo múltiple*) DATOS MUSE (*multiple subsampling encoding*)
mutilación *f* GEN *criptología* garbling
mutilado *adj* GEN garbled
mutilar *vt* GEN garble
MUX *abr* (*multiplexor*) CONMUT, DATOS, INFO, TEC RAD, TRANS MUX (*multiplexer*)
MV *abr* (*multivibrador*) ELECTRÓN MV (*multivibrator*)

N

n: **~-vías** *f pl* DATOS n-ways

N *abr* GEN (*número*) N (*number*), PRUEBA (*newton*) N (*newton*)

N/A *abr* (*inaplicable*) DATOS, TELEFON N/A (*not applicable*)

NACK *abr* (*reconocimiento negativo*) TEC RAD NACK (*negative acknowledgement*)

NAE *abr* (*extensión de dirección de red*) DATOS NAE (*network address extension*)

nailon *m* COMPON nylon

NAM *abr* (*número de abonado múltiple*) TELEFON MSN (*multiple subscriber number*)

nanosegundo *m* (*ns*) PRUEBA nanosecond (*ns*)

NAPI/TOA *abr* (*numeración y direccionamiento de indicador de plan/tipo de dirección*) DATOS NAPI/TOA (*numbering and addressing plan indicator/type of address*)

navegación *f* GEN navigation; **~ hipermediática** INFO hypermedia navigation; **~ por hipertexto** INFO hypertext navigation; **~ de largo recorrido** TEC RAD long-range navigation; **~ multimedia** INFO multimedia navigation; **~ de la Red** DATOS, INFO *Internet* netsurfing

navegador *m* INFO *Internet* surfer

navegante *m* INFO *Internet* navigator

navegar *vt* INFO *Internet* navigate, surf

NBP *abr* (*parámetro no básico*) DATOS NBP (*nonbasic parameter*)

NCL *abr* (*lenguaje de control de la red*) ELECTRÓN, INFO, TELEFON NCL (*network control language*)

NCSA *abr* (*Centro Nacional para Aplicaciones de Superinformática*) DATOS, INFO *Internet* NCSA (*National Center for Supercomputing Applications*)

ND *abr* (*fallo de la red*) DATOS network failure, ELECTRÓN ND (*network default*), TELEFON network default (*ND*), network failure

NDIS *abr* (*especificaciones del interfaz de arrastre de la red*) ELECTRÓN NDIS (*network drive interface specifications*)

necesidad: **~ de potencia** *f* ELECTRÓN power requirement; **~ de potencia de la portadora** *f* TRANS carrier power requirement

NEF *abr* (*trama no esperada*) TELEFON NEF (*not-expected frame*)

negación *f* ELECTRÓN, INFO, TRANS denial, negation

negar *vt* ELECTRÓN, INFO, TRANS *acceso* deny, negate

negativo: **~ de batería** *m* ELECTROTEC battery negative; **~ directo** *m* ELECTROTEC direct negative; **~ de disco compacto fotográfico** *m* INFO photo CD Master; **~ de distribución** *m* ELECTROTEC *suministro de energía* distribution negative; **~ de reinstrucción** *m* (*RTN*) DATOS, TELEFON retrain negative (*RTN*)

negentropía *f* DATOS negentropy

negociación *f* GEN negotiation; **~ de retardo de tránsito de extremo a extremo** (*EETDN*) DATOS, TELEFON end-to-end transit delay negotiation (*EETDN*)

negocio: **~ de línea única** *m* GEN single-line business

negrita *f* INFO *imprenta* bold

negro: **~ nominal** *m* ELECTROTEC nominal black

nemática: **~ de superalabeo** *f* TRANS supertwist nematic; **~ torcida** *f* ELECTRÓN twisted nematic; **~ torcida supermonocromática** *f* ELECTRÓN monochrome super-twist nematic

neopreno *m* COMPON neoprene

neper *m* (*np*) ELECTRÓN neper (*np*)

Net: **la ~** *f* DATOS, INFO the Net

netiquette *f* DATOS, INFO *Internet* netiquette

Netscape *m* DATOS, INFO *Internet* Netscape

neutralización *f* GEN counter-action, neutralization

neutro *adj* ELECTROTEC neutral

newsgroup *m* INFO newsgroup

newton *m* (*N*) PRUEBA newton (*N*)

NFM *abr* (*gestor de archivo de red*) DATOS, INFO NFM (*network file manager*)

nicromio *m* COMPON nichrome

nieve *f* TEC RAD snow

niobato: **~ de litio** *m* COMPON *química* lithium neobate

NIP *abr* (*interrupción de proceso negativa*) TELEFON PIN (*procedure interrupt negative*)

níquel *m* GEN nickel

nitruración *f* COMPON nitriding

nivel *m* GEN level; **~ de aceite** GEN oil level; **~ acústico** GEN sound level, noise level; **~ de base** CONMUT base level; **~ de blanco** INFO picture white, white level, TELEFON, TRANS white level; **~ de borrado** TELEFON blanking level; **~ de carga flotante** ELECTROTEC *baterías* float-charging level; ~ **de confianza** GEN confidence level; ~ **de control de los auriculares** TELEFON headphones monitoring level; ~ **de coordinación** DATOS coordination level; ~ **de la corriente de retención** ELECTRÓN *semiconductores* holding current level; ~ **de declinación** INFO decay level; ~ **de disparo** INFO trigger level; ~ **de dureza de radiación** COMPON radiation-hardness level; ~ **de eco residual** TELEFON residual echo level; ~ **de energía** COMPON *semiconductores* energy level; ~ **de entrada** DATOS *digitalizador* entry level; ~ **fantasma** TRANS phantom level; ~ **de Fermi** GEN *física* Fermi level; ~ **de graduación por teclado** INFO keyboard scaling level; ~ **de grises** TELEFON grey (*AmE* gray) level; ~ **de indentación** DATOS level of indentation; ~ **inicial** GEN starting level; ~ **de inspección** PRUEBA *calidad* inspection level; ~ **de intensidad** TRANS intensity level; ~ **de interferencia** ELECTROTEC, TEC RAD interference level; ~ **keranuico** TEC RAD keranuic level; ~ **de mal funcionamiento** PRUEBA malfunction level; ~ **de manipulación del tráfico** DATOS, TELEFON traffic-handling level; ~ **máximo permisible de interferencia** (*MPIL*) DATOS maximum permissible interference level (*MPIL*); ~ **medio** TEC RAD mid-level; ~ **de movilidad** TEC RAD level of mobility; ~ **de negro** INFO, TELEFON, TRANS black level; ~ **no afectado** TELEFON unaffected level; ~ **de ocupación** TELEFON traffic level; ~ **de polarización** INFO bias level; ~ **de potencia** INFO power level; ~ **de presentación** DATOS, TRANS presentation level; ~ **de prioridad** TELEFON priority level; ~ **de prueba adaptado** ELECTROTEC, PRUEBA terminated level; ~ **de realimentación** INFO feedback level; ~ **de recepción** INFO listening level; ~ **de red** DATOS, TELEFON, TRANS network layer; ~ **de referencia** GEN *acústica* background level; ~ **relativo** TELEFON relative level; ~ **relativo nominal** TELEFON nominal relative level; ~ **relativo de potencia real o aparente** TELEFON relative power level; ~ **de rendimiento** CONTROL performance index; ~ **de retardo** INFO delay level; ~ **de reverberación** INFO reverberation level; ~ **de reverbero** INFO reverb level; ~ **de riesgo** PRUEBA risk level; ~ **de ruido** GEN noise level; ~ **de ruido de circuito** TELEFON circuit noise level; ~ **de ruido percibido** TELEFON perceived noise level; ~ **de salida** ELECTRÓN, INFO output level; ~ **de señal** SEÑAL, TRANS signal level; ~ **de servicio** TRANS service layer; ~ **de sesión** DATOS, TRANS session level; ~ **de significación** PRUEBA significance level; ~ **de sonoridad** TELEFON *acústica* loudness level; ~ **de tensión** ELECTROTEC voltage level; ~ **total** INFO total level; ~ **de transferencia del registro** CONMUT register-transfer level; ~ **umbral del receptor** TEC RAD receiver threshold level; ~ **de zumbido** ELECTRÓN hum level

nivelación *f* GEN levelling (*AmE* leveling); ~ **de la amplitud** TRANS amplitude equalization

nivelar *vt* GEN level

nivelímetro *m* PRUEBA level meter

NL *abr* (*carácter de cambio de línea*) DATOS NL (*new-line character*)

NNI *abr* (*interfaz de nodo de la red*) TRANS NNI (*network node interface*)

NNTP *abr* (*protocolo de transferencia de noticias de la red*) DATOS *Internet* NNTP (*network news transfer protocol*)

NO: **~-Y** *m* (*NAND*) ELECTRÓN NOT-AND (*NAND*)

nodal *adj* INFO nodal

nodo *m* GEN node; ~ **de acceso** GEN access node; ~ **de acceso internacional** CONMUT, TELEFON international gateway node (*IGN*); ~ **de almacenamiento de información** TELEFON information storage node; ~ **de árbol** DATOS, INFO tree node; ~ **ascendente** TEC RAD ascending node; ~ **de base** DATOS, INFO root node; ~ **de la base de parte superior** INFO root top node; ~ **del blanco** INFO target node; ~ **compuesto** INFO composite node; ~ **de conmutación** CONMUT, SEÑAL switching node; ~ **de conmutación de circuito virtual** CONMUT, TRANS virtual-circuit switching node; ~ **de conmutación de paquetes** CONMUT packet-switching node; ~ **de control móvil** TELEFON mobile control node (*MCN*); ~ **de destino** INFO destination node; ~ **fuente** INFO source node; ~ **de gestión de la red** TELEFON network support node; ~ **de hipertexto** INFO hypertext node; ~ **de hoja** DATOS leaf node; ~ **receptor** INFO destination node; ~ **de la red** TRANS network node; ~ **de servicio** TELEFON service node; ~ **de servicio/transferencia** TELEFON service-transfer node; ~ **sin hijo** INFO childless node; ~ **terminal** INFO tip node, top node; ~ **de tránsito internacional** CONMUT, TRANS international transit node; ~ **de transmisión** TRANS transmission node

NOGAD *abr* (*ajustador de ganancia activado por el ruido*) ELECTRÓN NOGAD (*noise-operated gain adjusting device*)

nombre *m* GEN name; ~ **caracterizado relativo** DATOS relative distinguished name; ~

comercial GEN brand name; ~ **cualificado** DATOS qualified name; ~ **descriptivo** DATOS descriptive name; ~ **distintivo** DATOS distinguished name; ~ **de fichero lógico** (*LFN*) INFO logical file name (*LFN*); ~ **funcional** DATOS performance name; ~ **genérico** DATOS generic name; ~ **impreso** GEN printed name; ~ **de parámetro** TELEFON parameter name; ~ **primitivo** DATOS primitive name; ~ **de símbolo** GEN symbol name; ~ **de tecla** INFO key label; ~ **de trayectoria** INFO path name; ~ **de usuario** INFO user name

nominal *adj* ELECTROTEC rated

nominativo *adj* TEC RAD registered

norma *f* GEN norm, standard; ~ **de aseguramiento de la calidad** PRUEBA quality assurance ruling; ~ **de cifrado de datos** *m* (*DES*) DATOS data encryption standard (*DES*); ~ **de codificación** TRANS coding standard; ~ **para la comprobación de conformidad** DATOS, PRUEBA conformance-testing standard; ~ **DIN** GEN DIN standard; **de ~ doble** *adj* GEN dual-standard; ~ **de funcionamiento común** TEC RAD common operating standard; ~ **internacional** (*SI*) GEN international standard (*IS*); ~ **ISO** GEN ISO standard; ~ **de rendimiento** PRUEBA performance objective; ~ **de seguridad basada en la identidad** APLIC identity-based security policy; ~ **vinculada al acondicionamiento** COMPON *técnicas de montaje* construction-bound rule

Norma: ~ **Europea de Telecomunicaciones** *f* GEN European Telecommunications Standard

normalización *f* GEN normalization, standardization

normalizar *vt* GEN normalize, standardize

normas: ~ **básicas de codificación** *f pl* DATOS basic encoding rules (*BER*); ~ **de frecuencia atómica** *f pl* ELECTRÓN atomic frequency standards

normativo *adj* GEN normative

nota: ~ **dominante** *f* INFO keynote; ~ **de embarque** *f* GEN shipping note; ~ **de entrega** *f* GEN *aprovisionamiento y distribución* delivery note; ~ **de fondo** *f* INFO bottom note; ~ **al pie** *f* INFO footnote

notación: ~ **en árbol** *f* DATOS, INFO tree notation; ~ **combinada en árbol y tabular** *f* (*TTCN*) DATOS, INFO tree and tabular combined notation (*TTCN*); ~ **hexadecimal** *f* DATOS, INFO hexadecimal notation; ~ **normal de Backus** *f* DATOS Backus Normal Form

notificación *f* GEN notification; ~ **explícita de bloqueo regresivo** CONMUT, DATOS, TELEFON backward explicit congestion notification (*BECN*); ~ **de falta de entrega** DATOS, TELEFON nondelivery notification (*NDN*); ~ **de intercambio de datos electrónicos** TELEFON EDI notification (*EDIN*); ~ **interpersonal** (*IPN*) TELEFON, TELEFON interpersonal notification (*IPN*); ~ **de mensajería por voz** APLIC, DATOS, TELEFON voice messaging notification; ~ **negativa** DATOS, TELEFON negative notification (*NN*); ~ **de no recepción** (*NRN*) DATOS, TELEFON nonreceipt notification (*NRN*); ~ **de no receptor** DATOS, TELEFON nonrecipient notification; ~ **oral** (*NV*) TELEFON voice notification (*VN*); ~ **positiva** DATOS, TELEFON positive notification (*PN*); ~ **de recepción** (*RN*) DATOS, TELEFON receipt notification (*RN*); ~ **reenviada** (*FN*) DATOS, TELEFON forwarded notification (*FN*); ~ **de servicio** DATOS, TELEFON service notification; ◆ **otra** ~ (*ON*) DATOS other notification (*ON*)

novato *adj* GEN fledgling

np *abr* (*neper*) ELECTRÓN np (*neper*)

NPA *abr* (*área de plan de numeración*) TELEFON NPA (*numbering plan area*)

NPD *abr* (*dispositivo de protección de la red*) TELEFON NPD (*network protection device*)

NPI/TOA *abr* (*indicador de plan de numeración/TOA*) DATOS, TELEFON NPI/TOA (*numbering plan indicator/TOA*)

NRN *abr* (*notificación de no recepción*) DATOS, TELEFON NRN (*nonreceipt notification*)

NRT *abr* (*tiempo irreal*) DATOS, INFO, PRUEBA, TELEFON NRT (*nonreal time*)

NRZ *abr* (*sin retorno a cero*) DATOS, TELEFON NRZ (*nonreturn to zero*)

ns *abr* (*nanosegundo*) PRUEBA ns (*nanosecond*)

NSC-CIG-DTC *abr* (*CIG-DTC de orden de instalación no normalizada*) DATOS NSC-CIG-DTC (*nonstandard facility command CIG-DTC*)

NSF *abr* (*instalación no normalizada*) TELEFON NSF (*nonstandard facility*)

NSF-CSI-DIS *abr* (*CSI-DIS de instalación no normalizada*) DATOS NSF-CSI-DIS (*nonstandard facility CSI-DIS*)

NSS *abr* (*ajuste no normalizado*) DATOS, TELEFON NSS (*nonstandard setup*)

NT1 *abr* (*terminación digital de la red*) TELEFON NT1 (*network digital termination*)

NTL *abr* (*pérdida de transmisión normalizada*) TEC RAD NTL (*normalized transmission loss*)

NTN *abr* (*número del terminal de la red*) DATOS, TELEFON NTN (*network terminal number*)

NTU *abr* (*unidad terminal de la red*) TRANS NTU (*network terminating unit*)

núcleo *m* GEN nucleus, INFO kernel, nucleus; ~

atómico GEN *física* atomic nucleus; ~ **de ferrita** ELECTROTEC, INFO ferrite core; ~ **de hierro** ELECTRÓN *electromagnetismo* iron core; ~ **de polvo de hierro** ELECTRÓN *electromagnetismo* iron-dust core; ~ **de relé** COMPON relay core; ~ **del sistema operativo** INFO nucleus of operating system; ~ **toroidal** ELECTRÓN toroidal core

nueva: ~ **dirección** *f* CONMUT, TELEFON redirection; ~ **tecnología** *f* GEN new technology

nuevas: ~ **portadoras comunes** *f pl* GEN new common carriers (*NCC*)

nuevo: ~ **ensayo** *m* PRUEBA, TELEFON retest

NUI *abr* (*identificación del usuario de la red*) DATOS, TELEFON NUI (*network user identification*)

NUL *abr* (*carácter nulo*) DATOS NUL (*null character*)

nulo *adj* TEC RAD null

numeración *f* GEN numbering; ~ **y direccionamiento de indicador de plan/tipo de dirección** (*NAPI/TOA*) DATOS numbering and addressing plan indicator/type of address (*NAPI/TOA*); ~ **de escala** PRUEBA scale numbering; ~ **de páginas** INFO *procesamiento de texto* page numbering; ~ **de ruta** TELEFON route numbering; ~ **uniforme** TELEFON uniform numbering

numeral: ~ **binario** *m* DATOS, INFO binary numeral; ~ **hexadecimal** *m* DATOS, INFO hexadecimal numeral; ~ **octal** *m* TELEFON octal numeral

numéricamente *adv* GEN digitally, numerically

numérico *adj* GEN digital, numeric

numerización *f* GEN digitizing

número *m* (*N*) GEN number (*N*); ~ **de abonado** DATOS, TELEFON subscriber number; ~ **de abonado múltiple** (*NAM*) TELEFON multiple subscriber number (*MSN*); ~ **abreviado** TELEFON abbreviated number; ~ **de acceso** TELEFON access number; ~ **de acceso universal** TELEFON universal-access number; ~ **de aciertos por blanco** TEC RAD number of hits per target; ~ **averiado** DATOS number out of order; ~ **de banda** TELEFON band number; ~ **de base** INFO base number; ~ **de condiciones significativas** TELEFON number of significant conditions; ~ **de dato internacional** DATOS international data number; ~ **de designación** TELEFON designation number; ~ **de diagrama** GEN *dibujo técnico* diagram number; ~ **de día juliano** TEC RAD Julian day number; ~ **de dibujo** GEN *dibujo técnico* drawing number; ~ **de dientes** GEN number of teeth; ~ **de directorio** TELEFON directory number (*DN*); ~ **de directorio indirecto** TELEFON indirect directory number (*IDN*); ~ **disponible** TELEFON spare number; ~ **del documento** GEN document number; ~ **efectivo de bitios** DATOS effective number of bits; ~ **de elementos defectuosos** PRUEBA number of defectives; ~ **de equipo** TELEFON equipment number; ~ **de espiras** ELECTROTEC number of turns; ~ **de estación de barco** DATOS, TEC RAD, TELEFON ship station number; ~ **de extensión** CONMUT *PBX* extension number; ~ **en guía del abonado** TELEFON subscriber's directory number; ~ **hexadecimal** DATOS, INFO hexadecimal number; ~ **de identificación** INFO key number; ~ **de identificación de parámetro** TELEFON parameter identification number; ~ **incompleto del abonado que llama** TELEFON calling-party number incomplete; ~ **de indicación de la posición del móvil** TEC RAD roaming number; ~ **inexistente** TELEFON nonexistent number; ~ **internacional** TELEFON international number; ~ **irracional** GEN irrational number; ~ **de lectura máximo utilizable** INFO maximum usable read-number; ~ **de llamada gratuita** TELEFON toll-free number *AmE*, freephone number *BrE*; ~ **llamado** TELEFON called number; ~ **llamado original** TELEFON original called number; ~ **local** TELEFON local number; ~ **de localización automática de estación móvil** (*MSRN*) TELEFON mobile station roaming number (*MSRN*); ~ **de marcado directo** TELEFON through-dialling number; ~ **de memoria** PERIF *contestador* memory number; ~ **de muestra** PRUEBA, TRANS sample number; ~ **muestral promedio** (*ASN*) PRUEBA *calidad*, TRANS average sample number (*ASN*); ~ **de nivel** INFO rank number; ~ **no adjudicado** DATOS number not allocated; ~ **no asignado** TELEFON vacant number; ~ **no decimal** TELEFON nondecimal numeral; ~ **de onda** TRANS *óptica de fibras* wave number; ~ **de orden** CONMUT, DATOS, INFO sequence number (*SN*); ~ **personal de identidad** GEN personal identity number (*PIN*); ~ **de pista** INFO track number; ~ **de rechazo** PRUEBA nonacceptance number, rejection number; ~ **de reexpedición** CONMUT, TELEFON redirecting number, redirection number; ~ **de secuencia de comando** CONMUT command sequence number; ~ **de secuencia decreciente** INFO backward sequence number; ~ **de secuencia hacia atrás** DATOS backward sequence number; ~ **de secuencia hija** DATOS, INFO child-sequence number; ~ **de secuencia inverso** INFO backward sequence number; ~ **de secuencia de parámetro** CONMUT, DATOS, INFO parent sequence number; ~ **de secuencia de retorno** TELEFON backward sequence number; ~ **de sesión** INFO session number; ~ **de teléfono gratuito** TELEFON freephone number *BrE*, toll-free number *AmE*; ~ **de teléfono móvil** TELEFON mobile number; ~ **de télex nacional de abonado** TELEFON subscriber's national telex number; ~ **de terminal de destino** DATOS destination terminal number; ~ **de terminal de la red** (*NTN*) DATOS, TELEFON network terminal number (*NTN*); ~ **total de abonados** TEC RAD, TELEFON subscriber base; ~ **universal** TELEFON universal number; ~ **vacante** TELEFON vacant number; ~ **viejo** TELEFON old number

números: ~ **marcados** *m pl* TELEFON digits dialled (*AmE* dialed)

NV *abr* (*notificación oral*) TELEFON VN (*voice notification*)

O

O *abr* (*opcional*) GEN O (*optional*)
OAS *abr* (*onda acústica de superficie*) TEC RAD, TELEFON, TRANS SAW (*surface acoustic wave*)
objetivo: ~ **de almacenamiento** *m* ELECTRÓN storage target; ~ **de control** *m* INFO control objective; ~ **de diseño** *m* TELEFON design objective; ~ **de entrada en servicio** *m* TELEFON commissioning objective; ~ **de rendimiento** *m* TELEFON performance objective
objetivos: ~ **de control** *m pl* INFO control objectives
objeto *m* DATOS, INFO object; ~ **de apoyo de la gestión** DATOS management support object; ~ **de asociación simple** DATOS single-association object; ~ **corporal** INFO body object; ~ **gestionado** DATOS managed object; ~ **gestionado de capa N** DATOS N-layer managed object; ~ **gestionado por sistemas** DATOS systems-managed object; ~ **del servicio de aplicación** (*ASO*) DATOS, TELEFON application service object (*ASO*)
oblea *f* COMPON, ELECTRÓN wafer
oblicuidad *f* GEN skewness
obligación: ~ **a corto plazo** *f* APLIC short-term duty
obligatorio: **no** ~ *adj* GEN nonmandatory
OBO *abr* (*retroceso de salida*) TRANS OBO (*output back-off*)
obras: ~ **públicas** *f pl* GEN public works
observabilidad *f* GEN observability
observación *f* GEN observation; ~ **radioviento** TEC RAD rawin observation; ~ **de servicio** PRUEBA service observation; ~ **del tráfico** ELECTRÓN, PRUEBA, TELEFON traffic observation
observar *vt* GEN observe
obsoleto *adj* GEN obsolete
obstruido *adj* GEN clogged, obstructed
obtener *vt* TELEFON *persona por teléfono* obtain
obturador *m* TV shutter; ~ **de protección** ELECTRÓN protective shutter
obturante: ~ **B** *m* ELECTRÓN B sealant
OC *abr* (*código de organización*) ELECTRÓN OC (*organization code*)
octeto: ~ **indicador** *m* DATOS flag byte
octodo *m* ELECTRÓN octode
octópodo *adj* ELECTRÓN octopus
ocultación *f* TEC RAD concealment; ~ **de campo** DATOS field blanking
ocultamiento: ~ **compensado por el movimiento** *m* TV motion-compensated concealment; ~ **predictivo espacial** *m* INFO spatial predictive concealment; ~ **predictivo provisional** *m* INFO temporal predictive concealment
ocupación: ~ **de contador de retransmisión** *f* (*RCB*) TELEFON redrive counter busy (*RCB*); ~ **doble** *f* TELEFON double seizure; ~ **ficticia** *f* SEÑAL false seizure; ~ **de ruta** *f* TELEFON route seizure
ocupaciones: ~ **por circuito por hora** *f pl* (*SCH*) TELEFON seizures per circuit per hour (*SCH*)
ocupado *adj* GEN busy, occupied
ocupador: ~ **de espacio corporal** *m* INFO body placeholder
ocupancia: ~ **espectral** *f* TRANS spectral occupancy
ocupar *vt* TELEFON occupy
ocurrencia: ~ **de colisiones** *f* ELECTRÓN occurrence of collisions
ODA *abr* (*arquitectura de documento administrativo*) DATOS, INFO ODA (*office document architecture*)
ODFMA *abr* (*acceso múltiple con división de frecuencia ortogonal*) TEC RAD ODFMA (*orthogonal frequency-division multiple access*)
O-E *abr* (*óptico-eléctrico*) GEN O-E (*optical-electrical*)
oferta *f* GEN *línea de enlace* offer; ~ **de servicio** GEN service offering
oficial *adj* GEN official, recognized
oficina *f* GEN office; ~ **central de anotación** TELEFON central recording office; ~ **de conmutación de telefonía móvil** (*MTSO*) TEC RAD mobile telephone switching office (*MTSO*); ~ **de coordinación de publicidad** TEC RAD traffic bureau; ~ **de correos** TELEFON post office (*PO*); ~ **móvil** TEC RAD mobile office; ~ **paso a paso** TELEFON step-by-step office; ~ **principal** TEC RAD host office; ~ **pública de telefax** TELEFON public facsimile bureau; ~ **de supervisión del servicio** TELEFON service supervision desk; ~ **de**

telégrafos (*TO*) TELEFON telegraph office (*TO*)
oficinas: ~ **conectadas** *f pl* TELEFON linked offices
oficinista *m* GEN office clerk
o/g *abr* (*saliente*) GEN o/g (*outgoing*)
óhmetro *m* ELECTRÓN ohmmeter
ohmio *m* ELECTRÓN ohm
ojal *m* CLIMAT, ELECTROTEC eyelet
OM *abr* (*operaciones y mantenimiento*) PRUEBA, TELEFON OM (*operations and maintenance*)
OMG *abr* (*grupo de gestión de objeto*) GEN OMG (*object management group*)
omisión *f* DATOS, INFO, TEC RAD, TELEFON omission
omitido *adj* DATOS, INFO, TEC RAD, TELEFON omitted
omitir *vt* DATOS, INFO, TEC RAD, TELEFON omit
omnipresencia *f* ELECTRÓN omnipresence
ON *abr* (*otra notificación*) DATOS ON (*other notification*)
onda *f* GEN wave; ~ **acústica de superficie** (*OAS*) TRANS surface acoustic wave (*SAW*); ~ **de choque** GEN shock wave; ~ **compleja** TEC RAD complex wave; ~ **compuesta** TEC RAD composite wave; ~ **continua** TRANS continuous wave; ~ **correctora de la distorsión geométrica de la imagen** TV field keystone waveform; ~ **cuadrada** ELECTRÓN, SEÑAL square waveform; ~ **decamétrica** TRANS decametric wave; ~ **dirigida** TRANS guided wave; ~ **eléctrica transversal** TEC RAD, TRANS transverse electric wave; ~ **espacial** TEC RAD sky wave, space wave; ~ **estacionaria** TEC RAD, TRANS standing wave; ~ **estacionaria de tensión** (*VSW*) TEC RAD, TRANS voltage standing wave (*VSW*); ~ **hertziana** ELECTRÓN, TEC RAD hertzian wave; ~ **ionosférica** TEC RAD ionospheric wave; ~ **longitudinal** TEC RAD longitudinal wave; ~ **media** TEC RAD medium wave; ~ **métrica** TRANS metric wave; ~ **modulada** TRANS modulated wave; ~ **moduladora** TRANS modulating wave; ~ **piloto** TEC RAD, TRANS pilot wave; ~ **plana** TEC RAD plane wave; ~ **portadora** TRANS carrier wave; ~ **portadora completa** TEC RAD full-carrier wave; ~ **portadora de haz global** TRANS global beam-carrier wave; ~ **portadora de imagen** TV visual carrier wave; ~ **portadora reducida** TEC RAD reduced carrier wave; ~ **de presión** TEC RAD pressure wave; ~ **progresiva** ELECTRÓN progressive wave, TEC RAD, TRANS progressive wave, travelling (*AmE* traveling) wave; ~ **pulsátil** INFO, TELEFON, TRANS pulse wave; ~ **rectangular** TRANS rectangular wave, ELECTRÓN, SEÑAL square wave; ~ **rotacional** TEC RAD shear wave; ~ **sinusoidal** ELECTRÓN, INFO, TELEFON sine wave; ~ **sonora** GEN sound wave; ~ **de superficie magnetostática** (*MSSW*) ELECTRÓN magnetostatic surface wave (*MSSW*); ~ **transversal** TEC RAD, TRANS transverse wave; ~ **troposférica** TEC RAD tropospheric wave; ~ **ultracorta** TEC RAD ultrashort wave
ondámetro *m* PRUEBA, TEC RAD wavemeter
ondas: ~ **continuas moduladas** *f pl* (*MCW*) TEC RAD modulated continuous waves (*MCW*); ~ **sinusoidales** *f pl* COMPON singing
ondulación *f* ELECTRÓN ripple; ~ **de ganancia** ELECTRÓN gain ripple
ondulado *adj* COMPON corrugated, ELECTRÓN rippled
ondulante *adj* COMPON *superficie* undulating, ELECTRÓN rippling
ONS *abr* (*en escena*) TELEFON ONS (*on-scene*)
OOB *abr* (*fuera de banda*) ELECTRÓN, TEC RAD OOB (*out-of-band*)
OOK *abr* (*manipulación todo-nada*) SEÑAL, TELEFON OOK (*on-off keying*)
OOP *abr* (*programación orientada por objeto*) INFO OOP (*object-oriented programming*)
opacado *m* DATOS retouch
opacidad *f* TV opacity
opaco *adj* TV opaque
opción *f* GEN option; ~ **de comando** TELEFON command option; ~ **normalizada** TELEFON standardized option; ~ **de recorrido** TELEFON layout option; ~ **de usuario** TELEFON user option
opcional *adj* (*O*) DATOS optional (*O*)
opciones: ~ **de puenteado** *f pl* ELECTROTEC strapping options
operabilidad *f* TELEFON operational performance
operación *f* GEN operation; ~ **abstracta de extracción** DATOS fetch abstract-operation; ~ **abstracta de MS de registrador** DATOS register-MS abstract-operation; ~ **abstracta sumaria** DATOS summary abstract-operation; ~ **administración, mantenimiento y provisión** PRUEBA, TELEFON operation, administration, maintenance and provisioning (*OAMP*); ~ **de atenuación limitada** TRANS attenuation-limited operation; ~ **auxiliar** INFO auxiliary operation; ~ **de banda transversal** TRANS cross-band operation; ~ **básica** INFO basic operation; ~ **de búsqueda y rescate** TEC RAD search-and-rescue operation, SRR operation; ~ **de carga** INFO loading operation; ~ **centralizada** GEN centralized operation; ~ **de**

control INFO control operation; ~ **controlada por la voz** CONMUT, CONTROL voice-controlled operation; ~ **cuántica limitada** ELECTRÓN quantum-limited operation; ~ **de distorsión limitada** ELECTRÓN distortion-limited operation; ~ **emisorizada** SEÑAL, TELEFON senderized operation; ~ **exclusiva** ELECTRÓN NOT-IF-THEN operation; ~ **de extracto de presentación de mensajes** DATOS message-submission abstract-operation; ~ **en una frecuencia** SEÑAL, TEC RAD, TELEFON single-frequency operation; ~ **de gestión de sistemas** INFO systems management operation; ~ **de horizontal** CONMUT operation of horizontal, operation of selecting; ~ **de identidad** INFO identity operation; ~ **de interconexión de redes** TRANS internetworking; ~ **limitada por difracción** TRANS *óptica de fibras* diffraction-limited operation; ~ **limitada por el ruido cuántico** ELECTRÓN, TRANS quantum-noise-limited operation; ~ **local** CONTROL local operation; ~ **lógica** INFO logic operation; ~ **mayorante** ELECTRÓN majority operation; ~ **en modo de demora** GEN delay-mode operation; ~ **en modo de paquetes** TELEFON packet-mode operation; ~ **en modo de pérdida** TRANS loss-mode operation; ~ **NO** ELECTRÓN NOT operation; ~ **de no equivalencia** ELECTRÓN nonequivalence operation; ~ **de no identidad** ELECTRÓN nonidentity operation; ~ **NO-O** ELECTRÓN NOR operation; ~ **NO-Y** ELECTRÓN NAND operation; ~ **O-EXCLUSIVO** ELECTRÓN EXCLUSIVE-OR operation; ~ **OR** ELECTRÓN OR operation; ~ **en paralelo** CONMUT parallel operation; ~ **en un paso** INFO one-step operation; ~ **paso a paso** INFO step-by-step operation; ~ **por pasos** CONTROL step operation; ~ **de portadora única** INFO, TEC RAD, TRANS single-carrier operation; ~ **preparatoria** INFO housekeeping operation; ~ **remota** (*RO*) DATOS, PRUEBA remote operation (*RO*); ~ **de respondedor múltiple** TEC RAD multiple-transponder operation; ~ **simultánea** TELEFON simultaneous operation; ~ **SRR** TEC RAD SRR operation; ~ **subordinada** DATOS background operation; ~ **totalmente automática** TELEFON fully-automatic operation; ~ **umbral** ELECTRÓN, TEC RAD threshold operation; ~ **Y** ELECTRÓN AND operation

operaciones: ~ **encadenadas** *f pl* DATOS linked operations; ~ **y mantenimiento** *f pl* (*OM*) PRUEBA, TELEFON operations and maintenance (*OM*); ~ **preparatorias** *f pl* TELEFON preparatory operations

operador *m* CONMUT telephonist *BrE*, telephone operator *AmE*, INFO handler, TELEFON telephonist *BrE*, telephone operator *AmE*; ~ **auxiliar** TELEFON assistant operator; ~ **de cable** TRANS cable operator; ~ **de CATV** TV CATV operator; ~ **comercial** TV commercial operator; ~ **del cuadro conmutador** GEN switchboard operator; ~ **de Internet** DATOS Internet operator; ~ **jefe** TELEFON chief operator; ~ **por modulación** INFO modulating operator; ~ **principal** TELEFON controlling operator; ~ **público** DATOS, TELEFON public operator; ~ **de red de cable** GEN cable network operator; ~ **de registro** TELEFON recording operator; ~ **de servicio nocturno** CONMUT, TELEFON night-service operator; ~ **del sistema de información sobre abonados** DATOS directory system agent; ~ **de sistemas** GEN systems operator; ~ **de sonido** INFO sound operator; ~ **SUT** DATOS SUT operator; ~ **de telecomunicaciones** GEN telecommunications operator; ~ **de telefonía móvil** TEC RAD mobile telephone operating company; ~ **de televisión por cable** GEN cable television operator; ◆ ~ **no disponible** TEC RAD operator unavailable

operativo *adj* GEN operative; **no** ~ GEN disabled

optativo *adj* TEC RAD optional (*O*)

óptica *f* GEN optics; ~ **de fibras** TRANS fibre (*AmE* fiber) optics; ~ **geométrica** (*GO*) TRANS geometric optics (*GO*); ~ **integrada** TRANS integrated optics; ~ **de onda** ELECTRÓN, TRANS wave optics; ~ **de rayos** ELECTRÓN, TRANS ray optics

ópticamente: ~ **activo** *adv* COMPON, TRANS *óptica de fibras* optically active

óptico *adj* INFO, TV optical; ~**-eléctrico** (*O-E*) GEN optical-electrical (*O-E*)

optimización *f* GEN optimization

optimizar *vt* GEN optimize

optoacoplador *m* DATOS, ELECTRÓN, TRANS optocoupler

optoaislador *m* ELECTRÓN optoisolator

optoeléctrico *adj* TV optoelectric

optoelectrónica *f* ELECTRÓN optoelectronics, optronics

optoelectrónico *adj* ELECTRÓN optoelectronic

optomecánica *f* GEN optomechanics

optomecánico *adj* GEN optomechanical

optrónico *adj* ELECTRÓN optronic

OPX *abr* (*extensión fuera del edificio*) TELEFON OPX (*off-premises extension*)

O/R *abr* (*remitente/receptor*) DATOS, TELEFON O/R (*originator/recipient*)

órbita *f* TEC RAD orbit; ~ **de aparcamiento** TEC RAD parking orbit; ~ **de aureola** TEC RAD halo orbit; ~ **circular** TEC RAD circular orbit; ~

ecuatorial TEC RAD equatorial orbit; ~ **ecuatorial casi sincrónica** TEC RAD quasi-synchronous equatorial orbit; ~ **elíptica** TEC RAD elliptical orbit; ~ **de excentricidad** TEC RAD eccentric orbit; ~ **inclinada** TEC RAD inclined orbit; ~ **de noche** TEC RAD graveyard orbit; ~ **no perturbada** TEC RAD unperturbed orbit; ~ **de pista terrestre cerrada** TEC RAD closed earth-track orbit; ~ **temporal** TEC RAD parking orbit; ~ **terrestre cerrada** TEC RAD closed earth-orbit

orden[1]: **en ~ alfabético** *adv* GEN in alphabetical order; **en ~ decreciente** *adv* GEN in descending order; **por ~ numérico** *adv* GEN in numerical order

orden[2] *f* GEN order; ~ **de los bitios de transmisión** INFO bit-order of transmission; ~ **codificada** INFO coded order; ~ **de diversidad** TEC RAD order of diversity; ~ **ejecutar** INFO execute order; ~ **de equilibrio** DATOS matching order; ~ **de imprimir** INFO print command; ~ **de instalación no normalizada** DATOS, TELEFON nonstandard facility command (*NSC*); ~ **jerárquica** GEN ranking order; ~ **de mando de tránsito digital** GEN digital transit command; ~ **8 de paridad de bitio intercalada** DATOS bit interleaved parity order 8; ~ **de prioridad** GEN order of priority; ~ **programada** TELEFON programmed command; ~ **relativa** INFO relative order; ~ **de servidor** DATOS server command; ~ **de telemetría y determinación de distancias** (*TCR*) PRUEBA telemetry command and ranging (*TCR*); ~ **de visualización** INFO display order

ordenador *Esp m* (*cf computadora AmL*) INFO computer; ~ **agenda** INFO palmtop computer; ~ **central** DATOS, INFO, TELEFON host computer, mainframe computer, mainframe; ~ **central de videotexto** INFO, TELEFON, TV videotext host; ~ **de control** INFO controlling computer; ~ **con cuadro de pinzas** INFO clipboard computer; ~ **frontal** CONMUT, INFO front-end computer; ~ **híbrido** INFO hybrid computer; ~ **incorporado** INFO built-in computer; ~ **de mano** INFO hand-held computer; ~ **neural** INFO neural computer; ~ **personal** (*CI*) INFO personal computer (*PC*); ~ **con programación reducida** DATOS, INFO reduced-instruction-set computer (*RISC*); ~ **con programa de control de conmutador matricial** CONMUT matrix switch control software machine; ~ **con programa para la gestión de auxiliares** TELEFON auxiliaries management software machine; ~ **con programa de mensajes de operación y mantenimiento** CONTROL operation and maintenance messaging software machine; ~ **con programa de control del enlace por PCM** TELEFON PCM link control software machine; ~ **con programa para control del complejo de conmutación** CONMUT switching complex control software machine; ~ **solo** ELECTRÓN stand-alone

oreja: ~ **de anclaje** *f* ELECTRÓN outrigger

organigrama: ~ **de la estructura** *m* INFO structure flow chart; ~ **funcional** *m* INFO functional flow chart; ~ **de funcionamiento secuencial** *m* TELEFON sequence chart; ~ **de funciones** *m* GEN function flow chart; ~ **de programación** *m* INFO programming flow chart; ~ **del sistema** *m* INFO system flow chart

organismo: ~ **de certificación** *m* PRUEBA *calidad* certification body; ~ **encargado del registro** *m* DATOS registration authority

organización: ~ **de estructuración** *f* INFO structuration organization; ~ **de registradores** *f* CONMUT register organization; ~ **de radiodifusión** *f* TEC RAD broadcasting organization; ~ **internacional para la normalización** *f* GEN international organization for standardization; ~ **y mantenimiento** *f* TELEFON organization and maintenance

Organización: ~ **Árabe de Comunicaciones por Satélite** *f* TEC RAD Arab Satellite Communications Organization; ~ **Europea de Control de Calidad** *f* (*EOQC*) GEN European Organization for Quality Control (*EOQC*); ~ **Internacional de Normalización** *f* (*ISO*) GEN International Standards Organization (*ISO*); ~ **del Satélite de Telecomunicaciones Europeo** *f* TEC RAD European Telecommunications Satellite Organization

organizar *vt* GEN organize

órgano: ~ **de llamada** *m* TELEFON calling device; ~ **de mando** *m* CONTROL controlling element

orientabilidad: ~ **de la antena** *f* TEC RAD antenna steerability

orientación: ~ **de antena** *f* TEC RAD antenna steering; ~ **automática hacia un transmisor** *f* TEC RAD homing; ~ **del haz** *f* TEC RAD beam steering

orificio: ~ **de referencia** *m* TELEFON mouth reference point

origen *m* GEN source, INFO initial point; **de ~** *adj* TELEFON originating; ~ **del archivo** DATOS *Internet* file source; ~ **del documento** DATOS document source; ~ **del fallo** PRUEBA failure source, source of failure, source of fault; ~ **de**

la interferencia ELECTROTEC source of interference; ~ **de la llamada** TELEFON call origin; ~ **del mensaje** DATOS *Internet* origin of message, TRANS message source; ~ **de la señal de vídeo** TV video source

originar: ~ **en** *vt* ELECTRÓN, INFO originate in

orticonoscopio *m* ELECTRÓN orthicon; ~ **de imagen** ELECTRÓN image orthicon

oscilación *f* ELECTRÓN, ELECTROTEC oscillation, INFO oscillation, swing; ~ **amortiguada** ELECTRÓN damped oscillation; ~ **en diente de sierra** ELECTRÓN, TRANS sawtooth wave; ~ **forzada** ELECTROTEC forced oscillation; ~ **libre** ELECTRÓN free oscillation; ~ **parásita** ELECTRÓN parasitic oscillation

oscilador *m* GEN oscillator; ~ **en anillo** ELECTRÓN ring oscillator; ~ **de circuitos T en paralelo** ELECTRÓN parallel-tee oscillator; ~ **de cristal de regulación por tensión** (*VCXO*) ELECTRÓN voltage-controlled crystal oscillator (*VCXO*); ~ **de cuarzo** ELECTRÓN quartz oscillator; ~ **de diapasón** TEC RAD tuning fork oscillator drive; ~ **de doble T** COMPON, ELECTRÓN twin-T oscillator; ~ **estándar** COMPON, ELECTRÓN standard oscillator; ~ **excitado por choque** ELECTRÓN shock-excited oscillator, TEC RAD ringing oscillator, shock-excited oscillator; ~ **de frecuencia de batido** ELECTRÓN, TRANS beat-frequency oscillator (*BFO*); ~ **de funcionamiento libre** ELECTRÓN free-running oscillator; ~ **local** ELECTRÓN, TEC RAD local oscillator (*LO*); ~ **maestro estabilizado** (*SMO*) PRUEBA stabilized master oscillator (*SMO*); ~ **de mando** TEC RAD switch driver; ~ **de onda de retorno** TELEFON backward wave oscillator (*BWO*); ~ **de onda de seno o coseno** ELECTRÓN sine-cosine oscillator; ~ **paramétrico** ELECTRÓN parametric oscillator; ~ **de relajación** ELECTRÓN relaxation oscillator; ~ **de reloj** ELECTRÓN clock oscillator; ~ **transitrón** COMPON, ELECTRÓN transitron oscillator; ~ **de variación de fase** ELECTROTEC phase-shift oscillator; ~ **con voltaje controlado** ELECTRÓN voltage-controlled oscillator (*VCO*)

oscilar *vi* GEN oscillate

oscilatorio *adj* GEN oscillatory

oscilógrafo: ~ **catódico** *m* ELECTRÓN oscillograph tube

osciloscopio *m* ELECTRÓN oscilloscope

oscurecer *vt* TV darken

OSI *abr* (*interconexión de sistemas abiertas*) DATOS, INFO, TELEFON, TRANS OSI (*open systems interconnection*)

OTF *abr* (*frecuencia óptima de tráfico*) TEC RAD OTF (*optimum traffic frequency*)

otorgado *adj* TV awarded

oval *adj* GEN oval

ovalidad *f* GEN ovality

OWF *abr* (*frecuencia óptima de trabajo*) TEC RAD OWF (*optimum working frequency*)

oxidación *f* COMPON scaling; ~ **interna en fase de vapor** ELECTRÓN inside vapour (*AmE* vapor) phase oxidation (*IVPO*)

P

PA *abr* (*amplificador de potencia*) DATOS, INFO, TRANS PA (*power amplifier*)

PABX *abr* (*centralita privada automática conectada a la red pública*) APLIC PABX (*private automatic branch exchange*); ~ **de integración del servicio** *f* APLIC service integration PABX; ~ **integrada de voz-datos** *f* PERIF integrated voice-data PABX

PAD *abr* (*ensamblador-desensamblador de paquetes*) DATOS, TELEFON PAD (*packet assembler-disassembler*); ~ **de videotexto** *m* TV videotext PAD

padre *m* DATOS, INFO *nódulo* parent

página: ~ **anterior** *f* GEN preceding page; ~ **de bienvenida** *f* DATOS, INFO welcome page, welcoming page; ~ **de continuación** *f* GEN *documentación* continuation page; ~ **en pantalla** *f* DATOS, INFO, TV screen page; ~ **de teletexto** *f* TELEFON teletext page; ~ **tipo** *f* PERIF type page; ~ **de título** *f* INFO cover page; ~ **Web** *f* INFO Web page

paginación *f* GEN pagination

paginar *vt* GEN paginate

páginas: ~ **amarillas** *f pl* TELEFON yellow pages

pago: ~ **fácil** *m* GEN easy payment; ~ **por teléfono** *m* APLIC, TELEFON payment by telephone; ~ **por el uso compartido de circuitos** *m* TELEFON remuneration for shared use of circuits; ~ **por el uso exclusivo de circuitos** *m* TELEFON remuneration for exclusive use of circuits

país: ~ **de destino** *m* TELEFON country of destination; ~ **de origen** *m* TELEFON country of origin; ~ **terminal** *m* TELEFON terminal country; ~ **de tránsito** *m* TELEFON transit country

palabra *f* DATOS, INFO word; ~ **clave** DATOS, INFO keyword, codeword; ~ **codificada de control** DATOS, INFO control codeword; ~ **de código de control** DATOS, INFO control codeword; ~ **de control** DATOS, INFO control word; ~ **criptografiada** TELEFON encrypted speech; ~ **de instrucción** DATOS, INFO instruction word; ~ **de parámetro** DATOS, INFO parameter word; ~ **reservada** DATOS, INFO reserved word; ~ **de sincronización** DATOS syncword; ~ **telegráfica** TELEFON telegraph word; ~ **única de referencia** DATOS reference unique word

palabras: ~ **y frases de procedimiento** *f pl* TELEFON procedure words and phrases; ~ **por minuto** *f pl* TELEFON words per minute

palanca *f* GEN lever; ~ **de desembrague** TELEFON release lever; ~ **de freno** COMPON brake handle; ~ **de modulación** INFO modulation lever

paleta *f* INFO paddle; ~ **de madera** COMPON wooden paddle

palo *m* ELECTROTEC pole

PAMA *abr* (*acceso múltiple de dirección de impulso*) TEC RAD PAMA (*pulse-address multiple access*)

panel *m* GEN panel; ~ **de acoplamiento** PERIF jack panel, TEC RAD, TELEFON patch board, patch panel; ~ **de base** COMPON, ELECTRÓN base panel; ~ **de bastidor** TELEFON rack panel; ~ **de células solares** ELECTROTEC solar array; ~ **de conexiones** INFO wiring board; ~ **de conmutación** CONMUT, TELEFON switchboard; ~ **de conmutación dicordio** CONMUT double-cord switchboard; ~ **de control** CONTROL, INFO control panel; ~ **de descarga gaseosa** ELECTRÓN, INFO gas panel; ~ **deslizante** COMPON slide panel; ~ **de diodo electroluminiscente** ELECTRÓN light-emitting diode panel; ~ **empotrado** COMPON flush panel; ~ **de fondo** ELECTRÓN back panel; ~ **frontal** GEN front panel; ~ **impreso múltiple** ELECTRÓN multiple-printed panel; ~ **laminado** ELECTRÓN *circuitos impresos* laminate sheet; ~ **de lámparas** TELEFON *dispositivos indicadores* lamp display panel, lamp panel; ~ **lateral** TV side panel; ~ **de montaje empotrado** ELECTRÓN flush-mount panel; ~ **mural** ELECTRÓN bulkhead panel; ~ **de observación** CONTROL observation panel; ~ **de plasma** INFO plasma panel; ~ **posterior** INFO rear panel; ~ **principal** ELECTRÓN host panel; ~ **de proyección de LCD** ELECTRÓN LCD projection panel; ~ **de prueba de programas** CONMUT, INFO program test panel; ~ **solar** ELECTROTEC solar panel; ~ **superior** INFO top panel; ~ **de supervisión** CONTROL supervisory panel; ~ **de supervisión de ruta** TELEFON route-supervision panel; ~ **de teclas** PERIF, TELEFON push-button panel; ~ **de visualización** TELEFON *aparatos de indicación* display panel

pantalla *f* GEN screen; ~ **aislante** ELECTROTEC insulation screen, *cables* core screen; ~ **de conductor** TRANS *cables* conductor screen; ~ **de cristal líquido ferroeléctrico** (*FLCD*) INFO ferro-electric liquid crystal display (*FLCD*); ~ **dividida en cuatro** INFO quarter screen; ~ **eléctrica** ELECTROTEC electric screen; ~ **electroluminiscente de película delgada** (*ELCD*) COMPON, ELECTRÓN thin-film electroluminescent display (*TFEL*); ~ **grande** DATOS large screen display; ~ **iluminada** INFO lighted display; ~ **metalizada** ELECTRÓN metallized (*AmE* metalized) screen; ~ **de multifrecuencia** TV multifrequency screen; ~ **de plasma** ELECTRÓN plasma display, plasma panel, INFO plasma display; ~ **de rayos catódicos** ELECTRÓN cathode-ray screen; ~ **selectiva de frecuencias** TEC RAD frequency-selective sunshield; ~ **táctil** INFO touch screen; ~ **térmica** ELECTRÓN heat shield; ~ **de tubo de rayos catódicos** TV scope; ~ **visual** INFO, TV visual display unit; ~ **de visualización** INFO, PERIF, TV display screen; ◆ **a toda** ~ DATOS *Internet* full-screen

papel: ~ **continuo** *m* ELECTRÓN fan-fold sheets; ~ **cuadriculado** *m* GEN *dibujo técnico* diagram paper; ~ **milimetrado** *m* GEN millimetre (*AmE* millimeter) paper; ~ **de tamaño A4** *m* GEN *documentación* A4-sized paper

paquete *m* DATOS bundling, package, packet, INFO, TELEFON package, packet; ~ **aritmético de bitios concordantes** INFO bit-true arithmetic package; ~ **de circuito de línea** TELEFON line circuit pack; ~ **de contorno pequeño** DATOS small-outline package; ~ **de datos de origen autenticado** TEC RAD data origin authentication package; ~ **delgado de contorno reducido** DATOS thin small-outline package; ~ **de equipos compartidos** DATOS, INFO shareware package; ~ **de equipo de terminal** TELEFON terminal equipment package; ~ **estándar** DATOS standard package; ~ **de gráficos** TV graphics package; ~ **de programas de distribución gratuita** INFO freeware package; ~ **de software** DATOS, INFO software package; ~ **de software básico** INFO basic software package; ~ **de tratamiento de texto** DATOS, INFO word processing package; ~ **de unidades funcionales de gestión de sistemas** INFO systems management functional unit package

par *m* GEN pair; ~ **de cables** TRANS cable pair; ~ **de canales complementarios** TELEFON pair of complementary channels; ~ **de conductores** GEN pair of conductors; ~ **térmico** ELECTRÓN thermocouple; ~ **terminal** ELECTROTEC terminal pair; ~ **termoeléctrico** COMPON thermocouple; ~ **torcido descubierto** (*UTP*) ELECTRÓN, TRANS unshielded twisted pair (*UTP*)

PAR *abr* (*relación cresta-promedio*) GEN PAR (*peak-to-average ratio*)

parábola *f* TEC RAD, TV parabola

paraboloide *m* TEC RAD paraboloid

parada *f* ELECTROTEC shutdown, INFO standstill, stop, TV stop; ~ **del aparato** TELEFON device stop; ~ **de armadura de relé** ELECTROTEC relay-armature stop; ~ **de bucle** INFO loop stop; ~ **completa** INFO standstill; ~ **de cuadro** TV frame grabbing; ~ **de emergencia** GEN emergency stop; ~ **no programada** PRUEBA unscheduled stop; ~ **programada** PRUEBA scheduled stop

paradero *m* TEC RAD whereabouts

paradiafonía *f* TRANS near-end crosstalk

paralelismo *m* GEN parallelism; ~ **conjuntivo** DATOS AND parallelism

parametrización *f* GEN parameterization

parametrizar *vt* GEN parameterize

parámetro *m* GEN parameter; ~ **de categoría fuera de rango** DATOS parameter out-of-range class; ~ **de circuito** TELEFON network parameter; ~ **de codificación** INFO coding parameter; ~ **de contexto** DATOS context parameter; ~ **de control de ráfaga de referencia** DATOS controlling reference burst parameter; ~ **definido por nombre de** ~ INFO parameter-name-defined parameter; ~ **definido por la posición** TELEFON position-defined parameter; ~ **dinámico** INFO program-generated parameter; ~ **de dispersión del perfil** TRANS profile-dispersion parameter; ~ **de función** INFO function parameter; ~ **de funcionamiento** INFO, PRUEBA performance parameter; ~ **general** INFO general parameter; ~ **no básico** (*NBP*) DATOS nonbasic parameter (*NBP*); ~ **obligatorio** DATOS obligatory parameter; ~ **de perfil** TRANS profile parameter; ~ **real** DATOS actual parameter; ~ **de tolerancia** PRUEBA tolerance parameter; ~ **de velocidad** INFO rate parameter

pararrayos *m* COMPON *protección* surge diverter, ELECTRÓN surge arrester, ELECTROTEC, PRUEBA lightning arrester, *protección* lightning rod; ~ **atmosférico** ELECTROTEC, PRUEBA lightning conductor; ~ **de carbón** ELECTROTEC *protección* carbon lightning arrester; ~ **de gas rarificado** ELECTROTEC *protección* vacuum

lightning arrester; ~ **de tipo gaseoso** ELECTROTEC *protección* gas-type arrester
parcial *adj* DATOS, INFO, TELEFON partial
parcialmente: ~ **suprimido** *adj* TV partially suppressed
pareado: ~ **de bandas de frecuencia** *m* TEC RAD frequency band pairing
parentesco *m* INFO relationship
pares[1]: **de ~ múltiples** *adj* TV multipair
pares[2]: **por ~** *fra* GEN in pairs
paridad *f* DATOS, ELECTRÓN parity; ~ **intercalada de bitio** (*BIP*) DATOS bit interleaved parity (*BIP*); ~ **8 intercalada de bitio** DATOS bit interleaved parity 8; ~ **X intercalada de bitio** (*BIP-X*) DATOS bit interleaved parity X (*BIP-X*)
parlante *m* TELEFON *dispositivo* speaker
parpadeante *adj* INFO, PRUEBA, TELEFON blinking
parpadear *vi* INFO, PRUEBA, TELEFON *visualización gráfica* blink
parpadeo *m* TEC RAD glint, TV flicker, flickering
parrilla *f* GEN grid; ~ **de desionización** ELECTRÓN de-ionization grid
parte: ~ **de la aplicación de operaciones y mantenimiento** *f* PRUEBA, TELEFON operations and maintenance application part; ~ **constitutiva** *f* GEN constituent part; ~ **de datos del usuario** *f* DATOS data user part; ~ **del dominio inicial** *f* DATOS initial domain part (*IDP*); ~ **específica del dominio** *f* DATOS domain-specific part; ~ **exterior** *f* TELEFON outside party; ~ **final** *f* TELEFON *montaje de equipo* end part; ~ **oral de un mensaje** *f* DATOS, INFO voice body part; ~ **posterior** *f* GEN rear, rear side; ~ **de la transferencia de mensaje** *f* (*MTP*) DATOS, TELEFON message transfer part (*MTP*); ~ **del usuario de datos** *f* DATOS data user part; ~ **del usuario de la RDSI** *f* TELEFON ISDN user part; ~ **de vista panorámica** *f* DATOS overview part
partición *f* ELECTRÓN, INFO, TRANS partition; ~ **de datos** *m* INFO data partitioning; ~ **del modo** TRANS *óptica de fibras* mode partition
participante *m* GEN participant
participar *vt* GEN participate
partícula *f* GEN particle
partida: ~ **de entrega** *f* PRUEBA *calidad* delivery lot
pasador: ~ **de aletas** *m* COMPON split pin
PASC *abr* (*codificación de subbanda ajustable de precisión*) INFO PASC (*precision adaptive subband coding*)
pase: ~ **en cadena** *m* INFO chain play; ~ **de testigo** *m* TRANS token passing
pasillo *m* GEN *montaje de equipo* gangway; ~ **entre filas** TELEFON row aisle; ~ **lateral** TELEFON side aisle; ~ **principal** CONMUT main aisle
pasivación *f* COMPON passivation; ~ **de nitruro de silicio** COMPON silicium nitride passivation
paso *m* GEN stage, pitch; ~ **de abonado** CONMUT subscriber stage, subscriber switch; ~ **de ampliación** GEN extension stage; ~ **banda** INFO, TEC RAD, TELEFON, TRANS *filtros* pass band; ~ **de carácter** APLIC *reconocimiento* character pitch; ~ **de contador** GEN counter step, counting step; ~ **de frecuencia** TV frequency step; ~ **de hélice** TEC RAD helix pitch; ~ **helicoidad del cable** TRANS cable pitch; ~ **de potencia** TEC RAD power stage; ~ **de selección final** CONMUT final selection stage; ~ **de selector final** TELEFON final selector stage; ~ **T** CONMUT T stage; ~ **de la unidad de bloque** DATOS frame unit step
pasta: ~ **para rellenar** *f* ELECTRÓN filling compound; ~ **supresora de ruidos** *f* GEN deadener
pastilla: ~ **de silicio** *f* COMPON silicon wafer
patente: ~ **pendiente** *fra* GEN patent pending
patrón *m* GEN standard, template, pattern; ~ **de capacitancia** ELECTRÓN capacitance standard; ~ **de frecuencia** TEC RAD frequency standard; ~ **de índice refractivo de cuatro círculos concéntricos** TRANS four-concentric-circle refractive-index template; ~ **de inductancia** COMPON, ELECTRÓN standard inductor, *electromagnetismo* inductance standard; ~ **de referencia de antena** TEC RAD reference antenna pattern; ~ **de resistencia** ELECTROTEC resistance standard
PATS *abr* (*sucesión de pruebas abstractas parametrizadas*) DATOS, PRUEBA PATS (*parameterized abstract test suite*)
pausa: ~ **interdigital** *f* TELEFON interdigital pause
pauta: ~ **de certificación** *f* DATOS certification path
PBC *abr* (*centro de reserva de programas*) TV PBC (*programme booking centre*)
PBX *abr* (*central telefónica privada*) TELEFON PBX (*private branch exchange*); ~ **intercom** *m* TELEFON intercom-PBX; ~ **de RDSI de banda ancha** *f* CONMUT broadband ISDN PBX; ~ **terminal** *m* TELEFON terminating PBX
PC *abr* (*ordenador personal*) INFO PC (*personal computer*)
PCD *abr* (*disco compacto fotográfico*) DATOS, INFO PCD (*photo compact disc*)

PCI *abr* (*placa de circuito impreso*) COMPON, ELECTRÓN PCB (*printed circuit board*)

PCL *abr* (*lenguaje de órdenes de imprimir*) DATOS PCL (*print command language*)

PCM *abr* (*modulación por impulsos codificados*) DATOS, INFO, TELEFON, TRANS PCM (*pulse-code modulation*)

PCN *abr* (*red de comunicación personal*) DATOS, TEC RAD PCN (*personal communication network*)

PCO *abr* (*punto de control y observación*) DATOS PCO (*point of control and observation*)

PCS *abr* (*servicios de comunicaciones personales*) APLIC PCS (*personal communications services*)

PCTR *abr* (*informe de prueba de conformidad de protocolo*) DATOS, PRUEBA PCTR (*protocol conformance test report*)

PDF *abr* (*función de densidad de probabilidad*) DATOS PDF (*probability density function*)

PDL *abr* (*lenguaje de descripción de página*) DATOS, INFO PDL (*page description language*)

PDM *abr* (*modulación de impulsos en duración*) DATOS, TEC RAD, TELEFON PDM (*pulse duration modulation*), TRANS pulse duration modulation (*PDM*), pulse width modulation (*PWM*)

PE *abr* (*elemento pictórico*) INFO, TELEFON PE (*pictorial element*)

pedal *m* INFO pedal; ~ **de expresión** INFO expression pedal; ~ **de portamento** INFO portamento footswitch; ~ **de sostén** INFO sustain footswitch, sustain pedal

pedido: ~ **por correo** *m* GEN mail order; ~ **de modificación técnica** *m* (*ECO*) COMPON engineering change order (*ECO*); ~ **de servicio** *m* GEN request for service

pedir *vt* GEN demand, order; ◆ ~ **por interclasificación** INFO order by merging

pegar *vt* INFO paste

peinado *adj* ELECTRÓN combed

peine: ~ **alzador** *m* CONMUT lifting comb; ~ **de apoyo** *m* COMPON, TELEFON supporting comb; ~ **recogedor** *m* TRANS *cables* collecting comb

pelacables *m* COMPON wire strippers, *cuchilla* stripper

pelador: ~ **de fibras** *m* ELECTRÓN fibre (*AmE* fiber) stripper

película *f* GEN film; ~ **aislante** COMPON lapping film; ~ **hidrófuga** CLIMAT vapour (*AmE* vapor) barrier; ~ **metálica delgada** ELECTRÓN metallic thin film

peligroso *adj* GEN hazardous

penalización *f* TRANS penalty

pendiente *f* ELECTRÓN, INFO, TEC RAD slope; ~ **de ataque inicial** INFO initial attack slope; ~ **de atenuación** TEC RAD attenuation slope; ~ **de filtro** INFO filter slope; ~ **de ganancia** ELECTRÓN gain slope; ~ **de la meseta** TEC RAD plateau slope; ~ **de retardo de grupo** TRANS group-delay slope

penetrabilidad *f* GEN penetrability

penetrable *adj* GEN penetrable

penetración: ~ **de mercado** *f* GEN market penetration

péntodo *m* ELECTRÓN pentode

PEP *abr* (*pico de potencia de la envolvente*) POWER *de modulación* PEP (*peak envelope power*)

pequeño: ~ **centro de telecomunicaciones** *m* ELECTRÓN small telecommunications centre (*AmE* center)

percentil *m* GEN percentile

percibir *vt* GEN perceive

pérdida *f* GEN loss; ~ **por absorción** TEC RAD absorption loss; ~ **de acoplador** TRANS coupler loss; ~ **de acoplamiento** TRANS coupling loss; ~ **de acoplamiento antena-medio** TEC RAD antenna-to-medium coupling loss; ~ **de acoplamiento de polarización** TEC RAD polarization coupling loss; ~ **de acoplamiento terminal** TELEFON terminal coupling loss; ~ **del aislamiento** TRANS insulation loss; ~ **del alimentador** TRANS feeder loss; ~ **de alineación de multitrama** TRANS loss of multiframe alignment; ~ **angular** TEC RAD corner loss; ~ **del apuntador** TRANS loss of pointer (*LOP*); ~ **por apuntamiento** TEC RAD pointing loss; ~ **por apuntamiento de la antena** TEC RAD antenna-pointing loss; ~ **de atenuación** TEC RAD attenuation loss; ~ **básica del espacio libre** TEC RAD free-space basic loss; ~ **del cable** TRANS cable loss; ~ **del conector** TRANS *óptica de fibras* connector loss; ~ **de conversión transversal** (*TCL*) TELEFON transverse conversion loss (*TCL*); ~ **por corrientes parásitas** ELECTROTEC eddy current loss; ~ **de datos** DATOS data loss; ~ **por desalineación** TRANS misalignment loss; ~ **por desalineación angular** TRANS angular misalignment loss; ~ **por desplazamiento lateral** TRANS lateral offset loss; ~ **por desplazamiento transversal** TRANS transverse offset loss; ~ **por desviación longitudinal** TRANS longitudinal offset loss; ~ **por dispersión** TEC RAD, TRANS scattering loss, spreading loss; ~ **de estabilidad** TELEFON stability loss; ~ **estándar** TRANS *de cable coaxial* standard loss; ~ **de explotación** GEN operating loss; ~ **o ganancia compuestas** TELEFON composite loss or gain; ~ **por histéresis** ELECTRÓN hysteresis loss; ~ **por inserción** TRANS insertion loss; ~ **de inserción en el conector** TRANS *óptica de* *fibras* connector insertion loss; ~ **de inteligibilidad equivalente de referencia** TELEFON articulation reference equivalent loss; ~ **de intervalo** TRANS gap loss; ~ **intrínseca de empalme** TRANS intrinsic joint loss; ~ **intrínseca de una unión** TRANS *óptica de fibras* intrinsic junction loss; ~ **de macroflexión** TRANS macrobend loss; ~ **de microflexión** TRANS microbend loss; ~ **mínima admisible de transmisión básica** (*MPBTL*) DATOS minimum permissible basic transmission loss (*MPBTL*); ~ **de multitrama** TRANS loss of multiframe (*LOM*); ~ **nominal de transmisión entre** TRANS nominal transmission loss between; ~ **de penetración en edificios** TEC RAD building penetration loss; ~ **de potencia en estado de conducción** COMPON, ELECTRÓN on-state power loss; ~ **de reflexión** ELECTROTEC, TRANS reflection loss; ~ **de refracción** ELECTRÓN bending loss; ~ **de retorno de frecuencia única** SEÑAL, TEC RAD, TELEFON single-frequency return loss; ~ **de semibucle** TELEFON semiloop loss; ~ **de señal** TRANS loss of signal (*LOS*); ~ **de tensión** ELECTROTEC voltage loss; ~ **total** TEC RAD *de enlace* total loss; ~ **de trama** TELEFON, TRANS loss of frame (*LOF*); ~ **de transmisión** TELEFON, TRANS transmission loss; ~ **de transmisión básica** TEC RAD basic transmission loss; ~ **de transmisión básica en el espacio libre** TEC RAD free-space basic transmission loss; ~ **de transmisión en el espacio libre** TEC RAD free-space transmission loss; ~ **de transmisión normalizada** (*NTL*) TEC RAD normalized transmission loss (*NTL*); ~ **de transmisión de la trayectoria de rayos** TEC RAD ray-path transmission loss; ~ **de unión extrínseca** TRANS *óptica de fibras* extrinsic joint loss; ◆ **sin** ~ INFO, TELEFON lossless

perecedero: **no** ~ *adj* GEN durable

perfeccionamiento *m* GEN *dibujo tecnológico* further development

perfecto *adj* TELEFON *modulación, restitución, señal* perfect

perfil *m* GEN profile; ~ **del campo modal** TRANS *óptica de fibras* mode-field profile; ~ **de caracteres** APLIC character outline; ~ **ESI** ELECTRÓN, TRANS ESI profile; ~ **de índice escalonado** ELECTRÓN step-index profile; ~ **de índice escalonado equivalente** ELECTRÓN, TRANS *óptica de fibras* equivalent step-index profile; ~ **de índice gradual** ELECTRÓN, TRANS graded-index profile; ~ **de índice de ley potencial** ELECTRÓN, ELECTROTEC, TRANS *óptica de fibras* power law index profile; ~ **del índice de refracción** TRANS refractive-index profile; ~ **parabólico** TEC RAD, TRANS parabolic profile; ~ **del parámetro** PRUEBA parameter profile

perforación *f* GEN perforation; ~ **marginal** COMPON sprocket hole

perforado: **no** ~ *adj* GEN unperforated

perforador: ~ **inicial** *m* COMPON, ELECTRÓN spudger; ~ **de teclado** *m* PERIF keyboard perforator

perforadora: ~ **de tecla** *f* INFO keypunch

perforar *vt* COMPON drill

periapsis *f* TEC RAD periapsis

periférico *m* GEN peripheral; ~ **inteligente** PERIF intelligent peripheral

perigeo *m* TEC RAD perigee

perilla: ~ **de llamada** *f* COMPON push button; ~ **del rodillo** *f* TELEFON platen knob

periodicidad: ~ **de la distribución de propaganda** *f* TV leafleting periodicity; ~ **de la señal de llamada** *f* TELEFON ringing periodicity

periódico *adj* GEN periodic

periodo *m* GEN period; ~ **de alineamiento** TELEFON line-up period; ~ **de almacenamiento de baterías** ELECTROTEC battery storage time; ~ **anomalístico** TEC RAD anomalistic period; ~ **de bloqueo** ELECTRÓN off period; ~ **de carga** PERIF *contestador*, TELEFON charging period; ~ **de carga completo** PERIF *contestador telefónico automático* completed charging period; ~ **de conducción** ELECTRÓN *semiconductores* conduction period; ~ **de contingencia** PRUEBA contingency period; ~ **de ejecución** INFO run time; ~ **erróneo** DATOS erroneous period (*EP*); ~ **espacial** DATOS space period; ~ **experimental de extensión** PRUEBA extension test period; ~ **de fallos por desgaste** PRUEBA wear-out failure period; ~ **de fallos prematuros** PRUEBA early-failure period; ~ **de línea de exploración** TV scanning line period; ~ **de la llamada** TELEFON ringing period; ~ **de marcación** TELEFON dialling (*AmE* dialing) period; ~ **mínimo libre de errores** DATOS minimum error-free period; ~ **de orbitación** TEC RAD orbiting period; ~ **orbital** TEC RAD orbital period; ~ **preparatorio** TELEFON preparatory period; ~ **de registro** GEN recording interval, recording period; ~ **de repetición de los impulsos** (*PRP*) TELEFON pulse repetition period (*PRP*); ~ **de retención** TEC RAD retention period; ~ **de retención del módem** DATOS modem holdover period; ~ **de revolución** TEC RAD period of revolution; ~

sidéreo de revolución TEC RAD sidereal period of revolution; ~ **sidéreo de rotación** TEC RAD sidereal period of rotation; ~ **de tarificación** TELEFON metering period
peritelefonía *f* TELEFON peritelephony
permanencia: ~ **del color** *f* GEN colourfastness (*AmE* colorfastness)
permanente *adj* GEN permanent
permeabilidad *f* COMPON permeability; ~ **inicial** ELECTRÓN *electromagnetismo* initial permeability
permeable *adj* COMPON permeable
permeación *f* ELECTRÓN permeance
permeámetro *m* PRUEBA permeameter
permiso: ~ **de desviación** *m* PRUEBA *calidad* deviation permit
permitido: **no** ~ *adj* TELEFON not permitted
permitividad *f* ELECTRÓN permittivity; ~ **piezoeléctrica masiva** ELECTRÓN bulk piezoelectric permittivity
permuta *f* INFO flipping, swop, TV bartering
permutar *vt* INFO flip, swop, TV barter
perno: ~ **de gancho** *m* COMPON hook bolt; ~ **de montaje** *m* COMPON mounting screw
perpendicular *adj* GEN perpendicular
persistencia *f* TELEFON tailing
perturbación *f* GEN disturbance, *en el tubo de gas* breakdown; ~ **de baja frecuencia** INFO glitch; ~ **de un canal** ELECTRÓN spot jamming; ~ **con cintas reflectoras** TEC RAD window jamming; ~ **entre símbolos** (*ISI*) INFO, TRANS, TV intersymbol interference (*ISI*); ~ **de fondo** PERIF *contestador automático* background mush; ~ **de una frecuencia** TEC RAD spot jamming; ~ **ionosférica** TEC RAD ionospheric disturbance; ~ **del servicio** PRUEBA service incident; ~ **temporal** PRUEBA, TELEFON temporary perturbation; ~ **con tiras antirradar** TEC RAD chaff jamming
perturbador: ~ **reemisor** *m* ELECTRÓN repeater-jammer
perturbar *vt* GEN *geotecnia* disturb
perturbarse *vi* DATOS, ELECTRÓN, TEC RAD jam
perveancia *f* ELECTRÓN perveance
pestaña: ~ **ancha** *f* COMPON wide flange
pestillo *m* ELECTRÓN latch
petición: ~ **A-asociada** *f* (*AARQ*) DATOS, TELEFON A-associate request (*AARQ*); ~ **de actualización rápida** *f* DATOS fast update request; ~ **de actualización rápida de instrucción de vídeo** *f* (*VCU*) TV video command fast-update request (*VCU*); ~ **de carga** *f* DATOS charge request; ~ **de comentarios** *f* DATOS *Internet* request for comments; ~ **de conexión N** *f* DATOS N-connect request; ~ **de confirmación** *f* DATOS acknowledgement request; ~ **de desconexión** *f* TELEFON clearing request; ~ **de envío** *f* DATOS forwarding request; ~ **de imagen inmóvil de la videoseñal de mando** *f* TV video command freeze-picture request (*VCF*); ~ **de primitiva** *f* DATOS request primitive; ~ **de puesta a cero de DTE** *f* TELEFON DTE clear request; ~ **de tiempo de transmisión** *f* TRANS request transmission time; ~ **de vaciado** *f* INFO dump request
PETS *abr* (*sucesión de pruebas ejecutables parametrizadas*) DATOS, PRUEBA PETS (*parameterized executable test suite*)
PFN *abr* (*red de formación de impulsos*) ELECTRÓN PFN (*pulse-forming network*)
PFT *abr* (*transferencia de fallo en PBX*) TELEFON PFT (*PBX failure transfer*)
PGI *abr* (*identificador de grupo de parámetros*) INFO PGI (*parameter-group identifier*)
PH *abr* (*manipulador de paquetes de datos*) DATOS PH (*packet handler*)
PI *abr* (*protocolo de Internet*) DATOS, INFO IP (*Internet protocol*)
picadura *f* ELECTRÓN pinhole
picar *vt* INFO pick
pico *m* ELECTROTEC peak; ~ **de longitud de onda** ELECTROTEC wavelength peak; ~ **máximo de potencia** ELECTROTEC peak power; ~ **de potencia de la envolvente** (*PEP*) ELECTROTEC *de modulación* peak envelope power (*PEP*); ~ **de ruido** TRANS noise spike
picosegundo *m* (*ps*) PRUEBA picosecond (*ps*)
PICS *abr* (*sentencia de conformidad de instrumentación de protocolos*) DATOS PICS (*protocol implementation conformance statement*)
pictograma *m* ELECTRÓN pictogram
pie *m* INFO bottom; ~ **de rey** PRUEBA slide calliper (*AmE* caliper), sliding calliper (*AmE* caliper)
piedra: ~ **de toque** *f* DATOS cornerstone
pieza *f* GEN part; ~ **de inserción** COMPON inset; ~ **de montaje** COMPON fixing bracket; ~ **polar** ELECTROTEC pole piece, pole shoe; ~ **de recambio** COMPON spare part
piezas: ~ **de montaje** *f pl* COMPON mounting hardware
piezoelectricidad *f* ELECTRÓN piezoelectricity
piezoeléctrico *adj* ELECTRÓN piezoelectric
pigmento *m* COMPON pigment
pila *n* INFO stack; ~ **de discos** *f* DATOS disk array; ~ **fotovoltaica** *f* ELECTRÓN photovoltaic cell; ~ **galvánica** *f* ELECTROTEC galvanic cell; ~ **patrón** *f* ELECTRÓN standard cell; ~ **seca** *f* ELECTROTEC dry cell, INFO dry battery
pilar *m* GEN pillar

piloto *m* GEN pilot
PIM *abr* (*intermodulación pasiva*) TRANS PIM (*passive intermodulation*)
pincel *m* COMPON *herramientas* paintbrush
pintar: **~ con pistola** *vt* COMPON spray-coat
pintura *f* INFO paint; **~ con efecto de relieve** COMPON texture lacquer
pinza: **~ de batería** *f* ELECTROTEC battery clip; **~ de cocodrilo** *f* ELECTRÓN crocodile clip; **~ rápida estañada** *f* COMPON solder-plated quick clip
pinzas: **~ para cables** *f pl* COMPON wire tweezers; **~ para hilos** *f pl* COMPON picking pliers
PIP *abr* (*imagen dentro de una imagen*) INFO PIP (*picture in picture*)
pirata[1] *adj* INFO, TV pirate
pirata[2]: **~ informático** *m* INFO hacker
pirateado *adj* INFO, TV pirated
pirateo *m* INFO, TV pirating; **~ informático** INFO hacking
piratería *f* INFO, TV piracy; **~ informática** DATOS, INFO software piracy
piroelectricidad *f* ELECTRÓN, ELECTROTEC pyroelectricity
piso: **~ para cables** *m* GEN cable floor; **~ semiconductor** *m* PRUEBA semiconductive floor
pisón *m* COMPON stamp
pista *f* DATOS, INFO, TEC RAD, TV track; **~ de arrastre** COMPON sprocket track; **~ de auditoría** GEN audit trail; **~ del programa** TEC RAD programme (*AmE* program) track; **~ de sonido** INFO, PERIF, TV soundtrack; **~ de vídeo** INFO, TV video track; **~ de voz** INFO vocal track
pistas: **~ por pulgada** *f pl* (*PPP*) INFO tracks per inch (*TPI*)
pistola: **~ calefactora** *f* ELECTROTEC heater gun
pistón *m* GEN *bombas hidráulicas, prensas* plunger, TEC RAD piston; **~ de cortocircuito** ELECTRÓN, TEC RAD shorting plunger
pitido *m* INFO beep sound, bleep
PIU *abr* (*unidad enchufable*) DATOS, ELECTRÓN, INFO, TELEFON PIU (*plug-in unit*)
PIV *abr* (*cresta de tensión inversa*) ELECTROTEC PIV (*peak inverse voltage*)
pivote *m* COMPON pivot; **~ guía** COMPON, ELECTRÓN tooling pin; **~ horizontal** CONMUT horizontal pivot; **~ de válvula** ELECTRÓN valve spindle
pixel *m* DATOS, INFO, TRANS, TV pixel
pixelación *f* DATOS *de digitalizador*, INFO pixellation (*AmE* pixelation)
pixelización *f* DATOS *de digitalizador*, INFO pixellization (*AmE* pixelization)
PIXIT *abr* (*información extra de instrumentación de protocolos para comprobación*) DATOS, PRUEBA PIXIT (*protocol implementation extra information for testing*)
placa[1]: **sobre la ~** *adv* INFO on-board
placa[2] *f* ELECTRÓN *de hierro colado* plate; **~ de alma metálica** ELECTRÓN metal-core board; **~ de asiento** COMPON wallplate, ELECTRÓN *de montaje* base plate; **~ auxiliar** ELECTROTEC *circuitos impresos* auxiliary board; **~ a base de plantilla de cable** TRANS formboard plate; **~ de capacitor** ELECTROTEC capacitor plate; **~ ciega de ocultación** ELECTRÓN knockout blanking plate; **~ de circuito impreso** (*PCI*) COMPON, ELECTRÓN printed circuit board (*PCB*); **~ de clavijas** INFO pinboard; **~ de compresión de datos** DATOS data-compression board; **~ de conexión** ELECTRÓN product board; **~ de conexión de cable** (*CCB*) TRANS *normas de construcción* cable connection board (*CCB*); **~ de conexión a tierra** ELECTROTEC grounding plate *AmE*, earthing plate *BrE*; **~ de contactos** COMPON contact plate; **~ de distribución de energía** ELECTROTEC *suministro* power distribution board; **~ enchufable** ELECTRÓN plug-in board; **~ de extensión** ELECTRÓN *circuitos impresos* extension board; **~ de fijación** COMPON fastening plate, fixing plate, mounting plate; **~ frontal rectangular** ELECTRÓN rectangular faceplate; **~ giratoria** COMPON, INFO swivel; **~ de Hall** ELECTRÓN *semiconductores* Hall plate; **~ impresa** ELECTRÓN printed board; **~ indicadora** ELECTROTEC rating plate; **~ de intercambio** ELECTRÓN replacement board; **~ interfaz de línea** (*LIB*) CONMUT line interface board (*LIB*); **~ lateral** TELEFON side plate; **~ madre** INFO mother board; **~ de memoria** DATOS memory board; **~ de memoria de códigos** TEC RAD *comunicación móvil* code memory board; **~ de montaje con juego de resortes** COMPON spring-set mounting plate; **~ posterior** GEN rear plate; **~ secundaria** ELECTRÓN *circuitos impresos*, INFO daughterboard; **~ separadora** COMPON spacer plate; **~ de soporte** COMPON supporting plate; **~ de toma de tierra** ELECTROTEC earthing plate *BrE*, grounding plate *AmE*; **~ del vértice** TEC RAD *de reflector de antena* vertex plate
plan *m* GEN plan; **~ de Babcock** TEC RAD, TELEFON Babcock plan; **~ de Babcock doble** TELEFON double Babcock plan; **~ de canalización con dos frecuencias** TRANS two-

frequency channelling (*AmE* channeling) plan; ~ **comercial** GEN business plan; ~ **de emergencia** GEN contingency plan; ~ **estratégico** GEN strategic plan; ~ **de frecuencia** TRANS, TV frequency plan; ~ **de muestreo** PRUEBA sampling plan; ~ **de numeración** GEN numbering plan; ~ **reticular** TEC RAD lattice plan; ~ **de seguridad** GEN *de transmisiones* security policy; ~ **de servicio** TV service plan; ~ **simulador de tiempo de ráfaga** (*SBTP*) DATOS simulator burst time plan (*SBTP*); ~ **temporal condensado de retransmisión por ráfagas** (*RCTP*) DATOS retransmitted condensed burst time plan (*RCTP*); ~ **de temporización de ráfagas con TDMA** SEÑAL, TEC RAD, TELEFON TDMA burst time plan; ~ **de transmisión** TRANS transmission plan

plancha: ~ **de conexión a tierra** *f* TEC RAD, TRANS earthing mat *BrE*, grounding mat *AmE*; ~ **de fijación** *f* ELECTRÓN wallplate

planeamiento: ~ **por contingencia** *m* GEN contingency planning

planear *vt* GEN plan

planicidad: ~ **del contacto** *f* TEC RAD flatness of bearing

planificación *f* GEN planning; ~ **a largo plazo** GEN long-term planning; ~ **de la memoria anexa** INFO bump mapping; ~ **de la red** TELEFON network planning; ~ **de sistemas** INFO systems planning

plano[1] *adj* GEN flat

plano[2]: ~ **de bitio** *m* TELEFON bit plane; ~ **de cableado** *m* GEN *documentación* cabling plan; ~ **de colocación** *m* GEN *documentación* floor plan; ~ **de colocación de cuadros conmutadores** *m* TELEFON switchboard floor plan; ~ **de disposición** *m* GEN scheme plan; ~ **de enrollamiento** *m* COMPON wire-wrap plane; ~ **de entrefase** *m* GEN mounting drawing; ~ **espectral** *m* TV spectrum plan; ~ **de memoria** *m* INFO memory plane; ~ **orbital** *m* TEC RAD orbital plane; ~ **de polarización** *m* TEC RAD plane of polarization; ~ **de la red** *m* TELEFON network chart, network plan; ~ **de referencia** *m* ELECTRÓN reference plane

planta *f* GEN plant; ~ **de energía** ELECTROTEC power plant; ~ **exterior** TRANS outside plant; ~ **de suministro de energía** ELECTROTEC power supply station

plantilla *f* COMPON, INFO, TRANS template; ~ **de bifurcación** TRANS *cables* branch formboard; ~ **de cable** TRANS cable formboard, formboard; ~ **de corte de cable** TRANS *cables* cutting formboard; ~ **de cuatro círculos concéntricos de campo cercano** TRANS four-concentric-circle near-field template; ~ **para formar cables** TRANS *cables* cable-forming board; ~ **pulidora** ELECTRÓN polishing jig; ~ **removible de teclado** APLIC, PERIF key overlay; ~ **de trazado** COMPON scribing template

plasma *m* GEN plasma; ~ **lasérico** ELECTRÓN laser plasma

plástico[1] *adj* COMPON plastic

plástico[2]: ~ **de acetato de vinilo** *m* COMPON vinyl acetate plastic; ~ **armado** *m* COMPON reinforced plastic; ~ **blando** *m* COMPON nonrigid plastic; ~ **de cloruro de vinilo** *m* COMPON vinyl chloride plastic; ~ **éster** *m* COMPON ester plastic; ~ **fenólico** *m* COMPON phenolic plastic; ~ **laminado** *m* COMPON laminated plastic; ~ **poliamídico** *m* COMPON polyamide plastic; ~ **de policarbonato** *m* COMPON polycarbonate plastic; ~ **poliestérico** *m* COMPON polyester plastic; ~ **polietilénico** *m* COMPON polyethylene plastic; ~ **politénico** *m* COMPON polythene plastic; ~ **rígido** *m* COMPON rigid plastic; ~ **semirrígido** *m* COMPON semirigid plastic; ~ **termoendurecible** *m* COMPON thermosetting plastic; ~ **uretánico** *m* COMPON urethane plastic

plastificante *m* COMPON plasticizer

plataforma: ~ **de gestión de la red** *f* TRANS network-management platform

plateado *m* COMPON silver-plating; ~ **por electrólisis** COMPON silver-electroplating

plato: ~ **de alimentación frontal** *m* TEC RAD front-fed dish

plausibilidad *f* DATOS plausibility

plazo: ~ **de desarrollo** *m* INFO development time

plegadora *f* INFO folder

plegar *vt* INFO fold

pleno *adj* GEN full

plesiócrono *adj* TELEFON, TRANS plesiochronous

plomería *f* TEC RAD plumbing

plotter *m* DATOS, INFO, PERIF, PRUEBA, TELEFON plotter

PLP *abr* (*protocolo de la capa de paquetes*) DATOS, TELEFON PLP (*packet layer protocol*)

PLR *abr* (*repetidor de enlace de impulsos*) TELEFON PLR (*pulse link repeater*)

PLU *abr* (*unidad de sincronización de fase*) ELECTRÓN PLU (*phase lock unit*)

pluralismo *m* ELECTRÓN pluralism

PLV *abr* (*vídeo en fase de producción*) INFO PLV (*production level video*)

PMA *abr* (*zona de memoria de programas*) INFO PMA (*program memory area*)

POAS *abr* (*sistema de análisis de opción de pago*) APLIC POAS (*payment option analysis system*)
población *f* PRUEBA population; ~ **estándar** PRUEBA standard population
podar *vt* INFO prune
poder: ~ **cortante** *m* INFO *metales* bite; ~ **resolutivo** *m* ELECTRÓN, TEC RAD resolving power; ~ **de saturación** *m* ELECTRÓN saturation power
polar *adj* GEN polar
polaridad *f* GEN polarity; ~ **de modulación** TV modulation polarity; ~ **negativa** ELECTROTEC minus polarity, negative polarity
polarización *f* GEN bias, polarization; ~ **de antena** TEC RAD antenna polarization; ~ **circular** TRANS circular polarization; ~ **cruzada** TEC RAD cross-polarization; ~ **a derecha** TEC RAD right-hand polarization; ~ **eléctrica** ELECTROTEC electric polarization; ~ **horizontal** TV horizontal polarization; ~ **de la onda** ELECTRÓN, TEC RAD wave polarization; ~ **ortogonal** TEC RAD orthogonal polarization; ~ **en sentido antihorario** TEC RAD counterclockwise polarization *AmE*, anticlockwise polarization *BrE*; ~ **vertical** TV vertical polarization
polarizador *m* GEN polarizer; ~ **de onda guiada** ELECTRÓN guided-wave polarizer
polea: ~ **motriz** *f* GEN driving pulley
poliamida *f* COMPON polyamide
polibutileno: ~ **tereftalato** *m* ELECTRÓN polybutylene terephthalate
poliéster *m* COMPON polyester
poliestireno: ~ **expandido** *m* COMPON expanded polystyrene
polietileno *m* COMPON polyethylene
polifonía *f* INFO polyphony
polifónico *adj* INFO polyphonic
polígono: ~ **de frecuencias** *m* PRUEBA *estadística* frequency polygon; ~ **de frecuencias acumuladas** *m* PRUEBA *calidad* cumulative frequency polygon
polimerización *f* COMPON polymerization
polimerizar *vt* COMPON polymerize
polimezcla *f* INFO poly mix
politeno *m* COMPON polythene
política: ~ **de aprovisionamiento** *f* GEN procurement policy; ~ **arancelaria** *f* APLIC, TELEFON tariff policy; ~ **comercial** *f* GEN commercial policy; ~ **de precios** *f* APLIC pricing policy; ~ **de seguridad basada en normas** *f* DATOS rule-based security policy
poliuretano *m* COMPON polyurethane
polo: ~ **inductor** *m* ELECTROTEC field pole; ~ **negativo** *m* ELECTROTEC negative pole, negative terminal; ~ **positivo** *m* ELECTROTEC positive pole, positive terminal
ponderación *f* GEN weight, weighting; ~ **de ruido** TRANS noise weighting
ponderado *adj* GEN weighted; **no** ~ GEN unweighted
poner: ~ **a cero** *vt* ELECTRÓN, PRUEBA zero, zeroize; ~ **una conferencia a cobro revertido** *vt* TELEFON transfer the charges; ~ **en contacto** *vt* TELEFON put in touch; ~ **a descargar** *vt* ELECTROTEC *baterías* put on discharge; ~ **en espera** *vt* TELEFON put on hold; ~ **en funcionamiento** *vt* GEN put into operation; ~ **en órbita** *vt* TEC RAD inject into orbit, put into orbit; ~ **en servicio** *vt* GEN bring into service, put into service
porcentaje *m* GEN percentage; ~ **de abonados** TELEFON subscriber percentage; ~ **de error** DATOS, SEÑAL, TELEFON error rate; ~ **de impulsos** TELEFON pulse rate; ~ **de marcación** TELEFON marking percentage; ~ **de modulación** TRANS modulation percentage; ~ **de realimentación** ELECTRÓN feedback ratio; ~ **de recombinación** COMPON recombination rate; ~ **de reposición** COMPON resetting percentage
porosidad *f* COMPON porosity
poroso *adj* COMPON porous
portabilidad *f* INFO, TELEFON portability; ~ **de hipertexto** INFO hypertext portability
portada *f* INFO front page, title page
portador: ~ **minoritario** *m* ELECTRÓN minority carrier; ~ **de transferencia fiable** *m* ELECTRÓN reliable-transfer server; ~ **de transmisión** *m* TRANS, TV transmission bearer
portadora *f* GEN carrier; ~ **de carga** ELECTRÓN *semiconductores* charge carrier; ~ **de carga móvil** ELECTRÓN mobile charge carrier; ~ **de color** TRANS colour (*AmE* color) carrier; ~ **común de telecomunicaciones** TELEFON telecommunications common carrier; ~ **de datos** TRANS data carrier; ~ **de datos en conmutación por paquetes** TRANS packet-switched data carrier; ~ **digital T1** TRANS T1 digital carrier; ~ **incrementalmente coherente** (*ICC*) TV incrementally coherent carrier (*ICC*); ~ **incrementalmente relacionada** TV incrementally related carrier; ~ **modulada en frecuencia** TRANS frequency-modulated carrier; ~ **monocanal** TEC RAD single-channel carrier; ~ **de múltiples destinos** TRANS multidestination carrier; ~ **permanente** TRANS continuous carrier; ~ **de pulsos** TRANS pulse carrier; ~ **recuperada** TRANS recovered carrier; ~ **en relación armónica** TV harmonically

related carrier; ~ **de sonido** TV sound carrier; ~ **del sonido de la televisión** TV television sound carrier; ~ **T1** (*T1C*) TRANS T1 carrier (*T1C*); ~ **de tono** TEC RAD tone-coded carrier; ~ **de vídeo** TV video carrier

portaetiquetas *m* GEN *marcaje* label holder

portalámparas *m* ELECTROTEC lamp holder; ~ **de llave** PERIF key-and-lamp unit

portamacho *m* COMPON, TEC RAD *fundición* bearing

portaondas: ~ **de ranura** *m* (*SWG*) DATOS, TEC RAD, TRANS slot waveguide (*SWG*)

portátil *adj* GEN portable

portatubo *m* COMPON tube socket

portaválvulas *m* COMPON tube socket, ELECTRÓN valve socket

portero: ~ **automático** *m* APLIC, PERIF, PRUEBA entryphone; ~ **eléctrico** *m* PERIF door interphone; ~ **de vídeo** *m* PERIF door video

posibilidad: ~ **de compartir** *f* DATOS sharing option; ~ **de desconexión** *f* GEN switch-off capability; ~ **de memorización** *f* PRUEBA storability

posición *f* GEN position; ~ **A** TELEFON position A; ~ **auxiliar de marcha** APLIC auxiliary service position; ~ **de carácter** TELEFON character position; ~ **central** GEN centre (*AmE* center) position; ~ **en la cola** TELEFON queue place; ~ **de contestación** TELEFON answering position; ~ **de control** CONTROL control position; ~ **de conversación** TELEFON speaking position; ~ **de desconexión** COMPON off position; ~ **de dígito** INFO digit place, SEÑAL, TRANS digit position; ~ **de equilibrio** ELECTRÓN rest position; ~ **final** GEN limiting position; ~ **de flanco posterior** COMPON, TELEFON trailing-edge position; ~ **de funcionamiento** GEN operating position; ~ **fundamental** INFO initial position; ~ **de horizontal** COMPON, CONMUT selecting bar position; ~ **de los impulsos** TELEFON pulse position; ~ **inicial** GEN starting position; ~ **libre** TELEFON *equipo* vacant position; ~ **de llamada** TELEFON ringing position; ~ **media** GEN middle position; ~ **neutra** GEN mid-position, neutral position; ~ **orbital** TEC RAD orbital position; ~ **orbital nominal** TEC RAD nominal orbital position; ~ **de las placas** COMPON *técnicas de montaje* board position; ~ **de pruebas** PRUEBA, TELEFON testing position; ~ **de relés** TELEFON relay space; ~ **de retención bloqueada** ELECTRÓN latched detent position; ~ **de la ruta de señalización** SEÑAL signalling-route (*AmE* signaling-route) set; ~ **seleccionada** TELEFON selected position; ~ **de servicio nocturno** CONMUT night-service board; ~ **de símbolo** DATOS symbol location; ~ **telegráfica internacional** TELEFON international telegraph position; ~ **télex** TELEFON telex position; ~ **de télex internacional** TELEFON international telex position; ~ **de trabajo** GEN on position; ~ **vertical** GEN vertical position; ~ **del vínculo de señalización** SEÑAL signalling-link (*AmE* signaling-link) set

posicionado *m* TEC RAD *de la antena* positioning

posicionamiento *m* DATOS placing, positioning, TEC RAD *de la antena* positioning; ~ **inicial** INFO initial positioning

positivo: ~ **de batería** *m* ELECTROTEC battery positive; ~ **de distribución** *m* ELECTROTEC *suministro de energía* distribution positive; ~ **de reinstrucción** *m* (*RTP*) DATOS, TELEFON retrain positive (*RTP*)

positrón *m* GEN positron

posprocesador *m* INFO postprocessor

posproducción *f* TV postproduction

postámbulo *m* DATOS postamble

postamplificador *m* ELECTRÓN, ELECTROTEC post-amplifier

poste: ~ **de telégrafos** *m* TELEFON telegraph pole

potencia *f* GEN force, power, strength; ~ **admisible fuera de banda** TEC RAD permissible out-of-band power; ~ **admisible máxima** TRANS maximum admissible power; ~ **de campo en ausencia de absorción** TEC RAD unabsorbed field strength; ~ **de campo utilizable** TEC RAD usable field strength; ~ **efectiva radiada** (*erp*) TEC RAD effective radiated power (*erp*); ~ **efectiva radiada isotrópicamente** TEC RAD effective isotropically radiated power; ~ **efectiva radiada en monopolio** (*emrp*) TEC RAD effective monopole-radiated power (*emrp*); ~ **de entrada de rejilla** ELECTRÓN grid input power; ~ **de entrada de vídeo** DATOS, TV video input; ~ **de la envolvente pico equivalente de una señal telefónica** TELEFON equivalent peak envelope power of a telephone signal; ~ **equivalente radiada isotrópicamente** (*EIRP*) TEC RAD equivalent isotropically radiated power (*EIRP*); ~ **equivalente de ruido** ELECTRÓN, TRANS noise equivalent power (*NEP*); ~ **de excitación** ELECTRÓN firing power, TEC RAD driving power; ~ **de excitación de rejilla** ELECTRÓN grid driving power; ~ **de fuga total** ELECTRÓN, ELECTROTEC total leakage power; ~ **del láser** ELECTRÓN laser power; ~ **media** ELECTROTEC medium power, TEC RAD mean power, TRANS average power; ~ **óptima de salida** ELECTRÓN optimum output power; ~

de la portadora TEC RAD carrier power; ~ radiada ELECTRÓN radiated power; ~ radiante ELECTRÓN, TRANS radiant power; ~ reactiva ELECTROTEC reactive power; ~ de reserva ELECTROTEC reserve power; ~ de salida ELECTRÓN, ELECTROTEC, INFO, TELEFON output, output power; ~ de salida del emisor TRANS transmitter output power; ~ de salida de impulso de pico ELECTRÓN peak-pulse output power; ~ de salida del impulso TEC RAD pulse output power; ~ de señal SEÑAL signal power; ~ solar ELECTROTEC solar power; ~ sonora GEN sound power; ~ telefónica de referencia TELEFON reference telephonic power; ~ total anódica de entrada ELECTRÓN, ELECTROTEC total anode-power input, total plate-power input; ~ de transmisión del satélite TEC RAD satellite transmit power; ~ de la trayectoria de bajada TEC RAD down-path power

potencial *m* GEN potential; ~ **del electrodo** ELECTRÓN electrode potential

potenciómetro *m* ELECTRÓN potentiometer; ~ **de ajuste** COMPON, ELECTRÓN trimming potentiometer

pozo: ~ **de inspección** *m* GEN manhole

PPM *abr* (*modulación de impulsos en posición*) DATOS, TELEFON, TRANS PPM (*pulse position modulation*)

ppp *abr* (*puntos por pulgada*) DATOS dpi (*dots per inch*)

PPP *abr* DATOS (*protocolo de punto a punto*) *Internet* PPP (*point-to-point protocol*), INFO (*pistas por pulgada*) TPI (*tracks per inch*)

PPSDN *abr* (*red pública de datos de paquetes conmutados*) DATOS PPSDN (*public packet-switched data network*)

practicable *adj* INFO feasible

preacentuación *f* ELECTRÓN, TEC RAD pre-emphasis

preajustador: ~ **de datos** *m* INFO data presetter

preajustar *vt* GEN preset

preajuste *m* GEN presetting

prealimentar *vt* ELECTROTEC pre-energize

preámbulo *m* DATOS preamble

preamplificación *f* ELECTRÓN preamplification

preamplificador *m* ELECTRÓN preamplifier

preasignado *adj* GEN preassigned

precableado[1] *adj* ELECTRÓN prewired, TRANS precabled

precableado[2] *m* ELECTRÓN prewiring, TRANS precabling

precablear *vt* ELECTRON prewire, TRANS precable

precalentamiento *m* ELECTRÓN preheating

precargado *adj* ELECTRÓN preloaded

preceptivo *adj* GEN mandatory

precio: ~ **de abono** *m* INFO subscription price; ~ **base** *m* GEN *comercio* basic price; ~ **base de acceso** *m* TELEFON basic rate access (*BRA*); ~ **base de servicio** *m* APLIC basic rate service; **de ~ fijo** *adj* GEN fixed-price; ~ **del servicio** *m* INFO service rate; ~ **sobre demanda** *m* APLIC price on request; ~ **de suscripción** *m* DATOS, TEC RAD, TELEFON, TV subscription price; ~ **de venta** *m* GEN cost price

precisión *f* INFO accuracy, PRUEBA, TEC RAD accuracy, precision; ~ **de ajuste** GEN setting accuracy; ~ **de muestra** TELEFON sample precision; ~ **del patrón de frecuencia** TEC RAD frequency-standard accuracy; ~ **de la posición** GEN position correctness; ~ **de regulación** GEN regulation accuracy; ◆ **de ~ simple** INFO single-precision

precorrección *f* TELEFON precorrection

predecir *vt* GEN predict

predefinir *vt* GEN predefine

predeterminado *adj* GEN predetermined

predeterminar *vt* GEN predetermine

predicción *f* GEN prediction; ~ **de cobertura celular** TEC RAD cell coverage prediction; ~ **ionosférica** TEC RAD ionospheric prediction; ~ **típica** TELEFON typical prediction

predictor *m* INFO predictor

predivisión *f* PRUEBA prescaler

preensamblaje *m* GEN preassembly

preensamblar *vt* GEN preassemble

preenvejecimiento *m* PRUEBA preageing

preestablecido *adj* GEN prearranged

preestreno: ~ **de una película** *m* TV movie preview

preexistente *adj* DATOS *Internet* pre-existing

preferencia: ~ **ejecutiva** *f* TELEFON executive override

prefijo *m* GEN prefix; ~ **de acceso** TELEFON access code; ~ **de código de servicio** CONMUT, TELEFON service code prefix; ~ **internacional** TELEFON international prefix; ~ **de mensaje** TEC RAD prosign

preforma: ~ **de vidrio** *f* TRANS *óptica de fibras* glass preform

preformar *vt* GEN preform

preformateado *adj* DATOS preformatted

pregrabado *adj* TV prerecorded

pregrupo *m* TRANS pregroup

preguntas: ~ **formuladas con frecuencia** *f pl* (*FAQ*) DATOS *Internet* frequently-asked questions (*FAQ*)

preigualación *f* ELECTRÓN pre-equalization

preimpreso *adj* INFO, PERIF preprinted

premarcación *f* TELEFON *de cintas magnéticas* premarking
premasterización *f* INFO premastering
prensa: ~ **de tornillo** *f* COMPON screw clamp
prensado: ~ **posproducción** *m* INFO postproduction pressing
preparación *f* INFO, TEC RAD, TELEFON setup; ~ **de datos** INFO data preparation; ~ **de muestra** PRUEBA sample preparation
preparador: ~ **de MML** *m* TELEFON initiator of MML; ~ **de salida MML** *m* TELEFON initiator of MML output
preparar *vt* GEN prepare
preparativos *m pl* GEN preliminaries
preprocesador *m* INFO preprocessor
preselección *f* GEN presetting
preseleccionar *vt* GEN preset
preselector *m* TELEFON, TV preselector
presencia: ~ **obligatoria** *f* DATOS obligatory presence
presentación *f* TELEFON presentation; ~ **distancia-amplitud** TEC RAD range-amplitude display; ~ **distancia-demora** TEC RAD range-bearing display; ~ **de la identificación de la línea que llama** (*CLIP*) TELEFON calling-line identification presentation (*CLIP*); ~ **en modo normal** DATOS normal-mode presentation; ~ **en modo X.410-1984** DATOS X.410-1984 mode presentation; ~ **montada** GEN *dibujo técnico* assembled representation; ~ **de página** DATOS, INFO page layout; ~ **de sector** TEC RAD sector display; ~ **de un servicio** APLIC introduction of service; ~ **sintética** PERIF synthetic display; ~ **tipo A** TEC RAD type A display; ~ **visual del número llamado** TELEFON called-number display
presentar *vt* INFO exhibit
preservación: ~ **de datos** *f* INFO data saving
presión *f* GEN pressure; ~ **de aceite** GEN oil pressure; ~ **acústica** TELEFON sound pressure; ~ **del agua** GEN water pressure; ~ **atmosférica** GEN atmospheric pressure; ~ **de evaporación** GEN evaporation pressure; ~ **negativa** GEN negative pressure; ~ **positiva** GEN positive pressure
presostato *m* COMPON pressure switch
prestación: ~ **de servicio** *f* APLIC service provision
préstamo: ~ **circular** *m* INFO end-around borrow
presupuestar *vi* GEN budget
presupuesto: ~ **calórico** *m* ELECTRÓN heat budget
presurización: ~ **de cables** *f* TRANS cable pressurization
pretratamiento *m* INFO preprocessing
prevención: ~ **de colisión de llamadas** *f* TELEFON prevention of call collision; ~ **de desastres** *f* PRUEBA *seguridad* disaster prevention
prevenir *vt* GEN prevent
prever *vt* GEN forecast, foresee
previo *adj* GEN previous
previsible *adj* GEN *interrupción* foreseeable
PRI-EOM *abr* (*final de mensaje de interrupción de procedimiento*) DATOS, TELEFON PRI-EOM (*procedure interrupt EOM*)
PRI-EOP *abr* (*fin de programa de interrupción de procedimiento*) TELEFON PRI-EOP (*procedure interrupt EOP*)
primario *adj* INFO primary
primer: ~ **canal adyacente** *m* TEC RAD first-adjacent channel; ~ **plano** *m* TV foreground; ~ **plano de pantalla** *m* INFO display foreground; ~ **término** *m* TV foreground
primitivo *m* DATOS, INFO *estadística* primitive; ~ **de confirmación del servicio N** DATOS N-service confirm-primitive; ~ **de indicación relacionada** DATOS related indication primitive; ~ **de indicación del servicio N** DATOS N-service indication-primitive; ~ **de petición del servicio N** DATOS N-service request-primitive; ~ **de respuesta** *f* DATOS response primitive; ~ **de servicio** *f* TELEFON service primitive
PRI-MPS *abr* (*MPS de interrupción de procedimiento*) DATOS, TELEFON PRI-MPS (*procedure interrupt MPS*)
principio: ~ **del bloque funcional** *m* CONMUT building-block principle; ~ **de facturación** *m* APLIC charging principle; ~ **de funcionamiento** *m* GEN operating principle; ~ **de retorno** *m* ELECTROTEC *suministro* flyback principle
prioridad *f* GEN priority; ~ **MCS de transferencia de datos** TELEFON MCS data-transfer priority; ~ **de pérdida celular** TEC RAD cell loss priority (*CLP*); ~ **con tono avisador** TELEFON priority with notifying tone
prioritario *adj* GEN overriding
priorizar *vt* TELEFON *llamadas* pre-empt
prisionero *m* COMPON stud
prisma *m* ELECTRÓN prism; ~ **de desviación** TRANS deviation prism
privacidad *f* DATOS, PERIF, TELEFON privacy; ~ **de la voz** PERIF, TELEFON voice privacy
probabilidad *f* GEN probability; ~ **de aceptación** PRUEBA probability of acceptance; ~ **de avería** PRUEBA probability of failure; ~ **de completar un servicio con**

éxito TELEFON probability of successful service completion; ~ **de congestión** TELEFON probability of congestion; ~ **de detección** TEC RAD probability of detection; ~ **de error** DATOS, SEÑAL error probability; ~ **de escasez** PRUEBA probability of shortage; ~ **de falsa alarma** GEN false alarm probability; ~ **de indicación** TEC RAD probability of indication; ~ **de rechazo** PRUEBA probability of rejection; ~ **de retraso excedente** DATOS probability of excess delay

probado: **no** ~ *adj* INFO, PRUEBA untested

probador: ~ **de cableado** *m* PRUEBA wiring test set, wiring tester; ~ **de cables** *m* PRUEBA, TRANS cable tester; ~ **de circuitos** *m* PRUEBA circuit test set; ~ **de conexiones** *m* PRUEBA wiring test set, wiring tester; ~ **de eslabones** *m* CONMUT, PRUEBA link test equipment; ~ **de facturación** *m* APLIC, PRUEBA charging tester; ~ **inferior** *m* (*TL*) DATOS, PRUEBA lower tester (*LT*); ~ **múltiple de abonados** *m* TELEFON subscriber multiple test unit; ~ **de ocupaciones** *m* PRUEBA, TELEFON traffic tester; ~ **superior** *m* (*UT*) DATOS, PRUEBA upper tester (*UT*)

probar *vt* GEN test; ◆ ~ **por comparación** INFO benchtest

procedimiento *m* GEN procedure; ~ **de acceso a la línea** TELEFON line access procedure (*LAP*); ~ **de actualización** GEN updating procedure; ~ **de adquisición de modo mandado** DATOS gated mode acquisition procedure; ~ **de adquisición de modo búsqueda** DATOS search mode acquisition procedure; ~ **de alineamiento** PRUEBA, TRANS line-up procedure; ~ **de autentificación** SEÑAL authentication procedure; ~ **B de acceso por enlace** TELEFON link access procedure B; ~ **binario de búsqueda** DATOS binary search procedure; ~ **de control** TELEFON control procedure; ~ **de coordinación de pruebas** PRUEBA test coordination procedure; ~ **de declaración** TV declaration procedure; ~ **de desconexión** TELEFON disconnection procedure; ~ **de despejo** SEÑAL clearing procedure; ~ **de desvío** INFO bypass procedure; ~ **detallado** GEN detailed procedure; ~ **de edición** GEN editing procedure; ~ **de emergencia** GEN emergency procedure; ~ **de enlace único** (*SLP*) TELEFON single-link procedure (*SLP*); ~ **de entrada remoto** ELECTRÓN remote log-in; ~ **esquemático de prueba** PRUEBA outline test procedure; ~ **de muestreo** PRUEBA sampling procedure; ~ **multienlace** (*MLP*) DATOS multilink procedure (*MLP*); ~ **de nueva asignación** DATOS new assignment procedure; ~ **orientado a caracteres** INFO character-oriented procedure; ~ **de programación** DATOS, INFO programming procedure; ~ **de prueba de posición de ruta de señalización** SEÑAL signalling-route-set (*AmE* signaling-route-set) test procedure; ~ **de prueba de salto de ciclo** PRUEBA cycle-skipping test procedure; ~ **de reasignación** TEC RAD reassignment procedure; ~ **de regeneración** DATOS, INFO refreshment procedure; ~ **de selección de menú** TELEFON menu selection procedure; ~ **de sincronización del marco de recepción** (*RFS*) TRANS receive-frame synchronization procedure (*RFS*); ~ **SMA** DATOS SMA procedure; ~ **de tasación de la unidad de tráfico** TELEFON traffic-unit price procedure

procesado[1] *adj* DATOS, INFO processed

procesado[2]: ~ **de texto** *m* INFO text processing; ~ **de transacciones** *m* DATOS transaction processing (*TP*)

procesador *m* DATOS, INFO processor, processor unit; ~ **administrativo** INFO administrative processor; ~ **de aplicaciones** INFO applications processor; ~ **asociativo** INFO associative processor; ~ **de bloques** DATOS frame processor; ~ **central** CONMUT central processor; ~ **de centro de operaciones y mantenimiento** INFO operation and maintenance centre (*AmE* center) processor; ~ **de coma flotante** INFO floating-point processor; ~ **para compresión de datos** DATOS data-compression processor; ~ **de comunicaciones** CONMUT communications processor; ~ **de conmutación** CONMUT, DATOS switching processor; ~ **de conmutación de mensajes** CONMUT message-switching processor; ~ **de conmutación de paquetes** INFO packet-switching processor; ~ **de datos** DATOS, INFO data processor; ~ **especializado** DATOS back-end processor; ~ **de gráficos** INFO graphics processor; ~ **híbrido** CONMUT hybrid processor; ~ **de interconexión** CONMUT, DATOS, INFO, TELEFON, TV server; ~ **de lenguaje** INFO language processor; ~ **de mantenimiento** INFO maintenance processor; ~ **de matrices** DATOS, INFO array processor; ~ **de mensajes de interfaz** INFO interface message processor (*IMP*); ~ **periférico** INFO peripheral processor (*PP*); ~ **regional** CONMUT regional processor (*RP*); ~ **de reserva** CONMUT stand-by processor; ~ **de retardo** INFO delay processor; ~ **de señales** SEÑAL signal processor; ~ **del sistema de conmutación** CONMUT, TELEFON switching system processor; ~ **de textos** DATOS, INFO word processor

procesadores: ~ **que trabajan en paralelo** *m pl* CONMUT parallel-working processors

procesamiento *m* CONMUT, DATOS, INFO processing; ~ **adaptado a la firma** TEC RAD signature-adapted processing; ~ **por lotes** DATOS, INFO batch processing; ~ **monomediático** TV monomedia processing; ~ **múltiple** DATOS, INFO multiprocessing; ~ **de la palabra** APLIC, TELEFON speech processing; ~ **en paralelo** DATOS, ELECTRÓN parallel processing; ~ **posproducción** INFO postproduction processing; ~ **prioritario** INFO foreground processing; ~ **remoto** CONTROL, DATOS remote processing; ~ **de señales** SEÑAL signal processing; ~ **de señales digitales** SEÑAL digital signal processing; ~ **subordinado** INFO background processing; ~ **Y-O** DATOS AND-OR processing

procesar *vt* DATOS, INFO process

proceso *m* DATOS, INFO process; ~ **aditivo** ELECTRÓN additive process; ~ **de aplicación** DATOS, TELEFON application process; ~ **de aplicación de gestión de sistemas** INFO systems management application process; ~ **aplicado de aleatorización de bitios de datos** DATOS applied data-bit scrambling process; ~ **automático de datos** DATOS, INFO datamation; ~ **de conexión** TELEFON switching sequence; ~ **de conmutación** CONMUT, TELEFON switching process, switching sequence; ~ **en control** PRUEBA process in control; ~ **de datos descentralizado** INFO distributing data processing; ~ **de descodificación** INFO decoding process; ~ **de documentos** DATOS document processing; ~ **de extracción** TRANS drawing process; ~ **e intercambio de documentos** DATOS document processing and interchange; ~ **de muestreo** TRANS sampling process; ~ **de restablecimiento** DATOS recovery procedure; ~ **reversible** GEN reversible process; ~ **de salida** INFO output process; ~ **semiaditivo** ELECTRÓN semiadditive process; ~ **suspendido** TELEFON suspended process; ~ **sustractivo** ELECTRÓN subtractive process

producción *f* GEN production; ~ **de fallo** TELEFON failure occurrence; ~ **local** TV local production; ~ **media** ELECTRÓN, INFO average output; ~ **de pares** COMPON pair production; ~ **transversal** TEC RAD track production; ~ **de voz** INFO, TEC RAD, TELEFON speech production

producir *vt* GEN produce, TELEFON *llamada* originate; ◆ ~ **una interrupción** ELECTRÓN raise an interrupt

productividad *f* GEN productivity

producto *m* GEN product; ~ **acabado** GEN finished product; ~ **de conversión de frecuencia espuria** TEC RAD spurious frequency-conversion product; ~ **final** GEN end product; ~ **de intermodulación** TRANS intermodulation product; ~ **de intermodulación parásita** TEC RAD, TRANS spurious intermodulation product; ~ **de una mezcla espuria** TRANS spurious mixing product; ~ **de modulación** TELEFON modulation product; ~ **de programación** INFO software product

productos: ~ **y servicios** *m pl* APLIC products and services

proforma: ~ **MOCS** *f* DATOS MOCS proforma; ~ **PCIS** *f* DATOS PCIS proforma; ~ **PICS** *f* DATOS PICS proforma; ~ **PISC** *f* DATOS PISC proforma; ~ **PIXIT** *f* DATOS PIXIT proforma

profundidad *f* INFO depth; ~ **de campo** ELECTRÓN field depth; ~ **de color** DATOS depth of colour (*AmE* color); ~ **de desintonización** INFO detune depth; ~ **de modulación** TRANS modulation depth; ~ **de modulación de altura** INFO pitch-modulation depth; ~ **de modulación de amplitud** TRANS amplitude modulation depth; ~ **de penetración** TEC RAD penetration depth

programa *m* GEN programme (*AmE* program), INFO program; ~ **de adjudicación de trayectoria** TELEFON path allocation program (*AmE* program); ~ **de análisis selectivo** DATOS snapshot program; ~ **antivirus** DATOS, INFO anti-virus program; ~ **de autor** DATOS authoring software; ~ **de biblioteca** INFO library program; ~ **de cable** TELEFON cable programme (*AmE* program); ~ **cifrado** INFO coded program; ~ **de compilación** INFO compiling program; ~ **de compresión de datos** DATOS, INFO data compression software; ~ **de comprobación** INFO checking program; ~ **de computadora** *AmL* (*cf programa de ordenador Esp*) DATOS, INFO computer program; ~ **comunitario** TV community programme (*AmE* program); ~ **para conexiones enrolladas** COMPON wrapping programme (*AmE* program); ~ **de control** CONTROL, DATOS monitoring program, INFO control program; ~ **de corrección ortográfica** INFO spelling checker; ~ **de demostración** INFO demo program; ~ **de diagnóstico** INFO diagnostic program; ~ **educacional** DATOS, INFO educational software; ~ **de ejecución** TELEFON executive program (*AmE* program); ~ **de**

encaminamiento INFO, TELEFON routing program; ~ **ensamblador** INFO assembly program; ~ **de estación de control de funciones del núcleo** DATOS, INFO, TELEFON core functions control station software; ~ **de exploración** INFO scanning program; ~ **fuente** INFO source program; ~ **genérico** TELEFON generic program; ~ **de gestión de la memoria** INFO memory management program; ~ **independiente** TV independent programme (*AmE* program); ~ **de instalación** DATOS *Internet* installation program; ~ **de instrucción** DATOS, INFO tutorial; ~ **de intercalación** INFO collator; ~ **de llamada** INFO fetch program; ~ **macrogenerador** INFO macrogenerating program; ~ **de manipulación de tráfico** DATOS, TELEFON traffic-handling program; ~ **de mantenimiento** PRUEBA maintenance programme (*AmE* program); ~ **monitor** DATOS monitor program; ~ **de muestreo** PRUEBA sampling scheme; ~ **objeto** INFO object program, target programme (*AmE* program); ~ **de ordenador** *Esp* (*cf programa de computadora AmL*) DATOS, INFO computer program; ~ **de presentación en página** DATOS, INFO page layout software; ~ **de procesamiento múltiple** INFO multiprocessing program; ~ **propio** TV own programme (*AmE* program); ~ **para prueba de línea** TELEFON line test programme (*AmE* program); ~ **de prueba orbital** TEC RAD orbital test programme (*AmE* program); ~ **de prueba preliminar** PRUEBA preliminary test programme (*AmE* program); ~ **de prueba previa** PRUEBA preliminary test programme (*AmE* program); ~ **de pruebas** PRUEBA *batería de pruebas* testing programme (*AmE* program); ~ **de rastreo selectivo** INFO snapshot trace program; ~ **reintroducible** INFO re-entrant program; ~ **reubicable** INFO relocatable program; ~ **reutilizable** INFO reusable program; ~ **de salida** INFO output program; ~ **secuenciador** DATOS sequencer program; ~ **de servicio** INFO service program; ~ **de sonido internacional de destino múltiple** TV international multiple-destination sound-programme (*AmE* sound-program); ~ **supervisor** INFO supervisory program; ~ **de televisión** TV television programme (*AmE* program); ~ **de televisión por satélite** TV satellite telecast; ~ **de trabajo con costes anuales** GEN annual-costed work programme (*AmE* program); ~ **de unidad de intercambio** TELEFON interchange unit program; ~ **de utilidades** INFO utility program; ~ **para utilización de la antememoria del disco** DATOS, INFO disk caching software; ~ **por vía satélite** TV satellite programme (*AmE* program)

programable *adj* DATOS, INFO, PERIF programmable; **no** ~ DATOS, INFO, PERIF unprogrammable

programación *f* DATOS, INFO programming, TV programme (*AmE* program) schedule; ~ **heurística** DATOS heuristic programming; ~ **lógica** INFO logic programming; ~ **con números enteros** INFO integer programming; ~ **orientada por objeto** (*OOP*) INFO object-oriented programming (*OOP*); ~ **paralela** DATOS parallel programming; ~ **paramétrica** INFO parametric programming

programador *m* INFO programmer; ~ **de ritmo** INFO rhythm programmer; ~ **de sistemas** INFO systems programmer

programar *vt* INFO program

programas: ~ **de comunicaciones** *m pl* DATOS, INFO communication software; ~ **de distribución gratuita** *m pl* INFO freeware; ~ **de dominio público** *m pl* INFO public-domain software; ~ **de enseñanza** *m pl* INFO teachware; ~ **en estación** *m pl* DATOS, INFO on-station software; ~ **de gestión** *m pl* DATOS, INFO management software; ~ **integrados** *m pl* DATOS, INFO integrated software; ~ **de libre disposición** *m pl* DATOS freeware; ~ **de navegación** *m pl* DATOS, INFO *Internet* navigation software

progresión *f* GEN stepping; ~ **de la nota** INFO note progression

prólogo: ~ **de procedimiento** *m* TELEFON procedure prologue

prolongador: ~ **eléctrico** *m* INFO extension cable

promedio: ~ **límite de calidad de subidas** *m* (*AOQL*) PRUEBA average outgoing quality limit (*AOQL*); ~ **de proceso** *m* PRUEBA process average; ~ **de retardo por tiempo de espera** *m* INFO mean waiting time average delay; ~ **de uso diario** *m* PRUEBA average daily use

promotor *m* COMPON *plástico* promotor

propagación *f* ELECTRÓN propagation, spread; ~ **de la dispersión de precipitación** TEC RAD precipitation-scatter propagation; ~ **por dispersión ionosférica** TEC RAD ionospheric scatter propagation; ~ **en el espacio libre** TEC RAD free-space propagation; ~ **ionosférica** TEC RAD ionospheric propagation; ~ **de la luz** TRANS *óptica de fibras* light propagation; ~ **en modo de silbidos atmosféricos** TEC RAD whistler-mode propagation; ~ **normal** TEC RAD standard propagation; ~ **de ondas** ELECTRÓN, TEC RAD wave propagation; ~ **por ráfagas meteóricas** TEC RAD meteor-burst propagation; ~ **rectilínea directa** TEC RAD line-of-sight propagation; ~ **submarina** TEC RAD underwater propagation; ~ **subterránea** TEC RAD subterranean propagation; ~ **transhorizonte** TEC RAD transhorizon propagation; ~ **transionosférica** TEC RAD transionospheric propagation; ~ **de trayectoria múltiple** TEC RAD multiple-path propagation; ~ **por trayectoria múltiple** TEC RAD multipath propagation; ~ **troposférica** TEC RAD tropospheric propagation

propaganda: ~ **de buzón** *f* DATOS junk mail

propagar *vt* ELECTRÓN propagate

propicio: **poco** ~ *adj* INFO unfriendly

propiedad: ~ **intelectual** *f* GEN intellectual property

propiedades: ~ **de propagación** *f pl* TEC RAD propagation properties

propio: **en el** ~ **lugar** *adv* TELEFON on-premises

proporción *f* GEN proportion; ~ **de errores de una manipulación** TELEFON error-rate of keying; ~ **de errores residuales** (*RER*) DATOS, INFO, TELEFON residual error rate (*RER*); ~ **fraccionaria** INFO fractional ratio

proporcional *adj* GEN proportional

proporcionar *vt* GEN provide

propuesta *f* GEN bid

propuesto *adj* DATOS proposed

propulsión: **de** ~ **nuclear** *adj* GEN nuclear-powered

protección *f* GEN protection; ~ **de almacenamiento** DATOS storage protection; ~ **de archivos** INFO file protection; ~ **selectiva de campo** DATOS, TEC RAD selective field-protection; ~ **de cartucho** INFO cartridge save; ~ **de casete** INFO cassette save; ~ **catódica** COMPON cathodic protection; ~ **de la comunicación** DATOS communication protection; ~ **contra eco** TEC RAD echo protection; ~ **contra incendios** GEN *seguridad* fire protection; ~ **contra intromisión** TELEFON protection against intrusion; ~ **contra el polvo** ELECTRÓN dust cover; ~ **contra roedores** ELECTRÓN rodent protection; ~ **contra el ruido** ELECTRÓN jabber protection; ~ **contra sobrecarga** ELECTRÓN, ELECTROTEC overload protection, overpower protection; ~ **contra sobrecarga de retardo inverso** ELECTROTEC inverse time overload protection; ~ **contra sobretensión** ELECTRÓN, ELECTROTEC overvoltage protection; ~ **contra tirones** COMPON, ELECTROTEC strain relief; ~ **contra vandalismo** GEN vandal-proofing; ~ **de errores** DATOS error protection; ~ **de la memoria** INFO memory protection; ~ **del número de orden** CONMUT, DATOS, INFO sequence number protection (*SNP*)

protector: ~ **contra sobretensión transitoria** *m* ELECTROTEC surge voltage protector; ~ **de lente** *m* GEN lens protector

proteger *vt* GEN protect

protegido *adj* GEN protected; ~ **contra el polvo** GEN dustproof

protocolo *m* DATOS, INFO, TEC RAD protocol; ~ **de acceso al directorio** DATOS directory access protocol (*DAP*); ~ **de aplicación de gestión de sistemas** (*SMAP*) DATOS systems management application protocol (*SMAP*); ~ **de la capa de paquetes** (*PLP*) DATOS, TELEFON packet layer protocol (*PLP*); ~ **de capa de red orientado a la conexión** (*CONP*) TRANS connection-oriented network-layer protocol (*CONP*); ~ **de comunicación** DATOS *Internet* communication protocol; ~ **de control de transmisión** DATOS, INFO transmission control protocol; ~ **de copia de Unix-a-Unix** (*UUCP*) DATOS *Internet* Unix-to-Unix copy protocol (*UUCP*); ~ **de entrelazado** DATOS interworking protocol (*IWP*); ~ **de gestión** DATOS management protocol; ~ **de gestión de capa N** DATOS N-layer management protocol; ~ **de información de encaminamiento** (*RIP*) DATOS routing information protocol (*RIP*); ~ **de Internet** (*PI*) DATOS, INFO Internet protocol (*IP*); ~ **de mantenimiento de tabla de encaminamiento** (*RTMP*) INFO routing table maintenance protocol (*RTMP*); ~ **N** DATOS N-protocol; ~ **de operación de mantenimiento** DATOS, ELECTRÓN maintenance operation protocol; ~ **de par a par** TRANS peer-to-peer protocol; ~ **de punto a punto** (*PPP*) DATOS *Internet* point-to-point protocol (*PPP*); ~ **de la red** DATOS *Internet*, INFO network protocol; ~ **de red microcom** DATOS microcom network protocol; ~ **de reserva rápida** DATOS fast-reservation protocol (*FRP*); ~ **de señalización** SEÑAL signalling (*AmE* signaling) protocol; ~ **de servidor de información de área extendida** DATOS wide area information server protocol; ~ **de sesión** DATOS, TRANS session protocol; ~ **simple de gestión de red** (*SNMP*) DATOS simple network management protocol (*SNMP*); ~ **simple de transferencia de correo** (*SMTP*) DATOS *Internet* simple mail transfer protocol (*SMTP*); ~ **del sistema de información sobre abonados** DATOS directory system protocol; ~ **de terminal virtual de**

patrón internacional ELECTRÓN international standard virtual terminal protocol; ~ **de transferencia de archivos** (*FTP*) DATOS *Internet* file transfer protocol (*FTP*); ~ **de transferencia de hipertexto** (*HTTP*) DATOS, INFO *Internet* hypertext transfer protocol (*HTTP*); ~ **de transferencia de noticias de la red** (*NNTP*) DATOS *Internet* network news transfer protocol (*NNTP*); ~ **de enlace de datos** DATOS data-link protocol

prototipo *m* GEN prototype

protuberancia *f* COMPON *de pieza de metal* burr

proveedor *m* PRUEBA provider, vendor; ~ **de la conexión** DATOS *Internet* connection provider; ~ **de equipo** GEN equipment supplier; ~ **de MCS** DATOS, TELEFON MCS provider; ~ **de MCS terminal** DATOS, TELEFON top MCS provider; ~ **RTSE** DATOS, TELEFON RTSE provider; ~ **de servicio abstracto MS** DATOS, TELEFON MS abstract service provider; ~ **del servicio de mantenimiento** (*PSM*) PRUEBA maintenance service provider (*MSP*); ~ **del servicio de procesamiento de transacciones** (*TPSP*) DATOS transaction processing service provider (*TPSP*); ~ **de servicios ACSE** DATOS, TELEFON ACSE service provider; ~ **de servicios de comunicación avanzada** DATOS, TELEFON ACSE service provider; ~ **de servicios de la red** DATOS network service provider; ~ **del sistema** APLIC system provider

proveer *vt* GEN provide, supply, furnish

provisional *adj* GEN provisional, provisory

provisto: ~ **de** *adj* GEN equipped with; ~ **de fusibles** *adj* ELECTRÓN, ELECTROTEC fused

proyección: ~ **a la americana** *f* GEN *dibujo técnico* third-angle projection; ~ **de giros** *f* INFO drawmap; ~ **a la inglesa** *f* GEN *dibujo técnico* first-angle projection

proyectar *vt* GEN project

proyectista *m* GEN *persona* designer

proyecto *m* GEN design; ~ **de satélite comercial EHF** TRANS EHF satcom project

proyector *m* GEN projector; ~ **de vídeo** TV video projector

PRP *abr* (*periodo de repetición de los impulsos*) TELEFON PRP (*pulse repetition period*)

PRR *abr* (*frecuencia de repetición de impulsos*) ELECTRÓN, SEÑAL, TELEFON, TRANS PRF (*pulse repetition frequency*), PRR (*pulse repetition rate*)

PRT-ID *abr* (*identificador de protocolos*) DATOS PRT-ID (*protocol identifier*)

prueba *f* GEN test, trial; ~ **de aislamiento** ELECTROTEC leakage test; ~ **ambiental** PRUEBA environmental testing; ~ **de aptitud** DATOS capability test, PRUEBA compliance test; ~ **de arco del efecto corona** TEC RAD corona arc testing; ~ **de articulación silábica** PRUEBA, TELEFON syllable articulation test; ~ **asistida por ordenador** (*CAT*) INFO computer-aided testing (*CAT*); ~ **de banda base de telefonía** PRUEBA telephony baseband test; ~ **básica de interconexión** (*BIT*) DATOS basic interconnection test (*BIT*); ~ **en bucle** TRANS back-to-back test; ~ **de cableado** PRUEBA wiring test; ~ **de caída de voltaje** PRUEBA drop test; ~ **de campo** PRUEBA *seguridad de funcionamiento*, TEC RAD field test; ~ **de choque** PRUEBA shock test; ~ **climática** PRUEBA environmental testing; ~ **para comparación** INFO benchmark test; ~ **de comportamiento** DATOS behaviour (*AmE* behavior) test; ~ **de conexiones** COMPON, PRUEBA wiring check, wiring test; ~ **conjunta** PRUEBA joint testing; ~ **de conmutación** CONMUT, PRUEBA switching test; ~ **de continuidad de extremo a extremo** TELEFON end-to-end continuity test; ~ **de cortocircuito** ELECTRÓN short-circuit test; ~ **de descargas atmosféricas** ELECTROTEC lightning impulse test; ~ **de descubrimiento** PRUEBA unwrapping test; ~ **de desenrollamiento** PRUEBA unwrapping test; ~ **de determinación** PRUEBA *seguridad de funcionamiento* determination test; ~ **de duración** DATOS, PRUEBA life testing; ~ **de dureza Brinell** COMPON *resistencia* Brinell test; ~ **de escucha** TEC RAD *móvil* listening test; ~ **de eslabones** CONMUT link test; ~ **estadística** PRUEBA statistical test; ~ **de facturación** APLIC charging test; ~ **final** PRUEBA final test; ~ **de finalización** PRUEBA completion test; ~ **funcional** PRUEBA functional test; ~ **de funcionamiento** PRUEBA performance test; ~ **de impacto** COMPON impact test; ~ **de laboratorio** PRUEBA laboratory test; ~ **de lámparas** ELECTROTEC lamp test; ~ **a largo plazo** PRUEBA long-term test; ~ **de línea** TELEFON line test; ~ **por lotes** COMPON batch testing; ~ **OC** ELECTRÓN OC test; ~ **de onda sinusoidal** PRUEBA sine vibration test; ~ **paramétrica** PRUEBA parametric test; ~ **parcial** PRUEBA subtest; ~ **de pelado** COMPON, PRUEBA stripping test; ~ **piloto** PRUEBA beta trial; ~ **de las placas de circuito** PRUEBA *ensayos* board-level testing; ~ **preliminar** PERIF preprint; ~ **previa** INFO preprint, PRUEBA pretest; ~ **de puesta en servicio** PRUEBA commissioning test; ~ **de recepción** GEN acceptance test; ~ **de resistencia** PRUEBA strength test; ~ **de reso-**

lución de conformidad DATOS, PRUEBA conformance-resolution test; **~ de resonabilidad** INFO resonableness check; **~ RF** PRUEBA, TEC RAD RF test; **~ de ruido** TELEFON noise test; **~ de ruta** TELEFON route test; **~ de rutina** PRUEBA routine test; **~ de sacudida** COMPON *resistencia* bump test; **~ de satélite integrada** (*IST*) PRUEBA integrated satellite test (*IST*); **~ de secuencia de fase** ELECTROTEC phase-sequence test; **~ de selección** PRUEBA screening test; **~ de señal** PRUEBA, SEÑAL signal test; **~ de señalización** SEÑAL signalling (*AmE* signaling) test; **~ de señalización de línea** PRUEBA line signalling (*AmE* signaling) test; **~ sesgada** GEN *estadística* biased test; **~ de significación** PRUEBA significance test; **~ de sí o no** PRUEBA, TELEFON yes-or-no test; **~ del sistema** INFO, PRUEBA system testing; **~ de sistema integrado** PRUEBA integrated system test; **~ de tensión** ELECTROTEC voltage test; **~ de tracción** COMPON, PRUEBA pulling test, tensile test; **~ de transmisión** PRUEBA, TRANS transmission test; **~ truncada** PRUEBA *fiabilidad* censored test; **~ de validez** CONTROL, DATOS validity check; **~ de vibración** PRUEBA vibration test; ◆ **a ~ de errores de manipulación** GEN foolproof; **a ~ de explosiones** TRANS explosion-proof; **a ~ de fallos** PRUEBA no-failure; **a ~ de incendios** COMPON fireproof; **a ~ de manipulaciones imprudentes** PRUEBA tamperproof

pruebas: **~ conjuntas** *f pl* PRUEBA joint tests; **~ no destructivas** *f pl* DATOS, PRUEBA nondestructive testing; **~ de taller** *f pl* TRANS bench testing

PRV *abr* (*tensión máxima inversa*) ELECTROTEC PRV (*peak reverse voltage*), TELEFON PIV (*peak inverse voltage*), PRV (*peak reverse voltage*)

ps *abr* (*picosegundo*) PRUEBA ps (*picosecond*)

PSAP *abr* (*punto de acceso del servicio de presentación*) DATOS PSAP (*presentation-service access point*)

pseudoacontecimiento *m* DATOS pseudo-event

pseudoaleatorio *adj* DATOS, ELECTRÓN, TRANS, TV pseudorandom

pseudocódigo *m* INFO pseudocode

pseudoinstrucción *f* INFO pseudo-instruction

pseudorruido *m* DATOS, TRANS pseudonoise

psicrómetro *m* TELEFON psychrometer

PSM *abr* (*proveedor del servicio de mantenimiento*) PRUEBA MSP (*maintenance service provider*)

PSN *abr* (*red conmutada pública*) CONMUT, DATOS, TELEFON, TRANS PSN (*public switched network*)

PSS *abr* (*subsistema de potencia*) ELECTROTEC PSS (*power subsystem*)

PTM *abr* (*modulación de impulsos en el tiempo*) TELEFON, TRANS PTM (*pulse time modulation*)

publicación *f* GEN *documentación* issue

publicidad *f* GEN publicity

puente *m* DATOS, ELECTRÓN, INFO bridge; **~ de alimentación** ELECTROTEC feeding bridge; **~ de alumbrado** ELECTROTEC lighting fixture; **~ de casete** GEN cassette deck; **~ de conexión** COMPON, CONMUT, DATOS, ELECTRÓN, ELECTROTEC jumper; **~ de conferencia** APLIC conference bridge; **~ de continuidad** COMPON wire bonding; **~ de diodos** ELECTRÓN diode bridge; **~ de distorsión** ELECTRÓN distortion bridge; **~ elástico** ELECTRÓN compliant bridge; **~ de hilo desnudo** TELEFON bare-wire strap; **~ independiente** ELECTRÓN stand-alone bridge; **~ de medición** PRUEBA measuring bridge; **~ rectificador** ELECTROTEC rectifier bridge; **~ de soldadura** COMPON solder bridge; **~ de transmisión** TEC RAD, TRANS transmission bridge; **~ de Wheatstone** ELECTROTEC Wheatstone bridge; ◆ **de ~** ELECTROTEC, TELEFON bridged

puenteable *adj* COMPON, ELECTROTEC strappable

puenteado *adj* COMPON, ELECTROTEC strapped

puerta *f* ELECTRÓN gate; **~ de banda** TEC RAD range gate; **~ de canal** TELEFON, TRANS channel gate; **~ de circuito de transferencia de voz** TELEFON, TRANS voice-order wire port; **~ de comunicaciones** INFO communication port; **~ corredera** COMPON sliding door; **~ de entrada de línea** PERIF line-in port; **~ de equivalencia** ELECTRÓN IF-AND-ONLY-IF gate; **~ de impresora** INFO printer port; **~ del módem** INFO modem port; **~ de muestreo** TRANS sampling gate; **~ en nivel de aplicación** TELEFON application-level gateway; **~ NO-O** ELECTRÓN NOR gate; **~ NO** ELECTRÓN NOT gate; **~ O** ELECTRÓN OR gate; **~ O-INCLUSIVO** ELECTRÓN *elementos lógicos* INCLUSIVE-OR gate; **~ de protector** ELECTRÓN, INFO serving port; **~ de recuperación** DATOS retrieval port; **~ de salida** ELECTRÓN output gate, output port, INFO output port; **~ en serie** INFO series port; **~ de velocidad** ELECTRÓN velocity gate; **~ Y** ELECTRÓN AND gate

puerto: **~ de acceso** *m* GEN access port; **~ de datos** *m* DATOS data port; **~ para discado público** *m* TELEFON public dial-up port; **~ paralelo** *m* INFO parallel port; **~ de presentación indirecta** *m* DATOS indirect submission port

puesta: **~ a cero** *f* GEN resetting; **~ en fase de**

blanco *f* TELEFON phase-white; ~ **en fase para negro** *f* TELEFON phase-black; ~ **en función en fábrica** *f* INFO factory setting; ~ **en marcha** *f* TELEFON *instalaciones* commissioning; ~ **en órbita** *f* TEC RAD injection into orbit; ~ **en servicio** *f* GEN launching; ~ **en tierra del cable** *f* TRANS, PRUEBA cable landing; ~ **a tierra de protección** *f* ELECTROTEC, PRUEBA protective earthing (*AmE* grounding)

puesto[1]: ~ **a tierra** *adj* ELECTRÓN, ELECTROTEC earthed *BrE*, grounded *AmE*

puesto[2] *m* INFO booth, TEC RAD station; ~ **de cajero automático** DATOS, PERIF ATM workstation; ~ **de exposición** GEN *comercio* exhibition stand; ~ **de mantenimiento** TELEFON maintenance station; ~ **de marcación** INFO dial-up port; ~ **de observación del tráfico** TELEFON traffic observation desk; ~ **de operadora auxiliar** TELEFON auxiliary position; ~ **de operador auxiliar** TELEFON assistant operator position; ~ **de operador de tráfico** CONMUT, TELEFON traffic operator position; ◆ ~ **en cola de espera** TEC RAD, TELEFON, TRANS queueing

pulgada *f* GEN *medidas* inch

pulsación *f* GEN beat

pulsador: ~ **de báscula** *m* ELECTROTEC rocker button; ~ **de micrófono telefónico** *m* TEC RAD push-to-talk button; ~ **de toma de líneas** *m* PERIF line seizure button; ~ **virtual** *m* INFO virtual pushbutton

pulso: ~ **de borrado** *m* TELEFON blanking pulse; ~ **de la señal en curso** *m* TEC RAD course blip pulse; ~ **de sincronización** *m* GEN synchronization pulse

pulverización *f* COMPON spraying, INFO powdering

pulverizar *vt* COMPON spray, INFO powder

punta *f* COMPON, ELECTROTEC, TELEFON point; ~ **de clavija** TELEFON plug tip; ~ **máxima proyectada** ELECTROTEC projected peak point; ~ **de pico** COMPON peak point; ~ **de tensión** ELECTROTEC voltage spike; ~ **de trazar** COMPON *láseres* scriber

puntal: ~ **exterior** *m* TELEFON outer stanchion

puntas: ~ **de clavijas distribuidoras** *f pl* ELECTRÓN dispensing tips

punto *m* INFO, TV dot; ~ **de acceso** INFO, TEC RAD port; ~ **de acceso al circuito** TELEFON circuit access point; ~ **de acceso a Internet** DATOS Internet access point; ~ **de acceso a la línea** TELEFON line access point; ~ **de acceso al mantenimiento de la red** DATOS, TRANS network service access maintenance point; ~ **de acceso de la prueba de circuitos** TELEFON circuit-test access point; ~ **de acceso a servicio** DATOS service access point (*SAP*); ~ **de acceso al servicio de enlace de datos** (*DLSAP*) DATOS data-link service access point (*DLSAP*); ~ **de acceso del servicio MCS** (*MCSAP*) TELEFON MCS service access point (*MCSAP*); ~ **de acceso al servicio de la red** DATOS network service access point (*NSAP*); ~ **de acceso al servicio de sesión** (*SSAP*) DATOS, TRANS session service access point (*SSAP*); ~ **de acceso al servicio de transporte** DATOS, TELEFON, TRANS transport-service access point (*TSAP*); ~ **de acceso del servicio de presentación** TELEFON presentation-service access point (*PSAP*); ~ **de acceso a videotexto** TELEFON, TV videotext access point; ~ **de adorno** INFO bullet; ~ **de análisis de redes** TELEFON network analysis point; ~ **auxiliar de conmutación** CONMUT auxiliary switching point; ~ **de bifurcación** ELECTRÓN, INFO, TRANS, TV branching point, branch point; ~ **de cajero automático** INFO ATM workstation; ~ **de carga** INFO bias point; ~ **central** GEN centre (*AmE* center) point; ~ **cero virtual** TELEFON virtual zero time; ~ **de codo** ELECTRÓN knee point; ~ **de conexión** GEN connection point; ~ **de conexión de grupo directo** TELEFON through-group connection point; ~ **de conexión de grupo secundario directo** TELEFON through-supergroup connection point; ~ **de conexión de grupo terciario directo** TELEFON through-mastergroup connection point; ~ **de conexión de montaje de grupo secundario 15 directo** TELEFON through-15-supergroup assembly connection point; ~ **de conmutación** CONMUT, TELEFON switching point; ~ **de conmutación analógica virtual** CONMUT virtual analog switching point; ~ **de conmutación virtual** CONMUT virtual switching point; ~ **de control** CONTROL control point, INFO, TELEFON checkpoint; ~ **de control y observación** (*PCO*) DATOS point of control and observation (*PCO*); ~ **de control de recuperación** (*RCP*) TELEFON restoration control point (*RCP*); ~ **de cruce bifilar** CONMUT two-wire crosspoint; ~ **de cruce de diodos** CONMUT diode crosspoint; ~ **de cruce de lengüeta rem** CONMUT remreed crosspoint; ~ **de cruce metálico** CONMUT metallic crosspoint; ~ **de cruce RCS** COMPON, ELECTRÓN SCR crosspoint; ~ **de cruce de rectificador controlado por silicio** COMPON,

ELECTRÓN silicon-controlled rectifier crosspoint; ~ **de cruce del relé de láminas** COMPON, CONMUT, ELECTROTEC reed-relay crosspoint; ~ **de cruce de semiconductores** COMPON, CONMUT, ELECTRÓN semiconductor crosspoint; ~ **de cruce sostenido eléctricamente** CONMUT electrically held crosspoint; ~ **de Curie** INFO Curie point; ~ **de derivación** TRANS *óptica de fibras* drop point; ~ **destellante** INFO flashing dot; ~ **de destino** TELEFON destination point; ~ **de destino de la señalización** SEÑAL signalling (*AmE* signaling) destination point; ~ **de determinación de tarifa** TELEFON tariff determination point; ~ **de distribución** TRANS distribution point; ~ **de división del teclado** INFO keyboard split point; ~ **de ebullición** GEN *física* boiling point; ~ **de entrada** INFO entry point; ~ **de entrada al bastidor** TELEFON rack entry point; ~ **de equilibrio** GEN breakeven point; ~ **de espacio muestral** PRUEBA sample point; ~ **de exploración** INFO, TV scanning spot; ~ **de facturación** APLIC charging point; ~ **final de la vía de transmisión** DATOS, TRANS transmission path endpoint (*TPE*); ~ **fosforescente** INFO phosphor dot; ~ **de gestión de la red** TELEFON network management point; ~ **hendido** INFO split point; ~ **de imagen** ELECTRÓN, INFO, TV pel; ~ **de incorporación a subred** (*SNPA*) DATOS subnetwork point of attachment (*SNPA*); ~ **de indiferencia** PRUEBA indifference point, point of control; ~ **de inflexión** GEN point of inflexion; ~ **de información sobre la disponibilidad del sistema** TELEFON system-availability information point; ~ **de informe de averías** TELEFON *red* fault report point; ~ **de inserción** INFO insertion point; ~ **de interconexión del sistema** TELEFON system interconnection point; ~ **de interrupción** INFO breakpoint; ~ **de intersección bloqueado magnéticamente** CONMUT magnetically latched crosspoint; ~ **de irrupción** COMPON *tiristores* breakover point; ~ **máximo** GEN peak; ~ **de medición** PRUEBA measuring point; ~ **medio** GEN midpoint; ~ **medio de clase** GEN class midpoint; ~ **Morse** TELEFON Morse dot; ~ **N de acceso al servicio** DATOS N-service access point; ~ **neutro** ELECTROTEC neutral point; ~ **de operación** CONMUT operation point; ~ **de orden de servicio** CONMUT service command point; ~ **de origen de la señalización** SEÑAL signalling (*AmE* signaling) originating point; ~ **de prueba** CONMUT, PRUEBA, SEÑAL, TRANS testing point; ~ **de prueba de línea** TELEFON line test-point; ~ **de prueba de relé** COMPON relay test-point; ~ **de reentrada** INFO re-entry point; ~ **de referencia** DATOS reference point; ~ **de referencia auditiva** TELEFON ear reference point; ~ **de referencia QOS** (*QRP*) GEN QOS reference point (*QRP*); ~ **de referencia R** DATOS R-reference point; ~ **de referencia S** DATOS S-reference point; ~ **de referencia T** DATOS T-reference point; ~ **de referencia de transmisión** (*TRP*) TELEFON, TRANS transmission reference point (*TRP*); ~ **remoto de conmutación** CONMUT remote switching point; ~ **de retransmisión de mensajes** TRANS message relay; ~ **de riesgo del cliente** PRUEBA *calidad* consumer's risk point; ~ **de riesgo del productor** PRUEBA producer's risk point; ~ **sensor** PRUEBA sensing point; ~ **de señalización** SEÑAL, TELEFON signalling (*AmE* signaling) point; ~ **de sobrecarga** TELEFON overload point; ~ **de soldadura** COMPON soldering point; ~ **de subsatélite** TEC RAD subsatellite point; ~ **de sujeción** INFO anchor point, TEC RAD hold-down point; ~ **terminal** ELECTRÓN spot; ~ **de toma de energía** ELECTRÓN power point; ~ **de trabajo** ELECTRÓN operating point; ~ **de transferencia de la señalización** CONMUT, SEÑAL, TELEFON, TRANS signalling (*AmE* signaling) transfer point; ~ **de transmisión de señales** CONMUT, SEÑAL, TELEFON, TRANS signal transfer point; ~ **de valle** COMPON valley point; ~ **de venta** GEN point of sale, sales outlet; ~ **virtual de conexión internacional** TELEFON virtual international connection point; ~ **de voltaje mínimo** ELECTROTEC, TEC RAD point of minimum voltage

puntos: ~ **de conexión internacional de frecuencias de banda base** *m pl* TELEFON points of international connection at baseband frequencies; ~ **de entrada y salida de telefonía** *m pl* TELEFON points of telephony input and output; ~ **de mantenimiento de línea de transmisión internacional** *m pl* TELEFON international transmission-line maintenance points; ~ **por pulgada** *m pl* (*dpi*), (*ppp*) DATOS *digitalizador* dots per inch (*dpi*)

puntuación *f* GEN punctuation

puntuar *vt* GEN punctuate

punzón *m* TELEFON stylus

PUP *abr* (*unidad principal de procesamiento*) INFO PUP (*principal processing unit*)

pupinización *f* TELEFON, TRANS pupinization

pupinizado *adj* TELEFON, TRANS pupinized

pupinizar *vt* TELEFON, TRANS pupinize
pupitre: ~ **de mando** *m* TELEFON chief operator's desk, chief supervisor's desk; ~ **mezclador** *m* INFO, TEC RAD mixing console; ~ **de proceso a distancia** *m* DATOS remote console
puro *adj* PRUEBA true
PWM *abr* (*modulación de impulsos en duración*) DATOS, TEC RAD, TELEFON PDM (*pulse duration modulation*), TRANS pulse width modulation (*PWM*)

Q

QCIF *abr* (*formato intermedio común dividido en cuatro*) INFO QCIF (*quarter common intermediate format*)

qdu *abr* (*unidad de distorsión de la cuantificación*) TELEFON, TRANS qdu (*quantizing distortion unit*)

QFP *abr* (*encapsulado plano de cuadrete*) TRANS QFP (*quad flatpack*)

QIC *abr* (*cartucho de cuarto de pulgada*) TV QIC (*quarter-inch cartridge*)

QPSK: ~ **filtrado** *m* TEC RAD *satélite marítimo* filtered QPSK

QRP *abr* (*punto de referencia QOS*) GEN QRP (*QOS reference point*)

QOS *abr* (*calidad de servicio*) GEN QOS (*quality of service*)

quemador *m* GEN burner; ~ **oxiacetilénico** GEN oxyacetylene burner

quemadura: ~ **de la trama** *f* ELECTRÓN raster burn

quinteto *m* DATOS quintet

quíntuplo *m* GEN quintuple

quiosco *m* DATOS kiosk; ~ **de Internet** DATOS Internet kiosk; ~ **multimedia** TV multimedia kiosk

R

R *abr* (*reluctancia*) ELECTROTEC R (*reluctance*)

rabillo *m* GEN pigtail

radar *m* TEC RAD radar; ~ **aéreo de exploración lateral** (*SLAR*) TEC RAD side-looking airborne radar (*SLAR*), sideways-looking airborne radar; ~ **aerotransportado** TEC RAD airborne radar; ~ **de agilidad de frecuencias** TEC RAD frequency agility radar; ~ **altimétrico** TEC RAD height-finding radar; ~ **de aterrizaje** TEC RAD precision approach radar; ~ **biestático** TEC RAD bistatic radar; ~ **buscador de blanco** TEC RAD homing radar; ~ **de detección de superficie de aeropuerto** TEC RAD airport surface detectionequipment (*ASDE*); ~ **de dirección de tiro** TEC RAD fire control radar; ~ **de diversidad** TEC RAD diversity radar; ~ **Doppler** TEC RAD Doppler radar; ~ **Doppler de impulsos** TEC RAD pulse-Doppler radar; ~ **de exploración** TEC RAD search radar; ~ **de exploración lateral no coherente** TEC RAD noncoherent side-looking radar; ~ **de iluminación** TEC RAD illuminating radar; ~ **de impulsos** TEC RAD pulse radar; ~ **indicador de blancos móviles** TEC RAD moving-target indicator radar, MTI radar; ~ **indicador distancia-altitud** TEC RAD range-height indicator; ~ **meteorológico** TEC RAD weather radar; ~ **modulado en frecuencia** TEC RAD frequency-modulated radar; ~ **modulado por impulsos** TEC RAD pulse-modulated radar; ~ **monoestático** TEC RAD monostatic radar; ~ **de monoimpulsos** TEC RAD monopulse radar; ~ **de navegación** TEC RAD navigational radar; ~ **de onda continua** TEC RAD continuous-wave radar, CW radar; ~ **de precisión para la aproximación** TEC RAD precision approach radar; ~ **de precisión de aterrizaje** TEC RAD precision approach radar; ~ **de puerto** TEC RAD harbour (*AmE* harbor) radar; ~ **secundario** TEC RAD secondary radar; ~ **totalmente coherente** TEC RAD fully-coherent radar; ~ **de vigilancia** TEC RAD surveillance radar

radiación *f* ELECTRÓN, TRANS, TEC RAD radiation; ~ **de antena** TEC RAD antenna radiation; ~ **de armario** TEC RAD cabinet radiation; ~ **coherente** TEC RAD coherent radiation; ~ **electromagnética** ELECTRÓN electromagnetic radiation; ~ **espuria** TRANS spurious radiation; ~ **de fondo** TEC RAD background radiation; ~ **incoherente** TRANS incoherent radiation; ~ **infrarroja** ELECTRÓN infrared radiation; ~ **IR** ELECTRÓN IR radiation; ~ **del lóbulo lateral** TEC RAD sidelobe radiation; ~ **monocromática** TRANS monochromatic radiation; ~ **óptica** TRANS optical radiation; ~ **parásita** TEC RAD spurious radiation; ~ **de partículas** GEN particle radiation; ~ **de radiofrecuencia** GEN radiofrequency radiation; ~ **de receptor** TEC RAD receiver radiation; ~ **secundaria** TEC RAD secondary radiation; ~ **térmica** ELECTRÓN *física* heat radiation; ~ **ultravioleta** TRANS ultraviolet radiation, UV radiation; ~ **visible** ELECTRÓN, TRANS visible radiation

radiador *m* GEN radiator; ~ **isotrópico** TEC RAD isotropic radiator; ~ **lambertiano** TRANS *óptica de fibras* Lambertian radiator

radial *adj* TEC RAD radio

radiancia *f* ELECTRÓN, TRANS radiance; ~ **espectral** ELECTRÓN, TRANS spectral radiance

radiar *vt* ELECTRÓN, TRANS radiate

radiativo: **no** ~ *adj* GEN nonradiative

radical *adj* ELECTRÓN radical

radio[1] *m* GEN *del círculo de la base* radius; ~ **de curvatura** ELECTRÓN bend radius; ~ **de curvatura mínimo** TRANS minimum bend radius; ~ **efectivo de la tierra** TEC RAD effective radius of the earth; ~ **de estampación** COMPON, ELECTRÓN tooling radius; ~ **de plegado** TRANS bend radius

radio[2] *f* TEC RAD radio; ~ **celular digital** TEC RAD digital cellular radio; ~ **enlazada** ELECTROTEC, TEC RAD trunked radio; ~ **intercalado en circuito** TEC RAD radio in the loop; ~ **de interior** TEC RAD indoor radio; ~ **marítima** TEC RAD maritime radio; ~ **de microondas** TEC RAD microwave radio; ~ **móvil privada** TEC RAD private mobile radio

radioactivo *adj* GEN radioactive

radioaficionado *m* TEC RAD radio ham

radioalarma *f* PERIF radio alarm

radioaltímetro *m* TEC RAD, TELEFON radio altimeter

radioanemómetro *m* TEC RAD radio wind

radiobaliza *f* TEC RAD marker beacon; ~ **fija** TEC

RAD fixed marker beacon; ~ **intermedia** TEC RAD middle marker beacon; ~ **interna** TEC RAD inner marker beacon
radiobuscador *m* DATOS, TEC RAD, TELEFON, TRANS radiopager
radiobúsqueda *f* DATOS, TEC RAD, TELEFON, TRANS radiopaging
radiocompás *m* TEC RAD radio compass
radiocomunicación *f* TEC RAD radiocommunication; ~ **espacial** TEC RAD, TV space radiocommunication; ~ **por paquetes** TEC RAD packet radio; ~ **terrestre** TEC RAD terrestrial radiocommunication
radiodifusión *f* TEC RAD radio broadcasting; ~ **de audio digital** DATOS, TEC RAD, TV digital audio broadcasting; ~ **directa vía satélite** DATOS, TEC RAD, TV direct broadcasting by satellite (*DBS*); ~ **por satélite** DATOS, TEC RAD, TV satellite broadcasting service; ~ **sonora** TEC RAD sound broadcasting; ~ **televisiva** TV television broadcasting
radiodifusor *m* TV broadcaster
radioducto: ~ **troposférico** *m* TEC RAD tropospheric radioduct
radioemisión: ~ **en el exterior** *f* TEC RAD, TV outside broadcast
radioemisora *f* TEC RAD broadcasting station
radioenlace: ~ **por microondas** *m* TRANS microwave relay link
radiofaro *m* TEC RAD radio beacon; ~ **de exploración** TEC RAD sector-scanning beacon; ~ **generado desde tierra** TECRAD earth-generated beacon; ~ **indicador de posición para emergencias** TEC RAD emergency position-indicating radio beacon (*EPIRB*); ~ **omnidireccional de VHF** (*VOR*) TEC RAD VHF omnidirectional radio range (*VOR*); ~ **para radar** TEC RAD ramark
radiofrecuencia *f* (*RF*) TEC RAD radio frequency (*RF*)
radiogoniometría *f* TEC RAD radio direction-finding
radiogoniómetro *m* TEC RAD radio direction finder, radiogoniometer; ~ **Adcock equilibrado** TEC RAD balanced Adcock direction finder; ~ **de antena fija** TEC RAD fixed-antenna direction finder; ~ **manual** TEC RAD manual direction finder
radioguía *f* TEC RAD radio guidance
radiolocalización *f* TEC RAD radio determination, radio location
radiómetro *m* PRUEBA radiometer; ~ **neto** PRUEBA net radiometer
radionavegación *f* TEC RAD radio navigation
radiopropagación *f* TEC RAD radio propagation
radiorreceptor: ~ **de FM** *m* TEC RAD FM radio receiver
radiosidad *f* INFO radiosity
radiosonda *f* TEC RAD radiosonde, rawin, rawin flight
radiosondeo *m* TEC RAD rawin sounding
radiotelefonear *vt* TEC RAD, TELEFON radiophone
radiotelefonía *f* TEC RAD, TELEFON radiotelephony
radioteléfono *m* TEC RAD, TELEFON radiotelephone; ~ **móvil** TEC RAD, TELEFON *proceso de datos a distancia* mobile radio telephone
radiotelegrafía *f* TEC RAD, TELEFON radiotelegraphy
radioteleimpresora *f* TEC RAD radio teleprinter
radioteodolito *m* TEC RAD radio theodolite
radiotransmisión: ~ **en díplex** *f* DATOS diplex radio transmission
radiotransmisor *m* TEC RAD radio transmitter
radiovigilancia *f* TEC RAD radio survey
radomo *m* TEC RAD radar dome, *antena* radome; ~ **húmedo** TEC RAD wet radome
ráfaga: ~ **corta** *f* DATOS short burst; ~ **de impulsos** *f* DATOS, TRANS pulse burst; ~ **portadora de tráfico** *f* DATOS traffic-carrying burst; ~ **primaria de referencia** *f* DATOS primary reference burst; ~ **secundaria auxiliar de referencia** *f* DATOS backup reference secondary burst; ~ **de subráfaga única** *f* DATOS single sub-burst burst; ~ **de tono** *f* SEÑAL, TELEFON burst of tone
RAG *abr* (*repetir la llamada*) TELEFON RAG (*ring again*)
raíz: ~ **cuadrada de la suma de los cuadrados** *f* GEN root sum square; ~ **cuadrada del valor medio cuadrático** *f* GEN root-mean square value
RAM *abr* (*memoria de acceso aleatorio*) INFO RAM (*random access memory*); ~ **dinámica** *f* DATOS, INFO dynamic RAM; ~ **estática** *f* INFO static RAM (*SRAM*); ~ **de vídeo** *f* (*VRAM*) INFO video RAM (*VRAM*)
rama: ~ **B** *f* TELEFON B-leg; ~ **común** *f* ELECTROTEC mutual branch; ~ **local** *f* TELEFON leg
ramal: ~ **de cable** *m* TRANS cable branch
ramificación *f* INFO *de circuito* branch; ~ **de cable** TRANS cable branching; ~ **de forma de cable** TELEFON skinner; ~ **óptica** TRANS optical branching
rampa *f* TELEFON ramp
rango *m* GEN range
ranura *f* GEN slot; ~ **de extensión** INFO expan-

sion slot; ~ **longitudinal** TRANS longitudinal slot; ~ **para las monedas** PERIF *teléfonos de pago* coin slot; ~ **radiante** TEC RAD *antena* slot radiator; ~ **en la tapa portabobinas** ELECTRÓN slot in the hub; ~ **transversal** TRANS transverse slot; ◆ **en** ~ COMPON, ELECTRÓN slot-based

rápido *adj* INFO quick

rarefacción *f* TEC RAD rarefaction

rarefactor *m* ELECTRÓN *tubos electrónicos* getter

rascado *m* COMPON *pintura* stripping

rasponazos *m pl* COMPON, PRUEBA scratching

rastreabilidad *f* ELECTRÓN, PRUEBA traceability

rastreador *m* TEC RAD tracker; ~ **de señal** SEÑAL signal tracer

rastrear *vt* CONMUT, ELECTRÓN, PRUEBA trace, TEC RAD keep track of, trace

rastreo *m* CONMUT, ELECTRÓN, PRUEBA, TEC RAD tracing; ~, **telemetría y órdenes** (*TT & C*) TEC RAD tracking, telemetry and command (*TT & C*)

ratio *f* GEN ratio; ~ **de aceleración posdeflexión** ELECTRÓN post-deflection acceleration ratio; ~ **de compensación** ELECTROTEC stand-off ratio; ~ **de discriminación** PRUEBA *calidad* discrimination ratio; ~ **de energía por bitio de información/densidad de ruido** TRANS energy per information bit to noise density ratio; ~ **de equilibrio de impedancia** ELECTRÓN impedance balance ratio; ~ **de equilibrio de señal** SEÑAL *generadores de señal sinusoidal* signal balance ratio; ~ **de error binario residual** *m* (*RBER*) TEC RAD residual bit error ratio (*RBER*); ~ **de error de caracteres** INFO character error ratio; ~ **de errores** TELEFON error ratio; ~ **de expansión de ancho de banda** TEC RAD, TRANS bandwidth expansion ratio; ~ **del impulso** CONMUT pulse ratio; ~ **de luminancia** PRUEBA luminance ratio; ~ **de precipitación** TEC RAD precipitation rate; ~ **de protección** TEC RAD protection ratio; ~ **de rechazo de respuesta espuria** TEC RAD spurious response rejection ratio; ~ **respuesta-ocupación** TELEFON answer-seizure ratio; ~ **de supresión de frecuencia de imagen** TEC RAD image rejection ratio; ~ **de supresión de modo lateral** (*SMSR*) TRANS side-mode suppression ratio (*SMSR*); ~ **de uniformidad de la iluminancia** ELECTROTEC illuminance uniformity ratio

ratón *m* INFO mouse

raya: ~ **continua** *f* GEN *dibujo técnico* continuous line; ~ **espectral** *f* TRANS spectral line; ~ **Morse** *f* TELEFON Morse dash

rayado *m* ELECTRÓN *circuitos impresos* cross-hatching

rayo *m* GEN ray; ~ **centrado del satélite** TEC RAD satellite spot beam; ~ **de efecto túnel** TRANS tunnelling (*AmE* tunneling) ray; ~ **de luz** TRANS *óptica de fibras* light ray; ~ **meridional** TRANS meridional ray; ~ **oblicuo** TRANS skew ray; ~ **paraxial** ELECTRÓN, TRANS paraxial ray; ~ **Pedersen** TEC RAD Pedersen ray; ~ **refractado** TRANS refracted ray

razón: ~ **de aspecto del pixel** *f* INFO, TEC RAD, TELEFON pixel aspect ratio; ~ **de conversión transversal** *f* TELEFON transverse conversion ratio; ~ **de potencia de portadora deseada a portadora indeseada** *f* ELECTRÓN, SEÑAL, TEC RAD wanted-to-unwanted carrier power ratio; ~ **de reexpedición** *f* CONMUT redirecting reason; ~ **de reproducción** *f* TELEFON reproduction ratio

razonamiento *m* DATOS reasoning; ~ **analógico** DATOS analogical reasoning; ~ **aproximativo** DATOS approximate reasoning; ~ **autoepistemológico** DATOS auto-epistemic reasoning; ~ **automatizado** DATOS automated reasoning; ~ **basado en casos prácticos** DATOS case-based reasoning; ~ **basado en memoria** DATOS memory-based reasoning; ~ **cualitativo** DATOS qualitative reasoning; ~ **cuantitativo** DATOS quantitative reasoning; ~ **deductivo** DATOS deductive reasoning; ~ **formal** INFO formal reasoning; ~ **híbrido** ELECTRÓN hybrid reasoning; ~ **hipotético** GEN hypothetical reasoning; ~ **implícito** DATOS default reasoning; ~ **monótono** DATOS monotonic reasoning; ~ **plausible** DATOS plausible reasoning; ~ **probabilístico** DATOS probabilistic reasoning; ~ **temporal** DATOS temporal reasoning

RBE *abr* (*entrada por lotes a distancia*) DATOS RBE (*remote batch entry*)

RBER *abr* (*ratio de error de binario residual*) TEC RAD RBER (*residual bit error ratio*)

RBV *abr* (*vidicón con haz de retorno*) TELEFON RBV (*return-beam vidicon*)

RC *abr* (*resistencia-capacitancia*) ELECTRÓN RC (*resistance-capacitance*)

RCB *abr* (*ocupación de contador de retransmisión*) TELEFON RCB (*redrive counter busy*)

RCMSS *abr* (*sistema de vigilancia de barrido por control remoto*) CONTROL RCMSS (*remote-control monitoring scan system*)

RCP *abr* (*punto de control de recuperación*) TELEFON RCP (*restoration control point*)

RCS *abr* (*rectificador controlado de silicio*) COMPON, ELECTRÓN SCR (*silicon-controlled rectifier*)

RCTP *abr* (*plan temporal condensado de retrans-*

misión por ráfagas) DATOS RCTP (*retransmitted condensed burst time plan*)

rcvr *abr* (*receptor*) DATOS, PERIF, TEC RAD, TELEFON rcvr (*receiver*)

RD *abr* (*datos recibidos*) DATOS receive data (*RD*), received data

RDA *abr* (*acceso remoto a base de datos*) DATOS RDA (*remote database access*)

R-DAT *abr* (*cinta de audio digital de cabeza rotatoria*) DATOS, INFO R-DAT (*rotary-head digital audio tape*)

RDSI *abr* (*red digital de servicios integrados*) TELEFON, TRANS ISDN (*integrated services digital network*)

RDSI-BA *abr* (*ISDN de banda*) TRANS B-ISDN (*broadband ISDN*)

RDU *abr* (*unidad de datos remota*) DATOS RDU (*remote data unit*)

reacción: **~ anódica** *f* COMPON anodic reaction

reactancia *f* ELECTROTEC reactance; **~ de electrodo** ELECTRÓN electrode reactance; **~ monofásica** ELECTROTEC single-phase reactance

reactivación *f* DATOS reactivation, PRUEBA, TEC RAD, TELEFON recovery

reactivar *vt* DATOS *digitalizador* reactivate

reagrupar *vt* INFO regroup

reajustabilidad *f* TEC RAD resettability

reajustar *vt* GEN readjust; ◆ **~ el ritmo de** ELECTRÓN, TRANS retime

reajuste *m* GEN readjustment; **~ del equilibrio** GEN rebalancing; **~ del ritmo** ELECTRÓN, TRANS retiming

real *adj* ELECTRÓN *función lógica*, TEC RAD true

realce *m* INFO highlighting

realidad: **~ virtual** *f* (*VR*) DATOS, INFO virtual reality (*VR*)

realimentación *f* INFO feedback; **~ AGC** TRANS feedback AGC; **~ de aplicabilidad** DATOS *Internet* relevance feedback; **~ negativa** GEN negative feedback; **~ positiva** GEN positive feedback

realineación *f* GEN realignment

realinear *vt* GEN realign

realzado *m* ELECTRÓN *circuitos impresos*, INFO embossing

realzar *vt* GEN highlight

reanudación: **~ menor** *f* CONMUT minor restart; **~ menor del sistema** *f* CONMUT minor system restart

rearranque: **~ de emergencia** *m* TELEFON emergency restart

reasignación *f* GEN reallocation

reasignado *adj* GEN reallocated

reasignar *vt* TEC RAD reassign

rebabar *vt* COMPON, INFO *cortar* clip

rebaja *f* INFO, PRUEBA reduction

rebajar *vt* INFO *nivel freático* lower; **~ de grado** INFO demote

rebanador *m* COMPON slicer

rebarba *f* COMPON blanking burr

rebobinado: **~ rápido** *m* INFO fast forward playback; **~ rápido invertido** *m* INFO fast reverse playback

reborde: **~ de brida plana** *m* TEC RAD plain coupling flange

rebordeado *m* GEN crimping

rebotar *vi* INFO bounce

REC *abr* DATOS, SEÑAL, TELEFON (*reconocimiento de recibo*) ACK (*acknowledgement*), TEC RAD (*recibir*) REC (*receive*)

recalada: **~ por radio** *f* TEC RAD radio homing

recalibración *f* GEN recalibration

recalibrar *vt* GEN recalibrate

recarga *f* INFO rebooting, reloading

recargar *vt* INFO reboot, reload

recaudar *vt* GEN collect

recebado *m* DATOS rebooting

recepción *f* GEN receipt; **~ amplificada** PERIF amplified reception; **~ en blanco y negro** TELEFON black-and-white reception; **~ comunitaria** TEC RAD community reception; **~ dentro de un edificio** TEC RAD in-building reception; **~ de diversidad** TEC RAD diversity reception; **~ de diversidad de frecuencias** TEC RAD frequency-diversity reception; **~ de diversidad espacial** TEC RAD space-diversity reception; **~ de emisión por radio** TEC RAD broadcast reception; **~ homodina** TEC RAD homodyne reception; **~ individual** TEC RAD individual reception; **~ de portadora reacondicionada** GEN reconditioned carrier reception; **~ de radiofrecuencia sintonizada** TEC RAD TRF reception, tuned radio frequency reception; **~ de régimen estacionario** (*SSR*) TEC RAD steady-state reception (*SSR*); **~ superregenerativa** TEC RAD superregenerative reception

receptáculo *m* GEN receptacle; **~ para cable de CA** INFO AC power-cord receptacle; **~ de mesa** TEC RAD *móvil* desk cassette

receptor *m* (*rcvr, RX*) DATOS, PERIF, TEC RAD, TELEFON receiver (*rcvr, RX*); **~ de banda ancha** TEC RAD wideband receiver; **~ de banda estrecha** TEC RAD narrow-band receiver; **~ de canal** TRANS channel receiver; **~ de cilindro** TELEFON drum receiver; **~ de cobertura general** TEC RAD general-coverage receiver; **~ de código** TELEFON code receiver; **~ continuo** TELEFON continuous receiver; **~ por diferen-**

cial de torsión ELECTROTEC torque differential receiver; ~ **fibroóptico** TRANS fibre-optic (*AmE* fiber-optic) receiver; ~ **flotante** TEC RAD floating receiver; ~ **de FM** SEÑAL MF receiver; ~ **del generador de frecuencia** TELEFON frequency-generator receiver; ~ **de guardia** TEC RAD guard receiver, watch-keeping receiver; ~ **de identificación** TEC RAD responser; ~ **integrado PIN-FET** TELEFON, TRANS *óptica de fibras* PIN-FET integrated receiver; ~ **del intercambio** DATOS interchange recipient; ~ **de mensajes** TRANS message sink; ~ **de mensaje de tarifa** TELEFON tariff-message receiver; ~ **de módem** DATOS modem receiver; ~ **multifrecuencia** SEÑAL multifrequency receiver; ~ **N** DATOS N-recipient; ~ **portátil** TEC RAD hand-held receiver; ~ **regenerativo** TEC RAD regenerative receiver; ~ **con reloj** TEC RAD watch receiver; ~ **de señales** SEÑAL, TEC RAD signal receiver; ~ **de señales de frecuencia vocal** SEÑAL, TELEFON voice-frequency signalling (*AmE* signaling) receiver; ~ **de televisión** TV television receiver, television set; ~ **de televisión de pantalla grande** TV projection television receiver; ~**-transmisor asíncrono universal** (*UART*) TEC RAD universal asynchronous receiver-transmitter (*UART*)

rechazar *vt* INFO, PRUEBA reject

rechazo *m* DATOS repudiation, GEN rejection, PRUEBA nonacceptance, rejection, *inspecciones* reject; **no** ~ DATOS nonrepudiation; ~ **de FI** TEC RAD IF rejection; ~ **de frecuencia de imagen** TEC RAD image frequency rejection; ~ **de receptor** TEC RAD receiver rejection; ~ **selectivo** TELEFON selective reject; ~ **de trama** TELEFON frame reject

recibido: ~ **y atendido** *fra* TEC RAD, TELEFON wilco

recibidor *m* GEN recipient

recibir *vt* (*REC*) COMPON *corriente de retención* obtain, DATOS, SEÑAL, TELEFON receive (*REC*)

reciprocidad *f* ELECTRÓN reciprocity

recíproco *adj* GEN reciprocal

reclamación *f* GEN fault complaint

reclasificar *vt* INFO re-sort

recogida: ~ **de datos** *f* DATOS data collection

recolección: ~ **de noticias** *f* TEC RAD, TV news gathering

recombinación *f* COMPON, TELEFON recombination, recombining

recomendación: ~ **de servicio** *f* TELEFON service advice

reconexión *f* GEN reconnection

reconfiguración *f* INFO reconfiguration, reconfiguring

reconfigurar *vt* INFO reconfigure

reconocedor: ~ **de voz** *m* APLIC, INFO, TELEFON voice recognizer

reconocimiento: ~ **de caracteres** *m* INFO character recognition; ~ **de errores** *m* PRUEBA *seguridad del funcionamiento* error recognition; ~ **del fallo** *m* TELEFON failure recognition; ~ **global** *m* GEN global recognition; ~ **del hablante** *m* APLIC, INFO, TELEFON speaker recognition; ~ **del impulso** *m* SEÑAL, TRANS pulse recognition; ~ **multilingüe** *m* APLIC, INFO, TELEFON multilingual recognition; ~ **negativo** *m* (*NACK*) TEC RAD negative acknowledgement (*NACK*); ~ **no numerado** *m* TELEFON unnumbered acknowledgement; ~ **de patrones** *m* APLIC, CONTROL, DATOS, TRANS pattern recognition; ~ **de recibo** *m* (*REC*) DATOS, SEÑAL, TELEFON acknowledgement (*ACK*); ~ **de una sola voz** *m* APLIC, INFO, TELEFON single-voice speech recognition; ~ **técnico** *m* DATOS technical acknowledgement; ~ **de textos manuscritos** *m* APLIC handwriting recognition; ~ **de voz** *m* APLIC, INFO, TELEFON speech recognition, voice recognition

reconstrucción *f* DATOS reconstruction, INFO rebuilding, reconstruction; ~ **de imagen** TV image reconstruction

reconstruir *vt* DATOS reconstruct, INFO rebuild, reconstruct

recopilación: ~ **de datos** *f* DATOS data gathering; ~ **de noticias por vía satélite** *f* TEC RAD, TV satellite news gathering

recordatorio *m* GEN follow-up; ~ **temporizado** PERIF, TELEFON timed reminder

recorrer *vt* INFO travel

recorrido: ~ **de cables** *m* TRANS *tendido de cables* cable route; ~ **de circuito** *m* TRANS circuit route; ~ **inverso de bucle** *m* ELECTRÓN loopback

recortado *m* INFO scissoring

recortar *vt* ELECTRÓN, ELECTROTEC trim, INFO crop, trim

recorte *m* INFO, TELEFON, TRANS clipping; ~ **por láser** ELECTRÓN laser trimming; ~ **de la sección de entrada** DATOS front-end clipping

rectenna *f* TEC RAD rectenna

rectificación *f* ELECTROTEC rectification, GEN correction; ~ **de media onda** ELECTRÓN half-wave rectification

rectificador *m* ELECTROTEC rectifier; ~ **de carga** ELECTROTEC charging rectifier; ~ **conectado a una secuencia** COMPON, ELECTROTEC sequence-connected rectifier; ~ **controlado de silicio** (*RCS*) COMPON, ELECTRÓN silicon-controlled rectifier (*SCR*); ~ **de cristal** ELECTROTEC crystal rectifier; ~ **de elementos de regulación** ELECTROTEC end-cell rectifier; ~ **de media onda** ELECTRÓN half-wave rectifier; ~ **metálico** ELECTRÓN metallic rectifier; ~ **de selenio** COMPON, ELECTROTEC selenium rectifier; ~ **trifásico** COMPON, ELECTROTEC three-phase rectifier

rectificar *vt* GEN rectify

recuantificación *f* INFO requantization

recubrimiento: ~ **de faceta** *m* TRANS *óptica de fibras* facet coating; ~ **metálico** *m* TRANS metallic coating

recuento: ~ **de justificación de puntero** *m* DATOS pointer justification count (*PJC*); ~ **de ruido impulsivo** *m* TELEFON impulse noise count

recuperación *f* GEN restoration, recovery, retrieval; ~ **de archivos** DATOS file recovery; ~ **después de un desastre** DATOS disaster recovery; ~ **de diálogo** DATOS dialog recovery; ~ **de errores** DATOS error recovery; ~ **en hilera** DATOS in-line recovery; ~ **de mensajes** DATOS message retrieval; ~ **del ritmo** TELEFON, TRANS timing recovery; ~ **del sistema** DATOS system recovery; ~ **de la transacción** DATOS transaction recovery

recuperador *m* (*RG*) CONMUT, ELECTRÓN, SEÑAL, TELEFON regenerator (*RG*)

recuperar *vt* GEN recover, retrieve, recuperate

recurrencia *f* DATOS, INFO recursiveness

recursión *f* DATOS, INFO recursion

recursivo *adj* DATOS, INFO recursive

recurso *m* GEN resource; ~ **compartido** DATOS, INFO shared resource; ~ **local** DATOS local resource; ~ **OSI** DATOS, TELEFON, TRANS OSI resource; ~ **de programación** DATOS, INFO software resource

recursos: ~ **espectrales** *m pl* TEC RAD spectral resources, spectrum resources; ~ **de frecuencia** *m pl* TEC RAD frequency resources; ~ **de la red** *m pl* TELEFON network resources

red *f* GEN network; ~ **de abonados** TRANS customer network; ~ **de acceso** GEN access network; ~ **de acceso libre** TELEFON *de sincronización mutua* democratic network; ~ **de adaptación** ELECTRÓN matching network; ~ **aérea** ELECTROTEC, TRANS overhead network; ~ **alámbrica** TEC RAD, TELEFON wired network, TRANS hard-wired network; ~ **en anillo** DATOS, INFO, TRANS ring network, token ring; ~ **de antenas de radiación regresiva** TEC RAD backfire antenna network; ~ **en árbol** TRANS tree network; ~ **de área extendida** (*WAN*) DATOS, INFO, TELEFON, TRANS wide-area network (*WAN*); ~ **de área local** DATOS, INFO, TELEFON, TRANS local area network (*LAN*); ~ **de área metropolitana** DATOS, INFO, TELEFON, TRANS metropolitan area network (*MAN*); ~ **de bloqueo** CONMUT, TTRANSMIS blocking network; ~ **de CA** INFO AC mains; ~ **de cables** TRANS, TV cable network; ~ **cerrada de distribución de datos al usuario** TRANS closed user data distribution network; ~ **de clientes** TRANS client network; ~ **Clos** TELEFON, TRANS Clos network; ~ **combinatoria** TRANS combining network; ~ **de comercialización** GEN dealership network; ~ **compensadora** TEC RAD compensating network, TRANS weighting network; ~ **de computadoras** *AmL* (*cf red de ordenadores Esp*) INFO, TRANS computer network; ~ **de comunicación de datos** DATOS data communication network, TRANS data-communication network; ~ **de comunicaciones urbanas** TRANS local communications network; ~ **de comunicación personal** (*PCN*) DATOS, TEC RAD personal communication network (*PCN*); ~ **comunitaria** TELEFON, TV community network; ~ **de concentración** TRANS concentration network; ~ **concentradora del haz** TEC RAD beam-forming network; ~ **de conmutación** CONMUT, TELEFON switching network; ~ **de conmutación de abonados** CONMUT, TELEFON subscriber switching network; ~ **de conmutación de banda ancha** CONMUT, TRANS broadband switching network, wideband switching network; ~ **de conmutación de banda estrecha** CONMUT, TRANS narrow-band switching network; ~ **de conmutación de circuitos** TRANS circuit-switched network, circuit-switching network; ~ **de conmutación de paquetes** TRANS packet-switch network, packet-switching network; ~ **de conmutación de tráfico** TRANS message-switching network; ~ **conmutada pública** (*PSN*) DATOS, TELEFON, TRANS public switched network (*PSN*); ~ **de coordinación y control de sistema** (*SCCN*) PRUEBA system coordination and control network (*SCCN*); ~ **de datos** DATOS, TRANS data network; ~ **de datos compartidos** (*SDN*) DATOS shared-data network (*SDN*); ~ **de datos internacional de conmutación de paquetes** TRANS international packet-switched data network; ~ **de desacentuación** ELECTRÓN de-emphasis network; ~ **de destino** TRANS destination network; ~ **digital integrada** TELEFON, TRANS integrated digital network; ~ **digital de servicios integrados** (*RDSI*) TELEFON, TRANS integrated services digital network (*ISDN*); ~ **de diodos** ELECTRÓN diode network; ~ **de distribución** CONMUT, TRANS, TV distribution network; ~ **de distribución local** ELECTRÓN, INFO, TELEFON, TRANS local distribution network; ~ **de dos bornes** ELECTRÓN, ELECTROTEC, TRANS two-terminal network; ~ **eléctrica** ELECTROTEC mains; ~ **de empresa** TRANS company network, enterprise network; ~ **de enlace** INFO link network, TELEFON, TRANS junction network; ~ **de entrega** TV delivery network; ~ **espacio-tiempo-espacio** CONMUT, TRANS space-time-space network; ~ **en estrella** ELECTROTEC, TRANS, TV star network; ~ **de expansión** TRANS expansion network; ~ **experimental** TEC RAD experimental network; ~ **de fibra óptica** TRANS fibre-optic (*AmE* fiber-optic) network; ~ **de ficha circulante** DATOS, INFO, TRANS token ring network; ~ **de formación de impulsos** (*PFN*) ELECTRÓN pulse-forming network (*PFN*); ~ **de formación de señal** TELEFON signal-shaping network; ~ **formadora del haz** ELECTRÓN beam-forming network; ~ **de gestión** TRANS management network; ~ **de gestión de jerarquía digital síncrona** TRANS SDH management network (*SMN*); ~ **de gestión de telecomunicaciones** DATOS, TELEFON, TRANS telecommunications management network (*TMN*); ~ **independiente** TRANS independent network; ~ **de instalación del abonado** TRANS subscriber premises network (*SPN*); ~ **integrada** TRANS integrated network; ~ **integrada de transmisión y conmutación** TRANS integrated transmission and switching network; ~ **inteligente** (*IN*) APLIC, TRANS intelligent network (*IN*); ~ **de interconexión** ELECTROTEC, TRANS interconnection network; ~ **de interconexión de procesador múltiple** DATOS multiprocessor interconnection network; ~ **interna** TRANS internetwork; ~ **internacional** TRANS international network; ~ **IST** TRANS IST network; ~ **jerárquica** TELEFON *mutuamente sincronizado* hierarchical network; ~ **en L** ELECTROTEC L network; ~ **de larga distancia** TELEFON long-distance network; ~ **de líneas** TELEFON line network; ~ **de mallas** TRANS meshed network; ~ **metropolitana** TELEFON, TRANS city network, metropolitan network; ~ **de microceldas** TEC RAD microcell network; ~ **de modulación de impulsos en amplitud** TELEFON, TRANS pulse amplitude modulation network (*PAM network*); ~ **móvil de transmisión de datos** TEC RAD mobile data transmission network; ~ **multietapa** ELECTRÓN multistage network; ~ **multihaz de elementos en fase** (*MPA*) TEC RAD multibeam phased array (*MPA*); ~ **multiservicio** TRANS multiservice network; ~ **mundial** TELEFON worldwide network; ~ **nacional** DATOS domestic network; ~ **neural** INFO, TRANS neural network; ~ **no bloqueante de reordenamiento** TRANS rearrangeable non-blocking network; ~ **de no bloqueo** CONMUT, TRANS nonblocking network; ~ **no conductora** ELECTRÓN nonconductive pattern; ~ **oligárquica** TELEFON *sincronizada* oligarchic network; ~ **óptica pasiva** TELEFON, TRANS passive optical network (*PON*); ~ **de ordenadores** *Esp* (*cf red de computadoras AmL*) INFO, TRANS computer network; ~ **pasiva** ELECTROTEC passive network; ~ **pasiva de cuatro terminales** ELECTRÓN, ELECTROTEC, TRANS passive four-terminal network; ~ **de pequeñas celdas** TEC RAD small-cell network; ~ **plegada** TRANS folded network; ~ **poligonal** TELEFON polygon network; ~ **de preacentuación** ELECTRÓN pre-emphasis network; ~ **primaria** ELECTRÓN, TRANS backbone; ~ **principal** INFO, TRANS backbone network; ~ **privada virtual** INFO, TELEFON virtual private network; ~ **pública** TEC RAD, TRANS public network; ~ **pública de datos** (*RPD*) DATOS, TELEFON, TRANS public data network (*PDN*); ~ **pública de datos de paquetes conmutados** (*PPSDN*) DATOS public packet-switched data network (*PPSDN*); ~ **de puertas** ELECTRÓN *elementos lógicos* gate network; ~ **de puesta a tierra** ELECTROTEC earthing *AmE*, grounding *BrE*; ~ **de puntos nodales** INFO nodes network; ~ **de radiación transversal** TEC RAD broadside array; ~ **de radiodifusión** TEC RAD broadcast network; ~ **de radio enlazada** TEC RAD trunked radio network; ~ **de radio de sondeo** TEC RAD polling radio network; ~ **ramificada** DATOS, INFO, TRANS tree and branch network; ~ **de referencia** DATOS key net; ~ **reguladora** ELECTRÓN equalizing network; ~ **de rejilla** TRANS grid network; ~ **de retransmisión** TEC RAD relay network; ~ **de retropropagación** DATOS back-propagation network; ~ **rural** TELEFON, TRANS rural network; ~ **de ruta estrecha** TRANS thin-route network; ~ **de satélites en modo paquetes** TEC RAD, TELEFON packet satellite network; ~ **satélite de reutilización de frecuencias** TEC RAD, TELEFON frequency re-use satellite network; ~ **de seguridad** TRANS security network; ~ **de seguridad a distancia** TRANS remote security network; ~ **de selectores sin congestión** CONMUT, TRANS nonblocking switching network; ~ **semántica** DATOS, INFO semantic network; ~ **semiasociada** CONMUT semiassociated network; ~ **de señalización** CONMUT, SEÑAL, TRANS signalling (*AmE* signaling) network; ~ **de servicio público móvil** TEC RAD, TRANS public mobile network; ~ **de sincronización** TRANS synchronization network; ~ **de sincronización despótica** TELEFON despotic network; ~ **de sincronización mutua** TELEFON mutually synchronized network; ~ **de sistema de espera** TELEFON queueing network; ~ **STS** CONMUT, TRANS STS network; ~ **sub-Clos** TELEFON, TRANS sub-Clos network; ~ **en T** ELECTROTEC T network; ~ **de telecomunicaciones** GEN telecommunications network; ~ **de telecomunicaciones comerciales por satélite** TELEFON, TRANS satellite business telecommunications network; ~ **telefónica** TELEFON, TRANS telephone network, voice network; ~ **telefónica alámbrica** TELEFON wired telephone network; ~ **telefónica conmutada** CONMUT, DATOS, TELEFON, TRANS public switched telephone network (*PSTN*); ~ **telefónica de larga distancia** TELEFON long-distance telephone network; ~ **telefónica local** TELEFON, TRANS local telephone network; ~ **telefónica pública** TELEFON, TRANS public telephone network; ~ **telefónica tradicional** TELEFON, TRANS traditional telephone network; ~ **telegráfica** TELEFON, TRANS telegraph network; ~ **telegráfica pública** TELEFON, TRANS public telegraph network; ~ **de teleproceso** INFO, TRANS teleprocessing network; ~ **de televisión por cable** TV cable television network; ~ **de tierra** ELECTROTEC earthing network *BrE*, grounding network *AmE*; ~ **de transferencia de datos** DATOS, TRANS data-transport network; ~ **de transmisión** TRANS transmission network; ~ **de transmisión de paquetes de datos** DATOS, TRANS packet data transmission network; ~ **de transmisión de voz y datos** DATOS, TRANS speech-data network; ~ **transoceánica** TRANS overseas network; ~ **de transporte** TELEFON transport network; ~ **de transporte de datos** DATOS, TRANS data-transport network; ~ **TS** TELEFON TS network; ~ **UHF** TV UHF network; ~ **urbana** TELEFON, TRANS urban network; ~ **de valor añadido** DATOS, TELEFON, TRANS value-added network (*VAN*); ~ **VHF** TV VHF network; ~ **por vía satélite** TEC RAD, TV satellite network; ~ **de videocomunicación** TV videocommunication network; ~ **VSAT** DATOS, TRANS VSAT network

Red: la ~ *f* DATOS, INFO the Net; ~ **Europea de Investigación Académica** *f* TRANS European Academic Research Network

redes: ~ **de establecimiento de conexión** *f pl* TRANS connection building networks

redimensionamiento *m* INFO resizing

redimensionar *vt* INFO resize

redirección: ~ **y deflexión de llamadas de red interna** *f* (*ICRD*) TRANS internetwork call redirection and deflection (*ICRD*)

redireccionador *m* DATOS redirector

redireccionar: ~ **el tráfico** *vt* CONMUT, TELEFON redirect traffic

rediscar *vt* TELEFON redial

rediseñar *vt* GEN redesign

redisponer *vt* GEN rearrange

redisposición *f* GEN rearrangement; ~ **de evento simple** (*SEU*) PRUEBA single event upset (*SEU*)

redistribución: ~ **de llamadas internas** *f* TELEFON in-call rearrangement

reducción *f* INFO downsizing, zoom-out; ~ **de datos** DATOS, INFO data reduction; ~ **de escala** PRUEBA scaling down; ~ **de la gama de luminancia** TELEFON compression of the luminance range; ~ **de tarifa** APLIC reduction of tariff

reducir *vt* INFO downsize

reductor *m* GEN *atenuador* pad; ~ **de ruido** TRANS noise reducer

redundancia *f* GEN redundancy; ~ **de llamada** TEC RAD, TRANS ring redundancy; ~ **pasiva** PRUEBA, TELEFON stand-by redundancy; ~ **relativa** GEN relative redundancy

reembolsar *vt* GEN refund, reimburse

reembolso *m* GEN refund, reimbursement

reemitir *vi* SEÑAL outpulse

reemplazar: ~ **el receptor** *vt* TELEFON replace the receiver; ~ **regularmente** *vt* PRUEBA replace regularly

reemplazo *m* GEN *direcciones* displacement

reencaminamiento *m* CONMUT, DATOS, TELEFON *del tráfico* rerouting; ~ **controlado** TELEFON controlled rerouting; ~ **forzado** TELEFON forced rerouting

reencaminar *vt* CONMUT, DATOS, TELEFON *tráfico* reroute

reencendido *m* DATOS, ELECTRÓN, TEC RAD reignition

reensamblaje *m* TELEFON reassembling

reentrada *f* TELEFON re-entry

reentrante *adj* INFO re-entrant

reescribir *vt* DATOS rewrite
reescritura *f* DATOS rewrite
reevaluar *vt* GEN remeasure
reexpedir *vt* CONMUT, TEC RAD, TELEFON redirect
referencia *f* GEN reference, mark, legend, benchmark, TEC RAD referencing; ~ **de aplicación** GEN application reference; ~ **de comando** TELEFON command reference; ~ **del control de intercambios** DATOS interchange control reference; ~ **cruzada** GEN *documentación* cross-reference; ~ **por defecto** DATOS default reference; ~ **local** GEN local reference; ~ **de posición de vuelo** TEC RAD attitude reference; ~ **del recibidor** DATOS recipient reference; ~ **de restricciones** DATOS constraints reference; ~ **de sesión ampliada** TELEFON expanded session reference; ~ **de sesión básica** TELEFON basic session reference
referenciador: ~ **gramatical** *m* INFO grammatical tagger
refinación *f* DATOS, INFO refinement
refinar *vt* DATOS, INFO refine
reflectancia *f* ELECTRÓN, TEC RAD, TRANS, TV reflectance; ~ **remanente** TRANS residual reflectance
reflectometría: ~ **óptica en el dominio del tiempo** *f* TRANS optical time domain reflectometry
reflectómetro *m* ELECTRÓN, ELECTROTEC, PRUEBA, TRANS reflectometer; ~ **de dominio frecuencial** ELECTRÓN, ELECTROTEC, PRUEBA, TRANS frequency-domain reflectometer; ~ **óptico en el dominio del tiempo** ELECTRÓN, ELECTROTEC, PRUEBA, TRANS optical time domain reflectometer
reflector *m* GEN reflector; ~ **angular** TEC RAD corner reflector; ~ **lambertiano** TRANS *óptica de fibras* Lambertian reflector; ~ **parabólico inverso** TEC RAD back dish; ~ **poligonal** TEC RAD mesh reflector; ~ **de radar** TEC RAD radar reflector; ~ **de varilla** TEC RAD rod reflector
reflejar *vt* GEN reflect, INFO mirror
reflexión *f* GEN reflection; ~ **difusa** TRANS diffuse reflection; ~ **especular** TRANS specular reflection; ~ **de Fresnel** TRANS Fresnel reflection; ~ **interna total** ELECTRÓN, TRANS *óptica de fibras* total internal reflection; ~ **ionosférica** TEC RAD ionospheric reflection; ~ **M** TEC RAD M reflection; ~ **total** ELECTRÓN, TRANS total reflection; ~ **troposférica** TEC RAD tropospheric reflection
reflexiones: ~ **con trayectorias múltiples** *f pl* TEC RAD multipath reflections
reflexivo *adj* DATOS, INFO reflexive
reforzamiento *m* GEN reinforcement
reforzar *vt* GEN reinforce, ruggedize
refracción *f* TEC RAD, TRANS, TV refraction; ~ **atmosférica** TEC RAD atmospheric refraction; ~ **normal** TEC RAD standard refraction; ~ **óptica** TRANS optical refraction
refractividad *f* TEC RAD, TRANS, TV refractivity
refractómetro: ~ **de mano** *m* TRANS hand-held refractometer
refrigeración *f* GEN cooling; ~ **por agua** GEN water cooling; ~ **independiente** GEN separate cooling
refrigerante *m* GEN refrigerant
refrigerar *vt* GEN refrigerate
refuerzo *m* GEN enhancement; ~ **de la imagen** TV image enhancement
regeneración *f* GEN regeneration; ~ **de imagen** TV image refreshing; ~ **de impulsos** GEN pulse regeneration; ~ **de onda luminosa** DATOS light-wave regeneration; ~ **de la señal** SEÑAL signal regeneration
regenerador: ~ **de impulso de sincronización** *m* TV synchronizing pulse regenerator; ~ **de impulsos** *m* SEÑAL, TELEFON pulse regenerator
regenerar *vt* GEN regenerate
regenerativo: **no** ~ *adj* TRANS nonregenerative
régimen: ~ **de carga de aparatos** *m* TEC RAD apparatus charge rate; ~ **de exploración** *m* PERIF scanning rate; ~ **de señalización de datos** *m* DATOS, INFO, TELEFON, TRANS data signalling (*AmE* signaling) rate; ~ **de sonoridad objetiva de efecto local** *m* (*SOLR*) TELEFON side-tone objective loudness rating (*SOLR*)
región *f* GEN region; ~ **del alma** TRANS core area; ~ **de arrastre** INFO drag region; ~ **de campo lejano** TEC RAD, TRANS far-field region; ~ **de campo próximo** TEC RAD, TRANS near-field region; ~ **de carga espacial** COMPON, ELECTRÓN space-charge region; ~ **del colector** COMPON drain region, ELECTRÓN *transistores* collector region; ~ **de drenaje** COMPON drain region; ~ **F** TEC RAD F-region; ~ **de Fresnel** TEC RAD, TRANS Fresnel region; ~ **Geiger** TEC RAD, TRANS Geiger region; ~ **de origen** COMPON source region; ~ **de proporcionalidad limitada** TEC RAD region of limited proportionality; ~ **de puerta** ELECTRÓN *transistores* gate region
registrable *adj* INFO, TELEFON recordable
registrado *adj* TEC RAD *radiodifusión* recorded, TV in-register
registrador *m* APLIC recorder, CONMUT, DATOS register, INFO logger, PERIF recorder, TEC RAD register, TELEFON register, transcriber; ~ **de curvas X-Y** PRUEBA X-Y recorder; ~ **de datos**

DATOS data logger; ~ **de dos etapas** TELEFON two-state register; ~ **de impulsos** TELEFON, TRANS pulse recorder; ~ **ionosférico de incidencia oblicua** TEC RAD oblique-incidence ionospheric recorder; ~ **múltiple** PERIF multiple recorder; ~ **de nivel** PRUEBA, TRANS level recorder; ~ **de secuencia** INFO sequence recorder; ~ **de señal y estado** CONMUT signal and state recorder; ~ **traductor** CONMUT register translator; ~ **de tráfico** TELEFON traffic recorder

registrar *vt* GEN record; ◆ ~ **en un diario** GEN enter in a journal

registro *m* DATOS, ELECTRÓN registration, INFO logging, TEC RAD, TV registration; ~ **del abonado** CONMUT subscriber's store; ~ **de alineamiento** TRANS line-up record; ~ **de almacenamiento** TELEFON storage register; ~ **analógico** GEN analog recording; ~ **de bitio fuera de texto** (*ZBR*) INFO zone bit recording (*ZBR*); ~ **de cola** DATOS trailer record; ~ **de comparación de resultados** GEN result-comparing register; ~ **de conformidad** DATOS conformance log; ~ **de control** INFO control register; ~ **de corrección de impulsos** SEÑAL pulse-correcting register; ~ **cronológico de transacciones** DATOS transaction logging; ~ **de daño en diario** DATOS log-damage record; ~ **de desplazamiento de realimentación** CONMUT, DATOS, ELECTRÓN feedback shift register; ~ **en diario** DATOS log record; ~ **en diario de operaciones potencial** DATOS potential log record; ~ **de dirección de instrucción** INFO control instruction register, instruction address register; ~ **directo** INFO home record; ~ **de doble longitud** DATOS double-length register; ~ **electrónico** PERIF *contestador* memo recording; ~ **de extensión doble** INFO double-length register; ~ **de frecuencia** TEC RAD frequency record; ~ **heurístico en diario** DATOS log-heuristic record; ~ **de imágenes por teclado** INFO keyframing; ~ **impreso de duración y coste del servicio de llamadas** APLIC, TELEFON printed record of duration and charge of calls service; ~ **incremental** TV incremental recording; ~ **inmediato** INFO at-once recording; ~ **inmediato en el libro de operaciones** DATOS log-ready record; ~ **de instrucción** INFO instruction register; ~ **de llamadas de entrada** TELEFON registration of incoming calls; ~ **local** CONMUT local register; ~ **de localización** TEC RAD location register, location registration; ~ **de localizaciones móviles** TEC RAD mobile location registration; ~ **de longitud cuádruple** DATOS, INFO quadruple-length register; ~ **de memoria** DATOS, INFO memory storage; ~ **del mensaje** DATOS message register; ~ **de mensaje EDI** TELEFON EDI message store; ~ **de operaciones** DATOS, INFO, TEC RAD, TELEFON log; ~ **de pasos** INFO step recording; ~ **de perturbaciones** TELEFON disturbance recording; ~ **de posición inicial** (*HLR*) TEC RAD home location register (*HLR*); ~ **de posición visitada** TEC RAD visited location register; ~ **de retención** ELECTRÓN holding register; ~ **sin retorno a referencia** DATOS nonreturn-to-reference recording; ~ **del sistema** GEN system record; ~ **de sobregrabación** INFO overdub recording; ~ **sonoro** INFO, TEC RAD, TV sound recording; ~ **de traducción** TELEFON translation register; ~ **de tráfico** TELEFON traffic record, traffic recording; ~ **de traslado a diario** DATOS log-commit record; ~ **de vídeo** INFO, TV video recording; ~ **de la voz** TELEFON speech recording

regla: ~ **de acero** *f* PRUEBA iron rule

reglaje: ~ **de precisión** *m* GEN fine setting

reglas: ~ **de la documentación** *f pl* GEN documentation rules

regleta: ~ **de bifurcación** *f* COMPON, ELECTRÓN *dispositivos de conexión* branching block; ~ **de conexiones** *f* COMPON terminal strip, ELECTRÓN connection strip, tag strip; ~ **distribuidora** *f* ELECTROTEC fanning strip; ~ **de guíahilos** *f* COMPON wire guide strip; ~ **de indicadores visuales** *f* INFO, PRUEBA, TELEFON visual indicator strip; ~ **de lámparas** *f* ELECTROTEC lamp strip; ~ **mural de terminales** *f* ELECTROTEC wall terminal strip; ~ **de rotulación** *f* GEN *marcaje* designation strip, label strip

regrabable *adj* DATOS, INFO rerecordable

regrabar *vt* DATOS, INFO rerecord

regulable: ~ **en altura** *adj* GEN vertically adjustable

regulación *f* GEN regulation, INFO, TELEFON setting; ~ **automática de la atenuación** TRANS automatic attenuation regulation; ~ **de caudal** CONTROL, DATOS, TELEFON flow control; ~ **con control anticipante** CONTROL feedforward control; ~ **por defecto** INFO default setting; ~ **de fase** ELECTRÓN phase regulation; ~ **de frecuencia** ELECTROTEC, TRANS frequency regulation; ~ **de luminosidad** PERIF brightness control; ~ **de precisión** GEN fine positioning, fine regulation; ~ **de temperatura** CLIMAT temperature control; ~ **de tensión** ELECTRÓN voltage regulation; ~ **a valor constante** CONTROL control with fixed set-point; ~ **de voltaje** ELECTROTEC voltage regulation; ◆ **de ~ continua** GEN continuously variable

regulador *m* CONTROL, ELECTRÓN, INFO regulator; ~ **de batería** ELECTROTEC battery regulator; ~ **de gasto** CONTROL flow regulator; ~ **de velocidad** TELEFON speed regulator; ~ **de voltaje** ELECTRÓN, ELECTROTEC voltage regulator

regular *vt* GEN adjust, regulate

reimportar *vt* DATOS reimport

reiniciación *f* INFO reinitialization, reboot

reiniciar *vt* GEN *sistema* restart, INFO reboot, reinitialize

reintento *m* TELEFON reattempt; ~ **de llamada** TELEFON repeated call attempt

reinyección: ~ **de configuraciones** *f* ELECTRÓN configurations foldback

rejilla *f* GEN grille (*AmE* grill); ~ **de difracción** CLIMAT, TRANS *óptica* diffraction grating; ~ **de gráficos** INFO graphics screen; ~ **de pantalla** ELECTRÓN screen grid; ~ **de protección** GEN protective grating, protective grille (*AmE* grill); ~ **protectora** GEN protective grating, protective grille (*AmE* grill); ~ **supresora** COMPON, ELECTRÓN suppressor grid

relación *f* GEN ratio; ~ **abierta de datos digitales** DATOS unrestricted digital-data service; ~ **de ajuste** COMPON setting ratio; ~ **de aspecto** TEC RAD, TELEFON aspect ratio; ~ **de ataque** INFO attack rate; ~ **de compresión** TELEFON ratio of compression, TRANS compression ratio; ~ **de contraste** TV contrast ratio; ~ **de controles de acceso** DATOS, PRUEBA access control list; ~ **cresta-promedio** (*PAR*) GEN peak-to-average ratio (*PAR*); ~ **de datos de banda de voz** DATOS voice-band data ratio; ~ **de densidad onda portadora-ruido** TRANS carrier-to-noise density ratio; ~ **día-hora cargada** TELEFON day-to-busy-hour ratio; ~ **entre el diámetro del alma y el espesor del aislamiento** TRANS core-cladding ratio; ~ **de expansión** TELEFON ratio of expansion; ~ **fuente/dispositivo consumidor** TELEFON source/sink relationship; ~ **hombre-máquina** INFO man-machine relationship; ~ **de impulsividad** TEC RAD impulsiveness ratio; ~ **impulso/pausa** TELEFON pulse-no-pulse ratio; ~ **de incidentes** TELEFON trouble report; ~ **de interferencia de la onda portadora** TRANS carrier-interference ratio; ~ **intrínseca de cresta** ELECTROTEC intrinsic standoff ratio; ~ **de justificación** TELEFON, TRANS justification ratio; ~ **lógica de par a par** TRANS peer-to-peer logical relationship; ~ **onda portadora-densidad del ruido intermodulación** TRANS carrier-to-intermodulation noise density ratio; ~ **de ondas estacionarias de tensión** (*VSWR*) TEC RAD, TRANS voltage standing wave ratio (*VSWR*); ~ **portadora/interferencia** TRANS carrier-to-interference ratio (*C/I*); ~ **portadora/potencia de ruido de intermodulación** TRANS carrier-to-intermodulation noise power ratio; ~ **portadora/potencia de ruido térmico** TRANS carrier-to-thermal noise power ratio; ~ **portadora/potencia total de ruido** TRANS carrier-to-total-noise-power ratio; ~ **portadora/ruido térmico del receptor** TRANS carrier-to-receiver thermal noise ratio; ~ **de protección de videofrecuencia** TEC RAD, TV videofrequency protection ratio; ~ **de radiación anterior-posterior** TEC RAD front-to-back ratio; ~ **de reposición** COMPON resetting ratio; ~ **de ritmo** TRANS timing relationship; ~ **de ruido señal/rms ponderado** TRANS signal-to-weighted rms noise ratio; ~ **señal/diafonía** TRANS signal-to-crosstalk ratio; ~ **señal/interferencia** TEC RAD signal-interference ratio; ~ **señal/ruido** ELECTRÓN, PRUEBA, SEÑAL, TRANS signal-to-noise ratio (*SNR*); ~ **señal/ruido normalizada** TEC RAD normalized signal-to-noise ratio; ~ **de señalización** SEÑAL signalling (*AmE* signaling) relation; ~ **señal de videofrecuencia/interferencia** TEC RAD, TV videofrequency signal-to-interference ratio; ~ **de servicio** DATOS service relationship; ~ **SINAD** GEN SINAD ratio; ~ **de télex** TELEFON telex relation; ~ **de terminación** TELEFON completion ratio; ~ **de tráfico** TELEFON traffic relation; ~ **de transformación** COMPON ratio of transformation; ◆ **sin ~ con la distancia** GEN unrelated to distance

relaciones: ~ **entre pares** *f pl* DATOS peer relationships; ~ **de grupo** *f pl* DATOS group relationships; ~ **de reserva** *f* DATOS backup relationships

relé *m* GEN relay; ~ **accionado por la voz** COMPON, CONMUT voice-operated relay; ~ **de acción escalonada** COMPON stepping relay; ~ **de apertura** COMPON, ELECTRÓN break relay; ~ **de auxilio** TEC RAD Mayday relay; ~ **de avance** COMPON notching relay; ~ **base-móvil** TEC RAD base-to-mobile relay; ~ **bisector de impulsos** COMPON pulse-bisecting relay; ~ **de bloque** DATOS frame relay; ~ **de bloqueo** ELECTROTEC latching relay; ~ **de capa de red** TRANS network-layer relay (*NLR*); ~ **de celda** TRANS cell relay; ~ **celular** TEC RAD cell relay; ~ **de circuito cerrado** ELECTROTEC closed-circuit relay; ~ **de comparación de fase** COMPON phase comparator relay; ~ **de conmutación de antena** TEC RAD antenna switching relay; ~ **con contactos humedecidos en mercurio** ELECTRÓN mercury-wetted contact relay; ~ **de contactos de mercurio** ELECTRÓN mercury contact relay; ~ **de corte de señal de llamada** TELEFON ring-trip relay; ~ **de decena** COMPON, TELEFON tens relay; ~ **en derivación** COMPON shunt relay; ~ **direccional** COMPON directional relay; ~ **de distribución** COMPON distribution relay; ~ **divisor de impulsos** COMPON bisecting relay; ~ **de doble armadura** COMPON double-armature relay; ~ **de dos pasos** COMPON two-stage relay; ~ **de dos tiempos** COMPON two-step relay; ~ **del emisor** COMPON, TRANS transmitter relay; ~ **encapsulado** COMPON encased relay; ~ **enchufable** COMPON plug-in relay; ~ **humedecido en mercurio** ELECTRÓN mercury-wetted relay; ~ **de impulsos** COMPON pulse relay; ~ **del indicador de señal** PRUEBA, SEÑAL signal indicator relay; ~ **de inducción** ELECTROTEC induction relay; ~ **de láminas** COMPON, CONMUT, ELECTROTEC reed relay; ~ **de láminas resonantes** ELECTROTEC resonant-reed relay; ~ **de mercurio** ELECTRÓN mercury relay; ~ **de mínima** COMPON undercurrent relay, under-power relay, undervoltage relay; ~ **de muelle de hilo** CONMUT wire-spring relay; ~ **N** DATOS N-relay; ~ **nominal** TEC RAD home base, home base station; ~ **no polarizado** COMPON, ELECTROTEC nonpolarized relay; ~ **de ocupación** COMPON, TELEFON seizure relay; ~ **polarizado** COMPON, ELECTROTEC polarized relay; ~ **de porcentaje** COMPON percentage relay; ~ **de potencia** COMPON power relay; ~ **de protección** COMPON protective relay; ~ **de radar** TEC RAD radar relay; ~ **de protección diferencial** COMPON protection relay; ~ **de reactancia** COMPON reactance relay; ~ **de recepción de impulsos** COMPON pulse-receiving relay; ~ **de la red** DATOS network relay; ~ **de regulación** COMPON regulation relay; ~ **de remanencia** ELECTROTEC remanent relay; ~ **de reposición** COMPON resetting relay, restoring relay; ~ **de retorno** COMPON turning relay; ~ **de ruta** TELEFON route relay; ~ **de sobrecargas** COMPON overpower relay; ~ **de sobrecorriente** COMPON overpower relay; ~ **de sobretensión** COMPON overvoltage relay; ~ **de supervisión** COMPON, TELEFON supervisory relay; ~ **telegráfico** COMPON telegraph relay; ~ **de temporización** COMPON, ELECTRÓN timing relay; ~ **del temporizador** COMPON, ELECTRÓN timer relay; ~ **de tensión** COMPON, ELECTROTEC voltage relay; ~ **de tiempo no especificado** COMPON nonspecified time relay; ~ **de traducción** COMPON, TELEFON translation relay; ~ **de tren de impulsos** COMPON, TELEFON, TRANS pulse-train relay; ~ **de tres bobinas** COMPON three-coil relay; ~ **de velocidad de variación** COMPON rate-of-change relay

relevador: ~ **de radar** *m* TEC RAD radar relay

relevo: ~ **de ficha** *m* DATOS token passing

rellamada: ~ **automática** *f* TELEFON automatic recall; ~ **automática programada** *f* TELEFON automatic timed recall; ~ **a número memorizado** *f* TELEFON saved number recall; ~ **del registrador** *f* TELEFON, TRANS register recall

rellamar *vt* TELEFON call back

rellenado: ~ **con ceros** *m* INFO zeroization, zeroizing

rellenar: ~ **con ceros** *vt* INFO zero-fill

relleno *m* GEN filling; ~ **digital** DATOS digital filling, TRANS digital padding; ~ **del tráfico** CONMUT, DATOS traffic padding

reloj *m* GEN clock; ~ **atómico** GEN atomic clock; ~ **de corrimiento** INFO shift clock; ~ **de estándar terrestre** TRANS terrestrial standard clock; ~ **fechador descodificador** (*DTS*) INFO decoding time stamp (*DTS*); ~ **fechador de presentación** INFO presentation time stamp; ~ **del módem** DATOS modem clock; ~ **parlante** TELEFON *información telefónica de la hora* speaking clock; ~ **recuperado** DATOS recovered clock; ~ **de referencia** TRANS reference clock; ~ **de sincronización** INFO synchronization clock

reluctancia *f* (*R*) ELECTROTEC reluctance (*R*)

remache *m* PRUEBA rivet

remanencia *f* ELECTRÓN, ELECTROTEC remanence

remarcación: ~ **automática** *f* TELEFON automatic redialling

remarcado: ~ **de número memorizado** *m* TELEFON saved number redial

remesa *f* GEN consignment

remiendo *m* ELECTRÓN patching, INFO vamp

remitente *m* GEN originator

remitente/receptor *m* (*O/R*) DATOS, TELEFON originator/recipient (*O/R*)

remolque: ~ **para baterías** *m* ELECTROTEC battery trolley; ~ **portacables** *m* TRANS cable trailer

remontar *vi* INFO *en pantalla* pop up

remoto *adj* ELECTRÓN remote

REMSEVS *abr* (*sistema remoto de voz protegida*) TEC RAD REMSEVS (*remote secure voice system*)
rendimiento *m* DATOS, INFO, TEC RAD throughput (*AmE* thruput); **~ de antena** TEC RAD radiation efficiency; **~ del canal** PRUEBA channel performance; **~ de inyección** ELECTRÓN *semiconductores* injection ratio; **~ de pantalla** ELECTRÓN screen efficiency; **~ de la pastilla** DATOS chip yield; **~ de potencia de detector** COMPON *semiconductores* detector power efficiency; **~ de propagación** TELEFON propagation performance; **~ de tensión de detector** COMPON *semiconductores* detector voltage efficiency; **~ total** DATOS, INFO, TEC RAD throughput (*AmE* thruput)
rendir *vt* INFO render
renominar *vt* INFO rename
renovación *f* GEN renewal, renovation
renovar *vt* GEN renovate
renta *f* GEN rental
rentabilidad *f* GEN profitability
reóstato *m* COMPON, ELECTROTEC rheostat; **~ en derivación** COMPON shunt rheostat
repaginar *vt* INFO repaginate
reparación *f* GEN repair; **~ de avería** GEN fault clearance; **~ por intercambio** TELEFON swap repair
reparar: **~ un fallo** *vt* PRUEBA clear a fault
repartición: **~ del campo en la superficie de apertura** *f* TEC RAD aperture illumination
repartido *adj* GEN distributed
repartidor: **~ de enlace** *m* CONMUT, TRANS junction distribution frame (*JDF*); **~ de grupo secundario** *m* TELEFON, TRANS supergroup distribution frame; **~ de una sola cara** *m* CONMUT single-sided distribution frame
repartir *vt* GEN distribute
repaso *m* GEN overhaul
reperforador *m* TELEFON reperforator, receiving perforator
repertorio *m* INFO, TELEFON directory, repertoire; **~ de instrucciones** INFO instruction repertoire
repetencia *f* TRANS *óptica de fibras* repetency
repetibilidad *f* GEN repeatability
repetición: **~ automática de número** *f* TELEFON number redial, selected number redial; **~ de impulsos** *f* ELECTRÓN, SEÑAL, TELEFON, TRANS pulse repetition; **~ de llamada** *f* INFO, TELEFON recall; **~ de llamada de operador** *f* TELEFON attendant recall; **~ del reglaje** *f* TELEFON retiming; **~ rutinaria** *f* TELEFON routine repetition
repetidor *m* DATOS, ELECTRÓN repeater, transponder, SEÑAL repeater, TEC RAD transponder, TELEFON, TRANS, TV repeater; **~ bidireccional** TRANS two-way repeater; **~ bifilar** TRANS two-wire repeater; **~ de conferencias** TELEFON conference repeater; **~ de cuatro hilos** TRANS four-wire repeater; **~ de enlace de impulsos** (*PLR*) TELEFON pulse link repeater (*PLR*); **~ generador de impulsos** TELEFON regenerative pulse repeater; **~ de impulsos** GEN pulse repeater; **~ de líneas** TELEFON line repeater; **~ multipuerta Ethernet de alambre fino** DATOS, ELECTRÓN thin-wire Ethernet multiport repeater; **~ multipunto** DATOS multipoint repeater; **~ óptico sumergido** TRANS submerged optical repeater; **~ de radiodifusión** TELEFON broadcast repeater; **~ reactivo** TRANS regenerative repeater; **~ sumergido** TRANS submerged repeater; **~ telegráfico** TELEFON telegraph repeater; **~ telegráfico para conferencias** TELEFON conference telegraph repeater; **~ transparente** GEN transparent repeater; ◆ **sin ~** GEN repeaterless
repetir: **~ el reglaje de** *vt* TELEFON retime; **~ la llamada** *vt* (*RAG*) TELEFON ring again (*RAG*)
reponer *vt* COMPON, DATOS, INFO reset
reposición: **~ sin cargo** *f* APLIC replacement free of charge; **~ a cero** *f* ELECTRÓN, PRUEBA zeroing
reposo *m* GEN idleness
representación *f* GEN representation; **~ alfanumérica del abonado** TELEFON subscriber's alphanumerical display; **~ codificada** INFO coded representation; **~ en coma flotante** INFO floating-point representation; **~ esquemática** GEN schematic representation; **~ estructurada en árbol** DATOS, INFO tree-structured view; **~ gráfica** GEN graphic representation; **~ en hileras** GEN semiassembled representation; **~ independiente** GEN separated representation; **~ de límite** INFO boundary representation; **~ de línea única** GEN single-line representation; **~ visual de identificación de la línea que llama** TELEFON calling-line identification display (*CLID*); **~ visual por diodos emisores de luz** INFO LED display; **~ visual de número marcado** GEN dialled-number (*AmE* dialed-number) display
representante: **~ comercial** *m* GEN sales representative; **~ del servicio** *m* GEN service representative
representar: **~ por imágenes** *vt* DATOS, INFO, PERIF, TELEFON, TV image
reproducción *f* GEN reproduction, INFO, PERIF, TEC RAD playback; **~ a distancia** PERIF *de un*

contestador automático remote playback; ~ **de sonido** INFO, TEC RAD, TV sound reproduction; ~ **de vídeo sólo por software** INFO, TV software-only video playback
reproducibilidad *f* GEN reproducibility, INFO playability
reproducible *adj* GEN reproducible
reproducir *vt* GEN reproduce
reproductor *m* GEN player; ~ **de CD-ROM** INFO CD-ROM player; ~ **de cintas** INFO tape reproducer; ~ **de cintas de vídeo** (*VTP*) INFO, TV videotape player (*VTP*); ~ **de discos compactos fotográficos** INFO photo CD player
repuenteable *adj* ELECTROTEC restrappable
requisito *m* INFO, PRUEBA requirement
requisitos: ~ **para la conexión** *m pl* DATOS *Internet* connection requirements; ~ **del diseño** *m pl* GEN design requirements; ~ **de funcionamiento secuencial** *m pl* CONMUT sequence requirements
RER *abr* (*proporción de errores residuales*) DATOS, INFO, TELEFON RER (*residual error rate*)
rerradiación *f* TEC RAD reradiation
resalto: ~ **de fondo** *m* COMPON bottom stud
reseña: ~ **marítima** *f* TELEFON maritime account
reserva *f* GEN booking; ~ **de baterías** PERIF battery backup; ~ **documental** DATOS *Internet* documentary resource
reservado *adj* GEN bespoke
residente *adj* DATOS, INFO resident
residuo *m* GEN residue
resina: ~ **para soldar** *f* COMPON resin flux; ~ **de urea** *f* COMPON urea resin
resincronización *f* ELECTRÓN reclocking, resynchronization
resincronizar *vt* ELECTRÓN reclock, resynchronize
resistencia *f* COMPON tether, ELECTRÓN resistance, tether, INFO resistance; ~ **de aislamiento** ELECTROTEC insulation resistance; ~ **de antena** TEC RAD antenna resistance; ~ **aparente en estado conductor** COMPON, ELECTRÓN on-state slope resistance; ~ **de base** ELECTRÓN *semiconductors* base resistance; ~ **base-emisor** ELECTRÓN *semiconductores* base-emitter resistance; ~ **de base extrínseca** COMPON *semiconductores* extrinsic base resistance; ~ **del bobinado** ELECTROTEC winding resistance; ~-**capacitancia** (*RC*) ELECTRÓN resistance-capacitance (*RC*); ~ **de CC** ELECTROTEC DC resistance; ~ **del colector** ELECTRÓN *semiconductores* collector resistance; ~ **en derivación** COMPON, ELECTROTEC shunt resistance; ~ **al desgarro** COMPON tearing resistance; ~ **de dispersión** ELECTRÓN spreading resistance; ~ **eléctrica** ELECTRÓN electrical resistance; ~ **electródica** ELECTRÓN electrode resistance; ~ **a emisiones perturbadoras** TEC RAD jamming resistance; ~ **del emisor** COMPON *semiconductores* emitter resistance; ~ **de entrada** ELECTRÓN transistor input resistance; ~ **equivalente** ELECTROTEC equivalent resistance; ~ **de la flexibilidad** GEN flexibility strength; ~ **al fuego** COMPON *materiales* fire resistance; ~ **de fuga** ELECTROTEC leakage resistance; ~ **a la iluminación** ELECTRÓN resistance under illumination; ~ **del kevlar** ELECTRÓN kevlar strength; ~ **de línea** TRANS line resistance; ~ **de pérdida** ELECTROTEC loss resistance; ~ **de puerta** ELECTRÓN *tiristores* gate resistance; ~ **de quantum Hall** ELECTRÓN quantum Hall resistance; ~ **a la ruptura** COMPON, PRUEBA breaking strength; ~ **a la ruptura por fluencia** PRUEBA creep rupture stress; ~ **de saturación** ELECTRÓN saturation resistance; ~ **en serie** ELECTRÓN, ELECTROTEC series resistance; ~ **en serie de bucles** ELECTRÓN loop series resistance; ~ **en serie del colector** COMPON, ELECTRÓN *semiconductores* collector series resistance; ~ **en serie del emisor** COMPON, ELECTRÓN *semiconductores* emitter series resistance; ~ **a la tracción** GEN tensile strength; ~ **volúmica** ELECTROTEC volume resistance; ◆ **de ~ fija** ELECTRÓN fixed-resistance
resistividad *f* ELECTRÓN, ELECTROTEC resistivity; ~ **específica** ELECTRÓN, ELECTROTEC specific resistivity; ~ **de volumen** ELECTRÓN, ELECTROTEC volume resistivity
resistor *m* COMPON, ELECTRÓN, ELECTROTEC resistor; ~ **de activación** COMPON pull-up resistor; ~ **de adaptación** *f* COMPON, ELECTRÓN matching resistor; ~ **de ajuste** *f* COMPON, ELECTRÓN trimming resistor; ~ **apagachispas** COMPON, ELECTROTEC spark-quenching resistor; ~ **autorreguladora** *f* ELECTROTEC ballast resistor; ~ **de bajada** COMPON pull-down resistor; ~ **bobinado** ELECTROTEC wire-wound resistor; ~ **de capa delgada** COMPON, ELECTRÓN thin-film resistor; ~ **de capa de difusión** *f* ELECTRÓN *semiconductores* diffused resistor; ~ **de capa gruesa** COMPON, ELECTRÓN thick-film resistor; ~ **de carga anódica** COMPON, ELECTROTEC plate-load resistor; ~ **combinado** COMPON, ELECTRÓN composite resistor; ~ **de compensación** *f* COMPON, ELECTROTEC balancing resistor; ~ **compuesto** ELECTRÓN

composite resistor; ~ **dependiente de la tensión** *f* COMPON, ELECTROTEC voltage-dependent resistor; ~ **de derivación** COMPON, ELECTROTEC shunt resistor; ~ **de desacoplo** *f* COMPON, ELECTROTEC decoupling resistor; ~ **devanado** COMPON wire-wound resistor; ~ **por difusión** ELECTRÓN, ELECTROTEC diffuse resistor; ~ **de drenaje** COMPON, ELECTRÓN, ELECTROTEC bleeding resistor; ~ **de escape** COMPON, ELECTRÓN grid leak resistor; ~ **de implantación iónica** *f* COMPON, ELECTRÓN *semiconductores* ion-implanted resistor; ~ **normal** COMPON, ELECTROTEC standard resistor; ~ **de película de carbón** *f* COMPON, ELECTRÓN carbon-film resistor; ~ **pelicular** COMPON, ELECTROTEC film resistor; ~ **de placa** COMPON, ELECTROTEC plate resistor; ~ **de protección** *f* COMPON, ELECTROTEC protective resistor, protecting resistor; ~ **en serie** COMPON, ELECTRÓN, ELECTROTEC series resistor; ~ **terminal** COMPON, ELECTRÓN terminating resistor

resolución *f* GEN resolution; ~ **de ambigüedad de fase** TRANS phase ambiguity resolution; ~ **angular** TEC RAD angular resolution; ~ **de distancia** TEC RAD range resolution; ~ **de escala** PRUEBA scale discrimination; ~ **espacial** INFO, TELEFON spatial resolution; ~ **de interpolación** PERIF *digitalizador* interpolation resolution; ~ **máxima** TV maximum resolution; ~ **de la nota** INFO note resolution; ~ **de pantalla** INFO display resolution; ~ **de problemas** GEN problem solving; ~ **de velocidad** TEC RAD velocity resolution

resonador *m* GEN resonator; ~ **acústico** GEN acoustic resonator; ~ **de cavidad** ELECTRÓN, TRANS cavity resonator; ~ **Fabry-Pérot** TRANS *óptica de fibras* Fabry-Pérot resonator

resonancia *f* GEN resonance; ~ **ciclotrónica** ELECTRÓN cyclotronic resonance; ~ **paralela** ELECTRÓN, ELECTROTEC parallel resonance; ~ **en serie** ELECTRÓN, ELECTROTEC series resonance

resonar *vi* GEN resonate

resorte: ~ **bimetálico** *m* CONTROL *mecanismo regulador* bimetal spring; ~ **de lámina** *m* COMPON leaf spring; ~ **tensor** *m* GEN tension spring

respondedor *m* DATOS, ELECTRÓN, TELEFON, TEC RAD transponder, responsor; ~ **de asociación** DATOS association responder; ~ **de prueba de abonado** TELEFON *para líneas y aparatos del abonado* subscriber test responder

responsabilidad *f* GEN accountability; ~ **EDI** TELEFON EDI responsibility; ~ **legal por el producto** PRUEBA product liability; ~ **de la línea** APLIC *administración de empresa* line responsibility; ~ **del mensaje por intercambio de datos electrónicos** DATOS, TELEFON EDIM responsibility

responsable: ~ **de adoptar decisiones** *m* TV decision-maker

responsividad *f* ELECTRÓN, TRANS responsivity; ~ **espectral** TRANS spectral responsivity

respuesta *f* GEN response; ~ **de amplitud** ELECTRÓN, SEÑAL, TRANS amplitude response; ~ **de amplitud de frecuencia** TRANS amplitude-frequency response; ~ **asociada A** DATOS, TELEFON A-associate response (*AARE*); ~ **del cable** TRANS cable response; ~ **codificada** TELEFON code answer; ~ **de comparador** ELECTRÓN *impulsos* comparator response; ~ **a continuar para corregir** (*CTR*) DATOS response to continue to correct (*CTR*); ~ **entrecortada** TEC RAD chopped response; ~ **escalonada** CONTROL step response; ~ **espuria** TEC RAD spurious response; ~ **de fase** TRANS phase response; ~ **para finalizar la transmisión** (*ERR*) DATOS response for end of transmission (*ERR*); ~ **de frecuencia** TEC RAD, TRANS frequency response; ~ **a la función escalón** TELEFON step-function response; ~ **a la función en rampa** CONTROL ramp response; ~ **de impulso** TEC RAD, TELEFON pulse response; ~ **de impulso finito** (*RIF*) ELECTRÓN, TRANS finite impulse response (*FIR*); ~ **de impulso infinito** ELECTRÓN, TRANS infinite impulse response (*IIR*); ~ **de impulso unitario** CONTROL unit impulse response; ~ **a un impulso unitario** TELEFON direct pulse response; ~ **a la llamada por tonos** INFO touch response; ~ **mutilada** TEC RAD garbled reply; ~ **con onda rectangular** ELECTRÓN, SEÑAL, TRANS square-wave response; ~ **primitiva** DATOS primitive response; ~ **de teclado** INFO keyboard response; ~ **telefónica** TELEFON voice response; ◆ **sin** ~ TEC RAD on no reply

resquebrajamiento *m* TRANS cleaving

restablecer *vt* GEN restore

restablecimiento *m* GEN restoration; ~ **del enlace de señalización** SEÑAL signalling-link (*AmE* signaling-link) restoration; ~ **del reloj** TRANS clock recovery; ~ **de una señal** SEÑAL signal restoration; ~ **del servicio** GEN restoration of service; ~ **del vínculo de señalización** SEÑAL signalling-link (*AmE* signaling-link) restoration

restauración *f* GEN restore

restitución *f* ELECTRÓN, TELEFON restitution

restricción: ~ **en la dirección salida** *f* TELEFON restriction in the outgoing direction; ~ **de extracción** *f* DATOS fetch restriction; ~ **de identificación de la línea que llama** *f* TELEFON calling-line identification restriction (*CLIR*)

restringir *vt* TEC RAD, TELEFON restrict

resultado *m* GEN outcome, result; ~ **de la facturación** APLIC charging output; ~ **imprevisto** PRUEBA unforeseen outcome; ~ **imprevisto de prueba** PRUEBA unforeseen test outcome; ~ **observado de la prueba** DATOS, PRUEBA observed test outcome; ~ **de una petición** TELEFON request output; ~ **preliminar** PRUEBA preliminary result; ~ **previsto de la prueba** PRUEBA foreseen test outcome

retardador: ~ **de fraguado** *m* COMPON retarder

retardo *m* GEN delay; ~ **de conexión** TV connection delay; ~ **de conexión directa** TELEFON through-connection delay; ~ **de conmutación** CONMUT switching delay; ~ **de desconexión de llamada urbana** CONMUT exchange call-release delay; ~ **de encaminamiento** DATOS *Internet* routing delay; ~ **de envolvente** TEC RAD envelope delay; ~ **estadístico de la ignición** ELECTRÓN statistical delay of ignition; ~ **de fase** GEN phase delay; ~ **de grupo** TRANS group delay; ~ **máximo aceptable de tránsito** (*MATD*) DATOS maximum acceptable transit delay (*MATD*); ~ **de paquete** DATOS packet delay; ~ **posmarcado** TEC RAD, TELEFON post-dialling (*AmE* post-dialing) delay; ~ **de propagación** TEC RAD propagation delay; ~ **en la señal de línea** TELEFON dial tone delay; ~ **en el sonido de una película** INFO picture-sound lag; ~ **en el tono de marcar** TELEFON dial tone delay; ~ **de transferencia de la red** TELEFON network transfer delay; ~ **de tránsito acumulado** (*DTC*) DATOS cumulative transit delay (*CTD*); ~ **de tránsito del blanco** (*TTD*) DATOS target transit delay (*TTD*); ~ **de transmisión** DATOS, TELEFON, TRANS transmission delay; ~ **en la trayectoria eléctrica** TRANS electrical path delay

retén *m* CONMUT pawl

retención *f* ELECTRÓN retention, GEN confinement; ~ **del abonado que llama** TELEFON *contestador* calling-party hold; ~ **AFC** TEC RAD AFC hold-in; ~ **de nivel de negro** INFO, TELEFON, TRANS black-level retention

retener *vt* APLIC *una llamada* book

retentividad *f* ELECTRÓN retentivity

retículo *m* GEN grating

retirada *f* ELECTRÓN, ELECTROTEC removal, INFO withdrawal

retirar *vt* GEN remove, INFO withdraw; ◆ ~ **la cubierta** GEN remove the sheath; ~ **de servicio** TELEFON remove from service

retoque: ~ **de soldadura** *m* COMPON soldering touch-up

retorcer *vt* COMPON twine

retorcido *adj* COMPON, ELECTRÓN twisted

retorno *m* GEN return, TV flyback; ~ **a cero** ELECTRÓN, INFO, TELEFON return to zero; ~ **a la conexión normal** TELEFON changeback; ~ **de eco** DATOS echo return; ~ **inteligente** DATOS intelligent backtracking; ~ **de llamada** TELEFON ring-back; ~ **de mar** TEC RAD sea return; ~ **al servicio** PRUEBA return to service; ~ **de tierra del circuito de señal** SEÑAL, TRANS signal ground *AmE*, signal earth *BrE*; ◆ **sin** ~ **a cero** (*NRZ*) DATOS, TELEFON nonreturn to zero (*NRZ*)

retráctil *adj* DATOS retractable

retransmisión *f* TEC RAD, TELEFON, TV rebroadcasting, retransmission; ~ **por cinta** TELEFON tape relay; ~ **N** TELEFON N-relay

retransmisor *m* TEC RAD, TV rebroadcasting transmitter; ~ **de radiocomunicaciones** ELECTROTEC, TEC RAD radiocommunication relay

retransmitir *vt* TEC RAD, TELEFON, TV rebroadcast, retransmit

retraso *m* ELECTRÓN, INFO lag; ~ **de grupo** TEC RAD group retardation; ~ **de restitución** TELEFON restitution delay; ~ **técnico** PRUEBA technical delay

retroacción *f* GEN feedback

retroalimentación *f* TELEFON message feedback

retroceso *m* ELECTRÓN retrace, INFO backspace (*BS*), retrace, TV retrace; ~ **del estaño** ELECTRÓN *soldaduras* dewetting; ~ **de salida** (*OBO*) TRANS output back-off (*OBO*)

retrodifusión: ~ **indirecta** *f* ELECTROTEC, TEC RAD indirect backscatter

retrodispersión *f* ELECTROTEC, TEC RAD, TELEFON, TRANS backscattering; ~ **de corta distancia** ELECTROTEC, TEC RAD short-distance backscattering; ~ **de larga distancia** ELECTROTEC, TEC RAD long-distance backscattering

retroiluminado *adj* ELECTRÓN backlit

retroseguimiento *m* DATOS backtracking

reubicar *vt* DATOS, INFO, TEC RAD relocate

reutilizable *adj* INFO, TEC RAD reusable

reutilización: ~ **de canal de frecuencia** *f* TEC RAD reuse of frequency channel; ~ **de espectro**

f TEC RAD spectrum reuse; ~ **de frecuencia** *f* TEC RAD frequency reuse
revendedor *m* TEC RAD reseller
reverberación *f* ELECTRÓN, INFO, TELEFON reverberation; ~ **digital** ELECTRÓN, INFO digital reverb
revertir *vi* INFO revert
revestido *adj* COMPON sheathed, coated
revestimiento *m* COMPON sheathing, coating; ~ **adaptado** TRANS matched cladding; ~ **antideslumbrante** GEN anti-glare coating; ~ **antirreflectante** GEN anti-reflective coating; ~ **antirreflector** GEN anti-reflection coating; ~ **de doble capa** COMPON double-layer coating; ~ **interior** COMPON lining; ~ **de sílice** COMPON silica cladding; ~ **de TY** COMPON, ELECTROTEC TY wrap
revestir *vt* COMPON wrap
revisable *adj* DATOS revisable
revisión: ~ **del diseño** *f* PRUEBA design review; ~ **del documento** *f* GEN document revision
revista: ~ **crítica de estudio** *f* GEN critical design review
revólver *m* DATOS revolver
RF *abr* (*radiofrecuencia*) TV RF (*radio frequency*)
RFI *abr* (*interferencia de radiofrecuencia*) TEC RAD RFI (*radio-frequency interference*)
RFP *abr* (*solicitud de propuesta*) GEN RFP (*request for proposal*)
RFS *abr* (*procedimiento de sincronización del marco de recepción*) TRANS RFS (*receive-frame synchronization procedure*)
RG *abr* (*recuperador*) CONMUT, ELECTRÓN, SEÑAL, TELEFON RG (*regenerator*)
RGB *abr* (*rojo-verde-azul*) INFO, TV RGB (*red-green-blue*)
riel: ~ **de guía** *m* DATOS running bar
riesgo *m* GEN hazard, risk; ~ **del cliente** PRUEBA client risk; ~ **del consumidor** PRUEBA consumer risk; ~ **heurístico** DATOS heuristic hazard; ~ **del productor** PRUEBA producer's risk
RIF *abr* (*respuesta de impulso finito*) ELECTRÓN, TRANS FIR (*finite impulse response*)
RIN *abr* (*ruido de intensidad relativa*) ELECTRÓN RIN (*relative-intensity noise*)
riostra *f* GEN brace
RIP *abr* (*protocolo de información de encaminamiento*) DATOS RIP (*routing information protocol*)
riqueza: ~ **higrométrica** *f* TEC RAD mixing ratio
ritmo *m* INFO rhythm; ~ **de ampliación** TELEFON rate of extension; ~ **de anotación** INFO recording tempo; ~ **binario** DATOS *temporización de los bitios*, TRANS bit timing; ~ **de conmutación** PRUEBA switching rate; ~ **por defecto** INFO default tempo; ~ **diurno** GEN diurnal rhythm; ~ **del patrón** TRANS clock rate; ~ **de salto del reloj** TRANS clock-skip rate; ~ **de tambor** INFO drum rhythm
RLE *abr* (*codificación de longitud de tirada*) INFO RLE (*run-length encoding*)
RLRE *abr* (*APDU de liberación-respuesta-A*) DATOS RLRE (*A-release-response APDU*)
RLRQ *abr* (*APDU de liberación-petición-A*) DATOS RLRQ (*A-release-request APDU*)
RLT *abr* (*línea troncal de enlaces de desconexión*) TELEFON RLT (*release-link trunk*)
RN *abr* (*notificación de recepción*) DATOS, TELEFON RN (*receipt notification*)
RO *abr* (*operación remota*) DATOS, PRUEBA RO (*remote operation*)
robustez *f* GEN ruggedness
rodeo *m* TELEFON detour
rodete *m* TELEFON rotor
rodillo: ~ **prensador** *m* PERIF pinch roller
rodiodeposición *f* GEN rhodium plating
ROE *abr* (*coeficiente de onda estacionaria*) TEC RAD, TRANS SWR (*standing-wave ratio*)
roja: ~**-verde-azul** *adj* (*RGB*) INFO, TV red-green-blue (*RGB*)
rol: ~ **de director** *m* DATOS manager role
rollo: ~ **de papel** *m* PERIF reel of paper
ROLR *abr* (*valoración de sonoridad del objetivo de recepción*) TELEFON ROLR (*receive objective loudness rating*)
ROM *abr* (*memoria sólo de lectura*) INFO ROM (*read-only memory*)
romperse *v refl* GEN fail
roña *f* COMPON rust
rosca: ~ **fina** *f* GEN fine thread
rotulación: ~ **serigrafiada** *f* ELECTRÓN silk-screened lettering
rótulo *m* INFO nameplate; ~ **de marca de tictac** INFO tick-mark label
rotura *f* COMPON rupture, GEN breaking; ~ **de cable** TRANS cable break; ~ **de avalancha** ELECTRÓN avalanche breakdown
RPD *abr* (*red pública de datos*) DATOS, TELEFON, TRANS PDN (*public data network*)
RPE *abr* (*equipo periférico remoto*) INFO RPE (*remote peripheral equipment*)
RPF *abr* (*formato de publicación de referencia*) DATOS RPF (*reference publication format*)
RPOA *abr* (*agencia privada autorizada*) TELEFON RPOA (*recognized private operating agency*)
RSA *abr* (*algoritmo de Rivest, Shamir, Adleman*) TELEFON RSA (*Rivest, Shamir, Adleman algorithm*)
RSB *abr* (*banda lateral residual*) TEC RAD, TRANS,

TV RSB (*residual sideband*), VSB (*vestigial sideband*)

RSM *abr* (*control espectral a distancia*) TEC RAD RSM (*remote spectrum-monitoring*)

RST *abr* (*terminación de la sección del regenerador*) DATOS *de impulsos* RST (*regenerator-section termination*)

RTF *abr* (*formato de texto muy elaborado*) INFO RTF (*rich text format*)

RTG *abr* (*generador de temporización del regenerador*) DATOS RTG (*regenerator-timing generator*)

RTMP *abr* (*protocolo de mantenimiento de tabla de encaminamiento*) INFO RTMP (*routing table maintenance protocol*)

RTN *abr* (*negativo de reinstrucción*) DATOS, TELEFON RTN (*retrain negative*)

RTP *abr* (*positivo de reinstrucción*) DATOS, TELEFON RTP (*retrain positive*)

RTSI *abr* (*intercambio de estado de terminal de referencia*) TEC RAD RTSI (*reference terminal status interchange*)

rueda *f* INFO wheel; ~ **accionada por el pulgar** ELECTRÓN thumbwheel; ~ **de cadena** COMPON sprocket; ~ **de impresión** PERIF printwheel, *impresión* type wheel; ~ **motriz** GEN driving wheel

ruido *m* GEN noise; ~ **acústico** DATOS, TELEFON acoustic noise; ~ **aditivo** TELEFON additive noise; ~ **ambiente** GEN ambient noise, room noise; ~ **de ambiente** GEN room noise, ambient noise; ~ **atmosférico** GEN static; ~ **de avalancha** ELECTRÓN *semiconductores* avalanche noise; ~ **de banda ancha** TRANS wideband noise, broadband noise; ~ **de banda estrecha** TRANS narrow-band noise; ~ **básico** INFO, TELEFON basic noise; ~ **blanco** GEN uniform spectrum random noise; ~ **de cebado** ELECTRÓN primer noise; ~ **de circuito** INFO, TELEFON line noise, circuit noise; ~ **de componente alterna** ELECTRÓN hum; ~ **cuántico** ELECTRÓN, TRANS quantum noise; ~ **de cuantificación** ELECTRÓN, TRANS quantization noise, quantizing noise; ~ **errático triangular** TEC RAD triangular random noise; ~ **fluctuante** ELECTRÓN flicker noise; ~ **de fondo** GEN background noise; ~ **de fotones** ELECTRÓN, TRANS photon noise; ~ **de fritura** TELEFON frying; ~ **galáctico** TEC RAD galactic noise; ~ **gaussiano** TELEFON Gaussian noise; ~ **granular** ELECTRÓN, TELEFON granular noise, shot noise; ~ **impulsivo** TELEFON impulse noise; ~ **de intensidad relativa** (*RIN*) ELECTRÓN relative-intensity noise (*RIN*); ~ **de intermodulación** TEC RAD cross-modulation noise, TRANS intermodulation noise; ~ **intrínseco** TRANS intrinsic noise; ~ **iónico** ELECTRÓN ion noise; ~ **de mácula** TRANS speckle noise; ~ **microfónico** ELECTRÓN microphone noise, microphonic noise; ~ **modal** TRANS modal noise; ~ **en modulación de frecuencia** TRANS frequency-modulation noise; ~ **no gaussiano** TELEFON non-Gaussian noise; ~ **de origen humano** ELECTRÓN, TEC RAD man-made noise; ~ **parásito** TEC RAD interference noise; ~ **parásito por precipitación** TEC RAD precipitation static; ~ **de partición** ELECTRÓN partition noise; ~ **de partición de modo** TRANS mode-partition noise; ~ **producido por la señal** TRANS modulation noise; ~ **radiado** ELECTRÓN radiated noise; ~ **residual** INFO residual noise; ~ **retrasado** TEC RAD trailing noise; ~ **rosado** INFO pink noise; ~ **de selectores** CONMUT, TELEFON selector noise; ~ **telegráfico** TELEFON telegraph noise; ~ **de tubo** ELECTRÓN tube noise; ~ **1/f** ELECTRÓN noise 1/f

rumbo: ~ **observado** *m* TEC RAD observed bearing

ruptura *f* DATOS, INFO truncation; ~ **eléctrica** COMPON, ELECTRÓN *de semiconductores* electric breakdown; ~ **por fluencia** PRUEBA creep rupture; ~ **de red** PRUEBA network breakdown; ~ **secundaria** COMPON secondary breakdown

ruta *f* ELECTRÓN, TELEFON, TRANS route; ~ **alternativa** TELEFON alternative route; ~ **del cable** TRANS cable run; ~ **de datos** TV data path; ~ **de mensaje de señalización** SEÑAL signalling-message (*AmE* signaling-message) route; ~ **de postes** TRANS pole route; ~ **principal** TELEFON backbone route; ~ **de señalización** SEÑAL signalling (*AmE* signaling) route; ~ **de sustitución** TELEFON substitute route; ~ **del tráfico** TELEFON traffic route; ~ **de tránsito** TELEFON transit routing; ~ **transversal** TELEFON transverse route; ~ **virtual del cajero automático** APLIC ATM virtual path

rutas: ~ **de emergencia** *f pl* TELEFON emergency routes

rutina *f* DATOS, INFO routine; ~ **de ayuda operacional** INFO utility routine; ~ **de diagnóstico** INFO diagnostic routine; ~ **de inicialización** INFO initializing routine; ~ **de operación** TELEFON operational procedure; ~ **de vaciado** INFO dump routine

RX *abr* (*receptor*) DATOS, PERIF, TEC RAD, TELEFON RX (*receiver*), rcvr (*receiver*)

S

S1 *abr* (*dibit doble descifrado*) DATOS S1 (*unscrambled double dibit*)
SABD *abr* (*sistema de administración de base de datos*) DATOS, INFO DBMS (*database management system*)
SABM *abr* (*modo equilibrado asincrónico de puesta a uno*) TELEFON SABM (*set asynchronous balanced mode*)
SABME *abr* (*modo equilibrado asincrónico de puesta a uno extendido*) TELEFON SABME (*set asynchronous balanced mode extended*)
sacudida *f* DATOS jerk, ELECTROTEC shock; **~ eléctrica** ELECTROTEC electric shock
sal: **~ de Rochelle** *f* ELECTRÓN Rochelle salt
sala: **~ de baterías** *f* ELECTROTEC battery room; **~ de conmutación** *f* TELEFON switching room; **~ de control** *f* PRUEBA control room; **~ de máquinas** *f* ELECTROTEC power room, TELEFON engine room; **~ de servicio** *f* PRUEBA operations room
salida: **~ de los auriculares** *f* TELEFON headphones output; **~ de control de codificación** *f* DATOS coding control output; **~ de copia impresa** *f* INFO hard-copy output; **~ digital** *f* INFO digital output; **~ de distribución** *f* ELECTROTEC *suministro de energía* distribution output; **~ espontánea** *f* TELEFON spontaneous output; **~ de formato** *f* TELEFON format output; **~ de guiaje** *f* TELEFON guidance output; **~ impresa** *f* INFO hard copy; **~ de información única incorporada** *f* ELECTRÓN flush-mounted single information outlet; **~ de línea** *f* INFO line-out; **~ máxima** *f* ELECTRÓN fan-out; **~ de menú** *f* TELEFON menu output; **~ mono** *f* INFO, PERIF mono output; **~ mono/estéreo** *f* INFO, PERIF mono/stereo output; **~ multibitios** *f* DATOS multibit output; **~ negociada** *f* DATOS negotiated exit (*NE*); **~ de la nota** *f* INFO note output; **~ de nota simultánea** *f* INFO simultaneous note output; **~ de orientación** *f* TELEFON prompting output; **~ de rechazo** *f* TELEFON rejection output; **~ de reconocimiento de selección** *f* TELEFON selection acknowledgement output; **~ de respuesta** *f* TELEFON response output; **~ de señal de vídeo** *f* TV video outlet
saliente *adj* (*o/g*) GEN outgoing (*o/g*)
salir *vi* INFO quit, exit
salpicadura: **~ de estaño** *f* COMPON solder spatter, solder splash
salto *m* TEC RAD hop; **~ de ciclo** TRANS cycle skipping; **~ de fase** TEC RAD phase hopping; **~ longitudinal** TELEFON longitudinal judder; **~ de modos** PERIF, TRANS mode hopping, mode jumping; **~ de papel** INFO *de impresora* slew, PERIF, TELEFON paper skip, paper throw; **~ del ritmo de la portadora de reloj** TRANS clock-carrier cycle skipping; **~ de tensión** ELECTROTEC voltage jump; **~ transversal** TELEFON transverse judder
salvaguarda *f* DATOS, INFO save, safeguard; **~ del programa** INFO program save
salvapantallas *m* INFO screensaver
sangría *f* ELECTRÓN tap, INFO indent, TELEFON tap
SAR *abr* (*segmentación y rearmado*) DATOS SAR (*segmentation and reassembly*)
SAS *abr* (*estación de acoplamiento simple*) ELECTRÓN SAS (*single-attachment station*)
satélite *m* TEC RAD, TV satellite; **~ auxiliar** TEC RAD co-located satellite; **~ de camino principal** TEC RAD major-path satellite; **~ de comunicaciones civiles** TEC RAD civil communications satellite; **~ de comunicaciones de cuerpo estabilizado** TEC RAD body-stabilized communications satellite; **~ DBS** TV DBS satellite; **~ en estación** TEC RAD on-station satellite; **~ independiente** TEC RAD stand-alone satellite; **~ interferente** TEC RAD interfering satellite; **~ LEO** TEC RAD LEO satellite; **~ para llenado de espacios interlobulares** TEC RAD gap-filler satellite; **~ de mantenimiento de órbita** TEC RAD station-keeping satellite; **~ en órbita terrestre** TEC RAD earth-orbiting satellite; **~ en órbita terrestre media** TEC RAD medium earth-orbiting satellite; **~ de orientación comercial** TEC RAD business-oriented satellite; **~ con posición de vuelo estabilizada** TEC RAD attitude-stabilized satellite; **~ principal** TEC RAD host satellite; **~ radiofaro** TEC RAD beacon satellite; **~ de recopilación de datos** TEC RAD data-collection satellite; **~ reflector** TEC RAD reflecting satellite; **~ regenerativo** TEC RAD regenerative satellite; **~ relé de datos** TEC RAD data-relay satellite; **~ repetidor de datos** TEC RAD data-relay satellite; **~ de reserva** TEC

RAD spare satellite; ~ **en serie** TEC RAD tandem satellite; ~ **sincronizado** TEC RAD synchronized satellite; ~ **subsincrónico** TEC RAD subsynchronous satellite; ~ **supersincrónico** TEC RAD supersynchronous satellite; ~ **de tecnología de aplicación** TEC RAD application technology satellite; ~ **con tecnología de la comunicación** TEC RAD communications technology satellite; ~ **de telecomunicaciones** TEC RAD, TV telecommunications satellite; ~ **de teledetección** TEC RAD remote sensing satellite

SATS *abr* (*batería de pruebas abstractas seleccionadas*) DATOS, PRUEBA SATS (*selected abstract test suite*)

saturación *f* ELECTRÓN saturation

saturado *adj* ELECTRÓN saturated

SB *abr* (*banda lateral*) TEC RAD, TELEFON, TRANS SB (*sideband*)

SB1 *abr* (*binarios cifrados*) DATOS SB1 (*scrambled binary ones*)

SBS *abr* (*sistema comercial de satélite*) TEC RAD SBS (*satellite business system*)

SBTP *abr* (*plan simulador de tiempo de ráfaga*) DATOS SBTP (*simulator burst time plan*)

SBV *abr* (*videotexto basado en sintaxis*) TELEFON SBV (*syntax-based videotext*)

S/C *abr* (*cortocircuito*) ELECTRÓN, ELECTROTEC, INFO S/C (*short circuit*)

SCCN *abr* (*red de coordinación y control de sistema*) PRUEBA SCCN (*system coordination and control network*)

SCCP *abr* (*dispositivo de control de la conexión de señalización*) SEÑAL, TELEFON SCCP (*signalling connection control part*)

SCH *abr* (*ocupaciones por circuito por hora*) TELEFON SCH (*seizures per circuit per hour*)

SCPC *abr* (*canal único por portadora*) TEC RAD SCPC (*single channel per carrier*)

SCS *abr* (*enunciado de conformidad de sistema*) DATOS, PRUEBA SCS (*system-conformance statement*)

SCSI *abr* (*interfaz de pequeño sistema informático*) INFO SCSI (*small computer system interface*)

SCTR *abr* (*informe de prueba de conformidad de sistema*) DATOS, PRUEBA SCTR (*system conformance test report*)

SCU *abr* (*unidad de señal de control de sistema*) SEÑAL SCU (*system control signal unit*)

SDA *abr* (*analizador de distorsión de señal*) PRUEBA, SEÑAL SDA (*signal distortion analyser*)

SDL *abr* (*lenguaje de descripción y especificación*) DATOS, INFO SDL (*specification and description language*); ~ **básica** *f* DATOS, INFO basic SDL

SDLC *abr* (*procedimiento de control de transmisión síncrona*) DATOS SDLC (*synchronous data-link control*)

SDN *abr* (*red de datos compartidos*) DATOS SDN (*shared-data network*)

sección *f* GEN section; ~ **de adaptación** TEC RAD transforming section; ~ **de cable** TRANS cable section; ~ **de cable elemental** TRANS elementary cable section; ~ **canalizada** TRANS channelized section; ~ **de circuito de programa sonoro** TV sound programme (*AmE* program) circuit section; ~ **de circuito de televisión** TV television circuit section; ~ **elemental de repetidores** TELEFON elementary repeater section; ~ **elemental regenerada** TELEFON elementary regenerated section; ~ **de filtro** ELECTRÓN filter section; ~ **de grupo** TELEFON group section; ~ **de grupo cuaternario** TELEFON supermastergroup section; ~ **de grupo secundario** TELEFON supergroup section; ~ **de grupo terciario** TELEFON mastergroup section; ~ **internacional** TELEFON international section; ~ **de línea** TRANS line section; ~ **de montaje del grupo secundario quince** TELEFON fifteen-supergroup assembly section; ~ **multiplexada** TRANS channelized section, multiplex section; ~ **de plantilla de cable** TRANS *cables* form-board section; ~ **principal** TELEFON main section; ~ **principal internacional** TELEFON international main section; ~ **de prueba principal** TELEFON principal test section; ~ **del regenerador** DATOS, TELEFON *de impulsos* regenerator section (*RS*); ~ **de regenerador elemental** DATOS, TELEFON elementary regenerator section; ~ **de regulación de línea** TELEFON regulated line section; ~ **de repetidor elemental** TELEFON elementary repeater section; ~ **terminal nacional** TELEFON terminal national section; ~ **transversal** ELECTRÓN *de vigas* cross section; ~ **transversal del conductor** ELECTRÓN conductor area; ~ **transversal del hilo** ELECTROTEC wire area; ~ **unitaria** TV unit section

seccionador *m* ELECTRÓN *de señal* slicer, ELECTROTEC *conmutador* disconnector

secreto[1] *adj* GEN private, secret

secreto[2]: ~ **comercial** *m* GEN commercial secrecy

sector *m* GEN sector; ~ **asociado** TV associated sector; ~ **de audio CD-I** INFO CD-I audio sector; ~ **CD-I** INFO CD-I sector; ~ **de coordenadas** CONMUT crossbar switch; ~ **de crecimiento** GEN growth sector; ~ **de des-**

criptor de pista INFO track-descriptor sector; ~ **de interferencia** TEC RAD interference sector; ~ **de servicios** GEN service sector; ~ **de transacción** DATOS transaction branch; ~ **de video CD-I** INFO CD-I video sector

sectores: ~ **por pista** *m pl* (*spt*) INFO sectors per track (*spt*)

secuencia *f* GEN sequence; ~ **de aleatorización de bitios anteriormente transmitidos** TELEFON earlier-transmitted bits scrambling process; ~ **binaria** DATOS binary sequence; ~ **bipolar de impulsos** TELEFON bipolar sequence of pulses; ~ **de codificación** DATOS scrambling sequence; ~ **de comprobación** PRUEBA check sequence; ~ **de control alterna** TEC RAD alternate control sequence; ~ **de control de trama** (*FCS*) DATOS, TELEFON frame check sequence (*FCS*); ~ **desencadenada** DATOS unchained sequence; ~ **encadenada** DATOS chained sequence; ~ **de entrelazamiento** TV interlacing sequence; ~ **de escape** DATOS escape sequence; ~ **de fases** ELECTROTEC phase sequence; ~ **de formato de entrada de parámetro** INFO format parameter-entry sequence; ~ **de imágenes** TV image sequence; ~ **de impulsos** SEÑAL pulse sequence; ~ **de intercalación** INFO collating sequence; ~ **de introducción de bloque paramétrico** TELEFON parameter block introduction sequence; ~ **de llamada** TELEFON calling sequence; ~ **de marcación** CONMUT marking sequence; ~ **de medidas en caso de avería** PRUEBA collapse action sequence; ~ **de modo desconectado** TELEFON disconnected-mode frame; ~ **de operación en modo menú** TELEFON menu-mode operating sequence; ~ **de paquetes** DATOS, TELEFON packet sequencing; ~ **que opera en modo de continuación** TELEFON continuation-mode operating sequence; ~ **de recuperación de ritmo binario** DATOS, TEC RAD bit-timing recovery sequence; ~ **de referencia digital PCM** TELEFON PCM digital reference sequence; ~ **de señalización** SEÑAL signalling (*AmE* signaling) sequence; ~ **de tambor** INFO drum sequence; ~ **unipolar de impulsos** TELEFON unipolar sequence of pulses

secuenciador *m* GEN sequencer; ~ **de llamadas** TELEFON call sequencer

secuencial *adj* GEN sequential

sede: ~ **remota** *f* GEN remote site

segmentación *f* DATOS, INFO, TELEFON segmentation, segmenting; ~ **y rearmado** (*SAR*) DATOS segmentation and reassembly (*SAR*)

segmentado *adj* DATOS, INFO, TELEFON segmented

segmento *m* DATOS, INFO, TELEFON segment; ~ **espacial** TEC RAD space segment; ~ **ISA** DATOS, INFO ISA segment; ~ **MHD** DATOS, INFO MHD segment; ~ **de pantalla** INFO display segment; ~ **ST** DATOS ST segment; ~ **ST del título del conjunto transaccional** DATOS transactional set header in ST segment; ~ **UNA** DATOS, TELEFON UNA segment; ~ **UNB** DATOS, TELEFON UNB segment; ~ **UNH** DATOS, TELEFON UNH segment

seguidor: ~ **de cátodo** *m* ELECTRÓN cathode follower; ~ **del emisor** *m* COMPON *semiconductores* emitter follower

seguimiento *m* GEN tracking; ~ **automático** TEC RAD auto-track; ~ **del blanco** TEC RAD target tracking; ~ **continuo** TEC RAD continuous tracking; ~ **gradual** TEC RAD step tracking; ~ **y mando** (*T & C*) PRUEBA tracking and command (*T & C*); ~ **del programa** TEC RAD programme (*AmE* program) tracking; ~ **de satélites** TEC RAD satellite tracking

segunda: ~ **elección** *f* TELEFON second choice

segundo[1]: ~ **enlace de distribución** *m* TEC RAD second distribution link; ~ **trayecto de satélite** *m* TEC RAD second satellite hop

segundo[2] *m* GEN *tiempo* second; ~ **erróneo** DATOS, TEC RAD errored second (*ES*); ~ **no disponible** DATOS unavailable second (*UAS*); ~ **de pérdida de trama** TRANS frame loss second; ~ **de salto** PRUEBA leap second; ~ **sumamente impreciso** DATOS, TEC RAD severely-errored second (*SES*)

seguridad *f* GEN safety, security; ~ **de archivos** INFO file security; ~ **a distancia** TRANS, TV remote security; ~ **de funcionamiento** PRUEBA dependability, TELEFON operational security; ~ **industrial** CONTROL industrial safety; ~ **informática** (*COMPUSEC*) DATOS, INFO computer security (*COMPUSEC*); ~ **de la red** DATOS, INFO network security; ~ **del sistema de proceso de datos** DATOS, INFO data-processing system security; ~ **de telecomunicaciones** TELEFON telecommunications security

seguro *adj* GEN dependable, foolproof, safe, secure

seleccionar *vt* GEN choose

selectividad *f* TEC RAD, TELEFON selectivity; ~ **contra canales adyacentes** TEC RAD, TRANS adjacent channel selectivity; ~ **efectiva** TEC RAD effective selectivity; ~ **del filtro** GEN filter selectivity

selector *m* CONMUT, DATOS, TELEFON selector; ~

de campo de acción MCS TELEFON MCS-domain selector; ~ **de canales** TEC RAD, TV channel selector; ~ **de contactos de relé** CONMUT, TELEFON relay-type switch; ~ **de dirección N** DATOS N-address selector; ~ **de dos movimientos** TELEFON two-motion selector; ~ **EG** (*selector de generador de envolvente*) INFO EG select (*envelope-generator select*); ~ **de frecuencias** TRANS frequency selector; ~ **por frecuencia vocal** TELEFON voice dialler (*AmE* dialer); ~ **de generador de envolvente** (*selector EG*) INFO envelope-generator select (*EG select*); ~ **giratorio** CONMUT rotary selector; ~ **de línea** TELEFON final selector; ~ **paso a paso** CONMUT step-by-step selector; ~ **de programas** CONMUT, INFO program selector; ~ **remoto de abonado** TELEFON remote subscriber stage, remote subscriber switch; ~ **Strowger** TELEFON Strowger selector; ~ **de tensión** ELECTROTEC voltage selector; ~ **de trabajos** INFO job selector; ~ **unidireccional** CONMUT uniselector

sellado[1] *adj* COMPON, ELECTRÓN sealed

sellado[2] *m* COMPON *de grietas*, ELECTRÓN sealing; ~ **hermético** COMPON hermetic sealing

sello *m* COMPON, ELECTRÓN seal; ~ **de plomo** COMPON leaded seal

SEM *abr* (*microscopio electrónico de exploración*) ELECTRÓN SEM (*scanning electron microscope*)

semación *f* TELEFON semation

semáforo *m* ELECTRÓN, TELEFON semaphore

semántica *f* GEN semantics; ~ **operacional** DATOS, INFO operational semantics; ~ **selectiva** DATOS, INFO snapshot semantics

semator *m* TELEFON semator

semiautomático *adj* CONMUT, TELEFON semiautomatic

semiciclo *m* GEN half-cycle

semiconductor *m* COMPON, ELECTRÓN semiconductor; ~ **con aislante metálico** (*SIA*) COMPON, ELECTRÓN metal insulator semiconductor (*MIS*); ~ **compensado** COMPON, ELECTRÓN compensated semiconductor; ~ **extrínseco** COMPON, ELECTRÓN extrinsic semiconductor; ~ **intrínseco** COMPON, ELECTRÓN intrinsic semiconductor; ~ **metal-óxido** COMPON, ELECTRÓN metal-oxide semiconductor; ~ **de potencia** COMPON, ELECTRÓN power semiconductor; ~ **tipo I** COMPON, ELECTRÓN I-type semiconductor; ~ **de tipo N** COMPON, ELECTRÓN N-type semiconductor; ~ **de tipo P** COMPON, ELECTRÓN P-type semiconductor

semiduplex *adj* TELEFON semiduplex

semigráfico *adj* PRUEBA semigraphic

Seminario: ~ **Europeo para Sistemas Abiertos** *m* (*EWOS*) GEN European Workshop for Open Systems (*EWOS*)

semirremolque *m* ELECTRÓN, INFO piggyback

semirrestador *m* ELECTRÓN half-subtracter

semisumador *m* ELECTRÓN half-adder

semisupresor: ~ **de eco** *m* TELEFON echo half-suppressor

semitono *m* INFO semitone

señal *f* GEN signal; ~ **de abonado ocupado** SEÑAL subscriber-busy signal; ~ **de acordes** INFO chord intelligence; ~ **de activación del transmisor** PERIF, SEÑAL, TELEFON, TRANS transmitter turn-on signal; ~ **acústica** GEN acoustic signal, beep, bleep; ~ **de alineación de trama** SEÑAL, TELEFON, TRANS frame-alignment signal; ~ **de alineación de grupo de tramas** SEÑAL, TELEFON, TRANS bunched frame-alignment signal; ~ **de arranque de unidad elemental de tiempo** DATOS, SEÑAL wink start signal; ~ **de avería RPOA** TELEFON RPOA out-of-order signal; ~ **de aviso navegacional** TEC RAD navigation warning signal; ~ **de banda ancha** ELECTRÓN, SEÑAL, TRANS wideband signal; ~ **de banda estrecha** ELECTRÓN, SEÑAL, TRANS narrow-band signal; ~ **básica** INFO basic signal; ~ **binaria** DATOS binary signal; ~ **bipolar** SEÑAL, TRANS bipolar signal; ~ **bipolar modificada** SEÑAL, TRANS modified alternate mark inversion; ~ **blanca nominal** TELEFON nominal white signal; ~ **de bloqueo** TELEFON blocking signal; ~ **de brote de ondas de acceso** DATOS access burst signal; ~ **de cambio a cifras** TELEFON figure-shift signal; ~ **de cambio a letras** TELEFON letter shift signal; ~ **de cancelación completada** SEÑAL cancellation-completed signal; ~ **de carácter** SEÑAL character signal; ~ **de circuito liberado** SEÑAL circuit-released signal; ~ **de colgar** SEÑAL hang-up signal; ~ **de color** INFO colour (*AmE* color) signal; ~ **compuesta** ELECTRÓN, SEÑAL, TV composite signal; ~ **de conexión de bus** CONMUT bus connection signal; ~ **de confusión** SEÑAL confusion signal; ~ **de congestión del equipo de conmutación** CONMUT, SEÑAL switching-equipment congestion signal; ~ **de congestión del grupo de circuito** DATOS, SEÑAL circuit-group congestion signal; ~ **de congestión de la red** SEÑAL network congestion signal; ~ **de conmutación** CONMUT, SEÑAL changeover signal; ~ **de conmutación de punto de prueba e interregistro** CONMUT, SEÑAL testing point switching and interregister signalling (*AmE* signaling); ~ **de contestación** SEÑAL answer-back signal; ~ **de contestación** SEÑAL answer signal; ~ **continua** DATOS continuous signal; ~ **de continuidad** SEÑAL continuity signal; ~ **de control** INFO control signal; ~ **de control de errores** TEC RAD error-check signal; ~ **de control multibloque** SEÑAL multiblock monitoring signal; ~ **cuadrada** ELECTRÓN square signal; ~ **de datos** DATOS data signal; ~ **de desbloqueo** SEÑAL, TEC RAD release-guard signal, TELEFON release-guard signal, unblocking signal; ~ **de desconexión** DATOS release signal; ~ **de desconexión doble** TELEFON both-party release signal; ~ **deseada** ELECTRÓN, SEÑAL, TEC RAD wanted signal; ~ **de destello** DATOS, SEÑAL wink signal; ~ **de detección de la portadora** DATOS carrier-sense signal; ~ **detectable mínima** (*MDS*) TRANS minimum detectable signal (*MDS*); ~ **digital N-aria** SEÑAL N-ary digital signal; ~ **eléctrica de abonado ocupado** SEÑAL electrical subscriber-busy signal; ~ **de entrada de retorno** TELEFON backward-input signal; ~ **de error** CONTROL error signal; ~ **de espaciado** TELEFON blanking signal; ~ **de espectro extendido** TEC RAD spread-spectrum signal; ~ **espuria** SEÑAL spurious signal; ~ **de estado de bus** CONMUT bus state signal; ~ **de explorador** SEÑAL scanner signal; ~ **de fallo de la red** SEÑAL network failure signal; ~ **de falta de corriente DCE** SEÑAL DCE-power off signal; ~ **fantasma** TV ghost signal; ~ **de fase** TELEFON phasing signal; ~ **de FI** ELECTRÓN, TEC RAD IF signal; ~ **de fin de bloque** SEÑAL end-of-block signal; ~ **de fin de comunicación** DATOS end-of-communication signal; ~ **de fin de numeración** SEÑAL end-of-pulsing signal; ~ **de fin de página** DATOS end-of-page signal; ~ **de fin de selección del registro** TELEFON register end-of-selection signal; ~ **de frecuencia única** SEÑAL, TEC RAD, TELEFON single-frequency signal; ~ **de fuera de fase** GEN out-of-phase signal; ~ **de gestión** SEÑAL management signal; ~ **de habilitación** INFO enabling signal; ~ **de identificación de la línea que llama** SEÑAL calling-line identification signal; ~ **de identificación de línea llamada** SEÑAL called-line identification signal; ~ **de imagen** INFO, TV picture signal; ~ **de impulso** SEÑAL pulse signal; ~ **de indicación de alarma** (*AIS*) PRUEBA, TELEFON alarm indication signal (*AIS*); ~ **para informes programados** SEÑAL scheduled reporting signal; ~ **inhibidora** ELECTRÓN inhibiting signal; ~ **de interferencia** TRANS interfering signal; ~ **internacional de socorro** TEC RAD Mayday; ~ **de interrogación** TEC RAD interrogating signal; ~ **de intervalo** TEC RAD interval signal; ~ **de inversión de marcas alternas** SEÑAL, TELEFON, TRANS alternate mark inversion signal; ~ **de invitación a transmitir** SEÑAL, TELEFON proceed-to-send signal; ~ **de invitación a transmitir en tráfico de tránsito** SEÑAL, TELEFON transit proceed-to-send signal; ~ **KP** TELEFON KP signal; ~ **de liberación del registro** SEÑAL register-release signal, TELEFON register-disconnect signal, register-release signal, register-clear signal; ~ **de línea fuera de servicio** SEÑAL line out-of-service signal; ~ **lineal de corrección de pendiente de línea** TV line tilt correction; ~ **de llamada** SEÑAL, TELEFON calling signal, ringing signal; ~ **de llamada aceptada** SEÑAL call-accepted signal; ~ **de llamada interrumpida** SEÑAL, TELEFON interrupted ringing signal; ~ **de luminancia** TV luminance signal; ~ **luminosa** TELEFON *de dispositivos indicadores* lamp signal; ~ **luminosa en la pantalla** TEC RAD blip; ~ **de mando digital** (*DCS*) DATOS digital command signal (*DCS*); ~ **de marcación retardada** SEÑAL delay-dialling (*AmE* delay-dialing) signal; ~ **de mensaje rechazado** SEÑAL message refusal signal; ~ **de modulación** TRANS modulating signal; ~ **modulada** TRANS modulated signal; ~ **muestreada** SEÑAL, TRANS sampled signal; ~ **multidimensional** DATOS multidimensional signal; ~ **multipágina** (*MPS*) DATOS, TELEFON multipage signal (*MPS*); ~ **N-aria redundante** SEÑAL redundant n-ary signal; ~ **de negro nominal** TELEFON nominal black signal; ~ **de peligro** TEC RAD danger signal; ~ **no regulada** SEÑAL unnotched signal; ~ **de número cambiado** SEÑAL, TELEFON changed-number signal; ~ **de número no asignado** SEÑAL, TELEFON unallocated-number signal; ~ **de número ocupado** SEÑAL, TELEFON number-busy signal; ~ **de ocupación** SEÑAL, TELEFON seizing signal; ~ **de onda sinusoidal** ELECTRÓN, INFO, TELEFON sine-wave signal; ~ **de operador auxiliar** TELEFON assistant operator signal; ~ **ortogonal** TEC RAD orthogonal signal; ~ **parabólica de corrección de campo** TV field bend correction; ~ **perturbadora** ELECTROTEC perturbing signal; ~ **de petición de auxilio** PRUEBA distress signal; ~ **de petición de cancelación** SEÑAL cancellation-request signal; ~ **portadora** INFO, TRANS carrier signal; ~ **preparada en espera** SEÑAL stand-by ready signal; ~ **de prioridad** SEÑAL,

TRANS priority signal; ~ **de proseguir** TELEFON go-ahead signal; ~ **de prueba de inserción** SEÑAL insertion test signal; ~ **pseudoternaria** TELEFON pseudoternary signal; ~ **de puerta** DATOS, ELECTRÓN *elementos lógicos* gate signal, gating signal; ~ **de puerta de un solo símbolo** DATOS, ELECTRÓN one-symbol gate signal; ~ **de puesta a uno** CONTROL, SEÑAL set signal; ~ **de radar** TEC RAD radar signal; ~ **de radiodifusión** TEC RAD broadcasting signal; ~ **de radiofaro** SEÑAL beacon signal; ~ **de realimentación** CONTROL feedback signal; ~ **de reconocimiento** SEÑAL acknowledgement signal; ~ **de reconocimiento de bloqueo** SEÑAL blocking acknowledgement signal; ~ **de reconocimiento de ocupación** SEÑAL, TELEFON seizing acknowledgement signal; ~ **de reconocimiento multibloque** SEÑAL multiblock acknowledgement signal; ~ **de reconocimiento de preparado en espera** SEÑAL stand-by ready acknowledgement signal; ~ **de reconocimiento tiempo de espera-tiempo de reserva** SEÑAL stand-by-ready acknowledgement signal; ~ **de reconocomiento emitida por un circuito** SEÑAL circuit-released acknowledgement signal; ~ **de recuperación** GEN recovery signal; ~ **de referencia** CONTROL set point; ~ **de referencia de inserción** SEÑAL insertion reference signal; ~ **del registro** SEÑAL, TELEFON register signal; ~ **de registro aceptado** SEÑAL registration-accepted signal; ~ **de registro concluido** SEÑAL registration-completion signal; ~ **de reiteración de respuesta** SEÑAL reanswer signal; ~ **de relleno DX** SEÑAL DX-stuffing signal; ~ **de reposición a cero del registro** SEÑAL, TELEFON register-reset signal; ~ **de respuesta** SEÑAL, TELEFON answer signal; ~ **retenida** SEÑAL retained signal; ~ **de retorno** TELEFON backward signal; ~ **de ruido** ELECTROTEC noise signal; ~ **de salida** ELECTRÓN, INFO output signal; ~ **de selección** SEÑAL, TELEFON selection signal; ~ **sencilla** SEÑAL, TELEFON simple signal; ~ **del servicio** SEÑAL, TELEFON service signal; ~ **de servicio degradado** SEÑAL, TELEFON degraded-service signal; ~ **de sincronismo de color** INFO burst; ~ **de sincronismo de línea** TV line-synchronizing signal; ~ **de sincronización** INFO synchronizing signal, SEÑAL synchronization signal, synchronizing signal; ~ **de sincronización de octeto** SEÑAL octet timing signal; ~ **de sincronización vertical** TV field-synchronizing signal; ~ **sincronizada discontinuamente** INFO discretely-timed signal; ~ **de solicitud de redireccionamiento** SEÑAL redirection request signal; ~ **de solicitud de registro** SEÑAL registration request signal; ~ **de sonido** SEÑAL sound signal; ~ **de sonar** TEC RAD sonar signal; ~ **ST** SEÑAL ST signal; ~ **de tarificación** TELEFON metering signal; ~ **telegráfica** TELEFON telegraph signal; ~ **temporal discreta** TRANS discretely-timed signal; ~ **de temporización** SEÑAL timing signal; ~ **de terminal ocupada** SEÑAL, TELEFON terminal-engaged signal; ~ **de toma de línea** SEÑAL, TELEFON seizure signal; ~ **de toma de tránsito** SEÑAL, TELEFON transit seizure signal; ~ **total** SEÑAL aggregate signal; ~ **de trabajo** TEC RAD marking wave; ~ **de transferencia de carga de emergencia** SEÑAL emergency load-transfer signal; ~ **transferida por el abonado** SEÑAL subscriber-transferred signal; ~ **de tránsito de conexión directa** SEÑAL, TELEFON transit through-connect signal; ~ **trapezoidal de corrección de línea** TV line keystone correction; ~ **de urgencia** TEC RAD urgency signal; ~ **de video** INFO, SEÑAL, TEC RAD, TV video signal; ~ **de vídeo positiva** INFO positive video signal; ~ **de vídeo sintética** SEÑAL, TEC RAD synthetic video signal; ~ **visual** INFO, SEÑAL visual signal; ~ **vocal** SEÑAL, TELEFON speech signal; ~ **vocal simulada** SEÑAL, TELEFON simulated speech

señalador: ~ **personalizado** *m* DATOS, INFO personalized bookmark

señalamiento: ~ **erróneo** *m* TEC RAD mispointing

señales: ~ **de código de frecuencia vocal** *f pl* SEÑAL, TELEFON voice-frequency code signalling (*AmE* signaling); ~ **de gestión de la red** *f pl* SEÑAL network management signals; ~ **de gestión de la red de señalización** *f pl* SEÑAL signalling-network (*AmE* signaling-network) management signals; ~ **de mantenimiento de la red** *f pl* SEÑAL network maintenance signals; ~ **parásitas** *f pl* TEC RAD grass; ~ **parásitas en pantalla** *f pl* TEC RAD *en radar* clutter; ~ **de selección de la red** *f pl* SEÑAL network selection signals; ~ **transmitidas efectivamente en programa de sonido** *f pl* SEÑAL effectively transmitted signals in sound programme (*AmE* sound-program)

señalización *f* DATOS, INFO, SEÑAL, TEC RAD, TELEFON signalling (*AmE* signaling); ~ **anexa** GEN associated signalling (*AmE* signaling); ~ **asociada** SEÑAL associated signalling (*AmE* signaling); ~ **asociada al canal** SEÑAL, TELEFON, TRANS channel-associated signalling (*AmE* signaling) (*CAS*); ~ **auxiliar** TELEFON auxiliary signalling (*AmE* signaling); ~ **de bitios robados** DATOS, SEÑAL robbed-bit signalling (*AmE* signaling); ~ **de CA** ELECTROTEC, SEÑAL AC signalling (*AmE* signaling); ~ **de canal común separada** SEÑAL separate common-channel signalling (*AmE* signaling); ~ **500/20** TELEFON 500/20 signalling (*AmE* signaling); ~ **codificada dentro de banda** SEÑAL coded in-band signalling (*AmE* signaling); ~ **de código de frecuencia vocal** TELEFON voice-frequency code signalling (*AmE* signaling); ~ **concentrada** TRANS bunched signalling (*AmE* signaling); ~ **de conmutación e interregistro del punto de prueba** CONMUT, PRUEBA testing point switching and interregister signalling (*AmE* signaling); ~ **de control centralizada** SEÑAL centralized control signalling (*AmE* signaling); ~ **de control de la red** TELEFON network control signalling (*AmE* signaling); ~ **de corriente alterna** ELECTROTEC, SEÑAL alternating-current signalling (*AmE* signaling); ~ **de corriente continua** ELECTROTEC, SEÑAL DC signalling (*AmE* signaling); ~ **cuasiasociada** SEÑAL, TELEFON quasi-associated signalling (*AmE* signaling); ~ **decádica** SEÑAL decadic signalling (*AmE* signaling); ~ **con dos frecuencias** SEÑAL, TELEFON two-frequency signalling (*AmE* signaling); ~ **a 2600 Hz** SEÑAL signalling (*AmE* signaling) at 2600 Hz; ~ **E & M** TELEFON E & M signalling (*AmE* signaling); ~ **entre bornes** SEÑAL interhub signalling (*AmE* signaling); ~ **entre registradores** TELEFON register signalling (*AmE* signaling); ~ **del estado del abonado** TELEFON subscriber-state signalling (*AmE* signaling); ~ **de extremo a extremo** SEÑAL end-to-end signalling (*AmE* signaling); ~ **fuera de banda** SEÑAL out-band signalling (*AmE* signaling); ~ **fuera del intervalo** SEÑAL out-slot signalling (*AmE* signaling); ~ **de impulsos** TELEFON pulse signalling (*AmE* signaling); ~ **en el intervalo de tiempo** SEÑAL in-slot signalling (*AmE* signaling); ~ **de línea** SEÑAL line signalling (*AmE* signaling); ~ **de línea del punto de prueba** PRUEBA, SEÑAL testing point line signalling (*AmE* signaling); ~ **manual** GEN ring-down signalling (*AmE* signaling); ~ **de multifrecuencia** SEÑAL multifrequency signalling (*AmE* signaling); ~ **no asociada** SEÑAL nonassociated signalling (*AmE* signaling); ~ **de protocolo simple para red ATM** TRANS simple protocol for ATM network signalling (*AmE* signaling); ~ **del punto central** SEÑAL centre-point signalling (*AmE* center-point signaling) ; ~ **del punto medio** SEÑAL midpoint signalling (*AmE* signaling); ~ **del recorrido inverso de un bucle** GEN loopback signalling (*AmE* signaling); ~ **de registradores** SEÑAL, TELEFON register signalling (*AmE* signaling); ~ **de registro en bloque** SEÑAL, TELEFON en bloc register signalling (*AmE* signaling); ~ **de retención** SEÑAL hold signalling (*AmE* signaling); ~ **sección por sección** SEÑAL, TRANS link-by-link signalling (*AmE* signaling); ~ **de secuencia obligada** SEÑAL compelled signalling (*AmE* signaling); ~ **de tonalidad** SEÑAL, TELEFON tone signalling (*AmE* signaling); ~ **totalmente disociada** SEÑAL fully-dissociated signalling (*AmE* signaling); ~ **de usuario a usuario** SEÑAL, TEC RAD user-to-user signalling (*AmE* signaling) (*UUS*); ~ **por voz digitalizada** SEÑAL speech-digit signalling (*AmE* signaling)

señalizador *m* SEÑAL signal device

senda *f* CONMUT bypath

senoidal *adj* ELECTRÓN, SEÑAL sinusoidal

sensibilidad *f* GEN sensitivity; ~ **de la articulación** ELECTRÓN knee sensitivity; ~ **del controlador de intervalo** INFO breath-controller sensitivity; ~ **del dispositivo de llamada** TELEFON ringer sensitivity; ~ **de emisión** TELEFON *de circuito telefónico local* sending sensitivity; ~ **espectral relativa** ELECTRÓN relative spectral sensitivity; ~ **máxima** GEN maximum sensitivity; ~ **máxima utilizable** TEC RAD maximum usable sensitivity; ~ **de modulación de amplitud** INFO amplitude modulation sensitivity; ~ **de radiación** TEC RAD radiation sensitivity; ~ **radiante** ELECTRÓN radiant sensitivity; ~ **de recepción** TELEFON *de un sistema telefónico* receiving sensitivity; ~ **del receptor** TEC RAD receiver sensitivity, RX sens; ~ **de referencia** TEC RAD, TRANS reference sensibility, reference sensitivity; ~ **de la rueda de modulación** INFO modulation wheel sensitivity; ~ **de sincronización** GEN synchronization sensitivity; ~ **sintonizadora** ELECTRÓN tuning sensitivity; ~ **a la temperatura** CLIMAT temperature sensitivity; ~ **de velocidad de desconexión** INFO release-velocity sensitivity

sensibilizar *vt* GEN sensitize

sensible *adj* GEN sensitive; ~ **a descargas electrostáticas** PRUEBA electrostatic-discharge-sensitive

sensor *m* GEN sensor; ~ **de fase** TRANS phase sensor; ~ **de imagen** INFO, TV image sensor; ~ **de intensidad** TRANS intensity sensor; ~ **de microcurvaturas** TRANS microbend sensor; ~

pasivo TEC RAD passive sensor; ~ **de posición de vuelo** TEC RAD attitude sensor

sentencia *f* DATOS, INFO *programación* sentence, declaration; ~ **de asignación** DATOS assignment statement; ~ **concurrente** DATOS concurrent statement; ~ **de conformidad de instrumentación de protocolos** (*PICS*) DATOS protocol implementation conformance statement (*PICS*)

sentido *m* GEN direction; ~ **longitudinal** GEN longitudinal direction; ~ **de rotación** GEN direction of rotation; ~ **de torsión** GEN *de cables* direction of lay; ◆ **en ~ antihorario** GEN anticlockwise *BrE*, counterclockwise *AmE*; **en ~ horario** GEN clockwise

separación: ~ **de agujeros** *f* ELECTRÓN *circuitos impresos* hole spacing; ~ **de almacenes** *f* GEN magazine spacing; ~ **angular** *f* TEC RAD, TRANS angular separation; ~ **de bloques** *f* TELEFON block separation; ~ **de canal** *f* TEC RAD channel separation; ~ **entre bastidores** *f* TELEFON rack spacing; ~ **entre conductores** *f* ELECTRÓN *circuitos impresos* conductor spacing; ~ **entre letras** *f* DATOS interletter spacing; ~ **entre portadoras** *f* TRANS carrier separation; ~ **entre registros** *f* DATOS interrecord gap; ~ **de las espigas** *f* ELECTRÓN pin spacing; ~ **de frecuencia** *f* TEC RAD frequency separation; ~ **de frecuencia de precisión** *f* TEC RAD precision frequency offset; ~ **de la frecuencia normalizada** *f* TEC RAD normalized frequency offset; ~ **normalizada** *f* TEC RAD normalized offset; ~ **de órbita** *f* TEC RAD orbit gap; ~ **de las placas** *f* COMPON *técnicas de montaje* board spacing; ~ **de relés** *f* TELEFON relay spacing; ~ **de secciones** *f* DATOS, TELEFON section spacing

separador: ~ **amagnético de armadura** *m* ELECTRÓN nonmagnetic armature shim; ~ **de bloques** *m* TELEFON block separator; ~ **de cables** *m* TRANS cable spacer; ~ **de crominancia** *m* INFO chromakey; ~ **de elementos de datos** *m* DATOS data element separator; ~ **por elementos de los datos de componentes** *m* COMPON, DATOS component data element separator; ~ **de fibra No-NiKs** *m* ELECTRÓN No-NiKs fibre (*AmE* fiber) stripper; ~ **de grupo** *m* (*SG*) DATOS group separator (*GS*); ~ **de llave** *m* INFO key split; ~ **de módem** *m* DATOS modem interface; ~ **de modos** *m* INFO, TRANS mode stripper; ~ **de nombre de dominio** *m* DATOS domain name resolver (*DNR*); ~ **de solicitud de sistema** *m* DATOS, TELEFON facility request separator

separata *f* PERIF offprint

septeto *m* INFO septet

septum *m* TEC RAD septum

serial *adj* GEN serial

serie *f* GEN series; ~ **de acontecimientos** PRUEBA sequence of events; ~ **de etiquetas** ELECTRÓN tag rash; ~ **numérica** GEN number series; ~ **de pruebas de conformidad** DATOS, PRUEBA conformance test-suite; ~ **de símbolos** GEN symbol string; ~ **unitaria** DATOS unit string

servibilidad *f* APLIC, TELEFON serveability performance, serviceability performance

servicio *m* GEN service; ~ **abierto** TELEFON open service; ~ **de abonado** INFO, TELEFON subscriber service; ~ **abstracto MS** DATOS MS abstract service; ~ **de adaptación de la sintaxis** DATOS syntax-matching service; ~ **de agenda** TELEFON diary service; ~ **de alarma recordatorio** APLIC reminder alarm service; ~ **de alta calidad** TV premium service; ~ **alternativo de palabras** APLIC alternate speech service; ~ **de bajo nivel** PRUEBA lower-level service; ~ **básico** TV basic service; ~ **B-RDSI** (*servicio-B de red digital de servicios integrados*) APLIC, TELEFON B-ISDN service; **~-B de red digital de ~s integrados** (*servicio B-RDSI*) APLIC, TELEFON B-ISDN service; ~ **de buzón de mensajes** INFO mailbox messaging service; ~ **de captación de línea PBX** CONMUT PBX line-hunting service; ~ **centrex** TELEFON centrex service; ~ **de circuito virtual del canal B** TELEFON B-channel virtual-circuit service; ~ **al cliente** GEN customer service; ~ **complementario** APLIC, TELEFON complementary service; ~ **de conferencia múltiple** APLIC conference call service; ~ **conmutado** APLIC circuit-switched service; ~ **de contestador** APLIC, TELEFON answering service; ~ **de corriente portadora** APLIC, INFO, TEC RAD, TRANS bearer service; ~ **de datagrama** DATOS, SEÑAL, TELEFON datagram service; ~ **de demanda manual internacional** TELEFON international manual demand service; ~ **de despertador** TELEFON wake-up service; ~ **de desvío en caso de número ocupado** TELEFON diversion-if-number-busy service; ~ **directo domiciliario** TELEFON home-direct service; ~ **de documentación de prensa** INFO clip-art library; ~ **de emergencia** APLIC emergency service, TELEFON emergency attention, TV emergency service; ~ **de emisión doble** DATOS multicast service; ~ **especial** TELEFON special service; ~ **en estrella** PERIF *contestador automático* star service; ~ **de extremo a extremo** TELEFON end-to-end servicing; ~ **de gestión de sistemas** INFO systems management service; ~ **de identificación de llamadas maliciosas** TELEFON malicious-call identification service; ~ **independiente de portador** (*BIS*) TELEFON bearer independent service (*BIS*); ~ **de información** APLIC, TELEFON directory enquiries *BrE* (*DQ*), directory assistance *AmE*; ~ **de información sobre abonados** TELEFON directory enquiry service; ~ **iniciado por el proveedor** DATOS provider-initiated service; ~ **ininterrumpido** GEN uninterrupted duty; ~ **de interceptación** TELEFON intercept service; ~ **internacional público de telefax** TELEFON international public facsimile service; ~ **interurbano de mensajería** TELEFON message toll service; ~ **limitado de abonados móviles** TELEFON limited roaming telephone service; ~ **de llamada gratuita** TELEFON freephone service *BrE*, toll-free service *AmE*; ~ **de llamadas de conferencia** TELEFON conference call service; ~ **de llamadas de emergencia** APLIC, TELEFON emergency call service; ~ **de llamadas por especificación** APLIC custom calling service; ~ **marítimo móvil de satélite** TEC RAD maritime mobile-service satellite; ~ **marítimo móvil terrestre** TEC RAD maritime mobile terrestrial service; ~ **a la medida del cliente** GEN custom-designed service; ~ **de mensajería** GEN courier service; ~ **de mensajería por voz** TELEFON voice messaging service; ~ **meteorológico** APLIC, TELEFON weather line; ~ **móvil** GEN mobile service; ~ **móvil marítimo** TEC RAD, TELEFON maritime mobile service; ~ **móvil telefónico** TELEFON mobile telephone service; ~ **móvil terrestre de satélite** (*LMSS*) TEC RAD, TELEFON, TRANS land-mobile satellite service (*LMSS*); ~ **de múltiples portadores** TELEFON multibearer service; ~ **N** DATOS, TELEFON N-service; ~ **de nivel superior** APLIC higher-level service; ~ **nocturno** CONMUT night-service mode, DATOS night service; ~ **nocturno general** TELEFON universal night-service; ~ **de notificación de gestión** DATOS management notification service; ~ **no transparente del portador** APLIC, DATOS nontransparent bearer service; ~ **de número de acceso universal** TELEFON universal-access number service; ~ **de observación** APLIC, TELEFON observation service; ~ **de operaciones internacionales** APLIC international operations service; ~ **de operación de gestión** DATOS management operation service; ~ **portador de circuito virtual** DATOS, TRANS virtual-circuit bearer service; ~ **portador de circuito virtual RDSI** DATOS, TRANS ISDN virtual circuit bearer service; ~ **portador con conmutación de paquetes** APLIC packet-switched bearer service; ~ **portador de modos de circuito** APLIC circuit mode-bearer service; ~ **portador de modo trama** APLIC, TRANS frame-mode bearer service (*FMBS*); ~ **portador en modos por paquete** DATOS, TELEFON packet-mode bearer service; ~ **portador no restringido a 64 kbps** APLIC sixty-four kbps unrestricted bearer service; ~ **portador restringido a 64 kbps** APLIC sixty-four kbps restricted bearer service; ~ **portador de señalización al usuario** DATOS, SEÑAL user-signalling (*AmE* user-signaling) bearer service; ~ **PPSDN** DATOS PPSDN service; ~ **de presentación** DATOS presentation service; ~ **de prioridad** TELEFON priority facility; ~ **público de telegrafía** TELEFON, TRANS public telegram service; ~ **público de transmisión de datos** TRANS public data-transmission service; ~ **de radioaficionados** TEC RAD amateur radio service; ~ **de radiobúsqueda** TEC RAD radio-paging service; ~ **de radiocomunicaciones** TEC RAD radiocommunication service; ~ **de radiodifusión por satélite** TEC RAD broadcasting-satellite service; ~ **RDSI de banda ancha** APLIC, TELEFON broadband ISDN service; ~ **recordatorio** TELEFON reminder service; ~ **recuperable** TEC RAD restorable service; ~ **de recuperación** INFO retrieval service; ~ **de red** DATOS, TELEFON network service (*NS*); ~ **de red de compromisos, acuerdos y recuperación** (*CCR*) DATOS commitment, concurrency, and recovery (*CCR*); ~ **en red en modo de conexión** DATOS connection-mode network service; ~ **de red orientado a la comunicación** (*CONS*) APLIC connection-oriented mode network service (*CONS*); ~ **de red OSI** DATOS, TELEFON OSI network service; ~ **reforzado** APLIC enhanced service; ~ **de registro detallado de llamadas** APLIC call-detail recording service; ~ **de repetición de número** TELEFON number repetition service; ~ **repetido** TELEFON callback; ~ **restringido a 64 kbps** APLIC sixty-four kbps restricted service; ~ **retardado** GEN delay working; ~ **de satélite** TEC RAD, TV satellite service; ~ **de satélite de frecuencia normal** TEC RAD, TV standard-frequency satellite service; ~ **según tarifa** GEN fee-based service; ~ **de seguridad** GEN *de transmisiones* security service; **Servicio de Red Xerox** (*XNS*) DATOS Xerox Network Service (*XNS*); ~ **de sesión** DATOS, TRANS session service; ~ **sin cargo** TELEFON toll-free service *AmE*, freephone service *BrE*; ~ **de**

subred DATOS subnetwork service; ~ **suplementario** APLIC, TEC RAD, TELEFON supplementary service; ~ **de tarifa uniforme** APLIC flat-rate service; ~ **de telecomunicaciones** GEN telecommunications service; ~ **de telefax** DATOS, TELEFON facsimile service; ~ **de telefonía visual** APLIC, TEC RAD, TELEFON, TV visual telephone service; ~ **telefónico de área extendida** TELEFON wide-area telephone service (*WATS*); ~ **por teléfono** APLIC teleservice; ~ **de teléfono monedero** APLIC, TELEFON payphone service; ~ **de telefoto** TELEFON telephoto service; ~ **telegráfico** TELEFON telegraph service; ~ **telegráfico internacional de cuenta transferida** TELEFON international transferred-account telegraph service; ~ **de telemática** DATOS, INFO, TEC RAD telematics service; ~ **de teletexto** APLIC, TV teletext service; ~ **de teletipo** APLIC, PERIF, TELEFON teleprinter service, teletypewriter service *AmE*; ~ **de télex** APLIC, PERIF, TELEFON teleprinter service, teletypewriter service *AmE*; ~ **de transferencia de información no reconocido** APLIC unacknowledged information-transfer service (*UITS*); ~ **de enlace de datos** APLIC, DATOS, TRANS data-link service (*DLS*); ~ **de transmisión de datos conmutados por paquetes** DATOS, TRANS packet-switched data transmission service; ~ **de transmisión de datos por circuito arrendado** DATOS, TRANS leased-circuit data transmission service; ~ **de transmisión de datos por circuito conmutado** DATOS, TRANS circuit-switched data transmission service; ~ **de transmisión de la palabra** APLIC speech service; ~ **de transmisión de mensaje verbal** TELEFON, TRANS verbal-message transmission service; ~ **transparente del portador** APLIC, DATOS transparent bearer service; ~ **de transporte** DATOS, TELEFON, TRANS transport service; ~ **tripartito** APLIC, TELEFON three-party service; ~ **de usuario** TELEFON user facility, user service; ~ **al usuario con información grabada** TELEFON customer recorded-information service; ~ **de usuario optativo esencial** TELEFON essential optional user-facility; ~ **de valor añadido y datos** APLIC, DATOS value-added and data service; ~ **de videoconferencias** APLIC, INFO, TEC RAD, TV videoconferencing; ~ **de videoteléfono** APLIC, TEC RAD, TELEFON, TV videophone service; ~ **de videotexto** APLIC, INFO, TELEFON, TV videotext service

servicios: ~ **comerciales** *m pl* GEN business services; ~ **de comunicaciones personales** *m pl* (*PCS*) APLIC personal communications services (*PCS*); ~ **de comunicación** *m pl* APLIC communication services (*CS*); ~ **de emergencia** *m pl* APLIC emergency services; ~ **facturados** *m pl* TELEFON billed services; ~ **de información sobre abonados** *m pl* DATOS directory services; ~ **de misión principal** *m pl* PRUEBA main mission services; ~ **prestados** *m pl* TELEFON services rendered; ~ **públicos** *m pl* GEN public utilities; ~ **de radioteléfono móvil** *m pl* TELEFON mobile radio telephone services; ~ **de red de valor añadido** *m pl* (*SRVA*), (*VANS*) APLIC, TRANS, TV value-added network services (*VANS*)

servidor: ~ **activado por la voz** *m* CONTROL, PERIF, TELEFON voice command server; ~ **de archivos** *m* DATOS, INFO file server; ~ **Audiotex** *m* TELEFON Audiotex server; ~ **de comunicaciones** *m* DATOS, INFO communications server; ~ **de correo** *m* DATOS, INFO mail server; ~ **designado** *m* DATOS, INFO named server; ~ **de documento** *m* DATOS, INFO document server; ~ **de emisión doble** *m* DATOS multicast server; ~ **especializado** *m* DATOS back-end server; ~ **FTP** *m* DATOS FTP server; ~ **de impresora** *m* DATOS, INFO printer server; ~ **de información de área extendida** *m* (*WAIS*) DATOS, INFO wide-area information server (*WAIS*); ~ **local** *m* INFO local server; ~ **de mensaje vocal** *m* TELEFON voice message server; ~ **de noticias** *m* DATOS, INFO news server; ~ **de la red** *m* DATOS, INFO network server; ~ **SDA virtual** *m* TELEFON virtual SDA server; ~ **de telefax** *m* DATOS, TELEFON facsimile server; ~ **telemático** *m* DATOS, INFO telematic server; ~ **de videotexto** *m* INFO, TELEFON, TV videotext server; ~ **vocal** *m* DATOS, INFO voice server

servoamplificador *m* ELECTRÓN, ELECTROTEC servoamplifier

servomecanismo *m* CONTROL servomechanism

servosistema *m* CONTROL servosystem

sesgado *adj* ELECTROTEC, TEC RAD biased

sesgo: ~ **de altura de tono** *m* ELECTROTEC, TEC RAD pitch bias

sesión *f* GEN session

SETS *abr* (*batería de pruebas ejecutables seleccionadas*) DATOS, PRUEBA SETS (*selected executable test suite*)

SEU *abr* (*redisposición de evento simple*) PRUEBA SEU (*single event upset*)

SG *abr* (*separador de grupo*) DATOS GS (*group separator*)

SGML *abr* (*lenguaje estándar generalizado de

referencia) INFO SGML (*standard generalized markup language*)
sí *m* DATOS yes (*Y*)
SI *abr* (*norma internacional*) GEN IS (*international standard*)
SIA *abr* (*semiconductor con aislante metálico*) COMPON, ELECTRÓN MIS (*metal insulator semiconductor*)
SIG *abr* (*elemento de firma*) SEÑAL, TELEFON SIG (*signature element*)
signatario *m* TV signatory
signatura: ~ **del blanco** *f* TEC RAD target signature
significativo *adj* GEN significant
signo *m* GEN sign; ~ **constitutivo** INFO key signature
sílaba *f* GEN syllable
silenciador *m* TELEFON, TRANS silencer *BrE*, muffler *AmE*
silencio: ~ **de escucha** *m* TEC RAD listening silence
silencioso *adj* GEN silent
sílice *f* COMPON silica, silicon; ~ **vítrea** TRANS vitreous silica
siluminio *m* COMPON silumin
SIM *abr* (*módulo de identidad del abonado*) TEC RAD SIM (*subscriber identity module*)
simbiosis *f* ELECTRÓN symbiosis
símbolo *m* GEN symbol; ~ **de anotación** TELEFON annotation symbol; ~ **de decisión** TELEFON decision symbol; ~ **de diagrama de flujo** INFO flow-chart symbol; ~ **de estado** TELEFON state symbol; ~ **gráfico** DATOS graphic symbol, graphical symbol; ~ **lógico** DATOS, ELECTRÓN logic symbol, logical symbol; ~ **mnemotécnico** GEN mnemonic symbol; ~ **no terminal** PERIF nonterminal symbol; ~ **de salida** TELEFON output symbol; ~ **de seguimiento** INFO tracking symbol; ~ **de tarea** TELEFON task symbol
simetría *f* GEN symmetry
simétrico *adj* GEN symmetric, symmetrical
similitud *f* TELEFON commonality
simplex[1] *adj* TEC RAD, TELEFON simplex
simplex[2]: ~ **por canales conjugados** *m* TELEFON *dúplex* two-way simplex; ~ **con dos frecuencias** *m* TEC RAD, TELEFON two-frequency simplex
simulación *f* GEN simulation; ~ **de avería** PRUEBA fault simulation; ~ **del circuito** ELECTRÓN *semiconductores* circuit simulation; ~ **por computadora** *AmL* (*cf simulación por ordenador Esp*) INFO computer simulation; ~ **por ordenador** *Esp* (*cf simulación por computadora AmL*) INFO computer simulation; ~ **de pequeña constante de tiempo** INFO fast-time simulation
simulador *m* INFO, TELEFON simulator; ~ **de emisor de indicativo** TELEFON answer-back unit simulator; ~ **de tensión** ELECTROTEC voltage simulator
simular *vt* GEN simulate
sincrocapacitor *m* COMPON, ELECTRÓN synchro capacitor
sincrónico *adj* GEN synch
sincronismo *m* GEN synchronism
sincronización *f* GEN synchronization, synchronizing; ~ **asimétrica** TELEFON, TRANS asymmetric synchronization; ~ **automática** TRANS auto-synchronization; ~ **biterminal** TELEFON, TRANS double-ended synchronization; ~ **en bucle cerrado** CONTROL, ELECTROTEC, TRANS closed-loop synchronization; ~ **de cinta** INFO tape synchronization; ~ **de fase** ELECTRÓN phase locking; ~ **de imagen y sonido** TV image and sound synchronization; ~ **inicial** DATOS preamble; ~ **mutua** ELECTRÓN mutual synchronization; ~ **de la red** DATOS network locking; ~ **de reloj** INFO clock synchronization; ~ **por reloj** TRANS clock phasing; ~ **de teclas** INFO key sync, key synchronization; ~ **de trama** DATOS, TRANS frame synchronization
sincronizado *adj* GEN synchronized; **no** ~ GEN nonsynchronized
sincronizador: ~ **de ciclo** *m* DATOS, TRANS frame synchronizer
sincronizar *vt* GEN synchronize
síncrono *adj* GEN synchronous
sincrotransmisor *m* TEC RAD synchro transmitter
sinergético *adj* GEN synergetic
sinergía *f* GEN synergy
sinérgico *adj* GEN synergic
sinónimo *m* GEN synonym
sinopsis *f* DATOS, INFO synopsis
sintáctico *adj* DATOS, INFO, TELEFON syntactic
sintaxis *f* GEN syntax; ~ **concreta** DATOS, INFO concrete syntax; ~ **gráfica concreta** DATOS, INFO concrete graphical syntax; ~ **del nombre del archivo** DATOS, INFO, TELEFON file-name syntax; ~ **textual concreta** DATOS, INFO concrete textual syntax
sinterización *f* COMPON sintering
síntesis *f* GEN synthesis; ~ **FM** (*síntesis de modulación de frecuencia*) INFO FM synthesis (*frequency-modulation synthesis*); ~ **de modulación de frecuencia** (*síntesis FM*) INFO frequency-modulation synthesis (*FM synthesis*); ~ **de la señal** DATOS, SEÑAL signal

synthesis; ~ **de sonido** INFO sound synthesis; ~ **de sonidos vocales** APLIC, INFO, TELEFON speech synthesis; ~ **sustractiva** INFO subtractive synthesis; ~ **vectorial** INFO vector synthesis; ~ **de la voz** APLIC, INFO, TELEFON voice synthesis

sintetizador *m* ELECTRÓN, INFO, TELEFON synthesizer; ~ **de frecuencia** TEC RAD frequency synthesizer; ~ **de sonido** INFO sound synthesizer

sintetizar *vt* ELECTRÓN, INFO, TELEFON synthesize

síntoma: ~ **de avería** *m* TELEFON fault symptom; ~ **de fallo** *m* TELEFON failure symptom

sintonía: ~ **por permeabilidad** *f* TEC RAD permeability tuning

sintonización *f* ELECTRÓN, INFO, TEC RAD tuning; ~ **de antena** TEC RAD antenna tuning; ~ **de fase** ELECTRÓN phase tuning; ~ **de frecuencia** TEC RAD frequency tuning; ~ **de reluctancia** TEC RAD reluctance tuning

sintonizador *m* ELECTRÓN, INFO, TEC RAD tuner; ~ **de televisión** TV television tuner

sintonizar *vt* ELECTRÓN, INFO, TEC RAD tune

sirena *f* TELEFON siren

siseo *m* TELEFON hiss

sismorresistente *adj* PRUEBA earthquake-resistant

sistema *m* GEN system; ~ **abierto de gestión** DATOS management open system; ~ **abierto gestionado** DATOS managed open system; ~ **de acceso múltiple** DATOS multiaccess system; ~ **de acceso a la red** TRANS network-access system; ~ **accionado por energía eléctrica** TELEFON power-driven system; ~ **de accionamiento motorizado** TEC RAD motor-driven system; ~ **activado por la voz** CONTROL, PERIF, TELEFON voice command system; ~ **adaptativo** DATOS adaptive system; ~ **de administración de base de datos** (*SABD*) DATOS, INFO database management system (*DBMS*); ~ **de alarma por control remoto** CONTROL remote-control warning system; ~ **de alimentación por telefonía** TV telephony head-end; ~ **de almacenamiento** (*S/SYS*) DATOS storage system (*S/SYS*); ~ **de almacenamiento y transferencia** (*ST/SYS*) DATOS storage and transfer system (*ST/SYS*); ~ **ALOHA** TEC RAD, TELEFON ALOHA system; ~ **ALOHA de línea ranurada** TEC RAD, TELEFON slotted ALOHA system; ~ **de amplificación** INFO amplification system; ~ **de análisis de opción de pago** (*POAS*) APLIC payment option analysis system (*POAS*); ~ **de análisis de pequeña constante de tiempo** INFO fast-time analysis system; ~ **de ancho de banda eficiente** TRANS bandwidth-efficient system; ~ **de antenas de radiación longitudinal** TEC RAD endfire array antenna; ~ **de anuncio de mensajes** TELEFON message announcement system; ~ **de aplicación** CONMUT, TELEFON application system; ~ **de apoyo** TELEFON support system; ~ **de archivo de red** DATOS network file system; ~ **de área extendida** DATOS, TELEFON, TRANS wide-area system; ~ **de asignación de memoria** DATOS storage allocation system; ~ **de aterrizaje por instrumentos** (*ILS*) TEC RAD instrument landing system (*ILS*); ~ **automático paso a paso** TELEFON step-by-step automatic system; ~ **de autor** INFO authoring system; ~ **de ayuda a la decisión** DATOS decision support system; ~ **básico de entrada/salida** (*BIOS*) PRUEBA basic input/output system (*BIOS*); ~ **de batería central** TELEFON central-battery system; ~ **de batería de filtros** TEC RAD *de satélite marítimo* filter-bank system; ~ **de batería local** TELEFON local battery system; ~ **de baterías** ELECTRÓN, ELECTROTEC, INFO battery system; ~ **binario** INFO binary system; ~ **binario de tarifa** TELEFON binary tariff system; ~ **de boletíns electrónicos** (*BBS*) DATOS, INFO, TELEFON bulletin board system (*BBS*); ~ **de bus** CONMUT bus system; ~ **buscapersonas** TELEFON staff locater system; ~ **de búsqueda bidireccional** TEC RAD two-way paging system; ~ **de cableado BULL** (*BCS*) DATOS, ELECTRÓN, TRANS BULL cabling system (*BCS*); ~ **de cableado IBM** DATOS, ELECTRÓN, TRANS IBM cabling system; ~ **de cables** DATOS, ELECTRÓN, TRANS cable system; ~ **de canalización** GEN duct system; ~ **de carga continua** TRANS krarup system; ~ **con carga independiente** ELECTROTEC separate charging system; ~ **con carga separada** ELECTROTEC divided-battery system; ~ **de CATV** TV CATV system; ~ **centralizado** CONMUT centralized system; ~ **centralizado para la separación de tráfico** CONMUT centralized traffic division system; ~ **de certificación** PRUEBA certification system; ~ **de circuito único** GEN single-circuit system; ~ **de codificación de cuatro firmas** TEC RAD four-signature coding system; ~ **de codificación de mensajes** APLIC, DATOS, TEC RAD, TELEFON message handling system (*MHS*); ~ **coherente de articulación múltiple** TV multipivoted coherent system; ~ **de color Munsell** TV Munsell colour (*AmE* color) system; ~ **combinado local-interurbano** CONMUT combined local-toll system; ~ **combinado de pérdida y retardo** TELEFON combined loss and delay system; ~ **comercial** APLIC, TEC RAD, TRANS business system; ~ **comercial de satélite** (*SBS*) APLIC, TEC RAD, TRANS satellite business system (*SBS*); ~ **en comprobación** (*SUT*) DATOS system under test (*SUT*); ~ **de comunicaciones comerciales** APLIC, TEC RAD, TRANS business communication system; ~ **de conmutación PCM** CONMUT, TELEFON PCM switching system; ~ **de comunicación de datos** DATOS, INFO data communication system; ~ **de comunicación militar** APLIC, TEC RAD, TRANS military communication system; ~ **de comunicación por cable radiante** TRANS radiating-cable communication system; ~ **de conexión del abonado** PERIF, TELEFON subscriber connection system; ~ **de conmutación** CONMUT message-switching system, switching system, TELEFON switching system, TRANS message-switching system; ~ **de conmutación bifilar** CONMUT two-wire switching system; ~ **de conmutación bifilar doble** CONMUT four-wire switching system; ~ **de conmutación de circuitos** CONMUT circuit-switching system; ~ **de conmutación por cable alámbrico** CONMUT hard-wired programmable switching system; ~ **de conmutación de división espacial** CONMUT, TELEFON spaced-division switching system; ~ **de conmutación por división de frecuencias** CONMUT FDS system, frequency-division switching system; ~ **de conmutación de elementos** ELECTROTEC *baterías* cell switch system; ~ **de conmutación híbrido** CONMUT, TRANS hybrid switching system; ~ **de conmutación de múltiple tarifa** CONMUT multirate switching system; ~ **de conmutación multiservicio** CONMUT multiservice switching system; ~ **de conmutación de protección** CONMUT, TELEFON protection switching system; ~ **de conmutación del videófono** CONMUT videophone switching system; ~ **conmutador rotatorio** CONMUT rotary system; ~ **de construcción mecánica** GEN equipment practice; ~ **de continuidad absoluta** ELECTROTEC no-break system; ~ **continuo de silenciador de tono controlado** (*CTCSS*) TEC RAD continuous tone-controlled squelch system (*CTCSS*); ~ **de control** CONTROL control system, monitor system; ~ **de control de actitud y de balanceo** TEC RAD attitude and roll control system; ~ **de control en bucle cerrado** CONTROL closed-loop control system; ~ **de control centralizado** CONMUT centralized control system; ~ **de control de la central** TELEFON exchange control system; ~ **de control de comunicación integrada** TEC RAD integrated communication control system; ~ **de control de impulsos** CONMUT, TELEFON pulse control system; ~ **de control indirecto** CONTROL indirect control system; ~ **de control del marcador** CONMUT marker control system; ~ **de control de posición de vuelo y de órbita** TEC RAD attitude and orbit control system; ~ **de control de programa cableado** CONTROL wired-programme (*AmE* wired-program) control system, WPC system; ~ **de control por registradores** CONMUT, TELEFON register-controlled system; ~ **de control totalmente distribuido** CONMUT fully-distributed control system; ~ **de corrección de errores** DATOS, TELEFON error-correcting system; ~ **de demanda de repetición** TELEFON request-repeat system; ~ **dependiente de códigos** TELEFON code-dependent system; ~ **de desensamblado de paquete facsímil** DATOS facsimile packet disassembly facility; ~ **de detección de errores** TELEFON error-detecting system; ~ **de detección de la portadora** DATOS carrier-sense system; ~ **de distribución con guía de luz** ELECTRÓN lightguide distribution system; ~ **de distribución de cabina y columna** TRANS cab and pillar distribution system; ~ **de distribución de línea troncal múltiple** TV multiple-trunk distribution system; ~ **dividido funcionalmente** DATOS functionally-divided system; ~ **de división espacial** CONMUT, TELEFON space-division system, spaced-division system; ~ **de división de función** CONMUT function division system; ~ **de división de tráfico** CONMUT traffic division system; ~ **de dos hilos** CONMUT two-wire system; ~ **de end** DATOS end system; ~ **de energía eléctrica** ELECTROTEC electrical power system; ~ **de enlaces** TRANS link system; ~ **enlazado** TEC RAD trunked system; ~ **de ensamblado de paquete facsímil** DATOS facsimile packet assembly facility; ~ **de equiseñal acoplado** TEC RAD interlocking equisignal system; ~ **espacial** TEC RAD space system; ~ **de espectro extendido** TEC RAD spread-spectrum system; ~ **de espera** TELEFON delay system; ~ **de estudio virtual** INFO virtual-studio system; ~ **de evaluación de la calidad y la fiabilidad** PRUEBA reliability and quality measurement system; ~ **de expedición de enlaces** TEC RAD trunked dispatch system; ~ **experto** APLIC, DATOS, INFO expert system; ~ **de facturación simple** APLIC single-billing system; ~ **fuente**

CONMUT source system; ~ **de gestión** DATOS managing system; ~ **gestionado** DATOS managed system; ~ **de gestión del conocimiento** (*KMS*) INFO knowledge management system (*KMS*); ~ **en gran escala** DATOS large-scale system; ~ **hexadecimal** DATOS, INFO hexadecimal system; ~ **híbrido** CONMUT, TRANS hybrid system; ~ **de identificación de la línea que llama** TELEFON calling-line identification facility; ~ **impulsor de casete** INFO tape cassette drive system; ~ **independiente de códigos** TELEFON code-independent system; ~ **de indicativos de país** GEN country-code system; ~ **de información de estación de barco** SEÑAL, TEC RAD ship reporting system; ~ **informático** INFO computer system; ~ **integrado** TRANS integrated system; ~ **inteligente de enseñanza asistida** (*ITS*) INFO intelligent tutoring system (*ITS*); ~ **inteligente para vehículos en carretera** (*IVHS*) TEC RAD intelligent vehicle highway system (*IVHS*); ~ **integrado de voz y datos** DATOS, TELEFON voice and data integrated system; ~ **interactivo digital vídeo-audio** (*VADIS*) INFO video-audio digital interactive system (*VADIS*); ~ **de intercambio de texto para mensaje** DATOS message-oriented text interchange system (*MOTIS*); ~ **de interfono** TELEFON intercom system; ~ **de interrogación** INFO polling system; ~ **de inversor** ELECTROTEC inverter system; ~ **de límites cruzados** TEC RAD cross-border system; ~ **lineal de tiempo discreto** TEC RAD discrete time-linear system; ~ **de línea ranurada** TEC RAD, TELEFON slotted system; ~ **de líneas** TELEFON, TRANS line system; ~ **de llamada selectiva** TEC RAD, TELEFON selective calling system (*SELCAL*); ~ **de llamadas de socorro radiotelefónicas** TEC RAD distress radio call system; ~ **de llamada visual** SEÑAL, TEC RAD visual call system; ~ **de localización** TEC RAD localization system; ~ **de localización de vehículos** TEC RAD vehicle localization system; ~ **de localización global** (*GPS*) GEN global positioning system (*GPS*); ~ **localizador de doble frecuencia** TEC RAD two-frequency localizer system; ~ **de lógica cableada** CONTROL wired-logic system; ~ **lógico** INFO logic system; ~ **loran** TEC RAD loran; ~ **de manguitos termoencogibles** COMPON heat-shrinkable sleeving; ~ **del marcador** CONMUT marker system; ~ **marítimo** TELEFON maritime system; ~ **marítimo local** TELEFON maritime local system; ~ **marítimo de satélite** TEC RAD maritime satellite system; ~ **MATV** TV MATV system; ~ **de medición** PRUEBA measurement process; ~ **de mensajería interpersonal** (*IPMS*) APLIC, DATOS, TEC RAD, TELEFON interpersonal messaging system (*IPMS*); ~ **de mensajería por intercambio de datos electrónicos** APLIC, DATOS, TEC RAD, TELEFON EDI-messaging system (*EDIMS*); ~ **de mensajería por voz** APLIC, TELEFON voice messaging system; ~ **monoprocesador** CONMUT single-processor system; ~ **Morse de telegrafía** TELEFON Morse telegraphy; ~ **móvil marítimo de satélite** TELEFON maritime mobile satellite system; ~ **móvil terrestre** TEC RAD, TELEFON, TRANS land-mobile system; ~ **de muestreo** PRUEBA sampling system; ~ **multimicroprocesador** CONMUT multimicroprocessor system; ~ **de multinivel** TRANS multilevel system; ~ **multiprocesador** CONMUT multiprocessor system; ~ **de multiservicio vía satélite** TEC RAD, TV satellite multiservice system; ~ **multitonal** TELEFON multitone system; ~ **mundial para comunicaciones móviles** (*GSM*) TEC RAD global system for mobile communications (*GSM*); ~ **mundial de satélites de telecomunicaciones** TEC RAD, TELEFON global telecommunications satellite system; ~ **mundial de telecomunicaciones** (*GTS*) TEC RAD, TELEFON global telecommunications system (*GTS*); ~ **de navegación hiperbólica** TEC RAD hyperbolic navigation system; ~ **neutral aislado** ELECTROTEC isolated neutral system; ~ **no duplicado** TELEFON nonduplicated system; ~ **no jerárquico** TRANS nonhierarchical system; ~ **de numeración** TELEFON numbering system; ~ **de oficina integrado** PERIF integrated office system; ~ **ómnibus** TELEFON omnibus system; ~ **de onda portadora de uno más uno** TRANS one-plus-one carrier system; ~ **de operación de CD-ROM en tiempo real** (*CD-SOTR*) INFO CD-ROM real-time operating system (*CD-RTOS*); ~ **de operación en red** ELECTRÓN networking operating system; ~ **operativo** (*SO*) DATOS, INFO, PRUEBA operating system (*OS*); ~ **operativo de disco Microsoft®** INFO Microsoft disk operating system; ~ **operativo de red** DATOS network operating system; ~ **de orden y control** CONTROL command and control system; ~ **Ostwald de color** INFO Ostwald colour (*AmE* color) system; ~ **panel** CONMUT panel system; ~ **de parada de seguridad** CONTROL emergency shutdown system, safety shutdown system; ~ **paso a paso** CONMUT step-by-step system; ~ **PCM** TELEFON PCM system; ~ **de pérdidas** TELEFON loss system; ~

planar TRANS planar system; ~ **polifásico** ELECTRÓN polyphase system; ~ **principal** DATOS host system; ~ **de procesamiento de datos** DATOS data-processing system, data processing (*DP*); ~ **de procesamiento múltiple** DATOS multiprocessing system; ~ **de proceso de datos** INFO data-processing system; ~ **de programación** DATOS, INFO programming system; ~ **de punto a multipunto** TRANS point-to-multipoint system; ~ **de punto a punto** TRANS point-to-point system; ~ **de radiación dirigida** TEC RAD guided radiation system; ~ **de radio de banda ancha** TEC RAD broadband radio system; ~ **de radiobúsqueda** TEC RAD radiopaging system, staff locater system; ~ **de radiobúsqueda por vía satélite** TEC RAD satellite paging system; ~ **de radiocomunicaciones ferroviarias** TEC RAD railroad radio system *AmE*, railway radio system *BrE*; ~ **de radiocomunicación móvil** TEC RAD mobile radio system; ~ **de radio enlace** TEC RAD radio-relay system; ~ **de radioenlace transhorizonte** TEC RAD transhorizon radio relay system; ~ **de radio por estación de derivación** TEC RAD shunting-yard radio system; ~ **de radio para servicio de salvamento** TEC RAD rescue-service radio system; ~ **de radiotaxi** TEC RAD taxi radio system; ~ **de radio de un tren** TEC RAD train radio system; ~ **Rebecca-Eureka** TEC RAD Rebecca-Eureka system; ~ **de referencia** GEN reference system; ~ **de referencia en bucle** (*LRS*) TELEFON loop reporting system (*LRS*); ~ **de reflexión múltiple** TEC RAD multihop system; ~ **de refrigeración** CLIMAT cooling system; ~ **de regulación por contrarreacción** CONTROL, INFO feedback control system; ~ **de relé en dos frecuencias** TEC RAD two-frequency relais system; ~ **de relé de láminas** COMPON, CONMUT, ELECTROTEC reed-relay system; ~ **de relé de radio de banda ancha** TEC RAD wideband radio relay system; ~ **de relé satélite** TEC RAD satellite relay system; ~ **remoto de conmutación** CONMUT remote switching system; ~ **remoto de voz protegida** (*REMSEVS*) TEC RAD remote secure voice system (*REMSEVS*); ~ **de repetidores sin amplificador de antena** TEC RAD nonboosted antenna repeater system; ~ **de reserva** GEN stand-by system; ~ **respondedor** DATOS transponder system; ~ **de respuesta telefónica** TELEFON voice response system; ~ **de retransmisión** CONMUT, TELEFON relay system; ~ **de retransmisión en dos frecuencias** TELEFON two-frequency relay system; ~ **de rodaje** DATOS, INFO script system; ~ **de satélite** TEC RAD, TV satellite system; ~ **secuencial** DATOS sequencing system; ~ **de seguimiento** TEC RAD tracking system; ~ **de seguimiento remoto** PRUEBA remote monitoring system; ~ **de senda** CONMUT bypath system; ~ **sensor** GEN sensor system; ~ **de señales de tráfico** GEN traffic signal system; ~ **de señalización** SEÑAL signalling (*AmE* signaling) system; ~ **de señalización de línea** CONTROL, SEÑAL line signalling (*AmE* signaling) system; ~ **de señalización ferroviaria** CONTROL, SEÑAL railroad signaling system *AmE*, railway signalling system *BrE*; ~ **de separación de banda de contacto** ELECTRÓN clamp-band separation system; ~ **simplex por canales conjugados** TELEFON two-way simplex system; ~ **síncrono en paralelo** INFO parallel synchronous system; ~ **de sintonía silenciosa** TEC RAD tone squelch system; ~ **SMATV** TV SMATV system; ~ **de sonido** INFO sound system; ~ **de soporte de pie** ELECTRÓN floor-standing system; ~ **de suministro de electricidad** ELECTROTEC electricity supply system; ~ **táctico de navegación aérea** TEC RAD tactical air navigation; ~ **de tarificación** APLIC toll system; ~ **de telecomunicaciones** TEC RAD telecommunications system; ~ **de telecomunicaciones ferroviarias** TEC RAD railroad telecommunications system *AmE*, railway telecommunications system *BrE*; ~ **de telefonía por interconexión de procesos** TELEFON switching process interworking telephony event (*SPITE*); ~ **de telefonía móvil** TEC RAD mobile telephone system, mobile telephony system, TELEFON mobile telephone system; ~ **telefónico con teclado** PERIF, TELEFON key telephone system; ~ **telefónico para zonas rurales** TELEFON rural telephone system; ~ **telegráfico de corrección de errores** DATOS, TELEFON error-correcting telegraph system; ~ **telegráfico doble equilibrado en puente** TELEFON bridge duplex system; ~ **de televisión** TV television system; ~ **de televisión de antena colectiva vía satélite** TV satellite master antenna television system; ~ **de televisión por cable** TV cable television system; ~ **de temporizador** GEN timer system; ~ **de tierra colectivo** ELECTROTEC collective earth system *BrE*, collective ground system *AmE*; ~ **de toma de tierra colectivo** ELECTROTEC collective earth system *BrE*, collective ground system *AmE*; ~ **de tramas ranuradas** TEC RAD frame-slotted system; ~ **transaccional** DATOS transactional system; ~ **de transferencia de datos** DATOS data transfer system; ~ **de transferencia de mensajes** DATOS, TELEFON message transfer system (*MTS*); ~ **de transmisión bidireccional** TEC RAD, TELEFON full duplex; ~ **de transmisión de mensaje** DATOS message transmission system; ~ **de trayectoria de planeo de doble frecuencia** TEC RAD two-frequency glide path system; ~ **de tuberías** COMPON, ELECTRÓN, ELECTROTEC tubing; ~ **universal de telecomunicaciones móviles** (*UMTS*) TEC RAD universal mobile telecommunication system (*UMTS*); ~ **universal de transporte** (*UTS*) ELECTRÓN universal transport system (*UTS*); ~ **de comunicación de vehículos inteligentes** (*VICS*) TEC RAD vehicle intelligent communication system (*VICS*); ~ **de vía gradual** TEC RAD step track system; ~ **de vías troncales** TEC RAD trunking system; ~ **de videotexto** INFO, TELEFON, TV videotext system; ~ **de vigilancia de barrido por control remoto** (*RCMSS*) CONTROL remote-control monitoring scan system (*RCMSS*); ~ **de visualización de información operativa** TEC RAD operational information display system; ~ **de visualización del documento** DATOS document imaging system; ~ **Wheatstone** TELEFON Wheatstone system

Sistema: ~ **Europeo de Mensajes Radiofónicos** *m* TEC RAD European Radio Message System

sistemas: ~ **y procedimientos** *m pl* INFO systems and procedures

sistemática *f* PRUEBA systematics

sistematización *f* GEN systematization

sistematizar *vt* GEN systematize

sitio: ~ **de lanzamiento** *m* TEC RAD launch site

situación *f* TEC RAD situation; ~ **de búsqueda** TEC RAD search condition; ~ **de desconexión** ELECTROTEC power-off condition; ~ **de fallo** TELEFON failure state; ~ **jurídica** GEN legal status

situado *adj* TEC RAD situated

situar *vt* TEC RAD situate

skiatrón *m* ELECTRÓN skiatron

SLAR *abr* (*radar aéreo de exploración lateral*) TEC RAD SLAR (*side-looking airborne radar*)

SLDA *abr* (*amplificador de láser diódico semiconductor*) ELECTRÓN SLDA (*semiconductor laser diode amplifier*)

SLIC *abr* (*circuito interfaz de línea de abonado*) CONMUT SLIC (*subscriber line interface circuit*)

SLP *abr* (*procedimiento de enlace único*) TELEFON SLP (*single-link procedure*)

SLS *abr* (*supresión del lóbulo lateral*) TEC RAD SLS (*side-lobe suppression*)

SMA *abr* (*adquisición en modo de búsqueda*) DATOS SMA (*search mode acquisition*)

SMAP *abr* (*protocolo de aplicación de gestión de sistemas*) DATOS SMAP (*systems management application protocol*)

SMATV *abr* (*televisión de antena colectiva vía satélite*) TV SMATV (*satellite master antenna television*)

SME *abr* (*equipo de gestión de sistemas*) INFO SME (*systems management equipment*)

SMI *abr* (*estructura de información de la gestión*) DATOS SMI (*structure of management information*)

SMO *abr* (*oscilador maestro estabilizado*) PRUEBA SMO (*stabilized master oscillator*)

SMPTE *abr* (*Sociedad de Técnicos Cinematográficos*) TRANS SMPTE (*Society of Motion Picture Technicians*)

SMSR *abr* (*ratio de supresión de modo lateral*) TRANS SMSR (*side-mode suppression ratio*)

SMTP *abr* (*protocolo simple de transferencia de correo*) DATOS SMTP (*simple mail transfer protocol*)

SMU *abr* (*unidad de medición de fuente*) PRUEBA SMU (*source-measure unit*)

SNMP *abr* (*protocolo simple de gestión de red*) DATOS SNMP (*simple network management protocol*)

SNPA *abr* (*punto de incorporación a subred*) DATOS SNPA (*subnetwork point of attachment*)

SO *abr* (*sistema operativo*) DATOS, INFO, PRUEBA OS (*operating system*)

soberano *m* INFO ruler

sobrante *adj* INFO excess

sobrealcance *m* COMPON overreaching

sobrecarga *f* DATOS, TRANS overload; ~ **de corriente de encendido** ELECTRÓN heater surge current; ~ **DCME** TELEFON DCME overload; ~ **de transmisión** TRANS transmission overload

sobrecargado *adj* CONMUT, ELECTROTEC overloaded

sobrecargar *vt* CONMUT, ELECTROTEC overload; ◆ ~ **fuera de órbita** TEC RAD blast out of orbit

sobrecorriente *f* CONMUT overcurrent, ELECTROTEC overcurrent, surge current

sobrecresta *f* TELEFON overshoot

sobredesvanecimiento *m* TEC RAD overfading

sobredesviación *f* TEC RAD overdeviation

sobreexponer *vt* DATOS *digitalizador*, TV overexpose

sobregrabación *f* INFO overdub, overdubbing

sobremesa: **de** ~ *adj* INFO desktop

sobremodulación *f* ELECTRÓN overexcitation, overmodulation
sobrepasar *vt* GEN exceed
sobreponer *vt* INFO, TELEFON superimpose
sobreposición *f* INFO, TELEFON superimposing, superimposition, superposition
sobretensión *f* ELECTROTEC overvoltage, voltage surge; ~ **positiva** ELECTROTEC positive overvoltage
sobrevoltaje *m* ELECTRÓN, ELECTROTEC surge
socavadura *f* COMPON, ELECTRÓN undercut
sociedad *f* GEN partnership; ~ **de cartera** GEN holding company
soda: ~ **cáustica líquida** *f* COMPON *química* liquid caustic soda
sofométrico *adj* TELEFON psophometric
sofómetro *m* TELEFON psophometer
software: ~ **de base** *m* DATOS, INFO system software; ~ **de cliente** *m* DATOS, INFO client software; ~ **del conocimiento** *m* DATOS, INFO knowledgeware; ~ **didáctico** *m* DATOS, INFO courseware; ~ **de dominio público** *m* DATOS, INFO public-domain software; ~ **de emulación de terminal** *m* DATOS, INFO terminal emulation software; ~ **de orientación** *m* DATOS, INFO steering software; ~ **de servidor** *m* DATOS, INFO server software
SOIC *abr* (*circuito integrado de contorno pequeño*) ELECTRÓN SOIC (*small-outline integrated circuit*)
solapa: **de** ~ *adj* INFO lap-held, lap-size
solapado: ~ **de líneas** *m* TV pairing
solapamiento *m* INFO, TEC RAD overlap
solape *m* INFO, TEC RAD overlapping
soldabilidad *f* COMPON *con estaño*, ELECTRÓN, TRANS solderability
soldable *adj* COMPON *con estaño*, ELECTRÓN, TRANS solderable
soldadura *f* COMPON, TRANS soldering, welding; ~ **automática** ELECTRÓN machine soldering; ~ **capilar** COMPON capillary soldering; ~ **en fase de vapor** COMPON condensation soldering; ~ **por fase vapor** COMPON vapour-phase (*AmE* vapor-phase) soldering; ~ **por inmersión** ELECTRÓN dip soldering; ~ **por refusión de estaño** ELECTRÓN reflow soldering; ~ **simultánea** ELECTRÓN mass soldering
soldar *vt* COMPON, ELECTRÓN, TRANS solder, weld
solenoide *m* ELECTROTEC solenoid
solicitante *m* DATOS invoker, ELECTRÓN applicant, TEC RAD, TELEFON invoker; ~ **indeseado** DATOS unwelcome caller
solicitud: ~ **en bucle digital de comando de reversión de bucle** *f* TV loopback command digital loop request; ~ **en bucle sonoro de comando de reversión de bucle** *f* TV loopback command audio loop request; ~ **de capacidades de transacción** *f* DATOS transaction capabilities application part; ~ **de conexión** *f* DATOS connection request; ~ **para enviar** *f* DATOS request to send; ~ **primitiva** *f* DATOS primitive request; ~ **de prioridad** *f* TELEFON priority request; ~ **de propuesta** *f* (*RFP*) GEN request for proposal (*RFP*); ~ **de sistema** *f* DATOS, TELEFON facility request; ~ **en videobucle de comando de reversión de bucle** *f* TV loopback command video loop request
solidez *f* GEN solidity; ~ **del color** GEN colourfastness (*AmE* colorfastness)
SOLR *abr* (*régimen de sonoridad objetiva de efecto local*) TELEFON SOLR (*side-tone objective loudness rating*)
solución: ~ **de emergencia** *f* GEN emergency solution
SOM *abr* (*media de opinión*) INFO MOS (*mean opinion score*)
sombear *vt* INFO hatch
sombra *f* ELECTRÓN, INFO shade, shadow, TEC RAD shadow, shadowing, TV shade; ~ **de cuello** ELECTRÓN neck shadow; ~ **de Gouraud** TV Gouraud shading; ~ **de interpolación de los colores** INFO colour (*AmE* color) interpolation shading; ~ **poligonal** INFO polygonal shading; ~ **del radar** TEC RAD radar shadow; ~ **radioeléctrica** TEC RAD radio shadow
sombreado[1] *adj* INFO dimmed
sombreado[2] *m* ELECTRÓN, TV shading; ~ **de interpolación de intensidad** TV intensity interpolation shading; ~ **de interpolación de vector normal** INFO normal-vector interpolation shading; ~ **Phong** INFO Phong shading; ~ **de símbolo** GEN symbol striping
someter *vt* INFO submit
someterse: ~ **a** *vt* PRUEBA undergo
sonar *m* TEC RAD sonar
sonda *f* PRUEBA, TEC RAD probe; ~ **de conexión a tierra** ELECTROTEC grounding probe *AmE*, earthing probe *BrE*; ~ **espacial** TEC RAD, TV space probe; ~ **de Hall** ELECTRÓN *semiconductores* Hall probe; ~ **ionosférica** TEC RAD ionospheric recorder; ~ **lógica** ELECTRÓN *circuitos impresos* logic probe; ~ **termométrica** CONTROL thermometer probe
sondear *vt* DATOS, INFO, TEC RAD poll
sondeo: ~ **ionosférico** *m* TEC RAD ionospheric sounding
sónico *adj* TELEFON sonic
sonido: ~ **atonal** *m* INFO pitchless sound; ~ **de**

tránsito lento *m* INFO slow-attack sound; ~ **ululado incorporado** *m* ELECTRÓN built-in warble tone

sonómetro *m* PRUEBA sound-level meter

sonoridad *f* GEN loudness; ~ **de intensidad creciente** TEC RAD escalating loudness

sonorizado *adj* APLIC voiced

soplete *m* GEN *para herramientas* cutting torch

soporte *m* COMPON holder; ~ **angular para montaje en el suelo** COMPON floor bracket, floor mounting bracket; ~ **del bastidor** TELEFON rack support; ~ **de la batería** ELECTROTEC battery tray; ~ **de cable** TRANS *tendido de cables* cable bracket; ~ **del componente** ELECTRÓN component holder; ~ **de cubierta** TELEFON case support, cover support; ~ **de datos automatizados** CONTROL automated data medium; ~ **de elevación de acimut** TEC RAD azimuth elevation mount; ~ **elevador** CONMUT lifting bar; ~ **de fijación** COMPON fixing bracket; ~ **físico** COMPON hardware; ~ **legible por la máquina** CONTROL machine-readable medium; ~ **lógico** DATOS, INFO software; ~ **lógico inalterable** DATOS, INFO firmware; ~ **maestro de producción de imagen múltiple** TV multiple-image production master; ~ **de memoria** INFO storage medium; ~ **de montaje** COMPON mounting holder; ~ **de muelles** COMPON spring holder; ~ **de relé** TELEFON relay support; ~ **del teclado** INFO keyboard stand; ~ **virgen** DATOS virgin medium

soportes: ~ **de acceso embutidos** *m pl* ELECTRÓN recessed access brackets; ~ **aislantes** *m pl* ELECTRÓN, ELECTROTEC base materials; ~ **de base** *m* COMPON base materials

sordina: ~ **de la tapa microfónica** *f* PERIF, TELEFON *de contestador* mouthpiece muffler *AmE*, mouthpiece silencer *BrE*

sostén *m* INFO sustain

spt *abr* (*sectores por pista*) INFO spt (*sectors per track*)

SPTV *abr* (*televisión de imagen fija*) TV SPTV (*still-picture television*)

SQL *abr* (*lenguaje de interrogación estructurado*) DATOS SQL (*structured query language*)

SQS *abr* (*Asociación Suiza para Certificados de Garantía de Calidad*) GEN SQS (*Swiss Association for Quality Assurance Certificates*)

SREJ *abr* (*trama de rechazo selectivo*) TELEFON SREJ (*selective reject frame*)

SRR *abr* (*búsqueda y rescate*) RADIO SRR (*search and rescue*)

SRS *abr* (*dispersión Raman estimulada*) TEC RAD SRS (*stimulated Raman scattering*)

SRT *abr* (*transparente de sonido fundamental de la fuente*) ELECTRÓN SRT (*source rooting transparent*)

SRVA *abr* (*servicios de red de valor añadido*), (*VANS*) APLIC, TRANS, TV VANS (*value-added network services*)

SS *abr* (*espectro extendido*) TEC RAD SS (*spread spectrum*)

SSA *abr* (*montaje de sensor solar*) TEC RAD SSA (*sun sensor assembly*)

SSAP *abr* (*punto de acceso al servicio de sesión*) DATOS, TRANS SSAP (*session service access point*)

SSC *abr* (*centro de conmutación de satélite*) CONMUT, TEC RAD SSC (*satellite switching centre*)

SSM *abr* (*mensaje de segmento único*) DATOS SSM (*single-segment message*)

SSMI *abr* (*generador de imágenes de microondas con sensor especial*) ELECTRÓN SSMI (*special sensing microwave imager*)

SSO *abr* (*desconexión espuria*) PRUEBA SSO (*spurious switch-off*)

SSR *abr* (*recepción de régimen estacionario*) TEC RAD SSR (*steady-state reception*)

S/SYS *abr* (*sistema de almacenamiento*) DATOS S/SYS (*storage system*)

ST *abr* (*tipo de segmento*) DATOS, INFO ST (*segment type*)

STA *abr* (*algoritmo de árbol espaciador*) ELECTRÓN STA (*spanning-tree algorithm*)

STC *abr* (*atenuador*) CONTROL STC (*sensitivity time control*), INFO attenuator, fader, TRANS attenuator

STL *abr* (*enlace entre estudio y transmisor*) TEC RAD STL (*studio-transmitter link*)

ST/SYS *abr* (*sistema de almacenamiento y transferencia*) DATOS ST/SYS (*storage and transfer system*)

SU *abr* (*unidad de señal*) SEÑAL SU (*signal unit*)

SUB *abr* (*subdirección*) APLIC, DATOS, TELEFON SA (*subaddress*), SUB (*subaddressing*)

subactividad *f* GEN subactivity

subagenda *f* INFO subnotebook

subárbol *m* DATOS, TELEFON *de proveedor de MCS* subtree

subarmónica *f* ELECTRÓN subharmonic

subbanda *f* CONMUT, INFO, TEC RAD sub-band

subbastidor *m* TELEFON subrack

subcadena *f* DATOS substring

subcampo: ~ **inferior** *m* TEC RAD, TRANS lower subfield

subcapa: ~ **de convergencia** *f* DATOS convergence sublayer; ~ **de convergencia de transmisión** *f* DATOS, TRANS transmission

convergence sublayer; ~ **de segmentación y rearmado** *f* DATOS segmentation and reassembly sublayer
subconjunto *m* DATOS subassembly; ~ **de afinación de uniformidad** DATOS consistent-refinement subset; ~ **uniforme de partición** DATOS consistent partitioning subset
subcontratación *f* GEN subcontracting
subcontratista *m* GEN subcontractor
subcubo *m* TRANS sub-hub
subdirección *f* (*SUB*) APLIC, DATOS, TELEFON subaddress (*SA*), subaddressing (*SUB*); ~ **terminal** APLIC terminal subaddressing
subdirectorio *m* INFO subdirectory
subdominio: ~ **denominador** *m* DATOS naming subdomain
subestación: ~ **de pantalla táctil** *f* INFO touch-screen kiosk
subestructura: ~ **de canal** *f* DATOS channel substructure
subfunción *f* GEN subfunction
subgrupo *m* TELEFON, TRANS subgroup; ~ **de búsqueda** TELEFON search subgroup; ~ **de circuitos** TELEFON, TRANS circuit subgroup; ~ **racional** PRUEBA rational subgroup
subíndice *m* DATOS, INFO subscript
subir *vi* INFO move up
submenú *m* DATOS submenu
subminiatura *f* ELECTRÓN subminiature
submodulación *f* TRANS undermodulation
submuestreo *m* INFO, TELEFON subsampling
submultiplexar *vt* TRANS submultiplex
suboperación *f* GEN suboperation
subpoblación *f* PRUEBA subpopulation
subportadora *f* GEN subcarrier
subprograma *m* DATOS, INFO subprogram; ~ **cerrado** INFO closed subroutine
subred *f* DATOS, TRANS, TV subnetwork; ~ **de gestión de jerarquía digital síncrona** TRANS SDH management subnetwork (*SMS*)
subrefracción *f* TEC RAD subrefraction
subreflector *m* TEC RAD subreflector
subruta *f* TELEFON subroute
subrutina *f* INFO subroutine; ~ **de programa reintroducible** TELEFON re-enterable program subroutine; ~ **reintroducible** TELEFON re-enterable subroutine
subsidiario *adj* GEN subsidiary
subsiguiente *adj* GEN subsequent
subsistema *m* GEN subsystem; ~ **de acceso** TEC RAD access subsystem; ~ **de control de determinación de posición de vuelo** TEC RAD attitude determination control subsystem; ~ **de estación de base** TELEFON base station subsystem (*BSS*); ~ **funcional de acceso de abonados** TELEFON subscriber access functional subsystem; ~ **de gestión de la transacción** DATOS transaction management subsystem; ~ **de localización de vehículos** TEC RAD vehicle localization subsystem; ~ **OM** (*subsistema de operaciones y mantenimiento*) DATOS, PRUEBA, TELEFON OM subsystem (*operations and maintenance subsystem*); ~ **de operaciones y mantenimiento** (*subsistema OM*) DATOS, PRUEBA, TELEFON operations and maintenance subsystem (*OM subsystem*); ~ **de potencia** (*PSS*) ELECTROTEC power subsystem (*PSS*); ~ **de servicio** INFO housekeeping subsystem
substrato *m* COMPON, ELECTRÓN substrate; ~ **de arseniuro de galio** COMPON, ELECTRÓN gallium arsenide substrate
substructura: ~ **de bloques** *f* DATOS block substructure
subtipo *m* DATOS subtype
subtitulado *m* TV subtitling
subtítulo *m* INFO subheader, subheading, TV subtitle
subunidad *f* ELECTRÓN, TELEFON subunit
subvariable *f* INFO subvariable
succión: ~ **al vacío** *f* ELECTRÓN vacuum suction
sucesión: ~ **de acontecimientos** *f* GEN course of events; ~ **de ensayos ejecutables** *f* DATOS, PRUEBA executable test suite (*ETS*); ~ **genérica de ensayos** *f* DATOS, PRUEBA generic test suite; ~ **de pruebas abstractas normalizadas** *f* DATOS, PRUEBA standardized abstract test-suite; ~ **de pruebas abstractas parametrizadas** *f* (*PATS*) DATOS, PRUEBA parameterized abstract test suite (*PATS*); ~ **de pruebas ejecutables parametrizadas** *f* (*PETS*) DATOS, PRUEBA parameterized executable test suite (*PETS*)
suceso *m* DATOS, ELECTRÓN, INFO, PRUEBA, TEC RAD, TELEFON event; ~ **ionizante** ELECTRÓN, TEC RAD ionizing event; ~ **ionizante inicial** ELECTRÓN, TEC RAD initial ionizing event
sufijo: ~ **de mensaje** *m* INFO message suffix
sujeción: ~ **firme** *f* ELECTRÓN fast clamping
sujetador: ~ **de barra** *m* COMPON *normas de montaje* bar holder; ~ **de cubierta** *m* COMPON case holder, cover holder; ~ **pequeño de plástico rojo** *m* ELECTRÓN small red plastic clip
suma: ~ **de comprobación** *f* DATOS check sum
sumador: ~ **paralelo** *m* ELECTRÓN parallel adder
sumar *vt* GEN add
suministrador: ~ **de servicios CMISE** *m* DATOS, TELEFON CMISE service provider
suministrar *vt* GEN supply

suministro: ~ **de corriente interrumpible** *m* ELECTROTEC interruptible power supply; ~ **eléctrico por batería** *m* ELECTRÓN, ELECTROTEC battery power supply; ~ **de energía OEM** *m* ELECTROTEC OEM power supply; ~ **de portadora redundante** *m* TRANS redundant carrier-supply; ~ **de potencia** *m* ELECTROTEC power feed

superación: ~ **de señal de operador ocupado** *f* TELEFON attendant busy override

superar *vt* INFO outperform

superconductor *m* ELECTRÓN, TRANS superconductor

superficie: ~ **de almacenamiento** *f* ELECTRÓN storage surface; ~ **de apoyo** *f* COMPON bearing surface; ~ **en bruto** *f* ELECTRÓN plate finish; ~ **de contacto** *f* COMPON contact surface; ~ **pasivada** *f* COMPON passivated surface; ~ **polar** *f* TELEFON pole face; ~ **de referencia** *f* TELEFON, TRANS reference surface; ~ **rugosa** *f* GEN orange peel; ~ **de la señal** *f* INFO signal surfacc; ~ **de soldadura** *f* COMPON solder pad; ~ **de visualización** *f* INFO display surface

supergrupo *m* TELEFON, TRANS supergroup

superheterodino *adj* TEC RAD superheterodyne

superíndice *m* DATOS, INFO superscript

superior *adj* INFO upper

superluminiscencia *f* ELECTRÓN, TRANS superluminescence

superponer *vt* INFO, TELEFON superpose, superimpose

superposición: ~ **de vídeo** *f* TEC RAD, TV video mapping

superradiancia *f* TRANS superradiance

superred *f* INFO, TELEFON superlattice

superrefracción *f* TEC RAD superrefraction

supervisar *vt* GEN supervise

supervisión *f* GEN supervision; ~ **de analizador de espectro** PRUEBA, TEC RAD spectrum-analyzer monitoring; ~ **de cierre de ruta** TELEFON route lockout supervision; ~ **de circuito de extremo a extremo** PRUEBA end-to-end circuit supervision; ~ **continua** PRUEBA continuous supervision; ~ **a distancia** CONTROL remote supervision, PRUEBA, TEC RAD telemonitoring, TV remote supervision, remote survey; ~ **de estado** CONTROL, PRUEBA status monitoring; ~ **del funcionamiento** TELEFON performance monitoring; ~ **de fusibles** ELECTROTEC fuse supervision; ~ **de grupo** TELEFON group monitoring; ~ **de nivel** TEC RAD level monitoring; ~ **de ocupaciones** TELEFON seizure supervision; ~ **de la onda portadora** TRANS carrier-wave indication supervision; ~ **ordinaria** PRUEBA routine monitoring; ~ **de perturbaciones** TELEFON disturbance supervision; ~ **de la red** DATOS, TELEFON network supervision; ~ **de la red eléctrica** ELECTROTEC mains monitoring; ~ **de ruido** TELEFON noise monitoring, noise supervision; ~ **de ruta** TELEFON route supervision; ~ **del servicio** TELEFON service supervision; ~ **del sistema** INFO system monitor; ~ **del sistema de comunicaciones** PRUEBA communications system monitoring (*CSM*); ~ **de la televisión** PRUEBA, TV television monitoring; ~ **de tráfico** ELECTRÓN, PRUEBA, TELEFON traffic monitoring, traffic supervision

supervisor *m* CONTROL, INFO, PRUEBA, TELEFON supervisor; ~ **jefe** TELEFON chief supervisor; ~ **de la red** PRUEBA network supervisor; ~ **de la red eléctrica** CONTROL, ELECTROTEC mains monitor; ~ **de ruido** PRUEBA noise monitor; ~ **del sistema** CONTROL system monitor; ~ **de tráfico** ELECTRÓN, PRUEBA, TELEFON traffic monitor

suplementar *vt* GEN supplement

supraconductor *m* ELECTRÓN, TRANS supraconductor

suprafónico *adj* ELECTRÓN supraphonic

supresión *f* ELECTRÓN, TEC RAD, TELEFON suppression; ~ **de cuadratura** TRANS squareax; ~ **del eco** TRANS echo suppression; ~ **de haz** TEC RAD back-off; ~ **del lóbulo lateral** (*SLS*) TEC RAD side-lobe suppression (*SLS*); ~ **del lóbulo lateral del interrogador** (*ISLS*) TEC RAD interrogator side-lobe suppression (*ISLS*); ~ **de modo lateral** TRANS side-mode suppression; ~ **de perturbaciones** PRUEBA disturbance suppression; ~ **de la portadora** TRANS carrier rejection, carrier suppression; ~ **de reflejos** TEC RAD clutter suppression; ~ **de ruido** TEC RAD noise suppression, squelch

suprimido *adj* ELECTRÓN, TEC RAD, TELEFON suppressed

suprimir *vt* GEN delete; ◆ ~ **la fluctuación de** TRANS dejitter

surfear *vt* DATOS surf

susceptibilidad: ~ **de error local** *f* TEC RAD site-error susceptibility; ~ **de errores** *f* DATOS error susceptibility

suscribirse *v refl* TV subscribe

suscripción: ~ **básica** *f* TELEFON basic subscription; ~ **multizonal** *f* TEC RAD multi-zone subscription

suspensión *f* TELEFON cease

suspensor: ~ **de cable** *m* TRANS cable lasher

sustancia: ~ **ferromagnética** *f* GEN ferromagnetic substance
sustitución *f* INFO, TELEFON replacement; ~ **de tarjeta** PERIF *teléfonos de pago* card follow-on
sustituir *vt* INFO, TELEFON replace
sustractor *m* ELECTRÓN subtracter
SUT *abr* (*sistema en comprobación*) DATOS SUT (*system under test*)
SWG *abr* (*portaondas de ranura*) DATOS, TEC RAD, TRANS SWG (*slot waveguide*)

T

T: ~ **híbrida** *f* TEC RAD hybrid T; ~ **mágica** *f* TEC RAD magic T; ~ **en paralelo** *f* TEC RAD shunt T; ~ **en serie** *f* TEC RAD series T

tabla: ~ **de adjudicación de archivo** *f* (*FAT*) DATOS file allocation table (*FAT*); ~ **de barras** *f* GEN bar chart; ~ **de cableado** *m* GEN *documentación* cable-set table, cabling table; ~ **de códigos** *f* TELEFON code table; ~ **de colocación de hilos** *f* GEN running list; ~ **de conexiones** *f* COMPON wiring table; ~ **de consulta de colores** *f* (*CLUT*) INFO colour (*AmE* color) look-up table (*CLUT*); ~ **de contingencia** *f* GEN *estadística* contingency table; ~ **de conversión** *f* INFO conversion table; ~ **de decisión** *f* GEN decision table; ~ **de frecuencias** *f* TRANS frequency table; ~ **de funciones** *f* GEN function table; ~ **de restricción simple** *f* DATOS single-constraint table; ~ **de terminales** *f* ELECTROTEC, TELEFON terminal table; ~ **de verdad** *f* ELECTRÓN truth table

tablero *m* TEC RAD dashboard; ~ **de aglomerado** COMPON particle board; ~ **de compresión de imagen** DATOS, TV image-compression board; ~ **conmutador de centralita privada** TELEFON PBX switchboard; ~ **de distribución** ELECTROTEC *suministro de energía* distribution panel; ~ **impreso de dos caras** ELECTRÓN *circuitos impresos* double-sided printed board; ~ **de instrumentos** PRUEBA instrument panel; ~ **de llaves** PERIF key shelf; **sonoro** INFO sound board; ~ **de terminales** ELECTROTEC terminal board

tableta *f* INFO tablet

tablón: ~ **de anuncios** *m* TELEFON notice board

tabulación *f* INFO tabulation

tabulador *m* INFO tabulator

taburete *m* GEN stool

Tacán *m* TEC RAD Tacan

tacógrafo *m* GEN tachograph

táctil *adj* INFO tactile

taller: ~ **de reparaciones** *m* TELEFON repair shop

tamaño: ~ **A4** *m* GEN *de documentación* A4 size; ~ **básico** *m* GEN basic size; ~ **de canal** *m* TRANS channel size; ~ **del diagrama** *m* GEN *dibujo técnico* diagram size; ~ **de documento** *m* DATOS *Internet* document size; ~ **de fuente tipográfica** *m* INFO font size; ~ **medio de las rachas** *m* PRUEBA *calidad* average run length; ~ **del papel** *m* PERIF paper size; ~ **de punto** *m* TV spot size

tambor: ~ **de impresión** *m* PERIF, TELEFON print drum; ~ **de traza** *m* INFO drum plotter

tamiz *m* ELECTRÓN sieve

tamizado: ~ **de guíaondas** *m* TEC RAD, TRANS waveguide bolting

tamizar *vt* ELECTRÓN sift

tándem[1]: **en** ~ *adj* TELEFON tandem

tándem[2]: **en** ~ *adv* TELEFON tandem

tanque: ~ **de contacto** *m* ELECTROTEC contact bank; ~ **de expansión** *m* GEN *contenedores* expansion tank

tapa *f* COMPON, INFO lid; ~ **de auricular** PERIF, TELEFON earcap; ~ **deslizante** COMPON sliding lid; ~ **microfónica** PERIF transmitter cap; ~ **posterior** GEN rear cover; ~ **de recipiente** COMPON housing lid; ~ **roscada** COMPON screw cover

tapón: ~ **de presión** *m* GEN pressure plug; ~ **de tuerca** *m* COMPON screw cap

tarea *f* INFO subjob; ~ **de desintonización** INFO detune job; ~ **rutinaria** INFO routine task; ~ **secundaria** DATOS background task

tarifa *f* APLIC, TELEFON tariff; ~ **inicial** GEN *telefonía* access charge rate; ~ **de llamada** APLIC call charge; ~ **local** TELEFON local rate, local tariff; ~ **más elevada** GEN higher tariff; ~ **media** APLIC *tasa de conversación* medium rate; ~ **mínima** APLIC minimum charge; ~ **reducida** GEN lower tariff; ~ **reducida de llamadas** APLIC, TELEFON cheap-call rate

tarificación *f* TELEFON metering

tarjeta *f* GEN card; ~ **de alarma** APLIC alarm card; ~ **de apertura** GEN aperture card; ~ **de bus local** INFO local bus card; ~ **de cargo** TELEFON charge card; ~ **CGA** INFO CGA card; ~ **de circuitos integrados** *m* ELECTRÓN chip card; ~ **comercial** GEN business card; ~ **de comunicación** DATOS communication card; ~ **de control de comunicaciones** INFO communication control card; ~ **de datos** INFO data card; ~ **de disco rígido** INFO hard-disk card; ~ **electrónica enchufable** ELECTRÓN plug-in chip card; ~ **de expansión** INFO expansion card; ~ **flexible rígida** ELECTRÓN rigid flex card; ~ **inteligente** APLIC smartcard; ~ **de interco-**

nexión universal S TELEFON *red de transmisión digital de servicios integrados* S-universal interface card; ~ **de línea con ramificación de salida** ELECTRÓN fan-out line card; ~ **de llamada** TELEFON calling card; ~ **con memoria** INFO smartcard; ~ **de memoria** DATOS memory card; ~ **de módem** DATOS modem card; ~ **de la palanca de arranque** APLIC, TELEFON swipe card; ~ **de periférico** INFO peripheral card; ~ **prepagada** TELEFON prepaid card; ~ **de procesamiento** DATOS, INFO processing card; ~ **de procesamiento de voz** APLIC speech-processing card; ~ **de punto nodal** INFO nodecard; ~ **de tarificación** TELEFON toll ticket; ~ **telefónica** DATOS, TELEFON phonecard; ~ **de vídeo** INFO video card

tasa *f* GEN rate; ~ **básica de bitios** DATOS basic bit rate; ~ **de bitios erróneos** DATOS, INFO, TEC RAD bit error rate; ~ **de bits** DATOS, INFO, TELEFON, TRANS bit rate; ~ **de caracteres erróneos** DATOS, INFO character error rate; ~ **de carga** APLIC charge rate; ~**de error de bitio a plazo medio** INFO medium-term bit error rate; ~**de error por elemento unitario** TELEFON *modulación isócrona* unit-element error rate; ~ **de errores binarios** TEC RAD binary error rate; ~ **de errores en los elementos** TELEFON element error rate; ~ **de fallos** GEN failure rate, fault rate; ~ **de incremento** GEN rate of increase; ~ **de interrupción** TELEFON rate of interruption; ~ **de línea terrestre** TELEFON land-line charge; ~ **de llamadas del abonado** TELEFON subscriber calling rate; ~ **máxima** GEN *conversación telefónica* peak rate; ~ **de media de fallos** INFO, PRUEBA mean failure rate; ~ **de ocupación** GEN occupancy rate; ~ **de ocupación de canal** TEC RAD channel occupancy rate; ~ **de penetración** TV penetration rate; ~ **de relleno** GEN filling rate; ~ **de reparaciones** PRUEBA repair rate; ~ **de tarificación** TELEFON metering rate; ~ **terrestre** TEC RAD, TELEFON land-station charge; ~ **de tránsito** TELEFON transit share; ~ **de transmisión** GEN churn rate

tasable *adj* TELEFON subject to charge

tasación *f* GEN valuation; ~ **automática con tarjetas** TELEFON toll ticketing

taxonomía: ~ **confusa** *f* INFO tangled taxonomy

TCI-DCS *abr* (*DCS de identificación del abonado emisor*) DATOS, TELEFON TCI-DCS (*transmitting subscriber identification-DCS*)

TCL *abr* (*pérdida de conversión transversal*) TELEFON TCL (*transverse conversion loss*)

TCM *abr* (*modulación codificada en enrejado*) DATOS, TRANS TCM (*trellis-coded modulation*)

TCR *abr* (*orden de telemetría y determinación de distancias*) PRUEBA TCR (*telemetry command and ranging*)

TCRF *abr* (*función de tránsito relacionada con la conexión*) DATOS, TRANS TCRF (*transit connection-related function*)

Tcu *abr* (*tiempo de transferencia en una centralita*) CONMUT Tcu (*cross-office transfer time*)

TDC *abr* (*código télex de destino*) TELEFON TDC (*telex destination code*)

TDD *abr* (*aparato de telecomunicaciones para sordos*) GEN TDD (*telecommunications device for the deaf*)

TDM *abr* (*teclado de desplazamiento mínimo*) SEÑAL MSK (*minimum-shift keying*)

TDMA *abr* (*acceso múltiple por división de tiempo*) SIG, TEC RAD, TELEFON, TRANS TDMA (*time-division multiple access*); ~ **de ruta estrecha** *f* TEC RAD thin-route TDMA

TE *abr* (*equipo terminal*) DATOS, INFO, PERIF, TELEFON TE (*terminal equipment*)

TEC *abr* (*transistor de efecto de campo*) COMPON, ELECTRÓN *semiconductores* FET (*field-effect transistor*)

tecla: ~ **alfanumérica** *f* PERIF *teclados* data character key; ~ **de atajo** *f* INFO keyboard shortcut; ~ **blanda** *f* INFO soft key; ~ **de cifra** *f* INFO, TELEFON digit key; ~ **de continuación** *f* INFO continue key; ~ **de control de un solo toque** *f* INFO single-stroke control key; ~ **de emisión de impulsos** *f* TELEFON pulsing key; ~ **de función** *f* CONTROL, INFO function key; ~ **para llamadas externas** *f* TELEFON key to callouts; ~ **de mayúsculas** *f* INFO, PERIF shift key; ~ **de ocupación** *f* INFO break key; ~ **seguir** *f* INFO key follow; ~ **de selección directa** *f* TEC RAD key for direct selection, key for preset selection; ~ **de servicio de repetición** *f* INFO repeat action key

teclado *m* APLIC, INFO key set, keyboard; ~ **alfanumérico** INFO alphanumeric keyboard; ~ **de desplazamiento mínimo** (*TDM*) SEÑAL minimum-shift keying (*MSK*); ~ **en diente de sierra** PERIF sawtooth keyboard; ~ **de función** INFO function keyboard; ~ **inferior** INFO lower keyboard; ~ **múltiple** TEC RAD multiple-key set; ~ **numérico** APLIC, GEN keypad; ~ **piezosensible** INFO pressure-sensitive keyboard; ~ **de registro** INFO recording keyboard; ~ **para registro de imágenes** INFO keyframe; ~ **con seguro de cambio** INFO shift-lock keyboard; ~ **con seguro de inversión**

PERIF shift-lock keyboard; ~ **sensible a la velocidad** INFO velocity-sensitive keyboard; ~ **superior** INFO upper keyboard; ~ **telefónico** TELEFON push-button set

tecleado *m* APLIC, PERIF, TELEFON keying; ~ **de cambio de amplitud** (*ASK*) SEÑAL amplitude-shift keying (*ASK*); ~ **de cambio en la relación de fase de amplitud** (*APSK*) SEÑAL amplitude phase-shift keying (*APSK*)

teclear *vt* GEN *información, texto* key

tecleo: ~ **de cambio de fase de banda estrecha** *m* SEÑAL narrow-band phase-shift keying (*NBPSK*); ~ **de cambio de fase multinivel** *m* (*MPSK*) SEÑAL multilevel phase-shift keying (*MPSK*); ~ **de inversión de frecuencia con filtro gaussiano** *m* (*GFSK*) SEÑAL Gaussian-filtered frequency-shift keying (*GFSK*); ~ **neutro** *m* TELEFON neutral keying

técnica: ~ **de campo próximo refractado** *f* TRANS *de fibras* refracted near-field technique; ~ **de conmutación** *f* CONMUT switching technique, switching technology; ~ **de crisol doble** *f* TRANS *óptica de fibras* double-crucible technique; ~ **de descripción formal** *f* (*FDT*) DATOS formal description technique (*FDT*); ~ **intercambiadora de iones** *f* ELECTRÓN ion-exchange technique; ~ **de medición** *f* PRUEBA measurement technique; ~ **de modulación** *f* TEC RAD modulation technique; ~ **de perturbación** *f* TRANS perturbation technique; ~ **planar** *f* COMPON planar technique; ~ **de predistorsión** *f* ELECTRÓN predistortion technique; ~ **de prueba** *f* PRUEBA testing technique; ~ **de recuento de impulsos** *f* TEC RAD pulse-counting technique; ~ **de representación** *f* INFO rendering technique; ~ **de retrodispersión** *f* TELEFON, TRANS backscattering technique; ~ **de teletráfico** *f* TELEFON teletraffic engineering; ~ **de transmisión** *f* TRANS transmission technique; ~ **de ubicación cero** *f* TEC RAD null placing technique; ~ **de varilla en tubo** *f* ELECTRÓN, TRANS rod-in-tube technique

técnicas: ~ **de cambio de código** *f pl* TELEFON code-extension techniques; ~ **de control** *f pl* CONTROL *automático* control engineering

técnico: ~ **de apoyo de la red** *m* GEN network support engineer; ~ **electrónico** *m* GEN electronics technician; ~ **de estudios** *m* GEN development engineer; ~ **de laboratorio** *m* GEN laboratory technician; ~ **mecánico** *m* GEN engineering mechanic; ~ **de montaje** *m* GEN editor; ~ **de operaciones y mantenimiento** *m* TELEFON operation and maintenance engineer; ~ **proyectista** *m* GEN engineering draftsman; ~ **de servicio** *m* GEN service technician; ~ **de servicio de ordenadores personales** *m* GEN PC support engineer; ~ **de telecomunicaciones y equipos electrónicos** *m* GEN telecommunications and electronic equipment technician; ~ **de ventas** *m* GEN sales engineer

tecnología: ~ **aplicada** *f* ELECTRÓN, INFO know-how; ~ **avanzada** *f* GEN cutting-edge technology; ~ **de componentes** *f* ELECTRÓN *del diseño* component engineering, component technology; ~ **de corrientes fuertes** *f* ELECTROTEC power-current engineering; ~ **la información** *f* COMP information technology (*IT*); ~ **de paso triple** *f* DATOS *de digitalizador* triple-pass technology; ~ **de refrigeración** *f* CLIMAT refrigeration engineering; ~ **de voz** *f* APLIC, INFO, TEC RAD, TELEFON speech technology

tecnológico *adj* GEN technological

tejido: ~ **absorbente** *m* TEC RAD space cloth

tela *f* COMPON *materiales* fabric; ~ **sin pelusas** ELECTRÓN lint-free cloth

telaraña: ~ **mundial** *f* (*WWW*) INFO World Wide Web (*WWW*)

teleafiche *m* DATOS remote indication

telearranque *m* CONTROL remote start

teleautógrafo *m* DATOS, INFO, PERIF, TEC RAD, TV telewriting

telecomercialización *f* GEN telemarketing

telecompra *f* GEN armchair shopping, home shopping

telecomunicación *f* INFO, TEC RAD, TELEFON telecommunication; ~ **móvil por vía satélite** TEC RAD mobile satellite telecommunication; ~ **personal universal** (*UPT*) APLIC universal personal telecommunication (*UPT*)

telecomunicaciones *f pl* GEN telecommunications (*telecom*); ~ **de relevo** TELEFON relief telecommunications

Telecomunidad: ~ **de Asia y el Pacífico** *f* TEC RAD Asia-Pacific Telecommunity

teleconferencia *f* APLIC, TEC RAD, TRANS teleconference; ~ **de tres direcciones** APLIC, TELEFON three-way teleconference

teleconmutación *f* GEN telecommuting

teledetección *f* PRUEBA remote detection

teledirección *f* PRUEBA remote management

teleespectador *m* TV television viewer

telefax *m* INFO telefax, PERIF facsimile terminal, fax machine, TELEFON facsimile terminal, telefax; ~ **rápido** TELEFON fast facsimile; ~ **en redes privadas** TELEFON facsimile on private networks

telefonía *f* TELEFON telephony; ~ **de banda ancha** TELEFON wideband telephony; ~ **de banda normal** TELEFON normal-band telephony; ~ **por cable** SEÑAL cable telephony; ~ **de cuatro canales** TELEFON four-channel telephony; ~ **estereofónica de dos canales** TELEFON twin-channel stereo telephony; ~ **ferroviaria** TEC RAD, TELEFON railroad telephony *AmE*, railway telephony *BrE*; ~ **móvil** TEC RAD, TELEFON mobile telephony; ~ **periférica** TELEFON peripheral telephony; ~ **pública móvil** TELEFON public wireless telephony; ~ **rural** TELEFON rural telephony; ~ **visual** INFO, TEC RAD, TELEFON, TV visual telephony

teléfono *m* TELEFON telephone; ~ **activado por la voz** TELEFON voice-activated phone; ~ **amplificado** PERIF amplified telephone; ~ **autoalimentado** TELEFON sound-powered telephone; ~ **de botonera** TELEFON touch-tone telephone; ~ **de campaña** TELEFON field telephone; ~ **de coche** TELEFON car telephone, car phone; ~ **combinado** TEC RAD *móviles*, TELEFON combi-telephone; ~ **con contestador** PERIF, TELEFON answerphone; ~ **de disco** PERIF, TELEFON dial telephone; ~ **de disco dactilar** PERIF, TELEFON dial telephone set; ~ **de emergencia** PERIF, TELEFON emergency telephone, emergency telephone set; ~ **especializado** PERIF, TELEFON specialized telephone; ~ **de extensión** CONMUT *PBX*, TELEFON extension telephone set; ~ **con fax** PERIF, TELEFON fax phone; ~ **de ficha** TELEFON coin box; ~ **inalámbrico** TELEFON cordless phone; ~ **intercom** TELEFON intercom telephone; ~ **llamado** TELEFON called telephone; ~ **manos libres** TELEFON speakerphone; ~ **de mesa** TELEFON table instrument, table set; ~ **monedero** TELEFON payphone; **móvil** TEC RAD, TELEFON mobile telephone, mobile; ~ **multifrecuencia** TELEFON MF telephone, multifrequency telephone; ~ **de observación** PERIF, TELEFON observation telephone; ~ **de onda continua** TELEFON phone/cw; ~ **de pago previo** TELEFON coin-box set; ~ **de pago con tarjeta** TELEFON card-operated payphone, cardphone; ~ **de pared** TELEFON wall telephone, wall telephone set; ~ **personal** TELEFON personal telephone; ~ **portátil** TELEFON hand-portable telephone, transportable telephone; ~ **de previo pago** PERIF, TELEFON coin-operated telephone; ~ **principal del abonado** TELEFON subscriber's main station; ~ **público** PERIF, TELEFON public call office, public telephone, *accionado por monedas* coin-operated payphone; ~ **público inalámbrico** TELEFON public cordless telephone; ~ **que llama** TELEFON calling telephone; ~ **de recepción amplificada** TELEFON receive-amplified telephone; ~ **supletorio** TELEFON auxiliary telephone set; ~ **de tarjeta de crédito** TELEFON creditphone; ~ **de teclado** APLIC keyset telephone, PERIF keypad telephone, push-button set, push-button telephone, TELEFON key-set telephone, keypad telephone, push-button telephone; ~ **de uso privado** PERIF, TELEFON residential telephone; ~ **visual** APLIC, INFO, TEC RAD, TELEFON, TV visual telephone; ◆ **al** ~ TELEFON on the telephone; **por** ~ TELEFON over the telephone

telegestión *f* APLIC telemanagement

telegrafía *f* TEC RAD, TELEFON, TRANS telegraphy; ~ **de corrientes portadoras** TELEFON carrier-current telegraphy; ~ **dentro de la banda de voz** TELEFON intraband telegraphy; ~ **diplex de cuatro frecuencias** TELEFON four-frequency diplex telegraphy; ~ **escalonada** TELEFON echelon telegraphy; ~ **en facsímil** TELEFON facsimile telegraphy; ~ **facsimilar de documentos** TELEFON document facsimile telegraphy; ~ **facsimilar fotográfica** TELEFON photograph facsimile telegraphy; ~ **de frecuencia vocal** (*VFT*) TELEFON voice-frequency telegraphy (*VFT*); ~ **infratelefónica** TELEFON subtelephone telegraphy; ~ **mosaico** TELEFON mosaic telegraphy; ~ **sin hilos** TEC RAD, TELEFON wireless telegraphy; ~ **ultratelefónica** TELEFON, TRANS super-telephone telegraphy

telégrafo: ~ **alfabético** *m* DATOS, TELEFON alphabet telegraph; ~ **de frecuencia vocal** *m* TELEFON voice-frequency telegraph

telegrama *m* TELEFON telegram; ~ **facsimilar fotográfico** TELEFON photograph facsimile telegram; ~ **facsímil de documento** TELEFON document facsimile telegram; ~ **de servicio** PERIF, TELEFON service telegram

teleguía *f* TEC RAD teleguidance

teleimpresión *f* INFO remote printing, PERIF *periódicos, revistas* teleprinting

teleimpresor *m* APLIC, PERIF, TELEFON teleprinter *BrE*, teletypewriter *AmE*

teleinformática *f* DATOS, INFO, TEC RAD teleinformatics

telemantenimiento *m* DATOS, INFO, PRUEBA, TV remote maintenance

telemática *f* GEN telematics

telemedición *f* DATOS remote metering, telemetering, PRUEBA, TEC RAD telemetering

telemedida *f* PRUEBA remote measuring
telemetría *f* DATOS, PRUEBA telemetry, TEC RAD rangefinding, ranging, telemetry; **~ interna** DATOS housekeeping telemetry; **~ por láser** DATOS laser telemetry
telémetro: **~ por láser** *m* DATOS, PRUEBA laser rangefinder; **~ radárico** *m* TEC RAD radar rangefinder
telepago *m* TV remote payment
Telepoint *m* ELECTROTEC Telepoint
telepresencia *f* INFO telepresence
teleprocesamiento: **~ de datos** *m* DATOS, INFO, TEC RAD remote data processing
teleproceso *m* GEN teleprocessing; **~ por lotes** DATOS, INFO remote batch processing
teleprogramas: **~ de videotexto** *m pl* INFO, TELEFON, TV videotext telesoftware
telepuerta *f* GEN teleport
telepunto *m* TELEFON Telepoint
teleseguridad *f* APLIC telesecurity
teleselección: **~ por frecuencia vocal** *f* TELEFON voice dialling (*AmE* dialing)
teleselector: **~ por frecuencia vocal** *m* TELEFON voice dialler (*AmE* dialer)
teletexto *m* (*TTX*) APLIC, TV teletext (*TTX*)
teletipo *m* (*TTY*) INFO teletype (*TTY*), PERIF teletype (*TTY*), teletype machine, TEC RAD, TELEFON teletype (*TTY*)
teletrabajo *m* GEN teleworking
teletráfico *m* CONMUT, TELEFON teletraffic
teletransmisión *f* TRANS remote transmission
televía *f* INFO teletrack
televigilancia *f* CONTROL, TV remote surveillance
televisión *f* (*TV*) TEC RAD, TV television (*TV*); **~ de antena colectiva** (*MATV*) TV master antenna television (*MATV*); **~ de antena colectiva vía satélite** (*SMATV*) TV satellite master antenna television (*SMATV*); **~ con antena comunitaria** TV community-antenna television (*CATV*); **~ de antena principal** TV main-antenna television; **~ por cable** TV cable television; **~ por cable SCART inclinado** TV tilted cable television, SCART socket television; **~ de calidad actual** INFO existing-quality television; **~ de calidad mejorada** INFO enhanced-quality television; **~ en circuito cerrado** TV CCTV, closed-circuit television; **~ comercial** TV business television; **~ educativa** (*ITV*) TV instructional television (*ITV*); **~ de exploración lenta** TV slow-scan television; **~ de imagen fija** (*SPTV*) TV still-picture television (*SPTV*); **~ industrial** TV industrial television; **~ intensificada** TEC RAD enhanced television; **~ por satélite comercial** TV commercial satellite telecast
televisor: **~ multinorma** *m* TV multistandard television set
télex *m* (*TLX*) DATOS, TEC RAD, TELEFON telex (*TLX*); **~-positivo** DATOS, TEC RAD, TELEFON telex-plus
telnet *f* DATOS *Internet* telnet
TEM *abr* (*electromagnético transversal*) TRANS TEM (*transverse electromagnetic*)
temperatura: **~ ambiente** *f* CLIMAT room temperature, INFO ambient temperature; **~ de encapsulado** *f* ELECTRÓN *semiconductores* case temperature; **~ equivalente de ruido** *f* TEC RAD equivalent noise temperature; **~ de evaporación** *f* GEN *física* evaporation temperature; **~ de funcionamiento** *f* GEN operating temperature; **~ media de la placa Hall** *f* ELECTRÓN mean Hall-plate temperature; **~ de ruido** *f* TEC RAD noise temperature; **~ de ruido del enlace por satélite** *f* TEC RAD satellite-link noise temperature; **~ del ruido de origen celeste** *f* TEC RAD sky-noise temperature; **~ de ruido puntual** *f* TEC RAD spot-noise temperature; **~ de ruido de sistema** *f* TRANS system-noise temperature; **~ de unión** *f* COMPON *semiconductores* junction temperature
templado *m* ELECTRÓN tempering
temporización *f* CONTROL timing control; **~ de componente de señal** SEÑAL signal-element timing; **~ de control** TRANS control timing; **~ de la imagen** DATOS frame timing
temporizador: **~ controlador de secuencia** *m* ELECTRÓN watchdog timer; **~ de cuenta atrás** *m* PERIF *contestador telefónico automático* countdown timer; **~ de intervalos** *m* ELECTRÓN, ELECTROTEC interval timer; **~ supervisor** *m* CONMUT supervisory timer
temporizar: **~ el ciclo** *vt* DATOS time the cycling
tenazas: **~ de boca en T** *f pl* GEN *herramientas* angle-nose pliers
tendedor: **~ de cables** *m* TRANS cable layer
tendencia *f* GEN trend
tendencias: **~ principales** *f pl* GEN major tendencies
tendido: **~ ascendente de cables** *m* TRANS cable runway; **~ de cables** *m* GEN *ingeniería exterior de planta* cable laying
teñir *vt* COMPON stain
tensado *adj* COMPON tightened
tensar *vt* COMPON tighten
tensión *f* (*V*) ELECTROTEC voltage (*V*); **~ anódica directa de cresta** ELECTRÓN peak forward-anode voltage; **~ de ánodo negativa de pico** ELECTRÓN peak negative-anode voltage; **~ aplanada** ELECTROTEC filtered

voltage, smoothed voltage; ~ **de aumento lineal** ELECTROTEC linearly-increasing voltage; ~ **auxiliar** ELECTROTEC auxiliary voltage; ~ **base-emisor** ELECTRÓN *semiconductores* base-emitter voltage; ~ **básica de emisor** COMPON *semiconductores* emitter-base voltage; ~ **de bombeo** ELECTROTEC pumping voltage; ~ **en los bornes** ELECTROTEC terminal voltage; ~ **de calentamiento** ELECTRÓN filament voltage; ~ **de carga** ELECTROTEC *baterías* charging voltage; ~ **de cebado** ELECTRÓN primer voltage, striking voltage; ~ **de celda** ELECTROTEC *baterías* cell voltage; ~ **colector-base** ELECTRÓN *semiconductores* collector-base voltage; ~ **colector-emisor** ELECTRÓN *semiconductores* collector-emitter voltage; ~ **continua permanente en estado bloqueado** ELECTRÓN continuous direct off-state; ~ **de contorneamiento** ELECTROTEC flashover voltage; ~ **de control inducida** ELECTRÓN *semiconductores* induced control-voltage; ~ **de corte** ELECTRÓN *transistores* cutoff voltage; ~ **de corte de puerta** ELECTRÓN *tiristores* gate turn-off voltage; ~ **cuántica** GEN quantum voltage; ~ **de cuasicresta** TEC RAD *móvil aeronáutico* quasi-peak voltage; ~ **de descomposición** COMPON *electroquímica* decomposition voltage; ~ **en diente de sierra** ELECTRÓN sawtooth voltage; ~ **de diodo rectificador** ELECTRÓN probe voltage; ~ **disruptiva** ELECTROTEC disruptive voltage; ~ **de emisor** COMPON *semiconductores* emitter voltage; ~ **de encendido** ELECTRÓN heater voltage, ignition voltage, striking voltage; ~ **entre conductores** ELECTROTEC voltage between lines; ~ **en estado de conducción** COMPON, ELECTRÓN on-state voltage; ~ **de estricción** ELECTROTEC pinch-off voltage; ~ **fantasma** ELECTRÓN phantom voltage; ~ **de fase** ELECTROTEC phase voltage; ~ **flotante** COMPON *semiconductores* floating voltage; ~ **interferente** PRUEBA interference voltage, *EMC* disturbance voltage; ~ **inversa continua permanente** ELECTRÓN *diodos* continuous direct reverse-voltage; ~ **inversa de cresta no repetitiva** COMPON, TELEFON nonrepetitive peak reverse-voltage; ~ **inversa de emisor** COMPON *semiconductores* emitter cutoff voltage, emitter reverse-voltage; ~ **inversa inicial** ELECTRÓN, ELECTROTEC initial inverse-voltage; ~ **inversa media** ELECTRÓN, ELECTROTEC mean reverse-voltage; ~ **de irrupción** COMPON *tiristores* break-over voltage; ~ **de limitación de crestas** ELECTRÓN *semiconductores* clipping voltage; ~ **de llamada** TELEFON ringing voltage; ~ **de luminiscencia** ELECTRÓN glow voltage; ~ **de mantenimiento** ELECTRÓN maintaining voltage, sustaining voltage; ~ **máxima inversa** (*PRV*) ELECTROTEC peak reverse-voltage (*PRV*), TELEFON peak inverse-voltage (*PIV*), peak reverse-voltage (*PRV*); ~ **máxima proyectada** ELECTROTEC projected peak-point voltage; ~ **de no disparo de puerta** ELECTRÓN *tiristores* gate non-trigger voltage; ~ **no filtrada** ELECTROTEC unsmoothed voltage; ~ **de ondulación** DATOS ripple voltage; ~ **de perturbación** PRUEBA *EMC* perturbation voltage; ~ **de pico** ELECTROTEC peak-point current; ~ **de placa** ELECTRÓN plate voltage; ~ **de polarización de rejilla** ELECTRÓN grid-bias voltage; ~ **principal** ELECTROTEC main voltage, principal voltage; ~ **pulsatoria** ELECTROTEC pulsating voltage; ~ **de punto de valle** ELECTROTEC valley-point voltage; ~ **en rampa** ELECTROTEC linearly-increasing voltage, *de aumento lineal* ramp voltage; ~ **de reencendido** ELECTRÓN reignition voltage; ~ **de referencia** ELECTROTEC reference voltage; ~ **reflejada** TRANS reflected voltage; ~ **de régimen** ELECTROTEC voltage rating; ~ **de reposición** ELECTRÓN resetting voltage; ~ **residual** ELECTROTEC residual voltage; ~ **de ruptura por impulsos** ELECTRÓN *de protector*, PRUEBA impulse spark-over voltage; ~ **de señalización** ELECTROTEC, SEÑAL signalling (*AmE* signaling) voltage; ~ **de sincronismo** ELECTRÓN synchronism voltage; ~ **transversal** ELECTROTEC, TELEFON *de protector* transverse voltage; ~ **de umbral en estado conductor** COMPON, ELECTRÓN on-state threshold voltage; ~ **umbral de meseta** TEC RAD plateau-threshold voltage; ~ **en vacío** ELECTROTEC no-load power, no-load voltage; ~ **de valle** COMPON valley-point voltage

tensión/frecuencia *f* (*V/F*) ELECTRÓN, ELECTROTEC voltage/frequency (*V/F*)

tensor *m* COMPON guy

teorema: ~ **de reciprocidad** *m* ELECTROTEC reciprocity theorem; ~ **de Thévenin** *m* ELECTROTEC Thévenin's theorem

teoría: ~ **del autómata** *f* DATOS automata theory; ~ **del cálculo** *f* GEN computation theory; ~ **de la comunicación** *f* GEN communication theory; ~ **cuántica** *f* GEN quantum theory; ~ **de duda** *f* DATOS fuzzy theory; ~ **de nebulosa** *f* DATOS fuzzy theory; ~ **de la probabilidad** *f* GEN probability theory; ~ **secundaria** *f* TRANS background theory; ~ **del tráfico** *f* TELEFON traffic theory

tercero *m* TELEFON third party
térmico *adj* GEN thermal
terminación *f* GEN completion; ~ **1 de red digital de servicios integrados** TRANS B-ISDN network termination 1 (*NT1-LB*); ~ **2 de red digital de servicios integrados** TRANS B-ISDN network termination 2 (*NT2-LB*); ~ **de camino de menor orden** TRANS lower-order path termination (*LPT*); ~ **de colector** ELECTRÓN *semiconductores* collector termination; ~ **digital frontal** TELEFON end digital termination; ~ **digital de red** (*NT1*) TELEFON network digital termination (*NT1*); ~ **de línea** TELEFON, TRANS line termination; ~ **de llamada** TELEFON completion of call; ~ **de red** APLIC, TELEFON, TRANS network termination (*NT*); ~ **de la red B-RDSI** (*T-RDSI*) TRANS B-ISDN network termination (*NT-LB*); ~ **de sección** DATOS, TELEFON section termination; ~ **de la sección del regenerador** (*RST*) DATOS *de impulsos* regenerator-section termination (*RST*); ~ **de trayecto de orden superior** TRANS higher-order path termination (*HPT*)
terminador *m* COMPON terminating set, DATOS terminating unit, terminator, ELECTRÓN, TRANS terminating set; ~ **de bus** TELEFON bus network; ~ **de segmento** DATOS, INFO segment terminator
terminal *m* GEN terminal; ~ **de abonado** TELEFON, TV subscriber terminal; ~ **de ánodo** ELECTRÓN *semiconductores* anode terminal; ~ **de apertura muy pequeña** (*VSAT*) DATOS, TRANS, TV very small aperture terminal (*VSAT*); ~ **bancaria doméstica** APLIC home-banking terminal; ~ **de base** ELECTRÓN *semiconductores* base terminal; ~ **de bloque** TELEFON block terminal; ~ **de cable** TRANS *normas de construcción* cable outlet; ~ **de canal** TRANS channel terminal; ~ **de cátodo** ELECTRÓN *semiconductores* cathode terminal; ~ **de central telefónica** TELEFON exchange termination (*ET*); ~ **de circuito de datos** (*DCTE*) TRANS data-circuit terminating equipment (*DCTE*); ~ **de circuitos** TELEFON circuit terminal; ~ **de colector** ELECTRÓN *semiconductores* collector terminal; ~ **de comunicación de datos** DATOS data-communication terminal; ~ **de conexión** INFO connection terminal, connecting terminal; ~ **compartida** TELEFON terminal share; ~ **para conexión enrollada** COMPON wire-wrap terminal; ~ **de conmutación** CONMUT switching terminal; ~ **conmutada** DATOS dial-up terminal; ~ **de control** CONTROL control terminal; ~ **controlado** CONTROL controlled terminal; ~ **de datos** PERIF, TELEFON data terminal; ~ **de digitalización lenta** TELEFON slow-scan terminal; ~ **digital de línea** TELEFON line-digital terminal; ~ **digital principal** INFO host digital terminal; ~ **a distancia** PERIF remote terminal; ~ **de emisor** COMPON *semiconductores* emitter terminal; ~ **de estación de barco** TEC RAD shipboard terminal; ~ **de la frecuencia portadora** TRANS carrier-frequency terminal; ~ **de Hall** ELECTRÓN *semiconductores* Hall terminal; ~ **hombre-máquina** TELEFON man-machine terminal; ~ **impresor** TELEFON printing terminal; ~ **de línea** TELEFON line terminal; ~ **de llamada simultánea** TELEFON simultaneous-call terminal; ~ **de mensajes de mano** DATOS hand-held message terminal; ~ **en modo de paquetes** TELEFON packet-mode terminal; ~ **operativo remoto** PERIF remote-operating terminal; ~ **de parada de cuadro** TV frame grabber; ~ **de pantalla** INFO, TV screen terminal; ~ **de pinza** *f* ELECTROTEC terminal clip; ~ **portátil** TEC RAD hand-held terminal; ~ **positivo** ELECTROTEC positive terminal; ~ **de puerta** ELECTRÓN *tiristores* gate terminal; ~ **con pulsador** INFO push-button terminal; ~ **que llama** TELEFON calling terminal; ~ **receptor** DATOS destination terminal, TRANS receiving terminal; ~ **de recuperación de microformas** DATOS microform-retrieval terminal; ~ **remoto** PERIF remote terminal unit; ~ **sin procesador** DATOS dumb terminal; ~ **de telecomunicaciones** TEC RAD, TELEFON telecommunications terminal; ~ **telemática** DATOS telematic terminal; ~ **de teletexto** INFO, PERIF, TEC RAD teletext terminal; ~ **de télex** DATOS, TEC RAD, TELEFON telex terminal; ~ **de tornillo** ELECTROTEC terminal screw; ~ **de trabajo** INFO operating terminal; ~ **de tráfico** TEC RAD traffic terminal; ~ **TVRO** TV TVRO terminal; ~ **de usuario** TELEFON, TRANS, TV user terminal; ~ **de usuario local** PERIF local-user terminal (*LUT*); ~ **de videotexto** INFO, TELEFON, TV videotext terminal; ~ **virtual** DATOS, INFO, TELEFON virtual terminal; ~ **VSAT** TEC RAD VSAT terminal; ~ **WAN** INFO WAN terminal
terminar *vt* TELEFON, TRANS terminate
terminarse *v refl* TELEFON, TRANS terminate
término *m* GEN term; ~ **colectivo** GEN *lingüística* collective term
termistor *m* CLIMAT temperature sensor,

COMPON thermistor, CONTROL temperature sensor, ELECTRÓN thermistor
termodinámica *f* GEN thermodynamics
termodinámico *adj* GEN thermodynamic
termoelectricidad *f* GEN thermoelectricity
termoeléctrico *adj* GEN thermoelectric
termómetro: ~ **de depósito líquido** *m* ELECTRÓN wet-bulb thermometer; ~ **de resistencia** *m* PRUEBA resistance thermometer
termoplástico *m* GEN thermoplastic
termorretracción *f* ELECTRÓN heat shrinking
termostato *m* CLIMAT, COMPON, CONTROL thermostat; ~ **de operación** COMPON, CONTROL operation thermostat
terraja *f* COMPON screw top
terreno: ~ **mixto** *m* GEN mixed terrain
testigo *m* GEN *red en anillo* token; ~ **de asignación de mando multipunto** DATOS multipoint command assignment token; ~ **de bloqueo** TELEFON blocking lamp; ~ **de liberación de mando multipunto** DATOS multipoint command release token; ~ **de profundidad** PRUEBA *moldeo en tierra* depth gauge (*AmE* gage)
tetrodo *m* COMPON, ELECTRÓN tetrode
texel *m* INFO texel
texto *m* DATOS, INFO, TELEFON text; ~ **aclaratorio** TELEFON clarifying text; ~ **ampliado** INFO stretched text; ~ **en borrador** INFO draft text; ~ **cifrado** DATOS, TEC RAD, TELEFON, TV ciphertext; ~ **corrido** GEN running text
TFT *abr* (*transistor de capa delgada*) COMPON, ELECTRÓN TFT (*thin-film transistor*)
tiempo[1]: **en ~ real** *adv* DATOS, INFO, PRUEBA, TELEFON in real time
tiempo[2]: ~ **de acceso** *m* DATOS, INFO access time; ~ **de acumulación de portadores** *m* ELECTRÓN *semiconductores* carrier-storage time; ~ **en el aire** *m* TEC RAD airtime; ~ **de ajuste** *m* CONTROL setting time; ~ **de ascenso** *m* GEN rise time; ~ **de asignación de canal** *m* TRANS channel-allocation time; ~ **atómico internacional** *m* (*IAT*) TEC RAD international atomic time (*IAT*); ~ **de borrado mínimo utilizable** *m* ELECTRÓN minimum usable erasing time; ~ **de calentamiento de calefactor** *m* ELECTRÓN heater warm-up time; ~ **de carga** *m* ELECTROTEC charge time; ~ **de coherencia** *m* TRANS coherence time; ~ **de conducción** *m* ELECTRÓN *semiconductores* conduction time; ~ **de confirmación** *m* DATOS submission time; ~ **de conmutación** *m* CONMUT, PRUEBA, TELEFON switching time; ~ **de contestación** *m* TELEFON answering time; ~ **conveniente** *m* TEC RAD proper time; ~ **de convergencia** *m* TELEFON convergence time; ~ **de conversación** *m* TEC RAD, TELEFON talk time; ~ **de corrección de fallo** *m* PRUEBA *seguridad de funcionamiento* fault-correction time, TELEFON failure-correction time; ~ **de demora logística** *m* PRUEBA, TELEFON logistic-delay time; ~ **de descenso de impulso** *m* TELEFON pulse fall time; ~ **de desintegración** *m* INFO, TELEFON, TRANS *de impulso* decay time; ~ **de desvanecimiento de impulso** *m* GEN pulsed decay time; ~ **de diagnóstico** *m* PRUEBA diagnosis time; ~ **de diagnóstico de fallo** *m* TELEFON failure-diagnosis time; ~ **de digitalización de imagen** *m* PERIF, TV image-scanning time; ~ **de ejecución** *m* INFO execution time; ~ **en emisión** *m* TEC RAD on-air time; ~ **de encendido controlado por puerta** *m* ELECTRÓN *tiristores* gate-controlled turn-off time; ~ **de encendido de transmisor** *m* PERIF, TELEFON, TRANS transmitter turn-on time; ~ **de ensamblaje** *m* INFO assembling time; ~ **de envío de la parte de transferencia de mensaje** *m* TELEFON message transfer part sending time; ~ **de espera** *m* GEN holding time, lead time; ~ **de espera en cola** *m* TELEFON queueing time; ~ **de espera para justificación** *m* TRANS justification waiting time; ~ **de establecimiento** *m* GEN setup time; ~ **de establecimiento de conexión** *m* CONTROL, TELEFON setting-up time; ~ **de establecimiento de sintetizador** *m* TELEFON synthesizer settling time; ~ **de establecimiento relativo** *m* TEC RAD *de una señal telegráfica* relative build-up time; ~ **de excitación** *m* ELECTRÓN firing time; ~ **de exposición** *m* TV *fotografía* exposure time; ~ **de fallo** *m* PRUEBA fault time; ~ **de falsa alarma** *m* GEN false-alarm time; ~ **de funcionamiento-avería** *m* TELEFON up-down time; ~ **de funcionamiento requerido** *m* PRUEBA, TELEFON required time; ~ **de guardia** *m* TEC RAD guard time; ~ **de ignición** *m* ELECTRÓN ignition time; ~ **de iluminación** *m* TEC RAD illumination time; ~ **improductivo** *m* ELECTRÓN, PRUEBA downtime; ~ **de inactividad** *m* TELEFON nonoperating time; ~ **de indisponibilidad** *m* DATOS unavailability time (*UAT*), outage time, TELEFON down time; ~ **inefectivo en el aire** *m* TEC RAD ineffective airtime; ~ **de inicio total** *m* ELECTRÓN total starting tie; ~ **de inoperatividad** *m* PRUEBA nonoperating time; ~ **de instalación** *m* GEN *ingeniería de planta* installation time; ~ **de intercambio** *m* CONMUT exchange time, DATOS

interchange time; ~ **interdigital** *m* TELEFON interdigital time; ~ **de ionización** *m* ELECTRÓN ionization time; ~ **irreal** *m* (*NRT*) DATOS, INFO, PRUEBA, TELEFON nonreal time (*NRT*); ~ **de lanzamiento** *m* COMPON release time; ~ **libre** *m* PRUEBA *seguridad de funcionamiento*, TELEFON free time; ~ **de localización de avería** *m* PRUEBA *seguridad de funcionamiento* fault-localization time; ~ **de manejo de la parte de usuario de datos** *m* DATOS data-user-part handling time; ~ **de manejo por el usuario** *m* TELEFON user handling-time; ~ **de maniobra** *m* GEN operating time; ~ **de manipulación de la parte de datos del usuario** *m* SEÑAL data-user-part handling time; ~ **de mantenimiento** *m* PRUEBA, TELEFON maintenance time; ~ **de mantenimiento correctivo** *m* PRUEBA, TELEFON corrective maintenance time; ~ **de mantenimiento de la interrupción** *m* TELEFON break-in hangover time; ~ **de mantenimiento preventivo** *m* PRUEBA, TELEFON preventive maintenance time; ~ **de marcado** *m* TELEFON dialling (*AmE* dialing) time; ~ **máximo de retención** *m* ELECTRÓN maximum retention time; ~ **máximo utilizable de lectura** *m* INFO maximum usable reading time; ~ **máximo utilizable de visado** *m* TV maximum usable viewing time; ~ **medio antes del fallo** *m* PRUEBA mean time before failure; ~ **medio de disponibilidad** *m* PRUEBA mean uptime; ~ **medio entre fallos** *m* (*MTBF*) PRUEBA mean time between failures (*MTBF*); ~ **medio entre interrupciones** *m* (*MTBI*) INFO mean time between interruptions (*MTBI*); ~ **medio entre revisiones** *m* (*MTBO*) PRUEBA mean time between overhauls (*MTBO*); ~ **medio de espera** *m* INFO mean waiting time; ~ **medio hasta el primer fallo** *m* (*MTFF*) INFO mean time to first failure (*MTFF*); ~ **medio hasta el restablecimiento** *m* (*MTTR*) PRUEBA mean time to recovery (*MTTR*); ~ **medio hasta el restablecimiento del servicio** *m* (*MTSR*) PRUEBA mean time to service restoration (*MTSR*); ~ **medio de indisponibilidad** *m* PRUEBA mean downtime; ~ **medio de ocupación** *m* TELEFON mean holding-time per seizure; ~ **medio de operación** *m* TELEFON average operating time; ~ **medio de propagación en un solo sentido** *m* TELEFON mean one-way propagation time; ~ **medio de reparación** *m* PRUEBA mean repair time, mean time to repair; ~ **de memoria intermedia** *m* ELECTRÓN buffer time; ~ **de muestreo** *m* TRANS sampling time; ~ **no requerido** *m* PRUEBA nonrequired time; ~ **no requerido requerido** *m* PRUEBA, TELEFON required unrequired time; ~ **de ocupación** *m* GEN occupation time, seizure time; ~ **de operación de la supresión** *m* TELEFON suppression operate time; ~ **de orbitación** *m* TEC RAD orbiting time; ~ **perdido** *m* GEN lost time; ~ **de persistencia** *m* TEC RAD, TELEFON hangover time; ~ **de persistencia de la supresión** *m* TELEFON suppression hangover time; ~ **de portamento** *m* INFO portamento time; ~ **de posicionamiento** *m* DATOS, INFO positioning time; ~ **de precalentamiento** *m* ELECTRÓN preheating time; ~ **productivo de sistema** *m* GEN system production time; ~ **de programación programado** *m* TEC RAD scheduled operating time; ~ **de propagación** *m* TEC RAD, TRANS propagation time, TELEFON, TRANS transmission time; ~ **de propagación en el canal de datos** *m* TELEFON data-channel propagation time; ~ **de propagación de señal** *m* SEÑAL *en un medio* signal delay; ~ **de puesta en servicio** *m* GEN commissioning time; ~ **rápido** *m* INFO quicktime *m*; ~ **de reacción** *m* GEN reaction time; ~ **de reanudación** *m* INFO rerun time; ~ **de recepción de la parte de transferencia de mensaje** *m* INFO message transfer part receiving time; ~ **de recuperación** *m* TELEFON reframing time; ~ **de recuperación de la alineación de trama** *m* TELEFON, TRANS frame-alignment recovery time; ~ **de reemisión** *m* TEC RAD re-emission time; ~ **de reparación** *m* TELEFON repair time; ~ **de repetición de impulso** *m* TELEFON pulse-repetition time; ~ **en reserva** *m* TELEFON stand-by time; ~ **de respuesta** *m* GEN response time; ~ **de retardo** *m* GEN delay time; ~ **de retardo de HV** *m* TEC RAD HV delay time; ~ **de retención** *m* INFO hold time; ~ **de retención de un circuito internacional** *m* TELEFON holding-time of an international circuit; ~ **de retorno** *m* DATOS, INFO turnaround time; ~ **de retraso técnico** *m* PRUEBA technical-delay time; ~ **de reverberación** *m* TELEFON reverberation time; ~ **de reverbero** *m* TELEFON reverb time; ~ **de secado** *m* GEN drying time; ~ **de servicio** *m* APLIC service time; ~ **de soldadura** *m* COMPON soldering time; ~ **de subida de impulso** *m* TELEFON pulse-rise time; ~ **tasable** *m* APLIC, TELEFON chargeable duration; ~ **total de retención** *m* TELEFON total holding time; ~ **de transferencia de emisor** *m* TELEFON sender transfer time; ~ **de transferencia del mensaje en los puntos de transferencia de señalización** *m* TRANS message transfer time at signalling (*AmE* signaling) transfer points; ~ **de transferencia de receptor** *m* TELEFON receiver transfer time; ~ **de transferencia en una centralita** *m* (*Tcu*) CONMUT cross-office transfer time (*Tcu*); ~ **de tránsito** *m* ELECTROTEC, PRUEBA, TELEFON transit time; ~ **de tránsito de señal** *m* ELECTRÓN signal-transit time; ~ **de transmisión total** *m* TELEFON, TRANS total transmission time; ~ **útil** *m* GEN operable time, up-time; ~ **de utilización de máquina** *m* INFO running time

TIFF *abr* (*formato de archivo de imagen con nombre simbólico*) DATOS, INFO TIFF (*tagged-image file format*)

tilde *f* INFO tilde

timbre *m* INFO timbre, tone colour (*AmE* color), TELEFON ringer, timbre; ~ **electromagnético** TELEFON polarized electric bell; ~ **recordatorio** TELEFON reminder ring; ~ **supletorio** PERIF extension bell

timbres *m pl* PERIF bellsets

tinta: ~ **para imprimir** *f* PERIF printing ink

tintura *f* COMPON staining

tipo: ~ **de accionamiento automático** *m* DATOS auto-action type; ~ **de atributo** *m* DATOS attribute type; ~ **básico** *m* DATOS base type; ~ **de bloque** *m* DATOS *SDL* block type; ~ **de cadena de caracteres** *m* DATOS character-string type; ~ **de cadena de octetos** *m* DATOS octet-string type; ~ **de circuito de programa de sonido** *m* TELEFON type of sound-programme (*AmE* sound-program) circuit; ~ **de codificación** *m* TRANS coding type; ~ **de datos** *m* DATOS data type; ~ **de datos resumidos** *m* DATOS abstract data type; ~ **de dirección** *m* (*TOA*) DATOS, TELEFON type of address (*TOA*); ~ **de disyuntivo** *adj* ELECTRÓN ORed; ~ **de doble cadena** *m* DATOS bistring type; ~ **de documento** *m* GEN document type; ~ **de entidad N** *m* DATOS N-entity type; ~ **de entrada** *m* DATOS entry type; ~ **enumerado** *m* DATOS enumerated type; ~ **estructurado** *m* DATOS structured type; ~ **de modulación** *m* TRANS modulation type; ~ **de notificación** *m* DATOS notification type; ~ **de número** *m* TELEFON type of number (*TON*); ~ **de pista** *m* INFO track type; ~ **de proceso de aplicación** *m* DATOS application-process type; ~ **resaltado** *m* DATOS set-of type; ~ **de reverberación** *m* INFO reverberation type; ~ **de reverbero** *m* INFO reverb type; ~ **de secuencia** *m* DATOS sequence type; ~ **de segmento** *m* DATOS, INFO segment type (*ST*); ~ **de selección** *m* TELEFON selection type; ~ **de situación normal** *m* DATOS normal-situation class; ~ **de transmisión** *m* TELEFON type of transmission

tipografía *f* DATOS *Internet*, INFO typography

tipos: ~ **de información codificada entregados** *m pl* DATOS delivered encoded-information types; ~ **de información codificada originales** *m pl* DATOS original encoded-information types

tira *f* COMPON, CONMUT, ELECTRÓN, ELECTROTEC strap; ~ **de contacto elastomérica** COMPON elastomer contact-strip; ~ **de goma** INFO rubber banding; ~ **de rotulación** COMPON strip label

tiratrón *m* ELECTRÓN thyatron

tiristor *m* COMPON, ELECTRÓN, ELECTROTEC thyristor; ~ **bidireccional** COMPON triac; ~ **diodo bloqueado en inversa** COMPON, ELECTRÓN reverse-blocking diode thyristor; ~ **de potencia** COMPON, ELECTRÓN power thyristor; ~ **de puerta remota** ELECTRÓN remote-gate thyristor; ~ **triodo bidireccional** ELECTRÓN bidirectional triode thyristor

tiro *m* GEN *chimenea* draft

titilación *f* TEC RAD, TV scintillation

titular: ~ **de contrato de abono** *m* DATOS, INFO, TEC RAD, TELEFON, TV subscriber-agreement holder; ~ **principal de la cuenta** *m* GEN major account holder

título *m* DATOS, INFO title; ~ **de conjunto transaccional** DATOS transactional set header; ~ **de entidad N** DATOS N-entity title; ~ **de proceso de aplicación** DATOS application process title; ~ **de sistema** DATOS system title; ~ **de TPSU** DATOS TPSU title; ~ **de la unidad de aplicación** DATOS application-entity title

títulos *m pl* TV credits; ~ **de crédito de producción** TV production credits

TL *abr* (*probador inferior*) DATOS, PRUEBA LT (*lower tester*)

TLX *abr* (*télex*) DATOS, TEC RAD, TELEFON TLX (*telex*)

TLXAU *abr* (*unidad de acceso télex*) DATOS, TEC RAD, TELEFON TLXAU (*telex access unit*)

TM *abr* (*modo magnético transversal*) TEC RAD, TRANS TM (*transverse magnetic mode*)

TMUX-P *abr* (*transmultiplexor tipo P*) TELEFON TMUX-P (*type P transmultiplexer*)

TMUX-S *abr* (*transmultiplexor tipo S*) TELEFON TMUX-S (*type S transmultiplexer*)

TNC *abr* (*controlador de nodo terminal*) TRANS TNC (*terminal node controller*)

TO *abr* (*oficina de telégrafos*) TELEFON TO (*telegraph office*)

TOA *abr* (*tipo de dirección*) DATOS, TELEFON TOA (*type of address*)

tolerancia *f* GEN allowance, tolerance; ~ **de ajuste** GEN fitting tolerance; ~ **de atenuación** TRANS attenuation tolerance; ~ **del diámetro del núcleo** TRANS core diameter tolerance; ~ **de errores** DATOS, ELECTRÓN, INFO, PRUEBA fault tolerance; ~ **de frecuencia** TEC RAD frequency tolerance; ~ **impulsiva al ruido** TEC RAD impulsive noise tolerance; ~ **posicional** GEN positional tolerance

toma *f* APLIC socket, power outlet, TELEFON seizure; ~ **para auriculares** APLIC, PERIF headphones socket; ~ **de bobinado** ELECTROTEC winding tapping; ~ **de CA** INFO AC outlet, alternative-current power outlet; ~ **central** ELECTRÓN *de transformador* midpoint tap, *transformadores* centre (*AmE* center) tap; ~ **corriente mural** ELECTRÓN, ELECTROTEC, INFO, TELEFON wall outlet; ~ **de electricidad de CA** INFO AC power outlet; ~ **de huellas de identificación de señal** SEÑAL signal-identification fingerprinting; ~ **de línea de corriente alterna** INFO alternative-current mains wall socket; ~ **múltiple** CONMUT multiple seizure; ~ **de salida** INFO output socket; ~ **de tensión de batería** ELECTROTEC battery-voltage terminal; ~ **de tierra de bastidor** ELECTROTEC rack earth *BrE*, rack ground *AmE*; ~ **de tierra colectiva** ELECTROTEC collective earthing *BrE*, collective grounding *AmE*; ~ **de tierra independiente** ELECTRÓN, ELECTROTEC independent earth electrode *BrE*, independent ground electrode *AmE*; ~ **de tierra de protección** ELECTROTEC, PRUEBA protective earth *BrE*, protective ground *AmE*

tomacorriente *f* ELECTRÓN, INFO, TV *electricidad* socket; ~ **de la red de corriente alterna** TV alternative-current mains-socket wall

tonal *adj* INFO, PERIF, TELEFON tonal

tonalidad: ~ **de llamada** *f* TELEFON ring tone

tono *m* GEN tone; ~ **agradable** TELEFON comfort tone; ~ **de alerta para terceros** TELEFON third-party warning tone; ~ **anunciador de bloqueo de línea** TELEFON line-lockout warning tone; ~ **APB** TELEFON APB tone; ~ **ATB** TELEFON ATB tone, all trunks busy tone; ~ **de aviso** TELEFON warning tone; ~ **bajo** INFO bass tone; ~ **de congestión** TELEFON congestion tone, reorder tone; ~ **de contestación** (*ANS*) DATOS answer tone (*ANS*); ~ **por debajo de la banda** TEC RAD tone below band; ~ **de discado interno PABX** TELEFON PABX internal dial tone; ~ **por encima de la banda** TEC RAD tone above band; ~ **de fin de periodo** PERIF *teléfonos de monedas* end-of-period tone; ~ **de indicación negativo** TELEFON negative indication tone; ~ **de indicación positivo** TELEFON positive indication tone; ~ **de información especial** TELEFON special information tone; ~ **de intervención** TELEFON intrusion tone; ~ **de llamada** TELEFON ringing tone; ~ **de llamada especial** TELEFON special ringing tone; ~ **de marcar** TELEFON dial tone *AmE*, dialling (*AmE* dialing) tone, TELEFON dialling (*AmE* dialing) tone; ~ **de marcar especial** TELEFON special dial tone; ~ **de marcar secundario** TELEFON second dial tone; ~ **normal de teclado** INFO standard keyboard pitch; ~ **de ocupado rápido** TELEFON fast-busy tone; ~ **de prioridad** TELEFON priority tone; ~ **rastreable** ELECTRÓN traceable tone; ~ **de reconocimiento de teléfono monedero** TELEFON payphone recognition tone; ~ **de retención de llamada** TELEFON tone on hold; ~ **de retorno de llamada** TELEFON ring-back tone; ~ **de sobrecarga** TELEFON overflow tone; ~ **de sobrecontrol** TELEFON override tone; ~ **supervisor** CONMUT, CONTROL supervisory tone; ~ **de tictac** TELEFON tick tone; ~ **de transferencia** TELEFON barge-in tone; ~ **único** GEN single tone; ~ **de vibrador** TELEFON ticker tone; ~ **de zumbido** TELEFON buzzer tone, buzzing tone

tonos: ~ **bajos** *m pl* INFO bass tones

tope: ~ **de capacidad** *m* TEC RAD *antenas* capacity top; ~ **de disco** *m* PERIF, TELEFON *teléfonos* finger stop; ~ **de entrehierro** *m* COMPON residual stud; ~ **de impulso** *m* TELEFON pulse top

tópico *m* INFO topic

topología *f* ELECTRÓN topology; **de anillo** DATOS *red* ring topology; ~ **de árbol** DATOS, INFO tree topology; ~ **de bus** DATOS bus topology; ~ **en estrella** DATOS star topology; ~ **de red** DATOS, TELEFON network topology

torcedura *f* TRANS kink

torcimiento: ~ **característico** *m* DATOS, TELEFON characteristic distortion

tormenta: ~ **ionosférica** *f* TEC RAD ionospheric storm

tormentas: ~ **de radiodifusión** *f pl* ELECTRÓN broadcast storms

torneado: ~ **con torno revólver** *m* COMPON turret turning

tornear *vt* COMPON turn

tornillo *m* COMPON screw; COMPON screw bolt; ~ **de cabeza hexagonal** COMPON screw with hexagonal head; ~ **cautivo** COMPON captive screw; ~ **de fijación** COMPON fixing screw; ~ **para madera** COMPON woodscrew; ~ **de mano**

GEN hand vice (*AmE* vise); ~ **de mariposa** COMPON, ELECTRÓN *sin destornillador* thumbscrew; ~ **nivelador** ELECTRÓN jackscrew; ~ **sin engarce** ELECTRÓN crimpless screw

tornillos: ~ **de fijación** *m pl* COMPON mounting screws

torno *m* COMPON *de tornear* lathe; ~ **de cables** TRANS cable winch; ~ **revólver** COMPON turret lathe; ~ **para tendido de cables** TRANS *ingeniería exterior de planta* cable-laying winch

torón *m* COMPON, ELECTRÓN *cuerdas, cables* strand, TRANS *cables* conductor unit

torreta *f* ELECTRÓN turret

torsión *f* COMPON twisting

total: ~ **medio inspeccionado** *m* (*ATI*) PRUEBA *calidad* average total inspected (*ATI*)

TPA *abr* (*arquitectura de protocolo telemático*) TELEFON TPA (*telematic protocol architecture*)

TPASE *abr* (*elemento del servicio de aplicación de procesamiento de transacciones*) DATOS TPASE (*transaction processing application service element*)

TPIWF *abr* (*función de interfuncionamiento de paquetes de télex*) DATOS TPIWF (*telex packet interworking function*)

TPPM *abr* (*máquina de protocolo de procesamiento de transacciones*) DATOS TPPM (*transaction processing protocol machine*)

TPSP *abr* (*proveedor del servicio de procesamiento de transacciones*) DATOS TPSP (*transaction processing service provider*)

TPSU *abr* (*usuario del servicio de procesamiento de transacciones*) DATOS TPSU (*transaction processing service user*)

TPSUI *abr* (*invocación del usuario del servicio de procesamiento de transacciones*) DATOS TPSUI (*transaction processing service user invocation*)

TQM *abr* (*gestión de calidad total*) PRUEBA TQM (*total quality management*)

trabajo *m* INFO job; ~ **cooperativo** DATOS cooperative work; ~ **en modo de pérdida** TRANS loss-mode working; ~ **troncal automático** APLIC automatic trunk working

trabajoso *adj* GEN laborious

tracción *f* TRANS *de cables* pulling

traducción *f* DATOS, INFO, TEC RAD, TELEFON, TV translation

traducir *vt* DATOS, INFO, TEC RAD, TELEFON, TV translate

traductor *m* INFO *software* translator; ~ **de frecuencia** TEC RAD frequency translator; ~ **de lenguaje** INFO language translator; ~ **de nivel** ELECTRÓN level translator

tráfico *m* GEN traffic; ~ **comercial** TELEFON business traffic; ~ **compartido** TEC RAD traffic sharing; ~ **conducido** TELEFON traffic carried; ~ **conseguido** TEC RAD successful traffic; ~ **coordinado** TEC RAD coordinated traffic; ~ **de corta distancia** TELEFON short-distance traffic; ~ **de cresta** TELEFON peaked traffic; ~ **desaprovechado** TELEFON traffic lost, waste traffic; ~ **de desbordamiento** CONMUT, TELEFON overflow traffic; ~ **diario** TELEFON day traffic; ~ **efectivo** GEN effective traffic; ~ **entre PBX** TELEFON inter-PBX traffic; ~ **frustrado** TEC RAD, TELEFON unsuccessful traffic; ~ **ideal** TELEFON pure-chance traffic; ~ **interlocal** TELEFON interlocal traffic; ~ **de larga distancia** TELEFON intertoll traffic; ~ **de línea de interconexión** CONMUT, TELEFON, TRANS tie-line traffic; ~ **local** GEN local traffic; ~ **no selectivo** TEC RAD nonselective traffic; ~ **ofrecido** TELEFON traffic offered; ~ **originado por abonados** TELEFON subscriber-dialled (*AmE* subscriber-dialed) traffic; ~ **perdido** TELEFON lost traffic; ~ **Poisson** TELEFON Poisson traffic; ~ **de red entre ligero y medio** ELECTRÓN light-to-medium network traffic; ~ **a servicios especiales** TELEFON traffic to special services; ~ **suburbano** TELEFON short-haul traffic; ~ **de telecomunicaciones** CONMUT, TELEFON telecommunications traffic; ~ **de télex** TELEFON telex traffic; ~ **terminal** TELEFON terminating traffic; ~ **de tránsito** TELEFON transit traffic; ~ **transversal** TELEFON transverse traffic

trama *f* GEN line; ~ **de acuse de recepción no numerada** TELEFON unnumbered acknowledgement frame; ~ **de distribución combinada** CONMUT combined distribution frame (*CDF*); ~ **de distribución de grupo** TELEFON, TRANS group distribution frame; ~ **de distribución principal de repetición** TRANS main repeater distribution frame; ~ **de identificación de central** TELEFON exchange identification frame; ~ **no esperada** (*NEF*) TELEFON not-expected frame (*NEF*); ~ **de no preparado para recibir** TELEFON receive-not-ready frame; ~ **de preparado para recibir** TELEFON receive-ready frame; ~ **principal de distribución** TELEFON, TRANS main distribution frame (*MDF*); ~ **de rechazo selectivo** (*SREJ*) TELEFON selective-reject frame (*SREJ*); ~ **RNR** TELEFON RNR frame; ~ **de subdistribución** TRANS subdistribution frame; ~ **temporal** INFO, TEC RAD, TRANS timescale; ~ **temporal en sincronización** TEC RAD timescale in synchronization

tramar *vt* INFO rasterize

tramitación: ~ **de tráfico** *f* DATOS traffic handling
trámite: **en** ~ *fra* GEN in the pipeline
tramo *m* DATOS, INFO, TELEFON, TRANS *carretera* section; ~ **de cables** TRANS cable route; ~ **de canalización** GEN duct run
trampa *f* ELECTRÓN, INFO, PERIF, TEC RAD trap; ~ **de ondas** TEC RAD wave trap
tranferencia: ~ **de fallo en PBX** *f* (*PFT*) TELEFON PBX failure transfer (*PFT*)
transacción *f* GEN transaction; ~ **de distribución asistida por aplicación** DATOS application-supported distribution transaction; ~ **distribuida respaldada por el proveedor** DATOS provider-supported distributed transaction; ~ **por telefax** DATOS, TELEFON facsimile transaction
transadmitancia *f* ELECTRÓN transadmittance
transceptor *m* DATOS, TEC RAD transceiver; ~ **de comprobación de continuidad** TELEFON continuity-check transceiver; ~ **de mochila** TEC RAD man-pack transceiver
transcodificación *f* ELECTRÓN, PRUEBA, TELEFON, TRANS transcoding
transcodificador *m* ELECTRÓN, PRUEBA, TELEFON, TRANS transcoder
transcodificar *vt* ELECTRÓN, PRUEBA, TELEFON, TRANS transcode
transconductancia *f* ELECTRÓN transconductance
transductor *m* GEN transducer, transductor; ~ **interdigital** TRANS interdigital transducer (*IDT*); ~ **de medición** PRUEBA measuring transducer; ~ **en modo ortogonal** TEC RAD orthogonal mode transducer; ~ **ortomodal** TEC RAD orthomode transducer; ~ **pasivo** ELECTROTEC passive transducer; ~ **de señal eléctrica** ELECTROTEC electric signal transducer
transferencia *f* DATOS transfer, TEC RAD hand-off, hand-over, TELEFON barge-in, transfer; ~ **de archivos binarios** (*BFT*) TELEFON binary-file transfer (*BFT*); ~ **automática** TELEFON auto-transfer; ~ **de bloques** INFO block transfer; ~ **de categorías** TELEFON category transfer; ~ **de datos** DATOS, TRANS data transfer; ~ **de datos pedidos** DATOS, TELEFON request-data transfer; ~ **de datos solicitada** DATOS data-transfer requested; ~ **de documentos en masa** TELEFON document bulk transfer; ~ **entre células** TEC RAD *móvil de tierra* intercell hand-off; ~ **del fallo de potencia** TELEFON power-failure transfer; ~ **de ficheros** DATOS, INFO file transfer; ~ **de imagen** DATOS, TRANS image transfer; ~ **de llamada retenida** TEC RAD holding call transfer; ~ **y manipulación de documentos en masa** TELEFON document bulk transfer and manipulation; ~ **megatónica** DATOS mt transfer; ~ **de mensajes** DATOS, TELEFON message transfer (*MT*); ~ **de un ordenador a otro** DATOS, ELECTRÓN, INFO, TV downloading; ~ **radial** INFO radial transfer; ~ **de tecnología** INFO technology transfer; ~ **telegráfica** TELEFON telegraph transfer; ~ **telemática de fichero** TELEFON telematic file transfer; ~ **temporal** TELEFON temporary transfer; ~ **temporal del número propio** TELEFON diversion follow me
transferir *vt* DATOS transfer, TEC RAD hand off, hand over, TELEFON *llamadas* transfer
transformación *f* ELECTRÓN transformation, reshaping; ~ **elevadora** ELECTROTEC step-up transformation; ~ **de Fourier** (*FT*) PRUEBA Fourier transformation (*FT*); ~ **rápida de Fourier** (*TRF*) INFO fast Fourier transform (*FFT*); ~ **reductora** ELECTROTEC step-down transformation; ~ **de señales** ELECTRÓN, SEÑAL signal transformation
transformador *m* GEN converter; ~ **de absorción** ELECTROTEC draining transformer; ~ **de adaptación** ELECTRÓN matching transformer; ~ **de aislamiento** ELECTROTEC isolating transformer; ~ **automático** ELECTROTEC auto-transformer; ~ **de bobinado sencillo** ELECTROTEC single-winding transformer; ~ **con conexión Scott** COMPON, ELECTROTEC Scott-connected transformer; ~ **de corriente** ELECTROTEC mains transformer; ~ **de corriente de llamada** TELEFON ringing-current transformer; ~ **diferencial** ELECTRÓN hybrid transformer; ~ **de distribución** ELECTROTEC *suministro de energía* distribution transformer; ~ **elevador de voltaje** ELECTROTEC step-up transformer; ~ **de modo** TEC RAD mode changer, mode transformer; ~ **de potencia** ELECTROTEC power transformer; ~ **puente** ELECTRÓN bridge transformer; ~ **reductor** ELECTROTEC step-down transformer; ~ **de retorno** TV flyback transformer; ~ **de salida** ELECTROTEC output transformer; ~ **de sonda** TEC RAD probe transformer; ~ **de toma central** ELECTROTEC *suministro de corriente* centre-tapped (*AmE* center-tapped) transformer; ~ **de tono en voltaje** INFO pitch-to-voltage converter
transformar *vt* ELECTROTEC transform
transgresión: ~ **del código AMI** *f* TRANS AMI violation; ~ **de la inversión alternada de**

marcas *f* TELEFON alternate mark inversion violation

transición *f* TELEFON transition; ~ **progresiva** CONMUT progressive transition

transistor *m* COMPON, ELECTRÓN, ELECTROTEC transistor; ~ **bipolar** COMPON, ELECTRÓN bipolar transistor; ~ **bipolar de heterounión** (*HBT*) COMPON, ELECTRÓN heterojunction bipolar transistor (*HBT*); ~ **de capa delgada** (*TFT*) COMPON, ELECTRÓN thin-film transistor (*TFT*); ~ **de conducción en serie** COMPON, ELECTRÓN, ELECTROTEC series pass transistor; ~ **de contacto de punta** COMPON, ELECTRÓN point-contact transistor; ~ **controlado por tensión** ELECTRÓN voltage-controlled transistor; ~ **con efecto de campo de enlace** (*JFET*) COMPON junction field-effect transistor (*JFET*); ~ **de efecto de campo** (*TEC*) COMPON, ELECTRÓN *semiconductores* field-effect transistor (*FET*); ~ **epitaxial** COMPON *semiconductores* epitaxial transistor; ~ **de microaleación difusa** (*MADT*) ELECTRÓN micro-alloy diffused transistor (*MADT*); ~ **de paso** ELECTRÓN pass transistor; ~ **de pequeña señal** COMPON, ELECTRÓN small-signal transistor; ~ **planar** COMPON planar transistor; ~ **de potencia** COMPON power transistor; ~ **de una sola unión** (*UJT*) COMPON, ELECTRÓN unijunction transistor (*UJT*); ~ **de tetrodo** COMPON, ELECTRÓN tetrode transistor; ~ **tetrodo de efecto de campo** COMPON, ELECTRÓN tetrode field-effect transistor; ~ **triodo de efecto de campo** COMPON, ELECTRÓN triode field-effect transistor; ~ **de unión por crecimiento** ELECTRÓN *semiconductores* grown junction transistor; ~ **de uniones** COMPON, ELECTRÓN junction transistor; ~ **unipolar** COMPON, ELECTRÓN unipolar transistor

tránsito: ~ **transportado** *m* TELEFON transit carried

transitorio *adj* GEN transient

transitorios *m pl* TRANS transients

transitrón *m* COMPON, ELECTRÓN transitron

transliterar *vt* DATOS, INFO transliterate

transmisión *f* GEN forwarding; ~ **en alternativa** TELEFON alternate transmission; ~ **de banda ancha** TRANS broadband transmission, wideband transmission; ~ **de banda lateral única** TEC RAD, TRANS single sideband transmission; ~ **en banda lateral residual** TEC RAD, TRANS vestigial-sideband transmission; ~ **con bandas laterales independientes** TEC RAD, TRANS independent sideband transmission; ~ **de bloques** TRANS block transmission; ~ **de caracteres en serie** TRANS character-serial transmission; ~ **codificada** TRANS coded transmission; ~ **de código** TRANS code transmission; ~ **por correa** COMPON belt drive; ~ **de corriente doble** TRANS double-current transmission; ~ **por corriente portadora** TRANS carrier transmission; ~ **por corriente simple** ELECTROTEC single-current transmission; ~ **a corta distancia** TRANS short-haul transmission; ~ **de datos** DATOS data link, data transmission, datacom, TRANS data link, data transmission; ~ **de datos en banda ancha** DATOS, TRANS wideband data transmission; ~ **de datos en modo de conexión en red** DATOS, TRANS network-connection-mode data transmission; ~ **de datos en modo sin conexión en red** DATOS, TRANS network-connectionless-mode data transmission; ~ **de datos simétricos de mando multipunto** DATOS, TRANS multipoint command symmetrical data transmission; ~ **de datos de voz alternada** TRANS alternate voice-data transmission; ~ **por desplazamiento de fase cuaternaria** SEÑAL, TEC RAD phase-shaped QPSK; ~ **de documento** DATOS document transmission; ~ **de enlace ascendente** TEC RAD, TRANS uplink transmission; ~ **de fibra óptica** TRANS fibre-optic (*AmE* fiber-optic) transmission; ~ **de imagen** DATOS, TRANS image transmission; ~ **con incidencia oblicua** TEC RAD oblique-incidence transmission; ~ **de larga distancia** TELEFON long-haul transmission; ~ **de una llamada terminal** APLIC terminal call forwarding; ~ **de memoria virtual** TELEFON VM forwarding; ~ **de mensaje** DATOS pass-along message (*PAM*); ~ **de mensajes hablados** (*MVG*) APLIC, DATOS, TELEFON voice messaging (*VMG*); ~ **múltiple** TRANS multiple transmission; ~ **múltiple de televisión** TRANS multiple television transmission; ~ **N inalámbrica** DATOS, TRANS N connectionless-mode transmission; ~ **de la onda** TRANS wave transmission; ~ **por paquetes** TRANS packet transmission; ~ **en paralelo** INFO, TRANS parallel transmission; ~ **de portadora reducida** TEC RAD reduced-carrier transmission; ~ **de programa sonoro internacional** TV international sound-programme (*AmE* sound-program) transmission; ~ **de punto de prueba** PRUEBA testing-point transmission; ~ **rápida por desplazamiento de frecuencia** SEÑAL fast frequency-shift keying; ~ **de retorno** TEC RAD backward transmission; ~ **por satélite** TEC RAD, TRANS,

TV satellite transmission; ~ **secreta** TELEFON scrambled message; ~ **de secuencia de líneas** ELECTRÓN line-sequential drive; ~ **por secuencias de octetos** TRANS byte-serial transmission; ~ **en semiduplex** TRANS half-duplex transmission; ~ **de señales** CONMUT, SEÑAL, TELEFON, TRANS signal transfer; ~ **simple** TRANS simple transmission; ~ **sonora** APLIC, TRANS sound transmission; ~ **por telefax** TELEFON facsimile transmission; ~ **de televisión internacional** TV international television transmission; ~ **de televisión por vía satélite** TV satellite television broadcasting; ~ **de tono de información especial** TELEFON send special information tone; ~ **por trayectoria múltiple** DATOS multipath transmission; ~ **por vía satélite** TEC RAD, TV satellite broadcasting; ~ **de vídeo** TRANS, TV video transmission; ~ **de la voz** TRANS voice transmission

transmisiones: ~ **ocasionales** *f pl* TRANS occasional transmissions

transmisómetro *m* PRUEBA, TRANS transmissometer

transmisor *m* (*XMTR*) GEN transmitter, sender; ~ **automático Wheatstone** TELEFON Wheatstone automatic system; ~ **de canal** TRANS channel transmitter; ~ **de corto alcance** TEC RAD short-range transmitter; ~ **de datos codificados** DATOS coded-data transmitter; ~ **de imagen de televisión** *f* TV television-picture transmitter; ~ **interferente** TEC RAD interfering transmitter; ~ **de Lambert** ELECTRÓN Lambertian radiator; ~ **de láser** TRANS *óptica de fibras* laser transmitter; ~ **localizador de emergencia** TEC RAD emergency locater transmitter (*ELT*); ~ **de mando** TRANS control transmitter; ~ **de mensajes** PERIF message broadcaster; ~ **de módem** DATOS modem transmitter; ~ **múltiple** TEC RAD multiple transmitter; ~ **de numeración automático** TELEFON automatic numbering transmitter; ~ **de radiodifusión** TELEFON broadcast transmitter; ~ **de radiodifusión de sonido** TEC RAD sound broadcast transmitter; ~ **de radiofaro** TEC RAD beacon transmitter; ~-**receptor** TRANS transmitter-receiver; ~ **repetidor** TEC RAD relay transmitter; ~ **del sonido de la televisión** TV television sound transmitter; ~ **de tambor** TELEFON drum transmitter; ~ **de teclado** TELEFON keyboard transmitter; ~ **telegráfico** TELEFON telegraph transmitter; ~ **de teleseñales** APLIC telesignalling (*AmE* telesignaling) transmitter; ~ **de televisión** TV television transmitter

transmitancia: ~ **isócrona** *f* PRUEBA frequency response

transmitir *vt* DATOS, INFO, TEC RAD, TELEFON, TRANS send, transmit; ◆ ~ **por cable coaxial** COMPON, ELECTRÓN, INFO, TRANS pipe

transmultiplexor *m* ELECTRÓN, TELEFON, TRANS transmultiplexer (*TMUX*); ~ **tipo P** (*TMUX-P*) TELEFON type P transmultiplexer (*TMUX-P*); ~ **tipo S** (*TMUX-S*) TELEFON type S transmultiplexer (*TMUX-S*)

transordenador *m* DATOS, INFO transputer

transparencia *f* DATOS, TELEFON, TRANS *red* transparency; ~ **de código** TRANS code transparency; ~ **de velocidad de señalización de datos** DATOS, SEÑAL data signalling (*AmE* signaling) rate transparency

transparente: ~ **de sonido fundamental de la fuente** *f* (*SRT*) ELECTRÓN source rooting transparent (*SRT*)

transpondedor: ~ **interrogador** *m* TEC RAD interrogator transponder; ~ **de satélite** *m* TEC RAD satellite transponder

transponer *vt* ELECTRÓN, INFO, TELEFON, TRANS transpose

transportabilidad *f* PRUEBA transportability

transportador *m* DATOS handler

transportar *vt* GEN convey

transporte *m* DATOS, TELEFON, TV carrying

transposición *f* ELECTRÓN, INFO, TELEFON, TRANS transposition; ~ **de octava** INFO octave transpose

transposiciones: ~ **de hilos** *f pl* ELECTRÓN wire transpositions

traslación: ~ **de grupo** *f* TRANS group translation

traslador: ~ **de anillo** *m* TRANS ring translator

traspondedor: ~ **que trabaja en eclipse** *m* TEC RAD eclipse-working transponder

trasposición: ~ **de tecla** *f* INFO key transpose

tratamiento: ~ **anódico** *m* COMPON anodic treatment; ~ **de errores** *m* DATOS error handling; ~ **de imagen** *m* DATOS, TV image processing; ~ **de llamadas** *m* TELEFON call handling; ~ **de texto** *m* DATOS, INFO word processing

travesaño *m* COMPON *marco de fundición* stay; ~ **de bastidor** TELEFON rack beam; ~ **de tramo de cable** TRANS cable-run rung

traviesa: ~ **de separación** *f* COMPON, TELEFON *estructura* tie bar

trayecto: ~ **de bajada** *m* TV down path; ~ **de cables** *m* TRANS cable way; ~ **de canalización** *m* TRANS *tendido de cables* conduit route; ~ **de comunicación de la conexión** *m* TELEFON connection communication path; ~ **de**

conexión *m* CONMUT switching path; **~ de interferencia** *m* TEC RAD interference path; **~ de respondedor** *m* TEC RAD transponder hopping; **~ de señal** *m* SEÑAL signal path; **~ de subida** *m* TEC RAD up-path

trayectoria: **~ de ascensión** *f* TRANS climb path; **~ a-t-b** *f* TEC RAD *atenuación*, TELEFON path a-t-b; **~ de comprobación de seguridad** *f* DATOS *de transmisiones* security audit trail; **~ eléctrica** *f* TEC RAD electrical length; **~ de enlace** *f* INFO linkway; **~ de migración directa** *f* ELECTRÓN straightforward migration path; **~ de navegación** *f* INFO navigation path; **~ óptica** *f* TEC RAD line-of-sight; **~ óptima** *f* TRANS optimal path; **~ de planeo de referencia nula** *f* TEC RAD null-reference glide path; **~ de radio digital** *f* TEC RAD digital radiopath; **~ de transmisión por satélite** *f* TV satellite transmission path; **~ de transmisión por vía satélite** *f* TEC RAD, TRANS satellite transmission path; **~ de vuelo** *f* TEC RAD flight path

trazabilidad *f* CONMUT traceability

trazado *m* COMPON *con lásers* scribing, DATOS layout, INFO layout, plot; **~ de organigrama** GEN flowcharting

trazador: **~ de curvas X-Y** *m* PRUEBA X-Y plotter; **~ de gráficos** *m* INFO graph plotter; **~ de líneas primitivas** *m* INFO pitchbender

trazar *vt* COMPON *carpintería* scribe, INFO *grafo* plot

trazo: **~ estroboscópico** *m* TEC RAD strobe marker; **~ negro** *m* ELECTRÓN black stroke; **~ oblicuo** *m* INFO oblique stroke

TRC *abr* (*tubo de rayos catódicos*) COMPON, ELECTRÓN CRT (*cathode-ray tube*)

T-RDSI *abr* (*terminación de la red B-RDSI*) TRANS NT-LB (*B-ISDN network termination*)

trémolo *m* INFO tremolo

tren *m* TRANS, TV train; **~ de impulsos** TELEFON, TRANS, TV pulse train; **~ de ondas** ELECTRÓN, TEC RAD wave train; **~ de señales** SEÑAL signal train

trenza *f* ELECTRÓN, TRANS braid, braiding; **~ de fibra óptica** TRANS optical fibre (*AmE* fiber) pigtail; **~ metálica** ELECTRÓN metallic twist

trenzado[1] *adj* COMPON, ELECTRÓN stranded

trenzado[2]: **~ cruzado** *m* TRANS *cables* cross-stranding

trepidación: **~ transversal** *f* TRANS transverse judder

TRF *abr* (*transformación rápida de Fourier*) INFO FFT (*fast Fourier transform*)

triac *m* ELECTRÓN *semiconductores* triac

tributario[1] *adj* DATOS, TELEFON tributary (*TR*)

tributario[2] *m* DATOS, TELEFON tributary (*TR*)

triestado: **de ~** *adj* ELECTRÓN tristate

trigatrón *m* ELECTRÓN trigatron

Trinitrón *m* INFO Trinitron

trinquete: **~ de arrastre** *m* GEN driving pawl

triodo *m* COMPON, ELECTRÓN triode

trituración *f* GEN crushing

TRL *abr* (*atenuación de retorno transversal*) TELEFON TRL (*transverse return loss*)

trocear *vt* GEN *herramientas* cut off

trombón *m* COMPON trombone

tronco: **~ de cables** *m* TRANS cable trunk

tropiezo *m* INFO hitch

tropopausa *f* TEC RAD tropopause

troposfera *f* TEC RAD troposphere

TRP *abr* (*punto de referencia de transmisión*) TELEFON, TRANS TRP (*transmission reference point*)

TRT *abr* (*verificación de ruta de tráfico*) PRUEBA, TELEFON TRT (*traffic-route testing*)

trueque *m* TV barter

truncación *f* DATOS, INFO truncation

truncar *vt* DATOS, INFO truncate

TRV *abr* (*vehículo de retransmisión por cinta*) TV TRV (*tape-relay vehicle*)

TSDU *abr* (*unidad de datos del servicio de transporte*) DATOS, TELEFON, TRANS TSDU (*transport-service data unit*)

TSM *abr* (*gestor de servidor terminal*) DATOS TSM (*terminal server manager*)

TT & C *abr* (*rastreo, telemetría y órdenes*) TEC RAD TT & C (*tracking, telemetry and command*)

TTCN *abr* (*notación combinada en árbol y tabular*) DATOS, INFO TTCN (*tree and tabular combined notation*)

TTD *abr* (*retardo de tránsito del blanco*) DATOS TTD (*target transit delay*)

TTT *abr* (*línea central de pruebas de transmisión*) PRUEBA, TELEFON TTT (*transmission test trunk*)

TTU *abr* (*unidad de arrastre de la cinta*) APLIC, INFO, TV TTU (*tape transport unit*)

TTX *abr* (*teletexto*) APLIC, TV TTX (*teletext*)

TTXAU *abr* (*unidad de acceso a teletexto*) APLIC TTXAU (*teletext access unit*)

TTY *abr* (*teletipo*) INFO, PERIF, TEC RAD, TELEFON TTY (*teletype*)

tubo *m* ELECTRÓN, TRANS tube; **~ de almacenamiento de memoria** ELECTRÓN storage tube; **~ analizador de punto móvil** ELECTRÓN flying-spot scanner tube; **~ de cámara fotoconductor** TV photoconductive camera tube; **~ de cámara fotoemisor** ELECTRÓN photoemissive camera tube; **~ de cámara de imagen** ELECTRÓN image camera tube; **~ de cámara con tensión**

anódica estabilizada ELECTRÓN anode-potential-stabilized camera tube; ~ **de campo** ELECTROTEC tube of force; ~ **capilar de vidrio** ELECTRÓN glass capillary; ~ **de cátodo frío** ELECTRÓN cold-cathode tube; ~ **contador de aguja** TEC RAD needle counter tube; ~ **contador de corriente gaseosa** ELECTRÓN gas-flow counter tube; ~ **contador de Geiger-Müller** TEC RAD Geiger-Müller counter tube; ~ **contador de radiación** TEC RAD radiation counter tube; ~ **contador de pared delgada** TEC RAD thin-wall counter tube; ~ **controlado por carga espacial** COMPON space-charge-controlled tube; ~ **convertidor de frecuencia** ELECTRÓN frequency converter tube; ~ **de convertidor de imagen** ELECTRÓN, TV image converter tube; ~ **de descarga de arco controlado por rejilla** ELECTRÓN grid-controlled arc discharge tube; ~ **de descarga luminiscente** ELECTRÓN glow discharge tube; ~ **desfasador** ELECTRÓN phase-shifter tube; ~ **ER** ELECTRÓN, TEC RAD TR tube; ~ **estabilizador de tensión** ELECTRÓN, ELECTROTEC voltage-stabilizing tube; ~ **de fuerza** ELECTROTEC tube of force; ~ **de gas** ELECTRÓN gas tube; ~ **de imagen** ELECTRÓN picture tube; ~ **de impulsos electrónicos** COMPON, ELECTRÓN trigger tube; ~ **indicador luminiscente** ELECTRÓN glow indicator tube; ~ **indicador de neón** ELECTRÓN neon indicator tube; ~ **intensificador de imagen** ELECTRÓN image intensifier tube; ~ **lleno de gas** ELECTRÓN gas-filled tube; ~ **mezclador** ELECTRÓN mixer tube; ~ **multielectródico estabilizador de voltaje** ELECTRÓN multielectrode voltage-stabilizing tube; ~ **múltiple** ELECTRÓN multiple tube; ~ **de onda de carga espacial** COMPON, ELECTRÓN space-charge wave tube; ~ **de onda de retorno** TELEFON backward-wave tube; ~ **de ondas progresivas** ELECTRÓN, TRANS travelling-wave (*AmE* traveling-wave) tube; ~ **de osciloscopio** ELECTRÓN oscilloscope tube; ~ **de plasma generador de ruido** ELECTRÓN noise-generator plasma tube; ~ **de potencia** ELECTRÓN power tube, power valve; ~ **pre-TR** ELECTRÓN pre-TR tube; ~ **pretransmisor/receptor** ELECTRÓN pre-transmit-receive tube; ~ **de proyección** ELECTRÓN projection tube; ~ **de rayos catódicos** (*TRC*) COMPON, ELECTRÓN cathode-ray tube (*CRT*); ~ **de rayos catódicos de cañón múltiple** ELECTRÓN multiple-gun cathode-ray tube; ~ **de rayos catódicos para televisión** ELECTRÓN, TV television tube; ~ **rectificador** ELECTRÓN, ELECTROTEC rectifier tube, rectifying tube; ~ **de rectificador de cátodo líquido** ELECTRÓN pool-rectifier tube; ~ **regulador de tensión** ELECTRÓN, ELECTROTEC voltage-regulator tube; ~ **de tensión de referencia** COMPON, ELECTRÓN voltage-reference tube; ~ **unidad** ELECTRÓN unit tube; ~ **de vacío** ELECTRÓN vacuum tube; ~ **de vapor de mercurio** ELECTRÓN mercury-vapour (*AmE* mercury-vapor) tube

tuerca *f* GEN nut

turno: ~ **de exploración** *m* TELEFON scanning shift

T1C *abr* (*portadora T1*) TRANS T1C (*T1 carrier*)

TV *abr* (*televisión*) TEC RAD, TV TV (*television*); ~ **por cable** *f* TV cable TV; ~ **comercial** *f* TV business TV

T & C *abr* (*seguimiento y mando*) PRUEBA T & C (*tracking and command*)

U

U: **--Matic** *m* TV U-Matic

UART *abr* (*receptor-transmisor asíncrono universal*) TEC RAD UART (*universal asynchronous receiver-transmitter*)

ubicar: **~ erróneamente** *vt* INFO misplace

UCM *abr* (*gestor de comunicaciones de usuario*) INFO UCM (*user communications manager*)

UCP *abr* (*unidad central de proceso*) CONMUT, INFO, TELEFON CPU (*central processing unit*)

UDC *abr* (*conector de datos universal*) ELECTRÓN UDC (*universal data connector*)

UE *abr* (*elemento de usuario*) DATOS, TELEFON UE (*user element*)

UHF *abr* (*frecuencia ultra-alta*) TV UHF (*ultrahigh frequency*)

UIT *abr* (*Unión Internacional de Telecomunicación*) TELEP ITU (*International Telecommunication Union*)

UJT *abr* (*transistor de una sola unión*) COMPON, ELECTRÓN UJT (*unijunction transistor*)

Ultimedia *m* INFO Ultimedia

último: **~ en entrar, primero en salir** *fra* (*LIFO*) INFO last in, first out (*LIFO*)

ultrasónico *adj* TEC RAD ultrasonic

ultrasonido *m* TEC RAD ultrasound

ultravioleta *adj* (*UV*) ELECTRÓN, TRANS ultraviolet (*UV*)

ululación *f* GEN warble

umbral *m* ELECTRÓN, ELECTROTEC, TEC RAD, TRANS threshold; **~ de acción láser** ELECTRÓN, TRANS lasing threshold; **~ de audibilidad** TEC RAD threshold of audibility, TELEFON threshold of audibility, threshold of hearing; **~ auditivo** TELEFON auditory threshold; **~ de decisión** GEN decision threshold; **~ de descodificación** TEC RAD decoding threshold; **~ de despliegue** TEC RAD deployment threshold; **~ de detección** ELECTRÓN, TELEFON, TRANS detection threshold; **~ de discriminación** PRUEBA *metrología* discrimination threshold; **~ de Geiger** TEC RAD Geiger threshold; **~ mínimo de interferencia** TEC RAD minimum interference threshold; **~ de rentabilidad** GEN threshold of profitability; **~ de ruido** TELEFON noise threshold; **~ de la señal** SEÑAL signal threshold

UMTS *abr* (*sistema universal de telecomunicaciones móviles*) TEC RAD UMTS (*universal mobile telecommunication system*)

unidad *f* GEN unit; **~ de abonado** TV subscriber unit; **~ de acceso** GEN access unit; **~ de acceso controlado** (*CAU*) TELEFON controlled-access unit (*CAU*); **~ de acceso digital del abonado** TELEFON subscriber digital-access unit; **~ de acceso EDI** (*EDI-AU*) TELEFON EDI access unit (*EDI-AU*); **~ de acceso multiestación** (*MAU*) DATOS multistation access unit (*MAU*); **~ de acceso a teletexto** (*TTXAU*) APLIC teletext access unit (*TTXAU*); **~ de acceso télex** (*TLXAU*) DATOS, TEC RAD, TELEFON telex access unit (*TLXAU*); **~ de adaptación** GEN interfacing unit; **~ de adaptador de datos** INFO data adaptor unit; **~ administrativa** CONTROL administrative unit (*AU*); **~ de alimentación** ELECTROTEC power pack, *apoyo* power supply unit; **~ de alimentación eléctrica** ELECTROTEC *ondulador* power supply unit; **~ de almacenamiento y conversión** (*CSU*) DATOS conversion and storage unit (*CSU*); **~ de aplicación** DATOS application entity; **~ aritmética y lógica** (*ALU*) INFO arithmetic and logic unit (*ALU*); **~ de arranque** TELEFON pull-out unit; **~ de arrastre de la cinta** (*TTU*) APLIC, INFO, TV tape transport unit (*TTU*); **~ auxiliar de conmutación** CONMUT auxiliary switching unit; **~ base** TEC RAD base unit (*BU*); **~ básica computada** TELEFON basic metered unit; **~ binaria** DATOS, INFO, TRANS bit; **~ de cableado** COMPON, ELECTRÓN wiring unit, TRANS cabling unit; **~ de cableado multicapa** COMPON, ELECTRÓN multilayer wiring unit; **~ de canal** TRANS channel unit; **~ de casete** INFO cassette drive; **~ de CD-ROM** INFO CD-ROM drive; **~ central de conmutación** TELEFON central switching unit; **~ central de control** CONMUT central control unit; **~ central de proceso** (*UCP*) CONMUT, INFO, TELEFON central processing unit (*CPU*); **~ de coma flotante** (*FPU*) INFO floating-point unit (*FPU*); **~ de comando** CONTROL command unit; **~combinadora** ELECTRÓN combiner unit; **~ de concentración remota** TRANS remote concentration unit; **~ de conexión del abonado** PERIF, TELEFON subscriber connection unit; ~ **de conexión cruzada** TELEFON cross-connecting unit; ~ **de conexión de líneas** TRANS line connection unit; ~ **de conmutación y control** CONMUT switching and control unit; ~ **conmutadora** CONMUT switching unit; ~ **de contacto de relé** COMPON relay contact unit; ~ **de contadores** TELEFON meter unit; ~ **de contadores de cifra de mérito** PRUEBA figure of merit meter; ~ **de control** CONTROL control unit, monitoring unit; ~ **de control de batería** ELECTROTEC battery control unit; ~ **de control de energía** ELECTROTEC power control unit; ~ **de control de posición de vuelo** TEC RAD attitude control unit; ~ **de control remoto** CONTROL remote-control unit (*RCU*); ~ **de copia impresa** APLIC, PERIF hard-copy device; ~ **de corrección de impulsos** TRANS pulse-correction unit; ~ **de datos del servicio de la red** DATOS network service data unit; ~ **de datos** GEN item; ~ **de datos de interfaz MCS** DATOS MCS interface data unit; ~ **de datos del interfaz MCS** TELEFON MCS interface data unit; ~ **de datos con protocolo de subcapa de convergencia** DATOS convergence sublayer protocol data unit (*CSPDU*); ~ **de datos de protocolo** DATOS protocol data unit (*PDU*); ~ **de datos de protocolo N** DATOS *para técnicos* N-service data unit, TELEFON N-protocol data unit; ~ **de datos de protocolo de mensaje** (*MPDU*) DATOS message protocol data unit (*MPDU*); ~ **de datos de protocolo de presentación** DATOS presentation-protocol data unit; ~ **de datos del protocolo N** DATOS N-protocol data unit; ~ **de datos del protocolo MCS** TELEFON MCS protocol data unit; ~ **de datos del protocolo de aplicación** (*APDU*) DATOS, TELEFON application protocol data unit (*APDU*); ~ **de datos del protocolo de la red** DATOS network protocol data unit (*NPDU*); ~ **de datos del protocolo de sesión** DATOS, TRANS session protocol data unit (*SPDU*); ~ **de datos remota** (*RDU*) DATOS remote data unit (*RDU*); ~ **de datos de servicio AAL** TELEFON AAL service data unit (*AAL-SDU*); ~ **de datos de servicio** DATOS service data unit (*SDU*); ~ **de datos de servicio de presentación** DATOS presentation-service data unit; ~ **de datos de servicio de red relanzada** DATOS expedited network service data-unit; ~ **de datos del servicio N** TELEFON N-service data unit; ~ **de datos del servicio MCS** TELEFON MCS service data unit; ~ **de datos del servicio de cajeros automáticos** TELEFON ATM service data unit; ~ **de datos del servicio de transporte** (*TSDU*) DATOS, TELEFON, TRANS transport-service data unit (*TSDU*); ~ **de diagnóstico de control** PRUEBA control diagnostic unit; ~ **de disco** INFO disk drive, disk driver, floppy-disk drive; ~ **de disco duro** INFO hard drive; ~ **de discos** INFO disk-storage drive; ~ **de distorsión de la cuantificación** (*qdu*) TELEFON, TRANS quantizing distortion unit (*qdu*); ~ **de distribución y control** (*DCU*) CONTROL distribution and control unit (*DCU*); ~ **enchufable** (*PIU*) DATOS *digitalizador*, ELECTRÓN, INFO, TELEFON plug-in unit (*PIU*); ~ **de encriptación** DATOS, SEÑAL encryption unit; ~ **de energía del bastidor** ELECTROTEC, TELEFON rack power unit; ~ **de energía de reserva** ELECTROTEC stand-by power supply; ~ **de entrada** GEN front end; ~ **de exclusión** PRUEBA *seguridad de funcionamiento* disablement unit, exclusion unit; ~ **exploradora** CONMUT scanner unit; ~ **de facturación** APLIC charge unit; ~ **funcional** DATOS, PRUEBA functional unit (*FU*); ~ **funcional entrelazada** DATOS interworking functional unit (*IFU*); ~ **de fusible de batería** ELECTROTEC battery fuse unit; ~ **de gestión** DATOS management unit; ~ **de gestión de la memoria** INFO memory management unit; ~ **de grabación de respuestas** TELEFON answer-recording unit; ~ **de identificación** GEN *marcaje* label frame; ~ **de imagen** DATOS frame unit; ~ **independiente** ELECTRÓN stand-alone unit; ~ **indicadora** TELEFON indicating unit; ~ **de interfaz de comando** CONTROL command interface unit; ~ **de integración eléctrica** ELECTRÓN electrical integration unit; ~ **de interconexión de guía de luz** (*LIU*) ELECTRÓN lightguide interconnection unit (*LIU*); ~ **de interfuncionamiento** DATOS interworking unit; ~ **de jacks de servicio** COMPON, TELEFON service jack unit; ~ **de línea** TELEFON line unit; ~ **de línea urbana** CONMUT *PBX* exchange line unit; ~ **lógica** (*LU*) DATOS logic unit, logic unit; ~ **M** TEC RAD M-unit; ~ **de medición** PRUEBA unit of measurement; ~ **de medición de fuente** (*SMU*) PRUEBA source-measure unit (*SMU*); ~ **de memoria de datos** CONMUT data store unit; ~ **modular estándar** TELEFON standard modular unit; ~ **monitora** PRUEBA monitor unit; ~ **móvil** TELEFON mobile unit, TV roadside cabinet, streetside cabinet; ~ **de muestreo** PRUEBA sampling unit; ~ **N** TEC RAD N-unit; ~ **N de datos del protocolo** DATOS N-protocol data unit; ~ **no conforme** PRUEBA nonconforming unit; ~ **de organización** TELEFON organizational unit; ~ **portadora** TELEFON bearer unit; ~ **de potencia** ELECTROTEC power unit; ~ **de presentación** INFO presentation unit; ~ **principal de control** INFO main control unit; ~ **principal de procesamiento** (*PUP*) INFO principal processing unit (*PUP*); ~ **del procesador principal** INFO main processor unit; ~ **de procesador de señales** SEÑAL signal processor unit; ~ **de proceso de 32 bitios** INFO 32-bit processor unit; ~ **de protección contra sobretensión** ELECTRÓN overvoltage protection unit; ~ **de protocolo** DATOS protocol unit (*PU*); ~ **de receptor de antena** TEC RAD antenna-receiver unit; ~ **rectificadora** ELECTROTEC rectifier unit; ~ **rectificadora del tiristor** ELECTROTEC thyristor rectifier unit; ~ **de refrigeración** CLIMAT cooling unit; ~ **remota** CONMUT remote unit (*RU*); ~ **remota de conmutación** CONMUT remote switching unit (*RSU*); ~ **de reserva** GEN stand-by unit; ~ **de respuestas telefónicas** APLIC, TELEFON voice response unit; ~ **de retardo** ELECTRÓN *elementos lógicos* delay element; ~ **de salida** INFO output unit; ~ **selectora** COMPON, CONMUT selecting unit; ~ **sensora** PRUEBA sensing unit; ~ **de señal** (*SU*) SEÑAL signal unit (*SU*); ~ **de señal de categoría de enlace** SEÑAL link status signal unit; ~ **de señal de control de sistema** (*SCU*) SEÑAL system control signal unit (*SCU*); ~ **de señal inicial** (*ISU*) SEÑAL initial signal unit (*ISU*); ~ **de señalización del mensaje** DATOS, TELEFON message signalling (*AmE* signaling) unit; ~ **de señal de sincronización** SEÑAL synchronization signal unit; ~ **de señal de sincronización multibloque** (*MBS*) SEÑAL multiblock synchronization signal unit (*MBS*); ~ **de señal solitaria** (*LSU*) SEÑAL lone signal unit (*LSU*); ~ **de servicio** DATOS housekeeping unit; ~ **de servicio de justificación** TELEFON, TRANS justification service unit; ~ **de sincronización de ciclo** TRANS frame synchronization unit; ~ **de sincronización de fase** (*PLU*) ELECTRÓN phase lock unit (*PLU*); ~ **de sincronización del bloque** DATOS frame synchronization unit; ~ **de síntesis de la voz** APLIC, INFO, TELEFON voice synthesis unit; ~ **sintetizadora** INFO synthesizer unit; ~ **de sintonización de antena** (*ATU*) TEC RAD antenna tuning unit (*ATU*); ~ **suministrable** TELEFON deliverable unit; ~ **de suministro de energía** ELECTROTEC *rectificador* power supply unit; ~ **de supervisión** CONTROL, PRUEBA, TELEFON supervisory unit; ~ **supervisora de función** CONMUT function supervisory unit; ~ **telefónica protegida** TELEFON secured telephone unit; ~ **terminal de la red** (*NTU*) TRANS network terminating unit (*NTU*); ~ **terminal de vídeo** TV video display terminal *AmE*; ~ **de tiempo** GEN unit of time; ~ **de trama temporal** INFO, TEC RAD, TRANS timescale unit; ~ **de transferencia de datos del protocolo** DATOS transport protocol data unit (*TPDU*); ~ **de transporte** GEN transport unit; ~ **virtual** INFO virtual unit; ~ **de visualización** INFO, TV display unit

unidades: ~ **constitutivas** *f pl* COMPON *técnicas de montaje* constituent units; ~ **relacionadas de decibelios** *f pl* TELEFON dB-related units

unidireccional *adj* DATOS one-way, INFO, TEC RAD, TELEFON unidirectional

unidireccionalidad *f* INFO one-way directionality

unificación *f* ELECTRÓN unification

uniforme *adj* GEN regular; **no** ~ GEN nonuniform

uniformidad: ~ **de luminancia** *f* ELECTRÓN uniformity of luminance

unilateral *adj* GEN unilateral

unión *f* GEN junction; ~ **adaptada** TEC RAD matched junction; ~ **en árbol** DATOS, INFO tree attachment; ~ **de colector** ELECTRÓN *semiconductores* collector junction; ~ **descendente de programas** TEC RAD down-link of programmes; ~ **por difusión** COMPON diffused junction; ~ **del emisor** COMPON *semiconductores* emitter junction; ~ **local** TRANS local junction; ~ **multifibra** TRANS multifibre (*AmE* multifiber) joint; ~ **PN** ELECTRÓN PN junction; ~ **recta** COMPON, ELECTRÓN straight splice

Unión: ~ **Europea de Radiodifusión** *f* (*EURB*) TEC RAD European Union for Radio Broadcasting (*EURB*); ~ **Internacional de Telecomunicación** *f* (*UIT*) TELEP International Telecommunication Union (*ITU*); ~ **de Radiodifusión de Asia y el Pacífico** *f* (*ABU*) TEC RAD Asia-Pacific Broadcasting Union (*ABU*); ~ **de Radiodifusión de los Estados Árabes** *f* TEC RAD Arab States Broadcasting Union; ~ **de Radiodifusión Europea** *f* (*EBU*) TEC RAD European Broadcasting Union (*EBU*)

unir *vt* GEN join; ◆ ~ **a tope** TRANS butt-joint

universo *m* PRUEBA universe

univocal *m* TELEFON speech plus simplex

UNTDI *abr* (*intercambio de datos de comercio de las Naciones Unidas*) TELEFON UNTDI (*United Nations trade-data interchange*)

UPC *abr* (*control de parámetro de uso*) APLIC UPC (*usage parameter control*)

UPS *abr* (*alimentación eléctrica ininterrumpible*) ELECTROTEC UPS (*uninterruptible power supply*)

UPT *abr* (*telecomunicación personal universal*) APLIC UPT (*universal personal telecommunication*)

URE *abr* (*Unión de Radiodifusión Europea*) TEC RAD EBU (*European Broadcasting Union*)

URL *abr* (*localizador de recursos uniforme*) DATOS *Internet* URL (*uniform resource locator*)

usanza *f* TELEFON usage

usar *vt* GEN use

USB *abr* (*banda lateral superior*) TEC RAD, TRANS USB (*upper sideband*)

USB1 *abr* (*binarios no descifrados*) DATOS USB1 (*unscrambled binary ones*)

Usenet *f* DATOS *Internet* Usenet

uso: **~ compartido de línea** *m* DATOS line sharing; **~ de la corriente parásita simétrica** *m* ELECTRÓN push-on sneak current use; **~ y desgaste** *m* INFO wear and tear; **~ de exclusivo** *adj* TELEFON private; **~ fraudulento** *m* GEN fraudulent use; **~ futuro** *m* GEN future use; **~de privado** *adj* PERIF private

usuario *m* GEN user; **~ comercial** GEN business user; **~ en conjunto** TRANS joint user; **~ de DTAM** TELEFON DTAM user; **~ final** GEN end user; **~ de intercambio de datos electrónicos** DATOS, TELEFON EDI user; **~ de MCS** TELEFON MCS user; **~ de mensajería por intercambio de datos electrónicos** DATOS EDI-messaging user, TELEFON EDI-messaging user, EDIMG user; **~ de mensajería por voz** APLIC, DATOS, TELEFON voice messaging user; **~ de MIS** DATOS MIS user; **~ de MS** DATOS MS user; **~ de la red** DATOS *Internet* network user; **~ de la red de servicio que llama** DATOS calling network-service user; **~ de RTSE** DATOS RTSE user; **~ del servicio** DATOS, TELEFON service-user; **~ de servicio abstracto MS** DATOS MS abstract service user; **~ del servicio de calidad de transmisión de CMISE** APLIC, DATOS performing CMISE service user; **~ de servicio CMISE** DATOS CMISE service user; **~ del servicio CMISE** TELEFON CMISE service user; **~ del servicio de procesamiento de transacciones** (*TPSU*) DATOS transaction processing service user (*TPSU*); **~ de servicio de red llamado** DATOS called network service user, called NS user; **~ de servicio de red que llama** DATOS calling NS user; **~ de servicios ACSE** DATOS, TELEFON ACSE service user; **~ de servicios de comunicación avanzada** DATOS, TELEFON ACSE service user; **~ de los servicios de la red** DATOS network service user; **~ de sistema de correo por voz** APLIC, TELEFON voice-mail system user; **~ del sistema de mensajería interpersonal** DATOS interpersonal messaging system user; **~ de sistema de mensajería por voz** APLIC, DATOS, TELEFON voice messaging system user

UT *abr* (*probador superior*) DATOS, PRUEBA UT (*upper tester*)

útil *adj* GEN useful

utilidad *f* DATOS, INFO utility; **~ de explotación** GEN operating profit; **~ de la red** TELEFON network utility

utilizable: **no ~** *adj* GEN unusable

utilización: **~ de la antememoria** *f* DATOS, INFO caching; **~ de la antememoria del disco** *f* DATOS, INFO disk caching; **~ comercial** *f* GEN business application

utillaje: **~ para revelado de disco compacto fotográfico** *m* INFO photo CD access development toolkit

UTP *abr* (*par torcido descubierto*) ELECTRÓN, TRANS UTP (*unshielded twisted pair*)

UTS *abr* (*sistema universal de transporte*) ELECTRÓN UTS (*universal transport system*)

UUCP *abr* (*protocolo de copia de Unix-a-Unix*) DATOS *Internet* UUCP (*Unix-to-Unix copy protocol*)

UV *abr* (*ultravioleta*) ELECTRÓN, TRANS UV (*ultraviolet*)

UW *abr* (*palabra singular*) DATOS UW (*unique word*)

V

V *abr* (*tensión*) ELECTROTEC V (*voltage*)
vaciado: **~ de datos** *m* INFO data dump
vaciamiento: **~ de la memoria** *m* INFO storage dump
vaciar *vt* TEC RAD *satélite* offload
vacío[1] *adj* DATOS empty
vacío[2] *m* GEN vacuum, void
vacuna *f* DATOS, INFO *protección contra los virus* vaccine
VADIS *abr* (*sistema interactivo digital vídeo-audio*) INFO VADIS (*video-audio digital interactive system*)
valencia *f* GEN valence, valency
validación *f* DATOS validation
validar *vt* DATOS validate
validez *f* DATOS validity
válido *adj* GEN valid
valor *m* GEN value; **~ de ajuste** GEN setting value; **~ asignable por defecto** (*D*) DATOS defaultable (*D*); **~ de atenuación** TRANS attenuation value; **~ del atributo** DATOS attribute value; **~ de autentificación** DATOS authentication value; **~ básico** GEN basic value; **~ de causa** GEN cause value; **~ de comprobación criptográfico** DATOS cryptographic check value; **~ cuantificado** TRANS quantized value; **~ de los datos de presentación** DATOS presentation data value; **~ de decisión** TELEFON, TRANS *PCM* decision value; **~ de decisión virtual** TELEFON virtual decision value; **~ distinto de cero** INFO nonzero value; **~ esperado** GEN *estadística* expected value; **~ indicativo** GEN standard value; **~ inicial** GEN initial value; **~ límite** PRUEBA limit value, limiting value; **~ máximo** GEN peak value; **~ máximo del impulso** TELEFON pulse-peak value; **~ medido** PRUEBA measured value; **~ medio de atenuación** TRANS mean loss value; **~ modulado** DATOS modulated value; **~ nominal** GEN nominal value; **~ normalizado** TEC RAD normalized value; **~ observado** PRUEBA observed value; **~ por defecto** GEN default value; **~ óptimo** GEN optimum value; **~ ordinal** DATOS ordinal value; **~ de parámetro** PRUEBA, TELEFON parameter value (*PV*); **~ de puntero** CONMUT pointer value; **~ de referencia** PRUEBA reference value; **~ de régimen permanente** GEN steady-state value; **~ de reposición** COMPON resetting value; **~ rms** GEN rms value; **~ umbral** GEN threshold value; **~ verdadero aceptado de una cantidad** PRUEBA *calidad* conventional true value of a quantity
valoración *f* GEN valuation; **~ de calidad** PRUEBA quality assessment; **~ del proveedor** PRUEBA supplier valuation; **~ de sonoridad del objetivo de recepción** (*ROLR*) TELEFON receive objective loudness rating (*ROLR*)
valorar *vt* GEN valuate
válvula *f* COMPON, ELECTRÓN valve; **~ de bola** COMPON ball valve; **~ de cátodo frío** ELECTRÓN cold-cathode valve; **~ de contrapresión** COMPON back-pressure valve; **~ de corte remoto** ELECTRÓN remote cutoff tube; **~ de desagüe** COMPON drain valve; **~ de descarga luminiscente** ELECTRÓN glow discharge valve; **~ de escape** COMPON exhaust valve, reducing valve; **~ manorreductora** COMPON pressure-reducing valve; **~ de onda progresiva** ELECTRÓN travelling-wave (*AmE* traveling-wave) valve; **~ rectificadora** ELECTRÓN, ELECTROTEC rectifier valve, rectifying valve; **~ reguladora** CLIMAT *tiro de chimeneas* throttle valve; **~ de solenoide** COMPON solenoid valve
VAM *abr* (*método de acceso virtual*) CONMUT VAM (*virtual access method*)
vano *m* GEN span
VANS *abr* (*servicios de red de valor añadido*), (*VANS*) APLIC, TRANS, TV VANS (*value-added network services*)
vapor *m* GEN vapour (*AmE* vapor)
vaporización: **~ al vacío** *f* GEN vacuum vaporization
varactor *m* COMPON, ELECTRÓN varactor
variabilidad *f* ELECTRÓN variability
variable[1] *adj* GEN variable, changeable
variable[2] *f* GEN variable; **~ controlada** CONTROL controlled variable; **~ lógica** CONMUT switching variable; **~ de mandato** CONTROL command variable; **~ manipulada** COMPON, CONTROL manipulated variable; **~ de referencia** CONTROL reference variable; **~ de salida** CONTROL output variable
variación *f* GEN variation; **~ de altura** INFO pitch variation; **~ de ancho** INFO width variation; **~ diaria** GEN day-to-day variation; **~ diurna** GEN

diurnal variation; ~ **en un lote** PRUEBA *calidad* batch variation; ~ **de nivel** INFO level variation; ~ **sistemática** PRUEBA systematic variation; ~ **de tensión** ELECTROTEC, PRUEBA voltage variation

variante *f* GEN variant

varianza *f* GEN variance

varilla *f* ELECTRÓN *soldadura* rod; ~ **de medición** TELEFON measuring rod; ~ **de nivel de aceite** TELEFON *aparatos indicadores* dipstick; ~ **de soporte** COMPON supporting rod; ~ **de tierra** ELECTROTEC earthing rod *BrE*, grounding rod *AmE*

varioplex *m* ELECTRÓN varioplex

varistor *m* COMPON, ELECTRÓN varistor

vástago: ~ **de válvula** *m* COMPON valve spindle

vatímetro: ~ **de cabeza de torsión** *m* PRUEBA torsion-head wattmeter; ~ **de Thruline** *m* PRUEBA Thruline wattmeter; ~ **de Truline** *m* PRUEBA Truline wattmeter

vatio: ~**-hora** *m* GEN watt-hour

VBI *abr* (*intervalo de borrado vertical*) DATOS, TV VBI (*vertical blanking interval*)

Vbox *m* INFO Vbox

VCCRF *abr* (*función de canal virtual relacionada con la conexión*) DATOS, TRANS VCCRF (*virtual-channel connection-related function*)

VCL *abr* (*visualización en cristal líquido*) ELECTRÓN, INFO LCD (*liquid crystal display*)

VCP *abr* (*reproductor de cintas de vídeo*) INFO, TV VCP (*videocassette player*)

VCU *abr* (*petición de actualización rápida de instrucción de vídeo*) TV VCU (*video command fast-update request*)

VCXO *abr* (*oscilador de cristal de regulación por tensión*) ELECTRÓN VCXO (*voltage-controlled crystal oscillator*)

vecino *adj* GEN neighbouring (*AmE* neighboring)

vector *m* GEN vector; ~ **de crominancia** TV chrominance vector; ~ **eléctrico** TRANS *óptica de fibras* electric vector; ~ **mayoritario** ELECTRÓN majority carrier; ~ **de movimiento** INFO, TV motion vector; ~ **de movimiento de retorno** TV backward-motion vector; ~ **de reflexión** INFO reflection vector; ~ **de retorno** INFO backward-motion vector

vehículo: ~ **de centro de mensajes** *m* TRANS message-centre (*AmE* message-center) vehicle; ~ **espacial** *m* TEC RAD space vehicle; ~ **de retransmisión por cinta** *m* (*TRV*) TV tape-relay vehicle (*TRV*)

velocidad *f* GEN speed; ~ **de acercamiento** PRUEBA rate of approach; ~ **de amortiguamiento** INFO decay rate; ~ **aproximada de tráfico binario** DATOS gross bit rate; ~ **de arrastre** COMPON *semiconductores* drift velocity, TEC RAD drift rate; ~ **de ataque** INFO attack velocity; ~ **baja de tráfico binario** TRANS low bit rate; ~ **de banda** TRANS band rate; ~ **binaria** DATOS, INFO, TELEFON, TRANS binary rate; ~ **binaria equivalente** TELEFON equivalent bit rate; ~ **binaria de error** DATOS binary error rate; ~ **binaria de señalamiento en serie** SEÑAL binary serial signalling (*AmE* signaling) rate; ~ **de bloqueo** GEN blocking rate; ~ **del cable** TELEFON cable rate; ~ **de cambio de fase** ELECTRÓN phase-change velocity; ~ **del chip** DATOS chip rate; ~ **de codificación** DATOS coding rate; ~ **de conversación** TELEFON speaking rate; ~ **de desconexión** INFO release rate, release velocity; ~ **digital** TRANS digit rate; ~ **digital de línea** DATOS, TELEFON symbol rate, TRANS line digit rate, symbol rate; ~ **de escape** TEC RAD escape velocity; ~ **de exploración** TEC RAD scanning speed; ~ **de fase** TEC RAD phase speed, phase velocity; ~ **de funcionamiento libre** DATOS free-running speed; ~ **de generación** ELECTRÓN *semiconductores* generation rate; ~ **de graduación por teclado** INFO keyboard scaling rate; ~ **de grupo** TRANS envelope velocity, group velocity; ~ **de imagen** INFO, TV picture rate; ~ **de impresión** PERIF printing speed; ~ **de impulso** APLIC impulse rate; ~ **de información obligada** DATOS committed information rate (*CIR*); ~ **de lectura** ELECTRÓN reading speed; ~ **de lectura mínima utilizable** ELECTRÓN minimum usable reading speed; ~ **de línea en baudios** INFO *teleproceso* baud rate; ~ **máxima utilizable de escritura** ELECTRÓN maximum usable writing speed; ~ **media de cruce** GEN average crossing rate; ~ **media de reparación** INFO mean repair rate; ~ **de modulación** TELEFON modulation rate, TRANS modulation rate, modulation speed; ~ **de una partícula** TELEFON particle velocity; **de poca** ~ *adj* GEN low-speed; ~ **de precipitación** TEC RAD precipitation rate; ~ **de propagación** ELECTRÓN, TRANS propagation velocity; ~ **de regeneración** DATOS refresh rate, INFO refresh rate, regeneration rate; ~ **de registro de imágenes por teclado** INFO keyframe rate; ~ **de sensibilización** ELECTRÓN priming rate, priming speed; ~ **simbólica nominal** DATOS nominal symbol rate; ~ **del sonido** GEN speed of sound, velocity of sound; ~ **de tráfico binario de transmisión** TEC RAD, TRANS transmission bit rate; ~ **de transferencia de información del usuario** DATOS user information transfer rate; ~ **de transmisión** TRANS transmission speed; ~ **de transmisión de datos** DATOS data-transfer rate; ~ **de transmisión dividida** DATOS split baud rate

vendedor *m* GEN dealer

ventana *f* DATOS, INFO, TEC RAD, TELEFON, TRANS, TV window; ~ **de Hann** TV Hann window; ~ **de palabra singular** DATOS unique word window; ~ **de prueba espectral** TRANS spectral window; ~ **temporal** INFO temporary window; ~ **de transmisión** TRANS transmission window; ~ **UW** DATOS UW window; ~ **de velocidad** TEC RAD speed gate, velocity gate

ventanas: ~ **en cascada** *f pl* INFO cascading windows

ventanilla: ~ **de estado** *f* CONMUT *de la primera parte conectada*, TELEFON status field

ventas: ~ **por teléfono** *f pl* APLIC, TV telesales

ventilación *f* CLIMAT ventilation

ventilador *m* INFO fan

ver *vt* INFO view

veracidad *f* PRUEBA, TEC RAD truth

verdad *f* INFO truth

verdadero *adj* INFO true

veredicto: ~ **no concluyente** *m* DATOS inconclusive verdict

verificación *f* DATOS *de comando* verification, PRUEBA, TELEFON *de mensaje* verification, verifying; ~ **de atenuación** TRANS attenuation check; ~ **desde la centralita** CONMUT cross-office check, *centralitas* cross-exchange check; ~ **de componente redundante** ELECTROTEC redundant-component check; ~ **de la consistencia de un circuito** DATOS circuit-consistency verification; ~ **de errores** TEC RAD error check; ~ **de fallo** TELEFON failure verification; ~ **de funcionamiento** PRUEBA *seguridad de funcionamiento* function check-out; ~ **de información aérea** TRANS verification of overhead information; ~ **de paridad** DATOS, TELEFON parity check; ~ **de redundancia** DATOS redundancy check; ~ **de redundancia cíclica** (*CRC*) DATOS, TELEFON cyclic redundancy check (*CRC*); ~ **de redundancia longitudinal** DATOS longitudinal redundancy check; ~ **de ruta de tráfico** (*TRT*) PRUEBA, TELEFON traffic-route testing (*TRT*); ~ **por saltos** INFO, PRUEBA leapfrog test; ~ **transversal por redundancia** DATOS transverse redundancy check

verificador: ~ **de vía de tráfico** *m* PRUEBA, TELEFON traffic route tester

verificar *vt* GEN check

vernier *m* PRUEBA vernier

versátil *adj* GEN versatile

versión *f* DATOS, INFO, TELEFON version; ~ **por idioma** GEN *documentación* language version; ~ **sintáctica** DATOS, TELEFON syntax version

vertical[1] *adj* GEN *equipo* vertical

vertical[2] *f* GEN vertical

vértice *m* GEN vertex

V/F *abr* (*tensión/frecuencia*) ELECTRÓN, ELECTROTEC V/F (*voltage/frequency*)

VFT *abr* (*telegrafía de frecuencia vocal*) TELEFON VFT (*voice-frequency telegraphy*)

VGA *abr* (*matriz de vídeo gráfico*) INFO VGA (*video graphics array*)

VHF *abr* (*frecuencia muy alta*) INFO VHF (*very high frequency*)

vía *f* CONMUT, ELECTRÓN, INFO, TEC RAD, TELEFON, TRANS path; ~ **de acceso** INFO approach; ~ **de comunicación** DATOS *SDL* communication path; ~ **de conversación** TELEFON speech path; ~ **de la eficacia en función del coste** GEN cost-effective way; ~ **férrea troncal** TRANS main line; ~ **no traficable** TRANS untrafficable path; ~ **pública** GEN public thoroughfare; ~ **de retorno** TRANS, TV return path; ~ **de tarjetas** INFO card path; ~ **de transmisión** TEC RAD, TRANS transmission path; ~ **de transmisión aérea** TRANS, TELEFON overhead transmission path; ~ **de transmisión de banda ancha** TRANS broadband channel; ~ **de transmisión multipunto centralizada** DATOS centralized multipoint facility; ~ **de transmisión virtual** DATOS, TRANS virtual path (*VP*)

vías: ~ **de tráfico reintroducibles** *f pl* TELEFON re-entrant trunking

vibración *f* INFO jar, TEC RAD swing, TV flutter

vibrador *m* ELECTRÓN, INFO, TEC RAD vibrator

vibrar *vi* INFO chatter

vibrato: ~ **de retardo** *m* INFO delay vibrato; ~ **de la rueda de modulación** *m* INFO modulation wheel vibrato

VICS *abr* (*sistema vehículo de comunicación inteligente*) TEC RAD VICS (*vehicle intelligent communication system*)

vida *f* PRUEBA lifespan, *seguridad de funcionamiento* life; ~ **del fractil de P** PRUEBA P-fractile life; ~ **mecánica** PRUEBA mechanical life; ~ **media** PRUEBA mean life; ~ **de servicio** DATOS housekeeping lifetime; ~ **útil** PRUEBA wear-out life; ~ **útil en almacenamiento** PRUEBA shelf life

video *AmL ver vídeo Esp*

vídeo *Esp m* APLIC, INFO, TV video; ~ **de banda ancha** TV wideband video; ~ **combinado** TV composite video; ~ **en fase de producción** (*PLV*) INFO production level video (*PLV*); ~

independiente INFO, TV separated video (*S-video*); ~ **inverso** TV inverse video; ~ **a petición** (*VOD*) INFO, TV video-on-demand (*VOD*); ~ **de reflexión múltiple** TV multihop video; ~ **S** TV S-video; ~ **en tiempo real** INFO, TV real-time video

videocámara *m* TV video camera

videocasete *f* INFO, TV videocassette

videocompresión *f* DATOS, TV video compression

videocomunicación *f* TV videocommunication

videoconferencia *f* APLIC videoconference, INFO, TEC RAD videoconference, visioconference, TELEFON visioconference, TV confravision, videoconference, visioconference; ~ **de exploración lenta** TV slow-scan videoconferencing

videocorreo *m* INFO video mail

videodisco *m* TV videodisc (*AmE* videodisk), videodisk; ~ **compacto** TV video CD

videoenlace *m* TV video link

videofonía *f* TEC RAD, TELEFON, TV videophony; ~ **de imagen fija** TEC RAD, TELEFON, TV still-picture videophony

videofrecuencia *f* TRANS, TV videofrequency

videograbador *m* INFO, TV video recorder

videografía *f* INFO, TEC RAD, TV videography; ~ **para difusión de mensajes** APLIC broadcast videography; ~ **de radiodifusión** TV broadcast videography

videomensajería *f* INFO, TV videomessaging

videoprocesador *m* INFO video processor

videoservicios *m pl* TV video services

videoservidor *m* INFO video server

videotelefonía *f* TEC RAD, TELEFON, TV videotelephony; ~ **de imagen fija** TEC RAD, TELEFON, TV still-picture videotelephony

videoteléfono *m* APLIC videophone, videotelephone, viewphone, visiophone, picture phone

videotexto *m* INFO, TEC RAD, TV videotext; ~ **basado en sintaxis** (*SBV*) TELEFON syntax-based videotext (*SBV*); ~ **para difusión de mensajes** APLIC broadcast videotext; ~ **de radiodifusión** TV broadcast videotext

vidicón *m* ELECTRÓN vidicon; ~ **con haz de retorno** (*RBV*) TELEFON return-beam vidicon (*RBV*)

vidrio: ~ **reforzado** *m* COMPON reinforced glass

viga *f* COMPON joist; ~ **de recalzo** INFO, TELEFON needle

vigilancia *f* TEC RAD watching; ~ **de sistemas y de controles** PRUEBA control and system monitoring

vigilante: ~ **de cables** *m* ELECTRÓN wire-minder

VIM *abr* (*mensajería independiente del proveedor*) TEC RAD VIM (*vendor-independent messaging*)

vincular *vt* DATOS bind

vínculo: ~ **con el prestador de servicios** *m* TRANS service provider link (*SPL*)

vinilo *m* COMPON vinyl

VIO *abr* (*entrada/salida virtual*) INFO VIO (*virtual input/output*)

violación: ~ **bipolar** *f* TRANS bipolar violation; ~ **de código** *f* TRANS code violation (*CV*); ~ **de código de transmisión** *f* TELEFON, TRANS transmission code violation

violar *vt* PRUEBA tamper with

viraje *m* COMPON turning

virus *m* DATOS, INFO virus; ~ **informático** DATOS computer virus

visibilidad: ~ **bajo los ecos parásitos** *f* TEC RAD subclutter visibility

visión: ~ **asistida por computadora** *AmL f* (*cf visión asistida por ordenador Esp*) INFO computer-assisted vision; ~ **asistida por ordenador** *Esp f* (*cf visión asistida por computadora AmL*) INFO computer-assisted vision

visitante: ~ **no autorizado** *m* DATOS *Internet* unauthorized visitor

visor *m* INFO display window

vista: ~ **en corte** *f* GEN *dibujo técnico* cutaway view; ~ **desplezada** *f* GEN *dibujo técnico* exploded view; ~ **exterior** *f* GEN outside view; ~ **de fondo** *f* GEN *dibujo técnico* bottom view; ~ **interior** *f* GEN *dibujo técnico* inside view; ~ **normal** *f* INFO normal view

visto[1]: ~ **desde abajo** *adj* GEN viewed from below; ~ **desde adentro** *adj* GEN viewed from inside; ~ **desde atrás** *adj* GEN viewed from rear; ~ **de lado** *adj* GEN viewed from side; ~ **por el lado de los componentes** *adj* GEN viewed from component side; ~ **desde el lado de la soldadura** *adj* GEN viewed from soldering side

visto[2]: ~ **bueno** *m* TEC RAD OK signal

visualización *f* GEN display; ~ **de acimut estabilizado** TEC RAD azimuth-stabilized display; ~ **en bruto** TEC RAD raw video display; ~ **en cristal líquido** (*VCL*) ELECTRÓN, INFO liquid crystal display (*LCD*); ~ **de datos** INFO video data; ~ **de indicación multipunto** DATOS multipoint indication visualization; ~ **de llamadas en LCD** PERIF LCD calls display

visualizador: ~ **alfanumérico** *m* TELEFON, TV alphanumeric display; ~ **de caracteres** *m* ELECTRÓN *dispositivos indicadores* character display; ~ **de cifras** *m* TELEFON *aparatos indicativos* digit display; ~ **de conexión** *m*

PERIF power-on display; ~ **para descarga de gas** *m* ELECTRÓN, INFO gas discharge display; ~ **de estado de referencia** *m* TEC RAD reference-status display

visualizar *vt* GEN display

VITC *abr* (*código de tiempo de intervalo vertical*) DATOS, INFO VITC (*vertical interval time code*)

vivo *adj* ELECTROTEC, TV live

VLBI *abr* (*interferometría con línea de base muy larga*) PRUEBA VLBI (*very long baseline interferometry*)

VLSI *abr* (*integración a escala muy grande*) COMPON, ELECTRÓN, TELEFON, TRANS VLSI (*very large-scale integration*)

VM *abr* (*correspondencia telefónica*) APLIC, TELEFON VM (*voice mail*)

VMG *abr* (*transmisión de mensajes hablados*) APLIC, DATOS, TELEFON VMG (*voice messaging*)

VMGE *abr* (*entorno de mensajería por voz*) DATOS, TELEFON VMGE (*voice messaging environment*)

VMGS-MS *abr* (*memoria de mensaje de sistema de mensajería por voz*) DATOS VMGS-MS (*voice messaging system message store*)

VMGS-UA *abr* (*agente de usuario de sistema de mensajería por voz*) DATOS, TELEFON VMGS-UA (*voice messaging system user agent*)

VM-MS *abr* (*memoria de mensajería por voz*) APLIC, DATOS, TELEFON VM-MS (*voice messaging store*)

vocal[1]: **no ~** *adj* TEC RAD nonvoice

vocal[2] *f* GEN vowel

vocalizado *adj* INFO, TELEFON voiced

voces: **de ~ múltiples** *adj* GEN multivoice

vocoder *m* INFO, TEC RAD, TELEFON vocoder; ~ **de canales** TEC RAD channel vocoder; ~ **formante** TEC RAD formant vocoder; ~ **de tono excitado** TEC RAD pitch-excited vocoder

VOD *abr* (*vídeo a petición*) INFO, TV VOD (*video-on-demand*)

volatilidad *f* INFO volatility

volcado: ~ **en masa** *m* INFO bulk dump

voltaje: ~ **ánodo-cátodo** *m* ELECTRÓN *semiconductors* anode-cathode voltage; ~ **aplicado** *m* ELECTROTEC applied voltage; ~ **de avalancha** *m* ELECTRÓN *semiconductores* avalanche voltage; ~ **de batería** *m* ELECTRÓN, ELECTROTEC battery voltage; ~ **de CA** *m* INFO AC voltage, alternating current voltage; ~ **de CC** *m* ELECTROTEC DC voltage; ~ **de control** *m* ELECTROTEC control voltage; ~ **desactivado** *m* COMPON off-state voltage; ~ **de descarga disruptiva de CA** *m* ELECTRÓN, ELECTROTEC *de un protector* AC spark-over voltage; ~ **de descarga disruptiva de CC** *m* ELECTRÓN, ELECTROTEC DC spark-over voltage; ~ **de entrada** *m* ELECTRÓN, ELECTROTEC threshold voltage; ~ **de entrada de rejilla** *m* ELECTRÓN grid input voltage; ~ **de foco de imagen** *m* ELECTRÓN image focus voltage; ~ **de Hall** *m* ELECTRÓN *semiconductores* Hall voltage; ~ **de hélice** *m* ELECTRÓN helix voltage; ~ **impulsor de rejilla** *m* ELECTRÓN grid driving voltage; ~ **impulsor rejilla-cátodo** *m* ELECTRÓN grid-cathode driving voltage; ~ **inactivo** *m* ELECTRÓN off-state voltage; ~ **inducido** *m* ELECTROTEC induced voltage; ~ **máximo** *m* ELECTROTEC peak voltage; ~ **de modulación** *m* ELECTRÓN modulation voltage; ~ **nominal de descarga disruptiva de CC** *m* ELECTRÓN, ELECTROTEC *de protector* nominal DC spark-over voltage; ~ **de onda cuadrada** *m* ELECTRÓN, SEÑAL square-wave voltage; ~ **de pico** *m* COMPON peak point voltage; ~ **de régimen** *m* ELECTRÓN operating voltage; ~ **de ruptura** *m* ELECTRÓN breakdown voltage; ~ **de salida** *m* ELECTRÓN, INFO output voltage; ~ **de saturación** *m* ELECTRÓN saturation voltage

voltamperio *m* ELECTRÓN, ELECTROTEC volt-ampere

voltímetro: ~ **de cresta** *m* PRUEBA peak voltmeter; ~ **digital** *m* PRUEBA digital voltmeter; ~ **de hierro móvil** *m* PRUEBA iron-vane voltmeter, moving-vane voltmeter; ~ **selectivo** *m* PRUEBA frequency-selective voltmeter; ~ **sintonizable** *m* PRUEBA tunable voltmeter; ~ **de varias sensibilidades** *m* PRUEBA multirange voltmeter

voltio *m* ELECTRÓN, ELECTROTEC volt

volumen *m* GEN volume; ~ **de modo efectivo** TRANS effective mode volume; ~ **de modos** TRANS mode volume; ~ **de muestra** PRUEBA sample size; ~ **de tráfico** CONMUT traffic volume, DATOS traffic flow, TELEFON traffic flow, traffic volume

volúmenes: ~ **limítrofes** *m pl* INFO bounding volumes

volúmetro: ~ **VU** *m* PRUEBA VU-meter

voluminoso *adj* COMPON bulky

volver: ~ **a ejecutar** *vt* INFO, TELEFON rerun; ~ **a la posición de reposo** *vt* CONMUT return to home position

VOR *abr* (*radiofaro omnidireccional de VHF*) TEC RAD VOR (*VHF omnidirectional radio range*)

voxel *m* INFO, TV voxel

voz *f* INFO, PERIF voice, TELEFON speech, voice; ~ **/onda continua** TELEFON voice/cw; ~ **sintética** PERIF *de contestador automático* synthetic voice; ◆ **sin ~** INFO unvoiced

VPCRF *abr* (*función de trayectoria virtual relacionada con la conexión*) DATOS, TRANS VPCRF (*virtual-path connection-related function*)

VR *abr* (*realidad virtual*) DATOS, INFO VR (*virtual reality*)

VRAM *abr* (*RAM de vídeo*) INFO VRAM (*video RAM*)

VSAT *abr* (*terminal de apertura muy pequeña*) DATOS, TRANS, TV VSAT (*very small aperture terminal*)

VSB *abr* (*banda lateral residual*) TEC RAD, TRANS, TV VSB (*vestigial sideband*)

VSW *abr* (*onda estacionaria de tensión*) TEC RAD, TRANS VSW (*voltage standing wave*)

VSWR *abr* (*relación de ondas estacionarias de tensión*) TEC RAD, TRANS VSWR (*voltage standing wave ratio*)

VTP *abr* (*reproductor de cintas de vídeo*) INFO, TV VTP (*videotape player*)

vuelco *m* DATOS dumping, INFO dump, dumping; **~ de memoria** INFO memory dump; **~ de la memoria central** INFO core dump

vuelta: **~ atrás de la transacción** *f* DATOS transaction roll-back

WAIS *abr* (*servidor de información de área extendida*) DATOS *Internet* WAIS (*wide-area information server*)

WAN *abr* (*red de área extendida*) DATOS, INFO, TELEFON, TRANS WAN (*wide-area network*)

watímetro: **~ de paleta** *m* PRUEBA vane wattmeter

WDM *abr* (*multiplexión por división de longitud de onda*) TELEFON, TRANS WDM (*wavelength division multiplexing*)

Web: **el ~** *m* INFO the Web

Whois *m* DATOS *Internet* Whois

WIC *abr* (*centro de interconexión montable en pared*) ELECTRÓN WIC (*wall-mountable interconnect centre*)

WSC *abr* (*centro de empalmes montable en pared*) ELECTRÓN WSC (*wall-mountable splice centre*)

WWW *abr* (*telaraña mundial*) DATOS, INFO WWW (*World Wide Web*)

WYSIWYG (*lo que se ve es lo que se obtiene*) INFO WYSIWYG (*what you see is what you get*)

X Y Z

X *abr* (*supletorio*) GEN X (*extension*)
XD *abr* (*ex-directorio*) GEN XD (*ex-directory*)
XID *abr* (*central de identificación*) TELEFON XID (*exchange of identification*)
XNS *abr* (*Servicio de Red Xerox*) DATOS XNS (*Xerox Network Service*)

Y *n* INFO *operaciones no matemáticas* AND
yugo *m* COMPON, ELECTRÓN, TELEFON yoke
yunque *m* COMPON anvil
yuxtapuesto *adj* GEN juxtaposed

zanja *f* TEC RAD ditch
zapear *vt* INFO zap
ZBR *abr* (*registro de bitio fuera de texto*) INFO ZBR (*zone bit recording*)
zigzag *m* ELECTROTEC, INFO zigzag
zócalo: **~ de armario** *m* TRANS *técnicas constructivas* cabinet base; **~ de tubo** *m* COMPON valve socket
zona *f* DATOS, INFO area, TEC RAD zone; **~ de base** ELECTRÓN base region; **~ de caída** TEC RAD footprint; **~ de captación** TEC RAD, TV catchment area; **~ de central interurbana principal** TELEFON main trunk exchange area; **~ de central urbana** TELEFON local exchange area; **~ de cobertura de estación de base** TEC RAD *comunicación móvil* base area; **~ de conexión** TEC RAD home area; **~ de contacto** (*I/F*) GEN interface (*I/F*); **~ de control** INFO control area; **~ de cobertura de estación de base** TEC RAD *comunicación móvil* base coverage area; **~ cubierta** TEC RAD, TV coverage area; **~ de declaraciones** DATOS declarations part; **~ despejada** ELECTROTEC *circuitos impresos* clearance hole; **~ de espacio libre** TEC RAD *contestador* free-space range; **~ de estación de base** TEC RAD base station area; **~ de facturación** APLIC, TELEFON charging area; **~ de franquicia** TV franchise area; **~ de Fresnel** TRANS Fresnel zone; **~ de interacción** ELECTRÓN interaction region; **~ de interferencia** GEN nuisance area; **~ más clara** PERIF *escáner* clearest zone; **~ más oscura** TV *digitalizador* darkest zone; **~ de memoria de programas** (*PMA*) INFO program memory area (*PMA*); **~ meta** TEC RAD, TRANS target zone; **~ de montaje** GEN assembly area; **~ de propagación** GEN propagation area; **~ de pulsación** INFO clicking area; **~ de radiación** TEC RAD radiation zone; **~ del radiador** TEC RAD radiator area; **~ de servicio** GEN service area; **~ de sombra** TEC RAD shadow region, TRANS shadow region, shadow zone; **~ de tarifa** TELEFON tariff zone, tariff area; **~ telefónica de centro de conmutación de grupo** CONMUT group switching centre (*AmE* center) catchment area, group switching centre (*AmE* center) exchange area; **~ de texto** INFO text box; **~ de tolerancia** PRUEBA tolerance zone; **~ de tráfico** TEC RAD traffic area; **~ para transmisión** TEC RAD, TRANS zone for transmission
zoom *m* INFO zooming
zumbador *m* GEN buzzer
zumbido *m* ELECTRÓN hum; **~ de la red** ELECTROTEC mains hum

ENGLISH–SPANISH
INGLÉS–ESPAÑOL

A

A-ABORT: ~ **APDU** *n* DATA APDU *m* de ABORTO-A

AAL *abbr* (*ATM adaptation layer*) GEN AAL (*envoltura de adaptación de un ATM*); ~ **protocol control information** *n* (*AAL-PCI*) DATA, TELEP información *f* de control de protocolos AAL; ~ **service data unit** *n* (*AAL-SDU*) TELEP unidad *f* de datos de servicio AAL

AAL-PCI *abbr* (*AAL protocol control information*) DATA, TELEP información *f* de control de protocolos AAL

AAL-SDU *abbr* (*AAL service data unit*) TELEP unidad *f* de datos de servicio AAL

AARE *abbr* (*A-associate response*) DATA, TELEP AARE (*respuesta A-asociada*)

AARQ *abbr* (*A-associate request*) DATA, TELEP AARQ (*petición A-asociada*)

A-associate: ~ **request** *n* (*AARQ*) DATA, TELEP petición *f* A-asociada (*AARQ*); ~ **response** *n* (*AARE*) DATA, TELEP respuesta *f* A-asociada (*AARE*)

abbreviated: ~ **dialling** *BrE n* TELEP llamada *f* abreviada; ~ **number** *n* TELEP número *m* abreviado

absolute: ~ **address** *n* COMP dirección *f* absoluta; ~ **stability** *n* TELEP estabilidad *f* absoluta

absorption: ~ **factor** *n* RADIO factor *m* de absorción; ~ **loss** *n* RADIO pérdida *f* por absorción; ~ **spectrometry** *n* RADIO espectrometría *f* de absorción

absorptive: ~ **attenuator** *n* TRANS atenuador *m* absorbente

abstract *n* COMP, DATA abstracto *m*; ~ **data type** DATA tipo *m* de datos resumidos; ~ **test method** DATA metódo *m* de tipos resumidos

ABU *abbr* (*Asia-Pacific Broadcasting Union*) RADIO ABU (*Unión de Radiodifusión de Asia y el Pacífico*)

AC *abbr* (*alternating current*) GEN CA (*corriente alterna*); ~ **generator** *n* POWER alternador *m* CA generador *m* de CA; ~ **mains** *n* COMP red *f* de CA; ~ **mains cord** *n* COMP conductor *m* flexible de la red de CA; ~ **mains wall socket** *n* TV enchufe *m* mural de alimentación de CA; ~ **outlet** *n* COMP toma *f* de CA; ~ **power** *n* COMP electricidad *f* de CA; ~ **power adaptor** *n* COMP adaptador *m* de electricidad de CA; ~ **power-cord receptacle** *n* COMP receptáculo *m* para cable de CA; ~ **power outlet** *n* COMP toma *f* de electricidad de CA; ~ **power supply** *n* COMP fuente *f* de alimentación de CA; ~ **pulsing** *n* TELEP emisión *f* de impulsos de CA; ~ **signalling** *BrE n* POWER, SIG señalización *f* de CA; ~ **socket** *n* COMP enchufe *m* de CA; ~ **spark-over voltage** *n* ELECT, POWER *of a protector* voltaje *m* de descarga disruptiva de CA; ~ **voltage** *n* COMP voltaje *m* de CA

accelerator *n* ELECT, PARTS *plastics* acelerador *m*

acceptance: ~ **angle** *n* TRANS *fibre optics* ángulo *m* de recepción; ~ **test** *n* TEST prueba *f* de recepción

acceptor *n* DATA, ELECT *semiconductors* aceptor *m*

access *n* GEN acceso *m*; ~ **burst signal** DATA señal *f* de brote de ondas de acceso; ~ **channel** TELEP, TRANS, TV canal *m* de acceso; ~ **charge rate** GEN tarifa *f* inicial; ~ **circuit** TELEP circuito *m* de acceso; ~ **concentrator** TRANS concentrador *m* de acceso; ~ **connection element** (*ACE*) DATA, TELEP elemento *m* de conexión de acceso (*ACE*); ~ **control** DATA, TEST *security* control *m* de acceso; ~ **control list** DATA, TEST *security* relación *f* de controles de acceso; ~ **link** TELEP enlace *m* de acceso; ~ **matrix** TELEP matriz *f* de acceso; ~ **method** TELEP método *m* de acceso; ~ **network** GEN red *f* de acceso; ~ **node** GEN nodo *m* de acceso; ~ **number** TELEP número *m* de acceso; ~ **path** COMP, TRANS, TV camino *m* de acceso; ~ **port** GEN puerto *m* de acceso; ~ **subsystem** RADIO *radio paging* subsistema *m* de acceso; ~ **time** COMP, DATA tiempo *m* de acceso; ~ **unit** GEN unidad *f* de acceso

accountability *n* GEN responsabilidad *f*

accumulator *n* POWER *batteries* acumulador *m*

accuracy *n* COMP, RADIO, TEST *quality* precisión *f*

ACD *abbr* (*automatic call distribution*) TELEP distribución *f* automática de llamadas

ACE *abbr* (*access connection element*) DATA, TELEP ACE (*elemento de conexión de acceso*)

ACID *abbr* (*atomicity, consistency, isolation and durability*) DATA ACID (*atomicidad, consistencia, aislamiento y durabilidad*)

ACK *abbr* (*acknowledgement*) DATA, SIG, TELEP REC (*reconocimiento de recibo*)

acknowledge *vt* DATA, SIG, TELEP confirmar recepción
acknowledgement *n* (*ACK*) DATA, SIG, TELEP reconocimiento *m* de recibo (*REC*); ~ **request** DATA petición *f* de confirmación; ~ **signal** SIG señal *f* de reconocimiento
acoustic: ~ **antenna** *n* RADIO antena *f* acústica; ~ **coupler** *n* DATA, TELEP acoplador *m* acústico; ~ **coupling** *n* TELEP acoplamiento *m* acústico; ~ **delay line** *n* DATA *datacom* línea *f* de retardo acústico; ~ **noise** *n* DATA, TELEP ruido *m* acústico; ~ **resonator** *n* GEN resonador *m* acústico; ~ **signal** *n* GEN señal *f* acústica
acousto: ~**-electric effect** *n* ELECT efecto *m* electroacústico
ACPM *abbr* (*association control protocol machine*) DATA ACPM (*máquina de asociación de protocolo de control*)
ACSE *abbr* (*association control service element*) DATA, TELEP ACSE (*elemento para el control de asociación*); ~ **service provider** *n* DATA, TELEP proveedor *m* de servicios ACSE; ~ **service user** *n* DATA, TELEP usuario *m* de servicios ACSE
ACTE *abbr* (*Advisory Committee for Telecommunications Equipment*) TELEP ACTE (*Comité Asesor para Equipo de Telecomunicaciones*)
activation *n* GEN activación
active *adj* GEN activo; ~ **component** *n* ELECT componente *m* activo; ~ **double star** *n* TRANS estrella *f* doble activa; ~ **laser medium** *n* ELECT, TRANS medio *m* activo de laser; ~ **processor** *n* COMP procesador *m* activo; ~ **region** *n* TRANS zona *f* activa; ~ **skin** *n* RADIO película *f* activa; ~ **supervisor** *n* DATA supervisor *m* activo; ~ **threat** *n* DATA amenaza *f* activa
activity *n* GEN actividad *f*
actual: ~ **key** *n* COMP clave *f* real; ~ **parameter** *n* DATA *specification and description language* parámetro *m* real
actuator *n* RADIO accionador *m*
A-D *abbr* (*analog-to-digital*) GEN A-D (*analógico a digital*); ~ **converter** *n* (*ADC*) COMP, SIG, TELEP convertidor *m* A-D
adaptive: ~ **antenna** *n* RADIO antena *f* adaptativa; ~ **coding** *n* SIG codificación *f* por adaptación; ~ **filtering** *n* TRANS filtración *f* por adaptación; ~ **routing** *n* TELEP encaminamiento *m* adaptable; ~ **system** *n* DATA sistema *m* adaptativo
adaptor *n* GEN adaptador *m*
ADC *abbr* (*A-D converter*) COMP, SIG, TELEP convertidor *m* A-D
add *vt* GEN sumar
add-drop: ~ **multiplexer** *n* TRANS multiplexor *m* sumar-restar
adder *n* DATA circuito *m* sumador
add-in: ~ **board** *n* COMP cuadro *m* aditivo
additive: ~ **noise** *n* TELEP ruido *m* aditivo; ~ **process** *n* ELECT *printed circuits* proceso *m* aditivo
address *n* GEN dirección *f*; ~ **generation** COMP generación *f* de dirección; ~ **presentation restricted indicator** DATA indicador *m* restringido de presentación de dirección
addressing *n* COMP, DATA direccionamiento *m*
adhesion *n* ELECT *soldering* adhesión *f*
adjacent: ~ **channel** *n* RADIO, TELEP canal *m* adyacente; ~ **channel selectivity** *n* RADIO, TRANS selectividad *f* contra canales adyacentes
adjust *vt* GEN regular
adjustable *adj* GEN ajustable
adjustment *n* GEN ajuste *m*
administrative: ~ **processor** *n* COMP procesador *m* administrativo; ~ **unit** *n* (*AU*) CONTR unidad *f* administrativa; ~ **unit group** *n* CONTR grupo *m* de unidad administrativa
admittance *n* POWER admitancia *f*
ADPCM: ~ **decoder** *n* TELEP decodificador *m* ADPCM; ~ **encoder** *n* TELEP codificador *m* de señales ADPCM
Advisory: ~ **Committee for Telecommunications Equipment** *n* (*ACTE*) TELEP Comité *m* Asesor para Equipo de Telecomunicaciones (*ACTE*)
AE *abbr* (*aerial*) COMP, RADIO antena *f*
AEI *abbr* (*application-entity invocation*) DATA AEI (*invocación de la unidad de aplicación*)
AENOR *abbr* (*Spanish Association of Normalization and Certification*) GEN AENOR (*Asociación española de normalización y certificación*)
aerial *n* (*AE*) COMP, RADIO antena *f*; ~ **cable** TRANS cable *m* aéreo; ~ **mast** RADIO mástil *m* de antena
aerosol *n* RADIO aerosol *m*
AFC: ~ **hold-in** *n* RADIO retención *f* AFC; ~ **pull-in** *n* RADIO conexión *f* AFC
AFI *abbr* (*authority and format identifier*) DATA, SIG identificador *m* de autorización y formato
A4: ~ **size** *n* GEN *documentation* tamaño *m* A4; ~**-sized paper** *n* GEN *documentation* papel *m* de tamaño A4
AGC *abbr* (*automatic gain control*) ELECT, TELEP, TRANS AIS (*control automática de ganancia*)
ageing *n* GEN envejecimiento *m*
agent *n* DATA, TELEP agente *m*
aggregate: ~ **signal** *n* SIG señal *f* total

A-grade: ~ **contour** *n* RADIO contorno *m* de grado A
AI *abbr* (*artificial intelligence*) APPL, COMP, SWG AI (*inteligencia artificial*)
AIFF *abbr* (*Audio-Interchange File Format*) RADIO FFIA (*Formato de Fichero de Intercambio-Audio*)
airborne: ~ **radar** *n* RADIO radar *m* aerotransportado
airtime *n* RADIO tiempo *m* en el aire
AIS *abbr* (*alarm indication signal*) TELEP, TEST AIS (*señal de indicación de alarma*)
AJ *abbr* (*anti-jamming*) RADIO AJ (*antiinterferencias*)
alarm *n* DATA, TELEP, TEST alarma *f*; ~ **call** APPL llamada *f* de alarma; ~ **card** APPL tarjeta *f* de alarma; ~ **circuit** TELEP, TEST circuito *m* de alarma; ~ **function** TEST función *f* de alarma; ~ **indication signal** (*AIS*) TELEP, TEST señal *f* de indicación de alarma (*AIS*); ~ **setting** TEST ajuste *m* de alarma
A-law *n* TRANS ley *f* A
A-leg *n* TELEP cadena *f* regional en A
alerting *n* RADIO, TEST alerta *f*
algorithmic *adj* COMP algorítmico; ~ **language** *n* COMP lenguaje *m* algorítmico
alias[1] *adv* GEN alias
alias[2] *n* GEN alias *m*
aliasing *n* COMP, TRANS ajeno *m*
align *vt* GEN alinear
alignment *n* GEN alineamiento *m*; ~ **fault** TRANS error *m* de alineación
all: ~ **trunks busy tone** *n* (*ATB tone*) TELEP tono *m* ATB
allocate *vt* GEN asignar
allocation *n* GEN asignación *f*
allowance *n* GEN tolerancia *f*
alloy *n* PARTS aleación *f*
all-plastic: ~ **fibre** *BrE n* DATA, PARTS, TRANS fibra *f* de plástico
all-silica: ~ **fibre** *BrE n* DATA, PARTS, TRANS fibra *f* de sílice
ALOHA: ~ **system** *n* RADIO, TELEP sistema *m* ALOHA
alpha: ~ **dial** *n* COMP cuadrante *m* alfa; ~ **mosaic** *n* TEST mosaico *m* alfa; ~ **trial** *n* TEST ensayo *m* alfa
alphabet *n* DATA alfabeto *m*; ~ **code** COMP código *m* alfabético; ~ **key** COMP clave *f* alfabética; ~ **telegraph** DATA, TELEP telégrafo *m* alfabético
alphabetic *adj* COMP alfabético; ~ **code** *n* COMP código *m* alfabético; ~ **signal** *n* SIG señal *f* alfabética; ~ **telegraphy** *n* TELEP telegrafía alfabética
alphabetical *adj* GEN *linguistics* alfabético; ◆ **in ~ order** GEN en orden alfabético
alphabetization *n* GEN *linguistics* alfabetización *f*
alphabetize *vt* GEN *linguistics* alfabetizar
alphageometric *adj* TV alfageométrico; ~ **display** *n* TELEP dispositivo *m* visualizador
alphamosaic *adj* TV alfamosaico
alphanumeric: ~ **display** *n* TELEP, TV visualizador *m* alfanumérico; ~ **keyboard** *n* COMP teclado *m* alfanumérico; ~ **pager** *n* RADIO avisador *m* alfanumérico
alphaphotographic *adj* TV alfafotográfico
alt. *abbr* (*alternative*) DATA *Internet* alt. (*alternativa*)
alter[1]: ~ **operation** *n* COMP operación *m* modificar
alter[2] *vt* COMP alterar
alterable *adj* COMP alterable
alteration *n* COMP alteración
alternate: **~-access channel** *n* TELEP canal *m* de acceso en alternativa; ~ **bit inversion** *n* TRANS inversión *f* alterna de bitios; ~ **control sequence** *n* RADIO secuencia *f* de control alterna; ~ **digit inversion** *n* TRANS inversión *f* alterna de dígitos; ~ **final trunk group** *n* ELECT, SWG, TELEP grupo *m* alterno de líneas troncales finales; ~ **group** *n* COMP grupo *m* alterno; ~ **mark inversion** *n* (*AMI*) SIG, TELEP, TRANS inversión *f* alternada de marcas (*AMI*); ~ **mark inversion code** *n* SIG, TELEP, TRANS código *m* de inversión alternada de marcas; ~ **mark inversion signal** *n* SIG, TELEP, TRANS señal *f* de inversión alternada de marcas; ~ **mark inversion violation** *n* TELEP transgresión *f* de inversión alternada de marcas; ~ **route** *n* TELEP ruta *f* alternativa; ~ **route final trunk group** *n* ELECT, SWG grupo *m* de líneas troncales finales de ruta alternativa, TELEP grupos *m* de rutas alternativas de líneas troncales finales; ~ **speech service** *n* APPL *ISDN* servicio *m* alternativo de palabras; ~ **transmission** *n* TELEP transmisión *f* en alternativa; ~ **voice-data channel** *n* TRANS canal *m* de datos de voz alternado; ~ **voice-data circuit** *n* (*AVD*) TRANS circuito *m* de datos de voz alternado (*AVD*); ~ **voice-data transmission** *n* TRANS transmisión *f* de datos de voz alternada
alternated *adj* RADIO alternado
alternately *adv* GEN alternadamente
alternating: ~ **current** *n* (*AC*) GEN corriente *f* alterna (*CA*); ~ **current coupler** *n* POWER acoplador *m* de corriente alterna; ~ **current generator** *n* POWER alternador *m* de corriente

alterna; ~ **current pulsing** *n* TELEP emisión *f* de impulsos de corriente alterna; **~-current power supply** *n* COMP fuente *f* de corriente alterna; **~-current signalling** *BrE n* POWER, SIG señalización *f* de corriente alterna; **~-current source** *n* POWER generador *m* de corriente alterna; ~ **discharge current** *n* ELECT *of protector* corriente *f* alterna de descarga; ~ **voltage** *n* POWER voltaje *m* alterna

alternation: ~ **on inquiry** *n* TELEP consulta *f* alternativa

alternative[1]: ~ **current mains wall socket** *n* COMP toma *f* de línea de corriente alterna, TV tomacorriente *f* de la red de corriente alterna; ~ **current power cord receptacle** *n* COMP enchufe *m* para cable de corriente alterna; ~ **current power outlet** *n* COMP toma *f* de corriente alterna; ~ **current socket** *n* COMP clavija *f* bipolar para corriente alterna; ~ **current source** *n* POWER generador *m* de corriente alterna; ~ **route** *n* TELEP ruta *f* alternativa; ~ **routing** *n* GEN enrutamiento *m* alternativo; ~ **routing indicator** *n* TELEP indicador *m* de enrutamiento alternativo; ~ **routing of signalling** *BrE n* SIG enrutamiento *m* alternativo de señales; ~ **test method** *n* TEST *fibre optics* método *m* alternativo de pruebas

alternative[2] *n* (*alt.*) DATA *Internet* alternativa *f* (*alt.*)

alternative[3]: **with ~ handling** *phr* RADIO con tratamiento alternativo

altimeter *n* TELEP, TEST *indicating devices* altímetro *m*

altitude: ~ **of the apogee** *n* RADIO altura *f* del apogeo

ALU *abbr* (*arithmetic and logic unit*) COMP ALU (*unidad aritmética y lógica*)

aluminium *BrE* (*AmE* **aluminum**) *n* GEN aluminio *m*; ~ **alloy** PARTS aleación *f* de aluminio; ~ **bronze** PARTS bronce *m* de aluminio

AM *abbr* (*amplitude modulation*) COMP, ELECT, SIG, TRANS, TV modulación *f* de amplitud

amateur: ~ **radio service** *n* RADIO servicio *m* de radioaficionados

ambient: ~ **noise** *n* GEN *acoustics* ruido *m* ambiente; ~ **temperature** *n* COMP temperatura *f* ambiente

ambiguity *n* GEN ambigüedad *f*

American: ~ **National Standards Institute** *n* (*ANSI*) GEN Instituto *m* Nacional Americano de Normas (*ANSI*); ~ **National Standards Institute committee 12** *n* DATA Comité *m* 12 del Instituto Nacional Americano de Normas; ~ **Standard Code for Information Interchange** *n* (*ASCII*) COMP, DATA, TELEP Código *m* Estándar Americano para Intercambio de Información (*ASCII*)

AMI *abbr* (*alternate mark inversion*) SIG, TELEP, TRANS AMI (*inversión alternada de marcas*); ~ **code** *n* TRANS código *m* AMI; ~ **violation** *n* TRANS transgresión *f* del código AMI

aminoplastic *n* PARTS aminoplástico *m*

ammeter *n* TEST amperímetro *m*

amorphous: ~ **channel** *n* TRANS canal *m* amorfo

amortization *n* GEN amortización *f*

amperage *n* POWER amperaje *m*

ampere *n* POWER amperio *m*; ~ **rating** ELECT amperaje *m* de servicio

ampersand *n* COMP *typography* ampersand *m*

amplification *n* GEN amplificación *f*; ~ **stage** ELECT etapa *f* de amplificación; ~ **system** COMP sistema *m* de amplificación

amplified: ~ **circuit** *n* TRANS circuito *m* amplificado; ~ **handset** *n* PER, TELEP micrófono *m* con amplificador; ~ **reception** *n* PER *on answering machine* recepción *f* amplificada; ~ **telephone** *n* PER APP teléfono *m* amplificado

amplifier *n* GEN amplificador *m*; ~ **circuit** POWER circuito *m* amplificador; ~ **stage** ELECT, POWER fase *f* de amplificación

amplify *vt* GEN amplificar

amplitude *n* COMP, ELECT, RADIO, TELEP amplitud *f*; **~-corrected echo** TELEP eco *m* de amplitud corregida; ~ **curve** COMP curva *f* de amplitud; ~ **distortion** TRANS distorsión *f* de amplitud; ~ **equalization** COMP, TRANS nivelación *f* de amplitud; ~ **equalizer** COMP, TRANS ecualizador *m* de amplitud; ~ **error** COMP error *m* de amplitud; **~-frequency response** TRANS respuesta *f* de amplitud de frecuencia; ~ **keying** SIG manipulación *f* de amplitud; **~-modulated** *adj* TRANS modulado en amplitud; ~ **modulation** COMP, ELECT, SIG, TRANS modulación *f* de amplitud; ~ **modulation depth** TRANS profundidad *f* de modulación de amplitud; ~ **modulation sensitivity** COMP sensibilidad *f* de modulación de amplitud; ~ **phase-shift keying** (*APSK*) SIG tecleado *m* de cambio en la relación de fase de amplitud (*APSK*); **~-probability distribution** SIG distribución *f* de la probabilidad de amplitud; **~-quantized control** TELEP control *m* de amplitud cuantificada; **~-reflection factor** TRANS *fibre optics* factor *m* de reflexión de amplitud; ~ **response** ELECT, SIG, TRANS respuesta *f* de amplitud; **~-response limit** TRANS límite *m* de amplitud de respuesta;

~-shift keying (*ASK*) SIG tecleado *m* de cambio de amplitud (*ASK*)

anaerobic: ~ **adhesive** *n* ELECT adhesivo *m* anaeróbico

analog *adj* GEN analógico; ~ **adder** *n* ELECT adicionador *m* analógico; ~ **computer** *n* COMP calculadora *f* analógica; ~ **compact cassette** *n* TRANS casete *f* compacta analógica; ~ **circuit** *n* COMP, TRANS circuito *m* analógico; ~ **private wire** *n* TRANS circuito *m* analógico de uso privado; ~ **colour code** *n* TELEP código *m* analógico de color; ~ **control** *n* TELEP control *m* analógico; ~ **converter** *n* COMP, DATA, SIG, TELEP convertidor *m* analógico; **~-to-digital** *n* (*A-D*) COMP, SIG, TELEP analógico a digital (*A-D*); **~-to-digital converter** *n* COMP, SIG, TELEP convertidor *m* analógico-digital; ~ **data** *n pl* COMP datos *m pl* analógicos; ~ **device** *n* GEN dispositivo *m* analógico; **~-digital** COMP analógico-digital; ~ **transmission link** *n* TRANS enlace *m* de transmisión analógica; ~ **input** *n* COMP entradas *f pl* analógicas; ~ **filter** *n* COMP, TRANS filtro *m* analógico; ~ **interface** *n* TRANS interfaz *m* analógico; ~ **modulation** *n* COMP modulación *f* analógica; ~ **line board** *n* SWG placa *f* de circuitos de línea analógica; ~ **call processor** *n* COMP procesador *m* analógico de llamadas; ~ **recording** *n* GEN registro *m* analógico; ~ **repeater** *n* TELEP repetidor *m* analógico; ~ **display** *n* COMP representación *f* analógica; ~ **signal** *n* COMP, SIG señal *f* analógica; ~ **synthesizer** *n* COMP sintetizador *m* analógico; ~ **system** *n* GEN sistema *m* de conmutación analógica; ~ **sound** *n* COMP sonido *m* analógico; **~-to-digital** (*A/D*) GEN analógico a digital; ~ **transmission** *n* TRANS transmisión *f* analógica; ~ **video** *n* COMP video *m* analógico

analogical: ~ **learning** *n* COMP aprendizaje *m* analógico; ~ **reasoning** *n* DATA razonamiento *m* analógico

analyse *BrE* (*AmE* **analyze**) *vt* GEN analizar

analyser *BrE* (*AmE* **analyzer**) *n* ELECT, SIG analizador *m*

analysis *n* GEN análisis *m*; ~ **filterbank** COMP batería *f* de filtros para el análisis; ~ **meter** TELEP instrumento *m* de análisis; ~ **for number length** TELEP análisis *m* de longitud numérica; ~ **table** GEN *documentation* cuadro *m* de análisis

analyst *n* GEN analista *m*; **~-programmer** GEN analista-programador *m*

anamorphosis *n* GEN anamorfosis *m*

anchor[1] *n* COMP anclaje *m*; ~ **MSC** RADIO MSC *m* de sujeción; ~ **point** COMP punto *m* de sujeción

anchor[2] *vt* ELECT sujetar

anchoring: ~ **bracket** *n* TRANS *cable laying* grapa *f* de sujeción

ancillary *adj* COMP auxiliar; ~ **equipment** *n* PER APP equipo *m* auxiliar; ~ **unit** *n* COMP unidad *f* ancilar

AND: **not** ~ *n* ELECT NO-Y; ~ **circuit** *n* ELECT *logic elements* circuito *m* Y; ~ **gate** *n* ELECT *logic elements* puerta *f* Y; ~ **operation** *n* ELECT *logic elements* operación *f* Y; ~ **parallelism** *n* DATA paralelismo *m* conjuntivo

AND-OR: ~ **graph** *n* DATA gráfico *m* Y-O; ~ **processing** *n* DATA procesamiento *m* Y-O

anechoic: ~ **chamber** *n* TEST cámara *f* anecoica; ~ **room** *n* TEST cámara *f* anecoica

angel *n* RADIO *radar* eco *m* parásito

angle: ~ **of arrival** *n* TRANS ángulo *m* de llegada; ~ **of azimuth** *n* RADIO ángulo *m* acimutal; ~ **of departure** *n* TRANS ángulo *m* de salida; ~ **of depression** *n* RADIO ángulo *m* de depresión; ~ **diversity link** *n* DATA enlace *m* de diversidad angular; ~ **of elevation** *n* RADIO, TRANS ángulo *m* de elevación; ~ **encoder** *n* DATA codificador *m* angular; ~ **from the axis** *n* RADIO ángulo *m* respecto de eje; ~ **of incidence** *n* TRANS ángulo *m* de incidencia; ~ **modulation** *n* TRANS modulación *f* de ángulo; **~-nose pliers** *n* GEN *tools* tenazas *f pl* de boca en T; ~ **of rotation** *n* TRANS *fibre optics* ángulo *m* de giro; ~ **of sight** *n* RADIO ángulo *m* de visión; ~ **of site** *n* TRANS ángulo *m* de situación; ~ **of tilt** *n* RADIO ángulo *m* de picado

angular: ~ **displacement** *n* TELEP, TRANS *fibre optics* desplazamiento *m* angular; ~ **misalignment loss** *n* TRANS *fibre optics* pérdida *f* por desalineación angular; ~ **momentum** *n* RADIO momento *m* cinético; ~ **resolution** *n* RADIO resolución *f* angular; ~ **separation** *n* RADIO, TRANS separación *f* angular

anisochronous *adj* GEN anisócrono; ~ **data** *n pl* DATA datos *m pl* anisócronos; ~ **data transmission** *n* DATA, TRANS transmisión *f* de datos anisócronos

anisotropic *adj* GEN anisotrópico

annealed: ~ **wire** *n* PARTS *heat treatment* hilo *m* recocido

annotation *n* COMP, TELEP anotación *f*; ~ **button** COMP botón *m* de anotador; ~ **symbol** TELEP símbolo *m* de anotación

announcement: ~ **machine** *n* PER, TELEP máquina *f* de avisos; **~-only facility** *n* PER *on answering machine*, TELEP contestador *m* estático; ~ **trunk** *n* TELEP línea *f* troncal parlante; ~ **trunk group** *n* TELEP grupo *m* de líneas troncales parlantes

annual: ~ **charge ratio** *n* TELEP índice *m* de carga anual; ~-**costed work programme** *BrE n* GEN programa *m* de trabajo con costes anuales

annular: ~ **sector radiating line** *n* RADIO línea *f* radiante de sector anular

anode *n* ELECT ánodo *m*; ~ **battery** POWER batería *f* anódica; ~-**cathode characteristics** *n pl* ELECT *semiconductors* características *f pl* ánodo-cátodo; ~-**cathode voltage** ELECT *semiconductors* voltaje *m* ánodo-cátodo; ~ **current** ELECT corriente *f* de ánodo; ~-**potential-stabilized camera tube** ELECT tubo *m* de cámara con tensión anódica estabilizada; ~ **terminal** ELECT *semiconductors* terminal *m* de ánodo; ~ **voltage** ELECT tensión *f* de ánodo

anodic: ~ **reaction** *n* PARTS reacción *f* anódica; ~ **treatment** *n* PARTS tratamiento *m* anódico

anodize *vt* ELECT anodizar

anomalistic: ~ **period** *n* RADIO periodo *m* anomalístico

anomalous: ~ **dispersion** *n* TRANS *fibre optics* dispersión *f* anómala

ANS *abbr* (*answer tone*) DATA ANS (*tono de contestación*)

Ansaphone® *n* TELEP contestador *m* automático

ANSI *abbr* (*American National Standards Institute*) GEN ANSI (*Instituto Nacional Americano de Normas*)

answer *n* TELEP contestación *f*; ~-**back** DATA, TELEP emisión *f* de indicativo; ~-**back code** DATA, TELEP *teleprinter* código *m* identificador; ~-**back signal** SIG señal *f* de contestación; ~-**back unit** DATA emisor *m* automático de indicativo, TELEP emisor *m* de indicativo; ~-**back unit simulator** TELEP simulador *m* de emisor de indicativo; ~ **code** TELEP código *m* de respuesta; ~ **indication** TELEP indicación *f* de respuesta; ~ **jack** TELEP conjuntor *m* de contestación; ~ **lamp** TELEP lámpara *f* de respuesta; ~ **machine function** PER, TELEP función *f* de contestador automático; ~-**recording unit** TELEP unidad *f* de grabación de respuestas; ~-**seizure ratio** TELEP ratio *f* respuesta-ocupación; ~ **signal** SIG, TELEP señal *f* de respuesta; ~ **signal, charge** SIG señal *f* de contestación, con cargo; ~ **signal, no charge** SIG señal *f* de contestación, sin cargo; ~ **tone** (*ANS*) DATA tono *m* de contestación (*ANS*)

answering *n* DATA *datacom* respuesta *f*; ~ **delay** TELEP demora *f* en la respuesta; ~ **device with voice synthesizer** TELEP contestador *m* con sintetizador de voz; ~ **equipment** TELEP equipo *m* de respuesta; ~ **modem** DATA módem *m* de contestación; ~ **position** TELEP posición *f* de contestación; ~ **service** APPL, TELEP servicio *m* de contestador; ~ **time** TELEP tiempo *m* de contestación

answerphone *n* PER, TELEP teléfono *m* con contestador

antenna *n* RADIO antena *f*; ~ **building** RADIO construcción *f* de antena; ~ **butterfly** RADIO antena *f* en mariposa; ~ **configuration** RADIO configuración *f* de antena; ~ **current** RADIO corriente *f* de antena; ~ **curtain** RADIO antena *f* en cortina; ~ **directivity diagram** RADIO diagrama *m* de directividad de antena; ~ **directivity factor** RADIO factor *m* de direccionalidad de antena; ~ **discrimination** RADIO discriminación *f* de antena; ~ **drive motor** RADIO accionador *m* de antena; ~ **effect** RADIO efecto *m* de antena; ~ **effective area** RADIO área *f* efectiva de antena; ~ **factor** RADIO factor *m* de antena; ~ **farm** RADIO campo *m* de antenas; ~ **feed** RADIO alimentador *m* primario de antena; ~ **gain** RADIO ganancia *f* de antena; ~ **gain pattern** RADIO diagrama *m* de ganancia de antena; ~ **input impedance** RADIO impedancia *f* de entrada de antena; ~ **lead-in** RADIO acometida *f* de antena, TELEP bajada *f* de antena; ~-**loading coil** RADIO bobina *f* de carga de antena; ~-**to-medium coupling loss** RADIO pérdida *f* de acoplamiento antena-medio; ~-**pointing loss** RADIO pérdida *f* por apuntamiento de la antena; ~ **polarization** RADIO polarización *f* de antena; ~ **power gain** RADIO ganancia *f* de potencia de antena; ~ **radiation** RADIO radiación *f* de antena; ~ **radiation pattern** RADIO diagrama *m* direccional de antena; ~ **range** RADIO alcance *m* de antena; ~-**receiver unit** RADIO unidad *f* de receptor de antena; ~ **remote control** RADIO control *m* remoto de antena; ~ **resistance** RADIO resistencia *f* de antena; ~ **side lobe** RADIO lóbulo *m* lateral de antena; ~ **spacing** RADIO distanciamiento *m* de antena radiogoniométrica; ~ **spacing error** RADIO error *m* de distanciamiento de antena; ~ **steerability** RADIO orientabilidad *f* de la antena; ~ **steering** RADIO orientación *f* de antena; ~ **structure** RADIO estructura *f* de la antena; ~ **supporting structure** RADIO estructura *f* de soporte de la antena; ~ **swing-off** RADIO balanceo *m* de antena; ~ **switching circuit** RADIO circuito *m* de conmutación de antena; ~ **switching relay** RADIO relé *m* de conmutación de antena; ~ **TDRSS** RADIO antena *f* de TDRSS; ~ **tuning** RADIO sintonización *f* de antena; ~ **tuning unit** (*ATU*) RADIO unidad *f* de sintonización de antena (*ATU*); ~ **width** RADIO ancho *m* de antena

anti-aliasing[1] *adj* COMP, TRANS anti-ajeno; ~ **filter** *n* TRANS filtro *m* anti-ajeno

anti-aliasing[2] *n* COMP, TRANS antisolapamiento *m*

anticlockwise[1] *BrE adv* (*cf counterclockwise AmE*) GEN inversamente al sentido horario

anticlockwise[2]: ~ **polarization** *n* RADIO polarización *f* en sentido antihorario

anti-glare: ~ **filter** *n* DATA filtro *m* antirreflectante; ~ **coating** *n* GEN revestimiento *m* antideslumbrante

anti-jamming *n* (*AJ*) RADIO antiinterferencias *f* (*AJ*)

anti-magnetic: ~ **shield** *n* DATA escudo *m* antimagnético

anti-phase *n* POWER contrafase *f*

anti-radiation: ~ **filter** *n* DATA filtro *m* antirradiación

anti-reflection *adj* (*AR*) GEN antirreflectante (*AR*); ~ **coating** *n* GEN revestimiento *m* antirreflector

anti-reflective: ~ **coating** *n* GEN revestimiento *m* antirreflectante

anti-shock: ~ **device** *n* TELEP dispositivo *m* de protección contra golpes

anti-sidetone *adj* TELEP contra efectos locales; ~ **circuit** *n* TELEP circuito *m* contra efectos locales; ~ **device** *n* TELEP dispositivo *m* contra efectos locales

anti-static *adj* ELECT antiestático; ~ **carpet** *n* ELECT alfombra *f* antiestática; ~ **mat** *n* ELECT alfombrilla *f* antiestática; ~ **wristband** *n* ELECT muñequera *f* antiestática; ~ **plastic** *n* ELECT plástico *m* antiestático

anti-virus: ~ **program** *n* COMP, DATA programa *m* antivirus

anvil *n* PARTS yunque *m*

any: ~-**value expression** *n* DATA expresión *f* de valor indeterminado

AOQ *abbr* (*average outgoing quality*) TEST AOQ (*calidad media de salida*)

AOQL *abbr* (*average outgoing quality limit*) TEST AOQL (*promedio límite de calidad de subidas*)

APB *abbr* (*all paths busy*) TELEP todas vías ocupadas; ~ **tone** *n* TELEP tono *m* APB

APCI *abbr* (*application protocol control information*) DATA, TELEP APCI (*información de control del protocolo de aplicación*)

APD *abbr* (*avalanche photodiode*) DATA, PARTS, TRANS *fibre optics* APD (*fotodiodo de avalancha*)

APDU *abbr* (*application protocol data unit*) DATA, TELEP APDU (*unidad de datos del protocolo de aplicación*)

aperiodic *adj* GEN *physics* aperiódico; ~ **aerial** *n* RADIO antena *f* aperiódica; ~ **antenna** *n* RADIO antena *f* aperiódica

aperture *n* GEN apertura *f*; ~ **antenna** RADIO antena *f* de apertura; ~ **card** GEN *documentation* tarjeta *f* de apertura; ~ **illumination** RADIO repartición *f* del campo en la superficie de apertura

apex *n* GEN ápice *m*

API *abbr* (*application program interface*) COMP, DATA API (*interfaz de programa de aplicación*)

apoapsis *n* RADIO apoápside *m*

apogee *n* RADIO apogeo *m*

apparatus *n* ELECT aparato *m*; ~ **charge rate** RADIO régimen *m* de carga de aparatos

apparent: ~ **height** *n* RADIO altura *f* aparente

append *vt* COMP añadir

appliance *n* GEN artefacto *m*

applicable: **not** ~ *phr* (*N/A*) DATA, TELEP inaplicable (*N/A*)

applicant *n* ELECT solicitante *m*

application *n* GEN aplicación *f*; ~ **appliance label** APPL etiqueta *f* para la utilización de aplicaciones; ~ **association** DATA asociación *f* de aplicaciones; ~ **entity** GEN entidad *f* de aplicación; ~-**entity invocation** (*AEI*) DATA invocación *f* de la unidad de aplicación (*AEI*); ~-**entity invocation identifier** DATA identificador *m* de la invocación de la unidad de aplicación; ~-**entity qualifier** DATA cualificador *m* de la unidad de aplicación; ~-**entity title** DATA título *m* de la unidad de aplicación; ~ **function** DATA función *f* de aplicación; ~ **generator** DATA generador *m* de aplicación; ~ **layer** APPL *open systems interconnection*, DATA, TRANS capa *f* de aplicación; ~-**layer structure** TRANS estructura *f* de la capa de aplicación; ~-**level gateway** TELEP puerta *f* en nivel de aplicación; ~ **process** DATA, TELEP proceso *m* de aplicación; ~ **process invocation** DATA invocación *f* de proceso de aplicación; ~ **process invocation identifier** DATA identificador *m* de invocación de proceso de aplicación; ~ **process title** DATA título *m* del proceso de aplicación; ~ **process type** DATA tipo *m* del proceso de aplicación; ~ **program interface** (*API*) COMP, DATA interfaz *f* de programa de aplicación (*API*); ~ **protocol control information** (*APCI*) DATA, TELEP información *f* de control del protocolo de aplicación (*APCI*); ~ **protocol data unit** (*APDU*) DATA, TELEP unidad *f* de datos del protocolo de aplicación (*APDU*); **~ reference** GEN referencia *f* de aplicación; **~ service element** (*ASE*) DATA, TELEP elemento *m* del servicio de aplicación (*ASE*); **~ service object** (*ASO*) DATA, TELEP objeto *m* del servicio de aplicación (*ASO*); **~-specific integrated circuit** (*ASIC*) ELECT, PARTS circuito *m* integrado de aplicación específica (*CIAE*); **~-supported distribution transaction** DATA transacción *f* de distribución asistida por aplicación; **~ system** SWG, TELEP sistema *m* de aplicación; **~ technology satellite** RADIO satélite *m* de tecnología de aplicación

applications: **~ context** *n* GEN contexto *m* de aplicaciones; **~ processor** *n* COMP procesador *m* de aplicaciones

applied: **~ data bit** *n* TELEP bitio *m* de información aplicado; **~ data-bit scrambling process** *n* DATA proceso *m* aplicado de aleatorización de bitios de datos; **~ research** *n* TEST investigación *f* aplicada; **~ voltage** *n* POWER voltaje *m* aplicado

apply *vt* GEN aplicar

appraisal: **~ cost** *n* TEST *quality* coste (*AmL* costo) *m* de tasación

approach *n* COMP vía *f* de acceso

approved *adj* GEN aprobado; **~ connecting agent** *n* TELEP agente *m* de conexión aprobado

approximate: **~ reasoning** *n* DATA razonamiento *m* aproximativo

approximation: **~ error** *n* GEN error *m* de aproximación

APSK *abbr* (*amplitude phase-shift keying*) SIG APSK (*tecleado de cambio en la relación de fase de amplitud*)

AR *abbr* (*anti-reflection*) GEN AR (*antirreflectante*)

Arab: **~ Satellite Communications Organization** *n* (*ARABSAT*) RADIO Organización *f* Árabe de Comunicaciones por Satélite (*ARABSAT*); **~ States Broadcasting Union** *n* RADIO Unión *f* de Radiodifusión de los Estados Árabes

ARABSAT *abbr* (*Arab Satellite Communications Organization*) RADIO ARABSAT (*Organización Arabe de Comunicaciones por Satélite*)

aramid: **~ yarn** *n* ELECT hilo *m* aramid

arc: **~ current** *n* ELECT corriente *f* de arco

architecture *n* COMP arquitectura *f*

archivage *n* DATA archivado *m*

archive[1]: **~ material** *n* TV material *m* de archivo

archive[2] *vt* COMP, DATA archivar

archives *n pl* GEN archivos *m pl*

area *n* COMP, DATA zona *f*; **~ code** TELEP código *m* de zona; **~ of propagation** GEN *physics* área *f* de propagación; **~ of work** GEN área *f* de trabajo

A-release-request: **~ APDU** *n* (*RLRQ*) DATA APDU *m* de liberación-petición-A (*RLRQ*)

A-release-response: **~ APDU** *n* (*RLRE*) DATA APDU *m* de liberación-respuesta-A (*RLRE*)

argument *n* DATA argumento *m*; **~ of perigee** RADIO argumento *m* de perigeo

arithmetical: **~ average** *n* GEN *statistics* media *f* aritmética; **~ coding** *n* COMP, DATA codificación *f* aritmética; **~ delimiter** *n* COMP delimitador *m* aritmético; **~ expression** *n* COMP expresión *f* aritmética; **~ and logic unit** *n* (*ALU*) COMP unidad *f* aritmética y lógica (*ALU*); **~ mean** *n* GEN *statistics* media *f* aritmética; **~ weighted mean** *n* GEN *statistics* media *f* aritmética ponderada

armature: **~ stick** *n* ELECT adhesión *f* de armadura; **~ winding** *n* POWER arrollamiento *m* de inducido

armchair: **~ shopping** *n* GEN telecompra *f*

armour *BrE* (*AmE* **armor**) *n* GEN, TRANS *cables* blindaje *m*

armoured *BrE* (*AmE* **armored**) *adj* GEN armado; **~ cable** *n* COMP, ELECT, TRANS cable *m* armado

armouring *BrE* (*AmE* **armoring**) *n* TRANS *cables* blindaje *m*

arpeggiator *n* COMP arpegiador *m*

arpeggio *n* COMP arpegio *m*

arrange *vt* GEN disponer

arrangement *n* GEN disposición *f*

arranger *n* COMP clasificador *m*

array *n* GEN array *m*; **~ antenna** RADIO antena *f* matricial; **~ connection** ELECT conexión *f* de la matriz; **~ connectors** *n pl* ELECT conectores *m pl* de la matriz; **~ hinge** RADIO articulación *f* de antena múltiple; **~ processor** COMP, DATA procesador *m* de matrices

arrow *n* GEN aguja *f* de cadeneo

articulation *n* TELEP articulación *f*; **~ index** RADIO, TELEP, TRANS índice *m* de nitidez; **~ reference equivalent loss** TELEP pérdida *f* de inteligibilidad equivalente de referencia; **~ score** RADIO medida *f* de inteligibilidad

artificial *adj* COMP, DATA, ELECT artificial; **~ ageing** *n* TEST envejecimiento *m* artificial; **~ antenna** *n* RADIO antena *f* artificial; **~ intelligence** *n* (*AI*) APPL, COMP, SWG inteligencia *f* artificial (*AI*); **~ language** *n* COMP lenguaje *m* artificial; **~ line** *n* TRANS línea *f* artificial; **~ load** *n* RADIO carga *f* artificial; **~ noise** *n* TEST ruido *m* artificial; **~ satellite** *n* DATA, RADIO satélite *m* artificial; **~ traffic** *n* TELEP tráfico *m* artificial; **~ voice** *n* COMP voz *f* artificial

artwork: ~ **master** *n* POWER *printed circuits* matriz *f* de diseño
ascending: ~ **node** *n* RADIO nodo *m* ascendente
ascent: ~ **phase** *n* RADIO fase *f* ascendente
ASCII *abbr* (*American Standard Code for Information Interchange*) COMP, DATA, TELEP ASCII (*Código Estándar Americano para Intercambio de Información*); ~ **data** *n pl* COMP datos *m pl* ASCII
ASE *abbr* (*application service element*) DATA, TELEP ASE (*elemento del servicio de aplicación*)
Asia-Pacific: ~ **Broadcasting Union** *n* (*ABU*) RADIO Unión *f* de Radiodifusión de Asia y el Pacífico (*ABU*); ~ **Telecommunity** *n* RADIO Telecomunidad *f* de Asia y el Pacífico
ASIC *abbr* (*application-specific integrated circuit*) ELECT, PARTS CIAE (*circuito integrado de aplicación específica*)
ASK *abbr* (*amplitude-shift keying*) SIG ASK (*tecleado de cambio de amplitud*)
ASN *abbr* (*average sample number*) TEST *quality assurance* ASN (*número muestral promedio*)
ASO *abbr* (*application service object*) DATA, TELEP ASO (*objeto del servicio de aplicación*)
aspect: ~ **ratio** *n* RADIO, TELEP relación *f* de aspecto
asphalt *n* GEN *construction* asfalto *m*
assault: ~ **alarm device** *n* TEST *security* instalación *f* de alarma antirrobo; ~ **alarm installation** *n* TEST *security* instalación *f* de alarma antirrobo
assemble *vt* GEN montar
assembled: ~ **representation** *n* GEN *technical drawing* presentación *f* montada
assembler *n* COMP ensamblador *m*
assembling: ~ **time** *n* COMP tiempo *m* de ensamblaje
assembly *n* GEN construcción *f*, montaje *m*, ensemblaje *m*; ~ **area** GEN zona *f* de montaje; ~ **instructions** *n pl* GEN *documentation* instrucciones *f pl* para el montaje; ~ **language** COMP lenguaje *m* ensamblador; ~ **line** GEN línea *f* de montaje; ~ **program** COMP programa *m* ensamblador
asserted *adj* TELEP afirmado
assess *vt* GEN evaluar
assessment *n* GEN evaluación *f*; ~ **of distance** GEN evaluación *f* de la distancia
assign *vt* COMP designar
assigned: ~ **frequency** *n* RADIO frecuencia *f* asignada; ~-**frequency band** *n* RADIO banda *f* de frecuencia asignada
assignee *n* GEN designado *m*
assignment *n* GEN asignación *f*; ~ **channel** TRANS canal *m* de asignación; ~ **of frequencies** RADIO distribución *f* de frecuencias; ~ **map** DATA mapa *m* de asignaciones; ~ **message** DATA aviso *m* de asignaciones; ~ **of orbit** RADIO asignación *f* de órbita; ~ **specifications** *n pl* GEN *documentation* especificaciones *f pl* de destino; ~ **statement** DATA *specification and description language* sentencia *f* de asignación
assistant *n* GEN ayudante *m*; ~ **operator** TELEP operador *m* auxiliar; ~ **operator position** TELEP puesto *m* de operador auxiliar; ~ **operator signal** TELEP señal *f* de operador auxiliar
assisted *adj* COMP asistido
associated: ~ **connectors** *n pl* ELECT conectores *m pl* anexos; ~ **mode** *n* TELEP *of signalling* modo *m* asociado; ~ **sector** *n* TV sector *m* asociado; ~ **signalling** *BrE n* GEN señalización *f* anexa
association *n* GEN asociación *f*; ~ **area** DATA *specification and description language* área *f* de asociación; ~ **control protocol machine** (*ACPM*) DATA máquina *f* de asociación de protocolo de control (*ACPM*); ~ **control service element** (*ACSE*) DATA, TELEP elemento *m* para el control de asociación; ~-**initiating application entity** DATA entidad *f* de aplicación iniciadora de asociación; ~ **initiator** DATA iniciador *m* de asociaciones; ~ **responder** DATA respondedor *m* de asociación; ~-**responding application entity** DATA entidad *f* de aplicación respondedora de asociación
Association: ~ **Class** *n* DATA Clase *f* de Asociación
associative: ~ **processor** *n* COMP procesador *m* asociativo; ~ **storage** *n* COMP almacenamiento *m* asociativo
assumed: ~ **model** *n* GEN *statistics* modelo *m* hipotético
astable *adj* ELECT *logic elements* astable; ~ **circuit** *n* ELECT circuito *m* astable
astatic *adj* GEN *physics* astático
astrionics *n* ELECT astriónica *f*
asymmetric *adj* COMP asimétrico; ~-**ended synchronization** *n* TELEP, TRANS sincronización *f* asimétrica
asymmetrical: ~ **distortion** *n* ELECT, TELEP distorsión *f* asimétrica
asymptomatic: ~ **approximation** *n* GEN aproximación *f* asintomática
asymptotic: ~ **availability** *n* TELEP, TEST *dependability* disponibilidad *f* asintótica; ~ **availability/unavailability** *n* TELEP disponibilidad *f* /indisponibilidad *f* asintótica

asynchronous *adj* COMP, DATA, TRANS asíncrono; ~ **balanced mode** *n* TELEP modo *m* asíncrono equilibrado; ~ **circuit** *n* DATA circuito *m* asíncrono; ~ **computer** *n* COMP ordenador *m* asíncrono *Esp*, computador *m* asíncrono *AmL*, computadora *f* asíncrona *AmL*; ~ **counter** *n* TEST contador *m* asincrónico; ~ **data communication** *n* DATA comunicación *f* de datos asíncronos; ~ **data transmission** *n* DATA, TRANS transmisión *f* asincrónica de datos; ~ **data units** *n pl* DATA, ELECT unidades *f pl* de datos asíncronos; ~ **digital subscriber loop** *n* TRANS bucle *m* asincrónico de abonado digital; ~ **disconnected mode** *n* TELEP modo *m* asíncrono desconectado; ~ **format** *n* DATA formato *m* asíncrono; ~ **interface module** *n* ELECT módulo *m* de interfaz asincrónico; ~ **link** *n* DATA enlace *m* asincrónico; ~ **motor** *n* POWER motor *m* asíncrono; ~ **network** *n* TELEP red *f* asíncrona; ~ **operation** *n* TELEP operación *f* asincrónica; ~ **port** *n* DATA puerto *m* asíncrono; ~ **time-division multiplexing** *n* COMP multiplexión *f* asíncrona de división del tiempo, DATA multiplexión *f* asíncrona por distribución en el tiempo; ~ **transfer mode** *n* APPL, COMP, TELEP, TRANS modo *m* de transferencia asíncrona; ~ **transmission** *n* TELEP, TRANS transmisión *f* asíncrona

ATB: ~ **tone** *n* (*all trunks busy tone*) TELEP tono *m* ATB

ATI *abbr* (*average total inspected*) TEST *quality* ATI (*total medio inspeccionado*)

ATM *abbr* APPL, COMP (*asynchronous transfer mode*) ATM (*modo de transferencia asíncrona*), DATA, TELEP, TEST (*abstract data method*) ATM (*metódo de datos resumidos*); ~ **adaptation layer** *n* (*AAL*) GEN envoltura *f* de adaptación de un ATM (*AAL*); ~ **cell** *n* COMP, DATA célula *f* de cajero automático; ~ **connection** *n* COMP conexión *f* de cajero automático; ~ **service data unit** *n* TELEP unidad *f* de datos del servicio de cajeros automáticos; ~ **switch** *n* SWG conmutador *m* ATM; ~ **virtual path** *n* APPL ruta *f* virtual del cajero automático; ~ **workstation** *n* COMP, DATA, PER APP punto *m* de cajero automático, puesto *m* de cajero automático

atmospheric: ~ **noise** *n* RADIO ruido *m* atmosférico; ~ **pressure** *n* GEN *physics* presión *f* atmosférica; ~ **refraction** *n* RADIO refracción *f* atmosférica

atom *n* COMP átomo *m*

atomic: ~**-action data** *n pl* DATA datos *m pl* sobre acción atómica; ~ **clock** *n* GEN reloj *m* atómico; ~ **frequency standards** *n pl* ELECT normas *f pl* de frecuencia atómica; ~ **nucleus** *n* GEN *physics* núcleo *m* atómico; ~ **standard clock** *n* RADIO reloj *m* atómico; ~ **time scale** *n* RADIO escala *f* de tiempo atómica

atomicity *n* DATA atomicidad *f*; ~**, consistency, isolation and durability** (*ACID*) DATA atomicidad, *f* consistencia, aislamiento y durabilidad (*ACID*)

atonal *adj* COMP atonal

at-once: ~ **recording** *n* COMP registro *m* inmediato

attach[1]: ~ **construct** *n* DATA estructura *f* de conexión

attach[2] *vt* DATA *Internet* conectar

attachment *n* TELEP aditamento *m*

attack *n* COMP ataque *m*; ~ **rate** COMP relación *f* de ataque; ~ **velocity** COMP velocidad *f* de ataque

attend *vt* GEN asistir

attendant[1]: ~**-access code** *n* TELEP código *m* de acceso de operadora; ~ **busy override** *n* TELEP superación *f* de señal de operador ocupado; ~**-to-**~ **calling** *n* TELEP llamada *f* de operador a operador; ~ **conference** *n* TELEP conferencia *f* por operador; ~**-controlled conference** *n* TELEP conferencia *f* controlada por operador; ~ **override** *n* TELEP acceso *m* directo a operador; ~ **position** *n* TELEP puesto *m* de operador; ~ **recall** *n* TELEP repetición *f* de llamada de operador; ~**'s set** *n* TELEP aparato *m* de operador; ~**-status indicator** *n* TELEP indicador *m* de estado del operador

attendant[2] *n* TELEP asistente *m*

attenuate *vt* COMP, POWER, RADIO, TRANS atenuar

attenuation *n* COMP, POWER, RADIO, TRANS atenuación *f*; ~ **check** TRANS verificación *f* de atenuación; ~ **coefficient** RADIO, TRANS coeficiente *m* de atenuación; ~ **constant** RADIO, TRANS constante *f* de atenuación; ~ **control** TRANS control *m* de atenuación; ~ **deviation** TRANS desviación *f* de atenuación; ~ **distortion** TELEP, TRANS distorsión *f* de atenuación; ~ **factor** TRANS constante *f* de atenuación; ~**-limited** *adj* TRANS limitado por la atenuación; ~**-limited operation** TRANS operación *f* de atenuación limitada; ~ **loss** RADIO pérdida *f* de atenuación; ~ **measurement** TRANS medición *f* de atenuación; ~ **slope** RADIO pendiente *f* de atenuación; ~ **tolerance** TRANS tolerancia *f* de atenuación; ~ **value** TRANS valor *m* de atenuación

attenuator *n* COMP, TRANS atenuador *m*

attitude: ~ **control** *n* RADIO control *m* de actitud; ~ **control system** *n* RADIO cadena *f* de mando de vuelo; ~ **control unit** *n* RADIO unidad *f* de control de posición de vuelo; ~ **determination control subsystem** *n* RADIO subsistema *m* de control de determinación de posición de vuelo; ~ **and orbit control system** *n* RADIO sistema *m* de control de posición de vuelo y de órbita; ~ **reference** *n* RADIO referencia *f* de posición de vuelo; ~ **and roll control system** *n* RADIO sistema *m* de control de actitud y de balanceo; ~ **sensor** *n* RADIO sensor *m* de posición de vuelo; **~-stabilized satellite** *n* RADIO satélite *m* con posición de vuelo estabilizada

attribute *n* GEN atributo *m*; ~ **group** COMP grupo *m* de atributos; ~ **method** TEST método *m* atributivo; ~ **type** DATA tipo *m* de atributo; ~ **value** DATA valor *m* del atributo; **~-value assertion** (*AVA*) DATA afirmación *f* de valor de atributo (*AVA*)

ATU *abbr* (*antenna tuning unit*) RADIO ATU (*unidad de sintonización de antena*)

AU *abbr* (*administrative unit*) CONTR unidad *f* administrativa

audibility *n* GEN *acoustics* audibilidad *f*; ~ **range** TELEP gama *f* de audibilidad

audible *adj* GEN *acoustics* audible; ~ **alarm** *n* RADIO, TEST alarma *f* audible; ~ **indication** *n* SIG indicación *f* audible; ~ **signal** *n* GEN señal *f* audible; ~ **sound** *n* GEN tono *m* de llamada audible

audio *adj* GEN audio; ~ **access unit** *n* COMP unidad *f* acústica de acceso; ~ **bearer service** *n* APPL servicio *m* de portador; ~ **buffer** *n* COMP memoria *f* tampón para sonido; ~ **card** *n* COMP tarjeta *f* de audio; ~ **carrier** *n* TRANS portador *m* de audio; ~ **channel** *n* TRANS, TV canal *m* de sonido; ~ **compression** *n* COMP compresión *f* acústica; ~ **compression board** *n* DATA placa *f* de compresión de sonido; ~ **cue** *n* DATA señal *f* auditiva; **~-digital** COMP audiodigital; ~ **document** *n* DATA documento *m* sonoro; ~ **equipment** *n* COMP equipo *m* de sonido, TRANS equipo *m* de audio; ~ **frequency** *n* COMP, TELEP frecuencia *f* de sonido; **~-frequency circuit** *n* RADIO circuito *m* de audiofrecuencia; **~-frequency protection ratio** *n* RADIO relación *f* de protección de audiofrecuencia; **~-frequency signal-to-interference ratio** *n* RADIO relación *f* señal/interferencia de audiofrecuencia; ~ **in** *n* COMP audio *m* incluido; ~ **indicate active** *n* TELEP indicador *m* de audio activo; ~ **indicate muted** *n* TELEP indicador *m* de audio desconectado; ~ **input** *n* COMP entrada *f* del micro; ~ **loop** *n* TELEP bucle *m* de audio; **~-messaging interchange specifications** *n pl* TELEP especificaciones *f pl* para intercambio de mensajes sonoros; ~ **mixer** *n* COMP mezclador *m* de sonido; ~ **news-gathering** *n* RADIO recopilación *f* de noticias sonoras; ~ **out** *n* COMP audio *m* suprimido, ELECT audio *m* desconectado; ~ **output** *n* COMP, ELECT escucha; ~ **path** *n* TRANS camino *m* audio; ~ **recording** *n* COMP registro *m* sonoro; ~ **sequence** *n* COMP secuencia *f* sonora; ~ **signal** *n* SIG señal *f* acústica; ~ **signal on hold** *n* TELEP señal *f* de audio a la espera; ~ **splitter** *n* TRANS separador *m* de audio; ~ **system** *n* COMP, GEN sistema *m* de audio; ~ **tape** *n* COMP cinta *m* sonora; ~ **track** *n* COMP pista *f* de audio; **~-videotext** *n* APPL, ELECT audio videotexto

audiochannel *n* TRANS canal *m* de sonido

audioconference *n* RADIO audioconferencia *f*

Audio-Interchange: ~ **File Format** *n* COMP Formato *m* de Fichero para Intercambio de Datos Acústicos, DATA, *Internet* Formato *m* de Fichero de Intercambio-Audio

Audiotex: ~ **server** *n* TELEP servidor *m* Audiotex

audiovisual *adj* TV audiovisual

audit *n* GEN comprobación *f*; ~ **trail** GEN pista *f* de auditoría

auditory: ~ **threshold** *n* TELEP umbral *m* auditivo

augmented: ~ **baseline** *n* RADIO línea *f* de base aumentada

aural *adj* COMP aural; ~ **signal** *n* COMP señal *f* aural

authentication *n* DATA, SIG autentificación *f*; ~ **exchange** DATA, SIG intercambio *m* de autentificación; ~ **function** DATA función *f* de autentificación; ~ **information** DATA, SIG información *f* de autentificación; ~ **mechanism** DATA mecanismo *m* de autentificación; ~ **procedure** SIG procedimiento *m* de autentificación; ~ **value** DATA valor *m* de autentificación

author *n* GEN autor *m*; **~-created link** COMP enlace *m* creado por el autor

authoring: ~ **software** *n* DATA programa *m* de autor; ~ **system** *n* COMP sistema *m* de autor

authority: ~ **and format identifier** *n* (*AFI*) DATA, SIG identificador *m* de autorización y formato

authorization *n* GEN autorización *f*; ~ **code** TELEP código *m* de autorización; ~ **information** DATA, SIG información *f* de autorización; **~-information qualifier** DATA cualificación *f* de información para autorización, SIG cualificador *m* de autoriza-

ción de la información; ~ **key** TELEP clave *f* de autorización
authorize *vt* GEN autorizar
authorized *adj* GEN autorizado; ~ **dealer** *n* GEN agente *m* autorizado; ~ **banker's transfer** *n* GEN transferencia *f* bancaria autorizada
auto: ~ **bass chord** *n* COMP cuerda *f* de bajo automático; ~ **power off** *n* COMP, POWER desconexión *f* automática
auto-action *n* DATA accionamiento *m* automático; ~ **type** DATA tipo *m* de accionamiento automático
auto-alert *n* DATA autoalerta *f*
auto-arranger *n* COMP clasificador *m* automático
auto-bend *n* COMP doblado *m* automático
auto-correction *n* COMP *spellcheck* corrección *f* automática, autocorrección *f*, DATA, SIG, TELEP autocorrección *f*
auto-correlation *n* GEN autocorrelación *f*; ~ **detector** RADIO detector *m* de autocorrelación
autocue *BrE n* (*cf teleprompter AmE*) TV indicador *m* automático visual
auto-dialling *BrE* (*AmE* **auto-dialing**) *n* TELEP discado *m* automático
auto-epistemic: ~ **reasoning** *n* DATA razonamiento *m* autoepistemológico
auto-forward *n* DATA envío *m* automático
auto-installation *n* ELECT instalación *f* automática
auto-load *n* COMP carga *f* automática
auto-locate *n* COMP localización *f* automática
auto-manual: ~ **switchboard** *n* PER, SWG cuadro *m* conmutador automático-manual
automata: ~ **theory** *n* DATA teoría *f* del autómata
automate *vt* GEN automatizar
automated: ~ **data medium** *n* CONTR soporte *m* de datos automatizados; ~ **reasoning** *n* DATA razonamiento *m* automatizado
automatic *adj* GEN automático; ~ **accompaniment** *n* COMP acompañamiento *m* automático; ~ **acquisition** *n* RADIO adquisición *f* automática; ~ **ADC** *n* TELEP ADC *m* automático; ~ **aligning** *n* RADIO alineación *f* automática; ~ **alignment** *n* RADIO alineación *f* automática; ~ **answering** *n* DATA, TELEP respuesta *f* automática; ~ **arranger** *n* COMP arreglador *m* automático; ~ **alternative routing** *n* SIG, TELEP encaminamiento *m* alternativo automático; ~ **attempt** *n* TELEP prueba *f* automática; ~ **antenna-pattern recording** *n* RADIO registro *m* automático de diagrama de antena; ~ **attenuation regulation** *n* TRANS regulación *f* automática de la atenuación; ~ **banking** *n* APPL banca *f* automática; ~ **battery charging** *n* POWER carga *f* automática de batería; ~ **billing** *n* TELEP facturación *f* automática; ~ **booked call** *n* TELEP llamada *f* solicitada automática; ~ **bypass** *n* ELECT derivación *m* automática; ~ **call** *n* TELEP llamada *f* automática; ~ **call acceptance** *n* TELEP aceptación *f* de llamada automática; ~ **callback** *n* TELEP retrollamada *f* automática; ~ **call distributor** *n* TELEP distribuidor *m* automático de llamadas; ~ **call distribution** *n* (*ACD*) TELEP distribución *f* de llamada automática; ~ **calling** *n* TELEP llamada *f* automática; ~ **call queue overflow** *n* TELEP sobrecarga *f* en cola de llamadas automáticas; ~ **call transfer** *n* SIG transferencia *f* automática de llamadas; ~ **call transfer unit** *n* TELEP unidad *f* de transferencia de llamadas automáticas; ~ **call transmitter** *n* TELEP transmisor *m* de llamadas automáticas; ~ **call unit** *n* TELEP unidad *f* de llamadas automáticas; ~ **changeover** *n* TEST cambio *m* automático; ~ **channel-tuning** *n* RADIO sintonización *f* automática de canales; ~ **charging-control unit** *n* POWER unidad *f* automática de control de carga; ~ **charging equipment** *n* APPL equipo *m* de cobro automático; ~ **control** *n* CONTR control *m* automático; ~ **control system** *n* COMP, CONTR sistema *m* de control automático; ~ **correction** *n* COMP, DATA, SIG, TELEP corrección *f* automática; ~ **credit card service** *n* APPL servicio *m* automático de tarjetas de crédito; ~ **cutoff** *n* COMP cierre *m* automático; ~ **data processing** *n* COMP, DATA procesamiento *m* automático de datos; ~ **detection** *n* RADIO detección *f* automática; ~ **device** *n* GEN dispositivo *m* automático; ~ **dialler** *n* TELEP marcador *m* automático; ~ **dialling** *n* TELEP marcación *f* automática; ~ **direction finding** *n* RADIO goniometría *f* automática; ~ **error correction** *n* COMP, DATA, SIG, TELEP corrección *f* automática de errores; ~ **error detection** *n* RADIO, TELEP detección *f* automática de los errores; ~ **exchange** *n* TELEP centralita *f* automática; ~ **fall-back** *n* DATA rearranque *m* automático; ~ **frequency control** *n* ELECT, TELEP, TRANS control *m* automático de frecuencia; ~ **function supervision** *n* TELEP supervisión *f* de función automática; ~ **gain control** *n* (*AGC*) ELECT, TELEP, TRANS control *m* automático de ganancia (*AGC*); ~ **harmonization** *n* COMP armonización *f* automática; ~ **harmonizing** *n* COMP armonización *f* automática; ~ **identification of outward dialling** *BrE n* RADIO identificación *f* automática de discado de

salida, DATA identificación *f* automática de llamadas al exterior, TELEP identificación *f* automática de llamada hacia afuera; ~ **intercept system** *n* TELEP sistema *m* automático de interceptación; ~ **interrupt** *n* COMP interrupción *f* automática; ~ **laser shutdown** *n* TRANS desconexión *f* láser automática; ~ **level control** *n* COMP control *de* nivel automático; ~ **line management** *n* TRANS gestión *f* automática de línea; ~ **location registration of ships** *n* APPL localización *f* automática de matrículas de buques; ~ **measurement** *n* RADIO medición *m* automática; ~ **maritime radio-telex service** *n* TELEP servicio *m* automático de radiotelex marítimo; ~ **message accounting** *n* TELEP contabilización *f* automática de comunicaciones; ~ **message switching centre** *BrE* *n* TELEP centro *m* automático conmutador de mensajes; ~ **mobile telephone system** *n* RADIO, TELEP sistema *m* automático de telefonía móvil; ~ **numbering equipment** *n* TELEP equipo *m* de numeración automática; ~ **numbering transmitter** *n* TELEP transmisor *m* de numeración automático; ~ **observation** *n* TEST observación *f* automática; ~ **personal call** *n* TELEP comunicación *f* automática de persona a persona; ~ **playback** *n* COMP reproducción *f* automática; ~ **power control** *n* POWER control *m* automático de potencia; ~ **preselection** *n* TELEP preselección *f* automática; ~ **protection switching** *n* SWG conmutación *f* automática de protección; ~ **radio mute** *n* PER, RADIO desconexión *f* automática de la radio; ~ **recall** *n* TELEP rellamada *f* automática; ~ **redialling** *BrE* *n* TELEP remarcación *f* automática; ~ **rerouting** *n* TELEP reencaminamiento *m* automático; ~ **repeat attempt** *n* TELEP repetición *f* automática de llamada; ~ **repetition** *n* TELEP repetición *f* automática; ~ **restart unit** *n* CONTR unidad *f* de rearranque automático; ~ **reverse-charge call** *n* TELEP llamada *f* automática de cobro revertido; ~ **ring trip** *n* TELEP interrupción *f* automática de llamada; ~ **roaming** *n* RADIO recorrido *m* automático; ~ **rhythm** *n* COMP ritmo *m* automático; ~ **route selection** *n* TELEP selección *f* automática de ruta; ~ **route selection with queueing** *n* TELEP selección *f* automática de ruta con cola de espera; ~ **sequential connection** *n* TELEP conexión *f* secuencial automática; ~ **shut-off** *n* ELECT interrupción *f* automática; ~ **signalling** *BrE* *n* TELEP señalización *f* automática; ~ **soldering** *n* ELECT soldadura *f* automática; ~ **stop** *n* COMP parada *f* automática; ~ **switchboard** *n* PER, SWG central *f* telefónica automática; ~ **switching** *n* TV conmutación *f* automática; ~ **switching equipment** *n* PER, SWG equipo *m* de conmutación automática; ~ **telephone exchange** *n* SWG, TELEP central *f* telefónica automática; ~ **timed recall** *n* TELEP rellamada *f* automática programada; ~ **tracking** *n* RADIO seguimiento *m* automático; ~ **traffic rearrangement** *n* TRANS redistribución *f* automática del tráfico; ~ **transferred-charge call** *n* TELEP cargo *m* automático en cuenta de una llamada; ~ **transferred debiting of charges** *n* TELEP adeudos *m pl* cargados en cuenta automáticamente; ~ **transmission** *n* APPL transmisión *f* automática; ~ **transmission-measuring equipment** *n* SWG equipo *m* automático de medición de transmisión; ~ **transmitter** *n* TELEP transmisor *m* automático; ~ **trunk testing** *n* TELEP comprobación *f* automática de línea troncal; ~ **trunk working** *n* PER APP trabajo *m* troncal automático; ~ **typesetting** *n* APPL composición *f* automática de textos; ~ **updating** *n* PER APP actualización *f* automática; ~ **vehicle location** *n* APPL, RADIO localización *f* automática de vehículos; ~ **verbal announcement of charges service** *n* TELEP comunicación *f* verbal automática del servicio de cargo; ~ **volume control** *n* ELECT control *m* automático de volumen; ~ **wake-up** *n* TELEP despertador *m* automático

automatically *adv* GEN automáticamente

automation *n* CONTR automatización *f*

automatize *vt* CONTR automatizar

automize *vt* CONTR autonomizar

auto-mode *n* DATA modo *m* automático

auto-modulation *n* TRANS modulación *f* automática

autonomous *adj* COMP, TELEP autónomo; ~ **registration device** *n* TELEP dispositivo *m* autónomo de registro; ~ **registration update** *n* TELEP actualización *f* autónoma de registro

auto-play *n* COMP juego *m* automático

auto-stop *n* COMP desconexión *f* automática

auto-synchronization *n* TRANS sincronización *f* automática

auto-track *n* RADIO seguimiento *m* automático

auto-transfer *n* TELEP transferencia *f* automática

auto-transformer *n* POWER transformador *m* automático

AUTOVON: ~ **trunk access** *n* TELEP acceso *m* a la red interurbana AUTOVON

auxiliaries: ~ **control station** *n* TELEP estación *f* de control de operadoras auxiliares; ~ **management software machine** *n* TELEP ordenador *m* con programa para la gestión de auxiliares *Esp*, computador *m* con programa para la gestión de auxiliares *AmL*, computadora *f* con programa para la gestión de auxiliares *AmL*

auxiliary: ~ **board** *n* POWER *printed circuits* placa *f* auxiliar; ~ **circuit** *n* POWER circuito *m* auxiliar; ~ **conditions** *n pl* TELEP *of modulation* condiciones *f pl* secundarias; ~ **contact** *n* POWER contacto *m* auxiliar; ~ **equipment** *n* COMP equipo *m* auxiliar; ~ **equipment rack** *n* ELECT bastidor *m* auxiliar; ~ **management software machine** *n* TELEP ordenador *m* con programa para la gestión de supletorios; ~ **operation** *n* COMP operación *f* auxiliar; ~ **position** *n* TELEP puesto *m* de operadora auxiliar; ~ **service position** *n* APPL posición *f* auxiliar de marcha; ~ **services trunk group** *n* TELEP grupo *m* de líneas troncales de servicios auxiliares; ~ **set** *n* TELEP aparato *m* auxiliar; ~ **signalling** *BrE n* TELEP señalización *f* auxiliar; ~ **switch** *n* COMP, SWG interruptor *m* auxiliar; ~ **switching point** *n* SWG punto *m* auxiliar de conmutación; ~ **switching unit** *n* SWG unidad *f* auxiliar de conmutación; ~ **telephone set** *n* TELEP teléfono *m* supletorio; ~ **voltage** *n* POWER tensión *f* auxiliar; ~ **winding** *n* POWER bobinado *m* auxiliar

AVA *abbr* (*attribute-value assertion*) DATA AVA (*afirmación de valor de atributo*)

availability *n* GEN disponibilidad *f*; ~ **figure** SWG cifra *f* de disponibilidad; ~ **function** TEST función *f* de disponibilidad; ~ **of a leased circuit** TELEP disponibilidad *f* de circuito dedicado; ~ **measure** TEST medida *f* de disponibilidad; ~ **unavailability** TELEP disponibilidad *f* - indisponibilidad *f*

available *adj* GEN disponible; ~ **line** *n* TELEP línea *f* disponible; ~ **memory** *n* COMP memoria *f* disponible; ~ **power** *n* TELEP, TRANS potencia *f* disponible

avalanche *n* ELECT *semiconductors* avalancha *f*; ~ **breakdown** ELECT rotura *f* por avalancha; ~ **diode** ELECT diodo *m* de avalancha; ~ **effect** ELECT efecto *m* de avalancha; ~ **noise** ELECT ruido *m* de avalancha; ~ **photodiode** (*APD*) DATA, PARTS, TRANS *fibre optics* fotodiodo *m* de avalancha (*APD*); ~ **voltage** ELECT voltaje *m* de avalancha

AVD *abbr* (*alternate voice-data circuit*) TRANS AVD (*circuito de datos de voz alternado*)

average: ~ **amount of inspection** *n* TEST *quality* cantidad *f* media de inspección; ~ **bits per sample** *n* DATA media *f* de bits por muestra; ~ **busy hour in CCS** *n* TELEP hora *f* ocupada promedio en CCS; ~ **call duration** *n* TELEP duración *f* promedio de las llamadas; ~ **cladding diameter** *n* TRANS diámetro *m* medio de chapeado; ~ **core diameter** *n* TRANS diámetro *m* medio de alma; ~ **crossing rate** *n* GEN velocidad *f* media de cruce; ~ **daily use** *n* TEST promedio *m* de uso diario; ~ **operating time** *n* TELEP tiempo *m* medio de operación; ~ **outgoing quality** *n* (*AOQ*) TEST calidad *f* media de salida (*AOQ*); ~ **outgoing quality limit** *n* (*AOQL*) TEST promedio *m* límite de calidad de subidas (*AOQL*); ~ **output** *n* COMP, ELECT producción *f* media; ~ **power** *n* TRANS potencia *f* media; ~ **reference-surface diameter** *n* (*DRav*) TELEP diámetro *m* medio de superficie de referencia (*DRav*); ~ **run length** *n* TEST *quality* tamaño *m* medio de las rachas; ~ **sample number** *n* (*ASN*) TEST *quality* número *m* muestral promedio (*ASN*); ~ **total inspected** *n* (*ATI*) TEST *quality* total *m* medio inspeccionado (*ATI*); ~ **traffic intensity** *n* TELEP intensidad *f* media de tráfico

avoidance: ~ **angle** *n* RADIO ángulo *m* de evitación

awarded *adj* TV otorgado

axial *adj* COMP, DATA, ELECT, TRANS axial; ~ **interference microscopy** *n* ELECT, TRANS microscopía *f* por interferencia axial; ~ **lateral plasma deposition** *n* DATA, ELECT sedimentación *f* lateroaxial de plasma; ~ **propagation coefficient** *n* ELECT, RADIO, TRANS coeficiente *m* de propagación axial; ~ **ratio** *n* RADIO razón *f* axial, RADIO relación *f* axial; ~ **ray** *n* RADIO, TRANS rayo *m* axial; ~ **slab interferometry** *n* RADIO, TRANS interferometría *f* axial

axiom *n* DATA *specification and description language* axioma *m*

axis *n* COMP eje *m*

axle *n* PARTS eje *m*

azimuth *n* RADIO acimut *m*; ~ **antenna** RADIO antena *f* acimutal; ~ **elevation mount** RADIO soporte *m* de elevación de acimut; **~-speed indicator** RADIO indicador *m* de la velocidad acimutal; **~-stabilized display** RADIO visualización *f* de acimut estabilizado

B

B: ~ **answer signal** *n* TELEP señal *f* de respuesta B; **~-grade contour** *n* RADIO contorno *m* de grado B; **~-leg** *n* TELEP rama *f* B; ~ **sealant** *n* ELECT obturante *m* B; **~-stage materials** *n* PARTS *plastics* resina *f* en estado B; ~ **version** *n* TELEP versión *f* B

babble *n* TELEP diafonía *f* múltiple

Babcock: ~ **plan** *n* RADIO, TELEP plan *m* de Babcock

back: **~-to-~ parabolic antennas** *n pl* RADIO antenas *f pl* parabólicas adosadas; **~-to-~ scanners** *n pl* RADIO exploradores *m pl* en oposición; **~-to-~ test** *n* TRANS prueba *f* en bucle; ~ **dish** *n* RADIO reflector *m* parabólico inverso; ~ **echo** *n* RADIO eco *m* posterior; **~-end processor** *n* DATA procesador *m* especializado; **~-end server** *n* DATA servidor *m* especializado; ~ **lobe** *n* RADIO *aerials* lóbulo *m* posterior; **~-off** *n* RADIO supresión *f* de haz; ~ **pressure** *n* GEN *physics* contrapresión *f*; **~-pressure valve** *n* PARTS válvula *f* de contrapresión; **~-propagation network** *n* DATA red *f* de retropropagación; ~ **scatter** *n* POWER, RADIO, TELEP, TRANS energía *f* retrodispersada

backbone *n* ELECT, TRANS red *f* primaria; ~ **circuit** TRANS circuito *m* primario; ~ **flexibility** ELECT flexibilidad *f* de la red primaria; ~ **network** COMP, TRANS red *f* principal; ~ **route** TELEP ruta *f* principal; ~ **structure** TRANS estructura *f* de red primaria

backfire: ~ **antenna** *n* RADIO red *f* de antenas de radiación regresiva

background *n* GEN fondo *m*; ~ **level** GEN *acoustics* nivel *m* de referencia; ~ **mode of operation** RADIO modo *m* de operación de fondo; ~ **mush** PER *on answering machine* perturbación *f* de fondo; ~ **noise** GEN *acoustics* ruido *m* de fondo; ~ **operation** DATA operación *f* subordinada; ~ **processing** COMP procesamiento *m* subordinado; ~ **radiation** RADIO radiación *f* de fondo; ~ **signature** RADIO indicativo *m* de identificación; ~ **task** DATA tarea *f* secundaria; ~ **theory** TRANS teoría *f* secundaria

backing: ~ **tape** *n* COMP cinta *f* auxiliar

backlight *vt* COMP, ELECT iluminar a contraluz, retroiluminar

backlighting *n* COMP, ELECT iluminación *f* a contraluz, retroiluminación *f*

backlit *adj* COMP, ELECT iluminado a contraluz, retroiluminado

backlog *n* COMP acumulación *f* de trabajo

backplane *n* ELECT panel *m* de fondo

backscattering *n* POWER, RADIO, TELEP, TRANS *fibre optics* retrodispersión *f*; ~ **technique** TELEP, TRANS *fibre optics* técnica *f* de retrodispersión

backspace *n* (*BS*) COMP retroceso *m*; ~ **character** COMP carácter *m* de retroceso

backtracking *n* DATA retroseguimiento *m*

backup *adj* COMP auxiliar, de reserva; ~ **battery** *n* PER APP batería *f* de reserva; ~ **battery system** *n* COMP sistema *m* de batería de reserva; ~ **function** *n* GEN función *f* de reserva; ~ **information** *n* COMP información *f* de reserva; ~ **launch** *n* RADIO emisión *f* de reserva; ~ **launcher** *n* RADIO emisor *m* de reserva; ~ **mode** *n* TEST modo *m* de reserva; ~ **period** *n* TEST periodo *m* de reserva; ~ **reference secondary burst** *n* DATA ráfaga *f* secundaria auxiliar de referencia de reserva; ~ **relationships** *n pl* DATA relaciones *f pl* de reserva; ~ **power supply** *n* TELEP fuente *f* de energía de reserva; ~ **station** *n* RADIO estación *f* de reserva; ~ **store** *n* COMP memoria *f* de reserva; ~ **system** *n* DATA sistema *m* de copia de seguridad; ~ **supervisor** *n* TELEP supervisor *m* de reserva; ~ **tape** *n* COMP cinta *f* de seguridad, TEST cinta *f* de reserva; ~ **time** *n* COMP tiempo *m* de copia de seguridad

Backus: ~ **Normal Form** *n* DATA notación *f* normal de Backus

Backus-Naur: ~ **form** *n* (*BNF*) DATA impreso *m* Backus-Naur (*BNF*)

backward: ~ **alarm** *n* TEST alarma *f* de retorno; ~ **attenuation** *n* TRANS atenuación *f* de retorno; ~ **channel** *n* TELEP, TRANS canal *m* de retroceso; ~ **echo** *n* TELEP eco *m* de retorno; ~ **explicit congestion notification** *n* (*BECN*) DATA, SWG, TELEP notificación *f* explícita de bloqueo regresivo; **~-indicator bit** *n* COMP bitio *m* indicador inverso; **~-input signal** *n* TELEP señal *f* de entrada de retorno; **~-interworking telephony event** *n* (*BITE*) TELEP acontecimiento *m* de telefonía entrelazada hacia atrás (*BITE*); **~-motion vector** *n* COMP vector *m* de retorno, TV vector *m* de movimiento de retorno; **~ sequence number** *n* COMP número *m* de secuencia decreciente, número *m* de secuencia inverso, DATA número *m* de secuencia hacia atrás, TELEP número *m* de secuencia de retorno; **~ signal** *n* TELEP señal *f* de retorno; **~ transmission** *n* RADIO transmisión *f* de retorno; **~ wave oscillator** *n* (*BWO*) TELEP oscilador *m* de onda de retorno; **~ wave tube** *n* TELEP tubo *m* de onda de retorno

badge: **~ reader** *n* APPL lector *m* de identificadores

baffle *n* GEN *acoustics* desviación *f*

bakelite *n* PARTS *plastics* baquelita *f*

balance[1]: **~ level** *n* COMP nivel *m* de equilibrio; **~ ratio** *n* TELEP factor *m* de equilibrado de un radiogoniómetro; **~-return loss** *n* GEN balance *m* de situación; **~-return loss attenuation** *n* TRANS atenuación *f* de circuito compensado; **~ tester** *n* TRANS comprobador *m* de equilibrio; **~ winding** *n* POWER arrollamiento *m* de equilibrio

balance[2] *vt* GEN equilibrar

balanced *adj* COMP equilibrado, DATA simétrico, POWER equilibrado, RADIO, TRANS balanceado; **~ Adcock direction finder** *n* RADIO radiogoniómetro *m* Adcock equilibrado; **~ bridge interferometer switch** *n* TELEP interruptor *m* equilibrado de interferómetro de puente; **~ code** *n* TELEP código *m* simétrico; **~ error** *n* GEN error *m* compensado; **~ grading group** *n* TELEP grupo *m* de líneas equilibradas; **~ input** *n* COMP entrada *f* simétrica; **~ mixer** *n* COMP, RADIO mezclador *m* compensado; **~ network** *n* DATA red *f* equilibrada; **~ opposing winding** *n* POWER arrollamiento *m* bifilar; **~ output** *n* COMP salida *f* simétrica; **~-to-earth** POWER equilibrado a tierra; **~-to-unbalanced transformer** *n* POWER transformador *m* simétrico/asimétrico

balancing *n* GEN equilibrio *m*; **~ resistor** PARTS, POWER resistor *m* de compensación

ball: **~ bearing** *n* PARTS cojinete *m* de bolas; **~ valve** *n* PARTS válvula *f* de bola

ballast: **~ resistor** *n* POWER resistor *m* autorreguladora

balloon: **~ help** *n* COMP ayuda *f* especial

balun *n* ELECT, POWER balun *m*; **~ adaptor** ELECT adaptador *m* simétrico/asimétrico

banana: **~ jack** *n* ELECT *connecting device* clavija *f* de punta cónica; **~ plug** *n* ELECT *connecting device*, TEST clavija *f* de punta cónica

band *n* COMP, RADIO, TRANS banda *f*; **~ elimination filter** TRANS filtro *m* de bloqueo de banda; **~ number** TELEP número *m* de banda; **out-~ signalling** *BrE* SIG señalización *f* fuera de banda; **~ rate** TRANS velocidad *f* de banda; **~ rejection filter** TRANS filtro *m* de bloqueo de banda

bandpass: **~ filter** *n* (*BPF*) POWER, TRANS filtro *m* paso banda

bands *n pl* COMP bandas *f pl*

bandstop: **~ filter** *n* TRANS filtro *m* eliminador de banda

bandwidth *n* GEN ancho *m* de banda; **~ compression** DATA compresión *f* de ancho de banda; **~ on demand** COMP ancho *m* de banda sobre demanda; **~-efficient system** TRANS sistema *m* de ancho de banda eficiente; **~ expansion** RADIO, TRANS expansión *f* de ancho de banda; **~-expansion ratio** RADIO, TRANS ratio *f* de expansión de ancho de banda; **~-limited** *adj* TRANS limitado por ancho de banda; **~-limited operation** TRANS funcionamiento *m* en banda limitada

bank *n* COMP, DATA, TV bloque *m*; **~ contact** TELEP contacto *m* de banco; **~ of contacts** TELEP lámina *f* de contactos; **~ memory** COMP memoria *f* de banco

banknote: **~ sorting machine** *n* APPL máquina *f* clasificadora de billetes de banco

BAPTA *abbr* (*bearing and power transfer assembly*) POWER BAPTA (*conjunto de marcación y transferencia de energía*)

bar *n* TELEP, COMP, PARTS *equipment practice*, TELEP barra *f*; **~ chart** GEN *statistics* tabla *f* de barras; **~ code** APPL, DATA código *m* de barras; **~-code scanner** APPL, DATA lector *m* de código de barras; **~ graph** GEN *statistics* gráfico *m* de barras; **~ histogram** GEN histograma *m* de barras; **~ holder** PARTS *equipment practice* sujetador *m* de barra

bare: **~ wire** *n* ELECT, POWER *conductors* hilo *m* desnudo; **~-wire multiple** *n* TELEP múltiple *m* de hilo desnudo; **~-wire strap** *n* TELEP puente *m* de hilo desnudo; **~ wiring** *n* TELEP conexiones *f pl* con hilo desnudo

barge: **~-in** *n* TELEP transferencia *f*; **~-in tone** *n* TELEP tono *m* de transferencia

Bark *n* COMP Bark *m*

barred *adj* RADIO bloqueado, TELEP barrado, restringido; **~ code** *n* TELEP indicativo *m* de servicio restringido; **~ station** *n* RADIO estación *f* bloqueada

barrette: **~ file** *n* PARTS lima *f* de bonete

barrier: **~ layer** *n* ELECT *semiconductors* capa *f* de detención

barring *n* RADIO, TELEP bloqueo *m*

barter *n* TV trueque *m*

bartering *n* TV permuta *f*

base *n* ELECT *semiconductors*, PARTS *structure* base *f*; ~ **address** COMP direccionamiento *m* de base; ~ **area** RADIO *mobicom* zona *f* de cobertura de estación de base; ~ **charge** ELECT *semiconductors* carga *f* de base; ~ **coverage area** RADIO *mobile* zona *f* de cobertura de estación de base; ~ **current** ELECT *semiconductors* corriente *f* de base; ~**-emitter resistance** ELECT *semiconductors* resistencia *f* base-emisor; ~**-emitter voltage** ELECT *semiconductors* tensión *f* base-emisor; ~ **frequency** COMP frecuencia *f* de base; ~ **iron** PARTS hierro *m* de base, hierro *m* de fondo; ~ **level** SWG nivel *m* de base; ~ **line** COMP itinerario *m* básico; ~ **load** TELEP *power supply* carga *f* base; ~ **materials** *n pl* ELECT soportes *m pl* aislantes, PARTS soporte *m pl* de base, POWER soportes *m pl* aislantes; ~ **metal** ELECT *soldering* metal *m* inestable; ~**-to-mobile relay** RADIO relé *m* base-móvil; ~ **number** COMP número *m* de base; ~ **panel** ELECT, PARTS panel *m* de base; ~ **plate** ELECT, PARTS *structure* placa *f* de asiento; ~ **region** ELECT *semiconductors* zona *f* de base; ~ **resistance** ELECT *semiconductors* resistencia *f* de base; ~ **reverse current** ELECT *semiconductors* corriente *f* inversa de base; ~ **station** RADIO, TELEP estación *f* base; ~ **station area** RADIO zona *f* de estación de base; ~ **station controller** RADIO controlador *m* de la estación base; ~ **-station separation** RADIO *mobile* distancia *f* entre estaciones de base; ~ **station subsystem** (*BSS*) TELEP subsistema *m* de estación de base; ~ **support bracket** ELECT abrazadera *f* de anclaje de base; ~ **terminal** ELECT *semiconductors* terminal *m* de base; ~ **type** DATA tipo *m* básico; ~ **unit** (*BU*) RADIO unidad *f* base

baseband *n* RADIO, TRANS, TV banda *f* base; ~ **bandwidth** RADIO ancho *m* de banda de banda base, TRANS anchura *f* de banda base; ~ **equipment** RADIO equipamiento *m* de banda base; ~ **modem** DATA, TRANS módem *m* de banda base; ~ **response** TELEP respuesta *f* de banda base; ~ **response function** APPL, CONTR, DATA, TELEP, TRANS función *f* de respuesta de banda base; ~ **switch matrix** SWG matriz *f* de conexión de banda de base; ~ **transfer function** APPL, CONTR, DATA, TRANS función *f* de transferencia de banda base; ~ **transmission** DATA transmisión *f* de banda base, TELEP transmisión *f* en banda base

baseline *n* RADIO línea *f* de tierra

basic: ~ **access** *n* TELEP, TRANS acceso *m* básico; ~ **amplitude** *n* RADIO amplitud *f* básica; ~ **bit rate** *n* DATA tasa *f* básica de bitios; ~ **channel** *n* COMP canal *m* básico; ~ **character font** *n* DATA *Internet* familia *f* de caracteres básica; ~ **charge** *n* TELEP comisión *f* básica; ~ **data** *n pl* GEN datos *m pl* básicos; ~ **dimensions** *n pl* GEN dimensiones *f pl* básicas; ~ **element** *n* GEN elemento *m* básico; ~ **encoding rules** *n pl* (*BER*) COMP codificación *f* básica, DATA normas *f pl* básicas de codificación; ~ **equipment rack** *n* PARTS *equipment practice* bastidor *m* de equipo básico; ~ **error correction method** *n* TELEP método *m* básico de corrección de errores; ~ **frame alignment** *n* (*BFA*) TRANS alineación *f* básica de imagen; ~ **frequency** *n* ELECT frecuencia *f* de base; ~ **group** *n* TELEP, TRANS grupo *m* básico; ~ **input/output system** *n* (*BIOS*) TEST sistema *m* básico de entrada/salida (*BIOS*); ~ **interconnection test** *n* (*BIT*) DATA prueba *f* básica de interconexión (*BIT*); ~ **loop** *n* COMP bucle *m* básico; ~ **loss** *n* TRANS atenuación *f* básica; ~ **material** *n* COMP material *m* básico; ~ **metered unit** *n* TELEP unidad *f* básica computada; ~ **metering pulse** *n* TELEP impulso *m* básico de cómputo; ~ **MUF** *n* RADIO MUF *f* básica; ~ **noise** *n* COMP, TELEP ruido *m* básico; ~ **operation** *n* COMP operación *f* básica; ~ **price** *n* GEN *commerce* precio *m* base; ~ **rack** *n* PARTS *equipment practice* bastidor *m* de equipo básico; ~ **rate access** *n* (*BRA*) TELEP precio *m* base de acceso; ~ **rate interface** *n* DATA interfaz *f* de frecuencia básica, TELEP interfaz *f* de tarifa básica; ~ **rate service** *n* APPL precio *m* base de servicio; ~ **research** *n* TEST investigación *f* básica; ~ **SDL** *n* COMP, DATA SDL *f* básica; ~ **service** *n* TV servicio *m* básico; ~ **session reference** *n* TELEP referencia *f* de sesión básica; ~ **signal** *n* COMP señal *f* básica; ~ **size** *n* GEN tamaño *m* básico; ~**-size drawing** *n* GEN *technical drawing* dibujo *m* a escala; ~ **software package** *n* COMP paquete *m* de software básico; ~ **specifications** *n pl* GEN *commerce* especificaciones *f pl* básicas; ~ **subscription** *n* TELEP suscripción *f* básica; ~ **supergroup** *n* TELEP, TRANS grupo *m* secundario básico; ~ **transmission loss** *n* RADIO pérdida *f* de transmisión básica; ~ **value** *n* GEN valor *m* básico

basilar: ~ **membrane** *n* RADIO membrana *f* basilar

basket *n* DATA cesta *f*

bass *n* COMP bajo *m*; ~ **chord** COMP cuerda *f* de bajo; ~ **tone** COMP tono *m* bajo; ~ **tones** *n pl* COMP tonos *m pl* bajos

batch *n* COMP, TEST *quality* lote *m*; ~ **processing** COMP, DATA procesamiento *m* por lotes; ~ **testing** PARTS prueba *f* por lotes; ~ **variation** TEST *quality* variación *f* en un lote

battery *n* ELECT *combined cells*, POWER *single cell* batería *f*; ~ **backup** PER *on answering machine* reserva *f* de baterías; ~ **cabinets** *n pl* TRANS armarios *m pl* de acumuladores; ~ **cable** POWER cable *m* de batería; ~ **cart** POWER carro *m* portabaterías; ~ **case** POWER cubierta *f* de batería; ~ **cell** POWER elemento *m* de batería; ~ **cell container** POWER contenedor *m* de elemento de batería; ~ **charger** POWER cargador *m* de batería o pila; ~ **charging** POWER carga *f* de batería; ~ **clip** POWER pinza *f* de batería; ~ **condition** POWER estado *m* de batería; ~ **control unit** POWER unidad *f* de control de batería; ~ **eliminator** POWER eliminador *m* de batería; ~ **fuse** PARTS, POWER fusible *m* de batería, cebo *m* eléctrico de cantidad; ~ **monitor** POWER control *m* de batería; ~ **negative** POWER negativo *m* de batería; ~ **positive** POWER positivo *m* de batería; ~ **power supply** COMP fuente *f* de alimentación de batería, ELECT, POWER suministro *m* eléctrico por batería; ~ **rack** POWER bastidor *m* de batería; ~ **regulator** POWER regulador *m* de batería; ~ **room** POWER sala *f* de baterías; ~ **storage time** POWER periodo *m* de almacenamiento de baterías; ~ **system** COMP, ELECT, POWER sistema *m* de baterías; ~ **tray** POWER soporte *m* de la batería; ~ **trolley** POWER remolque *m* para baterías; ~ **voltage** ELECT, POWER voltaje *m* de batería; ~ **voltage terminal** POWER toma *f* de tensión de batería

baud: ~ **rate** *n* COMP velocidad *f* de línea en baudios

baudot: ~ **code** *n* COMP código *m* baudot

bay *n* TELEP módulo *m*

bayonet *n* ELECT, PARTS, POWER bayoneta *f*

BB *abbr* (*bulletin board*) COMP, DATA, TELEP BB (*boletín electrónico*)

BBD *abbr* (*bucket brigade device*) SWG *shift register* circuitos *m* de noria

BBS *abbr* (*bulletin board system*) COMP, DATA, TELEP BBS (*sistema de boletíns electrónicos*)

BC *abbr* (*broadcast*) GEN emisión *f*

B-channel *n* DATA, TRANS *information channel* canal *m* B; ~ **virtual-circuit service** TELEP servicio *m* de circuito virtual del canal B

BCS *abbr* (*BULL cabling system*) DATA BCS (*sistema de cableado BULL*)

beacon *n* RADIO faro *m*; ~ **satellite** RADIO satélite *m* radiofaro; ~ **signal** SIG señal *f* de radiofaro; ~ **transmitter** RADIO transmisor *m* de radiofaro

bead: ~ **of epoxy** *n* ELECT cuenta *f* de epoxia

beam *n* GEN haz *m*; ~ **aerial** RADIO antena *f* del haz; ~ **antenna** RADIO antena *f* del haz; ~ **area** RADIO ángulo *m* sólido del haz; ~ **axis** RADIO eje *m* del haz; ~ **centre** *BrE* RADIO centro *m* del haz; ~ **centre** *BrE* **gain** RADIO ganancia *f* del centro del haz; ~ **diameter** TRANS *fibre optics* diámetro *m* del haz; ~ **divergence** RADIO, TRANS *fibre optics* divergencia *f* del haz; ~ **edge** RADIO borde *m* del haz; ~ **forming** ELECT, RADIO modelación *f* del haz; ~**-forming network** ELECT red *f* formadora del haz, RADIO red *f* concentradora del haz; ~ **scanning** RADIO exploración *f* del haz; ~ **splitter** COMP, TRANS *fibre optics* divisor *m* del haz; ~ **steerability** RADIO direccionabilidad *f* del haz; ~ **steering** RADIO orientación *f* del haz; ~ **switching** RADIO conmutación *f* de haz; ~ **waveguide** RADIO guíaondas *m* del haz; ~ **waveguide feed** RADIO alimentación *f* del guíaondas del haz; ~ **width** RADIO, TRANS *fibre optics* amplitud *f* del haz

bear against *vt* PARTS descansar sobre

bear on *vt* PARTS apoyar sobre

bearer *n* TRANS *of circuits* montante *m*; ~ **cable** TRANS cable *m* sustentador; ~ **capability** TELEP capacidad *f* del portador; ~ **capacity** TRANS capacidad *f* del portador; ~ **channel** TELEP, TRANS canal *m* de corriente portadora; ~ **circuit** DATA circuito *m* portador; ~ **identification code** DATA, TELEP código *m* de identificación de portador; ~ **independent service** (*BIS*) TELEP servicio *m* independiente de portador (*BIS*); ~ **service** APPL, COMP, RADIO, TRANS servicio *m* de corriente portadora; ~ **unit** TELEP unidad *f* portadora

bearing *n* PARTS, RADIO *of antenna* portamacho *m*, *direction* dirección *f*; ~ **housing** PARTS caja *f* de cojinete; ~ **and power transfer assembly** (*BAPTA*) POWER conjunto *m* de marcación y transferencia de energía (*BAPTA*); ~ **seat** PARTS asiento *m* de rodamiento; ~ **shell** PARTS casquillo *m* de cojinete; ~ **surface** PARTS superficie *f* de apoyo; ◆ **take a** ~ RADIO determinar la dirección

beat *n* GEN *physics* pulsación *f*; ~ **frequency** ELECT, TRANS frecuencia *f* de batido; ~**-frequency oscillator** (*BFO*) ELECT, TRANS oscilador *m* de frecuencia de batido; ~ **indicator** COMP indicador *m* de batido

beats: ~ **per minute** *n pl* COMP batidos *m pl* por minuto

BECN *abbr* (*backward explicit congestion notification*) DATA, SWG, TELEP notificación *f* explícita de bloqueo regresivo

bedding *n* TRANS *cable laying* estratificación *f*
beep *n* RADIO señal *f* acústica; ~ **sound** COMP pitido *m*
beeper *n* COMP indicador *m* acústico
beginning: ~ **of call demand** *n* DATA demanda *f* de comienzo de llamada; ~ **of message** *n* (*BOM*) DATA comienzo *m* de mensaje; ~ **of pulse** *n* COMP comienzo *m* de la pulsación; ~**-of-tape marker** *n* COMP marca *f* de comienzo de cinta magnética
behaviour *BrE* (*AmE* **behavior**) *n* DATA *specification and description language* comportamiento *m*; ~ **line** DATA línea *f* de comportamiento; ~ **test** DATA prueba *f* de comportamiento; ~ **tree** DATA árbol *m* de comportamiento
behavioural: ~ **animation** *n* COMP animación *f* funcional
Beldfoil: ~ **aluminium-polyester shield** *BrE* *n* ELECT blindaje *m* de aluminio-poliéster de Beldfoil
Belleville: ~ **spring** *n* PARTS arandela *f* Belleville
bellsets *n pl* PER APP timbres *m pl*
belt: ~ **drive** *n* PARTS transmisión *f* por correa; ~ **insulation** *n* TRANS *cables* aislamiento *m* por cinturón; ~ **loop** *n* ELECT bucle *m* de cinta
belted: ~ **cable** *n* TRANS cable *m* encintado
benchmark: ~ **test** *n* COMP prueba *f* para comparación; ~ **testing** *n* COMP ensayo *m* para comparación
bend[1]: ~ **aerial** *n* RADIO antena *f* acodada; ~ **carbide** *n* ELECT carburo *m* metálico curvado; ~ **radius** *n* TRANS *cables* radio *m* de plegado, radio *m* de curvatura
bend[2] *vt* COMP, PARTS, RADIO, TRANS *cables* curvar
bender *n* PARTS curvadora *f*
bending: ~ **loss** *n* ELECT pérdida *f* de refracción, TRANS amortiguación *f* de flexión; ~ **strength** *n* PARTS resistencia *f* a la flexión; ~ **tool** *n* PARTS herramienta *f* dobladora
BER *abbr* (*basic encoding rules*) COMP codificación *f* básica, DATA normas *f pl* básicas de codificación
beta: ~ **trial** *n* TEST prueba *f* piloto
bevel *vt* PARTS biselar
bevelling *BrE* (*AmE* **beveling**) *n* TEST biselado *m*
Bézier: ~ **curve** *n* COMP curva *f* de Bézier
BFA *abbr* (*basic frame alignment*) TRANS alineación *f* básica de imagen
b-field: ~ **picture** *n* COMP, TV imagen *f* de campo-b
BFO *abbr* (*beat-frequency oscillator*) ELECT, TRANS oscilador *m* de frecuencia de batido
b-frame: ~ **picture** *n* COMP, TV imagen *f* de columna-b
BFT *abbr* (*binary file transfer*) TELEP BFT (*transferencia de archivos binarios*); ~ **file attributes** *n pl* TELEP atributos *m pl* de fichero BFT
bias *n* COMP, POWER, RADIO, TEST *quality* polarización *f*; ~ **control** COMP control *m* por polarización; ~ **current** ELECT corriente *f* de polarización; ~ **distortion** ELECT, TELEP distorsión *f* de desviación; ~ **level** COMP nivel *m* de polarización; ~ **point** COMP punto *m* de carga
biased *adj* POWER sesgado; ~ **sample** *n* TEST *quality* muestra *f* sesgada; ~ **test** *n* GEN *statistics* prueba *f* sesgada
biconic: ~ **connectors** *n pl* ELECT conectores *m pl* bicónicos
biconical: ~ **antenna** *n* RADIO antena *f* bicónica
bid *n* GEN propuesta *f*; ~ **for a line** TELEP intento *m* de ocupación de una línea
bidirectional *adj* GEN bidireccional; ~ **diode thyristor** *n* ELECT, PARTS tiristor *m* diodo bidireccional; ~**flow** *n* GEN flujo *m* bidireccional; ~ **network** *n* TV red *f* bidireccional; ~ **transistor** *n* ELECT transistor *m* bidireccional; ~ **triode thyristor** *n* ELECT, PARTS tiristor *m* triodo bidireccional
bidirectionally: ~**-predictive coded picture** *n* COMP imagen *f* codificada bidireccionalmente predictiva, TV imagen *f* codificada predictiva en ambos sentidos
bids: ~ **per circuit per hour** *n pl* TELEP intentos *m pl* de ocupación por circuito y por hora
bifilar: ~ **winding** *n* POWER devanado *m* bifilar
bilateral *adj* GEN bilateral; ~ **agreement** *n* GEN convenio *m* bilateral; ~ **closed user group** *n* DATA, TELEP grupo *m* de usuarios cerrado bilateral; ~ **closed user group call indicator** *n* TELEP indicador *m* de llamada de grupo de usuarios cerrado bilateral; ~ **closed user group with outgoing access** *n* DATA, TELEP grupo *m* de usuarios cerrado bilateral con acceso de salida; ~ **control** *n* TELEP, TRANS control *m* bilateral
bi-level: ~ **image** *n* TELEP imagen *f* a dos niveles
bilinear *adj* PER APP *scanner* bilineal
billed: ~ **services** *n pl* TELEP servicios *m pl* facturados
billing *n* GEN facturación *f*; ~ **address** TELEP dirección *f* de facturación; ~ **centre** *BrE* APPL centro *m* de facturación; ~ **information** TELEP información *f* de la facturación
bimetal *n* PARTS bimetal *m*; ~ **spring** CONTR *regulating devices* resorte *m* bimetálico; ~ **springset** CONTR *regulating devices* conjunto *m* de resorte bimetálico
binary: ~ **coding** *n* SIG codificación *f* binaria; ~

counter *n* ELECT *logic elements* contador *m* binario; ~ **data message** *n* TELEP mensaje *m* de datos binarios; ~ **digit** *n* TRANS dígito *m* binario; ~ **error rate** *n* COMP grado *m* del error binario, DATA velocidad *f* binaria de error, RADIO tasa *f* de errores binarios; ~ **figure** *n* COMP figura *f* binaria, TRANS cifra *f* binaria; ~ **file** *n* DATA fichero *m* binario, TELEP archivo *m* binario; ~ **file transfer** *n* (*BFT*) TELEP transferencia *f* de archivos binarios (*BFT*); ~ **image** *n* DATA imagen *f* binaria; ~ **modulation** *n* DATA modulación *f* binaria; ~ **multiplier** *n* DATA multiplicador *m* binario; ~ **numeral** *n* COMP, DATA numeral *m* binario; ~ **phase-shift keying** *n* (*BPSK*) SIG manipulación *f* por desviación de fase binaria; ~ **rate** *n* COMP, DATA, TELEP, TRANS velocidad *f* binaria; ~ **search** *n* COMP búsqueda *f* dicotómica; ~ **search procedure** *n* DATA procedimiento *m* binario de búsqueda; ~ **sequence** *n* DATA secuencia *f* binaria; ~ **serial signalling rate** *BrE n* SIG velocidad *f* binaria de señalamiento en serie; ~ **signal** *n* DATA señal *f* binaria; ~ **synchronous communication** *n* (*BSC*) DATA comunicación *f* binaria sincrónica; ~ **system** *n* COMP sistema *m* binario; ~ **tariff system** *n* TELEP sistema *m* binario de tarifa

bind *vt* DATA fijar, vincular

binder *n* GEN ligador *m*

binding *n* DATA *specification and description language* fijación *f*

binomial: ~ **distribution** *n* GEN *statistics* distribución *f* binomial

BIOS *abbr* (*basic input/output system*) TEST BIOS (*sistema básico de entrada/salida*)

BIP *abbr* (*bit interleaved parity*) DATA BIP (*paridad intercalada de bitio*)

biphase: ~ **code** *n* APPL código *m* bifásico

bipolar: ~ **circuit** *n* ELECT *logic elements* circuito *m* bipolar; ~ **code** *n* SIG código *m* bipolar; ~ **code with three-zero substitution** *n* (*B3ZS*) DATA, SIG, TELEP código *m* bipolar con sustitución de tres ceros (*B3ZS*); ~ **logic** *n* ELECT *logic elements* circuito *m* lógico bipolar; ~ **pulse** *n* TELEP impulso *m* bipolar; ~ **sequence of pulses** *n* TELEP secuencia *f* bipolar de impulsos; ~ **signal** *n* SIG, TRANS señal *f* bipolar; ~ **transistor** *n* ELECT, PARTS transistor *m* bipolar; ~ **violation** *n* TRANS violación *f* bipolar

BIP-X *abbr* (*bit interleaved parity X*) DATA BIP-X (*paridad X intercalada de bitio*)

birefringence *n* DATA, TRANS *fibre optics* birrefringencia *f*

BIS *abbr* (*bearer independent service*) TELEP BIS (*servicio independiente de portador*)

B-ISDN *abbr* (*broadband ISDN*) TRANS RDSI-BA (*ISDN de banda*); ~ **network termination** *n* (*NT-LB*) TRANS terminación *f* de la red B-RDSI (*T-RDSI*); ~ **network termination 1** *n* (*NT1-LB*) TRANS terminación *f* 1 de red digital de servicios integrados; ~ **network termination 2** *n* (*NT2-LB*) TRANS terminación *f* 2 de red digital de servicios integrados; ~ **service** *n* APPL servicio-B *m* de red digital de servicios integrados, TELEP servicio *m* B-RDSI; ~ **terminal adaptor** *n* (*TA-LB*) TELEP adaptador *m* terminal B de la red de transmisión digital de RDSI-BA; ~ **terminal equipment** *n* TELEP equipo *m* terminal B de la RDSI

bisecting: ~ **relay** *n* PARTS relé *m* divisor de impulsos

bistability *n* GEN biestabilidad *f*

bistatic: ~ **radar** *n* RADIO radar *m* biestático

bistring: ~ **type** *n* DATA tipo *m* de doble cadena

bit *n* COMP, DATA, TRANS unidad *f* binaria; ~-**by-**~ **encoding** DATA codificación *f* bit a bit; ~ **combination** COMP, TELEP combinación *f* binaria; ~ **disagreements** *n pl* DATA discordancias *f pl* de bitios; ~ **efficiency** DATA eficiencia *f* por bitio; ~ **error** TRANS error *m* de bitio; ~ **error code** COMP código *m* de error de bitios; ~ **error rate** COMP, DATA, RADIO tasa *f* de bitios erróneos; ~ **interleaved parity** (*BIP*) DATA paridad *f* intercalada de bitio (*BIP*); ~ **interleaved parity 8** DATA paridad *f* 8 intercalada de bitio; ~ **interleaved parity X** (*BIP-X*) DATA paridad *f* X intercalada de bitio (*BIP-X*); ~ **interleaved parity order 8** DATA orden *f* 8 de paridad de bitio intercalada; ~ **interleaving** DATA intercalación *f* de bitios; ~ **mapping** DATA configuración *f* de la unidad binaria; ~-**order of transmission** COMP orden *f* de los bitios de transmisión; ~ **plane** TELEP plano *m* de bitio; ~ **rate** COMP, DATA, TELEP, TRANS tasa *f* de bits; ~-**rate compression** DATA compresión *f* de velocidad binaria; ~ **sequence independence** COMP independencia *f* de la secuencia de bitios; ~ **stream** DATA flujo *m* de bitio; ~ **switch** SWG conmutador *m* de bit; ~ **timing** DATA, TRANS ritmo *m* binario; ~-**timing recovery sequence** DATA, RADIO secuencia *f* de recuperación de ritmo binario; ~-**true arithmetic package** COMP paquete *m* aritmético de bitios concordantes

BIT *abbr* (*basic interconnection test*) DATA BIT (*prueba básica de interconexión*)

bite *n* COMP poder *m* cortante

BITE *abbr* (*backward-interworking telephony event*) TELEP BITE (*acontecimiento de telefonía entrelazada hacia atrás*)
bi-timbral *adj* COMP bitímbrico
Bitnet *n* DATA *Internet* Bitnet *m*
bits: ~ **per colour** *BrE phr* DATA *scanner* bitios *m pl* por color; ~ **per inch** *phr* (*bpi*) COMP bitios *m pl* por pulgada (*bpp*); ~ **per second** *phr* (*bps*) DATA bitios *m pl* por segundo (*bps*)
black: ~**-and-white reception** *n* TELEP recepción *f* en blanco y negro; ~**-level retention** *n* COMP, TELEP, TRANS retención *f* de nivel de negro; ~ **stroke** *n* ELECT trazo *m* negro
blackout *n* RADIO apagón *m*
blade: ~ **contact** *n* ELECT contacto *m* de cuchilla, PARTS contacto *m* de espigas planas; ~ **contact unit** *n* ELECT, PARTS bloque *m* de contactos de patillas planas
blank: ~**-and-burst message** *n* DATA mensaje *m* de borrado y error; ~ **entry** *n* DATA entrada *f* en blanco; ~ **label** *n* ELECT etiqueta *f* en blanco; ~ **line** *n* TELEP línea *f* vacía
blanked: ~**-off channel** *n* TRANS canal *m* aislado
blanket: ~ **coverage** *n* RADIO cobertura *f* general
blanking *n* TELEP *video* borrado *m*; ~ **burr** PARTS rebarba *f*; ~ **level** TELEP *video* nivel *m* de borrado; ~ **pulse** TELEP *video* pulso *m* de borrado; ~ **signal** TELEP *video* señal *f* de borrado
blast: ~ **out of orbit** *vi* RADIO sobrecargar fuera de órbita
blaster *n* COMP dinamitero *m*
bleeding: ~ **resistor** *n* ELECT, PARTS, POWER resistor *m* de drenaje
bleep *n* COMP pitido *m*, RADIO señal *f* acústica
bleeper *n* RADIO mecanismo *m* de señal sonora
blind *adj* COMP ciego; ~ **carbon copy** *n* DATA copia *f* libre; ~ **copy** *n* DATA copia *f* libre; ~ **courtesy copy** *n* TELEP copia *f* libre sin cargo; ~ **region** *n* RADIO zona *f* ciega; ~ **speed** *n* RADIO velocidad *f* ocultamiento a la detección
blink *vi* COMP, TELEP, TEST *indicating devices* parpadear
blinking *adj* COMP, TELEP, TEST parpadeante
blip *n* RADIO señal *f* luminosa en la pantalla
block *n* GEN *supply and distribution* bloque *m*; ~**-acknowledged counter** TELEP contador *m* de bloques con acuse de recibo; ~**-cancel character** COMP carácter *m* de cancelación de bloque; ~ **check** COMP control *m* por bloques; ~ **code** COMP código *m* de bloques; ~ **coding** TRANS codificación *f* de bloque; ~ **companding** COMP compresión-expansión *f* de bloque; ~**-completed counter** TELEP contador *m* de bloques completos; ~ **container** POWER *batteries* recipiente *m* con compartimentos; ~ **diagram** GEN *documentation* diagrama *m* de conjuntos; ~**-dropping indicator** DATA indicador *m* de deposición de bloques; ~ **interleaving** RADIO intercalación *f* de bloques; ~ **of parameters** TELEP bloque *m* de coordenadas; ~ **printing offset litho** ELECT litografía *f* offset para impresión de bloques; ~ **quantization** TRANS cuantificación *f* de bloques; ~ **separation** TELEP separación *f* de bloques; ~ **separator** TELEP separador *m* de bloques; ~ **sort** COMP clasificación *f* de bloques; ~ **substructure** DATA *specification and description language* substructura *f* de bloques; ~ **terminal** TELEP terminal *m* de bloque; ~ **transfer** COMP transferencia *f* de bloques; ~ **transmission** TRANS transmisión *f* de bloques; ~**-tree diagram** DATA diagrama *m* de árbol en bloques; ~ **type** DATA *specification and description language* tipo *m* de bloque
blocked *adj* GEN bloqueado, TELEP agrupado
blocking *n* GEN bloqueo *m*; ~ **acknowledgement signal** SIG señal *f* de reconocimiento de bloqueo; ~ **alarm** TELEP avisador *m* de bloqueo; ~ **button** TELEP botón *m* de bloqueo; ~ **counter** TELEP contador *m* de bloqueo; ~ **direction** ELECT *semiconductors* dirección *f* de bloqueo; ~ **effect** GEN efecto *m* de bloqueo; ~ **grade of service** TELEP grado *m* de bloqueo del servicio; ~ **indication** TELEP indicación *f* de bloqueo; ~ **lamp** TELEP testigo *m* de bloqueo; ~ **network** SWG red *f* de bloqueo; ~ **rate** GEN velocidad *f* de bloqueo; ~ **signal** TELEP señal *f* de bloqueo
blow *vi* POWER *fuses* fundirse
blowing *n* ELECT, POWER *of fuse* fundición *f*
blown: ~ **fibre** *BrE n* TRANS fibra *f* soplada; ~ **fuse** *n* PARTS, POWER fusible *m* fundido
blurred *adj* COMP borroso
BNF *abbr* (*Backus-Naur form*) DATA BNF (*impreso Backus-Naur*)
board *n* COMP, PARTS, SWG *switchboard* cuadro *m*; ~ **compartment** PARTS *equipment practice* compartimiento *m* para placa de circuitos; ~ **extraction tool** TEST *tools* extractor *f* de placas; ~ **extractor** TEST *tools* extractor *f* de placas; ~ **frame** PARTS *equipment practice* bastidor *m* de placas; ~**-level testing** TEST prueba *f* de las placas de circuito; ~ **position** PARTS *equipment practice* posición *f* de las placas; ~ **spacing** PARTS *equipment practice* separación *f* de las placas
bobbin *n* POWER carrete *m*

Bode: ~ **diagram** *n* CONTR, POWER diagrama *m* de Bode
body *n* GEN cuerpo *m*; ~ **object** COMP objeto *m* corporal; ~ **placeholder** COMP ocupador *m* de espacio corporal; **~-stabilized communications satellite** RADIO satélite *m* de comunicaciones de cuerpo estabilizado
bold *n* COMP *printing* negrita *f*
BOM *abbr* (*beginning of message*) DATA comienzo *m* de mensaje
bonding *n* ELECT, POWER conexión *f* a tierra; ~ **sheet** ELECT, POWER lámina *f* de conexión
book *vt* APPL *call*, TELEP retener
bookmark *n* COMP marcador *m* de páginas Web, DATA, *Internet* dirección *f* URL
Boolean *adj* COMP, DATA *specification and description language* booleano
boost *vt* GEN *sales* aumentar
booster *n* COMP acelerador *m* intermedio; ~ **battery** POWER batería *f* de refuerzo; ~ **converter** POWER convertidor *m* de refuerzo
boot *vt* COMP arrancar
boot up *vt* COMP arrancar
booth *n* COMP puesto *m*
bootstrap *n* COMP autoelevación *f*; ~ **loader** COMP cargador *m* automático inicial
border *n* RADIO caballón *m*
bore *vt* PARTS mandrilar
boresight *n* RADIO alineación *f* óptica; ~ **facility** RADIO instrumental *m* para alineación óptica; ~ **method** RADIO método *m* de alineación óptica
boring *n* PARTS perforación *f*; ~ **bar** PARTS barra *f* taladradora; ~ **fixture** ELECT, PARTS montaje *m* de taladrado
both: **~-party release** *n* COMP, TELEP desconexión *f* doble; **~-party release signal** *n* TELEP señal *f* de desconexión doble; **~-way communication** *n* TELEP comunicación *f* en los dos sentidos; **~-way group** *n* TRANS grupo *m* bidireccional; **~-way line** *n* TRANS línea *f* bidireccional
bottleneck *n* ELECT atasco *m*
bottom *n* COMP pie *m*; ~ **field** TV *MPEG* campo *m* de fondo; ~ **note** COMP nota *f* de fondo; ~ **stud** PARTS resalto *m* de fondo; ~ **view** GEN *technical drawing* vista *f* de fondo
bounce *vi* COMP rebotar
bound: ~ **data** *n pl* DATA datos *m pl* editados; ~ **mode** *n* TRANS *fibre optics* modo *m* ligado
boundary *n* COMP, ELECT *semiconductors*, RADIO límite *m*; ~ **representation** COMP representación *f* de límite
bounding: ~ **volumes** *n pl* COMP volúmenes *m pl* limítrofes
bow *n* POWER *printed circuits* asa *f*; **~-tie antenna** RADIO antena *f* replegada
BPF *abbr* (*bandpass filter*) POWER, TRANS filtro *m* paso banda
bpi *abbr* (*bits per inch*) COMP bpp (*bitios por pulgada*)
bps *abbr* (*bits per second*) DATA bps (*bitios por segundo*)
BPSK *abbr* (*binary phase-shift keying*) SIG manipulación *f* por desviación de fase binaria
BRA *abbr* (*basic rate access*) TELEP precio *m* base de acceso
brace *n* GEN riostra *f*
bracket *n* GEN *linguistics* ménsula *f*
braid *n* ELECT, TRANS *cable* trenza *f*
braided: ~ **cable** *n* ELECT, PARTS, TRANS cable *m* trenzado
braiding *n* ELECT, PARTS, TRANS *cables* trenza *f*
brainstorming *n* DATA intercambio *m* de ideas
brake *n* GEN freno *m*; ~ **block** PARTS bloque *m* de frenado; ~ **fluid** PARTS líquido *m* para frenos; ~ **handle** PARTS palanca *f* de freno; ~ **lining** PARTS guarnición *f*
branch *n* COMP ramificación *f*; ~ **flip-flop** SWG basculador *m* de desviación; ~ **formboard** TRANS *cables* plantilla *f* de bifurcación; ~ **joint** TRANS *cables* conexión *f* en T; **~-off** TRANS bifurcación *f*; ~ **point** COMP, TRANS punto *m* de bifurcación
branching: ~ **block** *n* ELECT, PARTS *connecting devices* regleta *f* de bifurcación; ~ **bypass** *n* ELECT derivación *f* de bifurcación; ~ **jack** *n* ELECT *connecting devices*, PARTS conjuntor *m* general; ~ **point** *n* ELECT, TRANS, TV punto *m* de bifurcación
brand *n* GEN clase *f*; ~ **name** GEN nombre *m* comercial
brass *n* ELECT, PARTS bronce *m*
break *n* GEN derrame *m*; **~-in** TELEP entrada *f* forzada; ~ **contact group** SWG grupo *m* de contacto de ruptura; ~ **contact spring set** SWG juego *m* de resortes de contacto de ruptura; **~-in hangover time** TELEP tiempo *m* de mantenimiento de la interrupción; ~ **jack** SWG clavija *f* de ocupación; ~ **key** COMP tecla *f* de ocupación; **~-over voltage** PARTS *thyristors* tensión *f* de irrupción; ~ **pulse** TELEP impulso *m* de apertura; ~ **pulsing generator** TELEP generador *m* de impulsos de apertura; ~ **relay** ELECT, PARTS relé *m* de apertura
break in *vi* TELEP forzar la entrada
breakage *n* POWER *electric conductors* rotura *f*
breakdown *n* GEN *of order* perturbación *f*; ~

signal TELEP interrupción *f* de señal; ~ **voltage** ELECT voltaje *m* de ruptura
breakeven: ~ **point** *n* GEN punto *m* de equilibrio
breaking *n* GEN *of code* rotura *f*; ~ **length** PARTS, TEST longitud *f* de la ruptura; ~ **strength** PARTS, TEST resistencia *f* a la ruptura
breakout: ~ **box** *n* ELECT caja *f* de desconexión; ~ **cable** *n* ELECT cable *m* de desconexión; ~ **jacketing** *n* ELECT hermetización *f* de la desconexión
breakover: ~ **current** *n* POWER *thyristors* corriente *f* de irrupción; ~ **point** *n* PARTS *thyristors* punto *m* de irrupción
breakpoint *n* COMP punto *m* de interrupción
breakthrough *n* COMP, RADIO *land mobile* perforación *f*
breastplate: ~ **microphone** *n* PARTS micrófono *m* de solapa
breath *n* COMP intervalo *m*; ~ **control** COMP control *m* de intervalo; ~ **controller** COMP controlador *m* de intervalo; ~**-controller assignment** COMP asignación *f* de controlador de intervalo; ~**-controller range** COMP amplitud *f* de controlador de intervalo; ~**-controller sensitivity** COMP sensibilidad *f* del controlador de intervalo
Brewster: ~**'s angle** *n* TRANS *fibre optics* ángulo *m* de Brewster
bridge *n* COMP, DATA, ELECT puente *m*; ~ **circuit** ELECT, POWER circuito *m* en puente; ~ **connection** POWER acoplamiento *m* en derivación; ~ **control cord circuit** TELEP circuito *m* de verificación; ~ **converter** POWER convertidor *m* de puente; ~ **duplex system** TELEP sistema *m* telegráfico doble equilibrado en puente; ~ **router** DATA encaminador *m* en puente; ~ **splice** TRANS conexión *f* en puente; ~ **transformer** ELECT transformador *m* puente
bridged *adj* POWER, TELEP de puente
bright *adj* COMP luminoso, TRANS claro; ~**-field detector** *n* TRANS detector *m* de campo claro; ~**-field sensor** *n* TRANS sensor *m* de campo claro
brightening *n* COMP abrillantamiento *m*
brightness *n* COMP, TV luminosidad *f*; ~ **control** PER APP regulación *f* de luminosidad
brilliance *n* COMP luminosidad *f*; ~ **control** COMP, TV control *m* de luminosidad
Brillouin: ~ **spectrum** *n* DATA espectro *m* de Brillouin
Brinell: ~ **test** *n* PARTS *strength* prueba *f* de dureza Brinell
British: ~ **Standards Institution** *n* (*BSI*) GEN Instituto *m* Británico de Normas (*BSI*)
brittleness *n* PARTS *strength* fragilidad *f*
broadband *n* COMP, DATA, TELEP, TRANS banda *f* ancha; ~ **amplifier** TRANS amplificador *m* de banda ancha; ~ **channel** TRANS vía *f* de transmisión de banda ancha; ~ **crosspoint** SWG contacto *m* de cruce de banda ancha; ~ **ISDN** (*B-ISDN*) TRANS ISDN *m* de banda (*RDSI-BA*); ~ **ISDN PBX** SWG PBX *f* de RDSI de banda ancha; ~ **ISDN service** APPL, TELEP servicio *m* RDSI de banda ancha; ~ **noise** COMP ruido *m* de banda ancha; ~ **radio system** RADIO sistema *m* de radio de banda ancha; ~ **switch** SWG conmutador *m* de banda ancha; ~ **switching network** SWG, TRANS red *f* de conmutación de banda ancha; ~ **transmission** TRANS transmisión *f* de banda ancha
broadcast *n* (*BC*) GEN emisión *f*; ~ **channel** RADIO canal *m* de radiodifusión; ~ **message** RADIO mensaje *m* radiodifundido; ~ **network** RADIO red *f* de radiodifusión; ~ **programme** *BrE* **channel** RADIO canal *m* de programas de radiodifusión; ~ **reception** RADIO recepción *f* de emisión por radio; ~ **repeater** TELEP repetidor *m* de radiodifusión; ~ **signalling** *BrE* **virtual channel** (*BSVC*) TRANS canal *m* virtual de señalización de radiodifusión; ~ **station** RADIO estación *f* de radiodifusión; ~ **storms** *n pl* ELECT tormentas *f pl* de radiodifusión; ~ **transmitter** TELEP transmisor *m* de radiodifusión; ~ **videography** APPL videografía *f* para difusión de mensajes, TV videografía *f* de radiodifusión; ~ **videotext** APPL videotexto *m* para difusión de mensajes, TV videotexto *m* de radiodifusión
broadcaster *n* TV radiodifusor *m*
broadcasting *n* GEN emisión *f*; ~ **authority** RADIO autoridades *f pl* de la radiodifusión; ~ **band** RADIO banda *f* radioemisora; ~ **organization** RADIO organización *f* de radiodifusión; ~ **rights** *n pl* TV derechos *m pl* de radiodifusión; ~**-satellite service** RADIO servicio *m* de radiodifusión por satélite; ~**-satellite space station** RADIO estación *f* espacial de radiodifusión por satélite; ~ **schedule** RADIO horarios *m* de radiodifusión; ~ **signal** RADIO señal *f* de radiodifusión; ~ **spectrum** RADIO, TRANS espectro *m* de radiodifusión; ~ **station** RADIO radioemisora *f*
broadside: ~ **array** *n* RADIO red *f* de radiación transversal; ~ **array antenna** *n* RADIO antena *f* en array de radiación transversal
broken *adj* PARTS *conductor* averiado
broker *n* COMP comisionista *m*
B-router *n* DATA asignador *m* de ruta-B

browser *n* COMP *software* lector *m*, buscador *m*
browsing *n* COMP búsqueda *f*
brush *n* GEN escobilla *f*; ~ **discharge** POWER descarga *f* en penacho
BS *abbr* (*backspace*) COMP retroceso *m*
BSC *abbr* (*binary synchronous communication*) DATA comunicación *f* binaria sincrónica
BSI *abbr* (*British Standards Institution*) GEN BSI (*Instituto Británico de Normas*)
BSS *abbr* (*base station subsystem*) TELEP subsistema *m* de estación de base
BSVC *abbr* (*broadcast signalling virtual channel*) TRANS canal *m* virtual de señalización de radiodifusión
BT *abbr* (*document bulk transfer class*) TELEP BT (*clase de transferencia de documentos en masa*)
B3ZS *abbr* (*bipolar code with three-zero substitution*) DATA, SIG, TELEP B3ZS (*código bipolar con sustitución de tres ceros*)
BTM *abbr* (*document bulk transfer and manipulation class*) TELEP BTM (*clase de transferencia y manipulación de documentos en masa*)
B-tree *n* DATA árbol *m* B
B-type: ~ **conduction** *n* ELECT conducción *f* de tipo B
BU *abbr* (*base unit*) RADIO unidad *f* base
bucket: ~ **brigade device** *n* (*BBD*) SWG *shift register* circuitos *m pl* de noria
budget *vi* GEN presupuestar
buffer *n* COMP, TRANS separador *m*; ~ **amplifier** ELECT amplificador *m* compensador; ~ **battery** POWER batería *f* tampón; ~ **circuit** COMP, DATA circuito *m* de compensación; ~ **memory** COMP, DATA memoria *f* dinámica; ~ **state** ELECT estado *m* tampón; ~ **storage** COMP, DATA memoria *f* intermedia; ~ **time** ELECT tiempo *m* de memoria intermedia
buffering *n* DATA, ELECT almacenamiento *m* en memoria intermedia
bug *n* GEN defecto *m*
build in *vt* GEN integrar
builder *n* COMP constructor *m*
building: ~**-block principle** *n* SWG principio *m* del bloque funcional; ~ **element** *n* GEN *design technology* elemento *m* de construcción; ~ **module** *n* GEN *design technology* módulo *m* de construcción; ~ **penetration loss** *n* RADIO pérdida *f* de penetración en edificios
built: ~**-in check** *n* TEST comprobación *f* automática; ~**-in computer** *n* COMP ordenador *m* incorporado *Esp*, computador *m* incorporado *AmL*, computadora *f* incorporada *AmL*; ~**-in filter** *n* ELECT filtro *m* incorporado; ~**-in loudspeaker** *n* PER APP altavoz *m* incorporado; ~**-in speaker** *n* COMP altavoz *m* incorporado; ~**-in test equipment** *n* TELEP, TEST equipo *m* de prueba incorporado; ~**-in warble tone** *n* ELECT sonido *m* ululado incorporado; ~**-in warble-tone source** *n* ELECT fuente *f* de sonido ululado incorporado; ~ **slide** *n* COMP corredera *f* articulada
bulk *n* COMP volumen *m*; ~ **data** *n pl* COMP datos *m pl* en masa; ~ **density** GEN *supply & distribution* densidad *f* aparente; ~ **dump** COMP volcado *m* en masa; ~ **information** COMP información *f* en masa; ~ **modulus** PARTS *strength* módulo *m* de compresibilidad; ~ **piezoelectric permittivity** ELECT permitividad *f* piezoeléctrica masiva; ~ **sampling** TEST *quality* muestreo *m* a granel; ~ **storage** DATA almacenamiento *m* de gran capacidad; ~ **store** COMP memoria *f* de gran capacidad, memoria *f* de masa
bulkhead: ~ **feed-through adaptor** *n* ELECT adaptador *m* de tipo pasante de alimentación transversal; ~ **panel** *n* ELECT panel *m* mural
bulky *adj* PARTS voluminoso
BULL: ~ **cabling system** *n* (*BCS*) DATA sistema *m* de cableado BULL (*BCS*)
bullet *n* COMP punto *m* de adorno
bulletin *n* TELEP boletín *m*; ~ **board** (*BB*) COMP, DATA, TELEP boletín *m* electrónico (*BB*); ~ **board system** (*BBS*) COMP, DATA, TELEP sistema *m* de boletíns electrónicos (*BBS*)
bump *n* POWER *printed circuits* memoria *f* anexa; ~ **mapping** COMP planificación *f* de la memoria anexa; ~ **test** PARTS *strength* prueba *f* de sacudida
bunch *vt* TRANS agrupar
bunched: ~ **frame-alignment signal** *n* SIG, TRANS señal *f* de alineación de grupo de tramas; ~ **signalling** *BrE* *n* TRANS señalización *f* concentrada
bunching *n* TRANS *cables* agrupación *f*
bundle *vt* DATA concentrar
bundling *n* DATA paquete *m*
burglar: ~ **alarm** *n* TEST *security* alarma *f* contra robos
buried: ~ **cable** *n* GEN *outside plant engineering*, TRANS cable *m* enterrado
burn: ~**-in** *n* COMP, TEST *testing components* envejecimiento *m*
burner *n* GEN quemador *m*
burning *n* ELECT, TELEP *of contacts* calcinación *f*
burr *n* PARTS protuberancia *f*
Burrus: ~ **diode** *n* PARTS, TELEP, TRANS diodo *m* de Burrus
burst *n* COMP señal *f* de sincronismo de color; ~ **data section** DATA sector *m* de datos por

paquetes; ~ **duration** DATA duración *f* de la ráfaga; **~-envelope fall time** DATA tiempo *m* de secuencia de envolvente de ráfaga; **~-envelope rise time** DATA tiempo *m* de subida de envolvente de ráfaga; ~ **isochronous** TELEP isócrono *m* de la ráfaga; ~ **loss rate** DATA velocidad *f* de pérdida de la ráfaga; ~ **meteor communications** *n pl* TRANS comunicaciones *f pl* por ionización meteórica; ~ **mode** DATA, RADIO modo *m* de ráfaga; **~-mode link analyser** *BrE* DATA analizador *m* de eslabón del modo ráfaga; **~-mode tester** DATA comprobador *m* del modo ráfaga, TEST probador *m* por impulsos de ráfagas; ~ **noise** TRANS ruido *m* de interferencia; ~ **of tone** SIG, TELEP ráfaga *f* de tono; ~ **position control information** DATA información *f* por ráfagas del control de la posición; ~ **rate** DATA velocidad *f* de ráfaga; ~ **synchronization** DATA sincronización *f* de ráfagas; **~-time plan** DATA plan *m* de sincronización por ráfagas; **~-time plan change control** DATA control *m* del cambio del plan de sincronización por ráfagas; **~-timing acquired** DATA sincronización *f* por ráfagas adquirida; ~ **timing not acquired** DATA temporización *f* de ráfagas no adquirida; **~-timing synchronization** DATA sincronización *f* por ráfagas; ~ **transmission** RADIO, TRANS transmisión *f* en ráfagas; ◆ ~ **acquired** DATA captado por ráfagas

bus *n* COMP, POWER, TELEP barra *f* colectora; ~ **adaptation unit** SWG adaptador *m* de bus; ~ **arbitrator** COMP controlador *m* del bus; ~ **cable** PARTS *equipment practice* cable *m* de bus; ~ **cable chute** PARTS *equipment practice* canal *m* de cable de bus; ~ **configuration** COMP configuración *f* del sistema bus; ~ **connection signal** SWG señal *f* de conexión de bus; ~ **connection unit** SWG conector *m* de bus; ~ **driver** ELECT *logic elements* activador *m* de bus; ~ **network** TELEP terminador *m* de bus; ~ **state signal** SWG señal *f* de estado de bus; ~ **system** SWG sistema *m* de bus; ~ **topology** DATA topología *f* de bus

busbar *n* COMP, POWER, TELEP barra *f* de distribución

busied: ~ **out** *adj* TELEP bloqueado

business: ~ **application** *n* GEN utilización *f* comercial; ~ **card** *n* GEN tarjeta *f* comercial; ~ **communication system** *n* APPL, RADIO, TRANS sistema *m* de comunicaciones comerciales; ~ **communications** *n pl* GEN comunicaciones *f pl* comerciales; ~ **customer** *n* GEN cliente *m* comercial; **~-oriented satellite** *n* RADIO satélite *m* de orientación comercial; ~ **plan** *n* GEN plan *m* comercial; ~ **services** *n pl* GEN servicios *m pl* comerciales; ~ **subscriber** *n* RADIO abonado *m* comercial; ~ **system** *n* APPL, RADIO, TRANS sistema *m* comercial; ~ **television** *n* TV televisión *f* comercial; ~ **traffic** *n* TELEP tráfico *m* comercial; ~ **user** *n* GEN usuario *m* comercial

busy *adj* GEN ocupado, de máxima ocupación, de ocupación punta; ~ **hour** *n* TELEP hora *f* de carga; ~ **hour-to-day ratio** *n* TELEP razón *f* tráfico de hora punta al tráfico de 24 horas; ~ **lamp** *n* TELEP lámpara *f* dc ocupado; ~ **lamp field** *n* TELEP campo *m* de lámparas de ocupado; ~ **lamp panel** *n* TELEP cuadro *m* de lámparas de ocupado; ~ **number** *n* TELEP número *m* ocupado; ~ **occurrence** *n* TELEP situación *f* de ocupado; ~ **out** *n* TELEP interrupción *f* del servicio; ~ **override** *n* TELEP acceso *m* a una línea ocupada; ~ **period** *n* TELEP periodo *m* cargado; ~ **season** *n* TELEP temporada *f* de alta ocupación; ~ **signal** *n* GEN, TELEP señal *f* de ocupado; ~ **state** *n* TELEP estado *m* de ocupación; **~-state display** *n* TELEP cuadro *m* de lámparas de ocupado; ~ **status** *n* TELEP estado *m* de ocupación; ~ **study** *n* TELEP estudio *m* dc ocupación; ~ **test** *n* TELEP ensayo *m* de ocupación; ~ **tone** *n* TELEP señal *f* de ocupación

busy out *vt* SWG interrumpir el servicio

butt[1]: **~-configurated** *adj* ELECT configurado en tope; **~contact** *n* PARTS, SWG, TELEP contacto *m* de presión directa; **~joint** *n* GEN, TRANS junta *f* a tope

butt[2] *n* ELECT bisagra *f*

butted: **~-ribbon arrays** *n pl* ELECT grupos *m pl* de cinta empalmada

button *n* GEN botón *m*

buzzer *n* GEN zumbador *m*; ~ **tone** TELEP tono *m* de zumbido

buzzing: ~ **tone** *n* TELEP tono *m* de zumbido

BWO *abbr* (*backward wave oscillator*) TELEP oscilador *m* de onda de retorno

bypass *n* COMP derivación *f*; ~ **procedure** COMP procedimiento *m* de desvío

bypassing *n* POWER contorneo *m*

bypath *n* SWG senda *f*; ~ **system** SWG sistema *m* de senda

byte *n* COMP, DATA, ELECT, TELEP byte *m*; **~-alignment** COMP alineación *f* basada en octetos **~-serial transmission** TRANS transmisión *f* por secuencias de octetos; ~ **switch** SWG conmutador *m* de octetos

C

C: ~ **component** *n* DATA componente *m* C

C³ *abbr* (*C-cube*) COMP C³ (*cubo C*)

cab: ~ **and pillar distribution system** *n* TRANS sistema *m* de distribución de cabina y columna

cabinet *n* COMP, DATA, PARTS *equipment practice*, TRANS armario *m*; ~ **base** TRANS *equipment practice* zócalo *m* de armario; ~ **connection chute** SWG *equipment practice* canal *m* de conexión de armario; ~ **radiation** RADIO radiación *f* de armario

cable *n* GEN cable *m*; ~ **address** TELEP dirección *f* del cable; ~ **armour** *BrE* TRANS armadura *f* de cable; ~ **assembly** TRANS montaje *m* de cables; ~ **bend** TRANS curva *f* de cable; ~ **boat** TRANS buque *m* cablero; ~ **box** GEN caja *f* de empalme; ~ **bracket** TRANS *cable laying* soporte *m* de cable; ~ **branch** TRANS ramal *m* de cable; ~ **branching** TRANS ramificación *f* de cable; ~ **break** TRANS rotura *f* de cable; ~ **butt** TRANS extremo *m* de cable; ~ **cabinet** TRANS *cable laying* armario *m* de cables; ~ **cap** TRANS cubierta *f* de cable; ~ **capacity** TRANS capacidad *f* del cable; ~ **cellar** TRANS *cable laying* galería *f* de cables; ~ **chamber** TRANS cámara *f* de cables; ~ **chute** TRANS *cable laying* canal *m* de cables; ~ **clamp** ELECT, TRANS *cable laying* abrazadera *f* de cable; ~ **code** TELEP código *m* telegráfico; ~ **communications** *n pl* APPL comunicaciones *f pl* por cable; ~ **connection** TRANS conexión *f* de cables; ~ **connection board** (*CCB*) TRANS *equipment practice* placa *f* de conexión de cable (*CCB*); ~ **core** GEN alma *f* de cable; ~ **covering** TRANS cubierta *f* del cable; ~ **designation** TRANS designación *f* de cable; ~ **distribution** TRANS distribución *f* de cables; **~-distribution cabinet** POWER armario *m* de distribución de cables; **~-distribution diagram** TRANS *documentation* diagrama *m* de distribución de cables; **~-distribution head** TRANS cabeza *f* de cable, *cables* manga *f* de empalme de cables; ~ **distribution table** GEN *documentation* esquema *m* de distribución de cables; ~ **drawing** GEN *documentation* esquema *m* de cableado; ~ **drum** TRANS carrete *m* para enrollar cables; ~ **entry** TRANS entrada *f* de cables; ~ **factory** TRANS fábrica *f* de cables; ~ **floor** GEN piso *m* para cables; ~ **formboard** TRANS plantilla *f* para cables, *cables* horma *f* de cable; ~ **forming** TRANS cosido *m* de cable, formación *f* de cable; **~-forming board** TRANS plantilla *f* para formar cables; ~ **grip** GEN *outside plant engineering* fijador *m* de cable; ~ **group** GEN grupo *m* de cables; ~ **group list** TRANS *documentation* lista *f* de grupos de cables; ~ **head** TRANS extremo *m* de cable; ~ **holder** PARTS fijación *f* de cable; ~ **hole** TRANS agujero *m* de cable; ~ **inlet** TRANS entrada *f* de cables; ~ **jacket** TRANS cubierta *f* de cable; ~ **joining** TRANS empalme *m* de cables; ~ **joint** TRANS empalme *m* de cables; ~ **jointer** TRANS empalmador *m* de cables; ~ **lacing** TRANS costura *f* de cables; ~ **landing** TRANS puesta *f* en tierra del cable; ~ **lasher** TRANS suspensor *m* de cable; ~ **layer** TRANS tendedor *m* de cables; ~ **laying** TRANS *outside plant engineering* tendido *f* de cables; **~-laying ship** TRANS buque *m* para tendido de cables submarinos; **~-laying winch** TRANS *outside plant engineering* torno *m* para tendido de cables; ~ **layout** TRANS tendido *m* de cables; ~ **length** TRANS longitud *f* de cable; ~ **list** TRANS *documentation* lista *f* de cables; ~ **localization** TRANS localización *f* de cables; ~ **locater** TRANS *outside plant engineering* localizador *m* de cables; ~ **loss** TRANS *CATV* pérdida *f* del cable; ~ **lug** TRANS asiento *m* de cable; ~ **mounting drawing** TRANS *documentation* esquema *m* de montaje de cables; ~ **network** TRANS, TV red *f* de cables; ~ **network operating company** TV operador *m* de red de cable; ~ **network operator** TV operador *m* de red de cable; ~ **operator** TRANS operador *m* de cable; ~ **outlet** TRANS *equipment practice* terminal *m* de cable; ~ **pair** TRANS par *m* de cables; ~ **pitch** TRANS paso *m* helicoidad del cable; ~ **plough** *BrE* GEN *outside plant engineering* excavadora *f* de zanjas para cables; ~ **pressurization** TRANS presurización *f* de cables; ~ **pressurization equipment** GEN equipo *m* de presurización de cables; ~ **programme** *BrE* TELEP programa *m* de cable; ~ **pulling** GEN *outside plant engineering* tirando *m* de cables; ~ **quad** TRANS cuadrete *m* de cable; ~ **raceway** TRANS *cable laying* conducto *m* de cables; ~ **rack** TRANS *cable laying* bastidor *m* de cables; ~ **rate** TELEP velocidad *f* del cable;

~ **reel** TRANS carrete *m* de cable; ~ **response** TRANS respuesta *f* del cable; ~ **riser** TRANS *cable laying* conducto *m* ascendente de cables; ~ **route** TRANS *cable laying* recorrido *m* de cables, tramo *m* de cables; ~ **run** TRANS *cable laying* ruta *f* del cable; ~ **running** TRANS *cable laying* tendido *m* de cables; ~**-run rung** TRANS travesaño *m* de tramo de cable; ~ **runway** TRANS conducto *m* ascendente de cables, *cable laying* tendido *m* ascendente de cables; ~ **scheme** TRANS esquema *m* de cableado; ~ **section** TRANS *CATV* sección *f* de cable; ~ **set** TRANS cableado *m*; ~**-set table** GEN *documentation* tabla *f* de cableado; ~ **sheath** TRANS cubierta *f* de cable; ~ **shelf** TRANS bandeja *f* de cable; ~ **ship** TRANS buque *m* cablero; ~ **sleeve** TRANS cabeza *f* de cable, manguito *m* de empalme de cable; ~ **socket** TV enchufe *m* de anclaje; ~ **spacer** TRANS separador *m* de cables; ~ **splice** TRANS empalme *m* de cables; ~**-storage hold** TRANS bodega *f* de almacenamiento de cables; ~ **support** TRANS apoyo *m* de cable; ~ **system** TRANS sistema *m* de cables; ~ **tag** TRANS etiqueta *f* de cable; ~ **tank** TRANS bodega *f* para cable; ~ **telephony** SIG telefonía *f* por cable; ~ **television** TV televisión *f* por cable; ~ **television network** TV red *f* de televisión por cable; ~ **television operator** TV operador *m* de televisión por cable; ~ **television system** TV sistema *m* de televisión por cable; ~ **terminal drawing** TRANS *documentation* esquema *m* de terminal de cables; ~ **termination** TRANS extremo *m* de cable; ~ **tester** TEST, TRANS probador *m* de cables; ~ **ties** *n pl* PARTS anclajes *m pl* para cable; ~ **trailer** TRANS remolque *m* portacables; ~ **transfer** TELEP giro *m* cablegráfico; ~ **trimmer** TRANS ajustador *m* de cables; ~ **trough** TRANS canalización *f* para cables; ~ **trunk** TRANS tronco *m* de cables; ~ **TV** TV TV *f* por cable; ~ **vault** TRANS galería *f* de cables; ~ **way** TRANS trayecto *m* de cables; ~**-way drawing** GEN *documentation* esquema *m* del recorrido de cables; ~ **winch** TRANS torno *m* de cables

cabled: ~ **distribution** *n* TV distribución *f* cableada

cabling *n* PARTS, TRANS cableado *m*, cableaje *m*; ~ **duct** PARTS, TRANS conducto *m* de cableado; ~ **element** PARTS, TRANS elemento *m* de cableado; ~ **plan** GEN *documentation* plano *m* de cableado; ~ **table** GEN *documentation* gráfico *m* de cableado; ~ **unit** PARTS, TRANS unidad *f* de cableado

cache: ~ **memory** *n* COMP, DATA memoria *f* intermedia; ~ **storage** *n* COMP, DATA memoria *f* de acceso rápido; ~ **store** *n* COMP, DATA memoria *f* caché

caching *n* COMP, DATA utilización *f* de la antememoria

CAD *abbr* (*computer-aided design*) COMP CAD (*diseño asistido por ordenador, diseño asistido por computador*)

CAD/CAM *abbr* (*computer-aided design and computer-aided manufacturing*) COMP CAD/CAM (*diseño y fabricación asistidos por ordenador*); ~ **system engineer** *n* COMP ingeniero *m* de sistemas CAD/CAM

cadence *n* RADIO cadencia *f*

CAE *abbr* (*computer-aided engineering*) COMP CAE (*ingeniería asistida por ordenador*)

caesura *n* DATA cesura *f*

CAL *abbr* (*computer-assisted learning*) COMP CAL (*aprendizaje auxiliada por computadora*)

calculate *vt* GEN calcular

calculation *n* GEN cálculo *m*; ~ **error** GEN error *m* de cálculo

calibrate *vt* TEST calibrar

calibrated: ~ **attenuator** *n* TRANS atenuador *m* calibrado

calibration *n* TEST calibrado *m*; ~ **centre** *BrE* TEST centro *m* de calibrado; ~ **certificate** TEST certificado *m* de calibrado; ~ **error** TEST error *m* de calibrado

call[1] *n* GEN llamada *f*; ~ **alignment** COMP alineación *f* de llamadas; ~**-accepted condition** DATA, TELEP condición *f* de llamada aceptada; ~**-accepted message** DATA, TELEP mensaje *m* de llamada aceptada; ~**-accepted signal** SIG señal *f* de llamada aceptada; ~**-back queue** TELEP cola *f* de espera de retrodemanda; ~ **barring** DATA, TELEP interdicción *f* de llamadas; ~**-barring equipment** TELEP equipo *m* de interdicción de llamadas; ~**-binding posts** *n pl* TRANS terminales *m pl* de enlace; ~ **buffer** TELEP memoria *f* intermedia de llamadas; ~**-by-~ service** APPL servicio *m* llamada por llamada; ~ **charge** APPL tarifa *f* de llamada; ~**-charge context** TELEP contexto *m* de tarificación de llamadas; ~**-charge counter** APPL contador *m* tarificador de llamadas; ~**-charge equipment** APPL equipo *m* tarificador de llamadas; ~**-charge rate** APPL, TELEP tarifa *f* de llamada; ~**charging** APPL tarificación *f* de llamadas; ~**-charging analysis** APPL análisis *m* de tarificación de llamadas; ~ **charging and observation** TELEP tarificación *m* y observación de llamadas; ~**-charging equipment** APPL equipos *m pl* de tarificación del lamadas;

~clearing COMP desconexión *f* de llamada, TELEP desconexión *f* de llamadas; **~ collision** TELEP colisión *f* de llamadas; **~ collision at the DTE/DCE interface** TELEP colisión *f* de llamadas en el interfaz DTE/DCE; **~ congestion** TELEP congestión *f* de llamadas; **~ congestion loss** TELEP pérdida *f* por congestión de llamadas; **~ congestion probability of loss** TELEP probabilidad *f* de pérdida por congestión de llamadas; **~-connected signal** SIG señal *f* de conexión; **~ connection** TELEP conexión *f* de llamada; **~ control** TELEP control *m* de llamadas; **~ control agent** DATA agente *m* de control de llamadas; **~ control character** TELEP carácter *m* de control de llamadas; **~ control procedure** TELEP procedimiento *m* de control de llamada; **~ control signals** *n pl* SIG señales *f pl* de control de llamada; **~ counter** TELEP contador *m* de llamadas; **~ data recording** APPL registro *m* de llamadas; **~ deflection** TELEP desvío *f* de llamadas; **~ destination** TELEP destino *m* de las llamadas; **~-detail recording service** APPL servicio *m* de registro detallado de llamadas; **~-detail recording system** TRANS sistema *m* de registro detallado de llamadas; **~-detail report** TELEP informe *m* detallado de llamadas; **~ discrimination** TELEP discriminación *f* de llamadas; **~ display** TELEP indicador *m* de llamada; **~ distributor** TELEP distribuidor *m* de llamadas; **~ diversion** TELEP desvío *m* de llamadas; **~ diversion on busy** TELEP desvío *m* de llamada por ocupado; **~ diversion on no reply** TELEP desvío *m* de llamada por falta de respuesta; **~ diverter** TELEP desviador *m* de llamadas; **~ duration** PER, TELEP duración *f* de llamada; **~ establishment** TELEP establecimiento *m* de llamadas; **~ extending** TELEP conexión *f* de llamadas; **~ failure** TELEP fracaso *m* de la llamada; **~ failure signal** SIG señal *f* de fracaso de llamada; **~ fee** TELEP importe *m* de la llamada; **~ filtering** TELEP filtrado *m* de llamadas; **~ finder** TELEP buscador *m* de llamada; **~ flow** TELEP flujo *m* de llamadas; **~ for bids** GEN convocatoria *f* a licitación; **~ for offers** GEN solicitud *f* de ofertas; **~ forward** TELEP desvío *m* de llamadas; **~ forward all ~s** TELEP desvío *m* de todas las llamadas; **~ forward all ~s programmable** TELEP desvío *m* programable de todas las llamadas; **~ forward busy** TELEP desvío *m* de llamadas ocupado; **~ forward intra-group** TELEP desvío *m* de llamadas intragrupal; **~ forward no answer** TELEP desvío *m* de llamadas sin respuesta; **~ forward reason display** TELEP indicación *f* de motivo de desvío de llamada; **~ forward remote** TELEP desvío *m* de llamadas remoto; **~ forwarder** TELEP desviador *m* de llamadas; **~ forwarding** TELEP desvío *m* de llamadas; **~-forwarding all ~s** TELEP desvío *m* de todas las llamadas; **~-forwarding busy** TELEP desvío *m* de llamadas ocupado; **~-forwarding may occur indicator** DATA indicador *m* de posible desvío de llamada; **~-forwarding no answer** TELEP desvío *m* de llamadas sin respuesta; **~-forwarding remote** TELEP desvío *m* de llamadas remoto; **~ handling** TELEP tratamiento *m* de llamadas; **~-handling area** TELEP zona *f* de control de llamadas; **~-handling software machine** TELEP máquina *f* de software de control de llamadas; **~ hold** TELEP llamada *f* retenida; **~ holding** TELEP retención *f* de llamadas; **~-holding facility** TELEP función *f* de retención de llamadas; **~ identification** DATA, TELEP identificación *f* de llamada; **~-identification line** TELEP línea *f* de identificación de llamadas; **~ identifier** TELEP identificador *m* de llamadas; **~ identity** DATA, TELEP identidad *f* de llamada; **~ in progress** TELEP comunicación *f* en curso, llamada *f* en curso; **~-in-progress cost information** APPL información *f* del precio de la llamada en curso; **~-in-progress switching** RADIO conmutación *f* de llamada en curso; **~-in software** TELEP programa *m* de llamadas en curso; **~-indicating device** TELEP dispositivo *m* indicador de llamada; **~ indicator** TELEP indicador *m* de llamada; **~ information** TELEP información *f* de la llamada; **~-information service signal** SIG señal *f* de servicio de información de llamadas; **~ initiation** TELEP inicio *m* de la llamada; **~ initiator** DATA iniciador *m* de llamada; **~ intensity** TELEP intensidad *f* de llamadas; **~ intent** TELEP intento *m* de llamada; **~ interception** TELEP intercepción *f* de llamada; **~ log** RADIO, TELEP registro *m* de llamadas; **~ logging** RADIO, TELEP registro *m* del mensaje; **~ memory** RADIO memoria *f* de llamadas; **~ meter** TELEP medidor *m* de llamadas; **~ metering** TELEP medida *f* de llamadas; **~-metering equipment** APPL equipo *m* de medida de llamadas; **~-metering store** APPL memoria *f* de medida de llamadas; **~ minute counter** APPL contador *m* de llamadas por minutos, TELEP contador *m* de minutos de conversación; **~-modification completed message** TELEP mensaje *m* completo de modificación de llamadas; **~-modification reject message**

(*CMRJ*), TELEP mensaje *m* de rechazo de modificación de llamadas; **~-modification request** TELEP solicitud *f* de modificación de llamadas; **~ not accepted** TELEP llamada *f* no aceptada; **~ origin** TELEP origen *m* de llamada; **~ origination** RADIO origen *m* de llamadas; **~ originator** RADIO productor *m* de llamadas; **~ park** TELEP retención *f* de llamada; **~ phases** *n pl* TELEP etapas *f pl* de la llamada; **~ pick-up** TELEP captación *f* de llamadas, TELEP intervención *f* de llamadas; **~ pick-up group** TELEP grupo *m* de captación de llamadas; **~ processing** TELEP procesamiento *m* de llamadas; **~ processor** TELEP procesador *m* de llamadas; **~ progress** TELEP progresión *f* de la llamada; **~ progress signal** TELEP señal *f* de progresión de la llamada; **~ progress signal logical channel** TELEP canal *m* lógico de señal de progresión de la llamada; **~ progress tone** TELEP tono *m* de progresión de la llamada; **~ queue** TELEP cola *f* de llamadas; **~ queueing** TELEP colocación *f* en cola de llamadas; **~-queueing facility** TELEP sistema *m* de puesta en fila de espera de llamadas; **~-queueing system** TELEP sistema *m* de cola de espera de las llamadas; **~ recorder** TELEP registrador *m* de llamadas; **~ reference** TELEP referencia *f* de llamada; **~ register** TELEP registro *m* de llamadas; **~-rejected message** DATA, TELEP mensaje *m* de rechazo de llamada; **~-release time** TELEP tiempo *m* de desconexión de llamada; **~ released** DATA llamada *f* desconectada; **~ request** DATA solicitud *f* de llamada, TELEP petición *f* de llamadas; **~ request packet** TELEP paquete *m* de petición de llamadas; **~ request signal** SIG señal *f* de petición de llamada; **~ restriction** TELEP restricción *f* de llamadas; **~ retrieval** TELEP recuperación *f* de llamada; **~ routine sequencer** TELEP secuenciador *m* de rutina de llamada; **~ screening** TELEP filtrado *m* de lamadas; **~ security** TELEP seguridad *f* de las llamadas; **~ selection** TELEP selección *f* de las llamadas; **~-selector responder** TELEP contestador *m* de selector de llamada; **~ sender** TEST emisor *m* de llamada; **~ sequence** TELEP secuencia *f* de llamada; **~ sequencer** TELEP secuenciador *m* de llamadas; **~ setup** GEN establecimiento *m* de llamadas; **~ setup control signal** DATA señal *f* de control de establecimiento de llamada; **~ setup delay** TELEP retardo *m* en el establecimiento de llamadas; **~ setup packet** COMP paquete *m* de establecimiento de llamada; **~ setup phase** GEN fase *f* de establecimento de llamadas; **~ setup time** GEN tiempo *m* de establecimiento de llamadas; **~ sign** RADIO, TELEP distintivo *m* de llamada; **~ signal** TELEP señal *f* de llamada; **~ simulator** TELEP, TEST simulador *m* de llamadas; **~ store** SWG memoria *f* de llamadas; **~ string** TELEP cadena *f* de llamadas; **~ success rate** TEST índice *m* de éxito de llamadas; **~ trace** DATA, TELEP localizador *m* de llamada; **~ transfer** TELEP transferencia *f* de llamada; **~ transfer on "busy"** TELEP transferencia *f* de llamada por "ocupado"; **~ transfer on "ringing tone no reply"** TELEP transferencia *f* de llamada por "tono de sin respuesta"; **~waiting** TELEP llamada *f* en espera; **~-waiting indication** TELEP indicación *f* de llamada en espera; **~-waiting services** *n pl* TELEP servicios *m pl* de llamada en espera; **~-waiting tone** TELEP tono *m* de llamada en espera; **~ wire** TELEP hilo *m* de llamada

call[2] *vt* GEN llamar; ◆ **~ collect** *AmE* (*cf reverse the charges BrE*) APPL, TELEP hacer una llamada a cobro revertido; **~ over the intercom** TELEP llamar por el teléfono interno; **~ sb back** TELEP devolver la llamada a alguien; **~ toll-free** TELEP llamar sin cargo

call back *vt* TELEP rellamar

callback *n* TELEP servicio *m* repetido

called: **~ exchange** *n* TELEP central *f* de destino; **~-line charging** *n* APPL cargo *m* al abonado llamado; **~-line identification** *n* (*CDLI*) TELEP identificación *f* de la línea llamada; **~-line identification request indicator** *n* TELEP indicador *m* de petición de identificación de la línea llamada; **~-line identification signal** *n* SIG señal *f* de identificación de la línea llamada; **~-line identity** *n* TELEP identidad *f* de la línea llamada; **~-line identity indicator** *n* TELEP indicador *m* de identidad de la línea llamada; **~ N address** *n* DATA dirección *f* del número llamado; **~ network service user** *n* DATA usuario *m* de servicio de red llamado; **~ NS user** *n* DATA usuario *m* de servicio de red llamado; **~ number** *n* TELEP número *m* llamado; **~-number display** *n* TELEP presentación *f* visual del número llamado; **~-party release** *n* SWG, TELEP desconexión *f* por abonado llamado; **~-station identity** *n* DATA, RADIO, TELEP identidad *f* de la estación llamada; **~-subscriber identification** *n* (*CSI*) TELEP identificación *f* del abonado llamado; **~-subscriber identification-DIS** *n* (*CSI-DIS*) TELEP identificación *f* DIS del abonado llamado (*CSI-DIS*); **~-subscriber release** *n* SWG, TELEP desconexión *f* del abonado llamado, liberación *f* del abonado llamado; ~ **telephone** *n* TELEP teléfono *m* llamado; ~ **terminal** *n* TELEP terminal *m* llamada

caller *n* TELEP abonado *m* que llama; ~ **display** TELEP indicador *m* del abonado que llama

calling *n* DATA *datacom* marcación *f*, TELEP, *for bids* llamada *f*; ~ **card** TELEP tarjeta *f* de llamada; ~ **channel** RADIO *mobicom* canal *m* de llamadas; ~ **device** TELEP órgano *m* de llamada; ~ **indicator** PER APP indicador *m* de llamadas; ~ **into execution** TELEP *of subroutine* llamada *f* en curso; ~ **lamp** PER APP lámpara *f* de llamada; ~**-line identification** TELEP identificación *f* de la línea que llama; ~**-line identification display** (*CLID*) TELEP representación *f* visual de identificación de la línea que llama; ~**-line identification facility** TELEP sistema *m* de identificación de la línea que llama; ~**-line identification presentation** (*CLIP*) TELEP presentación *f* de la identificación de la línea que llama (*CLIP*); ~**-line identification restriction** (*CLIR*) TELEP restricción *f* de identificación de la línea que llama; ~**-line identification signal** SIG señal *f* de identificación de la línea que llama; ~**-line identity** TELEP identidad *f* de la línea que llama; ~**-line indicator** (*CLI*) DATA, TELEP indicador *m* de la línea que llama; ~ **modem** DATA módem *m* que llama; ~ **N address** DATA domicilio *m* del número que llama; ~ **network-service user** DATA usuario *m* de la red de servicio que llama; ~ **NS user** DATA usuario *m* de servicio de red que llama; ~**-number identification** DATA, RADIO, TELEP identificación *f* del número que llama; ~**-number indication** TELEP indicación *f* del número que llama; ~**-party address** TELEP domicilio *m* del abonado que llama; ~**-party category indicator** TELEP indicador *m* de categoría del abonado que llama; ~**-party category request indicator** TELEP indicador *m* solicitando la categoría del abonado que llama; ~**-party category response indicator** TELEP indicador *m* contestando la categoría del abonado que llama; ~**-party data** *n pl* TELEP datos *m pl* del abonado que llama; ~**-party hold** TELEP *on answering machine* retención *f* del abonado que llama; ~**-party number incomplete** TELEP número *m* incompleto del abonado que llama; ~**-party release** SWG, TELEP desconexión *f* por el abonado que llama; ~ **rate** TELEP densidad *f* de llamadas; ~ **sequence** TELEP secuencia *f* de llamada; ~ **signal** TELEP señal *f* de llamada; ~ **station** TELEP estación *f* que llama; ~**-subscriber release** SWG, TELEP desconexión *f* de abonado que llama; ~**-subscriber identification** (*CIG*) TELEP identificación *f* de abonado que llama (*CIG*); ~**-subscriber identification-DTC** (*CIG-DTC*) TELEP identificación *f* DTC del abonado que llama (*CIG-DTC*); ~**-subscriber release** SWG, TELEP desconexión *f* por el abonado que llama; ~ **telephone** TELEP teléfono *m* que llama; ~ **terminal** TELEP terminal *m* que llama; ~ **tree** DATA conjunto de circuitos de llamada

CAM *abbr* (*computer-aided manufacturing*) COMP CAM (*fabricación asistida por ordenador*)

CAMC *abbr* (*customer-access maintenance centre*) *BrE* TEST centro *m* de mantenimiento abierto al cliente

camera *n* GEN cámara *f*

camp: ~**-on** *n* TELEP conexión *f* en espera

can *n* ELECT *semiconductors* lata *f*

cancel *vt* GEN cancelar

cancellation *n* GEN cancelación *f*; ~**-completed signal** SIG señal *f* de cancelación completada; ~ **ratio** RADIO coeficiente *m* de cancelación; ~**-request signal** SIG señal *f* de petición de cancelación

candidate: ~ **MSC** *n* RADIO centro *m* candidato de conmutación de mensajes

canning *n* POWER envainado *m*

capability *n* TEST aptitud *f*, capacidad *f*; ~ **approval** TEST *quality* aprobación *f* de aptitud; ~ **test** DATA prueba *f* de aptitud

capacitance *n* ELECT capacitancia *f*; ~ **diode** ELECT *semiconductors*, PARTS diodo *m* de capacidad variable; ~ **standard** ELECT patrón *m* de capacitancia

capacitive *adj* ELECT capacitivo; ~ **coupling** *n* ELECT acoplamiento *m* capacitivo; ~ **load** *n* ELECT carga *f* capacitiva

capacitor *n* ELECT, POWER capacitor *m*; ~ **microphone** TELEP micrófono *m* electrostático; ~ **plate** POWER placa *f* de capacitor

capacity *n* GEN *quantity* capacidad *f*; ~ **decrease** DATA disminución *f* de capacidad; ~ **imbalance** TELEP desequilibrio *m* de capacidad; ~ **increase** DATA aumento *m* de capacidad; ~ **shortage** GEN escasez *f* de capacidad; ~ **top** RADIO *aerials* tope *m* de capacidad

capillary: ~ **soldering** *n* PARTS soldadura *f* capilar

capsule *n* GEN cápsula *f*

captive: ~ **screw** *n* PARTS tornillo *m* cautivo

capture *n* GEN captación *f*; ~ **area** RADIO área *f* de captación; ~ **effect** ELECT efecto *m* de captación; ~ **frequency** ELECT frecuencia *f* de captación

CAQ *abbr* (*computer-aided quality*) COMP CAQ (*calidad asistida por ordenador*)

car: ~ **installation kit** *n* RADIO *mobicom* equipo *m* para instalar en automóviles; ~ **phone** *n* DATA, RADIO, TELEP teléfono *m* de coche; ~ **telephone** *n* DATA, RADIO, TELEP teléfono *m* de coche

carbon: ~**-film resistor** *n* ELECT, PARTS resistor *m* de película de carbón; ~**-granule chamber** *n* PARTS *microphones* cámara *f* de gránulos de carbón; ~ **granules** *n pl* PARTS *microphones* gránulos *m pl* de carbón; ~ **lightning arrester** *n* POWER pararrayos *m* de carbón; ~ **microphone** *n* PARTS micrófono *m* de carbón; ~ **microphone current feed** *n* POWER, TELEP alimentación *f* de micrófono de carbón

carcinotron *n* ELECT carcinotrón *m*

card *n* GEN tarjeta *f*; ~ **edge** ELECT *printed circuits* borde *m* de la ficha; ~ **follow-on** PER APP *payphone* sustitución *f* de tarjeta; ~ **holder** TELEP caja *f* para fichas; ~ **hopper** COMP alimentador *m* de tarjetas; ~**-operated payphone** TELEP teléfono *m* de pago con tarjeta; ~ **path** COMP vía *f* de tarjetas; ~ **reader** APPL, COMP, PER APP, TV lector *m* de tarjetas

cardphone *n* TELEP teléfono *m* de pago con tarjeta

carriage: ~**-control tape** *n* PER APP *printers* cinta *f* de control de carro; ~ **return character** *n* GEN carácter *m* de retorno del carro

carrier *n* GEN portadora *f*; ~ **circuit** TRANS circuito *m* de ondas portadoras; ~ **component** ELECT componente *m* de ondas portadoras; ~**-current connection** TRANS conexión *f* por corrientes portadoras; ~**-current device** TRANS dispositivo *m* de corrientes portadoras; ~**-current telegraphy** TELEP telegrafía *f* de corrientes portadoras; ~ **detect** TRANS localización *f* de onda portadora; ~ **frequency** TRANS frecuencia *f* de ondas portadoras; ~**-frequency connection** TRANS conexión *f* en frecuencia portadora; ~**-frequency offset** TRANS desplazamiento *m* de frecuencia de la onda portadora; ~**-frequency terminal** TRANS terminal *m* de la frecuencia portadora; ~ **group alarm** TRANS alarma *f* del grupo de portadoras; ~**-interference ratio** TRANS relación *f* de interferencia de la onda portadora; ~**-link interference** TRANS interferencia *f* del enlace de la línea portadora; ~**-operated switch** RADIO conmutador *m* operado por onda portadora; ~ **power** RADIO potencia *f* de la portadora; ~ **power requirement** TRANS necesidad *f* de potencia de la portadora; ~**-to-receiver thermal noise ratio** TRANS relación *f* portadora/ruido térmico del receptor; ~ **rejection** TRANS supresión *f* de la portadora; ~**-sense multiple access** (*CSMA*) DATA acceso *m* múltiple de detección de la portadora; ~**-sense signal** DATA señal *f* de detección de la portadora; ~**-sense system** DATA sistema *m* de detección de la portadora; ~ **separation** TRANS separación *f* entre portadoras; ~ **signal** COMP, TRANS señal *f* portadora; ~ **storage** ELECT, PARTS *semiconductors* acumulación *f* de portadoras; ~ **storage time** ELECT, PARTS *semiconductors* tiempo *m* de acumulación de portadores; ~ **supply** TRANS generador *m* de portadoras; ~ **suppression** TRANS supresión *f* de la portadora; ~**-to-interference ratio** (*C/I*) TRANS relación *f* portadora/interferencia; ~**-to-intermodulation noise density ratio** TRANS relación *f* onda portadora-densidad del ruido intermodulación; ~**-to-intermodulation noise power ratio** TRANS relación *f* onda portadora/potencia de ruido de intermodulación; ~**-to-noise density ratio** TRANS relación *f* onda portadora-densidad del ruido; ~**-to-thermal noise power ratio** TRANS relación *f* portadora/potencia del ruido térmico; ~**-to-total-noise-power ratio** TRANS relación *f* portadora/potencia total del ruido; ~ **transmission** TRANS transmisión *f* por corriente portadora; ~ **wave** TRANS onda *f* portadora; ~**-wave indication supervision** TRANS supervisión *f* de la onda portadora

carry *vt* GEN transportar

carrying *n* DATA, TELEP, TV transporte *m*; ~ **case** GEN estuche *m* de protección; ~ **strap** GEN banda *f* protectora

cartridge *n* COMP, PARTS, POWER cartucho *m*; ~ **box** COMP cartuchera *f*; ~ **fuse** PARTS, POWER fusible *m* de cartucho; ~ **loading** COMP carga *f* de cartucho; ~ **save** COMP protección *f* de cartucho

CAS *abbr* (*channel-associated signalling*) *BrE* SIG, TELEP, TRANS señalización *f* asociada al canal

cascade *n* GEN cascada *f*; ~ **coupling** ELECT acoplamiento *m* capacitivo; ~ **load** ELECT carga *f* capacitiva

cascading: ~ **windows** *n pl* COMP ventanas *f pl* en cascada

case *n* GEN caja; ~**-based reasoning** DATA razonamiento *m* basado en casos prácticos; ~ **capacitance** ELECT capacitancia *f* del encapsulado; ~**-hardened steel** PARTS acero *m* de cementación; **~holder** PARTS sujetador *m* de cubierta; **~ study** GEN estudio *m* de caso práctico; **~ support** TELEP soporte *m* de cubierta; **~ temperature** ELECT temperatura *f* de encapsulado

CASE *abbr* (*computer-aided software engineering*) COMP ingeniería *f* de soporte lógico auxiliada por computador *AmL*, ingeniería *f* de soporte lógico auxiliada por computadora *AmL*, ingeniería *f* de soporte lógico auxiliada por ordenador *Esp*

cash: **~-box** *n* GEN caja *f*; **~ management** *n* GEN gestión *f* de tesorería

casing *n* COMP, PARTS, TELEP marco *m*

Cassegrain: **~ antenna** *n* RADIO antena *f* Cassegrain

cassette *n* GEN casete *f*; **~ deck** GEN puente *m* de casete; **~ drive** COMP unidad *f* de casete; **~ interface** COMP interfaz *f* de casete; **~ save** COMP protección *f* de casete

CAT *abbr* (*computer-aided testing*) COMP CAT (*prueba asistida por ordenador*)

catalogue *BrE* (*AmE* **catalog**) *n* GEN catálogo *m*

catastrophic: **~ degradation** *n* TELEP *fibre optics* degradación *f* catastrófica; **~ failure** *n* TEST fallo *m* catastrófico

catching: **~-up** *n* GEN actualización *f*

catchment: **~ area** *n* RADIO, TV zona *f* de captación

category *n* GEN categoría *f*, clase *f*; **~ analyser** *BrE* TELEP analizador *m* de categorías; **~ marking** TELEP marcado *m* de categorías; **~ store** TELEP memoria *f* de categorías; **~ transfer** TELEP transferencia *f* de categorías

catenate *vt* COMP catenular

cater for *vt* TELEP *consumer needs* atender a

cathode *n* ELECT, PARTS cátodo *m*; **~-covering current** ELECT corriente *f* de revestimiento catódico; **~ follower** ELECT seguidor *m* de cátodo; **~-ray direction finder** RADIO goniómetro *m* de rayos catódicos; **~-ray screen** ELECT pantalla *f* de rayos catódicos; **~-ray tube** (*CRT*) ELECT, PARTS tubo *m* de rayos catódicos (*TRC*); **~ terminal** ELECT *semiconductors* terminal *m* de cátodo

cathodic: **~ protection** *n* PARTS protección *f* catódica

CATV *abbr* (*community-antenna television*) TV televisión *f* con antena comunitaria; **~ operator** *n* TV operador *m* de CATV; **~ system** *n* TV sistema *m* de CATV

CAU *abbr* (*controlled-access unit*) TELEP CAU (*unidad de acceso controlado*)

caulk *vt* GEN calafatear

caulking *n* GEN calafateo *m*

causality *n* DATA causalidad *f*

cause: **~ of trouble** *n* TEST *dependability* causa *f* de la alteración; **~ value** *n* GEN valor *m* de causa

cavity *n* ELECT, TRANS *waveguides* cavidad *f*; **~ resonator** ELECT, TRANS *waveguides* resonador *m* de cavidad

CB *abbr* (*citizens' band*) RADIO banda *f* ciudadana

C-band *n* RADIO banda *f* C

CBO: **~ stream** *n* DATA corriente *f* CBO

CBT *abbr* (*computer-based training*) COMP CBT (*formación basada en ordenadores*)

CC *abbr* (*concatenated coding*) DATA código *m* encadenado

CCB *abbr* (*cable connection board*) TRANS CCB (*placa de conexión de cable*)

CCD *abbr* (*charge-coupled device*) ELECT DAC (*dispositivo acoplado por carga*); **~ detector** *n* PER APP *scanner* detector *m* de dispositivo acoplada por carga

CCITT: **~ high-level language** *n* TELEP lenguaje *m* CCITT de alto nivel

CCR *abbr* (*commitment, concurrency and recovery*) DATA CCR (*servicio de red de compromisos, acuerdos y recuperación*)

CCTV *abbr* (*closed-circuit television*) TV televisión *f* en circuito cerrado

C-cube *n* (C^3) COMP cubo *m* C (C^3)

CD *abbr* (*compact disc*) COMP CD (*disco compacto*); **~ reader** *n* COMP lector *m* de discos compactos

CDB *abbr* (*command-descriptor block*) DATA CDB (*bloque descriptor de orden*)

CDE *abbr* (*convolutionally and differentially encoded*) DATA CDE (*codificado convolucional y diferencialmente*)

cdf *abbr* (*cumulative distribution function*) DATA cdf (*función de distribución acumulativa*)

CD-I *abbr* (*compact disc interactive*) COMP CD-I (*disco compacto interactivo*); **~ audio sector** *n* COMP sector *m* de audio CD-I; **~ data** *n pl* COMP datos *m pl* CD-I; **~ sector** *n* COMP sector *m* CD-I; **~ video sector** *n* COMP sector *m* de video CD-I

CDLI *abbr* (*called-line identification*) DATA, TELEP identificación *f* de la línea llamada

CDM *abbr* (*companded delta modulation*) SIG, TRANS modulación *f* delta compandida

CDMA *abbr* (*code-division multiple access*) RADIO CDMA (*acceso múltiple de división de código*)

CD-ROM *abbr* (*compact disc read-only memory*) GEN CD-ROM (*disco compacto de memoria sólo de lectura*); **~ address** *n* COMP dirección *f*

de CD-ROM; ~ **drive** *n* COMP unidad *f* de CD-ROM; ~ **player** *n* COMP reproductor *m* de CD-ROM; ~ **real-time operating system** *n* (*CD-RTOS*) COMP sistema *m* de operación de CD-ROM en tiempo real (*CD-SOTR*)

CD-RTOS *abbr* (*CD-ROM real-time operating system*) COMP CD-SOTR (*sistema de operación de CD-ROM en tiempo real*)

CE *abbr* (*connection element*) DATA elemento *m* de conexiones

cease *n* TELEP *telephone line* suspensión *f*

ceased: ~ **subscriber** *n* TELEP abonado *m* dado de baja

cell *n* GEN célula *f*; ~ **border** RADIO límite *m* entre células; ~ **boundary** RADIO límite *m* de la célula; ~ **container** POWER *batteries* contenedor *m* de células; ~ **coverage prediction** RADIO predicción *f* de cobertura celular; ~-**interconnection bar** POWER *batteries* barra *f* de interconexión de células; ~ **loss priority** (*CLP*) RADIO prioridad *f* de pérdida celular; ~ **relay** RADIO relé *m* celular, TRANS relé *m* de célula; ~ **splitting** RADIO división *f* de células; ~ **switch** COMP, POWER *batteries* conmutador *m* de elementos; ~ **switching** COMP, POWER conmutación *f* de elementos; ~ **switch system** POWER *batteries* sistema *m* de conmutación de elementos; ~ **voltage** POWER *batteries* tensión *f* de célula; ~ **voltage equalizer** POWER *batteries* ecualizador *m* de tensión de células

cellular *adj* GEN celular; ~ **insulation** *n* TRANS aislamiento *m* celular; ~ **mobile exchange** *n* RADIO centralita *f* móvil celular; ~ **mobile telephone network** *n* RADIO red *f* de teléfonos celulares móviles, TELEP red *f* de telefonía celular móvil; ~ **network** *n* DATA, RADIO, TELEP red *f* celular; ~ **radio handset** *n* RADIO microteléfono *m* de radiocelular; ~ **radiotelephony** *n* RADIO telefonía *f* celular por radio; ~ **rexolene conductor** *n* ELECT conductor *m* celular de rexoleno; ~ **structure** *n* RADIO estructura *f* celular; ~ **system** *n* RADIO sistema *m* celular; ~ **technique** *n* RADIO técnica *f* celular

CELP: ~ **coder** *n* SIG codificador *m* CELP

CEN *abbr* (*European Committee for Standardization*) GEN CEN (*Comité Europeo para la Normalización*)

CENELEC *abbr* (*European Committee for Electrotechnical Standardization*) GEN CENELEC (*Comité Europeo para la Normalización Electrotécnica*)

censored: ~ **test** *n* TEST *dependability* prueba *f* truncada

center *AmE see centre BrE*

central: ~ **alarm coupler** *n* CONTR acoplador *m* central de alarma; ~ **battery** *n* TELEP batería *f* central; ~-**battery system** *n* TELEP sistema *m* de batería central; ~-**charging equipment** *n* APPL equipo *m* de tarificación central; ~ **conductor** *n* TRANS conductor *m* central; ~ **control unit** *n* SWG unidad *f* central de control; ~ **defence** *BrE n* TELEP defensa *f* central; ~ **exchange** *n* SWG central *f* telefónica; ~ **exchange switch** *n* SWG, TELEP central *f* telefónica; ~ **exchange trunk** *BrE n* (*cf central office trunk AmE*) SWG central *f* de comunicaciones; ~ **office trunk** *AmE n* (*cf central exchange trunk BrE*) SWG enlace *m* oficina central; ~ **processing unit** *n* (*CPU*) COMP, SWG, TELEP unidad *f* central de proceso (*UCP*); ~ **processor** *n* SWG procesador *m* central; ~ **pulse distributor** *n* SWG distribuidor *m* central de impulsos; ~ **remote monitoring station** *n* TEST estación *f* de seguimiento remoto de centrales; ~ **strand** *n* TRANS alma *f* de cable; ~ **strength member** *n* TRANS *cable* elemento *m* central de resistencia mecánica; ~ **suspension cable** *n* TRANS cable *m* de suspensión central; ~ **switching matrix** *n* SWG matriz *f* de conmutación central; ~ **switching unit** *n* TELEP unidad *f* central de conmutación

centralize *vt* GEN centralizar

centralized: ~ **billing** *n* APPL facturación *f* centralizada; ~ **control signalling** *BrE n* SIG señalización *f* de control centralizada; ~ **control system** *n* SWG sistema *m* de control centralizado; ~ **DSA** *n* DATA área *f* de servicio digital centralizada; ~ **maintenance** *n* GEN mantenimiento *m* centralizado; ~ **multipoint facility** *n* DATA vía *f* de transmisión multipunto centralizada, TELEP instalación *f* multipunto centralizada; ~ **operation** *n* GEN operación *f* centralizada; ~ **system** *n* SWG sistema *m* centralizado; ~ **traffic division system** *n* SWG sistema *m* centralizado para la separación de tráfico

centralograph *n* TELEP centralógrafo *m*

centre *BrE* (*AmE* **center**) *n* GEN centro *m*; ~ **frequency** TRANS frecuencia *f* nominal; ~ **line** RADIO eje *m* longitudinal; ~ **point** GEN punto *m* central; ~-**point signalling** SIG señalización *f* del punto central; ~ **position** GEN posición *f* central; ~ **tap** ELECT *transformers* toma *f* central; ~-**tapped transformer** POWER *supply* transformador *m* de toma central

centrex: ~ **service** *n* TELEP servicio *m* centrex; ~ **system** *n* TELEP sistema *m* Centrex

CEQ *abbr* (*customer equipment*) PER APP equipo *m* del abonado
ceramic: ~ **capacitor** *n* ELECT condensador *m* cerámico; ~ **encapsulation** *n* ELECT encapsulación *f* cerámica; ~ **encasement** *n* ELECT encapsulado *m* de cerámica; ~ **quad flatpack** *n* TRANS cápsula *f* planacerámica de cuadrete
certificate *n* GEN certificado *m*; ~ **of approval** GEN certificado *m* de aprobación; ~ **of compliance** TEST *quality* certificado *m* de aptitud; ~ **of conformance** TEST *quality* certificado *m* de conformidad; ~ **of conformity** TEST *quality* certificado *m* de conformidad
certification *n* GEN certificación *f*; ~ **authority** DATA autoridad *f* de certificación; ~ **body** TEST *quality* organismo *m* de certificación; ~ **path** DATA pauta *f* de certificación; ~ **system** TEST *quality* sistema *m* de certificación
CF *abbr* (*control function*) GEN CF (*función de control*)
CFM *abbr* (*companding and frequency modulation*) TRANS CFM (*modulación de compresión-expansión y de frecuencia*)
CFR *abbr* (*confirmation to receive*) DATA, TELEP CFR (*confirmación por recibir*)
CGA *abbr* (*colour graphics adaptor*) COMP CGA (*adaptador gráfico de colores*); ~ **card** *n* COMP tarjeta *f* CGA
chaff: ~ **echo** *n* RADIO eco *m* de tiras antirradar; ~ **jamming** *n* RADIO perturbación *f* con tiras antirradar
chain *n* GEN cadena *f*; ~ **clear** COMP desconexión *f* en cadena; ~ **directory** COMP directorio *m* en cadena; ~ **edit** COMP edición *f* en cadena; ~ **play** COMP pase *m* en cadena; ~ **sampling** TEST *quality* muestreo *m* en cadena; ~ **structure** COMP estructura *f* en cadena
chained: ~ **sequence** *n* DATA secuencia *f* encadenada
chaining: ~ **search** *n* COMP búsqueda *f* encadenada
change[1]: ~ **cause** *n* TEST causa *f* del cambio; ~ **channel signal** *n* RADIO señal *f* de cambio de canal; ~ **dump** *n* COMP vuelco *m* de cambios; ~ **in temperature** *n* GEN cambio *m* de temperatura; ~ **notice** *n* DATA notificación *f* de cambio; ~ **order** *n* GEN modificación *f* contractual; ~ **request** *n* DATA demanda *f* de modificación
change[2] *vt* GEN cambiar
changeable *adj* GEN variable
changeback *n* TELEP retorno *m* a la conexión normal; ~ **code** SIG código *m* de retorno a la conexión normal
changed: ~-**number signal** *n* SIG señal *f* de número cambiado
changeover *n* GEN conmutación *f*; ~ **signal** SIG señal *f* de conmutación; ~ **to standby** GEN conmutación *f* a modo de espera
changer *n* POWER cambiador *m*
channel *n* GEN canal *m*; ~ **allocation store** TRANS memoria *f* de asignación de canal; ~ **allocation time** TRANS tiempo *m* de asignación de canal; ~-**associated signalling** *BrE* (*CAS*) SIG, TELEP, TRANS señalización *f* asociada al canal; ~ **B** TELEP canal *m* B; ~ **capacity** TRANS capacidad *f* del canal; ~-**check test** TEST ensayo *m* de comprobación de canal; ~ **coder** RADIO *mobile* codificador *m* de canales; ~ **concentrator** TRANS concentrador *m* de canales; ~ **control** TRANS control *m* de canal; ~ **D** TELEP canal *m* D; ~ **dedication** TELEP asignación *f* de canal; ~ **efficiency** TRANS eficiencia *f* del canal; ~ **equipment** TRANS equipo *m* de canales; ~ **extension equipment** TRANS equipo *m* de extensión de canales; ~ **filtering** TV filtrado *m* de canal; ~ **gate** TELEP, TRANS puerta *f* de canal; ~ **loading** RADIO, TRANS carga *f* de canal; ~ **mapping list** TRANS lista *f* de configuración de canales; ~ **marking** RADIO *mobile* marcación *f* de canal; ~ **modem** TRANS módem *m* de canal; ~ **occupancy rate** RADIO tasa *f* de ocupación del canal; ~ **performance** TEST rendimiento *m* del canal; ~ **pool** TRANS agrupación *f* de canales; ~ **protocol machine** (*CPM*) DATA máquina *f* de protocolo de canal; ~ **receiver** TRANS receptor *m* de canal; ~ **selector** RADIO, TV selector *m* de canales; ~ **separation** RADIO separación *f* de canales; ~ **size** TRANS tamaño *m* del canal; ~ **spacing** RADIO espaciado *m* de canales; ~ **state** TRANS estado *m* del canal; ~ **substructure** DATA *specification and description language* subestructura *f* de canales; ~ **switch** RADIO, SWG conmutador *m* de canales; ~ **switching** RADIO, SWG conmutación *f* de canal; ~ **terminal** TRANS terminal *m* de canal; ~ **time slot** TELEP, TRANS intervalo *m* de tiempo de canal; ~ **transmitter** TRANS transmisor *m* de canal; ~ **unit** TRANS unidad *f* de canal; ~ **vocoder** RADIO vocoder *m* de canales
channeling *AmE see channelling BrE*
channelization *n* TRANS canalización *f*, multiplexaje *m*
channelized: ~ **section** *n* TRANS sección *f* canalizada, sección *f* multiplexada; ~ **stage**

of repeater *n* TRANS etapa *f* de repetidor canalizada

channelling *BrE* (*AmE* **channeling**) *n* TRANS canalización *f*

chaos *n* ELECT caos *m*

character *n* GEN carácter *m*; ~ **alignment** TELEP alineación *f* de caracteres; ~ **boundary** APPL límite *m* de carácter; ~ **check** TELEP comprobación *f* de caracteres; ~ **code** COMP código *m* de carácter; ~ **coupling switch** SWG conmutador *m* de acoplamiento de carácter; ~ **display** GEN *indicating devices* visualizador *m* de caracteres; ~ **error rate** COMP, DATA tasa *f* de caracteres erróneos; ~ **error ratio** COMP, DATA ratio *f* de error de caracteres, índice *m* de error de caracteres; ~ **font** COMP, DATA *Internet* familia *f* de caracteres; ~ **generation** COMP generación *f* de caracteres; ~-**generation logic** COMP lógica *f* de generación de caracteres; ~ **generator** COMP generador *m* de caracteres; ~ **instruction** COMP instrucción *f* de caracteres; ~-**oriented procedure** COMP procedimiento *m* orientado a caracteres; ~ **outline** APPL perfil *m* de caracteres; ~ **pitch** APPL paso *m* de carácter; ~ **position** TELEP posición *f* de carácter; ~ **reader** APPL lector *m* de caracteres; ~ **recognition** COMP reconocimiento *m* de caracteres; ~-**serial transmission** TRANS transmisión *f* de caracteres en serie; ~ **set** COMP juego *m* de caracteres; ~ **of signal** SIG carácter *m* de señal; ~ **signal** SIG señal *f* de carácter; ~ **skew** APPL inclinación *f* de un carácter; ~-**spacing reference line** COMP línea *f* de referencia de espaciado de caracteres; ~-**string type** DATA tipo *m* de cadena de caracteres; ~ **switch** SWG conmutador *m* de caracteres

characteristic[1]: ~ **distortion** *n* DATA, TELEP torcimiento *m* característico; ~ **distortion compensation** *n* DATA, TELEP compensación *f* de distorsión característica; ~ **frequency** *n* RADIO frecuencia *f* característica

characteristic[2] *n* GEN característica *f*

characters: ~ **per inch** *n pl* (*cpi*) COMP caracteres *m pl* por pulgada (*cpi*)

charge[1] *n* APPL, ELECT, POWER *batteries* carga *f*, TELEP, *billing* facturación *f*; ~ **advice** APPL aviso *m* de facturación; ~ **card** TELEP tarjeta *f* de cargo; ~ **carrier** ELECT *semiconductors* portadora *f* de carga; ~-**coupled device** (*CCD*) ELECT dispositivo *m* acoplado por carga (*DAC*); ~-**coupled device circuit** DATA circuito *m* con dispositivo de carga acoplada; ~ **cycle** POWER ciclo *m* de carga; ~ **injection device** POWER dispositivo *m* de inyección de carga; ~-**metering converter** APPL convertidor *m* del contador de carga; ~/**no** ~ **indicator** TELEP indicador *m* de carga/no carga; ~ **rate** APPL tasa *f* de carga; ~ **request** DATA petición *f* de carga; ~ **time** POWER tiempo *m* de carga; ~ **transfer device** ELECT dispositivo *m* de transferencia de carga; ~ **unit** APPL unidad *f* de facturación; ◆ **no** ~ TELEP sin cargo

charge[2] *vt* ELECT *battery*, TELEP cargar; ◆ ~ **a call to** APPL cargar una llamada a

chargeable *adj* APPL, TELEP facturable, tasable; ~ **duration** *n* APPL, TELEP tiempo *m* tasable; ~ **duration of telex call** *n* TELEP duración *f* facturable de comunicación por telex

charged: ~ **duration** *n* APPL, TELEP duración *f* facturada

charger *n* POWER *batteries* cargador *m*

charging *n* APPL *of batteries* carga *f*, TELEP, *billing* facturación *f*; ~ **analysis** APPL análisis *m* de facturación; ~ **area** APPL, TELEP zona *f* de facturación; ~ **centre** *BrE* APPL centro *m* de facturación; ~ **current** POWER *batteries* corriente *f* de carga; ~ **data** *n pl* APPL información *f* de la facturación; ~ **equipment** APPL equipo *m* de carga; ~ **error** APPL error *m* de facturación; ~ **impulse** APPL impulso *m* de carga; ~ **information** APPL información *f* de carga; ~ **interval** APPL intervalo *m* de facturación; ~ **method** APPL, TELEP método *m* de facturación; ~ **mode** APPL, TELEP modo *m* de facturación; ~ **output** APPL resultado *m* de la facturación; ~ **period** APPL, TELEP periodo *m* de carga; ~ **point** APPL punto *m* de facturación; ~ **principle** APPL principio *m* de facturación; ~ **in progress PE** TELEP EP *m* de carga en ejecución; ~-**recording centre** *BrE* TELEP centro *m* de registro de facturación; ~ **rectifier** POWER rectificador *m* de carga; ~ **set** POWER *batteries* cargador *m*; ~ **statistics** *n pl* APPL estadísticas *f pl* de facturación; ~ **store** APPL memoria *f* de facturación; ~ **test** APPL, TEST prueba *f* de facturación; ~ **tester** APPL, TEST probador *m* de facturación; ~ **unit** APPL, POWER *batteries* cargador *m* de batería; ~ **voltage** POWER *batteries* tensión *f* de carga

charstring *n* DATA cadena *f* de caracteres

chart *n* GEN gráfica *f*

chassis *n* GEN chasis *m*

chatter *vi* COMP vibrar

cheap: ~-**call rate** *n* APPL, TELEP tarifa *f* reducida de llamadas

Chebyshev: ~ **approximation** *n* TRANS aproximación *f* de Chebichef

check[1]: ~ **battery** *n* COMP batería *f* de comprobación; ~ **bit** *n* COMP, DATA clave *f* de control; ~ **box** *n* COMP casilla *f* de control; ~ **indicator** *n* COMP indicador *m* de control; ~ **loop** *n* TELEP bucle *m* de comprobación; ~**-number field** *n* SWG campo *m* del número de control; ~**-out equipment** *n* TEST equipo *m* de verificación; ~**-out time** *n* TEST, TELEP hora *f* de la verificación; ~ **print-out** *n* SWG impresión *f* de verificación; ~ **sequence** *n* TEST secuencia *f* de comprobación; ~ **sum** *n* DATA suma *f* de comprobación; ~**-symbol time slot** *n* TRANS intervalo *m* de tiempo de símbolo de verificación

check^2 *vt* GEN comprobar

checking: ~ **off** *n* TEST comprobación *f*; ~ **program** *n* COMP programa *m* de comprobación

checklist *n* TEST lista *f* de control

checkpoint *n* COMP, TELEP punto *m* de control

cheese: ~ **antenna** *n* RADIO antena *f* con parábola achatada

chemical: ~ **vapour** *BrE* **deposition** *n* ELECT deposición *f* de vapor químico

cheque: ~**-processing equipment** *BrE n* APPL equipo *m* de procesamiento de cheques

chief: ~ **operator** *n* TELEP operador *m* jefe; ~ **operator's desk** *n* TELEP pupitre *m* de mando; ~ **supervisor** *n* TELEP supervisor *m* jefe; ~ **supervisor's desk** *n* TELEP pupitre *m* de mando

child *n* COMP hijo *m*; ~ **entry** DATA entrada *f* hija; ~**-sequence number** COMP, DATA número *m* de la secuencia hija

childless: ~ **node** *n* COMP nodo *m* sin hijo

chip *n* COMP, DATA, ELECT chip *m*; ~ **card** ELECT circuito *m* estampado para plaquetas; ~ **rate** DATA velocidad *f* del chip; ~ **yield** DATA rendimiento *m* de la pastilla

Chireix: ~ **array** *n* RADIO matriz *f* de Chireix

chirping *n* RADIO, TRANS *fibre optics* chirrido *m*

chloroprene *n* GEN cloropreno *m*

choke *n* POWER bobina *f* de autoinducción

choking: ~ **coil** *n* POWER bobina *f* de bloqueo

choose *vt* GEN seleccionar

chopped: ~ **response** *n* RADIO respuesta *f* entrecortada

chopper *n* ELECT interruptor *m*; ~ **amplifier** ELECT amplificador *m* de interrupción

chord *n* COMP acorde *m*; ~ **intelligence** COMP señal *f* de acordes

chosen: ~ **path** *n* DATA *Internet* camino *m* elegido

chromakey *n* COMP separador *m* de crominancia

chromatic *adj* GEN cromático; ~ **balance** *n* TV balance *m* cromático; ~ **colour** *n* TV color *m* cromático; ~ **dispersion** *n* TRANS dispersión *f* cromática; ~ **distortion** *n* TV distorsión *f* cromática; ~ **space** *n* TV espacio *m* cromático

chromaticity *n* GEN cromaticidad *f*

chrominance *n* GEN *optics* cromeado *m*; ~ **axis** TV eje *m* de crominancia; ~ **component** TV componente *m* de crominancia; ~**-luminance delay inequality** TV desigualdad *f* de retardo de luminancia-crominancia; ~ **vector** TV vector *m* de crominancia

chronological: ~ **link** *n* COMP enlace *m* cronológico

chunk *n* COMP *of information* bloque *m*

churn: ~ **rate** *n* GEN tasa *f* de transmisión

chute *n* PARTS, SWG, TELEP, TRANS *cable laying* canal *m*

CI *abbr* (*concatenation indication*) ELECT indicación *f* de concatenación

C/I *abbr* (*carrier-to-interference ratio*) TRANS relación *f* portadora/interferencia

CIG *abbr* (*calling-subscriber identification*) DATA, TELEP CIG (*identificación del abonado que llama*)

cigar: ~ **aerial** *n* RADIO antena *f* en cigarro; ~ **antenna** *n* RADIO antena *f* en cigarro

CIG-DTC *abbr* (*calling-subscriber identification-DTC*) DATA, TELEP CIG-DTC (*identificación DTC de abonado que llama*)

CIM *abbr* (*computer-integrated manufacturing*) COMP CIM (*fabricación informatizada*)

CIME *abbr* (*customer installation maintenance entity*) APPL entidad *f* de mantenimiento de la instalación del cliente

cinching *n* COMP cincha *f*

ciphertext *n* DATA, RADIO, TELEP, TV texto *m* cifrado

CIR *abbr* (*committed information rate*) DATA velocidad *f* de información obligada

CIRC *abbr* (*cross-interleaved Reed-Solomon code*) COMP CIRC (*código Reed-Solomon intercruzado*)

circuit *n* GEN circuito *m*; ~ **access point** TELEP punto *m* de acceso al circuito; ~ **availability** RADIO, TELEP disponibilidad *f* de circuitos; ~ **board** COMP, ELECT cuadro *m* de circuito; ~ **breaker** ELECT, POWER interruptor *m* automático; ~ **closer** POWER contactor *m*; ~**-consistency verification** DATA verificación *f* de la consistencia de un circuito; ~ **control** GEN control *m* de circuito; ~ **control station** TELEP estación *f* de control de circuito; ~ **description** GEN descripción *f* de circuito; ~ **design** GEN *technical drawing* diseño *m* de circuito; ~ **diagram** GEN diagrama *m* de circuito; ~ **equipment** TELEP equipo *m* de circuito; ~ **group** GEN grupo *m* de circuitos;

~-group congestion signal DATA, SIG señal *f* de congestión del grupo de circuito; ~-group query message (*CQM*) DATA, SWG mensaje *m* de interrogación de un grupo de circuitos; ~-group query response message (*CQR*) DATA, SWG mensaje *m* de respuesta de un grupo de circuitos; ~ interface DATA interfaz *f* del circuito; ~ junctor TELEP conector *m* del circuito; ~ loudness rating SIG indicación *f* de sonoridad del circuito; ~ mode-bearer service APPL servicio *m* portador de modos de circuito; ~ noise GEN ruido *m* del circuito; ~ noise level TELEP nivel *m* de ruido de circuito; ~-released acknowledgement signal SIG señal *f* de reconocimiento emitida por un circuito; ~-released signal SIG señal *f* de circuito liberado; ~ route TRANS recorrido *m* de circuito; ~ simulation ELECT *semiconductors* simulación *f* del circuito; ~ stability TRANS estabilidad *f* del circuito; ~ state indicator GEN indicador *m* del estado del circuito; ~ state message GEN mensaje *m* de estado del circuito; ~ subcontrol station TELEP estación *f* de subcontrol del circuito; ~ subgroup GEN subgrupo *m* de circuitos; ~ switch SWG conmutador *m* de circuito; ~-switched connection DATA, SWG enlace *m* conmutado; ~-switched data transmission service DATA, TRANS servicio *m* de transmisión de datos por circuito conmutado; ~-switched exchange SWG central *f* de circuitos conmutados; ~-switched network TRANS red *f* de conmutación de circuitos; ~-switched service APPL servicio *m* conmutado; ~ switching GEN conmutación *f* de circuitos; ~-switching centre *BrE* SWG, TRANS centro *m* de conmutación de circuitos; ~-switching network SWG, TRANS red *f* de conmutación de circuitos; ~-switching system SWG sistema *m* de conmutación de circuitos; ~-switching unit SWG conmutador *m* de circuitos; ~ terminal TELEP terminal *m* de circuitos; ~-test access point TELEP punto *m* de acceso de la prueba de circuitos; ~ test set TEST probador *m* de circuitos; ~ and wiring diagram GEN diagrama *m* de circuito y cableado

circuitry *n* COMP conjunto *m* de circuitos, ELECT circuitería *f*

circular: ~ orbit *n* RADIO órbita *f* circular; ~ polarization *n* TRANS polarización *f* circular

circularize *vt* RADIO *orbit* circunvolucionar

circulator *n* TRANS *waveguide* circulador *m*

CISE *abbr* (*complex instruction set equipment*) DATA CISE (*equipo complejo de conjunto de instrucciones*)

citizens': ~ band *n* (*CB*) RADIO banda *f* ciudadana

city: ~ call *n* TELEP llamada *f* metropolitana; ~ network *n* TELEP red *f* metropolitana; ~ paging *n* RADIO buscapersonas *m* metropolitano

civil: ~ communications satellite *n* RADIO satélite *m* de comunicaciones civiles; ~ engineering *n* GEN ingeniería *f* civil

CL *abbr* (*connectionless*) GEN CL (*sin conexiones*)

cladding *n* GEN revestimiento *m*, chapado *m*; ~ centre *BrE* TRANS *optical fibre* centro *m* de chapado; ~ diameter TRANS *optical fibre* diámetro *m* de revestimiento; ~ mode TRANS *fibre optics* modo *m* de revestimiento; ~ mode stripper TRANS *fibre optics* eliminador *m* de modos de revestimiento; ~-surface diameter deviation TELEP desviación *f* de diámetro de superficie de revestimiento; ~ tolerance field TRANS *fibre optics* campo *m* de tolerancia del chapado

clamp *n* GEN abrazadera *f*; ~-band separation system ELECT sistema *m* de separación de banda de contacto; ~ diode ELECT diodo *m* fijador de nivel; ~ terminal ELECT collarín *m* de borne

clamping: ~ circuit *n* ELECT circuito *m* de fijación de amplitud; ~ diode *n* ELECT diodo *m* de fijación; ~ sleeve *n* ELECT manguito *m* de apriete

clarifying: ~ text *n* TELEP texto *m* aclaratorio

clarity *n* GEN claridad *f*

clashing *n* RADIO *land mobile* colisión *f*

class: ~ boundary *n* GEN *statistics* límite *m* de clase; ~ of document *n* GEN clase *f* de documento; ~ of emission *n* RADIO clase *f* de emisión; ~ of HDLC *n* TELEP clase *f* de HDLC; ~ interval *n* GEN *statistics* intervalo *m* de clase; ~ limits *n pl* GEN *statistics* límites *m pl* de clase; ~ marking *n* TELEP marcación *f* de categoría; ~-mark store *n* TELEP memoria *f* de categorías; ~ midpoint *n* GEN punto *m* medio de clase

classification *n* GEN clasificación *f*; ~ of failures TEST clasificación *f* de los fallos

classify *vt* GEN clasificar

clean[1]: ~ carrier *n* RADIO portadora *f* sin perturbaciones

clean[2] *vt* GEN limpiar

cleaning *n* GEN limpieza *f*; ~ head TELEP cabeza *f* de limpieza; ~ kit GEN equipo *m* de limpieza

clear[1]: ~ acrylic tube *n* ELECT tubo *m* acrílico claro; ~ area *n* APPL área *f* libre, COMP área *f* de exploración; ~-back condition *n* SIG desconectado; ~-backsignal *n* SIG, TELEP señal *f* de desconexión; ~ band *n* APPL área *f* reservada; ~ **channel capability** *n* APPL capacidad *f* de onda exclusiva, TELEP capacidad *f* de canal libre; ~ **confirmation** *n* TELEP confirmación *f* de liberación; ~**-forward signal** *n* SIG, TELEP señal *f* de desconexión; ~ **indication delay** *n* TELEP retardo *m* de indicación de liberación; ~ **message** *n* TEST mensaje *m* de liberación; ~ **operation** *n* COMP reposición *f* acero; ~ **request condition** *n* TELEP petición *f* de liberación; ~ **request packet** *n* DATA paquete *m* de solicitud de liberación; ~ **sixty-four service** *n* APPL servicio *m* sesenta y cuatro de desconexión; ~**-sky absorption loss** *n* RADIO pérdida *f* de absorción por cielo despejado; ~ **speech** *n* RADIO voz *f* clara; ~ **text** *n* DATA texto *m* despejado; ~ **to send** *adj* DATA listo para emitir; ~ **transmission** *n* TRANS transmisión *f* despejada; ~ **varnish** *n* PARTS barniz *m* transparente; ~ **weather attenuation** *n* RADIO amortiguación *f* de tiempo despejado

clear[2] *vt* COMP *screen* borrar; ◆ ~ **a fault** TELEP corregir un fallo, TEST reparar un fallo

clearance: ~ **hole** *n* POWER *printed circuits* zona *f* despejada

clearest: ~ **zone** *n* PER APP *scanner* zona *f* más clara

clearing *n* COMP, TELEP desconexión *f*; ~ **collision** TELEP choque *m* de desconexión; ~ **confirmation** TELEP confirmación *f* de desconexión; ~ **indicator** TELEP indicador *m* de fin de conversación; ~ **lamp** TELEP lámpara *f* de fin de conversación; ~ **message** TEST mensaje *m* de desconexión; ~ **procedure** SIG procedimiento *m* de despejo; ~ **request** TELEP demanda *f* de desconexión, petición *f* de desconexión

cleartext *n* DATA texto *m* legible

cleaved *adj* TRANS cortado

cleaver *n* TRANS instrumento *m* para resquebrajar

cleaving *n* TRANS resquebrajamiento *m*; ~ **tool** TRANS herramienta *f* de corte

clerical: ~ **costs** *n pl* GEN costes *m pl* (*AmL* costos) administrativos

CLI *abbr* (*calling-line indicator*) DATA, TELEP indicador *m* de la línea que llama

click[1] *n* COMP cliqueo; ~**-stop position** COMP posición *f* de clic de tenido

click[2] *vt* COMP cliquear

clicking: ~ **area** *n* COMP zona *f* de pulsación

CLID *abbr* (*calling-line identification display*) TELEP representación *f* visual de identificación de línea que llama

client: ~ **checklist** *n* DATA lista *f* de comprobación de clientes; ~**-installation maintenance equipment** *n* TEST equipo *m* de mantenimiento de instalación del cliente; ~ **network** *n* TRANS red *f* de clientes; ~ **risk** *n* TEST *quality* riesgo *m* del cliente; ~ **software** *n* COMP, DATA software *m* de cliente

client/server *n* COMP cliente/servidor *m*

climate: ~ **control equipment** *n* CLIM equipo *m* de control del clima

climb: ~ **path** *n* TRANS trayectoria *f* de ascensión

clinched: ~**-wire through connection** *n* ELECT, POWER *printed circuits* conexión *f* transversal de hilo remachado

clip[1]: ~**-art library** *n* COMP servicio *m* de documentación de prensa; ~**-on ammeter** *n* TEST amperímetro *m* de mordazas

clip[2] *vt* COMP, PARTS rebabar

CLIP *abbr* (*calling-line identification presentation*) TELEP CLIP (*presentación de la identificación de la línea que llama*)

clipboard *n* COMP cuadro *m* de pinzas; ~ **computer** COMP ordenador *m* con cuadro de pinzas *Esp*, computador *m* con cuadro de pinzas *AmL*, computadora *f* con cuadro de pinzas *AmL*

clipping *n* GEN recorte *m*; ~ **voltage** ELECT *semiconductors* tensión *f* de limitación de crestas

clique *n* GEN grupo *m* de personas con intereses comunes

CLIR *abbr* (*calling-line identification restriction*) TELEP restricción *f* de identificación de la línea que llama

clock *n* GEN reloj *m*; ~**-carrier cycle skipping** TRANS salto *m* del ritmo de la portadora de reloj; ~ **change** PER APP *on answering machine* cambio *m* del reloj; ~ **cycle** COMP ciclo *m* cronometrado; ~**-cycle skipping** TRANS salto *m* del ritmo; ~ **distribution** TRANS distribución *f* de señales de reloj; ~ **error** TRANS error *m* de reloj; ~ **jitter** ELECT fluctuación *f* de fase de reloj; ~ **oscillator** ELECT oscilador *m* de reloj; ~ **phasing** TRANS sincronización *f* por reloj; ~ **pulse** ELECT impulso *m* de sincronización; ~ **rate** TRANS ritmo *m* del patrón; ~**-rate distribution** TRANS distribución *f* de la cadencia; ~ **recovery** TRANS restablecimiento *m* del reloj; ~**-recovery circuit** ELECT circuito *m* de recuperación del reloj; ~**-skip rate** TRANS ritmo *m* de salto del reloj; ~ **synchronization** COMP sincronización *f* de reloj; ~ **time difference** GEN diferencia *f* en la hora del reloj

clockwise[1] *adv* GEN en sentido horario

clockwise[2]: ~ **polarization** *n* RADIO polarización *f* en sentido horario

clogged *adj* GEN obstruido
clone *n* GEN clon *m*
Clos: ~ **network** *n* TELEP, TRANS red *f* Clos
close[1]: ~-**talking microphone** *n* PARTS micrófono *m* de voz baja; ~-**up** *n* GEN primer *m* plano
close[2] *vt* GEN cerrar
closed: ~-**circuit current** *n* POWER corriente *f* en circuito cerrado; ~-**circuit relay** *n* POWER relé *m* de circuito cerrado; ~-**circuit television** *n* TV televisión *f* en circuito cerrado; ~-**circuit working** *n* TRANS funcionamiento *m* en circuito cerrado; ~ **coupler** *n* TELEP acoplador *m* cerrado; ~ **earth-orbit** *n* RADIO órbita *f* terrestre cerrada; ~ **earth-track orbit** *n* RADIO órbita *f* de pista terrestre cerrada; ~-**loop circuit** *n* ELECT circuito *m* de bucle cerrado; ~-**loop control** *n* CONTR, POWER funcionamiento *m* en bucle cerrado; ~-**loop control system** *n* CONTR, POWER sistema *m* de control en bucle cerrado; ~-**loop synchronization** *n* CONTR, POWER, TRANS sincronización *f* en bucle cerrado; ~ **subroutine** *n* COMP subprograma *m* cerrado; ~ **subscriber group** *n* DATA, TELEP grupo *m* cerrado de abonados; ~ **user data distribution network** *n* TRANS red *f* cerrada de distribución de datos al usuario; ~ **user group** *n* (*CUG*) DATA, TELEP grupo *m* cerrado de usuarios; ~ **user group call indicator** *n* DATA, TELEP indicador *m* de llamada a grupo de usuarios cerrado; ~ **user group with outgoing access** *n* DATA, TELEP grupo *m* cerrado de usuarios con acceso de salida
closure: ~ **gasket** *n* PARTS junta *f* hermética de cierre
cloud: ~ **return** *n* RADIO eco *m* de nubes
cloverleaf: ~ **antenna** *n* RADIO antena *f* de trébol
CLP *abbr* (*cell loss priority*) RADIO prioridad *f* de pérdida celular
CLR *abbr* (*circuit-loudness rating*) TELEP CLR (*indicador de sonoridad de un circuito*)
cluster *n* GEN *statistics* grupo *m*; ~ **of cells** RADIO conglomerado *m* de celdillas; ~ **link** COMP enlace *m* de grupos; ~ **sampling** GEN muestreo *m* por universos
CLUT *abbr* (*colour look-up table*) COMP CLUT (*tabla de consulta de colores*)
clutter *n* RADIO señales *f pl* parásitas en pantalla; ~ **attenuation** RADIO atenuación *f* de reflejos; ~ **suppression** RADIO supresión *f* de reflejos
CM *abbr* (*conditionally mandatory*) TELEP condicionalmente obligatorio
CMB *abbr* (*CRC message bloc*) TELEP CMB (*bloque de mensajes CRC*)
cmf *abbr* (*cyclomotive force*) RADIO cmf (*fuerza ciclomotora*)
CMI *abbr* (*code-mark inversion*) TRANS inversión *f* de la marca de código
CMIS *abbr* (*common management-information service*) DATA servicio *m* de información de gestión común
CMISE *abbr* (*common management-information service element*) DATA elemento *m* de servicio de información de gestión común; ~ **service provider** *n* DATA, TELEP suministrador *m* de servicios CMISE; ~ **service user** *n* DATA, TELEP usuario *m* del servicio CMISE
CMOS: ~ **crosspoint** *n* TELEP contacto *m* de cruce CMOS
CMRJ *abbr* (*call-modification reject message*) TELEP modificación *f* de rechazo de llamada mensaje
CMY: ~ **colour** *BrE* **model** *n* COMP modelo *m* de color CMY
C-n *abbr* (*container-n*) APPL *SDH*, DATA, TELEP contenedor-n *m*
coalesce *vi* COMP fusionarse
coarse: ~ **adjustment** *n* GEN ajuste *m* aproximado; ~ **jumper** *n* ELECT interconexión *f* aproximada; ~ **setting** *n* GEN ajuste *m* aproximado; ~ **sunsensor** *n* RADIO sensor *m* solar aproximado; ~ **switch** *n* ELECT conmutación *f* aproximada
coast: ~ **station** *n* RADIO, TELEP estación *f* costera; ~-**station identity** *n* RADIO, TELEP identidad *f* de estación costera
coastal: ~ **earth station** *n* RADIO estación *f* terrena costera; ~ **station** *n* RADIO, TELEP estación *f* costera; ~ **station identity** *n* RADIO, TELEP identidad *f* de la estación costera
coating *n* GEN capa *f*, PARTS revestimiento *m*
coax *n* TRANS cable *m* coaxial
coaxial *adj* TRANS coaxial; ~ **antenna** *n* RADIO antena *f* de alimentador coaxial; ~ **cable** *n* TRANS cable *m* coaxial; ~ **cable distributor** *n* TV distribuidor *m* de cable coaxial; ~ **jack** *n* ELECT clavija *f* coaxial; ~ **line** *n* TRANS línea *f* coaxial; ~ **line system** *n* TRANS sistema *m* de línea coaxial; ~ **network** *n* TRANS red *f* coaxial; ~ **plug** *n* ELECT enchufe *m* coaxial; ~ **push-pull connector** *n* TRANS conector *m* coaxial en contrafase; ~ **reed relay changeover switch** *n* SWG conmutador *m* coaxial de relé de lengüeta; ~ **switched jumper** *n* TRANS conector *m* coaxil con mutado; ~ **tube** *n* TRANS tubo *m* coaxial
co-channel *n* RADIO cocanal *m*; ~ **interference** RADIO, TRANS interferencia *f* del cocanal; ~ **operation** RADIO *mobile* explotación *f* en

cocanal; ~ **reuse distance** RADIO *land mobile* distancia *f* de reutilización en el mismo canal

code[1] *n* GEN código *m*; ~ **answer** TELEP respuesta *f* codificada; ~**-answering device** TELEP *ATME* contestador *m* de código; ~**-answering unit** TELEP *ATME* contestador *m* de código; ~ **block** DATA bloque *m* de código; ~ **character** TELEP carácter *m* de código; ~ **conversion** GEN conversión *f* de código; ~ **converter** GEN convertidor *m* de código; ~**-dependent system** TELEP sistema *m* dependiente de códigos; ~ **division** RADIO, TELEP división *f* de código; ~**-division multiple access** (*CDMA*) RADIO acceso *m* múltiple de división de código (*CDMA*); ~ **efficiency** SIG eficiencia *f* de un código; ~ **element** COMP clave *f*; ~ **of ethics** GEN código *m* de ética; ~ **extension** TELEP extensión *f* de código; ~ **extension techniques** *n pl* TELEP técnicas *f pl* de extensión de código; ~ **4B3T** TRANS código *m* 4B3T; ~ **generator** TELEP generador *m* de código; ~**-independent system** TELEP sistema *m* independiente de códigos; ~ **key** RADIO clave *f* de código; ~**-mark inversion** (*CMI*) TRANS inversión *f* de la marca de código; ~ **memory board** RADIO *mobicom* placa *f* de memoria de códigos; ~ **plug** RADIO *mobile* clavija *f* de código; ~ **receiver** TELEP receptor *m* de códigos; ~ **sender** TELEP emisor *m* de códigos; ~**-sender finder** TELEP buscador *m* de emisor de códigos; ~ **table** TELEP tabla *f* de códigos; ~ **transmission** TRANS transmisión *f* de códigos; ~ **transparency** TRANS transparencia *f* de código; ~ **violation** (*CV*) TRANS violación *f* de código

code[2] *vt* GEN codificar

codec *n* TRANS codificador/descodificador *m*

coded: ~ **character data element** *n* TELEP elemento *m* de datos de carácter codificado; ~ **character set** *n* DATA, TELEP conjunto *m* de caracteres codificados; ~**-data transmitter** *n* DATA transmisor *m* de datos codificados; ~**-image storage space** *n* COMP memoria *f* de almacenamiento de imágenes codificadas; ~ **in-band signalling** *BrE n* SIG señalización *f* codificada dentro de banda; ~ **MPEG frame** *n* COMP marco *m* MPEG codificado; ~ **order** *n* COMP orden *f* codificada; ~ **orthogonal frequency division multiplex** *n* (*COFDM*) TRANS múltiplex de división de frecuencia ortogonal codificada (*COFDM*); ~ **program** *n* COMP programa *m* cifrado; ~ **representation** *n* COMP representación *f* codificada; ~ **ringing current** *n* TELEP corriente *f* de llamada codificada; ~ **transmission** *n* TRANS transmisión *f* codificada

coder *n* COMP, TRANS codificador *m*

codeword *n* COMP, DATA palabra *f* clave

coding *n* GEN codificación *f*; ~ **circuit** SIG circuito *m* de codificación; ~ **control output** DATA salida *f* de control de codificación; ~ **information** COMP información *f* de codificación; ~ **law** TELEP, TRANS ley *f* de codificación; ~ **parameter** COMP parámetro *m* de codificación; ~ **process** GEN codificación *f*, proceso *m* de codificación; ~ **rate** DATA velocidad *f* de codificación; ~ **sheet** GEN hoja *f* de codificación; ~ **standard** GEN norma *f* de codificación; ~ **type** TRANS tipo *m* de codificación

coefficient *n* GEN coeficiente *m*; ~ **of linear expansion** GEN coeficiente *m* de expansión lineal

coercive: ~ **field** *n* ELECT campo *m* coercitivo

coercivity *n* ELECT coercividad *f*

COFDM *abbr* (*coded orthogonal frequency division multiplex*) TRANS COFDM (*múltiplex de división de frecuencia ortogonal codificada*)

cognitive *adj* GEN cognitivo; ~ **mapping** *n* COMP aplicación *f* cognitiva; ~ **modelling** *BrE n* GEN modelado *m* cognitivo

coherence *n* GEN coherencia *f*; ~ **area** TRANS área *f* de coherencia; ~ **bandwidth** RADIO *land mobile*, TRANS ancho *m* de banda de coherencia; ~ **degree** TRANS *fibre optics* grado *m* de coherencia; ~ **length** TRANS longitud *f* de coherencia; ~ **time** TRANS tiempo *m* de coherencia

coherent: ~ **area** *n* TRANS área *f* coherente; ~ **demodulation** *n* TRANS demodulación *f* coherente; ~ **detection** *n* TRANS detección *f* coherente; ~ **light** *n* TRANS luz *f* coherente; ~ **modulation** *n* TRANS modulación *f* coherente; ~ **oscillator** *n* ELECT oscilador *m* coherente; ~ **phase-shiftkeying** *n* SIG transmisión *f* por desplazamiento coherente de fase; ~**-pulse radar** *n* RADIO radar *m* de impulsos coherentes; ~ **radiation** *n* RADIO radiación *f* coherente; ~ **significant instant** *n* SIG, TELEP instante *m* significativo coherente, TEST instante *m* coherente significativo; ~ **transmission** *n* TRANS transmisión *f* coherente

cohesion *n* GEN cohesión *f*

coil[1] *n* POWER bobina *f*; ~ **former** POWER carrete *m*; ~ **loading** TRANS carga *f* por bobinas

coil[2]: ~**-load** *vi* TRANS cargar bobinas

coin: ~**-box telephone** *n* TELEP teléfono *m* de ficha; ~**-box line** *n* TELEP línea *f* de pago previo; ~**-box set** *n* TELEP teléfono *m* de pago previo; ~ **buffer** *n* TELEP memoria *f* intermedia de monedero; ~**-operated payphone** *n* TELEP teléfono *m* público de pago previo; ~**-operated telephone** *n* PER APP, TELEP teléfono *m* de pago previo; ~**-return cup** *n* PER APP, TELEP cajetín *m* de devolución de moneda; ~ **slot** *n* PER APP, TELEP ranura *f* para las monedas

coincidence *n* GEN coincidencia *f*; ~ **method of measurement** TEST método *m* de medición de coincidencias

coincident *adj* TELEP coincidente

cold: ~**-cathode tube** *n* ELECT tubo *m* de cátodo frío; ~**-cathode valve** *n* ELECT válvula *f* de cátodo frío

collaborative: ~ **computing** *n* COMP, DATA computación *f* colaboradora *AmL*, informática *f* en colaboración *Esp*

collapse *n* GEN derrumbe *m*; ~ **action sequence** TEST secuencia *f* de medidas en caso de avería

collate *vt* COMP, DATA intercalar, cotejar

collating: ~ **sequence** *n* COMP secuencia *f* de intercalación

collator *n* COMP programa *m* de intercalación

collect[1]: ~ **call** *AmE n* (*cf reverse-charge call BrE*) APPL, TELEP llamada *f* pagadera en destino; ~**-call acceptance** *AmE n* (*cf reverse-charge call acceptance BrE*) APPL aceptación *f* de cobro revertido, TELEP aceptación *f* de llamada a cobro revertido; ~**-call acceptance not subscribed signal** *AmE n* (*cf reverse-charge call acceptance not subscribed signal BrE*) SIG señal *f* de cobro revertido no aceptado, TELEP señal *f* no suscrita de aceptación de cobro revertido; ~**-call service** *AmE n* (*cf reverse-charge call service BrE*) TELEP servicio *m* de cobro revertido; ~**-calling acceptance** *AmE n* (*cf reverse-charge calling acceptance BrE*) APPL aceptación *f* de cobro revertido, TELEP aceptación *f* de llamada a cobro revertido; ~**-calling request indicator** *AmE n* (*cf reverse-charge calling request indicator BrE*) APPL, TELEP indicador *m* solicitado de cobro revertido

collect[2] *vt* GEN recaudar

collecting: ~ **comb** *n* TRANS *cables* peine *m* recogedor

collection: ~ **charge** *n* TELEP cargo *m* por cobro; ~ **of terms** *n* GEN *linguistics* colección *f* de términos

collective: ~ **aerial** *n* TV antena *f* colectiva; ~ **earthing** *BrE n* (*cf collective grounding AmE*) POWER toma *f* de tierra colectiva; ~ **earth system** *BrE n* (*cf collective ground system AmE*) POWER sistema *m* de tierra colectivo; ~ **grounding** *AmE n* (*cf collective earthing BrE*) POWER toma *f* de tierra colectiva; ~ **ground system** *AmE n* (*cf collective earth system BrE*) POWER sistema *m* de toma de tierra colectivo; ~ **term** *n* GEN *linguistics* término *m* colectivo

collector *n* GEN colector *m*; ~**-base voltage** ELECT *semiconductors* tensión *f* colector-base; ~ **current** ELECT *semiconductors* corriente *f* del colector; ~**-emitter voltage** ELECT *semiconductors* tensión *f* colector-emisor; ~ **junction** ELECT *semiconductors* unión *f* de colectores; ~ **region** ELECT *transistors* región *f* del colector; ~ **resistance** ELECT *semiconductors* resistencia *f* del colector; ~ **series resistance** ELECT, PARTS *semiconductors* resistencia *f* en serie del colector; ~ **terminal** ELECT *semiconductors* terminal *m* del colector; ~ **termination** ELECT *semiconductors* terminación *f* del colector

collimation *n* TRANS colimación *f*

collimator *n* TRANS colimador *m*

collision *n* GEN colisión *f*; ~ **detection** ELECT detección *f* de colisión

collocated: ~ **concentrator** *n* SWG concentrador *m* colocado

co-located: ~ **antenna** *n* RADIO antena *f* anexa; ~ **exchange concentrator** *n* TELEP concentrador *m* anexo a la central; ~ **satellite** *n* RADIO satélite *m* anexa

colon *n* COMP colón *m*

colophony *n* ELECT colofonia *f*

color *AmE see colour BrE*

coloration *n* GEN coloración *f*

colorfast *AmE see colourfast BrE*

colorfastness *AmE see colourfastness BrE*

colorimetric: ~ **function** *n* COMP función *f* colorimétrica

colorization *n* COMP coloración *f*

ColorSence® *n* COMP ColorSence *m*®

ColorSync® *n* COMP ColorSync *m*®

colour *BrE* (*AmE* **color**) *n* GEN color *m*; ~ **carrier** TRANS portadora *f* de color; ~ **code** GEN código *m* de color; ~ **component** TELEP componente *m* cromático; ~ **graphics adaptor** (*CGA*) COMP adaptador *m* gráfico de colores (*CGA*); ~ **image** TELEP imagen *f* cromática; ~ **image scanner** PER APP escáner *m* de imágenes en color; ~ **interpolation shading** COMP sombra *f* de interpolación de los colores; ~ **look-up table** (*CLUT*) COMP tabla *f* de consulta de colores (*CLUT*); ~ **muting** COMP amortiguación *f* de colores; ~ **scheme** COMP esquema *m* de colores; ~ **separation overlay** COMP capa *f* de separación de colores; ~ **signal** COMP señal *f* de color; ~ **slide** GEN diapositiva *f* en color; ~ **space** TV espacio *m* cromático; ~ **stimulus** COMP estímulo *m* de colores

coloured: ~ **noise** *BrE* (*AmE* **colored noise**) *n* RADIO ruido *m* disfrazado

colourfast *BrE* (*AmE* **colorfast**) *adj* GEN de color *m* sólido

colourfastness *BrE* (*AmE* **colorfastness**) *n* GEN solidez *f* del color

COLP *abbr* (*connected-line identification presentation*) TELEP presentación *f* de identificación de línea conectada

COLR *abbr* (*connected-line identification restriction*) TELEP restricción *f* de identificación de línea conectada

column *n* GEN columna *f*

COM *abbr* (*continuation of message*) DATA continuación *f* de mensaje

COMA *abbr* (*ML COM of A branch*) TELEP COMA (*ML COM de la ramificación A*)

COMB *abbr* (*ML COM of B branch*) TELEP COMB (*ML COM de ramificación B*)

combed *adj* ELECT peinado

combi: ~-**telephone** *n* RADIO *mobile*, TELEP teléfono *m* combinado

combinational *adj* GEN combinatorio; ~ **circuit** *n* ELECT circuito *m* combinacional

combinatorial: ~ **circuit** *n* ELECT circuito *m* combinatorio

combine *vt* ELECT combinar

combined: ~ **distribution frame** *n* (*CDF*) SWG trama *f* de distribución combinada; ~ **local/transit exchange** *n* SWG centralita *f* combinada local/tránsito; ~ **local-toll system** *n* SWG sistema *m* combinado local-interurbano; ~ **loss and delay system** *n* TELEP sistema *m* combinado de pérdida y retardo; ~ **signalling** *BrE* **sender and receiver PE** *n* SIG EP *m* emisor y receptor de señalización combinada; ~ **station** *n* COMP estación *f* mixta; ~ **transmitting/receiving station** *n* RADIO estación *f* mixta de transmisión/recepción

combiner *n* ELECT combinador *m*; ~ **unit** ELECT unidad *f* combinadora

combining: ~ **network** *n* TRANS red *f* combinatoria

come: ~ **on stream** *vi* GEN llegar a un funcionamiento normal

comet: ~ **tail** *n* COMP cabellera *f* de cometa; ~ **tail effect** *n* COMP efecto *m* de cabellera de cometa

comfort: ~ **tone** *n* TELEP tono *m* agradable

comma *n* COMP coma *f*

command: ~ **antenna** *n* RADIO antena *f* de mando; ~ **character** *n* COMP, DATA carácter *m* de mando; ~ **class** *n* TELEP clase *f* de comando; ~ **code** *n* DATA, TELEP código *m* de mandato; ~ **and control system** *n* CONTR sistema *m* de orden y control; ~-**descriptor block** *n* (*CDB*) DATA bloque *m* descriptor de orden (*CDB*); ~ **document start** *n* DATA inicio *m* de comando de documento; ~ **function** *n* DATA función *f* de orden; ~ **identifier** *n* TELEP identificador *m* de comando; ~ **interface unit** *n* CONTR unidad *f* de interfaz de comando; ~ **language** *n* COMP lenguaje *m* de órdenes; ~ **line** *n* COMP, DATA *Internet* línea *f* de comandos; ~ **option** *n* TELEP opción *f* de comando; ~ **reference** *n* TELEP referencia *f* de comando; ~ **sequence number** *n* SWG número *m* de secuencia de comando; ~ **unit** *n* CONTR unidad *f* de comando; ~ **variable** *n* CONTR variable *f* de mandato

command/response: ~ **bit** *n* (*C/R*) TELEP bitio *m* de comando/respuesta (*C/R*)

comment *n* GEN comentario *m*

commentary: ~ **channel** *n* TRANS canal *m* de comentario

commercial[1]: ~ **confidence** *n* GEN confianza *f* comercial; ~ **operator** *n* TV operador *m* comercial; ~ **policy** *n* GEN política *f* comercial; ~ **satellite telecast** *n* TV televisión *f* por satélite comercial; ~ **secrecy** *n* GEN secreto *m* comercial

commercial[2] *n* TV anuncio *m* publicitario

commission *vt* TELEP encomendar, autorizar

commissioned *adj* TELEP encomendado, autorizado; ~ **research** *n* TEST investigación *f* encomendada

commissioning *n* TELEP puesta *f* en marcha; ~ **objective** TELEP objetivo *m* de entrada en servicio; ~ **test** TEST prueba *f* de puesta en servicio; ~ **time** GEN tiempo *m* de puesta en servicio

commitment *n* GEN compromiso *m*; ~, **concurrency and recovery** (*CCR*) DATA servicio *m* de red de compromisos, acuerdos y recuperación (*CCR*); ~ **coordinator** DATA coordinador *m* de obligaciones

committed: ~ **information rate** *n* (*CIR*) DATA velocidad *f* de información obligada

common *adj* GEN común; ~ **aerial** *n* RADIO antena *f* común; ~ **air interface** *n* RADIO interfaz *f* aérea común; ~ **ASEs** *n pl* DATA ASEs *m pl* comunes; ~ **associated signalling** *BrE* *n* SIG señalización *f* asociada común; ~ **base** *n* ELECT base *f* común; ~ **base station working** *n* RADIO trabajo *m* con estación de base común; ~ **battery** *n* TELEP batería *f* central; ~ **branch** *n* TRANS rama *f* común; ~ **carrier** *n* GEN, TRANS portadora *f* común; ~ **channel exchange** *n* SWG intercambio *m* por canal común; ~ **channel exchange, first** *n* SWG central *f* telefónica de canal común, primera; ~ **channel exchange, last** *n* SWG central *f* telefónica de canal común, última; ~**-channel signalling** *BrE n* SIG señalización *f* de canal común; ~**-channel signalling** *BrE* **system defined by CCITT** *n* SIG sistema *m* de señalización de canal común definido por CCITT; ~ **collector** *n* ELECT colector *m* común; ~ **command language** *n* COMP lenguaje *m* común de órdenes; ~ **control channel** *n* CONTR, TRANS canal *m* de control común; ~ **control equipment** *n* SWG equipo *m* de control común; ~**-control switching system** *n* SWG sistema *m* de conmutación de control común; ~ **control system** *n* SWG sistema *m* de control común; ~ **emitter** *n* ELECT emisor *m* común; ~ **equipment** *n* SWG equipo *m* común; ~ **field** *n* COMP campo *m* común; ~ **highway** *n* SWG conductor *m* común; ~ **intermediate format** *n* COMP formato *m* intermedio común; ~ **management information protocol** *n* SWG protocolo *m* de información de gestión común; ~ **management-information protocol** *n* DATA protocolo *m* común de información sobre gestión; ~**-management information protocol data unit** *n* DATA unidad *f* de datos de protocolo de información de gestión común; ~ **management-information protocol machine** *n* DATA ordenador *m* común de protocolo de información sobre gestión *Esp*, computador *m* común de protocolo de información sobre gestión *AmL*, computadora *f* común de protocolo de información sobre gestión *AmL*; ~ **management-information service** *n* (*CMIS*) DATA servicio *m* común de información sobre gestión; ~ **management-information service element** *n* (*CMISE*) DATA elemento *m* común del servicio de información sobre gestión; ~ **memory** *n* COMP memoria *f* común; ~**-mode rejection ratio** *n* TELEP porcentaje *m* de rechazo de modo común; ~ **operating standard** *n* RADIO norma *f* de funcionamiento común; ~ **return** *n* GEN retorno *m* combinado; ~ **routing channel** *n* RADIO canal *m* de encaminamiento común; ~**-sense reasoning** *n* DATA razonamiento *m* de sentido común; ~ **store** *n* DATA memoria *f* común; ~ **textual grammar** *n* DATA gramática *f* textual común; ~ **traffic terminal equipment** *n* TELEP equipo *m* común de terminal de tráfico; ~ **user access** *n* DATA acceso *m* de usuario común; ~ **waveguide multiplexer** *n* RADIO multiplexor *m* guiaondas común

commonality *n* TELEP similitud *f*

communicate *vi* GEN comunicar

communication *n* GEN comunicación *f*; ~ **application** TELEP aplicación *f* de comunicación; ~ **capability** DATA función *f* de comunicación; ~ **card** DATA tarjeta *f* de comunicación; ~ **channel** TV canal *m* de comunicaciones; ~ **console** DATA consola *f* de comunicación; ~ **control card** COMP tarjeta *f* de control de comunicaciones; ~ **engineering** TELEP ingeniería *f* de telecomunicaciones; ~ **error** GEN error *m* de comunicación; ~ **path** DATA vía *f* de comunicación; ~ **port** COMP puerta *f* de comunicaciones; ~ **protection** DATA protección *f* de la comunicación; ~ **protocol** DATA *Internet* protocolo *m* de comunicación; ~ **services** *n pl* (*CS*) APPL servicios *m pl* de comunicación; ~ **software** COMP, DATA programas *m pl* de comunicaciones; ~ **station** RADIO estación *f* de comunicaciones; ~ **theory** GEN teoría *f* de la comunicación

communications: ~ **agreement** *n* GEN acuerdo *m* de comunicaciones; ~ **centre** *BrE n* TRANS centro *m* de comunicaciones; ~ **processor** *n* SWG procesador *m* de comunicaciones; ~ **research centre** *BrE n* GEN centro *m* de investigación de comunicaciones; ~ **server** *n* DATA servidor *m* de comunicaciones; ~ **spacecraft operation centre** *BrE n* RADIO centro *m* de operación de vehículo espacial de comunicaciones; ~ **system control** *n* (*CSC*) GEN control *m* del sistema de comunicaciones (*CSC*); ~ **system management** *n* GEN gestión *f* del sistema de comunicaciones; ~ **system monitoring** *n* (*CSM*) TEST supervisión *f* del sistema de comunicaciones; ~ **technology satellite** *n* RADIO satélite *m* con tecnología de la comunicación

communicator *n* TELEP comunicante *m*

community: ~ **aerial** *n* RADIO, TV antena *f* colectiva; ~ **antenna** *n* RADIO, TV antena *f* colectiva; ~**-antenna television** *n* (*CATV*) TV televisión *f* con antena comunitaria; ~ **dial office** *AmE n* SWG central *f* telefónica comunitaria; ~ **network** *n* TELEP, TV red *f* comunitaria; ~ **programme** *BrE n* TV programa *m* comunitario; ~ **reception** *n* RADIO recepción *f* comunitaria; ~ **of transponders** *n* RADIO comunidad *f* de radiofaros respondedores

commutation *n* GEN conmutación *f*

commutator *n* GEN conmutador *m*

compact[1] *adj* GEN compacto; ~ **constraint table** *n* DATA tabla *f* de vínculo compacto; ~ **disc** *n* COMP disco *m* compacto; ~ **disc file manager** *n* COMP gestor *m* de ficheros de disco compacto;

~ **disc interactive** *n* COMP disco *m* compacto interactivo; ~ **disc read-only memory** *n* (*CD-ROM*) COMP disco *m* compacto de memoria sólo de lectura (*CD-ROM*); ~ **disc-write once** *n* COMP disco *m* compacto de una sola escritura; ~ **test-case table** *n* DATA tabla *f* compacta de proceso de prueba
compact[2] *vt* GEN compactar
compacted: ~ **conductor** *n* TRANS *cables* conductor *m* compactado
compactness *n* GEN compacidad *f*
compand *vt* COMP, DATA, TELEP, TRANS comprimir-expandir
companded: ~ **delta modulation** *n* (*CDM*) SIG, TRANS modulación *f* delta compandida
companding: ~ **and frequency modulation** *n* (*CFM*) TRANS modulación *f* de compresión-expansión y de frecuencia (*CFM*)
company: ~ **code** *n* PER APP *PBX* código *m* de empresa; ~ **identity code** *n* PER APP *PBX* código *m* de identificación de la empresa; ~ **network** *n* TRANS red *f* de empresa
comparator *n* ELECT comparador *m*; ~ **response** ELECT respuesta *f* de comparador
comparing: ~ **element** *n* CONTR elemento *m* de comparación
comparison *n* GEN comparación *f*; ~ **circuit** ELECT circuito *m* de comparación; ~ **measurement** TEST medida *f* de comparación; ~ **method** TEST método *m* de comparación; ~ **method of measurement** TEST método *m* de medición por comparación
compartmental: ~ **model** *n* TEST modelo *m* compartimentado
compatibility *n* COMP, GEN compatibilidad *f*
compatible *adj* GEN compatible; ~ **computer** *n* COMP ordenador *m* compatible *Esp*, computador *m* compatible *AmL*, computadora *f* compatible *AmL*
compelled: ~ **cycle** *n* SIG ciclo *m* de secuencia obligada; ~ **multi-frequency code** *n* SIG código *m* multifrecuencia de secuencia obligada; ~ **signalling** *BrE* *n* SIG señalización *f* de secuencia obligada
compendium *n* GEN compendio *m*
compensate *vt* GEN compensar
compensated: ~ **semiconductor** *n* ELECT semiconductor *m* compensado
compensating: ~ **element** *n* CONTR elemento *m* compensador; ~ **network** *n* RADIO *aerials* red *f* compensadora
compensation *n* GEN compensación *f*; ~ **circuit** ELECT circuito *m* de compensación
competition *n* GEN competencia *f*
competitive *adj* GEN competitivo
competitiveness *n* GEN competitividad *f*
compilation *n* GEN compilación *f*
compile *vt* DATA *directory*, GEN compilar
compiler *n* COMP compilador *m*; ~ **generator** COMP generador *m* de compiladores
compiling: ~ **phase** *n* COMP fase *f* de compilación; ~ **program** *n* COMP programa *m* de compilación
complement *n* GEN complemento *m*
complementary: ~ **code** *n* TRANS código *m* complementario; ~ **method of measurement** *n* TEST método *m* complementario de medición; ~ **service** *n* APPL, TELEP servicio *m* complementario
complete[1] *adj* GEN completo; ~ **failure** *n* TEST fallo *m* total; ~ **valid input-signal set** *n* DATA juego *m* completo de señal de entrada válida
complete[2] *vt* GEN completar
completed: ~ **call** *n* TELEP llamada *f* terminada; ~ **call attempt** *n* RADIO, TELEP intento *m* de llamada terminado; ~ **charging period** *n* APPL periodo *m* de carga completo
completion *n* GEN terminación *f*; ~ **of call** TELEP terminación *f* de llamada; ~ **ratio** TELEP relación *f* de terminación; ~ **test** TEST prueba *f* de finalización
complex: ~ **instruction set equipment** *n* (*CISE*) DATA equipo *m* complejo de conjunto de instrucciones (*CISE*); ~ **wave** *n* RADIO onda *f* compleja
complexity *n* GEN complejidad *f*
compliance *n* GEN elasticidad *f*; ~ **test** TEST prueba *f* de aptitud
compliant: ~ **bridge** *n* ELECT puente *m* elástico
comply *vi* GEN *with requirements* cumplir
copolar: ~ **attenuation** *n* TRANS atenuación *f* copolar
component: ~ **data element separator** *n* DATA separador *m* por elementos de los datos de componentes; ~ **engineering** *n* GEN tecnología *f* de los componentes; ~ **front** *n* ELECT *printed circuits* frente *m* del componente; ~ **height** *n* ELECT *printed circuits* altura *f* del componente; ~ **holder** *n* ELECT soporte *m* del componente; ~ **hole** *n* ELECT *printed circuits* agujero *m* del componente; ~ **housing** *n* ELECT alojamiento *m* de componente; ~**-level testing** *n* TEST comprobación *f* de componentes; ~ **technology** *n* ELECT tecnología *f* de los componentes
composite: ~ **gain** *n* TELEP, TRANS ganancia *f* compuesta; ~ **loss** *n* TELEP, TRANS atenuación *f* compuesta; ~ **loss or gain** *n* TELEP, TRANS pérdida *f* o ganancia compuestas; ~ **node** *n* COMP nodo *m* compuesto; ~ **resistor** *n* ELECT,

PARTS resistor *m* compuesto, resistor *m* combinado; ~ **signal** *n* ELECT, SIG, TV señal *f* compuesta; ~ **signal coding** *n* TRANS codificación *f* de la señal compuesta; ~ **structure of grooved silicon chips** *n* ELECT estructura *f* combinada de las pastillas de sílice ranuradas; ~ **video** *n* TV vídeo *m* combinado; ~ **wave** *n* RADIO onda *f* compuesta; ~ **waveform** *n* ELECT forma *f* de onda compuesta

composition *n* GEN composición *f*

compound[1]: ~ **document architecture** *n* DATA arquitectura *f* compuesta de documento; ~ **parameter argument** *n* TELEP argumento *m* de parámetro compuesto; ~ **signal** *n* SIG señal *f* compuesta

compound[2] *n* GEN compuesto *m*

comprehensive *adj* GEN comprehensivo

compress *vt* DATA comprimir

compressed *adj* DATA compressed; ~ **data** *n pl* DATA datos *m pl* comprimidos; ~ **digital transmission** *n* TRANS transmisión *f* digital comprimida; ~ **mode** *n* DATA modo *m* comprimido; ~ **operation** *n* DATA operación *f* comprimida

compression *n* GEN compresión *f*; ~ **of the luminance range** TELEP reducción *f* de la gama de luminancia; ~ **ratio** TRANS relación *f* de compresión

compressor *n* GEN compresor *m*

compromise: ~ **equalization** *n* TRANS compensación *f* reveladora

compromising: ~ **emanations** *n pl* RADIO *security* emanaciones *f pl* comprometedoras

compulsory *adj* GEN obligatorio, forzoso; ~ **clearing request** *n* SIG demanda *f* obligatoria de fin de conversación

COMPUSEC *abbr* (*computer security*) COMP, DATA COMPUSEC (*seguridad informática*)

computation *n* COMP cómputo *m*, cálculo *m*; ~ **theory** COMP teoría *f* del cálculo

compute[1]: ~ **mode** *n* COMP modo *m* de cálculo

compute[2] *vt* COMP computar

computer *n* GEN ordenador *m Esp*, computador *m AmL*, computadora *f AmL*; ~**-aided design** (*CAD*) COMP diseño *m* asistido por ordenador *Esp* (*CAD*), diseño *m* asistido por computador *AmL*, diseño *m* asistido por computadora *AmL*; ~**-aided design and** ~**-aided manufacturing** (*CAD/CAM*) COMP diseño *m* y fabricación asistidos por ordenador *Esp* (*CAD/CAM*), diseño *m* y fabricación asistidos por computador *AmL* (*CAD/CAM*), diseño *m* y fabricación asistidos por computadora *AmL* (*CAD/CAM*); ~**-aided engineering** (*CAE*) COMP ingeniería *f* asistida por ordenador *Esp* (*CAE*), ingeniería *f* asistida por computador *AmL* (*CAE*), ingeniería *f* asistida por computadora *AmL* (*CAE*); ~**-aided logistics** COMP logística *f* asistida por ordenador *Esp*, logística *f* asistida por computador *AmL*, logística *f* asistida por computadora *AmL*; ~**-aided management** COMP gestión *f* informatizada; ~**-aided manufacturing** (*CAM*) COMP fabricación *f* asistida por ordenador *Esp* (*CAM*), fabricación *f* asistida por computador *AmL* (*CAM*), fabricación *f* asistida por computadora *AmL* (*CAM*); ~**-aided quality** (*CAQ*) COMP calidad *f* asistida por ordenador *Esp* (*CAQ*), calidad *f* asistida por computador *AmL* (*CAQ*), calidad *f* asistida por computadora *AmL* (*CAQ*); ~**-aided software engineering** (*CASE*) COMP ingeniería *f* de soporte lógico auxiliada por computadora *AmL*, ingeniería *f* de soporte lógico auxiliada por computador *AmL*, ingeniería *f* de soporte lógico auxiliada por ordenador *Esp*; ~**-aided testing** (*CAT*) COMP prueba *f* asistida por ordenador *Esp* (*CAT*), prueba *f* asistida por computador *AmL* (*CAT*), prueba *f* asistida por computadora *AmL* (*CAT*); ~**-assisted instruction** COMP aprendizaje *m* asistido por computadora *AmL*, aprendizaje *m* asistido por computador *AmL*, aprendizaje *m* asistido por ordenador *Esp*; ~**-assisted learning** (*CAL*) COMP aprendizaje *m* auxiliado por computadora *AmL* (*CAL*), aprendizaje *m* auxiliado por computador *AmL* (*CAL*), aprendizaje *m* auxiliado por ordenador *Esp* (*CAL*); ~**-assisted management** COMP gestión *f* asistida por ordenador *Esp*, gestión *f* asistida por computador *AmL*, gestión *f* asistida por computadora *AmL*; ~**-assisted vision** COMP visión *f* asistida por ordenador *Esp*, visión *f* asistida por computador *AmL*, visión *f* asistida por computadora *AmL*; ~ **bank** COMP banco *m* de memoria; ~**-based learning** COMP aprendizaje *m* basado en el ordenador *Esp*, enseñanza *f* basada en el computador *AmL*, enseñanza *f* basado en el computadora *AmL*; ~**-based training** (*CBT*) COMP formación *f* basada en ordenadores *Esp* (*CBT*), formación *f* basada en computadores *AmL* (*CBT*), formación *f* basada en computadoras *AmL* (*CBT*); ~ **circuitry** COMP circuitería *f* de ordenador *Esp*, circuitería *f* de computador *AmL*, circuitería *f* de computadora *AmL*; ~ **enthusiast** COMP entusiasta *m* de ordenador *Esp*, entusiasta *m* de computador *AmL*, entusiasta *m* de computadora *AmL*; ~ **fraud** COMP fraude *m* informático; ~ **graphics** *n pl*

COMP, TV gráficos *m pl* de computadora *AmL*, gráficos *m pl* de computador *AmL*, gráficos *m pl* de computadora *AmL*; ~ **imagery** COMP imaginería *f* informática; ~ **instruction** COMP instrucción *f* de máquina; ~**-integrated manufacturing** (*CIM*) COMP fabricación *f* informatizada (*CIM*); ~ **interface** COMP, DATA interfaz *f* de ordenador *Esp*, interfaz *f* de computador *AmL*, interfaz *f* de computadora *AmL*; ~ **link** TRANS conexión *f* con ordenador *Esp*, conexión *f* con computador *AmL*, conexión *f* con computadora *AmL*; ~ **network** COMP, DATA, TRANS red *f* de ordenadores *Esp*, red *f* de computadores *AmL*, red *f* de computadoras *AmL*; ~ **program** COMP, DATA programa *m* de ordenador *Esp*, programa *m* de computador *AmL*, programa *m* de computadora *AmL*; ~ **scientist** GEN especialista *m* en informática; ~ **security** (*COMPUSEC*) COMP, DATA seguridad *f* informática (*COMPUSEC*); ~ **simulation** COMP simulación *f* por ordenador *Esp*, simulación *f* por computador *AmL*, simulación *f* por computadora *AmL*; ~ **system** COMP sistema *m* informático; ~ **virus** COMP, DATA virus *m* informático

computerization *n* COMP informatización *f*

computerized *adj* COMP automatizado, informatizado *Esp*, computadorizado *AmL*; ~ **power supply** *n* POWER suministro *m* eléctrico automatizado; ~ **typesetting** *n* APPL composición *f* automatizada

computer-PBX: ~ **interface** *n* (*CPI*) COMP interfaz *f* ordenador-PBX *Esp*, interfaz *f* computador-PBX *AmL*, interfaz *f* computadora-PBX *AmL*

computing *n* COMP informática *f Esp*, computación *f AmL*, cálculo *m*; ~ **centre** *BrE* COMP centro *m* de cálculo

concatenated: ~ **coding** *n* (*CC*) DATA código *m* encadenado

concatenation *n* DATA concatenación *f*; ~ **indication** (*CI*) ELECT indicación *f* de concatenación

concealment *n* GEN ocultación *f*; ~ **method** COMP método *m* de ocultación

concentration *n* GEN concentración *f*; ~ **module** ELECT módulo *m* de concentración (*ASM*); ~ **network** TRANS red *f* de concentración; ~ **stage** TRANS etapa *f* de concentración; ~ **in a switching stage** TELEP concentración *f* en una etapa de conmutación

concentrator *n* ELECT, TRANS concentrador *m*

concentric *adj* GEN concéntrico; ~ **conductor** *n* TRANS conductor *m* concéntrico; ~ **neutral cable** *n* TRANS cable *m* neutral concéntrico; ~ **stranding** *n* TRANS cableado *m* concéntrico

concentricity *n* GEN concentricidad *f*

concept *n* GEN concepto *m*

conceptual *adj* GEN conceptual; ~ **clustering** *n* COMP agrupación *f* conceptual; ~ **primitive** *n* COMP primitivo *m* conceptual

conclude *vt* GEN concluir

concrete: ~ **grammar** *n* DATA gramática *f* concreta; ~ **graphical grammar** *n* DATA gramática *f* gráfica concreta; ~ **graphical syntax** *n* DATA sintaxis *f* gráfica concreta; ~ **syntax** *n* DATA sintaxis *f* concreta; ~ **textual syntax** *n* DATA sintaxis *f* textual concreta

concurrent: ~ **statements** *n pl* DATA sentencias *f pl* concurrentes

condensation: ~ **soldering** *n* PARTS soldadura *f* en fase de vapor

condense *vt* DATA, GEN, TELEP condensar

condensed *adj* DATA condensado; ~ **sub-burst data** *n pl* DATA datos *m pl* condensados en subráfagas

condenser *n* COMP capacitor *m*; ~ **microphone** TELEP micrófono *m* de condensador

condensing *n* DATA condensación *f*

condition *n* GEN condición *f*; ~ **A** TELEP condición *f* A; ~ **Z** TELEP condición *f* Z

conditional: ~ **access** *n* TV acceso *m* condicional; ~ **access control** *n* RADIO control *m* de acceso condicional; ~ **component** *n* DATA componente *m* condicional; ~ **expression** *n* DATA expresión *f* condicional; ~ **jump** *n* COMP salto *m* condicional; ~ **notification class of service** *n* DATA notificación *f* condicional de clase de servicio; ~ **parameter** *n* DATA parámetro *m* condicional; ~ **replenishment** *n* DATA reaprovisionamiento *m* condicional; ~ **requirement** *n* DATA requerimiento *m* condicional, DATA requisito *m* condicional; ~ **selection** *n* SWG selección *f* condicional

conditionally: ~ **mandatory** *adj* (*CM*) TELEP condicionalmente obligatorio

conductance *n* ELECT, POWER conductancia *f*

conducted: ~ **effect** *n* ELECT *EMC* efecto *m* propagado por conducción; ~ **spurious emission** *n* ELECT emisión *f* parásita conducida

conductibility *n* GEN conductibilidad *f*

conducting *adj* ELECT, POWER conductor; ~ **on** ELECT *transistors* en estado de conducción; ~ **direction** *n* ELECT dirección *f* de conducción; ~ **layer** *n* ELECT capa *f* conductora; ~ **state** *n* ELECT volumen *m* conductor

conduction *n* GEN conducción *f*; ~ **band** ELECT

banda *f* de conducción; ~ **electron** ELECT electrón *m* de conducción; ~ **period** ELECT periodo *m* de conducción; ~ **time** ELECT tiempo *m* de conducción; ~ **time control** ELECT control *m* del tiempo de conducción

conductive *adj* ELECT, POWER conductor; ~ **layer** *n* ELECT capa *f* conductora; ~ **pattern** *n* ELECT impresión *f* conductora

conductivity *n* GEN conductividad; ~ **modulation** ELECT modulación *f* de conductividad

conductor *n* GEN conductor *m*; ~ **area** ELECT sección *f* transversal del conductor; ~ **insulation** TRANS aislamiento *m* de conductor; ~ **screen** TRANS pantalla *f* de conductor; ~ **spacing** ELECT *printed circuits* separación *f* entre conductores; ~ **unit** TRANS torón *m*; ~ **width** ELECT *printed circuits* anchura *f* de conductor

conduit *n* ELECT, PARTS, TRANS conducto *m* portacables; ~ **route** TRANS *cable laying* trayecto *m* de canalización; ~ **run** TRANS *cable laying* tramo *m* de canalización; ~ **system** TRANS *cable laying* sistema *m* de canalización

cone: ~ **loudspeaker** *n* PARTS altavoz *m* de cono

conference *n* GEN conferencia *f*; ~ **bridge** APPL puente *m* de conferencia; ~ **call** APPL, TELEP comunicación *f* colectiva; ~ **call chairman** APPL director *m* de conferencia colectiva; ~ **call chairperson** APPL director *m* de conferencia múltiple; ~ **calling** APPL llamada *f* de conferencia múltiple; ~ **call service** APPL servicio *m* de conferencia múltiple, TELEP servicio *m* de llamadas de conferencia; ~ **circuit** TELEP circuito *m* de conferencias; ~ **repeater** TELEP repetidor *m* de conferencias; ~ **telegraph repeater** TELEP repetidor *m* telegráfico para conferencias

conferencing: ~ **configuration** *n* TV configuración *f* de la conferencia

confidence: ~ **coefficient** *n* GEN coeficiente *m* de seguridad; ~ **interval** *n* GEN intervalo *m* de confianza; ~ **level** *n* GEN nivel *m* de confianza; ~ **limit** *n* GEN límite *m* de confianza

confidential *adj* GEN confidencial

configuration *n* GEN configuración *f*; ~ **and data management** DATA configuración *f* y gestión de datos

configurations: ~ **foldback** *n* ELECT reinyección *f* de configuraciones

configure *vt* GEN configurar

confinement *n* GEN retención *f*

confirm *vt* GEN confirmar

confirmed: ~ **service** *n* APPL, DATA servicio *m* confirmado

confirmation *n* DATA, TELEP confirmación *f*; ~ **of receipt** (*COR*) DATA, TELEP confirmación *f* de recepción (*COR*); ~ **to receive** (*CFR*) DATA, TELEP confirmación *f* por recibir (*CFR*)

confluence *n* GEN confluencia *f*

conformal: ~ **antenna** *n* RADIO antena *f* conforme

conformance *n* GEN conformidad *f*; ~ **of devices** TELEP conformidad *f* de los dispositivos; ~ **log** DATA registro *m* de conformidad; ~**-resolution test** DATA, TEST prueba *f* de resolución de conformidad; ~ **testing** DATA, TEST comprobación *f* de conformidad; ~**-testing specification document** DATA documento *m* de especificación de comprobación de conformidad; ~**-testing specification manual** (*CTSM*) DATA manual *m* de especificación de comprobación de conformidad (*CTSM*); ~**-testing standard** DATA, TEST norma *f* para la comprobación de conformidad; ~ **test specification manual** DATA, TEST manual *m* de especificación de prueba de conformidad; ~ **test-suite** DATA, TEST serie *f* de pruebas de conformidad

conforming: ~ **implementation** *n* DATA ejecución *f* de conformidad

conformity *n* GEN conformidad *f*; ~ **specifications** *n pl* TEST especificaciones *f pl* de conformidad

confravision *n* TV videoconferencia *f*

confusion *n* GEN confusión *f*; ~ **jamming** RADIO emisión *f* perturbadora de confusión; ~ **message** DATA mensaje *m* de confusión; ~ **signal** SIG señal *f* de confusión

congested: ~**-route counter** *n* SWG, TELEP contador *m* de congestiones de ruta

congestion *n* SWG congestión *f*; ~ **indicator** RADIO indicador *m* de congestión; ~ **tone** TELEP tono *m* de congestión

conical *adj* GEN cónico; ~ **scan** *n* RADIO exploración *f* cónica; ~ **scanning with notation** *n* RADIO exploración *f* cónica con notación; ~ **scanning with rotation** *n* RADIO exploración *f* cónica con rotación

conjunction *n* ELECT *logic elements* unión *f*

connect *vt* GEN conectar; ◆ ~ **in cascade** ELECT conectar en cascada; ~ **in a loop** POWER conectar en un bucle; ~ **negative** POWER conectar negativo; ~ **in parallel** ELECT, POWER conectar en paralelo; ~ **in series** ELECT conectar en serie

connectability *n* TV conectabilidad *f*

Connect-Additional *n* TELEP conexión *f* adicional

connected *adj* GEN conectado; ~**-line**

identification *n* DATA, TELEP identificación *f* de línea conectada; **~-line identification presentation** *n* TELEP presentación *f* de identificación de línea conectada; **~-line identification restriction** *n* (*COLR*) TELEP restricción *f* de identificación de línea conectada; **~-line identity request indicator** *n* TELEP indicador *m* de solicitud de identidad de línea conectada; **~ switching path PE** *n* SWG elemento *m* de procesamiento con camino de conmutación conectado; **~ to** GEN conectado con; **~ to voltage** POWER conectado a voltaje

connecter *n* COMP conector *m*

connecting: **~ box** *n* TELEP caja *f* de distribución; **~ cable** *n* PARTS cable *m* de conexión; **~ device** *n* ELECT dispositivo *m* de conexión; **~ pin** *n* ELECT espiga *f* de conexión; **~ tag** *n* ELECT espiga *f* de conexión; **~ terminal** *n* COMP terminal *m* de conexión; **~ wire** *n* ELECT alambre *m* de conexión

Connect-Initial *n* TELEP conexión *f* inicial

connection *n* GEN conexión *f*; **~ box** TELEP, TV cuadro *m* de conexiones; **~ building network** TRANS red *f* de establecimiento de conexión; **~ cable** PARTS cable *m* de conexión; **~ centre** *BrE* TV centro *m* de conexión; **~ charge** TELEP cuota *f* de conexión, TV carga *f* de conexión; **~ communication path** TELEP trayecto *m* de comunicación de la conexión; **~ delay** TV retardo *m* de conexión; **~ diagram** GEN diagrama *m* de conexión; **~ disk** COMP, DATA *Internet* disco *m* de conexión; **~ element** (*CE*) DATA elemento *m* de conexiones; **~ equipment** GEN equipo *m* de conexión; **~ field** POWER campo *m* de conexión; **~ kit** DATA *Internet* equipo *m* de conexión; **~ matrix** SWG matriz *f* de conexión; **~ mode** DATA modo *m* de conexión; **~-mode network service** DATA servicio *m* en red en modo de conexión; **~-oriented mode network service** (*CONS*) APPL servicio *m* de red orientado a la comunicación (*CONS*); **~-oriented network-layer protocol** (*CONP*) TRANS protocolo *m* de capa de red orientado a la conexión (*CONP*); **~ overhead** DATA inicio *m* de la conexión; **~ point** GEN punto *m* de conexión; **~ in progress** TELEP conexión *f* en curso; **~ provider** DATA *Internet* proveedor *m* de la conexión; **~-related function** (*CRF*) DATA, TRANS función *f* relacionada con la conexión; **~ request** DATA solicitud *f* de conexión; **~ requirement** DATA *Internet* requisito *m* para la conexión; **~ in series** ELECT conexión *f* en serie; **~ strip** ELECT regleta *f* de conexiones; **~ terminal** COMP terminal *m* de conexión; **~ through an exchange** SWG conexión *f* a través de una centralita; **~ unit** GEN bloque *m* de conexión

connectionless *adj* (*CL*) GEN sin conexiones (*CL*); **~ bearer service** *n* APPL servicio *m* portador sin conexiones; **~ broadband data service** *n* APPL, DATA servicio *m* de datos de banda ancha sin conexiones; **~ convergence protocol** *n* DATA protocolo *m* de convergencia sin conexiones; **~ mode** *n* DATA modo *m* de desconexión; **~-mode network service** *n* APPL servicio *m* de reden el modo sin conexión; **~-mode presentation service** *n* DATA servicio *m* de presentación en modo de desconexión; **~-mode session service** *n* DATA servicio *m* de sesión en modo de desconexión; **~-mode transmission** *n* TRANS transmisión *f* en modo sin conexión; **~ network-layer protocol** *n* TRANS protocolo *m* de capa de red sin conexión; **~ network-layer service** *n* APPL servicio *m* de capa de red sin comunicación; **~ server** *n* DATA servidor *m* sin conexiones; **~ service function** *n* APPL función *f* de servicio sin conexiones

connector *n* GEN conector *m*; **~ insertion loss** TRANS *fibre optics* pérdida *f* de inserción en el conector; **~ loss** TRANS *fibre optics* pérdida *f* del conector; **out ~** GEN conector *m* de salida; **~ terminal** ELECT borne *m* de conexión

CONP *abbr* (*connection-oriented network-layer protocol*) TRANS CONP (*protocolo de capa de red orientado a la conexión*)

CONS *abbr* (*connection-oriented mode network service*) APPL CONS (*servicio de red orientado a la comunicación*)

consecutive *adj* GEN consecutivo; **~ error block** *n* DATA bloque *m* de errores consecutivos; **~ identical digits** *n pl* DATA dígitos *m pl* idénticos consecutivos; **~ number** *n* GEN número *m* consecutivo

consignment *n* GEN remesa *f*

consistency *n* GEN coherencia *f*

consistent: **~ partitioning subset** *n* DATA *specification and description language* subconjunto *m* uniforme de partición; **~-refinement subset** *n* DATA subconjunto *m* de afinación de uniformidad

console *n* GEN consola *f*

consolidated: **~ link-layer management message** *n* DATA mensaje *m* de gestión consolidada de enlace entre capas

consolidation *n* TRANS consolidación *f*

consonant *n* GEN consonante *f*

constant[1]: **~ angular velocity** *n* COMP velocidad *f* angular constante; **~ bit rate** *n* COMP, DATA tasa *f* de bits constante; **~ failure intensity**

period *n* TELEP periodo *m* de intensidad de fallo constante; ~ **linear velocity** *n* COMP velocidad *f* lineal constante; ~ **quality** *n* COMP calidad *f* constante; ~ **rate** *n* COMP, DATA velocidad *f* constante; ~ **shading** *n* COMP sombra *f* constante; ~**-voltage rectifier** *n* POWER rectificador *m* de tensión constante

constant² *n* GEN constante *m*

constituent: ~ **part** *n* GEN parte *f* constitutiva; ~ **unit** *n* PARTS unidad *f* constitutiva

constraint *n* GEN limitación *f*

constraints: ~ **part** *n* DATA componente *m* de restricciones; ~ **reference** *n* DATA referencia *f* de restricciones

construction *n* GEN construcción *f*; ~**-bound rule** PARTS *equipment practice* norma *f* vinculada al acondicionamiento; ~ **element** GEN elemento *m* de construcción; ~ **manager** GEN jefe *m* de obras

constructive: ~ **interference** *n* TRANS *fibre optics* interferencia *f* constructiva; ~ **solid geometry** *n* (*CSG*) COMP geometría *f* constructiva tridimensional (*CSG*)

consultation *n* GEN consulta *f*

consumer: ~**-friendliness** *n* GEN facilidad *f* de uso, facilidad *f* de manejo; ~ **market** *n* GEN mercado *m* de consumo; ~ **risk** *n* TEST *quality* riesgo *m* del consumidor; ~**'s risk point** *n* TEST *quality* punto *m* de riesgo del cliente

contact *n* GEN contacto *m*; ~ **bank** POWER tanque *m* de contacto; ~ **bar** POWER barra *f* de contactos; ~ **erosion** POWER erosión *f* de contactos; ~ **metallization** *BrE* PARTS *semiconductors* metalización *f* por contacto; ~ **method of measurement** TEST método *m* de medida por contacto; ~ **pin** ELECT clavija *f* de contacto; ~ **plate** CLIM *lighting* placa *f* de contactos; ~ **point** POWER, SWG *switchgear* contacto *m* del ruptor; ~**-spring adjuster** TEST doblador *m* de muelles de contacto; ~**-spring bender** TEST doblador *m* de muelles de contacto; ~ **surface** PARTS superficie *f* de contacto; ~ **unit** ELECT, SWG dispositivo *m* de contacto

contactor *n* ELECT, SWG contactor *m*

container *n* (*CT*) GEN contenedor *m*; ~**-n** (*C-n*) APPL, DATA, TELEP contendor *m* n

content: ~**-addressable memory** *n* COMP memoria *f* de contenido direccionable; ~**-addressable storage** *n* DATA almacenamiento *m* de contenido direccionable; ~**-addressed memory** *n* COMP memoria *f* de contenido direccionado; ~**-addressed storage** *n* COMP memoria *f* con contenido direccionado; ~ **length** *n* DATA amplitud *f* de contenido

contention *n* GEN capacidad; ~ **control** RADIO control *m* de contención

context *n* COMP, DATA contexto *m*; ~ **parameter** DATA parámetro *m* del contexto; ~ **search** COMP búsqueda *f* por contexto

contextual *adj* COMP, DATA contextual

continent: ~ **code** *n* TELEP *telex* código *m* continental

continental: ~ **circuit** *n* TELEP circuito *m* continental; ~ **connection** *n* TELEP conexión *f* continental; ~ **exchange** *n* SWG centralita *f* continental

contingency: ~ **period** *n* TEST periodo *m* de contingencia; ~ **plan** *n* GEN plan *m* de emergencia; ~ **planning** *n* GEN planeamiento *m* por contingencia; ~ **table** *n* GEN *statistics* tabla *f* de contingencia

continuation: ~ **character** *n* TELEP carácter *m* de continuación; ~ **of message** *n* (*COM*) DATA continuación *f* de mensaje; ~ **mode** *n* TELEP modo *m* de continuación; ~**-mode operating sequence** *n* TELEP secuencia *f* que opera en modo de continuación; ~ **page** *n* GEN *documentation* página *f* de continuación; ~ **rack** *n* PARTS bastidor *m* de continuación

continue: ~ **key** *n* COMP tecla *f* de continuación

continuity: ~**-check responder** *n* TELEP contestador *m* de comprobación de continuidad; ~**-check transceiver** *n* TELEP transceptor *m* de comprobación de continuidad; ~ **fault** *n* TEST fallo *m* en la continuidad; ~ **signal** *n* SIG señal *f* de continuidad

continuous *adj* COMP continuo; ~ **action** *n* CONTR acción *f* continua; ~ **bit oriented stream** *n* DATA corriente *f* continua orientada al bitio; ~ **carrier** *n* TRANS portadora *f* permanente; ~ **direct off-state** *n* ELECT tensión *f* continua permanente en estado bloqueado; ~ **direct reverse-voltage** *n* ELECT *diodes* tensión *f* inversa continua permanente; ~ **direct on-state current** *n* ELECT *thyristors* corriente *f* continua permanente en estado de conducción; ~ **form** *n* PER APP formulario *m* continuo; ~ **line** *n* GEN raya *f* contínua; ~ **loading** *n* GEN *cables* carga *f* continua; ~**-mode transmission monitor** *n* TEST monitor *m* de transmisión en modo continuo; ~ **multiplexing** *n* RADIO multiplexación *f* continua; ~ **receiver** *n* TELEP receptor *m* continuo; ~ **signal** *n* DATA señal *f* continua; ~ **supervision** *n* TEST supervisión *f* continua; ~ **tone-controlled squelch system** *n* (*CTCSS*) RADIO sistema *m* continuo de silenciador de tono controlado (*CTCSS*); ~ **tracking** *n* RADIO seguimiento *m* continuo; ~ **wave** *n* (*CW*) TRANS onda *f* continua; ~**-wave radar** *n* RADIO radar *m* de onda continua

continuously: ~ **variable** *adj* GEN de regulación continua

contour: ~ **correction** *n* COMP corrección *f* del contorno; ~ **extraction** *n* COMP extracción *f* del contorno

contouring *n* COMP contorneo *m*

contractor *n* GEN contratista *m*

contradirectional: ~ **interface** *n* TRANS interfaz *f* contradireccional

contrast *n* GEN contraste *m*; ~ **control** TV control *m* de contraste; ~ **filter** TV *photography* filtro *m* para contraste; ~ **range** TV *video* contraste *m* máximo; ~ **ratio** TV *video* relación *f* de contraste

contribution: ~ **link** *n* RADIO enlace *m* de contribución

control: ~**-and-delay channel** *n* TRANS canal *m* de control y retardo; ~ **area** *n* COMP zona *f* de control; ~ **cabinet** *n* CONTR armario *m* de controles; ~ **cable** *n* CONTR, TRANS cable *m* de mando; ~ **card** *n* CONTR ficha *f* de control; ~ **change** *n* CONTR cambio *m* de control; ~ **channel** *n* CONTR canal *m* de control; ~ **character** *n* COMP, TELEP carácter *m* de control; ~ **circuit** *n* CONTR circuito *m* de control; ~ **codeword** *n* COMP palabra *f* de código de control, DATA palabra *f* codificada de control; ~ **component** *n* CONTR componente *m* de control; ~ **crystal** *n* ELECT cristal *m* de control; ~ **data** *n pl* COMP datos *m pl* de control; ~ **desk** *n* CONTR, TV mesa *f* de control; ~ **device** *n* CONTR dispositivo *m* de control; ~ **diagnostic unit** *n* TEST unidad *f* de diagnóstico de control; ~ **element PE** *n* TELEP elemento *m* pictórico de elemento de control; ~ **engineering** *n* CONTR técnicas *f pl* de control; ~ **equipment** *n* CONTR equipo *m* de maniobra; ~ **with fixed set-point** *n* CONTR regulación *f* a valor constante; ~ **flow diagram** *n* DATA diagrama *m* de circulación de control; ~ **frame** *n* DATA marco *m* de control; ~ **frequency** *n* CONTR frecuencia *f* piloto; ~ **function** *n* (*CF*) GEN función *f* de control (*CF*); ~ **information** *n* COMP información *f* de control; ~ **instruction register** *n* COMP registro *m* de dirección de instrucción; ~ **knob** *n* GEN botón *m* de mando; ~ **language** *n* COMP, ELECT, TELEP lenguaje *m* de control; ~ **limits** *n pl* TEST *quality* límites *m pl* de control; ~ **loop** *n* CONTR bucle *m* de control; ~ **MCSAP** *n* TELEP MCSAP *m* de control; ~ **memory** *n* COMP memoria *f* de control; ~ **message** *n* COMP mensaje *m* de control; ~ **mode** *n* RADIO modo *m* de control; ~ **objectives** *n pl* COMP objetivos *m pl* de control; ~ **operation** *n* COMP, CONTR operación *f* de control; ~ **panel** *n* COMP, CONTR panel *m* de control; ~ **point** *n* CONTR punto *m* de control; ~ **position** *n* CONTR posición *f* de control; ~ **procedure** *n* CONTR, TELEP procedimiento *m* de control; ~ **program** *n* COMP programa *m* de control; ~ **program language** *n* COMP lenguaje *m* de programa de control; ~ **pulse** *n* CONTR impulso *m* de control; ~ **range** *n* CONTR campo *m* de control; ~ **register** *n* COMP registro *m* de control; ~ **room** *n* GEN sala *f* de control; ~ **signal** *n* COMP señal *f* de control; ~ **signalling** *BrE* **code** *n* SIG código *m* de señalamiento de control; ~ **statement** *n* COMP especificación *f* de control; ~ **station** *n* GEN estación *f* de control; ~**-station access multiplex** *n* TELEP multiplexor *m* de acceso de estación de control; ~ **storage** *n* COMP memoria *f* de control; ~ **structure** *n* CONTR estructura *f* de control; ~ **system** *n* CONTR sistema *m* de control; ~ **and system monitoring** *n* TEST vigilancia *f* de sistemas y de controles; ~ **terminal** *n* CONTR terminal *m* de control; ~ **timing** *n* TRANS temporización *f* de control; ~ **transmitter** *n* TRANS transmisor *m* de mando; ~ **unit** *n* CONTR unidad *f* de control; ~ **voltage** *n* POWER voltaje *m* de control; ~ **word** *n* COMP, DATA palabra *f* de control

controllability *n* GEN controlabilidad *f*

controlled: ~**-access unit** *n* (*CAU*) TELEP unidad *f* de acceso controlado (*CAU*); ~**-avalanche rectifier diode** *n* ELECT *semiconductors* diodo *m* rectificador de avalancha controlada; ~ **corrective maintenance** *n* TEST *dependability* mantenimiento *m* correctivo controlado; ~ **maintenance** *n* GEN mantenimiento *m* controlado; ~ **rerouting** *n* TELEP reencaminamiento *m* controlado; ~ **slip** *n* RADIO, TELEP deslizamiento *m* controlado; ~ **terminal** *n* CONTR terminal *m* controlado; ~ **variable** *n* CONTR variable *f* controlada

controller *n* COMP controlador *m*

controlling: ~ **computer** *n* COMP ordenador *m* de control *Esp*, computador *m* de control *AmL*, computadora *f* de control *AmL*; ~ **device** *n* CONTR mecanismo *m* de control; ~ **element** *n* CONTR órgano *m* de mando; ~ **equipment** *n* CONTR equipo *m* de control; ~ **exchange** *n* SWG centro *m* director; ~ **operator** *n* TELEP operadora *f* principal; ~ **reference burst**

parameter *n* DATA parámetro *m* de control de ráfaga de referencia
conurbation *n* GEN conurbación *f*
conventional: ~ **degree of distortion** *n* TELEP grado *m* aceptado de distorsión; ~ **true value of a quantity** *n* TEST *quality* valor *m* verdadero aceptado de una cantidad
convergence *n* DATA, TRANS convergencia *f*; ~ **sublayer** DATA subcapa *f* de convergencia; ~ **sublayer protocol data unit** (*CSPDU*) DATA unidad *f* de datos con protocolo de subcapa de convergencia; ~ **time** TELEP tiempo *m* de convergencia
conversation *n* GEN conversación *f*; ~ **impossible** RADIO conversación *f* imposible; ~ **time** RADIO, TELEP duración *f* de una conversación; ~ **time limiter** TELEP limitador *m* de conversaciones
conversational *adj* DATA conversacional; ~ **language** *n* COMP lenguaje *m* coloquial; ~ **mode** *n* COMP, TRANS modo *m* conversacional; ~ **service** *n* COMP servicio *m* de conversación; ~ **system** *n* COMP sistema *m* de conversación
conversion *n* GEN conversión *f*; ~ **to bipolar code** TRANS conversión *f* a código bipolar; ~ **and storage unit** (*CSU*) DATA unidad *f* de almacenamiento y conversión (*CSU*); ~ **table** COMP tabla *f* de conversión
convert *vt* COMP, DATA convertir; ◆ ~ **to bipolar code** TRANS convertir a código bipolar
converter *n* GEN transformador *m*; ~ **cabinet** POWER armario *m* del convertidor; ~ **module** ELECT módulo *m* conversor; ~ **rack** POWER bastidor *m* de convertidor
convertible *adj* GEN cambiable
convey *vt* GEN transportar
convolution: ~ **code** *n* COMP código *m* convolucional
convolutional: ~ **code** *n* COMP código *m* convolucional; ~ **coding** *n* COMP codificación *f* convolucional; ~ **encoder** *n* COMP, DATA codificador *m* convolucional
convolutionally: ~ **and differentially encoded** *n* (*CDE*) DATA codificado convolucional y diferencialmente (*CDE*)
convolutive: ~ **code** *n* COMP código *m* convolutivo
cool *vt* GEN enfriar
cooling *n* GEN refrigeración *f*; ~ **flange** PARTS aleta *f* de refrigeración; ~ **jacket** GEN camisa *f* refrigerante; ~ **system** CLIM sistema *m* de refrigeración; ~ **unit** CLIM unidad *f* de refrigeración
cooperating: ~ **DSA** *n* DATA DSA *f* cooperativa
cooperative: ~ **work** *n* DATA trabajo *m* cooperativo
coordinate[1]: ~ **clock** *n* RADIO reloj *m* coordinado; ~ **graphics** *n* COMP infografía *f* por coordenadas; ~ **time** *n* RADIO tiempo *m* coordinado
coordinate[2] *vt* GEN coordinar
coordinated: ~ **single-layer embedded test method** *n* TEST método *m* coordinado de comprobación interna de una capa; ~ **single-layer test method** *n* TEST método *m* coordinado de prueba de una capa; ~ **test method** *n* TEST método *m* de prueba coordinado; ~ **time scale** *n* RADIO escala *f* temporal coordinada; ~ **traffic** *n* RADIO tráfico *m* coordinado
coordination *n* GEN coordinación *f*; ~ **level** DATA nivel *m* de coordinación
co-polar *adj* RADIO co-polar
co-polarized *adj* RADIO co-polarizado
copper: ~ **alloy** *n* GEN aleación *f* de cobre; ~ **busbar** *n* POWER barra *f* colectora de cobre; ~ **cable** *n* TRANS cable *m* de cobre; ~ **cabling** *n* ELECT cableado *m* de cobre; ~ **conductor** *n* ELECT conductor *m* de cobre; ~ **distributed data interface** *n* DATA interfaz *f* de datos distribuido por cobre; ~ **wire** *n* GEN alambre *m* de cobre
coprocessor *n* COMP coprocesador *m*
coproduction *n* TV coproducción *f*
copy[1]: ~ **destination point** *n* COMP punto *m* de destino de la copia
copy[2] *vt* GEN copiar
COR *abbr* (*confirmation of receipt*) DATA, TELEP COR (*confirmación de recepción*)
cord *n* COMP, GEN, TELEP, TEST cordón; ~ **circuit** TELEP, TEST circuito *m* de cordón flexible; ~ **set** TRANS cordón *m* conector
cordless *adj* GEN sin cable, inalámbrico; ~ **handset** *n* PER APP microteléfono *m* inalámbrico, TELEP aparato *m* inalámbrico; ~ **PABX** *n* PER APP PABX *m* inalámbrico; ~ **phone** *n* PER APP teléfono *m* inalámbrico, TELEP teléfono *m* sin cordón; ~ **switchboard** *n* GEN, SWG, TELEP cuadro *m* de distribución sin hilos conductores; ~ **telephone** *n* PER, TELEP teléfono *m* inalámbrico; ~ **telephone set** *n* PER, TELEP aparato *m* telefónico inalámbrico; ~ **telephony** *n* RADIO telefonía *f* sin hilos
core *n* GEN alma *f*, núcleo *m*; ~ **activity** GEN actividad *f* de núcleo; ~ **area** TRANS *fibre optics* región *f* del alma; ~ **centre** *BrE* TRANS centro *m* del alma; **~-cladding concentricity error** TELEP, TRANS *fibre optics* error *m* de concentricidad en el revestimiento del alma; **~-cladding ratio** TRANS relación *f* entre el

diámetro del alma y el espesor del aislamiento; ~ **diameter** TELEP, TRANS diámetro *m* de núcleo; ~ **diameter deviation** TELEP, TRANS desviación *f* del diámetro del núcleo; ~ **diameter tolerance** TRANS tolerancia *f* del diámetro del núcleo; ~ **dump** COMP vuelco *m* de la memoria central; ~ **of exchange** DATA memoria *f* de intercambio; ~ **flush** COMP ajuste *m* de núcleo; ~ **functions control station software** COMP, DATA, TELEP programa *m* de estación de control de funciones del núcleo; ~ **image** COMP imagen *f* de la memoria; ~ **logic** COMP lógica *f* de núcleos; ~ **matrix** COMP matriz *f* de núcleos; ~ **mode** TRANS *fibre optics* modo *m* de núcleo; ~ **screen** POWER *cables* pantalla *f* aislante; ~ **storage** COMP almacenamiento *m* de núcleos; ~ **tolerance field** TRANS campo *m* de tolerancia del alma; ~ **wrap** PARTS *cable* aislamiento *m* de conductor

corkscrew: ~ **antenna** *n* RADIO antena *f* helicoidal

corner: ~ **antenna** *n* RADIO antena *f* de ángulo cóncavo; ~ **loss** *n* RADIO pérdida *f* angular; ~ **reflector** *n* RADIO reflector *m* angular

cornerstone *n* DATA piedra *f* de toque

cornucopia *n* RADIO cornucopia *f*

corona *n* GEN corona *f*; ~ **arc testing** RADIO prueba *f* de arco del efecto corona; ~ **effect** POWER efecto *m* corona

corporate *adj* GEN colectivo

corpus *n* COMP, DATA corpus *m*

correct *vt* GEN corregir

correctable: ~ **fault** *n* TEST defecto *m* corregible

corrected: ~ **equivalent resistance error** *n* TELEP error *m* corregido de resistencia equivalente; ~ **fault** *n* TEST fallo *m* corregido; ~ **reference equivalent** *n* (*CRE*) TELEP equivalente *m* de referencia corregido (*CRE*)

correction *n* GEN rectificación *f*; ~ **area** COMP área *f* de asociación; ~ **character** TELEP carácter *m* de corrección; ~ **factor** TEST *metrology* factor *m* de corrección; ~ **function** TEST *metrology* función *f* de corrección

corrective: ~ **maintenance** *n* TEST mantenimiento *m* correctivo; ~ **maintenance time** *n* TELEP, TEST tiempo *m* de mantenimiento correctivo; ~ **measure** *n* GEN medida *f* correctiva

correlation *n* GEN correlación *f*

correlative: ~ **phase-shift keying** *n* SIG manipulación *f* por desviación de fase correlativa

correlator *n* GEN correlador *m*

corridor *n* RADIO corredor *m*

corroded *adj* GEN corroído

corrosion *n* GEN corrosión *f*

corrugated *adj* GEN ondulado; ~ **horn feed** *n* RADIO excitador *m* de bocina ondulada

cosecant *n* RADIO cosecante *f*; ~-**squared antenna** RADIO antena *f* compensada

cosine: ~ **emission** *n* TRANS emisión *f* del coseno; ~ **emission law** *n* TRANS *fibre optics* ley *f* de emisión del coseno; ~ **pulse** *n* TRANS impulso *m* en coseno; ~-**squared pulse** *n* TRANS impulso *m* en coseno cuadrado

cost *n* GEN coste *m* (*AmL* costo); ~ **centre** GEN centro *m* de coste *Esp*, centro *m* de costo *AmL*, centro *m* de cálculo de costes *Esp*, centro *m* de cálculo de costos *AmL*; ~ **of ownership** GEN coste *m* total de posesión *Esp*, costo *m* total de posesión *AmL*; ~ **price** GEN precio *m* de venta; ~ **of production** GEN coste *m* de producción *Esp*, costo *m* de producción *AmL*

Costas: ~ **loop** *n* TELEP bucle *m* de Costas

count *vt* GEN contar

count down *vt* GEN contar hacia atrás

countdown *n* GEN cuenta *f* atrás; ~ **timer** PER APP *on answering machine* temporizador *m* de cuenta atrás

counter *n* GEN contador *m*; ~-**acting winding** POWER arrollamiento *m* antagonista; ~-**action** GEN neutralización *f*; ~ **cell** POWER elemento *m* de fuerza contraelectromotriz; ~-**electromotive force cell** POWER elemento *m* de fuerza contraelectromotriz; ~-**measure** GEN contramedida *f*; ~ **step** GEN paso *m* de contador

counteract *vt* GEN contrarrestar

counterbalance *n* GEN equilibrio *m*

counterclockwise[1] *AmE adv* (*cf anticlockwise BrE*) GEN en sentido antihorario

counterclockwise[2]: ~ **polarization** *AmE n* (*cf anticlockwise polarization BrE*) RADIO polarización *f* en sentido antihorario

counterpoise *n* GEN compensación por toma de tierra

counterpressure *n* GEN contrapresión *f*

countersinking *n* GEN avellanado *m*

counterweight *n* GEN contrapeso *m*

counting: ~ **circuit** *n* ELECT circuito *m* contador; ~ **device** *n* GEN contador *m*; ~ **function** *n* ELECT función *f* de conteo; ~ **step** *n* GEN paso *m* de contador

country: ~ **code** *n* DATA, TELEP código *m* de país; ~ **code and echo suppressor indicators** *n pl* TELEP indicadores *m pl* de código de país y de supresor de eco; ~ **code indicator** *n* TELEP indicador *m* de código de país; ~ **code system** *n* GEN sistema *m* de indicativos de país; ~ **of**

destination *n* TELEP país *m* de destino; **~ of origin** *n* TELEP país *m* de origen
couple *vt* COMP acoplar
coupled: **~ mode** *n* TRANS modo *m* acoplado
coupler *n* ELECT acoplador *m*; **~ loss** TRANS *fibre optics* pérdida *f* de acoplador
coupling *n* GEN acoplamiento *m*; **~ capacitor** ELECT, POWER condensador *m* de acoplamiento; **~ efficiency** TRANS *fibre optics* eficacia *f* del acoplamiento; **~ factor** TRANS *fibre optics* factor *m* de acoplamiento; **~ loss** TRANS *fibre optics* pérdida *f* de acoplamiento
courier *n* GEN estafeta *f*; **~ service** GEN servicio *m* de mensajería
course: **~ blip pulse** *n* RADIO pulso *m* de la señal en curso; **~ of events** *n* GEN sucesión *f* de acontecimientos
courseware *n* COMP, DATA software *m* didáctico
courtesy: **~ music** *n* TELEP música *f* ambiental
cover: **~ holder** *n* PARTS sujetador *m* de cubierta; **~ page** *n* COMP página *f* de título; **~ plate** *n* PARTS cubierta *f* protectora de tubos de rayos catódicos; **~ support** *n* TELEP soporte *m* de cubierta
coverage *n* RADIO, TV cobertura *f*; **~ area** RADIO, TV zona *f* cubierta; **~ diagram** RADIO, TV diagrama *m* de cobertura; **~ factor** RADIO, TV factor *m* de recubrimiento
covering *n* GEN cubierta *f*
CPE *abbr* (*customer premises equipment*) PER APP CPE (*equipo en locales del abonado*)
cpi *abbr* (*characters per inch*) COMP cpi (*caracteres por pulgada*)
CPL *abbr* (*combined programming language*) PERIF CPL (*lenguaje combinado de programación*); **~/1 language** *n* COMP lenguaje *m* CPL/1
CPM *abbr* (*channel protocol machine*) DATA máquina *f* de protocolo de canal
CPU *abbr* (*central processing unit*) COMP, SWG, TELEP UCP (*unidad central de proceso*)
CQM *abbr* (*circuit-group query message*) DATA, SWG mensaje *m* de interrogación de un grupo de circuitos
CQR *abbr* (*circuit-group query response message*) DATA, SWG mensaje *m* de respuesta de un grupo de circuitos
C/R *abbr* (*command/response bit*) TELEP C/R (*bitio de comando/respuesta*)
crack *n* GEN grieta *f*; **~ detector** TEST detector *m* de grietas
cracking *n* TEST fisura *f*; **~ test** TEST prueba *f* de agrietamiento
crackle *n* TELEP craquelado *m*
crackling: **~ noise** *n* TELEP crepitación *f*
cradle *n* GEN cuna *f*; **~ switch** PER APP, RADIO, TELEP interruptor *m* de gancho
crank *n* PER APP manivela *f*
CRC *abbr* (*cyclic redundancy check*) DATA, TELEP CRC (*verificación de redundancia cíclica*); **~ message bloc** *n* (*CMB*) TELEP bloque *m* de mensajes CRC (*CMB*)
CRE *abbr* (*corrected reference equivalent*) TELEP CRE (*equivalente de referencia corregido*)
create *vt* GEN crear
creation *n* GEN creación *f*; **~ date** COMP fecha *f* de creación; **~ time** DATA hora *f* de creación
credentials *n pl* DATA credenciales *f pl*
credit: **~ card call** *n* TELEP llamada *f* con tarjeta de crédito; **~ indicator** *n* TELEP indicador *m* de crédito; **~ rating** *n* GEN calificación *f* de obligaciones
creditphone *n* TELEP teléfono *m* de tarjeta de crédito
credits *n pl* TV títulos *m pl*
creep *n* TEST corrimiento *m*; **~ modulus** TEST límite *m* de fluencia; **~ rupture** TEST ruptura *f* por fluencia; **~ rupture stress** TEST resistencia *f* a la ruptura por fluencia
creepage *n* POWER contorneamiento *m*
creeping: **~ distance** *n* POWER distancia *f* de contorneo
crew: **~ factor** *n* TEST factor *m* del equipo de operadores
CRF *abbr* (*connection-related function*) DATA, TRANS función *f* relacionada con la conexión
crimp[1]: **~ ratchet tool frame** *n* ELECT alojamiento *m* de herramienta con bloqueo de engarce; **~ tool** *n* ELECT herramienta *f* para engarzar
crimp[2] *vt* ELECT aboquillar, engarzar
crimping *n* GEN rebordeado *m*
crimpless: **~ screw** *n* ELECT tornillo *m* sin engarce
crisis: **~ management** *n* GEN gestión *f* de la crisis
criterion *n* GEN constante *f* de comparación
critical: **~ angle** *n* TRANS ángulo *m* crítico; **~ band** *n* TELEP banda *f* crítica; **~ band rate** *n* APPL frecuencia *f* crítica de banda; **~ damping** *n* TELEP *acoustics* amortiguamiento *m* crítico; **~ design review** *n* GEN revista *f* crítica de estudio; **~ failure** *n* TEST fallo *m* crítico; **~ frequency** *n* TRANS frecuencia *f* crítica; **~ path** *n* GEN *project management* camino *m* crítico; **~-path analysis** *n* GEN *project management* análisis *m* de camino crítico; **~-path method** *n* GEN *project management* método *m* del camino crítico
crocodile: **~ clip** *n* ELECT pinza *f* de cocodrilo

crocus: ~ **cloth** *n* ELECT tela *f* fina de esmeril
crop *vt* COMP recortar
cross[1]: ~ **assembler** *n* COMP ensamblador *m* cruzado; ~-**band operation** *n* TRANS operación *f* de banda transversal; ~-**border system** *n* RADIO *land mobile* sistema *m* de límites cruzados; ~-**check** *n* COMP determinación *f* de sentido; ~ **chute** *n* PARTS canal *m* transversal; ~-**connect** *n* ELECT interconexión *f*; ~-**connect cabinet** *n* TELEP armario *m* interconectado; ~-**connecting unit** *n* TELEP unidad *f* de conexión cruzada; ~-**exchange check** *n* SWG *cross-office* verificación *f* desde la centralita; ~ **hair** *n* COMP cruz *f*; ~-**hatching** *n* ELECT rayado *m*; ~-**index** *n* GEN índice *m* cruzado; ~-**interleaved Reed-Solomon code** *n* (*CIRC*) COMP código *m* Reed-Solomon intercruzado (*CIRC*); ~-**media** *adj* ELECT de mediostransversales; ~-**modulation noise** *n* RADIO ruido *m* de intermodulación; ~-**module** *adj* ELECT de módulo transversal; ~-**net** *adj* TRANS de red transversal; ~-**office check** *n* SWG verificación *f* desde la centralita; ~-**office transfer time** *n* SWG tiempo *m* de transferencia en una centralita; ~-**point switch** *n* SWG conmutador *m* de punto de cruce; ~-**polar channel** *n* TRANS canal *m* de polarización cruzada; ~-**polar discrimination** *n* TRANS discriminación *f* de polarización cruzada; ~-**polar isolation** *n* TRANS aislamiento *m* de polarización cruzada; ~-**polarization** *n* RADIO, TRANS polarización *f* cruzada; ~-**polarization discrimination** *n* RADIO, TRANS discriminación *f* de polarización cruzada; ~-**polarization isolation** *n* RADIO aislamiento *m* de polarización cruzada; ~-**polarized** *adj* TRANS de polarización cruzada; ~-**polar performance** *n* TRANS funcionamiento *m* de polarización cruzada; ~-**reference** *n* GEN referencia *f* cruzada; ~ **section** *n* GEN sección *f* transversal; ~-**site link** *n* TRANS enlace *m* de sitio cruzado; ~ **state** *n* TRANS *fibre optics* estado *m* transversal; ~-**stranding** *n* TRANS *cables* trenzado *m* cruzado; ~-**strapping** *n* ELECT interconexión *f* cruzada; ~-**twisting** *n* TRANS conexión *f* por torsión cruzada
cross[2]: ~-**connect** *vt* ELECT interconectar
crossbar: ~ **switch** *n* SWG sector *m* de coordenadas
crossed: ~ **field** *n* TELEP campo *m* cruzado; ~-**field amplifier** *n* TRANS amplificador *m* de campo cruzado; ~-**slot antenna** *n* RADIO antena *f* de ranuras cruzadas
crossfade *n* TV fundido *m* encadenado
cross-office: ~ **transfer time** *n* (*Tcu*) SWG tiempo *m* de transferencia en una centralita (*Tcu*)
crossover: ~ **patch cable** *n* ELECT cable *m* de conexión de filtro
crosspoint *n* SWG contacto *m* de cruce
crosstalk *n* TRANS interferencia *f* entre los canales
CRT *abbr* (*cathode-ray tube*) ELECT, PARTS TRC (*tubo de rayos catódicos*)
crushing *n* GEN trituración *f*
cryogenic: ~ **current comparator** *n* ELECT comparador *m* de tensión criogénica; ~ **storage** *n* COMP memoria *f* criogénica
cryogenics *n* GEN criogenia *f*
cryotron *n* ELECT criotrón *m*
cryptanalysis *n* DATA, TELEP, TV criptoanálisis *f*
cryptogram *n* GEN criptograma *m*
cryptographic: ~ **check value** *n* DATA valor *m* de comprobación criptográfico
cryptography *n* DATA, TELEP, TV criptografía *f*
crystal[1]: ~ **diode** *n* POWER diodo *m* de cristal; ~ **filter** *n* RADIO filtro *m* de cristal; ~ **rectifier** *n* POWER rectificador *m* de cristal
crystal[2] *n* GEN cristal *m*
CS *abbr* (*communication services*) APPL servicios *m pl* de comunicación
CSC *abbr* (*communications system control*) GEN CSC (*control del sistema de comunicaciones*)
CSG *abbr* (*constructive solid geometry*) COMP CSG (*geometría constructiva tridimensional*)
CSI *abbr* (*called-subscriber identification*) TELEP identificación *f* del abonado llamado
CSI-DIS *abbr* (*called-subscriber identification-DIS*) TELEP CSI-DIS (*identificación DIS del abonado llamado*)
CSM *abbr* (*communications system monitoring*) TEST supervisión *f* del sistema de comunicaciones
CSPDU *abbr* (*convergence sublayer protocol data unit*) DATA unidad *f* de datos con protocolo de subcapa de convergencia
CSU *abbr* (*conversion and storage unit*) DATA CSU (*unidad de almacenamiento y conversión*)
CT *abbr* (*container*) GEN contenedor *m*
CTCSS *abbr* (*continuous tone-controlled squelch system*) RADIO CTCSS (*sistema continuo de silenciador de tono controlado*)
CTD *abbr* (*cumulative transit delay*) DATA DTC (*retardo de tránsito acumulado*)
CTR *abbr* (*response to continue to correct*) DATA CTR (*respuesta a continuar para corregir*)
CTSM *abbr* (*conformance-testing specification manual*) DATA CTSM (*manual de especificación de comprobación de conformidad*)

CUG *abbr* (*closed user group*) DATA, TELEP grupo *m* cerrado de usuarios

cumbersome *adj* GEN incómodo

cumulative: ~ **distribution function** *n* (*cdf*) DATA función *f* de distribución acumulativa (*cdf*); ~ **frequency polygon** *n* TEST *quality* polígono *m* de frecuencias acumuladas; ~ **leakage index** *n* TRANS índice *m* acumulado de fuga; ~ **sum chart** *n* TEST *quality* gráfico *m* de valores acumulados; ~ **transit delay** *n* (*CTD*) DATA retardo *m* de tránsito acumulado (*DTC*)

curbed: ~ **modulation** *n* TELEP modulación *f* fragmentada

Curie: ~ **point** *n* COMP punto *m* de Curie

current *n* GEN corriente *f*; ~ **amplification** POWER amplificación *f* de corriente; ~**-carrying capacity** POWER capacidad *f* de conducción de corriente; ~ **collector** POWER aparato *m* tomador de corriente; ~ **consumption** POWER potencia *f* absorbida; ~ **density** POWER densidad *f* de corriente; ~ **distribution** POWER distribución *f* de corriente; ~ **drain** POWER consumo *f* de corriente; ~ **drift** POWER deriva *f* de la corriente; ~ **feed** POWER alimentación *f* de corriente; ~**-feed relay** POWER relé *m* de alimetación de corriente; ~ **intensity** POWER intensidad *f* de corriente; ~ **limit level** POWER nivel *m* de limitación de corriente; ~ **limiter** POWER limitador *m* de corriente; ~ **limiting** POWER limitación *f* de corriente; ~ **load** POWER carga *f* de corriente; ~ **loop** POWER cresta *f* de la corriente; ~ **outlet** POWER toma *f* de corriente; ~ **relay** POWER relé *m* de corriente; ~ **return conductor** POWER conductor *m* de retorno del corriente; ~ **rush** POWER salto *m* de corriente; ~ **state of the art** GEN estado *m* actual de la tecnología; ~ **surge** POWER golpe *m* de corriente; ~ **tap** POWER toma *f* de corriente; ~ **version** COMP versión *f* básica

cursor *n* COMP, RADIO cursor *m*; ◆ ~ **off** COMP cursor detenido; ~ **on** COMP cursor activo

curvature *n* RADIO *of the earth* curvatura *f*

curve *n* GEN curva *f*; ~ **follower** COMP lectora *f* de curvas; ~ **generator** COMP generador *m* de curvas

curved: ~**-nose pliers** *n pl* GEN alicates *m pl* de punta angular

cushion *n* GEN colchón *m*

custom: ~ **calling service** *n* APPL servicio *m* de llamadas por especificación; ~**-designed service** *n* GEN servicio *m* a la medida del cliente

customer: ~**-access maintenance centre** *BrE n* (*CAMC*) TEST centro *m* de mantenimiento abierto al cliente; ~ **advisor** *n* GEN asesor *m* de la clientela; ~ **equipment** *n* (*CEQ*) PER APP equipo *m* del abonado; ~ **installation maintenance entity** *n* (*CIME*) APPL entidad *f* de mantenimiento de la instalación del cliente; ~ **network** *n* TRANS red *f* de abonados; ~ **premises equipment** *n* (*CPE*) PER APP equipo *m* en locales del abonado (*CPE*); ~ **recorded-information service** *n* TELEP servicio *m* al usuario con información grabada; ~ **service** *n* GEN servicio *m* al cliente

customize *vt* GEN adaptar especialmente

customized *adj* GEN adaptado especialmente; ~ **menu** *n* COMP menú *m* personalizado

customizing *n* GEN adaptación *f* especial

cut[1]: ~ **face** *n* TRANS superficie *f* del corte; ~**-off** *adj* RADIO bloqueado; ~**-out** *n* ELECT disyuntor *m*; ~**-out key** *n* TELEP conmutador *m* de bloqueo; ~**-over** *n* GEN, TELEP puesta *f* en servicio; ~**-sheet feeder** *n* COMP alimentador *m* de papel cortado

cut[2]: ~ **and paste** *vt* COMP, DATA cortar y pegar

cut back *vt* GEN desbarbar

cut off *vt* GEN trocear

cutaway: ~ **view** *n* GEN vista *f* en corte

cutback *n* TEST cese *m* repentino; ~ **technique** TRANS *fibre optics* método *m* de alteración

Cutler: ~ **feed** *n* RADIO alimentacion *f* Cutler

cutoff *n* GEN corte *m*; ~ **current** ELECT corriente *f* de corte; ~ **frequency** ELECT frecuencia *f* de cierre; ~ **switch** GEN conmutador *m* de corte; ~ **voltage** ELECT tensión *f* de corte; ~ **wavelength** RADIO, TRANS longitud *f* de onda de corte

cutter *n* GEN cortadora *f*

cutting *n* GEN corte *m*; ~**-edge technology** GEN tecnología *f* avanzada; ~ **formboard** GEN plantilla *f* de corte de cable; ~ **length** GEN longitud *f* de corte; ~ **pliers** *n pl* GEN alicates *m pl* de corte; ~ **and stripping tool** GEN herramienta *f* para cortar y pelar; ~ **tool** GEN herramienta *f* de corte; ~ **torch** GEN soplete *m*

CV *abbr* (*code violation*) TRANS violación *f* de código

CW *abbr* (*continuous wave*) RADIO onda *f* continua; ~ **radar** *n* RADIO radar *m* de onda continua

cyber *adj* DATA *Internet* ciber

cyberculture *n* DATA *Internet* cibercultura *f*

cybernaut *n* DATA *Internet* cibernauta *m*

cybernetics *n* DATA cibernética *f*

cyberspace *n* DATA *Internet* ciberespacio *m*

cycle *n* GEN ciclo *m*; ~ **skipping** TRANS salto *m* de ciclo; ~**-skipping test procedure** TEST procedimiento *m* de prueba de salto de ciclo

cyclic: ~ **alarm** *n* TEST alarma *f* cíclica; ~ **block**

code *n* DATA código *m* cíclico de bloque; ~ **code** *n* DATA código *m* cíclico; ~ **distortion** *n* TELEP distorsión *f* cíclica; ~ **redundancy check** *n* (*CRC*) DATA, TELEP verificación *f* de redundancia cíclica (*CRC*); ~ **redundancy code** *n* DATA, TELEP código *m* de redundancia cíclica

cyclical *adj* GEN cíclico

cycling *n* TEST funcionamiento *m* cíclico

cyclomotive: ~ **force** *n* (*cmf*) RADIO fuerza *f* ciclomotora (*cmf*)

cyclotronic: ~ **resonance** *n* ELECT resonancia *f* ciclotrónica

cylindrical: ~ **reflector antenna** *n* RADIO antena *f* cilíndrica de reflector

cypher *n* GEN cifra *f*

cyphering *n* GEN codificación *f*

D

D *abbr* (*defaultable*) DATA D (*valor asignable por defecto*)

D-A *abbr* (*digital-to-analog*) GEN D-A (*digital a analógico*); **~ converter** *n* GEN convertidor *m* D-A

daisy[1]: **~-chaining** *n* COMP, ELECT conexión en batería

daisy[2]: **~-chain** *vt* COMP, ELECT conectar en batería

damage *n* GEN avería *f*; **~ recovery** ELECT arreglo *m* de avería

damaged *adj* TRANS *cable* deteriorado

damp[1]: **~ heat testing** *n* TEST prueba *f* de calor húmedo; **~ point** *n* TRANS punto *m* de amortiguación

damp[2] *vt* TRANS amortiguar

damped: **~ oscillation** *n* ELECT oscilación *f* amortiguada

damping *n* TRANS amortiguamiento *m*

danger: **~ signal** *n* TEST señal *f* de peligro

DAP *abbr* (*directory access protocol*) DATA protocolo *m* de acceso al directorio

dark: **~ current** *n* ELECT corriente *f* residual; **~ field** *n* TRANS campo *m* oscuro; **~-field detector** *n* TRANS *fibre optics* detector *m* de campo oscuro

darken *vt* TV oscurecer

darkest: **~ zone** *n* TV *scanner* zona *f* más oscura

Darlington: **~ amplifier** *n* ELECT amplificador *m* Darlington; **~ circuit** *n* ELECT circuito *m* Darlington

dash *n* GEN guión *m*

dashboard *n* RADIO tablero *m*

data *n pl* GEN datos *m pl*; **~ above voice** *n* RADIO datos *m pl* sobre voz; **~ acquisition** *n* DATA adquisición *f* de datos; **~ adaptor unit** *n* COMP unidad *f* de adaptador de datos; **~-addressable memory** *n* COMP memoria *f* direccionable de datos; **~ amount** *n* DATA cantidad *f* de datos; **~ backup** *n* DATA copia *f* de seguridad de datos; **~ bank** *n* DATA banco *m* de datos; **~ block** *n* DATA bloque *m* de datos; **~ buffering** *n* DATA memorización *f* provisional de datos; **~ carrier** *n* TRANS portadora *f* de datos; **~-carrier detection** *n* (*DCD*) TRANS detección *f* de portadora de datos; **~-carrier failure detector** *n* TRANS detector *m* de fallo en la portadora de datos; **~ cassette recorder** *n* (*DCR*) PER APP grabadora *f* de datos de casete (*DCR*); **~ channel** *n* DATA, TELEP canal *m* de datos; **~-channel failure detector** *n* TELEP detector *m* de fallo en el canal de datos; **~-channel propagation time** *n* TELEP tiempo *m* de propagación en el canal de datos; **~-channel received-line signal detector** *n* DATA, TRANS detector *m* de señal de línea recibida de canal de datos; **~ character key** *n* PER APP *keyboards* tecla *f* alfanumérica; **~ circuit** *n* DATA, TRANS circuito *m* de datos; **~ circuit terminal equipment** *n* DATA, PER APP equipo *m* terminal del circuito de datos; **~ circuit terminating equipment** *n* (*DCTE*) TRANS terminal *m* de circuito de datos (*DCTE*); **~ code** *n* DATA código *m* de datos; **~ collection** *n* DATA recogida *f* de datos; **~-collection centre** *BrE* *n* DATA centro *m* de recogida de datos; **~-collection equipment** *n* DATA equipo *m* de recogida de datos; **~-collection satellite** *n* RADIO satélite *m* de recopilación de datos; **~ communication** *n* DATA, TRANS comunicación *f* de datos; **~-communication channel** *n* DATA, TRANS canal *m* de comunicación de datos; **~-communication network** *n* DATA, TRANS red *f* de comunicación de datos; **~-communications controller** *n* DATA controlador *m* de comunicaciones de datos; **~-communications equipment** *n* PER APP equipo *m* de communicación de datos; **~-communication system** *n* DATA, COMP sistema *m* de comunicación de datos; **~-communication terminal** *n* DATA terminal *m* de comunicación de datos; **~ compatibility** *n* DATA, COMP compatibilidad *f* de datos; **~ compression** *n* COMP, DATA *remote processing* compresión *f* de datos; **~-compression board** *n* DATA *remote processing* placa *f* de compresión de datos; **~-compression processor** *n* DATA, COMP *remote processing* procesador *m* para compresión de datos; **~-compression software** *n* DATA, COMP *remote processing* programa *m* de compresión de datos; **~ concentrator** *n* DATA concentrador *m* de datos; **~ connection** *n* DATA conexión *f* de datos; **~ conversion** *n* DATA, COMP conversión *f* de datos; **~ converter** *n* DATA, COMP convertidor *m* de datos; **~ country code** *n* DATA código *m* de país de datos; **~-description**

language *n* DATA, COMP lenguaje *m* de descripción de datos; ~ **distribution** *n* DATA, TELEP distribución *f* de datos; ~ **dump** *n* COMP vaciado *m* de datos; ~ **editing** *n* COMP edición *f* de información, DATA edición *f* de datos; ~**-element separator** *n* DATA separador *m* de elementos de datos; ~**-encryption standard** *n* (*DES*) DATA norma *f* de cifrado de datos (*DES*); ~ **entry** *n* COMP introducción *f* de datos; ~ **exchange file** *n* COMP fichero *m* de intercambio de datos; ~**-flow analysis** *n* DATA análisis *m* de flujo de datos; ~**-flow chart** *n* DATA, COMP diagrama *m* de flujo de datos; ~**-flow diagram** *n* DATA, COMP diagrama *m* de flujo de datos; ~ **gathering** *n* DATA recopilación *f* de datos; ~ **group** *n* RADIO grupo *m* de datos; ~ **handling** *n* DATA manejo *m* de datos; ~ **independence** *n* DATA, COMP independencia *f* de datos; ~ **integrity** *n* DATA integridad *f* de datos; ~ **interchange** *n* DATA, COMP intercambio *m* de datos; ~ **link** *n* DATA, TRANS enlace *f* de datos; ~**-link connection** *n* (*DLC*) DATA, TRANS conexión *f* de enlace de datos; ~**-link connection identifier** *n* (*DLCI*) DATA, TRANS identificador *m* de enlace de datos; ~**-link control** *n* DATA, TRANS control *m* de enlace de datos; ~**-link layer** *n* DATA, TRANS capa *f* de enlace de datos; ~**-link protocol** *n* DATA, TRANS protocolo *m* de enlace de datos; ~**-link service** *n* (*DLS*) APPL, DATA, TRANS servicio *m* de enlace de datos; ~**-link service access point** *n* (*DLSAP*) APPL, DATA, TRANS punto *m* de acceso al servicio de enlace de datos (*DLSAP*); ~**-link specialized access** *n* DATA, TRANS acceso *m* especializado del enlace de datos; ~**-link switching** *n* DATA, TRANS, ELECT conmutación *f* de enlace de datos; ~ **logger** *n* DATA registrador *m* de datos; ~ **loss** *n* DATA pérdida *f* de datos; ~ **management** *n* (*DM*) COMP, DATA gestión *f* de datos; ~**-manipulation language** *n* DATA lenguaje *m* de manipulación de datos; ~ **modelling** *BrE n* DATA modelización *f* de datos; ~ **modifier** *n* DATA, COMP modificador *m* de datos; ~ **modulation** *n* DATA, COMP modulación *f* de datos; ~ **multiplexer** *n* COMP, TELEP multiplexador *m* de datos; ~ **network** *n* DATA, TELEP, TRANS red *f* de datos; ~**-network identification code** *n* (*DNIC*) DATA, TELEP, TRANS código *m* de identificación de la red de datos; ~**-origin authentication** *n* DATA autentificación *f* del origen de datos, autentificación *f* del origen de la información, RADIO paquete *m* de datos de origen autenticado; ~**-origin authentication package** *n* DATA paquete *m* de datos de origen autenticado; ~**packet** *n* DATA paquete *m* de datos; ~ **packet link** *n* DATA, TRANS enlace *m* de paquetes de datos; ~ **partitioning** *n* COMP partición *f* de datos; ~ **path** *n* TV ruta *f* de datos; ~**-permissible** *adj* TRANS admisible para los datos; ~**-permissive** *adj* TRANS permitido para los datos; ~ **phase** *n* TRANS fase *f* de datos; ~ **port** *n* DATA puerto *m* de datos; ~ **preparation** *n* COMP, DATA preparación *f* de datos; ~ **presetter** *n* COMP preajustador *m* de datos; ~**-private wire** *n* TRANS circuito *m* de uso privado de datos; ~ **processing** *n* (*DP*) COMP, DATA sistema *m* de procesamiento de datos; ~**-processing centre** *BrE n* COMP, DATA centro *m* de procesamiento de datos, centro *m* de cálculo, centro *m* de tratamiento de datos; ~**-processing connection** *n* COMP, DATA conexión *f* para procesamiento de datos; ~**-processing system** *n* COMP sistema *m* de proceso de datos, DATA sistema *m* de procesamiento de datos; ~**-processing system security** *n* COMP, DATA seguridad *f* del sistema de proceso de datos; ~**-processing terminal equipment** *n* DATA, PER APP equipo *m* terminal para el proceso de datos; ~ **processor** *n* COMP, DATA procesador *m* de datos; ~ **quantization** *n* TELEP cuantificación *f* de datos; ~ **recording** *n* DATA grabación *f* de datos; ~ **reduction** *n* COMP, DATA reducción *f* de datos; ~**-relay satellite** *n* RADIO satélite *m* repetidor de datos; ~ **saving** *n* COMP preservación *f* de datos; ~**-set identifier** *n* (*DSI*) DATA identificador *m* de conjunto de datos; ~ **sheet** *n* GEN *documentation* cuadro *m* de características, hoja *f* de especificaciones; ~ **signal** *n* DATA señal *f* de datos; ~ **signalling** *BrE n* DATA, SIG señalización *f* de datos; ~**-signal quality detection** *n* DATA, SIG detección *f* de calidad de señal de datos; ~ **sink** *n* DATA colector *m* de datos; ~ **store** *n* SWG almacén *m* de datos; ~**-store block** *n* SWG *code* bloque *m* de almacén de datos; ~**-store channel** *n* SWG *code* canal *m* de almacén de datos; ~**-store element** *n* DATA elemento *m* de almacén de datos; ~**-store handler** *n* SWG manipulador *m* de almacén de datos; ~**-store multiplexer** *n* SWG *code* multiplexador *m* de almacén de datos; ~**-store unit** *n* SWG unidad *f* de almacén de datos; ~ **stream** *n* DATA flujo *m* de datos; ~ **switch** *n* DATA, ELECT, SWG conmutador *m* de datos; ~ **switching** *n* DATA, ELECT, SWG conmutación *f* de datos; ~**-switching exchange** *n* DATA, SWG central *f* de conmutación de datos; ~ **terminal** *n* PER APP, TELEP terminal *m* de datos; **~-terminal equipment** *n* (*DTE*) DATA, PER APP, TELEP equipo *m* terminal de datos (*DTE*); **~-terminal-equipment address** *n* DATA, PER APP, TELEP dirección *f* de terminal de datos; **~ transfer** *n* DATA, TRANS transferencia *f* de datos; **~-transfer phase** *n* DATA, TELEP fase *f* de transferencia de datos; **~-transfer rate** *n* DATA velocidad *f* de transmisión de datos; **~ transfer requested** *n* DATA transferencia *f* de datos solicitada; **~-transfer system** *n* DATA sistema *m* de transferencia de datos; **~ transmission** *n* DATA, TRANS transmisión *f* de datos; **~-transmission balloon** *n* DATA globo *m* de transmisión de datos; **~-transport network** *n* DATA, TRANS red *f* de transporte de datos, red *f* de transferencia de datos; **~ type** *n* DATA tipo *m* de datos; **~ under voice** *n* (*DUV*) RADIO datos *m pl* bajo voz (*DUV*); **~ unit** *n* RADIO unidad *f* de datos; **~-user part** *n* DATA parte *f* del usuario de datos; **~-user-part handling time** *n* DATA, SIG tiempo *m* de manejo de la parte de usuario de datos; **~-wrap plug** *n* ELECT conexión *f* arrollada para datos

database *n* COMP, DATA base *f* de datos; **~ language** COMP, DATA lenguaje *m* de base de datos; **~ management** COMP, DATA administración *f* de base de datos; **~-management system** (*DBMS*) COMP, DATA sistema *m* de administración de base de datos (*SABD*); **~ manager** COMP, DATA gestor *m* de base de datos

datacom *n* DATA transmisión *f* de datos, TRANS comunicación *f* de datos

datagram *n* DATA, SIG, TELEP datagrama *m*; **~ delivery data confirmation** DATA, SIG, TELEP confirmación *f* de los datos enviados por datagrama; **~ non-delivery indication** DATA, SIG, TELEP indicación *f* de ausencia de envío del datagrama; **~ service** DATA, SIG, TELEP servicio *m* de datagrama

datamation *n* COMP, DATA proceso *m* automático de datos

date *n* GEN fecha *f*; **~ line** RADIO línea *f* de cambio de fecha; **~ of manufacture** GEN fecha *f* de fabricación; **~-time stamping** PER APP *on answering machine* especificación *f* de fecha y hora; **~ and time of transmission** DATA fecha *f* y hora de transmisión

daughterboard *n* COMP, ELECT *printed circuits* placa *f* secundaria

daughters *n pl* COMP hijas *f pl*

day: **~-to-busy-hour ratio** *n* TELEP relación *f* día-hora cargada; **~-to-~ variation** *n* GEN variación *f* diaria; **~ traffic** *n* TELEP tráfico *m* diario

dB *abbr* (*decibel*) GEN dB (*decibelio*); **~-related units** *n pl* TELEP unidades *f pl* relacionadas de decibelios

D-bit: **~ modification facility** *n* COMP instalación *f* de modificación D-bitio

dBm *abbr* (*decibels above one milliwatt*) TELEP dBm (*decibelios por encima de un milivatio*)

DBMS *abbr* (*database management system*) COMP, DATA SABD (*sistema de administración de base de datos*)

DBS: **~ satellite** *n* (*direct broadcasting by satellite*) TV satélite *m* DBS (*radiofusión directa vía satélite*)

DC *abbr* (*direct current*) POWER CC (*corriente continua*); **~ bus** *n* POWER barra *f* colectora de CC; **~ busbar** *n* POWER barra *f* de distribución de CC; **~ clamp diode** *n* RADIO diodo *m* de fijación por CC; **~ component** *n* TELEP componente *m* de CC; **~ generator** *n* POWER generador *m* de CC; **~ power** *n* COMP electricidad *f* de CC; **~ resistance** *n* POWER resistencia *f* de CC; **~ restorer diode** *n* RADIO diodo *m* restaurador de CC; **~ signalling** *BrE n* POWER, SIG señalización *f* de CC; **~ spark-over voltage** *n* ELECT, POWER voltaje *m* de descarga disruptiva de CC; **~ voltage** *n* ELECT, POWER voltaje *m* de CC

DCA *abbr* (*document content architecture*) DATA arquitectura *f* del contenido del documento

DCC/DNIC: **~ indicator** *n* SIG indicador *m* DCC/CNIC

DCD *abbr* (*data-carrier detection*) DATA detección *f* del portador de datos

DC-DC: **~ converter** *n* DATA convertidor *m* CD-CD, POWER convertidor *m* CA-CC, convertidor *m* CC-CA

DCE *abbr* (*despun control electronics*) ELECTRÓN DCE (*conjunto electrónico de rotación*); **~ clear indication** *n* TELEP indicación *f* de borrado de DCE; **~-power off signal** *n* SIG señal *f* de falta de corriente DCE; **~ source** *n* TELEP fuente *f* DCE; **~ waiting** *n* TELEP espera *f* de DCE

D-channel *n* DATA, TELEP *ISDN* canal *m* D; **~ virtual circuit** DATA, TELEP *ISDN* circuito *m* virtual canal D

DCME *abbr* (*digital circuit multiplication equipment*) TELEP DCME (*equipo de multiplicación del circuito digital*); **~ frame** *n* TELEP cuadro *m* DCME; **~ function** *n* TELEP función *f* DCME; **~ gain** *n* TELEP amplificador *m* DCME; **~ overload** *n* TELEP sobrecarga *f* DCME; **~ overload mode** *n* TELEP modo *m* de sobrecarga DCME

DCN *abbr* (*disconnect*) GEN DCN (*desconectar*)
DCR *abbr* (*data cassette recorder*) PER APP DCR (*grabadora de datos de casete*)
DCS *abbr* (*digital command signal*) SIG DCS (*señal de mando digital*)
DCTE *abbr* (*data circuit terminating equipment*) TRANS DCTE (*terminal de circuito de datos*)
DCU *abbr* (*distribution and control unit*) CONTR DCU (*unidad de distribución y control*)
DCX *abbr* (*multipage PCX file*) DATA DCX (*archivo PCX multipágina*)
DD *abbr* (*double density*) COMP DD (*doble densidad*)
DDA *abbr* (*defined-domain attribute*) TELEP DDA (*atributo de campo definido*)
deactivate *vt* GEN desactivar
deactivation *n* GEN desactivación *f*
dead *adj* GEN muerto; **~centre** *BrE n* COMP punto *m* muerto; **~room** *n* TEST cámara *f* anecoica; **~time** *n* CONTR tiempo *m* muerto
deadener *n* GEN *ergonomics* pasta *f* supresora de ruidos
deadline *n* GEN fin *m* de plazo
dealer *n* GEN vendedor *m*
deal with *vt* COMP ocuparse de
dealership: **~ network** *n* GEN red *f* de comercialización
de-averaging *n* GEN despromediación *f*
deblock *vt* RADIO, TELEP desbloquear
deblocking *n* RADIO, TELEP desbloqueo *m*
debug *vt* DATA depurar
debunch *vt* ELECT desagrupar
decade: **~ counter** *n* ELECT *logic elements* contador *m* de décadas; **~ resistance box** *n* POWER caja *f* de resistencia de décadas
decadic: **~ signalling** *BrE n* SIG señalización *f* decádica
decametric: **~ wave** *n* TRANS onda *f* decamétrica
decay *n* GEN amortiguamiento *m*; **~ level** COMP nivel *m* de declinación; **~ rate** COMP velocidad *f* de amortiguamiento; **~ time** COMP, TELEP *pulses*, TRANS tiempo *m* de desintegración
decentralize *vt* GEN descentralizar
decentralized *adj* GEN descentralizado; **~-control signalling** *BrE n* SIG señalización de control descentralizado; **~ operation** *n* GEN operación *f* descentralizada; **~ system** *n* SWG sistema *m* descentralizado
deception: **~ jamming** *n* RADIO interferencia *f* de diversión
decibel *n* (*dB*) GEN decibelio *m* (*dB*)
decibels: **~ above one milliwatt** *n pl* (*dBm*) TELEP decibelios *m pl* por encima de un milivatio (*dBm*)
decimal *adj* GEN decimal
decipher *vt* GEN descodificar, descifrar
deciphering *n* GEN descodificación *f*, desciframiento *m*
decision: **~ circuit** *n* TELEP circuito *m* de decisión; **~ content** *n* COMP contenido *m* de decisión; **~ feedback equalizer** *n* ELECT ecualizador *m* con realimentador de decisión; **~ function** *n* TEST *quality* criterio *m* de decisión; **~ instruction** *n* COMP instrucción *f* de decisión; **~-support system** *n* DATA sistema *m* de ayuda a la decisión; **~ symbol** *n* TELEP símbolo *m* de decisión; **~ table** *n* GEN tabla *f* de decisión; **~ threshold** *n* GEN umbral *m* de decisión; **~ tree** *n* DATA árbol *m* de decisión; **~ value** *n* TELEP, TRANS valor *m* de decisión
declaration *n* COMP sentencia *f*, DATA declaración *f*; **~ procedure** TV procedimiento *m* de declaración
declarations: **~ part** *n* DATA zona *f* de declaraciones
decode *vt* GEN decodificar, descifrar
decoded: **~ stream** *n* COMP flujo *m* descodificado
decoder *n* GEN descodificador *m*
decoding *n* GEN descifrado *m*, desciframiento *m*; **~ margin** RADIO margen *m* de descodificación; **~ process** COMP proceso *m* de descodificación; **~ threshold** RADIO umbral *m* de descodificación; **~ time stamp** (*DTS*) COMP reloj *m* fechador descodificador (*DTS*)
decomposition *n* PARTS descomposición *f*; **~ voltage** PARTS tensión *f* de descomposición
decompress *vt* COMP, DATA descomprimir
decompressed: **~ data** *n pl* COMP, DATA datos *m pl* descomprimidos
decompression *n* COMP, DATA descompresión *f*
deconvolution *n* GEN eliminación *f* de espiras
decorrelation *n* GEN descorrelación *f*
decoupling *n* ELECT, POWER desacoplamiento *m*; **~ capacitor** ELECT, POWER condensador *m* de desacoplo; **~ resistor** PARTS, POWER resistor *m* de desacoplo
decrease[1] *n* GEN decremento *m*
decrease[2] *vt* GEN disminuir; ◆ **~ load** TELEP descongestionar
decrement *n* COMP disminución *f*
decryption *n* GEN descriptación *f*
dedicated: **~ channel** *n* TRANS canal *m* dedicado; **~ frequency** *n* TRANS frecuencia *f* dedicada; **~ line** *n* DATA, TELEP, TRANS línea *f* dedicada; **~ port** *n* COMP, TRANS puerto *m* dedicado; **~ signalling** *BrE* **channel** *n* SIG canal *m* de señalización dedicado; **~ terminal** *n* TV

terminal *m* dedicado; ~ **terrestrial circuit** *n* TRANS circuito *m* terrestre dedicado
deducted *adj* GEN deducido
deductive: ~ **reasoning** *n* DATA razonamiento *m* deductivo
de-emphasis *n* ELECT desacentuación *f*; ~ **network** ELECT red *f* de desacentuación
deep: ~ **space** *n* RADIO espacio *m* remoto
de facto: ~ **standard** *n* DATA estándar *m* de hecho
default *n* COMP valor *m* por defecto; ~ **behaviour** *BrE* DATA comportamiento *m* por defecto; ~ **button** COMP botón *m* por defecto; ~ **context** DATA contexto *m* por defecto; ~ **group** DATA grupo *m* por defecto; ~ **group inference** DATA inferencia *f* de grupo por defecto; ~ **identification** DATA identificación *f* por defecto; ~ **identifier** DATA identificador *m* por defecto; ~ **library** DATA biblioteca *f* implícita; ~ **reasoning** DATA razonamiento *m* implícito; ~ **reference** DATA referencia *f* por defecto; ~ **setting** COMP regulación *f* por defecto; ~ **tempo** COMP ritmo *m* por defecto; ~ **value** GEN valor *m* por omisión
defaultable *n* (*D*) DATA valor *m* asignable por defecto (*D*)
defect *n* GEN defecto *m*; **~-counting method** TEST método *m* de recuento de defectos
defective *adj* GEN defectuoso
deferrals *n pl* ELECT comunicaciones *f pl* diferidas
deferred: ~ **addressing** *n* COMP direccionamiento *m* diferido; ~ **transmission** *n* GEN transmisión *f* diferida
defined: **~-context set** *n* DATA conjunto *m* de contexto definido; **~-domain attribute** *n* (*DDA*) TELEP atributo *m* de campo definido (*DDA*)
definition *n* DATA definición *f*
deflection *n* GEN desviación *f*; ~ **method of measurement** TEST método *m* de medición por desviación; ~ **yoke** ELECT, PARTS bobina *f* de yugo de desviación
deflector *n* RADIO deflector *m*
deframing *n* DATA *scanner* desalineación *f* de trama
defrost *vt* GEN descongelar
defrosting *n* GEN descongelación *f*
defruiter *n* RADIO supresor de señales de retorno asíncronas
degas *vt* GEN desgasificar
degassing *n* GEN desgasificación *f*
degradation *n* GEN degradación *f*; ~ **figure** RADIO factor *m* de degradación
degraded: ~ **minute** *n* DATA, RADIO minuto *m* degradado; **~-service signal** *n* SIG señal *f* de servicio degradado
degree *n* GEN *physics* grado *m*; ~ **of coherence** TRANS grado *m* de coherencia; ~ **of distortion in service** TELEP grado *m* de distorsión en el servicio; ~ **of early anisochronous parallel distortion** TELEP grado *m* de distorsión paralela anisócrona prematura; ~ **of freedom** COMP grado *m* de libertad; ~ **of gross start-stop distortion** TELEP grado *m* de distorsión arranque-parada bruta; ~ **of inherent distortion** TELEP grado *m* de distorsión inherente; ~ **of isochronous distortion** TELEP grado *m* de distorsión isócrona; ~ **of late anisochronous parallel distortion** TELEP grado *m* de distorsión paralela anisócrona tardía; ~ **of modulation** TRANS grado *m* de modulación; ~ **of standardized test distortion** TELEP grado *m* de distorsión de la prueba normalizada; ~ **of start-stop distortion** TELEP grado *m* de distorsión arranque-parada; ~ **of synchronous start-stop distortion** TELEP grado *m* de distorsión arranque-parada sincrónica; ~ **of urgency** GEN grado *m* de urgencia
degressive *adj* GEN degresivo
de-ice *vt* GEN descongelar
de-icing *n* GEN deshielo *m*, descongelación *f*
de-ionization *n* ELECT desionización *f*; ~ **grid** ELECT parrilla *f* de desionización
dejitter *vt* TRANS suprimir la fluctuación de
delamination *n* ELECT delaminación *f*
delay *n* GEN demora *f*; ~ **circuit** ELECT circuito *m* de retardo; **~-dialling** *BrE* SIG señal *f* de marcación retardada; ~ **effect** COMP efecto *m* de retardo; ~ **element** ELECT *logic elements* unidad *f* de retardo; ~ **equalizer** COMP igualador *m* de retardos; ~ **lens** RADIO lente *f* de retardo; ~ **level** COMP nivel *m* de retardo; ~ **line** DATA línea *f* de retardo; **~-lock loop** ELECT bucle *m* de bloqueo de demora; **~-mode operation** GEN operación *f* en modo de demora; ~ **processor** COMP procesador *m* de retardo; ~ **slug** PARTS *relays* anillo *m* de retardo; ~ **system** TELEP sistema *m* de espera; ~ **time** GEN tiempo *m* de retardo; ~ **vibrato** COMP vibrato *m* de retardo; ~ **working** GEN servicio *m* retardado
delayed: ~ **action** *n* GEN acción *f* retardada, acción *f* diferida; ~ **delivery** *n* TELEP entrega *f* aplazada
delete[1]: ~ **abstract-operation** *n* DATA operación *f* abstracta de borrado; ~ **channel** *n* COMP canal *m* de borrado
delete[2] *vt* GEN borrar, anular

deleted: **~-time sending** *n* DATA envío *m* de tiempo suprimido
deleting *n* GEN borrado *m*, anulación *f*
deletion *n* GEN anulación *f*, borrado *m*; **~ character** COMP carácter *m* de borrado
delimiter *n* GEN delimitador *m*
deliver *vt* GEN distribuir, entregar
deliverable: **~ unit** *n* TELEP unidad *f* suministrable
delivered: **~ encoded-information types** *n pl* DATA tipos *m pl* de información codificada entregados; **~-message entry** *n* DATA entrada *f* de mensaje entregado; **~-report entry** *n* DATA entrada *f* de informe entregado
delivery *n* GEN entrega *f*; **~ confirmation** TELEP confirmación *f* de entrega; **~ date** GEN fecha *f* de entrega; **~ inspection** TEST inspección *f* de entrega; **~ lot** TEST partida *f* de entrega; **~ network** TV red *f* de entrega; **~ note** GEN nota *f* de entrega
de-load *vt* POWER *cables* despupinizar
delrine: **~ rod** *n* PARTS varilla *f* de delrin
delta *n* ELECT *magnetic cell* delta *m*; **~ modulation** SIG, TRANS modulación *f* triangular; **~ modulation PSK-2** SIG modulación *f* delta PSK-2
demagnetization *n* ELECT desmagnetización *f*
demagnetize *vt* ELECT desmagnetizar
demagnetizing *n* ELECT desmagnetización *f*
demand[1]: **on ~** *adv* TV por encargo
demand[2]: **~-assigned multiple-access system** *n* RADIO sistema *m* de acceso múltiple asignado por demanda; **~-assigned SCPC system** *n* RADIO sistema *m* SCPC asignado por demanda; **~-assigned switching and signalling** *BrE n* SIG conmutación *f* y señalización asignadas por demanda; **~-assigned switching and signalling** *BrE* **subsystem** *n* SIG subsistema de conmutación y señalización asignado por demanda; **~ assignment** *n* RADIO asignación *f* por demanda; **~-assignment circuit** *n* ELECT circuito *m* de asignación por demanda; **~-assignment multiple access** *n* SIG acceso *m* múltiple de asignación por demanda; **~-assignment signalling** *BrE n* SIG señalización de asignación por demanda; **~ operating** *n* TELEP operación *f* por demanda; **~ service** *n* GEN servicio *m* de llamadas sin espera; **~ working** *n* TELEP servicio *m* inmediato
demand[3] *vt* COMP pedir
demo *n* COMP demostración *f*; **~ button** COMP botón *m* de demostración; **~ program** COMP programa *m* de demostración; **~ song** COMP canción *f* de propaganda
democratic: **~ network** *n* TELEP *mutually synchronized* red *f* de acceso libre
demodulate *vt* COMP, ELECT, TRANS desmodular
demodulation *n* COMP, ELECT, TRANS desmodulación *f*
demodulator *n* COMP, ELECT, TRANS desmodulador *m*
demote *vt* COMP rebajar de grado
demount *vt* GEN desmontar
demultiplexer *n* COMP, RADIO, TRANS desmultiplexor *m*
demultiplexing *n* COMP, RADIO, TRANS desmultiplexado *m*
denial *n* COMP negación *f*
denomination *n* GEN denominación *f*
deny *vt* COMP denegar, TRANS, *access* negar
depacketizer *n* DATA, TELEP despaquetificador *m*
dependability *n* TEST seguridad *f* de funcionamiento
dependable *adj* GEN seguro
dependent: **~ conformance** *n* DATA conformidad *f* subordinada; **~ exchange** *n* SWG, TELEP central *f* telefónica dependiente; **~ repeater station** *n* TELEP estación *f* repetidora subordinada; **~ station** *n* TELEP estación *f* alimentada a distancia
depletion: **~ layer** *n* PARTS *semiconductors* capa *f* de transición; **~-mode operation** *n* PARTS *semiconductors* funcionamiento *m* en modo de agotamiento
deployment *n* RADIO despliegue *m*; **~ threshold** RADIO umbral *m* de despliegue
depointing *n* RADIO desorientación *f*
depolarization *n* RADIO despolarización *f*
deposition *n* COMP depósito *m*, GEN deposición *f*, TV depósito *m*
depress *vt* COMP deprimir
depressed: **~ cladding** *n* TRANS *fibre optics* chapado *m* deprimido
depth *n* GEN profundidad *f*; **~ of colour** *BrE* DATA profundidad *f* de color; **~ gauge** *BrE* TEST testigo *m* de profundidad, calibrador *m* de profundidad
dequantization *n* COMP decuantización *f*
deregistration *n* RADIO baja *f*
deregulation *n* GEN *economics* desregulación *f*
derivation: **~ path** *n* DATA camino *m* de derivación
derivative *n* GEN *mathematics* derivado *m*
derive *vt* GEN *mathematics* derivar
DES *abbr* (*data encryption standard*) DATA DES (*norma de cifrado de datos*)
descending *adj* GEN decreciente; **~ node** *n*

RADIO nodo *m* descendente; ◆ **in ~ order** GEN en orden decreciente
descramble *vt* GEN desencriptar
descrambler *n* GEN desencriptador
descreening *n* DATA *scanner* desfiltrado *m*
description *n* GEN descripción *f*; **~ language** DATA lenguaje *m* de descripción
descriptive: **~ name** *n* DATA nombre *m* descriptivo
desensitization *n* ELECT desensibilización *f*
design *n* GEN diseño *m*, proyecto *m*; **~ aid** COMP asistencia *f* a diseño; **~ failure** TEST *dependability* defecto *m* de cálculo; **~ fault** TEST *dependability* fallo *m* de diseño; **~ modification** GEN modificación *f* del diseño; **~ objective** TELEP objetivo *m* de diseño; **~ requirements** *n pl* GEN requisitos *m pl* del diseño; **~ review** TEST *quality* revisión *f* del diseño
designate *vt* GEN designar
designated: **~ frequency** *n* TRANS frecuencia *f* designada
designating: **~ tape** *n* GEN cinta *f* de designación
designation *n* GEN designación *f*; **~ drawing** GEN *documentation* dibujo *m* funcional; **~ frame** GEN *marking* marco *m* de rotulación; **~ number** TELEP número *m* de designación; **~ strip** GEN *marking* regleta *f* dc rotulación
designer *n* GEN proyectista *m*
desired: **~ spectrum** *n* RADIO espectro *m* deseado
desk: **~ cassette** *n* RADIO *mobile* receptáculo *m* de mesa; **~ checking** *n* COMP comprobación *f* simulada, control *m* de programación; **~ microphone** *n* TELEP micrófono *m* de mesa
desktop *adj* COMP de sobremesa; **~ computer** *n* COMP computador *m* de sobremesa *AmL*, computadora *f* de sobremesa *AmL*, ordenador *m* de sobremesa *Esp*; **~ connectivity** *n* ELECT conectividad *f* de sobremesa; **~ publishing** *n* GEN autoedición *f* de textos; **~ publishing system** *n* GEN sistema *m* de autoedición de sobremesa; **~ video** *n* COMP, DATA vídeo *m* de sobremesa *Esp*, video *m* de sobremesa *AmL*
despotic: **~ network** *n* TELEP *synchronized* red *f* de sincronización despótica
destaticize *vt* COMP convertir de paralelo a serie
destination *n* GEN destino *m*; **~ address** TELEP dirección *f* de destino; **~ details** *n pl* DATA *Internet* detalles *m pl* de destino; **~ equipment** COMP equipo *m* receptor; **~ exchange** TELEP central *f* de destino; **~ identifier** DATA identificador *m* de destino; **~ indicator** TELEP indicador *m* de destino; **~ link** COMP enlace *m* receptor; **~ memory** COMP memoria *f* de receptor; **~ network** TRANS red *f* de destino; **~ node** COMP nodo *m* de destino, nodo *m* receptor; **~ office** TELEP central *f* de destino; **~ point** TELEP punto *m* de destino; **~ point code** DATA, TELEP código *m* de punto de destino; **~ terminal** DATA terminal *m* receptor; **~ terminal number** DATA número *m* de terminal de destino
destructive: **~ interference** *n* TRANS *fibre optics* interferencia *f* destructiva; **~ read** *n* COMP lectura *f* destructiva; **~ testing** *n* TEST ensayo *m* destructivo
detachable *adj* GEN desmontable
detachment *n* GEN disgregación *f*
detail: **~ drawing** *n* GEN *documentation* dibujo *m* detallado
detailed: **~ billing** *n* APPL facturación *f* detallada; **~ diagnosis** *n* GEN diagnóstico *m* detallado; **~ procedure** *n* GEN procedimiento *m* detallado
detect *vt* GEN detectar
detectable: **~ element** *n* COMP elemento *m* detectable; **~ group** *n* COMP grupo *m* detectable
detection *n* GEN detección *f*; **~ equipment** RADIO equipo *m* de detección; **~ of marker** DATA detección *f* de marcador; **~ threshold** ELECT, TELEP, TRANS umbral *m* de detección
detectivity *n* ELECT, TEST, TRANS *fibre optics* detectividad *f*
detector *n* GEN detector *m*; **~ power efficiency** PARTS *semiconductors* rendimiento *m* de potencia de detector; **~ voltage efficiency** PARTS *semiconductors* rendimiento *m* de tensión de detector
deter *vt* COMP disuadir
detergent *n* GEN detergente *m*
deterioration: **~ of quality** *n* TEST deterioro *m* de calidad
determination: **~ test** *n* TEST *dependability* prueba *f* de determinación
deterrent *n* COMP factor *m* disuasorio
detour *n* TELEP rodeo *m*
detune[1]: **~ depth** *n* COMP profundidad *f* de desintonización; **~ job** *n* COMP tarea *f* de desintonización
detune[2] *vt* COMP desintonizar
detuning *n* COMP desintonización *f*
deuterium *n* GEN deuterio *m*
deuteron *n* GEN deuterón *m*
develop *vt* GEN desarrollar
development *n* GEN desarrollo *m*; **~ engineer** GEN técnico *m* de estudios; **~ time** COMP plazo *m* de desarrollo
deviation *n* GEN desviación *f*; **~ measurement** TEST medida *f* de desviación; **~ permit** TEST

quality permiso *m* de desviación; ~ **prism** TRANS prisma *m* de desviación

device *n* GEN dispositivo *m*; ~ **adaptor** COMP adaptador *m* de aparato; ~ **adaptor interface** COMP interfaz *f* de adaptador de aparato; ~ **control** COMP, DATA, ELECT control *m* de unidad; ~ **controller** COMP, DATA, ELECT controlador *m* de dispositivo; ~ **description** TELEP descripción *f* de aparato; ~ **driver** DATA controlador *m* de dispositivo; ~ **start** TELEP arranque *m* del aparato; ~ **stop** TELEP parada *f* del aparato; ~ **under test** TEST aparato *m* en prueba; ◆ ~ **waiting** TELEP aparato *m* en espera

devise *vt* COMP concebir

dewetting *n* ELECT *soldering* retroceso *m* de estaño

DFB: ~ **laser diode** *n* (*distributed feedback laser diode*) TRANS *fibre optics* diodo *m* láser DFB (*diodo láser de realimentación distribuida*)

diac *n* ELECT diodo *m* bidireccional

diagnose *vt* TEST diagnosticar

diagnosis *n* GEN diagnóstico *m*; ~ **time** TEST tiempo *m* de diagnóstico

diagnostic: ~ **aid** *n* TEST ayuda *f* a diagnóstico; ~ **code** *n* TELEP código *m* de diagnóstico; ~ **element** *n* TELEP elemento *m* de diagnóstico; ~ **program** *n* COMP programa *m* de diagnóstico; ~ **routine** *n* COMP rutina *f* de diagnóstico

diagonal: ~ **cutting pliers** *n pl* GEN alicates *m pl* de corte diagonal

diagram *n* GEN diagrama *m*; ~ **issue** GEN edición *f* de diagramas; ~ **number** GEN *technical drawing* número *m* de diagrama; ~ **paper** GEN *technical drawing* papel *m* cuadriculado; ~ **size** GEN *technical drawing* tamaño *m* del diagrama

dial[1] *n* COMP, PER APP, TELEP disco *m*; ~ **disk** *AmE* (*cf finger wheel BrE*) PER APP, TELEP disco *m* dactilar, disco *m* marcador; ~ **in**/~ **out** ELECT llamada *f* entrante/llamada *f* saliente; ~ **pulse** TELEP impulso *m* de marcación por disco telefónico; ~ **switching** COMP conmutación *f* de dial; ~ **telephone** PER APP, TELEP teléfono *m* de disco; ~ **telephone set** PER APP, TELEP teléfono *m* de disco dactilar; ~ **tone** *AmE* DATA, TELEP tono *m* de marcar; ~ **tone delay** TELEP retardo *m* en la señal de línea, retardo *m* en el tono de marcar; ~**-up** DATA *Internet*, ELECT marcación *f* manual; ~**-up connection** DATA *Internet* conexión *f* conmutada; ~**-up line** DATA línea *f* conmutada; ~**-up modem** DATA módem *m* conmutado; ~**-up port** COMP puesto *m* de marcación; ~**-up terminal** DATA terminal *m* conmutado

dial[2] *vt* TELEP marcar

dial up *vt* TELEP marcar manualmente

dialled: ~**-number display** *BrE* *n* GEN representación *f* visual de número marcado

dialler *BrE* (*AmE* **dialer**) *n* TELEP disco *m* dactilar

dialling *BrE* (*AmE* **dialing**) *n* TELEP marcación *f*; ~ **code** TELEP código *m* de marcación; ~ **error** TELEP error *m* de marcación; ~ **period** TELEP periodo *m* de marcación; ~ **time** TELEP tiempo *m* de marcado; ~ **tone** DATA, TELEP tono *m* de marcar

dialog *n* DATA, TV diálogo *m*; ~ **recovery** DATA recuperación *f* de diálogo

diaphragm *n* PARTS diafragma *m*

diary *n* GEN diario *m*; ~ **call** GEN llamada *f* de agenda; ~ **service** TELEP servicio *m* de agenda

dichotomizing: ~ **search** *n* COMP búsqueda *f* dicotómica

dichroic: ~ **filter** *n* ELECT, TRANS filtro *m* dicroico; ~ **mirror** *n* TRANS *fibre optics* espejo *m* dicroico

die[1]: ~**-cast** *adj* PARTS fundido a presión

die[2] *n* PARTS *semiconductors* hilera *f* de estirar; ~ **set** ELECT, PARTS juego *m* de matrices

die away *vi* POWER desvanecerse

die cast *vt* PARTS fundir a presión

die down *vi* POWER apagarse

dielectric *adj* GEN dieléctrico; ~ **antenna** *n* RADIO antena *f* dieléctrica; ~ **breakdown** *n* ELECT perforación *f* dieléctrica; ~ **charge** *n* ELECT carga *f* dieléctrica; ~ **constant** *n* POWER constante *f* dieléctrica; ~ **lens** *n* RADIO lente *f* dieléctrica; ~**-loss measurement** *n* DATA medición *f* de pérdida dieléctrica; ~ **radiator** *n* RADIO antena *f* dieléctrica; ~ **resonator** *n* ELECT resonador *m* dieléctrico; ~ **rod radiator** *n* RADIO antena *f* dieléctrica de varilla; ~ **strength** *n* POWER resistencia *f* dieléctrica; ~ **strength test** *n* POWER prueba *f* de resistencia dieléctrica

diesel: ~ **engine** *n* GEN motor *m* diesel; ~ **generator set** *n* POWER grupo *m* electrógeno diesel

difference *n* GEN diferencia *f*; ~ **of capacity** TELEP diferencia *f* de capacidad

differential *adj* GEN diferencial; ~ **amplifier** *n* TRANS amplificador *m* diferencial; ~ **circuit** *n* TRANS circuito *m* diferencial; ~ **delay time** *n* TRANS tiempo *m* de retardo diferencial; ~ **echo suppressor** *n* RADIO, TELEP supresor *m* de eco diferencial; ~ **encoding** *n* DATA codificación *f* diferencial; ~ **equation** *n* GEN ecuación *f* diferencial; ~ **interpolation** *n* COMP interpolación *f* diferencial; ~ **loop modulation** *n* COMP modulación *f* en bucle diferencial; ~ **measurement method** *n* TEST método *m* de medición diferencial; ~ **modal attenuation** *n* TRANS atenuación *f* diferencial entre modos; ~ **mode attenuation** *n* TRANS atenuación *f* en modo diferencial; ~ **pressure** *n* GEN presión *f* diferencial; ~ **pulsation-coding modulation** *n* TRANS modulación *f* diferencial de codificación de la pulsación; ~ **pulse-code modulation** *n* COMP modulación *f* diferencial por código de impulsos, TELEP modulación *f* diferencial de codificación del impulso; ~ **quantum efficiency** *n* TEST rendimiento *m* cuántico diferencial; ~ **relay** *n* PARTS relé *m* diferencial; ~ **resistance** *n* PARTS resistencia *f* diferencial; ~ **sensitivity** *n* TELEP sensibilidad *f* diferencial; ~ **threshold** *n* TELEP umbral *m* diferencial; ~ **transformer** *n* POWER transformador *m* diferencial

differentiate *vt* GEN diferenciar

differentiating: ~ **circuit** *n* POWER circuito *m* diferenciador

differentiation *n* TELEP diferenciación *f*

diffraction *n* TRANS *fibre optics* difracción *f*; ~ **grating** CLIM, TRANS *fibre optics* rejilla *f* de difracción; ~**-limited operation** TRANS *fibre optics* operación *f* limitada por difracción; ~ **pattern** TRANS *fibre optics* configuración *f* de difracción

diffuse[1]: ~ **reflection** *n* COMP, RADIO reflexión *f* difusa; ~ **reflection coefficient** *n* RADIO coeficiente *m* de reflexión difusa

diffuse[2] *vt* GEN *chemistry* difundir

diffused: ~ **junction** *n* PARTS *semiconductors* unión *f* por difusión; ~ **resistor** *n* ELECT *semiconductors* resistor *m* por difusión

diffusion *n* GEN difusión *f*; ~ **capacitance** PARTS *semiconductors* capacitancia *f* de difusión; ~ **constant** PARTS *semiconductors* constante *f* de difusión; ~ **length** PARTS *semiconductors* extensión *f* de difusión

digit *n* DATA, TELEP, TRANS dígito *m*; ~ **display** TELEP visualizador *m* de cifras; ~ **key** TELEP tecla *f* de cifra; ~ **place** COMP posición *f* de dígito; ~ **plate** PER APP disco *m* de cifras; ~ **position** SIG, TRANS posición *f* de dígito; ~ **rate** TRANS velocidad *f* digital; ~ **sequence integrity** TELEP integridad *f* de la secuencia numérica; ~ **storing** TELEP almacenamiento *m* de cifras; ~ **time slot** TELEP intervalo *m* de tiempo de un solo dígito

digital *adj* GEN digital, numérico; ~ **announcement** *n* PER APP anuncio *m* digital; ~ **announcement machine** *n* TELEP máquina *f* de anuncios digitales; ~ **audio** *n* COMP audio *m* digital; ~ **audio broadcast** *n* COMP radiodifusión *f* de audio digital; ~ **audio broadcasting** *n* TV transmisión *f* en audio digital; ~ **audio reproduction** *n* COMP reproducción *f* en audio digital; ~ **audio stationary head** *n* COMP cabeza *f* estacionaria de audio digital; ~ **audiotape** *n* COMP, TV cinta *f* de audio digital; ~ **block** *n* DATA bloque *m* digital; ~ **carrier module** *n* TRANS módulo *m* de portadora digital; ~ **cassette** *n* COMP casete *f* digital; ~ **cellular radio** *n* RADIO radio *f* celular digital; ~ **cellular system** *n* RADIO sistema *m* celular digital; ~ **circuit** *n* TELEP, TRANS circuito *m* digital; ~ **circuit multiplication equipment** *n* TELEP, TRANS equipo *m* de multiplicación del circuito digital; ~ **circuit multiplication gain** *n* TELEP, TRANS ganancia *f* de multiplicación del circuito digital; ~**-circuit multiplication system** *n* TELEP, TRANS sistema *m* de multiplicación de circuito digital; ~ **circuitry** *n* COMP circuitería *f* digital; ~ **coding** *n* DATA codificación *f* digital; ~ **command signal** *n* (*DCS*) CONTR, DATA señal *f* de mando digital (*DCS*); ~ **compact cassette** *n* COMP casete *f* compacta digital; ~ **compression** *n* TV compresión *f* digital; ~ **concentrator** *n* TELEP concentrador *m* digital; ~ **connection** *n* GEN conexión *f* digital; ~ **connectivity** *n* TRANS conectividad *f* digital; ~ **control** *n* COMP, TELEP control *m* digital; ~ **control unit** *n* TELEP unidad *f* de control digital; ~**-controlled oscillator** *n* COMP oscilador *m* de control digital; ~ **converter** *n* GEN convertidor *m* digital; ~ **cross-connect** *n* SWG interconexión *m* digital; ~ **cross-connect equipment** *n* SWG equipo *m* de interconexión digital; ~ **data** *n pl* COMP, DATA datos *f* digitales; ~ **data communications message protocol** *n* DATA protocolo *m* de mensaje de comunicación de datos digitales; ~ **delay line** *n* COMP línea *f* de retardo digital; ~ **demultiplexer** *n* TELEP demultiplexor *m* digital; ~ **display** *n* COMP elemento *m* de visualización digital; ~ **distribution frame** *n* SWG repartidor *m* digital; ~ **document interchange format** *n* DATA formato *m* digital de intercambio de documento; ~ **document interchange standard** *n* DATA norma *f* digital de intercambio de documento; ~ **dynamic amplifier** *n* COMP amplificador *m* dinámico digital; ~ **dynamic filter** *n* COMP filtro *m* dinámico digital; ~ **editing** *n* COMP edición *f* digital; ~ **encoder-decoder** *n* COMP, TRANS codificador-descodificador *m* digital; ~ **encoding** *n* COMP, DATA, TRANS codificación *f* digital; ~ **engineering** *n* TELEP ingeniería *f* digital; ~

error *n* TELEP error *m* digital; ~ **European cordless telephone service** *n* POWER servicio *m* de teléfono inalámbrico europeo digital; ~ **exchange** *n* SWG central *f* telefónica digital; ~ **feedback** *n* ELECT realimentación *f* digital; ~ **filling** *n* DATA relleno *m* digital; ~ **filter** *n* COMP, DATA, TRANS filtro *m* digital; ~ **filtering** *n* COMP, DATA, TRANS filtrado *m* digital; ~ **FM tone generator** *n* COMP generador *m* digital de tono FM; ~ **frame structure** *n* DATA estructura *f* de tramas digitales; ~ **frequency generator** *n* COMP generador *m* de frecuencia digital; ~ **frequency synthesis** *n* TELEP síntesis *f* de frecuencia digital; ~ **harmonics generator** *n* COMP generador *m* de armonía digital; ~ **hierarchy** *n* TELEP, TRANS jerarquía *f* digital; ~ **IC** *n* ELECT circuito *m* integrado digital; ~ **identification frame** *n* DATA trama *f* de identificación digital; ~ **impulse** *n* COMP impulso *m* digital; ~ **input** *n* COMP entrada *f* digital; ~ **integrated circuit** *n* ELECT circuito *m* integrado digital; ~ **interface** *n* DATA interconexión *f* digital; ~ **interference** *n* TRANS interferencia *f* digital; ~ **lineboard** *n* SWG tarjeta *f* de línea digital; ~ **linelink** *n* TRANS enlace *m* de línea digital; ~ **linepath** *n* TRANS trayecto *m* de línea digital; ~ **line section** *n* TELEP sección *f* de línea digital; ~ **line system** *n* TELEP sistema *m* de línea digital; ~ **loop** *n* TRANS bucle *m* digital; ~ **loss** *n* TRANS pérdida *f* digital; ~ **main network switching centre** *BrE* *n* SWG centro *m* de conmutación digital de la red principal; ~ **mastering** *n* COMP masterización *f* digital; ~ **micro-mirror device** *n* COMP dispositivo *m* de micro espejo digital; ~ **mobile system** *n* TELEP sistema *m* móvil digital; ~ **modulation** *n* COMP, TRANS modulación *f* digital; ~ **modulation link** *n* COMP, TRANS enlace *m* de modulación digital; ~ **modulation system** *n* COMP, TRANS sistema *m* de modulación digital; ~ **multi-junctor** *n* TELEP multijuntor *m* digital; ~ **multiplex equipment** *n* TELEP multiplexor *m* digital, TRANS equipo *m* múltiplex digital; ~ **multiplex hierarchy** *n* TELEP, TRANS jerarquía *f* de multiplexión digital; ~ **multiplexer** *n* TRANS multiplexor *m* digital; ~ **musical instrument** *n* COMP instrumento *m* musical digital; ~ **network** *n* TRANS red *f* digital; ~ **noise** *n* DATA ruido *m* digital; ~ **non-interpolated channel** *n* TRANS canal *m* digital no interpolado; ~ **output** *n* COMP salida *f* digital; ~ **pad** *n* TRANS cuaderno *m* digital; ~ **paper storage** *n* DATA almacenamiento *m* de papel digital; ~ **path** *n* TRANS trayecto *m* digital; ~ **phase-shift modulation** *n* TRANS modulación *f* digital del cambio de fase; ~ **processing** *n* DATA procesamiento *m* digital; ~ **pseudo-noise sequence** *n* TRANS secuencia *f* de pseudorruidos digitales; ~ **publishing** *n* APPL, COMP, DATA edición *f* digital; ~ **pulse stream** *n* TELEP cadena *f* de impulsos digitales; ~ **quality** *n* RADIO calidad *f* digital; ~ **radio broadcasting** *n* RADIO transmisión *f* de radio digital; ~ **radio concentrator** *n* RADIO concentrador *m* de radio digital; ~ **radiopath** *n* TELEP trayectoria *m* de radio digital; ~ **radio relay for synchronous hierarchy** *n* RADIO relé *m* de radio digital para jerarquía síncrona; ~ **radio section** *n* TELEP sección *f* de radio digital; ~ **radio system** *n* DATA, RADIO sistema *m* de radio digital; ~ **radio telephony** *n* POWER telefonía *f* de radio digital; ~ **readout** *n* COMP lectura *f* digital, TEST presentador *m* visual digital; ~ **recorder** *n* COMP, PER APP, TEST registrador *m* digital; ~ **recording** *n* COMP, PER APP, TEST registro *m* digital; ~ **reverb** *n* COMP reverberación *f* digital; ~ **reverberation** *n* COMP reverberación *f* digital; ~ **rhythm programmer** *n* COMP programador *m* de ritmo digital; ~ **satellite concentrator** *n* RADIO concentrador *m* satélite digital; ~ **scale** *n* TEST escala *f* digital; ~ **section** *n* TRANS sección *f* digital; ~ **selective call** *n* RADIO llamada *f* selectiva digital; ~ **sequence** *n* COMP secuencia *f* digital; ~ **signal** *n* GEN señal *f* digital; ~ **signal processing** *n* COMP procesamiento *m* de señal digital; ~ **signal-processing chip** *n* DATA microcircuito *m* de procesamiento de señal digital; ~ **signal processor** *n* COMP, DATA procesador *m* de señal digital; ~ **signature** *n* DATA firma *f* digital; ~ **soundtrack** *n* COMP banda *f* sonora digital; ~ **speech compression** *n* TELEP, TRANS compresión *f* de voz digital; ~ **speech interpolation** *n* TELEP, TRANS interpolación *f* vocal digital; ~ **still image** *n* COMP imagen *f* fija digital; ~ **storage** *n* COMP, DATA almacenamiento *m* digital; ~ **storage media** *n pl* COMP soportes *m pl* de memorización digital; ~ **stream** *n* DATA cadena *f* digital; ~ **subscriber access unit** *n* TRANS unidad *f* digital de acceso al abonado; ~ **sum** *n* TELEP suma *f* digital; **~-sum variation** *n* TELEP variación *f* de suma digital; ~ **surround processing** *n* COMP procesamiento *m* perimétrico digital; ~ **switch** *n* SWG, TELEP, TRANS conmutador *m* digital; ~ **switching** *n* SWG, TELEP, TRANS conmutación *f* digital; ~ **switching centre** *BrE* *n* SWG centro *m* de conmutación digital; ~ **switching element** *n* SWG elemento *m* de conmutación digital; ~ **switching matrix** *n*

SWG matriz *f* de conmutación digital; ~ **switching network** *n* SWG, TRANS red *f* de conmutación digital; ~ **switching system** *n* SWG sistema *m* de conmutación digital; ~ **system** *n* COMP sistema *m* digital; ~ **talking machine** *n* PER APP máquina *f* parlante digital; ~ **technology** *n* GEN tecnología *f* digital; ~ **telephone** *n* SWG, TELEP teléfono *m* digital; ~**-to-analog conversion** *n* GEN conversión *f* digital-analógica; ~**-to-analog converter** *n* GEN conversor *m* digital-analógico; ~ **transit command** *n* SIG orden *f* de mando de tránsito digital; ~ **transmission** *n* TRANS transmisión *f* digital; ~ **transmission link** *n* TRANS enlace *m* de transmisión digital; ~**-transmit command** *n* DATA mando *m* de transmisión digital; ~ **transmitters** *n pl* DATA transmisores *m pl* digitales; ~ **trunk interface** *n* SIG interconexión *f* troncal digital; ~ **video** *n* COMP vídeo *m* digital; ~ **video editing** *n* COMP edición *f* digital de vídeo; **interactive ~ video** *n* COMP vídeo *m* digital interactivo; ~ **video tape recorder** *n* COMP videograbadora *f* digital; ~ **voltmeter** *n* TEST voltímetro *m* digital

digitalization *n* GEN digitalización *f*

digitalize *vt* GEN digitalizar

digitalized *adj* GEN digitalizado

digitally[1] *adv* GEN digitalmente, numéricamente; ◆ ~**-mastered** COMP masterizado digitalmente

digitally[2]: ~**-compensated crystal oscillator** *n* ELECT oscilador *m* de cristal digitalmente compensado

digitization *n* GEN digitalización *f*

digitize *vt* GEN digitalizar

digitized *adj* GEN digitalizado; ~ **data** *n pl* DATA información *f* digitalizada; ~ **image** *n* DATA, TV imagen *f* digitalizada; ~ **speech** *n* TELEP voz *f* digitalizada

digitizing *n* GEN numerización *f*

digits: ~ **dialled** *BrE n pl* TELEP números *m pl* marcados

dihedral: ~ **antenna** *n* TEST antena *f* de diedro

dimensional: ~ **compatibility** *n* GEN compatibilidad *f* dimensional; ~ **draft** *n* GEN *technical drawing* croquis *m* a escala; ~ **drawing** *n* GEN *technical drawing* dibujo *m* a escala; ~ **stability** *n* PARTS estabilidad *f* dimensional

dimensioning *n* GEN acotación *f*

dimmed *adj* COMP sombreado; ~ **icon** *n* COMP icono *m* sombreado

DIN *abbr* (*Deutsche Institut für Normung*) GEN DIN (*Instituto Alemán para la Normalización*); ~ **connector** *n* COMP conector *m* DIN; ~ **jack** *n* COMP clavija *f* DIN; ~ **plug** *n* COMP clavija *f* DIN; ~ **standard** *n* GEN norma *f* DIN

diode *n* GEN diodo *m*; ~ **array** ELECT array *m* de diodos; ~ **bridge** ELECT puente *m* de diodos; ~ **circuit** ELECT circuito *m* de diodos; ~ **crosspoint** SWG punto *m* de cruce de diodos; ~ **laser** ELECT, PARTS *semiconductors*, TRANS *fibre optics* láser *m* de diodo; ~ **network** ELECT red *f* de diodos; ~ **photodetector** DATA, PARTS, TRANS fotodetector *m* de diodos; ~ **tree** ELECT árbol *m* de diodos

dip: ~ **soldering** *n* ELECT soldadura *f* por inmersión

diplex *n* PARTS, RADIO, TELEP, TRANMIS díplex *m*; ~ **radio transmission** DATA radiotransmisión *f* en díplex

diplexer *n* PARTS, RADIO, TELEP, TRANMIS diplexor *m*

dipole *n* RADIO dipolo *m*; ~ **aerial** RADIO antena *f* dipolo; ~ **antenna** RADIO antena *f* dipolo

dipping: ~ **lacquer** *n* GEN *paints* laca *f* de inmersión; ~ **varnish** *n* GEN *paints* barniz *m* de inmersión

dipstick *n* TELEP varilla *f* de nivel de aceite

Dirac: ~ **function** *n* TELEP *pulses* función *f* de Dirac

direct[1]: ~ **to disk** *adv* COMP directo al disco

direct[2]: ~ **access** *n* COMP, DATA, PER APP acceso *m* directo; ~ **access memory** *n* COMP, DATA memoria *f* de acceso directo; ~ **access service** *n* COMP servicio *m* de acceso directo; ~ **addressing** *n* COMP direccionamiento *m* directo; ~ **beam** *n* RADIO haz *m* electromagnético directo; ~ **broadcasting by satellite** *n* (*DBS*) DATA, RADIO, TV radiodifusión *f* directa vía satélite; ~**-buried cable** *n* ELECT cable *m* enterrado de punto a punto; ~ **call** *n* DATA, TELEP llamada *f* directa; ~ **carrier** *n* TRANS portadora *f* directa; ~ **comparison** *n* TEST medición *f* por comparación directa; ~ **control** *n* COMP, SWG control *m* directo; ~ **control system** *n* COMP, SWG sistema *m* de control directo; ~ **current** *n* GEN corriente *f* continua; ~ **current converter** *n* POWER convertidor *m* de corriente continua; ~ **current coupler** *n* POWER acoplador *m* de corriente continua; ~ **current signalling** *BrE n* POWER, SIG señalización *f* por corriente continua; ~ **current transmission** *n* TRANS transmisión *f* por corriente continua; ~**-dial call** *n* TELEP llamada *f* automática; ~ **dialling-in** *BrE n* DATA, TELEP llamada automática; ~ **digital frequency synthesis** *n* ELECT síntesis *f* de frecuencia digital directa; ~ **digital input** *n* COMP entrada *f* digital directa; ~ **distance**

dialling *BrE* *n* TELEP marcado *m* directo a distancia; ~ **distance dialling** *BrE* **access code** *n* TELEP código *m* de acceso para marcado directo a distancia; ~ **distribution** *n* RADIO distribución *f* directa; ~-**driven selector** *n* SWG selector *m* controlado directamente; ~ **feed** *n* RADIO alimentación *f* directa; ~ **function** *n* COMP función *f* directa; ~ **input dialling** *BrE* *n* TELEP marcado *m* de entrada directa; ~-**insert subroutine** *n* COMP subrutina *f* de inserción directa; ~ **inward dialling** *BrE* *n* TELEP marcación *f* directa entrante; ~ **line** *n* SWG, TRANS línea *f* directa; ~ **memory** *n* COMP, PER APP memoria *f* directa; ~ **memory access** *n* (*DMA*) COMP acceso *m* directo a la memoria (*DMA*); ~ **method of measurement** *n* TEST método *m* de medición directo; ~ **negative** *n* POWER negativo *m* directo; ~ **orbit** *n* RADIO órbita *f* directa; ~ **outward dialling** *BrE* *n* TELEP marcación *f* directa saliente; ~ **positive** *n* POWER positivo *m* directo; ~ **pulse response** *n* TELEP respuesta *f* a un impulso unitario; ~ **routing** *n* TELEP encaminamiento *m* directo; ~ **satellite broadcasting** *n* TV radiodifusión *f* por satélite directa; ~ **select** *n* COMP selección *f* directa; ~ **sequence** *n* GEN secuencia *f* directa; ~-**sequence spread spectrum** *n* RADIO espectro *m* extendido de secuencia directa; ~ **service circuit** *n* TELEP circuito *m* de servicio directo; ~-**transit country** *n* TELEP país *m* de tránsito directo; ~ **voltage** *n* POWER tensión *f* continua; ~ **wiring** *n* PARTS cableado *m* de punto a punto

direct[3] *vt* TELEP dirigir

directed: ~-**beam scanner** *n* COMP digitalizador *m* de haz dirigido; ~ **connection** *n* TELEP conexión *f* orientada; ~ **scan** *n* COMP digitalización *f* dirigida

direction *n* GEN dirección *f*; ~ **finder** RADIO radiogoniómetro *m*; ~ **finding** RADIO radiogoniometría *f*; ~ **of lay** GEN *cables* sentido *m* de torsión; ~ **of rotation** GEN sentido *m* de rotación; ~ **of speech** TELEP dirección *f* de la voz

directional: ~ **aerial** *n* RADIO antena *f* dirigida, antena *f* direccional; ~ **antenna** *n* RADIO antena *f* dirigida, antena *f* direccional; ~ **beam** *n* RADIO haz *m* direccional; ~ **coupler** *n* DATA, SWG, TRANS *fibre optics* acoplador *m* direccional; ~ **coupler optical switch** *n* SWG conmutador *m* óptico de acoplador direccional; ~ **division multiplexing** *n* TRANS multiplexación *f* de división direccional; ~ **filter** *n* ELECT filtro *m* direccional; ~ **light source** *n* COMP fuente *f* de luz direccional; ~ **microphone** *n* TELEP micrófono *m* direccional; ~ **relay** *n* PARTS relé *m* direccional; ~ **sample** *n* TEST *quality* muestra *f* orientada

directive[1]: ~ **antenna** *n* DATA antena *f* directiva; ~ **gain** *n* RADIO ganancia *f* directiva

directive[2] *n* COMP, DATA directriz *f*

directivity *n* RADIO directividad *f*

directly: ~-**heated cathode** *n* ELECT, PARTS cátodo *m* de caldeo directo; ~-**powered repeater station** *n* TRANS estación *f* repetidora autónoma

directory *n* GEN directorio *m*, repertorio *m*; ~ **access protocol** (*DAP*) DATA protocolo *m* de acceso al directorio; ~ **assistance** APPL, TELEP información *f* del directorio; ~ **enquiries** *BrE* (*DQ*) APPL, TELEP servicio *m* de información; ~ **enquiry service** TELEP servicio *m* de información sobre abonados; ~ **entry** DATA entrada *f* de directorio; ~ **information** *AmE* APPL, TELEP información *f* sobre abonados; ~ **information tree** (*DIT*) DATA, TELEP árbol *m* de información de directorio; ~ **number** (*DN*) TELEP número *m* de directorio; ~ **services** *n pl* DATA servicios *m pl* de información sobre abonados; ~ **store** TELEP almacenamiento *m* en directorio; ~ **system agent** DATA operador *m* del sistema de información sobre abonados; ~ **system protocol** DATA protocolo *m* del sistema de información sobre abonados; ~ **user agent** (*DUA*) APPL, DATA apoderado *m* de usuarios de directorio (*DUA*)

disable *vt* GEN desactivar

disabled *adj* GEN desactivado; ~ **channel** *n* RADIO canal *m* inhabilitado; ~ **state** *n* TEST estado *m* no operativo; ~ **time** *n* TEST tiempo *m* no operativo

disablement: ~ **unit** *n* TEST *dependability* unidad *f* de exclusión

disabling *n* COMP desactivación *f*

disadvantage *n* GEN desventaja *f*

disassemble *vt* GEN desensamblar

disaster: ~ **prevention** *n* TEST *security* prevención *f* de desastres; ~ **recovery** *n* DATA recuperación *f* después de un desastre

disc: ~ **antenna** *BrE* *n* RADIO antena *f* de disco

discharge *n* GEN descarga *f*; ~ **cycle** POWER ciclo *m* de descarga

disclosure *n* COMP revelación *f*

discolour *BrE* (*AmE* **discolor**) *vt* GEN cambiar de color

discolouration *BrE* (*AmE* **discoloration**) *n* GEN alteración *f* de color

disconnect[1]: ~ **frame** *n* TELEP secuencia *f* de desconexión; ~ **signal** *n* DATA, SIG señal *f* de desconexión

disconnect[2] *vt* (*DCN*) GEN desconectar (*DCN*)
disconnected: ~ **mode** *n* TELEP modo *m* desconectado; ~**-mode frame** *n* TELEP secuencia *f* de modo desconectado
disconnection *n* GEN desconexión *f*; ~ **procedure** TELEP procedimiento *m* de desconexión
disconnector *n* POWER *switchgear* seccionador *m*
discontinue *vt* GEN *project* interrumpir
discontinuous: ~ **selection** *n* COMP selección *f* discontinua
discrete *adj* PARTS discreto; ~ **channel** *n* TRANS canal *m* discreto; ~ **circuit** *n* ELECT circuito *m* discreto; ~ **component** *n* PARTS componente *m* discreto; ~ **cosine transform** *n* COMP, DATA transformación *f* cosenoidal discreta; ~ **cosine transform coding** *n* DATA codificación *f* de transformación cosenoidal discreta; ~ **data collector** *n* COMP compilador *m* de datos no integrados; ~ **Fourier transform** *n* DATA transformación *f* discreta Fourier; ~ **system** *n* DATA sistema *m* discreto; ~**-time deterministic signal** *n* TRANS señal *f* determinista de tiempo discreto; ~**-time linear system** *n* TRANS sistema *m* lineal de tiempo discreto; ~**-time system** *n* DATA sistema *m* de tiempo discreto; ~ **value** *n* GEN valor *m* discreto; ~ **variable** *n* GEN variable *f* discreta
discretely: ~**-timed signal** *n* COMP señal *f* sincronizada discontinuamente, TRANS señal *f* temporal discreta
discretionary *adj* COMP discrecional
discriminating: ~ **satellite exchange** *n* SWG central *f* telefónica discriminante de satélite
discrimination *n* DATA, ELECT, SWG, TEST discriminación *f*; ~ **instruction** COMP instrucción *f* de discriminación; ~ **ratio** TEST *quality* ratio *f* de discriminación; ~ **threshold** TEST *metrology* umbral *m* de discriminación
discriminator *n* DATA, ELECT, SWG, TEST discriminador *m*; ~ **input object** DATA ejecución *f* de entrada del discriminador
disengage *vt* GEN desembragar
dish *n* RADIO plato *m*; ~ **aerial** RADIO antena *f* parabólica, antena *f* de plato; ~ **antenna** RADIO antena *f* parabólica, antena *f* de plato
disjunction *n* ELECT *logic elements* disgregación *f*
disk: ~ **antenna** *AmE* (*see disc antenna BrE*); ~ **array** *n* DATA pila *f* de discos; ~ **box** *n* COMP caja *f* de discos; ~ **caching** *n* COMP, DATA utilización *f* de la antememoria del disco; ~ **caching software** *n* COMP, DATA programa *m* para utilización de la antememoria del disco; ~ **change** *n* COMP cambio *m* de disco; ~ **drive** *n* COMP, DATA unidad *f* de disco; ~**-drive controller** *n* COMP, DATA controlador *m* de la unidad de disco; ~ **driver** *n* COMP unidad *f* de disco; ~ **memory** *n* COMP memoria *f* de disco; ~ **slot** *n* COMP disquetera *f*; ~ **space** *n* COMP espacio *m* en disco; ~ **storage** *n* COMP almacenamiento *m* en disco magnético; ~**-storage drive** *n* COMP unidad *f* de discos; ~ **store** *n* COMP almacenamiento *m* en disco
diskette *n* COMP disquete *m*
dismantle *vt* GEN desarmar
dismount *vt* GEN desmontar
disparity *n* TELEP disparidad *f*
dispatoh[1]: ~**-received file** *n* DATA archivo *m* recibido por distribución; ~ **service** *n* TRANS servicio *m* de distribución; ~ **system** *n* TRANS sistema *m* de despacho
dispatch[2] *vt* GEN despachar
dispatcher *n* GEN despachador *m*
dispensing: ~ **tips** *n pl* ELECT puntas *f pl* de clavijas distribuidoras
dispersed *adj* GEN disperso
dispersion *n* GEN dispersión *f*; ~**-flattened fibre** *BrE* TRANS fibra *f* con dispersión nivelada
displacement *n* GEN desplazamiento *m*; ~ **amplitude** ELECT amplitud *f* de desplazamiento; ~ **from centre** *BrE* **frequency** RADIO desplazamiento *m* respecto a la frecuencia central
display[1] *n* GEN visualización *f*; ~ **background** COMP fondo *m* de pantalla; ~ **board** DATA cuadrante *m* de visualización; ~ **console** COMP, TV consola *f* de visualización; ~ **device** COMP, DATA dispositivo *m* de visualización; ~ **element** COMP elemento *m* infográfico; ~ **field** COMP campo *m* de representación visual; ~ **foreground** COMP primer plano *m* de pantalla; ~ **group** COMP grupo *m* de visualización; ~ **light** COMP lámpara *f* de señalización; ~ **order** COMP orden *f* de visualización; ~ **paging** RADIO buscapersonas *m* con pantalla; ~ **panel** TELEP panel *m* de visualización; ~ **resolution** COMP resolución *f* de pantalla; ~ **screen** COMP, PER APP, TV pantalla *f* de visualización; ~ **segment** COMP segmento *m* de pantalla; ~ **surface** COMP superficie *f* de visualización; ~ **unit** COMP, TV unidad *f* de visualización; ~ **window** COMP visor *m*
display[2] *vt* GEN visualizar
disposition: ~ **of wires** *n* PARTS disposición *f* de cables
dispute *n* GEN conflicto *m*
disputed: ~ **bill** *n* GEN factura *f* no conforme
disrupt *vt* GEN interrumpir
disruptive: ~ **breakdown** *n* GEN *electricity*

descarga *f* disruptiva; ~ **voltage** *n* POWER tensión *f* disruptiva
dissipate *vt* POWER disipar
dissipation: ~ **factor** *n* POWER factor *m* de disipación
dissolved: ~ **image** *n* COMP imagen *f* desvanecida
distance *n* GEN distancia *f*; ~ **learning** TV educación *f* a distancia
distant: ~ **field** *n* GEN campo *m* distante; ~ **light source** *n* COMP fuente *f* de luz distante
distinguish *vt* GEN distinguir
distinguished: ~ **name** *n* DATA nombre *m* distintivo
distort *vt* GEN distorsionar
distortion *n* GEN distorsión *f*; ~ **bridge** ELECT puente *m* de distorsión; ~ **factor** ELECT factor *m* de distorsión; ~**-limited** *adj* TRANS limitado por la distorsión; ~**-limited operation** ELECT operación *f* limitada por la distorsión
distress: ~ **alerting** *n* RADIO señal *f* de socorro; ~ **radio call system** *n* RADIO sistema *m* de llamadas de socorro radiotelefónicas; ~ **signal** *n* TEST *security* señal *f* de petición de auxilio
distribute *vt* GEN distribuir, repartir
distributed *adj* GEN distribuido, repartido; ~ **application** *n* COMP aplicación *f* distribuida; ~ **architecture** *n* SWG arquitectura *f* distribuida; ~ **Bragg reflector laser diode** *n* PARTS diodo *m* láser de reflector Bragg distribuido; ~ **computing** *n* COMP, DATA cálculo *m* repartido; ~**-computing environment** *n* COMP entorno *m* informático descentralizado; ~ **control** *n* DATA, SWG control *m* distribuido; ~ **control system** *n* SWG sistema *m* de control distribuido; ~ **data processing** *n* DATA informática *f* distribuida; ~ **feedback** *n* DATA, ELECT, TRANS realimentación *f* distribuida; ~ **feedback laser** *n* DATA, ELECT, TRANS láser *m* de realimentación distribuida; ~ **feedback laser diode** *n* TRANS diodo *m* láser de realimentación distribuida; ~ **file** *n* DATA archivo *m* distribuido; ~ **file server** *n* DATA servidor *m* de archivo distribuido; ~ **frame-alignment signal** *n* TRANS señal *f* de alineación de trama distribuida; ~**-information processing task** *n* DATA tarea *f* de proceso de información distribuida; ~**-light source** *n* COMP fuente *f* de luz repartida; ~ **multi-antenna** *n* RADIO antena *f* multicanal distribuida; ~ **network** *n* COMP, TV red *f* distribuida; ~ **office support system** *n* COMP sistema *m* descentralizado de apoyo para oficina, DATA sistema *m* distribuido de apoyo de oficina; ~ **PBX** *n* PER APP central *f* telefónica privada distribuida; ~ **processes** *n pl* COMP procesos *m pl* descentralizados; ~ **processing** *n* DATA proceso *m* distribuido; ~**-queue double system** *n* DATA sistema *m* doble de cola distribuida; ~**-queue dual bus** *n* COMP bus *m* dual de cola repartida; ~ **queueing** *n* ELECT colas *f pl* distribuidas; ~**-sequence controller** *n* COMP controlador *m* de secuencia repartida; ~ **signalling** *BrE n* TRANS señalización *f* distribuida; ~ **single layer** *n* DATA capa *f* única distribuida; ~ **single-layer embedded test method** *n* TEST método *m* de prueba integrada en capa única distribuida; ~ **single-layer test method** *n* TEST método *m* de prueba de capa única distribuida; ~ **system** *n* COMP sistema *m* distribuido; ~**-system architecture** *n* COMP arquitectura *f* de sistema descentralizado; ~ **test method** *n* TEST método *m* de prueba distribuido; ~ **transaction** *n* DATA transacción *f* distribuida
distributing: ~ **data processing** *n* COMP proceso *m* de datos descentralizado
distribution *n* GEN distribución *f*, repartición *f*; ~ **block** ELECT *connecting devices* bloque *m* de distribución; ~ **box** POWER, TRANS caja *f* de distribución; ~ **busbar** POWER barra *f* de distribución; ~ **cabinet** TRANS armario *m* de distribución; ~ **cable** POWER cable *m* de distribución; ~ **centre** *BrE* POWER, TV centro *m* de distribución; ~ **channel** DATA *scanner* canal *m* de distribución; ~ **and control unit** (*DCU*) CONTR unidad *f* de distribución y control (*DCU*); ~ **frame** TRANS marco *m* repartidor; ~ **function** GEN *statistics* función *f* de distribución; ~ **fuse** PARTS fusible *m* de distribución; ~ **link** RADIO enlace *m* de distribución; ~ **list** (*DL*) COMP *e-mail*, DATA, TELEP lista *f* de distribución; ~ **module** POWER módulo *m* de distribución; ~ **negative** POWER negativo *m* de distribución; ~ **network** SWG, TRANS, TV red *f* de distribución; ~ **output** POWER salida *f* de distribución; ~ **panel** POWER tablero *m* de distribución; ~ **plug** POWER clavija *f* de distribución; ~ **point** TRANS punto *m* de distribución; ~ **positive** POWER positivo *m* de distribución; ~ **rack** POWER bastidor *m* de distribución; ~ **relay** PARTS relé *m* de distribución; ~ **relay set** SWG juego *m* de relés de distribución; ~ **stage** TRANS etapa *f* de distribución; ~ **terminal** POWER bloque *m* de distribución; ~ **of traffic** TELEP distribución *f* de tráfico; ~ **transformer** POWER transformador *m* de distribución
distributor *n* DATA, TELEP distribuidor *m*
disturb *vt* GEN perturbar

disturbance *n* GEN perturbación *f*; ~ **counter** TELEP contador *m* de perturbaciones; ~**-free** *adj* TELEP sin interferencias; ~ **recording** TELEP registro *m* de perturbaciones; ~ **supervision** TELEP supervisión *f* de perturbaciones; ~ **suppression** TEST supresión *f* de perturbaciones; ~ **voltage** TEST tensión *f* interferente

DIT *abbr* (*directory information tree*) DATA, TELEP árbol *m* de información de directorio

ditch *n* RADIO zanja *f*

dithering *n* COMP motivo *m* de tramado

diurnal: ~ **rhythm** *n* GEN *physiology* ritmo *m* diurno; ~ **variation** *n* GEN variación *f* diurna

diversification *n* GEN diversificación *f*

diversion: ~ **on busy** *n* TELEP desvío *m* en caso de ocupado; ~ **of call** *n* TELEP desvío *m* de llamada; ~ **follow me** *n* TELEP transferencia *f* temporal del número propio; ~**-if-number-busy service** *n* TELEP servicio *m* de desvío en caso de número ocupado; ~ **on no reply** *n* TELEP desvío *m* cuando no hay contestación

diversity *n* GEN diversidad *f*; ~ **radar** RADIO radar *m* de diversidad; ~ **reception** RADIO recepción *f* de diversidad

divert *vt* TELEP desviar

divert to *vt* COMP desviar hacia

divert towards *vt* COMP desviar hacia

divide *vt* GEN dividir; ◆ ~ **into syllables** COMP dividir en sílabas

divided: ~**-battery system** *n* POWER sistema *m* con carga separada

divider *n* ELECT divisor *m*

divisibility *n* GEN divisibilidad *f*

divisible *adj* GEN divisible

division *n* GEN división *f*; ~ **into syllables** APPL *speech processing* división *f* en sílabas

divisor *n* GEN *mathematics* divisor *m*

DL *abbr* (*distribution list*) COMP *e-mail*, DATA, TELEP lista *f* de distribución

DLC *abbr* (*data-link connection*) DATA, TRANS conexión *f* de transmisión de datos

DLCI *abbr* (*data-link connection identifier*) DATA, TRANS identificador *m* de conexión de transmisión de datos

DLS *abbr* (*data-link service*) APPL, DATA, TRANS servicio *m* de transmisión de datos

DLSAP *abbr* (*data-link service access point*) APPL, DATA, TRANS DLSAP (*punto de acceso al servicio de transmisión de datos*)

DM *abbr* (*data management*) COMP, DATA gestión *f* de datos

DMA *abbr* (*direct memory access*) COMP DMA (*acceso directo a la memoria*)

DN *abbr* (*directory number*) TELEP número *m* de directorio

DNIC *abbr* (*data-network identification code*) DATA, TELEP, TRANS código *m* de identificación de la red de datos

DNR *abbr* (*domain name resolver*) DATA *Internet* separador *m* de nombre de dominio

DNTX: ~ **code** *n* SIG código *m* DNTX

do: ~**-not-disturb service** *n* TELEP servicio *m* no molestar; ~**-not-transmit code** *n* SIG código *m* no transmitir

docking: ~ **station** *n* ELECT estación *f* de acoplamiento

document *n* GEN documento *m*; ~ **bulk transfer** TELEP transferencia *f* de documentos en masa; ~ **bulk transfer and manipulation** TELEP transferencia *f* y manipulación de documentos en masa; ~ **bulk transfer and manipulation class** (*BTM*) TELEP clase *f* de transferencia y manipulación de documentos en masa (*BTM*); ~ **bulk transfer class** (*BT*) TELEP clase *f* de transferencia de documentos en masa (*BT*); ~ **content architecture** (*DCA*) DATA arquitectura *f* del contenido del documento; ~ **description language** DATA lenguaje *m* de descripción de documento; ~ **facsimile telegram** TELEP telegrama *m* facsímil de documento; ~ **facsimile telegraphy** TELEP telegrafía *f* facsimilar de documentos; ~ **file** DATA *Internet* archivo *m* de documentos; ~ **format** DATA *Internet* formato *m* de documento; ~ **heading** GEN encabezamiento *m* de documento; ~ **imaging system** DATA sistema *m* de visualización de documentos; ~ **interchange format** DATA formato *m* de intercambio de documentos; ~ **issue** GEN emisión *f* de un documento; ~ **list** GEN lista *f* de documentos; ~ **loader** DATA *scanner* cargador *m* de documentos, TELEP manipulador de documentos; ~ **number** GEN número *m* de documento; ~ **processing** DATA proceso *m* de documentos; ~ **processing and interchange** DATA proceso *m* e intercambio de documentos; ~ **revision** GEN revisión *f* de documento; ~ **server** DATA *Internet* servidor *m* de documentos; ~ **size** DATA *Internet* tamaño *m* de documento; ~ **source** DATA origen *m* de documento; ~ **structure** GEN estructura *f* de documento; ~ **transmission** DATA transmisión *f* de documentos; ~ **type** GEN tipo *m* de documento

documentary: ~ **resource** *n* DATA *Internet* reserva *f* documental

documentation *n* GEN documentación *f*; ~ **rules** *n pl* GEN reglas *f pl* de la documentación

dog *n* PARTS garra *f*
domain *n* DATA campo *m*; **~-defined attribute** TELEP atributo *m* definido por campo; **~ expert** COMP experto *m* de campo; **~ name resolver** (*DNR*) DATA *Internet* separador *m* de nombre de dominio; **~-specific part** DATA parte *f* específica de dominio
domestic: **~ network** *n* DATA red *f* nacional
domotics *n pl* APPL domótica *f*
donor *n* PARTS *semiconductors* donador *m*
door: **~ interphone** *n* PER APP portero *m* eléctrico; **~ video** *n* PER APP portero *m* de vídeo
dopant *n* PARTS adulterante *m*
Doppler: **~ effect** *n* GEN efecto *m* Doppler; **~ radar** *n* RADIO radar *m* Doppler
dormant *adj* COMP durmiente
dosage *n* GEN dosis *f*
dose *vt* GEN dosificar
dot *n* GEN punto *m*; **~ diagram** GEN *statistics* diagrama *m* de puntos; **~ frequency** TV *video* frecuencia *f* de exploración; **~-interlace scanning** TV *video* exploración *f* por puntos sucesivos; **~-interlace sweep** TV *video* barrido *m* de puntos entrelazados; **~-matrix printer** COMP, PER APP impresora *f* de puntos
dots: **~ per inch** *n pl* (*dpi*) DATA *scanner* puntos *m pl* por pulgada (*dpi*, *ppp*)
dotted: **~ line** *n* GEN *technical drawing* línea *f* de puntos
double: **~-armature relay** *n* PARTS relé *m* de doble armadura; **~ Babcock plan** *n* TELEP plan *m* de Babcock doble; **~-base diode** *n* PARTS *semiconductors* diodo *m* de doble base; **~-break contact** *n* PARTS *electric contacts* contacto *m* bipolar; **~ call** *n* TELEP llamada *f* doble; **~ check** *n* COMP doble control *m*; **~-circuit line** *n* APPL *outside plant engineering* línea *f* doble; **~ click** *n* COMP, DATA doble click *m*; **~ conductor** *n* GEN conductor *m* doble; **~-conversion converter** *n* TRANS convertidor *m* de conversión doble; **~-cord switchboard** *n* SWG panel *m* de conmutación dicordio; **~-crucible technique** *n* TRANS *fibre optics* técnica *f* de crisol doble; **~ current** *n* TELEP *telex*, TRANS corriente *f* doble; **~-current code** *n* TELEP, TRANS código *m* de corriente doble; **~-current transmission** *n* TRANS transmisión *f* de corriente doble; **~-current working** *n* TELEP funcionamiento *m* con ambas corrientes; **~ density** *n* (*DD*) COMP doble densidad *f* (*DD*); **~-ended control** *n* TRANS control *m* biterminal; **~-ended synchronization** *n* TELEP, TRANS sincronización *f* biterminal; **~ error** *n* TELEP error *m* doble; **~ heterojunction** *n* TRANS *fibre optics* heterounión *f* doble; **~-layer coating** *n* PARTS revestimiento *m* de doble capa; **~-length register** *n* COMP registro *m* de extensión doble, DATA registro *m* de doble longitud; **~-make contact** *n* PARTS contacto *m* de doble cierre; **~-phantom balanced telegraph circuit** *n* TELEP circuito *m* telegráfico balanceado superfantasma; **~-phantom circuit** *n* TRANS circuito *m* superfantasma; **~-reflector antenna** *n* RADIO antena *f* de doble reflector; **~ seizure** *n* TELEP ocupación *f* doble; **~ sideband** *n* (*DSB*) RADIO, TRANS banda *f* lateral doble (*DSB*); **~-sided distribution frame** *n* TRANS marco *m* repartidor de dos caras; **~-sided printed board** *n* ELECT tablero *m* impreso de dos caras; **~-sided rack** *n* TELEP bastidor *m* de dos caras; **~ solidus** *n* GEN *symbols* barra *f* doble
doubler *n* ELECT doblador *m* de voltaje
down: **~ converter** *n* RADIO convertidor *m* de RF a IF; **~ lead** *n* RADIO *aerials* bajante *m*, TELEP bajada *f*; **~ link** *n* RADIO, TELEP enlace *m* descendente; **~-link beam** *n* RADIO haz *m* de enlace descendente; **~-link feeder link** *n* RADIO enlace *m* de alimentador de enlace descendente; **~-link of programmes** *BrE n* RADIO unión *f* descendente de programas; **~ path** *n* TV trayecto *m* de bajada; **~-path power** *n* RADIO potencia *f* de la trayectoria de bajada; **~-range station** *n* TRANS estación *f* de recorrido descendente; **~-state** *n* TELEP estado *m* no operativo; **~ time** *n* TELEP tiempo *m* de indisponibilidad
downbeat *n* COMP compás *m* descendente
downgrade *vt* GEN degradar
downline: **~-loadable** *adj* ELECT cargable por teleproceso
download[1] *n* DATA descarga *f*
download[2] *vt* DATA descargar
downloadable *adj* DATA cargable por teleproceso
downloading *n* DATA transferencia *f* de un ordenador a otro
downsize *vt* COMP reducir
downsizing *n* COMP reducción *f*
downtime *n* ELECT, TEST *dependability* tiempo *m* improductivo
downward: **~ compatibility** *n* COMP compatibilidad *f* hacia abajo
DP *abbr* (*data processing*) COMP, DATA sistema *m* de procesamiento de datos
dpi *abbr* (*dots per inch*) DATA *scanner* dpi (*puntos por pulgada*)
DQ *BrE abbr* (*directory enquiries*) APPL, TELEP servicio *m* de información
draft *n* GEN *technical drawing* tiro *m*; **~ international standard** GEN borrador *m* de norma

internacional; ~ **project** GEN anteproyecto *m*; ~ **to scale** GEN *technical drawing* borrador *m* a escala; ~ **sketch** GEN *technical drawing* borrador *m* de esquema; ~ **text** COMP texto *m* en borrador

drag[1] *n* RADIO draga; ~ **region** COMP región *f* de arrastre

drag[2]: ~ **and drop** *vt* COMP arrastrar y eliminar

dragging *n* GEN arrastre *m* de imagen

drain: ~ **region** *n* PARTS *transistors* región *f* del colector, región *f* de drenaje; ~ **valve** *n* PARTS válvula *f* de desagüe; ~ **wire** *n* GEN *cables* alambre *m* de drenaje

drainage: ~ **coil** *n* POWER bobina *f* de drenaje

draining: ~ **transformer** *n* POWER transformador *m* de absorción

DRav *abbr* (*average reference-surface diameter*) TELEP DRav (*diámetro medio de superficie de referencia*)

drawing: ~ **number** *n* GEN *technical drawing* número *m* de dibujo; ~ **process** *n* TRANS *fibre optics* proceso *m* de extracción; ~ **to scale** *n* GEN *technical drawing* dibujo *m* a escala

drawmap *n* COMP proyección *f* de giros

DRCS *abbr* (*dynamically-redefinable characters*) COMP DRCS (*caracteres dinámicamente redefinibles*)

drift *n* GEN deriva *f*; ~ **compensation** TELEP compensación *f* de arrastre; ~ **failure** TEST fallo *m* progresivo; ~ **rate** RADIO velocidad *f* de arrastre; ~ **velocity** PARTS *semiconductors* velocidad *f* de arrastre

driftpin *n* GEN mandril *m*

drill *vt* PARTS perforar

drive *n* COMP mecanismo *m*; ~ **distance** ELECT distancia *f* de arrastre; ~ **fit** GEN ajuste *m* forzado

driven: ~ **element** *n* RADIO elemento *m* excitado

driver *n* GEN impulsor *m*; ~ **stage** ELECT etapa *f* excitadora

driving: ~ **belt** *n* PARTS correa *f* de transmisión; ~ **circuit** *n* POWER circuito *m* excitador; ~ **current** *n* POWER corriente *f* de excitación; ~ **motor** *n* GEN motor *m* de impulsión; ~ **pawl** *n* GEN trinquete *m* de arrastre; ~ **power** *n* RADIO potencia *f* de excitación; ~ **pulley** *n* GEN polea *f* motriz; ~ **stage** *n* ELECT etapa *f* de excitación; ~ **wheel** *n* GEN rueda *f* motriz

drop: **~-down menu** *n* COMP menú *m* desplegable; ~ **and insert** *n* TRANS derivación *f* e inserción; ~ **point** *n* TRANS *fibre optics* punto *m* de derivación; ~ **test** *n* TEST prueba *f* de caída de voltaje

drop out *vi* TELEP desaparecer

drum: ~ **factor** *n* TELEP factor *m* de cilindro; ~ **plotter** *n* COMP tambor *m* de traza; ~ **receiver** *n* TELEP receptor *m* de cilindro; ~ **rhythm** *n* COMP ritmo *m* de tambor; ~ **sequence** *n* COMP secuencia *f* de tambor; ~ **store** *n* SWG almacenamiento *m* de tambor; ~ **transmitter** *n* TELEP transmisor *m* de tambor

dry: ~ **battery** *n* COMP pila *f* seca; ~ **cell** *n* POWER pila *f* seca; **~-cell battery** *n* POWER batería *f* de célula seca

drying: ~ **time** *n* GEN tiempo *m* de secado

DSB *abbr* (*double sideband*) RADIO, TRANS DSB (*banda lateral doble*)

DSI *abbr* (*data set identifier*) DATA identificador *m* de conjunto de datos

D-star *n* TEST estrella *f* D

DTAM: ~ **user** *n* TELEP usuario *m* de DTAM

DTE *abbr* (*data terminal equipment*) DATA, PER APP, TELEP DTE (*equipo terminal de datos*); ~ **busy** *n* TELEP equipo *m* terminal de datos ocupado; ~ **clear request** *n* TELEP petición *f* de puesta a cero de DTE; **~-inactive registration facility** *n* TELEP instalación *f* de registro de DTE inactiva; ~ **waiting** *n* TELEP espera *f* de DTE; ◆ **~-controlled not-ready** TELEP controlado por DTE no disponible; **~-uncontrolled not-ready** TELEP no controlado por DTE no disponible

DTMF *abbr* (*digital tone multi-frequency*) DATA DTMF (*multifrecuencia de tono dual*)

DTS *abbr* (*decoding time stamp*) COMP DTS (*reloj fechador descodificador*)

DUA *abbr* (*directory user agent*) APPL, DATA DUA (*apoderado de usuarios de directorio*)

dual *adj* COMP doble; **~-attaching copper station** *n* ELECT estación *f* de cobre de doble conexión; **~-attaching fibre station** *n* ELECT estación *f* de fibra de doble conexión; ~ **attachment** *n* ELECT accesorio *m* dual; **~-beam antenna** *n* DATA antena *f* de doble haz; **~-beam oscilloscope** *n* ELECT osciloscopio *m* de doble haz; **~-carrier algorithm** *n* COMP, TRANS algoritmo *m* de portadora doble; **~-carrier transmission** *n* TRANS transmisión *f* de portadora doble; **~-channel mode** *n* COMP modo *m* de canal doble; **~-channel visual direction finding** *n* RADIO ubicación *f* visual de dirección de canal doble; **~-conversion receiver** *n* RADIO receptor *m* de conversión doble; **~-frequency channel** *n* RADIO canal *m* de doble frecuencia; ~ **front end** *n* RADIO sección *f* de entrada doble; ~ **homing** *n* ELECT reposición *f* doble; **~-in-line package** *n* DATA empaque *m* dual en línea; **~-launch system** *n* RADIO sistema *m* de emisión dual; **~-level plan** *n* GEN plan *m* de doble nivel; ~ **path** *n* TRANS camino *m* doble; ~**-polarized antenna** *n* RADIO antena *f* de doble polarización; ~**-processor load-sharing system** *n* SWG sistema *m* de reparto de la carga de doble procesador; ~**-processor system** *n* SWG sistema *m* de doble procesador; ~**-purpose connector** *n* ELECT conector *m* de finalidad doble; ~**-purpose wire insertion cap** *n* ELECT caperuza *f* de inserción de cable de doble finalidad; ~**-ring** ELECT de anillo doble; ~ **ring of trees** *n* ELECT doble *m* anillo de árboles; ~ **seizure** *n* TELEP ocupación *f* doble; ~**-signalling telephone** *BrE n* PER APP, TELEP teléfono *m* de doble señalización; ~**-standard** GEN de norma doble; ~ **telephone numbers** *n pl* TELEP números *m pl* telefónicos dobles; ~**-tone multifrequency** *n* (*DTMF*) TELEP multifrecuencia *f* de doble tono (*DTMF*); ~**-tone multifrequency dialling** *BrE n* TELEP discado *m* de multifrecuencia de doble tono; ~**-tone multifrequency signalling** *BrE n* TELEP señalización de multifrecuencia de doble tono

dub *vt* GEN doblar

dubbing *n* GEN doblaje *m*

duct *n* ELECT, PARTS, TRANS conducto *m*; ~ **cable** GEN cable *m* para canalizaciones; ~ **height** RADIO altura *f* de conducto portacables; ~ **intensity** RADIO intensidad *f* de envolvente; ~ **route** GEN *cable laying* conducto *m* multitubular; ~ **run** GEN *cable laying* tramo *m* de canalización; ~ **system** GEN *cable laying* sistema *m* de canalización; ~ **thickness** RADIO ancho *m* de conducto

ducting *n* RADIO canalización *f*; ~ **layer** RADIO capa *f* de canalización

dumb: ~ **terminal** *n* DATA terminal *m* sin procesador

dummy[1] *adj* COMP ficticio, RADIO artificial, TELEP falso; ~ **antenna** *n* RADIO antena *f* artificial; ~ **bit** *n* DATA bitio *m* ficticio; ~ **capacitance** *n* TRANS capacitancia *f* figurada; ~ **fibre** *BrE n* TRANS fibra *f* ficticia; ~ **front** *n* TELEP panel *m* falso; ~ **load** *n* RADIO carga *f* ficticia; ~ **panel** *n* TELEP panel *m* falso; ~ **strip** *n* PARTS regleta *f* ciega, regleta *f* falsa; ~ **unit** *n* TRANS unidad *f* ficticia

dummy[2] *n* GEN boceto *m* inicial

dump *n* COMP vuelco *m*; ~ **request** COMP petición *f* de vaciado; ~ **routine** COMP rutina *f* de vaciado

dumping *n* COMP, DATA vuelco *m*

duplex *adj* GEN dúplex; ~**-channel group system** *n* RADIO sistema *m* de grupo de canal dúplex; ~ **line** *n* RADIO, TRANS línea *f* dúplex; ~ **traffic** *n* RADIO tráfico *m* dúplex; ~ **working** *n* RADIO, TELEP funcionamiento *m* en dúplex

duplexer *n* RADIO, SWG, TRANS duplexor *m*

duplicate[1]: ~ **equipment** *n* TELEP equipo *m* duplicado; ~ **operation** *n* TELEP operación *f* duplicada

duplicate[2] *vt* GEN duplicar

duplication *n* GEN duplicación *f*

durability *n* GEN durabilidad *f*

durable *adj* GEN no perecedero

duration *n* GEN duración *f*

dust: ~ **cover** *n* ELECT protección *f* contra el polvo, PARTS guardapolvos *m*; ~**-free room** *n* TEST habitación *f* libre de polvo

dustproof *adj* GEN protegido contra el polvo

dustproofing: ~ **kit** *n* ELECT equipo *m* protegido contra el polvo

duty: ~ **cycle** *n* POWER, RADIO ciclo *m* de trabajo

DUV *abbr* (*data under voice*) RADIO DUV (*datos bajo voz*)

DX: ~**-stuffing signal** *n* SIG señal *f* de relleno DX

dynamic *adj* GEN dinámico; ~ **allocation** *n* COMP asignación *f* dinámica; ~ **allocation of frequency** *n* GEN adjudicación *f* dinámica de frecuencia; ~ **buffering** *n* COMP tamponamiento *m* dinámico; ~ **chaining** *n* DATA encadenamiento *m* dinámico; ~ **channel allocation** *n* RADIO adjudicación *f* de canal dinámico; ~ **characteristic** *n* COMP característica *f* dinámica; ~ **compression** *n* COMP compresión *f* dinámica; ~ **conformance requirement** *n* DATA requisito *m* de conformidad dinámico; ~ **data exchange** *n* DATA intercambio *m* dinámico de datos; ~ **keyboard** *n* COMP teclado *m* dinámico; ~**-link library** *n* TELEP biblioteca *f* de enlace dinámico; ~ **load control** *n* SWG control *m* de carga dinámica; ~ **measurement** *n* TEST medición *f* dinámica; ~ **noise** *n* TELEP ruido *m* dinámico; ~ **part** *n* DATA parte *f* dinámica; ~ **RAM** *n* COMP, DATA RAM *f* dinámica; ~ **range** *n* TEST gama *f* dinámica; ~ **relocation** *n* COMP reubicación *f* dinámica; ~ **resource allocation** *n* COMP asignación *f* dinámica de recursos; ~ **response** *n* GEN respuesta *f* dinámica; ~ **stability** *n* GEN estabilidad *f* dinámica; ~ **storage** *n* COMP memoria *f* dinámica; ~ **storage allocation** *n* COMP asignación *f* de memoria dinámica

dynamically: ~**-redefinable characters** *n pl* (*DRCS*) COMP caracteres *m pl* dinámicamente redefinibles (*DRCS*)

E

E: ~ **mode** *n* RADIO modo *m* E

E & M: ~ **signalling** *BrE n* TELEP señalización *f* E & M

ear: ~ **reference point** *n* TELEP punto *m* de referencia auditiva

earcap *n* PER APP, TELEP tapa *f* de auricular

earlier: ~**-transmitted bits** *n pl* TRANS bitios *m pl* transmitidos anteriormente; ~**-transmitted bits scrambling process** *n* TELEP secuencia *f* de aleatorización de bitios anteriormente transmitidos

early: ~ **failure** *n* TEST fallo *m* inicial; ~**-failure period** *n* TEST periodo *m* de fallos prematuros

earphone *n* PER APP, TELEP auricular *m*

earpiece *n* PER APP, TELEP auricular *m*

earth *BrE n* (*cf ground AmE*) POWER masa *f*; ~ **acquisition** RADIO adquisición *f* de tierra; ~ **bar** ELECT, POWER barra *f* de tierra; ~ **button** SWG botón *m* de puesta a tierra; ~ **cable** POWER cable *m* de masa; ~ **clamp** POWER terminal *m* de puesta a tierra; ~ **collector** POWER colector *m* de tierra; ~ **collector bar** POWER barra *f* colectora de tierra; ~ **conductor** POWER conductor *m* de tierra; ~ **connection** POWER conexión *f* a tierra; ~ **electrode** POWER electrodo *m* de masa; ~ **electrode installation** POWER instalación *f* de electrodo de tierra; ~ **electrode resistance** POWER resistencia *f* de electrodo de tierra; ~**-facing panel** RADIO panel *m* orientado hacia la tierra; ~ **fault** POWER pérdida *f* a tierra; ~**-fault circuit breaker** POWER interruptor *m* accionado por corriente de pérdida a tierra; ~**-fault protection** POWER protección *f* contra fuga a tierra; ~**-fault relay** ELECT, PARTS relé *m* de avería por puesta a tierra; ~**-generated beacon** RADIO radiofaro *m* generado desde tierra; ~**-generated pilot** RADIO piloto *m* generado desde tierra; ~ **lead** POWER cable *m* a tierra; ~ **leakage** POWER fuga *f* a tierra; ~ **leakage current** POWER corriente *f* de fuga a tierra; ~ **leakage guard** POWER monitor *m* de fuga a tierra; ~ **loop** ELECT bucle *m* de masa; ~**-orbiting artificial satellite** RADIO satélite *m* artifical en órbita terrestre; ~ **satellite** RADIO satélite *m* en órbita terrestre; ~**-oriented geo-synchronous satellite** RADIO satélite *m* geosincrónico orientado desde tierra; ~ **plane** POWER plano *m* de tierra; ~ **plane antenna** RADIO antena *f* de plano de tierra; ~**plate** POWER toma *f* de tierra; ~ **potential** POWER potencial *m* terrestre; ~ **potential difference** POWER diferencia *f* de potencial terrestre; ~ **radial termination** POWER terminación *f* radial terrestre; ~ **resistance** POWER resistencia *f* de tierra; ~ **return** SIG, TRANS retorno *m* por tierra; ~**-return circuit** TRANS circuito *m* con vuelta por tierra; ~**-return double phantom circuit** TELEP circuito *m* superfantasma con vuelta por tierra; ~**-return phantom circuit** TELEP circuito *m* fantasma con vuelta por tierra; ~ **ring collector** POWER colector *m* anular de tierra; ~ **sensor** RADIO sensor *m* terrestre; ~ **shadow** RADIO sombra *f* de la tierra; ~ **station** GEN estación *f* terrestre; ~ **station record** RADIO registro *m* de estación terrena; ~ **station verification and assistance** RADIO verificación *f* y asistencia de estación terrena; ~**-sun sensor** RADIO sensor *m* tierra-sol; ~**-synchronous satellite** RADIO satélite *m* sincrónico terrestre; ~ **system** POWER, RADIO sistema *m* de tierra; ~ **terminal** POWER borne *m* de puesta a tierra; ~**-to-orbit shuttle** GEN lanzadera *f* Tierra-Orbita; ~ **wire** POWER hilo *m* de puesta a tierra

earthed *BrE adj* (*cf grounded AmE*) ELECT, POWER puesto a tierra; ~ **emitter circuit** *n* ELECT circuito *m* emisor puesto a tierra; ~ **emitter stage** *n* ELECT fase *f* de emisor puesto a tierra; ~**-grid amplifier** *n* ELECT amplificador *m* de rejilla puesta a tierra

earthing *BrE n* (*cf grounding AmE*) ELECT, POWER conexión *f* a tierra; ~ **arrangement** POWER conexión *f* a tierra; ~ **button** SWG *PBX* botón *m* de puesta a tierra; ~ **conductor** POWER conductor *m* de tierra; ~ **indent** ELECT muesca *f* para toma de tierra; ~ **mat** RADIO, TRANS plancha *f* de conexión a tierra; ~ **network** POWER red *f* de tierra; ~ **plate** POWER placa *f* de toma de tierra; ~ **probe** POWER diodo *m* rectificador de toma de tierra; ~ **rod** POWER varilla *f* de tierra; ~ **switch** POWER conmutador *m* de puesta a tierra

earthquake: ~**-resistant** *adj* TEST *construction* sismorresistente

easy: ~ **access** *n* COMP acceso *m* fácil; ~

connection *n* GEN conexión *f* fácil; ~ **payment** *n* GEN pago *m* fácil

eavesdropping *n* COMP, RADIO, TELEP escucha *f*

EB *abbr* (*erroneous block*) DATA EB (*bloque erróneo*)

ebonite *n* PARTS ebonita *f*

EBU *abbr* (*European Broadcasting Union*) RADIO EBU (*Unión de Radiodifusión Europea*)

EC *abbr* (*European Community*) GEN CE (*Comunidad Europea*)

ECC *abbr* (*error-correction code*) GEN código *m* corrector de errores

eccentric: ~ **orbit** *n* RADIO órbita *f* de excentricidad

ECD *abbr* (*error-control device*) DATA dispositivo *m* de control de errores

echelon: ~ **telegraphy** *n* TELEP telegrafía *f* escalonada

echo *n* GEN eco *m*; ~ **balance return loss** TELEP atenuación *f* de equilibrado de eco; ~ **box** RADIO caja *f* de ecos; ~ **cancellation** ELECT, RADIO, TRANS cancelador *f* de eco, eliminación *f* de ecos; ~ **canceller** *BrE* ELECT, RADIO, TRANS compensador *m* de ecos, eliminador *m* de ecos; ~ **cancelling** *BrE* ELECT, RADIO, TRANS cancelación *f* de eco; ~ **control** RADIO, TELEP control *m* de eco; ~**-control device** RADIO, TELEP aparato *m* de control de eco; ~ **curve** TELEP curva *f* de eco; ~ **effect** COMP, TELEP efecto *m* de eco; ~**-half-suppressor** TELEP semisupresor *m* de eco; ~ **loss** TELEP atenuación *f* de eco; ~ **protection** RADIO protección *f* contra eco; ~ **return** DATA retorno *m* de eco; ~ **suppression** TRANS supresión *f* de eco; ~**-suppressor indicator** TELEP indicador *m* del supresor de eco

Echo: ~ **Doppler indicator** *n* (*EDI*) RADIO indicador *m* de ecos por efecto Doppler (*EDI*)

echoless: ~ **chamber** *n* TEST cámara *f* sin reverberaciones

echometric: ~ **measurement** *n* TELEP medición *f* ecométrica

echoplex: ~ **mode** *n* TELEP modo *m* echoplex

ECL *abbr* (*emitter-coupled logic*) DATA, ELECT *logic elements* CLEA (*circuito lógico de emisor acoplado*)

eclipse: ~ **capability** *n* RADIO capacidad *f* de eclipse; ~ **interference** *n* RADIO interferencia *f* de eclipse; ~**-working transponder** *n* RADIO traspondedor *m* que trabaja en eclipse

ECMA *abbr* (*European Computer Manufacturers Association*) COMP ECMA (*Asociación Europea de Fabricantes de Ordenadores*)

ECO *abbr* (*engineering change order*) PARTS ECO (*pedido de modificación técnica*)

economic: ~ **standard antenna** *n* RADIO antena *f* estándar económica

economy: ~ **of scale** *n* GEN economía *f* de escala

ECPT *abbr* (*European Conference of Postal and Telecommunications Administrations*) GEN ECPT (*Conferencia Europea de Administración Postal y Telecomunicaciones*)

eddy: ~ **current** *n* POWER corriente *f* parásita; ~ **current loss** *n* POWER pérdida *f* por corrientes parásitas

edge: ~**-board contact** *n* ELECT contacto *m* de borde; ~**-of-coverage gain** *n* RADIO ganancia *f* de borde de cobertura; ~ **curl** *n* COMP enroscamiento *m* de borde; ~ **detection** *n* DATA detección *f* de borde; ~ **distance** *n* ELECT *printed circuits* distancia *f* a los bordes; ~**-emitting** *adj* TRANS de emisión de borde; ~**-emitting diode** *n* ELECT, TRANS *fibre optics* diodo *m* de emisión de borde; ~**-emitting LED** *n* TRANS *fibre optics* diodo *m* electroluminiscente con emisión de borde; ~**-emitting light-emitting diode** *n* ELECT, TRANS diodo *m* fotoemisor de emisión marginal; ~ **flare** *n* COMP distorsión *f* de contraste en los bordes; ~ **warping** *n* ELECT *printed circuits* alabeo *m* de borde

EDI *abbr* (*Echo Doppler indicator*) RADIO EDI (*indicador de ecos por efecto Doppler*); ~ **access unit** *n* (*EDI-AU*) TELEP unidad *f* de acceso EDI (*EDI-AU*); ~**-forwarding** *n* DATA, TELEP envío *m* por EDI; ~ **message** *n* (*EDIM*) DATA, TELEP mensaje *m* por intercambio de datos electrónicos; ~ **message store** *n* DATA memoria *f* de mensaje EDI, TELEP registro *m* de mensaje EDI; ~ **messaging** *n* (*EDIMG*) DATA, TELEP mensajería *f* por intercambio de datos electrónicos; ~**-messaging environment** *n* (*EDIME*) APPL, DATA, RADIO, TELEP entorno *m* de mensajería por intercambio de datos electrónicos; ~**-messaging system** *n* (*EDIMS*) APPL, DATA, RADIO, TELEP sistema *m* de mensajería por intercambio de datos electrónicos; ~**-messaging user** *n* APPL, DATA, RADIO, TELEP usuario *m* de mensajería por intercambio de datos electrónicos; ~ **notification** *n* (*EDIN*) TELEP notificación *f* de intercambio de datos electrónicos; ~ **responsibility** *n* DATA, TELEP responsabilidad *f* de EDI; ~ **user** *n* DATA, TELEP usuario *m* de intercambio de datos electrónicos; ~ **user agent** *n* (*EDI-UA*) DATA, TELEP agente *m* de usuario de EDI

EDI-AU *abbr* (*EDI access unit*) TELEP EDI-AU (*unidad de acceso EDI*)

EDIM *abbr* (*EDI message*) DATA, TELEP mensaje

m por intercambio de datos electrónicos; ~ **responsibility** *n* DATA, TELEP responsabilidad *f* del mensaje por intercambio de datos electrónicos

EDIME *abbr* (*EDI-messaging environment*) APPL, DATA, RADIO, TELEP entorno *m* de mensajería por intercambio de datos electrónicos

EDIMG *abbr* (*EDI messaging*) DATA, TELEP mensajería *f* por intercambio de datos electrónicos; ~ **user** *n* DATA, TELEP usuario *m* de mensajería por intercambio de datos electrónicos

EDIMS *abbr* (*EDI-messaging system*) APPL, DATA, RADIO, TELEP sistema *m* de mensajería por intercambio de datos electrónicos

EDIN *abbr* (*EDI notification*) TELEP notificación *f* de intercambio de datos electrónicos

edit[1]: ~ **buffer** *n* COMP memoria *f* intermedia de edición; ~ **buffer initialization** *n* COMP inicialización *f* de la memoria intermedia de edición; ~ **button** *n* COMP botón *m* de edición; ~ **decision list** *n* COMP lista *f* de decisión de edición; ~ **parameter** *n* COMP parámetro *m* de edición; ~ **recall** *n* COMP cancelación *f* de edición

edit[2] *vt* GEN editar

editing: ~ **procedure** *n* GEN procedimiento *m* de edición

edition *n* GEN edición *f*

editor *n* GEN técnico *m* de montaje, editor *m*

EDI-UA *abbr* (*EDI user agent*) DATA, TELEP agente *m* de usuario de EDI

EDP *abbr* (*electronic data processing*) DATA procesamiento *m* de datos electrónicos; ~ **analyst** *n* GEN analista *m* de procesamiento de datos electrónicos

educational: ~ **software** *n* COMP, DATA programa *m* educacional

EETDN *abbr* (*end-to-end transit delay negotiation*) DATA, TELEP EETDN (*negociación de retardo de tránsito de extremo a extremo*)

EF *abbr* (*entrance facility*) ELECT EF (*instalación de entrada*)

effect *n* GEN efecto *m*; ~ **memory** COMP memoria *f* de efecto

effective: ~ **address** *n* COMP direccional *m* efectivo; ~ **antenna height** *n* RADIO altura *f* efectiva de antena; ~ **call attempt** *n* RADIO intento *m* efectivo de llamada; ~ **capacity** *n* TELEP capacidad *f* efectiva; ~ **Earth-radius factor** *n* RADIO factor *m* efectivo de radio terrestre; ~ **gain** *n* TRANS ganancia *f* efectiva; ~ **height** *n* RADIO altura *f* efectiva; ~ **height above average terrain** *n* RADIO altura *f* efectiva sobre la media del terreno; ~ **input admittance** *n* ELECT, POWER admitancia *f* efectiva de entrada; ~ **input capacitance** *n* POWER capacitancia *f* efectiva de entrada; ~ **input impedance** *n* POWER impedancia *f* efectiva de entrada; ~ **instruction** *n* COMP instrucción *f* efectiva; ~ **isotropically radiated power** *n* RADIO potencia *f* efectiva radiada isotrópicamente; ~ **loss** *n* TRANS atenuación *f* compuesta; ~ **mode volume** *n* TRANS volumen *m* de modo efectivo; ~ **monopole-radiated power** *n* (*emrp*) RADIO potencia *f* efectiva radiada en monopolio (*emrp*); ~ **number of bits** *n* DATA número *m* efectivo de bitios; ~ **output admittance** *n* POWER admitancia *f* efectiva de salida; ~ **output capacitance** *n* POWER capacitancia *f* efectiva de salida; ~ **output impedance** *n* POWER impedancia *f* efectiva de salida; ~ **overall noise band** *n* TRANS banda *f* de ruido general efectivo; ~ **radiated power** *n* (*e.r.p.*) RADIO potencia *f* efectiva radiada (*e.r.p.*); ~ **radius of the Earth** *n* RADIO radio *m* efectivo de la tierra; ~ **selectivity** *n* RADIO selectividad *f* efectiva; ~ **traffic** *n* GEN tráfico *m* efectivo

effectively: ~ **transmitted signals in sound program** *n pl* SIG señales *f pl* transmitidas efectivamente en programa de sonido

effectiveness *n* GEN efectividad *f*

effector *n* COMP efector *m*

efficiency *n* GEN eficiencia; ~ **factor in time** TELEP factor *m* de eficiencia en el tiempo; ~ **ratio** TELEP ratio *f* de eficiencia

E-field: ~ **antenna** *n* RADIO antena *f* de campo E

EFIR *abbr* (*equiripple finite-impulse response filter*) ELECT EFIR (*filtro de respuesta de impulso finito igualmente ondulado*)

EG: ~ **select** *n* COMP selector *m* EG

EGA *abbr* (*enhanced graphics adaptor*) COMP EGA (*adaptador de gráficos destacados*)

egg: ~-**box lens** *n* RADIO lente *f* de caja oval

EHF: ~ **satcom project** *n* TRANS proyecto *m* de satélite comercial EHF

EID *abbr* (*escape in data*) DATA EID (*cambio de código en los datos*)

eight: ~ **to fourteen modulation** *n* (*ETF*) COMP modulación *f* ocho a catorce (*ETF*); ~ **to ten modulation** *n* (*ETM*) COMP modulación *f* ocho a diez (*ETM*)

EIRP *abbr* (*equivalent isotropically-radiated power*) RADIO EIRP (*potencia equivalente radiada isotrópicamente*)

eject[1]: ~ **button** *n* COMP, DATA botón *m* de expulsión

eject[2] *vt* COMP, DATA *disk* expulsar

elasticity *n* GEN elasticidad *f*

elastomer: ~ **contact-strip** *n* PARTS tira *f* de contacto elastomérica
elastometer *n* TEST elastómetro *m*
electret *n* PARTS, POWER electreto *m*; ~ **microphone** TELEP micrófono *m* de electreto
electric: ~ **arc** *n* POWER arco *m* eléctrico; ~ **breakdown** *n* ELECT, PARTS *semiconductors* ruptura *f* eléctrica; ~ **circuit** *n* POWER circuito *m* eléctrico; ~ **cord** *n* COMP cable *m* eléctrico; ~ **current** *n* POWER corriente *f* eléctrica; ~ **field** *n* ELECT campo *m* eléctrico; ~ **measuring instrument** *n* TEST instrumento *m* de medición eléctrica; ~ **migration** *n* POWER migración *f* eléctrica; ~ **polarization** *n* POWER polarización *f* eléctrica; ~ **power** *n* POWER energía *f* eléctrica; ~ **power station** *n* POWER central *f* eléctrica; ~ **screen** *n* POWER pantalla *f* eléctrica; ~ **shock** *n* POWER sacudida *f* eléctrica; ~ **signal transducer** *n* POWER transductor *m* de señal eléctrica; ~ **vector** *n* TRANS *fibre optics* vector *m* eléctrico
electrical: ~ **admittance** *n* ELECT admitancia *f* eléctrica; ~ **capacitance** *n* ELECT capacitancia *f* eléctrica; ~ **charge** *n* ELECT carga *f* eléctrica; ~ **circuit** *n* ELECT circuito *m* eléctrico; ~ **conductance** *n* ELECT conductancia *f* eléctrica; ~ **data** *n pl* TRANS *fibre optics* datos *m pl* eléctricos; ~ **engineer** *n* GEN electrotécnico *m*; ~ **engineering** *n* GEN electrotecnia *f*; ~ **equipment** *n* DATA equipo *m* eléctrico; ~ **fitter** *n* GEN montador *m* electricista; ~ **fitting** *n* PARTS accesorio *m* eléctrico; ~ **integration unit** *n* ELECT unidad *f* de integración eléctrica; ~ **interference** *n* COMP interferencia *f* eléctrica; ~ **length** *n* RADIO trayectoria *f* eléctrica; ~ **model** *n* ELECT modelo *m* eléctrico; ~ **path delay** *n* TRANS retardo *m* en la trayectoria eléctrica; ~ **path equalization** *n* TRANS compensación *f* en la trayectoria eléctrica; ~ **path length** *n* TRANS longitud *f* del trayecto eléctrico; ~ **power** *n* POWER energía *f* eléctrica; ~ **power system** *n* POWER sistema *m* de energía eléctrica; ~ **resistance** *n* ELECT resistencia *f* eléctrica; ~ **subscriber-busy signal** *n* SIG señal *f* eléctrica de abonado ocupado; ~ **tape** *n* POWER cinta *f* aislante; ~ **wavelength** *n* RADIO longitud *f* de onda eléctrica; ~ **zero** *n* TEST *measuring device* cero *m* eléctrico
electrically: ~**-erasable programmable read-only memory** *n* ELECT memoria *f* sólo de lectura programable borrable por medios eléctricos; ~**-held crosspoint** *n* SWG punto *m* de cruce sostenido eléctricamente; ~**-programmable read-only memory** *n* (*EPROM*) COMP, ELECT memoria *f* sólo de lectura programable eléctricamente (*EPROM*)
electrician *n* GEN electricista *m*
electricity: ~ **supply system** *n* POWER sistema *m* de suministro de electricidad
electrification *n* TEST electrificación *f*
electro: ~**-optic** *adj* ELECT, SWG, TRANS electroóptico; ~**-optical** *adj* ELECT, SWG, TRANS *fibre optics* electroóptico; ~**-optic effect** *n* TRANS efecto *m* electroóptico; ~**-optic switch** *n* SWG conmutador *m* electroóptico
electroacoustic *adj* TELEP electroacústico
electroacoustics *n* TELEP electroacústica *f*
electrochemical: ~ **capacitor** *n* PARTS condensador *m* electroquímico; ~ **dissociation** *n* GEN disociación *f* electroquímica
electrochemistry *n* GEN electroquímica *f*
electrode *n* ELECT, POWER electrodo *m*; ~ **admittance** ELECT admitancia *f* de electrodo; ~ **capacitance** ELECT capacitancia *f* electródica; ~ **conductance** ELECT conductancia *f* de electrodo; ~ **configuration** ELECT configuración *f* de electrodo; ~ **dark current** ELECT corriente *f* oscura electródica; ~ **dissipation** ELECT disipación *f* de electrodo; ~ **impedance** ELECT impedancia *f* de electrodo; ~ **potential** ELECT potencial *m* de electrodo; ~ **reactance** ELECT reactancia *f* de electrodo; ~ **resistance** ELECT resistencia *f* electródica
electrodynamic *adj* ELECT, TELEP *electromagnetism* electrodinámico; ~ **microphone** *n* TELEP micrófono *m* electrodinámico
electroforming *n* ELECT *chemistry* electroplastia *f*
electrographic: ~ **printer** *n* PER APP impresora *f* electrográfica
electroluminescence *n* ELECT, TRANS *fibre optics* electroluminiscencia *f*
electrolyte *n* ELECT electrólito *m*
electrolytic: ~ **capacitor** *n* PARTS condensador *m* electrolítico; ~ **cell** *n* ELECT *chemistry* cuba *f* electrolítica, célula *f* electrolítica; ~ **copper** *n* GEN cobre *m* electrolítico; ~ **dissociation** *n* GEN *chemistry* disociación *f* electrolítica; ~**-dissociation constant** *n* GEN constante *f* de disociación electrolítica
electromagnet *n* POWER electroimán *m*
electromagnetic *adj* GEN electromagnético; ~ **beam** *n* RADIO haz *m* electromagnético; ~ **compatibility** *n* ELECT, TEST, TV compatibilidad *f* electromagnética; ~ **disturbance** *n* ELECT, RADIO, TEST perturbación *f* electromagnética; ~ **environment** *n* TEST entorno *m* electromagnético; ~ **field** *n* ELECT campo *m* electromagnético; ~ **force** *n* ELECT, POWER

fuerza *f* electromagnética; ~ **interference** *n* ELECT, TEST interferencia *f* electromagnética; ~ **microphone** *n* TELEP micrófono *m* electromagnético; ~ **mode** *n* TRANS modalidad *f* electromagnética; ~ **pulse** *n* ELECT, RADIO impulso *m* electromagnético; ~ **radiation** *n* ELECT, TRANS radiación *f* electromagnética; ~ **relay** *n* PARTS relé *m* electromagnético; ~ **remote sensing** *n* TEST detección *f* remota electromagnética; ~ **wave** *n* DATA, ELECT onda *f* electromagnética; ~ **wave propagation** *n* ELECT, RADIO propagación *f* de las ondas electromagnéticas

electromechanical *adj* GEN electromecánico; ~ **relay** *n* PARTS relé *m* electromecánico; ~ **technician** *n* GEN técnico *m* electromecánico

electromechanics *n* GEN electromecánica *f*

electromotive: ~ **force** *n* (*EMF*) ELECT, POWER fuerza *f* electromotriz (*EMF*)

electron *n* GEN electrón *m*; ~ **beam** ELECT haz *m* de electrones; ~ **charge** ELECT carga *f* de electrón; ~ **collision** ELECT colisión *f* electrónica; ~ **conduction** ELECT, PARTS conducción *f* electrónica; ~ **current** ELECT, POWER corriente *f* electrónica; ~ **flow** ELECT, POWER flujo *m* electrónico; ~ **gas** GEN gas *m* de electrones; ~ **mass** GEN masa *f* del electrón; ~ **microscope** ELECT microscopio *m* electrónico; ~ **microscopy** ELECT microscopía *f* electrónica; ~ **shell** GEN capa *f* de electrón; ~ **volt** GEN electronvoltio *m*

electronic *adj* GEN electrónico; ~ **address** *n* DATA dirección *f* de correo electrónico; ~ **automatic exchange** *n* SWG central *f* electrónica automática; ~ **book** *n* DATA libro *m* electrónico; ~ **business set** *n* TELEP conjunto *m* empresarial electrónico; ~ **call-signal generator** *n* PER APP generador *m* electrónico de señal de llamada; ~ **chip** *n* ELECT placa *f* electrónica; ~ **circuit** *n* ELECT circuito *m* electrónico; ~ **circuitry** *n* ELECT circuitos *m pl* electrónicos; ~ **clock** *n* ELECT reloj *m* electrónico; ~ **component** *n* COMP, ELECT, PARTS componente *m* electrónico; ~ **countermeasure** *n* RADIO contramedida *f* electrónica; ~ **crosspoint** *n* SWG punto *m* de contacto electrónico; ~ **data interchange** *n* COMP, DATA, TELEP intercambio *m* de datos electrónicos; ~ **data processing** *n* COMP, DATA tratamiento *m* de datos electrónico; ~ **design automation** *n* PARTS automatización *f* del diseño electrónico; ~ **design interchange format** *n* DATA formato *m* de intercambio de diseño electrónico; ~ **directory** *n* DATA, TELEP directorio *m* electrónico; ~ **document interchange** *n* COMP, DATA, TELEP intercambio *m* de documentos electrónicos; ~ **document management** *n* DATA gestión *f* de documentos electrónicos; ~ **engineering** *n* GEN ingeniería *f* electrónica; ~ **equipment** *n* GEN equipo *m* electrónico; ~ **equipment technician** *n* GEN técnico *m* en equipo electrónico; ~ **exchange** *n* COMP, SWG central *f* electrónica; ~ **funds transfer** *n* APPL, DATA, TELEP transferencia *f* electrónica de fondos; ~ **funds transfer at point of sale** *n* APPL transferencia *f* electrónica de fondos en punto de venta; ~ **key** *n* DATA manipulador *m* electrónico; ~ **key system** *n* PER APP sistema *m* de llave electrónica; ~ **lock** *n* RADIO cerradura *f* electrónica; ~ **mail** *n* APPL, COMP, DATA correo *m* electrónico; ~ **mailbox** *n* APPL, COMP, DATA buzón *m* de correo electrónico; ~ **mailbox service** *n* COMP servicio *m* de buzón electrónico; ~ **message** *n* APPL, DATA mensaje *m* electrónico; ~ **message service** *n* APPL, COMP, DATA servicio *m* de mensajería electrónica, servicio *m* de mensaje electrónico, servicio *m* de mensajes electrónicos; ~ **message switch** *n* SWG conmutador *m* electrónico de mensajes; ~ **message system** *n* APPL, COMP, DATA sistema *m* electrónico de mensajes; ~ **messaging** *n* APPL, COMP, DATA mensajería *f* electrónica; ~ **module** *n* COMP módulo *m* electrónico; ~ **music** *n* COMP música *f* electrónica; ~ **news gathering** *n* GEN recolección *f* electrónica de noticias; ~ **payment** *n* APPL pago *m* electrónico; ~ **payment terminal** *n* APPL terminal *m* de pago electrónico; ~ **pen** *n* DATA lapiz *m* electrónico; ~ **piano** *n* PER APP piano *m* electrónico; ~ **publishing** *n* APPL, COMP, DATA publicación *f* electrónica; ~ **satellite concentrator** *n* RADIO concentrador *m* de satélite electrónico; ~ **storyboard** *n* COMP guión *m* visualizado electrónico; ~ **supervision** *n* APPL supervisión *f* electrónica; ~ **switch** *n* COMP, DATA, SWG conmutador *m* electrónico; ~ **switching** *n* COMP, DATA, SWG conmutación *f* electrónica; ~ **switching system** *n* SWG sistema *m* de conmutación electrónico; ~ **technology** *n* COMP tecnología *f* electrónica; ~ **telephone set** *n* PER APP, TELEP teléfono *m* electrónico; ~ **time meter** *n* TEST medidor *m* electrónico de tiempo; ~ **tube** *n* ELECT tubo *m* electrónico; ~ **base** *n* ELECT, PARTS base *f* de tubo electrónico; ~ **tuning** *n* ELECT sintonización *f* electrónica; ~ **valve** *n* ELECT válvula *f* electrónica

electronically: ~**-despun antenna** *n* RADIO antena *f* de rotación inversa electrónica

electronics *n* GEN electrónica *f*; ~ **technician** GEN técnico *m* electrónico
electrostatic *adj* GEN electrostático; ~ **discharge** *n* ELECT, TEST descarga *f* electrostática; ~**-discharge-sensitive** TEST sensible a descargas electrostáticas; ~ **microphone** *n* PARTS, TELEP micrófono *m* electrostático; ~ **precautions** *n pl* ELECT precauciones *f pl* electrostáticas; ~ **printer** *n* PER APP impresora *f* electrostática; ~ **relay** *n* PARTS relé *m* electrostático
electrostatics *n* GEN electrostática *f*
element *n* GEN elemento *m*; ~ **error rate** TELEP tasa *f* de errores en los elementos
elementary: ~ **cable section** *n* TELEP sección *f* de cable elemental; ~ **echo** *n* TELEP eco *m* elemental; ~ **regenerated section** *n* TELEP sección *f* elemental regenerada; ~ **regenerator section** *n* TELEP sección *f* de regenerador elemental; ~ **repeater section** *n* TELEP sección *f* de repetidor elemental; ~ **stream** *n* COMP cadena *f* elemental
elevated: ~ **duct** *n* RADIO conducto *m* elevado
elevation *n* GEN elevación *f*; ~ **angle** RADIO ángulo *m* de elevación
ellipse *n* GEN elipse *f*
elliptic *adj* GEN elíptico
elliptical *adj* GEN elíptico; ~ **orbit** *n* RADIO órbita *f* elíptica; ~ **polarization** *n* RADIO polarización *f* elíptica
ELT *abbr* (*emergency locater transmitter*) RADIO transmisor *m* localizador de emergencia
e-mail *abbr* (*electronic mail*) APPL, COMP, DATA correo *m* electrónico; ~ **address** *n* DATA *Internet* dirección *f* de correo electrónica; ~ **message** *n* DATA *Internet* mensaje *m* de correo electrónico; ~ **message store** *n* DATA *Internet* almacenamiento *m* de mensaje de correo electrónico
embed *vt* COMP, DATA, ELECT empotrar, integrar
embedded: ~ **computer** *n* COMP ordenador *m* especializado integrado en un equipo *Esp*, computador *m* especializado integrado en un equipo *AmL*, computadora *f* especializada integrada en un equipo *AmL*; ~ **control channel** *n* DATA canal *m* de control embebido; ~ **operations channel** *n* (*EOC*) DATA canal *m* operativo embebido; ~ **testing** *n* DATA comprobación *f* integrada
embedding *n* COMP, DATA, ELECT integración *f*
embellish *vt* DATA *Web page* adornar
emboss *vt* COMP, ELECT *printed circuits* realzar, PARTS, *plastics* estampar
embossing *n* COMP, ELECT *printed circuits* realzado *m*, PARTS, *plastics* estampación *f*
emergency *n* GEN emergencia *f*; ~ **action** TELEP medidas *f pl* de urgencia; ~ **attention** TELEP *fault maintenance* servicio *m* de emergencia; ~ **brake** GEN freno *m* de emergencia; ~ **call** APPL, PER APP *on answering machine* llamada *f* de emergencia; ~ **call box** TELEP buzón *m* para llamadas de emergencia; ~ **call service** APPL, TELEP servicio *m* de llamadas de emergencia; ~ **call station** PER APP estación *f* para llamadas de socorro; ~ **changeover** TELEP conmutación *f* de emergencia; ~ **installation** APPL instalación *f* de emergencia; ~ **lighting** POWER alumbrado *m* de emergencia; ~ **load-transfer signal** SIG señal *f* de transferencia de carga de emergencia; ~ **locater transmitter** (*ELT*) RADIO transmisor *m* localizador de emergencia; ~ **position-indicating radio beacon** (*EPIRB*) RADIO radiofaro *m* indicador de posición para emergencias; ~ **position-indicating radio beacon station** RADIO estación *f* de radiofaro indicador de posición para emergencias; ~ **power supply** DATA fuente *f* de alimentación de emergencia; ~ **procedure** GEN procedimiento *m* de emergencia; ~ **restart** TELEP rearranque *m* de emergencia; ~ **route** TELEP ruta *f* de emergencia; ~ **service** APPL, TV servicio *m* de emergencia; ~ **shutdown system** CONTR sistema *m* de parada de seguridad; ~ **solution** GEN solución *f* de emergencia; ~ **stop** CONTR parada *f* de emergencia; ~ **telephone** PER APP, TELEP teléfono *m* de emergencia; ~ **telephone set** PER APP, TELEP teléfono *m* de emergencia; ~ **telephone station** PER APP, TELEP estación *f* telefónica para emergencias; ~ **transmitter** RADIO emisor *m* de socorro; ~ **trunk group** GEN grupo *m* de líneas troncales de emergencia
EMF *abbr* (*electromotive force*) ELECT, POWER EMF (*fuerza electromotriz*)
emission *n* GEN emisión *f*
emissivity *n* ELECT, RADIO, TRANS *fibre optics* emisividad *f*
emit *vt* GEN emitir
emittance *n* ELECT, RADIO, TRANS emitancia *f*
emitter *n* GEN emisor *m*; ~**-base voltage** PARTS *semiconductors* tensión *f* básica de emisor; ~**-coupled logic** (*ECL*) DATA, ELECT *logic elements* circuito *m* lógico de emisor acoplado (*CLEA*); ~ **current** ELECT *semiconductors* corriente *f* de emisor; ~ **cutoff current** ELECT *semiconductors* corriente *f* residual de emisor; ~ **cutoff voltage** PARTS *semiconductors* tensión *f* inversa de emisor; ~ **depletion layer** PARTS *semiconductors* capa *f* agotada de emisor; ~ **depletion layer capacitance** PARTS

semiconductors capacitancia *f* de la capa agotada de emisor; ~ **efficiency** PARTS *semiconductors* eficiencia *f* de emisor; ~ **follower** PARTS *semiconductors* seguidor *m* de emisor; ~ **junction** PARTS *semiconductors* unión *f* de emisor; ~ **resistance** PARTS *semiconductors* resistencia *f* de emisor; ~ **reverse voltage** PARTS *semiconductors* tensión *f* inversa de emisor; ~ **series resistance** ELECT, PARTS *semiconductors* resistencia *f* en serie de emisor; ~ **terminal** PARTS *semiconductors* terminal *m* de emisor; ~ **voltage** PARTS *semiconductors* tensión *f* de emisor

emphasis *n* COMP énfasis *m*

empty *adj* GEN vacío; ~ **space** *n* TELEP espacio *m* vacío

emrp *abbr* (*effective monopole-radiated power*) RADIO emrp (*potencia efectiva radiada en monopolio*)

emulate *vt* COMP, ELECT emular

emulation *n* COMP, ELECT emulación *f*

emulator *n* COMP, ELECT emulador *m*

EN *abbr* (*Euronorm*) GEN EN (*Euronorma*)

enable[1]: ~ **current** *n* POWER corriente *f* de activación; ~ **dependency** *n* ELECT dependencia *f* de validación; ~ **input** *n* ELECT entrada *f* de habilitación; ~ **pulse** *n* PARTS impulso *m* autorizado

enable[2] *vt* COMP, RADIO autorizar

enabling *n* COMP, RADIO autorización *f*; ~ **signal** COMP señal *f* de habilitación

enamel *n* PARTS esmalte *m*

enamelled *BrE* (*AmE* **enameled**) *adj* GEN esmaltado; ~ **wire** *BrE n* GEN *electric conductors* hilo *m* esmaltado

en bloc: ~ **register signalling** *BrE n* SIG señalización *f* de registro en bloque

encapsulated *adj* GEN encapsulado; ~ **relay** *n* PARTS relé *m* encapsulado

encapsulating *n* GEN encapsulamiento *m*

encapsulation *n* GEN encapsulación *f*

encased: ~ **relay** *n* PARTS relé *m* encapsulado

encipherment *n* DATA, SIG cifrado *m*

enclose *vt* GEN adjuntar

enclosed: ~ **fuse-link** *n* POWER fusible *m* de fusión cerrada

enclosure *n* GEN documento *m* adjunto

encode *vt* GEN codificar

encoded *adj* GEN codificado; ~ **relay** *n* PARTS relé *m* encapsulado

encoder *n* GEN codificador *m*

encoding *n* GEN codificación *f*; ~ **law** TELEP, TRANS ley *f* de codificación

encrypt *vt* GEN *cryptology* cifrar

encrypted: ~ **speech** *n* TELEP palabra *f* criptografiada

encrypting: ~ **equipment** *n* GEN *cryptology* equipo *m* de encriptación

encryption *n* GEN encriptación *f*; ~ **control channel** TRANS canal *m* de control de cifrado; ~ **key distribution** RADIO distribución *f* de clave de encriptación; ~ **unit** DATA, SIG unidad *f* de encriptación

encryptor *n* GEN encriptador *m*

end: ~**-around borrow** *n* COMP préstamo *m* circular; ~**-around carry** *n* COMP acarreo *m* circular; ~**-around shift** *n* COMP desplazamiento *m* lógico; ~**-of-block signal** *n* SIG señal *f* de fin de bloque; ~ **cell** *n* POWER *batteries* elemento *m* de regulación; ~**-cell rectifier** *n* POWER rectificador *m* de elementos de regulación; ~ **of communication** *n* DATA, TELEP fin *m* de comunicación; ~**-of-communication signal** *n* DATA, TELEP señal *f* de fin de comunicación; ~ **cutter** *n* GEN alicates *m pl* de corte frontal; ~ **of dialog** *n* TELEP fin *m* de diálogo; ~ **digital termination** *n* TELEP terminación *f* digital frontal; ~**-to-~ circuit supervision** *n* TEST supervisión *f* de circuito de extremo a extremo; ~**-to-~ communication** *n* TELEP comunicación *f* de extremo a extremo; ~**-to-~ continuity test** *n* TELEP prueba *f* de continuidad de extremo a extremo; ~**-to-~ digital connection** *n* ELECT conexión *f* digital de extremo a extremo; ~**-to-~ encipherment** *n* DATA cifrado *m* de extremo a extremo; ~**-to-~ information indicator** *n* DATA indicador *m* de datos de extremo a extremo; ~**-to-~ method indicator** *n* DATA indicador *m* de método dc extremo a extremo; ~**-to-~ servicing** *n* TELEP servicio *m* de extremo a extremo; ~**-to-~ signalling** *BrE n* SIG señalización *f* de extremo a extremo; ~**-to-~ test call** *n* TELEP llamada *f* de prueba de extremo a extremo; ~**-to-~ transit delay negotiation** *n* (*EETDN*) DATA, TELEP negociación *f* de retardo de tránsito de extremo a extremo (*EETDN*); ~**-of-life power** *n* POWER fin *m* de energía de vida; ~ **of line** *n* PER APP *printers* fin *m* de línea; ~ **mark** *n* COMP marca *f* final; ~ **of message** *n* (*EOM*) DATA, TELEP fin *m* de mensaje (*EOM*); ~ **of output** *n* DATA, TELEP fin *m* de salida; ~**-of-page signal** *n* DATA señal *f* de fin de página; ~ **part** *n* TELEP *equipment practice* parte *f* final; ~**-of-period tone** *n* PER APP *payphones* tono *m* de fin de periodo; ~ **of procedure** *n* (*EOP*) DATA, TELEP fin *m* de procedimiento (*EOP*); ~ **product** *n* GEN producto *m* final; ~ **of pulse** *n* COMP fin *m* de impulso; ~**-of-pulsing signal** *n* SIG señal *f* de fin de numeración; ~ **of retransmission EOM** *n* (*EOR-EOM*) DATA, TELEP fin *m* de retransmisión EOM (*EOR-EOM*); ~ **of retransmission EOP** *n* (*EOR-EOP*) DATA, TELEP fin *m* de retransmisión EOP (*EOR-EOP*); ~ **of retransmission MPS** *n* (*EOR-MPS*) DATA, TELEP fin *m* de retransmisión MPS (*EOR-MPS*); ~ **of retransmission NULL** *n* (*EOR-NULL*) DATA, TELEP fin *m* de retransmisión NULL (*EOR-NULL*); ~ **of retransmission PRI-EOM** *n* (*EOR-PRI-EOM*) DATA, TELEP fin *m* de retransmisión PRI-EOM (*EOR-PRI-EOM*); ~ **of retransmission PRI-EOP** *n* (*EOR-PRI-EOP*) DATA, TELEP fin *m* de retransmisión PRI-EOP (*EOR-PRI-EOP*); ~ **of retransmission PRI-MPS** *n* (*EOR-PRI-MPS*) DATA, TELEP fin *m* de retransmisión PRI-MPS (*EOR-PRI-MPS*); ~ **of selection** *n* (*EOS*) DATA, TELEP fin *m* de selección (*EOS*); ~ **statement** *n* TELEP declaración *f* final; ~ **system** *n* DATA sistema *m* de end; ~ **of transaction** *n* TELEP fin *m* de movimiento; ~ **of transmission** *n* (*EOT*) DATA fin *m* de transmisión (*EOT*); ~ **user** *n* GEN usuario *m* final; **~-user computing** *n* COMP, DATA informática *f* de usuario final

endfire: ~ **array antenna** *n* RADIO sistema *m* de antenas de radiación longitudinal

endless: ~ **loop** *n* COMP bucle *m* sin fin; ~ **repeat** *n* COMP repetición *f* sin fin

endurance: ~ **test** *n* TEST *dependability* prueba *f* de duración

energizing: ~ **circuit** *n* POWER circuito *m* de excitación

energy *n* GEN energía *f*; ~ **absorption** ELECT absorción *f* de carga; ~ **band** PARTS *semiconductors* banda *f* de energía; ~ **converter** POWER conversor *m* de energía; ~ **density** GEN densidad *f* de energía; ~ **dispersal** RADIO dispersión *f* de energía; ~ **gap** PARTS *semiconductors* intervalo *m* de energía; ~ **level** PARTS *semiconductors* nivel *m* de energía; **~-to-noise density** TRANS densidad *f* de energía/ruido; ~ **per information bit to noise density ratio** TRANS ratio *f* de energía por bitio de información/densidad de ruido; ~ **saving** GEN ahorro *m* de energía

engage *vt* GEN encajar

engaged *adj* GEN ocupado; ~ **signal** *n* TELEP señal *f* de ocupado; ~ **state display** *n* TELEP señal *f* visual de ocupado; ~ **tone** *BrE n* TELEP tono *m* de ocupado

engine: ~ **room** *n* TELEP sala *f* de máquinas

engineer *n* TELEP ingeniero *m*

engineering: ~ **change order** *n* (*ECO*) PARTS pedido *m* de modificación técnica (*ECO*); ~ **data** *n pl* GEN *design technology* datos *m pl* técnicos; ~ **draftsman** *n* GEN técnico *m* proyectista; ~ **fitter** *n* GEN montador *m* técnico; ~ **instructions** *n pl* GEN instrucciones *f pl* técnicas; ~ **mechanic** *n* GEN técnico *m* mecánico; ~ **order wire** *n* (*EOW*) TELEP línea *f* de órdenes de ingeniería; ~ **service channel** *n* TRANS canal *m* de servicio técnico; ~ **service circuit** *n* TRANS circuito *m* de servicio técnico

engrave *vt* GEN grabar

engraving *n* GEN grabado *m*

enhance *vt* GEN aumentar

enhanced: ~ **graphics adaptor** *n* (*EGA*) COMP adaptador *m* de gráficos destacados (*EGA*); ~ **MAC** *n* TV MAC *m* aumentado; **~-quality television** *n* COMP televisión *f* de calidad mejorada; ~ **service** *n* APPL servicio *m* reforzado; ~ **television** *n* TV televisión *f* intensificada

enhancement *n* GEN refuerzo *m*; **~-mode operation** PARTS funcionamiento *m* en modo de enriquecimiento

ensemble: ~ **activity** *n* GEN actividad *f* colectiva

en spirale *adj* ELECT, POWER en espiral

ensure *vt* GEN asegurar

enter[1]: ~ **compressed mode** *n* DATA modo *m* comprimido de entrada; ~ **transparent mode** *n* DATA modo *m* transparente de entrada

enter[2] *vt* GEN introducir; ◆ ~ **in a journal** GEN registrar en un diario

enterprise: ~ **network** *n* TRANS red *f* de empresa; **~-systems connection** *n* (*ESCON*) COMP *IBM* conexión *f* de sistemas de empresa (*ESCON*)

entertainment: ~ **broadcasting** *n* RADIO, TV difusión *f* de programas de entretenimiento

enthalpy *n* GEN entalpia *f*

entity *n* TEST *quality* entidad *f*

entrance: ~ **facility** *n* (*EF*) ELECT instalación *f* de entrada (*EF*)

entropy *n* GEN entropía *f*; ~ **coding** SIG codificación *f* de entropía

entry *n* GEN *documentation* introducción *f*, *point of access* entrada *f*; ~ **information** DATA información *f* sobre la entrada; ~ **level** DATA *scanner* nivel *m* de entrada; ~ **phone** PER APP portero *m* automático; ~ **point** COMP punto *m* de entrada; ~ **status** DATA condición *f* de entrada; ~ **type** DATA tipo *m* de entrada

entryphone *n* GEN portero *m* automático

enumerated: ~ **type** *n* DATA tipo *m* enumerado

envelope *n* COMP, DATA, SWG envoltura *f*; ~ **curve** COMP curva *f* envolvente; ~ **generator** COMP generador *m* de onda envolvente; **~-gen-**

erator select COMP selector *m* de generador de envolvente; **~ switch** SWG conmutador *m* de grupo; **~ velocity** TRANS velocidad *f* de grupo
environment *n* COMP, DATA *Internet* entorno *m*; **~ mapping** COMP cartografía *f* medioambiental
environmental: **~ endurance** *n* CLIM *engineering* resistencia *f* al ambiente; **~ factor** *n* CLIM *engineering* factor *m* ambiental; **~ specification** *n* GEN especificación *f* ambiental; **~ stress** *n* CLIM *engineering* esfuerzo *m* ambiental; **~ test** *n* GEN ensayo *m* ambiental; **~ testing** *n* CLIM prueba *f* climática, TEST prueba *f* ambiental
EOC *abbr* (*embedded operations channel*) DATA canal *m* operativo embebido
EOM *abbr* (*end of message*) DATA, TELEP EOM (*fin de mensaje*)
EOP *abbr* (*end of procedure*) DATA, TELEP EOP (*fin de procedimiento*)
EOQC *abbr* (*European Organization for Quality Control*) GEN EOQC (*Organización Europea de Control de Calidad*)
EOR-EOM *abbr* (*end of retransmission EOM*) DATA, TELEP EOR-EOM (*fin de retransmisión EOM*)
EOR-EOP *abbr* (*end of retransmission EOP*) DATA, TELEP EOR-EOP (*fin de retransmisión EOP*)
EOR-MPS *abbr* (*end of retransmission MPS*) DATA, TELEP EOR-MPS (*fin de retransmisión MPS*)
EOR-NULL *abbr* (*end of retransmission NULL*) DATA, TELEP EOR-NULL (*fin de retransmisión NULL*)
EOR-PRI-EOM *abbr* (*end of retransmission PRI-EOM*) DATA, TELEP EOR-PRI-EOM (*fin de retransmisión PRI-EOM*)
EOR-PRI-EOP *abbr* (*end of retransmission PRI-EOP*) DATA, TELEP EOR-PRI-EOP (*fin de retransmisión PRI-EOP*)
EOR-PRI-MPS *abbr* (*end of retransmission PRI-MPS*) DATA, TELEP EOR-PRI-MPS (*fin de retransmisión PRI-MPS*)
EOS *abbr* (*end of selection*) DATA, TELEP EOS (*fin de selección*)
EOT *abbr* (*end of transmission*) DATA EOT (*fin de transmisión*)
EOW *abbr* (*engineering order wire*) TELEP línea *f* de órdenes de ingeniería
EP *abbr* (*erroneous period*) DATA periodo *m* erróneo
EPIRB *abbr* (*emergency position-indicating radio beacon*) RADIO radiofaro *m* indicador de posición para emergencias
epitaxial: **~ layer** *n* PARTS capa *f* epitaxial; **~ transistor** *n* PARTS *semiconductors* transistor *m* epitaxial
epitaxy *n* PARTS *semiconductors* epitaxia *f*
E-plane: **~ lens** *n* RADIO lente *f* plana E
EPLD *abbr* (*erasable programmable logic device*) ELECT EPLD (*elemento lógico programable y borrable*)
epoxide *n* PARTS *plastics* epóxido *m*
epoxy *adj* PARTS epoxi; **~ bead** *n* ELECT cuenta *f* de epoxia; **~ buffer** *n* TRANS buffer *m* de epoxi; **~ plastic** *n* PARTS plástico *m* epoxi; **~ putty** *n* ELECT masilla *f* epoxi; **~ resin** *n* PARTS resina *f* epoxi
EPROM *abbr* (*electrically-programmable read-only memory*) COMP, ELECT EPROM (*memoria sólo de lectura programable eléctricamente*)
EQL *abbr* (*equalizer*) GEN EQL (*ecualizador*)
equal: **~-length code** *n* DATA, TELEP código *m* de igual longitud; **~-tempered scale** *n* COMP escala *f* de temple constante
equalization *n* GEN igualación *f*, ecualización *f*
equalize *vt* GEN igualar, ecualizar
equalizer *n* (*EQL*) GEN ecualizador *m* (*EQL*); **~ circuit** TRANS circuito *m* de ecualizador
equalizing: **~ charge** *n* POWER *power supply* carga *f* compensadora; **~ network** *n* ELECT red *f* reguladora; **~ pulse** *n* TV *video* impulso *m* de igualación
equate *vt* COMP igualar
equate to *vt* COMP igualar con
equatorial: **~ orbit** *n* RADIO órbita *f* ecuatorial
equidistant *adj* GEN equidistante
equilibrium: **~ length** *n* TRANS *fibre optics* longitud *f* de equilibrio; **~-mode distribution** *n* TRANS distribución *f* en modalidad de equilibrio; **~ radiation pattern** *n* TRANS diagrama *m* de radiación de equilibrio
equip *vt* GEN equipar
equipment *n* GEN equipo *m*; **~ assembler** GEN montador *m* de equipo; **~ availability** TELEP disponibilidad *f* de equipo; **~ disabled** GEN equipo *m* fuera de servicio; **~ failure** COMP avería *f* de máquina; **~ fuse** PARTS, POWER fusible *m* de aparato; **~ installation** COMP instalación *f* de equipo; **~ number** TELEP número *m* de equipo; **~ practice** GEN sistema *m* de construcción mecánica; **~ supplier** GEN proveedor *m* de equipo
equipped *adj* TRANS completo; **~ with** GEN provisto de
equiripple: **~ finite-impulse response filter** *n* (*EFIR*) ELECT filtro *m* de respuesta de impulso finito igualmente ondulado (*EFIR*)
equivalence *n* GEN equivalencia *f*

equivalent *n* GEN equivalencia *f*; ~ **articulation loss** TELEP atenuación *f* de nitidez equivalente; ~ **binary content** TELEP contenido *m* binario equivalente; ~ **bit rate** TELEP velocidad *f* binaria equivalente; ~ **isotropically radiated power** (*EIRP*) RADIO potencia *f* equivalente radiada isotrópicamente (*EIRP*); ~ **noise temperature** RADIO temperatura *f* equivalente de ruido; ~ **peak envelope power of a telephone signal** TELEP potencia *f* equivalente de la envolvente pico de una señal telefónica; ~ **random circuit group** TELEP grupo *m* equivalente de circuitos aleatorios; ~ **random traffic intensity** TELEP intensidad *f* equivalente de tráfico aleatorio; ~ **resistance** POWER resistencia *f* equivalente; ~ **resistance error** TELEP error *m* de resistencia equivalente; ~ **step index** (*ESI*) TELEP, TRANS índice *m* escalonado equivalente; ~ **step index fibre** *BrE* TRANS fibra *f* de índice escalonado equivalente; ~**-step-index profile** ELECT, TRANS *fibre optics* perfil *m* de índice escalonado equivalente; ~ **thermal network** PARTS *semiconductors* circuito *m* térmico equivalente

erasable: ~ **programmable logic device** *n* (*EPLD*) ELECT elemento *m* lógico programable y borrable (*EPLD*); ~ **store** *n* COMP almacenamiento *m* borrable

erase *vt* GEN borrar

erasing *n* COMP borrado *m*; ~ **rate** COMP índice *m* de borrado

erasure *n* PER APP, RADIO borrado *m*; ~ **of messages** PER APP *on answering machine* borrado *m* de mensajes

erbium: ~**-doped fibre** *BrE* **amplification** *n* TRANS amplificación *f* de fibra de erbio impurificada; ~**-doped fibre** *BrE* **amplifier** *n* TRANS amplificador *m* de fibra de erbio impurificada

erect *adj* TELEP *equipment practice* vertical

ergonomic *adj* COMP ergonómico

ergonomics *n* COMP ergonomía *f*

Ericsson: ~ **transport network architecture** *n* (*ETNA*) TRANS arquitectura *f* de la red de transporte Ericsson (*ETNA*)

Erlang *n* ELECT, TELEP Erlangio *m*

erosion *n* PARTS erosión *f*

ERR *abbr* (*response for end of transmission*) DATA ERR (*respuesta para finalizar la transmisión*)

erroneous: ~ **bit** *n* COMP bitio *m* erróneo; ~ **block** *n* (*EB*) DATA bloque *m* erróneo (*EB*); ~ **period** *n* (*EP*) DATA periodo *m* erróneo

error *n* GEN error *m*; ~ **analysis** DATA análisis *m* de errores; ~ **burst-correcting capability** DATA capacidad *f* de corrección de errores por ráfagas; ~ **in calculation** GEN error *m* de cálculo; ~ **catalogue** TELEP *O & M* catálogo *m* de fallos; ~ **check** RADIO *maritime* verificación *f* de errores; ~**-check character** RADIO *maritime* carácter *m* de control de errores; ~**-check signal** RADIO *maritime* señal *f* de control de errores; ~ **code** TELEP código *m* de error; ~ **constant** CONTR constante *f* de error; ~**-control device** (*ECD*) DATA dispositivo *m* de control de errores; ~**-control loop** TELEP bucle *m* de control de errores; ~**-correcting code** DATA, SIG, TELEP código *m* de corrección de errores; ~**-correcting system** DATA, TELEP sistema *m* de corrección de errores; ~**-correcting telegraph system** DATA, TELEP sistema *m* telegráfico de corrección de errores; ~ **correction** GEN corrección *f* de errores; ~**-correction code** (*ECC*) GEN código *m* corrector de errores; ~ **correction coding** DATA codificación *f* de corrección de errores; ~ **correction mode** TELEP modo *m* corrección de errores; ~ **counter** TELEP contador *m* de fallos; ~ **density** DATA densidad *f* de errores; ~**-detecting code** COMP, DATA, TELEP código *m* de detección de errores (*EDC*); ~**-detecting system** TELEP sistema *m* de detección de errores; ~ **detection** COMP, DATA, TELEP detección *f* de errores; ~**-detection coding** DATA codificación *f* para la detección de errores; ~ **detector** COMP, DATA, TELEP detector *m* de errores; ~ **diagnosis** DATA diagnóstico *m* de error; ~**-free decisecond** TELEP decisegundo *m* sin error; ~ **handling** DATA tratamiento *m* de errores; ~ **of judgment** GEN estimación *f* incorrecta; ~ **marking** TELEP *O & M* marcación *f* de error; ~ **message** DATA mensaje *m* de error; ~ **of method** TEST *metrology* error *m* de método; ~ **multiplication** TELEP multiplicación *f* de errores; ~**-multiplication factor** TELEP factor *m* de multiplicación de errores; ~ **pattern** DATA curva *f* de error; ~ **print-out** SWG impresión *f* de errores; ~ **probability** DATA, SIG probabilidad *f* de error; ~ **protection** DATA protección *f* contra errores; ~**-protection code** DATA código *m* de protección contra errores; ~ **pulse** DATA impulso *m* de error; ~ **rate** DATA, SIG, TELEP *telegraph communication* porcentaje *m* de error; ~**-rate of keying** TELEP proporción *f* de errores de manipulación; ~**-rate measurement** DATA medición *f* de índice de errores; ~**-rate monitor** TELEP monitor *m* de proporción de errores; ~ **ratio** TELEP ratio *f* de errores; ~ **recognition** TEST *dependability* reconocimiento *m* de errores; ~ **recovery**

DATA recuperación *f* de errores; ~ **report** TELEP *O & M* informe *m* de errores; ~ **resilience** COMP flexibilidad *f* de reacción al error; ~ **signal** CONTR señal *f* de error; ~ **spread** TELEP extensión *f* de error; ~ **susceptibility** DATA susceptibilidad *f* de errores

errored: ~ **second** *n* (*ES*) DATA, RADIO segundo *m* erróneo

e.r.p. *abbr* (*effective radiated power*) RADIO e.r.p. (*potencia efectiva radiada*)

ES *abbr* (*errored second*) DATA, RADIO segundo *m* erróneo

ESA *abbr* (*European Space Agency*) RADIO ESA (*Agencia Espacial Europea*)

ESC *abbr* (*escape*) COMP, DATA, TELEP ESC (*escape*)

escalating: ~ **loudness** *n* RADIO sonoridad *f* de intensidad creciente

escape *n* (*ESC*) COMP, DATA, TELEP escape *m* (*ESC*); ~ **character** COMP, DATA carácter *m* de escape; ~ **code** DATA, TELEP cambio *m* de código; ~ **in data** (*EID*) DATA cambio *m* de código en los datos (*EID*); ~ **indication** TELEP indicación *f* de escape; ~ **sequence** DATA secuencia *f* de escape; ~ **velocity** RADIO velocidad *f* de escape

ESCON *abbr* (*enterprise-systems connection*) COMP *IBM* ESCON (*conexión de sistemas de empresa*)

ESI *abbr* (*equivalent step index*) TELEP, TRANS índice *m* escalonado equivalente; ~ **profile** *n* ELECT, TRANS *fibre optics* perfil *m* ESI; ~ **refractive-index difference** *n* TELEP diferencia *f* de índice de refracción ESI

esoteric *adj* DATA *Internet* confidencial; ~ **marking** *n* DATA marcado *m* confidencial

essential: ~ **optional user-facility** *n* TELEP servicio *m* de usuario optativo esencial

establish *vt* GEN establecer; ◆ ~ **a direct link with** TELEP establecer una comunicación directa con

ester: ~ **plastic** *n* PARTS plástico *m* éster

estimate *n* GEN estimación *f*

estimated *adj* GEN estimado; ~ **value** *n* TEST valor *m* estimado

estimation *n* GEN estimación *f*; ~ **error** GEN *statistics* error *m* de estimación

e/t: ~ **switch** *n* (*eleven/twelve switch*) RADIO *gigahertz* conmutador *m* e/t (*conmutador once/doce*)

ET *abbr* (*exchange termination*) TELEP terminal *m* de central telefónica

ETC *abbr* (*exchange terminal circuit*) SWG ETC (*circuito terminal de central*)

etch *vt* GEN grabar

etching *n* GEN grabación *f*

ETF *abbr* (*eight to fourteen modulation*) COMP ETF (*modulación ocho a catorce*)

Ethernet *n* DATA Ethernet *m*; ~ **card** DATA tarjeta *f* Ethernet

ethyl: ~ **cellulose** *n* PARTS etilcelulosa *f*

ETM *abbr* (*eight to ten modulation*) COMP ETM (*modulación ocho a diez*)

ETNA *abbr* (*Ericsson transport network architecture*) TRANS ETNA (*arquitectura de la red de transporte Ericsson*)

ETS *abbr* (*executable test suite*) DATA, TEST sucesión *f* de ensayos ejecutables

ETSI *abbr* (*European Telecommunications Standards Institute*) GEN Instituto *m* Europeo de Normas de Telecomunicaciones

EURB *abbr* (*European Union for Radio Broadcasting*) RADIO EURB (*Unión Europea de Radiodifusión*)

eurobeam: ~ **transmit antenna** *n* RADIO antena *f* de transmisión de eurohaz

Euronorm *n* (*EN*) GEN Euronorma *f* (*EN*)

European: ~ **Academic Research Network** *n* TRANS Red *f* Europea de Investigación Académica; ~ **Broadcasting Union** *n* (*EBU*) RADIO Unión *f* de Radiodifusión Europea (*EBU*); ~ **Committee for Electrotechnical Standardization** *n* (*CENELEC*) GEN Comité *m* Europeo para la Normalización Electrotécnica (*CENELEC*); ~ **Committee for Standardization** *n* (*CEN*) GEN Comité *m* Europeo para la Normalización (*CEN*); ~ **Community** *n* (*EC*) RADIO Comunidad *f* Europea (*CE*); ~ **Computer Manufacturers Association** *n* (*ECMA*) COMP Asociación *f* Europea de Fabricantes de Ordenadores (*ECMA*); ~ **Conference of Postal and Telecommunications Administrations** *n* (*ECPT*) GEN Conferencia *f* Europea de Administración Postal y Telecomunicaciones (*ECPT*); ~ **Institute of Research and Strategic Studies** *n* GEN Instituto *m* Europeo de Investigación y Estudios Estratégicos; ~ **Organization for Quality Control** *n* (*EOQC*) GEN Organización *f* Europea de Control de Calidad (*EOQC*); ~ **Radio Message System** *n* RADIO Sistema *m* Europeo de Mensajes Radiofónicos; ~ **Space Agency** *n* (*ESA*) RADIO Agencia *f* Espacial Europea (*ESA*); ~ **Telecommunications Standard** *n* GEN Norma *f* Europea de Telecomunicaciones; ~ **Telecommunications Satellite Organization** *n* RADIO Organización *f* del Satélite de Telecomunicaciones Europeo; ~ **Telecommunications Standards Institute** *n* (*ETSI*) GEN Instituto *m* Europeo de Normas de Telecomunicaciones; ~ **Union for Radio Broadcasting** *n* (*EURB*) RADIO Unión *f* Europea de Radiodifusión (*EURB*); ~ **Workshop for Open Systems** *n* (*EWOS*) GEN Seminario *m* Europeo para Sistemas Abiertos (*EWOS*)

eutectic: ~ **alloy** *n* PARTS aleación *f* eutéctica

evaluate *vt* GEN evaluar

evaluation *n* GEN evaluación *f*

evanescent: ~ **field** *n* ELECT, TRANS *fibre optics* campo *m* evanescente

evaporation *n* GEN evaporación *f*; ~ **pressure** GEN presión *f* de evaporación; ~ **temperature** GEN temperatura *f* de evaporación

even *adj* GEN *mathematics* llano; ~ **note** *n* COMP nota *f* constante; ~ **parity check** *n* COMP, DATA, TELEP comprobación *f* de paridad

event *n* GEN suceso *m*; ~**-forwarding discriminator** DATA discriminadador *m* de progresión de eventos; ~ **presentation restriction indicator** DATA indicador *m* de restricción en la presentación de eventos

EWOS *abbr* (*European Workshop for Open Systems*) GEN EWOS (*Seminario Europeo para Sistemas Abiertos*)

EX *abbr* (*extinction ratio*) ELECT, TRANS *fibre optics* coeficiente *m* de extinción

examination *n* GEN examen *m*, investigación *f*

examine *vt* GEN examinar

excavation: ~ **damage** *n* APPL *outside plant engineering* daño *m* por excavación

excavator: ~ **damage** *n* APPL *outside plant engineering* daño *m* por excavadora

exceed *vt* GEN sobrepasar

excess *adj* COMP sobrante; ~ **attenuation** *n* TRANS exceso *m* de atenuación; ~ **current** *n* POWER sobre corriente *f*; ~ **solder** *n* ELECT exceso *m* de soldadura; ~ **voltage** *n* POWER sobre tensión *f*

exchange *n* GEN central *f*; ~ **assembly drawing** GEN *documentation* dibujo *m* de conjunto de la central; ~ **battery** POWER *power supply* batería *f* de central; ~ **call-release delay** SWG retardo *m* de desconexión de llamada urbana; ~ **code** GEN código *m* de central; ~ **concentrator** SWG concentrador *m* de central; ~ **control system** TELEP sistema *m* de control de la central; ~ **diagram** POWER, SWG *documentation* diagrama *m* de central; ~ **identification** TELEP identificación *f* de central; ~ **identification frame** TELEP trama *f* de identificación de central; ~ **of identification** (*XID*) TELEP central *f* de identificación (*XID*); ~ **installation drawing** GEN *documentation* dibujo *m* de instalación de la central; ~ **library** GEN *documentation* biblioteca *f* de la central; ~ **line** GEN línea *f* de la central; ~ **line junction set** SWG *PBX* juego *m* de relés de línea urbana; ~ **line unit** SWG *PBX* unidad *f* de línea urbana; ~ **manual** GEN *documentation* manual *m* de la central; ~ **mechanical structure** TELEP *equipment practice* estructura *f* mecánica de centrales; ~ **method** TELEP método *m* de la central; ~ **switchboard** SWG central *f* telefónica; ~ **terminal circuit** (*ETC*) SWG circuito *m* terminal de central (*ETC*); ~ **termination** (*ET*) TELEP terminal *m* de central telefónica; ~ **time** SWG tiempo *m* de intercambio

exchangeability *n* GEN conmutabilidad *f*

exchangeable *adj* GEN conmutable

excitation *n* ELECT, TELEP excitación *f*

excite *vt* ELECT, TELEP excitar

exclude *vt* GEN excluir

excluded *adj* GEN excluido

exclusion *n* GEN exclusión *f*; ~ **unit** TEST *dependability* unidad *f* de exclusión

exclusive: ~ **external line** *n* SWG *PBX*, TELEP línea *f* externa privada; ~ **rights** *n pl* TV derechos *m pl* de exclusividad

EXCLUSIVE-OR: ~ **element** *n* ELECT circuito *m* O-EXCLUSIVO; ~ **operation** *n* ELECT operación *f* O-EXCLUSIVO

ex-directory *adj* (*XD*) GEN ex-directorio (*XD*)

executable: ~ **memory image** *n* TELEP imagen *f* de memoria ejecutable; ~ **test case** *n* DATA, TEST juego *m* de ensayo ejecutable; ~ **test case error** *n* DATA, TEST error *m* en juego de ensayo ejecutable; ~ **test suite** *n* (*ETS*) DATA, TEST sucesión *f* de ensayos ejecutables

execute[1]: ~ **button** *n* COMP botón *m* ejecutar; ~ **order** *n* COMP orden *f* ejecutar; ~ **phase** *n* COMP fase *f* de ejecución; ~ **statement** *n* COMP instrucción *f* de ejecución

execute[2] *vt* COMP ejecutar

execution: ~ **character** *n* TELEP carácter *m* de ejecución; ~ **time** *n* COMP tiempo *m* de ejecución

executive: ~ **override** *n* TELEP preferencia *f* ejecutiva; ~ **program** *n* TELEP programa *m* de ejecución

exhaust: ~ **gas** *n* CLIM gas *m* de escape; ~ **valve** *n* PARTS válvula *f* de escape

exhibit *vt* COMP presentar

exhibition *n* GEN exposición *f*; ~ **stand** GEN puesto *m* de exposición

existing: ~**-quality television** *n* COMP televisión *f* de calidad actual

exit: ~ **button** *n* COMP botón *m* salir

exocentric: ~ **angle** *n* RADIO ángulo *m* exocéntrico
expand *vt* COMP, TELEP ampliar, extender
expandability *n* COMP, TELEP ampliabilidad *f*, extensibilidad *f*
expandable *adj* COMP, TELEP extensible
expanded: ~ **memory** *n* COMP, DATA memoria *f* expandida; ~ **polystyrene** *n* PARTS poliestireno *m* expandido; ~ **polystyrene foam** *n* PARTS espuma *f* de poliestireno expandido; ~ **session reference** *n* TELEP referencia *f* de sesión ampliada; ~ **sweep** *n* RADIO exploración *f* ampliada
expander *n* COMP extensor *m*
expansion *n* GEN expansión *f*; ~ **band** RADIO banda *f* de expansión; ~ **button** COMP botón *m* de ampliación; ~ **card** COMP tarjeta *f* de expansión; ~ **force** GEN *physics* fuerza *f* de expansión; ~ **joint** PARTS *construction* junta *f* de dilatación; ~ **network** TRANS red *f* de expansión; ~ **slot** COMP ranura *f* de extensión; ~ **stage** TRANS etapa *f* de expansión; ~ **tank** GEN *containers* tanque *m* de expansión; ~ **window** TRANS *fibre optics* margen *m* de alargamiento
expectation *n* GEN *statistics* expectativa *f*
expected: ~ **value** *n* GEN *statistics* valor *m* esperado
expedited: ~ **network service data-unit** *n* DATA unidad *f* de datos de servicio de red relanzada
experiment: ~ **design** *n* GEN *statistics* diseño *m* de experimentos
experimental: ~ **network** *n* RADIO red *f* experimental
expert: ~ **system** *n* APPL *artificial intelligence*, COMP, DATA sistema *m* experto
expiration: ~ **of timer** *n* RADIO *maritime* caducidad *f* del temporizador
expire *vi* GEN expirar
explicit *adj* GEN explícito; ~ **congestion notification** *n* DATA, SWG, TELEP notificación *f* explícita de congestión; ~ **link** *n* COMP enlace *m* explícito
exploded: ~ **view** *n* GEN *technical drawing* vista *f* despiezada
explosion: ~-**proof** *adj* TRANS a prueba de explosiones
exponent *n* GEN *mathematics* exponente *m*
exponential *adj* GEN exponencial; ~ **curve** *n* COMP curva *f* exponencial; ~ **distribution** *n* GEN distribución *f* exponencial; ~ **equation** *n* GEN ecuación *f* exponencial; ~ **function** *n* GEN función *f* exponencial
export[1] *n* COMP exportación *f*
export[2] *vt* COMP *data* exportar
expose *vt* TV *photography* exponer
exposure *n* TV *photography* exposición *f*; ~ **limits** *n pl* GEN límites *m pl* de exposición; ~ **meter** TV *photography* fotómetro *m*; ~ **time** TV *photography* tiempo *m* de exposición
expression *n* GEN expresión *f*; ~ **control** COMP control *m* de expresión; ~ **pedal** COMP pedal *m* de expresión
extend *vt* GEN aumentar, ampliar
extended *adj* GEN ampliado, aumentado; ~ **application-layer structure** *n* TRANS estructura *f* de capa de aplicación ampliada; ~ **bandwidth system** *n* RADIO sistema *m* de ancho de banda ampliado; ~ **binary coded decimal interchange code** *n* DATA, TELEP código *m* decimal codificado en binario ampliado; ~ **BNF** *n* DATA BNF *f* ampliada; ~ **eclipse use** *n* RADIO uso *m* de eclipse ampliado; ~ **industry standard architecture** *n* COMP arquitectura *f* normalizada muy difundida en el sector; ~ **memory** *n* COMP, DATA memoria *f* extendida; ~ **range** *n* TV gama *f* extensa
extendibility *n* COMP, TELEP ampliabilidad *f*
extensibility *n* COMP, TELEP extensibilidad *f*
extension *n* (*X*) GEN extensión *f* (*X*); ~ **bell** PER APP timbre *m* supletorio; ~ **board** ELECT *printed circuits* placa *f* de extensión; ~ **cable** GEN prolongador *m* eléctrico; ~ **circuit** TELEP circuito *m* de extensión; ~ **cord** POWER cordón *m* de extensión; ~ **line** DATA, SWG *PBX* línea *f* de extensión; ~ **mode** TELEP modo *m* de ampliación; ~ **module** TELEP módulo *m* de ampliación; ~ **number** SWG *PBX* número *m* de extensión; ~ **rack** TELEP *equipment practice* bastidor *m* de ampliación; ~ **set** SWG *PBX* estación *f* supletoria, TELEP extensión *f* supletoria; ~ **socket** GEN ficha *f* para extensión; ~ **stage** GEN paso *m* de ampliación; ~ **step** TELEP etapa *f* de ampliación; ~ **store** COMP memoria *f* de extensión; ~ **telephone set** SWG *PBX*, TELEP teléfono *m* de extensión; ~ **test period** TEST periodo *m* experimental de extensión
external *adj* GEN externo; ~ **access equipment** *n* TELEP equipo *m* de acceso externo; ~ **alarm** *n* RADIO alarma *f* externa; ~ **blocking** *n* SIG bloqueo *m* exterior; ~ **cable** *n* ELECT cable *m* exterior; ~ **cabling** *n* ELECT cableado *m* exterior; ~ **calls barred** *n* SWG prohibición *f* de llamadas externas; ~ **calls barred extension** *n* TELEP extensión *f* no operativa para llamadas externas; ~ **clock** *n* COMP reloj *m* externo; ~ **damage** *n* GEN daño *m* externo; ~

device *n* COMP dispositivo *m* externo; ~ **disabled time** *n* TEST tiempo *m* no operativo externo; ~ **equipment** *n* GEN equipo *m* externo; ~ **failure cost** *n* TEST coste *m* (*AmL* costo) por fallos externos; ~ **interrupt signal** *n* SWG señal *f* externa de interrupción; ~ **keyboard** *n* COMP teclado *m* externo; ~ **layer** *n* ELECT capa *f* exterior; ~ **loss time** *n* TEST tiempo *m* no operativo externo; ~ **memory** *n* COMP memoria *f* externa; ~ **photoelectric effect** *n* ELECT, TRANS efecto *m* fotoeléctrico externo; ~ **plant centre** *BrE* *n* GEN centro *m* de instalaciones exteriores; ~ **program parameter** *n* COMP parámetro *m* externo de programa; ~ **supervision station** *n* TELEP estación *f* de supervisión exterior; ~ **type** *n* DATA tipo *m* externo

extinction: ~ **ratio** *n* (*EX*) ELECT, TRANS *fibre optics* coeficiente *m* de extinción

extra *adj* GEN extra

extract[1]: ~ **instruction** *n* COMP instrucción *f* de extracción

extract[2] *vt* COMP extraer

extraneous *adj* COMP, POWER, TELEP extraño; ~ **noise** *n* TELEP ruido *m* parásito; ~ **voltage** *n* POWER tensión *f* extraña

extrapolate *vt* COMP *mathematics* extrapolar

extrapolation *n* COMP *mathematics* extrapolación *f*

extrinsic: ~ **base resistance** *n* PARTS *semiconductors* resistencia *f* de base extrínseca; ~ **conduction** *n* ELECT, PARTS *semiconductors* conducción *f* extrínseca; ~ **joint loss** *n* TRANS *fibre optics* pérdida *f* de unión extrínseca; ~ **semiconductor** *n* ELECT, PARTS semiconductor *m* extrínseco

extrude *vt* GEN *cables* extruir

extruded: ~ **solid insulation** *n* GEN *cables* aislamiento *m* sólido extruido

extrusion *n* GEN *of cables* extrusión *f*

eye: ~ **diagram** *n* ELECT, TRANS diagrama *m* de ojo; ~ **opening** *n* TRANS apertura *f* de ojo; ~ **pattern** *n* ELECT diagrama *m* en ojo, TRANS diagrama *m* de ojos; ~ **width** *n* RADIO amplitud *f* de ojo

eyedropper *n* COMP cuentagotas *m*

eyelet *n* CLIM, ELECT, POWER *lighting* ojal *m*; ~**-bond strap** ELECT ojal metálico adherente de remache hueco

EyePhone *n* COMP EyePhone *m*

eyeshape: ~ **pattern** *n* ELECT, TRANS diagrama *m* en ojo

F

F *abbr* (*farad*) ELECT, POWER F (*faradio*)
fabric *n* PARTS tela *f*
Fabry-Pérot: ~ **laser diode** *n* (*FP laser diode*) TRANS *fibre optics* diodo *m* láser Fabry-Pérot (*diodo láser FP*); ~ **resonator** *n* (*FP resonator*) TRANS *fibre optics* resonador *m* Fabry-Pérot (*resonador FP*)
faceplate *n* ELECT chapa *f* frontal
facet *n* TRANS *fibre optics* faceta *f*; ~ **coating** TRANS *fibre optics* recubrimiento *m* de faceta; ~ **damage** TRANS *fibre optics* daño *m* en la faceta
facial: ~ **animation** *n* COMP animación *f* facial; ~ **warping** *n* COMP deformación *f* facial
facilities: ~ **management** *n* (*FM*) APPL gestión *f* de instalaciones, COMP gerencia *f* de informática (*FM*)
facility *n* GEN instalación *f*; ~ **cancellation request accepted message** DATA, TELEP mensaje *m* aceptado de solicitud de cancelación de sistema; ~ **cancellation request message** TELEP mensaje *m* de solicitud de cancelación de sistema; ~ **cancellation request rejected message** TELEP mensaje *m* rechazado de solicitud de cancelación de sistema; ~ **indicator** DATA indicador *m* de capacidad; ~ **registration request accepted message** DATA, TELEP mensaje *m* aceptado de solicitud de registro de sistema; ~ **registration request message** TELEP mensaje *m* de solicitud de registro de sistema; ~ **registration request rejected message** TELEP mensaje *m* rechazado de solicitud de registro de sistema; ~**-rejected message** (*FRJ*) DATA mensaje *m* de sistema rechazado; ~ **request** DATA, TELEP solicitud *f* de sistema; ~**-request message** (*FAR*) DATA, TELEP mensaje *m* de solicitud de sistema; ~ **request separator** DATA, TELEP separador *m* de solicitud de sistema
facsimile *n* GEN facsímil *m*; ~ **apparatus** TELEP aparato *m* facsímil; ~ **compression** TELEP compresión *f* facsímil; ~ **information field** (*FIF*) COMP, DATA campo *m* de información facsímil (*FIF*); ~ **interworking function** TELEP función *f* de interoperación de facsímil; ~**-line manager** DATA gestor *m* de línea de fax; ~ **machine** PER APP, TELEP máquina *f* de facsímil; ~ **message** TELEP mensaje *m* de facsímil; ~ **packet assembly facility** DATA sistema *m* de ensamblado de paquete facsímil; ~ **packet assembly-disassembly** DATA ensamblado-desensamblado *m* de paquete facsímil; ~ **packet disassembly facility** DATA sistema *m* de desensamblado de paquete facsímil; ~ **on private networks** TELEP telefax *m* en redes privadas; ~ **server** DATA, TELEP servidor *m* de telefax; ~ **service** DATA, TELEP servicio *m* de telefax; ~ **telegraphy** TELEP telegrafía *f* en facsímil; ~ **terminal** DATA, PER APP, TELEP telefax *m*; ~ **transaction** DATA, TELEP transacción *f* por telefax; ~ **transmission** DATA, TELEP transmisión *f* por telefax
factor *n* GEN factor *m*; ~ **of cooperation** TELEP factor *m* de cooperación
factorial *n* COMP *mathematics*, TEST factorial *m*; ~ **experiment** TEST experimento *m* factorial
factoring *n* COMP *mathematics*, TEST factoraje *m*
factory: ~**-preset** *adj* COMP preajustado en fábrica; ~**-programmed** *adj* COMP programado en fábrica; ~ **setting** *n* COMP puesta *f* en función en fábrica
facts: ~ **base** *n* COMP base *f* de hechos
factual: ~ **basis** *n* ELECT base *f* factual
fade *n* COMP, TV desvanecimiento *m*; ~**-in** COMP, TV aparición gradual de la imagen; ~**-out** COMP, TV desaparición gradual de la imagen
fader *n* COMP atenuador *m* (*STC*)
fading *n* COMP, RADIO, TV atenuación *f*, desvanecimiento *m*; ~ **rate** RADIO índice *m* de desvanecimiento
fail[1]: ~**-safe** *adj* TELEP, TEST sin riesgo de fallo, seguro en caso de fallo; ~**-safe concept** *n* TEST concepto *m* seguro en caso de fallos; ~ **verdict** *n* DATA diagnóstico *m* de fallo
fail[2] *vi* GEN romperse; ◆ ~ **to appear** GEN no aparecer
failure *n* GEN fallo *m*; ~ **alarm** TELEP alarma *f* de averías; ~ **alarm circuit** TELEP circuito *m* de alarma de averías; ~ **alarm lamp** TEST lámpara *f* de alarma de averías; ~ **analysis** TEST análisis *m* de averías; ~ **catalogue** TELEP catálogo *m* de fallos; ~ **cause** TEST *dependability* causa *f* de fallo; ~ **consequence diagram** GEN *documentation* diagrama *m* de consecuencia de fallo; ~**-correction time** TELEP tiempo *m* de corrección de fallo; ~ **counter** TELEP contador *m* de fallos; ~ **criterion** TEST *dependability*

criterio *m* de fallo; ~ **criticality** TEST *dependability* criticidad *f* de fallo; ~ **diagnosis time** TELEP tiempo *m* de diagnóstico de fallo; ~ **distribution** TELEP *O & M* distribución *f* de fallos; ~ **indication** TELEP *O & M* indicación *f* de fallo; ~ **intensity** TEST *dependability* intensidad *f* de fallo; ~ **marking** TELEP *O & M* marcación *f* de error; ~ **mechanism** TEST *dependability* mecanismo *m* de fallo; ~ **mode** TELEP, TEST *dependability* modo *m* de fallo; ~ **occurrence** TELEP producción *f* de fallo; ~ **rate** GEN tasa *f* de fallos; ~ **recognition** TELEP reconocimiento *m* de fallo; ~ **source** TEST origen *m* de fallo; ~ **state** TELEP situación *f* de fallo; ~ **symptom** TELEP *O & M* síntoma *m* de fallo; ~ **verification** TELEP verificación *f* de fallo

fall: ~**-back circuit** *n* TEST circuito *m* auxiliar; ~**-back rate** *n* ELECT índice *m* de reserva; ~**-back relationship** *n* DATA relación *f* de reserva; ~ **time** *n* COMP, ELECT, TELEP intervalo *m* de caída

falling: ~ **exponential curve** *n* COMP curva *f* exponencial descendente; ~ **linear curve** *n* COMP curva *f* lineal descendente

false: ~ **alarm** *n* GEN falsa alarma *f*; ~**-alarm probability** *n* GEN probabilidad *f* de falsa alarma; ~**-alarm time** *n* GEN tiempo *m* de falsa alarma; ~ **call** *n* TELEP llamada *f* falsa; ~ **calling rate** *n* TELEP índice *m* de llamadas falsas; ~ **echo** *n* RADIO eco *m* falso; ~ **seizure** *n* SIG ocupación *f* ficticia; ~ **start** *n* TEST falso arranque *m*

fan *n* COMP ventilador *m*; ~**-in** ELECT número de señales que entran; ~**-beam antenna** RADIO antena *f* en abanico; ~**-fold sheets** *n pl* GEN papel *m* continuo; ~**-out** ELECT salida *f* máxima; ~**-out cable** ELECT cable *m* de multiplicación de salidas; ~**-out device** ELECT dispositivo *m* de multiplicación de salidas; ~**-out jacketing** ELECT envuelta *f* con ramificación de salida; ~**-out line card** ELECT tarjeta *f* de línea con ramificación de salida; ~**-shaped beam** POWER haz *m* en abanico

fanning *n* ELECT distribución *f*; ~ **strip** POWER regleta *f* distribuidora

FAQ *abbr* (*frequently-asked questions*) DATA *Internet* FAQ (*preguntas formuladas con frecuencia*)

far: ~**-end block error** *n* DATA error *m* de bloqueo distante; ~**-end crosstalk** *n* RADIO telediafonía *f*; ~**-end receive failure** *n* SIG fallo *m* de recepción en diafonía; ~ **field** *n* GEN campo *m* lejano; ~**-field analysis** *n* GEN análisis *m* en campo lejano; ~**-field diffraction pattern** *n* RADIO, TRANS diagrama *m* de difracción de campo lejano; ~**-field radiation pattern** *n* RADIO, TRANS diagrama *m* de radiación de campo lejano; ~**-field region** *n* GEN región *f* de campo lejano

FAR *abbr* (*facility request message*) DATA, TELEP capacidad *f* de petición de mensajes

farad *n* (*F*) ELECT, POWER faradio *m* (*F*)

Faraday: ~ **effect** *n* TRANS *fibre optics* efecto *m* Faraday; ~**'s law** *n* POWER ley *f* de Faraday

fast: ~ **access** *n* COMP acceso *m* rápido; ~**-access memory** *n* COMP memoria *f* de acceso rápido; ~**-access storage** *n* COMP memoria *f* con acceso rápido; ~**-busy tone** *n* TELEP tono *m* de ocupado rápido; ~ **circuit switch** *n* SWG conmutador *m* rápido de circuitos; ~ **clamping** *n* ELECT sujeción *f* firme; ~ **facsimile** *n* TELEP telefax *m* rápido; ~**-forward playback** *n* COMP rebobinado *m* rápido; ~ **Fourier transform** *n* (*FFT*) COMP transformación *f* rápida de Fourier (*TRF*); ~ **frequency-shift keying** *n* SIG transmisión *f* rápida por desplazamiento de frecuencia; ~ **packet switch** *n* (*FPS*) DATA, SWG, TRANS conmutador *m* rápido de paquetes; ~ **packet switching** *n* DATA, SWG, TRANS conmutación *f* rápida de paquetes; ~**-reservation protocol** *n* (*FRP*) DATA protocolo *m* de reserva rápida; ~**-reverse playback** *n* COMP rebobinado *m* rápido invertido; ~**-select acceptance** *n* TELEP aceptación *f* de selección rápida; ~ **storage** *n* DATA almacenamiento *m* rápido; ~**-time analysis system** *n* COMP sistema *m* de análisis de pequeña constante de tiempo; ~**-time simulation** *n* COMP simulación *f* de pequeña constante de tiempo; ~ **update** *n* DATA actualización *f* rápida; ~**-update request** *n* DATA petición *f* de actualización rápida

fasteners *n pl* ELECT fijaciones *f pl*

fastening: ~ **plate** *n* PARTS placa *f* de fijación

FAT *abbr* (*file allocation table*) DATA FAT (*tabla de adjudicación de archivo*)

fatigue *n* PARTS fatiga *f*; ~ **test** TEST ensayo *m* de fatiga

fault *n* GEN avería *f*, fallo *m*; ~ **alarm circuit** TELEP circuito *m* de alarma de fallos; ~ **alarm lamp** TEST lámpara *f* indicadora de avería; ~ **analysis** TEST *dependability* análisis *m* de averías; ~ **centre** *BrE* TELEP centro *m* de reclamaciones; ~ **clearance** GEN reparación *f* de avería; ~**-cleared message** TEST mensaje *m* de fallo eliminado; ~ **code** TELEP código *m* de avería; ~ **complaint service** TELEP servicio *m* de reclamaciones; ~ **condition** TELEP condición *f* de falla; ~ **correction** TEST *dependability* corrección *f* de fallos; **~-correction time** TEST *dependability* tiempo *m* de corrección de fallos; **~ current** ELECT corriente *f* por avería; **~ detection** TELEP, TEST detección *f* de averías; **~ diagnosis** TEST *dependability* diagnosis *f* de averías; **~ display** TEST indicador *m* de averías; **~-free** *adj* TEST sin averías; **~ indication** TELEP *O & M* indicación *f* de averías; **~ isolation** TELEP *O & M* localización *f* de averías; **~ localization** TELEP, TEST *dependability* localización *f* de averías; **~-localization time** TELEP, TEST *dependability* tiempo *m* de localización de averías; **~ location** DATA detección *f* de averías; **~ maintenance** TEST reparación *f* de avería; **~ marking** TELEP *O & M* marcado *m* de averías; **~ message** DATA mensaje *m* de fallo; **~ mode** TEST *dependability* modo *m* de fallo; **~-mode analysis** (*FMA*) TEST *dependability* análisis *m* de modo de fallo (*FMA*); **~ pattern** TELEP *O & M* cuadro *m* descriptivo de fallo; **~ print-out** SWG impresión *f* de error; **~ rate** GEN tasa *f* de fallos; **~ reception centre** *BrE* (*FRC*) TEST centro *m* de recepción de fallos; **~ report** GEN informe *m* de averías; **~ reporting** POWER aviso *m* de averías; **~-report point** TELEP *network* punto *m* de informe de averías; **~ simulation** TEST simulación *f* de avería; **~ state** TELEP estado *m* de avería; **~ symptom** TELEP *O & M* síntoma *m* de avería; **~ syndrome** TELEP *O & M* cuadro *m* descriptivo de fallo; **~ time** TEST tiempo *m* de fallo; **~ tolerance** GEN tolerancia *f* de errores; **~-tolerant computing** COMP, DATA informática *f* con tolerancia de fallos; **~ tracing** TELEP *general software name* localización *f* de avería; **~-tracing procedure** TELEP *O & M* método *m* de búsqueda de averías; **~ tree** TEST *dependability* árbol *m* de fallos; **~-tree analysis** DATA, TEST *dependability* análisis *m* de árbol de fallos

faulty *adj* GEN averiado; **~ call** *n* TELEP, TEST llamada *f* errónea; **~ connection** *n* TELEP, TEST conexión *f* errónea; **~ link information** *n* TELEP información *f* de enlace averiado; **~ working** *n* TEST funcionamiento *m* defectuoso

favour *BrE* (*AmE* **favor**) *n* ELECT favor *m*

fax[1] *n* PER APP fax *m*; **~ answerphone** PER APP, TELEP contestador *m* con fax; **~ machine** TELEP telefax *m*, aparato *m* de fax; **~ modem** DATA *Internet* fax *m* módem; **~ phone** PER APP, TELEP teléfono *m* con fax; **~ switch** DATA conmutador *m* de fax

fax[2] *vt* GEN faxear, enviar por fax

FBI *abbr* (*field blanking interval*) DATA, TV FBI (*intervalo de supresión de campo*)

FC *abbr* (*functional class*) TELEP FC (*clase funcional*)

FCA *abbr* (*functional class A*) TELEP FCA (*clase funcional A*)

FCB *abbr* (*functional class B*) TELEP FCB (*clase funcional B*)

FCC *abbr* (*Federal Communication Commission*) DATA FCC (*Comisión Federal de Comunicación*)

FCS *abbr* (*frame check sequence*) DATA, TELEP FCS (*secuencia de control de trama*)

FD *abbr* (*functional description*) GEN FD (*descripción funcional*)

FDM *abbr* (*frequency-division multiplexing*) TELEP, TRANS MDF (*multiplexado por división de frecuencias*)

FDMA *abbr* (*frequency-division multiple access*) SIG, SWG, TELEP, TRANS AMDF (*acceso múltiple por división de frecuencias*)

FDS: **~ system** *n* SWG sistema *m* de conmutación por división de frecuencias

FDT *abbr* (*formal description technique*) DATA FDT (*técnica de descripción formal*)

FE *abbr* (*functional entity*) DATA entidad *f* funcional

FEA *abbr* (*functional entity action*) DATA acción *f* de entidad funcional

feasibility: **~ study** *n* TEST *systems development* estudio *m* de factibilidad

feasible *adj* TEST factible, practicable

feature *n* GEN dispositivo *m*

fed: **~ with holding current** *phr* PARTS alimentado con corriente de retención

Federal: **~ Communication Commission** *n* (*FCC*) DATA Comisión *f* Federal de Comunicación (*FCC*)

fee *n* GEN cuota *f*; **~-based service** GEN servicio *m* según tarifa; **~ digits** *n pl* TELEP cifras *f pl* de determinación de tarifa

feed[1] *n* GEN alimentación; **~ assembly** ELECT ensamblaje *m* de alimentación; **~-forward control** CONTR control *m* por alimentación anticipada; **~ hole** COMP, DATA ranura *f* de alimentación; **~ horn** ELECT bocina *f* excitadora; **~ input** ELECT entrada *f* de alimentación; **~ loss** TRANS pérdida *f* de alimentación; **~ mechanism** GEN mecanismo *m* de alimentación; **~ pitch** DATA paso *m* de arrastre; **~ system** ELECT sistema *m* de alimentación; **~ system bandwidth** RADIO, TRANS anchura *f* de banda del sistema de alimentación; **~ track** COMP vía *f* de arrastre, DATA pista *f* de arrastre

feed[2] *vt* GEN alimentar

feed back *vt* COMP realimentar

feedback *n* GEN retroacción *f*; ~ **AGC** TRANS realimentación *f* AGC; ~ **closed loop** ELECT bucle *m* de bucle de realimentación; ~ **control** COMP, CONTR control *m* en lazo cerrado; ~**-control system** COMP, CONTR sistema *m* de regulación por contrarreacción; ~ **decoding** ELECT descodificación *f* de la realimentación; ~ **instability** GEN inestabilidad *f* de la realimentación; ~ **level** COMP nivel *m* de realimentación; ~ **loop** COMP, ELECT, RADIO circuito *m* cerrado de realimentación; ~ **ratio** ELECT porcentaje *m* de realimentación; ~ **shift register** DATA, ELECT, SWG registro *m* de desplazamiento de realimentación; ~ **signal** CONTR señal *f* de realimentación

feeder *n* COMP *paper*, RADIO *aerials*, TRANS aparato *m* alimentador; ~ **link** RADIO, TRANS enlace *m* de alimentador; ~**-link beam area** RADIO ángulo *m* sólido del haz del enlace del alimentador; ~**-link service area** RADIO área *f* de servicio del enlace del alimentador; ~ **loss** TRANS pérdida *f* del alimentador

feedforward: ~ **AGC** *n* TRANS AGC *m* con corrección anticipante; ~ **amplifier** *n* ELECT amplificador *m* con corrección anticipante; ~ **control** *n* CONTR regulación *f* con control anticipante

feeding *n* COMP, PARTS, POWER *supply* avance *m*; ~ **bridge** POWER puente *m* de alimentación; ~ **cable** TRANS cable *m* alimentador; ~ **current** POWER corriente *f* de alimentación; ~ **device** COMP dispositivo *m* de alimentación; ~ **fuse** PARTS, POWER fusible *m* de alimentación; ~ **loop** POWER *supply* bucle *m* de alimentación

feedthrough: ~ **capacitor** *n* POWER capacitor *m* pasante; ~ **pad** *n* POWER almohadilla *f* pasante

feeler: ~ **gauge** *BrE n* TEST *metrology* calibre *m* de espesor

female: ~-~ **adaptor** *n* COMP, ELECT adaptador *m* hembra-hembra; ~ **plug** *n* COMP, ELECT, TEST enchufe *m* hembra

Fermi: ~ **level** *n* GEN *physics* nivel *m* de Fermi

ferrite: ~ **antenna** *n* RADIO antena *f* de ferrita, antena *f* magnética; ~ **core** *n* COMP núcleo *m* de ferrita; ~**-core matrix** *n* COMP matriz *f* de núcleo de ferrita; ~**-core storage** *n* COMP memoria *f* de núcleos de ferrita; ~ **matrix** *n* COMP matriz *f* de ferrita; ~ **storage** *n* COMP memoria *f* de núcleos magnéticos

ferro: ~**-electric liquid-crystal display** *n* (*FLCD*) COMP pantalla *f* de cristal líquido ferroeléctrico (*FLCD*)

ferroalloy *n* GEN ferroaleación *f*

ferrocobalt *n* PARTS ferrocobalto *m*

ferromagnetic: ~ **substance** *n* GEN sustancia *f* ferromagnética

ferromanganese *n* PARTS ferromanganeso *m*

ferronickel *n* COMPON ferroníquel *m*

ferrosilicon *n* COMPON ferrosilicio *m*

ferrule *n* PARTS, TRANS *fibre optics* casquillo *m*

FET *abbr* (*field-effect transistor*) ELECT *semiconductors*, PARTS TEC (*transistor de efecto de campo*)

fetch: ~ **abstract-operation** *n* DATA operación *f* abstracta de extracción; ~ **program** *n* COMP programa *m* de llamada; ~ **restriction** *n* DATA restricción *f* de extracción

FFD *abbr* (*freeform deformation*) COMP FFD (*deformación de forma libre*)

FFT *abbr* (*fast Fourier transform*) COMP TRF (*transformación rápida de Fourier*)

FI *abbr* (*format identifier*) POWER, TELEP FI (*identificador de formato*)

fibre *BrE* (*AmE* **fiber**) *n* GEN fibra *f*; ~ **axis** TRANS eje *m* de fibra; ~ **buffer** TRANS amortiguador *m* de fibra; ~ **bundle** TRANS haz *m* de fibras; ~ **cable plant** TRANS fábrica *f* de cables de fibra; ~**-distributed data interface** DATA, TRANS interfaz *f* de datos distribuidos por fibra; ~ **-end face** TRANS faz *f* de fibra; ~ **geometry** TRANS geometría *f* de fibras; ~ **jacket** PARTS, TRANS camisa *f* de fibra; ~ **to the office** (*FTTO*) TRANS fibra *f* hasta la oficina; ~**-optic backbone** ELECT cable *m* principal fibroóptico; ~**-optic bridge-repeater link** ELECT enlace *m* de repetidor en puente de fibra óptica; ~**-optic cladding** TRANS chapado *m* de fibra óptica; ~**-optic link** POWER enlace *m* fibroóptico; ~**-optic loop** TRANS bucle *m* fibroóptico; ~**-optic network** TRANS red *f* de fibra óptica; ~**-optic receiver** TRANS receptor *m* fibroóptico; ~ **optics** TRANS óptica *f* de fibras; ~**-optic splice** TRANS empalme *m* de fibra óptica; ~**-optic terminal device** TRANS dispositivo *m* terminal de fibra óptica; ~**-optic transmission** TRANS transmisión *f* de fibra óptica; ~ **scattering** TRANS dispersión *f* por fibra; ~ **stripper** ELECT pelador *m* de fibras; ~ **to the building** (*FTTB*) TRANS fibra *f* hasta el edificio; ~ **to the curb** (*FTTC*) TRANS fibra *f* para pozo de registro de canalización; ~ **to the floor** (*FTTF*) TRANS fibra *f* al suelo (*FTTF*); ~ **to the home** (*FTTH*) TRANS fibra *f* hasta el domicilio

fibreglass *BrE* (*AmE* **fiberglass**) *n* ELECT fibra *f* de vidrio

fidelity *n* ELECT, RADIO *receivers* fidelidad *f*, exactitud *f* de reproducción

Fidonet *n* DATA *Internet* Fidonet *m*

field[1] *n* GEN campo *m*, TV imagen *f*; ~ **of activities** GEN campo *m* de actividades; ~ **bend correction** TV señal *f* parabólica de corrección de campo; ~ **blanking** DATA ocultación *f* de campo; ~**-blanking interval** (*FBI*) DATA, TV intervalo *m* de supresión de campo (*FBI*); ~ **coil** RADIO bobina *f* de campo; ~ **depth** ELECT profundidad *f* de campo; ~ **effect** DATA, ELECT efecto *m* de campo; ~**-effect integrated circuit** DATA circuito *m* integrado con efecto de campo; ~**-effect transistor** (*FET*) ELECT *semiconductors*, PARTS transistor *m* de efecto de campo (*TEC*); ~ **emission** ELECT emisión *f* de campo; ~ **frequency** TV frecuencia *f* de imagen; ~ **indicator** TELEP indicador *m* de campo; ~ **intensity** TEST intensidad *f* de campo; ~**-intensity meter** TEST medidor *m* de campo; ~ **keystone correction** TV corrección *f* de la distorsión trapecial del barrido vertical; ~ **keystone waveform** TV onda *f* correctora de la distorsión geométrica de la imagen; ~ **of knowledge** GEN campo *m* de conocimiento; ~**-length indicator** TELEP indicador *m* de longitud de campo; ~ **maintenance** TEST *dependability* mantenimiento *m* local; ~ **measurement** TEST medida *f* sobre el terreno; ~ **mesh** ELECT malla *f* de campo; ~ **pole** POWER polo *m* inductor; ~ **purpose** TEST finalidad *f* de campo; ~ **scan** COMP exploración *f* de campo; ~ **strength** POWER, RADIO, TEST intensidad *f* de campo; ~**-strength measurement** POWER medición *f* de la intensidad de campo; ~**-strength meter** TEST medidor *m* de intensidad de campo; ~ **sweep** TV barrido *m* vertical; ~**-synchronizing signal** TV señal *f* de sincronización vertical; ~ **sync pulse** TV impulso *m* de sincronización vertical; ~ **telephone** TELEP teléfono *m* de campaña; ~ **test** RADIO, TEST *dependability* prueba *f* de campo; ~**-tilt correction** TV corrección *f* de declive en tiempo de campo; ~**-time luminance distortion** COMP distorsión *f* lumínica en tiempo de campo; ~**-time waveform distortion** DATA distorsión *f* de forma de onda en tiempo de campo; ~ **trial** RADIO, TEST ensayo *m* de campo; ~ **of view** (*FOV*) RADIO campo *m* visual (*FOV*)

field[2]: ~**-test** *vt* GEN probar en campo

FIF *abbr* (*facsimile information field*) COMP, DATA FIF (*campo de información facsímil*)

fifteen: ~**-supergroup assembly section** *n* TELEP sección *f* de montaje de grupo secundario quince; ~**-supergroup assembly link** *n* TELEP enlace *m* de ensamblado de grupo secundario quince

figurative: ~ **constant** *n* COMP constante *f* figurada

figure *n* GEN *documentation* cifra *f*; ~**-of-eight diagram** RADIO diagrama *m* en ocho; ~ **of merit** RADIO, TEST coeficiente *m* de calidad; ~**-of-merit meter** TEST unidad *f* de contadores de cifra de mérito; ~ **shift** PER APP, TELEP *telex* cambio *m* a cifras; ~**-shift signal** TELEP señal *f* de cambio a cifras

filament *n* ELECT, PARTS *plastics* filamento *m*; ~ **current** ELECT corriente *f* de filamento; ~ **starting current** ELECT corriente *f* inicial de filamento; ~ **surge current** ELECT corriente *f* inicial de filamento; ~ **voltage** ELECT tensión *f* de filamento

file *n* GEN fichero *m*; ~ **access** COMP, DATA acceso *m* a archivos; ~ **allocation table** (*FAT*) DATA tabla *f* de adjudicación de archivo (*FAT*); ~ **attribute** TELEP atributo *m* de archivo; ~ **backup** COMP copia *f* de seguridad de un fichero; ~ **compression** DATA *Internet* compresión *f* de fichero; ~ **conversion** DATA conversión *f* de fichero; ~ **decompression** DATA *Internet* descompresión *f* de fichero; ~ **divider** TV divisor *m* de archivo; ~ **of addresses** TELEP fichero *m* de direcciones de archivos; ~ **layout** DATA diseño *m* de fichero; ~ **maintenance** COMP mantenimiento *m* de archivos; ~**-name syntax** DATA, TELEP sintaxis *f* del nombre del archivo; ~ **protection** COMP protección *f* de archivos; ~ **recovery** DATA *Internet* recuperación *f* de archivos; ~ **search** DATA búsqueda *f* de archivo; ~ **security** COMP seguridad *f* de archivos; ~ **server** COMP, DATA, TEST servidor *m* de archivos; ~ **source** DATA *Internet* origen *m* de archivo; ~ **transfer** COMP, DATA transferencia *f* de ficheros; ~ **transfer access and management** (*FTAM*) COMP *remote data processing*, DATA, TELEP acceso *m* y gestión de transferencia de archivos (*FTAM*), acceso *m* y gestión de transferencia de ficheros; ~ **transfer access and management service element** (*FTAMSE*) DATA elemento *m* del servicio de acceso y gestión de transferencia de ficheros; ~ **transfer protocol** (*FTP*) DATA *Internet* protocolo *m* de transferencia de archivos (*FTP*)

fill[1]: ~ **character** *n* DATA carácter *m* de relleno; ~**-in** *n* COMP relleno; ~**-in signal unit** *n* TELEP unidad *f* de señal de relleno

fill[2] *vt* GEN llenar

filled: ~ **cable** *n* GEN cable *m* relleno

filler *n* GEN *cables* material *m* de relleno; ~ **character** DATA carácter *m* de relleno

filling *n* GEN relleno *m*; ~ **coefficient** DATA coeficiente *m* de relleno; ~ **compound** ELECT pasta *f* para rellenar; ~ **rate** GEN tasa *f* de relleno

film *n* GEN película *f*; ~ **resistor** PARTS, POWER resistor *m* pelicular

filter[1]: ~ **amplifier** *n* ELECT amplificador *m* de filtro; ~-**bank system** *n* RADIO *maritime satellite* sistema *m* de batería de filtros; ~ **capacitor** *n* PARTS, POWER condensador *m* de filtro; ~ **coupler** *n* ELECT acoplador *m* de filtro; ~ **cutoff** *n* COMP corte *m* por filtración; ~ **flank** *n* ELECT flanco *m* de filtro; ~ **fuse** *n* PARTS, POWER fusible *m* de filtro; ~ **item** *n* DATA elemento *m* de filtro; ~ **roll-off** *n* ELECT atenuación *f* progresiva de filtro; ~-**and-sample detector** *n* ELECT detector *m* de filtrado y muestreo; ~ **section** *n* ELECT sección *f* de filtro; ~ **selectivity** *n* GEN selectividad *f* de filtro; ~ **set** *n* DATA *Internet* conjunto *m* de filtros; ~ **slope** *n* COMP pendiente *f* de filtro

filter[2] *vt* GEN filtrar

filterbank *n* COMP batería *f* de filtros

filtered: ~ **QPSK** *n* RADIO *maritime satellite* QPSK *m* filtrado; ~ **voltage** *n* POWER tensión *f* aplanada

filtering *n* GEN filtración *f*

final: ~ **amplifier** *n* ELECT amplificador *m* final; ~ **bit** *n* TELEP bitio *m* final; ~ **character** *n* DATA carácter *m* final; ~-**circuit group** *n* TELEP grupo *m* final de circuitos; ~-**controlling element** *n* CONTR elemento *m* de control final; ~ **inspection** *n* TEST *quality* inspección *f* final; ~-**selection stage** *n* SWG paso *m* de selección final; ~ **selector** *n* TELEP selector *m* de línea; ~-**selector stage** *n* TELEP paso *m* de selector final; ~ **stage** *n* RADIO, TELEP etapa *f* de salida; ~ **state** *n* DATA estado *m* final; ~ **test** *n* TEST prueba *f* final

find *vt* COMP, DATA encontrar

find/replace *vt* COMP, DATA encontrar/sustituir, buscar/reemplazar

fine: ~ **adjustment** *n* GEN ajuste *m* preciso; ~ **diagnosis** *n* GEN diagnóstico *m* detallado; ~ **positioning** *n* GEN regulación *f* de precisión; ~ **regulation** *n* GEN regulación *f* de precisión; ~ **setting** *n* GEN reglaje *m* de precisión; ~ **switching** *n* COMP conmutación *f* de precisión; ~ **thread** *n* GEN rosca *f* fina; ~ **tuning** *n* GEN ajuste *m* fino de sintonización; ~-**wire fuse** *n* PARTS, POWER fusible *m* de hilo delgado

finger *n* DATA *Internet* dedo *m*; ~ **hole** PER APP, TELEP agujero *m* de disco dactilar; ~ **stop** PER APP, TELEP tope *m* de disco; ~ **wheel** *BrE* (*cf dial disk AmE*) PER APP, TELEP disco *m* dactilar, disco *m* marcador

fingering *n* COMP digitación *f*

finish *n* COMP acabado *m*

finish off *vt* COMP completar

finished: ~ **product** *n* GEN producto *m* acabado

finite: ~ **automation** *n* DATA automatización *f* finita; ~ **field** *n* SIG campo *m* finito; ~ **impulse response** *n* (*FIR*) ELECT, TRANS respuesta *f* de impulso finito (*RIF*); ~ **impulse response filter** *n* ELECT, TRANS filtro *m* de respuesta de impulso finito; ~ **state** *n* DATA, SIG estado *m* finito; ~-**state vector quantization** *n* SIG cuantización *f* de vector de estado finito

FIR *abbr* (*finite impulse response*) ELECT, TRANS RIF (*respuesta de impulso finito*); ~ **filter** *n* ELECT, TRANS filtro *m* FIR

fire: ~-**control radar** *n* RADIO radar *m* de dirección de tiro; ~ **detector** *n* APPL *safety*, PARTS detector *m* de incendios; ~ **protection** *n* GEN *safety* protección *f* contra incendios; ~ **resistance** *n* PARTS *materials* resistencia *f* al fuego; ~ **valve** *n* PARTS *safety* compuerta *f* cortafuegos

fireproof *adj* PARTS a prueba de incendios

firing *n* ELECT alimentación *f*; ~ **power** ELECT potencia *f* de excitación; ~ **time** ELECT tiempo *m* de excitación

firmware *n* COMP soporte *m* lógico inalterable

first: ~-**adjacent channel** *n* RADIO primer canal adyacente *m*; ~-**angle projection** *n* GEN *technical drawing* proyección *f* a la inglesa; ~ **breakdown** *n* ELECT *semiconductors* disrupción *f* primaria; ~-**choice circuit group** *n* ELECT, SWG, TELEP grupo *m* de circuitos preferente; ~-**choice group** *n* ELECT, SWG, TELEP grupo *m* preferente; ~ **cost** *n* GEN *economics* coste *m* (*AmL* costo) original; ~-**data multiplexer** *n* DATA, TRANS multiplexor *m* de primer dato; ~-**line maintenance** *n* TEST mantenimiento *m* de primera línea; ~-**order multiplexer** *n* DATA, TRANS multiplexor *m* de primera orden; ~-**party release** *n* COMP, TELEP desconexión *f* simple; ~-**subscriber release** *n* COMP, TELEP desconexión *f* del primer abonado; ~ **upright** *n* TELEP *equipment practice* montante *m* inicial

fishbone: ~ **antenna** *n* RADIO antena *f* de varillas telescópicas

fishplate *n* PARTS brida *f*

fit *vt* GEN ajustar

fitted: ~ **with** *adj* GEN equipado con

fitter *n* GEN instalador *m*

fitting *n* GEN accesorio *m*; ~ **tolerance** GEN tolerancia *f* de ajuste

fittings *n pl* PARTS herrajes *m pl*

five: ~-**level RTTY** *n* TELEP RTTY *m* de cinco niveles; ~-**point impairment scale** *n* TEST escala *f* de reducción de cinco puntos; ~-**unit code** *n* TELEP código *m* de cinco unidades

fix *vt* GEN fijar

fixed: ~-**antenna direction finder** *n* RADIO radiogoniómetro *m* de antena fija; ~ **assets** *n pl* GEN capital *m* fijo; ~ **attenuator** *n* TRANS atenuador *m* permanente; ~ **bias** *n* ELECT polarización *f* fija; ~ **blade connector** *n* ELECT, PARTS toma *f* de espigas; ~ **charge** *n* TELEP cuota *f* fija; ~ **costs** *n* GEN costes *m pl* (*AmL* costos) fijos; ~-**destination call services** *n pl* TELEP servicios *m pl* de llamada de destino fijo; ~ **direction finder** *n* RADIO goniómetro *m* fijo; ~ **fork connector** *n* ELECT, PARTS toma *f* de contactos de horquilla; ~ **frequency** *n* RADIO frecuencia *f* fija; ~ **light** *n* TV luz *f* fija; ~ **network** *n* TRANS red *f* fija; ~ **night-service** *n* TELEP servicio *m* nocturno fijo; ~ **packet recording** *n* COMP registro *m* de paquete fijo; ~ **pin connector** *n* ELECT, PARTS toma *f* de espigas; ~ **point** *n* GEN punto *m* fijo; ~ **point method** *n* COMP método *m* de la coma fija; ~ **price** *n* GEN precio *m* fijo; ~ **radio beacon** *n* RADIO radiobaliza *f* fija; ~-**resolution progressive mode** *n* COMP modo *m* progresivo de resolución fija; ~ **segmentation** *n* COMP segmentación *f* fija; ~ **sequencer** *n* DATA secuenciador *m* fijo; ~ **store** *n* COMP memoria *f* inalterable

fixing: ~ **bracket** *n* PARTS pieza *f* de montaje, soporte *m* de fijación; ~ **plate** *n* PARTS placa *f* de fijación; ~ **ring** *n* PARTS anillo *m* de fijación; ~ **screw** *n* PARTS tornillo *m* de fijación

flag *n* DATA etiqueta *f*; ~ **bit** DATA bit *m* indicador; ~ **byte** DATA octeto *m* indicador

flap: ~ **attenuator** *n* RADIO atenuador *m* de aleta

flare: ~ **antenna** *n* RADIO antena *f* de bocina

flash *n* ELECT destello; ~-**arc** ELECT arco *m* de flameo; ~ **time** PER APP tiempo *m* de destello

flashing: ~ **cursor** *n* COMP cursor *m* destellante; ~ **dot** *n* COMP punto *m* destellante; ~ **light** *n* TEST, TV *optics* luz *f* intermitente; ~ **rate** *n* POWER frecuencia *f* de destellos

flashover *n* POWER contorneamiento *m*; ~ **clearance** POWER distancia *f* de seguridad; ~ **voltage** POWER tensión *f* de contorneamiento

flat *adj* GEN plano; ~ **antenna** *n* RADIO antena *f* plana; ~ **battery** *n* POWER batería *f* plana; ~-**bed plotter** *n* PER APP, TELEP trazador *m* plano; ~-**bed scanner** *n* PER APP, TELEP escáner *m* plano; ~-**bed transmitter** *n* PER APP, TRANS transmisor *m* de campo plano; ~ **cable** *n* ELECT cable *m* cinta; ~ **conductor cable** *n* ELECT cable *m* de conductores planos; ~ **fading** *n* RADIO desvanecimiento *m* plano; ~-**fading margin** *n* RADIO margen *m* de tonalidad plana; ~ **frequency characteristic** *n* ELECT respuesta *f* de frecuencia plana; ~ **leakage power** *n* ELECT potencia *f* transmitida en régimen permanente; ~ **level** *n* COMP nivel *m* plano; ~ **line** *n* RADIO línea *f* no resonante; ~-**nose pliers** *n pl* GEN alicates *m pl* de boca plana; ~-**pack** *n* POWER componente *m* plano; ~ **panel display** *n* PER APP pantalla *f* plana; ~ **price** *n* GEN precio *m* neto; ~ **rate** *n* APPL tarifa *f* uniforme; ~-**rate service** *n* APPL servicio *m* de tarifa uniforme; ~ **response** *n* ELECT respuesta *f* plana; ~ **screen** *n* TV pantalla *f* plana; ~ **shading** *n* TV tonalidad *f* uniforme; ~ **structure** *n* GEN estructura *f* uniforme; ~-**top antenna** *n* RADIO antena *f* de hilos horizontales, antena *f* en hoja; ~-**topped pulse** *n* GEN impulso *m* de cúspide plana

flatness: ~ **of bearing** *n* RADIO planicidad *f* de contacto

flatpack *n* TRANS encapsulado *m*

flaw *n* GEN defecto *m*

F-layer *n* COMP, RADIO capa *f* F

FLCD *abbr* (*ferro-electric liquid-crystal display*) COMP FLCD (*pantalla de cristal líquido ferroeléctrico*)

fledgling *adj* GEN novato

flex *n* ELECT conductor *m* flexible; ~-**ferrule** ELECT conductor *m* flexible de casquillo

flexibility *n* GEN flexibilidad *f*; ~ **strength** GEN resistencia *f* de la flexibilidad

flexible *adj* GEN flexible; ~ **access switch** *n* SWG conmutador *m* de acceso flexible; ~ **cable** *n* POWER cable *m* flexible; ~ **conductor** *n* GEN conductor *m* flexible; ~ **disk** *n* COMP disco *m* flexible; ~ **media stack** *n* ELECT pila *f* de medios flexibles; ~ **night-service** *n* TELEP servicio *m* nocturno flexible; ~ **printed board** *n* ELECT placa *f* impresa flexible; ~ **screwdriver** *n* GEN destornillador *m* flexible; ~ **shaft** *n* PARTS eje *m* flexible; ~ **waveguide** *n* DATA, RADIO, TRANS guíaondas *m* flexible

flexstrip *n* POWER cinta *f* de conductores flexibles

flicker *n* TV parpadeo *m*; ~ **noise** ELECT ruido *m* fluctuante

flickering *n* TV parpadeo *m*; ~ **light** TV luz *f* parpadeante

flight: ~ **path** *n* RADIO trayectoria *f* de vuelo;

~-path angle *n* RADIO ángulo *m* de trayectoria de vuelo

flip: **~ chip** *n* ELECT pastilla *f* dispersora; **~-flop** *n* ELECT báscula *f*

flip *vt* COMP bascular, permutar

float: **~-charging current** *n* POWER *batteries* corriente *f* de carga flotante; **~-charging level** *n* POWER *batteries* nivel *m* de carga flotante

floating: **~ battery** *n* POWER batería *f* de carga equilibrada; **~ charge** *n* POWER *batteries* carga *f* flotante; **~ drift-tube klystron** *n* ELECT clistrón *m* de variación equilibrada; **~ point** *n* COMP coma *f* flotante; **~-point hardware** *n* COMP equipo *m* de coma flotante; **~-point processor** *n* COMP procesador *m* de coma flotante; **~-point representation** *n* COMP representación *f* en coma flotante; **~-point unit** *n* (*FPU*) COMP unidad *f* de coma flotante (*FPU*); **~ receiver** *n* RADIO receptor *m* flotante; **~ voltage** *n* PARTS *semiconductors* tensión *f* flotante

floor: **~ bar** *n* PARTS barra *f* de suelo; **~ bracket** *n* PARTS soporte *m* angular para montaje en el suelo; **~ load** *n* GEN *construction* carga *f* sobre el suelo; **~ mounting** *n* GEN montaje *m* en el suelo; **~ mounting bracket** *n* PARTS soporte *m* angular para montaje en el suelo; **~ plan** *n* GEN *documentation* plano *m* de colocación; **~ rail** *n* PARTS barra *f* de suelo; **~-standing system** *n* ELECT sistema *m* de soporte de pie

floppy: **~ disk** *n* COMP disco *m* flexible; **~-disk drive** *n* COMP unidad *f* de disco; **~-disk reader** *n* COMP lector *m* de discos flexibles

flow[1]: **~ chart** *n* GEN diagrama *m* de flujo; **~-chart symbol** *n* COMP símbolo *m* de diagrama de flujo; **~ control** *n* CONTR, DATA, TELEP regulación *f* de caudal; **~ diagram** *n* GEN diagrama *m* de flujo; **~ direction** *n* GEN dirección *f* de flujo; **~ indicator** *n* TELEP indicador *m* de flujo; **~ line** *n* GEN *flowcharts* línea *f* de flujo; **~ melting** *n* ELECT *printed circuits* fusión *f* térmica; **~ meter** *n* CONTR, TEST flujómetro *m*; **~ regulator** *n* CONTR regulador *m* de gasto; **~ stress** *n* GEN *strength* esfuerzo *m* de fluencia

flow[2] *n* GEN colada *f*

flow through *vt* POWER pasar

flowcharting *n* GEN trazado *m* de organigrama

FLS *abbr* (*free-line signal*) TELEP lámpara *f* de línea libre

fluctuate *vi* GEN fluctuar

fluctuation *n* GEN fluctuación *f*

fluorescence *n* RADIO, TRANS *fibre optics* fluorescencia *f*

fluorescent: **~ lamp** *n* ELECT lámpara *f* fluorescente

flush: **~ antenna** *n* RADIO antena *f* empotrada; **~ conductor** *n* ELECT conductor *m* nivelado; **~-mounted single information outlet** *n* ELECT salida *f* de información única incorporada; **~ mounting** *n* GEN montaje *m* embutido; **~-mount panel** *n* ELECT panel *m* de montaje empotrado; **~ panel** *n* PARTS panel *m* empotrado

flutter *n* TV vibración *f*

flux *n* GEN fundente *m*; **~ density** ELECT densidad *f* de flujo; **~ pattern** ELECT espectro *m* de flujo

fluxmeter *n* CONTROL, TEST flujómetro *m*

fly: **~-back transformer** *n* TV transformador *m* de retorno

flyback *n* POWER, TV retorno *m*; **~ principle** POWER *supply* principio *m* de retorno; **~ transformer** TV transformador *m* de retorno

flying: **~ lead** *n* ELECT *components* aleta *f* adelantada; **~-lead component** *n* ELECT componente *m* soldado; **~-spot scanner tube** *n* ELECT tubo *m* analizador de punto móvil

FM *abbr* (*facilities management*) APPL gestión *f* de instalaciones, COMP FM (*gerencia de informática*); **~ expander** *n* COMP extensor *m* FM; **~ radio receiver** *n* RADIO radiorreceptor *m* de FM; **~ stack** *n* COMP cabeza *f* de lectura de FM; **~ synthesis** *n* COMP síntesis *f* FM

FMA *abbr* (*fault-mode analysis*) TEST *dependability* FMA (*análisis de modo de fallo*)

FMBS *abbr* (*frame-mode bearer service*) APPL, TRANS servicio *m* portador de modo trama

FN *abbr* (*forwarded notification*) DATA, TELEP FN (*notificación reenviada*)

F/O: **~ link** *n* POWER, TRANS enlace *m* F/O

focal: **~-fed antenna** *n* RADIO antena *f* de alimentación focal

focused *adj* GEN enfocado

focusing *n* ELECT enfoque *m*; **~ coil** ELECT bobina *f* de enfoque; **~ electrode** ELECT electrodo *m* de enfoque; **~ magnet** ELECT electroimán *m* de concentración

fog: **~ effect** *n* COMP efecto *m* de niebla

foil *n* PARTS, POWER hoja *f* delgada de metal

fold[1]: **~-out array** *n* RADIO matriz *f* desplegada

fold[2] *vt* COMP plegar

folded: **~-dipole antenna** *n* RADIO antena *f* dipolo doblado; **~ network** *n* TRANS red *f* plegada

folder *n* COMP plegadora *f*

follow: **~-me call forward** *n* TELEP desvío *m* de llamada al número propio; **~-me diversion** *n*

TELEP transferencia *f* temporal del número propio; **~-on call** *n* ELECT, PER APP llamada *f* sucesiva; **~-on instructions** *n pl* GEN instrucciones *f pl* sucesivas; **~-on satellite** *n* RADIO satélite *m* sucesivo; **~-on service advice** *n* TELEP asesoramiento *m* de servicio sucesivo; **~-up** *n* GEN recordatorio *m*; **~-up control** *n* CONTR control *m* de seguimiento

follow up *vt* GEN seguir hasta la terminación

font *n* COMP, DATA juego *m* de caracteres; **~ manager** COMP gestor *m* de tipos; **~ size** COMP tamaño *m* de fuente tipográfica; **~ style** COMP estilo *m* de fuente tipográfica

foolproof *adj* GEN seguro, a prueba de errores de manipulación

footcandle: **~ meter** *n* TEST medidor *m* de bujía-pie

footnote *n* COMP nota *f* al pie

footprint *n* RADIO zona *f* de caída

force *n* GEN potencia *f*

forced: **~ oscillation** *n* POWER oscilación *f* forzada; **~ rerouting** *n* TELEP reencaminamiento *m* forzado

forecast *vt* GEN prever

foreground *n* TV primer término *m*, primer plano *m*; **~ processing** COMP procesamiento *m* prioritario

foresee *vt* GEN prever

foreseeable *adj* GEN previsible

foreseen: **~ test outcome** *n* TEST resultado *m* previsto de la prueba

fork: **~ contact** *n* ELECT contacto *m* de horquilla; **~-contact unit** *n* ELECT bloque *m* de contactos de horquilla

forked: **~ working** *n* TELEP comunicación *f* bifurcada

form *n* GEN impreso *m*; **~-board section** TRANS sección *f* de plantilla de cable; **~ factor** TEST factor *m* de forma; **~ feed** COMP, TELEP avance *m* de hoja; **~ heading** GEN *documentation* encabezamiento *m* de formulario

formal: **~ description technique** *n* (*FDT*) DATA técnica *f* de descripción formal (*FDT*); **~ logic** *n* COMP lógica *f* formal; **~ reasoning** *n* COMP razonamiento *m* formal; **~ specification** *n* DATA especificación *f* formal

formant *n* TELEP *acoustics* formante *m*; **~ vocoder** RADIO *maritime satellite* vocoder *m* formante

format: **~ control** *n* COMP control *m* de formato; **~ effector** *n* DATA, TELEP carácter *m* determinante de formato; **~ identifier** *n* (*FI*) POWER, TELEP identificador *m* de formato (*FI*); **~ output** *n* TELEP salida *f* de formato; **~ parameter-entry sequence** *n* COMP secuencia *f* de formato de entrada de parámetro; **~ parameter input** *n* COMP entrada *f* de parámetro de formato; **~ tape** *n* DATA banda *f* piloto

formatting *n* COMP, DATA formateado *m*

formboard *n* GEN plantilla *f* de cable; **~ base** TRANS base *f* de plantilla de cable; **~ designation** TRANS designación *f* de plantilla de cable; **~ peg** TRANS espiga *f* de plantilla de cable; **~ plate** TRANS placa *f* a base de plantilla de cable; **~ specification** TRANS especificación *f* de plantilla de cable

formula *n* COMP fórmula *f*

fortuitous: **~ distortion** *n* TELEP distorsión *f* de línea

forum *n* DATA *Internet* foro *m*

forward[1]: **~ bias** *n* ELECT polarización *f* directa; **~ biasing** *n* ELECT polarización *f* directa; **~ chaining** *n* DATA encadenado *m* progresivo; **~ channel** *n* DATA, TELEP, TRANS canal *m* de emisión; **~ characteristic** *n* ELECT característica *f* directa; **~ current** *n* ELECT, POWER corriente *f* directa; **~ current transfer ratio** *n* ELECT relación *f* de transferencia de corriente directa; **~ direction** *n* ELECT sentido *m* directo; **~ echo** *n* TELEP eco *m* directo; **~ error correction** *n* GEN corrección *f* de errores hacia adelante; **~ error correction codec** *n* COMP, DATA, RADIO codificador-descodificador *m* de corrección de errores hacia adelante; **~ error correction coding** *n* COMP, DATA, RADIO codificación *f* de corrección de errores hacia adelante; **~ explicit congestion notification** *n* DATA, SWG, TELEP notificación *f* de congestión explícita hacia adelante; **~ hold** *n* SWG mantenimiento *m* por delante; **~ indicator bit** *n* TELEP bitio *m* indicador directo; **~-input signal** *n* SIG señal *f* de entrada hacia adelante; **~ interworking telephony event** *n* TELEP suceso *m* de telefonía con interconexión directa; **~-motion vector** *n* COMP, TV vector *m* de movimiento progresivo; **~ power dissipation** *n* ELECT pérdida *f* de potencia directa; **~ power loss** *n* ELECT pérdida *f* de potencia directa; **~ prediction** *n* TV predicción *f* hacia adelante; **~ recall signal** *n* TELEP señal *f* de llamada hacia delante; **~ resistance** *n* ELECT resistencia *f* a la penetración; **~ round-the-world echo** *n* RADIO eco *m* directo de circunvalación; **~ scattering** *n* RADIO propagación *f* transhorizonte; **~ sequence number** *n* DATA, SWG, TELEP número *m* de secuencia progresiva; **~ signal** *n* POWER señal *f* hacia adelante; **~ signalling** *BrE* *n* TELEP señalización *f* hacia adelante; **~ slope resistance** *n* ELECT, PARTS resistencia *f* directa aparente; ~ **solar array** *n* RADIO panel *m* de células solares frontal; ~ **transfer facility** *n* DATA sistema *m* de transferencia progresiva, TELEP equipo *m* de transferencia directa; ~ **transfer signal** *n* DATA, SIG señal *f* de transferencia directa; ~ **voltage** *n* ELECT tensión *f* directa; ~ **voltage drop** *n* ELECT caída *f* de tensión directa

forward[2] *vt* GEN enviar

forwarded: ~ **notification** *n* (*FN*) DATA, TELEP notificación *f* reenviada (*FN*)

forwarding *n* GEN transmisión *f*; ~ **request** DATA petición *f* de envío

four: ~-**channel telephony** *n* TELEP telefonía *f* de cuatro canales; ~-**concentric-circle near-field template** *n* TRANS plantilla *f* de cuatro círculos concéntricos de campo cercano; ~-**concentric-circle refractive-index template** *n* TRANS patrón *m* de índice refractivo de cuatro círculos concéntricos; ~-**frequency diplex telegraphy** *n* TELEP telegrafía *f* diplex de cuatro frecuencias; ~-**pole filter** *n* ELECT filtro *m* de cuatro polos; ~-**signature coding system** *n* RADIO sistema *m* de codificación de cuatro firmas; ~-**terminal network** *n* ELECT, POWER, TRANS cuadripolo *m*; ~-**wire cable** *n* ELECT cable *m* tetrafilar; ~-**wire chain** *n* TELEP cadena *f* tetrafilar; ~-**wire circuit** *n* TELEP circuito *m* de cuatro hilos conductores; ~-**wire crosspoint** *n* SWG contacto *m* de cruce bifilar doble; ~-**wire repeater** *n* TRANS repetidor *m* de cuatro hilos; ~-**wire switch** *n* SWG conmutador *m* bifilar doble; ~-**wire switching** *n* SWG conmutación *f* bifilar doble; ~-**wire switching system** *n* SWG sistema *m* de conmutación bifilar doble; ~-**wire type circuit** *n* TELEP circuito *m* de tipo tetrafilar

Fourier: ~ **transformation** *n* (*FT*) TEST transformación *f* de Fourier (*FT*); ~ **transform infrared spectography** *n* (*FTIR*) TEST espectroscopía *f* infrarroja transformada de Fourier (*FTIR*)

FOV *abbr* (*field of view*) RADIO FOV (*campo visual*)

FP: ~ **laser diode** *n* (*Fabry-Pérot laser diode*) TRANS *fibre optics* diodo *m* láser FP (*diodo láser Fabry-Pérot*); ~ **resonator** *n* (*Fabry-Pérot resonator*) TRANS *fibre optics* resonador *m* FP (*resonador Fabry-Pérot*)

FPS *abbr* (*fast-packet switch*) DATA, SWG, TRANS conmutador *m* rápido de paquetes

FPU *abbr* (*floating-point unit*) COMP FPU (*unidad de coma flotante*)

fractal: ~ **dimension** *n* COMP dimensión *f* fractal

fractional: ~ **ratio** *n* COMP proporción *f* fraccionaria

fracture *n* PARTS *strength* fractura *f*

fragment *vt* GEN fragmentar

frame *n* GEN imagen *f*; ~ **acquisition and synchronization** TRANS adquisición *f* y sincronización de trama; ~ **alignment** TRANS alineación *f* de trama; ~-**alignment recovery time** TELEP, TRANS tiempo *m* de recuperación de la alineación de trama; ~-**alignment signal** TELEP, TRANS señal *f* de alineación de trama; ~-**alignment state** TRANS estado *m* de alineación de trama; ~-**alignment time slot** TELEP, TRANS intervalo *m* de alineación de trama; ~-**check sequence** (*FCS*) DATA, TELEP secuencia *f* de control de trama (*FCS*); ~ **code** DATA código *m* frame; ~ **distributor** TRANS distribuidor *m* de tramas; ~ **format** TV formato *m* de cuadro; ~ **frequency** DATA, TV frecuencia *f* de imágenes; ~ **generator** TRANS generador *m* de tramas; ~ **grabber** COMP dispositivo *m* de registro de imágenes, TV terminal *m* de parada de cuadro; ~ **grabbing** TV parada *f* de cuadro; ~ **length** DATA longitud *f* de bloque; ~-**loss second** TRANS segundo *m* de pérdida de trama; ~-**management message** DATA mensaje *m* de gestión de bloque; ~-**mode bearer service** (*FMBS*) APPL, TRANS servicio *m* portador del modo trama; ~ **overhead** DATA bloque *m* superior; ~ **pattern** DATA configuración *f* de bloque; ~ **processor** DATA procesador *m* de bloques; ~ **rate** DATA frecuencia *f* de imagen; ~ **reject** TELEP rechazo *m* de trama; ~-**reject** ~ (*FRMR*) TELEP cuadro *m* de rechazo de trama (*FRMR*); ~ **relay** DATA relé *m* de bloque; ~ **slip** TV deslizamiento *m* de imagen; ~-**slotted system** RADIO *land mobile* sistema *m* de tramas ranuradas; ~ **structure** TRANS estructura *f* de trama; ~ **synchronization** DATA, TRANS sincronización *f* de trama; ~-**synchronization unit** DATA unidad *f* de sincronización de bloque, TRANS unidad *f* de sincronización de ciclo; ~ **synchronizer** DATA, TRANS sincronizador *m* de ciclo; ~ **timing** DATA temporización *f* de imagen; ~ **unit** DATA unidad *f* de imagen; ~-**unit step** DATA paso *m* de la unidad de bloque

framework *n* COMP, PARTS armazón *m*; ~ **agreement** GEN acuerdo *m* marco

framing *n* DATA ajuste *m* de imagen; ~ **bit** TELEP bitio *m* delimitador; ~ **information** TRANS información *f* de alineación de trama; ~ **pattern** DATA configuración *f* de bloque; ~ **pulse** RADIO impulso *m* de referencia

franchise: ~ **area** *n* TV zona *f* de franquicia
fraud *n* GEN fraude *m*
fraudulent: ~ **access attempt** *n* RADIO intento *m* de acceso fraudulento; ~ **use** *n* GEN uso *m* fraudulento
Fraunhofer: ~ **diffraction pattern** *n* TRANS diagrama *m* de difracción de Fraunhofer
frayed: ~ **wiring** *n* ELECT cable *m* deshilachado
FRC *BrE abbr* (*fault reception centre*) TEST centro *m* de recepción de fallos
free[1] *adj* GEN libre; **~-blade connector** *n* ELECT clavija *f* de espigas, conector *m* de espigas, PARTS clavija *f* de espigas; ~ **competition** *n* GEN libre competencia *f*; ~ **connector** *n* ELECT, PARTS clavija *f*; ~ **convection** *n* CLIM convección *f* libre; **~-field room** *n* TEST *acoustics* cámara *f* anecoica; **~-fork connector** *n* ELECT, PARTS clavija *f* de horquilla; ~ **form** *n* COMP forma *f* libre; ~ **format** *n* DATA formato *m* libre; **~-line signal** *n* (*FLS*) TELEP lámpara *f* de línea libre; ~ **oscillation** *n* ELECT oscilación *f* libre; **~-pin connector** *n* ELECT clavija *f* de espigas, conector *m* de espigas redondas, PARTS conector *m* de espigas redondas; **~-running frequency** *n* ELECT frecuencia *f* de marcha continua; **~-running oscillator** *n* ELECT oscilador *m* de funcionamiento libre; **~-running oscillator frequency** *n* ELECT frecuencia *f* de oscilación libre; **~-running speed** *n* DATA velocidad *f* de funcionamiento libre; **~-sleeve connector** *n* ELECT, PARTS clavija *f* de manguito; **~-space antenna** *n* RADIO antena *f* en el espacio libre; **~-space attenuation** *n* RADIO, TRANS atenuación *f* en el espacio libre; **~-space basic loss** *n* RADIO pérdida *f* básica de espacio libre; **~-space basic transmission loss** *n* RADIO pérdida *f* de transmisión básica en el espacio libre; **~-space conditions** *n pl* RADIO condiciones *f pl* de espacio libre; **~-space propagation** *n* RADIO propagación *f* en el espacio libre; **~-space range** *n* RADIO *answering machine* zona *f* de espacio libre; **~-space transmission loss** *n* RADIO pérdida *f* de transmisión en el espacio libre; ~ **state** *n* TEST *dependability* estado *m* no requerido; ~ **time** *n* TELEP, TEST *dependability* tiempo *m* libre
free[2] *vt* RADIO despejar
freeform: ~ **deformation** *n* (*FFD*) COMP deformación *f* de forma libre (*FFD*)
freephone: ~ **call** *BrE n* (*cf toll-free call AmE*) TELEP llamada *f* gratuita; ~ **number** *BrE n* (*cf toll-free number AmE*) TELEP número *m* de teléfono gratuito; ~ **service** *BrE n* (*cf toll-free service AmE*) TELEP servicio *m* de llamada gratuita
freeware *n* COMP programas *m pl* de distribución gratuita, DATA programas *m pl* de libre disposición; ~ **package** COMP paquete *m* de programas de distribución gratuita
freeze: **~-frame** *n* TV cuadro *m* bloqueado; **~-out** *n* DATA, TRANS bloqueo *m* momentáneo; **~-out fraction** *n* DATA, TRANS fracción *f* de bloqueo momentáneo; **~-picture** *n* TV imagen *f* congelada; **~-picture request** *n* TV petición *f* de congelación de la imagen; ~ **update** *n* TV actualización *f* congelada
freeze *vt* COMP fijar
F-region *n* RADIO región *f* F
frequency *n* GEN frecuencia *f*; **~-agility pager** RADIO buscapersonas *m* de agilidad de frecuencias; **~-agility radar** RADIO radar *m* de agilidad de frecuencias; ~ **alignment** RADIO alineación *f* de frecuencias; ~ **allocation** RADIO distribución *f* de frecuencias; ~ **allotment** RADIO adjudicación *f* de frecuencias; ~ **assignment** RADIO asignación *f* de frecuencias; ~ **band** TRANS banda *f* de frecuencias; **~-band pairing** RADIO pareado *m* de bandas de frecuencia; ~ **changer** POWER cambiador *m* de frecuencia; **~-change signalling** *BrE* RADIO modulación *f* de frecuencia; ~ **changing** RADIO cambio *m* de frecuencia; ~ **channel** RADIO canal *m* de frecuencias; ~ **characteristic** RADIO característica *f* de frecuencia; ~ **coherence** RADIO coherencia *f* de la frecuencia; ~ **control** POWER, TRANS control *m* de frecuencias; ~ **conversion** RADIO, TRANS, TV conversión *f* de frecuencia; ~ **converter** GEN convertidor *m* de frecuencia; **~-converter tube** ELECT tubo *m* convertidor de frecuencia; ~ **counter** TEST contador *m* de frecuencia; ~ **curve** GEN *statistics* curva *f* de frecuencias; ~ **demodulation** TRANS demodulación *f* de frecuencia; ~ **demodulator** TRANS demodulador *m* de frecuencia; ~ **density** GEN *statistics* densidad *f* de frecuencia; ~ **departure** RADIO, TRANS desviación *f* de frecuencia; **~-derived channel** RADIO canal *m* derivado; ~ **deviation** RADIO, TRANS desviación *f* de frecuencia; **~-deviation meter** TEST indicador *m* de desviación de frecuencia; ~ **difference** RADIO diferencia *f* de frecuencia; ~ **displacement** TRANS desplazamiento *m* en frecuencia; ~ **distribution** GEN *statistics* distribución *f* de frecuencias; ~ **diversity** RADIO, TRANS diversidad *f* de frecuencias; **~-diversity reception** RADIO recepción *f* de

diversidad de frecuencias; ~ **divider** TRANS divisor *m* de frecuencias; ~ **division** TRANS división *f* de frecuencias; **~-division duplexing** TRANS conmutación *f* por división de frecuencias; **~-division multiplex** TRANS múltiplex *m* por división de frecuencias; **~-division multiple access** (*FDMA*) SIG, SWG, TELEP, TRANS acceso *m* múltiple por división de frecuencias (*AMDF*); **~-division multiplex** TELEP, TRANS múltiplex *m* por división de frecuencias; **~-division multiplexing** (*FDM*) TELEP, TRANS multiplexado *m* por división de frecuencias (*MDF*); **~-division switching** TELEP conmutación *f* por división de frecuencias; **~-division switching system** SWG sistema *m* de conmutación por división de frecuencias; ~ **domain** GEN *statistics*, TEC RAD, TRANS ámbito *m* de frecuencia; **~-domain analysis** DATA análisis *m* de dominio de frecuencia; **~-domain characteristic** RADIO característica *f* de dominio frecuencial; **~-domain reflectometer** ELECT, POWER, TEST, TRANS reflectómetro *m* de dominio frecuencial; ~ **doubler** TRANS duplicador *m* de frecuencia; ~ **down-conversion** TRANS conversión *f* de frecuencia descendente; ~ **drift** ELECT, RADIO, TRANS deriva *f* de frecuencia; ~ **encoding** RADIO codificación *f* de frecuencia; **~-exchange signalling** *BrE* SIG modulación *f* por mutación de frecuencia; ~ **filter** ELECT, RADIO filtro *m* de frecuencias; **~-generating stage** TRANS fase *f* de generación de frecuencia; ~ **generator** ELECT, TELEP generador *m* de frecuencia; **~-generator receiver** TELEP receptor *m* de generador de frecuencia; ~ **hopping** RADIO *land mobile* desplazamiento *m* de frecuencia; **~-hopping spread spectrum** RADIO ensanchamiento *m* del espectro del salto de frecuencia; ~ **instability** RADIO inestabilidad *f* de frecuencia; ~ **inversion** RADIO *land mobile* inversión *f* de frecuencia; ~ **lock** RADIO fijador *m* de frecuencia; ~ **management** RADIO gestión *f* de frecuencia; ~ **measurement** TEST medición *f* de frecuencia; ~ **meter** TEST frecuencímetro *m*; **~-modulated carrier** TRANS portadora *f* modulada en frecuencia; **~-modulated radar** RADIO radar *m* modulado en frecuencia; ~ **modulation** RADIO, TRANS, TV modulación *f* de frecuencia; **~-modulation noise** TRANS ruido *m* en modulación de frecuencia; **~-modulation synthesis** COMP síntesis *f* de modulación de frecuencia; ~ **monitor** RADIO monitor *m* de frecuencia; ~ **multiplexing** TELEP, TRANS multiplexado *m* de frecuencia; ~ **multiplier** TRANS multiplicador *m* de frecuencia; ~ **offset** RADIO, TRANS dispersión *f* de frecuencias; ~ **plan** TRANS, TV plan *m* de frecuencia; ~ **polygon** PARTS *statistics* polígono *m* de frecuencia; **~-prediction chart** RADIO diagrama *m* de predicción de frecuencia; **~-pulling effect** ELECT efecto *m* de arrastre de las frecuencias; ~ **range** RADIO, TRANS gama *f* de frecuencias; ~ **ratio** TRANS índice *m* de frecuencia; ~ **record** RADIO registro *m* de frecuencia; ~ **regulation** POWER, TRANS regulación *f* de frecuencia; ~ **resources** *n pl* RADIO recursos *m pl* de frecuencia; ~ **response** RADIO, TRANS respuesta *f* de frecuencia; **~-response curve** ELECT, RADIO curva *f* de réplica de frecuencia; ~ **re-use** RADIO reutilización *f* de frecuencia; **~-re-use satellite network** RADIO, TELEP red *f* satélite de reutilización de frecuencia; ~ **scan** RADIO, TRANS digitalización *f* de frecuencia; ~ **scanner** TRANS explorador *m* de frecuencias; **~-selective ringing** TELEP llamada *f* selectiva; **~-selective sunshield** RADIO pantalla *f* selectiva; **~-selective voltmeter** TEST voltímetro *m* selectivo; ~ **selector** TRANS selector *m* de frecuencias; ~ **separation** RADIO separación *f* de frecuencia; ~ **shift** TRANS desplazamiento *m* de frecuencia; **~-shift keying** (*FSK*) SIG manipulación *f* por desplazamiento de frecuencia (*MDF*); ~ **spacing** RADIO espaciamiento *m* de frecuencias; ~ **spectrum** RADIO espectro *m* de frecuencias; ~ **stability** RADIO estabilidad *f* de frecuencia; ~ **staggering** RADIO escalonamiento *m* de frecuencias; ~ **standard** RADIO patrón *m* de frecuencia; **~-standard accuracy** RADIO precisión *f* de patrón de frecuencia; ~ **step** TV paso *m* de frecuencia; ~ **sweep** RADIO, TRANS barrido *m* de frecuencia; ~ **swing** GEN desvío *m* de frecuencia; ~ **synthesizer** RADIO sintetizador *m* de frecuencia; ~ **table** TRANS tabla *f* de frecuencias; ~ **tolerance** RADIO tolerancia *f* de frecuencia; ~ **tracking** RADIO ajuste *m* de frecuencias; ~ **translation** RADIO, TRANS conversión *f* de frecuencia; ~ **translator** RADIO traductor *m* de frecuencia; ~ **tuning** RADIO sintonización *f* de frecuencia; **~-tuning range** ELECT banda *f* de sintonización de frecuencias; ~ **uncertainty** TELEP inestabilidad *f* de frecuencia; **~-uncertainty band** TRANS banda *f* de incertidumbre de frecuencias

frequently: **~-asked questions** *n pl* (*FAQ*) DATA *Internet* preguntas *f pl* formuladas con frecuencia (*FAQ*)

Fresnel: **~ diffraction pattern** *n* TRANS diagrama *m* de difracción de Fresnel; **~ reflection** *n* TRANS reflexión *f* de Fresnel; **~ reflection method** *n* TRANS método *m* de reflexión de Fresnel; **~ region** *n* RADIO, TRANS región *f* de Fresnel; **~ zone** *n* TRANS zona *f* de Fresnel; **~ zone blockage** *n* TRANS blocaje *m* de zona de Fresnel

fringe: **~ effect** *n* POWER efecto *m* ondulatorio

fringing *n* POWER alteración *f* cromática de extremos

FRJ *abbr* (*facility-rejected message*) DATA mensaje *m* de sistema rechazado

FRMR *abbr* (*frame reject frame*) TELEP FRMR (*cuadro de rechazo de trama*)

front: **~-to-back ratio** *n* RADIO relación *f* de radiación anterior-posterior; **~ end** *n* GEN unidad *f* de entrada; **~-end clipping** *n* DATA recorte *m* de la sección de entrada; **~-end computer** *n* COMP, SWG ordenador *m* frontal *Esp*, computador *m* frontal *AmL*, computadora *f* frontal *AmL*; **~-end equipment** *n* RADIO equipo *m* frontal; **~-fed dish** *n* RADIO plato *m* de alimentación frontal; **~-fed reflector antenna** *n* RADIO antena *f* reflectora de alimentación frontal; **~ feed** *n* GEN alimentación *f* frontal; **~ loading** *n* GEN carga *f* frontal; **~ page** *n* COMP portada *f*; **~ panel** *n* GEN panel *m* frontal

FRP *abbr* (*fast-reservation protocol*) DATA protocolo *m* de reserva rápida

frying *n* TELEP ruido *m* de fritura

FS *abbr* (*further study*) DATA FS (*estudio posterior*)

FSK *abbr* (*frequency shift keying*) SIG MDF (*manipulación por desplazamiento de frecuencia*)

FT *abbr* (*Fourier transformation*) TEST FT (*transformación de Fourier*)

FTAM *abbr* (*file transfer access and management*) COMP *remote data processing*, DATA, TELEP acceso *m* y gestión de transferencia de archivos (*FTAM*)

FTAMSE *abbr* (*file transfer access and management service element*) DATA elemento *m* del servicio de acceso y gestión de transferencia de ficheros

FTIR *abbr* (*Fourier transform infrared spectography*) TEST FTIR (*espectroscopía infrarroja transformada de Fourier*)

FTP *abbr* (*file-transfer protocol*) DATA *Internet* FTP (*protocolo de transferencia de archivos*); **~ server** *n* DATA *Internet* servidor *m* FTP

FTTB *abbr* (*fibre to the building*) TRANS fibra *f* hasta el edificio

FTTC *abbr* (*fibre to the curb*) TRANS fibra *f* para pozo de registro de canalización

FTTF *abbr* (*fibre to the floor*) TRANS FTTF (*fibra para lecho de canalización*)

FTTH *abbr* (*fibre to the home*) TRANS fibra *f* hasta el domicilio

FTTO *abbr* (*fibre to the office*) TRANS fibra *f* hasta la oficina

FU *abbr* (*functional unit*) DATA, TEST *presentation* unidad *f* funcional

fulfil: **~ requirements** *BrE vt* GEN cumplir los requisitos

full *adj* GEN completo, lleno, pleno; **~ adder** *n* ELECT sumador; **~ availability** *n* TELEP accesibilidad *f* completa; **~ break-in** *n* TELEP interposición *f* plena; **~ break-in operate time** *n* RADIO, TELEP tiempo *m* de operación de interposición plena; **~ carrier** *n* RADIO onda *f* portadora completa; **~-carrier emission** *n* RADIO emisión *f* de onda portadora completa; **~-carrier wave** *n* RADIO onda *f* portadora completa; **~ digital processing** *n* COMP procesamiento *m* digital pleno; **~ duplex** *n* TRANS sistema *m* de transmisión bidireccional; **~-duplex operation** *n* TRANS trabajo *m* en duplex integral; **~-duration half-maximum** *n* DATA medio *m* máximo de duración completa; **~ echo suppressor** *n* RADIO, TELEP supresor *m* de eco completo; **~-float system** *n* POWER sistema *m* de distribución en paralelo; **~ load** *n* POWER carga *f* plena; **~-motion video** *n* COMP, TV vídeo *m* de movimiento total; **~-motion videoconferencing** *n* COMP, TV videoconferencia *f* de movimiento total; **~ rate** *n* APPL precio *m* sin descuento; **~-rate channel** *n* TRANS canal *m* de plena velocidad; **~ refund** *n* GEN reembolso *m* total; **~-scale test** *n* TEST prueba *f* a escala real; **~ screen** *n* COMP pantalla *f* llena; **~-steering capability** *n* RADIO capacidad *f* plena de maniobra; **~ subtracter** *n* ELECT substractor *m* completo; **~ sunlight period** *n* RADIO periodo *m* de pleno sol; **~ text** *n* COMP texto *m* integral; **~ view** *n* COMP panorámica *f*; **~-wave dipole** *n* RADIO dipolo *m* de onda completa; **~-wave rectification** *n* ELECT rectificación *f* de onda completa; **~-wave rectifier** *n* ELECT rectificador *m* de onda completa; **~-width half-maximum** *n* DATA, ELECT, TRANS medio *m* máximo de ancho completo

fully: **~-automatic operation** *n* TELEP operación *f* totalmente automática; **~-charged battery** *n* POWER batería *f* de carga plena; **~-coherent radar** *n* RADIO radar *m* totalmente coherente; **~-dissociated mode of operation** *n* TELEP modo *m* de operación totalmente disociada; **~-dissociated signalling** *BrE n* SIG señalización *f* totalmente disociada; **~-distributed control system** *n* SWG sistema *m* de control totalmente distribuido; **~-loaded** *adj* GEN completamente cargado; **~-provided circuit group** *n* TELEP grupo *m* de circuitos totalmente abastecidos; **~-restricted station** *n* SWG estación *f* totalmente restringida

function *n* GEN función *f*; **~-affecting maintenance** TELEP, TEST *dependability* mantenimiento *m* relativo a la función; **~ block** SWG bloque *m* de funciones; **~ change** GEN modificación *f* de función; **~ check-out** TEST *dependability* verificación *f* de funcionamiento; **~ data** *n pl* DATA datos *m pl* de función; **~-degrading failure** TELEP fallo *m* que deteriora la función; **~-degrading fault** TEST *dependability* fallo *m* que deteriora la función; **~-degrading maintenance** TELEP mantenimiento *m* que deteriora la función; **~ description** GEN *documentation* descripción *f* de función; **~ designation** GEN designación *f* de función; **~-disturbing fault** TEST *dependability* fallo *m* que perturba la función; **~-division system** SWG sistema *m* de división de función; **~-division system architecture** SWG arquitectura *f* de sistema de división de función; **~-edit buffer** COMP memoria *f* intermedia de edición; **~ flow chart** GEN organigrama *m* de funciones; **~ generator** COMP generador *m* de función; **~ identification** TELEP identificación *f* de función; **~ key** COMP, CONTR *equipment* tecla *f* de función; **~ keyboard** COMP teclado *m* de función; **~ modification** GEN modificación *f* de función; **~ module** SWG módulo *m* de función; **~ parameter** COMP parámetro *m* de función; **~-permitting failure** TELEP fallo *m* que no afecta a la función; **~-permitting maintenance** TELEP *dependability*, TEST mantenimiento *m* que no afecta a la función; **~-preventing failure** TELEP, TEST fallo *m* que impide la función; **~-preventing fault** TEST *dependability* fallo *m* que impide la función; **~-preventing maintenance** TELEP, TEST *dependability* mantenimiento *m* que impide la función; **~ specification and design** SWG especificación *f* y diseño de función; **~ supervisory unit** SWG *O & M* unidad *f* supervisora de función; **~ switch** GEN conmutador *m* funcional; **~ table** GEN tabla *f* de funciones

functional: **~ block** *n* TELEP bloque *m* funcional; **~-block description** *n* TELEP descripción *f* de bloque funcional; **~-block specification** *n* TELEP especificación *f* de bloque funcional; **~ characteristic** *n* GEN característica *f* funcional; **~ class** *n* (*FC*) TELEP clase *f* funcional (*FC*); **~ class A** *n* (*FCA*) TELEP clase *f* funcional A (*FCA*); **~ class B** *n* (*FCB*) TELEP clase *f* funcional B (*FCB*); **~ description** *n* (*FD*) GEN *in SDL* descripción *f* funcional (*FD*); **~ entity** *n* (*FE*) DATA entidad *f* funcional; **~ entity action** *n* (*FEA*) DATA acción *f* de entidad funcional; **~ feature** *n* GEN característica *f* funcional; **~ group header** *n* DATA encabezamiento *m* de grupo funcional; **~ group leader** *n* DATA líder *m* de grupo funcional; **~ mode** *n* TELEP modo *m* funcional, TEST modo *m* de funcionamiento; **~ specification** *n* GEN especificación *f* de funcionamiento; **~ stress** *n* TEST *dependability* esfuerzo *m* funcional; **~ test** *n* TEST prueba *f* funcional; **~ testing** *n* TEST comprobación *f* de funcionamiento; **~ unit** *n* (*FU*) DATA, TEST *presentation* unidad *f* funcional

functionality *n* GEN funcionalidad *f*

functionally: **~-divided system** *n* DATA sistema *m* dividido funcionalmente

functor *n* COMP elemento *m* lógico

fundamental *adj* GEN fundamental; **~ component** *n* TEST componente *m* fundamental; **~ feature** *n* GEN característica *f* fundamental; **~ mode** *n* TRANS modo *m* fundamental; **~ sound** *n* TELEP nota *f* fundamental

fundamentals *n pl* GEN fundamentos *m pl*

funnel *n* ELECT conducto *m*; **~ mouthpiece** TELEP boquilla *f* de embudo

furnish *vt* GEN proveer

further: **~ development** *n* GEN *design technology* perfeccionamiento *m*; **~ study** *n* (*FS*) DATA estudio *m* posterior (*FS*)

fuse *n* ELECT, PARTS, POWER fusible *m*; **~ alarm** TEST *electricity* alarma *f* de fusible; **~ box** POWER caja *f* de fusibles; **~ holder** POWER cuadro *m* de fusibles; **~ link** POWER elemento *m* de sustitución; **~ panel** POWER cuadro *m* de fusibles; **~-protected** *adj* ELECT, POWER protegido por fusible; **~ strip** POWER lámina *f* de fusible; **~ supervision** POWER supervisión *f* de fusibles; **~ switch** POWER conmutador *m* de fusibles; **~-tapered fibre** *BrE* TRANS fibra *f* de fusible progresivo; **~ unit** POWER conjunto *m* de fusible; **~ wire** POWER hilo *m* fusible

fused *adj* ELECT, POWER provisto de fusibles; **~-fibre** *BrE* **splice** *n* TRANS empalme *m* por fusión de fibra; **~ joint** *n* TRANS empalme *m* soldado; **~ quartz** *n* TRANS cuarzo *m* fundido; **~ silica** *n* TRANS sílice *f* fundida; **~ transistor** *n* ELECT transistor *m* por fusión

fusing *n* ELECT *printed circuits*, TRANS *fibre optics* fusión *f*

fusion: **~ splice** *n* TRANS *fibre optics* empalme *m* por fusión

future: **~ reference picture** *n* COMP, TV imagen *f* de referencia futura; **~ use** *n* GEN uso *m* futuro

fuzziness *n* TV *scanner* borrosidad *f*

fuzzy: **~ theory** *n* DATA teoría *f* de duda, teoría *f* de nebulosa

G

GaAs *abbr* (*gallium arsenide*) PARTS GaAs (*arseniuro de galio*)
gage *AmE see gauge BrE*
gain *n* GEN ganancia *f*; **~ calibration error** ELECT error *m* de calibrado de ganancia; **~ control** ELECT control *m* de ganancia; **~ degradation** RADIO degradación *f* de ganancia; **~ inequality** ELECT desigualdad *f* de ganancia; **~ ripple** ELECT ondulación *f* de ganancia; **~ scale** ELECT escala *f* de ganancia; **~ slope** ELECT pendiente *f* de ganancia
galactic: **~ noise** *n* RADIO ruido *m* galáctico
gallium: **~ arsenide** *n* (*GaAs*) PARTS arseniuro *m* de galio (*GaAs*); **~ arsenide substrate** *n* ELECT, PARTS *semiconductors* substrato *m* de arseniuro de galio
galvanic: **~ cell** *n* POWER pila *f* galvánica
galvanometer *n* TEST galvanómetro *m*
gamma *n* ELECT, TV gamma *f*; **~ curve** ELECT, PER APP *scanner*, TV *camera* curva *f* gamma
gamut *n* COMP, TV *chromaticity* gama *f*
gang: **~ mountable adaptor** *n* ELECT adaptador *m* múltiple montable
gap *n* GEN intervalo *m*; **~ character** DATA carácter *m* de separación; **~ coding** TRANS codificación *f* por intervalos; **~-filler satellite** RADIO satélite *m* para llenado de espacios interlobulares; **~ loss** TRANS *fibre optics* pérdida *f* de intervalo
garble *vt* GEN mutilar
garbled *adj* GEN mutilado; **~ reply** *n* RADIO respuesta *f* mutilada
garbling *n* GEN *cryptology* mutilación *f*
gas: **~-content factor** *n* ELECT factor *m* de contenido de gas; **~ current** *n* ELECT corriente *f* de gas; **~ discharge** *n* ELECT descarga *f* de gas; **~ discharge display** *n* COMP, ELECT visualizador *m* para descarga de gas; **~-filled phototube** *n* ELECT fototubo *m* lleno de gas; **~-filled tube** *n* ELECT tubo *m* lleno de gas; **~-flow counter tube** *n* TEST tubo *m* contador de corriente gaseosa; **~ multiplication** *n* ELECT multiplicación *f* debida al gas; **~ multiplication factor** *n* ELECT factor *m* de multiplicación de gas; **~ panel** *n* COMP, ELECT panel *m* de descarga gaseosa; **~ sample counter tube** *n* TEST contador *m* para muestras gaseosas; **~ tube** *n* ELECT tubo *m* de gas; **~-type arrester** *n* POWER *protection* pararrayos *m* de tipo gaseoso
gasket *n* PARTS junta *f* de estanqueidad; **~ ring** PARTS anillo *m* obturador
gate *n* ELECT *semiconductors* puerta *f*; **~ array** ELECT *logic elements* conjunto *m* de puertas; **~ array circuit** ELECT *logic elements* circuito *m* prefundido; **~ circuit** ELECT *logic elements* circuito *m* puerta; **~ condition** ELECT *logic elements* condición *f* para activación de puerta; **~-controlled turn-off time** ELECT *thyristors* tiempo *m* de corte controlado por puerta; **~ cutoff current** ELECT *thyristors* corriente *f* residual de puerta; **~ network** ELECT *logic elements* red *f* de puertas; **~ non-trigger current** ELECT *thyristors* corriente *f* de no disparo de puerta; **~ non-trigger voltage** ELECT *thyristors* tensión *f* de no disparo de puerta; **~ pulse** ELECT *logic elements* impulso *m* de puerta; **~ region** ELECT *transistors* región *f* de puerta; **~ resistance** ELECT resistencia *f* de puerta; **~ signal** ELECT *logic elements* señal *f* de puerta; **~ terminal** ELECT terminal *m* de puerta; **~ turn-off current** ELECT *thyristors* corriente *f* de corte de puerta; **~ turn-off voltage** ELECT *thyristors* tensión *f* de corte de puerta; **~ turn-on current** ELECT *thyristors* corriente *f* de encendido de puerta; **~ turn-on voltage** ELECT *thyristors* tensión *f* de encendido de puerta
gated: **~ amplifier** *n* ELECT amplificador *m* de desbloqueo periódico; **~ mode acquisition procedure** *n* DATA procedimiento *m* de adquisición de modo mandado
gateway *n* RADIO, SWG, TRANS pasillo *m*; **~ exchange** SWG central *f* de pasillo; **~ MSC** SWG MSC *m* de pasillo; **~ station** RADIO estación *f* de pasillo
gating: **~ signal** *n* RADIO señal *f* de puerta
gauge[1] *BrE* (*AmE* **gage**) *n* TEST calibrador *m*, indicador *m*, medidor *m*, *measure* norma *f* de medida
gauge[2] *BrE* (*AmE* **gage**) *vt* TEST calibrar, medir
gauging *BrE* (*AmE* **gaging**) *n* TEST calibración *f*, medición *f*
Gauss: **~'s law** *n* TEST *statistics* ley *f* de Gauss
Gaussian: **~ beam** *n* TRANS haz *m* gaussiano; **~ channel** *n* DATA canal *m* gaussiano; **~ curve** *n*

TEST *statistics* curva *f* gaussiana; ~ **distribution** *n* TEST *statistics* distribución *f* gaussiana; **~-filtered frequency-shift keying** *n* (*GFSK*) SIG tecleo *m* de inversión de frecuencia con filtro gaussiano (*GFSK*); **~-filtered minimum shift keying** *n* (*GMSK*) SIG codificación *f* por mínimo desplazamiento de fase por filtrado gaussiano; ~ **noise** *n* SIG, TELEP, TRANS ruido *m* gaussiano; ~ **pulse** *n* SIG, TELEP, TRANS impulso *m* gaussiano

Gb *abbr* (*gigabyte*) COMP, DATA Gb (*gigabitio*)

gc *abbr* (*gigacycle*) TEST gc (*gigaciclo*)

GDI *abbr* (*graphic device interface*) COMP GDI (*interfaz de estación de presentación visual*)

Geiger: ~ **counter** *n* TEST contador *m* Geiger; ~ **region** *n* RADIO, TRANS región *f* Geiger; ~ **threshold** *n* RADIO umbral *m* de Geiger

Geiger-Müller: ~ **counter** *n* TEST contador *m* Geiger-Müller; ~ **counter tube** *n* TEST tubo *m* contador de Geiger-Müller

gel *n* PARTS *plastics* gel *m*; ~ **battery** POWER batería *f* de gel

general: ~ **attribute** *n* DATA atributo *m* general; ~ **auto-action** *n* DATA acción *f* automática general; ~ **call** *n* TELEP llamada *f* general; ~ **circuit interface** *n* COMP interfaz *f* de circuito general; **~-coverage receiver** *n* RADIO receptor *m* de cobertura general; ~ **localization** *n* TEST localización *f* general; ~ **model** *n* TELEP modelo *m* general; ~ **parameter** *n* COMP parámetro *m* general; **~-purpose interface** *n* COMP interfaz *f* universal; **~-purpose interface bus** *n* (*GPIB*) DATA conductor *m* común de acoplamiento mutuo universal (*GPIB*); ~ **recorded announcement** *n* TELEP anuncio *m* general grabado; ~ **supervisory station** *n* TELEP estación *f* supervisora general

generalized: ~ **phase control** *n* TEST control *m* de fase generalizado

generated: ~ **address** *n* COMP dirección *f* calculada

generation *n* GEN generación *f*; ~ **rate** ELECT *semiconductors* velocidad *f* de generación

generator *n* GEN generador *m*

generic: ~ **address** *n* DATA dirección *f* genérica; ~ **definition** *n* DATA definición *f* genérica; ~ **flow control** *n* (*GFC*) CONTR, TRANS control *m* genérico del flujo; ~ **name** *n* DATA nombre *m* genérico; ~ **program** *n* TELEP programa *m* genérico; ~ **test case** *n* TEST juego *m* de ensayo genérico; ~ **test suite** *n* TEST sucesión *f* genérica de ensayos

genlocking *n* TV intersincronización *f*

genuine: ~ **receive point** *n* RADIO punto *m* de recepción genuino

geocentric: ~ **angle** *n* RADIO ángulo *m* geocéntrico

geographically: **~-distributed exchange** *n* SWG central *f* distribuida geográficamente

geometric: ~ **distortion** *n* TV distorsión *f* geométrica; ~ **mean** *n* TEST media *f* geométrica; ~ **optics** *n pl* (*GO*) TRANS óptica *f* geométrica (*GO*); ~ **sample** *n* TEST muestra *f* geométrica

geostationary *adj* RADIO geoestacionario; ~ **fixed satellite** *n* RADIO satélite *m* fijo geoestacionario; ~ **orbit** *n* RADIO órbita *f* geoestacionaria; ~ **satellite** *n* RADIO satélite *m* geoestacionario; ~ **satellite orbit** *n* RADIO órbita *f* de satélite geoestacionario

geosynchronous *adj* RADIO geosincrónico

getter *n* ELECT rarefactor *m*

GFC *abbr* (*generic flow control*) CONTR, TRANS control *m* genérico del flujo

GFSK *abbr* (*Gaussian-filtered frequency-shift keying*) SIG GFSK (*tecleo de inversión de frecuencia con filtro gaussiano*)

ghost *n* TV imagen *f* fantasma; ~ **call** TRANS llamada *f* fantasma; ~ **signal** TV señal *f* fantasma

GHz *abbr* (*gigahertz*) GEN GHz (*gigahertzio*)

GI *abbr* (*group identifier*) TELEP GI (*identificador de grupo*)

GIF *abbr* (*graphics interchange format*) DATA *Internet* GIF (*formato de interfaz de gráficos*)

gigabit *n* DATA gigabit *m*

gigabyte *n* (*Gb*) COMP, DATA gigabitio *m* (*Gb*)

gigacycle *n* (*gc*) TEST gigaciclo *m* (*gc*)

gigahertz *n* (*GHz*) GEN gigahertzio *m* (*GHz*)

given: ~ **code** *n* DATA código *m* especificado

GL *abbr* (*group length*) TELEP GL (*longitud de grupo*)

glare *n* TV brillo *m*

glass *n* PARTS vidrio *m*, cristal *m*; ~ **capillary** PARTS tubo *m* capilar de vidrio; ~ **fibre** *BrE* PARTS, TRANS fibra *f* de vidrio; ~ **preform** PARTS *fibre optics* preforma *f* de vidrio; ~ **tube fuse** ELECT, PARTS, POWER fusible *m* en tubo de vidrio

glide *n* GEN barrido *m* de frecuencia; ~ **effect** COMP efecto *m* de barrido

glint *n* RADIO parpadeo *m*

glitch *n* COMP perturbación *f* de baja frecuencia

global: ~ **beam** *n* TRANS haz *m* global; ~ **beam-carrier wave** *n* TRANS onda *f* portadora de haz global; ~ **call** *n* RADIO llamada *f* global; ~ **coverage** *n* RADIO cobertura *f* global; ~ **network addressing domain** *n* DATA dominio *m* mundial de dirección de red; ~ **positioning system** *n* (*GPS*) RADIO sistema *m* de localización global (*GPS*); ~ **recognition** *n* GEN reconocimiento *m* global; ~ **reference burst distribution** *n* DATA distribución *f* mundial por ráfaga de referencia; ~ **system for mobile communications** *n* (*GSM*) RADIO sistema *m* mundial para comunicaciones móviles (*GSM*); ~ **telecommunications satellite system** *n* RADIO, TELEP sistema *m* mundial de satélites de telecomunicaciones; ~ **telecommunications system** *n* (*GTS*) GEN sistema *m* mundial de telecomunicaciones (*GTS*)

glow: ~ **current of protector** *n* ELECT corriente *f* luminiscente de protector; ~ **discharge** *n* ELECT descarga *f* luminiscente; ~ **discharge lamp** *n* ELECT lámpara *f* de descarga luminiscente; ~ **discharge tube** *n* ELECT tubo *m* de descarga luminiscente; ~ **discharge valve** *n* ELECT válvula *f* de descarga luminiscente; ~ **indicator tube** *n* ELECT tubo *m* indicador luminiscente; ~ **lamp** *n* ELECT lámpara *f* de efluvios; ~ **voltage** *n* ELECT tensión *f* de luminiscencia

GMSK *abbr* (*Gaussian-filtered minimum shift keying*) SIG codificación *f* por mínimo desplazamiento de fase por filtrado gaussiano

GMT *abbr* (*Greenwich Mean Time*) GEN GMT (*hora media de Greenwich*)

go[1]: ~**-ahead signal** *n* TELEP *telex* señal *f* de télex; ~**-to-return crosstalk** *n* TRANS diafonía *f* entre las dos direcciones

go[2]: ~ **high** *vt* ELECT ascender; ~ **low** *vt* ELECT descender; ~ **on-hook** *vt* TELEP colgar

GO *abbr* (*geometric optics*) TRANS GO (*óptica geométrica*)

Golay: ~ **coding** *n* DATA codificación *f* de Golay

gold: ~**-bonded diode** *n* ELECT, PARTS *semiconductors* diodo *m* de punta de oro

gooseneck: ~ **feed** *n* RADIO alimentador *m* de cuello de cisne

gopher *n* COMP, DATA *Internet* gopher *m*

Gouraud: ~ **shading** *n* TV sombra *f* de Gouraud

GPIB *abbr* (*general-purpose interface bus*) DATA GPIB (*conductor común de acoplamiento mutuo universal*)

GPS *abbr* (*global positioning system*) RADIO GPS (*sistema de localización global*)

grab *n* COMP agarradera *f*

grabber *n* COMP acaparador *m*

graceful: ~ **degradation** *n* RADIO, TRANS degradación *f* con garbo

grade *n* GEN grado *m*; ~ **of service** TEST grado *m* de servicio

graded: ~ **index** *n* ELECT, TRANS índice *m* gradual; ~**-index fibre** *BrE n* ELECT, PARTS, TRANS fibra *f* de índice gradual; ~**-index profile** *n* ELECT, TRANS perfil *m* de índice gradual

grading *n* TELEP granulometría *f*; ~**-coupling loss cable** TRANS cable *m* de pérdidas de acoplamiento de mejora; ~**-index fibre** *BrE* ELECT, PARTS, TRANS fibra *f* de índice graduado, fibra *f* de índice nivelador

gradual *adj* GEN gradual; ~ **degradation** *n* RADIO, TRANS degradación *f* gradual; ~ **failure** *n* TEST fallo *m* gradual; ~ **phasing-in** *n* GEN asimilación *f* progresiva

graduate *vt* TEST graduar

graduation *n* TEST graduación *f*

grammatical: ~ **tagger** *n* COMP referenciador *m* gramatical

grandfathered: ~ **station** *n* RADIO estación *f* con registros de reserva

grant *vt* APPL *licence* conceder

granular: ~ **noise** *n* TELEP ruido *m* granular

granularity *n* TV granularidad *f*

granule *n* TV gránulo *m*

graph *n* GEN gráfico *m*; ~ **plotter** COMP trazador *m* de gráficos

graphic: ~ **character** *n* DATA carácter *m* gráfico; ~ **code extension** *n* TELEP extensión *f* de código gráfico; ~ **device interface** *n* (*GDI*) COMP interfaz *f* de estación de presentación visual (*GDI*); ~ **programming aid** *n* COMP auxiliar *m* para programación gráfica; ~ **representation** *n* GEN representación *f* gráfica; ~ **symbol** *n* DATA símbolo *m* gráfico

graphical: ~ **interface** *n* COMP interfaz *f* gráfica; ~ **symbol** *n* DATA símbolo *m* gráfico; ~ **user interface** *n* (*GUI*) COMP interfaz *f* de usuario gráfica (*GUI*)

graphics *n pl* COMP, DATA, TV gráficos *m pl*; ~ **database** *n* DATA base *f* de datos gráficos; ~ **interchange format** *n* (*GIF*) DATA *Internet* formato *m* de interfaz de gráficos (*GIF*); ~ **mode** *n* COMP modo *m* gráfico; ~ **package** *n* COMP paquete *m* de gráficos; ~ **processor** *n* COMP procesador *m* de gráficos; ~ **screen** *n* COMP rejilla *f* de gráficos

grass *n* RADIO señales *f pl* parásitas

graveyard: ~ **orbit** *n* RADIO órbita *f* de noche

gravity: ~ **gradient** *n* RADIO gradiente *m* de gravedad

gray *AmE see grey BrE*

Gray: ~ **code** *n* DATA código *m* de Gray

Greenwich: ~ **Mean Time** *n* (*GMT*) GEN hora *f* media de Greenwich (*GMT*)

Gregorian: ~ **reflector aerial** *n* RADIO antena *f*

de reflector Gregoriana; ~ **reflector antenna** *n* RADIO antena *f* de reflector Gregoriana
grey: ~ **level** *n* TELEP nivel *m* de grises; ~ **scale** *n* COMP, PER APP escala *f* de grises
grid *n* GEN parrilla *f*, rejilla *f*; ~ **amplitude modulation** ELECT modulación *f* de amplitud de rejilla; ~ **bias voltage** ELECT tensión *f* de polarización de rejilla; ~**-cathode driving voltage** ELECT voltaje *m* impulsor rejilla-cátodo; ~**-controlled arc discharge tube** ELECT tubo *m* de descarga de arco controlado por rejilla; ~ **current** ELECT corriente *f* de polarización de rejilla; ~ **data message** DATA mensaje *m* de datos en cuadrícula; ~ **driving power** ELECT potencia *f* de excitación de rejilla; ~ **driving voltage** ELECT voltaje *m* impulsor de rejilla; ~ **input power** ELECT potencia *f* de entrada de rejilla; ~ **input voltage** ELECT voltaje *m* de entrada de rejilla; ~ **leak resistor** ELECT, PARTS resistor *m* de escape; ~ **network** TRANS red *f* de rejilla
gridded: ~ **reflector aerial** *n* RADIO antena *f* reflectora cuadriculada; ~ **reflector antenna** *n* RADIO antena *f* reflectora cuadriculada
grille *BrE* (*AmE* **grill**) *n* GEN rejilla *f*
grip *n* COMP asa *f*
grommet *n* ELECT, PARTS arandela *f* aislante
groove *n* GEN hendidura *f*
grooved: ~ **cable** *n* PARTS, TRANS *fibre optics* alambre *m* acanalado
gross: ~ **bit rate** *n* DATA velocidad *f* aproximada de tráfico binario
ground *AmE n* (*cf earth BrE*) ELECT, POWER masa *f*, tierra *f*; ~ **address** GEN dirección *f* terrestre; ~ **bar** ELECT, POWER barra *f* de tierra; ~**-based duct** RADIO conducto *m* superficial; ~**-based pilot** RADIO piloto *m* en tierra; ~ **button** SWG botón *m* de tierra; ~ **cable** POWER cable *m* de descarga a tierra; ~ **clamp** POWER abrazadera *f* de toma de masa; ~**clutter** RADIO ecos *m pl* de tierra; ~ **clutter coefficient** RADIO coeficiente *m* de ecos de tierra; ~ **collector** POWER colector *m* de tierra; ~ **collector bar** POWER barra *f* colectora de tierra; ~ **command** RADIO orden *f* de tierra; ~ **conductor** POWER hilo *m* de tierra; ~ **connection** POWER conexión *f* a tierra; ~ **control** RADIO control *m* del terreno; ~ **data** *n pl* RADIO datos *m pl* de tierra; ~ **electrode** POWER electrodo *m* de tierra; ~ **electrode installation** POWER instalación *f* de electrodo de tierra; ~ **electrode resistance** POWER resistencia *f* del electrodo de tierra; ~ **fault** POWER pérdida *f* a tierra; ~**-fault circuit breaker** POWER interruptor *m* de circuito accionado por corriente de pérdida; ~**-fault protection** POWER protección *f* contra fuga a tierra; ~**-fault relay** ELECT, PARTS relé *m* accionado por corriente de pérdida a tierra; ~**-generated pilot** RADIO piloto *m* generado en tierra; ~ **junction** ELECT *semiconductors* conexión *f* a tierra; ~ **lead** POWER cable *m* a tierra; ~ **leakage** POWER fuga *f* a tierra; ~ **leakage current** POWER corriente *f* de pérdida a tierra; ~ **leakage guard** POWER monitor *m* de fuga a tierra; ~ **loop** ELECT bucle *m* de tierra; ~ **movement radar** RADIO radar *m* indicador de movimiento en tierra; ~ **network** TRANS red *f* de tierra; ~ **noise** ELECT ruido *m* de fondo; ~ **noise level** ELECT nivel *m* de ruido de fondo; ~ **plane** POWER plano *m* de base; ~ **plane antenna** RADIO antena *f* con polarización horizontal; ~ **plate** POWER placa *f* de tierra; ~ **potential** POWER potencial *m* de tierra; ~ **potential difference** POWER diferencia *f* de potencial de tierra; ~ **radial termination** POWER terminación *f* radial de tierra; ~ **resistance** POWER resistencia *f* de tierra; ~ **resolution** ELECT resolución *f* de tierra; ~ **return** RADIO, SIG, TRANS vuelta *f* por tierra; ~**-return circuit** TRANS circuito *m* con vuelta por tierra; ~**-return double phantom circuit** TELEP circuito *m* superfantasma con retorno por tierra; ~ **ring collector** POWER colector *m* de anillo de masa; ~ **segment** TRANS segmento *m* de tierra; ~ **station** RADIO estación *f* terrestre; ~ **support equipment** RADIO equipo *m* de apoyo en tierra; ~**-synchronous satellite** RADIO satélite *m* sincrónico por tierra; ~ **system** POWER, RADIO sistema *m* de antena-tierra; ~ **terminal** POWER, RADIO borne *m* de contacto a tierra; ~ **test** TRANS ensayo *m* en tierra; ~ **translating equipment** TRANS equipo *m* traductor en tierra; ~ **wave** RADIO onda *f* de tierra; ~ **wire** POWER conductor *m* de tierra
grounded *AmE adj* (*cf earthed BrE*) ELECT, POWER puesto a tierra; ~ **emitter circuit** *n* ELECT circuito *m* con emisor a tierra; ~ **emitter stage** *n* ELECT etapa *f* con emisor a tierra; ~**-grid amplifier** *n* ELECT amplificador *m* con rejilla a masa
grounding *AmE n* (*cf earthing BrE*) ELECT, POWER conexión *f* a tierra, puesta *f* a tierra; ~ **arrangement** POWER instalación *f* con puesta a tierra; ~ **button** SWG *PBX* botón *m* de tierra; ~ **conductor** POWER conductor *m* de puesta a tierra; ~ **indent** ELECT muesca *f* de puesta a tierra; ~ **mat** RADIO, TRANS emparrillado *m* de puesta a tierra; ~ **network** POWER red *f* de puesta a tierra; ~ **plate** POWER placa *f* de conexión a tierra; ~ **probe** POWER sonda *f* de

conexión a tierra; ~ **rod** POWER varilla *f* de tierra; ~ **switch** POWER conmutador *m* de puesta a tierra

group *n* GEN grupo *m*; ~ **call** TELEP llamada *f* de grupo; ~ **call identity** TELEP identidad *f* de grupo llamante; ~ **controller** DATA controlador *m* de grupo; ~ **delay** TRANS retardo *m* de grupo; ~**-delay distortion** TRANS distorsión *f* de retardo de grupo; ~**-delay equalization** TRANS igualación *f* de retardo de grupo; ~**-delay mask** TRANS máscara *f* de retardo de grupo; ~**-delay slope** TRANS pendiente *f* de retardo de grupo; ~ **distribution frame** TELEP, TRANS trama *f* de distribución de grupo; ~ **identifier** (*GI*) TELEP identificador *m* de grupo (*GI*); ~ **index** TRANS *fibre optics* índice *m* de grupo; ~ **key** ELECT clave *f* de grupo; ~ **length** (*GL*) TELEP longitud *f* de grupo (*GL*); ~ **link** TELEP conexión *f* en grupo; ~ **listening** TELEP escucha *f* en grupo; ~ **mark** DATA marca *f* de grupo; ~ **of matrix links** TELEP grupo *m* de enlaces de matrices; ~ **monitoring** TELEP supervisión *f* de grupo; ~ **number call** TELEP llamada *f* a número de grupo; ~**-pilot failure** TRANS fallo *m* en onda piloto de grupo; ~**-pilot generation** TRANS generación *f* de onda piloto de grupo; ~ **relationships** *n pl* DATA relaciones *f pl* de grupo; ~ **retardation** RADIO retraso *m* de grupo; ~ **section** TELEP sección *f* de grupo; ~ **separator** (*GS*) DATA separador *m* de grupo (*SG*); ~ **switching** SWG conmutación *f* de grupo; ~ **switching centre** *BrE* (*GSC*) SWG centro *m* de conmutación de grupo; ~ **switching centre** *BrE* **catchment area** SWG zona *f* telefónica de centro de conmutación de grupo; ~ **switching centre** *BrE* **exchange area** SWG zona *f* telefónica de centro de conmutación de grupo; ~ **testing** TEST comprobación *f* de grupo; ~ **translation** TRANS traslación *f* de grupo; ~ **velocity** TRANS velocidad *f* de grupo; ~**-velocity dispersion** (*GVD*) TRANS dispersión *f* de velocidad de grupo (*GVD*)

grouping *n* GEN agrupación *f*, acoplamiento *m*; ~ **diagram** GEN *documentation* esquema *m* de agrupación

groupware *n* COMP equipo *m* de grupo

grown: ~ **junction transistor** *n* ELECT *semiconductors* transistor *m* de unión por crecimiento

growth: ~ **capability** *n* GEN capacidad *f* de crecimiento; ~ **rate** *n* GEN índice *m* de crecimiento; ~ **sector** *n* GEN sector *m* de crecimiento

GS *abbr* (*group separator*) DATA SG (*separador de grupo*)

GSC *BrE abbr* (*group switching centre*) SWG centro *m* de conmutación de grupo

GSM *abbr* (*global system for mobile communications*) RADIO GSM (*sistema mundial para comunicaciones móviles*)

GTS *abbr* (*global telecommunications system*) GEN GTS (*sistema mundial de telecomunicaciones*)

guaranteed *adj* APPL, PER APP garantizado

guarantor: ~ **administration** *n* APPL administración *f* de garante

guard: ~ **band** *n* RADIO banda *f* de protección; ~ **circuit** *n* TELEP circuito *m* de seguridad; ~ **receiver** *n* RADIO receptor *m* de guardia; ~ **ring** *n* TELEP anillo *m* de guardia; ~ **space** *n* RADIO *land mobile* espacio *m* de protección; ~ **time** *n* RADIO tiempo *m* de guardia

GUI *abbr* (*graphical user interface*) COMP GUI (*interfaz de usuario gráfica*)

guidance: ~ **output** *n* TELEP salida *f* de guiaje

guide *n* GEN guía *f*; ~ **characteristic wave impedance** ELECT, RADIO impedancia *f* de onda característica; ~ **finger** TELEP dedo *m* de guía; ~ **spool** TELEP carrete *m* de guía; ~ **wavelength** RADIO, TRANS longitud *f* de onda en la guía

guided: ~ **mode** *n* TRANS *fibre optics* modo *m* guiado; ~ **radiation system** *n* RADIO *land mobile* sistema *m* de radiación dirigida; ~ **wave** *n* TRANS *fibre optics* onda *f* dirigida; ~**-wave demultiplexer** *n* (*GWD*) TRANS desmultiplexor *m* de onda guiada (*GWD*); ~**-wave polarizer** *n* ELECT polarizador *m* de onda guiada

guillotine: ~ **attenuator** *n* RADIO atenuador *m* de guillotina

guy *n* PARTS tensor *m*; ~ **attachment** PARTS hierro *m* tensor de soporte; ~ **wire** PARTS, RADIO cable *m* de retenida

GVD *abbr* (*group velocity dispersion*) TRANS GVD (*dispersión de velocidad de grupo*)

GWD *abbr* (*guided-wave demultiplexer*) TRANS GWD (*desmultiplexor de onda guiada*)

gyrator *n* RADIO girador *m*

gyrofrequency *n* RADIO girofrecuencia *f*

gyroscopic: ~ **drift** *n* RADIO deriva *f* giroscópica

gyrotron *n* ELECT girotrón *m*

H

H *abbr* (*henry*) ELECT H (*henrio*); **~ bend** *n* RADIO codo *m* H; **~ corner** *n* RADIO codo *m* H; **~ line** *n* RADIO línea *f* H; **~ mode** *n* RADIO modo *m* H

hacker *n* COMP pirata *m* informático

hacking *n* COMP pirateo *m* informático

halation *n* TV halo *m*

half[1]: **~-adder** *n* ELECT semisumador *m*; **~-cycle** *n* POWER semiciclo *m*; **~-duplex circuit** *n* TELEP circuito *m* semiduplex; **~-duplex connection** *n* TELEP conexión *f* semiduplex; **~-duplex transmission** *n* TRANS transmisión *f* en semiduplex; **~-duplex transmission module** *n* (*HDTM*) TRANS módulo *m* de transmisión en semiduplex (*HDTM*); **~-lattice filter** *n* ELECT, TRANS filtro *m* de semicelosía; **~-line pulse** *n* TV impulso *m* de semilínea; **~-power bandwidth** *n* RADIO, TRANS ancho *m* de banda de mitad de potencia; **~-power beamwidth** *n* RADIO, TRANS ancho *m* de haz de mitad de potencia; **~-rate channel** *n* TRANS canal *m* de velocidad mitad; **~-rate FEC decoder** *n* SIG descodificador *m* FEC de velocidad mitad; **~-size plug** *n* PARTS clavija *f* media; **~-subtracter** *n* ELECT semirrestador *m*; **~-tone distortion** *n* TV distorsión *f* de medias tintas; **~-toning** *n* TELEP gradaciones *f* intermedias; **~-wave dipole** *n* RADIO dipolo *m* de semionda; **~-wave rectification** *n* ELECT rectificación *f* de media onda; **~-wave rectifier** *n* ELECT rectificador *m* de media onda

half[2]: **~-wave rectify** *vt* ELECT rectificar media onda

Hall: **~ angle** *n* ELECT ángulo *m* de Hall; **~ coefficient** *n* ELECT coeficiente *m* de Hall; **~ effect** *n* ELECT efecto *m* Hall; **~ effect device** *n* ELECT elemento *m* de efecto Hall; **~ generator** *n* ELECT generador *m* Hall; **~ integrated circuit** *n* ELECT circuito *m* integrado Hall; **~ modulator** *n* ELECT modulador *m* Hall; **~ multiplier** *n* ELECT multiplicador *m* Hall; **~ plate** *n* ELECT placa *f* de Hall; **~ probe** *n* ELECT sonda *f* de Hall; **~ terminal** *n* ELECT terminal *m* de Hall; **~ voltage** *n* ELECT voltaje *m* de Hall

hallway: **~ telephone** *n* APPL, PER APP *security* portero *m* automático

halo *n* TV halo *m*; **~ orbit** RADIO órbita *f* de aureola

halogen: **~-free** *adj* PARTS sin halógenos; **~-quenched counter tube** *n* ELECT tubo *m* contador con halógeno

halt[1]: **~ indicator** *n* COMP indicador *m* de parada; **~ instruction** *n* COMP instrucción *f* de parada

halt[2] *vt* COMP detener

ham *n* RADIO radioaficionado *m*

Hamming: **~ code** *n* DATA código *m* de Hamming; **~ distance** *n* COMP distancia *f* de Hamming

hand: **~ control** *n* COMP mando *m* manual; **~-held computer** *n* COMP ordenador *m* de mano *Esp*, computador *m* de mano *AmL*, computadora *f* de mano *AmL*; **~-held message terminal** *n* PER APP terminal *m* de mensajes de mano; **~-held receiver** *n* RADIO receptor *m* portátil; **~-held refractometer** *n* TRANS *optics* refractómetro *m* de mano; **~-held scanner** *n* PER APP digitalizador *m* manual; **~-held terminal** *n* RADIO terminal *m* portátil; **~-off** *n* RADIO transferencia *f*; **~-over** *n* RADIO transferencia *f*; **~-portable telephone** *n* TELEP teléfono *m* portátil; **~ vice** *BrE* *n* PARTS tornillo *m* de mano

hand off *AmE* *vt* (*cf hand over BrE*) RADIO transferir

hand over *BrE* *vt* (*cf hand off AmE*) RADIO transferir

handle *vt* GEN *control by hand* manejar

handler *n* COMP operador *m*, DATA transportador *m*, SWG manipulador *m*

handling *n* GEN manipulación *f*

hands: **~-free** *adj* GEN manos libres; **~-free answerback** *n* TELEP respuesta *f* manos libres; **~-free conversation** *n* APPL, PER APP conversación *f* manos libres; **~-free operation** *n* RADIO operación *f* manos libres; **~-free tapedeck** *n* GEN mecanismo *m* impulsor de cinta manos libres; **~-free telephone set** *n* PER APP, TELEP teléfono *m* manos libres; **~-free terminal** *n* APPL, PER APP terminal *m* manos libres; **~-on experience** *n* GEN experiencia *f* práctica

handset *n* PER APP, TELEP aparato *m* telefónico; **~ cord** PER APP, TELEP cordón *m* del aparato telefónico; **~ handle** PER APP, TELEP mango *m* del aparato telefónico; **~ telephone** PER APP, TELEP microteléfono *m*

handshake *n* COMP, DATA, TRANS entrada *f* en comunicación, saludo *m* inicial

handwriting: ~ **recognition** *n* APPL reconocimiento *m* de textos manuscritos

hanger *n* PARTS soporte *m*

hanging: ~ **up** *n* TELEP corte *m*

hang on *vi* TELEP mantener la comunicación

hangover: ~ **time** *n* RADIO, TELEP tiempo *m* de persistencia

hang-up: ~ **signal** *n* SIG señal *f* de colgar

hang up *vi* TELEP colgar

Hann: ~ **window** *n* TV ventana *f* de Hann

harbor *AmE see harbour BrE*

harbour: ~ **radar** *BrE n* RADIO radar *m* de puerto

hard: ~ **copy** *n* COMP, PER APP copia *f* impresa, salida *f* impresa; ~**-copy device** *n* PER APP unidad *f* de copia impresa; ~**-copy image** *n* PER APP imagen *f* de copia impresa; ~**-copy output** *n* PER APP salida *f* de copia impresa; ~ **decision** *n* ELECT decisión *f* relacionada con la circuitería; ~**-decision decoder** *n* SIG descodificador *m* de decisiones relacionadas con la circuitería; ~**-decision decoding** *n* SIG descodificación *f* de decisiones relacionadas con la circuitería; ~ **disk** *n* COMP disco *m* duro, disco *m* rígido; ~**-disk card** *n* COMP tarjeta *f* de disco rígido; ~**-disk interface** *n* COMP interfaz *f* de disco rígido; ~ **drive** *n* COMP unidad *f* de disco duro; ~ **magnetic material** *n* ELECT, PARTS material *m* de fuerte remanencia magnética; ~ **shorting** *n* POWER cortocircuito *m* de equipo; ~**-wired network** *n* TRANS red *f* alámbrica; ~**-wired programmable switching system** *n* SWG sistema *m* de conmutación por cable alámbrico

hardening: ~ **agent** *n* PARTS *plastics* endurecedor *m*

hardware *n* COMP hardware *m*, soporte *m* físico; ~ **description language** (*HDL*) COMP lenguaje *m* de descripción del soporte físico (*HDL*)

harmful: ~ **interference** *n* RADIO interferencia *f* perjudicial; ~ **out-of-band components** *n pl* TELEP componentes *m pl* fuera de banda perjudiciales; ~ **out-of-band components direct through-connection** *n* TELEP conexión *f* directa de componentes fuera de banda perjudiciales

harmless: ~ **out-of-band components** *n pl* TELEP componentes *m pl* fuera de banda inofensivos

harmonic *adj* GEN armónico; ~ **addition** *n* TRANS adición *f* armónica; ~ **analysis** *n* ELECT análisis *m* armónico; ~ **component** *n* TRANS componente *m* armónico; ~ **content** *n* COMP residuo *m* armónico; ~ **distortion** *n* TRANS distorsión *f* armónica; ~ **emission** *n* RADIO emisión *f* armónica; ~ **generator** *n* TRANS generador *m* armónico; ~ **interference** *n* RADIO interferencia *f* por armónicas; ~ **oscillation** *n* ELECT, TRANS oscilación *f* armónica; ~ **progression** *n* TRANS progresión *f* armónica; ~ **spectrum** *n* TRANS espectro *m* armónico; ~ **structure** *n* TRANS estructura *f* armónica

harmonically: ~ **related carrier** *n* TV portadora *f* en relación armónica

harness *n* ELECT, PARTS arnés *m*, conexionado *m* preformado

Hartley *n* ELECT hartleyo *m*

hash: ~ **coding** *n* COMP, DATA direccionamiento *m* calculado; ~ **function** *n* COMP, DATA función *f* de creación de códigos de comprobación

hashing *n* COMP, DATA método *m* de claves

hatch *vt* COMP *shade* sombrear

hazard *n* GEN riesgo *m*; ~ **analysis** TEST análisis *m* de riesgos

hazardous *adj* GEN peligroso; ~ **state** *n* TEST estado *m* peligroso

HBT *abbr* (*heterojunction bipolar transistor*) ELECT, PARTS HBT (*transistor bipolar de heterounión*)

HDBC *abbr* (*heavy duty building cable*) ELECT, PARTS HDBC (*cable de construcción para grandes amperajes*)

HDL *abbr* (*hardware description language*) COMP HDL (*lenguaje de descripción del soporte físico*)

HDR *abbr* (*header*) COMP, DATA HDR (*cabezal de tubos*)

HDTM *abbr* (*half-duplex transmission module*) TRANS HDTM (*módulo de transmisión en semiduplex*)

head *n* COMP *disk drive* cabeza *f*; ~ **gap** COMP entrehierro *m* de la cabeza; ~ **of queue** SWG cabeza *f* de cola

header *n* (*HDR*) COMP, DATA cabezal *m* de tubos (*HDR*); ~ **card** COMP ficha *f* de cabecera; ~ **error control** (*HEC*) DATA control *m* de error de cabecera; ~ **label** COMP etiqueta *f* de cabecera

heading *n* DATA encabezamiento *m*

headphone: ~ **monitoring** *n* TELEP control *m* de los auriculares; ~ **monitoring level** *n* TELEP nivel *m* de control de los auriculares

headphones *n pl* PER APP, TELEP auriculares *m pl*; ~ **output** *n* TELEP salida *f* de los auriculares; ~ **socket** *n* PER APP toma *f* para auriculares

headset *n* TELEP casco *m* con auriculares; ~ **jack** PARTS, PER APP jack *m* del casco

hearing: ~**-aid compatible** *adj* PER APP *answering machine* compatible con aparato auditivo

heart *n* COMP centro *m*

heat *n* GEN calor *m*; ~ **budget** ELECT presupuesto *m* calórico; ~ **coil** POWER bobina *f* térmica; ~ **concentrator** COMP concentrador *m* de calor; ~ **detector** TV detector *m* de calor; ~ **dissipation** ELECT disipación *f* de calor; ~ **dissipator** ELECT disipador *m* de calor; ~ **radiation** ELECT radiación *f* térmica; ~ **shield** ELECT pantalla *f* térmica; ~**-shrinkable sleeve** PARTS manguito *m* termoencogible; ~**-shrinkable sleeving** PARTS sistema *m* de manguitos termoencogibles; ~ **shrinking** ELECT, PARTS termorretracción *f*; ~ **sink** ELECT disipador *m* térmico; ~ **strip** PARTS banda *f* térmica; ~ **trace cable** PARTS cable *m* termosensor

heater *n* ELECT calentador *m*; ~**-cathode insulation current** ELECT corriente *f* de aislamiento entre filamento calefactor-cátodo; ~ **current** ELECT corriente *f* de filamento; ~ **gun** POWER pistola *f* calefactora; ~ **starting current** ELECT corriente *f* de encendido; ~ **surge current** ELECT sobrecarga *f* de corriente de encendido; ~ **voltage** ELECT tensión *f* de encendido; ~ **warm-up time** ELECT tiempo *m* de calentamiento del calefactor

heating *n* GEN calefacción *f*; ~ **cable** PARTS, POWER cable *m* de calefacción; ~ **fault** POWER fallo *m* de calentamiento

heavy: ~ **current** *n* POWER corriente *f* fuerte; ~**-current engineering** *n* POWER ingeniería *f* de la corriente fuerte; ~**-duty building cable** *n* (*HDBC*) ELECT, PARTS cable *m* de construcción para grandes amperajes (*HDBC*)

HEC *abbr* (*header error control*) DATA control *m* de error de cabecera

height *n* GEN altura *f*; ~ **of arch** TEST altura *f* de arco; ~**-finding radar** RADIO radar *m* altimétrico; ~ **indicator** RADIO altímetro *m* de sonda; ~**-position indicator** RADIO indicador *m* de altitud y posición

helical *adj* GEN helicoidal; ~ **antenna** *n* RADIO antena *f* helicoidal; ~ **potentiometer** *n* POWER potenciómetro *m* helicoidal

helix *n* GEN hélice *f*; ~ **aerial** RADIO antena *f* helicoidal; ~ **antenna** RADIO antena *f* helicoidal; ~ **pitch** RADIO paso *m* de hélice; ~ **spiral wrap** ELECT arrollamiento *m* espiral helicoidal; ~ **voltage** ELECT voltaje *m* de hélice

help: ~ **file** *n* COMP fichero *m* de ayuda

helpware *n* COMP, DATA equipo *m* de ayuda

hemispherical: ~ **coverage** *n* RADIO cobertura *f* hemisférica

henry *n* (*H*) ELECT henrio *m* (*H*)

heptode *n* ELECT heptodo *m*

hermetic *adj* PARTS hermético; ~ **sealing** *n* PARTS sellado *m* hermético

hermetically: ~ **sealed** *adj* PARTS sellado herméticamente

hertz *n* (*Hz*) GEN hertzio *m* (*Hz*)

hertzian: ~ **wave** *n* ELECT, RADIO onda *f* hertziana

heterochronous *adj* TRANS heterócrono

heterodyne *adj* RADIO heterodino; ~ **detection** *n* ELECT detección *f* heterodina; ~ **frequency** *n* RADIO frecuencia *f* heterodina; ~ **frequency meter** *n* TEST frecuencímetro *m* heterodino; ~ **reception** *n* RADIO recepción *f* heterodina

heterogeneity *n* GEN heterogeneidad *f*

heterogeneous *adj* GEN heterogéneo; ~ **multiplex** *n* TELEP múltiplex *m* heterogéneo

heterojunction *n* ELECT, PARTS, TRANS heterounión *f*; ~ **bipolar transistor** (*HBT*) ELECT, PARTS transistor *m* bipolar de heterounión (*HBT*)

heuristic: ~ **decision** *n* DATA decisión *f* heurística; ~ **hazard** *n* DATA riesgo *m* heurístico; ~ **method** *n* DATA método *m* heurístico; ~ **mix** *n* DATA combinación *f* heurística; ~ **programming** *n* DATA programación *f* heurística; ~ **search** *n* DATA investigación *f* heurística

heuristics *n* DATA heurística *f*

hexadecimal: ~ **digit** *n* COMP, DATA dígito *m* hexadecimal; ~ **notation** *n* COMP, DATA notación *f* hexadecimal; ~ **number** *n* COMP, DATA número *m* hexadecimal; ~ **numeral** *n* COMP, DATA número *m* hexadecimal; ~ **system** *n* COMP, DATA sistema *m* hexadecimal

hexode *n* ELECT hexodo *m*

hidden: ~ **line** *n* COMP línea *f* oculta; ~**-line removal** *n* COMP eliminación *f* de línea oculta

hide *vt* COMP ocultar

hierarchical *adj* GEN jerárquico; ~ **data structure** *n* COMP, DATA estructura *f* jerárquica de datos; ~ **link** *n* COMP enlace *m* jerárquico; ~ **multiplexer** *n* TELEP multiplexor *m* jerárquico; ~ **network** *n* TELEP *mutually synchronized* red *f* jerárquica; ~ **progressive mode** *n* COMP, TELEP modo *m* jerárquico progresivo; ~ **routing** *n* TELEP encaminamiento *m* jerárquico; ~ **system** *n* GEN sistema *m* jerárquico

hierarchy *n* GEN jerarquía *f*

high *adj* GEN alto; ~ **BER alarm** *n* DATA alarma *f* máxima BER; ~ **bit-rate ATM network mixer** *n* SWG mezclador *m* de redes ATM de alta velocidad binaria; ~ **bit-rate private switch** *n* SWG conmutador *m* privado de alta velocidad binaria; ~**-capacity communication satellite** *n* RADIO satélite *m* de comunicación de gran capacidad; ~ **data rate** *n* DATA alta *f* velocidad de datos; ~ **definition** *n* TV alta *f* definición; ~**-definition digital video narrow band emission** *n* TV emisión *f* en banda estrecha de alta definición; ~**-definition image** *n* TV imagen *f* de alta definición; ~**-definition television** *n* TV televisión *f* de alta definición; ~**-density binary code** *n* COMP, ELECT código *m* binario de alta densidad; ~**-density bipolar 2 code** *n* DATA, SIG, TELEP código *m* 2 bipolar de alta densidad; ~**-density bipolar 3 code** *n* DATA, SIG, TELEP código *m* 3 bipolar de alta densidad; ~ **electron mobility transistor** *n* ELECT transistor *m* de alta movilidad electrónica; ~**-fidelity** GEN de alta fidelidad; ~**-frequency** POWER de alta frecuencia; **very ~ frequency** *n* (*VHF*) POWER frecuencia muy alta (*VHF*); ~**-impedance** ELECT de elevada impedancia; ~**-performance** GEN de alto rendimiento; ~**-power** POWER de alta potencia; ~**-priority** GEN de alta prioridad; ~**-resistance** ELECT, POWER de alta resistencia; ~**-resolution** (*hi-res*) COMP, TV de alta resolución (*hi-res*); ~**-risk** GEN de alto riesgo; ~**-security** GEN de alta seguridad; ~**-voltage** POWER de alto voltaje

higher: ~ **frequency** *n* RADIO frecuencia *f* más alta; ~**-level function** *n* (*HLF*) TRANS función *f* de nivel superior; ~**-level service** *n* APPL servicio *m* de nivel superior; ~**-order path adaptation** *n* TRANS adaptación *f* de trayecto de orden superior; ~**-order path connection** *n* (*HPC*) TRANS conexión *f* de trayecto de orden superior; ~**-order path termination** *n* (*HPT*) TRANS terminación *f* de trayecto de orden superior; ~ **tariff** *n* APPL tarifa *f* más elevada

highlight *vt* GEN realzar

higlighting *n* COMP realce *m*

hi-res *abbr* (*high-resolution*) COMP, TV hi-res (*de alta resolución*)

hiss *n* TELEP siseo *m*

histogram *n* TEST histograma *m*

hit *n* RADIO acierto *m*

hitch *n* COMP tropiezo *m*

hits: ~ **per scan** *n pl* RADIO aciertos *m pl* por exploración; ~ **per target** *n pl* RADIO aciertos *m pl* por blanco

HLF *abbr* (*higher-level function*) TRANS función *f* de nivel superior

HLR *abbr* (*home location register*) RADIO HLR (*registro de posición inicial*)

hoghorn *n* RADIO antena *f* de embudo

hogtrough: ~ **aerial** *n* RADIO antena *f* de artesa; ~ **antenna** *n* RADIO antena *f* de artesa

hold[1]: ~ **current** *n* ELECT corriente *f* de retención; ~**-down point** *n* RADIO punto *m* de sujeción; ~ **facility** *n* COMP capacidad *f* de retención; ~ **signalling** *BrE* *n* SIG señalización *f* de retención; ~ **time** *n* COMP tiempo *m* de retención; ~**-up alarm installation** *n* APPL, TEST *security* instalación *f* de alarma de parada

hold[2]: **on** ~ *phr* TELEP en espera; ~ **the line** *phr* TELEP retener la línea

hold in *vi* POWER contenerse

holding *n* GEN retención *f*; ~ **call transfer** RADIO transferencia *f* de llamada retenida; ~ **circuit** ELECT circuito *m* de mantenimiento de nivel; ~ **coil** POWER bobina *f* de retención; ~ **contact** POWER contacto *m* de retención; ~ **current** ELECT corriente *f* de retención; ~ **current level** ELECT nivel *m* de la corriente de retención; ~ **range** ELECT campo *m* de retención; ~ **register** ELECT registro *m* de retención; ~ **time** SWG, TELEP tiempo *m* de retención; ~**-time distribution** SWG, TELEP distribución *f* del tiempo de retención; ~**-time of an international circuit** SWG, TELEP tiempo *m* de retención de un circuito internacional; ~ **winding** PARTS, POWER arrollamiento *m* de retención

hole *n* ELECT, PARTS *printed circuits* agujero *m*, perforación *f*; ~ **drawing** ELECT, PARTS *printed circuits* dibujo *m* de perforación; ~ **mounting** ELECT, PARTS *printed circuits* montaje *m* en agujero; ~ **pattern** ELECT, PARTS *printed circuits* configuración *f* de perforaciones; ~**-plating thickness** ELECT, PARTS *printed circuits* espesor *m* del revestimiento electrolítico del agujero; ~ **spacing** ELECT, PARTS *printed circuits* separación *f* de agujeros

holed *adj* ELECT, PARTS *printed circuits* agujereado

Hollerith: ~ **code** *n* COMP código *m* Hollerith

hollow: ~ **conductor** *n* PARTS, TRANS *cables* conductor *m* hueco

holograph *n* DATA, ELECT, SWG holografía *f*

holographic: ~ **exchange** *n* SWG central *f* holográfica; ~ **optical element** *n* DATA elemento *m* óptico holográfico; ~ **storage** *n* DATA memoria *f* holográfica

holography *n* DATA, ELECT, SWG holografía *f*

home: ~ **area** *n* RADIO zona *f* de conexión; ~**-banking terminal** *n* APPL terminal *m* bancaria doméstica; ~ **base** *n* RADIO relé *m* nominal; ~ **base station** *n* RADIO relé *m* nominal; ~ **computing** *n* COMP informática *f* doméstica; ~**-direct service** *n* TELEP servicio *m* directo domiciliario; ~ **exchange** *n* SWG central *f* primaria; ~ **location** *n* RADIO posición *f* inicial; ~ **location register** *n*

(*HLR*) RADIO registro *m* de posición inicial (*HLR*); ~ **position** *n* COMP posición *f* de reposo; ~ **record** *n* COMP registro *m* directo; ~ **shopping** *n* APPL telecompra *f*; ~ **switching centre** *BrE n* SWG centro *m* de conmutación de posición inicial

homing *n* RADIO orientación *f* automática hacia un transmisor; ~ **radar** RADIO radar *m* buscador de blanco

homochronous *adj* TRANS homócrono

homodyne: ~ **detection** *n* TRANS detección *f* homodina; ~ **reception** *n* RADIO recepción *f* homodina

homogeneity *n* GEN homogeneidad *f*

homogeneous *adj* GEN homogéneo; ~**cladding** *n* TRANS chapeado *m* homogéneo; ~ **multiplex** *n* TELEP múltiplex *m* homogéneo; ~ **section** *n* RADIO, TELEP sección *f* homogénea

homogenize *vt* GEN homogeneizar

homojunction *n* ELECT, PARTS, TRANS homounión *f*

hood *n* ELECT, PARTS campana *f*

hook *n* TELEP gancho *m*; ~ **bolt** PARTS perno *m* de gancho; ~ **flash** TELEP aviso *m* de conexión; ~ **switch** PER APP conmutador *m* de horquilla; ~-**up** COMP conexión *f*; ~-**up wire** PARTS alambre *m* para conexiones; ◆ **off the** ~ RADIO, TELEP descolgado; **on the** ~ TELEP colgado

hook up *vt* COMP, TELEP interconectar

hop *n* RADIO salto *m*

hopping: ~ **switch** *n* RADIO conmutador *m* de saltos

horizontal: ~ **aerial** *n* RADIO antena *f* horizontal; ~ **antenna** *n* RADIO antena *f* horizontal; ~ **blanking interval** *n* TV intervalo *m* de supresión de la imagen; ~ **directivity pattern** *n* RADIO configuración *f* de directividad horizontal; ~ **line frequency** *n* TV frecuencia *f* de línea horizontal; ~ **pivot** *n* PARTS pivote *m* horizontal; ~ **polarization** *n* TV polarización *f* horizontal; ~ **sweep** *n* TV barrido *m* horizontal; ~ **sweep frequency** *n* TV frecuencia *f* de barrido horizontal; ~ **tabulation character** *n* (*HT*) DATA carácter *m* de tabulación horizontal

horn *n* RADIO bocina *f*; ~ **aerial** RADIO antena *f* de bocina; ~ **antenna** RADIO antena *f* de bocina

host *n* COMP anfitrión *m*; ~ **computer** COMP ordenador *m* central *Esp*, computador *m* central *AmL*, computadora *f* central *AmL*; ~ **digital terminal** COMP terminal *m* digital principal; ~ **exchange** RADIO, SWG central *f* primaria; ~ **office** RADIO oficina *f* principal; ~ **panel** ELECT panel *m* principal; ~ **satellite** RADIO satélite *m* principal; ~ **station** RADIO estación *f* principal; ~ **system** DATA sistema *m* principal

hot: ~ **billing** *n* RADIO facturación *f* inmediata; ~ **carrier diode** *n* ELECT diodo *m* portador con gran actividad; ~ **cathode** *n* ELECT cátodo *m* caliente; ~-**cathode stepping tube** *n* ELECT tubo *m* gradual de cátodo caliente; ~ **conditions** *n pl* ELECT situación *f* de emergencia; ~ **line** *n* TELEP línea *f* de prioridad; ~-**line mode of operation** *n* TELEP modo *m* de operación sin marcación; ~ **satellite** *n* RADIO satélite *m* de funcionamiento inmediato; ~ **spot** *n* COMP punto *m* caliente; ~**stand-by** *n* TEST reserva *f* para funcionamiento inmediato; ~ **stand-by system** *n* TEST sistema *m* auxiliar para urgencias; ~ **swapping** *n* ELECT intercambio *m* de memoria activa; ~-**wire ammeter** *n* TEST amperímetro *m* térmico; ~ **zone** *n* COMP zona *f* caliente

hotel: ~ **service extension** *n* SWG, TELEP *PBX* extensión *f* de servicio hotelero

hour: ~ **meter** *n* APPL, TEST contador *m* de horas de funcionamiento

household: ~ **electricity** *n* POWER electricidad *f* para consumo doméstico

housekeeping: ~ **bit** *n* DATA bitio *m* de servicio; ~ **function** *n* DATA función *f* de servicio; ~ **information** *n* DATA información *f* preparatoria; ~ **lifetime** *n* DATA vida *f* de servicio; ~ **operation** *n* COMP operación *f* preparatoria; ~ **routine** *n* COMP rutina *f* de iniciación; ~ **subsystem** *n* COMP subsistema *m* de servicio; ~ **telemetry** *n* DATA telemetría *f* interna; ~ **unit** *n* DATA unidad *f* de servicio

housing *n* PARTS envoltura *f*; ~ **lid** PARTS tapa *f* de recipiente

how: ~ **do you read me?** *phr* RADIO ¿cómo me recibe?

howl *vi* TELEP, TRANS aullar

howling *n* TELEP, TRANS aullido *m*

HPC *abbr* (*higher-order path connection*) TRANS conexión *f* de trayecto de orden superior

HPT *abbr* (*higher-order path termination*) TRANS terminación *f* de trayecto de orden superior

HRC *abbr* (*hypothetical reference circuit*) RADIO, TELEP, TRANS HRC (*circuito de referencia hipotético*)

HRDP *abbr* (*hypothetical reference digital path*) RADIO, TELEP, TRANS HRDP (*camino digital de referencia hipotética*)

HT *abbr* (*horizontal tabulation character*) DATA carácter *m* de tabulación horizontal

HTML *abbr* (*hypertext markup language*) COMP,

DATA *Internet* HTML (*lenguaje de referencia de hipertexto*)
HTTP *abbr* (*hypertext transfer protocol*) COMP, DATA *Internet* HTTP (*protocolo de transferencia de hipertexto*)
hub *n* COMP *of disk* anillo *m* central, GEN centro *m*, cubo *m*; ~ **station** RADIO estación *f* de cubo
hue *n* TV matiz *m*
hum *n* ELECT zumbido *m*; ~ **level** ELECT nivel *m* de zumbido
human: ~ **error** *n* TEST error *m* humano
hundreds: ~ **digit** *n* COMP dígito *m* de centena
HV *abbr* (*high voltage*) ELECT, POWER HV (*alta tensión*); ~ **delay time** *n* RADIO tiempo *m* de retardo de HV
hybrid *n* GEN híbrido *m*; ~ **circuit** ELECT circuito *m* híbrido; ~ **circulator** RADIO circulador *m* híbrido; ~ **coil** POWER bobina *f* híbrida; ~ **computer** COMP computador *m* híbrido *AmL*, computadora *m* híbrida *AmL*, ordenador *m* híbrido *Esp*; ~ **connector** TRANS *fibre optics* conector *m* híbrido; ~ **integrated circuit** ELECT circuito *m* integrado híbrido; ~ **junction** RADIO conexión *f* híbrida; ~ **mode** ELECT modo *m* híbrido; ~ **processor** SWG procesador *m* híbrido; ~ **reasoning** ELECT razonamiento *m* híbrido; ~ **ring** RADIO anillo *m* híbrido; ~ **spread spectrum** RADIO espectro *m* de difusión híbrido; ~ **switch** SWG conmutador *m* híbrido; ~ **switching system** SWG, TRANS sistema *m* de conmutación híbrido; ~ **system** SWG, TRANS sistema *m* híbrido; ~ **T** RADIO t *f* híbrida; ~ **transformer** ELECT transformador *m* diferencial
hydrometeor *n* RADIO hidrometeoro *m*
hydrophone *n* PER APP hidrófono *m*
hyperbolic: ~ **navigation system** *n* RADIO sistema *m* de navegación hiperbólica
hyperbook *n* COMP hiperlibro *m*
hypercourse *n* COMP hipercurso *m*
hyperdata *n pl* DATA hiperdatos *m pl*
hyperdocument *n* COMP hiperdocumento *m*
hypergroup *n* TRANS hipergrupo *m*
hypermedia *adj* COMP, DATA *Internet* hipermedia
hyperobject *n* COMP hiperobjeto *m*
hypertext *n* COMP, DATA hipertexto *m*; ~ **application** COMP aplicación *f* de hipertexto; ~ **conversion** COMP conversión *f* de hipertexto; ~ **document** COMP documento *m* de hipertexto; ~ **link** COMP enlace *m* con hipertexto; ~ **markup language** (*HTML*) COMP, DATA *Internet* lenguaje *m* de referencia de hipertexto (*HTML*); ~ **navigation** COMP navegación *f* por hipertexto; ~ **navigation path** COMP camino *m* de navegación por hipertexto; ~ **node** COMP nodo *m* de hipertexto; ~ **portability** COMP portabilidad *f* de hipertexto; ~ **transfer protocol** (*HTTP*) COMP, DATA *Internet* protocolo *m* de transferencia de hipertexto (*HTTP*)
hypertrail *n* COMP hiperpista *f*
hypervisor *n* TELEP hipervisor *m*
hyphen *n* COMP *word processing* guión *m*
hyphenate *vt* COMP *word processing* escribir con guión
hypothesis *n* GEN hipótesis *f*
hypothetical: ~ **reasoning** *n* GEN razonamiento *m* hipotético; ~ **reference circuit** *n* (*HRC*) RADIO, TELEP, TRANS circuito *m* de referencia hipotético (*HRC*); ~ **reference circuit for sound-programme** *BrE* **transmissions** *n* RADIO circuito *m* de referencia hipotética para transmisión de programas de sonido; ~ **reference circuit for telephony** *n* TELEP circuito *m* de referencia hipotética para telefonía; ~ **reference connection** *n* TELEP conexión *f* de referencia hipotética; ~ **reference digital path** *n* (*HRDP*) RADIO, TELEP, TRANS camino *m* digital de referencia hipotética (*HRDP*)
hysteresis *n* ELECT histéresis *f*; ~ **curve** ELECT curva *f* de histéresis; ~ **cycle** ELECT ciclo *m* de histéresis; ~ **loss** ELECT pérdida *f* por histéresis
Hz *abbr* (*hertz*) GEN Hz (*hertzio*)

I

I *abbr* (*implemented*) GEN I (*instalado*)

IA5 *abbr* (*international alphabet no. 5*) DATA, TELEP IA5 (*alfabeto internacional nº 5*)

IAT *abbr* (*international atomic time*) RADIO IAT (*tiempo atómico internacional*)

I/C *abbr* (*incoming*) GEN I/C (*entrante*)

ICC *abbr* (*incrementally coherent carrier*) TV ICC (*portadora incrementalmente coherente*)

ICCM *abbr* (*interworking by call-control mapping*) TELEP ICCM (*interfuncionamiento por diagrama de control de llamadas*)

ICD *abbr* (*international code designator*) DATA ICD (*designador de código internacional*)

ICE *abbr* (*interface configuration environment*) TELEP ICE (*entorno de configuración del interfaz*)

icon *n* COMP icono *m*; ~ **interface** COMP interfaz *f* icónica

iconoscope *n* ELECT iconoscopio *m*

ICRD *abbr* (*internetwork call redirection and deflection*) TRANS ICRD (*redirección y deflexión de llamadas de red interna*)

ICS *abbr* (*implementation conformance statement*) DATA ICS (*declaración de conformidad de la instalación*)

ID *abbr* (*identification*) GEN identificación *f*

IDA *abbr* (*integrated digital access*) TRANS IDA (*acceso digital integrado*)

IDD *BrE abbr* (*international direct dialling*) TELEP marcado *m* directo internacional

IDDD *BrE abbr* (*international direct distance dialling*) TELEP marcado *m* directo internacional a distancia

ideal: ~ **instants** *n pl* TELEP instantes *m pl* ideales; ~ **noise diode** *n* ELECT diodo *m* de ruido ideal

identification *n* (*ID*) GEN identificación *f*; ~ **code** DATA, TELEP código *m* de identificación; **~-code qualifier** DATA cualificador *m* de código de identificación; ~ **friend or foe** (*IFF*) RADIO identificación *f* de naves propias (*IFF*); ~ **line** DATA línea *f* de identificación

identifier *n* GEN identificador *m*; ~ **circuit** ELECT circuito *m* identificador

identity: **~-based security policy** *n* APPL norma *f* de seguridad basada en la identidad; ~ **card** *n* APPL *security* documento *m* de identidad; ~ **element** *n* ELECT *logic elements* elemento *m* de identidad, TELEP elemento *m* identificador; ~ **gate** *n* ELECT *logic elements* elemento *m* de identidad; ~ **operation** *n* COMP operación *f* de identidad

IDI *abbr* (*initial domain identifier*) DATA identificador *m* de dominio inicial

idle *adj* GEN libre; ~ **channel** *n* RADIO canal *m* en reposo; **~-channel noise** *n* TRANS ruido *m* de canal inactivo; ~ **circuit** *n* TRANS circuito *m* en reposo; ~ **condition** *n* ELECT, POWER, TELEP, TRANS estado *m* de reposo; ~ **line** *n* TELEP línea *f* libre; ~ **period** *n* ELECT período *m* inactivo; ~ **position** *n* ELECT posición *f* de reposo; ~ **state** *n* SWG, TEST estado *m* libre; ~ **state test** *n* TEST prueba *f* de estado no requerido; ~ **testing state** *n* DATA estado *m* de comprobación no requerida; ~ **time** *n* TEST tiempo *m* de inactividad; ~ **voice channel** *n* RADIO canal *m* de voz inactivo; ~ **working channel** *n* RADIO canal *m* operativo libre

idleness *n* GEN reposo *m*

idler: ~ **frequency** *n* ELECT frecuencia *f* lenta

IDN *abbr* (*indirect directory number*) TELEP número *m* de directorio indirecto

IDP *abbr* (*initial domain part*) DATA parte *f* del dominio inicial

IDSE *abbr* (*international data switching exchange*) DATA, SWG central *f* de conmutación para mensajes internacionales

IDT *abbr* (*interdigital transducer*) TRANS transductor *m* interdigital

IEE *BrE abbr* (*Institution of Electrical Engineers*) ELECT Instituto de Ingenieros Eléctricos

IEEE *AmE abbr* (*Institute of Electrical and Electronics Engineers*) ELECT Instituto de Ingenieros Eléctricos y Electrónicos

IF *abbr* (*intermediate frequency*) RADIO, TRANS FI (*frecuencia intermedia*); ~ **amplification** *n* RADIO, TRANS amplificación *f* de FI; ~ **amplifier** *n* RADIO, TRANS amplificador *m* de FI; ~ **filter** *n* RADIO, TRANS filtro *m* de FI; ~ **rejection** *n* RADIO, TRANS rechazo *m* de FI; ~ **signal** *n* RADIO, TRANS señal *f* de FI; ~ **stage** *n* RADIO, TRANS momento *m* de FI

I/F *abbr* (*interface*) GEN I/F (*interfaz*)

IF-AND-ONLY-IF: ~ **element** *n* ELECT *logic elements* elemento *m* de equivalencia; ~ **gate**

n ELECT puerta *f* de equivalencia; ~ **operation** *n* ELECT *logic elements* equivalencia *f* lógica

IFF *abbr* (*identification friend or foe*) RADIO IFF (*identificación de naves propias*)

i-field: ~ **picture** *n* COMP, TV imagen *f* de campo-i

IFLU *abbr* (*initial full line-up*) TEST IFLU (*ajuste inicial pleno*)

I-frame: ~ **counter** *n* TELEP contador *m* de cuadro I

IF-THEN: ~ **element** *n* ELECT *logic elements* circuito *m* SI-ENTONCES

IFU *abbr* (*interworking functional unit*) DATA unidad *f* funcional entrelazada

IG *abbr* (*interpolation gain*) TRANS ganancia *f* de interpolación

IGES *abbr* (*initial graphics exchange specification*) COMP IGES (*especificación inicial de intercambio de gráficos*)

IGN *abbr* (*international gateway node*) SWG, TELEP nodo *m* de acceso internacional

ignition *n* ELECT ignición *f*; ~ **time** ELECT tiempo *m* de ignición; ~ **voltage** ELECT tensión *f* de encendido

ignitron *n* ELECT ignitrón *m*

IIR *abbr* (*infinite impulse response*) ELECT, TRANS respuesta *f* de impulso infinito

ILD *abbr* (*injection laser diode*) ELECT *semiconductors* diodo *m* láser de inyección

illuminance: ~ **meter** *n* TEST medidor *m* de iluminancia; ~ **range** *n* POWER campo *m* de iluminancia; ~ **uniformity ratio** *n* POWER ratio *f* de uniformidad de la iluminancia

illuminate *vt* COMP, ELECT, POWER iluminar

illuminated *adj* COMP, ELECT, POWER iluminado; ~ **display** *n* RADIO pantalla *f* iluminada; ~ **keyboard** *n* RADIO teclado *m* iluminado

illuminating: ~ **device** *n* RADIO dispositivo *m* de iluminación; ~ **radar** *n* RADIO radar *m* de iluminación

illumination: ~ **time** *n* RADIO tiempo *m* de iluminación

ILS *abbr* (*instrument landing system*) RADIO ILS (*sistema de aterrizaje por instrumentos*)

IM *abbr* (*interface module*) GEN módulo *m* de interfaz

image *n* GEN imagen *f*; ~ **analyser** *BrE* TV analizador *m* de imagen; ~ **analysis** TV análisis *m* de imagen; ~ **and sound synchronization** TV sincronización *f* de imagen y sonido; ~ **attenuation coefficient** TRANS coeficiente *m* de atenuación de imagen; ~ **attenuation constant** TRANS constante *f* de atenuación de imagen; ~ **bank** TV banco *m* de imágenes; ~ **camera tube** ELECT tubo *m* de cámara de imagen; ~ **coding** DATA codificación *f* de imagen; ~ **compression** DATA, TV compresión *f* de imagen; ~-**compression board** DATA, TV tablero *m* de compresión de imagen; ~-**compression manager** DATA, TV gestor *m* de compresión de imagen; ~ **conversion** TV conversión *f* de imagen; ~ **converter tube** ELECT, TV tubo *m* convertidor de imagen; ~ **data** *n pl* DATA, TV datos *m pl* gráficos; ~ **enhancement** TV refuerzo *m* de la imagen; ~ **file** DATA fichero *m* de imagen; ~ **focus voltage** ELECT voltaje *m* de foco de imagen; ~ **frequency** RADIO frecuencia *f* de imagen; ~ **frequency rejection** RADIO rechazo *m* de frecuencia de imagen; ~ **graphics** TV gráfico *m* de imágenes; ~ **iconoscope** ELECT iconoscopio *m* de imagen; ~ **impedance** ELECT impedancia *f* de imagen; ~ **intensifier tube** ELECT tubo *m* intensificador de imagen; ~ **orthicon** ELECT orticonoscopio *m* de imagen; ~ **parameter filter** ELECT filtro *m* de bandas sin atenuación de imagen; ~ **phase-change coefficient** TRANS coeficiente de desfase *m* de imágenes; ~ **processing** DATA, TV tratamiento *m* de imagen, procesamiento *m* de imagen; ~-**processing equipment** DATA, TV equipo *m* de procesamiento de la imagen; ~ **quality** TV calidad *f* de la imagen; ~ **reconstruction** TV reconstrucción *f* de imagen; ~ **refreshing** TV regeneración *f* de imagen; ~ **rejection ratio** RADIO ratio *f* de supresión de frecuencia de imagen; ~ **retention fault** TV fallo *m* de retención de imagen; ~ **scanner** PER APP, TV digitalizador *m* de imagen; ~ **scanning time** PER APP, TV tiempo *m* de digitalización de imagen; ~ **sensor** PER APP, TV sensor *m* de imagen; ~ **sequence** TV secuencia *f* de imágenes; ~ **space** COMP espacio *m* de imagen; ~ **storage and retrieval** DATA almacenamiento *m* y recuperación de imagen; ~-**storage space** COMP espacio *m* de memoria de imagen; ~ **transfer** DATA, TRANS transferencia *f* de imagen; ~ **transfer coefficient** DATA, TRANS coeficiente *m* de transferencia de imagen; ~ **transfer constant** DATA, TRANS constante *f* de transferencia de imagen; ~ **transmission** DATA, TRANS transmisión *f* de imagen

imbalance *n* COMP, POWER desequilibrio *m*

imitate *vt* SIG imitar

imitative: ~ **deception** *n* ELECT falsa *f* imitación

immediate: ~ **addressing** *n* COMP direccionamiento *m* inmediato; ~ **dialling** *BrE* *n* TELEP marcado *m* inmediato

immersed: ~ **gun** *n* ELECT cañón *m* electrónico sumergido

immunity: ~ **to interference** *n* RADIO *mobile* inmunidad *f* a las interferencias; ~ **to noise** *n* TRANS inmunidad *f* al ruido
immutable *adj* GEN inmutable
IMP *abbr* (*interface message processor*) COMP procesador *m* de mensajes de interfaz
impact: ~ **printer** *n* COMP, PER APP impresora *f* de impacto; ~ **switch** *n* POWER conmutador *m* por inercia; ~ **test** *n* PARTS, TEST prueba *f* de impacto
impairment *n* GEN desperfecto *m*
impedance *n* COMP, ELECT, POWER, RADIO impedancia *f*; ~ **balance** ELECT equilibrio *m* de impedancia; ~ **balance ratio** ELECT ratio *f* de equilibrio de impedancia; ~ **bond** POWER conexión *f* inductiva; ~ **imbalance** ELECT desequilibrio *m* de impedancia; ~ **matching** ELECT adaptación *f* de impedancias
impermeability *n* GEN impermeabilidad *f*
impermeable *adj* GEN impermeable
impervious *adj* GEN impermeable
imperviousness *n* GEN hermeticidad *f*
impetus *n* ELECT impulso *m*
implantation *n* ELECT acoplamiento *m*
implanted *adj* ELECT acoplado
implanter *n* ELECT implantador *m*
implement *vt* GEN implementar
implementation *n* GEN implementación *f*; ~ **conformance statement** (*ICS*) DATA declaración *f* de conformidad de la instalación (*ICS*); ~ **under test** (*IUT*) TEST instalación *f* en prueba (*IUT*)
implemented *adj* (*I*) GEN instalado (*I*)
implementor *n* DATA instalador *m*
implicit *adj* DATA implícito
implied: ~ **addressing** *n* COMP direccionamiento *m* implícito
imply *vt* COMP implicar
import[1] *n* DATA importación *f*
import[2] *vt* DATA importar
imported *adj* DATA importado
impregnate *vt* PARTS impregnar
impregnating: ~ **agent** *n* PARTS impregnador *m*
impregnation *n* PARTS impregnación *f*
impulse *n* GEN impulso *m*; ~ **discharge current** ELECT *of protector* corriente *f* de descarga de impulsos; ~ **excitation** ELECT excitación *f* por impulsos; ~ **noise** TELEP ruido *m* impulsivo; ~ **noise count** TELEP recuento *m* de ruido impulsivo; ~ **rate** APPL velocidad *f* de impulso; ~ **response** TELEP respuesta *f* de impulso; ~ **spark-over voltage** ELECT *of protector* tensión *f* de ruptura por impulsos; ~ **spark-over voltage-time curve** ELECT *of protector* curva *f* de duración de tensión de ruptura de impulsos
impulsing *n* TELEP impulsión *f*
impulsive: ~ **noise tolerance** *n* RADIO tolerancia *f* impulsiva al ruido
impulsiveness: ~ **ratio** *n* RADIO *air safety* relación *f* de impulsividad
impurity *n* ELECT, PARTS impureza *f*
IN *abbr* (*intelligent network*) APPL, TRANS IN (*red inteligente*)
inaccuracy *n* TEST imprecisión *f*, inexactitud *f*
inactive *adj* GEN inactivo; ~ **character** *n* TELEP carácter *m* desactivado; ~ **node** *n* TRANS nodo *m* inactivo; ~ **signalling link** *BrE n* SIG enlace *m* de señalización inactivo
inaugurate *vt* RADIO inaugurar
inauguration *n* RADIO inauguración *f*
in-band *adj* SIG, TRANS en banda; ~ **carrier** *n* TRANS portadora *f* en banda; ~ **data** *n pl* DATA datos *m pl* en banda; ~ **signalling** *BrE n* SIG señalización en banda
in-between *n* TV inserto *m*
inboard: ~ **grommet** *n* ELECT, PARTS arandela *f* aislante interior
inbound: ~ **communication** *n* RADIO, TRANS comunicación *f* de entrada
in-building: ~ **coverage** *n* RADIO cobertura *f* dentro de un edificio; ~ **reception** *n* RADIO recepción *f* dentro de un edificio
in-call: ~ **rearrangement** *n* TELEP redistribución *f* de llamadas internas
inch *n* GEN *measurement* pulgada *f*
incidence *n* TEST incidencia *f*
incision *n* GEN incisión *f*
inclination *n* RADIO inclinación *f*
inclined: ~ **orbit** *n* RADIO órbita *f* inclinada
inclusion *n* ELECT *printed circuits* inclusión *f*
inclusive: ~ **price** *n* GEN precio *m* incluidos todos los gastos
INCLUSIVE-OR *n* ELECT disyuntivo *m* O-INCLUSIVO; ~ **circuit** ELECT circuito *m* O-INCLUSIVO; ~ **element** ELECT *logic elements* elemento *m* O-INCLUSIVO; ~ **gate** ELECT *logic elements* puerta *f* O-INCLUSIVO
incoherence *n* TRANS incoherencia *f*
incoherent *adj* TRANS incoherente; ~ **light** *n* TRANS luz *f* incoherente; ~ **radiation** *n* TRANS radiación *f* incoherente
incoming *adj* (*I/C*) GEN entrante (*I/C*); ~ **access** *n* DATA acceso *m* de entrada; ~ **call** *n* DATA, TELEP llamada *f* entrante; ~-**call barring** *n* TELEP bloqueo *m* de comunicaciones de llegada; ~ **call identification** *n* DATA, RADIO, TELEP identificación *f* de llamada entrante; ~ **call log** *n* TELEP registro *m* de llamadas

entrantes; ~ **modification** *n* TELEP modificación *f* de llamada entrante; ~ **call packet** *n* DATA, TELEP paquete *m* de llamadas entrantes; ~ **calls** *n pl* TELEP llamadas *f pl* entrantes; ~ **calls barred** *n* DATA, TELEP llamadas *f pl* entrantes prohibidas; ~**-calls-barred line** *n* TELEP, TRANS línea *f* con bloqueo de llamadas entrantes; ~ **circuit** *n* TELEP circuito *m* entrante; ~ **group** *n* TRANS grupo *m* entrante; ~ **inspection** *n* TEST control *m* de recepción; ~ **international R2 register** *n* TELEP registro *m* R2 de llamadas internacionales entrantes; ~ **line** *n* TRANS línea *f* entrante; ~ **matrix link** *n* TELEP enlace *m* de matriz entrante; ~**-only line** *n* TELEP, TRANS línea *f* exclusiva para llamadas entrantes; ~ **procedure** *n* RADIO procedimiento *m* entrante; ~ **R2 register** *n* TELEP registro *m* R2 de llamadas entrantes; ~ **register** *n* TELEP registro *m* entrante; ~ **response delay** *n* TELEP retardo *m* de respuesta entrante; ~ **signal** *n* SIG, TRANS señal *f* de llegada; ~ **surge** *n* ELECT sacudida *f* eléctrica de llegada; ~ **traffic** *n* TELEP tráfico *m* entrante; ~ **trunk** *n* TELEP enlace *m* de llegada; ~ **trunk circuit** *n* TELEP, TRANS circuito *m* de enlace entrante

incompatible *adj* DATA incompatible

inconclusive *adj* DATA no decisivo; ~ **verdict** *n* DATA veredicto *m* no concluyente

in-connector *n* RADIO conector *m* de entrada

inconsistency *n* COMP inconsistencia *f*

inconsistent *adj* COMP inconsistente

incorrect *adj* TELEP *modulation, restitution, signal* incorrecto

increment *n* GEN incremento *m*

incremental: ~ **charge advice** *n* TELEP aviso *m* de carga incremental; ~ **power gain** *n* ELECT ganancia *f* de potencia incremental; ~ **recording** *n* TV registro *m* incremental

incrementally: ~ **coherent carrier** *n* (*ICC*) TV portadora *f* incrementalmente coherente (*ICC*); ~ **related carrier** *n* TV portadora *f* incrementalmente relacionada

indent *n* COMP *word processing* sangría *f*

indentation *n* ELECT, PARTS *printed circuits* huella *f*, muesca *f*

independent: ~ **alarm** *n* PER APP *on answering machine* alarma *f* independiente; ~ **earth electrode** *BrE n* (*cf ground electrode AmE*) ELECT, POWER toma *f* de tierra independiente; ~ **ground electrode** *AmE n* (*cf earth electrode BrE*) ELECT, POWER toma *f* de tierra independiente; ~ **network** *n* TRANS red *f* independiente; ~ **programme** *BrE n* TV programa *m* independiente; ~ **sideband** *n* (*ISB*) RADIO banda *f* lateral independiente (*ISB*); ~ **sideband transmission** *n* RADIO, TRANS transmisión *f* con bandas laterales independientes

index *n* GEN índice *m*; ~ **dip** TRANS gozo *m* de índice; ~**-matching material** TRANS material *m* adaptador de índices; ~ **of cooperation** RADIO, TELEP módulo *m* de cooperación; ~ **of refraction** TRANS índice *m* de refracción; ~ **profile** ELECT, TRANS perfil *m* de índice; ~ **register** COMP registro *m* de índice; ~ **word** COMP, DATA palabra *f* índice

indexed: ~ **address** *n* COMP dirección *f* indexada

indexing *n* GEN indexación *f*

indicate *vt* GEN indicar

indicating: ~ **element** *n* TELEP elemento *m* de marcación; ~ **lamp** *n* TELEP lámpara *f* indicadora; ~**-measuring instrument** *n* TEST instrumento *m* de medición indicador; ~ **unit** *n* TELEP unidad *f* indicadora

indication *n* GEN indicación *f*; ~ **of fracture** TEST indicación *f* de rotura; ~ **primitive** DATA función *f* primitiva de indicación

indicator *n* GEN indicador *m*; ~ **lamp** TEST lámpara *f* indicadora

indifference: ~ **point** *n* TEST punto *m* de indiferencia

indirect: ~ **address** *n* COMP dirección *f* indirecta; ~ **addressing** *n* COMP direccionamiento *m* indirecto; ~ **backscatter** *n* POWER, RADIO retrodifusión *f* indirecta; ~ **control system** *n* CONTR sistema *m* de control indirecto; ~ **directory number** *n* (*IDN*) TELEP número *m* de directorio indirecto; ~ **distribution** *n* RADIO distribución *f* indirecta; ~ **memory** *n* PER APP *on answering machine* memoria *f* indirecta; ~ **method of measurement** *n* TEST método *m* indirecto de medición; ~ **strike** *n* POWER descarga *f* atmosférica indirecta; ~ **submission port** *n* DATA puerto *m* de presentación indirecta

indirectly: ~**-heated cathode** *n* ELECT cátodo *m* equipotencial

individual: ~ **aerial** *n* TV antena *f* individual; ~ **antenna** *n* TV antena *f* individual; ~ **charging** *n* APPL facturación *f* individual; ~ **current feed** *n* POWER alimentación *f* individual; ~ **distortion** *n* TELEP distorsión *f* individual; ~ **fault** *n* TEST fallo *m* individual; ~ **reception** *n* RADIO recepción *f* individual; ~ **store** *n* DATA almacén *m* individual

indoor: ~ **radio** *n* RADIO radio *f* de interior

induce *vt* COMP inducir

induced: ~ **control-voltage** *n* ELECT *semiconductors* tensión *f* de control inducida;

~ **current** *n* POWER corriente *f* inducida; ~ **voltage** *n* POWER voltaje *m* inducido

inductance *n* ELECT, POWER inductancia *f*; ~ **capacitance** ELECT capacitancia *f* inductiva; ~ **capacity** ELECT capacidad *f* de inductancia; ~ **coil** POWER bobina *f* de inductancia; ~ **standard** ELECT patrón *m* de inductancia

induction *n* ELECT, POWER inducción *f*; ~ **coil** ELECT bobina *f* de inducción; ~ **field** ELECT campo *m* inductor; ~ **meter** POWER contador *m* de inducción; ~ **relay** POWER relé *m* de inducción

inductive *adj* ELECT, POWER inductivo; ~ **circuit** *n* ELECT circuito *m* inductivo; ~ **load** *n* ELECT carga *f* inductiva; ~ **parallel tuning** *n* ELECT sintonización *f* paralela inductiva; ~ **reasoning** *n* DATA razonamiento *m* inductivo; ~ **residual voltage** *n* ELECT voltaje *m* residual inductivo

inductor *n* ELECT, PARTS, POWER inductor *m*

industrial: ~ **design** *n* APPL diseño *m* industrial; ~ **designer** *n* APPL diseñador *m* industrial; ~ **environment** *n* GEN entorno *m* industrial; ~ **espionage** *n* GEN *security* espionaje *m* industrial; ~ **safety** *n* CONTR seguridad *f* industrial; ~ **television** *n* TV televisión *f* industrial

industry: ~ **standard architecture** *n* (*ISA*) COMP, DATA arquitectura *f* industrial estándar (*ISA*)

ineffective: ~ **airtime** *n* RADIO tiempo *m* inefectivo en el aire; ~ **call** *n* TELEP llamada *f* inefectiva

inequality *n* GEN desigualdad *f*

inertia *n* POWER, TEST inercia *f*; ~ **damper** POWER amortiguador *m* de inercia; ~ **error** TEST error *m* debido a la inercia

inference: ~ **engine** *n* APPL *artificial intelligence* dispositivo *m* de inferencia

infinite: ~ **impulse response** *n* (*IIR*) ELECT, TRANS respuesta *f* de impulso infinito; ~ **impulse response filter** *n* ELECT, TRANS filtro *m* de respuesta de impulso infinito

infinitely: ~ **variable** *adj* CONTR infinitamente variable

in-flight *adj* RADIO a bordo; ~ **telephone service** *n* RADIO servicio *m* telefónico a bordo

influence *n* GEN influencia *f*; ~ **quantity** TEST *metrology* magnitud *f* influyente

influencing: ~ **factor** *n* TEST factor *m* influyente

infometrics *n* DATA infométrica *f*

information *n* APPL, GEN, TELEP información; ~ **base** DATA base *f* de información; ~ **base type** DATA tipo *m* de base de información; ~ **bearer channel** TELEP canal *m* portador de información; ~ **bit** DATA, TELEP bit *m* de información; ~ **channel** TELEP, TV canal *m* de información; ~ **content** DATA contenido *m* informático; ~ **element** TELEP elemento *m* de información; ~ **exchange** DATA intercambio *m* de información; ~ **exchange protocol** COMP protocolo *m* de intercambio de información; ~ **facility** TELEP servicio *m* de información; ~ **feedback system** DATA, TELEP sistema *m* de realimentación de la información; ~ **frame** TELEP cuadro *m* de información; ~ **highway** TRANS infovía; ~ **inquiry facility** TELEP servicio *m* de consulta de información; ~ **interchange** TELEP intercambio *m* de información; ~ **measure** GEN medida *f* de información; ~ **message** GEN mensaje *m* de información; ~ **network** DATA red *f* de información; ~ **processing** TELEP tratamiento *m* de la información; ~ **processing system** COMP, DATA, TELEP sistema *m* de proceso de la información; ~ **provider** DATA proveedor *m* de información; ~ **pulse** RADIO impulso *m* de información; ~ **rate per character** DATA velocidad *f* de información por carácter; ~ **receiver station** DATA estación *f* receptora de información; ~ **request message** DATA, TELEP mensaje *m* de petición de información; ~ **retrieval** COMP, DATA recuperación *f* de información; ~ **retrieval centre** *BrE* DATA centro de recuperación de la información; ~ **retrieval system** DATA sistema *m* de recuperación de información; ~ **security** TELEP seguridad *f* de la información; ~**-sending station** DATA estación *f* de envío de información; ~ **separator** DATA, ELECT separador *m* de información; ~**-separator character** DATA carácter *m* separador de la información; ~ **server** DATA servidor *m* de información; ~ **storage node** TELEP nodo *m* de almacenamiento de información; ~ **superhighway** TRANS autopista *f* de la información; ~ **system** DATA sistema *m* de información; ~ **technology** (*IT*) COMP, DATA tecnología *f* de la información; ~ **theory** GEN teoría *f* de la información; ~ **transfer** TELEP transferencia *f* de informaciones; ~ **transfer rate** DATA velocidad *f* de transferencia de la información; ~ **type** COMP, DATA tipo *m* de información; ~ **unit** TELEP unidad *f* de información

in-frame: ~ **coding** *n* DATA codificación *f* dentro de la trama

infrared *adj* (*IR*) GEN infrarrojo (*IR*); ~ **detector** *n* TRANS detector *m* de infrarrojos; ~ **light link**

n TRANS enlace *m* de luz infrarroja; ~ **radiation** *n* ELECT radiación *f* infrarroja
infrasound *n* COMP infrasonido *m*
infrastructure *n* GEN infraestructura *f*
infringement *n* GEN contravención *f*
inhibit *vt* ELECT inhibir
inhibiting: ~ **pulse** *n* ELECT impulso *m* inhibidor; ~ **signal** *n* ELECT señal *f* inhibidora
inhibition *n* ELECT inhibición *f*
initial: ~ **acceptance** *n* TEST aceptación *f* inicial; ~ **address message** *n* SIG, SWG, TELEP mensaje *m* inicial de dirección; ~ **alignment** *n* TELEP alineación *f* inicial; ~ **attack slope** *n* COMP pendiente *f* de ataque inicial; ~ **charge** *n* TELEP carga *f* inicial; ~ **condition** *n* ELECT condición *f* inicial; ~ **condition mode** *n* COMP modo *m* de estado inicial; ~ **connection charge** *n* APPL cargo *m* de conexión inicial; ~ **deployment** *n* GEN despliegue *m* inicial; ~ **domain identifier** *n* (*IDI*) DATA identificador *m* de dominio inicial; ~ **domain part** *n* (*IDP*) DATA parte *f* del dominio inicial; ~ **full line-up** *n* (*IFLU*) TEST ajuste *m* inicial plena (*IFLU*); ~ **graphics exchange specification** *n* (*IGES*) COMP especificación *f* inicial de intercambio de gráficos (*IGES*); ~ **inverse voltage** *n* ELECT, POWER tensión *f* inversa inicial; ~ **ionizing event** *n* RADIO suceso *m* ionizante inicial; ~ **metering** *n* APPL medición *f* inicial; ~ **permeability** *n* ELECT permeabilidad *f* inicial; ~ **point** *n* COMP origen *m*; ~ **position** *n* COMP posición *f* fundamental; ~ **positioning** *n* COMP posicionamiento *m* inicial; ~ **signal unit** *n* (*ISU*) SIG unidad *f* de señal inicial (*ISU*); ~ **state** *n* DATA, ELECT estado *m* inicial; ~ **TC** *n* TELEP código *m* interurbano inicial; ~ **testing state** *n* TEST estado *m* de comprobación inicial; ~ **value** *n* TEST valor *m* inicial
initialization *n* GEN inicialización *f*
initialize *vt* GEN inicializar
initializing: ~ **routine** *n* COMP rutina *f* de inicialización
initiate *vt* GEN iniciar
initiation *n* TELEP iniciación *f*
initiator: ~ **of man-machine language** *n* TELEP preparador *m* de lenguaje hombre-máquina; ~ **of MML output** *n* TELEP preparador *m* de salida MML
inject: ~ **into orbit** *vt* RADIO poner en órbita
injected *adj* TV inyectado; ~**-beam magnetron** *n* ELECT, RADIO magnetrón *m* de haz inyectado
injection: ~ **into orbit** *n* RADIO puesta *f* en órbita; ~ **laser diode** *n* (*ILD*) ELECT *semiconductors* diodo *m* láser de inyección; ~**-locked laser** *n* ELECT láser *m* sincronizado por inyección; ~ **ratio** *n* ELECT *semiconductors* rendimiento *m* de inyección
inkjet: ~ **printer** *n* COMP, PER APP impresora *f* de chorro de tinta
inland: ~ **call** *n* TELEP llamada *f* nacional
inlet: ~ **access** *n* DATA acceso *m* de entrada
in-line: ~ **instruction** *n* COMP instrucción *f* en hilera; ~ **recovery** *n* DATA recuperación *f* en hilera
inner: ~ **conductor** *n* TELEP, TRANS conductor *m* interior; ~ **duct** *n* ELECT, PARTS conducto *m* interior; ~ **marker beacon** *n* RADIO radiobaliza *f* interna; ~ **shell electron** *n* ELECT electrón *m* interno
inoperable *adj* COMP, GEN desactivado
inoperability *n* COMP, GEN desactivación *f*
inoperative *adj* GEN inactivo; ~ **reference station code** *n* RADIO código *m* de estación de referencia inactivo
in-process: ~ **inspection** *n* TEST inspección *f* durante la instalación; ~ **testing** *n* TEST comprobación *f* durante la instalación
input[1] *n* COMP entrada *f*; ~ **admittance** ELECT admitancia *f* de entrada; ~ **amplifier** ELECT amplificador *m* de alimentación; ~ **back-off** RADIO despojamiento *m* de entrada; ~ **capacitance** ELECT capacidad *f* de entrada; ~ **channel** ELECT canal *m* de entrada; ~ **circuit** ELECT circuito *m* de entrada; ~ **current** ELECT corriente *f* de entrada; ~ **data** *n pl* DATA datos *m pl* de entrada; ~ **device** COMP, DATA dispositivo *m* de introducción de datos; ~ **error** DATA error *m* de entrada; ~ **filter** DATA filtro *m* de entrada; ~ **gate** ELECT circuito *m* de entrada; ~**-gate signal** ELECT señal *f* de puerta de entrada; ~ **impedance** COMP, ELECT impedancia *f* de entrada; ~ **indicator** COMP indicador *m* de entrada; ~ **jitter** TELEP fluctuación *f* de entrada; ~ **level** ELECT nivel *m* de entrada; ~ **message acknowledgement** DATA reconocimiento *m* de mensaje de entrada; ~**-output device** TELEP unidad *f* de entrada-salida; ~ **port** COMP puerta *f* de entrada; ~ **power** ELECT potencia *f* de entrada; ~ **power flux density** RADIO densidad *f* de flujo de potencia de entrada; ~ **process** COMP proceso *m* de entrada; ~ **pulse** COMP impulso *m* de entrada; ~ **resistance** ELECT resistencia *f* de entrada; ~ **sensibility** TRANS sensibilidad *f* de entrada; ~ **signal** ELECT señal *f* de entrada; ~**-signal level-range** TRANS amplitud *f* de nivel de la señal de entrada; ~ **socket** COMP toma *f* de entrada; ~ **stage** ELECT, RADIO fase *f* de entrada; ~ **symbol** TELEP símbolo *m* de entrada; ~ **transaction accepted for delivery**

DATA transacción *f* de entrada aceptada para ejecución; ~ **transaction rejected** DATA transacción *f* de entrada rechazada; ~ **transformer** POWER transformador *m* de entrada; ~ **unit** COMP unidad *f* de entrada; ~ **validation** DATA validación *m* de entrada; ~ **variable** CONTR variable *f* de entrada; ~ **voltage** ELECT voltaje *m* de entrada

input[2] *vt* COMP entrar

input/output *n* (*I/O*) COMP, ELECT, TELEP entrada/salida *f* (*E/S*); ~ **bus** SWG bus *m* de entrada/salida; ~ **channel** COMP canal *m* de entrada/salida; ~ **controller** COMP controlador *m* entrada/salida; ~ **device** COMP dispositivo *m* de entrada/salida; ~ **group** SWG grupo *m* entrada/salida; ~ **process** COMP proceso *m* entrada/salida; ~ **program** DATA programa *m* entrada/salida; ~ **subsystem** SWG subsistema *m* entrada/salida; ~ **unit** COMP unidad *f* de entrada/salida

inquiry: ~ **reciprocation** *n* TELEP consulta *f* alternativa; ~ **station** *n* DATA estación *f* de consulta

in-register *phr* TV registrado

inrush: ~ **current** *n* ELECT corriente *f* de entrada

insert *n* PARTS inserto *m*

insertion *n* GEN inserción *f*; ~ **gain** TRANS ganancia *f* de inserción; ~ **loss** TRANS pérdida *f* por inserción; ~ **loss pad** PARTS, TRANS almohadilla *f* de pérdida de inserción; ~ **point** COMP punto *m* de inserción; ~ **reference signal** SIG señal *f* de referencia de inserción; ~ **test signal** SIG señal *f* de prueba de inserción; ~ **tool** GEN instrumento *m* de inserción

inset *n* PARTS pieza *f* de inserción; ~ **picture** COMP imagen *f* de detalle

inside: ~ **call** *n* TELEP llamada *f* interna; ~ **vapour** *BrE* **phase oxidation** *n* (*IVPO*) ELECT oxidación *f* interior en fase de vapor, oxidación *f* interna en fase de vapor; ~ **view** *n* GEN *technical drawing* vista *f* interior

in-slot: ~ **signalling** *BrE n* SIG señalización *f* en el intervalo de tiempo

inspect *vt* TEST inspeccionar

inspection *n* TEST inspección *f*; ~ **by attributes** TEST inspección *f* por atributos; ~ **by variables** TEST inspección *f* por variables; ~ **diagram** TEST diagrama *m* de inspección; ~ **lamp** TEST lámpara *f* de mano, lámpara *f* portátil; ~ **level** TEST nivel *m* de inspección; ~ **specification** TEST especificación *f* de inspección

instability *n* ELECT, RADIO inestabilidad *f*

install *vt* GEN instalar

installation *n* GEN instalación *f*; ~ **address** TELEP dirección *f* de la instalación; ~ **call forwarding** TELEP *ISDN* envío *m* de llamada de instalación; ~ **data** *n pl* DATA datos *m pl* de instalación; ~ **date** GEN *plant engineering* fecha *f* de instalación; ~ **diagram** GEN *documentation* diagrama *m* de instalación; ~ **documentation** GEN documentación *f* de instalación; ~ **drawing** GEN *documentation* dibujo *m* de instalación; ~ **kit** PARTS equipo *m* de instalación; ~ **library** GEN *documentation* biblioteca *f* de instalación; ~ **manual** GEN manual *m* de instalación; ~ **program** COMP programa *m* de instalación; ~ **site** GEN lugar *m* de instalación; ~ **time** GEN *plant engineering* tiempo *m* de instalación

installer *n* GEN instalador *m*

instant *n* GEN instante *m*

instantaneous *adj* GEN instantáneo; ~ **access** *n* TRANS acceso *m* instantáneo; ~ **automatic gain control** *n* ELECT, TELEP, TRANS control *m* automático de ganancia instantánea; ~ **availability** *n* TELEP disponibilidad *f* instantánea; ~ **availability/unavailability** *n* TELEP disponibilidad/nodisponibilidad *f* instantánea; ~ **failure intensity** *n* TEST intensidad *f* instantánea de fallo; ~ **failure rate** *n* TEST tasa *f* de fallos instantánea; ~ **frequency** *n* TRANS frecuencia *f* instantánea; ~ **power** *n* POWER potencia *f* instantánea; ~ **repair rate** *n* TELEP tasa *f* de reparaciones instantánea; ~ **value** *n* GEN valor *m* instantáneo

Institute: ~ **of Electrical and Electronics Engineers** *n* (*IEEE*) ELECT Instituto de Ingenieros Eléctricos y Electrónicos

instruction *n* GEN instrucción *f*; ~ **address** COMP dirección *f* de instrucción; ~ **address register** COMP registro *m* de dirección de instrucción; ~ **booklet** GEN folleto *m* de instrucciones; ~ **manual** GEN manual *m* de instrucciones; ~ **modifier** COMP modificador *m* de instrucciones; ~ **register** COMP registro *m* de instrucción; ~ **repertoire** COMP repertorio *m* de instrucciones; ~ **set** COMP juego *m* de instrucciones; ~ **word** COMP, DATA palabra *f* de instrucción

instructional: ~ **television** *n* (*ITV*) TV televisión *f* educativa (*ITV*)

instrument *n* GEN instrumento *m*; ~ **constant** TEST constante *f* de un instrumento de medición; ~ **cord** PER APP cordón *m* de instrumento; ~ **landing system** (*ILS*) RADIO sistema *m* de aterrizaje por instrumentos (*ILS*); ~ **panel** TEST tablero *m* de instrumentos

instrumental: ~ **error** *n* TEST error *m* instrumental

insulant *n* ELECT, PARTS, POWER aislante *m*

insulate *vt* ELECT, PARTS, POWER aislar
insulated: ~ **cable** *n* ELECT, PARTS, POWER cable *m* aislado; ~ **conductor** *n* ELECT, PARTS, POWER conductor *m* aislado
insulating *adj* ELECT, PARTS, POWER aislante; ~ **bead** *n* ELECT cuenta *f* aislante; ~ **foil** *n* POWER hoja *f* aislante; ~ **lacquer** *n* ELECT barniz *m* aislante; ~ **layer** *n* POWER capa *f* aislante; ~ **material** *n* ELECT, PARTS, POWER material *m* aislante; ~ **plate** *n* POWER placa *f* aislante; ~ **property** *n* GEN propiedad *f* aislante; ~ **sheet** *n* POWER lámina *f* aislante; ~ **spacer** *n* POWER separador *m* aislante; ~ **strip** *n* POWER regleta *f* aislante; ~ **tape** *n* POWER cinta *f* aislante; ~ **tube** *n* POWER tubo *m* aislante; ~ **varnish** *n* ELECT barniz *m* aislante; ~ **washer** *n* POWER arandela *f* aislante
insulation *n* ELECT, PARTS, POWER aislamiento *m*; ~ **displacement termination** ELECT corrección *f* de desplazamiento del aislamiento; ~ **fault** POWER defecto *m* de aislamiento; ~ **gap** POWER distancia *f* aislante; ~ **loss** TRANS pérdida *f* del aislamiento; ~ **measurement** POWER medición *f* de aislamiento; ~ **resistance** POWER resistencia *f* de aislamiento; ~ **screen** POWER pantalla *f* aislante
insulator *n* ELECT, PARTS, POWER aislante *m*
integer: ~ **programming** *n* COMP programación *f* con números enteros
integral[1]: ~ **aerial** *n* RADIO antena *f* integral; ~ **antenna** *n* RADIO antena *f* integral; ~ **sampling** *n* TRANS muestreo *m* integral
integral[2] *n* GEN *mathematics* integral *f*
integrate *vt* GEN integrar
integrated: ~ **access** *n* TELEP acceso *m* integrado; ~ **adaptor** *n* COMP adaptador *m* integrado; ~ **circuit** *n* ELECT, PARTS circuito *m* integrado; ~ **circuit panel assembly** *n* ELECT, PARTS montaje *m* de panel de circuito integrado; ~ **communication control system** *n* RADIO sistema *m* de control de comunicación integrada; ~ **data processing** *n* DATA informática *f* integrada; ~ **digital access** *n* (*IDA*) TRANS acceso *m* digital integrado (*IDA*); ~ **digital exchange** *n* SWG central *f* digital integrada; ~ **digital network** *n* TRANS red *f* digital integrada; ~ **digital services exchange** *n* SWG central *f* de servicios digitales integrados; ~ **electronics** *n* ELECT electrónica *f* de integración; ~ **microcircuit** *n* ELECT microcircuito *m* integrado; ~ **network** *n* TRANS red *f* integrada; ~ **office system** *n* PER APP sistema *m* de oficina integrado; ~ **optical circuit** *n* (*IOC*) TRANS circuito *m* óptico integrado; ~ **optical switch** *n* SWG conmutador *m* óptico integrado; ~ **optical switching matrix** *n* SWG matriz *f* de conmutación óptica integrada; ~ **optics** *n* TRANS óptica *f* integrada; ~ **optoelectronic circuit** *n* ELECT circuito *m* opticoelectrónico integrado; ~ **satellite test** *n* (*IST*) TEST prueba *f* de satélite integrada (*IST*); ~ **services digital network** *n* (*ISDN*) APPL, TELEP, TRANS red *f* digital de servicios integrados (*RDSI*); ~ **software** *n* COMP, DATA programas *m pl* integrados; ~ **system** *n* TRANS sistema *m* integrado; ~ **system test** *n* TEST prueba *f* de sistema integrado; ~ **transmission and switching network** *n* SWG, TRANS red *f* integrada de transmisión y conmutación; ~ **voice-data PABX** *n* TELEP PABX *f* integrada de voz-datos; ~ **voice-data switch** *n* SWG conmutador *m* de voz-datos integrados
integrating: ~ **circuit** *n* ELECT circuito *m* integrador; ~**-measuring instrument** *n* TEST instrumento *m* de integración-medición
integration *n* GEN integración *f*
integrator *n* DATA, ELECT, RADIO integrador *m*
integrity *n* DATA, TELEP, TRANS integridad *f*
intelligent: ~ **backtracking** *n* DATA retorno *m* inteligente; ~ **building** *n* APPL edificio *m* inteligente; ~ **computer interface** *n* COMP interfaz *f* de ordenador inteligente *Esp*, interfaz *f* de computador inteligente *AmL*, interfaz *f* de computadora inteligente *AmL*; ~ **network** *n* (*IN*) APPL, TRANS red *f* inteligente (*IN*); ~ **optical modem** *n* PER APP módem *m* óptico inteligente; ~ **peripheral** *n* PER APP periférico *m* inteligente; ~ **tutoring system** *n* (*ITS*) COMP sistema *m* inteligente de enseñanza asistida (*ITS*); ~ **vehicle highway system** *n* (*IVHS*) RADIO sistema *m* inteligente para vehículos en carretera (*IVHS*)
intelligible: ~ **crosstalk component** *n* TELEP componente *m* de diafonía inteligible
intensity *n* GEN intensidad *f*; ~ **control** TV control *m* de intensidad; ~ **interpolation shading** TV sombreado *m* de interpolación de intensidad; ~ **level** TRANS nivel *m* de intensidad; ~ **sensor** TRANS *fibre optics* sensor *m* de intensidad
interaction *n* GEN interacción *f*; ~ **gap** POWER intervalo *m* de interacción; ~ **loss** TRANS atenuación *f* por interacción; ~ **region** ELECT zona *f* de interacción
interactive *adj* GEN interactivo; ~ **area** *n* TRANS área *f* interactiva; ~ **audiovisual** *n* TV audiovisual *m* interactivo; ~ **computer graphics** *n* COMP, TV gráficos *m pl* de ordenador interactivo; ~ **medium** *n* COMP medio *m*

interactivo; ~ **mode** *n* COMP modo *m* interactivo; **~-mode operating sequence** *n* TELEP secuencia *f* de operación en modo interactivo; ~ **multimedia** *n* COMP sistema *m* multimedia interactivo; ~ **network** *n* TRANS red *f* interactiva; ~ **primary link** *n* DATA enlace *m* primario interactivo; ~ **service** *n* TV servicio *m* interactivo; ~ **system** *n* TV sistema *m* interactivo; ~ **television** *n* TV televisión *f* interactiva; ~ **terminal** *n* TV terminal *m* interactivo; ~ **TV** *n* TV TV *f* interactiva; ~ **video** *n* TV vídeo *m* interactivo; ~ **videodisk** *n* COMP videodisco *m* interactivo; ~ **videography** *n* RADIO, TV videografía *f* interactiva; ~ **videotext** *n* TV videotext *m* interactivo; ~ **voice data servers** *n pl* TELEP servidores *m pl* de datos acústicos interactivos; ~ **voice response** *n* TELEP respuesta *f* vocal interactiva; ~ **voice service** *n* APPL servicio *m* de voz interactivo

interactivity *n* GEN interactividad *f*

interband *adj* TELEP, TV intercalado

interblock: ~ **gap** *n* DATA intervalo *m* entre bloques

intercell: ~ **hand-off** *n* RADIO *land mobile* transferencia *f* entre células; ~ **switching** *n* RADIO *land mobile* conmutación *f* entre células

intercept: ~ **announcer** *n* TELEP avisador *m* de interceptación; ~ **receiver** *n* ELECT localizador *m* de emisoras; ~ **service** *n* TELEP servicio *m* de interceptación; ~ **trunk** *n* TELEP circuito *m* de intercepción troncal

interception: ~ **of calls** *n* TELEP interceptación *f* de llamadas; ~ **equipment** *n* TELEP equipo *m* de interceptación

interchange *n* GEN intercambio *m*; ~ **circuit** TRANS circuito *m* de intercambio; ~ **control reference** DATA referencia *f* del control de intercambios; ~ **date** DATA fecha *f* de intercambio; ~ **header** DATA cabecera *f* de intercambio; ~ **receiver identification** DATA, TELEP identificación *f* del receptor del intercambio; ~ **recipient** DATA receptor *m* del intercambio; ~ **sender** DATA emisor *m* de intercambio; ~ **sender identification** DATA, TELEP identificación *f* del emisor del intercambio; ~ **time** DATA tiempo *m* de intercambio; ~ **trunk** *BrE* (*cf intraoffice trunk AmE*) TELEP línea *f* interurbana de intercambio; ~ **trunk group** *BrE* (*cf intraoffice trunk group AmE*) TELEP grupo *m* de líneas interurbanas de intercambio; ~ **unit program** TELEP programa *m* de unidad de intercambio

interchangeability *n* GEN intercambiabilidad *f*

interchangeable *adj* GEN intercambiable; ~ **connectors** *n pl* TRANS conectores *m pl* intercambiables

intercharacter: ~ **rest condition** *n* TELEP condición *f* de reposo entre caracteres

intercity *adj* TELEP interurbano

intercom *n* (*intercommunication*) TELEP intercomunicación *f*; ~ **system** *n* TELEP sistema *m* de interfono; ~ **telephone** *n* TELEP teléfono *m* intercom

intercommunication *n* (*intercom*) TELEP intercomunicación *f*; ~ **system** TELEP interfono *m*; ~ **telephone set** PER APP, TELEP aparato *m* de intercomunicación telefónica

interconnect *vt* GEN interconectar

interconnection *n* GEN interconexión *f*; ~ **circuit** TELEP circuito *m* de interconexión; ~ **equipment** TELEP equipo *m* de interconexión; ~ **network** POWER, TRANS red *f* de interconexión

intercontinental: ~ **circuit** *n* TELEP circuito *m* intercontinental; ~ **connection** *n* TELEP conexión *f* intercontinental; ~ **transit circuit** *n* TELEP circuito *m* de tránsito intercontinental; ~ **transit exchange** *n* TELEP central *f* de tránsito intercontinental

intercontrol: ~ **station multiplex** *n* TELEP múltiplex *m* de estación de intercontrol

intercouple *vt* PER APP convertir información en paralelo a información en serie

interdigital: ~ **filter** *n* TRANS filtro *m* interdigital; ~ **pause** *n* TELEP pausa *f* interdigital; ~ **time** *n* TELEP tiempo *m* interdigital; ~ **transducer** *n* (*IDT*) TRANS transductor *m* interdigital

interelectrode: ~ **capacitance** *n* ELECT capacitancia *f* interelectródica

interexchange *n* TELEP intercambio *m* entre centrales; ~ **trunk** *BrE* (*cf interoffice trunk AmE*) TELEP línea *f* interurbana entre centrales; ~ **trunk group** *BrE* (*cf interoffice trunk group AmE*) TELEP grupo *m* de líneas interurbanas entre centrales

interface *n* (*I/F*) GEN interfaz *f* (*I/F*); ~ **adaptor** COMP, TELEP adaptador *m* de interfaz; ~ **circuit** ELECT circuito *m* de enlace; ~ **coder** TRANS codificador *m* de interfaz; ~ **configuration environment** (*ICE*) TELEP entorno *m* de configuración del interfaz (*ICE*); ~ **connection** COMP conexión *f* interfacial; ~ **converter** DATA convertidor *m* interfacial; ~ **decoder** TRANS descodificador *m* interfacial; ~ **message processor** (*IMP*) COMP procesador *m* de mensajes de interfaz; ~ **module** (*IM*) GEN módulo *m* de interfaz

interfacial: ~ **connection** *n* ELECT conexión *f* interfacial

interfacility: ~ **links** *n pl* TRANS enlaces *m pl* entre soportes de transmisión
interfacing: ~ **unit** *n* GEN unidad *f* de adaptación
interference *n* GEN interferencia *f*; ~ **canceller** *BrE* RADIO eliminador *m* de interferencias; ~ **eliminator** POWER eliminador *m* de interferencias; ~ **filter** ELECT, TRANS filtro *m* eliminador de interferencias; ~**-free coverage area** RADIO área *f* de cobertura libre de interferencias; ~ **fringe pattern** TRANS *fibre optics* franjas *f pl* de interferencia; ~ **immunity** RADIO inmunidad *f* a las interferencias; ~ **level** POWER, RADIO nivel *m* de interferencia; ~ **level contribution** RADIO contribución *f* del nivel de interferencia; ~ **noise** RADIO ruido *m* parásito; ~ **path** RADIO trayecto *m* de interferencia; ~ **pulse** ELECT impulso *m* interferente; ~ **sector** RADIO sector *m* de interferencia; ~ **suppression** ELECT, RADIO, TRANS eliminación *f* de interferencias; ~ **suppression equipment** ELECT, RADIO, TRANS equipo *m* eliminador de interferencia; ~ **voltage** TEST tensión *f* interferente
interferer *n* GEN factor *m* de interferencia
interfering: ~ **channel** *n* TRANS canal *m* interferente; ~ **power flux** *n* TRANS flujo *m* de potencia interferente; ~ **satellite** *n* RADIO satélite *m* interferente; ~ **signal** *n* TRANS señal *f* de interferencia; ~ **source** *n* RADIO fuente *f* interferente; ~ **transmitter** *n* RADIO transmisor *m* interferente
interferometer *n* TEST interferómetro *m*
interflow *n* TELEP interflujo *m*; ~ **key** TELEP clave *f* de interflujo
interframe: ~ **coding** *n* DATA codificación *f* intertrama; ~ **compression** *n* TV compresión *f* entre imágenes
interhub: ~ **signalling** *BrE n* SIG señalización *f* entre bornes
interlace *vt* RADIO, TELEP, TRANS, TV entrelazar
interlaced *adj* RADIO, TELEP, TRANS, TV entrelazado
interlacing *n* RADIO, TELEP, TRANS, TV entrelazamiento *m*; ~ **sequence** TV secuencia *f* de entrelazamiento
interlayer: ~ **connection** *n* ELECT, PARTS *printed circuits* conexión *f* entre capas
interleave *vt* COMP, RADIO, TELEP, TRANS intercalar
interleaved *adj* COMP, RADIO, TELEP, TRANS intercalado; ~ **carriers** *n pl* TRANS portadoras *f pl* intercaladas; ~ **channel coding** *n* TRANS codificación *f* de canal por división de tiempo; ~ **polarization** *n* RADIO polarización *f* intercalada
interleaver *n* COMP, RADIO, TELEP, TRANS intercalador *m*
interleaving *n* COMP, RADIO, TELEP, TRANS intercalación *f*
interletter: ~ **spacing** *n* DATA separación *f* entre letras
interline: ~ **spacing** *n* DATA espaciado *m* interlineal
interlocal: ~ **traffic** *n* TELEP tráfico *m* interlocal; ~ **trunk** *n* TELEP línea *f* troncal interlocal
interlock: ~ **code** *n* DATA, TELEP código *m* de bloqueo; ~ **switch** *n* RADIO conmutador *m* de enclavamiento
interlocking: ~ **equisignal system** *n* RADIO sistema *m* de equiseñal acoplado
intermarker: ~ **group trunk** *n* TELEP línea *f* troncal de grupo intermarcador
intermateable: ~ **connector** *n* PARTS *fibre optics* conector *m* acoplable
intermediate: ~ **crossconnect** *n* SIG interconexión *f* intermedia; ~ **data rate** *n* DATA velocidad *f* de flujo de datos intermedios; ~ **distribution frame** *n* SWG, TELEP trama *f* de distribución intermedia; ~ **equipment** *n* COMP, TELEP equipo *m* intermedio; ~ **frame** *n* TV imagen *f* intermedia; ~ **frequency** *n* TRANS frecuencia *f* intermedia; ~ **frequency amplification** *n* RADIO, TRANS amplificación *m* en frecuencia intermedia; ~ **frequency amplifier** *n* RADIO, TRANS amplificador *m* de frecuencia intermedia; ~ **frequency filter** *n* RADIO, TRANS filtro *m* de frecuencia intermedia; ~ **frequency rejection** *n* RADIO supresión *f* de frecuencia intermedia; ~ **frequency rejection ratio** *n* RADIO ratio *f* de rechazo de interferencias; ~ **frequency signal** *n* ELECT, RADIO señal *f* de frecuencia intermedia; ~ **frequency stage** *n* ELECT, RADIO fase *f* de frecuenciaintermedia; ~ **high usage trunk** *n* TELEP línea *f* troncal intermedia de uso elevado; ~ **reference system** *n* TELEP sistema *m* de referencia intermedio; ~ **repeater** *n* TRANS repetidor *m* intermedio; ~ **result** *n* GEN resultado *m* intermedio; ~ **satellite band** *n* RADIO banda *f* satélite intermedia; ~ **system** *n* DATA, RADIO sistema *m* intermedio; ~ **trunk channel** *n* TRANS canal *m* de línea troncal intermedio
intermittence *n* GEN intermitencia *f*
intermittent: ~ **failure** *n* TEST fallo *m* intermitente; ~ **fault** *n* TEST fallo *m* intermitente
intermodal: ~ **distortion** *n* TRANS *fibre optics* distorsión *f* intermodal

intermodulation *n* TRANS intermodulación *f*; ~ **attenuation** TRANS atenuación *f* de intermodulación; ~ **component** TRANS componente *m* de intermodulación; ~ **distortion** TRANS distorsión *f* de intermodulación; ~ **noise** TRANS ruido *m* de intermodulación; ~ **product** TRANS producto *m* de intermodulación

internal *adj* GEN interno; ~ **blocking** *n* SWG bloqueo *m* interno; ~ **cable** *n* ELECT cable *m* interno; ~ **circuitry** *n* ELECT circuitería *f* interna; ~ **clock** *n* PER APP reloj *m* interno; ~ **connecter** *n* ELECT conector *m* interno; ~ **damage** *n* GEN, TEST daño *m* interno; ~ **disabled time** *n* TEST tiempo *m* no operativo interno; ~ **downtime** *n* TELEP tiempo *m* no operativo interno; ~ **extension** *n* PER, TELEP extensión *f* interna; ~ **external downtime** *n* TELEP tiempo *m* no operativo internoexterno; ~ **failure** *n* TEST fallo *m* interno; ~ **gas pressure cable** *n* GEN cable *m* con gas a presión interior; ~ **input signal** *n* ELECT señal *f* de entrada interna; ~ **label** *n* DATA etiqueta *f* interna; ~ **layer** *n* ELECT capa *f* interior; ~**-memory capacity** *n* COMP capacidad *f* de memoria interna; ~ **PCM link** *n* TELEP enlace *m* PCM interno; ~ **photoelectric effect** *n* ELECT, TRANS efecto *m* fotoeléctrico interno; ~ **reference point** *n* ELECT punto *m* de referencia interna; ~ **resistance** *n* ELECT resistencia *f* interna; ~ **signal** *n* SIG señal *f* interna; ~ **speaker** *n* PER APP altavoz *m* interno; ~ **storage** *n* DATA almacenamiento *m* interno; ~ **traffic** *n* TELEP tráfico *m* interno; ~ **wiring** *n* ELECT cableado *m* interno

international: ~ **alphabet no. 5** *n* (*IA5*) DATA, TELEP alfabeto *m* internacional nº 5 (*IA5*); ~ **atomic time** *n* (*IAT*) RADIO tiempo *m* atómico internacional (*IAT*); ~ **automatic circuit** *n* TELEP circuito *m* automático internacional; ~ **chain** *n* TELEP cadena *f* internacional; ~ **circuit** *n* TELEP circuito *m* internacional; ~ **code designator** *n* (*ICD*) DATA designador *m* de código internacional (*ICD*); ~ **communications** *n* TELEP comunicaciones *f pl* internacionales; ~ **connection** *n* TELEP conexión *f* internacional; ~ **data number** *n* DATA número *m* de dato internacional; ~ **data number format** *n* DATA formato *m* internacional de numeración de datos; ~ **data switching exchange** *n* (*IDSE*) DATA, SWG central *f* de conmutación para mensajes internacionales; ~ **direct dialling** *BrE n* (*IDD*) TELEP marcado *m* directo internacional; ~ **direct distance dialling** *BrE n* (*IDDD*) TELEP marcado *m* directo internacional a distancia; ~ **exchange** *n* SWG, TELEP central *f* internacional; ~ **gateway exchange** *n* SWG, TELEP central *f* de acceso internacional; ~ **gateway node** *n* (*IGN*) SWG, TELEP nodo *m* de acceso internacional; ~ **leased circuit** *n* TELEP circuito *m* internacional dedicado; ~ **leased group link** *n* TELEP enlace *m* de grupo especializado internacional; ~ **line** *n* TELEP línea *f* internacional; ~ **link** *n* TELEP enlace *m* internacional; ~ **main section** *n* TELEP sección *f* principal internacional; ~ **manual demand service** *n* TELEP servicio *m* de demanda manual internacional; ~ **Morse code** *n* SIG código *m* Morse internacional; ~ **multiple-destination television circuit** *n* TV circuito *m* internacional de TV de destino múltiple; ~ **multiple-destination television connection** *n* TV conexión *f* internacional de TV de destino múltiple; ~ **multiple-destination television link** *n* TV enlace *m* internacional de TV de destino múltiple; ~ **network** *n* TELEP, TRANS red *f* internacional; ~ **network management** *n* TELEP, TRANS gestión *f* de red internacional; ~ **number** *n* TELEP número *m* internacional; ~ **operations service** *n* APPL servicio *m* de operaciones internacionales; ~ **packet-switched data network** *n* SWG, TRANS red *f* de datos internacional de conmutación de paquetes; ~ **packet-switching gateway exchange** *n* SWG, TRANS central *f* internacional de acceso a la conmutación de paquetes; ~ **prefix** *n* TELEP prefijo *m* internacional; ~ **public facsimile service** *n* TELEP servicio *m* internacional público de telefax; ~ **reference alphabet** *n* (*IRA*) TELEP alfabeto *m* de referencia internacional (*IRA*); ~ **registration authority** *n* DATA autoridad *f* de registro internacional; ~ **section** *n* TELEP sección *f* internacional; ~ **sound-programme centre** (*AmE* international sound-program center) *n* (*ISPC*) TV centro *m* de programas de sonido internacional (*ISPC*); ~ **sound-programme** *BrE* **circuit** *n* TV circuito *m* de programa sonoro internacional; ~ **sound-programme** *BrE* **connection** *n* TV conexión *f* de programa sonoro internacional; ~ **sound-programme** *BrE* **link** *n* TV enlace *m* de programa sonoro internacional; ~ **sound-programme** *BrE* **transmission** *n* TV transmisión *f* de programa sonoro internacional; ~ **standard** *n* (*IS*) GEN norma *f* internacional (*SI*); ~ **standard virtual terminal protocol** *n* ELECT protocolo *m* de terminal virtual de patrón internacional; ~ **subscriber dialling** *BrE n* TELEP abonado *m* a línea telefónica internacional; ~ **switching centre** *BrE* *n* SWG, TELEP centro *m* de conmutación internacional; ~ **switching maintenance centre** *BrE* *n* (*ISMC*) SWG, TEST centro *m* de mantenimiento de conmutación internacional (*ISMC*); ~ **telegraph alphabet no. 2** *n* (*ITA2*) TELEP alfabeto *m* telegráfico internacional nº 2 (*ITA2*); ~ **telegraph position** *n* TELEP posición *f* telegráfica internacional; ~ **telephone circuit** *n* TELEP circuito *m* telefónico internacional; ~ **telephone connection** *n* TELEP conexión *f* telefónica internacional; ~ **television centre** *BrE* *n* TV centro *m* de televisión internacional; ~ **television circuit** *n* TV circuito *m* de televisión internacional; ~ **television connection** *n* TV conexión *f* de televisión internacional; ~ **television link** *n* TV enlace *m* de televisión internacional; ~ **television programme centre** (*AmE* international television program center) *n* TV centro *m* de programas de televisión internacional; ~ **television transmission** *n* TV transmisión *f* de televisión internacional; ~ **telex position** *n* TELEP posición *f* de télex internacional; ~ **transferred-account telegraph service** *n* TELEP servicio *m* telegráfico internacional de cuenta transferida; ~ **transit exchange** *n* (*INTTR*) SWG, TELEP central *f* de tránsito internacional (*INTTR*); ~ **transit node** *n* SWG, TRANS nodo *m* de tránsito internacional; ~ **transmission-maintenance centre** *BrE* *n* (*ITMC*) TEST, TRANS centro *m* internacional de mantenimiento de la transmisión (*ITMC*); ~ **X.121 format** *n* DATA formato *m* internacional X.121

International: ~ **Organization for Standardization** *n* (*ISO*) GEN Organización *f* Internacional de Normalización (*ISO*); ~ **Telecommunication Union** *n* (*UIT*) TELEP Unión *f* Internacional de Telecomunicación (*ITU*)

Internet *n* COMP, DATA Internet *m*; ~ **access point** DATA punto *m* de acceso a Internet; ~ **kiosk** DATA quiosco *m* de Internet; ~ **operator** DATA operador *m* de Internet; ~ **packet exchange** (*IPX*) DATA, SWG intercambio *m* de paquetes de Internet (*IPX*); ~ **protocol** (*IP*) COMP, DATA protocolo *m* de Internet (*PI*)

internetwork *n* TRANS red *f* interna; ~ **call redirection and deflection** (*ICRD*) TRANS redirección *f* y deflexión de llamadas de red interna (*ICRD*); ~ **protocol exchange** DATA intercambio *m* de protocolo de red interna

internetworking *n* TRANS operación *f* de interconexión de redes

interoffice *adj* TELEP interoficinal; ~ **trunk** *AmE* *n* (*cf* *interexchange trunk BrE*) TELEP línea *f* interurbana entre centrales; ~ **trunk group** *AmE* *n* (*cf* *interexchange trunk group BrE*) TELEP grupo *m* de líneas interurbanas entre centrales

interoperability *n* GEN interoperabilidad *f*

inter-PABX: ~ **tie circuit** *n* TRANS circuito *m* de enlace entre centralitas telefónicas automáticas

inter-PBX: ~ **traffic** *n* TELEP tráfico *m* entre PBX

interpersonal: ~ **messaging** *n* TELEP mensajería *f* interpersonal; ~ **messaging environment** *n* (*IPME*) TELEP entorno *m* de mensajería interpersonal (*IPME*); ~ **messaging system** *n* (*IPMS*) TELEP sistema *m* de mensajería interpersonal (*IPMS*); ~ **messaging system user** *n* TELEP usuario *m* del sistema de mensajería interpersonal; ~ **messaging system user agent** *n* TELEP agente *m* de usuario del sistema de mensajería interpersonal; ~ **notification** *n* (*IPN*) TELEP notificación *f* interpersonal (*IPN*)

interphone *n* PER APP interfono *m*; ~ **equipment** PER APP equipo *m* de interfono

interplay *n* GEN efecto *m* recíproco

interpolate *vt* PER APP, TEST, TRANS interpolar

interpolated *adj* PER APP, TEST, TRANS interpolado

interpolation *n* PER APP, TEST, TRANS interpolación *f*; ~ **error** TEST *metrology* error *m* de interpolación; ~ **gain** (*IG*) TRANS ganancia *f* de interpolación; ~ **pool** TRANS fondo *m* común de interpolación; ~ **resolution** PER APP *scanner* resolución *f* de interpolación; ~ **of speech signals** TRANS interpolación *f* de señales vocales

interpolator *n* PER APP, TEST, TRANS interpolador *m*

interpose *vt* TELEP interponer

interposition *n* TELEP interposición *f*; ~ **trunk** TELEP enlace *m* entre posiciones; ~ **trunk group** TELEP grupo *m* de circuitos de transferencia

interpret *vt* GEN interpretar

interpretability *n* GEN interpretabilidad *f*

interpretable *adj* GEN interpretable

interpretation *n* GEN interpretación *f*

interpreter *n* COMP intérprete *m*

interpretive: ~ **code** *n* COMP código *m* interpretativo

interprocessor: ~ **link** *n* COMP enlace *m* entre procesadores

interrecord: ~ **gap** *n* DATA separación *f* entre registros

interrogate *vt* GEN interrogar

interrogating: ~ **pulse** *n* RADIO impulso *m* de interrogación; ~ **signal** *n* RADIO señal *f* de interrogación

interrogation *n* GEN interrogación *f*; ~ **signal generator** (*ISG*) RADIO generador *m* de señal de interrogación (*ISG*)
interrogator *n* RADIO interrogador *m*; ~ **side-lobe suppression** (*ISLS*) RADIO supresión *f* del lóbulo lateral del interrogador (*ISLS*); ~ **transponder** RADIO transpondedor *m* interrogador
interrupt[1] *n* COMP, DATA, SWG, TELEP interrupción; ~ **feedback** TRANS realimentación *f* de interrupción; ~ **indicator** COMP indicador *m* de interrupción
interrupt[2] *vt* COMP, DATA, SWG, TELEP interrumpir
interrupted: ~ **ringing** *n* TELEP llamada *f* intermitente; ~ **ringing signal** *n* TELEP señal *f* de llamada interrumpida
interrupter *n* GEN interruptor *m*
interruptible: ~ **power supply** *n* POWER suministro *m* de corriente interrumpible
interrupting: ~ **capacity** *n* POWER capacidad *f* interruptora
interruption *n* GEN interrupción *f*; ~ **of a call in progress** TELEP interrupción *f* de una llamada en curso; ~ **control** TELEP control *m* de interrupción; ~ **of transmission** TRANS interrupción *f* de la transmisión; ~ **of transmission service** TELEP interrupción *f* del servicio de la transmisión
intersatellite: ~ **link** *n* (*ISL*) RADIO enlace *m* entre satélites (*ISL*)
interscan *n* RADIO barrido *m* intermedio
interspersed: ~ **frequency** *n* RADIO frecuencia *f* intercalada
interstage: ~ **coupling** *n* ELECT acoplamiento *m* interetápico
interswitchboard: ~ **line** *n* TRANS línea *f* entre cuadros conmutadores
intersymbol: ~ **interference** *n* (*ISI*) COMP, TRANS, TV perturbación *f* entre símbolos (*ISI*)
intertandem: ~ **trunk** *n* TELEP línea *f* troncal entre tándems; ~ **trunk group** *n* TELEP grupo *m* de líneas troncales entre tándems
intertoll: ~ **primary trunk group** *n* TELEP grupo *m* de líneas troncales primarias de larga distancia; ~ **traffic** *n* TELEP tráfico *m* de larga distancia; ~ **trunk** *n* TELEP enlace *m* de larga distancia; ~ **trunk group** *n* TELEP grupo *m* troncal de larga distancia
interval *n* GEN intervalo *m*; ~ **signal** RADIO señal *f* de intervalo; ~ **timer** ELECT, POWER temporizador *m* de intervalos
interwork *n* GEN interfuncionamiento; ~ **description** GEN descripción *f* de interfuncionamiento; ~ **diagram** GEN esquema *m* de interfuncionamiento
interwork *vi* GEN cooperar
interworkability *n* GEN interfuncionabilidad *f*
interworking *n* GEN interfuncionamiento *m*; ~ **between networks** TELEP interfuncionamiento *m* entre redes; ~ **by call-control mapping** (*ICCM*) TELEP interfuncionamiento *m* por diagrama de control de llamadas (*ICCM*); ~ **by port access** (*IPA*) DATA interfuncionamiento *m* mediante acceso de puerta (*IPA*); ~ **function** TELEP función *f* de interoperación; ~ **functional unit** (*IFU*) DATA unidad *f* funcional entrelazada; ~ **protocol** (*IWP*) DATA protocolo *m* de entrelazado; ~ **unit** DATA unidad *f* de interfuncionamiento
interzonal *adj* ELECT, RADIO interzonal
interzone: ~ **interference** *n* RADIO interferencia *f* interzonal
intraband: ~ **telegraphy** *n* TELEP telegrafía *f* dentro de la banda de voz
intracoded: ~ **picture** *n* COMP, TV imagen *f* intracodificada
intracoding *n* COMP, TV intracodificación *f*
intraframe *n* TV cuadro *m* interno; ~ **coding** TV codificación *f* dentro del cuadro; ~ **compression** TV compresión *f* dentro del cuadro
intramodal: ~ **distortion** *n* TRANS distorsión *f* intramodal
intraoffice: ~ **junctor circuit** *n* SWG circuito *m* de conexión dentro de la oficina; ~ **trunk** *AmE n* (*cf interchange trunk BrE*) TELEP línea *f* interurbana de intercambio; ~ **trunk group** *AmE n* (*cf interchange trunk group BrE*) TELEP grupo *m* de líneas interurbanas de intercambio
intrinsic: ~ **conduction** *n* ELECT, PARTS *semiconductors* conducción *f* intrínseca; ~ **error** *n* TEST error *m* intrínseco; ~ **joint loss** *n* TRANS pérdida *f* intrínseca de empalme; ~ **junction loss** *n* TRANS *fibre optics* pérdida *f* intrínseca de una unión; ~ **layer** *n* ELECT, PARTS *semiconductors* capa *f* intrínseca; ~ **noise** *n* TRANS *fibre optics* ruido *m* intrínseco; ~ **semiconductor** *n* ELECT, PARTS semiconductor *m* intrínseco; ~ **standoff ratio** *n* POWER relación *f* intrínseca de cresta
introduction: ~ **of service** *n* APPL presentación *f* de un servicio
intruder: ~**-presence detector** *n* APPL detector *m* de presencia de intrusos
intrusion *n* ELECT, TELEP intrusión *f*, intervención *f*; ~ **tone** TELEP tono *m* de intervención
INTTR *abbr* (*international transit exchange*) SWG, TELEP INTTR (*central de tránsito internacional*)

invalid *adj* GEN inválido; ~ **cell** *n* COMP celda *f* noválida; ~ **command** *n* COMP comando *m* noválido; ~ **key** *n* COMP clave *f* noválida; ~ **MCSPDU** *n* TELEP MCSPDU *m* noválido; ~ **sequence** *n* DATA secuencia *f* noválida; ~ **test event** *n* DATA, TEST evento *m* de prueba noválido
inventory *n* GEN inventario *m*
inverse: ~ **electrode current** *n* ELECT corriente *f* inversa de electrodo; ~ **function** *n* GEN función *f* inversa; ~ **impedance** *n* ELECT impedancia *f* inversa; ~ **solidus** *n* COMP *typesetting* barra *f* oblicua invertida; ~ **time overload protection** *n* POWER protección *f* contra sobrecarga de retardo inverso; ~ **video** *n* TV vídeo *m* inverso *Esp*, video *m* inverso *AmL*
inversion *n* GEN inversión *f*; ~ **layer** TRANS estratificador *m* de inversión
invert *vt* GEN invertir
inverted: ~ **background** *n* TV fondo *m* invertido
inverter *n* ELECT, POWER inversor *m*, convertidor *m*; ~ **circuit** ELECT circuito *m* de inversor; ~ **system** POWER sistema *m* de inversor
inverting: ~ **amplifier** *n* POWER amplificador *m* de inversión
investigate *vt* TEST investigar
investigation *n* TEST investigación *f*
invitation: ~ **to transmit** *n* TRANS invitación *f* a transmitir; **~-to-transmit polling** *n* TRANS interrogación de invitación *f* a transmitir
invoke *vt* DATA, RADIO, TELEP invocar
invoked: ~ **group** *n* DATA grupo *m* invocado
invoker *n* DATA, RADIO, TELEP solicitante *m*
I/O *abbr* (*input/output*) COMP, ELECT, TELEP E/S (*entrada/salida*); ~ **channel** *n* COMP canal *m* de entrada/salida; ~ **controller** *n* COMP controlador *m* entrada/salida; ~ **device** *n* COMP, TELEP dispositivo *m* de entrada/salida
IOC *abbr* (*integrated optical circuit*) TRANS circuito *m* óptico integrado
ion *n* ELECT ion *m*; ~ **burn** TV mancha *f* iónica; ~ **concentrate** ELECT concentración *f* iónica; ~ **deposition printer** PER APP impresora *f* por deposición iónica; ~ **exchange** ELECT intercambio *m* iónico; ~ **exchanger** ELECT intercambiador *m* de iones; **~-exchange technique** ELECT técnica *f* intercambiadora de iones; ~ **implantation** ELECT, PARTS *semiconductors* implantación *f* de iones; **~-implanted resistor** ELECT, PARTS *semiconductors* resistor *m* de implantación iónica; **~-matter interaction** ELECT interacción *f* de materia iónica; ~ **migration** ELECT migración *f* de los iones; ~ **noise** ELECT ruido *m* iónico; ~ **trap** ELECT captador *m* de iones
ionic: ~ **current** *n* ELECT corriente *f* iónica; ~ **mobility** *n* ELECT movilidad *f* iónica
ionization *n* ELECT, RADIO ionización *f*; ~ **energy** ELECT energía *f* de ionización; ~ **time** ELECT tiempo *m* de ionización
ionizing: ~ **event** *n* ELECT suceso *m* ionizante
ionogram *n* RADIO ionograma *m*
ionoscope *n* ELECT ionoscopio *m*
ionosonde *n* RADIO ionosonda *f*
ionosphere *n* RADIO ionosfera *f*
ionospheric: ~ **cross-modulation** *n* RADIO intermodulación *f* ionosférica; ~ **disturbance** *n* RADIO perturbación *f* ionosférica; ~ **layer** *n* RADIO capa *f* ionosférica; ~ **prediction** *n* RADIO predicción *f* ionosférica; ~ **propagation** *n* RADIO propagación *f* ionosférica; ~ **recorder** *n* RADIO sonda *f* ionosférica; ~ **reflection** *n* RADIO reflexión *f* ionosférica; ~ **scatter** *n* RADIO dispersión *f* ionosférica; ~ **scatter propagation** *n* RADIO propagación *f* por dispersión ionosférica; ~ **sounding** *n* RADIO sondeo *m* ionosférico; ~ **storm** *n* RADIO tormenta *f* ionosférica; ~ **wave** *n* RADIO onda *f* ionosférica
IP *abbr* (*Internet protocol*) COMP, DATA PI (*protocolo de Internet*)
IPA *abbr* (*interworking by port access*) DATA IPA (*interfuncionamiento mediante acceso de puerta*)
IPME *abbr* (*interpersonal messaging environment*) TELEP IPME (*entorno de mensajería interpersonal*)
IPMS *abbr* (*interpersonal messaging system*) TELEP IPMS (*sistema de mensajería interpersonal*)
IPN *abbr* (*interpersonal notification*) TELEP IPN (*notificación interpersonal*)
IPX *abbr* (*Internet packet exchange*) DATA, SWG IPX (*intercambio de paquetes de Internet*)
IR *abbr* (*infrared*) GEN IR (*infrarrojo*); ~ **radiation** *n* ELECT radiación *f* IR
IRA *abbr* (*international reference alphabet*) TELEP IRA (*alfabeto de referencia internacional*)
iron: ~ **core** *n* ELECT, PARTS *electromagnetism* núcleo *m* de hierro; **~-core coil** *n* ELECT, PARTS *electromagnetism* bobina *f* de núcleo de hierro; **~-dust core** *n* ELECT, PARTS *electromagnetism* núcleo *m* de polvo de hierro; ~ **rule** *n* TEST regla *f* de acero; **~-vane voltmeter** *n* TEST voltímetro *m* de hierro móvil
irradiance *n* ELECT, POWER, TRANS irradiancia *f*
irradiation *n* ELECT, POWER, TRANS irradiación *f*;

~-saturation current ELECT, POWER corriente *f* de irradiación-saturación

irrational: **~ number** *n* GEN *mathematics* número *m* irracional

irrecoverable: **~ error** *n* COMP error *m* irrecuperable

irregularity: **~ reflection coefficient** *n* TELEP coeficiente *m* de reflexión de irregularidad

IS *abbr* (*international standard*) GEN SI (*norma internacional*)

ISA *abbr* (*industry standard architecture*) COMP, DATA ISA (*arquitectura industrial estándar*); **~ segment** *n* COMP, DATA segmento *m* ISA

ISB *abbr* (*independent sideband*) RADIO ISB (*banda lateral independiente*)

ISDN *abbr* (*integrated services digital network*) APPL, TELEP, TRANS RDSI (*red digital de servicios integrados*); **~ access** *n* TELEP acceso *m* a la RDSI; **~ adaptor** *n* TELEP adaptador *m* para RDSI; **~ environment** *n* DATA entorno *m* RDSI; **~ exchange** *n* SWG central *f* de la RDSI; **~ primary-rate access** *n* TELEP acceso *m* en velocidad primaria a la RDSI; **~ switch** *n* SWG conmutador *m* de la RDSI; **~ user part** *n* TELEP parte *f* del usuario de la RDSI; **~ virtual circuit bearer service** *n* DATA servicio *m* portador de circuito virtual RDSI

ISG *abbr* (*interrogation signal generator*) RADIO ISG (*generador de señal de interrogación*)

ISI *abbr* (*intersymbol interference*) COMP, TRANS, TV ISI (*perturbación entre símbolos*)

ISL *abbr* (*intersatellite link*) RADIO ISL (*enlace entre satélites*)

ISLS *abbr* (*interrogator side-lobe suppression*) RADIO ISLS (*supresión del lóbulo lateral del interrogador*)

ISMC *abbr* (*international switching maintenance centre*) SWG ISMC (*centro de mantenimiento de conmutación internacional*)

ISO *abbr* (*International Organization for Standardization*) GEN ISO (*Organización Internacional de Normalización*); **~ 7-bit and 8-bit coded character sets** *n pl* TELEP conjuntos *m pl* de caracteres codificados ISO de 7 y 8 bitios; **~ standard** *n* GEN norma *f* ISO

isochronous *adj* TRANS isócrono; **~ data** *n pl* DATA datos *m pl* isócronos; **~ data transmission** *n* DATA transmisión *f* isócrona de datos

isolated: **~ neutral system** *n* POWER sistema *m* neutral aislado

isolating: **~ transformer** *n* POWER transformador *m* de aislamiento

isolation *n* ELECT, PARTS, POWER aislamiento *m*

isolator *n* ELECT, PARTS, POWER aislante *m*

isophon *n* TELEP *acoustics* contorno *m* de sonoridad

isotropic *adj* ELECT, RADIO isotrópico; **~ aerial** *n* RADIO antena *f* isotrópica; **~ gain** *n* RADIO ganancia *f* isotrópica; **~ radiator** *n* RADIO radiador *m* isotrópico

ISPC *abbr* (*international sound-programme centre*) TV ISPC (*centro de programas sonoros internacional*)

IST *abbr* (*integrated satellite test*) TEST IST (*prueba de satélite integrada*); **~ network** *n* TRANS red *f* IST

ISU *abbr* (*initial signal unit*) SIG ISU (*unidad de señal inicial*)

IT *abbr* (*information technology*) COMP tecnología *f* de la información

ITA2 *abbr* (*international telegraph alphabet no. 2*) TELEP ITA2 (*alfabeto telegráfico internacional nº 2*)

itemized: **~ billing** *n* APPL facturación *f* por conceptos; **~ statement** *n* APPL declaración *f* detallada

itinerant *adj* RADIO itinerante

ITMC *abbr* (*international transmission-maintenance centre*) TEST, TRANS ITMC (*centro internacional de mantenimiento de la transmisión*)

ITS *abbr* (*intelligent tutoring system*) COMP ITS (*sistema inteligente de enseñanza asistida*)

ITU *abbr* (*International Telecommunication Union*) TELEP UIT (*Unión Internacional de Telecomunicación*)

ITV *abbr* (*instructional television*) TV ITV (*televisión educativa*)

I-type: **~ semiconductor** *n* ELECT, PARTS semiconductor *m* tipo I

IUT *abbr* (*implementation under test*) TEST IUT (*instalación en prueba*)

IVHS *abbr* (*intelligent vehicle highway system*) RADIO IVHS (*sistema inteligente para vehículos en carretera*)

IVPO *abbr* (*inside vapour phase oxidation*) ELECT oxidación *f* interior en fase de vapor, oxidación *f* interna en fase de vapor

IWP *abbr* (*interworking protocol*) DATA protocolo *m* de entrelazado

J

jabber: ~ **protection** *n* ELECT protección *f* contra el ruido
jack *n* PARTS, PER APP clavija *f* de conexión; ~ **panel** PER APP panel *m* de acoplamiento
jacket *n* ELECT, PARTS, TRANS camisa *f*
jacketing *n* ELECT, PARTS, TRANS envuelta *f*
jackscrew *n* ELECT, PARTS tornillo *m* nivelador
jam[1] *vt* DATA, ELECT, RADIO agarrotar, perturbar
jam[2] *vi* DATA, ELECT, RADIO agarrotarse, perturbarse
jammed *adj* DATA, ELECT, RADIO agarrotado, perturbado
jamming *n* DATA, ELECT, RADIO agarrotación *f*, perturbación *f*; ~ **resistance** RADIO resistencia *f* a emisiones perturbadoras; ~ **transmitter** RADIO emisor *m* interferente intencional
JDF *abbr* (*junction distribution frame*) SWG, TRANS repartidor *m* de enlace
jelly: ~-**filled cable** *n* PARTS, TRANS cable *m* relleno de gelatina
jewel: ~ **bearing** *n* PARTS, POWER cojinete *m* de piedra dura; ~ **mount** *n* PARTS, POWER montura *f* de piedra dura
JFET *abbr* (*junction field effect transistor*) ELECT, PARTS JFET (*transistor con efecto de campo de enlace*)
JFIF *abbr* (*JPEG file interchange format*) COMP, DATA JFIF (*formato de intercambio de ficheros JPEG*)
jig *n* PARTS dispositivo *m* de fijación
jitter *n* COMP, RADIO, TRANS, TV inestabilidad *f* de la imagen
job *n* COMP trabajo *m*; ~ **division** COMP división *f* de trabajos; ~ **execution** COMP ejecución *f* del trabajo; ~-**recovery control file** COMP fichero *m* de control de recuperación de trabajos; ~-**recovery file** COMP fichero *m* de recuperación de trabajos; ~ **selector** COMP selector *m* de trabajos; ~ **stream** COMP flujo *m* de trabajos
Johnson: ~ **counter** *n* ELECT contador *m* de Johnson
joint[1] *adj* TEST conjunto; ~ **testing** *n* TEST prueba *f* conjunta; ~ **tests** *n pl* TEST pruebas *f pl* conjuntas; ~ **user** *n* TRANS usuario *m* en conjunto
joint[2] *n* PARTS empalme *m*
jointer *n* PARTS juntera *f*
jointing *n* PARTS empalme *m*; ~ **sleeve** PART manguito *m* de empalme
joist *n* PARTS viga *f*
joker *n* COMP carácter *m* de sustitución
Joule: ~ **effect** *n* ELECT, POWER efecto *m* Joul ~'**s law** *n* ELECT, POWER ley *f* de Joule
joystick *n* COMP, PER APP joystick *m*
JPEG: ~ **file interchange format** *n* (*JFIF*) COM DATA formato *m* de intercambio de ficherc JPEG (*JFIF*)
JSM *abbr* (*jumper storage module*) ELECT JSM (*módulo de almacenamiento del puente d conexión*)
judgment: ~ **error** *n* TEST error *m* de apreciació
Julian: ~ **date** *n* RADIO fecha *f* según calendario juliano; ~ **day number** *n* RADI número *m* de día según el calendario juliano
jumbler: ~ **cable** *n* PARTS cable *m* del mezcladc
jump *n* COMP bifurcación *f*
jumper *n* ELECT, PARTS, POWER, SWG puente *m* d conexión; ~ **king** GEN acoplamiento maestro; ~ **ring** SWG anillo *m* de conexión; **storage module** (*JSM*) ELECT módulo *m* d almacenamiento del puente de conexió (*JSM*); ~ **wire** PARTS cable *m* d acoplamiento; ~-**wire eyelet** PARTS anillo guíahilos; ~-**wire guide** PARTS guíahilos *m*
jumper up *vt* ELECT puentear
jumpering *n* ELECT, PARTS, POWER, SW interconexión *f*
jumping: ~ **allotter** *n* TELEP distribuidor aleatorio
junction *n* GEN unión *f*; ~ **box** ELECT, PART POWER, TRANS *cables* caja *f* de derivación; **capacitance** ELECT, PARTS *semiconductor* capacitancia *f* en una unión; ~ **capacito** ELECT, PARTS condensador *m* en una unión d semiconductor; ~ **circuit** TELEP circuito *m* d enlace; ~ **diode** ELECT, PARTS diodo *m* d unión; ~ **distribution frame** (*JDF*) SWC TRANS repartidor *m* de enlace; ~ **field effec transistor** (*JFET*) ELECT, PARTS transistor con efecto de campo de enlace (*JFET*); ~ **lin** TELEP línea *f* de unión; ~-**line relay set** SW juego *m* de relés de línea de enlace; ~ **networ** TELEP, TRANS red *f* de enlace; ~ **tanden exchange** SWG, TELEP central *f* telefónic intermedia de enlace, central *f* tándem d

enlace; ~ **temperature** ELECT, PARTS *semiconductors* temperatura *f* de unión; ~ **transistor** ELECT, PARTS transistor *m* de uniones

nctor *n* GEN conjuntor *m*; ~ **block** SWG bloque *m* de circuitos troncales

nk: ~ **mail** *n* DATA propaganda *f* de buzón

stifiable: ~ **digit** *n* TELEP, TRANS dígito *m* justificable; ~ **digit time slot** *n* TELEP, TRANS intervalo *m* de tiempo de dígito justificable

justification *n* COMP, TELEP, TRANS justificación *f*; ~ **ratio** TELEP, TRANS relación *f* de justificación; ~ **service digit** TELEP, TRANS dígito *m* de servicio de justificación; ~ **service unit** TELEP, TRANS unidad *f* de servicio de justificación; ~ **waiting time** TRANS tiempo *m* de espera para justificación

justify *vt* COMP, TELEP, TRANS justificar

justifying: ~ **digit** *n* TELEP, TRANS dígito *m* de justificación

K

kaledia *n* COMP kaledia *f*
Karnaugh: ~ **map** *n* GEN mapa *m* de Karnaugh
kc *abbr* (*kilocycle*) TEST kc (*kilociclo*)
keep[1]: ~**-alive circuit** *n* ELECT, POWER circuito *m* de acción residual; ~**-alive electrode** *n* ELECT, POWER electrodo *m* de entretenimiento
keep[2]: ~ **track of** *phr* RADIO rastrear
Kelvin: ~ **effect** *n* POWER efecto *m* Kelvin
keranuic: ~ **level** *n* RADIO nivel *m* keranuico
kernel *n* COMP núcleo *m*
Kerr: ~ **effect** *n* TRANS *fibre optics* efecto *m* Kerr
kevlar: ~ **skin** *n* RADIO membrana *f* del kevlar; ~ **strength** *n* ELECT resistencia *f* del kevlar
key: ~**-and-lamp unit** *n* PER APP portalámparas *m* de llave; ~ **assignment** *n* COMP asignación *f* de clave; ~ **click** *n* RADIO, TELEP chasquido *m* de manipulación; ~ **for direct selection** *n* RADIO tecla *f* de selección directa; ~ **down** *n* TELEP manipulador *m* cerrado; ~ **factor** *n* TV factor *m* clave; ~ **filter** *n* COMP filtro *m* de llave; ~ **follow** *n* COMP tecla *f* seguir; ~ **groove** *n* GEN muesca *f* posicionadora; ~ **label** *n* COMP nombre *m* de tecla; ~ **lock** *n* COMP cerradura *f* de llave; ~ **management** *n* DATA, TELEP gestión *f* por teclado; ~ **matrix** *n* COMP matriz *f* de referencia; ~ **net** *n* DATA red *f* de referencia; ~ **number** *n* COMP número *m* de identificación; ~ **off** *n* COMP manipulador *m* desconectado; ~ **overlay** *n* PER APP plantilla *f* removible de teclado; ~**-per-line console** *BrE n* (*cf key-per-trunk console AmE*) PER APP consola *f* con una clave por línea; ~**-per-trunk console** *AmE n* (*cf key-per-line console BrE*) PER APP consola *f* con una clave por línea principal; ~ **for preset selection** *n* RADIO tecla *f* de selección directa; ~ **release** *n* COMP descarga *f* por medio de tecla; ~ **shelf** *n* PER APP tablero *m* de llaves; ~ **signature** *n* COMP signo *m* constitutivo; ~ **split** *n* COMP separador *m* de llave; ~ **stream** *n* APPL, PER APP cadena *f* de referencia; ~ **synchronization** *n* COMP sincronización *f* de teclas; ~ **system** *n* RADIO centralita *f* local de marcado directo; ~ **telephone set** *n* PER APP, TELEP equipo *m* telefónico con teclado; ~ **telephone system** *n* PER APP, TELEP sistema *m* telefónico con teclado; ~ **to callouts** *n* TELEP tecla *f* para llamadas externas; ~ **touch force** *n* COMP fuerza *f* de pulsión de la tecla; ~ **transpose** *n* COMP trasposición *f* de tecla
key in *vt* GEN introducir información mediante el teclado
keyboard *n* COMP, PER APP teclado *m*; ~ **amplifie[r]** COMP amplificador *m* de teclado; ~ **controlle[r]** COMP controlador *m* de teclado; ~ **matrix** COM[P] matriz *f* de teclado; ~ **perforator** PER AP[P] perforador *m* de teclado; ~ **range** COMP líne[a] *f* de teclados; ~ **response** COMP respuesta *f* d[e] teclado; ~ **scaling** COMP graduación *f* po[r] teclado; ~ **scaling level** COMP nivel *m* d[e] graduación por teclado; ~ **scaling rate** COM[P] velocidad *f* de graduación por teclado; ~ **sender** PER APP manipulador *m* dactilográfico; ~ **shortcut** COMP tecla *f* de atajo; ~ **split point** COMP punto *m* de división del teclado; ~ **stand** COMP soporte *m* de teclado; ~ **transmitter** TELEP transmisor *m* de teclado
keyboarding *n* COMP, DATA captura *f* de dato[s] con teclado, mecanografiado *m*
keyed *adj* COMP, ELECT tecleado
keyer *n* COMP, TELEP manipulador *m*, teclista *m*
keyframe *n* COMP teclado *m* para registro de imágenes; ~ **animation** COMP animación *f* po[r] teclado de registro de imágenes; ~ **rate** COM[P] velocidad *f* de registro de imágenes por teclad[o]
keyframing *n* COMP registro *m* de imágenes po[r] teclado
keying *n* APPL, PER, TELEP manipulación *f*, tecleado *m*; ~ **error** APPL, PER, TELEP error *m* de tecleado
keynote *n* COMP nota *f* dominante; ~ **field** RADI[O] campo *m* de dominante
keypad *n* COMP, PER APP, TELEP teclado *m* numérico; ~ **telephone** PER APP, TELE[P] teléfono *m* de teclado; ~ **telephone set** PE[R] APP, TELEP aparato *m* telefónico de teclado
keypunch *n* COMP perforadora *f* de tecla
keysender *n* TELEP emisor *m* de llamadas
keyset *n* PER APP teclado *m*, código *m* de teclado; ~ **telephone** PER APP, TELEP teléfono *m* de teclado
keyswitch *n* SWG conmutador *m* de tipo telefónico
keyway *n* GEN muesca *f* posicionadora
keyword *n* COMP, DATA palabra *f* clave; ~ **access**

COMP acceso *m* por palabra clave; ~ **link** COMP enlace *m* por palabra clave; ~ **list** COMP lista *f* de palabras clave

kHz *abbr* (*kilohertz*) RADIO kHz (*kilohertzio*)

kill *vt* COMP suprimir

kilo: **~-instructions per second** *n pl* (*KIPS*) COMP kiloinstrucciones *f pl* por segundo (*KIPS*); **~-operations per second** *n pl* (*KOPS*) COMP kilooperaciones *f pl* por segundo (*KOPS*)

kilobaud *n* TRANS kilobaudio *m*

kilocycle *n* (*kc*) TEST kilociclo *m* (*kc*)

kilohertz *n* (*kHz*) RADIO kilohertzio *m* (*kHz*)

kilostream: ~ **circuit** *n* APPL circuito *m* de kiloflujo

kink *n* TRANS *fibre optics* torcedura *f*

kiosk *n* DATA quiosco *m*

KIPS *abbr* (*kilo-instructions per second*) COMP KIPS (*kiloinstrucciones por segundo*)

Kirchhoff: **~'s laws** *n pl* ELECT, POWER leyes *f pl* de Kirchhoff

klystron *n* ELECT, RADIO klistron *m*

KMS *abbr* (*knowledge management system*) COMP KMS (*sistema de gestión del conocimiento*)

knee: **~-luminous flux** *n* ELECT flujo *m* luminoso acodado; ~ **point** *n* ELECT punto *m* de codo; ~ **sensitivity** *n* ELECT sensibilidad *f* de la articulación

knife: ~ **switch** *n* POWER interruptor *m* de cuchilla

knockout *n* ELECT eyector *m*; ~ **blanking plate** ELECT placa *f* ciega de ocultación

knowledge: ~ **acquisition** *n* APPL, COMP *artificial intelligence* adquisición *f* de conocimiento; ~ **base** *n* APPL, COMP *artificial intelligence* base *f* de conocimiento; ~ **chunk** *n* COMP bloque *m* de conocimiento; ~ **engineer** *n* COMP ingeniero *m* del conocimiento; ~ **engineering** *n* COMP ingeniería *f* del conocimiento; ~ **management system** *n* (*KMS*) COMP sistema *m* de gestión del conocimiento (*KMS*)

knowledgeware *n* COMP, DATA software *m* del conocimiento

KOPS *abbr* (*kilo-operations per second*) COMP KOPS (*kilooperaciones por segundo*)

KP: ~ **signal** *n* TELEP señal *f* KP

Kr *abbr* (*krypton*) PARTS Kr (*criptón*)

krarup: ~ **cable** *n* PARTS, TRANS cable *m* krarupizado; ~ **loading** *n* TRANS krarupización *f*; ~ **system** *n* TRANS sistema *m* de carga continua

krarupization *n* TRANS krarupización *f*

krarupize *vt* TRANS *cables* krarupizar

krypton *n* (*Kr*) PARTS criptón *m* (*Kr*)

L

L: ~ **network** *n* POWER red *f* en L

LA *abbr* (*local application*) TELEP LA (*aplicación local*)

label *n* GEN *marking* etiqueta *f*; ~ **frame** GEN unidad *f* de identificación; ~ **holder** GEN portaetiquetas *m*; ~ **printer** PER APP impresora *f* de etiquetas

labelling *BrE* (*AmE* **labeling**) *n* GEN etiquetado *m*

laboratory *n* GEN laboratorio *m*; ~ **technician** GEN técnico *m* de laboratorio; ~ **test** TEST prueba *f* de laboratorio; ~ **test conditions** *n pl* TEST condiciones *f pl* de laboratorio

lace *vt* PARTS *cables* coser con tiretas

laced: ~ **cable fan** *n* PARTS abanico *m* de cable cosido

lacquer *n* PARTS laca *f*

ladder: ~**-type filter** *n* RADIO filtro *m* en escalera

lag *n* COMP, ELECT retraso *m*

lagging: ~ **phase shift** *n* POWER desplazamiento *m* de fase retardada

Lambert: ~**'s cosine law** *n* ELECT, TRANS ley *f* de coseno de Lambert

Lambertian: ~ **emission** *n* ELECT, TRANS emisión *f* lambertiana; ~ **radiator** *n* ELECT, TRANS *fibre optics* transmisor *m* de Lambert, radiador *m* lambertiano; ~ **reflector** *n* ELECT, TRANS reflector *m* lambertiano; ~ **source** *n* ELECT, TRANS generador *m* de Lambert, generador *m* lambertiano

laminate[1] *n* PARTS material *m* laminar; ~ **sheet** ELECT, PARTS panel *m* laminado

laminate[2] *vt* PARTS laminar

laminated: ~ **plastic** *n* PARTS plástico *m* laminado

laminating *n* PARTS laminación *f*

lamination *n* PARTS laminación *f*

lamp: ~ **base** *n* POWER casquillo *m* de lámpara; ~ **cap** *n* POWER casquillo *m* de lámpara; ~ **display panel** *n* PER APP, TELEP panel *m* de lámparas; ~ **extractor** *n* GEN arrancador *m* de lámpara; ~ **holder** *n* POWER portalámparas *m*; ~ **lens** *n* POWER lente *f* de lámpara; ~ **panel** *n* PER APP, TELEP panel *m* de lámparas; ~ **signal** *n* PER APP, TELEP *indicating device* señal *f* luminosa; ~**-sleeve pliers** *n pl* PARTS alicates *m pl* para retirar lámparas; ~ **strip** *n* POWER regleta *f* de lámparas; ~ **test** *n* POWER, TEST prueba *f* de lámparas

LAN *abbr* (*local area network*) COMP, DATA, TELEP red *f* de área local; ~ **manager** *n* COMP, DATA, TELEP gestor *m* de red de área local

land *n* ELECT, PARTS isla *f*; ~ **cable** TRANS cable *m* terrestre; ~ **line** TELEP, TRANS línea *f* terrestre; ~**-line charge** TELEP tasa *f* de línea terrestre; ~**-mobile satellite service** (*LMSS*) RADIO, TELEP, TRANS servicio *m* móvil terrestre de satélite (*LMSS*); ~**-mobile station** (*LMS*) RADIO, TELEP estación *f* móvil terrestre; ~**-mobile system** RADIO, TELEP, TRANS sistema *m* móvil terrestre; ~ **return** RADIO eco *m* del terreno; ~ **station** RADIO, TELEP estación *f* terrestre; ~**-station charge** RADIO, TELEP tasa *f* terrestre

landless: ~ **hole** *n* ELECT, PARTS *printed circuits* agujero *m* sin nudo

lane *n* RADIO calle *f*; ~ **identification** RADIO identificación *f* de calle

language *n* GEN lenguaje *m*; ~ **digit** TELEP cifra *f* de idioma; ~ **processor** COMP procesador *m* de lenguaje; ~ **translator** COMP traductor *m* de lenguaje; ~ **version** GEN *documentation* versión *f* por idioma

lanthanide: ~**-doped fibre** *BrE n* PARTS, TRANS fibra *f* dopada con lantánido

LAP *abbr* (*line access procedure*) TELEP procedimiento *m* de acceso a la línea

lapel: ~ **microphone** *n* TELEP micrófono *m* de solapa

Laplace: ~**'s law** *n* GEN *in mathematics* ley *f* de Laplace

lapper *n* POWER encintadora *f*

lapping *n* PARTS *of cables* lapeado *m*; ~ **film** PARTS película *f* aislante

laptop[1] *adj* COMP de falda, portátil; ~ **computer** *n* COMP computador *m* portátil *AmL*, computadora *f* portátil *AmL*, ordenador *m* portátil *Esp*

laptop[2] *n* COMP computador *m* portátil *AmL*, computadora *f* portátil *AmL*, ordenador *m* portátil *Esp*

large: ~**-core multiple-mode optic fibre** *BrE n* PARTS, TRANS fibra *f* óptica multimodo de núcleo grande; ~**-scale integration** *n* (*LSI*) ELECT, TELEP, TRANS integración *f* a gran escala (*IGE*); ~**-scale integration circuit** *n* ELECT, TELEP, TRANS circuito *m* de integración a gran escala; **~-scale system** *n* DATA sistema *m* en gran escala; **~ screen display** *n* PER APP pantalla *f* grande; **very ~-scale integration** *n* (*VLSI*) ELECT, TELEP, TRANS integración *f* a escala muy grande (*VLSI*); **very ~-scale integration circuit** *n* ELECT, TELEP, TRANS circuito *m* de integración a escala muy grande

laryngophone *n* RADIO, TELEP laringófono *m*

laser *n* ELECT, PER APP, TEST, TRANS láser *m*; **~ attenuator** ELECT atenuador *m* lasérico; **~ beam** ELECT haz *m* de láser; **~ beam scanning** COMP digitalización *f* por haz lasérico; **~ diode** (*LD*) ELECT, TRANS *fibre optics* diodo *m* láser (*LD*); **~ medium** ELECT, TRANS *fibre optics* medio *m* láser; **~ plasma** ELECT plasma *m* lasérico; **~ power** ELECT potencia *f* del láser; **~ printer** COMP, PER APP impresora *f* láser; **~ printing** COMP impresión *f* láser; **~ rangefinder** TEST telémetro *m* por láser; **~ telemetry** TEST telemetría *f* por láser; **~ transmitter** TRANS *fibre optics* transmisor *m* de láser; **~ trimming** ELECT recorte *m* por láser

lasing: **~ threshold** *n* ELECT, TRANS *fibre optics* umbral *m* de acción láser

last[1]: **~-choice circuit group** *n* ELECT, SWG, TELEP grupo *m* de circuitos de última alternativa; **~-choice group** *n* ELECT, SWG, TELEP grupo *m* de última alternativa; **~ known location** *n* RADIO última *f* posición conocida; **~ number called redial** *n* TELEP repetición *f* automática de último número llamado; **~ number dialled** *BrE* **recall** *n* TELEP rellamada del último número marcado; **~ number recall** *n* TELEP rellamada *f* de último número; **~ number redial** *n* TELEP repetición *f* automática del último número; **~ number redial key** *n* PER APP clave *f* de repetición automática de último número; **~-party release** *n* COMP, TELEP desconexión *f* doble; **~-party release signal** *n* TELEP señal *f* de desconexión doble; **~ try** *n* TELEP último *m* ensayo

last[2]: **~ in, first out** *phr* (*LIFO*) COMP último en entrar, primero en salir (*LIFO*)

latch *n* ELECT *logic elements* pestillo *m*; **~ circuit** ELECT *logic elements* circuito *m* de enclavamiento; **~ lock** ELECT cerrojo *m*; **~-up** PARTS *semiconductors* cierre *m*

latched: **~ detent position** *n* ELECT posición *f* de retención bloqueada

latching: **~ current** *n* POWER *thyristors* corriente *f* de enganche; **~ relay** *n* POWER relé *m* de bloqueo

latency *n* COMP latencia *f*

lateral: **~ deviation** *n* RADIO desviación *f* lateral; **~ displacement** *n* RADIO desplazamiento *m* lateral; **~ offset loss** *n* TRANS pérdida *f* por desplazamiento lateral; **~ plasma deposition** *n* ELECT deposición *f* lateral de plasma

lathe *n* PARTS torno *m*

lattice *n* ELECT, RADIO malla *f*; **~ network** POWER circuito *m* en celosía; **~ plan** RADIO *land mobile* plan *m* reticular

launch: **~ angle** *n* TRANS *fibre optics* ángulo *m* de alimentación; **~ complex** *n* RADIO complejo *m* de lanzamiento; **~ numerical aperture** *n* (*LNA*) ELECT, TRANS apertura *f* numérica de lanzamiento; **~ rocket** *n* RADIO cohete *m* de lanzamiento; **~ site** *n* RADIO sitio *m* de lanzamiento

launching: **~ fibre** *BrE* *n* TRANS fibra *f* de lanzamiento

law *n* GEN ley *f*

lay out *vt* COMP diagramar

layer[1] *n* GEN capa *f*; **~ of cabling elements** PARTS capa *f* de elementos de cableado; **~ diameter** PARTS *cables* diámetro *m* de capa; **~ drawing** ELECT, PARTS *printed circuits* dibujo *m* de capas; **~ management** DATA gestión *f* en niveles; **~ management entity** (*LME*) DATA entidad *f* para la gestión de capas (*LME*); **~ of units** PARTS *cables* capa *f* de unidades

layer[2] *vt* TRANS disponer en capas

layered: **~ architecture** *n* COMP arquitectura *f* estratiforme

layering *n* TRANS disposición *f* en capas

layout *n* COMP trazado *m*; **~ character** COMP, DATA carácter *m* de formato; **~ drawing** GEN *documentation* dibujo *m* de implantación; **~ option** TELEP opción *f* de recorrido; **~ sketch** ELECT *printed circuits* esquema *m* de disposición

LBA *abbr* (*logical block address*) COMP LBA (*dirección de bloque lógico*)

LCA *abbr* (*local calling area*) TELEP área *f* de comunicación urbana

LCC *abbr* (*leadless chip-carrier*) ELECT LCC (*cápsula sin patillas*)

LCD *abbr* (*liquid crystal display*) ELECT VCL (*visualización en cristal líquido*); **~ calls display** *n* PER APP *on answering machine* visualización *f* de llamadas en LCD; **~ projection panel** *n* ELECT panel *m* de proyección de LCD

LCS *abbr* (*large capacity storage*) PER APP gran capacidad de almacenamiento; **~ printer** *n* PER APP impresora *f* con gran capacidad de almacenamiento

LD *abbr* (*laser diode*) ELECT, TRANS *fibre optics* LD (*diodo láser*)

LE *abbr* (*local exchange*) SWG, TELEP central *f* local

lead[1] *n* GEN *cable* cable *m*; **~-in area** COMP área *f* de entrada; **~-in insulator** POWER aislador *m* pasamuros; **~ time** GEN tiempo *m* de espera

lead[2] *n* PARTS *metal* plomo *m*; **~ accumulator** POWER *batteries* acumulador *m* de plomo; **~ seal** PARTS junta *f* hermética de plomo; **~ sheath** PARTS, POWER cubierta *f* de plomo; **~-sheathed cable** PARTS, POWER cable *m* con cubierta de plomo; **~ sheathing** PARTS, POWER cubierta *f* de plomo; **~ sleeve** PARTS, POWER manguito *m* de plomo; **~ storage battery** POWER acumulador *m* de plomo

leaded: **~ chip-carrier** *n* ELECT cápsula *f* con patillas; **~ seal** *n* PARTS sello *m* de plomo

leading: **~ edge** *n* GEN, PARTS, TELEP borde *m* de entrada; **~-edge position** *n* TELEP posición *f* en flanco anterior; **~-edge technology** *n* GEN tecnología *f* avanzada; **~ phase shift** *n* POWER desplazamiento *m* en fase adelantada

leadless: **~ chip-carrier** *n* (*LCC*) ELECT cápsula *f* sin patillas (*LCC*)

leaf *n* DATA nodo *m* del árbol; **~ node** DATA nodo *m* de hoja; **~ spring** PARTS resorte *m* de lámina

leak *n* COMP, TELEP fuga *f*; **~ time** TELEP duración *f* de la fuga

leakage *n* ELECT, POWER fuga *f*; **~ current** ELECT corriente *f* de fuga; **~ field** ELECT campo *m* de fugas; **~ resistance** POWER resistencia *f* de fuga; **~ test** POWER, TEST prueba *f* de aislamiento

leap: **~ second** *n* TEST segundo *m* de salto

leapfrog: **~ test** *n* COMP, TEST verificación *f* por saltos

leased: **~ circuit** *n* TELEP circuito *m* arrendado; **~-circuit data transmission service** *n* TRANS servicio *m* de transmisión de datos por circuito arrendado; **~ line** *n* (*LS*) TELEP, TRANS línea *f* alquilada (*LS*)

least: **~ significant bit** *n* (*LSB*) DATA bit *m* menos significativo (*BMS*)

lecture: **~ call** *n* TELEP conferencia *f*

LED *abbr* (*light-emitting diode*) ELECT diodo *m* electroluminiscente, diodo *m* emisor de luz; **~ display** *n* PER APP representación *f* visual por diodos emisores de luz; **~ printer** *n* PER APP impresora *f* de LED; **~ read-out** *n* COMP lectura *f* por diodo electroluminiscente

leg *n* TELEP rama *f* local

legend *n* GEN *marking* referencia *f*

legible *adj* GEN legible

length *n* GEN longitud *f*; **~ of call** TELEP duración *f* de la llamada; **~ of fracture** PARTS, TEST *strength* extensión *f* de la fractura; **~ indicator** (*LI*) DATA indicador *m* de longitud; **~-indicator field** DATA campo *m* indicador de longitud; **~ of kevlar** ELECT longitud *f* del kevlar; **~ of lay** PARTS *cables* longitud *f* de paso de hélice

lens *n* COMP, TV cristal *m*, lente *f*; **~ aerial** RADIO antena *f* de lente; **~ antenna** RADIO antena *f* de lente; **~ pliers** *n pl* PARTS *tools* alicates *m pl* para retirar lentes; **~ protector** GEN protector *m* de lente

LEO: **~ satellite** *n* RADIO satélite *m* LEO

letter: **~ quality** *n* COMP calidad *f* de letra; **~ shift** *n* TELEP cambio *m* a letras; **~ shift signal** *n* TELEP señal *f* de cambio a letras

letterhead *n* GEN *documentation* encabezamiento *m* de carta

level *n* GEN nivel *m*; **~ attenuator** COMP atenuador *m* de nivel; **~ control** COMP control *m* de nivel; **~ hunt** TELEP exploración *f* de nivel; **~ of indentation** DATA nivel *m* de indentación; **~ measuring set** TRANS hipsómetro *m*; **~ meter** TEST nivelímetro *m*; **~ of mobility** RADIO nivel *m* de movilidad; **~ monitor** RADIO, TRANS monitor *m* de nivel; **~ monitoring** RADIO, TRANS supervisión *f* de nivel; **~ recorder** TEST, TRANS registrador *m* de nivel; **~-selecting bar** SWG barra *f* de selección de nivel; **~-selecting horizontal bar** SWG horizontal *f* de nivel; **~ shifting** TELEP defasaje *m* de nivel; **~ translator** ELECT traductor *m* de nivel; **~ variation** COMP variación *f* de nivel

levelling *BrE* (*AmE* **leveling**) *n* GEN nivelación *f*

lexical: **~ access** *n* GEN acceso *m* léxico

LFA *abbr* (*loss of frame alignment*) TELEP, TRANS alineación *f* de pérdida de trama

LFN *abbr* (*logical file name*) COMP, DATA LFN (*nombre de fichero lógico*)

LI *abbr* (*length indicator*) DATA indicador *m* de longitud

liaison *n* GEN enlace *m*

LIB *abbr* (*line interface board*) SWG LIB (*placa interfaz de línea*)

liberalization *n* TV liberalización *f*

library *n* GEN biblioteca *f*; **~ of data** DATA biblioteca *f* de datos; **~ program** COMP programa *m* de biblioteca

LIC *abbr* (*line interface circuit*) SWG LIC (*circuito de interconexión de línea*)

license *vt* APPL conceder licencia a

lid *n* COMP, PARTS tapa *f*

life *n* TEST *dependability* vida *f*; **~ cycle cost** APPL *quality* coste *m* (*AmL* costo) completo de utilización; **~ expectancy** APPL esperanza *f* de vida; **~-length distribution function** TEST *dependability* función *f* de distribución del

tiempo de vida; ~ **testing** TEST prueba *f* de duración
lifespan *n* TEST *dependability* vida *f*
lifetime *n* TEST *dependability* duración *f* de vida; ~ **expectancy** APPL esperanza *f* de vida
LIFO *abbr* (*last in, first out*) COMP LIFO (*último en entrar, primero en salir*)
lifted: ~ **chip** *n* COMP microplaca *f* elevada; ~ **land** *n* ELECT, PARTS *printed circuits* levantamiento *m* de la isla
lifting: ~ **bar** *n* SWG soporte *m* elevador; ~ **comb** *n* SWG peine *m* alzador; ~ **spring** *n* SWG muelle *m* alzador
light *n* GEN luz *f*; ~ **beam** ELECT, TRANS haz *m* luminoso; ~ **current** ELECT, POWER, TRANS *fibre optics* corriente *f* luminosa; ~-**current engineering** POWER ingeniería *f* de corriente fotoeléctrica; ~ **detector** ELECT, TV detector *m* luminoso; ~ **emitter** TV cuerpo *m* fotoemisor; ~-**emitting diode** (*LED*) ELECT diodo *m* electroluminiscente, diodo *m* emisor de luz; ~-**emitting diode panel** ELECT panel *m* de diodo electroluminiscente; ~ **indicator** GEN indicador *m* luminoso; ~ **interferometer** TEST interferómetro *m* luminoso; ~ **microscope** TV microscopio *m* óptico; ~ **output** TEST emisión *f* luminosa; ~ **pen** COMP, PER APP lápiz *m* fotosensible; ~-**pen detection** COMP detección *f* por lápiz fotosensible; ~ **propagation** TRANS *fibre optics* propagación *f* de la luz; ~ **pulse** TV impulso *m* luminoso; ~ **ray** TRANS *fibre optics* rayo *m* de luz; ~-**signal transfer characteristic** ELECT característica *f* de transferencia de señal luminosa; ~ **source** TV fuente *f* de luz; ~-**to-medium network traffic** TRANS tráfico *m* de red entre ligero y medio; ~-**wave regeneration** ELECT regeneración *f* de onda luminosa; ~-**wave stream** ELECT corriente *f* de ondas lumínicas
lighted: ~ **display** *n* COMP pantalla *f* iluminada
lightguide *n* ELECT, TRANS guía *f* de luz; ~ **building cable** PARTS cable *m* de construcción con guía de luz; ~ **distribution system** ELECT sistema *m* de distribución con guía de luz; ~ **express entry** (*LXE*) ELECT entrada *f* expresa para guía de luz (*LXE*); ~ **interconnection unit** (*LIU*) ELECT unidad *f* de interconexión de guía de luz (*LIU*)
lighting *n* POWER, TV iluminación *f*; ~ **fixture** POWER puente *m* de alumbrado; ~ **model** COMP modelo *m* de iluminación
lightness *n* COMP claridad *f*
lightning: ~ **arrester** *n* POWER pararrayos *m*; ~ **conductor** *n* POWER pararrayos *m* atmosférico; ~ **impulse test** *n* POWER, TEST prueba *f* de descargas atmosféricas; ~ **rod** *n* POWER *protection* pararrayos *m*
lightweight: ~ **headset** *n* TELEP auricular *m* ligero; ~ **telephone headset** *n* TELEP auricular *m* telefónico ligero
limit[1] *n* COMP, ELECT, RADIO límite *m*; ~ **check** CONTR control *m* de tolerancias; ~ **switch** POWER interruptor *m* de fin de carrera; ~ **value** TEST valor *m* límite
limit[2] *vt* DATA, TEST limitar
limitation *n* GEN limitación *f*
limited: ~ **access** *n* TV acceso *m* restringido; ~-**access channel** *n* TV canal *m* de acceso restringido; ~ **availability** *n* SWG accesibilidad *f* limitada; ~ **bandwidth** *n* TRANS ancho *m* de banda limitado; ~ **multilevel image** *n* TELEP imagen *f* multinivel limitada; ~ **roaming telephone service** *n* TELEP servicio *m* limitado de abonados móviles; ~-**waiting queue** *n* TELEP cola *f* de espera limitada
limiter *n* ELECT, POWER limitador *m*
limiting: ~ **current** *n* POWER corriente *f* de limitación; ~ **position** *n* GEN posición *f* final; ~ **state** *n* TEST estado *m* límite; ~ **value** *n* TEST valor *m* límite
line *n* GEN línea *f*, trama *f*; ~ **access point** TELEP punto *m* de acceso a la línea; ~ **access procedure** (*LAP*) TELEP procedimiento *m* de acceso a la línea; ~ **amplifier** TV amplificador *m* de línea; ~ **analyser** *BrE* TEST analizador *m* de línea; ~ **balance** TRANS equilibrio *m* de línea; ~ **bend correction** TV corrección *f* de compensación de línea; ~ **blanking interval** TV intervalo *m* de borrado de línea; ~ **of centre** *BrE* RADIO eje *m*; ~ **circuit** SWG, TELEP circuito *m* de línea; ~ **circuit pack** TELEP paquete *m* de circuito de línea; ~ **code** DATA, TELEP, TRANS código *m* en línea; ~ **concentrator** TELEP, TRANS concentrador *m* de líneas; ~ **connection unit** TRANS unidad *f* de conexión de líneas; ~ **connector cord** COMP cable *m* de conectador de línea; ~ **cord** COMP cordón *m* de alimentación, ELECT, POWER cordón *m* de conexión a la red; ~ **current** COMP corriente *f* de línea; ~ **decoder** TRANS descodificador *m* de línea; ~ **digital terminal** TELEP terminal *m* digital de línea; ~ **digit rate** TRANS velocidad *f* digital de línea; ~ **drawing** DATA dibujo *m* lineal; ~ **equipment** TRANS equipo *m* de línea; ~ **fault** TELEP fallo *m* de línea; ~ **feed** COMP, TELEP alimentación *f* de la línea; ~-**feed character** DATA carácter *m* de alimentación de línea; ~ **filter** COMP, POWER filtro *m* antiparasitario; ~ **finder** SWG, TELEP, TRANS buscador *m* de llamada; ~-**finder stage** SWG,

TELEP, TRANS fase *f* de buscador de llamada; ~ **frequency** COMP, TV frecuencia *f* de línea; ~ **frequency generator** TV generador *m* de la frecuencia de línea; ~ **group** SWG, TRANS grupo *m* de líneas; ~ **hunting** TELEP búsqueda *f* de línea; ~ **input** COMP entrada *f* de línea; ~ **interface board** (*LIB*) SWG placa *f* interfaz de línea (*LIB*); ~ **interface circuit** (*LIC*) SWG circuito *m* de interconexión de línea (*LIC*); ~ **interface module** SWG módulo *m* de interfaz de líneas; ~ **keystone correction** TV señal *f* trapezoidal de corrección de línea; ~ **keystone waveform** TV compensador *m* de distorsión trapezoidal; ~ **lamp** TELEP lámpara *f* de llamada; ~ **lengthener** RADIO extensor *m* de guíaondas; ~ **link** TELEP enlace *m* en línea; ~ **load control** TELEP control *m* de carga de línea; ~ **lockout** TELEP bloqueo *m* de línea; **~-lockout warning tone** TELEP tono *m* anunciador de bloqueo de línea; ~ **of maintenance** TELEP línea *f* de mantenimiento; ~ **module** SWG módulo *m* de líneas; ~ **module equipment** SWG equipo *m* modular de línea; ~ **network** TELEP red *f* de líneas; ~ **noise** COMP ruido *m* de circuito; ~ **non-linearity** RADIO no linealidad *f* de la línea; **~-out** COMP salida *f* de línea; ~ **output** COMP base *f* de tiempo de línea; ~ **out-of-service signal** SIG señal *f* de línea fuera de servicio; ~ **polarity inversion** TRANS inversión *f* de polaridad de línea; ~ **polarity reversal** TRANS inversión *f* de polaridad de línea; **~-in port** PER APP puerta *f* de entrada de línea; ~ **printer** PER APP impresora *f* de línea; ~ **printing** PER APP impresión *f* línea por línea; ~ **rental** APPL alquiler *m* de líneas; ~ **repeater** TELEP repetidor *m* de líneas; ~ **resistance** TRANS resistencia *f* de línea; ~ **reversal** TELEP inversión *f* de líneas; ~ **scan** COMP, RADIO exploración *f* de línea; ~ **scanning** COMP, RADIO exploración *f* de línea; ~ **section** TRANS sección *f* de línea; ~ **seizure button** PER APP pulsador *m* de toma de líneas; **~-sequential drive** ELECT transmisión *f* de secuencia de líneas; ~ **sharing** DATA uso *m* compartido de línea; ~ **shift** TELEP cambio *m* de línea; ~ **of sight** RADIO línea *f* de visión, trayectoria *f* óptica, TV eje *m* de visación; **~-of-sight propagation** RADIO propagación *f* rectilínea directa; **~-of-sight range** RADIO alcance *m* de visibilidad óptica; ~ **signalling** *BrE* SIG señalización *f* de línea; ~ **signalling** *BrE* **code** (*LSC*) SIG código *m* de señalización de línea (*LSC*); ~ **signalling** *BrE* **equipment** SIG equipo *m* de señalización de línea; ~ **signalling** *BrE* **system** SIG sistema *m* de señalización de línea; ~ **signalling** *BrE* **test** SIG, TEST prueba *f* de señalización de línea; ~ **simulator** TRANS línea *f* artificial; ~ **slip** TV deslizamiento *m* de línea; ~ **space** COMP *word processing* espacio *m* interlineal; ~ **spectrum** ELECT, TRANS *fibre optics* espectro *m* de líneas; **~-stabilized oscillator drive** ELECT excitación *f* de oscilador estabilizado por línea; ~ **stretcher** RADIO extensor *m* de línea; ~ **sweep** TV barrido *m* lineal; ~ **switch** SWG conmutador *m* de línea; ~ **switching** SWG conmutación *f* de línea; **~-synchronizing signal** TV señal *f* de sincronismo de línea; ~ **system** TELEP, TRANS sistema *m* de líneas; ~ **tapping** ELECT derivación *f* de línea; ~ **terminal** TELEP, TRANS terminal *m* de línea; **~-terminal station** TELEP estación *f* terminal de línea; **~-terminating equipment** TELEP equipo *m* de terminación de línea; **~-terminating magazine** TRANS almacén *m* de terminación de línea; ~ **termination** TELEP, TRANS terminación *f* de línea; ~ **test** TEST prueba *f* de línea; ~ **tester** TEST aparato *m* para probar líneas eléctricas; ~ **test-point** TEST punto *m* de prueba de línea; ~ **test programme** *BrE* TEST programa *m* para prueba de línea; ~ **test set** TEST equipo *m* para prueba de línea; ~ **tilt correction** TV señal *f* corrección de pendiente de línea; **~-time luminance distortion** COMP distorsión *f* de luminancia en tiempo de línea; ~ **unit** TELEP unidad *f* de línea; **~-up** TRANS alineamiento *m*; **~-up instructions** *n pl* TRANS instrucciones *f pl* de alineamiento; **~-up period** TRANS periodo *m* de alineamiento; **~-up procedure** TRANS procedimiento *m* de alineamiento; **~-up record** TRANS registro *m* de alineamiento; ~ **width** ELECT *printed circuits* ancho *m* de línea

linear *adj* GEN lineal; ~ **amplifier** *n* TRANS amplificador *m* lineal; ~ **amplitude distortion** *n* TRANS distorsión *f* de amplitud lineal; ~ **analog control** *n* TELEP, TRANS control *m* analógico lineal; ~ **approximation** *n* ELECT aproximación *f* lineal; ~ **bit density** *n* COMP densidad *f* de registro lineal; ~ **block code** *n* SIG código *m* de bloque lineal; ~ **channel** *n* TRANS canal *m* lineal; ~ **circuit** *n* DATA, TRANS circuito *m* lineal; ~ **code** *n* DATA código *m* lineal; ~ **control** *n* COMP control *m* lineal; ~ **curve** *n* COMP curva *f* lineal; ~ **detection** *n* GEN detección *f* lineal; ~ **detector** *n* ELECT detector *m* lineal; **~-digital voice scrambler** *n* TELEP escamoteador *m* lineal digital de voz; ~ **distortion** *n* TRANS distorsión *f* lineal; ~ **expansion** *n* GEN expansión *f* lineal; ~ **expansion coefficient** *n* ELECT, GEN, PARTS coeficiente *m* de expansión lineal; ~ **filter** *n* DATA, TRANS filtro *m* lineal; ~ **filtering** *n* TRANS filtrado *m* lineal; ~ **integrated circuit** *n* DATA, ELECT circuito *m* integrado lineal; ~ **interpolation** *n* TRANS interpolación *f* lineal; ~ **joint** *n* GEN empalme *m* longitudinal; ~ **measure** *n* TEST medida *f* delongitud; ~ **oscillation** *n* TRANS oscilación *f* lineal; ~ **path** *n* COMP trayectoria *f* lineal; ~ **polarization** *n* POWER, RADIO, TRANS polarización *f* lineal; ~ **power amplifier** *n* TRANS amplificador *m* de potencia lineal; ~ **predicting coding** *n* DATA codificación *f* predictivalineal; ~ **predicting coding vocoder** *n* RADIO codificación *f* predictiva vocoder lineal; ~ **predictive vocoder** *n* TELEP vocoder *m* de predicción lineal; ~ **rackspace** *n* ELECT separación *f* lineal de bastidores; ~ **reading** *n* COMP lectura *f* lineal; ~ **receiver** *n* RADIO receptor *m* lineal; ~ **relation** *n* GEN relación *f* lineal; ~ **relationship** *n* GEN relación *f* lineal; ~ **scale** *n* TEST escala *f* lineal; ~**-scaling calculation** *n* DATA cálculo *m* deescalalíneal; ~ **slider** *n* COMP deslizador *m* lineal; ~ **synthesizer** *n* COMP sintetizador *m* lineal; ~ **system** *n* CONTR sistema *m* lineal; ~ **video** *n* COMP vídeo *m* lineal

linearity *n* GEN linealidad *f*; ~ **error** PARTS *semiconductors* error *m* de linealidad

linearizer *n* TRANS linealizador *m*

linearly: ~**-increasing voltage** *n* POWER tensión *f* de aumento lineal, tensión *f* en rampa; ~ **polarized** *adj* (*LP*) TRANS linealmente polarizado; ~**-polarized mode** *n* TRANS modo *m* de polarización lineal, modo *m* linealmente polarizado

liner *n* ELECT, PARTS revestimiento *m*

lines: ~ **per screen** *n pl* TV líneas *f pl* por pantalla

lingering: ~ **effect** *n* COMP efecto *m* persistente

linguistic: ~ **engineering** *n* COMP ingeniería *f* lingüística

lining *n* PARTS revestimiento *m* interior

link *n* GEN enlace *m*; ~ **access procedure B** TELEP procedimiento *m* B de acceso por enlace; ~**-by-**~ **encipherment** DATA cifrado *m* enlace por enlace; ~**-by-**~ **signalling** *BrE* SIG, TRANS señalización *f* sección por sección; ~**-by-**~ **traffic routing** TELEP enrutamiento *m* del tráfico enlace por enlace; ~ **coupling** RADIO acoplamiento *m* de eslabón; ~ **encryption** DATA cifrado *m* de enlaces; ~ **file** COMP fichero *m* de enlace; ~**-handling module** COMP, DATA *datacom* módulo *m* de tratamiento de enlaces; ~ **icon** COMP icono *m* de enlace; ~ **layer** TRANS capa *f* de enlace; ~ **map** COMP mapa *m* de enlaces; ~ **margin** TRANS margen *m* de enlace; ~ **marker** COMP marcador *m* de enlace; ~ **network** COMP red *f* de enlace; ~ **power budget** RADIO *mobile*, TRANS cálculo *m* de la potencia de enlace; ~ **status signal unit** SIG unidad *f* de señal de categoría de enlace; ~ **system** TRANS sistema *m* de enlaces; ~ **test** SWG, TEST prueba *f* de eslabones; ~ **test equipment** SWG, TEST probador *m* de eslabones

link up *vt* DATA, TRANS enlazar; ◆ ~ **with** DATA, TRANS establecer conexión con

linkage *n* COMP enlazamiento *m*; ~ **editor** COMP editor *m* de enlace

linked: ~ **offices** *n pl* TELEP oficinas *f pl* conectadas; ~ **operations** *n pl* DATA operaciones *f pl* encadenadas

linker *n* COMP enlazador *m*

linking: ~ **clusters of balun adaptors** *n pl* ELECT agrupaciones *f pl* de enlace de adaptadores balun

links: ~ **menu** *n* COMP menú *m* de enlaces

linkway *n* COMP trayectoria *f* de enlace

lipping *n* COMP formación *f* de reborde

LIPS *abbr* (*logical inferences per second*) TEST LIPS (*inferencias lógicas por segundo*)

liquid *n* GEN líquido *m*; ~ **caustic soda** PARTS soda *f* cáustica líquida; ~ **crystal display** (*LCD*) ELECT visualización *f* en cristal líquido (*VCL*); ~ **crystals** *n pl* ELECT, PARTS cristales *m pl* líquidos; ~**-level indicator** TELEP indicador *m* de nivel de líquidos; ~**-sample counter** RADIO contador *m* de muestras líquidas

LI/S *abbr* (*logical inferences per second*) TEST LI/S (*inferencias lógicas por segundo*)

list *vt* COMP, DATA enumerar

listed *adj* COMP, DATA enumerado

listen *vi* TELEP escuchar

listener: ~ **echo** *n* TELEP eco *m* en el receptor; ~ **echo loss** *n* TELEP atenuación *f* del eco en el receptor

listening *n* GEN escucha *f*; ~**-in** RADIO, TELEP, TEST escucha *f*; ~ **jack** PARTS, TELEP jack *m* de escucha; ~ **level** COMP nivel *m* de recepción; ~ **silence** RADIO silencio *m* de escucha; ~ **test** RADIO, TEST prueba *f* de escucha

listing *n* COMP, DATA listado *m*

lit *adj* COMP, ELECT, TELEP *display* iluminado

literal *n* COMP literal *m*

lithium *n* PARTS litio *m*; ~ **neobate** PARTS niobato *m* de litio

LIU *abbr* (*lightguide interconnection unit*) ELECT LIU (*unidad de interconexión de guía de luz*)

live *adj* POWER, TV vivo; ~ **conversation** *n* DATA conversación *f* directa; ~ **running** *n* COMP ejecución *f* activa; **~-situation traffic planning** *n* TELEP planificación *f* de tráfico de situación real; **~ traffic test** *n* TEST prueba *f* de tráfico real; **~ transmission** *n* TV emisión *f* en directo; **~ TV** *n* GEN TV *f* en directo; **~ video coverage** *n* TV cobertura *f* de vídeo en directo

LLC *abbr* (*logical link control*) CONTR control *m* lógico de enlace

LME *abbr* (*layer management entity*) DATA LME (*entidad para la gestión de capas*)

LMS *abbr* (*land-mobile station*) RADIO, TELEP estación *f* móvil terrestre

LMSS *abbr* (*land-mobile satellite service*) RADIO, TELEP, TRANS LMSS (*servicio móvil terrestre de satélite*)

lm/W *abbr* (*lumen per watt*) TEST lm/W (*lumen por watt*)

LNA *abbr* (*launch numerical aperture*) ELECT, TRANS apertura *f* numérica de lanzamiento

LO *abbr* (*local oscillator*) ELECT, RADIO oscilador *m* local

load[1] *n* COMP carga *f*; **~ balancing** DATA equilibrio *m* de carga; **~ capacitance** ELECT capacitancia *f* de carga; **~ capacity** TRANS capacidad *f* de carga; **~ circuit** ELECT circuito *m* en carga; **~ current** POWER corriente *f* de carga; **~ curve diagram** POWER diagrama *m* de carga; **~ cycle** POWER ciclo *m* de funcionamiento; **~ distribution** DATA, POWER distribución *f* de carga; **~ factor** POWER densidad *f* de ocupación; **~ function** POWER diagrama *m* de carga; **~ host** ELECT central *f* de carga; **~ impedance** ELECT impedancia *f* de carga; **~ number** TELEP número *m* de circuito de carga; **~ observation** TELEP observación *f* de la carga; **~ power** ELECT tensión *f* de carga; **~ range** POWER margen *m* de carga; **~ resistor** PARTS, POWER resistor *m* de carga; **~ sharing** DATA, TELEP carga *f* compartida; **~-sharing system** ELECT sistema *m* de reparto de la carga; **~ tester** TELEP probador *m* de ocupaciones; **~ transfer** ELECT transferencia *f* de cargas; **~-transfer acknowledgement signal** TELEP señal *f* de recepción de transferencia de cargas; **~-transfer signal** SIG señal *f* de transferencia de cargas

load[2] *vt* GEN cargar; ◆ **~ continuously** TRANS *cables* krarupizar

loaded *adj* GEN cargado

loading *n* GEN carga *f*; **~ cartridge** COMP cartucho *m* de carga; **~ coil** TRANS bobina *f* de carga; **~-coil case** TRANS caja *f* de bobinas de carga; **~-coil pot** TRANS caja *f* de bobinas de carga; **~ error** COMP error *m* de carga; **~ operation** COMP operación *f* de carga

loadsharing *n* TELEP carga *f* compartida

lobe *n* ELECT, RADIO lóbulo *m*; **~ attachment module** ELECT módulo *m* de fijación de lóbulo; **~ length** ELECT longitud *f* de lóbulo

local: **~ action** *n* POWER acción *f* local; **~ application** *n* (*LA*) TELEP aplicación *f* local (*LA*); **~ area network** *n* (*LAN*) COMP, DATA, TELEP, TRANS red *f* de área local; **~ authority** *n* TV administración *f* local; **~ battery** *n* TELEP batería *f* local; **~ battery system** *n* TELEP sistema *m* de batería local; **~ bus** *n* COMP bus *m* local; **~ bus card** *n* COMP tarjeta *f* de bus local; **~ call** *n* TELEP llamada *f* local; **~ calling area** *n* (*LCA*) TELEP área *f* de comunicación urbana; **~ central office** *n* TELEP central *f* telefónica urbana; **~ channel** *n* TV canal *m* local; **~ charge-rate call** *n* TELEP llamada *f* de tarifa urbana; **~ charge-rate trunk call** *n* TELEP llamada *f* de línea troncal de tarifa urbana; **~ communications network** *n* TRANS red *f* de comunicaciones urbanas; **~ control** *n* CONTR, TRANS mando *m* local; **~ copy** *n* TRANS copia *f* local; **~ digital concentrator** *n* TELEP concentrador *m* digital local; **~ distribution cable** *n* TRANS cable *m* de distribución local; **~ distribution network** *n* COMP, ELECT, TELEP, TRANS red *f* de distribución local; **~ end** *n* TELEP conjunto *m* terminal; **~ exchange** *n* (*LE*) SWG, TELEP central *f* local; **~ exchange area** *n* TELEP zona *f* de central urbana; **~ function capabilities** *n pl* SWG capacidad *f* de las funciones locales; **~ index** *n* TELEP índice *m* local; **~ junction** *n* TRANS unión *f* local; **~ lighting** *n* GEN iluminación *f* localizada; **~ line** *n* TELEP línea *f* local; **~-line concentrator** *n* TRANS distribuidor *m* de líneas urbanas; **~ loop** *n* RADIO, TELEP bucle *m* local; **~ maintenance** *n* GEN mantenimiento *m* local; **~ mode** *n* TELEP modo *m* local; **~ network interconnect** *n* ELECT interconexión *f* con red telefónica local; **~ number** *n* TELEP número *m* local; **~ operation** *n* CONTR operación *f* local; **~ oscillator** *n* (*LO*) ELECT, RADIO oscilador *m* local; **~ production** *n* TV producción *f* local; **~ rate** *n* TELEP tarifa *f* local; **~ record** *n* TELEP control *m* local; **~ reference** *n* GEN referencia *f* local; **~ register** *n* SWG registro *m* local; **~ resource** *n* DATA recurso *m* local; **~ single-layer embedded test method** *n* TEST método *m* de prueba incorporado de capa única local; **~ studio** *n* TV estudio *m* local; **~ switch** *n* SWG, TELEP conmutador *m* local; **~ tariff** *n* TELEP tarifa *f* local; **~ telephone network** *n* TELEP, TRANS red *f* telefónica local; **~ test method** *n* TEST método *m* de prueba local; **~ traffic** *n* GEN tráfico *m* local; ~ **tree** *n* DATA árbol *m* local; ~ **user terminal** *n* (*LUT*) PER APP terminal *m* de usuario local

localization *n* GEN localización *f*; ~ **system** RADIO sistema *m* de localización

locally: ~ **controlled** *adj* CONTR, TRANS controlado localmente

locate *vt* GEN localizar

locater *n* COMP localizador *m* de coordenadas

locating *n* GEN localización *f*; ~ **hole** ELECT *printed circuits* agujero *m* de posicionamiento; ~ **lamp** TELEP *indicating device* lámpara *f* de posición; ~ **notch** ELECT *printed circuits* muesca *f* de posicionamiento

location *n* GEN *position* emplazamiento *m*, *finding* localización *f*; ~ **area** RADIO, TELEP área *f* de localización; ~ **area identification** RADIO, TELEP identificación *f* de área de localización; ~ **cancellation** RADIO cancelación *f* de localización; ~ **deregistration** RADIO eliminación *f* de registro de localización; ~ **information** RADIO información *f* de localización; ~ **management** RADIO gestión *f* de emplazamiento de posición; ~ **register** RADIO registro *m* de localización; ~ **registration** RADIO registro *m* de localización; ~ **updating** RADIO actualización *f* de localización

lock *n* GEN bloqueo *m*; ~**-in range** RADIO gama *f* de señales de sincronización

lock on *vt* ELECT, RADIO *channel* engancharsea

lock out *vt* COMP cerrar

locked: ~ **file** *n* COMP fichero *m* cerrado; ~ **loop** *n* ELECT bucle *m* enganchado

locking: ~ **clip** *n* ELECT clip *m* de retención; ~ **device** *n* PER APP dispositivo *m* de parada; ~ **lever key** *n* GEN conmutador *m* de palanca con retención; ~ **mechanism** *n* COMP mecanismo *m* de bloqueo; ~ **ring** *n* ELECT, GEN anillo *m* de fijación; ~ **spring** *n* PARTS muelle *m* de bloqueo; ~**-type button** *n* COMP botón *m* con retención

locknut *n* PARTS contratuerca *f*

lockout *n* TELEP bloqueo *m*

LOF *abbr* (*loss of frame*) TELEP, TRANS pérdida *f* de trama

log[1] *abbr* (*logarithm*) RADIO log (*logaritmo*); ~**-normal fading** *n* RADIO *mobile* atenuación *f* logarítmica normal; ~**-normal shadowing** *n* RADIO distribución *f* logarítmica normal; ~**-periodic antenna** *n* RADIO antena *f* logarítmica

log[2] *n* GEN registro *m* de operaciones; ~**-commit record** DATA registro *m* de traslado a diario; ~**-damage record** DATA registro *m* de daño en diario; ~**-in** COMP, RADIO iniciación *f* de registro; ~**-in dialogue** *BrE* ELECT diálogo *m* de apertura; ~**-heuristic record** DATA registro *m* heurístico en diario; ~**-ready record** DATA registro *m* inmediato en el libro de operaciones; ~ **record** DATA registro *m* en diario

logarithm *n* (*log*) RADIO logaritmo *m* (*log*)

logarithmic: ~ **gain** *n* CONTR *automatic control* ganancia *f* logarítmica

logatom *n* RADIO, TELEP *acoustics* logátomo *m*; ~ **articulation index** RADIO, TELEP índice *m* de nitidez en logátomos

logbook *n* GEN libro *m* de guardia

logger *n* COMP registrador *m*

logging *n* COMP registro *m*

logic *n* GEN lógica *f*; ~ **analyser** *BrE* DATA analizador *m* lógico, analizador *m* de lógica; ~ **array** DATA conjunto *m* de unidades lógicas; ~ **circuit** ELECT circuito *m* lógico; ~ **design** GEN diseño *m* lógico; ~ **device** ELECT dispositivo *m* lógico; ~ **diagram** GEN diagrama *m* lógico; ~ **element** ELECT elemento *m* lógico; ~ **filtering** COMP filtrado *m* lógico; ~ **instruction** COMP instrucción *f* lógica; ~ **operation** COMP operación *f* lógica; ~ **probe** ELECT *printed circuits* sonda *f* lógica; ~ **programming** COMP programación *f* lógica; ~ **shift** COMP desplazamiento *m* lógico; ~ **state** ELECT *logic elements* estado *m* lógico; ~ **symbol** DATA, ELECT símbolo *m* lógico; ~ **system** COMP sistema *m* lógico; ~ **unit** (*LU*) COMP, DATA, RADIO, TELEP unidad *f* lógica (*LU*)

logical: ~ **block address** *n* (*LBA*) COMP dirección *f* de bloque lógico (*LBA*); ~ **channel** *n* TELEP canal *m* lógico; ~ **comparison** *n* COMP comparación *f* lógica; ~ **file** *n* DATA archivo *m* lógico; ~ **file name** *n* (*LFN*) COMP, DATA nombre *m* de fichero lógico (*LFN*); ~ **inferences per second** *n* (*LIPS*), (*LI/S*) TEST inferencias *f pl* lógicas por segundo (*LIPS*, *LI/S*); ~ **link control** *n* (*LLC*) CONTR control *m* lógico de enlace; ~ **symbol** *n* DATA, ELECT símbolo *m* lógico

log in *vi* COMP conectarse al sistema, iniciar la sesión

log off *vi* COMP desconectarse del sistema, terminar la sesión

log on *vi* COMP conectarse al sistema, iniciar la sesión

log out *vi* COMP desconectarse del sistema, terminar la sesión

logistic: ~ **delay** *n* TELEP, TEST demora *f* logística; ~ **delay time** *n* TELEP, TEST tiempo *m* de demora logística

logistics *n* GEN logística *f*

LOM *abbr* (*loss of multiframe*) TELEP, TRANS pérdida *f* de multitrama

lone: ~ **signal unit** *n* (*LSU*) SIG unidad *f* de señal solitaria (*LSU*)

long: ~**-address acceptance** *n* DATA aceptación *f* de direcciones extensas; ~ **circuit** *n* TELEP circuito *m* largo; ~**-distance backscatter** *n* POWER, RADIO retrodispersión *f* de larga distancia; ~**-distance call** *n* TELEP llamada *f* de larga distancia; ~**-distance connection** *n* TELEP conexión *f* de larga distancia; ~**-distance dual-homing link** *n* ELECT enlace *m* de doble búsqueda a larga distancia; ~**-distance line** *n* TELEP, TRANS línea *f* de larga distancia; ~**-distance network** *n* TELEP red *f* de larga distancia; ~**-distance telephone network** *n* TELEP red *f* telefónica de larga distancia; ~**-haul transmission** *n* TRANS transmisión *f* de larga distancia; ~**-life battery** *n* POWER batería *f* de larga duración; ~**-line equipment** *n* TRANS equipo *m* de línea larga; ~**-nose pliers** *n pl* GEN *tools* alicates *m pl* de punta larga; ~**-path bearing** *n* RADIO cojinete *m* de largo recorrido; ~**-range navigation** *n* RADIO navegación *f* de largo recorrido; ~**-term bit error rate** *n* DATA índice *m* de error de bitios en periodos largos; ~**-term planning** *n* GEN planificación *f* a largo plazo; ~**-term test** *n* TEST prueba *f* a largo plazo; **very** ~ **baseline interferometry** *n* (*VLBI*) TEST interferometría *f* con línea de base muy larga (*VLBI*)

longevity *n* DATA, PARTS, TEST *dependability* longevidad *f*

longitudinal: ~ **balance** *n* POWER equilibrio *m* longitudinal; ~ **direction** *n* GEN sentido *m* longitudinal; ~ **displacement** *n* GEN desplazamiento *m* longitudinal; ~ **judder** *n* TELEP salto *m* longitudinal; ~ **offset loss** *n* TRANS *fibre optics* pérdida *f* por desviación longitudinal; ~ **redundancy check** *n* DATA verificación *f* de redundancia longitudinal; ~ **slot** *n* TRANS ranura *f* longitudinal; ~ **time code** *n* (*LTC*) COMP código *m* temporal longitudinal (*LTC*); ~ **wave** *n* TEC RAD onda *f* longitudinal

look: ~**-ahead carry** *n* ELECT acarreo *m* de lectura anticipada

loop: ~ **aerial** *n* RADIO antena *f* de cuadro; ~ **alignment error** *n* TEST error *m* instrumental; ~ **amplifier** *n* TEST amplificador *m* en bucle; ~ **antenna** *n* RADIO antena *f* de cuadro; ~ **circuit** *n* POWER circuito *m* de bucle; ~ **dialling** *BrE n* PER APP, TELEP marcación *f* en bucle; ~**-disconnect dialling** *BrE n* PER APP, TELEP *of answering machine* marcado *m* de desconexión en bucle; ~**-disconnect mode** *n* PER APP, TELEP *of answering machine* modo *m* de desconexión en bucle; ~ **extender** *n* TELEP extensor *m* del bucle; ~ **reporting system** *n* (*LRS*) TELEP sistema *m* de referencia en bucle (*LRS*); ~ **series resistance** *n* ELECT resistencia *f* en serie de bucles; ~ **start** *n* TELEP arranque *m* en bucle; ~**-start trunk** *n* TELEP línea *f* troncal de arranque en bucle; ~ **station connector** *n* ELECT conector *m* de estación en bucle; ~ **stop** *n* COMP parada *f* de bucle; ~ **wiring concentrator** *n* (*LWC*) ELECT concentrador *m* de cableado en bucle (*LWC*)

loopback *n* ELECT recorrido *m* inverso de bucle, TRANS circuito *m* cerrado; ~ **command** TV mando *m* de desandar un bucle; ~ **command audio loop request** TV solicitud *f* en bucle sonoro de comando de reversión de bucle; ~ **command digital loop request** TV solicitud *f* en bucle digital de comando de reversión de bucle; ~ **command off** TV eliminación *f* de orden de desandar un bucle; ~ **command video loop request** TV solicitud *f* en videobucle de comando de reversión de bucle; ~ **configuration** ELECT configuración *f* del recorrido inverso de un bucle; ~ **signalling** *BrE* SIG señalización *f* del recorrido inverso de un bucle; ~ **test method** TEST método *m* de prueba del recorrido inverso de un bucle

looped *adj* COMP en bucle

looping *n* COMP enlace *m*

loose: ~ **cable-structure** *n* TRANS estructura *f* de cable suelto; ~ **connection** *n* POWER conexión *f* floja; ~ **contact** *n* POWER contacto *m* flojo; ~**-structure cable** *n* TRANS cable *m* de estructura suelta; ~**-tube cable** *n* TRANS cable *m* de tubo amplio

LOP *abbr* (*loss of pointer*) TRANS pérdida *f* del apuntador

loran *n* (*long-range navigation*) RADIO sistema *m* loran

LOS *abbr* (*loss of signal*) TRANS pérdida *f* de señal

loss *n* GEN pérdida *f*; ~ **around a corner** TRANS atenuación *f* por recodo; ~ **factor** POWER factor *m* de atenuación; ~ **of frame** (*LOF*) TELEP, TRANS pérdida *f* de trama; ~ **of frame alignment** (*LFA*) TELEP, TRANS alineación *f* de pérdida de trama; ~**-of-frame-alignment detector** TELEP, TRANS detector *m* de pérdida de alineación de trama; ~**-mode operation** TRANS operación *f* en modo de pérdida; ~**-mode working** TRANS trabajo *m* en modo de pérdida; ~ **of multiframe** (*LOM*) TELEP, TRANS pérdida *f* de multitrama; ~ **of multi-**

frame alignment TELEP, TRANS pérdida *f* de alineación de multitrama; ~ **of pointer** (*LOP*) TRANS pérdida *f* del apuntador; ~ **relative to free space** RADIO atenuación *f* debida a espacio abierto; ~ **resistance** POWER resistencia *f* de pérdida; ~ **of signal** (*LOS*) TRANS pérdida *f* de señal; ~ **system** TELEP sistema *m* de pérdidas

lossless *adj* COMP, TELEP sin pérdida

lossy *adj* COMP, TELEP disipativo

lost: ~ **call** *n* SWG, TELEP, TRANS llamada *f* perdida; ~ **call attempt** *n* TELEP intento *m* de llamada perdido; ~ **time** *n* GEN tiempo *m* perdido; ~ **traffic** *n* TELEP tráfico *m* perdido

lot *n* TEST *quality* lote *m*

loudness *n* GEN sonoridad *f*; ~ **contour** TELEP contorno *m* de sonoridad; ~ **level** TELEP nivel *m* de sonoridad; ~ **scale** TELEP escala *f* de sonoridad

loudspeaker *n* COMP, RADIO, TELEP altavoz *m*; ~ **paging** RADIO búsqueda *f* de personas por altavoz

low *adj* GEN bajo; ~ **bit data communication** *n* DATA, RADIO comunicación *f* de datos de poca significación; ~ **bit rate** *n* TRANS velocidad *f* baja de tráfico binario; ~ **dissipation** *n* ELECT disipación *f* baja; ~ **earth orbit** *n* RADIO órbita *f* terrestre baja; ~ **earth orbit satellite** *n* RADIO satélite *m* de órbita terrestre baja; ~ **frequency** *n* COMP, RADIO baja *f* frecuencia; ~**-frequency amplifier** *n* ELECT amplificador *m* de baja frecuencia; ~**-frequency oscillator** *n* COMP oscilador *m* de baja frecuencia; ~**-frequency sweep** *n* COMP barrido *m* de baja frecuencia; ~**-grade code** *n* RADIO código *m* de bajo índice; ~ **insertion loss** *n* TRANS bajas *f pl* pérdidas de inserción; ~ **insulation loss** *n* TELEP pérdida *f* escasa de aislamiento; ~**-layer compatibility** *n* DATA, SWG, TELEP compatibilidad *f* de capa baja; ~**-layer compatibility information element** *n* TELEP elemento *m* de información con compatibilidad de capa baja; ~ **level** *n* ELECT bajo nivel; ~**-level language** *n* COMP lenguaje *m* de bajo nivel; ~**-level modulation** *n* GEN modulación *f* de bajo nivel; ~**-level signalling device** *BrE n* GEN dispositivo de señalización de bajo nivel; ~ **loss** *n* TELEP bajas *f pl* pérdidas; ~**-loss connector** *n* TRANS conector *m* con baja atenuación; ~**-loss fibre connector** *BrE n* TRANS conector de fibra con baja atenuación; ~ **medium** *n* COMP medio bajo; ~**-noise amplifier** *n* ELECT, RADIO, TRANS amplificador *m* con bajo nivel de ruidos; ~**-noise block** *n* TV bloque *m* de bajo nivel de ruido; ~ **note** *n* COMP nota *f* grave; ~ **orbit** *n* RADIO órbita *f* baja; ~**-passband** *n* TRANS banda *f* de pasobajo; ~**-pass filter** *n* POWER, TRANS filtro *m* de paso bajo; ~**-performance equipment** *n* COMP equipo *m* de bajo rendimiento; ~**-power distress transmitter** *n* RADIO transmisor *m* de socorro de baja potencia; ~**-power flux density** *n* RADIO densidad *f* de flujo de baja potencia; ~**-power laser diode** *n* ELECT diodo *m* láser de baja potencia; ~**-power transistor** *n* PARTS transistor *m* de baja potencia; ~ **power transmitter** *n* RADIO transmisor *m* de baja potencia; ~**-power transmitter** *n* RADIO transmisor *m* de baja potencia; ~ **priority** *n* GEN escasa *f* prioridad; ~**-range note** *n* COMP nota *f* de corto alcance; ~**-rate encoding** *n* DATA codificación *f* a baja velocidad; ~ **resistance** *n* GEN baja *f* resistencia; ~**-resistance distribution** *n* POWER distribución *f* de baja resistencia; ~**-resistance** ELECT, POWER de baja resistencia; ~ **ring to ground** *n* TEST anillo *m* bajo a tierra; ~**-speed** GEN de poca velocidad; ~ **speed** *n* GEN baja *f* velocidad; ~ **tip to ground** *n* TEST punta *f* baja a tierra; ~ **tip to ring** *n* TEST punta *f* baja a anillo; ~**-to-high** ELECT bajo a alto; ~**-velocity camera tube** *n* ELECT tubo *m* de cámara de poca velocidad; ~**-velocity scanning** *n* ELECT exploración *f* de poca velocidad; ~ **voltage** *n* POWER bajo voltaje; ~**-wattage resistor** *n* PARTS, POWER resistor *m* de bajo vataje; **very** ~**-frequency wave** *n* PARTS onda *f* de muy baja frecuencia; ~**-voltage** ELECT, POWER de baja tensión

lower: ~ **case** *n* GEN minúscula *f*; ~**-frequency** *n* RADIO baja frecuencia *f*; ~ **keyboard** *n* COMP teclado *m* inferior; ~**-level service** *n* TEST servicio *m* de bajo nivel; ~ **limit** *n* GEN límite *m* inferior; ~**-order path adaptation** *n* (*LPA*) TRANS adaptación *f* de ruta de orden inferior; ~**-order path connection** *n* (*LPC*) TRANS ruta *f* de conexión de orden inferior; ~**-order path termination** *n* (*LPT*) TRANS terminación *f* de ruta de orden inferior; ~ **quality of service** *n* TEST calidad *f* inferior de servicio; ~ **quartile** *n* GEN *statistics* cuartil *m* inferior; ~ **sideband** *n* TRANS banda *f* lateral inferior; ~ **subfield** *n* RADIO, TRANS subcampo *m* inferior; ~ **tariff** *n* APPL tarifa *f* reducida; ~ **tester** *n* (*LT*) DATA, TEST probador *m* inferior (*TL*)

lowest: ~ **usable frequency** *n* (*LUF*) RADIO frecuencia *f* mínima utilizable (*LUF*)

LP *abbr* (*linearly polarized*) TRANS linealmente polarizado

LPA *abbr* (*lower-order path adaptation*) TRANS adaptación *f* de ruta de orden inferior

LPC *abbr* (*lower-order path connection*) TRANS ruta *f* de conexión de orden inferior

LPT *abbr* (*lower-order path termination*) TRANS terminación *f* de ruta de orden inferior

LRS *abbr* (*loop reporting system*) TELEP LRS (*sistema de referencia en bucle*)

LS *abbr* (*leased line*) TELEP, TRANS LS (*línea alquilada*)

LSB *abbr* (*least significant bit*) DATA BMS (*bit menos significativo*)

LSC *abbr* (*line signalling code*) SIG LSC (*código de señalización de línea*)

L-section: ~ **filter** *n* ELECT filtro *m* con bobina de entrada

LSI *abbr* (*large-scale integration*) ELECT, TELEP, TRANS IGE (*integración a gran escala*); ~ **circuit** *n* ELECT, TELEP, TRANS circuito *m* IGE

LSU *abbr* (*lone signal unit*) SIG LSU (*unidad de señal solitaria*)

LT *abbr* (*lower tester*) DATA, TEST TL (*probador inferior*)

LTC *abbr* (*longitudinal time code*) COMP LTC (*código temporal longitudinal*)

LU *abbr* (*logic unit*) COMP, DATA, TELEP LU (*unidad lógica*)

LUF *abbr* (*lowest usable frequency*) RADIO LUF (*frecuencia mínima utilizable*)

luggable *adj* COMP conectable

lumen *n* TEST lumen *m*; ~ **maintenance** TV constancia *f* del flujo luminoso; ~ **per watt** (*lm*/*W*) TEST lumen *m* por watt (*lm*/*W*)

luminance *n* COMP, ELECT, TV luminancia *f*; ~ **axis** COMP eje *m* de luminancia; ~ **compensation** TELEP compensación *f* de luminancia; ~ **component** COMP componente *m* de luminancia; ~ **contrast** TV contraste *m* de luminancia; ~ **factor** COMP factor *m* de luminancia; ~ **meter** TEST medidor *m* de luminancia; ~ **ratio** TEST ratio *f* de luminancia; ~ **signal** TV señal *f* de luminancia

luminescence *n* ELECT, TRANS luminiscencia *f*

luminescent *adj* ELECT, TRANS luminiscente; ~ **screen** *n* ELECT, TV pantalla *f* luminiscente

luminous *adj* GEN luminoso; ~ **button** *n* CONTR botón *m* luminoso; ~ **efficacity** *n* GEN rendimiento *m* luminoso; ~ **efficacy of a source of light** *n* DATA rendimiento *m* luminoso de una fuente de luz; ~ **emittance** *n* (*M*) TRANS *from point of surface* emitancia *f* luminosa (*M*);; ~**-flux threshold** *n* ELECT umbral *m* de flujo luminoso

LUT *abbr* (*local user terminal*) PER APP terminal *m* de usuario local

LWC *abbr* (*loop wiring concentrator*) ELECT LWC (*concentrador de cableado en bucle*)

LXE *abbr* (*lightguide express entry*) ELECT LXE (*entrada expresa para guía de luz*)

M

M *abbr* (*multivalued*) DATA, RADIO M (*polivalente*); **~ component** *n* DATA componente *m* M; **~ reflection** *n* RADIO reflexión *f* M; **~ unit** *n* RADIO unidad *f* M

mA *abbr* (*milliamp*) ELECT mA (*miliamperio*)

MA *abbr* (*medium adaptor*) ELECT, TELEP adaptador *m* intermedio

MAC *abbr* (*medium access control*) CONTR control *m* de acceso medio

MACF *abbr* (*multiple association control function*) DATA MACF (*función de control de asociación múltiple*)

machine: **~ code** *n* COMP código *m* de máquina; **~ language** *n* COMP lenguaje *m* de máquina; **~-readable medium** *n* CONTR soporte *m* legible por máquina; **~ soldering** *n* ELECT soldadura *f* automática

machining *n* GEN mecanización *f*

macro *n* COMP, DATA macro *m*

macrobend: **~ loss** *n* TRANS pérdida *f* de macroflexión

macrobending *n* TRANMIS macroflexión *f*

macroblock *n* COMP macrobloque *m*

macrocell *n* RADIO macrocélula *f*

macrogenerating: **~ program** *n* COMP programa *m* macrogenerador

macrogenerator *n* COMP macrogenerador *m*

macroinstruction *n* COMP, DATA macroinstrucción *f*

macrosegmentation *n* RADIO macrosegmentación *f*

MADT *abbr* (*micro-alloy diffused transistor*) ELECT MADT (*transistor de microaleación difusa*)

MAF *abbr* (*management application function*) DATA función *f* de gestión de la aplicación

magazine *n* GEN almacén *m*; **~ spacing** GEN separación *f* de almacenes

magic: **~ T** *n* RADIO T *f* mágica

magnet *n* ELECT imán *m*; **~ coil** ELECT bobina *f* de electroimán; **~ coupling** ELECT acoplamiento *m* magnético

magnetic *adj* DATA, ELECT magnético; **~ blowout** *n* ELECT soplador *m* magnético; **~ bubble memory** *n* DATA memoria *f* de burbuja magnética; **~ card** *n* DATA tarjeta *f* magnética; **~-card storage** *n* DATA memoria *f* de tarjetas magnéticas; **~-card unit** *n* DATA unidad *f* de tarjetas magnéticas; **~ circuit** *n* ELECT circuito *m* magnético; **~ clutch** *n* PARTS embrague *m* magnético; **~ core** *n* ELECT núcleo *m* magnético; **~-core storage** *n* DATA memoria *f* de núcleos magnéticos; **~ deflection** *n* ELECT deflexión *f* magnética; **~-delay line** *n* ELECT línea *f* de retardo magnética; **~ disk** *n* COMP, DATA disco *m* magnético; **~-disk storage** *n* COMP, DATA memoria *f* de discos magnéticos; **~ drum** *n* DATA tambor *m* magnético; **~ field** *n* ELECT campo *m* magnético; **~-field intensity** *n* ELECT intensidad *f* del campo magnético; **~-flux density** *n* ELECT densidad *f* del flujo magnético; **~ focusing** *n* ELECT concentración *f* magnética; **~ head** *n* ELECT cabeza *f* magnética; **~ hysteresis** *n* ELECT histéresis *f* magnética; **~ induction** *n* ELECT inducción *f* magnética; **~-induction current loop** *n* ELECT bucle *m* de corriente de inducción magnética; **~ ink character recognition** *n* ELECT reconocimiento *m* de caracteres en tinta magnética; **~ media** *n* COMP soporte *m* magnético; **~-potential difference** *n* ELECT diferencia *f* de potencialmagnético; **~ recording** *n* ELECT registro *m* magnético; **~-resonance spectroscopy** *n* ELECT espectroscopia *f* de resonancia magnética; **~ sensitivity** *n* ELECT sensibilidad *f* magnética; **~ shield** *n* ELECT pantalla *f* magnética; **~ storage** *n* DATA memoria *f* magnética; **~ storm** *n* RADIO tormenta *f* magnética; **~ strip** *n* ELECT, TELEP banda *f* magnética; **~-strip card** *n* ELECT tarjeta *f* de banda magnética; **~ support** *n* COMP soporte *m* magnético; **~ tape** *n* DATA cinta *f* magnética; **~-tape cartridge** *n* DATA casete *m* de cinta magnética; **~-tape drive** *n* TV unidad *f* de cinta magnética; **~-tape leader** *n* DATA comienzo *m* de cinta magnética; **~-tape storage** *n* DATA memoria *f* de cinta magnética; **~-tape trailer** *n* DATA fin *m* de cinta; **~ track** *n* DATA pista *f* magnética; **~ valve** *n* CONTR válvula *f* magnética

magnetically: **~ latched crosspoint** *n* SWG punto *m* de intersección bloqueado magnéticamente

magnetics *n* ELECT magnetismo *m*

magnetization *n* ELECT magnetización *f*

magnetizing: **~ current** *n* ELECT corriente *f*

magnetizante; ~ **field** *n* ELECT campo *m* magnetizante; ~ **inductance** *n* ELECT inductancia *f* magenetizante

magneto: ~**-optic** *adj* (*MO*) COMP, ELECT magnetoóptico (*MO*)

magnetoconductivity *n* ELECT magnetoconductividad *f*

magnetoresistance *n* ELECT magnetoresistencia *f*

magnetoscope *n* ELECT magnetoscopio *m*

magnetostatic: ~ **surface wave** *n* (*MSSW*) ELECT onda *f* de superficie magnetostática (*MSSW*)

magnetostriction *n* ELECT magnetostricción *f*

magnetostrictive: ~ **effect** *n* ELECT efecto *m* magnetostrictivo

magnetron *n* ELECT, RADIO magnetrón *m*; ~ **injection gun** ELECT cañón *m* de inyección magnetrónica

magnifying: ~ **glass** *n* GEN lupa *f* de aumento

magnitude *n* GEN magnitud *f*

mail *n* COMP correo *m*; ~ **order** GEN pedido *m* por correo; ~ **server** COMP, DATA servidor *m* de correo; ~ **shot** GEN campaña *f* agresiva por correo

mailbox *n* DATA buzón *m*; ~ **messaging service** COMP servicio *m* de buzón de mensajes

mailing: ~ **list** *n* DATA lista *f* de direcciones

main: ~ **aisle** *n* SWG pasillo *m* principal; ~**-antenna television** *n* TV televisión *f* de antena principal; ~ **bus** *n* POWER bus *m* principal; ~ **busbar** *n* POWER barra *f* colectora principal; ~ **chute** *n* SWG canal *m* principal; ~ **circuit breaker** *n* POWER interruptor *m* principal; ~ **contractor** *n* GEN contratante *m* principal; ~ **control station** *n* TELEP estación *f* principal de control; ~ **control unit** *n* COMP unidad *f* principal de control; ~ **current** *n* POWER corriente *f* principal; ~ **distribution frame** *n* (*MDF*) TELEP, TRANS trama *f* principal de distribución; ~ **entry** *n* DATA entrada *f* principal; ~ **event loop** *n* COMP bucle *m* de evento principal; ~ **exchange** *n* SWG central *f* principal; ~ **fuse** *n* PARTS, POWER fusible *m* principal; ~ **gap** *n* ELECT espacio *m* principal de descarga; ~ **international switching centre** *BrE* *n* SWG centro *m* principal de distribución internacional; ~ **international-trunk switching centre** *BrE* *n* SWG central *f* principal de línea troncal internacional; ~ **line** *n* TRANS vía *f* férrea troncal; ~ **lobe** *n* RADIO lóbulo *m* principal; ~ **mission services** *n pl* TEST servicios *m pl* de misión principal; ~ **multiplex coupler** *n* TELEP acoplador *m* múltiplex principal; ~ **processor unit** *n* COMP unidad *f* de procesador principal; ~ **repeater distribution frame** *n* TRANS trama *f* de distribución principal de repetición; ~ **repeater station** *n* TELEP estación *f* repetidora principal; ~ **section** *n* TELEP sección *f* principal; ~ **storage** *n* DATA memoria *f* central; ~ **switch** *n* POWER conmutador *m* principal; ~ **telephone line** *n* TELEP, TRANS línea *f* telefónica principal; ~ **trunk exchange area** *n* TELEP zona *f* de central interurbana principal; ~ **trunk switching centre** *BrE* *n* SWG central *f* interurbana principal; ~ **voltage** *n* POWER tensión *f* principal

mainframe *n* COMP, DATA, TELEP gran ordenador *m* *Esp*, gran computador *m* *AmL*, gran computadora *f* *AmL*, ordenador *m* central *Esp*, computador *m* central *AmL*, computadora *f* central *AmL*; ~ **computer** COMP, DATA, TELEP gran ordenador *m* *Esp*, gran computador *m* *AmL*, gran computadora *f* *AmL*, ordenador *m* central *Esp*, computador *m* central *AmL*, computadora *f* central *AmL*

mains *n* POWER *network* red *f* eléctrica; ~ **connection** POWER conexión *f* a la red; ~ **failure** POWER corte *m* de suministro eléctrica; ~**-failure alarm** POWER alarma *f* de corte de suministro eléctrico; ~ **frequency** POWER frecuencia *f* de la red; ~ **fuse** PARTS, POWER fusible *m* de red; ~ **hum** POWER zumbido *m* de la red; ~ **monitor** CONTR, POWER supervisor *m* de la red eléctrica; ~ **monitoring** POWER supervisión *f* de la red eléctrica; ~ **supply** POWER conexión *f* a la red; ~ **switch** POWER interruptor *m* de red; ~ **transformer** POWER transformador *m* de corriente

maintain *vt* GEN mantener

maintainability *n* GEN mantenibilidad *f*; ~ **function** TEST función *f* de mantenibilidad

maintaining: ~ **voltage** *n* ELECT tensión *f* de mantenimiento

maintenance *n* GEN mantenimiento *m*; ~ **analysis** TEST análisis *m* de mantenimiento; ~ **cell description** (*MCD*) DATA descripción *f* de célula de mantenimiento; ~ **channel** TRANS canal *m* de mantenimiento; ~ **entity** DATA entidad *f* de mantenimiento; ~ **function** GEN función *f* de mantenimiento; ~ **instructions** GEN instrucciones *f pl* de mantenimiento; ~ **interface** TELEP interfaz *f* de mantenimiento; ~ **interval** TEST intervalo *m* de mantenimiento; ~ **man-hours** TEST horas-hombre *f pl* de mantenimiento; ~ **operation protocol** DATA, ELECT protocolo *m* de operación de mantenimiento; ~ **painting** GEN pintura *f* de mantenimiento; ~ **processor** COMP procesador *m* de mantenimiento; ~ **programme** *BrE* TEST programa *m* de mantenimiento; ~ **service provider** (*MSP*) TEST proveedor *m* del servicio de mantenimiento (*PSM*); ~ **staff** GEN equipo *m* de mantenimiento; ~ **station** TELEP puesto *m* de mantenimiento; ~ **support logistics** TEST logística *f* de apoyo al mantenimiento; ~ **support performance** TELEP actuación *f* de apoyo del mantenimiento; ~ **time** TELEP, TEST tiempo *m* de mantenimiento

major: ~ **alarm** *n* TEST alarma *f* importante; ~ **failure** *n* TEST fallo *m* mayor; **~-path concept** *n* RADIO concepto *m* de camino principal; **~-path satellite** *n* RADIO satélite *m* de camino principal; ~ **tendencies** *n pl* GEN tendencias *f pl* principales

majority: ~ **carrier** *n* ELECT vector *m* mayoritario; ~ **decision logic** *n* ELECT lógica *f* de decisión mayoritaria; ~ **element** *n* ELECT elemento *m* mayoritario; ~ **gate** *n* ELECT circuito *m* perta de mayoría; ~ **operation** *n* ELECT operación *f* mayorante

make: **~-before-break contact** *n* ELECT contacto *m* de conmutación sin interrupción; ~ **pulse** *n* TELEP impulso *m* de cierre

maker *n* GEN constructor *m*

maladjustment *n* GEN ajuste *m* incorrecto

male: ~ **plug** *n* COMP, ELECT, TEST enchufe *m* macho

malfunction *n* TEST mal funcionamiento *m*; ~ **level** TEST nivel *m* de mal funcionamiento

malicious: ~ **call** *n* TELEP llamada *f* maliciosa; **~-call identification** *n* TELEP identificación *f* de llamadas maliciosas; **~-call identification service** *n* TELEP servicio *m* de identificación de llamadas maliciosas; **~-call tracing** *n* TELEP localización *f* de llamadas maliciosas

man: **~-machine communications** *n pl* (*MMC*) COMP comunicaciones *f pl* hombre-máquina (*MMC*); **~-machine dialog** *n* COMP diálogo *m* hombre-máquina; **~-machine interface** *n* COMP interfaz *f* hombre-máquina; **~-machine language** *n* (*MML*) COMP lenguaje *m* hombre-máquina (*MML*); **~-machine relationship** *n* COMP relación *f* hombre-máquina; **~-machine terminal** *n* TELEP terminal *m* hombre-máquina; **~-made noise** *n* ELECT, RADIO ruido *m* de origen humano; **~-pack transceiver** *n* RADIO transceptor *m* de mochila

MAN *abbr* (*metropolitan area network*) TRANS red *f* de área metropolitana

manageability *n* GEN manejabilidad *f*

manageable *adj* GEN manejable

managed: ~ **object** *n* DATA objeto *m* gestionado; **~-object boundary** *n* DATA límite *m* de objeto gestionado; **~-object class** *n* (*MOC*) DATA clase *f* de objeto gestionado; **~-object conformance statement** *n* (*MOCS*) DATA declaración *f* de conformidad del objeto gestionado (*MOCS*); ~ **open system** *n* DATA sistema *m* abierto gestionado; ~ **system** *n* DATA sistema *m* gestionado

management *n* GEN gestión *f*; ~ **account** GEN cuenta *f* de gestión; ~ **application function** (*MAF*) DATA función *f* de aplicación de gestión; ~ **centre** *BrE* TEST centro *m* de gestión; ~ **data** *n pl* DATA datos *m pl* de gestión; ~ **database** DATA base *f* de datos de gestión; ~ **domain** DATA, TELEP dominio *m* de gestión; ~ **information base** (*MIB*) DATA base *f* de información para dirección de empresas; ~ **information system** (*MIS*) DATA sistema *m* de información para gestión (*MIS*); ~ **interaction** DATA interacción *f* de gestión; ~ **network** TRANS red *f* de gestión; ~ **notification service** DATA servicio *m* de notificación de gestión; ~ **open system** DATA sistema *m* abierto de gestión; ~ **operation service** DATA servicio *m* de operación de gestión; ~ **protocol** DATA protocolo *m* de gestión; ~ **signal** SIG señal *f* de gestión; ~ **software** COMP, DATA programas *m pl* de gestión; ~ **support object** DATA objeto *m* de apoyo de la gestión; ~ **unit** DATA unidad *f* de gestión

manager *n* GEN director *m*

managing: ~ **system** *n* DATA sistema *m* de gestión

Manchester: ~ **data** *n pl* COMP datos *m pl* Manchester

mandatory *adj* GEN obligatorio; ~ **component** *n* DATA componente *m* obligatorio; ~ **requirement** *n* DATA requisito *m* obligatorio

manhole *n* GEN pozo *m* de inspección

manipulated: ~ **variable** *n* CONTR, PARTS variable *f* manipulada

manipulation: ~ **detection** *n* CONTR, DATA detección *f* por manipulación

manometer *n* TEST manómetro *m*

manual[1]: ~ **answering** *n* TELEP respuesta *f* manual; ~ **attempt** *n* TELEP prueba *f* manual; ~ **calling** *n* TELEP llamada *f* manual; **~-changeover acknowledgement signal** *n* SIG señal *f* de reconocimiento de conmutación manual; **~-changeover signal** *n* SIG señal *f* de conmutación manual; ~ **control** *n* CONTR control *m* manual; ~ **direction finder** *n* RADIO radiogoniómetro *m* manual; ~ **marking** *n* DATA marcado *m* manual; ~ **observation** *n* TEST observación *f* manual; ~ **operation** *n* GEN funcionamiento *m* manual; ~ **position** *n* TELEP posición *f* manual; ~ **regulation** *n* GEN regulación *f* manual; ~ **repatching** *n* TEST resintonización *f* manual; ~ **ringing** *n* TELEP llamada *f* manual; ~ **setting** *n* DATA, GEN ajuste *m* manual; ~ **switching system** *n* SWG sistema *m* de conmutación manual

manual[2] *n* GEN manual *m*

manually: ~**-controlled** *adj* CONTR controlado manualmente

manufacturing *n* GEN fabricación *f*; ~ **defect** TEST defecto *m* de fabricación; ~ **failure** TEST fallo *m* de fabricación; ~ **fault** TEST fallo *m* de fabricación

many: ~**-valued logic** *n* ELECT lógica *f* plurívoca

map[1] *n* COMP mapa *m*

map[2] *vt* COMP trazar un mapa, TELEP levantar planos

mapping *n* COMP encuadre *m*

MAR *abbr* (*miscellaneous apparatus rack*) TELEP MAR (*bastidor para aparatos diversos*)

margin *n* GEN margen *m*; ~ **of error** GEN margen *m* de error

marginal *adj* GEN marginal; ~ **check** *n* TEST verificación *f* marginal; ~ **test** *n* TEST prueba *f* marginal

maritime: ~ **account** *n* TELEP reseña *f* marítima; ~ **centre** *BrE n* TELEP centro *m* marítimo; ~ **centre** *BrE* **shore station** *n* TELEP estación *f* costera de un centro marítimo; ~ **local system** *n* TELEP sistema *m* marítimo local; ~ **mobile satellite system** *n* TELEP sistema *m* móvil marítimo de satélite; ~ **mobile service** *n* RADIO, TELEP servicio *m* móvil marítimo; ~ **mobile-service satellite** *n* RADIO servicio *m* marítimo móvil de satélite; ~ **mobile terrestrial service** *n* RADIO servicio *m* marítimo móvil terrestre; ~ **radio** *n* RADIO radio *f* marítima; ~ **satellite circuit** *n* RADIO circuito *m* marítimo de satélite; ~ **satellite switching centre** *BrE n* RADIO, SWG central *f* telefónica marítima por satélite; ~ **satellite system** *n* RADIO sistema *m* marítimo de satélite; ~ **switching centre** *BrE n* SWG, TELEP centro *m* de conmutación marítima; ~ **system** *n* TELEP sistema *m* marítimo; ~ **terrestrial circuit** *n* TELEP circuito *m* marítimo terrestre

mark[1] *n* TEST referencia; ~ **period** TELEP período *m* de marcado; ~ **scanning** DATA exploración *f* de marcas; ~ **scanning equipment** DATA equipo *m* de lectura óptica de marcas; ~ **sensing** GEN detección *f* de marcas

mark[2] *vt* GEN marcar

marked: ~ **idle channel** *n* SWG canal *m* libre marcado

marker *n* GEN marcador *m*; ~ **beacon** RADIO radiobaliza *f*; ~ **control system** SWG sistema *m* de control del marcador; ~ **system** SWG sistema *m* del marcador

market: ~ **demand** *n* GEN demanda *f* de mercado; ~ **growth** *n* GEN crecimiento *m* de mercado; ~ **penetration** *n* GEN penetración *f* de mercado; ~ **survey** *n* GEN estudio *m* de mercado

marketing: ~ **manager** *n* GEN director *m* comercial de mercadotecnia; ~ **mix** *n* GEN mixtura *f* de comercialización

marking *n* GEN marcación *f*; ~ **percentage** TELEP porcentaje *m* marcación; ~ **plate** PARTS chapa *f* de marcado; ~ **sequence** SWG secuencia *f* de marcación; ~ **wave** RADIO señal *f* de trabajo

marshal: ~ **document** *n* SIG especificación *f* de trayecto de señales

MASE *abbr* (*message administration service element*) DATA MASE (*elemento de servicio de administración de mensajes*)

maser *n* ELECT máser *m*

mask *n* ELECT, RADIO, TELEP, TRANS rejilla *f*

masking *n* ELECT, RADIO, TELEP, TRANS enmascaramiento *m*; ~ **curve** TELEP curva *f* de enmascaramiento

masquerade *n* DATA enmascaramiento *m*

mass *n* GEN masa *f*; ~ **soldering** ELECT soldadura *f* simultánea; ~ **storage** DATA almacenamiento *m* de gran capacidad

mast *n* RADIO mástil *m*

master *n* GEN maestro *m*; ~ **antenna television** (*MATV*) TV televisión *f* de antena colectiva; ~ **antenna television system** TV sistema *m* de televisión de antena colectiva (*MATV*); ~ **clear** COMP borrado *m* maestro; ~ **clock** ELECT reloj *m* maestro; ~**-clock pulse** ELECT impulso *m* de reloj maestro; ~ **data** *n pl* DATA datos *m pl* principales; ~ **document** COMP documento *m* original; ~ **oscillator** ELECT oscilador *m* maestro; ~ **oscillator frequency** ELECT frecuencia *f* de oscilador maestro; ~ **processor** COMP procesador *m* maestro; ~ **programme** *BrE* TV programa *m* principal; ~ **register** DATA registro *m* maestro; ~ **set** PER, TELEP teléfono *m* principal; ~**-slave control** CONTR control *m* principal-subordinado; ~**-slave flip-flop** ELECT biestable *m* ordenador-seguidor; ~ **station** RADIO puesto *m* principal; ~ **tape** TV cinta *f* maestra; ~ **time-plan** RADIO cronograma *m* principal; ~ **volume control** PER APP control *m* maestro del volumen; ~ **volume slider** COMP regulador *m* principal de volumen

mastergroup *n* TELEP, TRANS grupo *m* terciario; ~ **link** TELEP enlace *m* en grupo terciario; ~ **section** TELEP sección *f* de grupo terciario

matched: ~ **cladding** *n* TRANS revestimiento *m* adaptado; ~ **junction** *n* RADIO unión *f* adaptada; ~ **load** *n* RADIO carga *f* equilibrada; ~ **termination** *n* RADIO extremidad *f* no reflejante

matching *n* COMP adaptación *f*, DATA equilibrio *m*, ELECT adaptación *f*; ~ **device** ELECT dispositivo *m* de adaptación; ~ **impedance** ELECT impedancia *f* de adaptación; ~ **network** ELECT red *f* de adaptación; ~ **order** DATA orden *f* de equilibrio; ~ **resistor** ELECT, PARTS resistor *m* de adaptación; ~ **transformer** ELECT transformador *m* de adaptación

MATD *abbr* (*maximum acceptable transit delay*) DATA MATD (*retardo máximo aceptable de tránsito*)

mate *vt* ELECT casar

material *n* GEN material *m*; ~ **dispersion** TRANS dispersión *f* de material; ~ **dispersion coefficient** TRANS coeficiente *m* de dispersión del material; ~ **scattering** RADIO dispersión *f* del material; ~ **specification** GEN especificación *f* de materiales; ~ **thickness** PARTS espesor *m* del material

mating: ~ **connectors** *n pl* PARTS conectores *m pl* complementarios

matrix *n* GEN matriz *f*; ~ **circuit** ELECT circuito *m* matricial; ~ **configuration** ELECT configuración *f* matriz; ~ **control station** TELEP estación *f* matriz de control; ~ **detector** PER APP *scanner* detector *m* matriz; ~ **display** ELECT display *m* matricial; ~ **link** TELEP enlace *m* matricial; ~-**link interface** TELEP interfaz *f* matriz-enlace; ~ **printer** COMP, PER APP impresora *f* matricial; ~ **remover** ELECT eliminador *m* de matriz; ~ **storage** COMP memoria *f* matricial; ~ **switch** SWG conmutador *m* matricial; ~ **switch control software machine** SWG ordenador *m* con programa de control de conmutador matricial *Esp*, computador *m* con programa de control de conmutador matricial *AmL*, computadora *f* con programa de control de conmutador matricial *AmL*

matter *n* GEN materia *f*

MATV *abbr* (*master antenna television*) TV MATV; ~ **system** *n* TV sistema *m* MATV

MAU *abbr* (*multistation access unit*) DATA MAU (*unidad de acceso multiestación*)

maximization *n* GEN maximización *f*

maximize *vt* GEN maximizar

maximum *adj* GEN máximo; ~ **acceptable transit delay** *n* (*MATD*) DATA retardo *m* máximo aceptable de tránsito (*MATD*); ~ **admissible power** *n* TRANS potencia *f* admisible máxima; ~ **announcement duration** *n* PER APP *on answering machine* duración *f* máxima de aviso; ~ **justification rate** *n* TELEP índice *m* máximo de justificación; ~ **length** *n* TV longitud *f* máxima; ~ **likelihood decoder** *n* RADIO descodificador *m* de máxima probabilidad; ~ **memory capacity** *n* PER APP *on answering machine* capacidad *f* máxima de memoria; ~ **message duration** *n* PER APP *on answering machine* duración *f* máxima del mensaje; ~ **non-operate current** *n* ELECT corriente *f* de inactividad máxima; ~ **permissible interference level** *n* (*MPIL*) DATA nivel *m* máximo permisible de interferencia (*MPIL*); ~ **relative time interval error** *n* (*MRTIE*) DATA error *m* máximo de intervalo relativo; ~ **resolution** *n* TV resolución *f* máxima; ~ **retention time** *n* ELECT tiempo *m* máximo de retención; ~ **sensitivity** *n* GEN sensibilidad *f* máxima; ~ **theoretical numerical aperture** *n* ELECT, TRANS máxima *f* apertura numérica teórica; ~ **time interval error** *n* (*MTIE*) DATA error *m* de intervalo de tiempo máximo; ~ **usable frequency** *n* (*MUF*) RADIO frecuencia *f* máxima utilizable (*MUF*; ~ **usable reading time** *n* COMP tiempo *m* máximo utilizable de lectura; ~ **usable read-number** *n* COMP número *m* de lectura máximo utilizable; ~ **usable sensitivity** *n* RADIO sensibilidad *f* máxima utilizable; ~ **usable viewing time** *n* TV tiempo *m* máximo utilizable de visado; ~ **usable writing speed** *n* ELECT velocidad *f* máxima utilizable de escritura

Mayday: ~ **relay** *n* RADIO relé *m* de auxilio

Mb *abbr* (*megabyte*) COMP, DATA Mb (*megabyte*)

M-bit *n* TELEP megabitio *m*

Mbps *abbr* (*millions of bits per second*) COMP Mbps (*millones de bitios por segundo*)

MBS *abbr* (*multiblock synchronization signal unit*) SIG MBS (*unidad de señal de sincronización multibloque*)

Mbyte *abbr* (*megabyte*) COMP, DATA Mbyte (*megabyte*)

MCD *abbr* (*maintenance cell description*) DATA descripción *f* de célula de mantenimiento

MCF *abbr* (*message communication function*) DATA función *f* de comunicación de mensajes

MCI *abbr* (*media control interface*) COMP MCI (*interfaz de control de medios*)

MCN *abbr* (*mobile control node*) TELEP nodo *m* de control móvil

MCS *abbr* (*multipoint communications system*)

TELEP MCS (*sistema de comunicaciones multipunto*); ~ **attachment** *n* TELEP accesorio *m* MCS; ~ **channel** *n* TELEP canal *m* MCS; ~ **connection** *n* TELEP conexión *f* MCS; ~ **data-transfer priority** *n* TELEP prioridad *f* MCS de transferencia de datos; ~ **domain** *n* TELEP campo *m* de acción MCS; ~**-domain selector** *n* TELEP selector *m* de campo de acción MCS; ~ **interface data unit** *n* DATA, TELEP unidad *f* de datos del interfaz MCS; ~ **private channel** *n* TELEP canal *m* privado MCS; ~ **private channel manager** *n* TELEP gestor *m* de canal privado MCS; ~ **protocol data unit** *n* DATA, TELEP unidad *f* de datos del protocolo MCS; ~ **provider** *n* DATA, TELEP proveedor *m* de MCS; ~ **service access point** *n* (*MCSAP*) TELEP punto *m* de acceso del servicio MCS (*MCSAP*); ~ **service data unit** *n* DATA, TELEP unidad *f* de datos del servicio MCS; ~ **user** *n* TELEP usuario *m* de MCS; ~ **user identification** *n* DATA, TELEP identificación *f* de usuario del MCS

MCSAP *abbr* (*MCS service access point*) TELEFON MCSAP (*punto de accesso del servicio MCS*)

M curve *n* RADIO curva *f* M

MCW *abbr* (*modulated continuous waves*) RADIO MCW (*ondas continuas moduladas*)

MD *abbr* (*mediation device*) COMP dispositivo *m* de mediación

m-derived: ~ **filter** *n* RADIO filtro *m* m-derivado

MDF *abbr* (*main distribution frame*) TELEP, TRANS trama *f* principal de distribución

MDS *abbr* (*minimum detectable signal*) TRANS MDS (*señal detectable mínima*)

MDSE *abbr* (*message delivery service element*) DATA MDSE (*elemento de servicio de entrega de mensajes*)

mean: ~ **availability** *n* TELEP disponibilidad *f* media; ~ **availability/unavailability** *n* TELEP disponibilidad/indisponibilidad *f* media; ~ **busy hour** *n* TELEP hora *f* cargada media; ~ **deviation** *n* TEST desviación *f* media; ~ **downtime** *n* TEST tiempo *m* medio de indisponibilidad; ~ **entropy per character** *n* DATA entropía *f* media por carácter; ~ **error** *n* TEST error *m* medio; ~ **failure intensity** *n* COMP, TEST intensidad *f* media de fallos; ~ **failure rate** *n* COMP, TEST tasa *f* media de fallos; ~ **forward current** *n* ELECT, POWER corriente *f* directa media; ~ **Hall-plate temperature** *n* ELECT temperatura *f* media de la placa Hall; ~ **holding time per seizure** *n* TELEP tiempo *m* medio de ocupación; ~ **life** *n* TEST vida *f* media; ~ **loss value** *n* TRANS valor *m* medio de atenuación; ~ **maintenance man-hours** *n* TEST media *f* de horas-hombre de mantenimiento; ~ **one-way propagation time** *n* TELEP tiempo *m* medio de propagación en un solo sentido; ~ **opinion score** *n* (*MOS*) COMP media *f* de opinión (*SOM*); ~ **power** *n* RADIO potencia *f* media; ~ **repair rate** *n* COMP velocidad *f* media de reparación; ~ **repair time** *n* TEST tiempo *m* medio de reparación; ~ **re-use distance** *n* RADIO distancia *f* media de reutilización; ~ **reverse voltage** *n* ELECT, POWER tensión *f* inversa media; ~ **time before failure** *n* TEST tiempo *m* medio antes del fallo; ~ **time between failures** *n* (*MTBF*) TEST tiempo *m* medio entre fallos (*MTBF*); ~ **time between interruptions** *n* (*MTBI*) COMP tiempo *m* medio entre interrupciones (*MTBI*); ~ **time between overhauls** *n* (*MTBO*) TEST tiempo *m* medio entre revisiones (*MTBO*); ~ **time to first failure** *n* (*MTFF*) COMP tiempo *m* medio hasta el primer fallo (*MTFF*); ~ **time to recovery** *n* (*MTTR*) TEST tiempo *m* medio hasta el restablecimiento (*MTTR*); ~ **time to repair** *n* TEST tiempo *m* medio de reparación; ~ **time to service restoration** *n* (*MTSR*) TEST tiempo *m* medio hasta el restablecimiento del servicio (*MTSR*); ~ **unavailability** *n* TEST indisponibilidad *f* media; ~ **uptime** *n* TEST tiempo *m* medio de disponibilidad; ~ **waiting time** *n* COMP tiempo *m* medio de espera; ~ **waiting-time average delay** *n* COMP promedio *m* de retardo por tiempo de espera

meaningful: ~ **information** *n* DATA información *f* significativa

means: ~ **of testing** *n* TEST instrumentos *m pl* de comprobación

measure *vt* GEN medir

measured: ~ **value** *n* TEST valor *m* medido

measurement *n* GEN medición *f*; ~ **error** TEST error *m* de medición; ~ **method** TEST método *m* de medición; ~ **process** TEST sistema *m* de medición; ~ **technique** TEST técnica *f* de medición

measures *n pl* GEN medidas *f pl*

measuring: ~ **amplifier** *n* ELECT amplificador *m* de medición; ~ **bridge** *n* TEST puente *m* de medición; ~ **circuit** *n* TEST circuito *m* de medición; ~ **detector** *n* TEST detector *m* de medición; ~ **equipment** *n* TEST equipo *m* de medición; ~ **instrument** *n* TEST instrumento *m* de medición; ~ **loop** *n* TEST bucle *m* de medición; ~ **plug** *n* TEST clavija *f* para medición; ~ **point** *n* TEST punto *m* de medición; ~ **range** *n* TEST campo *m* de

medición; ~ **rod** *n* TELEP varilla *f* de medición; ~ **transducer** *n* TEST transductor *m* de medición
mechanical: ~ **contact reading** *n* TV lectura *f* de contacto mecánico; ~ **design** *n* GEN diseño *m* mecánico; ~ **engineering** *n* GEN ingeniería *f* mecánica; ~ **life** *n* TEST vida *f* mecánica; ~ **optical switch** *n* SWG dispositivo *m* de conmutación óptico mecánico; ~ **splice** *n* TRANS empalme *m* mecánico; ~ **spot displacement** *n* ELECT desplazamiento *m* mecánico de punto; ~ **spot misalignment** *n* ELECT desalineación *f* mecánica de punto; ~ **structure** *n* GEN estructura *f* mecánica; ~ **tuning range** *n* ELECT alcance *m* de sintonización mecánica; ~ **zero** *n* TEST cero *m* mecánico
mechanics *n* GEN mecánica *f*
mechanism *n* GEN mecanismo *m*
media *n pl* GEN medios *m pl*; ~ **control interface** *n* (*MCI*) COMP interfaz *f* de control de medios (*MCI*); ~ **conversion** *n* DATA cambio *m* de soportes; ~ **interface connector** *n* PARTS conector *m* de interfaz de medios
median *n* GEN mediana *f*
mediation: ~ **device** *n* (*MD*) COMP dispositivo *m* de mediación; ~ **function** *n* (*MF*) COMP función *f* de mediación
medical: ~ **computer graphics** *n* COMP gráficos *m* médicos informatizados
medium[1] *adj* GEN medio; ~ **access control** *n* (*MAC*) CONTR control *m* de acceso medio; ~ **adaptor** *n* (*MA*) ELECT, TELEP adaptador *m* intermedio; ~ **earth-orbiting satellite** *n* RADIO satélite *m* en órbita terrestre media; ~ **power** *n* POWER potencia *f* media; ~**-power flux density** *n* RADIO densidad *f* de flujo de potencia media; ~ **range** *n* RADIO alcance *m* medio; ~ **rate** *n* APPL *call charge* tarifa *f* media; ~**-scale integration** *n* ELECT, PARTS, TELEP, TRANS integración *f* de mediana escala; ~**-term bit error rate** *n* COMP tasa *f* de error de bitio a plazo medio; ~ **wave** *n* RADIO onda *f* media
medium[2] *n* GEN medio *m*
meet[1]: ~**-me bridge** *n* APPL puente *m* de encuentro; ~**-me call** *n* TELEP llamada *f* de conferencia múltiple; ~**-me conference** *n* APPL conferencia *f* múltiple
meet[2] *vt* GEN cumplir; ◆ ~ **the deadline** GEN cumplir con el plazo fijado; ~ **the needs of** GEN cubrir las necesidades de; ~ **the requirements** GEN cumplir con los requisitos; ~ **the specifications** GEN adecuarse a las especificaciones
meg: ~ **of memory** *n* DATA megaherzio *m* de memoria
megabyte *n* (*Mb*) COMP, DATA megabyte *m* (*Mb*)
megaflops *n* (*one million floating-point operations per second*) COMP megaflops *f*
megalips *n* (*one million logical inferences per second*) COMP megalips *m*
megastream: ~ **circuit** *n* TRANS circuito *m* de megaflujo
Megger® *n* TEST megaóhmetro *m*
megohmmeter *n* TEST megaóhmetro *m*
membrane *n* PARTS membrana *f*; ~ **keyswitch** ELECT interruptor *m* de membrana; ~ **switch** ELECT interruptor *m* de membrana
memo: ~ **recording** *n* PER APP *on answering machine* registro *m* electrónico
memorandum: ~ **of understanding** *n* (*MOU*) GEN memorándum *m* de acuerdo (*MOU*)
memory *n* COMP, DATA memoria *f*; ~ **access mode** COMP modo *m* de acceso a memoria; ~ **architecture** COMP arquitectura *f* de memoria; ~ **bank** COMP banco *m* de memoria; ~**-based reasoning** DATA razonamiento *m* basado en memoria; ~ **board** DATA placa *f* de memoria; ~ **card** DATA tarjeta *f* de memoria; ~ **circuit** ELECT circuito *m* de memoria; ~ **cleaning** COMP borrado *m* de memoria; ~ **configuration** COMP configuración *f* de memoria; ~**-control logic** COMP lógica *f* de control de memoria; ~ **dump** COMP vuelco *m* de memoria; ~ **erasure** COMP borrado *m* de memoria; ~ **expansion capacity** COMP capacidad *f* de ampliación de la memoria; ~ **extender** COMP ampliador *m* de memoria; ~ **management program** COMP programa *m* de gestión de la memoria; ~ **management unit** COMP unidad *f* de gestión de la memoria; ~ **manager** COMP gestor *m* de memoria; ~ **number** PER APP *on answering machine* número *m* de memoria; ~ **plane** COMP plano *m* de memoria; ~ **printout** COMP impresión *f* del contenido de la memoria; ~ **protection** COMP protección *f* de la memoria; ~ **shortage** COMP insuficiencia *f* de la memoria; ~ **space** COMP espacio *m* de memoria; ~ **storage** COMP, DATA registro *m* de memoria
menu *n* COMP menú *m*; ~ **bar** COMP barra *f* de menú; ~ **mode** TELEP modo *m* de menú; ~**-mode operating sequence** TELEP secuencia *f* de operación en modo menú; ~ **output** TELEP salida *f* de menú; ~ **selection procedure** TELEP procedimiento *m* de selección de menú
mercury: ~ **contact** *n* ELECT contacto *m* de mercurio; ~ **contact relay** *n* ELECT relé *m* de contactos de mercurio; ~ **relay** *n* ELECT relé *m* de mercurio; ~**-vapour** *BrE* **tube** *n* ELECT tubo *m* de vapor de mercurio; ~**-wetted contact**

relay *n* ELECT relé *m* con contactos humedecidos en mercurio; **~-wetted relay** *n* ELECT relé *m* humedecido en mercurio; **~-wetting** *n* ELECT humedecimiento *m* en mercurio
merge[1] *n* COMP fusión *f*
merge[2] *vt* COMP fusionar
meridional: **~ ray** *n* TRANS rayo *m* meridional
mesh *n* ELECT, RADIO malla *f*; **~ analysis** ELECT análisis *m* granulométrico; **~ effect** RADIO efecto *m* poligonal; **~ electrode** ELECT electrodo *m* poligonal; **~ reflector** RADIO reflector *m* poligonal
meshed: **~ network** *n* TRANS red *f* de mallas
meshing *n* COMP engranado *m*
mesochronous *adj* TELEP, TRANS mesócrono
mesomorph: **~ state** *n* ELECT estado *m* mesomorfo
message *n* GEN mensaje *m*; **~ administration service element** (*MASE*) DATA elemento *m* de servicio de administración de mensajes (*MASE*); **~ announcement system** TELEP sistema *m* de anuncio de mensajes; **~ beginning** PER APP comienzo *m* de mensaje; **~ broadcaster** PER APP transmisor *m* de mensajes; **~-centre** *BrE* **vehicle** TRANS vehículo *m* de centro de mensajes; **~ communication function** (*MCF*) DATA función *f* de comunicación de mensajes; **~ confirmation** DATA confirmación *f* de mensajes; **~ delivery service element** (*MDSE*) DATA elemento *m* de servicio de entrega de mensajes (*MDSE*); **~ discrimination** COMP discriminación *f* de mensaje; **~ display pager** RADIO buscapersonas *m* con visualización de mensaje; **~ distribution** TELEP distribución *f* de mensajes; **~ feedback** TELEP retroalimentación *f* de mensajes; **~ handling** APPL, TELEP manipulación *f* de mensajes; **~ handling environment** (*MHE*) APPL, DATA, RADIO, TELEP entorno *m* de manipulación del tráfico de mensajes (*MHE*); **~ handling system** (*MHS*) APPL, DATA, RADIO, TELEP sistema *m* de codificación de mensajes; **~ header** DATA cabecera *f* de mensaje; **~ mode** DATA modo *m* de mensaje; **~-oriented text interchange system** (*MOTIS*) DATA sistema *m* de intercambio de texto para mensajes; **~ pager** GEN buscapersonas *m* con mensajes; **~ protocol data unit** (*MPDU*) DATA unidad *f* de datos de protocolo de mensajes (*MPDU*); **~-received counter** PER APP *on answering machine* contador *m* de mensajes recibidos; **~-received indicator** PER APP *on answering machine* indicador *m* de mensaje recibido; **~ refusal signal** SIG señal *f* de mensaje rechazado; **~ register** DATA registro *m* del mensaje; **~ relay** TRANS punto *m* de retransmisión de mensajes; **~ retrieval** DATA recuperación *f* de mensajes; **~ retrieval service element** (*MRSE*) DATA elemento *m* del servicio de recuperación de mensajes (*MRSE*); **~ route** TRANS encaminamiento *m* del mensaje; **~ signalling** *BrE* **unit** DATA, TELEP unidad *f* de señalización del mensaje; **~ sink** TRANS receptor *m* de mensajes; **~ source** TRANS origen *m* del mensaje; **~ store** DATA, TELEP memoria *f* de mensajes; **~ storing** DATA memorización *f* de mensajes; **~-submission abstract operation** DATA operación *f* de extracto de presentación de mensajes; **~ submission service element** (*MSSE*) DATA elemento *m* de servicio de presentación de mensajes (*MSSE*); **~ suffix** COMP sufijo *m* de mensaje; **~ switch** SWG conmutador *m* de mensajes; **~ switching** SWG conmutación *f* de mensajes; **~-switching centre** *BrE* SWG centro *m* de conmutación de mensajes; **~-switching network** TRANS red *f* de conmutación de tráfico; **~-switching processor** SWG procesador *m* de conmutación de mensajes; **~-switching system** SWG, TRANS sistema *m* de conmutación de mensajes; **~ toll service** TELEP servicio *m* interurbano de mensajería; **~ transfer** (*MT*) DATA, TELEP transferencia *f* de mensajes; **~ transfer agent** (*MTA*) DATA, TELEP agente *m* de transferencia de mensajes; **~ transfer part** (*MTP*) DATA, TELEP parte *f* de la transferencia de mensaje (*MTP*); **~ transfer part receiving time** COMP tiempo *m* de recepción de la parte de transferencia de mensaje; **~ transfer part sending time** TELEP tiempo *m* de envío de la parte de transferencia de mensaje; **~ transfer service element** (*MTSE*) DATA elemento *m* del servicio de transferencia de mensajes (*MTSE*); **~ transfer system** (*MTS*) DATA, TELEP sistema *m* de transferencia de mensajes; **~ transfer time at signalling** *BrE* **transfer points** TRANS tiempo *m* de transferencia del mensaje en los puntos de transferencia de señalización; **~ transmission system** DATA sistema *m* de transmisión de mensajes; **~ waiting indication** TELEP indicación *f* de mensaje en espera
messaging: **~ service** *n* APPL servicio *m* de mensajería
messenger: **~ cable** *n* TRANS cable *m* portador; **~ call** *n* TELEP conversación *f* con aviso de llamada; **~ wire** *n* TRANS cable *m* portador
metal: **~-clad base material** *n* ELECT material *m* a base de revestimiento metálico; **~-clad**

conductor *n* ELECT conductor *m* con revestimiento metálico; ~ **conductor cable** *n* TRANS cable *m* de conductor metálico; **~-core board** *n* ELECT placa *f* de alma metálica; **~-film capacitor** *n* ELECT capacitor *m* de película metálica; ~ **insulator semiconductor** *n* (*MIS*) ELECT semiconductor *m* con aislante metálico (*SIA*); **~-organic chemical vapour** *BrE* **deposition** *n* ELECT, TRANS deposición *f* de vapor químico órgano-metálico; **~-organic vapour-phase** *BrE* **epitaxy** *n* ELECT epitaxia *f* de fase de vapor órgano-metálico; **~-oxide semiconductor** *n* ELECT semiconductor *m* metal-óxido; ~ **twinax plug** *n* ELECT clavija *f* metálica de doble eje

metalanguage *n* TELEP metalenguaje *m*

metallic: ~ **circuit** *n* TELEP circuito *m* metálico; ~ **coating** *n* TRANS recubrimiento *m* metálico; ~ **conductor cable** *n* TRANS cable *m* de conductores metálicos; ~ **crosspoint** *n* SWG punto *m* de cruce metálico; ~ **rectifier** *n* ELECT rectificador *m* metálico; ~ **thin film** *n* ELECT película *f* metálica delgada; ~ **twist** *n* ELECT trenza *f* metálica

metallization *BrE* (*AmE* **metalization**) *n* ELECT metalización *f*

metallized *BrE* (*AmE* **metalized**) *adj* GEN metalizado; **~-film capacitor** *n* ELECT capacitor *m* de película metalizada; **~-paper capacitor** *n* (*capacitor MP*) POWER capacitor *m* de papel metalizado; ~ **screen** *n* ELECT pantalla *f* metalizada

metareasoning *n* COMP metarrazonamiento *m*

metastable *adj* ELECT metaestable

meteor: **~-burst communication** *n* RADIO comunicación *f* por ráfagas meteóricas; **~-burst propagation** *n* RADIO propagación *f* por ráfagas meteóricas

meteorological *adj* CLIM meteorológico

meter *n* GEN medidor *m*; ~ **account save** TELEP almacenamiento *m* de la lectura del contador; ~ **changeover clock** POWER conmutador *m* de tiempo; ~ **indication** TELEP indicación *f* del contador; ~ **pulse** TELEP impulso *m* de cómputo; ~ **pulse generation** TELEP generación *f* de impulsos de cómputo; ~ **pulse generator** TELEP generador *m* de impulsos de cómputo; ~ **pulse interval** TELEP intervalo *m* de impulsos de cómputo; **~-pulse sender** TELEP emisor *m* de impulsos de cómputo; ~ **rack** TELEP bastidor *m* de contadores; ~ **unit** TELEP unidad *f* de contadores; ~ **value indication** TELEP indicación *f* del contador; ~ **wire** TELEP hilo *m* de cómputo

metering *n* TELEP tarificación *f*; ~ **period** TELEP periodo *m* de tarificación; ~ **rate** TELEP tasa *f* de tarificación; ~ **signal** TELEP señal *f* de tarificación

method *n* GEN método *m*

methodic *adj* GEN metódico

metric: ~ **wave** *n* TRANS onda *f* métrica

metrology *n* GEN metrología *f*

metropolitan: ~ **area network** *n* (*MAN*) TRANS *remote data processing* red *f* de área metropolitana; ~ **network** *n* TRANS red *f* metropolitana; ~ **switch** *n* SWG conmutador *m* metropolitano

MF *abbr* COMP (*mediation function*) función *f* de mediación, TELEP (*medium frequency*) FM (*frecuencia media*), (*multifrequency*) MF (*multifrecuencia*); ~ **generator** *n* SIG, TELEP generador *m* de FM; ~ **mode** *n* PER APP *on answer machine* modo *m* OM; ~ **receiver** *n* SIG receptor *m* de FM; ~ **sender-receiver** *n* SIG, TELEP emisor-receptor *m* multifrecuencia; ~ **telephone** *n* TELEP teléfono *m* multifrecuencia

MFA *abbr* (*multiframe alignment*) TRANS alineamiento *m* de múltiples bastidores

M-field *abbr* (*mode-of-operation field*) PER APP campo-M *m* (*campo de modo de operación*)

MHD: ~ **segment** *n* COMP, DATA segmento *m* MHD

MHE *abbr* (*message handling environment*) APPL, DATA, RADIO, TELEP MHE (*entorno de manipulación del tráfico de mensajes*)

MHS *abbr* (*message handling system*) APPL, DATA, RADIO, TELEP sistema *m* de codificación de mensajes

MIB *abbr* (*management information base*) DATA base *f* de información para dirección de empresas

mica: ~ **capacitor** *n* ELECT, PARTS condensador *m* de mica; ~ **washer** *n* ELECT arandela *f* de mica

micro-: **~alloy diffused transistor** *n* (*MADT*) ELECT transistor *m* de microaleación difusa (*MADT*); **~assembly** *n* ELECT microconjunto *m*; **~image** *n* DATA microimagen *f*; **~inch** *n* GEN micropulgada *f*

microbend: ~ **loss** *n* TRANS pérdida *f* de microflexión; ~ **sensor** *n* TRANS sensor *m* de microflexión

microbending *n* TRANS microflexión *f*

microcassette *n* PER APP *for answer machine* microcasete *f*

microcell: ~ **network** *n* RADIO red *f* de microceldas

microchip *n* ELECT microchip *m*

microcircuit *n* ELECT microcircuito *m*

microcom: ~ **network protocol** *n* DATA protocolo *m* de red microcom
microcomputer *n* COMP microordenador *m Esp*, microcomputador *m AmL*, microcomputadora *f AmL*
microelectronics *n* ELECT microelectrónica *f*
microfiche *n* GEN microficha *f*
microfilm *n* GEN micropelícula *f*
microform *n* DATA microforma *f*; ~ **reader** PER APP lector *m* de microformas; ~ **reader/printer** PER APP lector/impresor *m* de microformas; ~**-retrieval terminal** DATA terminal *m* de recuperación de microformas
microphone *n* GEN, PER APP, TELEP micrófono *m*; ~ **current feed** POWER, TELEP alimentación *f* de micrófono; ~ **cutoff** PER APP *of answer machine* bloqueo *m* de micrófono; ~ **input** ELECT entrada *f* de micrófono; ~ **noise** ELECT ruido *m* microfónico
microphonic: ~ **effect** *n* ELECT, TELEP efecto *m* microfónico; ~ **noise** *n* ELECT, TELEP ruido *m* microfónico
microphonism *n* ELECT, TELEP efecto *m* microfónico
microprocessor *n* COMP microprocesador *m*; ~ **control** SWG control *m* del microprocesador
microprogramming *n* COMP microprogramación *f*
microribbon *n* ELECT microcinta *f*
microsecond *n* GEN microsecundo *m*
microsegmentation *n* RADIO microsegmentación *f*
microslot *n* TRANS microranura *f*
Microsoft®: ~ **disk operating system** *n* COMP sistema *m* operativo de disco Microsoft®
microstripper *n* ELECT microseparador *m*
microswitch *n* ELECT, SWG microinterruptor *m*
microtonal *adj* COMP microtonal
microtune *n* ELECT microsintonía *f*
microtuning *n* ELECT microsintonía *f*
microwave *n* RADIO microonda *f*; ~ **absorption** RADIO absorción *f* de microondas; ~ **aerial** RADIO antena *f* de microondas; ~ **amplifier** RADIO amplificador *m* de microondas; ~ **antenna** RADIO antena *f* de microondas; ~ **attenuator** RADIO atenuador *m* de microondas; ~ **beam** RADIO haz *m* de microondas; ~ **circulator** RADIO circulador *m* de microondas; ~ **communications** *n pl* RADIO comunicaciones *f pl* por microondas; ~ **detector** RADIO detector *m* de microondas; ~ **engineering** RADIO técnica *f* de microondas; ~ **frequency** RADIO, TV frecuencia *f* de microondas; ~ **generation** RADIO generación *f* de microondas; ~ **integrated circuit** ELECT circuito *m* integrado de microondas; ~ **landing system** RADIO sistema *m* de aterrizaje por microondas; ~ **link** RADIO enlace *m* por microondas; ~ **link analyser** *BrE* TEST analizador *m* de enlace por microondas; ~ **modulator** RADIO modulador *m* de microondas; ~ **optical duplex antenna link** RADIO enlace *m* de antena dúplex por haz hertziano de trayectoria óptima; ~ **phase changer** RADIO cambiador *m* de fase del haz hertziano; ~ **radio** RADIO radio *f* de microondas; ~ **radio system** RADIO sistema *m* radioeléctrico por microondas; ~ **relay link** *n pl* TRANS radioenlace *m* por microondas; ~ **resonator** RADIO resonador *m* de microondas; ~ **signal** RADIO señal *f* de microondas; ~ **switch sequence** SWG secuencia *f* de conmutación por microondas; ~ **system** RADIO sistema *m* de microondas; ~ **tower** RADIO torre *f* de microondas
mid-: ~**level** *n* RADIO nivel *m* medio; ~**position** *n* GEN posición *f* media; ~**range** *n* GEN alcance *m* medio; ~**range speaker** *n* ELECT altavoz *m* para el registro medio
MID *abbr* (*multiplexing identification*) TRANS MID (*identificación de multiplexión*)
middle: ~ **marker beacon** *n* RADIO radiobaliza *f* intermedia; ~ **position** *n* GEN posición *f* media
midpoint *n* GEN punto *m* medio; ~ **signalling** *BrE* SIG señalización *f* del punto medio; ~ **tap** ELECT *of transformer* toma *f* central
migrate *vi* GEN migrar
migration *n* GEN migración *f*
mike *n* GEN micrófono *m*
military: ~ **communication system** *n* APPL, RADIO, TRANS sistema *m* de comunicación militar
Miller: ~ **effect** *n* ELECT efecto *m* Miller
milliamp *n* (*mA*) ELECT miliamperio *m* (*mA*)
milliken: ~ **conductor** *n* ELECT conductor *m* segmentado
millimetre *BrE* (*AmE* **millimeter**) *n* GEN milímetro *m*; ~ **paper** GEN papel *m* milimetrado
millimetric: ~ **monolithic integrated circuit** *n* ELECT circuito *m* integrado monolítico milimétrico
millions: ~ **of bits per second** *n pl* (*Mbps*) COMP millones *m pl* de bitios por segundo (*Mbps*)
millisecond *n* GEN milisegundo *m*
mineral: ~ **insulation** *n* ELECT aislamiento *m* mineral; ~ **oil** *n* PARTS aceite *m* mineral
mini: ~ **disk** *n* COMP minidisco *m*; ~ **floppy** *n* COMP minidisco *m* flexible; ~ **jack** *n* PER APP miniclavija *m*; ~**-star configuration** *n* TV

configuración *f* en miniestrella; **~-wrapping** *n* ELECT minienrollado *m*
miniature: **~ fuse** *n* PARTS, POWER fusible *m* miniatura
miniaturization *n* GEN miniaturización *f*
miniaturize *vt* GEN miniaturizar
miniaturized *adj* GEN miniaturizado
minigauge: **~ loop** *BrE n* TELEP bucle *m* de minicalibre
minimize *vt* COMP minimizar
minimizing *n* COMP minimización *f*
minimum: **~ acceptable interval** *n* TELEP intervalo *m* mínimo aceptable; **~ bend radius** *n* TRANS radio *m* de curvatura mínimo; **~ cathode current** *n* ELECT corriente *f* catódica mínima; **~ charge** *n* APPL, TELEP tarifa *f* mínima; **~ clearing** *n* RADIO mejora *f* del mínimo; **~ detectable signal** *n* (*MDS*) TRANS señal *f* detectable mínima (*MDS*); **~ error-free period** *n* DATA periodo *m* mínimo libre de errores; **~ interference threshold** *n* RADIO umbral *m* mínimo de interferencia; **~ interval** *n* TELEP intervalo *m* mínimo; **~ net loss** *n* TRANS equivalente *m* mínimo admisible; **~ operate current** *n* ELECT corriente *f* operatoria mínima; **~ payable** *n* APPL mínimo *m* cotizable; **~ permissible basic transmission loss** *n* (*MPBTL*) DATA pérdida *f* mínima admisible de transmisión básica (*MPBTL*); **~-phase filter** *n* TRANS filtro *m* de fase mínima; **~-shift keying** *n* (*MSK*) SIG teclado *m* de desplazamiento mínimo (*TDM*); **~-standard antenna** *n* RADIO antena *f* de estándar mínimo; **~ usable erasing time** *n* ELECT tiempo *m* de borrado mínimo utilizable; **~ usable field strength** *n* RADIO intensidad *f* mínima de campo utilizable; **~ usable power flux-density** *n* RADIO densidad *f* mínima de flujo energético utilizable; **~ usable reading speed** *n* ELECT velocidad *f* de lectura mínima utilizable
minor: **~ alarm** *n* TEST alarma *f* menor; **~ defect** *n* TEST defecto *m* menor; **~ restart** *n* SWG reanudación *f* menor; **~ system restart** *n* SWG reanudación *f* menor del sistema
minority: **~ carrier** *n* ELECT portador *m* minoritario
minus: **~ polarity** *n* POWER polaridad *f* negativa
mirror[1] *n* DATA, TRANS espejo; **~ site** DATA sitio *m* especular; **~ surface** TRANS superficie *f* especular; **~ writing** DATA escritura *f* especular
mirror[2] *vt* COMP reflejar
MIS *abbr* DATA (*management information system*) MIS (*sistema de información para gestión*), ELECT (*metal insulator semiconductor*) SIA (*semiconductor con aislante metálico*); **~ user** *n* DATA usuario *m* de MIS
misalignment: **~ loss** *n* TRANS *fibre optics* pérdida *f* por desalineación
miscalculation *n* GEN error *m* de cálculo
miscellaneous: **~ apparatus rack** *n* (*MAR*) TELEP bastidor *m* para aparatos diversos (*MAR*)
misfire *n* POWER fallo *m* de encendido
mishandling: **~ failure** *n* TEST fallo *m* por manipulación indebida
misjudgment *n* GEN estimación *f* incorrecta
mismatch *n* GEN discordancia *f*
misplace *vt* COMP ubicar erróneamente
mispointing *n* RADIO señalamiento *m* erróneo
misprint *n* GEN errata *f*
misread *n* COMP error *m* de lectura
misspelling *n* GEN falta *f* de ortografía
mistake *n* GEN error *m*
mistracking *n* COMP error *m* de seguimiento
misuse: **~ failure** *n* TEST fallo *m* por uso indebido
mix *vt* GEN mezclar
mixed: **~ analog-digital channel** *n* TELEP canal *m* mixto análogo-digital; **~ lines** *n pl* TELEP líneas *f pl* mixtas; **~ mode** *n* GEN modo *m* combinado; **~-mode dialling** *BrE n* PER APP *of answer machine* marcado *m* en modo combinado; **~ terrain** *n* GEN terreno *m* mixto
mixer *n* COMP mezclador *m*; **~ stage** RADIO etapa *f* mezcladora; **~ tube** ELECT tubo *m* mezclador
mixing: **~ console** *n* COMP, RADIO pupitre *m* mezclador; **~ desk** *n* COMP mesa *f* mezcladora; **~ ratio** *n* RADIO riqueza *f* higrométrica; **~ tray** *n* ELECT bandeja *f* mezcladora
MJD *abbr* (*modified Julian date*) RADIO MJD (*fecha del calendario juliano modificada*)
MLI *abbr* (*multilayer isolation*) ELECT MLI (*aislamiento multicapa*)
MLM *abbr* (*multilongitudinal mode*) TELEP, TRANS modo *m* multilongitudinal
MLP *abbr* (*multilink procedure*) DATA MLP (*procedimiento multienlace*)
MMC *abbr* (*man-machine communications*) COMP MMC (*comunicaciones hombre-máquina*)
MMJ *abbr* (*modified modular jack*) ELECT, PARTS MMJ (*jack modular modificado*)
MML *abbr* (*man-machine language*) COMP MML (*lenguaje hombre-máquina*)
MMP *abbr* (*modified modular plug*) ELECT MMP (*clavija modular modificada*)
mnemonic *n* TELEP ayuda *f* mnemotécnica; **~ symbol** GEN símbolo *m* mnemotécnico
MO *abbr* (*magneto-optic*) COMP, ELECT MO

(*magnetoóptico*); ~ **disk** *n* COMP, DATA, ELECT disco *m* MO

mobile *n* TELEP teléfono *m* móvil; ~ **charge carrier** ELECT portadora *f* de carga móvil; ~ **communications** *n pl* RADIO comunicaciones *f pl* móviles; ~ **control node** (*MCN*) TELEP nodo *m* de control móvil; ~ **data processing terminal equipment** RADIO equipo *m* terminal móvil de procesamiento de datos; ~ **data transmission network** RADIO red *f* móvil de transmisión de datos; ~ **installation** RADIO instalación *f* móvil; ~ **location registration** RADIO registro *m* de localizaciones móviles; ~ **number** TELEP número *m* de teléfono móvil; ~ **office** RADIO oficina *f* móvil; ~ **phone** RADIO, TELEP telefono *m* móvil; ~ **radio fleet** RADIO flotilla *f* de radiocomunicación móvil; ~ **radio station** RADIO estación *f* móvil radioeléctrica; ~ **radio system** RADIO sistema *m* de radiocomunicación móvil; ~ **radio telephone** RADIO, TELEP *remote data processing* radioteléfono *m* móvil; ~ **radio telephone services** *n pl* TELEP servicios *m pl* de radioteléfono móvil; ~ **satellite telecommunication** RADIO telecomunicación *f* móvil por vía satélite; ~ **service** RADIO servicio *m* móvil; ~ **services switching centre** *BrE* SWG, TELEP centro *m* de conmutación de servicios móviles; ~ **station** (*MS*) RADIO, TELEP estación *f* móvil; ~**-station charge** TELEP carga *f* de teléfono móvil; ~**-station roaming number** (*MSRN*) TELEP número *m* de localización automática de estación móvil (*MSRN*); ~ **subscriber identity** TELEP identidad *f* de abonado al servicio móvil; ~ **switching centre** *BrE* (*MSC*) SWG, TELEP centro *m* de conmutación móvil (*MSC*); ~ **telephone** RADIO, TELEP teléfono *m* móvil; ~ **telephone operating company** RADIO, TELEP operador *m* de telefonía móvil; ~ **telephone service** RADIO, TELEP servicio *m* móvil telefónico; ~ **telephone station** RADIO, TELEP teléfono *m* móvil; ~ **telephone switching office** (*MTSO*) RADIO, TELEP oficina *f* de conmutación de telefonía móvil (*MTSO*); ~ **telephone system** RADIO, TELEP sistema *m* de telefonía móvil; ~ **telephony** RADIO, TELEP telefonía *f* móvil; ~ **telephony system** RADIO, TELEP sistema *m* de telefonía móvil; ~**-to-base relay** RADIO estación *f* repetidora para enlace móvil-base; ~ **unit** TELEP unidad *f* móvil

MOC *abbr* (*managed-object class*) DATA clase *f* de objeto gestionado

mock-up *n* GEN maqueta *f*

MOCS *abbr* (*managed-object conformance statement*) DATA MOCS (*declaración de conformidad del objeto gestionado*); ~ **proforma** *n* DATA proforma *f* MOCS

modal: ~ **distortion** *n* TRANS distorsión *f* modal; ~ **noise** *n* TRANS ruido *m* modal

mode *n* GEN modo *m*; ~ **changer** RADIO transformador *m* de modo; ~ **coupling** TRANS acoplamiento *m* de modos; ~ **dependency** DATA dependencia *f* de modo de funcionamiento; ~ **distortion** TRANS distorsión *f* de modos; ~ **distribution** TRANS distribución *f* de modos; ~ **field** TRANS *fibre optics* campo *m* modal; ~**-field diameter** TRANS *fibre optics* diámetro *m* del campo modal; ~**-field profile** TRANS *fibre optics* perfil *m* del campo modal; ~ **filter** TRANS filtro *m* de función; ~ **hopping** PER APP, TRANS salto *m* de modos; ~ **jumping** PER APP, TRANS salto *m* de modos; ~ **mixer** TRANS mezclador *m* de modos; ~ **mixing** TRANS mezcla *f* de modos; ~ **of mounting** GEN forma *f* de montaje; ~ **of operation** GEN modo *m* de funcionamiento; ~**-of-operation field** (*M field*) PER APP campo *m* de modo de operación (*campo-M*); ~ **of oscillation** RADIO modo *m* de oscilación; ~ **partition** TRANS *fibre optics* partición *f* de modo; ~**-partition noise** TRANS *fibre optics* ruido *m* de partición de modo; ~ **scrambler** TRANS *fibre optics* codificador *m* de modos; ~**-selective coupler** TRANS acoplador *m* de selección de modo; ~ **stripper** TRANS separador *m* de modos; ~ **transformer** RADIO transformador *m* de modo; ~ **volume** TRANS volumen *m* de modos

model *n* GEN modelo *m*

modelling *BrE* (*AmE* **modeling**) *n* GEN modelación *f*

modem *n* (*modulator/demodulator*) GEN módem *m* (*modulador/desmodulador*); ~ **card** DATA tarjeta *f* de módem; ~ **clock** DATA reloj *m* del módem; ~ **control** DATA control *m* del módem; ~ **control key** DATA clave *f* de control del módem; ~ **eliminator** DATA eliminador *m* del módem; ~ **holdover period** DATA periodo *m* de retención del módem; ~ **interface** DATA separador *m* de módem; ~ **performance** PER APP funcionamiento *m* de módem; ~ **port** COMP puerta *f* del módem; ~ **protection equipment** DATA equipo *m* de protección del módem; ~ **radio** RADIO módem *m*; ~ **receiver** DATA receptor *m* de módem; ~ **tester** TEST analizador *m* de módem; ~ **transmitter** DATA transmisor *m* de módem; ~ **turnaround** DATA ejecución *f* del módem

modernization *n* GEN modernización *f*

modifiable *adj* GEN modificable

modification *n* GEN modificación *f*; **~ indicator** DATA indicador *m* de modificaciones
modified: **~ alternate mark inversion** *n* SIG, TRANS señal *f* bipolar modificada; **~ constraint** *n* DATA condicionamiento *m* modificado; **~ Julian date** *n* (*MJD*) RADIO fecha *f* de calendario juliano modificada (*MJD*); **~ Julian day** *n* RADIO día *m* de calendario juliano modificado; **~ modular jack** *n* (*MMJ*) ELECT jack *m* modular modificado (*MMJ*); **~ modular plug** *n* (*MMP*) ELECT clavija *f* modular modificada (*MMP*); **~ refractive index** *n* RADIO índice *m* de refracción modificado
modifier *n* COMP modificador *m*
modify *vt* GEN modificar
moding *n* ELECT gama *f* de modos
modular *adj* GEN modular; **~ construction** *n* GEN construcción *f* modular; **~ jack** *n* ELECT jack *m* modular; **~ patching** *n* ELECT parcheo *m* modular; **~ plug blade** *n* ELECT espiga *f* de clavija modular; **~ principle** *n* TELEP principio *m* modular; **~ spacing** *n* TELEP separación *f* modular; **~ system** *n* TELEP sistema *m* modular
modularity *n* GEN modularidad *f*
modularization *n* GEN modularización *f*
modulate *vt* GEN modular
modulated: **~ continuous waves** *n pl* (*MCW*) RADIO ondas *f pl* continuas moduladas (*MCW*); **~ signal** *n* TRANS señal *f* modulada; **~ tariff time slot** *n* DATA intervalo *m* de tiempo de tarifa modulada; **~ value** *n* DATA valor *m* modulado; **~ wave** *n* RADIO, TRANS onda *f* modulada
modulating: **~ action** *n* CONTR acción *f* por modulación; **~ frequency** *n* TELEP, TRANS frecuencia *f* de modulación; **~ operator** *n* COMP operador *m* por modulación; **~ signal** *n* TRANS señal *f* de modulación; **~ wave** *n* TRANS onda *f* moduladora
modulation *n* GEN modulación *f*; **~ by pulses** TRANS modulación *f* por impulsos; **~ coherence** TELEP coherencia *f* de modulación; **~ depth** TRANS profundidad *f* de modulación; **~ element** TELEP, TRANS elemento *m* de modulación; **~ factor** TRANS factor *m* de modulación; **~ with fixed reference** TELEP modulación *f* con referencia fija; **~ frequency** TELEP, TRANS frecuencia *f* de modulación; **~ generator** TRANS generador *m* de modulación; **~ index** RADIO índice *m* de modulación; **~ lever** COMP palanca *f* de modulación; **~ percentage** TRANS porcentaje *m* de modulación; **~ polarity** TV polaridad *f* de modulación; **~ product** TELEP producto *m* de modulación; **~ of pulses** TELEP modulación *f* de impulsos; **~ rate** TELEP, TRANS velocidad *f* de modulación; **~ restitution signal element** SIG elemento *m* señalizador de restitución de modulación; **~ speed** TRANS velocidad *f* de modulación; **~ switch** RADIO conmutador *m* de modulación; **~ technique** RADIO técnica *f* de modulación; **~ transfer function** ELECT función *f* de transferencia de modulación; **~ type** TRANS tipo *m* de modulación; **~ voltage** ELECT voltaje *m* de modulación; **~-wheel assignment** COMP asignación *f* de rueda de modulación; **~ wheel range** COMP alcance *m* de rueda de modulación; **~ wheel sensitivity** COMP sensibilidad *f* de la rueda de modulación; **~ wheel vibrato** COMP vibrato *m* de la rueda de modulación
modulator *n* GEN modulador *m*; **~ charger** ELECT cargador *m* del modulador; **~/demodulator** (*modem*) COMP, DATA, TRANS modulador/desmodulador *m* (*módem*)
module *n* GEN módulo *m*
modulo: **~-n counter** *n* ELECT contador *m* módulo-n
modulus: **~ of elasticity** *n* PARTS módulo *m* de elasticidad
moiré *n* TV muaré *m*
moisture *n* GEN humedad *f*; **~-absorbing capacity** GEN capacidad *f* de absorción de humedad; **~ content** GEN contenido *m* de humedad; **~ resistance** GEN humidifugacia *f*
mold *AmE see mould BrE*
molecular: **~ beam** *n* TV haz *m* molecular; **~ electronics** *n* ELECT electrónica *f* molecular
moment *n* GEN momento *m*
momentary *adj* GEN momentáneo
monitor *n* GEN monitor *m*; **~ program** DATA programa *m* monitor; **~ system** CONTR sistema *m* de control; **~ unit** CONTR unidad *f* monitora
monitoring *n* GEN control *m*; **~ equipment** CONTR, GEN equipo *m* de control; **~ mode** RADIO modo *m* de control; **~ program** CONTR, DATA programa *m* de control; **~ unit** CONTR unidad *f* de control
mono: **~ input** *n* COMP, PER APP entrada *f* mono; **~ output** *n* COMP, PER APP salida *f* mono
monocellular *adj* RADIO monocelular
monochromatic *adj* TRANS monocromático
monochromator *n* TRANS monocromador *m*
monochrome *adj* GEN monocromo; **~ supertwist nematic** *n* ELECT nemática *f* torcida supermonocromática; **~ television** *n* TV televisión *f* monocromática
monolithic *adj* ELECT monolítico; **~ circuit** *n* ELECT circuito *m* monolítico; **~ filter** *n* ELECT filtro *m* monolítico; **~ integrated circuit** *n*

ELECT, PARTS circuito *m* integrado monolítico; ~ **microwave integrated circuit** *n* ELECT circuito *m* integrado monolítico de microondas

monologue: ~ **interaction** *n* DATA interacción *f* en monólogo

monomedia: ~ **communication** *n* TV comunicación *f* monomediática; ~ **processing** *n* TV procesamiento *m* monomediático

monomode *adj* ELECT monomodo

monophonic *adj* RADIO monofónico; ~ **signal** *n* RADIO señal *f* monofónica

monophony *n* RADIO monofonía *f*

monopole: ~ **antenna** *n* RADIO antena *f* monopolo

monopulse: ~ **radar** *n* RADIO radar *m* de monoimpulsos

monoscope *n* ELECT monoscopio *m*

monostable *adj* ELECT monoestable

monostatic: ~ **radar** *n* RADIO radar *m* monoestático

mono/stereo: ~ **output** *n* COMP, PER APP, RADIO salida *f* mono/estéreo

monotonic: ~ **reasoning** *n* DATA razonamiento *m* monótono

morphing *n* COMP morfeado *m*

Morse: ~ **code** *n* TELEP código *m* Morse; ~ **dash** *n* TELEP raya *f* Morse; ~ **dot** *n* TELEP punto *m* Morse; ~ **key** *n* TELEP manipulador *m* Morse; ~ **space** *n* TELEP espacio *m* Morse; ~ **telegraphy** *n* TELEP telegrafía *f* Morse

MOS *abbr* (*mean opinion score*) COMP SOM (*media de opinión*)

mosaic *n* ELECT, TEST mosaico *m*; ~ **effect** COMP efecto *m* mosaico; ~ **telegraphy** TELEP telegrafía *f* mosaico

mother: ~ **board** *n* COMP placa *f* madre

motion: ~ **blur** *n* TV imagen *f* borrosa en movimiento; **~-compensated concealment** *n* TV ocultamiento *m* compensado por el movimiento; ~ **compensation** *n* TV compensación *f* de movimiento; ~ **estimation** *n* TV estimación *f* del movimiento; ~ **picture experts group algorithm** *n* DATA algoritmo *m* de grupo de expertos en cinematografía; ~ **picture experts group channel** *n* TV canal *m* de grupo de expertos en cinematografía; ~ **vector** *n* COMP, TV vector *m* de movimiento

MOTIS *abbr* (*message-oriented text interchange system*) DATA sistema *m* de intercambio de texto para mensajes

motor: **~-driven system** *n* RADIO sistema *m* de accionamiento motorizado; ~ **protector switch** *n* POWER interruptor *m* del protector de motor

MOU *abbr* (*memorandum of understanding*) GEN MOU (*memorándum de acuerdo*)

mould *BrE* (*AmE* **mold**) *n* GEN molde *m*

moulded *BrE* (*AmE* **molded**) *adj* GEN moldeado; ~ **composition resistor** *n* ELECT, PARTS resistor *m* de composición moldeada; ~ **plastic** *n* ELECT plástico *m* moldeado

mount *vt* GEN montar

mounting *n* GEN montaje *m*; ~ **bar** PARTS barra *f* de montaje; ~ **base** PARTS base *f* de montaje; ~ **bracket** PARTS consola *f* para montaje; ~ **cord** ELECT cuerda *f* de montaje; ~ **cut-out** ELECT desconexión *f* de montaje; ~ **drawing** GEN plano *m* de entrefase; ~ **frame** PARTS marco *m* de montaje; ~ **hardware** PARTS piezas *f pl* de montaje; ~ **holder** PARTS soporte *m* de montaje; ~ **hole** GEN agujero *m* de fijación; ~ **instructions** *n pl* GEN instrucciones *f pl* de montaje; ~ **mode** TELEP modo *m* de montaje; ~ **plate** PARTS placa *f* de fijación; ~ **screw** PARTS perno *m* de montaje, tornillo *m* de fijación

mouse *n* COMP ratón *m*

mouth: ~ **reference point** *n* TELEP orificio *m* de referencia

mouthpiece *n* PER APP, TELEP boca *f*; ~ **muffler** *AmE* (*cf mouthpiece silencer BrE*) PER APP *of answer machine*, TELEP sordina *f* de la tapa microfónica; ~ **silencer** *BrE* (*cf mouthpiece muffler AmE*) PER APP *of answer machine*, TELEP sordina *f* de la tapa microfónica

move *vt* GEN mover

move down *vt* COMP bajar

move up *vt* COMP subir

movie *n* TV película *f*; ~ **preview** TV preestreno *m* de una película

moving: **~-coil instrument** *n* POWER instrumento *m* de bobina móvil; **~-coil permanent magnet instrument** *n* POWER instrumento *m* de imán permanente con bobina móvil; ~ **fibre** *BrE* **switch** *n* SWG conmutador *m* móvil aislado con fibra; **~-iron instrument** *n* POWER instrumento *m* electromagnético; **~-target indicator radar** *n* (*MTI radar*) RADIO radar *m* indicador de blancos móviles; **~-vane voltmeter** *n* TEST voltímetro *m* de hierro móvil

MP: ~ **capacitor** *n* (*metallized paper capacitor*) POWER capacitor *m* MP (*capacitor de papel metalizado*)

MPA *abbr* (*multibeam phased array*) RADIO MPA (*red multihaz de elementos en fase*)

MPBTL *abbr* (*minimum permissible basic transmission loss*) DATA MPBTL (*pérdida mínima admisible de transmisión básica*)

MPDU *abbr* (*message protocol data unit*) DATA

MPDU (*unidad de datos de protocolo de mensaje*)

MPEG: ~ **algorithm** *n* DATA algoritmo *m* MPEG; ~ **channel** *n* TV canal *m* MPEG

MPIL *abbr* (*maximum permissible interference level*) DATA MPIL (*nivel máximo permisible de interferencia*)

MPS *abbr* (*multipage signal*) DATA, TELEP MPS (*señal multipágina*)

MPSK *abbr* (*multilevel phase-shift keying*) SIG MPSK (*tecleo de cambio de fase multinivel*)

MRSE *abbr* (*message retrieval service element*) DATA MRSE (*elemento del servicio de recuperación de mensajes*)

MRTIE *abbr* (*maximum relative time interval error*) DATA error máximo de intervalo relativo

MS *abbr* (*mobile station*) RADIO, TELEP MS (*estación móvil*); ~ **abstract service** *n* DATA, TELEP servicio *m* abstracto MS; ~ **abstract service provider** *n* DATA, TELEP proveedor *m* de servicio abstracto MS; ~ **abstract service user** *n* DATA, TELEP usuario *m* de servicio abstracto MS; ~ **stereo** *n* TV estéreo *m* MS; ~ **user** *n* DATA usuario *m* de MS

MSC *abbr* (*mobile switching centre*) SWG MSC (*centro de conmutación móvil*)

MSI *abbr* (*multisignal interconnect*) DATA IMS (*interconexión multiseñal*)

MSK *abbr* (*minimum-shift keying*) SIG TDM (*teclado de desplazamiento mínimo*)

MSN *abbr* (*multiple subscriber number*) TELEP número *m* de abona (*número de abonado múltiple, NAM*)

MSP *abbr* (*maintenance service provider*) TEST PSM (*proveedor del servicio de mantenimiento*)

MSRN *abbr* (*mobile station roaming number*) TELEP MSRN (*número de localización automática de estación móvil*)

MSSC *abbr* (*maritime satellite switching centre*) RADIO MSSC (*centro marítimo de conmutación por satélite*)

MSSE *abbr* (*message submission service element*) DATA MSSE (*elemento de servicio de presentación de mensaje*)

MSSS *abbr* (*multisatellite support centre*) RADIO MSSS (*centro de apoyo multisatélite*)

MSSW *abbr* (*magnetostatic surface wave*) ELECT MSSW (*onda de superficie magnetostática*)

mt: ~ **transfer** *n* DATA transferencia *f* megatónica

MT *abbr* (*message transfer*) DATA, TELEP transferencia *f* de mensajes

MTA *abbr* (*message transfer agent*) DATA, TELEP agente *m* de transferencia de mensajes

MTBF *abbr* (*mean time between failures*) TEST MTBF (*tiempo medio entre fallos*)

MTBI *abbr* (*mean time between interruptions*) COMP MTBI (*tiempo medio entre interrupciones*)

MTBO *abbr* (*mean time between overhauls*) TEST MTBO (*tiempo medio entre revisiones*)

MTFF *abbr* (*mean time to first failure*) COMP MTFF (*tiempo medio hasta el primer fallo*)

MTI: ~ **radar** *n* (*moving target indicator radar*) RADIO radar *m* indicador de blancos móviles

MTIE *abbr* (*maximum time interval error*) DATA error de intervalo de tiempo máximo

MTP *abbr* (*message transfer part*) DATA, TELEP MTP (*parte de la transferencia de mensaje*)

MTPI *abbr* (*multiplexer timing physical interface*) TRANS interfaz *f* física para sincronismo del multiplexor

MTS *abbr* (*message transfer system*) DATA, TELEP sistema de transferencia de mensajes

MTSE *abbr* (*message transfer service element*) DATA, TELEP MTSE (*elemento del servicio de transferencia de mensajes*)

MTSO *abbr* (*mobile telephone switching office*) RADIO, TELEP MTSO (*oficina de conmutación de telefonía móvil*)

MTSR *abbr* (*mean time to service restoration*) TEST MTSR (*tiempo medio hasta el restablecimiento del servicio*)

MTTR *abbr* (*mean time to recovery*) TEST MTTR (*tiempo medio hasta el restablecimiento*)

mu: ~ **factor** *n* POWER factor *m* de amplificación; ~ **law codec** *n* TRANS códice *m* legal mu

MUF *abbr* (*maximum usable frequency*) RADIO MUF (*frecuencia máxima utilizable*)

muffler *AmE n* (*cf silencer BrE*) TELEP, TRANS silenciador *m*

muldex *n* TELEP muldex *m*

multiaccess: ~ **system** *n* DATA sistema *m* de acceso múltiple

multiaddress: ~ **calling** *n* TELEP llamada *f* de dirección múltiple

multiaddressing *n* DATA multidireccionamiento *m*

multiarea *adj* RADIO multiárea

multiband: ~ **filter** *n* TRANS filtro *m* multibanda

multibeam: ~ **phased array** *n* (*MPA*) RADIO red *f* multihaz de elementos en fase (*MPA*)

multibearer: ~ **service** *n* TELEP *ISDN* servicio *m* de múltiples portadores

multibit: ~ **input** *n* DATA entrada *f* multibitios; ~ **output** *n* DATA salida *f* multibitios

multiblock *adj* TELEP multibloque; ~ **acknowledgement signal** *n* SIG señal *f* de reconocimiento multibloque; ~ **monitoring signal** *n* SIG señal *f* de control multibloque; ~ **synchronization signal unit** *n* (*MBS*) SIG unidad *f* de señal de sincronización multibloque (*MBS*)

multicast: ~ **call** *n* DATA llamada *f* de emisión doble; ~ **group** *n* DATA grupo *m* de emisión doble; ~ **server** *n* DATA servidor *m* de emisión doble; ~ **service** *n* DATA servicio *m* de emisión doble; ~ **service model** *n* DATA modelo *m* de servicio de emisión doble

multicavity: ~ **klystron** *n* ELECT klistron *m* de cavidad múltiple

multichannel *adj* TRANS multicanal; ~ **audio-digital interface** *n* COMP interfaz *m* audio-digital multicanal; ~ **filter** *n* TRANS filtro *m* multicanal; ~ **recording** *n* TV grabación *f* multicanal; ~ **rms deviation** *n* TRANS desviación *f* rms multicanal

multiconductor: ~ **cable** *n* TRANS cable *m* de múltiples conductores

multicyclic: ~ **mode** *n* DATA modo *m* multicíclico

multidestination: ~ **carrier** *n* TRANS portadora *f* de múltiples destinos; ~ **mode** *n* DATA modo *m* de múltiples destinos

multidimensional: ~ **data analysis** *n* DATA análisis *m* multidimensional de datos; ~ **digital filter** *n* DATA filtro *m* digital multidimensional; ~ **filtering** *n* ELECT filtrado *m* multidimensional; ~ **signal** *n* DATA, SIG señal *f* multidimensional

multielectrode: ~ **voltage-stabilizing tube** *n* ELECT tubo *m* multielectródico estabilizador de voltaje

multielement: ~ **service charge** *n* APPL carga *f* de servicio de elementos múltiples

multifaceted *adj* GEN multifacetado

multifibre *BrE* (*AmE* **multifiber**) *adj* TRANS multifibro; ~ **cable** *n* TRANS cable *m* multifibro; ~ **connector** *n* TRANS conector *m* multifibro; ~ **joint** *n* TRANS unión *f* multifibra

multiframe *n* DATA, TELEP, TRANS multitrama *f*; ~ **alignment** (*MFA*) TRANS alineamiento *m* de múltiples bastidores; ~ **alignment code** TRANS código *m* de múltiples bastidores; ~ **buffering** DATA almacenamiento *m* intermedio de multitrama; ~ **structure** DATA estructura *f* de multitrama

multifrequency *n* SIG, TELEP, TRANS multifrecuencia *f*; ~ **antenna** RADIO antena *f* multifrecuencia; ~ **generator** SIG, TELEP generador *m* multifrecuencia; ~ **receiver** SIG, TELEP receptor *m* multifrecuencia; ~ **screen** TV pantalla *f* de multifrecuencia; ~ **sender-receiver** SIG, TELEP emisor-receptor *m* multifrecuencia; ~ **signalling** *BrE* SIG señalización *f* de multifrecuencia; ~ **telephone** TELEP *signalling mode* teléfono *m* multifrecuencia

multifunction *adj* GEN de función múltiple, multifunción; ~ **key** *n* COMP tecla *f* de función múltiple; ~ **printer** *n* PER APP impresora *f* de función múltiple

multihop: ~ **satellite link** *n* RADIO enlace *m* de satélite de reflexión múltiple; ~ **system** *n* RADIO sistema *m* de reflexión múltiple; ~ **video** *n* TV vídeo *m* de reflexión múltiple

multilateral *adj* GEN multilateral

multilayer: ~ **CD** *n* DATA CD *m* multicapa; ~ **compact disc** *n* DATA disco *m* compacto multicapa; ~ **isolation** *n* (*MLI*) ELECT aislamiento *m* multicapa (*MLI*); ~ **optical compact disc** *n* DATA disco *m* compacto óptico multicapa; ~ **printed circuit** *n* ELECT circuito *m* impreso multicapa; ~ **wiring unit** *n* ELECT, PARTS unidad *f* de cableado multicapa

multilevel: ~ **continuous sampling** *n* TEST muestreo *m* continuo multinivel; ~ **image** *n* TELEP imagen *f* multinivel; ~ **modulation** *n* RADIO modulación *f* multinivel; ~ **phase-shift keying** *n* (*MPSK*) SIG tecleo *m* de cambio de fase multinivel (*MPSK*); ~ **system** *n* TRANS sistema *m* multinivel

multiline *adj* GEN multilineal; ~ **representation** *n* GEN representación *f* multilineal

multilingual: ~ **recognition** *n* APPL, COMP, TELEP reconocimiento *m* multilingüe

multilink: ~ **connection** *n* RADIO conexión *f* multienlace; ~ **procedure** *n* (*MLP*) DATA procedimiento *m* multienlace (*MLP*)

multilobe: ~ **aerial** *n* RADIO antena *f* multilobular

multilongitudinal: ~ **mode** *n* (*MLM*) TELEP, TRANS modo *m* multilongitudinal

multimedia *adj* COMP, TV multimedia; ~ **and hypermedia information coding experts group** *n* COMP grupo *m* de expertos en codificación de información multimedia e hipermedia; ~ **authoring system** *n* COMP sistema *m* de autoría de multimedia; ~ **capabilities** *n pl* COMP posibilidades *f pl* multimedia; ~ **conferencing** *n* TV conferencias *f pl* por multimedia; ~ **configuration** *n* COMP configuración *f* multimedia; ~ **database** *n* COMP, DATA base *m* de datos multimedia; ~ **document** *n* DATA documento *m* multimedia; ~ **editing** *n* TV edición *f* multimedia; ~ **file server** *n* COMP, DATA servidor *m* de archivo multimedia; ~ **kiosk** *n* TV quiosco *m* multimedia; ~ **messaging** *n* COMP mensajería *f* multimedia; ~ **navigation** *n* COMP navegación *f* multimedia;

~ **network** *n* TRANS red *f* multimedia; ~ **PC** *n* COMP ordenador *m* personal multimedia *Esp*, computador *m* personal multimedia *AmL*, computadora *f* personal multimedia *AmL*; ~ **personal computer** *n* COMP ordenador *m* personal multimedia *Esp*, computador *m* personal multimedia *AmL*, computadora *f* personal multimedia *AmL*; ~ **processing** *n* COMP procesamiento *m* en multimedia; ~ **server** *n* COMP, DATA servidor *m* de multimedia; ~ **service** *n* TV servicio *m* multimedia; ~ **station** *n* TV estación *f* multimedia; ~ **synchronization** *n* DATA sincronización *f* multimedia; ~ **system** *n* TV sistema *m* multimedia; ~ **terminal** *n* TV terminal *m* multimedia; ~ **transmission** *n* DATA, TRANS transmisión *f* multimedia; ~ **user interface** *n* DATA interfaz *m* de usuario demultimedia; ~ **workstation** *n* COMP, DATA, PER APP estación *f* de trabajo multimedia

multimeter *n* TEST multímetro *m*

multimetering *n* TELEP cómputo *m* múltiple

multimicroprocessor: ~ **system** *n* SWG sistema *m* multimicroprocesador

multimode: ~ **distortion** *n* TRANS distorsión *f* multimodal; ~ **fibre** *BrE n* DATA, PARTS, TRANS fibra *f* multimodo; ~ **laser** *n* ELECT, TRANS láser *m* multimodo

multipage: ~ **PCX file** *n* (*DCX*) DATA archivo *m* PCX multipágina (*DCX*); ~ **signal** *n* (*MPS*) DATA, TELEP señal *f* multipágina (*MPS*)

multipair *adj* TV de pares múltiples; ~ **cable** *n* TRANS, TV cable *m* de pares múltiples

multipath: ~ **fading** *n* RADIO desvanecimiento *m* por multitrayectoria; ~ **interference** *n* RADIO interferencia *f* de trayectos múltiples; ~ **propagation** *n* RADIO propagación *f* por trayectoria múltiple; ~ **reflections** *n* RADIO reflexiones *f pl* con trayectorias múltiples; ~ **time dispersion** *n* RADIO dispersión *f* de tiempo por trayectos múltiples; ~ **transmission** *n* DATA, TRANS transmisión *f* por trayectoria múltiple

multipin: ~ **connector** *n* ELECT, PARTS conector *m* de contactos múltiples

multipivoted: ~ **coherent system** *n* TV sistema *m* coherente de articulación múltiple

multiple *adj* GEN múltiple; ~ **access** *n* RADIO acceso *m* múltiple; ~ **association control function** *n* (*MACF*) DATA función *f* de control de asociación múltiple (*MACF*); ~ **beam** *n* RADIO multihaz *m*; ~-**beam antenna** *n* RADIO antena *f* multihaz; ~ **cable** *n* SWG cable *m* múltiple; ~-**code access** *n* DATA acceso *m* de código múltiple; ~ **columns** *n pl* DATA *Internet* columnas *f pl* múltiples; ~ **container** *n* POWER *for batteries* recipiente *m* con compartimentos; ~ **diffraction** *n* TRANS difracción *f* múltiple; ~-**duct block** *n* PARTS, TRANS bloque *m* de conducto múltiple; ~-**filter chain** *n* TRANS cadena *f* de filtro múltiple; ~ **frame** *n* SWG cuadro *m* múltiple; ~ **frequency** *n* GEN frecuencia *f* múltiple; ~-**frequency encoder** *n* TRANS codificador *m* de múltiples frecuencias; ~-**gun cathode-ray tube** *n* ELECT tubo *m* de rayos catódicos de cañón múltiple; ~ **image** *n* TV efecto *m* eco; ~-**image production master** *n* TV soporte *m* maestro de producción de imagen múltiple; ~-**key set** *n* RADIO teclado *m* múltiple; ~ **lines at the same address** *n pl* TELEP líneas *f pl* múltiples a la misma dirección; ~ **memory** *n* RADIO memoria *f* múltiple; ~-**mode connector** *n* TRANS conector *m* de modo múltiple; ~-**mode modem** *n* PER APP, TRANS módem *m* de modo múltiple; ~ **modulation** *n* TRANS modulación *f* múltiple; ~-**path propagation** *n* RADIO propagación *f* de trayectoria múltiple; ~ **pattern** *n* ELECT impresión *f* múltiple; ~-**printed panel** *n* ELECT panel *m* impreso múltiple; ~ **quantum well** *n* ELECT efecto *m* cuántico múltiple; ~ **recorder** *n* PER APP registrador *m* múltiple; ~-**reflector antenna** *n* RADIO antena *f* de reflector múltiple; ~ **sampling** *n* TRANS muestreo *m* múltiple; ~ **seizure** *n* SWG toma *f* múltiple; ~-**server queue** *n* COMP, DATA, PER APP cola *f* en servidor múltiple; ~ **subsampling encoding** *n* (*MUSE*) DATA codificación *f* por submuestreo múltiple (*MUSE*); ~ **subscriber number** *n* (*MSN*) TELEP número *m* de abonado múltiple (*número de abona*, *NAM*); ~ **switchboard** *n* SWG cuadro *m* múltiple; ~ **television transmission** *n* TRANS, TV transmisión *f* múltiple de televisión; ~ **transmission** *n* TRANS, TV transmisión *f* múltiple; ~ **transmitter** *n* RADIO, TRANS, TV transmisor *m* múltiple; ~-**transmitter distributor** *n* TELEP distribuidor *m* de transmisor múltiple; ~-**transponder operation** *n* RADIO operación *f* de respondedor múltiple; ~-**trunk distribution system** *n* TV sistema *m* de distribución de línea troncal múltiple; ~ **tube** *n* ELECT tubo *m* múltiple; ~ **twin cable** *n* TRANS cable *m* de pares combinados

multiplex *n* COMP, DATA, RADIO, SWG, TRANS múltiplex *m*; ~ **aggregate bit rate** TELEP velocidad *f* de tráfico binario agregado múltiplex; ~ **equipment** TELEP, TRANS equipo *m* múltiplex; ~ **interface** TELEP interfaz *m* múltiplex; ~ **link** TELEP enlace *m* múltiplex; ~ **section** TRANS sección *f* de múltiplex; ~**-section alarm indication signal** TRANS señal *f* deindicación de alarma de la sección de múltiplex; ~**-section overhead** TRANS cabecera *f* de sección de múltiplex; ~**-section protection** TRANS protección *f* de sección de múltiplex; ~**-section termination** TRANS final *m* de sección de múltiplex; ~ **space switch** SWG conmutación *f* de espacio múltiplex; ~**-timing generator** SWG generador *m* de temporización de múltiplex

multiplexed: ~ **analog components** *n* DATA, TV componentes *m* análogos multiplexados

multiplexer *n* (*MUX*) COMP, DATA, RADIO, SWG, TRANS multiplexor *m* (*MUX*); ~ **channel** TRANS canal *m* multiplexor; ~ **timing physical interface** (*MTPI*) TRANS interfaz *f* física para sincronismo del multiplexor; ~ **timing source** TRANS fuente *f* de señales de sincronismo de multiplexor

multiplexing *n* COMP, DATA, RADIO, SWG, TRANS multiplexión *f*; ~ **identification** (*MID*) TRANS identificación *f* de multiplexión (*MID*)

multiplicator *n* ELECT, TELEP multiplicador *m*

multiplied *adj* ELECT, TELEP multiplicado

multiplier *n* ELECT, TELEP multiplicador *m*

multipoint *adj* GEN multipunto; ~ **command assignment token** *n* DATA testigo *m* de asignación de mando multipunto; ~ **command conference** *n* DATA conferencia *f* de mando multipunto; ~ **command negating MCS** *n* DATA MCS *m* anulación de mando multipunto; ~ **command release token** *n* DATA testigo *m* de liberación de mando multipunto; ~ **command symmetrical data transmission** *n* DATA, TRANS transmisión *f* de datos simétricos de mando multipunto; ~ **command token claim** *n* DATA demanda *f* de testigo de mando multipunto; ~ **communication service** *n* TELEP servicio *m* de comunicaciones multipunto; ~ **communications system** *n* TRANS sistema *m* de comunicaciones multipunto; ~ **connection** *n* DATA conexión *f* multipunto; ~ **connector** *n* ELECT, PARTS conector *m* multipunto; ~ **indication loop** *n* DATA bucle *m* de indicación multipunto; ~ **indication secondary status** *n* DATA estado *m* secundario de indicación multipunto; ~ **indication visualization** *n* DATA visualización *f* de indicación multipunto; ~ **indication zero communication** *n* DATA indicación *f* multipunto de comunicación cero; ~ **repeater** *n* TRANS repetidor *m* multipunto

multipolar *adj* POWER multipolar

multipole *adj* POWER multipolar

multipolling *n* COMP interrogación *f* múltiple

multiprocessing *n* COMP, DATA procesamiento *m* múltiple; ~ **program** COMP programa *m* de procesamiento múltiple; ~ **system** DATA sistema *m* de procesamiento múltiple

multiprocessor *n* COMP, DATA multiprocesador *m*; ~ **interconnection network** DATA red *f* de interconexión de procesador múltiple; ~ **station bus** TELEP bus *m* de estación multiprocesadora; ~ **system** SWG sistema *m* multiprocesador

multiprogramming *n* COMP multiprogramación *f*

multiprotocol: ~ **routing** *n* COMP, DATA, TELEP, TRANS encaminamiento *m* multiprotocolario encaminamiento *m* multiprotocolo; ~ **signalling** *BrE* **coupler** *n* SIG acoplador *m* de señalización multiprotocolaria

multipurpose *adj* GEN polivalente (*M*); ~ **key** *n* COMP tecla *f* de finalidad múltiple; ~ **wiring system** *n* ELECT sistema *m* de conexiones polivalente

multirange: ~ **grommet** *n* ELECT anillo *m* protector de varias escalas; ~ **voltmeter** *n* TEST voltímetro *m* de varias sensibilidades

multirate: ~ **filtering** *n* TRANS filtración *f* de varias velocidades; ~ **switching system** *n* SWG sistema *m* de conmutación de tarifa múltiple

multiray: ~ **channel** *n* RADIO canal *m* multihaz; ~ **diversity** *n* RADIO diversidad *f* multihaz

multiregional *adj* GEN multiregional

multirelay *adj* RADIO multirelé

multisampling *n* TRANS muestreo *m* múltiple

multisatellite: ~ **link** *n* RADIO, TV enlace *m* multisatélite; ~ **support centre** *BrE n* (*MSSS*) RADIO centro *m* de apoyo multisatélite (*MSSS*)

multiscan: ~ **monitor** *n* TV monitor *m* de digitalización múltiple

multiscreen: ~ **configuration** *n* TV configuración *f* de pantallas múltiples

multiservice: ~ **automatic exchange** *n* SWG centralita *f* automática multiservicio; ~ **network** *n* TRANS red *f* multiservicio; ~ **switching system** *n* SWG sistema *m* de conmutación multiservicio

multisession *n* COMP multisesión *f*

multisignal: ~ **interconnect** *n* (*MSI*) DATA interconexión *f* multiseñal (*IMS*)

multisite *adj* RADIO multilocal

multislot: ~ **connection** *n* TELEP conexión *f* de ranura múltiple

multispot: ~ **antenna** *n* RADIO antena *f* de puntos múltiples

multistage: ~ **circuit** *n* ELECT circuito *m* multietapa; ~ **network** *n* ELECT red *f* multietapa; ~ **sampling** *n* TEST muestreo *m* multietapa

multistandard: ~ **television set** *n* TV televisor *m* multinorma

multistate: ~ **modulation** *n* RADIO modulación *f* transicional

multistation: ~ **access unit** *n* (*MAU*) DATA unidad *f* de acceso multiestación (*MAU*); ~ **teletex terminal installation** *n* TELEP instalación *f* de terminal teletexto multiestación

multistep: ~ **action** *n* CONTR acción *f* a niveles múltiples

multistrand: ~ **wire** *n* ELECT, PARTS, TRANS conductor *m* de hilos múltiples

multistranded *adj* ELECT, PARTS, TRANS dotado de conductor de hilos múltiples

multisynchronous *adj* TV multisincrónico

multitask *adj* COMP, DATA multitarea

multitasking *n* COMP, DATA multitarea *f*

multiterminal: ~ **service circuit** *n* TELEP circuito *m* de servicio de terminal múltiple

multitimbral *adj* COMP multitímbrico

multitone: ~ **system** *n* TELEP sistema *m* multitonal

multitrack *adj* DATA multipista; ~ **recorder** *n* DATA registrador *m* multipista; ~ **sequencer** *n* DATA secuenciador *m* multipista

multiunit: ~ **message** *n* TELEP mensaje *m* de unidades múltiples

multiuse: ~ **communication** *n* TELEP comunicación *f* multiuso

multiuser *adj* GEN multiusuario; ~ **system** *n* DATA sistema *m* multiusuario

multivalued *adj* (*M*) DATA polivalente (*M*); ~ **attribute** *n* DATA atributo *m* polivalente

multivariable: ~ **control** *n* TEST control *m* multivariable

multivendor *n* GEN multiproveedor *m*; ~ **host support** GEN logística *f* de sistema central de multiproveedores

multivibrator *n* (*MV*) ELECT multivibrador *m* (*MV*)

multivoice *adj* GEN de voces múltiples; ~ **speech recognition** *n* APPL, COMP, TELEP reconocimiento *m* de la voz de hablantes múltiples

multiwire: ~ **conductor** *n* ELECT, PARTS conductor *m* de hilos múltiples

multizone: ~ **subscription** *n* RADIO suscripción *f* multizonal

Munsell: ~ **colour** *BrE* **system** *n* TV sistema *m* de color Munsell

MUSE *abbr* (*multiple subsampling encoding*) DATA MUSE (*codificación por submuestreo múltiple*)

mute *adj* COMP mudo, TV sin sonido; ~ **button** *n* PER APP botón *m* de silenciamiento

mutilate *vt* TELEP estropear

mutilation *n* TELEP deterioro *m*

muting *n* RADIO amortiguación *f*; ~ **device** RADIO dispositivo *m* silenciador

mutual: ~ **branch** *n* POWER rama *f* común; ~ **conductance** *n* ELECT conductancia *f* mutua; ~ **coupling** *n* ELECT acoplamiento *m* mutuo; ~ **impedance** *n* ELECT impedancia *f* mutua; ~ **inductance** *n* ELECT inductancia *f* mutua; ~ **induction** *n* ELECT inducción *f* mutua; ~ **synchronization** *n* ELECT sincronización *f* mutua

mutually: ~ **exclusive attribute group** *n* COMP grupo *m* de atributos mutuamente excluyentes; ~ **synchronized network** *n* TELEP red *f* de sincronización mutua

MUX *abbr* (*multiplexer*) COMP, DATA, RADIO, SWG, TRANS MUX (*multiplexor*)

MV *abbr* (*multivibrator*) ELECT MV (*multivibrador*)

mylar: ~ **capacitor** *n* ELECT capacitor *m* mylar

N

N *abbr* (*number*) GEN N (*número*), TEST (*newton*) N (*newton*); ~ **address** *n* DATA dirección *f* N; ~**-address selector** *n* DATA selector *m* de dirección N; ~**-ary digital signal** *n* SIG señal *f* digital N-aria ~ **association** *n* DATA asociación *f* N; ~ **connection** *n* TELEP conexión *f* N; ~**-connection end point identifier** *n* DATA, TELEP identificador *m* de punto final de conexión N; ~ **connectionless-mode transmission** *n* DATA, TRANS transmisión *f* N inalámbrica; ~ **connect request** *n* DATA petición *f* de conexión N; ~**-directory function** *n* DATA función *f* de directorio N; ~ **disconnect indication** *n* DATA indicación *f* de desconectar N; ~ **entity** *n* DATA entidad *f* N; ~**-entity invocation** *n* DATA invocación *f* de entidad N; ~**-entity title** *n* DATA título *m* de entidad N; ~**-entity type** *n* DATA tipo *m* de entidad N; ~ **facility** *n* TELEP instalación *f* N; ~ **function** *n* DATA función *f* N; ~ **initiator** *n* DATA iniciador *m* N; ~ **layer** *n* DATA, ELECT capa *f* N; ~**-layer managed object** *n* DATA objeto *m* gestionado de capa N; ~**-layer management protocol** *n* DATA protocolo *m* de gestión de capa N; ~**-level code** *n* DATA código *m* de nivel N; ~ **protocol** *n* DATA, TELEP protocolo *m* N; ~**-protocol addressing information** *n* DATA, TELEP información *f* sobre direccionamiento de protocolo N; ~**-protocol control information** *n* DATA, TELEP información *f* de control de protocolo N; ~**-protocol data unit** *n* DATA, TELEP unidad *f* de datos de protocolo N; ~ **recipient** *n* DATA receptor *m* N; ~ **relay** *n* DATA relé *m* N; ~ **service** *n* DATA, TELEP servicio *m* N; ~**-service access point** *n* DATA, TELEP punto *m* de acceso al servicio N; ~**-service access-point address** *n* DATA, TELEP dirección *f* del punto de acceso al servicio N; ~**-service confirm-primitive** *n* DATA, TELEP primitivo *m* de confirmación del servicio N; ~**-service data unit** *n* DATA, TELEP unidad *f* de datos del servicio N; ~**-service indication-primitive** *n* DATA, TELEP primitivo *m* de indicación del servicio N; ~**-service request-primitive** *n* DATA, TELEP primitivo *m* de petición del servicio N; ~**-type conduction** *n* ELECT *semiconductors*, PARTS conducción *f* de tipo N; ~**-type semiconductor** *n* ELECT, PARTS semiconductor *m* de tipo N; ~**-user data** *n pl* DATA, TELEP datos *m pl* de usuario N

N/A *abbr* (*not applicable*) GEN N/A (*inaplicable*)

NACK *abbr* (*negative acknowledgement*) RADIO NACK (*reconocimiento negativo*)

NAD *abbr* (*noise amplitude distribution*) TRANS distribución *f* de amplitud de ruido

NAE *abbr* (*network address extension*) DATA NAE (*extensión de dirección de red*)

name *n* GEN nombre *m*

named: ~ **server** *n* DATA servidor *m* designado

nameplate *n* COMP rótulo *m*

naming: ~ **authority** *n* DATA autoridad *f* denominadora; ~ **domain** *n* DATA dominio *m* denominador; ~ **subdomain** *n* DATA subdominio *m* denominador; ~ **tree** *n* DATA árbol *m* denominador

NAND *abbr* (*NOT-AND*) ELECT (*NO-Y*); ~ **circuit** *n* ELECT circuito *m* NO-Y; ~ **gate** *n* ELECT circuito *m* NO-Y; ~ **operation** *n* ELECT operación *f* NO-Y

NANDed *adj* ELECT NANDed

nanosecond *n* (*ns*) TEST nanosegundo *m* (*ns*)

NAPI/TOA *abbr* (*numbering and addressing plan indicator/type of address*) DATA NAPI/TOA (*numeración y direccionamiento de indicador de plan/tipo de dirección*)

narrow: ~ **band** *n* RADIO, SIG, TRANS banda *f* estrecha; ~**-band frequency modulation** *n* (*NBFM*) RADIO, TRANS *of maritime satellite* modulación *f* de frecuencia de banda estrecha (*MFBE*); ~**-band noise** *n* RADIO, TRANS ruido *m* de banda estrecha; ~**-band phase-shift keying** *n* (*NBPSK*) SIG tecleo *m* de cambio de fase de banda estrecha; ~**-band receiver** *n* RADIO, TRANS receptor *m* de banda estrecha; ~**-band signal** *n* ELECT, RADIO, SIG, TRANS señal *f* de banda estrecha; ~**-band switch** *n* SWG, TRANS conmutador *m* de banda estrecha; ~**-band switching network** *n* SWG, TRANS red *f* de conmutación de banda estrecha; ~**-band voice modulation** *n* (*NBVM*) TRANS modulación *f* vocal de banda estrecha; ~**-beam transmit antenna** *n* RADIO, TRANS antena *f* de transmisión de banda estrecha; ~ **flange** *n* ELECT brida *f* estrecha; ~ **shift** *n* RADIO desplazamiento *m* estrecho

national: ~ **call indicator** *n* TELEP indicador *m* de llamada nacional; ~ **circuit** *n* TELEP circuito *m* nacional; ~ **code** *n* DATA, TELEP código *m* nacional; ~ **control centre** *BrE n* TEST centronacional decontrol; ~ **coverage** *n* RADIO cobertura *f* nacional; ~ **destination code** *n* DATA, TELEP código *m* nacional de destino; ~ **exchange** *n* SWG centralita *f* de ámbito nacional; ~ **extension** *n* TELEP extensión *f* nacional; ~ **identification digit** *n* TELEP dígito *m* de identificación nacional; ~ **indicator** *n* TELEP indicador *m* nacional; ~ **line** *n* TELEP línea *f* nacional; ~ **main section** *n* TELEP sección *f* principalnacional; **~-network congestion signal** *n* SIG señal *f* decongestión de la red nacional; ~ **number** *n* DATA, TELEP número *m* nacional; ~ **section** *n* TELEP sección *f* nacional; ~ **significant number** *n* TELEP número *m* nacionalsignificativo; ~ **sound-programme centre** *BrE n* RADIO centro nacional de programas sonoros; ~ **system** *n* TELEP sistema *m* nacional; ~ **television centre** *BrE n* TV centro nacional de televisión; ~ **usage** *n* TELEP usanza *f* del pais; ~ **variant** *n* TEST variante *f* nacional

National: ~ **Center for Supercomputing Applications** *AmE n* (*NCSA*) DATA Centro Nacional para Aplicaciones de Superinformática (*NCSA*)

nationwide *adj* GEN de ámbito nacional; ~ **paging** *n* TRANS paginación *f* de ámbito nacional

natural: ~ **colour** *BrE n* GEN color *m* natural; ~ **frequency** *n* POWER frecuencia *f* propia; ~ **language** *n* DATA lenguaje *m* natural; **~-language analysis** *n* DATA análisis *m* de lenguaje natural; **~-language generation** *n* DATA generación *f* de lenguaje natural; ~ **language interface** *n* DATA interfaz *m* de lenguaje natural; **~-language processing** *n* COMP, DATA procesamiento *m* de lenguaje natural; ~ **noise** *n* ELECT ruido *m* natural; ~ **oscillator** *n* ELECT oscilador *m* natural

nature: **~-of-circuit indicator** *n* TELEP indicador *m* de tipo de circuito

navigate *vt* COMP *Internet* navegar

navigation *n* COMP *Internet* navegación *f*; ~ **button** COMP *Internet* botón *m* de navegación; ~ **path** COMP *Internet* trayectoria *f* de navegación; ~ **software** COMP *Internet*, DATA programas *m pl* de navegación; ~ **warning signal** RADIO señal *f* de aviso navegacional

navigational: ~ **radar** *n* RADIO radar *m* de navegación

navigator *n* COMP *Internet* navegante *m*

NBFM *abbr* (*narrow-band frequency modulation*) RADIO, TRANS *of maritime satellite* MFBE (*modulación de frecuencia de banda estrecha*)

NBP *abbr* (*nonbasic parameter*) DATA NBP (*parámetro no básico*)

NBPSK *abbr* (*narrow-band phase-shift keying*) SIG tecleo *m* de cambio de fase de banda estrecha

NBVM *abbr* (*narrow-band voice modulation*) TRANS modulación *f* vocal de banda estrecha

NCC *abbr* (*new common carriers*) GEN nuevas portadoras *f pl* comunes

NCL *abbr* (*network control language*) ELECT, TELEP NCL (*lenguaje de control de la red*)

NCSA *abbr* (*National Center for Supercomputing Applications*) DATA *Internet* NCSA (*Centro Nacional para Aplicaciones de Superinformática*)

ND *abbr* (*network default*) ELECT, TELEP ND (*fallo de la red*)

NDF *abbr* (*new data flag*) DATA bandera *f* de datos nuevos

NDIS *abbr* (*network drive interface specifications*) ELECT NDIS (*especificaciones del interfaz de arrastre de la red*)

NDN *abbr* (*nondelivery notification*) DATA, TELEP notificación *f* de falta de entrega

NE *abbr* (*negociated exit*) DATA salida *f* negociada

near: **~-Earth space** *n* RADIO espacio *m* próximo a la Tierra; **~-end crosstalk** *n* TRANS paradiafonía *f*; ~ **field** *n* RADIO, TRANS *fibre optics* campo *m* próximo; **~-field analysis** *n* RADIO, TRANS *fibre optics* análisis *m* de campo próximo; **~-field diffraction pattern** *n* RADIO, TRANS *fibre optics* diagrama *m* de difracción de campo próximo; **~-field pattern** *n* RADIO, TRANS *fibre optics* diagrama *m* de campo próximo; **~-field radiation pattern** *n* RADIO, TRANS *fibre optics* diagrama *m* de radiación de campo próximo; **~-field region** *n* RADIO, TRANS *fibre optics* región *f* de campo próximo; **~-field scanning** *n* RADIO, TRANS *fibre optics* exploración *f* de un campo próximo

necessary: ~ **bandwidth** *n* RADIO, TRANS ancho *m* de banda necesario

neck *n* ELECT collarín *m*; ~ **shadow** ELECT sombra *f* de cuello

needle *n* COMP, TELEP viga *f* de recalzo; ~ **counter tube** RADIO tubo *m* contador de aguja; ~ **deflection** TELEP *of indicating device* deflexión *f* de aguja; ~ **file** PARTS lima *f* de aguja

NEF *abbr* (*not-expected frame*) TELEP NEF (*trama no esperada*)

negate *vt* COMP, ELECT anular

negation *n* COMP, ELECT negación *f*, anulación *f*

negative: ~ **acknowledge character** *n* TELEP carácter *m* de reconocimiento negativo; ~ **acknowledgement** *n* (*NACK*) RADIO reconocimiento *m* negativo (*NACK*); ~ **amplitude modulation** *n* ELECT, TV modulación *f* negativa de amplitud; ~ **battery** *n* POWER batería *f* negativa; ~ **effect** *n* COMP efecto *m* negativo; ~ **feedback** *n* CONTR, DATA, ELECT, POWER, RADIO realimentación *f* negativa; ~**-feedback amplifier** *n* ELECT amplificador *m* de realimentación negativa; ~ **frequency modulation** *n* ELECT modulación *f* negativa de frecuencia; ~ **indication tone** *n* TELEP tono *m* de indicación negativo; ~ **justification** *n* TELEP, TRANS justificación *f* negativa; ~ **notification** *n* (*NN*) DATA, TELEP notificación *f* negativa; ~ **pattern** *n* ELECT impresión *f* negativa; ~ **polarity** *n* POWER polaridad *f* negativa; ~ **pole** *n* POWER polo *m* negativo; ~ **pressure** *n* GEN presión *f* negativa; ~ **pulse** *n* ELECT impulso *m* negativo; ~ **terminal** *n* POWER *of battery* polo *m* negativo; ~ **tilt** *n* TELEP inclinación *f* negativa; ~ **wire** *n* TELEP conductor *m* negativo

negentropy *n* DATA entropía *f* negativa, negentropía *f*

negotiated: ~ **exit** *n* (*NE*) DATA salida *f* negociada

negotiation *n* GEN negociación *f*

neighbouring *BrE adj* GEN vecino

nematic: ~ **liquid crystals** *n pl* ELECT cristales *m pl* líquidos nemáticos; ~ **phase** *n* ELECT fase *f* nemática

neon: ~ **indicator tube** *n* ELECT tubo *m* indicador de neón

neoprene *n* PARTS neopreno *m*

NEP *abbr* (*noise equivalent power*) ELECT, TRANS *fibre optics* potencia *f* equivalente de ruido

neper *n* (*np*) ELECT neper *m* (*np*)

NERF *abbr* (*network element function*) TRANS función *f* del elemento de red

nest *vt* COMP *subroutine*, TEST anidar

nested: ~ **sampling** *n* TEST muestreo *m* polietápico

Net: **the** ~ *n* COMP, DATA la Red *f*, la Net *f*

netiquette *n* COMP, DATA *Internet* netiquette *f*

Netscape *n* COMP, DATA *Internet* Netscape *m*

netsurfing *n* COMP, DATA *Internet* navegación *f* de la Red

network *n* GEN red *f*; ~ **access** DATA, RADIO, TRANS acceso *m* a la red; ~ **access component** DATA, TELEP componente *m* de acceso a la red; ~**-access system** TRANS sistema *m* de acceso a la red; ~ **address** DATA dirección *f* de red; ~ **address extension** (*NAE*) DATA extensión *f* de dirección de red (*NAE*); ~ **addressing** DATA direccionamiento *m* de red; ~**-addressing authority** DATA autoridad *f* de direccionamiento de la red; ~**-addressing domain** DATA dominio *m* de direccionamiento de la red; ~ **administrator** COMP administrador *m* de red; ~ **analyser** *BrE* TEST analizador *m* de redes; ~ **analysis point** TELEP punto *m* de análisis de red; ~ **breakdown** TEST ruptura *f* de red; ~ **cable** TRANS cable *m* de red; ~ **chart** TELEP plano *m* de la red; ~ **cluster** TELEP conglomerado *m* de redes; ~ **configuration** TELEP configuración *f* de la red; ~ **congestion** TELEP, TRANS congestión *f* de red; ~ **congestion signal** SIG señal *f* de congestión de la red; ~ **connection** DATA, TELEP conexión *f* a la red; ~ **connection fee** TELEP cuota *f* de conexión a la red; ~**-connectionless-mode data transmission** DATA, TRANS transmisión *f* de datos en modo sin conexión en red; ~**-connection-mode data transmission** DATA, TRANS transmisión *f* de datos en modo de conexión en red; ~ **control centre** *BrE* TEST centro *m* de control de la red; ~ **control channel** TRANS canal *m* de control de la red; ~ **control language** (*NCL*) ELECT, TELEP lenguaje *m* de control de la red (*NCL*); ~ **control phase** ELECT, TELEP fase *f* de control de la red; ~ **control signalling** *BrE* TELEP señalización *f* de control de la red; ~ **default** (*ND*) ELECT, TELEP fallo *m* de la red (*ND*); ~ **diagram** TELEP diagrama *m* de la red; ~ **digital termination** (*NT1*) TELEP terminación *f* digital de la red (*NT1*); ~ **discard indicator** TEST indicador *m* de abandono de la red; ~ **disruption** ELECT interrupción *f* de la red; ~ **drive interface specifications** *n pl* (*NDIS*) ELECT especificaciones *f pl* del interfaz de arrastre de la red (*NDIS*); ~ **element** TRANS elemento *m* de red; ~ **element function** (*NERF*) TRANS función *f* del elemento de red; ~ **engineering** TELEP ingeniería *f* de red; ~ **entity** DATA entidad *f* de red; ~ **establishment agreement** TV acuerdo *m* de creación de una red; ~ **failure** DATA, TELEP fallo *m* de la red (*ND*); ~ **failure signal** SIG señal *f* de fallo de la red; ~ **fault in local loop signal** SIG fallo *m* de la red en señal de bucle local; ~ **file manager** (*NFM*) COMP gestor *m* de archivo de red (*NFM*); ~ **file system** DATA sistema *m* de archivo de red; ~ **gateway** DATA dispositivo *m* de acceso a la red; ~ **gateway element** TRANS elemento *m* de acceso a la red; ~ **head** GEN cabeza *f* de red; ~ **head-end** TV final *m* de cabeza de red; ~ **identity** TELEP identidad *f* de la red; ~ **infrastructure** TRANS infraestructura *f* de la red; ~ **interconnection** DATA *Internet* interconexión *f* de redes; ~ **interface** COMP, DATA, TELEP interfaz *f* de red; ~ **layer** DATA, TELEP, TRANS nivel *m* de red; ~**-layer relay** (*NLR*) TELEP, TRANS relé *m* de capa de red; ~ **locking** DATA sincronización *f* de la red; ~ **maintenance** GEN mantenimiento *m* de la red; ~ **maintenance signals** *n pl* SIG señales *f pl* de mantenimiento de la red; ~ **management** GEN administración *f* de red; ~ **management application service element** (*NM-ASE*) DATA elemento *m* de red del servicio de aplicación de gestión; ~ **management architecture** TRANS arquitectura *f* de gestión de la red; ~ **management centre** *BrE* (*NMC*) TEST centro *m* de gestión de la red; ~ **management data** *n pl* COMP datos *m pl* de gestión de la red; ~ **management function** TRANS función *f* de gestión de la red; ~ **management platform** TRANS plataforma *f* de gestión de la red; ~ **management point** TELEP punto *m* de gestión de la red; ~ **management signals** *n pl* SIG señales *f pl* de gestión de la red; ~ **map** SWG *exchange data store*, TELEP mapa *m* de la red; ~ **news transfer protocol** (*NNTP*) DATA protocolo *m* de transferencia de noticias de la red (*NNTP*); ~ **node** TRANS nodo *m* de red; ~ **node interface** (*NNI*) TRANS interfaz *f* del nodo de red (*NNI*); ~ **nonaddressing equipment** TELEP equipo *m* de no direccionamiento de la red; ~ **operating agreement** TV acuerdo *m* de explotación de red; ~ **operating company** TELEP empresa *f* operadora de red; ~ **operating system** DATA sistema *m* operativo de red; ~ **operation and maintenance centre** *BrE* TEST centro *m* de explotación y mantenimiento de red; ~ **operator** TELEP empresa *f* operadora de red; ~**-operators' maintenance channel** (*NOMC*) TRANS canal *m* de mantenimiento de operadores de red; ~ **parameter** TELEP parámetro *m* de circuito; ~ **performance** (*NP*) TEST características *f pl* de la red; ~ **plan** TELEP plano *m* de la red; ~ **planning** TELEP planificación *f* de la red; ~ **protection device** (*NPD*) TELEP dispositivo *m* de protección de la red (*NPD*); ~ **protocol** COMP, DATA *Internet* protocolo *m* de la red; ~ **protocol address information** DATA información *f* sobre dirección de protocolo de la red; ~ **protocol data unit** (*NPDU*) DATA unidad *f* de datos del protocolo de la red; ~ **relay** DATA relé *m* de la red; ~ **resources** TELEP recursos *m pl* de la red; ~ **routing** DATA encaminamiento *m* en la red; ~ **security** DATA seguridad *f* de la red; ~ **selection signals** *n pl* SIG señales *f pl* de selección de la red; ~ **server** DATA servidor *m* de la red; ~ **service** (*NS*) DATA, TELEP mantenimiento *m* de la red; ~ **service access component** DATA, TELEP componente *m* de acceso a los servicios de la red; ~ **service access maintenance point** TELEP, TRANS punto *m* de acceso al mantenimiento de la red; ~ **service access point** (*NSAP*) DATA punto *m* de acceso al servicio de la red; ~**-service access-point address** DATA dirección *f* del punto de acceso al servicio de la red; ~ **service data unit** DATA unidad *f* de datos del servicio de la red; ~ **service provider** DATA proveedor *m* de servicios de la red; ~ **service user** DATA usuario *m* de servicios de la red; ~ **service utilization component** DATA, TELEP componente *m* de utilización de los servicios de la red; ~ **structure** GEN estructura *f* de la red; ~ **supervision** DATA, TELEP supervisión *f* de la red; ~ **supervision and management** GEN gestión *f* y supervisión de la red; ~ **supervisor** GEN supervisor *m* de la red; ~ **support engineer** GEN técnico *m* de apoyo de la red; ~ **terminal equipment** (*NTE*) TELEP, TRANS equipo *m* terminal de la red; ~ **terminal number** (*NTN*) DATA, TELEP número *m* del terminal de la red (*NTN*); ~ **terminating unit** (*NTU*) TRANS unidad *f* terminal de la red (*NTU*); ~ **termination** (*NT*) APPL, TELEP, TRANS terminación *f* de la red; ~ **topology** DATA, TELEP topología *f* de red; ~ **transfer delay** TELEP retardo *m* de transferencia de la red; ~ **user** DATA *Internet* usuario *m* de la red; ~ **user identification** (*NUI*) DATA, TELEP identificación *f* del usuario de la red (*NUI*); ~ **utility** TELEP utilidad *f* de la red; ~ **utility field** TELEP campo *m* de utilidad de la red; ~ **utilization component** DATA, TELEP componente *m* de utilización de la red

networking *n* TELEP integración *f* en red; ~ **operating system** ELECT sistema *m* de operación en red

neural: ~ **computer** *n* COMP ordenador *m* neural *Esp*, computador *m* neural *AmL*, computadora *f* neural *AmL*; ~ **network** *n* COMP, TRANS red *f* neural

neutral *adj* POWER neutro; ~ **bar** *n* POWER barra *f* neutra; ~ **conductor** *n* ELECT conductor *m* neutral; ~ **keying** *n* TELEP tecleo *m* neutro; ~ **point** *n* POWER punto *m* neutro; ~ **position** *n* GEN posición *f* neutra; ~ **terminal** *n* POWER borne *m* neutro; ~ **wire** *n* GEN alambre *m* neutro

neutralization *n* ELECT neutralización *f*

neutralizing: ~ **capacitor** *n* ELECT condensador *m* de neutralización

new: ~ **assignment message** *n* DATA mensaje de nueva asignación; ~ **common carriers** *n* (*NCC*) GEN nuevas portadoras *f pl* comunes; ~ **data flag** *n* (*NDF*) DATA bandera *f* de datos nuevos; **~-line character** *n* (*NL*) DATA carácter *m* de cambio de línea (*NL*); ~ **technology** *n* GEN nueva tecnología

news: ~ **gathering** *n* RADIO, TV recolección *f* de noticias; ~ **server** *n* DATA *Internet* servidor *m* de noticias

newsletter *n* COMP boletín *m*

newsgroup *n* COMP newsgroup *m*

newspaper *n* DATA diario *m*

newton *n* (*N*) TEST newton *m* (*N*)

NFM *abbr* (*network file manager*) COMP NFM (*gestor de archivo de red*)

niche: ~ **market** *n* GEN mercado *m* de mercado; ~ **marketing** *n* GEN comercialización *f* por nichos

nichrome *n* PARTS nicromio *m*

night: ~ **effect** *n* RADIO efecto *m* de noche; ~ **error** *n* RADIO error *m* de noche; ~ **service** *n* DATA servicio *m* nocturno; **~-service board** *n* SWG posición *f* de servicio nocturno; **~-service connection** *n* SWG conexión *f* con el servicio nocturno; **~-service key** *n* SWG conmutador *m* de servicio nocturno; **~-service mode** *n* SWG servicio *m* nocturno; **~-service operator** *n* SWG, TELEP operador *m* de servicio nocturno

nickel *n* GEN níquel *m*

nitriding *n* PARTS nitruración *f*; ~ **steel** PARTS acero *m* de nitruración

NL *abbr* (*new-line character*) DATA NL (*carácter de cambio de línea*)

NLR *abbr* (*network-layer relay*) TRANS relé *m* de capa de red

NM-ASE *abbr* (*network management application service element*) DATA elemento *m* de red del servicio de aplicación de gestión

NMC *BrE abbr* (*network management centre*) TEST centro *m* de gestión de red

NN *abbr* (*negative notification*) DATA, TELEP notificación *f* negativa

NNI *abbr* (*network node interface*) TRANS NNI (*interfaz de nodo de la red*)

NNTP *abbr* (*network news transfer protocol*) DATA *Internet* NNTP (*protocolo de transferencia de noticias de la red*)

no.: ~ **seven exchange** *n* SWG central *f* n° 7

no-break: ~ **system** *n* POWER sistema *m* de continuidad absoluta

nodal *adj* COMP nodal; ~ **analysis** *n* TRANS análisis *m* nodal; ~ **period** *n* RADIO período *m* nodal; ~ **point** *n* TELEP punto *m* nodal

node *n* GEN nodo *m*; ~ **attribute** COMP atributo *m* nodal

nodecard *n* COMP tarjeta *f* de punto nodal

nodes: ~ **network** *n* COMP red *f* de puntos nodales

no-failure *adj* TEST a prueba de fallos

NOGAD *abbr* (*noise-operated gain adjusting device*) ELECT NOGAD (*ajustador de ganancia activado por el ruido*)

noise *n* GEN ruido *m*; ~ **allocation** TRANS atribución *f* de ruido; ~ **amplitude distribution** (*NAD*) TRANS distribución *f* de amplitud de ruido; ~ **cancellation** RADIO eliminación *f* de interferencias; **~-cancelling** *BrE* **microphone** TELEP micrófono *m* antirruidos; ~ **current** TELEP corriente *f* perturbadora; ~ **detector** TELEP detector *m* de ruidos; ~ **eliminator** POWER eliminador *m* de ruidos; ~ **equivalent power** (*NEP*) ELECT, TRANS *fibre optics* potencia *f* equivalente de ruido; ~ **factor** RADIO, TEST, TRANS factor *m* de ruido; ~ **field** TRANS campo *m* perturbador; ~ **figure** RADIO, TEST, TRANS coeficiente *m* de ruido; **~-generating device** COMP, TRANS generador *m* de ruido; ~ **generator** COMP, TRANS generador *m* de ruido; **~-generator diode** ELECT diodo *m* generador de ruido; **~-generator plasma tube** ELECT tubo *m* de plasma generador de ruido; ~ **indicator** TELEP indicador *m* de ruido; ~ **level** GEN nivel *m* de ruido; ~ **level control** TELEP control *m* de nivel de ruido; ~ **limit** RADIO límite *m* de ruido; ~ **masking** TRANS enmascaramiento *m* de ruido; ~ **measurement** TELEP medición *f* de ruido; **~-measuring amplifier** TELEP amplificador *m* para medición de ruido; **~-measuring device** TELEP medidor *m* de ruido; **~-measuring set** TEST equipo *m* de medición de ruido; ~ **meter** TELEP medidor *m* de ruido; ~ **monitor** TEST supervisor *m* de ruido; ~ **monitoring** TELEP supervisión *f* de ruido; ~ **1/f** ELECT ruido *m* 1/f; **~-operated gain adjusting device** (*NOGAD*) ELECT ajustador *m* de ganancia activado por el ruido (*NOGAD*); ~ **reducer** TRANS reductor *m* de ruido; ~ **signal** POWER señal *f* de ruido; ~ **source** POWER fuente *f* de ruido; ~ **spike** TRANS pico *m* de ruido; ~ **supervision** TELEP supervisión *f* de ruido; ~ **suppression** RADIO supresión *f* de ruido; ~ **suppression component** TEST componente *m* eliminador de interferencias; ~ **suppressor device** TEST dispositivo *m* eliminador de interferencias; ~ **temperature** RADIO temperatura *f* de ruido; ~

test TELEP, TEST prueba *f* de ruido; ~ **threshold** TELEP umbral *m* de ruido; ~**-weighing switch** ELECT conmutador *m* de ponderación del ruido; ~ **weighting** TRANS ponderación *f* de ruido

noiseless *adj* COMP silencioso

no-load: ~ **current** *n* POWER corriente *f* en vacío; ~ **operation** *n* GEN funcionamiento *m* sin carga; ~ **power** *n* POWER tensión *f* en vacío; ~ **voltage** *n* POWER tensión *f* en vacío

NOMC *abbr* (*network-operators' maintenance channel*) TRANS canal *m* de mantenimiento de operadores de red

nominal: ~ **alternating discharge current** *n* POWER corriente *f* nominal de descarga alterna; ~ **bandwidth** *n* COMP, TV ancho *m* de banda nominal; ~ **black** *n* POWER negro *m* nominal; ~ **black signal** *n* TELEP señal *f* de negro nominal; ~ **channel width** *n* TV ancho *m* de canal nominal; ~ **coverage area** *n* RADIO área *f* de cobertura nominal; ~ **DC spark-over voltage** *n* ELECT, POWER *of protector* voltaje *m* nominal de descarga disruptiva de CC; ~ **impulse discharge current** *n* POWER corriente *f* de choque nominal; ~ **justification rate** *n* TELEP índice *m* de justificación nominal; ~ **load** *n* ELECT carga *f* nominal; ~ **orbital position** *n* RADIO posición *f* orbital nominal; ~ **relative level** *n* TELEP nivel *m* relativo nominal; ~ **symbol rate** *n* DATA velocidad *f* simbólica nominal; ~ **transmission loss** *n* TRANS *of 4-wire circuit* pérdida *f* nominal de transmisión; ~ **value** *n* GEN valor *m* nominal; ~ **white** *n* TELEP blanco *m* nominal; ~ **white signal** *n* TELEP señal *f* blanca nominal

nonacceptance *n* TEST rechazo *m*; ~ **number** TEST número *m* de rechazo

nonadjacent *adj* RADIO no adyacente

nonapproved *adj* PER APP *apparatus* no aprobado

nonarmoured *BrE* (*AmE* **nonarmored**) *adj* ELECT no blindado

nonassociated: ~ **mode** *n* TELEP modo *m* no asociado; ~ **mode of operation** *n* TELEP modo *m* de operación no asociado; ~ **mode of signalling** *BrE n* SIG modo *m* de señalización no asociado; ~ **signalling** *BrE n* SIG señalización *f* no asociada

nonbasic: ~ **parameter** *n* (*NBP*) DATA parámetro *m* no básico (*NBP*)

nonbinary: ~ **code** *n* DATA código *m* no binario

nonblocking: ~ **network** *n* SWG, TRANS red *f* de no bloqueo; ~ **switching network** *n* SWG, TRANS red *f* de selectores sin congestión

nonboosted: ~ **antenna repeater system** *n* RADIO sistema *m* de repetidores sin amplificador de antena

noncircularity *n* TELEP, TRANS no circularidad *f*; ~ **of cladding** TELEP, TRANS no circularidad *f* del revestimiento; ~ **of core** TELEP, TRANS no circularidad *f* del núcleo; ~ **of reference surface** TELEP no circularidad *f* de la superficie de referencia

noncoherent: ~ **demodulation** *n* DATA desmodulación *f* incoherente; ~ **side-looking radar** *n* RADIO radar *m* de exploración lateral no coherente; ~ **swept-tone modulation** *n* TRANS modulación *f* no coherente por barrido de frecuencia

nonconducting *adj* POWER adiatérmico

nonconductive: ~ **pattern** *n* ELECT red *f* no conductora

nonconfirmed *adj* DATA no confirmado

nonconforming: ~ **items** *n pl* TEST artículos *m pl* no conformes; ~ **unit** *n* TEST unidad *f* no conforme

nonconformity *n* TEST disconformidad *f*

nonconjunction *n* ELECT conjunción *f* negativa

nonconnected *adj* DATA *to Internet* no conectado

noncorrectable: ~ **fault** *n* TEST fallo *m* no reparable

nondecimal: ~ **numeral** *n* TELEP número *m* no decimal

nondedicated: ~ **signalling** *BrE* **channel** *n* TRANS canal *m* de señalización no dedicado

nondelivery: ~ **notification** *n* (*NDN*) DATA, TELEP notificación *f* de falta de entrega

nondestructive: ~ **testing** *n* DATA, TEST pruebas *f pl* no destructivas

nondialled: ~ **connection** *BrE n* APPL, TELEP conexión *f* sin marcación

nondissipative: ~ **stub** *n* RADIO brazo *m* de reactancia

nondroppable: ~ **blocks** *n pl* DATA bloques *m pl* no separables

nonduplicated: ~ **equipment** *n* TELEP equipo *m* no duplicado; ~ **system** *n* TELEP sistema *m* no duplicado; ~ **working mode** *n* TELEP modo *m* de trabajo no duplicado

nonequilibrium: ~**-mode distribution** *n* TRANS distribución *f* en modo de desequilibrio

nonequivalence *n* ELECT *of logic elements* no equivalencia *f*; ~ **operation** ELECT operación *f* de no equivalencia

nonerasable: ~ **memory** *n* COMP memoria *f* indeleble; ~ **storage** *n* COMP almacenamiento *m* indeleble

nonexistent: ~ **number** *n* TELEP número *m* inexistente

nonfading *adj* GEN sin desvanecimiento
nonferrous: ~ **metal** *n* PARTS metal *m* no ferroso
non-Gaussian: ~ **noise** *n* TELEP ruido *m* no gaussiano
nonhalogen *adj* ELECT no halogénico
nonhazardous: ~ **state** *n* TEST estado *m* no peligroso
nonhierarchical: ~ **system** *n* TRANS sistema *m* no jerárquico
nonidentity: ~ **operation** *n* ELECT operación *f* de no identidad
No-NiKs: ~ **fibre** *BrE* **stripper** *n* ELECT separador *m* de fibra No-NiKs
nonimpact: ~ **printer** *n* DATA, TELEP impresora *f* sin impacto
noninductive: ~ **circuit** *n* POWER circuito *m* no inductivo
noninterchangeable *adj* GEN no intercambiable
noninterlaced *adj* COMP no entrelazado
noninterpolated: ~ **channel** *n* TRANS canal *m* no interpolado
non-intra: ~ **coding** *n* COMP codificación *f* non-intra
nonlinear *adj* GEN no lineal; ~ **amplification** *n* TRANS amplificación *f* no lineal; ~ **circuit** *n* TRANS circuito *m* no lineal; ~ **control system** *n* DATA sistema *m* de control no lineal; ~ **digital speech signals** *n pl* TELEP señales *f pl* de telefonía digital no lineal; ~ **distortion** *n* ELECT, GEN, SIG, TRANS distorsión *f* no lineal; ~ **effect** *n* ELECT, TRANS efecto *m* no lineal; ~ **filter** *n* DATA, TRANS filtro *m* no lineal; ~ **filtering** *n* TRANS filtrado *m* no lineal; ~ **interpolation** *n* TRANS interpolación *f* no lineal; ~ **optics** *n* DATA óptica *f* no lineal; ~ **oscillation** *n* TRANS oscilación *f* no lineal; ~ **processing loss** *n* TELEP pérdida *f* de procesamiento no lineal; ~ **recursive filter** *n* DATA, TRANS filtro *m* recursivo no lineal; ~ **resistance** *n* ELECT resistencia *f* no lineal; ~ **scale** *n* TEST escala *f* no lineal; ~ **scattering** *n* RADIO, TRANS dispersión *f* no lineal; ~ **system** *n* DATA sistema *m* no lineal
nonlinearity *n* GEN no linealidad *f*
nonloaded: ~ **cable** *n* TELEP cable *m* no cargado
nonmagnetic: ~ **armature shim** *n* ELECT separador *m* amagnético de armadura
nonmandatory *adj* GEN no obligatorio
nonmetallic: ~ **sheathed cable** *n* GEN cable *m* con revestimiento no metálico
nonmode: ~ **of signalling** *BrE n* SIG modo *m* negativo de señalización
non-network: ~ **interface** *n* TELEP interfaz *f* no de red
nonnumerical *adj* GEN no numérico
nonoperating: ~ **state** *n* TELEP, TEST estado *m* no operativo; ~ **time** *n* TELEP, TEST tiempo *m* de inactividad, tiempo *m* de inoperatividad
nonparametric *adj* GEN aparamétrico
nonpolarized *adj* POWER impolarizado, no polarizado; ~ **relay** *n* PARTS, POWER relé *m* no polarizado
nonprecision: ~ **frequency offset** *n* RADIO desnivel *m* de frecuencia no preciso
nonradiative *adj* GEN no radiativo
nonreal: ~ **time** *adj* (*NRT*) COMP, DATA, TELEP, TEST tiempo *m* irreal (*NRT*)
nonreceipt: ~ **notification** *n* (*NRN*) DATA, TELEP notificación *f* de no recepción (*NRN*)
nonrecipient: ~ **notification** *n* DATA, TELEP notificación *f* de no receptor
nonreciprocal: ~ **circuit** *n* ELECT circuito *m* no recíproco; ~ **waveguide** *n* DATA, RADIO, TRANS guía *f* de ondas no recíproca
nonrecurrent: ~ **charge** *n* TELEP cuota *f* de conexión; ~ **fee** *n* TELEP cuota *f* única
nonrecurring: ~ **pulse** *n* GEN impulso *m* aperiódico
nonrecursive: ~ **filter** *n* DATA, TRANS filtro *m* no recurrente
nonregenerative *adj* TRANS no regenerativo
nonregistration *n* RADIO falta *f* de registro
nonrelevant: ~ **failure** *n* TEST fallo *m* no significativo
nonrepetitive: ~ **peak reverse-voltage** *n* PARTS, TELEP tensión *f* inversa de cresta no repetitiva
nonrequired: ~ **time** *n* TEST tiempo *m* no requerido
nonreturn: ~**-to-reference recording** *n* DATA registro *m* sin retorno a referencia; ~ **to zero** *adj* (*NRZ*) DATA, TELEP, TRANS sin retorno a cero (*NRZ*)
nonreversible *adj* GEN irreversible
nonrigid: ~ **plastic** *n* PARTS plástico *m* blando
nonselective: ~ **traffic** *n* RADIO tráfico *m* no selectivo
non-self-maintained: ~ **discharge** *n* ELECT descarga *f* sin mantenimiento automático
non-sine-wave: ~ **waveform** *n* COMP forma *f* de onda no sinusoidal
nonspecified: ~ **time relay** *n* PARTS relé *m* de tiempo no especificado
nonstandard: ~ **facilities equipment** *n* TELEP equipo *m* de instalaciones no normalizadas; ~ **facility** *n* (*NSF*) TELEP instalación *f* no normalizada (*NSF*); ~ **facility CSI-DIS** *n* (*NSF-CSI-DIS*) DATA CSI-DIS *m* de instalación no normalizada (*NSF-CSI-DIS*); ~ **facility command** *n* (*NSC*) DATA, TELEP orden *f* de instalación no normalizada; ~ **facility command CIG-DTC** *n* (*NSC-CIG-DTC*) DATA

CIG-DTC *m* de orden de instalación no normalizada (*NSC-CIG-DTC*); ~ **setup** *n* (*NSS*) DATA, TELEP ajuste *m* no normalizado (*NSS*)

nonstructured *adj* COMP inestructurado

nonsupervized: ~ **location** *n* ELECT localización *f* no supervisada

nonsynchronized *adj* GEN no sincronizado; ~ **network** *n* TELEP red *f* no sincronizada

nonterminal: ~ **symbol** *n* TELEP símbolo *m* no terminal

nontransparent: ~ **bearer service** *n* APPL, DATA servicio *m* no transparente del portador

nonuniform *adj* GEN no uniforme; ~ **encoding** *n* TELEP, TRANS codificación *f* no uniforme; ~ **frequency response** *n* TRANS distorsión *f* de atenuación; ~ **quantizing** *n* TELEP, TRANS cuantificación *f* no uniforme

nonuseful: ~ **DC component** *n* ELECT componente *m* de CC inútil

nonverbal: ~ **communication** *n* DATA comunicación *f* no verbal

nonvoice *adj* RADIO no vocal

nonvolatile: ~ **memory** *n* COMP, DATA memoria *f* estable; ~ **storage** *n* COMP, DATA almacenamiento *m* estable

nonzero: ~ **value** *n* COMP valor *m* distinto de cero

NOR: ~ **circuit** *n* ELECT circuito *m* NO-O; ~ **element** *n* ELECT circuito *m* NO-O; ~ **gate** *n* ELECT puerta *f* NO-O; ~ **operation** *n* ELECT operación *f* NO-O

norm *n* GEN norma *f*

normal: ~**-band telephony** *n* TELEP telefonía *f* de banda normal; ~ **charging** *n* POWER carga *f* de operación; ~ **disconnected mode** *n* DATA modo *m* desconectado normal; ~ **distribution** *n* GEN distribución *f* normal; ~ **flow direction** *n* GEN dirección *f* normal de flujo; ~ **inspection** *n* TEST inspección *f* normal; ~ **lighting** *n* GEN iluminación *f* general; ~ **mode** *n* DATA modo *m* normal; ~**-mode presentation** *n* DATA presentación *f* en modo normal; ~**-response mode** *n* DATA modo *m* de respuesta normal; ~ **routing** *n* TELEP encaminamiento *m* normal; ~ **routing of signalling** *BrE* *n* SIG encaminamiento *m* normal de la señalización; ~**-situation class** *n* DATA tipo *m* de situación normal; ~**-vector interpolation shading** *n* COMP sombreado *m* de interpolación de vector normal; ~ **view** *n* COMP vista *f* normal

normalization *n* GEN normalización *f*

normalize *vt* GEN normalizar

normalized: ~ **detectivity** *n* ELECT, TEST, TRANS detectividad *f* normalizada; ~ **form** *n* COMP forma *f* normalizada; ~ **frequency** *n* RADIO, TRANS frecuencia *f* normalizada; ~ **frequency departure** *n* RADIO, TRANS desajuste *m* de frecuencia normalizado; ~ **frequency difference** *n* RADIO, TRANS diferencia *f* de frecuencia normalizada; ~ **frequency drift** *n* RADIO, TRANS deriva *f* de frecuencia normalizada; ~ **frequency offset** *n* RADIO, TRANS separación *f* de la frecuencia normalizada; ~ **offset** *n* RADIO, TRANS separación *f* normalizada; ~ **signal-to-noise ratio** *n* RADIO, TRANS relación *f* señal/ruido normalizada; ~ **transmission loss** *n* (*NTL*) RADIO, TRANS pérdida *f* de transmisión normalizada (*NTL*); ~ **value** *n* RADIO valor *m* normalizado

normative *adj* GEN normativo; ~ **references** *n pl* TELEP referencias *f pl* normativas

nose: ~ **bandwidth** *n* ELECT ancho *m* de banda de punta

NOT: ~ **circuit** *n* ELECT circuito *m* NO; ~ **element** *n* ELECT circuito *m* inversor; ~ **gate** *n* ELECT puerta *f* NO; ~ **operation** *n* ELECT operación *f* NO

notation *n* GEN anotación *f*

notch *n* ELECT, RADIO, TRANS entalladura *f*; ~ **filter** ELECT, TRANS filtro *m* de entalladura

notching: ~ **relay** *n* PARTS relé *m* de avance

note *n* COMP *remark* observación *f*; ~ **output** COMP salida *f* de la nota; ~ **progression** COMP progresión *f* de la nota; ~ **resolution** COMP resolución *f* de la nota; ~ **search** COMP búsqueda *f* de nota

notebook *n* COMP agenda *f*; ~ **computer** COMP agenda *f* electrónica

notepad *n* COMP anotador *m*; ~ **computer** COMP agenda *f* electrónica

not-expected: ~ **frame** *n* (*NEF*) TELEP trama *f* no esperada (*NEF*)

notice: ~ **to airmen** *n* RADIO aviso *m* a los aviadores; ~ **board** *n* TELEP tablón *m* de anuncios

notification *n* GEN notificación *f*; **no** ~ **class** DATA *of service* clase *f* sin notificación; ~ **type** DATA tipo *m* de notificación

NOT-IF-THEN: ~ **element** *n* ELECT circuito *m* exclusivo; ~ **operation** *n* ELECT operación *f* exclusiva

nozzle *n* RADIO boquilla *f*

np *abbr* (*neper*) ELECT np (*neper*)

NPA *abbr* (*numbering plan area*) TELEP NPA (*área de plan de numeración*)

NPD *abbr* (*network protection device*) TELEP NPD (*dispositivo de protección de la red*)

NPDU *abbr* (*network protocol data unit*) DATA unidad *f* de datos del protocolo de la red

NPI *abbr* (*null pointer indication*) DATA indicación *f* de cero con aguja

NPI/TOA *abbr* (*numbering plan indicator/TOA*) DATA, TELEP NPI/TOA (*indicador de plan de numeración/TOA*)

n-plus-one: ~ **address instruction** *n* COMP instrucción *f* de dirección n-más-una

NPO: ~**-type capacitor** *n* POWER capacitor *m* de tipo NPO

N relay *n* TELEP retransmisión *f* N

NRN *abbr* (*nonreceipt notification*) DATA, TELEP NRN (*notificación de no recepción*)

NRT *abbr* (*nonreal time*) COMP, DATA, TELEP, TEST NRT (*tiempo irreal*)

NRZ *abbr* (*nonreturn to zero*) DATA, TELEP NRZ (*sin retorno a cero*); ~ **code** *n* DATA, TRANS código *m* NRZ

ns *abbr* (*nanosecond*) TEST ns (*nanosegundo*)

NS *abbr* (*network service*) DATA, TELEP servicio *m* de la red

NSAP *abbr* (*network service access point*) DATA punto *m* de acceso al servicio de la red

NSC *abbr* (*nonstandard facility command*) DATA, TELEP orden *f* de instalación no normalizada

NSC-CIG-DTC *abbr* (*nonstandard facility command CIG-DTC*) DATA NSC-CIG-DTC (*CIG-DTC de orden de instalación no normalizada*)

NSF *abbr* (*nonstandard facility*) TELEP NSF (*instalación no normalizada*)

NSF-CSI-DIS *abbr* (*nonstandard facility CSI-DIS*) DATA NSF-CSI-DIS (*CSI-DIS de instalación no normalizada*)

NSS *abbr* (*nonstandard setup*) DATA, TELEP NSS (*ajuste no normalizado*)

NT *abbr* (*network termination*) APPL, TELEP, TRANS terminación *f* de la red

NT1 *abbr* (*network digital termination*) TELEP NT1 (*terminación digital de la red*)

NT1-LB *abbr* (*B-ISDN network termination 1*) TRANS terminación *f* 1 de red digital de servicios integrados

NT2-LB *abbr* (*B-ISDN network termination 2*) TRANS terminación *f* 2 de red digital de servicios integrados

NTE *abbr* (*network terminal equipment*) TELEP, TRANS equipo *m* terminal de la red

nth: ~ **choice group** *n* TELEP grupo *m* enésimo de elección; ~ **order harmonic distortion** *n* TELEP distorsión *f* armónica de enésimo orden

NTL *abbr* (*normalized transmission loss*) RADIO NTL (*pérdida de transmisión normalizada*)

NT-LB *abbr* (*B-ISDN network termination*) TRANS T-RDSI (*terminación de la red B-RDSI*)

NTN *abbr* (*network terminal number*) DATA, TELEP NTN (*número del terminal de la red*)

NTU *abbr* (*network terminating unit*) TRANS NTU (*unidad terminal de la red*)

nuance *n* COMP matiz *m*; ~ **of darkness** DATA, PER APP *of scanner* grado *m* de oscuridad; ~ **of lightness** DATA, PER APP *scanner* grado *m* de luminosidad

nuclear: ~**-powered** *adj* GEN de propulsión nuclear

nucleus *n* COMP núcleo *m*; ~ **of operating system** COMP núcleo *m* del sistema operativo

NUI *abbr* (*network user identification*) DATA, TELEP NUI (*identificación del usuario de la red*)

nuisance: ~ **area** *n* GEN zona *f* de interferencia

NUL *abbr* (*null character*) DATA NUL (*carácter nulo*)

null *adj* GEN nulo; ~ **AM-modulation method** *n* RADIO método *m* de modulación AM nulo; ~ **character** *n* (*NUL*) DATA carácter *m* nulo (*NUL*); ~**-data check** *n* COMP, DATA control *m* de datos nulos; ~ **detector** *n* RADIO detector *m* de cero; ~ **meter** *n* TEST indicador *m* de cero; ~ **method** *n* TEST método *m* de cero; ~ **modem** *n* DATA módem *m* nulo; ~ **placing technique** *n* RADIO técnica *f* de ubicación cero; ~ **pointer indication** *n* (*NPI*) DATA indicación *f* de cero con aguja; ~**-reference glide path** *n* RADIO trayectoria *f* de planeo de referencia nula; ~ **string** *n* DATA cadena *f* nula

number *n* (*N*) GEN número *m* (*N*); ~ **of bit disagreements** DATA desacuerdos *m pl* en el número de bitios; ~**-busy signal** SIG, TELEP señal *f* de número ocupado; ~ **of defectives** TEST número *m* de elementos defectuosos; ~ **not allocated** DATA número *m* no adjudicado; ~ **out of order** DATA número *m* averiado; ~ **plan identification** DATA, TELEP identificación *f* de plano numérico; ~ **redial** TELEP repetición *f* automática de número; ~ **repetition service** TELEP servicio *m* de repetición de número; ~ **series** GEN serie *f* numérica; ~ **of significant conditions** TELEP número *m* de condiciones significativas; ~ **of teeth** GEN número *m* de dientes; ~ **of turns** POWER número *m* de espiras

numbering *n* GEN numeración *f*; ~ **area** TELEP área *f* de numeración; ~ **drawing** GEN dibujo *m* de numeración; ~ **plan** GEN plan *m* de numeración; ~ **plan area** (*NPA*) TELEP área *f* de plan de numeración (*NPA*); ~ **plan identification** DATA, TELEP identificación *f* de plan de numeración; ~ **plan identifier** DATA identificador *m* de plan de numeración; ~ **plan indicator** DATA, TELEP indicador *m* de plan de numeración; ~ **plan indicator/TOA** (*NPI/*

TOA) DATA, TELEP indicador *m* de plan de numeración/TOA (*NPI/TOA*); ~ **plan interworking** DATA, TELEP interfuncionamiento *m* del plan de numeración; ~ **scheme** DATA, TELEP plan *m* de numeración; ~ **system** TELEP sistema *m* de numeración

numeric *adj* GEN numérico; ~ **coded character set** *n* DATA subconjunto *m* de caracteres numéricos codificados; ~ **data** *n pl* COMP, DATA datos *m pl* numéricos; ~ **keypad** *n* COMP teclado *m* numérico; ~ **pad** *n* COMP teclado *m* numérico; ~ **word** *n* COMP, DATA palabra *f* numérica

numerical[1]: ~ **aperture** *n* DATA, ELECT, TRANS apertura *f* numérica; ~ **code** *n* DATA código *m* numérico; ~ **control** *n* COMP control *m* numérico; ~ **data** *n pl* COMP datos *m pl* numéricos; ~ **database** *n* DATA base *f* de datos numéricos; ~ **display** *n* COMP presentación *f* numérica, RADIO información *f* numérica; ~ **display paging** *n* RADIO paginación *f* de presentación numérica; ~ **horizontal** *n* SWG horizontal *m* numérico; ~ **message** *n* RADIO mensaje *m* numérico; ~ **order** *n* GEN orden *m* numérico; ~ **selecting bar** *n* SWG barra *f* de selección numérica; ~ **sequence** *n* GEN secuencia *f* numérica

numerical[2]: **in ~ order** *phr* GEN por orden numérico

numerically *adv* GEN numéricamente

nut *n* GEN tuerca *f*

nylon *n* PARTS nailon *m*; ~ **binder** ELECT aglomerante *m* de nailon

Nyquist: ~ **diagram** *n* CONTR diagrama *m* de Nyquist; ~ **sampling** *n* COMP muestreo *m* de Nyquist

O

O *abbr* (*optional*) GEN O (*opcional*); ~ **component** *n* DATA componente *m* O
OAMP *abbr* (*operation, administration, maintenance and provisioning*) TELEP, TEST operación *f* administración, mantenimiento y provisión
object *n* COMP, DATA objeto *m*; ~ **class** DATA clase *f* de objeto; ~ **identifier** DATA identificador *m* de objeto; ~ **language** COMP lenguaje *m* objeto; ~ **management** COMP gestión *f* de objeto; ~ **management group** (*OMG*) GEN grupo *m* de gestión de objeto (*OMG*); ~ **module** COMP módulo *m* objeto; ~**-oriented programming** (*OOP*) COMP programación *f* orientada por objeto (*OOP*); ~ **program** COMP programa *m* de objeto; ~ **space** COMP espacio *m* objeto; ~ **study** GEN estudio *m* objeto
objective: ~ **loudness rating** *n* TELEP medida *f* de sonoridad objetiva
obligatory: ~ **parameter** *n* DATA parámetro *m* obligatorio; ~ **presence** *n* DATA presencia *f* obligatoria
oblique: ~**-incidence ionospheric recorder** *n* RADIO registrador *m* ionosférico de incidencia oblicua; ~**-incidence transmission** *n* RADIO transmisión *f* con incidencia oblicua; ~ **stroke** *n* COMP trazo *m* oblicuo
OBO *abbr* (*output back-off*) TRANS OBO (*retroceso de salida*)
observability *n* GEN observabilidad *f*
observation *n* GEN observación *f*; ~ **desk** CONTR mesa *f* de observación; ~ **error** TEST error *m* de observación; ~ **panel** CONTR panel *m* de observación; ~ **service** APPL, TELEP servicio *m* de observación; ~ **telephone** PER APP, TELEP teléfono *m* de observación
observe[1]: ~ **agent key** *n* TELEP clave *f* de observación de agente
observe[2] *vt* GEN observar
observed: ~ **bearing** *n* RADIO rumbo *m* observado; ~ **deviation** *n* TEST desviación *f* observada; ~ **dimension** *n* TEST dimensión *f* observada; ~ **test outcome** *n* DATA, TEST resultado *m* observado de la prueba; ~ **value** *n* TEST valor *m* observado
obsolete *adj* GEN obsoleto
obstacle: ~ **effect** *n* TELEP efecto *m* de un obstáculo; ~ **gain** *n* RADIO ganancia *f* impeditiva
obtain *vt* PARTS *holding current* recibir, TELEP obtener
OC *abbr* (*organization code*) ELECT OC (*código de organización*); ~ **curve** *n* GEN curva *f* OC; ~ **test** *n* ELECT prueba *f* OC
occasional: ~ **transmissions** *n pl* TRANS transmisiones *f pl* ocasionales
occupancy: ~ **rate** *n* GEN tasa *f* de ocupación
occupation: ~ **time** *n* GEN tiempo *m* de ocupación
occupied *adj* GEN ocupado; ~ **band** *n* RADIO banda *f* ocupada; ~ **bandwidth** *n* RADIO, TRANS ancho *m* de banda ocupado; ~ **spectrum space** *n* RADIO, TRANS zona *f* de espectro ocupado
occupy *vt* TELEP ocupar
occurrence: ~ **of collisions** *n* ELECT ocurrencia *f* de colisiones
octal: ~ **numeral** *n* TELEP numeral *m* octal
octantal: ~ **error** *n* RADIO error *m* octantal
octave: ~ **shift** *n* COMP desplazamiento *m* de octava; ~ **transpose** *n* COMP transposición *f* de octava
octet: ~ **alignment** *n* TELEP alineación *f* de octeto; ~**-string type** *n* DATA tipo *m* de cadena de octetos; ~ **timing signal** *n* SIG señal *f* de sincronización de octeto
octode *n* ELECT octodo *m*
octopus *adj* ELECT octópodo
ODA *abbr* (*office document architecture*) COMP, DATA ODA (*arquitectura de documento administrativo*)
odd *adj* GEN impar
ODFMA *abbr* (*orthogonal frequency-division multiple access*) RADIO ODFMA (*acceso múltiple con división de frecuencia ortogonal*)
O-E *abbr* (*optical-electrical*) GEN O-E (*óptico-eléctrico*)
OEM: ~ **power supply** *n* POWER suministro *m* de energía OEM
off-air *adj* RADIO que no transmite
off-axis *adj* RADIO descentrado del eje
off-beam *adv* RADIO fuera de haz
offbeat *adj* COMP excéntrico
off-duty: ~ **state marking** *n* TELEP marcación *f* de ausencia de operador

off-hook *adj* TELEP descolgado; ~ **queue** *n* TELEP cola *f* de contestación; ~ **queue offer tone** *n* TELEP tono *m* de oferta de cola de contestación

off-hooking *n* COMP, DATA, PER APP, RADIO, TELEP descolgamiento *m*

offer *n* GEN oferta *f*

office *n* GEN oficina *f*; ~ **automation** GEN automatización *f* de oficina; ~ **clerk** GEN oficinista *m*; ~ **directory** DATA guía *f* administrativa; ~ **document architecture** (*ODA*) COMP, DATA arquitectura *f* de documento administrativo (*ODA*); ~ **document interchange format** DATA formato *m* de intercambio de documentos administrativos; ~ **environment** GEN entorno *m* administrativo

off-line[1] *adv* COMP, DATA, TELEP fuera de línea

off-line[2]: ~ **storage** *n* COMP almacenamiento *m* fuera de línea; ~ **working** *n* TELEP funcionamiento *m* fuera de línea

offload *vt* RADIO *satellite* vaciar

off-net: ~ **station** *n* TELEP, TV estación *f* fuera de red

off-premises: ~ **extension** *n* (*OPX*) TELEP extensión *f* fuera del edificio (*OPX*)

offprint *n* PER APP separata *f*

offset *adj* GEN desplazado; ~ **current** *n* ELECT corriente *f* de desnivel; ~ **dish** *n* RADIO antena *f* parabólica excéntrica; ~ **feed** *n* RADIO alimentador *m* primario desplazado; ~ **Gregory antenna** *n* RADIO antena *f* Gregory excéntrica; ~ **key protection** *n* ELECT protección *f* con clave desplazada; ~ **parabolic antenna** *n* RADIO antena *f* parabólica excéntrica; ~ **quaternary phase shift keying** *n* DATA, SIG conmutación *f* de fase cuaternaria desplazada; ~ **screwdriver** *n* GEN destornillador *m* de mango acodado; ~ **voltage** *n* ELECT tensión *f* de desplazamiento

off-site: ~ **maintenance** *n* TEST mantenimiento *m* en el exterior

off-state: ~ **current** *n* ELECT corriente *f* desactivada; ~ **voltage** *n* ELECT voltaje *m* inactivo, PARTS voltaje *m* desactivado

off-time *adj* COMP, RADIO inactivo

o/g *abbr* (*outgoing*) GEN o/g (*saliente*)

OHA *abbr* (*overhead access*) DATA acceso *m* aereo

ohm *n* ELECT ohmio *m*

ohmic: ~ **contact** *n* PARTS contacto *m* óhmico

ohmmeter *n* ELECT óhmetro *m*

oil: ~ **can** *n* GEN aceitera *f*; ~**-filled cable** *n* GEN cable *m* de aceite fluido; ~**-filled capacitor** *n* ELECT capacitor *m* de aceite; ~ **gun** *n* GEN inyector *m* de aceite; ~ **level** *n* GEN nivel *m* de aceite; ~ **level gauge** *BrE n* TELEP indicador *m* de nivel de aceite; ~ **mist lubrication** *n* GEN lubricación *f* por vapor de aceite; ~ **nipple** *n* GEN aceitera *f*; ~ **pressure** *n* GEN presión *f* de aceite; ~ **pressure gauge** *BrE n* TELEP manómetro *m* de aceite

OK: ~ **signal** *n* RADIO visto *m* bueno

old: ~ **code** *n* TELEP código *m* antiguo; ~ **number** *n* TELEP número *m* viejo

oligarchic: ~ **network** *n* TELEP *synchronized* red *f* oligárquica

OM *abbr* (*operations and maintenance*) TELEP, TEST OM (*operaciones y mantenimiento*); ~ **subsystem** *n* TELEP, TEST subsistema *m* OM

OMG *abbr* (*object management group*) GEN OMG (*grupo de gestión de objeto*)

omit *vt* GEN omitir

omitted *adj* GEN omitido

omnibus: ~ **service circuit** *n* TELEP circuito *m* de servicio ómnibus; ~ **system** *n* TELEP sistema *m* ómnibus

omnidirectional: ~ **aerial** *n* RADIO antena *f* omnidireccional; ~ **antenna** *n* RADIO antena *f* omnidireccional

omnipresence *n* ELECT omnipresencia *f*

on *adj* ELECT conectado; ~ **position** *n* GEN posición *f* de trabajo

ON *abbr* (*other notification*) DATA ON (*otra notificación*)

on-air: ~ **time** *n* RADIO tiempo *m* en emisión

on-axis: ~ **gain** *n* RADIO ganancia *f* en el eje

on-board *adj* COMP de abordo, sobre la placa; ~ **apparatus** *n* RADIO instrumentos *m pl* de abordo; ~ **beacon** *n* RADIO baliza *f* de abordo; ~ **equipment** *n* RADIO equipo *m* de abordo; ~ **processing** *n* COMP procesamiento *m* abordo; ~ **subscriber** *n* RADIO abonado *m* abordo

on-busy *adj* TELEP ocupado

on-demand *adj* TV por encargo; ~ **service** *n* TELEP servicio *m* por encargo

one: ~ **address instruction** *n* COMP instrucción *f* de una sola dirección; ~**-ahead addressing** *n* COMP direccionamiento *m* progresivo automático; ~**-plus-**~ **address instruction** *n* COMP instrucción *f* de una más una direcciones; ~**-plus-**~ **carrier system** *n* TRANS sistema *m* de onda portadora de uno más uno; ~ **out of five decode** *n* ELECT descodificación *f* uno de cinco; ~**-piece cover** *n* TELEP cubierta *f* de una sola pieza; ~**-shot multivibrator** *n* ELECT multivibrador *m* monoestable; ~**-step operation** *n* COMP operación *f* en un paso; ~**-stop shopping** *n* GEN compra *f* concentrada; ~**-symbol gate signal** *n* DATA señal *f* de puerta de un solo símbolo; ~**-unit message** *n* TELEP mensaje *m* de una unidad;

~-**way** *adj* COMP, DATA, RADIO, SWG, TELEP, TV unidireccional; ~-**way communication** *n* RADIO, SWG, TELEP, TV comunicación *f* unidireccional; ~-**way directionality** *n* COMP unidireccionalidad *f*; ~-**way link** *n* COMP malla *f* unidireccional

ongoing *adj* GEN en curso

on-hold *adj* TELEP a la espera

on-hook *adj* COMP, TELEP colgado; ~ **condition** *n* TELEP condición *f* en reposo; ~ **dialling** *BrE n* GEN, RADIO, TELEP marcación *f* en reposo; ~ **signal** *n* TELEP señal *f* de término de conversación

on-line *adj* COMP, DATA, ELECT, TELEP en línea; ~ **catalogue** *n* DATA catálogo *m* en línea; ~ **data service** *n* COMP servicio *m* de datos en línea; ~ **database** *n* DATA base *f* de datos en línea; ~ **full-text document retrieval** *n* DATA recuperación *f* de documento de texto puro en línea; ~ **information retrieval** *n* DATA recuperación *f* de información en línea; ~ **information service** *n* DATA servicio *m* de información en línea; ~ **newspaper** *n* DATA periódico *m* en línea; ~ **optical amplification** *n* TRANS amplificación *f* óptica en línea; ~ **processing** *n* DATA procesamiento *m* en línea; ~ **registration facility** *n* TELEP función *f* de registro en línea; ~ **service** *n* DATA servicio *m* en línea; ~ **telematic service** *n* DATA servicio *m* telemático en línea; ~ **transaction processing** *n* DATA procesamiento *m* de transacción en línea

only-route: ~ **circuit group** *n* TELEP haz *m* de circuitos de ruta única

on-net: ~ **to off-net call** *n* TRANS llamada *f* desde la red hacia fuera

on-off: ~ **function** *n* PER APP *of answering machine* función *f* conexión-desconexión; ~ **keying** *n* (*OOK*) SIG, TELEP manipulación *f* todo-nada (*OOK*); ~ **switch** *n* COMP interruptor *m* de red

on-premises *adv* TELEP en el propio lugar

ONS *abbr* (*on-scene*) TELEP ONS (*en escena*)

on-scene *adj* (*ONS*) TELEP en escena (*ONS*)

on-site: ~ **acceptance** *n* RADIO aceptación *f* in situ; ~ **maintenance** *n* TEST mantenimiento *m* in situ

on-state: ~ **current** *n* ELECT, PARTS corriente *f* en estado de conducción; ~ **power loss** *n* ELECT, PARTS pérdida *f* de potencia en estado de conducción; ~ **slope resistance** *n* ELECT, PARTS resistencia *f* aparente en estado conductor; ~ **threshold voltage** *n* ELECT, PARTS tensión *f* de umbral en estado conductor; ~ **voltage** *n* ELECT, PARTS tensión *f* en estado de conducción

on-station: ~ **satellite** *n* RADIO satélite *m* en estación; ~ **software** *n* COMP, DATA programas *m pl* en estación

on-the-job: ~ **training** *n* GEN formación *f* en el puesto de trabajo

OOB *abbr* (*out-of-band*) ELECT, RADIO OOB (*fuera de banda*)

OOK *abbr* (*on-off keying*) SIG, TELEP OOK (*manipulación todo-nada*)

OOP *abbr* (*object-oriented programming*) COMP OOP (*programación orientada por objeto*)

opacity *n* TV opacidad *f*

opaque *adj* TV opaco

open[1] *adj* GEN abierto; ~ **battery** *n* POWER batería *f* abierta; ~ **circuit** *n* POWER circuito *m* abierto; ~-**circuit admittance** *n* POWER admitancia *f* de circuito abierto; ~-**circuit current** *n* POWER corriente *f* de circuito abierto; ~-**circuit impedance** *n* POWER impedancia *f* de circuito abierto; ~-**circuit test** *n* ELECT prueba *f* en circuito abierto; ~-**circuit voltage** *n* POWER tensión *f* de circuito abierto; ~-**circuit working** *n* ELECT, TELEP funcionamiento *m* en circuito abierto; ~ **collaboration environment** *n* COMP entorno *m* de colaboración abierta; ~ **collector-gate** *n* ELECT puerta *f* de colector abierto; ~ **data link interface** *n* ELECT interfaz *m* abierto de enlace de datos; ~ **distribution processing** *n* COMP procesamiento *m* de distribución abierta; ~ **document architecture** *n* DATA arquitectura *f* de documento abierto; ~ **document interchange format** *n* DATA formato *m* abierto de intercambio de documento; ~ **forum** *n* DATA foro *m* abierto; ~ **group** *n* DATA grupo *m* abierto; ~ **line** *n* POWER línea *f* abierta; ~ **loop** *n* COMP, POWER, TRANS bucle *m* abierto; ~-**loop acquisition** *n* CONTR, POWER adquisición *f* en circuito abierto; ~-**loop control** *n* CONTR, POWER control *m* de bucle abierto; ~-**loop gain** *n* CONTR, POWER ganancia *f* de bucle abierto; ~-**loop loss** *n* CONTR, POWER pérdida *f* de bucle abierto; ~-**loop synchronization** *n* CONTR, POWER, TRANS sincronización *f* de bucle abierto; ~ **messaging interface** *n* DATA interfaz *m* abierto de mensajería; ~ **network architecture** *n* DATA arquitectura *f* de red abierta; ~ **network provision** *n* TRANS provisión *f* de la red abierta; ~ **parenthesis** *n* GEN paréntesis *m* de apertura; ~ **position** *n* SWG posición *f* abierta; ~ **shortest path first** *n* ELECT el camino *m* abierto más corto primero; ~ **software foundation** *n* DATA soporte *m* de programas abiertos; ~ **standards computing platform** *n* DATA plataforma *f* abierta de informatización de normas; ~ **subroutine** *n* COMP subprograma *f* abierto; ~ **system** *n* DATA, TELEP, TRANS sistema *m* abierto; ~ **systems architecture** *n* TELEP, TRANS arquitectura *f* de sistemas abiertos; ~ **systems interconnection** *n* COMP, DATA, TELEP, TRANS interconexión *f* de sistemas abiertos; ~ **systems interconnection environment** *n* DATA entorno *m* abierto de interconexión de sistemas; ~ **systems interconnection layer** *n* COMP capa *f* de interconexión de sistemas abiertos; ~ **systems interconnection network layer** *n* DATA capa *f* de red deinterconexión de sistemas abiertos; ~ **systems interconnection network service** *n* TELEP servicio *m* abierto de red de interconexión desistemas; ~ **systems interconnection resource** *n* DATA, TELEP, TRANS recurso *m* de interconexión de sistemas abiertos; ~**-systems interconnection standard** *n* ELECT, TRANS norma *f* de interconexión de sistemas abiertos; ~ **wire** *n* GEN alambre *m* desnudo; ~**-wire line** *n* GEN, TELEP línea *f* de hilos desnudos

open[2] *vt* GEN abrir

opening *n* GEN *transistors* apertura *f*

operable: ~ **time** *n* GEN tiempo *m* útil

operator *n* GEN, RADIO, SWG, TELEP, TV operador *m*; ~**-assisted call** GEN, TELEP llamada *f* con intervención de operador; ~ **authorization** TELEP autorización *f* de operador; ~ **centre** *BrE* TELEP centro *m* de operadores; ~ **command** DATA instrucción *f* de operador; ~**-connected call** TELEP llamada *f* conectada por operador; ~**'s console** TELEP consola *f* de operador; ~ **control panel** COMP panel *m* de control; ~**-extended call** TELEP llamada *f* con extensión por operador; ~ **factor** RADIO factor *m* de operador; ~ **junctor trunk group** TELEP bloque *m* de circuitos troncales de operador; ~ **office** TELEP oficina *f* de operador; ~ **office trunk group** TELEP circuitos *m pl* troncales de la oficina de operador; ~ **output level** COMP, ELECT nivel *m* de salida de operador; ~ **position** GEN, TELEP puesto *m* deoperador; ~ **select** COMP selección *f* de operador; ~ **service centre** *BrE* TELEP centro de trabajo de operador; ~**'s set** TELEP teléfono *m* de operador; ~**'s switching set** SWG unidad *f* de dicordio; ~ **system** TELEP sistema *m* de operador; ~**'s telephone** PER, TELEP teléfono *m* de operador; ◆ ~ **disable** COMP operador fuero de servicio; ~ **unavailable** RADIO operador *m* no disponible; ~ **enable** COMP operador en servicio

operate[1]: ~ **condition** *n* GEN posición *f* de trabajo; ~ **current** *n* POWER corriente *f* de excitación; ~ **time** *n* PARTS tiempo *m* de funcionamiento

operate[2] *vt* TELEP *equipment* manejar

operate[3] *vi* GEN funcionar

operated: ~ **position** *n* GEN posición *f* de funcionamiento

operating: ~ **authorization** *n* TV autorización *f* de conexión; ~ **bandwidth** *n* RADIO, TRANS ancho *m* de banda de trabajo; ~ **centre** *BrE n* TELEP centro *m* de operación; ~ **channel** *n* RADIO canal *m* de trabajo; ~ **characteristic** *n* TELEP característica *f* funcional; ~ **characteristic curve** *n* GEN curva *f* característica operativa; ~ **company** *n* GEN compañía *f* operadora; ~ **conditions** *n pl* ELECT, TELEP condiciones *f pl* de servicio; ~ **current** *n* POWER corriente *f* de servicio; ~ **curve** *n* ELECT curva *f* de funcionamiento; ~ **device** *n* CONTR dispositivo *m* de maniobra; ~ **directions** *n pl* GEN instrucciones *f pl* de operación; ~ **environment** *n* DATA configuración *f* operativa; ~ **expense** *n* GEN coste *m* (*AmL* costo) de explotación; ~ **frequency** *n* RADIO frecuencia *f* de trabajo; ~ **instructions** *n pl* GEN instrucciones *f pl* para el uso; ~ **licence** *BrE n* GEN licencia *f* de explotación; ~ **line** *n* ELECT línea *f* de servicio; ~ **loss** *n* GEN pérdida *f* de explotación; ~ **point** *n* ELECT punto *m* de trabajo; ~ **position** *n* GEN posición *f* de funcionamiento; ~ **principle** *n* GEN principio *m* de funcionamiento; ~ **profit** *n* GEN utilidad *f* de explotación; ~ **revenues** *n pl* GEN ingresos *m pl* de explotación; ~ **spring** *n* GEN muelle *m* impulsor; ~ **state** *n* TEST estado *m* de funcionamiento; ~ **statistics** *n pl* TELEP estadística *f* de funcionamiento; ~ **system** *n* (*OS*) COMP, DATA, TELEP, TEST sistema *m* operativo (*SO*); ~ **system function** *n* (*OSF*) COMP, DATA, TELEP, TEST función *f* del sistema operativo (*FSO*); ~ **temperature** *n* GEN temperatura *f* de funcionamiento; ~ **terminal** *n* COMP terminal *m* de trabajo; ~ **time** *n* GEN tiempo *m* de maniobra; ~ **voltage** *n* ELECT voltaje *m* de régimen

operation *n* GEN funcionamiento *m*, operación *f*; ~, **administration and maintenance centre** *BrE* TELEP, TEST centro *m* de operaciones, administración y mantenimiento; ~, **administration, maintenance and provisioning** (*OAMP*) TELEP, TEST operación *f* administración, mantenimiento y provisión; ~ **class** DATA clase *f* de operaciones; ~ **code** COMP código *m* de operación; ~ **directory** COMP directorio *m* de operación; ~ **error** COMP error *m* de funcionamiento; ~ **of horizontal** SWG operación *f* de horizontal; ~ **interruption** TELEP interrupción *f* de la operación; ~ **and maintenance centre** *BrE* TELEP, TEST centro *m* de operaciones y mantenimiento; ~ **and maintenance centre** *BrE* **processor** COMP, TELEP, TEST procesador *m* de centro de operaciones y mantenimiento; ~ **and maintenance engineer** TELEP, TEST técnico *m* de operaciones y mantenimiento; ~ **and maintenance manual** GEN manual *m* de operación y mantenimiento; ~ **and maintenance messaging software machine** CONTR ordenador *m* con programa de mensajes de operación y mantenimiento; ~ **manual** GEN manual *m* de funcionamiento; ~ **point** SWG punto *m* de operación; ~ **thermostat** CONTR, PARTS termostato *m* de operación

operational: ~ **amplifier** *n* ELECT, TRANS amplificador *m* operacional; ~ **availability factor** *n* TEST factor *m* de disponibilidad operativa; ~ **centre** *BrE n* TELEP centro *m* operativo; ~ **conditions** *n pl* ELECT condiciones *f pl* de funcionamiento, TELEP condiciones *f pl* de servicio; ~ **data** *n pl* TELEP datos *m pl* de explotación; ~ **description** *n* TELEP descripción *f* de funcionamiento; ~ **experience** *n* TELEP experiencia *f* de trabajo; ~ **information display system** *n* RADIO sistema *m* de visualización de información operativa; ~ **lifetime** *n* TELEP, TEST duración *f* de la vida operativa; ~ **mode** *n* TELEP modo *m* de operación; ~ **MUF** *n* RADIO MUF *f* operativa; ~ **performance** *n* TELEP operabilidad *f*; ~ **procedure** *n* TELEP rutina *f* de operación; ~ **security** *n* TELEP seguridad *f* de funcionamiento; ~ **semantics** *n* COMP, DATA semántica *f* operacional

operations: ~ **centre** *BrE n* GEN centro *m* de explotaciones, centro *m* de operaciones; ~ **chart** *n* DATA diagrama *m* de operaciones; ~ **control centre** *BrE n* TEST centro *m* de control de operaciones; ~ **and maintenance** *n pl* (*OM*) TELEP, TEST operaciones *f pl* y mantenimiento (*OM*); ~ **and maintenance application part** *n* TELEP, TEST parte *f* de la aplicación de operaciones y mantenimiento; ~ **and maintenance subsystem** *n* DATA, TELEP, TEST subsistema *m* de operaciones y mantenimiento; ~ **room** *n* TEST sala *f* de servicio; ~ **support** *n* TELEP soporte *m* de operaciones; ~ **support node** *n* TELEP nodo *m* de soporte de operaciones; ~ **systems computing architecture** *n* COMP arquitectura *f* informática de los sistemas de operaciones

operative *adj* GEN operativo

optic: ~ **axis** *n* TRANS eje *m* óptico

optical *adj* COMP, TV óptico; ~ **aberration** *n* TRANS aberración *f* óptica; ~ **activity** *n* TRANS actividad *f* óptica; ~ **amplifier** *n* DATA, TRANS amplificador *m* óptico; ~ **attenuator** *n* TRANS atenuador *m* óptico; ~ **axis** *n* TRANS eje *m* óptico; ~ **bar-code reader** *n* DATA lector *m* óptico de códigos de barras; ~ **beam** *n* ELECT haz *m* óptico; ~ **bench** *n* TRANS banco *m* óptico; ~ **bistability** *n* TRANS biestabilidad *f* óptica; ~ **branching** *n* TRANS ramificación *f* óptica; ~ **cable** *n* TRANS cable *m* óptico; ~ **cable assembly** *n* TRANS montaje *m* de cables ópticos; ~ **cavity** *n* TRANS cavidad *f* óptica; ~ **centre** *BrE n* SWG centro *m* óptico; ~ **character recognition** *n* COMP, DATA reconocimiento *m* óptico de caracteres; ~ **character-recognition software** *n* COMP, DATA programa *m* de reconocimiento óptico de caracteres; ~ **characteristic** *n* TRANS característica *f* óptica; ~ **circulator** *n* TRANS circulador *m* óptico; ~ **coherence** *n* TRANS coherencia *f* óptica; ~ **combiner** *n* TRANS combinador *m* óptico; ~ **communications** *n pl* TRANS comunicaciones *f pl* ópticas; ~ **control boards** *n pl* TRANS paneles *m pl* de control óptico; ~ **cord** *n* TRANS cable *m* óptico; ~ **coupler** *n* DATA, ELECT, TRANS acoplador *m* óptico; ~-**data bus** *n* TRANS bus *m* de datos ópticos; ~ **demultiplexer** *n* TV desmultiplexor *m* óptico; ~ **detector** *n* TRANS detector *m* óptico; ~ **disk** *n* DATA, TV disco *m* óptico; ~ **disk drive** *n* DATA unidad *f* de disco óptico; ~ **distribution** *n* TRANS distribución *f* óptica; ~-**electrical** (*O-E*) GEN óptico-eléctrico (*O-E*); ~ **exchange** *n* SWG, TRANS central *f* óptica; ~ **fibre** *BrE n* TRANS, COMP, DATA, PARTS, TRANS fibra *f* óptica; ~ **fibre** *BrE* **cable** *n* TRANS cable *m* de fibra óptica; ~ **fibre** *BrE* **connector** *n* TRANS conector *m* de fibra óptica; ~ **fibre** *BrE* **coupler** *n* DATA, TRANS acoplador *m* de fibra óptica; ~ **fibre** *BrE* **frame** *n* TRANS trama *f* de fibra óptica; ~-**fibre** *BrE* **line system** *n* TRANS sistema *m* de línea de fibra óptica; ~ **fibre** *BrE* **link** *n* TRANS enlace *m* de fibra óptica; ~ **fibre** *BrE* **pigtail** *n* TRANS trenza *f* de fibra óptica; ~ **fibre** *BrE* **sensor** *n* TRANS sensor *m* de fibra óptica; ~ **fibre** *BrE* **splice** *n* TRANS unión *f* de fibra óptica; ~ **fibre** *BrE* **transmission** *n* TRANS transmisión de fibra óptica; ~ **filter** *n* TRANS filtro *m* óptico; ~ **head** *n* DATA cabeza *f* óptica;

~ **image** *n* TV imagen *f* óptica; ~ **interface** *n* TRANS interfaz *m* óptico; ~ **interference** *n* TRANS interferencia *f* óptica; ~ **isolator** *n* ELECT, TRANS aislador *m* óptico; ~**-line terminating equipment** *n* TRANS equipo *m* determinación de línea óptica; ~ **link** *n* TRANS enlace *m* óptico; ~ **loss** *n* DATA, TRANS pérdida *f* óptica; ~ **media** *n* TV medios *m pl* ópticos; ~ **modulation** *n* DATA modulación *f* óptica; ~ **modulation index** *n* TRANS índice *m* de modulación óptica; ~ **modulator** *n* TRANS modulador *m* óptico; ~ **mouse** *n* COMP ratón *m* óptico; ~ **multiplexer** *n* TV multiplexor *m* óptico; ~ **network unit** *n* TRANS unidad *f* de red óptica; ~ **noise** *n* DATA ruido *m* óptico; ~ **noise measurement** *n* DATA medición *f* del ruido óptico; ~ **parametric oscillator** *n* ELECT oscilador *m* paramétrico óptico; ~ **path** *n* TRANS trayectoria *f* óptica; ~ **path length** *n* TRANS longitud *f* de la trayectoria óptica; ~ **pencil** *n* APPL, COMP lápiz *m* óptico; ~ **phase-locked loop** *n* ELECT bucle *m* sincronizado por fase óptica; ~ **polarization** *n* TRANS polarización *f* óptica; ~ **power** *n* TRANS potencia *f* óptica; ~ **power meter** *n* TRANS medidor *m* de potencia óptica; ~ **power output** *n* TRANS potencia *f* de salida óptica; ~ **processing** *n* TRANS procesamiento *m* óptico; ~ **radiation** *n* TRANS radiación *f* óptica; ~ **reader** *n* DATA, TV lector *m* óptico; ~ **reading** *n* DATA, TV lectura *f* óptica; ~ **receiver** *n* TRANS receptor *m* óptico; ~ **recording** *n* TRANS registro *m* óptico; ~ **refraction** *n* TRANS refracción *f* óptica; ~ **regenerative repeater** *n* TRANS repetidor *m* regenerativo óptico; ~ **regenerator** *n* TRANS regenerador *m* óptico; ~ **repeater** *n* TRANS repetidor *m* óptico; ~ **resolution** *n* COMP resolución *f* óptica; ~ **resonator** *n* TRANS resonador *m* óptico; ~ **return loss** *n* TELEP, TRANS pérdidas *f pl* de retorno ópticas; ~ **scanner** *n* DATA explorador *m* óptico; ~ **shaft encoder** *n* ELECT codificador *m* de eje óptico; ~ **signal** *n* TRANS señal *f* óptica; ~ **spectrum** *n* TRANS espectro *m* óptico; ~ **splice** *n* TRANS junta *f* óptica; ~ **switch** *n* SWG, TRANS conmutador *m* óptico; ~ **switching** *n* SWG, TRANS conmutación *f* óptica; ~ **switching crosspoint** *n* SWG contactos *m pl* de conmutación óptica; ~ **switching matrix** *n* SWG matriz *f* de conmutación óptica; ~ **switching network** *n* SWG, TRANS red *f* de conmutación óptica; ~ **switching system** *n* SWG sistema *m* de conmutación óptica; ~ **tap** *n* TRANS toma *f* óptica; ~ **tape storage** *n* DATA memoria *f* de cinta óptica; ~ **tapping** *n* TRANS sistema *m* de conexión óptica; ~ **thickness** *n* TRANS espesor *m* óptico; ~ **time domain reflectometer** *n* ELECT, TRANS reflectómetro *m* óptico en el dominio del tiempo; ~ **time domain reflectometry** *n* ELECT, POWER, TEST, TRANS reflectometría *f* óptica en el dominio del tiempo; ~ **transistor** *n* ELECT transistor *m* óptico; ~ **transmission** *n* TRANS transmisión *f* óptica; ~ **transmitter** *n* TRANS transmisor *m* óptico; ~ **tuning** *n* TRANS sintonización *f* óptica; ~ **typefont** *n* DATA tipo *m* de caracteres ópticos; ~ **waveguide** *n* DATA, RADIO, TRANS guiaondas *m* óptico

optically: ~ **active** *adv* PARTS, TRANS ópticamente activo

optics *n* GEN óptica *f*

optimal: ~ **control** *n* CONTR control *m* óptimo; ~ **path** *n* TRANS trayectoria *f* óptima; ~ **sampling** *n* TRANS muestreo *m* óptimo

optimization *n* GEN optimización *f*

optimize *vt* GEN optimizar

optimum: ~ **gain** *n* ELECT ganancia *f* óptima; ~ **load** *n* RADIO carga *f* óptima; ~ **mechanical tuning range** *n* ELECT alcance *m* óptimo de sintonización mecánica; ~ **output power** *n* ELECT potencia *f* óptima de salida; ~ **scale range** *n* TEST amplitud *f* óptima de escala; ~ **traffic frequency** *n* (*OTF*) RADIO frecuencia *f* óptima de tráfico (*OTF*); ~ **value** *n* GEN valor *m* óptimo; ~ **working frequency** *n* (*OWF*) RADIO frecuencia *f* óptima de trabajo (*OWF*)

option *n* GEN opción *f*

optional *adj* (*O*) GEN opcional (*O*); ~ **component** *n* DATA componente *m* opcional; ~ **network-specific digit** *n* TELEP dígito *m* opcional específico de la red; ~ **pause instruction** *n* COMP instrucción *f* de pausa optativa; ~ **presence** *n* DATA presencia *f* opcional; ~ **requirement** *n* DATA demanda *f* opcional; ~ **service** *n* TV servicio *m* opcional; ~ **stop instruction** *n* COMP instrucción *f* de parada facultativa; ~ **user-facility** *n* TELEP servicio *m* de usuario facultativo

optionally: ~**-droppable blocks** *n* DATA bloques *m* adicionables optativos

optocoupler *n* DATA, ELECT, TRANS optoacoplador *m*

optoelectric *adj* TV optoeléctrico

optoelectronic *adj* ELECT optoelectrónico; ~ **component** *n* PARTS componente *m* optoelectrónico; ~ **crosspoint** *n* SWG punto *m* de cruce optoelectrónico; ~ **equipment** *n* ELECT equipo *m* optoelectrónico; ~ **isolator** *n* ELECT aislador *m* optoelectrónico; ~ **receiver** *n* TRANS receptor *m* optoelectrónico; ~ **switching**

matrix *n* SWG matriz *f* de conmutación optoelectrónica
optoelectronics *n* ELECT optoelectrónica *f*
optoisolator *n* ELECT optoaislador *m*
optomechanical *adj* GEN optomecánico
optomechanics *n* GEN optomecánica *f*
optronic *adj* ELECT optrónico
optronics *n* ELECT optoelectrónica *f*
OPX *abbr* (*off-premises extension*) TELEP OPX (*extensión fuera del edificio*)
OR: ~ **circuit** *n* ELECT circuito *m* O; ~ **circuit of N signals** *n* COMP circuito *m* O de señales N; ~ **element** *n* ELECT elemento *m* O; ~ **gate** *n* ELECT puerta *f* O; ~ **operation** *n* ELECT operación *f* O
O/R *abbr* (*originator/recipient*) DATA, TELEP O/R (*remitente/receptor*)
orange: ~ **book** *n* COMP libro *m* naranja; ~ **filter** *n* DATA *of scanner* filtro *m* de superficie rugosa; ~ **peel** *n* GEN superficie *f* rugosa
orbit *n* RADIO órbita *f*; ~ **determination** RADIO determinación *f* de órbita; ~ **drift** RADIO deriva *f* de órbita; ~ **gap** RADIO separación *f* de órbita; ~ **insertion** RADIO inserción *f* en órbita
orbital: ~ **decay** *n* RADIO decadencia *f* orbital; ~ **element** *n* RADIO elemento *m* orbital; ~ **guidance** *n* RADIO guía *f* orbital; ~ **inclination control** *n* RADIO control *m* de inclinación orbital; ~ **location** *n* RADIO emplazamiento *m* orbital; ~ **period** *n* RADIO periodo *m* orbital; ~ **plane** *n* RADIO plano *m* orbital; ~ **position** *n* RADIO posición *f* orbital; ~ **slot** *n* RADIO canal *m* orbital; ~ **test programme** *BrE* *n* RADIO programa *m* de prueba orbital
orbiting: ~ **period** *n* RADIO periodo *m* de orbitación; ~ **time** *n* RADIO tiempo *m* de orbitación
orchestration *n* COMP instrumentación *f*
order[1]: ~ **of diversity** *n* RADIO orden *f* de diversidad; ~ **of priority** *n* GEN orden *f* de prioridad; ~ **statistic** *n* GEN estadístico *m* de orden; ~ **taking** *n* GEN aceptación *f* de pedido; ~ **wire** *n* TRANS circuito *m* de intercomunicación
order[2] *vt* GEN *services* pedir; ◆ ~ **a call** TELEP encargar una llamada; ~ **by merging** COMP pedir por interclasificación
ordinal: ~ **date** *n* GEN fecha *f* ordinal; ~ **value** *n* DATA valor *m* ordinal
ordinary *adj* GEN común; ~ **both-way line** *n* TRANS línea *f* bidireccional ordinaria; ~ **line** *n* TRANS línea *f* ordinaria; ~ **network** *n* TRANS red *f* ordinaria; ~ **private telex call** *n* TELEP llamada *f* ordinaria privada de télex; ~ **wave** *n* RADIO onda *f* fundamental
ORed *adj* ELECT de tipo *m* disyuntivo
organization: ~ **and maintenance** *n* TELEP organización *f* y mantenimiento
organizational: ~ **link** *n* COMP enlace *m* organizacional; ~ **unit** *n* TELEP unidad *f* de organización
organize *vt* GEN organizar
orientation: ~ **control** *n* CONTR control *m* de la orientación
origin: ~ **indicator** *n* TELEP indicador *m* de origen; ~ **of message** *n* DATA origen *m* del mensaje
original: ~ **called number** *n* TELEP número *m* llamado original; ~ **document** *n* COMP, DATA documento *m* de origen; ~ **encoded-information types** *n pl* DATA tipos *m pl* de información codificada originales; ~ **inspection** *n* TEST inspección *f* en primera presentación; ~ **production master** *n* ELECT master *m* de producción original; ~ **redirection reason indicator** *n* TELEP indicador *m* de razón de reexpedición original
originate *vt* TELEP *call* producir
originate in *vt* COMP, ELECT originar en
originated: ~ **point code** *n* TELEP código *m* de punto de origen
originating *adj* TELEP de origen; ~ **calling position** *n* TELEP posición *f* de origen de la llamada; ~ **device** *n* TELEP aparato *m* de origen; ~ **exchange** *n* SWG central *f* de procedencia; ~ **feature code** *n* TELEP prefijo *m* de origen; ~ **junctor** *n* PARTS, SWG conjuntor; ~ **network identification request indicator** *n* TELEP indicador *m* de petición de identificacion de la red de origen; ~ **point** *n* SIG punto *m* de origen; ~ **point code** *n* DATA, TELEP código *m* de punto de origen; ~ **register** *n* SWG registro *m* de origen; ~ **traffic** *n* TELEP tráfico *m* de origen
originator *n* GEN remitente *m*
originator/recipient *n* (*O/R*) DATA, TELEP remitente/receptor *m* (*O/R*)
O-ring *n* PARTS anillo *m* en O
orphan *n* COMP línea *f* suelta
orthicon *n* ELECT orticonoscopio *m*
orthogonal: ~ **co-channel** *n* RADIO canal *m* colateral ortogonal; ~ **frequency-division multiple access** *n* (*ODFMA*) RADIO acceso *m* múltiple con división de frecuencia ortogonal (*ODFMA*); ~ **mode transducer** *n* RADIO transductor *m* en modo ortogonal; ~ **polarization** *n* RADIO polarización *f* ortogonal; ~ **signal** *n* RADIO señal *f* ortogonal
orthomode: ~ **transducer** *n* RADIO transductor *m* ortomodal

OS *abbr* (*operating system*) COMP, DATA, TELEP, TEST SO (*sistema operativo*)
oscillate *vi* GEN oscilar
oscillation *n* GEN oscilación *f*; ~ **circuit** COMP, ELECT, POWER, RADIO circuito *m* oscilante; ~ **frequency** COMP, ELECT, POWER, RADIO frecuencia *f* de oscilación
oscillator *n* GEN oscilador *m*; ~ **frequency** COMP, ELECT, POWER, RADIO frecuencia *f* de oscilación
oscillatory *adj* GEN oscilatorio, *circuit* oscilante
oscillograph: ~ **tube** *n* ELECT oscilógrafo *m* catódico
oscilloscope *n* ELECT osciloscopio *m*; ~ **tube** ELECT tubo *m* de osciloscopio
OSF *abbr* (*operating system function*) COMP, DATA, TELEP, TEST FSO (*función del sistema operativo*)
OSI *abbr* (*open systems interconnection*) COMP, DATA, TELEP, TRANS OSI (*interconexión de sistemas abiertas*); ~ **environment** *n* COMP, DATA entorno *m* OSI; ~ **layer** *n* COMP, DATA capa *f* OSI; ~ **model** *n* COMP, DATA modelo *m* OSI; ~ **model layer** *n* COMP, DATA capa *f* de modelo OSI; ~ **network layer** *n* COMP, DATA, TELEP capa *f* de red OSI; ~ **network service** *n* COMP, DATA, TELEP servicio *m* de red OSI; ~ **resource** *n* DATA, TELEP, TRANS recurso *m* OSI; ~ **transport layer** *n* COMP, DATA, TELEP capa *f* de transporte OSI
Ostwald: ~ **colour** *BrE* **system** *n* COMP sistema *m* Ostwald de color
OTF *abbr* (*optimum traffic frequency*) RADIO OTF (*frecuencia óptima de tráfico*)
other: ~ **notification** *phr* (*ON*) DATA otra notificación (*ON*)
outage *n* TELEP estado *m* no operativo, TEST estado *m* no operativo, indisponibilidad *f*; ~ **time** TEST tiempo *m* de indisponibilidad
out: ~ **of date** *adj* GEN fuera de fecha; ~ **of order** *adj* TEST fuera de servicio; ~ **of range** *adj* RADIO fuera de alcance; ~ **of reach** *adj* RADIO fuera de alcance; ~ **of service** *adj* POWER fuera de servicio
outboard: ~ **grommet** *n* ELECT arandela *f* aislante exterior
outbound: ~ **communication** *n* RADIO, TRANS comunicación *f* saliente
outcome *n* GEN resultado *m*
outdoor: ~ **aerial** *n* RADIO antena *f* exterior
outer: ~ **conductor** *n* TELEP, TRANS conductor *m* exterior; ~ **diameter** *n* GEN diámetro *m* exterior; ~ **electron** *n* ELECT electrón *m* periférico; ~ **rim** *n* COMP borde *m* exterior; ~ **stanchion** *n* TELEP puntal *m* exterior; ~ **vapour** *BrE* **deposition** *n* PARTS deposición *f* de vapor exterior; ~ **vertical iron** *n* TELEP hierro *m* vertical exterior
outgoing *adj* (*o/g*) GEN saliente (*o/g*); ~ **access** *n* DATA acceso *m* de salida; ~ **call** *n* TELEP llamada *f* de salida; ~ **calls barred extension** *n* SWG, TELEP extensión *f* restringida para llamadas de salida; ~ **calls barred line** *n* TELEP línea *f* restringida para llamadas de salida; ~ **channel** *n* TRANS canal *m* de salida; ~ **circuit** *n* ELECT circuito *m* saliente; ~ **group** *n* TELEP grupo *m* saliente; ~ **international R2 register** *n* TELEP registro *m* R2 de salida internacional; ~ **line** *n* TELEP línea *f* de salida; ~ **matrix link** *n* TELEP enlace *m* de matriz de salida; ~ **message** *n* PER APP mensaje *m* saliente; ~ **only line** *n* TELEP línea *f* sólo de salida; ~ **procedure** *n* TELEP procedimiento *m* saliente; ~ **R2 register** *n* TELEP registro *m* R2 de salida; ~ **traffic** *n* TELEP tráfico *m* saliente; ~ **trunk** *n* SWG troncal *f* de salida, TELEP troncal *f* saliente; ~ **trunk circuit** *n* SWG, TELEP, TRANS circuito *m* de troncal de salida; ~ **trunk queuing** *n* SWG cola *f* en troncal de salida, TELEP cola *f* en troncal saliente; ~ **wave** *n* ELECT onda *f* emitida; ◆ ~ **calls barred** DATA, TELEP llamadas de salida bloqueadas
outgrowth *n* ELECT excrecencia *f*
outlet *n* GEN *for sales* salida *f*; ~ **access** SWG acceso *m* de salida
outline *n* GEN boceto *m*; ~ **agreement** GEN acuerdo *m* de boceto; ~ **drawing** ELECT croquis *m*; ~ **font** COMP fuente *f* tipográfica esquemática; ~ **test procedure** TEST procedimiento *m* esquemático de prueba
out-of-action *adj* TELEP fuera de servicio
out-of-band *adj* (*OOB*) ELECT, RADIO fuera de banda (*OOB*); ~ **emission** *n* RADIO emisión *f* fuera de banda; ~ **power** *n* RADIO potencia *f* fuera de banda; ~ **spectrum** *n* RADIO espectro *m* fuera de banda
out-of-channel: ~ **filtering** *n* TRANS filtración *f* fuera de canal
out-of-frame: ~ **alignment time** *n* TELEP tiempo *m* de alineación fuera de cuadro; ~ **second** *n* DATA segundo *m* fuera de cuadro
out-of-phase: ~ **signal** *n* GEN señal *f* de fuera de fase
out-of-sync: ~ **error** *n* TRANS error *m* de fuera de sincronismo
outperform *vt* COMP superar
outpulse *vi* SIG reemitir
output *n* COMP, ELECT, POWER *of a four terminal network*, TELEP potencia *f* de salida; ~ **admittance** ELECT, POWER admitancia *f* de salida; ~ **angle** ELECT, TRANS ángulo *m* de

salida; ~ **assignment** COMP asignación *f* de salida; ~ **back-off** (*OBO*) TRANS retroceso *m* de salida (*OBO*); ~ **capacitance** ELECT capacitancia *f* de salida; ~ **channel** COMP canal *m* de salida; ~ **characteristics** *n pl* ELECT características *f pl* de salida; ~ **circuit** ELECT circuito *m* de salida; ~ **current** ELECT corriente *f* de salida; ~ **data** *n pl* DATA datos *m pl* de salida; ~ **device** COMP dispositivo *m* de salida; ~ **divergence** ELECT divergencia *f* de salida; ~ **gate** ELECT puerta *f* de salida; ~ **impedance** COMP, ELECT impedancia *f* de salida; ~ **indicator** COMP indicador *m* de salida; ~ **jack** COMP, PARTS jack *m* de salida; ~ **level** COMP, ELECT nivel *m* de salida; ~**-level attenuation** COMP, ELECT atenuación *f* de nivel de salida; ~**-level attenuator** COMP, ELECT atenuador *m* de nivel de salida; ~ **load** POWER carga *f* de salida; ~ **outside dialogue** TELEP diálogo *m* exterior de salida; ~ **port** COMP, ELECT puerta *f* de salida; ~ **power** ELECT potencia *f* de salida; ~ **process** COMP proceso *m* de salida; ~ **program** COMP programa *m* de salida; ~ **pulse** COMP, ELECT impulso *m* de salida; ~ **row** ELECT fila *f* de salida; ~ **signal** COMP, ELECT señal *f* de salida; ~ **socket** COMP, ELECT toma *f* de salida; ~ **stage** ELECT, RADIO etapa *f* de salida; ~ **symbol** TELEP símbolo *m* de salida; ~ **terminal** POWER borne *f* de salida; ~ **transformer** POWER transformador *m* de salida; ~ **unit** COMP unidad *f* de salida; ~ **variable** CONTR variable *f* de salida; ~ **voltage** COMP, ELECT voltaje *m* de salida

outrigger *n* ELECT oreja *f* de anclaje

outside: ~ **broadcast** *n* RADIO, TV radioemisión *f* en el exterior; ~ **location** *n* TV emplazamiento *m* exterior; ~ **party** *n* TELEP parte *f* exterior; ~ **plant** *n* TRANS planta *f* exterior; ~ **view** *n* GEN vista *f* exterior

out-slot: ~ **signalling** *BrE n* SIG señalización *f* fuera del intervalo

oval *adj* GEN oval

ovality *n* GEN ovalidad *f*

overall *adj* GEN general; ~ **adjacent channel protection margin** *n* RADIO margen *m* global de protección de canal adyacente; ~ **braid screen** *n* ELECT blindaje *m* de malla total; ~ **carrier-to-interference ratio** *n* RADIO relación *f* total portadora-interferencia; ~ **co-channel protection margin** *n* RADIO margen *m* total de protección en el mismo canal; ~ **equivalent protection margin** *n* RADIO margen *m* total de protección equivalente; ~ **objective loudness rating** *n* TELEP régimen *m* global de sonoridad objetiva; ~ **pitch** *n* COMP tono *m* general; ~ **second adjacent channel protection margin** *n* RADIO margen *m* total de protección de segundo canal adyacente

overcoverage *n* RADIO cobertura *f* superpuesta

overcurrent *n* POWER, SWG sobrecorriente *f*; ~ **factor** POWER, SWG índice *m* de sobrecorriente; ~ **relay** PARTS relé *m* de sobrecorriente

overdeviation *n* RADIO sobredesviación *f*

overdub *n* COMP sobregrabación *f*; ~ **recording** COMP registro *m* de sobregrabación

overdubbing *n* COMP sobregrabación *f*

overexcitation *n* ELECT sobremodulación *f*

overexpose *vt* TV, DATA *scanner* sobreexponer

overfading *n* RADIO sobredesvanecimiento *m*

overflow *n* GEN desbordamiento *m*, sobrecarga *f*; ~ **accept** SWG, TELEP aceptación *f* de sobrecarga; ~ **check** COMP comprobación *f* de sobrecarga; ~ **indicator** COMP indicador *m* de desbordamiento; ~ **meter** TELEP contador *m* de sobrecarga; ~ **tone** TELEP tono *m* de sobrecarga; ~ **traffic** SWG, TELEP tráfico *m* de desbordamiento

overhang *n* ELECT bovedilla *f*

overhaul *n* GEN repaso *m*

overhead: ~ **access** *n* (*OHA*) DATA acceso *m* aéreo; ~ **bit** *n* DATA, TELEP bit *m* de rebose; ~ **cable** *n* POWER, TRANS cable *m* aéreo; ~ **cabling** *n* POWER, TRANS cableado *m* aéreo; ~ **line** *n* POWER, TRANS línea *f* aérea; ~ **network** *n* POWER, TRANS red *f* aérea; ~ **span** *n* POWER, TRANS amplitud *f* aérea

overlap *n* TELEP solapamiento *m*; ~ **angle** ELECT ángulo *m* de superposición; ~ **area** RADIO área *f* de superposición; ~ **operation** TELEP operación *f* superpuesta

overlapping *n* COMP, RADIO solape *m*

overlay *n* COMP capa *f* superpuesta

overload[1] *n* POWER, SWG sobrecarga *f*; ~ **channel** TRANS canal *m* de sobrecarga; ~ **class field** TELEP campo *m* de clase de sobrecarga; ~ **control** TELEP control *m* de sobrecarga; ~ **forward current** ELECT, POWER corriente *f* directa de sobrecarga; ~ **point** TELEP punto *m* de sobrecarga; ~ **protection** ELECT, POWER protección *f* contra sobrecarga

overload[2] *vt* POWER, SWG sobrecargar

overloaded *adj* POWER, SWG sobrecargado

overloading *n* POWER, SWG sobrecarga *f*

overmodulation *n* ELECT sobremodulación *f*

overpower: ~ **protection** *n* ELECT, POWER protección *f* contra sobrecarga; ~ **relay** *n* PARTS relé *m* de sobrecargas

overreaching *n* PARTS sobrealcance *m*

override[1]: ~ **circuit** *n* TEST circuito *m* de transferencia de mando; ~ **feature** *n* TELEP característica *f* de sobrecontrol; ~ **tone** *n* TELEP tono *m* de sobrecontrol

override[2] *vt* DATA, TELEP cancelar el efecto de

overriding *adj* GEN prioritario

overrun *vi* COMP excederse

overseas: ~ **network** *n* TRANS red *f* transoceánica

oversee *vt* TV inspeccionar

overshoot *n* TELEP sobrecresta *f*

overtone *n* COMP armónico *m* superior

overview: ~ **part** *n* DATA parte *f* de vista panorámica

overvoltage *n* ELECT, POWER sobretensión *f*; ~ **alarm** ELECT, POWER alarma *f* de sobretensión; ~ **protection** ELECT, POWER protección *f* contra sobretensión; ~ **protection unit** ELECT, POWER unidad *f* de protección contra sobretensión; ~ **relay** ELECT, PARTS, POWER relé *m* de sobretensión

OWF *abbr* (*optimum working frequency*) RADIO OWF (*frecuencia óptima de trabajo*)

own: ~**-exchange supervisory circuit** *n* SWG circuito *m* de supervisión de central propia; ~ **programme** *BrE n* TV programa *m* propio

oxide: ~ **capacitor** *n* ELECT, PARTS condensador *m* de óxido; ~**-coated cathode** *n* ELECT, PARTS cátodo *m* con capa de óxido

oxyacetylene: ~ **burner** *n* GEN quemador *m* oxiacetilénico

P

PA *abbr* (*power amplifier*) COMP, DATA, TRANS PA (*amplificador de potencia*)

PABX *abbr* (*private automatic branch exchange*) TELEP PABX (*centralita privada automática conectada a la red pública*); ~ **internal dial tone** *n* TELEP tono *m* de discado interno PABX

pace *n* GEN marcha *f*

pack *vt* GEN empaquetar

package *n* COMP, DATA paquete *m*

packaging *n* GEN *of services* envasado *m*

packet *n* DATA, TELEP paquete *m*; ~ **assembler-disassembler** (*PAD*) DATA, TELEP ensamblador-desensamblador *m* de paquetes (*PAD*); ~ **assembly-disassembly facility** DATA, TELEP instalación *f* de montaje-desmontaje de paquete de datos; ~ **data transmission network** DATA, TRANS red *f* de transmisión de paquetes de datos; ~ **delay** DATA, TRANS retardo *m* de paquete; ~ **disassembly** DATA, TELEFON desmontaje *m* de paquetes de datos; ~ **format** DATA, TELEP formato *m* del paquete de datos; ~ **handler** (*PH*) DATA manipulador *m* de paquetes de datos (*PH*); ~ **layer** DATA capa *f* de paquetes de datos, TELEP categorías *f* para la conmutación de paquetes; ~ **layer protocol** (*PLP*) DATA, TELEP protocolo *m* de la capa de paquetes (*PLP*); ~ **layer reference event** DATA, TELEP evento *m* de referencia de la capa de paquetes; **~-mode bearer service** DATA, TELEP servicio *m* portador en modos por paquete; **~-mode operation** DATA, TELEP operación *f* en modo paquetes; **~-mode terminal** DATA, TELEP terminal *m* en modo de paquetes; ~ **multiplexing** DATA, RADIO multiplexión *f* del paquete de datos; ~ **radio** RADIO radiocomunicación *f* por paquetes; ~ **satellite network** RADIO, TELEP red *f* de satélites en modo paquetes; ~ **sequencing** DATA, TELEP secuenciado *m* de paquetes; ~ **switch** APPL, DATA, SWG, TELEP, TRANS conmutador *m* de paquete; **~-switched bearer service** APPL, DATA, SWG, TELEP, TRANS servicio *m* portador con conmutación de paquetes; **~-switched data carrier** DATA, SWG, TELEP, TRANS portadora *f* de datos en conmutación por paquetes; **~-switched data transmission service** DATA, SWG, TELEP, TRANS servicio *m* de transmisión de datos conmutados por paquetes; ~ **switching** (*PS*) APPL, DATA, SWG, TELEP, TRANS conmutación *f* por paquetes; **~-switching exchange** (*PSE*) DATA, SWG, TELEP, TRANS central *f* de conmutación de paquetes; **~-switching network** DATA, SWG, TELEP, TRANS red *f* de conmutación de paquetes; **~-switching node** DATA, SWG, TELEP, TRANS nodo *m* de conmutación de paquetes; **~-switching processor** COMP, DATA, SWG, TELEP, TRANS procesador *m* de conmutación de paquetes; **~-switch network** DATA, SWG, TELEP, TRANS red *f* de conmutación de paquetes; ~ **transmission** DATA, SWG, TELEP, TRANS transmisión *f* por paquetes

packetizer: **~-depacketizer** *n* TRANS empaquetador-desempaquetador *m*

packing: ~ **density** *n* DATA, ELECT densidad *f* de grabación; ~ **fraction** *n* DATA fracción *f* de empaquetado

pad *n* GEN reductor *m*

PAD *abbr* (*packet assembler-disassembler*) DATA, TELEP PAD (*ensamblador-desensamblador de paquetes*)

padding *n* COMP, DATA relleno *m*

paddle *n* COMP paleta *f*; ~ **board** POWER cuadro *m* de paletas; **~-shaped beam** RADIO haz *m* en forma de paleta

page[1]: ~ **description language** *n* (*PDL*) COMP, DATA lenguaje *m* de descripción de página (*PDL*); ~ **frame** *n* COMP, DATA marco *m* de página; ~ **layout** *n* COMP, DATA presentación *f* de página; ~ **layout software** *n* COMP, DATA programa *m* de presentación en página; ~ **marker** *n* COMP, DATA marcador *m* de página; ~ **numbering** *n* COMP, DATA numeración *f* de páginas; ~ **printer** *n* COMP, DATA, TELEP impresora *f* de páginas

page[2] *vt* GEN, RADIO, TELEP paginar

pager *n* GEN localizador *m*

paginate *vt* GEN paginar

paging *n* COMP, DATA, GEN, RADIO, TELEP paginación *f*; ~ **channel** TRANS canal *m* de paginación; ~ **service** APPL, RADIO, TELEP servicio *m* de paginación; ~ **system** RADIO sistema *m* busca personas, TELEP sistema *m* de radiobúsqueda

pagination *n* GEN paginación *f*

paid: **~-service advice** *n* TELEP aviso *m* de servicio tasado
paint[1] *n* COMP pintura *f*
paint[2] *vt* COMP pintar
paintbrush *n* COMP, TEST pincel *m*
pair *n* GEN par *m*; **~ of complementary channels** TELEP par *m* de canales complementarios; **~ of conductors** GEN par *m* de conductores; **~ generation** PARTS generación *f* de pares; **~ production** PARTS producción *f* de pares
paired: **~ cable** *n* GEN cable *m* de pares; **~-disparity code** *n* TELEP código *m* de disparidad emparejada
pairing *n* TV solapado *m* de líneas
pairs: **in ~** *phr* GEN por pares
palette *n* DATA caja *f* de pinturas
palmtop: **~ computer** *n* COMP ordenador *m* agenda *Esp*, computador *m* agenda *AmL*, computadora *f* agenda *AmL*
PAM *abbr* (*pass-along message*) DATA transmisión *f* de mensaje; **~ network** *n* TELEP, TRANS red *f* con modulación de amplitud de impulsos
PAMA *abbr* (*pulse-address multiple access*) RADIO PAMA (*acceso múltiple de dirección de impulso*)
pan[1] *n* COMP cuba *f* de amalgamación; **~ control** COMP control *m* panorámico
pan[2] *vi* COMP panoramizar
panel *n* GEN panel *m*; **~ plating** ELECT metalización *f* total del panel; **~ system** SWG sistema *m* panel
panoramic: **~ space** *n* DATA espacio *m* panorámico
Pantone: **~ scale** *n* ELECT escala *f* de Pantone
paper: **~ capacitor** *n* POWER condensador *m* de papel; **~ directory** *n* TELEP directorio *m* de papel; **~ feed** *n* PER APP alimentación *f* de papel; **~-feed unit** *n* PER APP alimentador *m* de papel; **~ format** *n* PER APP formato *m* del papel; **~-insulated mass-impregnated cable** *n* TRANS cable *m* con aislamiento de papel impregnado; **~-insulated non-draining cable** *n* TRANS cable *m* de material no absorbente; **~ laminate** *n* PARTS laminado *m* de papel; **~-out switch** *n* PER APP indicador *m* de falta de papel; **~ size** *n* PER APP tamaño *m* del papel; **~ skip** *n* PER APP, TELEP salto *m* de papel; **~ throw** *n* PER APP, TELEP salto *m* de papel; **~ tray** *n* PER APP bandeja *f* de papel; **~ winder** *n* PER APP, TELEP arrollador *m* de papel; **~-winder bail** *n* PER APP, TELEP impulsor *m* del arrollador de papel
PAR *abbr* (*peak-to-average ratio*) GEN PAR (*relación cresta-promedio*)
parabola *n* RADIO, TV parábola *f*
parabolic: **~ aerial** *n* RADIO antena *f* parabólica; **~ antenna** *n* RADIO antena *f* parabólica; **~ cylinder antenna** *n* RADIO antena *f* de cilindro parabólico; **~ profile** *n* RADIO, TRANS perfil *m* parabólico; **~ reflector antenna** *n* RADIO antena *f* de reflector parabólico
paraboloid *n* RADIO paraboloide *m*
parallax: **~ error** *n* TEST error *m* de paralaje
parallel: **~ adder** *n* ELECT sumador *m* paralelo; **~ architecture** *n* TRANS arquitectura *f* en paralelo; **~ carry** *n* COMP arrastre *m* en paralelo; **~ circuit** *n* ELECT circuito *m* en paralelo; **~ clocking arrangement** *n* ELECT acuerdo *m* de sincronización en paralelo; **~ connection** *n* POWER conexión *f* en paralelo; **~ cooling** *n* CLIM enfriamiento *m* en paralelo; **~ displacement** *n* TRANS desplazamiento *m* en paralelo; **~ interface** *n* DATA interfaz *f* en paralelo; **~ jack** *n* ELECT, PARTS *connecting device* jack *m* en paralelo; **~ operation** *n* SWG operación *f* en paralelo; **~-plate lens** *n* RADIO lente *f* de placas paralelas; **~ pliers** *n pl* GEN alicates *m pl* paralelos; **~ port** *n* COMP puerto *m* paralelo; **~ processing** *n* DATA, ELECT procesamiento *m* en paralelo; **~ programming** *n* DATA programación *f* paralela; **~ resonance** *n* ELECT, POWER resonancia *f* paralela; **~ resonant circuit** *n* ELECT, POWER circuito *m* de resonancia paralelo; **~-to-serial converter** *n* ELECT, TELEP convertidor *m* paralelo-serie; **~ structure** *n* TEST estructura *f* paralela; **~ synchronous system** *n* COMP sistema *m* síncrono en paralelo; **~ tasking** *n* ELECT asignación *f* de tareas paralelas; **~-tee oscillator** *n* ELECT oscilador *m* de circuitos T en paralelo; **~ transmission** *n* COMP, TRANS transmisión *f* en paralelo; **~-working processors** *n pl* SWG procesadores *m pl* que trabajan en paralelo
parallel up *vt* POWER acoplar en derivación
parallelism *n* GEN paralelismo *m*
parameter *n* GEN parámetro *m*; **~ assign** COMP, TELEP asignación *f* de parámetro; **~ block** COMP, TELEP bloque *m* paramétrico; **~ block introduction sequence** COMP, TELEP secuencia *f* de introducción de bloque paramétrico; **~ block request indication** COMP, TELEP indicación *f* de petición de bloque paramétrico; **~-group identifier** (*PGI*) COMP, TELEP identificador *m* de grupo de parámetros (*PGI*); **~ identification** COMP, TELEP identificación *f* de parámetros; **~ length** COMP, TELEP longitud *f* de parámetro; **~**

name COMP, TELEP nombre *m* de parámetro; ~ **out-of-range class** DATA parámetro *m* de categoría fuera de rango; **~-name-defined** ~ COMP, TELEP parámetro *m* definido por nombre de parámetro *m*; ~ **profile** COMP, TELEP, TEST perfil *m* del parámetro; ~ **value** (*PV*) COMP, TELEP, TEST valor *m* de parámetro; ~ **word** COMP, DATA, TELEP palabra *f* de parámetro

parameterization *n* GEN parametrización *f*

parameterize *vt* GEN parametrizar

parameterized: ~ **abstract test case** *n* DATA, TEST caso *m* de prueba abstracta parametrizada; ~ **abstract test suite** *n* (*PATS*) DATA, TEST sucesión *f* de pruebas abstractas parametrizadas (*PATS*); ~ **executable test case** *n* DATA, TEST caso *m* de prueba ejecutable parametrizada; ~ **executable test suite** *n* (*PETS*) DATA, TEST sucesión *f* de pruebas ejecutables parametrizadas (*PETS*)

parametric: ~ **amplifier** *n* (*paramp*) TRANS amplificador *m* paramétrico; ~ **oscillator** *n* ELECT oscilador *m* paramétrico; ~ **programming** *n* COMP programación *f* paramétrica; ~ **test** *n* TEST prueba *f* paramétrica

paramp *abbr* (*parametric amplifier*) TRANS amplificador *m* paramétrico

paraphase: ~ **amplifier** *n* ELECT amplificador *m* parafásico

parasitic: ~ **coupling** *n* ELECT acoplamiento *m* parásito; ~ **element** *n* RADIO elemento *m* parásito; ~ **error** *n* TEST error *m* parásito; ~ **oscillation** *n* ELECT oscilación *f* parásita

paraxial: ~ **ray** *n* ELECT, TRANS rayo *m* paraxial

parent *n* COMP, DATA padre *m*; ~ **company** GEN casa *f* matriz; ~ **entry** DATA entrada *f* de origen

Pareto: ~ **analysis** *n* TEST análisis *m* de Pareto

parity *n* DATA, ELECT paridad *f*; ~ **bit** DATA, RADIO bit *m* de paridad; ~ **check** DATA, TELEP verificación *f* de paridad; ~ **digit** COMP dígito *m* de paridad; ~ **function** TELEP función *f* de paridad; **~-generating circuit** SWG circuito *m* generador de paridad

park *vt* SWG, TELEP *call* aparcar

parked *adj* SWG, TELEP *call* aparcado

parking: ~ **orbit** *n* RADIO órbita *f* de aparcamiento, órbita *f* temporal

part *n* GEN pieza *f*; ~ **sample** TEST muestra *f* parcial; **~-time leased circuit** TRANS circuito *m* arrendado a tiempo parcial

partial *adj* COMP, DATA, TELEP parcial; ~ **attribute request** *n* DATA petición *f* de atributo parcial; ~ **break-in** *n* TELEP irrupción *f* parcial; ~ **break-in echo suppressor** *n* RADIO, TELEP supresor *m* de eco de irrupción parcial; ~ **break-in operate time** *n* RADIO, TELEP tiempo *m* de operación de irrupción parcial; ~ **call barring** *n* TELEP bloqueo *m* parcial de llamada; ~ **coherence** *n* TRANS coherencia *f* parcial; ~ **dial** *n* TELEP marcado *m* parcial; ~ **failure** *n* TEST fallo *m* parcial; **~-page request** *n* DATA, TELEP petición *f* de página parcial; **~-page signal EOM** *n* DATA, TELEP señal *f* EOM de página parcial; **~-page signal EOP** *n* DATA, TELEP señal *f* EOP de página parcial; **~-page signal MPS** *n* DATA, TELEP señal *f* MPS de página parcial; **~-page signal NULL** *n* DATA, TELEP señal *f* NULL de página parcial; **~-page signal PRI-EOM** *n* DATA, TELEP señal *f* PRI-EOM de página parcial; **~-page signal PRI-EOP** *n* DATA, TELEP señal *f* PRI-EOP de página parcial; **~-page signal PRI-MPS** *n* DATA, TELEP señal *f* PRI-MPS de página parcial; ~ **redundance** *n* TEST redundancia *f* parcial; ~ **refund** *n* TELEP reembolso *m* parcial; ~ **response code** *n* DATA código *m* de respuesta parcial; ~ **restoring time** *n* TRANS tiempo *m* de recuperación parcial; ~ **test** *n* TEST prueba *f* parcial; ~ **tone reversal** *n* TELEP inversión *f* parcial de matices

partially: ~ **suppressed** *adj* TV parcialmente suprimido

participant *n* GEN participante *m*

participate *vi* GEN participar

particle *n* GEN partícula *f*; ~ **board** PARTS tablero *m* de aglomerado; ~ **radiation** GEN radiación *f* de partículas; ~ **shredding** COMP fragmentación *f* de partículas; ~ **velocity** TELEP velocidad *f* de una partícula

partition *n* COMP, ELECT, TRANS partición *f*; ~ **noise** ELECT ruido *m* de partición

partitioning *n* COMP, ELECT, TRANS *of channel* partición *f*

partnership *n* GEN sociedad *f*

party: ~ **address** *n* DATA dirección *f* compartida; ~ **line** *n* TELEP línea *f* compartida

PASC *abbr* (*precision adaptive sub-band coding*) COMP PASC (*codificación de subbanda ajustable de precisión*)

pass *n* RADIO *of satellite* paso *m*; **~-along message** (*PAM*) DATA transmisión *f* de mensaje; ~ **band** COMP, RADIO, TELEP, TRANS paso *m* banda; **~-band attenuation** COMP, RADIO, TELEP, TRANS atenuación *f* de banda de paso; ~ **transistor** ELECT transistor *m* de paso; ~ **verdict** DATA decisión *f* de paso

passing: ~ **contact** *n* PARTS contacto *m* de impulsión

pass on *vt* RADIO, TRANS, TV trasladar

passivated: ~ **surface** *n* PARTS superficie *f* pasivada

passivation *n* PARTS pasivación *f*

passive: ~ **alerting** *n* RADIO alerta *f* pasiva; ~ **bus** *n* GEN bus *m* pasivo; ~ **circuit** *n* ELECT circuito *m* pasivo; ~ **component** *n* ELECT, PARTS componente *m* pasivo; ~ **double star** *n* TRANS doble estrella *f* pasiva; ~ **filter** *n* ELECT filtro *m* pasivo; ~ **four-terminal network** *n* ELECT, POWER, TRANS red *f* pasiva de cuatro terminales; ~ **intermodulation** *n* (*PIM*) TRANS intermodulación *f* pasiva (*PIM*); ~ **maintenance** *n* TEST mantenimiento *m* pasivo; ~ **matrix** *n* COMP matriz *f* pasiva; ~ **network** *n* POWER red *f* pasiva; ~ **optical network** *n* (*PON*) TELEP, TRANS red *f* óptica pasiva; ~ **sensor** *n* RADIO sensor *m* pasivo; ~ **threat** *n* DATA amenaza *f* pasiva; ~ **transducer** *n* POWER transductor *m* pasivo

password *n* COMP, DATA, PER APP *of answering machine* contraseña *f*

past: ~ **reference picture** *n* COMP, TV imagen *f* de referencia pasada

paste *vt* COMP pegar

patch[1]: ~ **board** *n* RADIO, TELEP panel *m* de acoplamiento; ~ **cord** *n* ELECT, TRANS cable *m* de conexión; ~ **panel** *n* RADIO, TELEP panel *m* de acoplamiento

patch[2] *vt* COMP conectar

patching *n* ELECT remiendo *m*

patent: ~ **pending** *phr* GEN patente pendiente

path *n* COMP, ELECT, RADIO, SWG, TELEP, TRANS vía *f*; ~ **allocation** SWG adjudicación *f* de trayectoria; ~ **allocation program** TELEP programa *m* de adjudicación de trayectoria; ~ **attenuation** RADIO, TELEP atenuación *f* de trayectoria; ~ **memory** SWG memoria *f* de la vía de transmisión; ~ **name** COMP nombre *m* de trayectoria; ~ **overhead** SWG, TELEP vía *f* de transmisión aérea

patrol: ~ **inspection** *n* TEST inspección *f* volante

PATS *abbr* (*parameterized abstract test suite*) DATA, TEST PATS (*sucesión de pruebas abstractas parametrizadas*)

pattern *n* APPL *of artificial intelligence*, COMP, ELECT, RADIO, TEST, TRANS estructura *f*; ~ **correspondence index** SWG índice *m* de correspondencia de formas; ~ **generator** TV generador *m* de imagen de prueba; ~ **mapping** COMP encuadre *m* de modelo; ~ **matching** APPL *of artificial intelligence*, COMP concordancia *f* con un modelo; ~ **plating** ELECT metalización *f* selectiva; ~ **recognition** APPL, CONTR, DATA, TRANS reconocimiento *m* de patrones

pause: ~ **instruction** *n* COMP instrucción *f* de pausa

pawl *n* SWG retén *m*

pay: ~-**by-use basis** *n* TV base *f* de pago por uso; ~ **channel** *n* TV canal *m* de pago; ~ **console** *n* PER APP impresora *f* de nóminas; ~-**per-view** *n* TV pago *m* por vista; ~-**per-view television** *n* TV televisión *m* por abono; ~ **station** *n* PER, TELEP teléfono *m* público; ~ **telephone** *n* PER, TELEP teléfono *m* monedero; ~ **television** *n* TV televisión *f* para abonados; ~ **television network** *n* TV red *f* de televisión de pago; ~ **tone** *n* TELEP tono *m* de pago

pay back *vt* GEN devolver, reembolsar

payload *n* GEN carga *f* útil

payment: ~ **by telephone** *n* APPL, TELEP pago *m* por teléfono; ~ **option analysis system** *n* (*POAS*) APPL sistema *m* de análisis de opción de pago (*POAS*)

payphone *n* PER APP, TELEP teléfono *m* monedero; ~ **recognition tone** TELEP tono *m* de reconocimiento de teléfono monedero; ~ **service** APPL, TELEP servicio *m* de teléfono monedero

PBC *abbr* (*programme booking centre*) TV PBC (*centro de reserva de programas*)

PBX *abbr* (*private branch exchange*) TELEP PBX (*central telefónica privada*); ~ **call** *n* TELEP llamada *f* PBX; ~ **failure transfer** *n* (*PFT*) TELEP transferencia *f* de fallo en PBX (*PFT*); ~ **group** *n* TELEP grupo *m* PBX; ~ **hunting** *n* TELEP búsqueda *f* por PBX; ~ **line-hunting service** *n* SWG, TELEP servicio *m* de captación de línea PBX; ~ **subscriber** *n* TELEP abonado *m* a PBX; ~ **switchboard** *n* TELEP tablero *m* conmutador de la centralita privada; ~ **trunk circuit** *n* TELEP circuito *m* troncal de PBX

PC *abbr* (*personal computer*) COMP PC (*ordenador personal*); ~ **support engineer** *n* GEN técnico *m* de servicio de ordenadores personales

PCB *abbr* (*printed circuit board*) ELECT, PARTS PCI (*placa de circuito impreso*)

PCD *abbr* (*photo compact disc*) COMP, DATA PCD (*disco compacto fotográfico*); ~ **writer** *n* COMP, DATA editor *m* de textos PCD

PCI *abbr* (*peripheral component interconnect*) COMP interconexión *f* de componente periférico

PCIS: ~ **proforma** *n* DATA proforma *f* PCIS

PCL *abbr* (*print command language*) COMP, DATA PCL (*lenguaje de órdenes de imprimir*)

PCM *abbr* (*pulse-code modulation*) COMP, DATA, TELEP, TRANS PCM (*modulación por impulsos codificados*); ~ **binary code** *n* DATA, ELECT,

TELEP código *m* binario PCM; ~ **digital reference sequence** *n* TELEP secuencia *f* de referencia digital PCM; ~ **link** *n* TELEP enlace *m* por PCM; ~ **link control software machine** *n* TELEP ordenador *m* con programas de control del enlace por PCM *Esp*, computador *m* con programas de control del enlace por PCM *AmL*, computadora *f* con programas de control del enlace por PCM *AmL*; ~ **multiplex equipment** *n* TELEP equipo *m* múltiplex de PCM, TRANS equipo *m* múltiplex PCM; ~ **switching system** *n* TELEP sistema *m* de conmutación PCM; ~ **system** *n* TELEP sistema *m* PCM; ~ **waveform** *n* COMP forma *f* de onda PCM

PCN *abbr* (*personal communication network*) DATA, RADIO PCN (*red de comunicación personal*)

PCO *abbr* (*point of control and observation*) DATA PCO (*punto de control y observación*)

PCS *abbr* (*personal communications services*) APPL PCS (*servicios de comunicaciones personales*); ~**-fibre** *BrE n* DATA, PARTS, TRANS fibra *f* PCS

PCTR *abbr* (*protocol conformance test report*) DATA, TEST PCTR (*informe de prueba de conformidad de protocolo*)

PDF *abbr* (*probability density function*) DATA PDF (*función de densidad de probabilidad*)

PDL *abbr* (*page description language*) COMP, DATA PDL (*lenguaje de descripción de página*)

PDM *abbr* (*pulse duration modulation*) DATA, ELECT, RADIO, TELEP, TRANS PDM (*modulación de impulsos en duración*)

PDN *abbr* (*public data network*) DATA, TELEP, TRANS RPD (*red pública de datos*)

PDU *abbr* (*protocol data unit*) DATA unidad *f* de datos de protocolo

PE *abbr* COMP (*pictorial element*) PE (*elemento pictórico*), TELEP (*peripheral exchange*) CP (*central periférica*), (*primary exchange*) CP (*central primaria*), (*public exchange*) CP (*central pública*)

peak *n* GEN punto *m* máximo; ~ **amplitude** TELEP *of elementary echo* amplitud *f* máxima; ~**-to-average ratio** (*PAR*) GEN relación *f* cresta-promedio (*PAR*); ~ **detector** GEN detector *m* de cresta; ~ **envelope power** (*PEP*) POWER pico *m* de potencia de la envolvente (*PEP*); ~ **forward-anode voltage** ELECT tensión *f* anódica directa de cresta; ~ **frequency deviation** RADIO, TRANS desviación *f* de frecuencia máxima; ~ **hour** TELEP hora *f* de máxima audiencia; ~**-intensity wavelength** RADIO, TRANS longitud *f* de onda de intensidad máxima; ~ **inverse voltage** (*PIV*) POWER cresta *f* de tensión inversa (*PIV*), TELEP tensión *f* máxima inversa (*PRV*); ~ **limiting** TELEP, TRANS limitación *f* de cresta; ~ **negative-anode voltage** ELECT tensión *f* de ánodo negativa de pico; ~**-to-~ amplitude** GEN amplitud *f* de cresta a cresta; ~ **point** PARTS punta *f* de pico; ~ **point current** PARTS, POWER tensión *f* de pico; ~ **point voltage** PARTS, POWER voltaje *m* de pico; ~ **power** POWER pico *m* máximo de potencia; ~**-pulse output power** ELECT potencia *f* de salida de impulso de pico; ~ **rate** GEN tasa *f* máxima; ~ **reading instrument** TEST instrumento *m* indicador de cresta; ~ **reverse voltage** (*PRV*) POWER, TELEP tensión *f* máxima inversa (*PRV*); ~ **traffic** TELEP cresta *f* de tráfico; ~ **value** GEN valor *m* máximo; ~ **voltage** POWER voltaje *m* máximo; ~ **voltmeter** TEST voltímetro *m* de cresta

peaked: ~ **traffic** *n* TELEP tráfico *m* de cresta

peakedness: ~ **factor** *n* TELEP factor *m* de agudeza

peaking: ~ **coil** *n* POWER bobina *f* correctora

pedal *n* COMP pedal *m*

Pedersen: ~ **ray** *n* RADIO rayo *m* Pedersen

peer *n* COMP, DATA par *m*; ~ **entity** DATA, RADIO entidad *f* par; ~**-entity authentication** DATA autentificación *f* por entidades de par; ~**-identity authentication** DATA autentificación *f* de identidad de par; ~**-to-~ logical relationship** TRANS relación *f* lógica de par a par; ~**-to-~ protocol** TRANS protocolo *m* de par a par; ~ **relationships** *n pl* DATA relaciones *f pl* entre pares

pel *n* COMP, ELECT, RADIO, TELEP, TV punto *m* de imagen; ~ **aspect ratio** COMP, ELECT, RADIO, TELEP, TV formato *m* del punto de imagen

pen *n* COMP estilete *m*

pencil: ~ **beam** *n* RADIO haz *m* en pincel; ~**-beam aerial** *n* RADIO antena *f* de haz en pincel; ~**-beam antenna** *n* RADIO antena *f* de haz en pincel

penalty *n* TRANS penalización *f*

penetrability *n* GEN penetrabilidad *f*

penetrable *adj* GEN penetrable

penetration: ~ **depth** *n* RADIO profundidad *f* de penetración; ~ **rate** *n* TV tasa *f* de penetración

pentode *n* ELECT péntodo *m*

PEP *abbr* (*peak envelope power*) POWER PEP (*pico de potencia de la envolvente*)

per: **on a ~-call basis** *adv* RADIO por llamadas

perceive *vt* GEN percibir

perceived: ~ **noise level** *n* TELEP nivel *m* de ruido percibido

percentage *n* TEST *of nonconforming items* porcentaje *m*; ~ **overflow** TELEP desbordamiento *m* porcentual; ~ **relay** PARTS relé *m* de porcentaje
percentile *n* GEN percentil *m*
perfect *adj* TELEP *modulation, restitution, signal* perfecto
perforated: ~ **tape** *n* TELEP cinta *f* perforada
perforation *n* GEN perforación *f*
perform *vt* COMP ejecutar
performance *n* GEN funcionamiento *m*; ~ **cartridge** COMP cartucho *m* de prestaciones; ~ **characteristic** TEST característica *f* de rendimiento; ~ **combination** COMP combinación *f* de prestaciones; ~ **concept** TEST concepto *m* relativo a las prestaciones; ~ **data** *n pl* COMP, DATA datos *m pl* sobre la calidad de servicio; ~ **effect** COMP efecto *m* del funcionamiento; ~ **index** CONTR nivel *m* de rendimiento; ~ **memory** COMP memoria *f* de funcionamiento; ~ **monitoring** TELEP supervisión *f* del funcionamiento; ~ **name** DATA nombre *m* funcional; ~ **objective** TELEP objetivo *m* de rendimiento, TEST norma *f* de rendimiento; ~ **parameter** COMP, TEST parámetro *m* de funcionamiento; ~ **test** TEST prueba *f* de funcionamiento
performer *n* DATA ejecutante *m*
performing: ~ **CMISE service user** *n* APPL, DATA usuario *m* del servicio de calidad de transmisión de CMISE
periapsis *n* RADIO periapsis *f*
perigee *n* RADIO perigeo *m*
period *n* GEN periodo *m*; ~ **of revolution** RADIO periodo *m* de revolución
periodic *adj* GEN periódico; ~ **frame** *n* COMP imagen *f* intermitente; ~ **quantity** *n* GEN cantidad *f* periódica; ~ **signal** *n* ELECT señal *f* periódica; ~ **systematic sampling** *n* TEST muestreo *m* sistemático periódico
peripheral *n* GEN periférico *m*; ~ **card** COMP tarjeta *f* de periférico; ~ **component interconnect** (*PCI*) COMP interconexión *f* de componente periférico; ~ **controller** COMP, DATA, ELECT controlador *m* periférico; ~ **device** COMP, DATA, PER APP dispositivo *m* periférico; ~ **device sharing** COMP, DATA, PER APP compartición *f* de dispositivo periférico; ~ **equipment** GEN equipo *m* periférico; ~ **exchange** TELEP central *f* periférica (*CP*); ~ **management** PER APP gestión *f* periférica; ~ **module** PER APP módulo *m* periférico; ~ **processor** (*PP*) COMP, PER APP procesador *m* periférico; ~ **telephony** TELEP telefonía *f* periférica
periscope: ~ **antenna** *n* RADIO antena *f* periscópica
peritelephony *n* TELEP peritelefonía *f*
permanent *adj* GEN permanente; ~ **circuit** *n* TELEP circuito *m* permanente; ~ **condition** *n* TELEP estado *m* permanente; ~ **connection** *n* DATA conexión *f* permanente; ~ **echo** *n* RADIO eco *m* estacionario, eco *m* permanente; ~ **failure** *n* TELEP avería *f* permanente; ~ **line** *n* TELEP línea *f* permanente; ~ **link** *n* GEN enlace *m* permanente; ~ **magnet** *n* ELECT, PARTS imán *m* permanente; ~ **signal** *n* GEN, TELEP, TRANS señal *f* permanente; ~**-signal trunk group** *n* TRANS grupo *m* troncal de señal permanente; ~ **subscriber number** *n* TELEP número *m* de abonado permanente; ~ **virtual circuit** *n* DATA, TRANS circuito *m* virtual permanente; ~ **wiring** *n* TELEP instalación *f* definitiva
permanently: ~ **connected** *adj* GEN conectado permanentemente
permeability *n* PARTS permeabilidad *f*; ~ **tuning** RADIO sintonía *f* por permeabilidad
permeable *adj* PARTS permeable
permeameter *n* TEST permeámetro *m*
permeance *n* ELECT permeación *f*
permissible: ~ **interference** *n* RADIO, TRANS interferencia *f* admisible; ~ **out-of-band power** *n* RADIO potencia *f* admisible fuera de banda; ~ **out-of-band spectrum** *n* RADIO espectro *m* admisible fuera de banda
permissive: ~ **circuit** *n* DATA circuito *m* permisivo
permittivity *n* ELECT permitividad *f*
perpendicular *adj* GEN perpendicular
person: ~**-to-**~ **call** *n* TELEP llamada *f* de persona a persona
personal: ~ **call** *n* TELEP comunicación *f* personal; ~ **communication network** *n* (*PCN*) DATA, RADIO red *f* de comunicación personal (*PCN*); ~ **communications** *n pl* TRANS comunicaciones *f pl* personales; ~ **communications services** *n pl* (*PCS*) APPL servicios *m pl* de comunicaciones personales (*PCS*); ~ **computer** *n* (*PC*) COMP ordenador *m* personal *Esp* (*PC*), computador *m* personal *AmL* (*PC*), computadora *f* personal *AmL* (*PC*); ~ **data** *n pl* GEN datos *m pl* personales; ~ **digital assistant** *n* COMP auxiliar *m* personal digital; ~ **icon** *n* DATA icono *m* personal; ~ **identity number** *n* GEN número *m* personal de identidad; ~ **telephone** *n* TELEP teléfono *m* personal
personalized: ~ **bookmark** *n* COMP, DATA señalador *m* personalizado
perturbation: ~ **technique** *n* TRANS técnica *f* de

perturbación; ~ **voltage** *n* TEST tensión *f* de perturbación

perturbing: ~ **effect** *n* GEN efecto *m* perturbador; ~ **signal** *n* POWER señal *f* perturbadora

perveance *n* ELECT perveancia *f*

PETS *abbr* (*parameterized executable test suite*) DATA, TEST PETS (*sucesión de pruebas ejecutables parametrizadas*)

PFM *abbr* (*pulse frequency modulation*) TELEP modulación *f* de frecuencia de impulsos

PFN *abbr* (*pulse-forming network*) ELECT PFN (*red de formación de impulsos*)

P-fractile: ~ **life** *n* TEST vida *f* del fractil de P

PFT *abbr* (*PBX failure transfer*) TELEP PFT (*transferencia de fallo en PBX*)

PGI *abbr* (*parameter group identifier*) COMP PGI (*identificador de grupo de parámetros*)

PH *abbr* (*packet handler*) DATA PH (*manipulador de paquetes de datos*)

phantastron *n* ELECT fantastrón *m*

phantom: ~ **circuit** *n* TELEP, TRANS circuito *m* fantasma; ~ **level** *n* TRANS nivel *m* fantasma; ~ **telegraph circuit** *n* TELEP circuito *m* telegráfico fantasma; ~ **voltage** *n* ELECT tensión *f* fantasma

phantoming *n* TELEP, TRANS fantomización *f*

phase *n* GEN fase *f*; ~ **advance** TRANS avance *m* de fase; ~ **alignment** ELECT alineamiento *m* de fase; ~ **alternate line** TV línea *f* con alternación de fase; ~ **ambiguity resolution** TRANS resolución *f* de ambigüedad de fase; ~**-amplitude characteristic** ELECT, TRANS característica *f* de amplitud de fase; ~**-amplitude distortion** ELECT, TRANS distorsión *f* de amplitud de fase; ~ **angle** COMP, POWER ángulo *m* de fase; ~**-black** TELEP fase *f* de negro; ~ **change** ELECT, POWER, RADIO, SIG, TRANS cambio *m* de fase; ~**-change coefficient** GEN coeficiente *m* de cambio de fase; ~**-change disk** DATA disco *m* de cambio de fase; ~**-change printer** DATA impresora *f* de cambio de fase; ~ **changer** ELECT, SIG cambiador *m* de fase; ~**-change velocity** ELECT velocidad *f* de cambio de fase; ~ **coefficient** GEN coeficiente *m* de fase; ~ **coherence** RADIO, TRANS coherencia *f* de fase; ~ **comparator relay** PARTS relé *m* de comparación de fase; ~ **compensation** ELECT, POWER compensación *f* de fase; ~ **conductor** POWER conductor *m* de fase; ~ **constant** GEN constante *f* de fase; ~ **control** ELECT control *m* de fase; ~**-corrected echo** TELEP eco *m* con corrección de fase; ~**-corrected horn** RADIO bocina *f* con corrección de fase; ~ **delay** GEN retardo *m* de fase; ~ **demodulation** ELECT desmodulación *f* de fase; ~ **detector** ELECT detector *m* de fase; ~ **deviation** GEN desviación *f* de fase; ~ **difference** ELECT, POWER diferencia *f* de fase; ~ **displacement** ELECT, POWER defase *f*; ~ **distortion** ELECT, TRANS distorsión *f* de fase; ~ **encoding** DATA codificación *f* de fase; ~ **equalization** GEN compensación *f* de fase; ~ **equalizer** GEN compensador *m* de fase; ~**-frequency distortion** ELECT, TRANS distorsión *f* de fase de frecuencia; ~ **hopping** RADIO salto *m* de fase; ~ **inversion** POWER, TELEP inversión *f* de fase; ~**-inversion modulation** POWER, TELEP modulación *f* por inversión de fase; ~ **inverter** POWER, TELEP inversor *m* de fase; ~**-locked filter** ELECT filtro *m* de fase sincronizada; ~**-locked loop** (*PLL*) ELECT lazo *m* de sincronización de fase; ~ **locking** ELECT sincronización *f* de fase; ~ **lock unit** (*PLU*) ELECT unidad *f* de sincronización de fase (*PLU*); ~ **meter** POWER medidor *m* de fase; ~ **modulation** (*PM*) GEN modulación *f* de fase (*MF*); ~**-modulation recording** DATA grabación *f* por modulación de fase; ~ **modulator** TRANS modulador *m* de fase; ~ **monitor** POWER monitor *m* de fase; ~ **plot** ELECT gráfico *m* de fase; ~ **regulation** ELECT regulación *f* de fase; ~ **response** TRANS respuesta *f* de fase; ~ **sensor** TRANS sensor *m* de fase; ~ **sequence** POWER secuencia *f* de fases; ~**-sequence indicator** POWER indicador *m* de secuencia de fases; ~**-sequence test** POWER prueba *f* de secuencia de fase; ~**-shaped QPSK** RADIO, SIG transmisión *f* por desplazamiento de fase cuaternaria; ~**-shaped quaternary ~-shift keying** RADIO, SIG manipulación *f* por secuencia de fase de cuaternaria de haz perfilado; ~ **shift** ELECT, POWER, RADIO, SIG, TRANS desplazamiento *m* de fase, desfasaje *m*; ~**-shift distortion** ELECT, POWER, RADIO, SIG, TRANS distorsión *f* de desplazamiento de fase; ~ **shifter** ELECT, POWER, RADIO, SIG, TRANS desfasador *m*; ~**-shifter tube** ELECT, POWER, RADIO, SIG, TRANS tubo *m* desfasador; ~**-shifting circuit** ELECT, POWER, RADIO, SIG, TRANS circuito *m* desfasador; ~**-shifting effect** COMP, ELECT, POWER, RADIO, SIG, TRANS efecto *m* de desfase; ~**-shift keying** (*PSK*) ELECT, POWER, RADIO, SIG, TRANS conmutación *f* de desplazamiento de fase (*CDF*); ~**-shift keying 2 modulation** ELECT, POWER, RADIO, SIG, TRANS modulación *f* 2 de manipulación por desplazamiento de fase; ~**-shift oscillator** ELECT, POWER, RADIO, SIG, TRANS oscilador *m* de variación de fase; ~ **speed** RADIO velocidad *f* de fase; ~ **stability** ELECT estabilidad *f* de fase; ~ **tuning** ELECT sintonización *f* de fase; ~ **velocity** RADIO velocidad *f* de fase; ~ **voltage** POWER tensión *f* de fase; ~-**white** TELEP puesta *f* en fase de blanco; ~ **winding** PARTS devanado *m* de fase

phase in *vt* GEN, RADIO introducir progresivamente

phase out *vt* GEN eliminar progresivamente

phased: ~-**array antenna** *n* RADIO antena *f* de elementos en fase

phasing *n* TV fasaje *m*; ~-**in** RADIO introducción *f* progresiva; ~-**out** RADIO eliminación *f* progresiva; ~ **signal** TELEP señal *f* de fase; ~-**type effect** COMP efecto *m* de tipo ajuste

phenolic: ~ **plastic** *n* PARTS plástico *m* fenólico

phon *n* TELEP fonio *m*

phone[1]: ~ **box** *BrE n* (*cf phone booth AmE*) TELEP cabina *f* telefónica; ~ **booth** *AmE n* (*cf phone box BrE*) TELEP cabina *f* telefónica; ~ **call** *n* TELEP llamada *f* telefónica; ~ **conference** *n* APPL, TELEP conferencia *f* telefónica; ~ **conferencing** *n* APPL, TELEP conferencia *f* telefónica; ~-**home feature** *n* DATA, TELEP dispositivo *m* para llamar a casa; ~-**in poll** *n* TELEP registro *m* telefónico; ~-**in programme** *BrE n* RADIO, TELEP programa *f* de llamadas; ~ **patch** *n* RADIO, TELEP acoplador *m* telefónico

phone[2] *vt* TELEP llamar por teléfono

phone back *vi* TELEP devolver la llamada

phonecard *n* DATA, TELEP tarjeta *f* telefónica

phones: ~ **jack** *n* PER APP cuadro *m* de audífonos

phonetic: ~ **alphabet** *n* RADIO alfabeto *m* fonético

Phong: ~ **shading** *n* COMP sombreado *m* Phong

phonic: ~ **motor** *n* RADIO motor *m* fónico

phoning: ~ **comfort** *n* RADIO, TELEP comodidad *f* telefónica

phosphor: ~ **dot** *n* COMP punto *m* fosforescente

phosphorescence *n* ELECT, TV fosforescencia *f*

photo: ~ **CD** *n* COMP, DATA disco *m* compacto fotográfico (*PCD*); ~-**CD access** *n* COMP, DATA acceso *m* a disco compacto fotográfico; ~-**CD access development toolkit** *n* COMP, DATA utillaje *m* para revelado de disco compacto fotográfico; ~-**CD catalogue** *BrE n* COMP, DATA catálogo *m* de disco compacto fotográfico; ~-**CD master** *n* COMP, DATA negativo *m* de disco compacto fotográfico; ~-**CD player** *n* COMP, DATA reproductor *m* de discos compactos fotográficos; ~-**CD portfolio** *n* COMP, DATA álbum *m* de discos compactos fotográficos; ~-**CD writer** *n* COMP, DATA impresor *m* de disco compacto fotográfico; ~ **compact disc** *n* (*PCD*) COMP, DATA disco *m* compacto fotográfico (*PCD*); ~ **imaging workstation** *n* COMP estación *f* de trabajo de imágenes fotográficas

photocathode *n* ELECT fotocátodo *m*

photocomposition *n* TV fotocomposición *f*

photoconductive: ~ **camera tube** *n* TV tubo *m* de cámara fotoconductor; ~ **cell** *n* ELECT célula *f* fotoconductiva; ~ **effect** *n* ELECT efecto *m* fotoconductor

photoconductivity *n* ELECT, TRANS fotoconductividad *f*

photocoupler *n* ELECT fotoacoplador *m*

photocurrent *n* ELECT, POWER, TRANS corriente *f* fotoeléctrica

photodetector *n* TRANS fotodetector *m*

photodiode *n* DATA, PARTS, TRANS fotodiodo *m*

photoelectric *adj* ELECT fotoeléctrico; ~ **cell** *n* ELECT célula *f* fotoeléctrica; ~ **current** *n* ELECT, POWER, TRANS corriente *f* fotoeléctrica; ~ **effect** *n* ELECT, TRANS efecto *m* fotoeléctrico; ~ **emission** *n* ELECT, TRANS emisión *f* fotoeléctrica

photoemission *n* ELECT, TRANS fotoemisión *f*

photoemissive: ~ **camera tube** *n* ELECT tubo *m* de cámara fotoemisor; ~ **effect** *n* ELECT, TRANS efecto *m* fotoemisivo

photograph: ~ **facsimile telegram** *n* TELEP telegrama *m* facsimilar fotográfico; ~ **facsimile telegraphy** *n* TELEP telegrafía *f* facsimilar fotográfica

photographic: ~ **coding** *n* COMP, TELEP codificación *f* fotográfica

photoisolator *n* ELECT fotoaislante *m*

photolithography *n* COMP fotolitografía *f*

photoluminescence *n* TRANS fotoluminescencia *f*

photomask *n* ELECT fotomáscara *f*

photomultiplier *n* ELECT fotomultiplicador *m*

photon *n* TRANS fotón *m*; ~ **noise** ELECT, TRANS ruido *m* de fotones

photonics *n* TRANS fotónica *f*

photorealistic *adj* COMP fotorrealista

photosensitive: ~ **cell** *n* DATA *of scanner*, ELECT célula *f* fotosensible

phototransistor *n* PARTS, TRANS fototransistor *m*

photovideotext *n* DATA fotovideotexto *m*

photovoltaic: ~ **cell** *n* ELECT pila *f* fotovoltaica, PARTS célula *f* fotovoltaica; ~ **effect** *n* ELECT efecto *m* fotovoltaico

phrase *n* COMP, DATA frase *f*

phrasing *n* COMP, DATA fraseo *m*

physical *adj* GEN físico; ~ **address** *n* TELEP domicilio *m* real; ~ **connection** *n* TRANS conexión *f* física; ~ **delivery** *n* DATA, TELEP entrega *f* física; **~-delivery access unit** *n* DATA, TELEP unidad *f* de acceso a la entrega física; **~-delivery system** *n* DATA, TELEP sistema *m* de entrega física; **~ installation** *n* GEN instalación *f* física; **~ interface** *n* DATA interfaz *m* físico; **~ layer** *n* DATA, TRANS capa *f* física; **~-layer operations and maintenance cell** *n* RADIO célula *f* de operaciones y mantenimiento de la capa física; **~-layer service** *n* TELEP, TEST servicio *m* de la capa física; **~ line** *n* TRANS línea *f* física; **~ medium** *n* TRANS medio *m* físico; **~-medium sublayer** *n* TRANS subcapa *f* del medio físico; **~ optics** *n* TRANS óptica *f* física; **~ security** *n* DATA, TEST seguridad *f* física; **~ service** *n* APPL, TELEP servicio *m* físico; **~-service access point** *n* DATA punto *m* de acceso al servicio físico

pick *vt* COMP picar

pick up *vt* COMP, RADIO, TELEP *sound* captar

picking: **~ pliers** *n pl* PARTS pinzas *f pl* para hilos; **~ up** *n* TELEP *of sound* captación *f*

pickling *n* GEN limpieza de metales

pick-up: **~ facility** *n* TELEP función *f* de captación

picosecond *n* (*ps*) TEST picosegundo *m* (*ps*)

PICS *abbr* (*protocol implementation conformance statement*) DATA PICS (*sentencia de conformidad de instrumentación de protocolos*); **~ proforma** *n* DATA proforma *f* PICS

pictogram *n* ELECT pictograma *m*

pictorial: **~ diagram** *n* GEN diagrama *m* pictórico; **~ element** *n* (*PE*) COMP, TELEP elemento *m* pictórico (*PE*); **~ icon** *n* COMP icono *m* pictórico

picture *n* GEN imagen *f*; **~ blemish** TV imperfección *f* pictórica; **~ channel** TV canal *m* de imagen; **~ coding** COMP, TELEP codificación *f* pictórica; **~ definition** TV definición *f* de la imagen; **~ element** COMP, ELECT, TV elemento *m* de imagen; **~ frequency** COMP, TV frecuencia *f* de imagen; **~ phone** PER APP, TELEP videoteléfono *m*; **~ in ~** (*PIP*) COMP, TV imagen *f* dentro de una imagen *f* (*PIP*); **~ quality** COMP, TV calidad *f* de la imagen; **~ rate** COMP, TV velocidad *f* de imagen; **~ signal** COMP, TV señal *f* de imagen; **~-sound desynchronization** COMP, TV desincronización *f* de sonido de película; **~-sound lag** COMP, TV retardo *m* en el sonido de una película; **~ store** CONTR memoria *f* de imagen; **~ tube** ELECT tubo *m* de imagen; **~ white** COMP nivel *m* de blanco

pie: **~ chart** *n* COMP gráfico *m* de sectores

piece: **~ part** *n* PARTS pieza *f*

Pierce: **~ gun** *n* ELECT cañón *m* electrónico de Pierce

piezoelectric *adj* ELECT piezoeléctrico; **~ crystal** *n* ELECT cristal *m* piezoeléctrico; **~ depolarization** *n* ELECT despolarización *f* piezoeléctrica; **~ transducer** *n* ELECT, PARTS transductor *m* piezoeléctrico

piezoelectricity *n* ELECT piezoelectricidad *f*

piggyback *n* COMP, ELECT semirremolque *m*

pigment *n* PARTS pigmento *m*

pigtail *n* GEN rabillo *m*

pillar *n* GEN pilar *m*

pillow: **~ loudspeaker** *n* TELEP altavoz *m* de cabecera

pilot *adj* GEN piloto; **~ carrier** *n* TRANS portadora *f* piloto; **~ frequency** *n* TRANS frecuencia *f* piloto; **~ interference** *n* TRANS interferencia *f* piloto; **~ lamp** *n* COMP, PER, RADIO, TELEP lámpara *f* piloto; **~ light** *n* COMP lámpara *f* piloto, PER, RADIO luz *f* piloto, TELEP luz *f* indicadora; **~ operation** *n* TELEP operación *f* experimental; **~ signal** *n* RADIO señal *f* piloto; **~ stud** *n* PARTS perno *m* de guía; **~ supervision** *n* TRANS supervisión *f* de piloto; **~ system** *n* RADIO sistema *m* experimental; **~ tone** *n* TRANS onda *f* piloto; **~-tone monitoring** *n* TRANS control *m* de señal de identificación; **~-tone system** *n* RADIO sistema *m* de señal de identificación; **~ trial** *n* TEST, TRANS prueba *f* piloto; **~ wave** *n* RADIO onda *f* piloto

PIM *abbr* (*passive intermodulation*) TRANS PIM (*intermodulación pasiva*)

pin *n* ELECT, PARTS, PER APP aguja *f*; **~ column** ELECT columna *f* de espigas; **~ contact** ELECT, PARTS contacto *m* de espigas; **~ contact unit** ELECT, PARTS *connecting device* bloque *m* de contactos de espigas; **~ feed** PER APP alimentación *f* por pernos; **~-grid array** ELECT matriz *f* de rejilla de espigas; **~-out** ELECT extensión *f* con rodillo; **~-to-~ wiring** ELECT, PARTS cableado *m* de espiga a espiga; **~ row** ELECT fila *f* de espigas; **~ spacing** ELECT separación *f* de las espigas

PIN *abbr* (*procedure interrupt negative*) DATA, TELEP, TRANS NIP (*interrupción de proceso negativa*)

pinboard *n* COMP placa *f* de clavijas

pinch: **~ effect** *n* PARTS efecto *m* de estricción; **~-off voltage** *n* POWER tensión *f* de estricción; **~ roller** *n* PER APP rodillo *m* prensador

pincushion: **~ distortion** *n* TV distorsión *f* en cojín

PIN-FET: **~ integrated receiver** *n* TELEP, TRANS receptor *m* integrado PIN-FET

pinhole *n* ELECT picadura *f*
pink: ~ **noise** *n* COMP ruido *m* rosado
pinpoint *vt* COMP detectar
pinpointing *n* COMP detección *f*
PIP *abbr* (*picture in picture*) COMP, TV PIP (*imagen dentro de una imagen*)
pipe[1] *n* COMP, ELECT, PARTS, TRANS tubo *m*, tubería *f*; ~ **cable** GEN cable *m* tubular
pipe[2] *vt* COMP, ELECT, PARTS, TRANS transmitir por cable coaxial
pipeline: **in the** ~ *phr* GEN en trámite
piracy *n* COMP, TV piratería *f*
pirate[1] *adj* COMP, TV pirata
pirate[2] *vt* COMP, TV piratear
pirated *adj* COMP, TV pirateado
pirating *n* COMP, TV pirateo *m*
PISC: ~ **proforma** *n* DATA proforma *f* PISC
piston *n* COMPON émbolo *m*, pistón *m*; ~ **attenuator** RADIO atenuador *m* de pistón
pit *n* COMP, PARTS foso *m*
pitch *n* GEN paso *m*; ~ **bias** RADIO sesgo *m* de altura de tono; ~ **circle** GEN círculo *m* primitivo; ~ **diameter** GEN diámetro *m* primitivo; ~**-excited vocoder** RADIO vocoder *m* de tono excitado; ~**-to-MIDI converter** COMP convertidor *m* de tono a MIDI; ~ **modulation** COMP modulación *f* de altura; ~**-modulation depth** COMP profundidad *f* de modulación de altura; ~ **scale** TELEP escala *f* primitiva; ~ **variation** COMP variación *f* de altura; ~**-to-voltage converter** COMP transformador *m* de tono en voltaje
pitchbend *n* COMP curva *f* primitiva; ~ **effect** COMP efecto *m* de curva primitiva; ~ **range** COMP amplitud *f* de curva primitiva
pitchbender *n* COMP trazador *m* de líneas primitivas
pitchless: ~ **sound** *n* COMP sonido *m* atonal
PIU *abbr* (*plug-in unit*) COMP, DATA *scanner*, ELECT, TELEP PIU (*unidad enchufable*)
PIV *abbr* (*peak inverse voltage*) POWER PIV (*cresta de tensión inversa*), TELEP PRV (*tensión máxima inversa*)
pivot *n* PARTS pivote *m*
pixel *n* COMP, DATOS, TRANS, TV pixel *m*; ~ **aspect ratio** COMP, RADIO, TELEP razón *f* de aspecto del pixel; ~ **map** COMP configuración *f* de pixel
pixelation *n* COMP, DATA *of scanner* pixelación *f*
pixellization *BrE n* COMP, DATA *of scanner* pixelización *f*
PIXIT *abbr* (*protocol implementation extra information for testing*) DATA PIXIT (*información extra de instrumentación de protocolos para comprobación*); ~ **proforma** *n* DATA proforma *f* PIXIT
PJC *abbr* (*pointer justification count*) DATA recuento *m* de justificación de puntero
PJE *abbr* (*pointer justification event*) DATA evento *m* de justificación de puntero
place *n* COMP posición *f*; ~ **of work** GEN lugar *m* de trabajo
placement *n* GEN colocación *f*
placing *n* GEN colocación *f*
plain: ~ **conductor** *n* GEN conductor *m* de metal desnudo; ~ **connector** *n* RADIO brida *f* plana; ~ **coupling flange** *n* RADIO reborde *m* de brida plana; ~ **language** *n* COMP lenguaje *m* claro; ~ **style** *n* COMP estilo *m* claro
plan *vt* GEN planear
planar: ~**-array antenna** *n* RADIO antena *f* de red planar de transductores; ~ **diode** *n* PARTS diodo *m* planar; ~ **system** *n* TRANS sistema *m* planar; ~ **technique** *n* PARTS técnica *f* planar; ~ **transistor** *n* PARTS transistor *m* planar
plane *n* COMP cepillo *m*; ~ **of polarization** RADIO plano *m* de polarización; ~ **washer** GEN arandela *f* plana; ~ **wave** RADIO onda *f* plana
planning *n* GEN planificación *f*
plant *n* GEN planta *f*; ~ **engineering** GEN ingeniería *f* de planta
plasma *n* GEN plasma *m*; ~**-addressed liquid crystal** ELECT cristal *m* líquido direccionado al plasma; ~ **diode** ELECT diodo *m* de plasma; ~ **display** COMP, ELECT pantalla *f* de plasma; ~ **panel** COMP panel *m* de plasma, ELECT pantalla *f* de plasma
plastic *adj* PARTS plástico; ~**-bending section modulus** *n* PARTS modulo *m* de la sección para flexión del plástico; ~**-clad silica** *n* TELEP sílice *m* con cubierta de plástico; ~**-clad silica fibre** *BrE n* DATA, PARTS, TRANS fibra *f* de sílice con revestimiento plástico; ~ **coating** *n* TRANS cobertura *f* plástica; ~ **deformation** *n* PARTS deformación *f* plástica; ~ **effect** *n* COMP efecto *m* plástico; ~**-encapsulated microcircuit** *n* ELECT microcircuito *m* encapsulado en plástico; ~ **encapsulation** *n* ELECT encapsulado *m* de plástico; ~ **fibre** *BrE n* DATA, PARTS, TRANS fibra *f* plástica; ~ **film** *n* PARTS lámina *f* de plástico; ~**-film capacitor** *n* POWER condensador *m* de película de plástico; ~ **foil** *n* PARTS película *f* de plástico; ~**-foil capacitor** *n* POWER condensador *m* de película de plástico; ~ **hose** *n* PARTS manguera *f* de plástico; ~**-insulated cable** *n* TRANS cable *m* aislado con plástico; ~ **material** *n* PARTS material *m* plástico; ~ **moulding** *BrE n* PARTS moldeado *m* plástico; ~ **rivet** *n* ELECT remache

m de plástico; ~ **strain** *n* PARTS deformación *f* plástica permanente; ~-**tipped** ELECT con cabeza de plástico; ~ **tubing** *n* PARTS manguera *f* de plástico

plasticizer *n* PARTS plastificante *m*

plate *n* ELECT placa *f*; ~ **battery** POWER batería *f* de placa; ~ **block** PER APP bloque *m* de placa; ~ **current** ELECT corriente *f* de placa; ~ **finish** ELECT superficie *f* en bruto; ~-**load resistor** PARTS, POWER resistor *m* de carga anódica; ~ **resistor** PARTS, POWER resistor *m* de placa; ~ **voltage** ELECT tensión *f* de placa

plateau *n* RADIO meseta *f*; ~ **slope** RADIO pendiente *f* de la meseta; ~-**threshold voltage** RADIO tensión *f* umbral de la meseta

platen: ~ **guide** *n* TELEP guía *f* de rodillo; ~ **knob** *n* TELEP perilla *f* del rodillo

platform *n* ELECT, TRANS andén *m*

plating *n* GEN fabricación *f* de chapas; ~ **bar** ELECT barra *f* de metalización; ~-**up** ELECT electrodeposición *f*

platter *n* COMP disco *m*

plausibility *n* DATA plausibilidad *f*

plausible: ~ **reasoning** *n* DATA razonamiento *m* plausible

play[1]: ~-**back** *n* COMP *audio*, RADIO reproducción *f*

play *vt* COMP *program, macro*, CONTR, DATA, PARTS, TELEP, TV reproducir

playability *n* COMP reproducibilidad *f*

playback *n* COMP *program, macro*, PER APP *of answering machine*, RADIO, TV lectura *f*, reproducción *f*; ~ **head** COMP cabeza *f* reproductora

player *n* GEN reproductor *m*

pleated: ~-**foil covered** *adj* TRANS cubierto de hoja plisada

plenum: ~-**rated cable** *n* ELECT cable *m* de distribución nominal; ~-**use area** *n* ELECT área *f* de uso pleno

plesiochronous *adj* TELEP, TRANS plesiócrono; ~ **digital hierarchy** *n* TELEP, TRANS jerarquía *f* digitalplesiocrónica; ~-**line terminal** *n* TRANS terminal *m* de línea plesiócrona; ~ **transmission equipment** *n* TRANS equipo *m* de transmisión plesiócrono

PLL *abbr* (*phase-locked loop*) ELECT lazo *m* de sincronización de fase

PL/M: ~ **language** *n* COMP lenguaje *m* PL/M

plot[1] *n* COMP trazado *m*

plot[2] *vt* COMP *graph* trazar

plotter *n* GEN plotter *m*

PLP *abbr* (*packet layer protocol*) DATA, TELEP PLP (*protocolo de la capa de paquetes*)

PLR *abbr* (*pulse link repeater*) TELEP PLR (*repetidor de enlace de impulsos*)

PLU *abbr* (*phase lock unit*) ELECT PLU (*unidad de sincronización de fase*)

plug *n* GEN conector *m* macho; ~-**and-jack connection** POWER conexión *f* mediante clavija y jack; ~ **bar** PARTS, POWER barra *f* de clavijas; ~-**in board** ELECT, PARTS placa *f* enchufable; ~-**in cable** ELECT, PARTS cable *m* enchufable; ~-**in chip card** ELECT, PARTS tarjeta *f* electrónica enchufable; ~-**in connector** ELECT, PARTS conector *m* enchufable; ~-**in module** COMP, ELECT, PARTS módulo *m* enchufable; ~-**in relay** PARTS relé *m* enchufable; ~-**in unit** (*PIU*) COMP, DATA *scanner*, ELECT, TELEP unidad *f* enchufable (*PIU*); ~ **pin** ELECT *connecting device*, PARTS espiga *f* de conector; ~-**sleeve indicator** TELEP manguito *m* indicador, indicador *m* de base de clavija; ~ **tip** TELEP punta *f* de clavija

plugged: ~ **in** *adj* GEN enchufado

plugging: ~ **chart** *n* COMP diagrama *m* de tablero

plumbing *n* RADIO plomería *f*

plunger *n* GEN pistón *m*; ~ **key** GEN conmutador *m* de botón

pluralism *n* ELECT pluralismo *m*

PLV *abbr* (*production level video*) COMP PLV (*vídeo en fase de producción*)

plywood *n* PARTS madera *f* contrachapada

PM *abbr* (*phase modulation*) GEN MF (*modulación de fase*)

PMA *abbr* (*program memory area*) COMP PMA (*zona de memoria de programas*)

PN *abbr* (*positive notification*) DATA, TELEP notificación *f* positiva; ~ **junction** *n* ELECT unión *f* PN

PO *abbr* (*post office*) TELEP oficina *f* de correos

POAS *abbr* (*payment option analysis system*) APPL POAS (*sistema de análisis de opción de pago*)

Pockels: ~ **effect** *n* TRANS efecto *m* Pockels

pocket *adj* COMP, RADIO de bolsillo; ~ **calculator** *n* GEN calculadora *f* de bolsillo; ~ **computer** *n* COMP ordenador *m* de bolsillo *Esp*, computador *m* de bolsillo *AmL*, computadora *f* de bolsillo *AmL*; ~ **receiver** *n* RADIO receptor *m* de bolsillo; ~-**sized** COMP, RADIO de bolsillo; ~ **telephone** *n* RADIO, TELEP teléfono *m* de bolsillo; ~ **terminal** *n* RADIO terminal *m* de bolsillo

point *n* COMP, TELEP punta *f*; ~ **code** DATA, TELEP código *m* de puntos fijos; ~ **contact** PARTS contacto *m* puntual; ~-**contact diode** PARTS diodo *m* de contacto dc punta; ~-**contact transistor** ELECT, PARTS transistor *m* de contacto de punta; ~ **of control** DATA, TEST punto *m* de indiferencia; ~ **of control and observation** (*PCO*) DATA, TEST punto *m* de control y observación (*PCO*); ~ **of inflexion** GEN punto *m* de inflexión; ~ **of minimum voltage** POWER, RADIO punto *m* de voltaje mínimo; **~-to-area communication** RADIO, TRANS comunicación *f* de punto a área; **~-to-~ communication** RADIO, TRANS comunicación *f* punto a punto; **~-to-multipoint communication** RADIO, TRANS comunicación *f* de punto a multipunto; **~-to-multipoint connection** APPL conexión *f* de punto a multipunto; **~-to-multipoint system** TRANS sistema *m* de punto a multipunto; **~-to-~ protocol** (*PPP*) DATA protocolo *m* de punto a punto (*PPP*); **~-to-~ system** TRANS sistema *m* de punto a punto; **~-to-~ wiring** PARTS cableado *m* de punto a punto; ~ **of sale** GEN punto *m* de venta; ~ **station** TRANS estación *f* puntual; ~ **in time** GEN momento *m*

pointer *n* COMP, DATA, TELEP aguja *f*; ~ **centring** *BrE* **error** RADIO error *m* de excentricidad; ~ **connection** SWG conexión *f* de indicador; ~ **deflection** TELEP *of indicating device* desviación *f* de la aguja; ~ **instrument** TELEP aparato *m* de aguja; ~ **justification count** (*PJC*) DATA recuento *m* de justificación de puntero; ~ **justification event** (*PJE*) DATA evento *m* de justificación de puntero; ~ **value** SWG valor *m* de puntero

pointing: ~ **angle** *n* RADIO ángulo *m* apuntador; ~ **device** *n* COMP dispositivo *m* apuntador; ~ **loss** *n* RADIO pérdida *f* por apuntamiento

points: ~ **of international connection at baseband frequencies** *n pl* TELEP puntos *m pl* de conexión internacional de frecuencias de banda base; ~ **of telephony input and output** *n pl* TELEP puntos *m pl* de entrada y salida de telefonía

Poisson: ~ **traffic** *n* TELEP tráfico *m* Poisson

polar *adj* GEN polar; ~ **keying** *n* TELEP manipulación *f* polar; ~ **modulation system** *n* RADIO sistema *m* de modulación polar; ~ **orbit** *n* RADIO órbita *f* polar; ~ **relay** *n* POWER relé *m* polarizado

polarity *n* GEN polaridad *f*; ~ **indicator** POWER indicador *m* de polaridad; ~ **inversion** POWER inversión *f* de polaridad; ~ **reversal** ELECT, POWER inversión *f* de polaridad; ~ **signalling** *BrE* SIG señalización de polaridad

polarization *n* GEN polarización *f*; ~ **controller** ELECT controlador *m* de polarización; ~ **coupling loss** RADIO pérdida *f* de acoplamiento de polarización; ~ **current** POWER corriente *f* de polarización; ~ **error** RADIO error *m* de polarización

polarized: **~-control mode** *n* DATA modo *m* de control polarizado; ~ **electric bell** *n* TELEP timbre *m* electromagnético; ~ **relay** *n* PARTS, POWER relé *m* polarizado

polarizer *n* GEN polarizador *m*

polarizing: ~ **isolation** *n* RADIO, TRANS aislamiento *m* polarizante; ~ **slot** *n* ELECT muesca *f* de polaridad

pole *n* POWER palo *m*; ~ **changer** POWER inversor *m* de polaridad; ~ **changing** POWER inversión *f* de polaridad; ~ **face** POWER, TELEP superficie *f* polar; ~ **inverter** POWER inversor *m* de polaridad; ~ **piece** POWER pieza *f* polar; ~ **reversal** POWER inversión *f* de polaridad; ~ **route** POWER, TRANS ruta *f* de postes; ~ **shoe** POWER pieza *f* polar

polishing: ~ **jig** *n* ELECT plantilla *f* pulidora; ~ **tool** *n* GEN bruñidor *m*

poll *vt* COMP interrogar, sondear, DATA, RADIO sondear

polling *n* GEN interrogación *f*; ~ **key** COMP clave *f* de interrogación; ~ **radio network** RADIO red *f* de radio de sondeo; ~ **system** COMP sistema *m* de interrogación

poly: ~ **mix** *n* COMP polimezcla *f*

polyamide *n* PARTS poliamida *f*; ~ **plastic** PARTS plástico *m* poliamídico

polybutylene: ~ **terephthalate** *n* ELECT polibutileno *m* tereftalato

polycarbonate: ~ **plastic** *n* PARTS plástico *m* de policarbonato

polyester *n* PARTS poliéster *m*; ~ **plastic** PARTS plástico *m* poliestérico

polyethylene *n* PARTS polietileno *m*; ~ **plastic** PARTS plástico *m* polietilénico

polygon: ~ **network** *n* TELEP red *f* poligonal

polygonal: ~ **shading** *n* COMP sombra *f* poligonal

polymerization *n* PARTS polimerización *f*

polymerize *vt* PARTS polimerizar

polyolefin: ~ **insulation** *n* ELECT aislamiento *m* de poliolefina

polyphase: ~ **system** *n* ELECT sistema *m* polifásico

polyphonic *adj* COMP polifónico; ~ **play** *n* COMP juego *m* polifónico; ~ **pressure** *n* COMP presión *f* polifónica; **~-pressure sensitivity** *n* COMP sensibilidad *f* a la presión polifónica

polyphony *n* COMP polifonía *f*

polyrod *n* RADIO antena *f* dieléctrica de varillas

polythene *n* PARTS politeno *m*; ~ **plastic** PARTS plástico *m* politénico

polyurethane *n* PARTS poliuretano *m*

polyvinyl: ~ **chloride** *n* ELECT cloruro *m* de polivinilo

PON *abbr* (*passive optical network*) TELEP, TRANS red *f* óptica pasiva

pool *n* TRANS *of channels* mancomunidad *f*; ~ **cathode** ELECT cátodo *m* líquido; ~ **rectifier tube** ELECT tubo *m* de rectificador de cátodo líquido

poor: ~ **installation** *n* TELEP instalación *f* deficiente; ~ **maintenance** *n* TELEP mantenimiento *m* deficiente

pop up *vi* COMP *on screen* remontar

population *n* TEST población *f*

porosity *n* PARTS porosidad *f*

porous *adj* PARTS poroso

port *n* COMP, RADIO punto *m* de acceso

portability *n* GEN portabilidad *f*

portable *adj* GEN portátil; ~ **computer** *n* COMP ordenador *m* portátil *Esp*, computador *m* portátil *AmL*, computadora *f* portátil *AmL*; ~ **document file** *n* DATA archivo *m* de documento transferible; ~ **operating-system interface** *n* DATA interfaz *m* de sistema operativo transferible; ~ **printer** *n* COMP, DATA, PER APP impresora *f* portátil; ~ **receiver** *n* RADIO receptor *m* portátil

position *n* GEN posición *f*; ~ **A** TELEP posición *f* A; ~ **correctness** GEN precisión *f* de la posición; ~**-defined parameter** TELEP parámetro *m* definido por la posición; ~ **relay set** SWG juego *m* de relés de posición; ~ **set** TELEP equipo *m* de posición de operadora

positional: ~ **tolerance** *n* GEN tolerancia *f* posicional

positioning *n* DATA, RADIO *of antenna* posicionamiento *m*; ~ **time** COMP, DATA tiempo *m* de posicionamiento

positive: ~ **amplitude modulation** *n* ELECT, TV modulación *f* de amplitud positiva; ~ **bar** *n* POWER barra *f* positiva; ~ **battery** *n* POWER batería *f* positiva; ~ **electron** *n* GEN electrón *m* positivo; ~ **emitter-coupled logic** *n* COMP lógica *f* acoplada de emisor positivo; ~ **feedback** *n* COMP, ELECT, RADIO realimentación *f* positiva; ~ **glow** *n* ELECT luz *f* anódica; ~ **indication tone** *n* TELEP tono *m* de indicación positivo; ~ **justification** *n* TELEP, TRANS justificación *f* positiva; ~**-negative justification** *n* TELEP, TRANS justificación *f* positiva-negativa; ~ **notification** *n* (*PN*) DATA, TELEP notificación *f* positiva; ~ **overvoltage** *n* POWER sobretensión *f* positiva; ~ **pole** *n* POWER polo *m* positivo; ~ **pressure** *n* GEN presión *f* positiva; ~ **terminal** *n* POWER terminal *m* positivo, *of battery* borne *m* positivo; ~ **video signal** *n* COMP señal *f* de vídeo positiva; ~ **wire** *n* TELEP hilo *m* positivo

positron *n* GEN positrón *m*

post: ~**-amplifier** *n* ELECT, POWER postamplificador *m*; ~**-amplifier loss** *n* ELECT, POWER atenuación *f* postamplificador; ~**-deflection acceleration factor** *n* ELECT factor *m* de aceleración posdeflexión; ~**-deflection acceleration ratio** *n* ELECT ratio *f* de aceleración posdeflexión; ~**-dialling** *BrE* **delay** *n* RADIO, TELEP retardo *m* posmarcado; ~ **office** *n* (*PO*) TELEP oficina *f* de correos (*PO*)

postamble *n* DATA postámbulo *m*

postcode *BrE n* (*cf zip code AmE*) COMP código *m* postal

posteriation *n* COMP distribución *f* de carteles

postprocessor *n* COMP posprocesador *m*

postproduction *n* TV posproducción *f*; ~ **pressing** COMP prensado *m* posproducción; ~ **processing** COMP procesamiento *m* posproducción

pot *n* ELECT célula *f* electrolítica

potential *n* GEN potencial *m*; ~ **barrier** PARTS barrera *f* de potencial; ~ **difference** POWER diferencia *f* de potencial; ~ **function** POWER función *f* potencial; ~ **gradient** POWER gradiente *m* de potencial; ~ **log record** DATA registro *m* en diario de operaciones potencial; ~ **market** GEN mercado *m* potencial

potentiometer *n* ELECT potenciómetro *m*

potting: ~ **boot** *n* GEN manguito *m* del tubo de impregnación

powdering *n* COMP pulverización *f*

power *n* GEN energía *f* eléctrica; ~ **amplification** COMP, DATA, TRANS amplificación *f* de potencia; ~ **amplifier** (*PA*) COMP, DATA, TRANS amplificador *m* de potencia (*PA*); ~ **budget** POWER, TRANS balance *m* de potencia; ~ **busbar** POWER barra *f* colectora para fuerza; ~ **cable** COMP, DATA, ELECT, POWER, TELEP cable *m* alimentador; ~ **cable chute** COMP, DATA, ELECT, POWER, TELEP canal *m* para cable de fuerza; ~ **control bus** POWER bus *m* de control de energía; ~ **control unit** POWER unidad *f* de control de energía; ~ **converter** COMP, DATA convertidor *m* de alimentación; ~ **cord** COMP, ELECT cordón *m* de alimentación, POWER cable *m* de alimentación; ~ **cube** RADIO bloque *m* de energía; ~ **current** POWER corriente *f* de gran amperaje; ~**-current engineering** POWER tecnología *f* de las corrientes fuertes; ~ **cut** POWER corte *m* en el suministro de corriente; ~ **density** POWER densidad *f* de potencia; ~**-on display** PER APP visualizador *m* de conexión; ~ **dissipation** PARTS, POWER disipación *f* de potencia; ~ **distribution** POWER distribución *f* de fuerza; ~ **distribution alarm** POWER alarma *f* de distribución de potencia; ~ **distribution board** POWER placa *f* de distribución de energía; ~ **distribution diagram** POWER esquema *m* de distribución de potencia; ~ **divider** POWER, RADIO divisor *m* de potencia; ~ **drain** POWER consumo *m* de potencia; ~-**driven system** TELEP sistema *m* accionado por energía eléctrica; ~ **electronics** POWER electrónica *f* de potencia; ~ **engineering** POWER ingeniería *f* de potencia; ~ **equipment** POWER equipo *m* de potencia; ~ **factor** POWER factor *m* de potencia; ~ **failure** COMP, POWER, TELEP fallo *m* de potencia; ~ **failure feature** COMP, POWER, TELEP, TELEP característica *f* del fallo de potencia; ~ **failure transfer** COMP, POWER, TELEP, TELEP transferencia *f* del fallo de potencia; ~ **feed** POWER suministro *m* de potencia; ~-**feeding repeater station** POWER estación *f* repetidora de suministro de potencia, TRANS estación *f* repetidora autoalimentada; ~-**feeding station** POWER estación *f* autoalimentada, TRANS estación *f* alimentadora; ~ **fuse** PARTS fusible *m* de potencia, POWER fusible *m* de línea de fuerza; ~ **gain** ELECT ganancia *f* de potencia; ~ **indicator** ELECT indicador *m* de potencia; ~ **law index fibre** *BrE* ELECT, POWER, TRANS fibra *f* de índice de ley potencial; ~ **law index profile** ELECT, POWER, TRANS perfil *m* de índice de ley potencial; ~ **level** COMP nivel *m* de potencia; ~ **limit** ELECT límite *m* de potencia; ~ **line** POWER línea *f* de transporte de energía; ~ **meter** TEST ergómetro *m*, medidor *m* de potencia; ~-**off alarm** POWER alarma *f* de desconexión; ~-**off condition** POWER situación *f* de desconexión; ~ **on/off switch** PARTS interruptor *m* de encendido/apagado; ~ **outage** POWER interrupción *f* de potencia; ~ **pack** POWER unidad *f* de alimentación; ~ **plant** POWER planta *f* de energía; ~ **point** POWER punto *m* de toma de energía; ~ **rack** POWER bastidor *m* de equipo de fuerza; ~ **relay** PARTS relé *m* de potencia; ~ **requirement** ELECT necesidad *f* de potencia; ~ **room** POWER sala *f* de máquinas; ~ **semiconductor** PARTS semiconductor *m* de potencia; ~ **source** POWER fuente *f* de energía; ~ **stage** RADIO paso *m* de potencia; ~ **subsystem** (*PSS*) POWER subsistema *m* de potencia (*PSS*); ~ **supply** ELECT, POWER fuente *f* de alimentación; ~ **supply equipment** ELECT, POWER equipo *m* de suministro eléctrico; ~ **supply station** ELECT, POWER planta *f* de suministro de energía; ~ **supply unit** ELECT, POWER *backup* unidad *f* de alimentación, *inverter* unidad *f* de alimentación eléctrica, *rectifier* unidad *f* de suministro de energía; ~ **switch** PARTS interruptor *m* de potencia; ~-**test bus** POWER bus *m* de prueba de potencia; ~ **thyristor** ELECT, PARTS tiristor *m* de potencia; ~ **transformer** POWER transformador *m* de potencia; ~ **transistor** PARTS transistor *m* de potencia; ~ **transmission line** POWER línea *f* de transporte de energía; ~ **tube** ELECT tubo *m* de potencia; ~ **unit** POWER unidad *f* de potencia; ~ **valve** ELECT tubo *m* de potencia

power down *vt* ELECT cortar la energía de

power up *vt* ELECT energizar

powered *adj* ELECT motorizado, POWER impulsado; ~-**down** ELECT con disminución de potencia; ~-**up** ELECT con aumento de potencia

PP *abbr* (*peripheral processor*) COMP, PER APP procesador *m* periférico

PPM *abbr* (*pulse position modulation*) DATA, TELEP, TRANS PPM (*modulación de impulsos en posición*)

PPP *abbr* (*point-to-point protocol*) DATA PPP (*protocolo de punto a punto*); ~ **Internet connection** *n* DATA conexión *f* PPP de Internet

pps *abbr* (*pulses per second*) RADIO, TELEP impulsos *m pl* por segundo

PPSDN *abbr* (*public packet-switched data network*) DATA PPSDN (*red pública de datos de paquetes conmutados*); ~ **service** *n* DATA servicio *m* PPSDN

practice: ~ **factor** *n* TELEP factor *m* de equipo

PRB: ~ **code** *n* DATA código *m* PRB

preageing *n* TEST preenvejecimiento *m*

preamble *n* DATA preámbulo *m*, sincronización *f* inicial

preamplification *n* ELECT preamplificación *f*

preamplifier *n* ELECT preamplificador *m*

preanalysis *n* TELEP análisis *m* preliminar

prearranged *adj* GEN preestablecido

preassemble *vt* GEN preensamblar

preassembled: ~ **cables** *n pl* TRANS cables *m pl* preensamblados

preassembly: ~ **kit** *n* GEN conjunto *m* para ensamblar

preassigned *adj* GEN preasignado; ~ **noninterpolated channel** *n* TRANS canal *m* nointerpolado preasignado; ~ **value** *n* TEST valor *m* preasignado

preavis: ~ **call** *n* TELEP llamada *f* con aviso previo; ~ **charge** *n* TELEP cargo *m* por preaviso; ~ **fee** *n* TELEP tasa *f* de preaviso

precable *vt* TRANS precablear

precabled *adj* TRANS precableado
precabling *n* TRANS precableado *m*
preceding: ~ **page** *n* GEN página *f* anterior
precious: ~ **metal** *n* PARTS metal *m* fino, metal *m* precioso
precipitation: ~ **rate** *n* RADIO ratio *f* de precipitación, velocidad *f* de precipitación; ~ **return** *n* RADIO econo *m* de precipitación; ~ **scatter** *n* RADIO dispersión *f* de precipitación; ~**-scatter propagation** *n* RADIO propagación *f* de la dispersión de precipitación; ~ **static** *n* GEN ruido *m* parásito por precipitación
precision *n* RADIO, TEST precisión *f*; ~ **adaptive sub-band coding** (*PASC*) COMP codificación *f* de subbanda ajustable de precisión (*PASC*); ~ **approach radar** RADIO radar *m* de precisión para la aproximación; ~ **framing** DATA *of scanner* alineación *f* de precisión; ~ **frequency offset** RADIO separación *f* de frecuencia de precisión; ~ **instrument** TEST instrumento *m* de precisión; ~ **mechanic** GEN mecánica *f* de precisión; ~ **radar approach** RADIO aproximación *f* por radar de precisión
precorrection *n* TELEP precorrección *f*
predefine *vt* GEN predefinir
predefined: ~ **marker** *n* DATA marcador *m* predefinido
predetermine *vt* GEN predeterminar
predetermined *adj* GEN predeterminado
predict *vt* GEN predecir
prediction *n* GEN predicción *f*; ~ **error** COMP error *m* de predicción
predictive: ~**-coded picture** *n* COMP, DATA, TV imagen *f* predictiva codificada; ~ **coding** *n* COMP, DATA, TV codificación *f* predictiva
predictor *n* COMP predictor *m*
predistortion: ~ **technique** *n* ELECT técnica *f* de predistorsión
pre-emphasis *n* ELECT, RADIO preacentuación *f*; ~ **network** ELECT, RADIO red *f* de preacentuación
pre-empt *vt* TELEP *call* priorizar
pre-energize *vt* POWER prealimentar
pre-equalization *n* ELECT corrección *f* previa, preigualación *f*
pre-existing *adj* DATA preexistente
prefix *n* GEN prefijo *m*
preform *vt* GEN preformar
preformatted *adj Internet* preformateado
preformed: ~ **fibre** *BrE n* TRANS fibra *f* premoldeada
pregroup *n* TRANS pregrupo *m*
preheating *n* ELECT precalentamiento *m*; ~ **time** ELECT tiempo *m* de precalentamiento
prejudice *n* ELECT daño *m*
preliminaries *n pl* GEN preparativos *m pl*
preliminary: ~ **result** *n* TEST resultado *m* preliminar; ~ **test** *n* TEST ensayo *m* preliminar; ~ **test programme** *BrE n* TEST programa *m* de prueba preliminar
preloaded *adj* ELECT precargado
premarking *n* TELEP *of magnetic tape* premarcación *f*
premastering *n* COMP premasterización *f*
premature: ~ **release** *n* COMP, TELEP desconexión *f* prematura
premises: **on the** ~ *adv* GEN en el establecimiento
premium: ~ **channel** *n* TV canal *m* de alta calidad, canal *m* de calidad superior; ~ **service** *n* TV servicio *m* de alta calidad
prepaid: ~ **card** *n* TELEP tarjeta *f* prepagada
preparatory: ~ **operations** *n pl* TELEP operaciones *f pl* preparatorias; ~ **period** *n* TELEP periodo *m* preparatorio
prepare *vt* GEN preparar
prepreg *n* ELECT hoja *f* preimpregnada
preprint *n* COMP, PER APP prueba *f* previa, prueba *f* preliminar
preprinted *adj* COMP, PER APP preimpreso
preprocessing *n* COMP pretratamiento *m*
preprocessor *n* COMP preprocesador *m*
preprogrammed: ~ **line** *n* TELEP línea *f* preprogramada
prerecorded *adj* TV pregrabado
prescaler *n* TEST predivisión *f*
preseizure: ~ **dialling** *BrE n* TELEP discado *m* de precaptación de línea, marcación *f* de toma previa
preselection *n* TELEP, TV preselección *f*
preselector *n* TELEP, TV preselector *m*
presentation *n* DATA, TELEP presentación *f*; ~ **address** DATA dirección *f* de presentación; ~ **connection** DATA conexión *f* de presentación; ~ **context** DATA contexto *m* de la presentación; ~ **control function** TELEP función *f* de control de la presentación; ~ **data value** DATA valor *m* de los datos de presentación; ~ **indicator** DATA indicador *m* de presentación; ~ **layer** DATA, TRANS nivel *m* de presentación; ~**-protocol data unit** DATA unidad *f* de datos de protocolo de presentación; ~ **service** DATA servicio *m* de presentación; ~**-service access point** (*PSAP*) DATA, TELEP punto *m* de acceso del servicio de presentación (*PSAP*); ~**-service data unit** DATA, TELEP unidad *f* de datos de servicio de presentación; ~ **time stamp** COMP reloj *m* fechador de presentación; ~ **unit** COMP unidad *f* de presentación
preset[1]: ~ **memory** *n* COMP memoria *f*

preestablecida; ~ **read only memory** *n* COMP memoria *f* preestablecida sólo de lectura; ~ **search** *n* COMP búsqueda *f* preestablecida; ~ **value** *n* GEN valor *m* prefijado; ~ **voice** *n* COMP voz *f* prefijada

preset[2] *vt* GEN preajustar, preseleccionar

presetting *n* GEN preselección *f*, preajuste *m*

pressing: ~ **master** *n* COMP master *m* de estampación

pressure *n* GEN presión *f*; ~ **contact** PARTS contacto *m* de presión; ~ **drop** GEN caída *f* de presión; ~ **increase** GEN aumento *m* de presión; ~ **plug** GEN tapón *m* de presión; ~**-reducing valve** PARTS válvula *f* manorreductora; ~**-sensitive keyboard** COMP teclado *m* piezosensible; ~ **test** GEN ensayo *m* manométrico; ~ **wave** RADIO onda *f* de presión

pressurized: ~ **cable** *n* GEN cable *m* presurizado

prestretched: ~ **cold-shrink tubing** *n* ELECT conducto *m* preestirado encogido en frío

pretest *n* TEST prueba *f* previa

pre-TR: ~ **tube** *n* ELECT tubo *m* pre-TR

pre-transmit: ~**-receive tube** *n* ELECT tubo *m* pretransmisor/receptor

prevent *vt* GEN prevenir

prevention: ~ **of call collision** *n* TELEP prevención *f* de colisión de llamadas; ~ **cost** *n* TEST coste *m* (*AmL* costo) de prevención

preventive: ~ **cyclic retransmission error control method** *n* TRANS método *m* preventivo de control de errores de transmisión cíclicos; ~ **inspection** *n* TEST inspección *f* preventiva; ~ **maintenance** *n* GEN mantenimiento *m* preventivo; ~ **maintenance time** *n* TELEP, TEST tiempo *m* de mantenimiento preventivo

preview *n* COMP, DATA *of scanner* avance *m*; ~ **channel** TV canal *m* de retransmisión de prueba

previous *adj* GEN previo

prewire *vt* ELECT precablear

prewired: ~ **connector** *n* ELECT conector *m* precableado

prewiring *n* ELECT precableado *m*

PRF *abbr* (*pulse repetition frequency*) ELECT, SIG, TELEP, TRANS PRR (*frecuencia de repetición de impulsos*)

price: ~ **fixing** *n* APPL estabilización *f* de los precios, fijación *f* de los precios; ~ **on request** *n* APPL precio *m* sobre demanda

pricing: ~ **policy** *n* APPL política *f* de precios; ~ **scheme** *n* APPL esquema *m* de precios

PRI-EOM *abbr* (*procedure interrupt EOM*) DATA, TELEP PRI-EOM (*final de mensaje de interrupción de procedimiento*)

PRI-EOP *abbr* (*procedure interrupt EOP*) DATA final *m* de programa de interrupción de procedimiento, TELEP PRI-EOP (*fin de programa de interrupción de procedimiento*)

primary *adj* COMP primario; ~ **access** *n* TELEP acceso *m* primario; ~ **battery** *n* POWER batería *f* primaria; ~ **block** *n* TELEP bloque *m* primario; ~ **body** *n* RADIO cuerpo *m* primario; ~ **centre** *BrE n* TELEP centro *m* primario; ~ **clock** *n* RADIO reloj *m* primario; ~ **coating** *n* TRANS revestimiento *m* primario; ~**-colour signal** *BrE n* COMP señal *f* de colorprimario; ~ **distribution link** *n* RADIO enlace *m* de distribución primario; ~ **document** *n* GEN documento *m* original; ~**-electron emission** *n* ELECT emisión *f* de electrón primario; ~ **exchange** *n* (*PE*) TELEP central *f* primaria (*CP*); ~ **failure** *n* TEST fallo *m* primario; ~ **flow direction** *n* GEN dirección *f* primaria del flujo; ~ **frequency standard** *n* RADIO estándar *m* primario de frecuencia; ~ **function** *n* DATA función *f* primaria; ~ **grade of reception quality** *n* RADIO grado *m* primario de calidad de recepción; ~ **translating equipment** *n* TELEP equipo *m* de translación de grupo primario; ~ **high-usage trunk group** *n* TELEP grupo *m* de enlace primario de explotación intensa; ~ **interlock trunk** *n* TELEP enlace *m* primario interconectado; ~ **intertoll trunk group** *n* TELEP grupo *m* de enlace interurbano primario; ~ **link** *n* COMP, TRANS enlace *m* primario; ~ **network** *n* TELEP red *f* primaria; ~ **radar** *n* RADIO radar *m* primario; ~ **rate access** *n* TELEP acceso *m* derivado primario; ~ **rate interface** *n* DATA, TELEP interfaz *f* derivada primaria; ~ **rate interface ISDN** *n* DATA, TELEP ISDN *m* de interfaz derivada primaria; ~ **reference burst** *n* DATA ráfaga *f* primaria de referencia; ~ **reference burst code** *n* DATA código *m* de ráfaga de referencia primaria; ~ **reference station** *n* RADIO estación *f* de referencia primaria; ~ **reference terminal** *n* RADIO terminal *m* de referencia primaria; ~ **relay** *n* PARTS relé *m* primario; ~ **route** *n* TELEP vía *f* primaria; ~ **source** *n* TV fuente *f* primaria; ~ **station** *n* DATA estación *f* primaria; ~ **storage** *n* DATA almacenamiento *m* principal; ~ **terminal** *n* TELEP borne *m* primario; ~ **winding** *n* PARTS devanado *m* primario

prime *vt* GEN imprimar

primer *n* GEN imprimador *m*; ~ **electrode** ELECT electrodo *m* cebador; ~ **ignition** ELECT encendido *m* por cebador; ~ **noise** ELECT ruido *m* de cebado; ~ **voltage** ELECT tensión *f* de cebado

priming *n* GEN imprimación *f*; ~ **rate** ELECT

velocidad *f* de sensibilización; ~ **speed** ELECT velocidad *f* de sensibilización

primitive *n* COMP, DATA primitivo *m*; ~ **confirm** DATA, TELEP confirmación *f* primaria; ~ **indication** DATA indicación *f* primaria; ~ **name** DATA nombre *m* primitivo; ~ **request** DATA solicitud *f* primitiva; ~ **response** DATA respuesta *f* primitiva

PRI-MPS *abbr* (*procedure interrupt MPS*) DATA, TELEP PRI-MPS (*MPS de interrupción de procedimiento*)

principal: ~ **processing unit** *n* (*PUP*) COMP unidad *f* principal de procesamiento (*PUP*); ~ **test section** *n* TELEP sección *f* de prueba principal; ~ **voltage** *n* POWER tensión *f* principal

principle: ~ **of operation** *n* TELEP principio *m* de funcionamiento

print[1]: ~ **bar** *n* TELEP barra *f* de impresión; ~ **barrel** *n* PER APP cilindro *m* de impresión; ~ **capacity** *n* PER APP capacidad *f* de impresión; ~ **command** *n* COMP orden *f* de imprimir; ~ **command language** *n* (*PCL*) COMP, DATA lenguaje *m* de órdenes de imprimir (*PCL*); **~-control character** *n* DATA carácter *m* de control de impresión; ~ **drum** *n* PER APP, TELEP tambor *m* de impresión; **~-out** *n* GEN copia *f*; ~ **settings** *n pl* COMP ajustes *m pl* de impresión; **~-through** *n* COMP, DATA distorsión por capa adyacente

print[2] *vt* GEN imprimir

printable: ~ **area** *n* COMP área *f* de impresión, área *f* imprimible

printed: ~ **board** *n* ELECT placa *f* impresa; **~-board assembly** *n* ELECT conjunto *m* impreso; ~ **circuit** *n* ELECT, PARTS circuito *m* impreso; ~ **circuit board** *n* (*PCB*) ELECT, PARTS placa *f* de circuito impreso (*PCI*); ~ **component** *n* ELECT componente *m* impreso; ~ **conductor** *n* ELECT conductor *m* estampado; ~ **contact** *n* ELECT contacto *m* impreso; ~ **edge-board contact** *n* ELECT contacto *m* impreso de borde de placa; ~ **edge-connector** *n* ELECT conector *m* de borde impreso; ~ **form** *n* GEN formulario *m* impreso; ~ **name** *n* GEN *in block capitals* nombre *m* impreso; ~ **record of duration and charge of calls service** *n* APPL, TELEP registro *m* impreso de duración y coste (*AmL* costo) del servicio de llamadas; ~ **wiring** *n* ELECT cableado *m* impreso, PARTS conexionado *m* impreso

printer *n* GEN impresora *f*; ~ **controller** COMP, DATA controlador *m* de impresora; ~ **driver** COMP, DATA gestor *m* de impresión; ~ **port** COMP puerta *f* de impresora; ~ **server** COMP, DATA servidor *m* de impresora

printing *n* APPL, COMP, DATA, ELECT, PER APP, TELEP impresión *f*; ~ **block** PER APP bloque *m* de impresión; ~ **ink** PER APP tinta *f* para imprimir; ~ **speed** PER APP velocidad *f* de impresión; ~ **terminal** TELEP terminal *m* impresor; ~ **track area** ELECT área *f* de pista de impresión

printwheel *n* PER APP margarita *f* de impresión, rueda *f* de impresión

priority *n* GEN prioridad *f*; ~ **call** TELEP llamada *f* con prioridad; ~ **classification** TELEP clasificación *f* prioritaria; ~ **disconnection** TELEP desconexión *f* de prioridad; ~ **extension** TELEP extensión *f* prioritaria; ~ **facility** TELEP servicio *m* de prioridad; **~-indication bit** DATA bitio *m* de indicación de prioridad; ~ **level** TELEP nivel *m* de prioridad; **~-marked line** TELEP línea *f* marcada como prioritaria; ~ **marking** TELEP marcación *f* de prioridad; ~ **with notifying tone** TELEP prioridad *f* con tono avisador; ~ **request** TELEP solicitud *f* de prioridad; ~ **signal** SIG, TELEP señal *f* de prioridad; ~ **subscriber** TELEP abonado *m* con prioridad; ~ **tone** TELEP tono *m* de prioridad

prism *n* ELECT prisma *m*

privacy *n* DATA, PER APP, TELEP privacidad *f*; ~ **button** DATA, PER APP, TELEP botón *m* de comunicación secreta; ~ **circuit** DATA, TELEP circuito *m* secreto; ~ **connection** DATA, TELEP conexión *f* secreta; ~ **release** COMP, DATA, TELEP desconexión *f* de circuito secreto

private *adj* DATA secreto, PER APP de uso privado, TELEP de uso exclusivo; ~ **automatic branch exchange** *n* APPL, SWG, TELEP centralita *f* privada automática conectada a la red pública; ~ **automatic exchange** *n* SWG, TELEP centralita *f* automática; ~ **branch exchange** *n* SWG, TELEP central *f* telefónica privada; ~ **branch network** *n* TELEP, TRANS red *f* de centralita privada; ~ **circuit** *n* TRANS circuito de uso privado; ~ **data-network identification code** *n* DATA código *m* de identificación de red de datos privada; ~ **dial-up port** *n* SWG puerto *m* de marcación privado; ~ **exchange** *n* TELEP central *f* privada; ~ **installation** *n* TELEP instalación *f* privada; ~ **integrated-services network** *n* TELEP, TRANS red *f* de servicios integrados privada; ~ **local system** *n* RADIO sistema *m* local privado; ~ **management domain** *n* DATA, TELEP, TEST ámbito *m* de gestión privada; ~ **manual branch exchange** *n* SWG, TELEP central *f* privada manual; ~ **manual**

exchange *n* SWG, TELEP central *f* manual privada; ~ **mobile radio** *n* RADIO radio *f* móvil privada; ~ **mobile-radio network** *n* RADIO red *f* privada de radio móvil; ~ **network** *n* SIG, TELEP, TRANS red *f* privada; ~ **network exchange** *n* SWG, TELEP central *f* de red privada; ~**-number ringing service** *n* SIG servicio *m* de llamada de número privada; ~ **numbering plan** *n* TELEP plan *m* de numeración privado; ~ **subscriber installation** *n* DATA instalación *f* propiedad del abonado, TELEP instalación *f* privada del abonado; ~ **system** *n* PER APP sistema *m* privado; ~ **telephone network** *n* TELEP, TRANS red *f* telefónica privada; ~ **use** *n* RADIO uso *m* privado; ~ **wire** *n* TRANS circuito *m* privado

probabilistic: ~ **reasoning** *n* DATA razonamiento *m* probabilístico

probability *n* GEN probabilidad *f*; ~ **of acceptance** TEST probabilidad *f* de aceptación; ~ **of congestion** TELEP probabilidad *f* de congestión; ~ **density** GEN densidad *f* de probabilidad; ~ **density function** (*PDF*) DATA función *f* de densidad de probabilidad (*PDF*); ~ **of detection** RADIO probabilidad *f* de detección; ~ **distribution** GEN distribución *f* de probabilidad; ~ **of excess delay** DATA probabilidad *f* de retraso excedente; ~ **of failure** TEST probabilidad *f* de avería; ~ **of indication** RADIO probabilidad *f* de indicación; ~ **of rejection** TEST probabilidad *f* de rechazo; ~ **of shortage** TEST probabilidad *f* de escasez; ~ **of successful service completion** TELEP probabilidad *f* de completar un servicio con éxito; ~ **theory** GEN teoría *f* de la probabilidad

probe *n* ELECT, RADIO, TEST sonda *f*; ~ **current** ELECT corriente *f* de sonda; ~ **microphone** TELEP micrófono *m* sonda; ~ **transformer** RADIO transformador *m* de sonda; ~ **voltage** ELECT tensión *f* de diodo rectificador

problem: ~ **solving** *n* GEN resolución *f* de problemas

procedural: ~ **error** *n* DATA error *m* de procedimiento

procedure *n* GEN procedimiento *m*; ~ **call** RADIO llamada *f* de procedimiento; ~ **epilogue** TELEP epílogo *m* de procedimiento; ~ **interrupt EOM** (*PRI-EOM*) DATA, TELEP final *m* de mensaje de interrupción de procedimiento (*PRI-EOM*); ~ **interrupt EOP** (*PRI-EOP*) DATA final *m* de programa de interrupción de procedimiento, TELEP fin *m* de programa de interrupción de procedimiento (*PRI-EOP*); ~ **interrupt MPS** (*PRI-MPS*) DATA, TELEP MPS *m* de interrupción de procedimiento (*PRI-MPS*); ~ **interrupt negative** (*PIN*) DATA interrupción *f* de procedimiento negativa (*NIP*); ~ **interrupt positive** DATA interrupción *f* de procedimiento positiva, TELEP interrupción *f* de proceso positiva, TRANS interrupción *f* del proceso positiva; ~ **prologue** TELEP prólogo *m* de procedimiento; ~ **words and phrases** TELEP palabras *f* y frases *f* de procedimiento

proceed: ~**-to-send signal** *n* TELEP señal *f* de invitación a transmitir

process[1] *n* COMP, DATA, SWG proceso *m*; ~ **average** TEST promedio *m* de proceso; ~ **capability** TEST capacidad *f* de proceso; ~ **control** CONTR, TEST control *m* de proceso; ~ **in control** TEST proceso *m* en control; ~ **description** TELEP descripción *f* de proceso; ~ **inspection** TEST inspección *f* del proceso; ~ **quality control** TEST control *m* de calidad de proceso

process[2] *vt* COMP, DATA, SWG procesar

processed *adj* COMP, DATA procesado

processing *n* COMP, DATA, SWG procesamiento *m*; ~ **card** COMP, DATA tarjeta *f* de procesamiento; ~ **delay** COMP, DATA demora *f* de procesamiento; ~ **power** COMP, DATA capacidad *f* de procesamiento; ~ **priority code** COMP, DATA código *m* de prioridad de procesamiento

processor *n* COMP, DATA procesador *m*; ~ **architecture** COMP, DATA arquitectura *f* de procesador; ~ **outage** TELEP inoperatividad *f* del procesador; ~ **unit** COMP, DATA procesador *m*

procurement: ~ **policy** *n* GEN política *f* de aprovisionamiento

produce *vt* GEN producir

producer *n* GEN fabricante *m*, productor *m*; ~**'s quality cost** TEST coste *m* (*AmL* costo) de la calidad del productor; ~**'s risk** TEST riesgo *m* del productor; ~**'s risk point** TEST punto *m* de riesgo del productor

product *n* GEN producto *m*; ~ **awareness** GEN conocimiento *m* del producto; ~ **board** ELECT placa *f* de conexión; ~ **liability** TEST responsabilidad *f* legal por el producto; ~ **line** GEN línea *f* de productos; ~ **specification** GEN especificación *f* de fabricación

production *n* GEN producción *f*; ~ **cost** GEN coste *m* (*AmL* costo) de producción; ~ **credits** *n pl* TV títulos *m pl* de crédito de la producción; ~ **level video** (*PLV*) COMP vídeo *m* en fase de producción (*PLV*); ~ **master** ELECT cliché *m* de producción; ~ **permit** TEST licencia *f* de producción

productivity *n* GEN productividad *f*
products: **~ and services** *n pl* APPL productos *m pl* y servicios *m pl*
professional: **~ classification** *n* TELEP clasificación *f* profesional; **~-level** *adj* PER APP en el ámbito profesional; **~ office system** *n* DATA sistema *m* profesional para oficinas
profile *n* GEN perfil *m*; **~ dispersion** TRANS dispersión *f* de perfil; **~-dispersion parameter** TRANS parámetro *m* de dispersión de perfil; **~ parameter** TRANS parámetro *m* de perfil
profitability *n* GEN rentabilidad *f*
profitable *adj* GEN lucrativo, rentable
program[1] *AmE see programme BrE*
program[2]: **~ cartridge** *n* COMP cartucho *m* de programa; **~ counter** *n* COMP contador *m* de instrucciones; **~ debugging** *n* DATA depuración *f* de programa; **~ flip-flop switch** *n* COMP, SWG conmutador *m* basculante de programas; **~ generator** *n* COMP generador *m* de programas; **~ library** *n* COMP biblioteca *f* de programas; **~ memory area** *n* (*PMA*) COMP zona *f* de memoria de programas (*PMA*); **~ package** *n* COMP, SWG conjunto *m* de programas; **~ save** *n* COMP salvaguarda *f* del programa; **~ selector** *n* COMP, SWG selector *m* de programas; **~ store** *n* COMP, SWG memoria *f* de programa; **~ store block** *n* COMP, SWG bloque *m* de memoria de programa; **~ store bus** *n* COMP, SWG bus *m* de memoria de programa; **~ store switch** *n* COMP, SWG conmutador *m* de memoria del programa; **~ store unit** *n* COMP, SWG memoria *f* del programa; **~ switch** *n* COMP, SWG conmutador *m* de programa; **~ test panel** *n* COMP panel *m* de prueba de programas, SWG panel *m* de pruebas del programa
program[3] *vt* COMP programar
programmable *adj* COMP, DATA, PER APP *answering machine* programable; **~ announcements** *n pl* TELEP anuncios *m pl* programables; **~ array logic** *n* COMP lógica *f* de matriz programable; **~ communication interface** *n* TELEP interfaz *m* de comunicación programable; **~ control** *n* SWG control *m* programable; **~ device** *n* ELECT dispositivo *m* programable; **~ keyboard** *n* COMP teclado *m* programable; **~ logic circuit** *n* ELECT circuito *m* lógico programable; **~ logic controller** *n* DATA controlador *m* lógico programable; **~ sequencer** *n* SWG secuenciador *m* programable; **~ unijunction transistor** *n* POWER transistor *m* programable de doble base
programme *BrE* (*AmE* program) *n* GEN programa *m*; **~ availability** TV disponibilidad *f* de programas; **~ booking centre** *BrE* (*PBC*) TV centro *m* de reserva de programas (*PBC*); **~ distribution** TV distribución *f* de programas; **~ schedule** TV programación *f*; **~ sound circuit** TRANS circuito *m* de sonido de programa; **~ track** RADIO pista *f* del programa; **~ tracking** RADIO seguimiento *m* del programa
programmed: **~ command** *n* TELEP orden *f* programada; **~ control** *n* CONTR control *m* programado; **~ sequence controller** *n* COMP controlador *m* de secuencias programado
programmer *n* COMP programador *m*
programming *n* COMP, DATA programación *f*; **~ error** COMP, DATA error *m* de programación; **~ flow chart** COMP organigrama *m* de programación; **~ interface** COMP, DATA interfaz *f* de programación; **~ method** COMP, DATA método *m* de programación; **~ procedure** COMP, DATA procedimiento *m* de programación; **~ system** COMP, DATA sistema *m* de programación
progressive: **~ behaviour** *BrE* *n* COMP, TELEP comportamiento *m* progresivo; **~ coding** *n* COMP, TELEP codificación *f* progresiva; **~-compatible-sequential coding** *n* TELEP codificación *f* secuencial compatible progresiva; **~ scanning** *n* COMP exploración *f* progresiva; **~ switching magazine** *n* SWG almacén *m* de conmutación progresiva; **~ transition** *n* SWG transición *f* progresiva; **~ wave** *n* ELECT, TRANS onda *f* progresiva
project *vt* GEN proyectar
projected: **~ peak point** *n* POWER punta *f* máxima proyectada; **~ peak point voltage** *n* POWER tensión *f* máxima proyectada
projection: **~ television receiver** *n* TV receptor *m* de televisión de pantalla grande; **~ tube** *n* ELECT tubo *m* de proyección
projector *n* GEN proyector *m*
promotor *n* PARTS *of plastics* promotor *m*
prompt *n* COMP apremio *m*; **~ line** COMP línea *f* de petición de orden
prompting *n* TELEP indicación *f*; **~ output** TELEP salida *f* de orientación
proof *n* PER APP prueba *f*; **~ test** TRANS ensayo *m* no destructivo
proofread *vi* PER APP corregir pruebas
proofreading *n* PER APP corrección *f* de pruebas
propagate *vt* ELECT propagar
propagation *n* ELECT propagación *f*; **~ area** GEN zona *f* de propagación; **~ coefficient** ELECT, RADIO, TRANS coeficiente *m* de propagación; **~ constant** ELECT, RADIO, TRANS constante *f* de propagación; **~ delay** RADIO retardo *m* de

propagación; ~ **error** RADIO error *m* de propagación; ~ **mode** TRANS modo *m* de propagación; ~ **performance** TELEP rendimiento *m* de propagación; ~ **properties** *n pl* RADIO propiedades *f pl* de propagación; ~ **time** RADIO, TRANS tiempo *m* de propagación; ~ **time equalization** RADIO, TRANS igualación *f* del tiempo de propagación; ~ **velocity** ELECT, TRANS velocidad *f* de propagación

proper: ~ **time** *n* RADIO tiempo *m* conveniente

proportion *n* COMP, TELEP *of lost calls*, TEST *of nonconforming items* proporción *f*

proportional *adj* COMP, TELEP, TEST proporcional; ~ **action** *n* CONTR acción *f* proporcional; ~ **counter tube** *n* RADIO tubo *m* contador proporcional; ~ **jitter** *n* RADIO fluctuación *f* proporcional de fase; ~ **region** *n* RADIO región *f* proporcional

proposed *adj* DATA propuesto

prosign *n* RADIO prefijo *m* de mensaje

prosodic: ~ **analysis** *n* DATA análisis *m* prosódico

protect *vt* GEN proteger

protecting: ~ **resistor** *n* POWER resistor *m* de protección

protection *n* GEN protección *f*; ~ **against intrusion** TELEP protección *f* contra intromisión; ~ **cable** TRANS cable *m* de protección; ~ **circuit** TEST circuito *m* de protección; ~ **equipment** POWER equipo *m* de protección; ~ **margin** RADIO margen *m* de protección; ~ **ratio** RADIO ratio *f* de protección; ~ **relay** PARTS relé *m* de protección diferencial; ~ **switching** SWG, TELEP conmutación *f* de protección; ~ **switching system** SWG, TELEP sistema *m* de conmutación de protección

protective: ~ **carrying case** *n* GEN cubierta *f* portadora de protección; ~ **case** *n* PARTS caja *f* de protección; ~ **cover** *n* PARTS cubierta *f* protectora; ~ **earth** *BrE n* (*cf protective ground AmE*) POWER, TEST toma *f* de tierra de protección; ~ **earthing** *BrE n* (*cf protective grounding AmE*) POWER, TEST puesta *f* a tierra de protección; ~ **grating** *n* GEN rejilla *f* protectora; ~ **grille** *n* GEN rejilla *f* de protección; ~ **ground** *AmE n* (*cf protective earth BrE*) POWER, TEST toma *f* de tierra de protección; ~ **grounding** *AmE n* (*cf protective earthing BrE*) POWER, TEST toma *f* de tierra de protección; ~ **plastic coating** *n* COMP cubierta *f* protectora de plástico; ~ **relay** *n* PARTS relé *m* de protección; ~ **resistor** *n* PARTS resistor *m* de protección; ~ **sheath** *n* ELECT camisa *f* de protección; ~ **shutter** *n* ELECT obturador *m* de protección

protector *n* POWER disyuntor *m*

protocol *n* COMP, DATA, RADIO *of land mobile* protocolo *m*; ~ **analysis** DATA análisis *m* de protocolos; ~ **class** DATA clase *f* de protocolo; ~ **conformance test report** (*PCTR*) DATA, TEST informe *m* de prueba de conformidad de protocolo (*PCTR*); ~ **control indicator** DATA indicador *m* de control del protocolo; ~ **control information** DATA información *f* para el control del protocolo; ~ **converter** DATA convertidor *m* de protocolo; ~ **data unit** (*PDU*) DATA unidad *f* de datos de protocolo; ~ **discriminator** DATA discriminador *m* de protocolo; ~ **emulator** DATA emulador *m* de protocolos; ~ **error** DATA, TELEP error *m* de protocolo; ~ **error class** DATA, TELEP clase *f* de error del protocolo; ~ **handling software machine** DATA, TELEP máquina *f* de software de gestión de protocolos; ~ **identifier** (*PRT-ID*) DATA identificador *m* de protocolos (*PRT-ID*); ~ **identifier field** DATA campo *m* de identificador de protocolos; ~ **implementation conformance statement** (*PICS*) DATA sentencia *f* de conformidad de instrumentación de protocolos (*PICS*); ~ **implementation extra information for testing** (*PIXIT*) DATA información *f* extra de instrumentación de protocolos para comprobación (*PIXIT*); ~ **machine** DATA máquina *f* de protocolos; ~ **reference model** DATA modelo *m* de referencia de protocolos; ~ **unit** (*PU*) DATA unidad *f* de protocolo

prototype *n* GEN prototipo *m*

provide *vt* GEN proporcionar

provider *n* GEN *device* proveedor *m*; **~-initiated service** DATA servicio *m* iniciado por el proveedor; **~-supported distributed transaction** DATA transacción *f* distribuida respaldada por el proveedor

proving: ~ **stand** *n* TEST banco *m* de pruebas

provisional *adj* GEN provisional

provisory *adj* GEN provisional

PRP *abbr* (*pulse repetition period*) TELEP PRP (*periodo de repetición de los impulsos*)

PRR *abbr* (*pulse repetition rate*) ELECT, SIG, TELEP, TRANS PRR (*frecuencia de repetición de impulsos*)

PRT-ID *abbr* (*protocol identifier*) DATA PRT-ID (*identificador de protocolos*)

prune *vt* COMP podar

PRV *abbr* (*peak reverse voltage*) POWER, TELEP PRV (*tensión máxima inversa*)

ps *abbr* (*picosecond*) TEST ps (*picosegundo*)

PSAP *abbr* (*presentation-service access point*)

DATA, TELEP PSAP (*punto de acceso del servicio de presentación*)
PSE *abbr* (*packet-switching exchange*) SWG, TELEP central *f* de conmutación de paquetes
pseudocode *n* COMP pseudocódigo *m*
pseudo-event *n* DATA pseudoacontecimiento *m*
pseudo-instruction *n* COMP pseudoinstrucción *f*
pseudonoise *n* DATA, TRANS pseudorruido *m*; ~ **modulation** DATA, TRANS modulación *f* por pseudorruido
pseudorandom *adj* DATA, ELECT, TRANS, TV pseudoaleatorio; ~ **binary sequence** *n* DATA secuencia *f* binaria pseudoaleatoria; ~ **bit** *n* DATA bitio *m* pseudoaleatorio; ~ **noise** *n* DATA, TRANS ruido *m* pseudoaleatorio; ~ **noise code** *n* DATA, TRANS código *m* de ruido pseudoaleatorio; ~ **sequence generator** *n* DATA generador *m* de secuencia pseudoaleatoria; ~ **signal** *n* ELECT señal *f* pseudoaleatoria
pseudoternary: ~ **signal** *n* TELEP señal *f* pseudoternaria
PSK *abbr* (*phase-shift keying*) RADIO, SIG CDF (*conmutación de retardo de fase*)
PSN *abbr* (*public switched network*) DATA, SWG, TELEP, TRANS PSN (*red conmutada pública*)
psophometer *n* TELEP sofómetro *m*
psophometric *adj* TELEP sofométrico; ~ **electromotive force** *n* TELEP fuerza *f* electromotriz psofométrica; ~ **emf** *n* TELEP fem *f* sofométrica; ~ **power** *n* TELEP potencia *f* sofométrica; ~ **voltage** *n* TELEP tensión *f* sofométrica
PSS *abbr* (*power subsystem*) POWER PSS (*subsistema de potencia*)
PSTN *abbr* (*public switched telephone network*) DATA, SWG, TELEP, TRANS red *f* telefónica conmutada
psychoacoustic: ~ **model** *n* COMP modelo *m* psicoacústico
psychrometer *n* TELEP psicrómetro *m*
PTM *abbr* (*pulse time modulation*) TELEP, TRANS PTM (*modulación de impulsos en el tiempo*)
P-type: ~ **semiconductor** *n* ELECT, PARTS semiconductor *m* de tipo p
PU *abbr* (*protocol unit*) DATA unidad *f* de protocolo
public: ~**-address equipment** *n* RADIO equipo *m* de megafonía; ~ **call box** *n* TELEP cabina *f* telefónica pública; ~ **call office** *n* TELEP teléfono *m* público; ~ **cordless telephone** *n* TELEP teléfono *m* público inalámbrico; ~ **data network** *n* (*PDN*) DATA, TELEP, TRANS red *f* pública de datos (*RPD*); ~ **data-transmission service** *n* TRANS servicio *m* público de transmisión de datos; ~ **dial-up port** *n* TELEP puerto *m* para discado público; ~ **domain** *n* COMP, DATA dominio *m* público; ~**-domain software** *n* COMP programas *m pl* de dominio público, DATA software *m* de dominio público; ~ **exchange** *n* (*PE*) TELEP central *f* pública (*CP*); ~ **facsimile bureau** *n* TELEP oficina *f* pública de telefax; ~ **facsimile station** *n* TELEP centro *m* público de telefax; ~ **key** *n* RADIO clave *f* pública; ~**-key encryption** *n* RADIO encriptación *f* de clave pública; ~ **line** *n* TELEP línea *f* pública; ~ **mobile network** *n* RADIO, TRANS red *f* de servicio público móvil; ~ **network** *n* RADIO, TRANS red *f* pública; ~ **operator** *n* DATA, TELEP operador *m* público; ~ **packet-switched data network** *n* (*PPSDN*) DATA red *f* pública de datos de paquetes conmutados (*PPSDN*); ~ **relations** *n* (*PR*) TEST relaciones *f pl* públicas (*PR*); ~ **switched network** *n* (*PSN*) DATA, SWG, TELEP, TRANS red *f* conmutada pública (*PSN*); ~ **switched telephone network** *n* (*PSTN*) DATA, SWG, TELEP, TRANS red *f* telefónica conmutada; ~ **telegram service** *n* TELEP, TRANS servicio *m* público de telegrafía; ~ **telegraph network** *n* TELEP, TRANS red *f* telegráfica pública; ~ **telephone** *n* TELEP teléfono *m* público; ~ **telephone booth** *n* TELEP cabina *f* telefónica pública; ~ **telephone exchange** *n* SWG, TELEP central *f* telefónica pública; ~ **telephone network** *n* TELEP, TRANS red *f* telefónica pública; ~ **thoroughfare** *n* GEN vía *f* pública; ~ **utilities** *n pl* GEN servicios *m pl* públicos; ~ **wireless telephony** *n* TELEP telefonía *f* pública móvil; ~ **works** *n pl* GEN obras *f pl* públicas
publicity *n* GEN publicidad *f*; ~ **campaign** GEN campaña *f* de publicidad
publish *vt* GEN editar
publisher *n* COMP, DATA editor *m*
publishing *n* GEN *of book* edición *f*
pull: ~**-down resistor** *n* PARTS resistor *m* de bajada; ~**-off strength** *n* ELECT fuerza *f* de arranque; ~**-out unit** *n* TELEP unidad *f* de arranque; ~**-relief clamp** *n* PARTS grapa *f* guardacable; ~**-up resistor** *n* PARTS resistor *m* de activación
pulled: ~ **surface** *n* PARTS superficie *f* rugosa
pulling *n* TRANS *of cable* tracción *f*, TV alargamiento *m* de la imagen; ~ **figure** ELECT factor *m* de arrastre; ~ **force** TEST fuerza *f* de tracción; ~ **test** PARTS, TEST prueba *f* de tracción
pulsating: ~ **current** *n* POWER corriente *f* pulsatoria; ~ **voltage** *n* POWER tensión *f* pulsatoria
pulse *n* GEN impulso *m*; ~**-address multiple access** (*PAMA*) RADIO acceso *m* múltiple de

dirección de impulso (*PAMA*); ~ **amplitude** TELEP, TRANS amplitud *f* de impulso; ~ **amplitude modulation** TELEP, TRANS modulación *f* de impulsos en amplitud; ~ **amplitude modulation network** TELEP, TRANS red *f* de modulación de impulsos en amplitud; ~ **base** TELEP base *f* de impulsos; ~**-bisecting relay** PARTS relé *m* bisector de impulsos; ~ **broadening** ELECT, TELEP, TRANS ensanchamiento *m* de pulsos; ~ **burst** DATA, TRANS ráfaga *f* de impulsos; ~ **cam disc** *BrE* PER APP leva *f* de impulsos; ~ **carrier** TRANS portadora *f* de pulsos; ~**-code modulation** (*PCM*) COMP, DATA, TELEP, TRANS modulación *f* por impulsos codificados (*PCM*); ~ **coding** DATA, TRANS codificación *f* de impulsos; ~ **compression** ELECT compresión *f* de impulso; ~ **control system** SWG, TELEP sistema *m* de control de impulsos; ~**-correcting register** SIG, TELEP, TRANS registro *m* de correción de impulsos; ~ **correction** SIG, TELEP, TRANS corrección *f* de impulsos; ~**-correction circuit** SIG, TELEP, TRANS circuito *m* de corrección de impulsos; ~**-correction unit** SIG, TELEP, TRANS unidad *f* de corrección de impulsos; ~ **corrector** SIG, TELEP, TRANS corrector *m* de impulso; ~ **counter** ELECT, RADIO, SIG contador *m* de impulsos; ~**-counting technique** ELECT, RADIO, SIG técnica *f* de recuento de impulsos; ~ **decay time** ELECT, TELEP duración *f* del amortiguamiento del impulso; ~ **detection** SIG detección *f* de impulso; ~ **disc** *BrE* PER APP leva *f* de impulsos; ~ **dispersion** ELECT, TELEP, TRANS dispersión *f* de impulsos; ~ **distortion** TELEP distorsión *f* de impulsos; ~**-Doppler radar** RADIO radar *m* Doppler de impulsos; ~ **droop** TELEP inclinación *f* del techo del impulso; ~ **duration** DATA, ELECT, RADIO, TELEP, TRANS duración *f* del impulso; ~ **duration modulation** (*PDM*) DATA, ELECT, RADIO, TELEP, TRANS modulación *f* de impulsos en duración (*PDM*, *PWM*); ~ **duty factor** TELEP coeficiente *m* de impulsos; ~ **echo attenuation** TELEP atenuación *f* del eco del impulso; ~ **echo meter** TELEP ecómetro *m* de impulsos; ~ **echo method** TRANS método *m* del eco de impulsos; ~ **echo return loss** TELEP atenuación *f* de reflexión del eco del impulso; ~ **edge** PARTS, TELEP flanco *m* de impulso; ~ **fall time** TELEP tiempo *m* de descenso del impulso; ~**-forming line** RADIO línea *f* conformadora de impulsos; ~**-forming network** (*PFN*) ELECT red *f* de formación de impulsos (*PFN*); ~ **frequency modulation** (*PFM*) TELEP modulación *f* de frecuencia de impulsos; ~ **generator** ELECT generador *m* de impulsos; ~ **interlacing** TELEP, TRANS entrelazamiento *m* de impulsos; ~ **interleaving** TELEP, TRANS intercalación *f* de impulsos; ~ **interval** TELEP, TRANS intervalo *m* de impulso; ~**-interval modulation** TELEP modulación *f* de pulsos en intervalo; ~ **leading edge** ELECT borde *m* frontal del impulso; ~ **length** DATA, ELECT, RADIO, TELEP, TRANS duración *f* del impulso; ~**-length discriminator** RADIO discriminador *m* de duración de impulsos; ~ **link repeater** (*PLR*) TELEP repetidor *m* de enlace de impulsos (*PLR*); ~ **magnitude** TELEP amplitud *f* de impulso; ~ **mask** TELEP, TRANS máscara *f* del impulso; ~ **meter** TELEP contador *m* de impulsos; ~**-modulated radar** RADIO radar *m* modulado por impulsos; ~ **modulation** COMP, TELEP, TRANS modulación *f* del impulso; ~ **output power** RADIO potencia *f* de salida del impulso; ~**-peak value** TELEP valor *m* máximo del impulso; ~ **position** TELEP posición *f* de los impulsos; ~ **position modulation** (*PPM*) DATA, TELEP, TRANS modulación *f* de impulsos en posición (*PPM*); ~**-no-~ ratio** TELEP relación *f* impulso/pausa; ~ **radar** RADIO radar *m* de impulsos; ~ **rate** TELEP porcentaje *m* de impulsos; ~ **ratio** SWG ratio *f* del impulso; ~**-receiving relay** PARTS relé *m* de recepción de impulsos; ~ **recognition** SIG, TRANS reconocimiento *m* del impulso; ~ **recorder** TELEP, TRANS registrador *m* de impulsos; ~ **recurrence frequency** ELECT, SIG, TELEP, TRANS cadencia *f* de impulsos; ~ **regeneration** GEN regeneración *f* de impulsos; ~ **regenerator** SIG, TELEP regenerador *m* de impulsos; ~ **relay** PARTS relé *m* de impulsos; ~ **repeater** GEN repetidor *m* de impulsos; ~ **repetition** ELECT, SIG, TRANS repetición *f* de impulsos; ~ **repetition frequency** (*PRF*) ELECT, SIG, TELEP, TRANS frecuencia *f* de repetición de impulsos (*PRR*); ~ **repetition period** (*PRP*) TELEP periodo *m* de repetición de los impulsos (*PRP*); ~ **repetition rate** (*PRR*) ELECT, SIG, TELEP, TRANS frecuencia *f* de repetición de impulsos (*PRR*); ~ **repetition time** TELEP tiempo *m* de repetición del impulso; ~ **response** RADIO respuesta *f* de impulso, TELEP respuesta *f* del impulso; ~ **rise time** TELEP tiempo *m* de subida del impulso; ~ **sequence** SIG secuencia *f* de impulsos; ~ **shaper** TELEP conformador *m* de impulsos; ~**-shaping circuit** ELECT circuito *m* configurador de impulsos; ~ **signal** SIG señal *f* de impulso; ~ **signalling** *BrE* TELEP señalización *f* de impulsos; ~ **spacing** TELEP, TRANS intervalo *m* de impulsos; ~ **spectrum** TELEP, TRANS espectro *m* de impulso; ~ **spike** TELEP impulso *m* parásito; ~ **spreading** ELECT, TELEP, TRANS *fibre optics* dispersión *f* de impulsos; ~ **stretcher** TELEP, TRANS dilatador *m* de impulsos; ~ **stuffing** TELEP, TRANS justificación *f*; ~ **time modulation** (*PTM*) TELEP, TRANS modulación *f* de impulsos en el tiempo (*PTM*); ~ **top** TELEP tope *m* del impulso; ~ **trailing edge** ELECT flanco *m* posterior de impulso; ~ **train** TELEP, TRANS tren *m* de impulsos; ~**-train relay** PARTS, TELEP, TRANS relé *m* de tren de impulsos; ~**-train spectrum** TELEP espectro *m* del tren de impulsos; ~ **wave** COMP, TELEP, TRANS onda *f* pulsátil; ~ **widening** ELECT, TELEP, TRANS ensanchamiento *m* de impulso; ~ **width** (*PW*) DATA, ELECT, RADIO, TELEP, TRANS ancho *m* de pulso; ~ **width modulation** (*PWM*) DATA, ELECT, RADIO, TELEP, TRANS modulación *f* de impulsos en duración (*PDM*, *PWM*)

pulsed: ~ **beacon** *n* RADIO baliza *f* por impulsos; ~ **carrier modulation** *n* TELEP modulación *f* de un portador de impulsos; ~ **decay time** *n* ELECT, TELEP tiempo *m* de desvanecimiento del impulso; ~ **magnetron** *n* ELECT, RADIO magnetrón *m* de impulsos

pulser *n* COMP, DATA, TELEP, TRANS generador *m* de impulsos

pulses: ~ **per second** *n* (*pps*) RADIO, TELEP impulsos *m pl* por segundo

pulsing *n* TELEP pulsación *f*; ~ **contact** TELEP contacto *m* de emisión de impulsos; ~ **key** TELEP tecla *f* de emisión de impulsos

pump[1] *n* POWER bomba *f*

pump[2] *vt* POWER bombear

pumping: ~ **voltage** *n* POWER tensión *f* de bombeo

punch: ~**-down block** *n* ELECT bloque *m* de perforaciones; ~**-downs** *n pl* ELECT perforaciones *f pl*; ~**-in** *n* COMP introducción *f*; ~**-out** *n* COMP perforación *f*; ~ **point** *n* COMP punzón *m* de perforación; ~**-through voltage** *n* POWER tensión *f* de perforación

punch in *vt* COMP introducir

punch out *vt* COMP cortar

punched: ~ **tape** *n* TELEP cinta *f* perforada

punctuate *vt* GEN puntuar

punctuation *n* GEN puntuación *f*

puncture *n* POWER perforación *f*

punctured: ~ **convolutional code** *n* COMP código *m* convolucional disruptivo

PUP *abbr* (*principal processing unit*) COMP PUP (*unidad principal de procesamiento*)

pupinization *n* TELEP, TRANS pupinización *f*

pupinize *vt* TELEP, TRANS pupinizar

pupinized *adj* TELEP, TRANS pupinizado

pure: ~**-chance traffic** *n* TELEP tráfico *m* ideal

purely: ~**-resistive circuit** *n* POWER circuito *m* puramente resistivo

push: ~ **button** *n* COMP botón *m* pulsador, PER APP, TELEP botón *m* de llamada; ~**-button dial** *n* PER APP, TELEP discado *m* de marcación con pulsador; ~**-button dialling** *BrE n* PER APP, TELEP marcación *f* por teclas; ~**-button panel** *n* PER APP, TELEP panel *m* de teclas; ~**-button set** *n* PER APP teléfono *m* de teclado, TELEP teclado *m* telefónico; ~**-button switch** *n* GEN interruptor *m* de botón de presión; ~**-button telephone** *n* PER APP, TELEP teléfono *m* de teclado; ~**-button terminal** *n* COMP terminal *m* con pulsador; ~**-down list** *n* COMP, DATA lista *f* de desplazamiento descendente; ~**-down storage** *n* DATA memoria *f* de lectura inversa; ~**-pull** *n* ELECT, POWER contrafase *f*; ~**-pull amplifier** *n* ELECT, POWER amplificador *m* en contrafase; ~**-pull arrangement** *n* ELECT, POWER montaje *m* en contrafase; ~**-pull circuit** *n* ELECT, POWER circuito *m* simétrico; ~**-pull operation** *n* ELECT, POWER funcionamiento *m* en contrafase; ~**-on sneak current use** *n* ELECT uso *m* de la corriente parásita simétrica; ~**-to-talk button** *n* RADIO pulsador *m* de micrófono telefónico; ~**-up list** *n* COMP, DATA lista *f* de desplazamiento ascendente

pushing: ~ **figure** *n* ELECT factor *m* de empuje

put: ~ **on charge** *vt* POWER *battery* poner a cargar; ~ **on discharge** *vt* POWER *battery* poner a descargar; ~ **on file** *vt* TELEP *telex* archivar; ~ **on hold** *vt* TELEP *call, caller* poner en espera; ~ **into operation** *vt* GEN poner en funcionamiento; ~ **into orbit** *vt* RADIO poner en órbita; ~ **into service** *vt* GEN poner en servicio; ~ **in touch with** *vt* TELEP poner en contacto con

put through *vt* SWG, TELEP conectar; ◆ ~ **a call** TELEP establecer una comunicación; ~ **to** TELEP conectar directamente

putty[1] *n* PARTS masilla *f*

putty[2] *vt* PARTS enmasillar

PV *abbr* (*parameter value*) TELEP, TEST valor *m* de parámetro

PVC: ~**-jacketed** *adj* ELECT, PARTS forrado de PVC

PW *abbr* (*pulse width*) DATA, ELECT, RADIO, TELEP, TRANS ancho *m* de pulso

PWM *abbr* (*pulse width modulation*) DATA, ELECT, RADIO, TELEP, TRANS PDM (*modulación de impulsos en duración*, *PWM*)

pyroelectricity *n* ELECT, POWER piroelectricidad *f*

Q

Q: ~ **factor** *n* POWER factor *m* Q

QCIF *abbr* (*quarter common intermediate format*) COMP QCIF (*formato intermedio común dividido en cuatro*)

qdu *abbr* (*quantizing distortion unit*) TELEP, TRANS qdu (*unidad de distorsión de la cuantificación*)

QED *abbr* (*quantum electrodynamics*) ELECT EDQ (*electrodinámica cuántica*)

QFP *abbr* (*quad flatpack*) TRANS QFP (*encapsulado plano de cuadrete*)

QIC *abbr* (*quarter-inch cartridge*) TV QIC (*cartucho de cuarto de pulgada*)

Q-information *n* TEST *on quality* información *f* Q

Q-meter *n* TEST cúmetro *m*

QOS *abbr* (*quality of service*) GEN QOS (*calidad de servicio*); ~ **reference point** *n* (*QRP*) GEN punto *m* de referencia QOS (*QRP*)

QRP *abbr* (*QOS reference point*) GEN QRP (*punto de referencia QOS*)

quad *n* GEN cuadrete *m*; ~ **cable** TRANS cable *m* cuádruple de pares; ~ **flatpack** (*QFP*) TRANS encapsulado *m* plano de cuadrete (*QFP*); ~**-pair cable** TRANS cable *m* de circuitos de cuatro parejas

quadplex *n* ELECT cuádruplex *m*

quadrantal: ~ **error** *n* RADIO error *m* cuadrantal

quadrature: **in** ~ *adj* POWER en cuadratura; ~ **amplitude modulation** *n* DATA, TRANS modulación *f* de amplitud encuadratura; ~ **amplitude shift keying** *n* RADIO desplazamiento *m* de amplitud encuadratura; ~ **component** *n* RADIO componente *m* encuadratura; ~ **mirror filter** *n* TRANS filtro *m* de espejo encuadratura; ~ **partial-response system** *n* DATA sistema *m* de respuesta parcial encuadratura; ~ **phase-shift keying** *n* RADIO, SIG desplazamiento *f* de fase encuadratura

quadriphase: ~ **shift keying** *n* RADIO, SIG desplazamiento *m* en cuadrifase

quadruple: ~**-length register** *n* COMP, DATA registro *m* de longitud cuádruple

qualification: ~ **approval** *n* GEN homologación *f*; ~ **test** *n* GEN ensayo *m* de homologación

qualified: ~ **event** *n* DATA evento *m* cualificado; ~ **name** *n* DATA nombre *m* cualificado

qualifier *n* DATA calificador *m*

qualimetry *n* TEST calimetría *f*

qualitative: ~ **reasoning** *n* DATA razonamiento *m* cualitativo

quality: ~ **assessment** *n* APPL, TEST valoración *f* de calidad; ~ **assurance** *n* APPL, TEST aseguramiento *m* de calidad; ~ **assurance office** *n* APPL, TEST oficina *f* de aseguramiento de calidad; ~ **assurance ruling** *n* TEST norma *f* de aseguramiento de calidad; ~ **audit** *n* APPL, TEST auditoría *f* de calidad; ~ **certificate** *n* TEST certificado *m* de calidad; ~ **chain** *n* APPL cadena *f* de calidad; ~ **circle** *n* TEST círculo *m* de calidad; ~ **control** *n* APPL, TEST control *m* de calidad; ~ **control planning** *n* TEST planificación *f* del control de calidad; ~ **control training** *n* GEN capacitación *f* en el control de calidad; ~ **cost** *n* TEST coste *m* (*AmL* costo) de calidad; ~ **cost accounting** *n* TEST contabilidad *f* del coste de calidad; ~ **degradation** *n* TEST degradación *f* de calidad; ~ **engineering** *n* TEST ingeniería *f* de calidad; ~ **factor** *n* POWER factor *m* de calidad; ~ **improvement** *n* TEST mejoramiento *m* de calidad; ~ **indicator** *n* GEN indicador *m* de calidad; ~ **information conference** *n* APPL conferencia *f* de información sobre calidad; ~ **level** *n* TEST nivel *m* de calidad; ~ **loop** *n* TEST bucle *m* de calidad; ~ **management** *n* APPL, TEST dirección *f* de calidad; ~ **manual** *n* TEST manual *m* de calidad; ~ **measure** *n* TEST evaluación *f* de calidad; ~ **plan** *n* TEST plan *m* de calidad; ~ **policy** *n* TEST política *f* de calidad; ~ **programme** *BrE n* TEST programa de calidad; ~ **requirement** *n* TEST requisitos *m pl* de calidad; ~ **of service** *n* (*QOS*) GEN calidad *f* de calidad (*QOS*); ~ **spiral** *n* TEST espiral *f* de lacalidad; ~ **surveillance** *n* TEST supervisión *f* de calidad; ~ **system** *n* TEST sistema *m* de calidad; ~ **system review** *n* TEST revisión *f* de sistema de calidad

quantification *n* GEN cuantificación *f*

quantify *vt* GEN cuantificar

quantitative: ~ **reasoning** *n* DATA razonamiento *m* cuantitativo

quantization *n* TRANS cuantificación *f*; ~ **error** TRANS error *m* de cuantificación; ~ **matrix** COMP matriz *f* de cuantificación; ~ **noise** ELECT, TRANS ruido *m* de cuantificación; ~ **step** TRANS escalón *m* de la cuantificación

quantize *vt* TRANS cuantificar
quantized: ~ **pulse modulation** *n* TRANS modulación *f* de impulsos cuantificados; ~ **value** *n* TRANS valor *m* cuantificado
quantizer *n* TRANS cuantificador *m*
quantizing *n* TRANS cuantificación *f*; ~ **distortion** TELEP, TRANS distorsión *f* de la cuantificación; ~ **distortion power** TELEP, TRANS capacidad *f* de distorsión de la cuantificación; ~ **distortion unit** (*qdu*) TELEP, TRANS unidad *f* de distorsión de la cuantificación (*qdu*); ~ **interval** TELEP intervalo *m* de cuantificación; ~ **noise** TRANS ruido *m* de cuantificación
quantum: ~ **efficiency** *n* ELECT, TRANS eficiencia *f* cuántico; ~**-efficiency spectral characteristic** *n* ELECT, TRANS característica *f* espectral de la eficiencia cuántica; ~ **electrodynamics** *n* (*QED*) ELECT electrodinámica *f* cuántica (*EDQ*); ~ **electronics** *n* ELECT electrónica *f* cuántica; ~ **Hall effect** *n* ELECT efecto *m* de quantum Hall; ~ **Hall resistance** *n* ELECT resistencia *f* de quantum Hall; ~**-limited operation** *n* ELECT operación *f* cuántica limitada; ~ **mechanics** *n* GEN mecánica *f* cuántica; ~ **noise** *n* ELECT, TRANS ruido *m* cuántico; ~**-noise-limited operation** *n* ELECT, TRANS operación *f* limitada por el ruido cuántico; ~ **theory** *n* GEN teoría *f* cuántica; ~ **voltage** *n* GEN tensión *f* cuántica
quarter: ~ **common intermediate format** *n* (*QCIF*) COMP formato intermedio común dividido en cuatro (*QCIF*); ~**-inch cartridge** *n* (*QIC*) TV cartucho *m* de cuarto de pulgada (*QIC*); ~ **screen** *n* COMP pantalla *f* dividida en cuatro; ~ **turn** *n* ELECT cuarto *m* de vuelta; ~**-wave whip antenna** *n* RADIO antena *f* de látigo de cuarto de onda
quartet *n* DATA cuarteto *m*
quartile *n* GEN cuartil *m*
quartz: ~ **filter** *n* ELECT filtro *m* de cuarzo; ~ **oscillator** *n* ELECT oscilador *m* de cuarzo
quasi: ~**-associated mode** *n* TELEP modo *m* cuasiasociado; ~**-associated mode of operation** *n* TELEP modo *m* de operación cuasiasociado; ~**-associated mode of signalling** *BrE* *n* SIG, TELEP modo *m* de señalización cuasiasociado; ~**-associated signalling** *BrE* *n* SIG, TELEP señalización *f* cuasiasociada; ~**-impulsive interference** *n* RADIO interferencia *f* cuasiimpulsora; ~**-omnidirectional antenna** *n* RADIO antena *f* cuasiomnidireccional; ~**-peak voltage** *n* RADIO tensión *f* de cuasicresta; ~**-synchronous equatorial orbit** *n* RADIO órbita *f* ecuatorial casi sincrónica
quaternary: ~ **phase-shift keying** *n* RADIO, SIG conmutación *f* de variación de fase cuaternaria
quench: ~ **frequency** *n* RADIO frecuencia *f* de corte
query: ~ **language** *n* DATA lenguaje *m* de consulta
queue *n* GEN cola *f*; ~ **control** TELEP control *m* de espera; ~ **file** TELEP archivo *m* de espera; ~ **place** TELEP posición *f* en la cola
queueing *n* RADIO, TELEP, TRANS puesto *m* en cola de espera; ~ **delay** TELEP demora *f* en la cola de espera; ~ **device** TELEP dispositivo *m* de puesta en cola; ~ **equipment** TELEP equipo *m* de cola de espera; ~ **network** TELEP red *f* de sistema de espera; ~ **time** TELEP tiempo *m* de espera en cola
quick *adj* COMP rápido; ~**-access memory** *n* COMP memoria *f* de acceso rápido; ~**-acting fuse** *n* PARTS, POWER fusible *m* ultrarrápido; ~ **charge** *n* POWER carga *f* rápida; ~ **load** *n* COMP carga *f* rápida; ~ **playback** *n* COMP reproducción *f* rápida
quicktime *n* COMP tiempo *m* rápido
quiescent: ~ **carrier modulation** *n* POWER modulación *f* de portadora suprimida; ~ **state** *n* POWER estado *m* de reposo
quintet *n* DATA quinteto *m*
quintuple *n* GEN quíntuplo *m*
quit *vi* COMP salir
quiz *n* COMP interrogatorio *m*
quotation *n* GEN *of price* cotización *f*; ~ **marks** *n pl* GEN comillas *f*

R

R[1] *abbr* (*reluctance*) POWER R (*reluctancia*)

R[2]: ~ **or T pads** *n* TELEP *in telephone extension* almohadillas *f* R o T

raceway: ~ **cover** *n* ELECT guarnición *f* de conducto

raceways *n pl* ELECT conductos *m pl*

rack *n* COMP, POWER *for electronic equipment*, TELEP cremallera *f*; ~ **beam** TELEP travesaño *m* de bastidor; ~ **cabling** TELEP cableado *m* de bastidor; ~ **earth** *BrE* (*cf rack ground AmE*) POWER toma *f* de tierra de bastidor; ~ **entry point** TELEP punto *m* de entrada al bastidor; ~ **ground** *AmE* (*cf rack earth BrE*) POWER toma *f* de tierra del bastidor; ~ **panel** TELEP panel *m* de bastidor; ~ **power** POWER, TELEP energía *f* del bastidor; ~ **power module** POWER, TELEP módulo *m* de energía del bastidor; ~ **power unit** POWER, TELEP unidad *f* de energía del bastidor; ~ **row** TELEP fila *f* de bastidores; ~ **spacing** TELEP separación *f* entre bastidores; ~ **stanchion** TELEP montante *m* del bastidor; ~ **support** TELEP soporte *m* del bastidor; ~ **upright** TELEP montante *m* del bastidor

racon *n* RADIO faro *m*

radar *n* RADIO radar *m*; ~ **aerial** RADIO antena *f* radárica; ~ **altitude control mode** RADIO modo *m* de control de altitud determinada por radar; ~ **angel** RADIO ángel *m*; ~ **antenna** RADIO antena *f* de radar; ~ **beacon** RADIO baliza *f* de radar; ~ **camouflage** RADIO enmascaramiento *m* antirradar; ~ **cross-section** RADIO área *f* de eco; ~ **dome** RADIO radomo *m*; ~ **duplexer** RADIO duplexador *m* de radar; ~ **echo** RADIO eco *m* de radar; ~ **equation** RADIO ecuación *f* del radar; ~ **head** RADIO equipo *m* de radar; ~ **horizon** RADIO horizonte *m* del radar; ~ **range** RADIO alcance *m* límite de radar; ~ **rangefinder** RADIO telémetro *m* radárico; ~ **reflector** RADIO reflector *m* de radar; ~ **reflector buoy** RADIO boya *f* reflectora de radar; ~ **relay** RADIO relé *m* de radar; ~ **set** RADIO equipo *m* de radar; ~ **shadow** RADIO sombra *f* del radar; ~ **signal** RADIO señal *f* de radar; ~**-warning receiver** RADIO detector *m* de radar

radial: ~ **acceleration** *n* COMP aceleración *f* radial; ~**-field cable** *n* GEN cable *m* de campo radial; ~ **leads** *n pl* POWER conductores *m pl* radiales; ~ **transfer** *n* COMP transferencia *f* radial

radiance *n* ELECT, TRANS radiancia *f*

radiant: ~ **efficiency** *n* ELECT, TRANS eficiencia *f* radiante; ~ **emittance** *n* ELECT, TRANS emitancia *f* radiante; ~ **energy** *n* ELECT, TRANS energía *f* radiante; ~ **exitance** *n* ELECT, TRANS exitancia *f* radiante; ~ **flux** *n* ELECT, TRANS flujo *m* energético; ~ **intensity** *n* ELECT, TRANS intensidad *f* radiante; ~ **power** *n* ELECT, TRANS potencia *f* radiante; ~ **sensitivity** *n* ELECT, TRANS sensibilidad *f* radiante

radiate *vt* ELECT, TRANS radiar

radiated: ~ **interference** *n* ELECT interferencia *f* radiada; ~ **noise** *n* ELECT ruido *m* radiado; ~ **power** *n* ELECT potencia *f* radiada

radiating: ~**-cable communication system** *n* TRANS sistema *m* de comunicación por cable radiante; ~ **device** *n* RADIO dispositivo *m* radiante; ~ **element** *n* RADIO elemento *m* radiante; ~ **horn** *n* RADIO bocina *f* radiante

radiation *n* ELECT, RADIO, TRANS radiación *f*; ~ **angle** ELECT, TRANS ángulo *m* de radiación; ~ **counter tube** RADIO tubo *m* contador de radiación; ~ **efficiency** RADIO *of antenna* rendimiento *m* de antena; ~**-hardness** PARTS dureza *f* de radiación; ~ **hardness level** PARTS nivel *m* de dureza de radiación; ~ **mode** ELECT, TRANS modo *m* de radiación; ~ **pattern** RADIO, TRANS diagrama *m* de radiación; ~ **sensitivity** RADIO sensibilidad *f* de radiación; ~ **zone** RADIO zona *f* de radiación

radiator: ~ **area** *n* RADIO zona *f* del radiador

radical *adj* ELECT radical

radio *adj* RADIO radial; ~ **alarm** *n* PER APP radioalarma *f*; ~ **altimeter** *n* RADIO, TELEP radioaltímetro *m*; ~ **amateur** *n* RADIO radioaficionado *m*; ~ **astronomical antenna** *n* RADIO antena *f* astronómica de radio; ~ **base station** *n* RADIO estación *f* base de radio; ~ **base station connection unit** *n* RADIO unidad *f* de conexión de estación base de radio; ~ **beacon** *n* RADIO radiofaro; ~ **blackout** *n* RADIO atenuación *f* de las señales radio; ~ **broadcaster** *n* RADIO emisora *f* de radio; ~ **broadcasting** *n* RADIO radiodifusión; ~ **button** *n* COMP botón *m* de radio; ~ **channel** *n* RADIO canal *m* de radio; ~ **channel spacing** *n* RADIO separación *f* de canales de radio; ~ **circuit** *n* RADIO circuito *m* radio eléctrico; ~ **common carriers** *n pl* RADIO servicios *m pl* públicos de radio; ~ **compass** *n* RADIO radiocompás; ~ **congestion** *n* RADIO sobrecarga *f* de radio; ~ **control** *n* RADIO control *m* de radio; ~ **control system** *n* RADIO sistema *m* de radio control; ~ **coverage** *n* RADIO cobertura *f* de radio; ~ **determination** *n* RADIO radiolocalización; ~**-determination satellite service** *n* RADIO servicio *m* de satélite para radiolocalización; ~ **direction-finding** *n* RADIO radiogoniometría; ~**-equipment maintenance** *n* POWER, RADIO mantenimiento *m* de equipos de radio; ~ **exchange** *n* RADIO central *f* retransmisora de radio; ~ **facsimile** *n* RADIO, TELEP facsímil *m* de radio; ~ **fade** *n* RADIO desvanecimiento *m* radioeléctrico; ~ **fade-out** *n* RADIO desvanecimiento *m* radioléctrico; ~ **frequency** *n* (*RF*) ELECT, RADIO radiofrecuencia *f* (*RF*); ~**-frequency cable** *n* RADIO cable *m* de radiofrencuencia; ~**-frequency channel** *n* RADIO canal *m* de radiofrecuencia; ~**-frequency disturbance** *n* RADIO perturbación *m* de radiofrecuencia; ~**-frequency input power** *n* RADIO potencia *f* de entrada de radiofrecuencia; ~**-frequency interference** *n* (*RFI*) RADIO interferencia *f* de radiofrecuencia (*RFI*); ~**-frequency protection ratio** *n* RADIO ratio *f* de protección de radiofrecuencia; ~**-frequency radiation** *n* RADIO radiación *f* de radiofrecuencia; ~**-frequency scanner** *n* RADIO digitalizador *m* de radiofrecuencia; ~**-frequency signal-to-signal interference ratio** *n* RADIO ratio *f* de interferencia señal/señal de radiofrecuencia; ~ **guidance** *n* RADIO radioguía *f*; ~ **ham** *n* RADIO radioaficionado *m*; ~ **homing** *n* RADIO recalada *f* por radio; ~ **horizon** *n* RADIO horizonte *m* radioeléctrico; ~**-horizon distance** *n* RADIO distancia *f* del horizonte radioeléctrico; ~ **in the loop** *n* RADIO radio *f* intercalado en circuito; ~ **influence voltage** *n* DATA tensión *f* de influencia radioeléctrica; ~ **jamming** *n* RADIO emisión *f* de ondas perturbadoras; ~ **link** *n* RADIO enlace *m* de microondas; ~ **location** *n* RADIO radiolocalización; ~**-location satellite service** *n* RADIO servicio *m* de satélite para radiolocalización; ~ **mast** *n* RADIO torre *f* de antena; ~ **navigation** *n* RADIO radionavegación; ~ **network** *n* RADIO red *f* de emisoras; ~ **noise** *n* RADIO ruido *m* radioeléctrico; ~ **path** *n* RADIO trayecto *m* radioeléctrico; ~ **programme** *BrE n* RADIO programa de radio; ~ **propagation** *n* RADIO radiopropagación; ~**-propagation prediction** *n* RADIO previsión *f* de propagación radioléctrica; ~ **range station** *n* RADIO estación *f* para determinar el alcance; ~ **rangefinding** *n* RADIO telemetría *f* radioeléctrica; ~ **receiver** *n* PER, RADIO, TELEP aparato *m* de radio; ~**-relay station** *n* RADIO estación *f* de radio enlace; ~**-relay system** *n* RADIO, TRANS sistema *m* de radio enlace; ~ **route** *n* RADIO ruta *f* radioeléctrica; ~ **shadow** *n* RADIO sombra *f* radioeléctrica; ~ **silence** *n* RADIO silencio *m* de radio; ~ **spectrum** *n* ELECT espectro *m* radioeléctrico; ~ **station** *n* RADIO estación *f* deradio; ~ **subsystem** *n* RADIO subsistema *m* radioeléctrico; ~ **survey** *n* RADIO radiovigilancia *f*; ~ **system** *n* RADIO red *f* radioeléctrica; ~ **teleprinter** *n* RADIO radioteleimpresora *f*; ~ **theodolite** *n* RADIO radioteodolito; ~ **tracking** *n* RADIO seguimiento *m* radioeléctrico; ~ **transmitter** *n* RADIO radiotransmisor *m*; ~ **unit** *n* RADIO unidad *f* de radio; ~ **warning** *n* RADIO radiovigilancia *f*; ~ **wave** *n* ELECT, RADIO onda *f* radioeléctrica; ~ **wind** *n* RADIO radioanemómetro *m*

radioactive *adj* GEN radioactivo

radiocommunication *n* RADIO radiocomunicación *f*; ~ **relay** POWER, RADIO retransmisor *m* de radiocomunicaciones; ~ **service** RADIO servicio *m* de radiocomunicaciones

radiogoniometer *n* RADIO radiogoniómetro *m*

radiographic: ~ **contrast** *n* RADIO contraste *m* radiográfico

radiopager *n* DATA, RADIO, TELEP, TRANS radiobuscador *m*

radiopaging *n* DATA, RADIO, TELEP, TRANS radiobúsqueda *f*; ~ **service** DATA, RADIO, TELEP, TRANS servicio *m* de radiobúsqueda; ~ **system** DATA, RADIO, TELEP, TRANS sistema *m* de radiobúsqueda

radiophone *vt* RADIO, TELEP radiotelefonear

radiosity *n* COMP radiosidad *f*

radiosonde *n* RADIO radiosonda *f*

radiotelephone *n* RADIO, TELEP radioteléfono *m*

radiotelephony *n* RADIO, TELEP radiotelefonía *f*

radius *n* GEN *of root circle* radio *m*

radome *n* RADIO radomo *m*

RAG *abbr* (*ring again*) TELEP RAG (*repita la llamada*)

RAI *abbr* (*remote-alarm indication*) TEST indicación *f* de alarma remota

railings *n pl* ELECT empalizada *f*

railroad: ~ **radio system** *AmE n* (*cf railway radio system BrE*) RADIO sistema *m* de radiocomunicaciones ferroviarias; ~ **signaling system** *AmE n* (*cf railway signalling system BrE*) CONTR, SIG sistema *m* de señalización ferroviaria; ~ **telecommunications system** *AmE n* (*cf railway telecommunications system BrE*) RADIO sistema *m* de telecomunicaciones ferroviarias; ~ **telephony** *AmE n* (*cf railway telephony BrE*) RADIO, TELEP telefonía *f* ferroviaria

railway: ~ **radio system** *BrE n* (*cf railroad radio system AmE*) RADIO sistema *m* de radiocomunicaciones ferroviarias; ~ **signalling system** *BrE n* (*cf railroad signaling system AmE*) CONTR, SIG sistema *m* de señalización ferroviaria; ~ **telecommunications system** *BrE n* (*cf railroad telecommunications system AmE*) RADIO sistema *m* de telecomunicaciones ferroviarias; ~ **telephony** *BrE n* (*cf railroad telephony AmE*) RADIO, TELEP telefonía *f* ferroviaria

rain: ~ **cell** *n* RADIO célula *f* de lluvia; ~ **clutter** *n* RADIO *on radar* ecos *m pl* de lluvia; ~ **fading** *n* RADIO desvanecimiento *m* por lluvia; ~ **model** *n* RADIO modelo *m* de lluvia

rainfall: ~ **rate** *n* RADIO intensidad *f* de lluvia

rainproof *adj* PARTS imbrífugo

raise: ~ **an interrupt** *vt* ELECT producir una interrupción

raised: ~ **cosine filtering** *n* GEN filtro *m* de coseno elevado

RAM *abbr* (*random access memory*) COMP RAM (*memoria de acceso aleatorio*); ~ **cartridge** *n* DATA cartucho *m* de RAM; ~ **disk** *n* DATA disco *m* de RAM

Raman: ~ **scattering** *n* TRANS dispersión *f* de Raman

ramark *n* RADIO radiofaro *m* para radar

ramp *n* TELEP rampa *f*; ~ **function** TELEP función *f* en rampa; ~ **response** CONTR respuesta *f* a la función en rampa; ~ **voltage** POWER tensión *f* en rampa

RAN *abbr* (*recorded announcement*) GEN mensaje *m* grabado

R & D *abbr* (*research and development*) TEST I & R (*investigación y desarrollo*)

random *adj* GEN aleatorio; ~ **access** *n* COMP acceso *m* aleatorio; ~ **access device** *n* COMP mecanismo *m* de acceso aleatorio; ~ **access memory** *n* (*RAM*) COMP memoria *f* de acceso aleatorio (*RAM*); ~**allotment** *n* TELEP selección *f* aleatoria; ~ **allotter** *n* TELEP distribuidor *m* aleatorio; ~ **block filemanager** *n* COMP gestor *m* aleatorio de ficheros en bloque; ~ **cause** *n* TEST causa *f* aleatoria; ~**error** *n* DATA, TELEP, TEST error *m* aleatorio; ~**excitation** *n* ELECT excitación *f* aleatoria; ~**factor** *n* COMP factor *m* aleatorio; ~**failure** *n* TEST fallo *m* aleatorio; ~ **field** *n* RADIO campo *m* de propiedades estadísticas; ~ **memory access** *n* COMP acceso *m* aleatorio a la memoria; ~ **multipleaccess** *n* COMP acceso *m* múltiple aleatorio; ~**metering** *n* TELEP cómputo *m* aleatorio; ~**noise** *n* ELECT, RADIO, TELEP ruido *m* aleatorio; ~ **numberelement** *n* TELEP elemento *m* numérico aleatorio; ~ **numbergeneration** *n* TELEP generación *f* numérica aleatoria; ~**process** *n* ELECT proceso *m* aleatorio; ~**pulse** *n* ELECT impulso *m* fortuito; ~ **repetitivemetering** *n* TELEP cómputo *m* repetitivo aleatorio; ~**sampling** *n* ELECT muestreo *m* aleatorio; ~**selection** *n* TELEP selección *f* al azar; ~**signal** *n* ELECT señal *f* fortuita; ~**traffic** *n* TELEP tráfico *m* aleatorio; ~**value** *n* COMP valor *m* aleatorio; ~**variable** *n* GEN variable *f* estadística

randomization *n* GEN aleatoriedad *f*

randomize *vt* GEN *pitch* aleatorizar

range *n* GEN rango *m*; ~ **of adjustment** GEN margen *m* de ajuste; ~**-amplitude display** RADIO presentación *f* distancia-amplitud; ~ **of audibility** TELEP intervalo *m* audible; ~**-bearing display** RADIO presentación *f* distancia-demora; ~ **of bearings** RADIO gama *f* azimutal; ~ **chart** TEST gráfico *m* de recorridos; ~ **discrimination** RADIO definición *f* en distancia; ~**-finding antenna spacing** RADIO separación *f* de antena radiogonométrica; ~ **gate** RADIO puerta *f* de banda; ~**-height indicator** RADIO radar *m* integrador distancia-altitud; ~ **marker** RADIO marcador *m* de alcance; ~ **of regulation** GEN margen *m* de regulación; ~ **resolution** RADIO resolución *f* de distancia; ~ **search** RADIO exploración *f* en distancia

rangefinding *n* RADIO telemetría *f*

ranging *n* RADIO telemetría *f*

rank *n* COMP número *m* de nivel; ~ **of selectors** TELEP juego *m* de selectores

ranking: ~ **order** *n* GEN orden *f* jerárquico

rapid: ~**-access loop** *n* DATA bucle *m* de acceso rápido; ~ **charge** *n* POWER *of battery* carga *f* rápida; ~ **mode** *n* DATA *of scanner* modo *m* rápido

rare: ~ **gas** *n* CLIM gas *m* raro; ~ **gas tube** *n* TEST tubo *m* de gas raro

rarefaction *n* RADIO rarefacción *f*

raster *n* ELECT, RADIO trama *f*; ~ **burn** ELECT quemadura *f* de la trama

rasterize *vt* COMP tramar

ratchet: ~ **control** *n* ELECT mando *m* de trinquete
rate *n* GEN tasa *f*; ~ **of approach** TEST velocidad *f* de acercamiento; ~**-of-change relay** PARTS relé *m* de velocidad de variación; ~ **of extension** TELEP ritmo *m* de ampliación; ~ **of increase** GEN tasa *f* de incremento; ~ **of interruption** TELEP tasa *f* de interrupción; ~ **one half convolutional coding** DATA codificación *f* convolucional de media unidad de velocidad; ~ **1/2 Golay code** DATA código *m* de Golay de media unidad de velocidad; ~ **parameter** COMP parámetro *m* de velocidad; ~ **timeband** TELEP banda *f* horaria de velocidad
rated *adj* POWER nominal; ~ **burden** *n* PARTS régimen *m* nominal; ~ **current** *n* POWER corriente *f* de régimen; ~ **frequency** *n* POWER frecuencia *f* de régimen; ~ **output level** *n* COMP, ELECT nivel *m* de salida nominal; ~ **output power** *n* POWER potencia *f* de salida nominal; ~**power** *n* POWER potencia *f* nominal; ~ **value** *n* POWER valor *m* asignado; ~ **voltage** *n* POWER voltaje *m* de régimen
rating *n* POWER, TELEP, TEST, TRANS clasificación *f*; ~ **plate** POWER placa *f* indicadora
ratio *n* GEN relación *f*, ratio *f*; ~ **circuit** ELECT circuito *m* de relación; ~ **of compression** TELEP relación *f* de compresión; ~ **of expansion** TELEP relación *f* de expansión; ~ **of transformation** PARTS relación *f* de transformación
ratioless: ~ **circuit** *n* ELECT circuito *m* sin relación
rational: ~ **subgroup** *n* TEST subgrupo *m* racional
raw: ~ **data** *n pl* DATA datos *m pl* no analizados; ~**-data counter** *n* DATA contador *m* de datos en bruto; ~ **video display** *n* RADIO visualización *f* en bruto
rawin *n* RADIO radiosonda *f*; ~ **flight** RADIO radiosonda *f*; ~ **observation** RADIO observación *f* radioviento; ~ **sounding** RADIO radiosondeo *m*
ray *n* GEN rayo *m*; ~ **optics** ELECT, TRANS óptica *f* de rayos; ~**-path transmission loss** RADIO pérdida *f* de transmisión de la trayectoria de rayos; ~ **tracing** COMP delineamiento *m* de rayos
Rayleigh: ~ **channel** *n* DATA, TELEP, TRANS canal *m* de Rayleigh; ~ **fading** *n* DATA, TELEP, TRANS desvanecimiento *m* de Rayleigh; ~ **scattering** *n* DATA, TELEP, TRANS dispersión *f* de Rayleigh
RBE *abbr* (*remote batch entry*) DATA RBE (*entrada por lotes a distancia*)
RBER *abbr* (*residual bit error ratio*) RADIO RBER (*ratio de error de binario residual*)
RBV *abbr* (*return-beam vidicon*) TELEP RBV (*vidicón con haz de retorno*)
RC *abbr* (*resistance-capacitance*) ELECT RC (*resistencia-capacitancia*); ~ **coupling** *n* ELECT acoplamiento *m* RC
RCB *abbr* (*redrive counter busy*) TELEP RCB (*ocupación de contador de retransmisión*)
RCD: ~ **circuit** *n* ELECT circuito *m* RCD
RCMSS *abbr* (*remote-control monitoring scan system*) CONTR RCMSS (*sistema de vigilancia de barrido por control remoto*)
RCP *abbr* (*restoration control point*) TELEP RCP (*punto de control de recuperación*)
RCTP *abbr* (*retransmitted condensed burst time plan*) DATA RCTP (*plan temporal condensado de retransmisión por ráfagas*)
RCU *abbr* (*remote-control unit*) CONTR unidad *f* de control remoto
rcvr *abbr* (*receiver*) DATA, PER APP, RADIO, TELEP rcvr (*receptor*)
RD *abbr* (*receive data*) DATA RD (*datos recibidos*)
RDA *abbr* (*remote database access*) DATA RDA (*acceso remoto a base de datos*)
R-DAT *abbr* (*rotary-head digital audio tape*) COMP, DATA R-DAT (*cinta de audio digital de cabeza rotatoria*)
RDB *abbr* (*relational database*) DATA base *f* de datos relacional
RDU *abbr* (*remote data unit*) DATA RDU (*unidad de datos remota*)
reach *n* TELEP *subscriber* alcance *m*
reactance *n* POWER reactancia *f*; ~ **relay** PARTS, POWER relé *m* de reactancia
reaction: ~ **time** *n* GEN tiempo *m* de reacción
reactivate *vt* DATA *scanner* reactivar
reactivation *n* DATA reactivación *f*
reactive: ~ **attenuator** *n* RADIO atenuador *m* reactivo; ~ **current** *n* POWER corriente *f* reactiva; ~ **load** *n* ELECT carga *f* reactiva; ~ **power** *n* POWER potencia *f* reactiva
read[1]: ~**-back check** *n* COMP control *m* por relectura; ~**-only memory** *n* (*ROM*) COMP, GEN memoria *f* sólo de lectura (*ROM*); ~**-out** *n* COMP, GEN extracción *f* mediante lectura; ~**-write head** *n* PARTS cabezal *m* de lectura-escritura; ~**-write memory** *n* COMP, DATA memoria *f* delectura-escritura
read[2] *vt* GEN leer
reader *n* GEN lector *m*
reading *n* GEN lectura *f*; ~ **error** COMP error *m* de lectura; ~ **gun** ELECT cañón *m* de lectura; ~ **speed** ELECT velocidad *f* de lectura; ~ **wire** TELEP hilo *m* de lectura

readjust *vt* GEN reajustar
readjustment *n* GEN reajuste *m*
read-out *n* COMP lectura *f* de salida
ready: ~ **for data** *adj* TRANS listo para datos; ~ **indication** *n* TELEP indicación *f* de disponibilidad; ~ **indicator** *n* TELEP indicador *m* de disponibilidad; ~ **on-line** *adj* DATA disponible en línea; ~ **for sending** *adj* DATA listo para enviar; ~ **stand by** *adj* DATA disponible en espera; ~ **to activate** *adj* GEN *symbol* listo para activar; ~**-to-change request** *n* DATA solicitud *f* de listo para cambiar; ~**-to-commit state** *n* DATA estado *m* de listo para grabar
real[1]: **in** ~ **time** *adv* COMP, DATA, TELEP, TEST en tiempo real
real[2]: ~ **address** *n* COMP direccional *m* real; ~ **end system** *n* DATA sistema *m* final real; ~ **file store** *n* DATA almacenamiento *m* de fichero real; ~ **open system** *n* DATA sistema *m* abierto real; ~ **storage** *n* DATA memoria *f* real; ~ **subnetwork** *n* DATA subred *f* real; ~ **system** *n* DATA sistema *m* real; ~ **time** *n* COMP, DATA tiempo *m* real; ~**-time control** *n* COMP control *m* en tiempo real; ~**-time conversion facility** *n* DATA servicio *m* de conversión en tiempo real; ~**-time editing** *n* COMP edición *f* en tiempo real; ~**-time effect** *n* COMP efecto *m* en tiempo real; ~**-time expert system** *n* SWG sistema *m* experto en tiempo real; ~**-time recording** *n* COMP grabación *f* en tiempo real; ~**-time transmission** *n* COMP transmisión *f* en tiempo real; ~**-time video** *n* COMP, TV vídeo *m* en tiempo real; ~**-time writing** *n* COMP escritura *f* en tiempo real; ~ **type** *n* DATA tipo *m* real
realign *vt* GEN realinear
realignment *n* GEN realineación *f*
reallocated *adj* GEN reasignado
reallocation *n* GEN reasignación *f*
ream *vt* PARTS avellanar
reanswer: ~ **signal** *n* SIG señal *f* de reiteración de respuesta
rear *n* GEN parte *f* posterior; ~ **cover** GEN tapa *f* posterior; ~ **edge** GEN borde *m* posterior; ~ **panel** COMP panel *m* posterior; ~ **plate** GEN placa *f* posterior; ~ **row** TELEP fila *f* posterior; ~ **side** GEN parte *f* posterior
rearrange *vt* GEN redisponer
rearrangeable: ~ **non-blocking network** *n* TRANS red *f* no bloqueante de reordenamiento
rearrangement *n* GEN redisposición *f*, reordenamiento *m*
reasoning *n* DATA razonamiento *m*
reassembling *n* TELEP reensamblaje *m*
reassign *n* RADIO reasignar
reassigned *adj* RADIO reasignado
reassignment: ~ **procedure** *n* RADIO procedimiento *m* de reasignación
reattempt *n* TELEP reintento *m*
rebalancing *n* GEN reajuste *m* del equilibrio
Rebecca-Eureka: ~ **system** *n* RADIO sistema *m* Rebecca-Eureka
reboot *vt* COMP reiniciar
rebooting *n* COMP reiniciación *f*, recarga *f*, DATA recebado *m*
rebroadcast[1] *n* RADIO, TRANS, TV retransmisión *f*
rebroadcast[2] *vt* RADIO, TRANS, TV retransmitir
rebroadcasting *n* RADIO, TRANS, TV retransmisión *f*; ~ **transmitter** RADIO, TRANS, TV retransmisor *m*
rebuild *vt* COMP reconstruir
rebuilding *n* COMP reconstrucción *f*
REC *abbr* (*receive*) DATA REC (*recibir*)
rec. *abbr* (*recreational*) DATA entr. (*de entretenimiento*)
recalibrate *n* GEN recalibrar
recalibration *n* GEN recalibración *f*
recall *n* COMP, TELEP repetición *f* de llamada; ~ **button** TELEP botón *m* de códigos; ~**-edit buffer** COMP búfer *m* de edición para rellamada
receipt *n* GEN recepción *f*; ~ **notification** (*RN*) DATA, TELEP notificación *f* de recepción (*RN*)
receive[1]: ~**-amplified handset telephone** *n* TELEP aparato *m* telefónico de recepción amplificada; ~**-amplified telephone** *n* TELEP teléfono *m* de recepción amplificada; ~ **antenna** *n* RADIO antena *f* receptora; ~ **buffer** *n* COMP búfer *m* de recepción; ~ **channel** *n* COMP, TRANS canal *m* de recepción; ~ **characteristic** *n* RADIO característica *f* de recepción; ~ **conditions** *n* COMP condiciones *f pl* de recepción; ~ **data** *n pl* (*RD*) DATA datos *m pl* recibidos (*RD*); ~**-data lamp** *n* TELEP lámpara *f* de recepción de datos; ~ **direction** *n* TEST, TRANS dirección *f* de recepción; ~ **echo loss** *n* TELEP amortiguación *f* de eco de recepción; ~ **end echo** *n* TELEP eco *m* final de recepción; ~ **event** *n* DATA acaecimiento *m* de recepción; ~ **fibre-optic** *BrE* **terminal device** *BrE n* TRANS dispositivo *m* terminal de recepción de fibra óptica; ~**-frame synchronization procedure** *n* (*RFS*) TRANS procedimiento *m* de sincronización del marco de recepción (*RFS*); ~ **gain** *n* RADIO ganancia *f* de recepción; ~ **lamp** *n* TELEP lámpara *f* de recepción; ~ **loss** *n* TELEP debilitación *f* en recepción; ~ **machine** *n* TRANS máquina *f* de recepción; ~**-not-ready frame** *n* TELEP trama *f* de no preparado para recibir; ~ **objective loudness rating** *n* (*ROLR*)

TELEP valoración *f* de sonoridad del objetivo de recepción (*ROLR*); **~-ready frame** *n* TELEP trama *f* de preparado para recibir
receive[2] *vt* (*REC*) DATA recibir (*REC*)
received: **~ data** *n pl* DATA datos *m pl* recibidos (*RD*)
receiver *n* (*rcvr*) DATA, PER APP, RADIO, TELEP receptor *m* (*rcvr*); **~ cap** PER APP, TELEP auricular *m*; **~ case** PER APP, TELEP caja *f* de auricular; **~ distributor** RADIO distribuidor *m* de recepción; **~ diversity** RADIO diversidad *f* de recepción; **~ headset** PER APP, TELEP auriculares *m pl*; **~ insertion** PER APP, SIG, TELEP inserción *f* de receptor; **~ multicoupler** RADIO multiacoplador *m* de escucha; **~-noise figure** RADIO cifra *f* de ruido del receptor; **~ radiation** RADIO radiación *f* de receptor; **~ rejection** RADIO rechazo *m* de receptor; **~ sensitivity** (*RX sens*) RADIO sensibilidad *f* del receptor; **~ station** TV estación *f* de recepción; **~ threshold level** RADIO nivel *m* umbral del receptor; **~ transfer time** TELEP tiempo *m* de transferencia del receptor
receiving: **~ antenna** *n* RADIO antena *f* de recepción; **~-application entity** *n* DATA dispositivo *m* para aplicación de recepción; **~ channel** *n* COMP canal *m* de recepción; **~ device** *n* COMP aparato *m* receptor, TELEP aparato receptor; **~ inspection** *n* TEST *quality* inspección *f* de recepción; **~ mode** *n* RADIO modo *m* de recepción; **~ perforator** *n* TELEP reperforador *m*; **~ sensitivity** *n* TELEP *of telephone system* sensibilidad *f* de recepción; **~ terminal** *n* TRANS terminal *m* receptor
receptacle *n* GEN receptáculo *m*
reception: **~ area** *n* RADIO área *f* de recepción; **~ channel** *n* COMP canal *m* de recepción; **~ conditions** *n* COMP condiciones *f pl* de recepción; **~ congestion** *n* TELEP congestión *f* de la recepción; **~ definition** *n* TELEP definición *f* de recepción
recess *n* GEN entrante *m* de pared
recessed: **~ access brackets** *n* ELECT soportes *m pl* de acceso embutidos; **~ mounting** *n* GEN montaje *m* embutido
recipient *n* GEN recibidor *m*; **~ identification code** DATA, TELEP código *m* de identificación del recibidor; **~ reference** DATA referencia *f* del recibidor; **~-reference qualifier** DATA calificador *m* de la referencia del receptor; **~'s reference password** COMP, DATA clave *f* de referencia del receptor
reciprocal *adj* GEN recíproco; **~ bearing** *n* RADIO marcación *f* recíproca; **~ circuit** *n* ELECT circuito *m* recíproco; **~ mixing** *n* RADIO mezcla *f* recíproca
reciprocity *n* ELECT reciprocidad *f*; **~ theorem** POWER teorema *m* de la reciprocidad
reclock *vt* ELECT resincronizar
reclocking *n* ELECT resincronización *f*
recognition *n* TELEP *of signal* reconocimiento *m*; **~ circuit** ELECT circuito *m* de reconocimiento
recognized *adj* GEN autorizado, oficial; **~ dealer** *n* GEN concesionario *m* oficial; **~ operating agency** *n* APPL, DATA agencia *f* oficial; **~ private operating agency** *n* (*RPOA*) APPL, DATA agencia *f* privada autorizada (*RPOA*); **~ private operation** *n* APPL, DATA representación *f* autorizada
recombination *n* PARTS recombinación *f*; **~ rate** PARTS porcentaje *m* de recombinación
recombining *n* TELEP recombinación *f*
recommendation: **~ indicator** *n* DATA indicador *m* de recomendación
reconditioned: **~ carrier reception** *n* GEN recepción *f* de portadora reacondicionada
reconfiguration *n* COMP reconfiguración *f*
reconfigure *vt* COMP reconfigurar
reconfiguring *n* COMP reconfiguración *f*
reconnection *n* GEN reconexión *f*
reconstruct *vt* COMP, DATA reconstruir
reconstructed: **~ picture** *n* COMP imagen *f* reconstruida; **~ sample** *n* TELEP muestra *f* reconstruida
reconstruction *n* COMP, DATA reconstrucción *f*
record[1]: **~ call** *n* TELEP llamada *f* de anotación; **~ cancel** *n* COMP anulación *f* de registro; **~ desk** *n* TELEP mesa *f* de anotaciones; **~ line** *n* TELEP línea de anotación; **~-line relay set** *n* TELEP juego *f* de relés de línea de anotación; **~ stand-by** *n* COMP reserva *f* de registro; **~ tone** *n* TELEP tono *m* de registro
record[2] *vt* GEN *documentation* registrar
recordable *adj* COMP, TELEP registrable
recorded *adj* RADIO *broadcast* registrado, TELEP anotado, TV grabado; **~ announcement** *n* (*RAN*) GEN mensaje *m* grabado; **~ announcement machine** *n* PER, TELEP máquina *f* de anuncios registrados; **~ announcement trunk** *n* TELEP centralita *f* de mensajes grabados; **~ broadcast** *n* RADIO, TV emisión diferida; **~ message** *n* GEN, TELEP mensaje *m* grabado; **~ public information service** *n* TELEP servicio *m* público de información registrada; **~ service** *n* TELEP servicio *m* grabado
recorder *n* APPL, PER APP registrador *m*
recording *n* PER APP, TELEP grabación *f*; **~ centre** *BrE* SWG, TELEP centro *m* de grabación; **~ channel** TELEP canal *m* de registro; **~-com-**

pleting trunk TELEP línea *f* telefónica; **~-completing trunk group** TELEP grupo *m* de enlace de llamada de la anotadora y de salida; **~ density** DATA densidad *f* de grabación; **~ head** GEN cabeza *f* de registro; **~ instrument** TEST instrumento *m* de registro; **~ interval** GEN periodo *m* de registro; **~ keyboard** COMP teclado *m* de registro; **~ operator** TELEP operador *m* de registro; **~ period** GEN periodo *m* de registro; **~ tempo** COMP ritmo *m* de anotación

recover *vt* COMP recuperar

recovered: **~ carrier** *n* TRANS portadora *f* recuperada; **~ clock** *n* DATA reloj *m* recuperado

recovery *n* DATA, RADIO, TELEP, TEST reactivación *f*; **~ log** DATA diario *m* de recuperaciones; **~ procedure** DATA proceso *m* de restablecimiento; **~ signal** GEN señal *f* de recuperación; **~ time** DATA, ELECT, PARTS duración *f* de restablecimiento

recreational *adj* (*rec.*) DATA de entretenimiento (*entr.*)

rectangular: **~ faceplate** *n* ELECT placa *f* frontal rectangular; **~ pulse** *n* TELEP impulso *m* rectangular; **~ wave** *n* TRANS onda *f* rectangular

rectenna *n* RADIO rectenna *f*

rectification *n* POWER rectificación *f*

rectifier *n* POWER rectificador *m*; **~ bridge** POWER puente *m* rectificador; **~ meter** POWER, TEST contador *m* de rectificador; **~ tube** ELECT, POWER tubo *m* rectificador; **~ unit** POWER unidad *f* rectificadora; **~ valve** ELECT, POWER válvula *f* rectificadora

rectify *vt* GEN rectificar

rectifying: **~ antenna** *n* RADIO antena *f* rectificadora; **~ device** *n* POWER dispositivo *m* rectificador; **~ tube** *n* ELECT, POWER tubo *m* rectificador; **~ valve** *n* ELECT, POWER válvula *f* rectificadora

rectilinear: **~ antenna** *n* RADIO antena *f* rectilínea

recuperate *vt* RADIO recuperar

recuperation *n* RADIO recuperación *f*

recursion *n* COMP, DATA recursión *f*

recursive *adj* COMP, DATA recursivo; **~ digital filter** *n* DATA filtro *m* digital recurrente; **~ filter** *n* DATA, TRANS filtro *m* recursivo; **~ function** *n* COMP, DATA función *f* recursiva; **~ neurons** *n pl* TRANS neuronas *f pl* recurrentes

recursiveness *n* COMP, DATA recurrencia *f*

red: **~ book** *n* COMP libro *m* rojo; **~-green-blue** *adj* TV rojo-verde-azul; **~ pass filter** *n* ELECT filtro *m* intermedio del rojo

redesign *vt* GEN rediseñar

redial[1]: **~ button** *n* PER APP botón *m* de rediscado

redial[2] *vt* TELEP rediscar

redirect *vt* RADIO *vehicle*, SWG, TELEP reexpedir; ◆ **~ traffic** SWG, TELEP redireccionar el tráfico

redirected: **~ call** *n* TELEP llamada *f* redireccionada; **~ call indicator** *n* TELEP indicador *m* de llamada redireccionada; **~ call signal** *n* SIG señal *f* de llamada redireccionada; **~ traffic** *n* TELEP tráfico *m* redireccionado

redirecting: **~ counter** *n* SWG contador *m* de reexpedición; **~ indicator** *n* SWG indicador *m* de reexpedición; **~ number** *n* SWG, TELEP número *m* de reexpedición; **~ reason** *n* SWG razón *f* de reexpedición

redirection *n* SWG, TELEP nueva dirección; **~ address** TELEP dirección *f* redireccionada; **~ address indicator** TELEP indicador *m* de llamada redireccionada; **~ number** SWG número *m* de reexpedición; **~ request signal** SIG señal *f* de solicitud de redireccionamiento

redirector *n* DATA redireccionador *m*

redrive: **~ counter** *n* TELEP contador *m* de retransmisión; **~ counter busy** *n* (*RCB*) TELEP ocupación *f* de contador de retransmisión (*RCB*)

reduce *vt* COMP, TEST adelgazar

reduced: **~-carrier emission** *n* RADIO emisión *f* de portadora reducida; **~-carrier transmission** *n* RADIO transmisión *f* de portadora reducida; **~ carrier wave** *n* RADIO onda *f* portadora reducida; **~ error** *n* TEST error *m* reducido; **~ inspection** *n* TEST inspección *f* reducida; **~-instruction-set computer** *n* (*RISC*) COMP, DATA ordenador *m* con programación reducida *Esp*, computador *m* con programación reducida *AmL*, computadora *f* con programación reducida *AmL*; **~ model** *n* GEN modelo *m* reducido; **~ rate** *n* APPL tarifa *f* reducida; **~ tariff** *n* APPL tarifa *f* reducida

reducing: **~ valve** *n* PARTS válvula *f* de escape

reduction *n* COMP, TEST *of sample* rebaja *f*; **~ of tariff** APPL reducción *f* de tarifa

redundancy *n* GEN redundancia *f*; **~ check** DATA verificación *f* de redundancia; **~ switching** TEST conmutación *f* por redundancia

redundant: **~ array of independent disks** *n* COMP agrupación *f* redundante de discos independientes; **~ carrier-supply** *n* TRANS suministro *m* de portadora redundante; **~ code** *n* SIG código *m* redundante; **~-component check** *n* POWER verificación *f* de componente redundante; **~ n-ary signal** *n* SIG señal *f* N-aria redundante

reed *n* PARTS lengüeta *f*; **~ contact** PARTS

contacto *m* de láminas; ~ **element** PARTS elemento *m* de lengüeta; ~ **frequency meter** TEST medidor *m* de frecuencia de lámina; ~ **relay** PARTS, POWER, SWG relé *m* de láminas; ~**-relay crosspoint** PARTS, POWER, SWG punto *m* de cruce del relé de láminas; ~**-relay electronic exchange** PARTS, POWER, SWG central *f* electrónica de relé de láminas; ~**-relay system** PARTS, POWER, SWG sistema *m* de relé de láminas

Reed-Solomon: ~ **code** *n* COMP código *m* Reed-Solomon

reel *n* DATA carrete *m*; ~ **hub** TELEP cubo *m* portabobina; ~ **of paper** PER APP rollo *m* de papel

re-emission: ~ **time** *n* RADIO tiempo *m* de reemisión

re-enterable: ~ **program** *n* COMP, TELEP programa *m* reintroducible; ~ **program subroutine** *n* COMP, TELEP subrutina *f* de programa reintroducible; ~ **subroutine** *n* TELEP subrutina *f* reintroducible

re-entrant: ~ **trunking** *n* TELEP vías *f pl* de tráfico reintroducibles

re-entry *n* TELEP reentrada *f*; ~ **point** COMP punto *m* de reentrada

reference: ~ **antenna pattern** *n* RADIO patrón *m* de referencia de antena; ~ **atmosphere for refraction** *n* RADIO atmósfera *f* de referencia para refracción; ~ **clock** *n* TRANS reloj *m* de referencia; ~ **conditions** *n pl* TEST condiciones *f pl* de referencia; ~ **designation** *n* COMP designación *f* de referencia; ~ **equivalent** *n* TELEP equivalente *m* de referencia; ~ **frequency** *n* RADIO, TRANS frecuencia *f* de referencia; ~ **guide** *n* COMP guía *f* de referencia; ~ **installation** *n* TELEP instalación *f* de referencia; ~ **level** *n* COMP nivel *m* de referencia; ~ **manual** *n* GEN manual *m* de referencia; ~ **model** *n* DATA modelo *m* de referencia; ~ **picture** *n* COMP, TV imagen *f* de referencia; ~**-pilot generation** *n* RADIO, TRANS generación *f* de onda de referencia; ~ **plane** *n* ELECT plano *m* de referencia; ~ **point** *n* DATA punto *m* de referencia; ~ **publication format** *n* (*RPF*) DATA formato de publicación de referencia (*RPF*); ~ **range** *n* TEST margen *m* de referencia; ~ **routing** *n* SWG encaminamiento *m* de referencia; ~ **sensibility** *n* RADIO, TRANS sensibilidad *f* de referencia; ~ **sensitivity** *n* RADIO, TRANS sensibilidad *f* de referencia; ~ **standardized abstract test-suite** *n* DATA, TEST batería *f* de pruebas abstractas estandarizadas de referencia; ~**-status display** *n* RADIO visualizador *m* de estado de referencia; ~ **surface** *n* TELEP, TRANS superficie *f* de referencia; ~**-surface centre** *BrE n* TELEP, TRANS centro *m* de superficie de referencia; ~**-surface diameter** *n* TELEP, TRANS diámetro *m* de superficie de referencia; ~**-surface diameter deviation** *n* TELEP, TRANS desviación *f* de diámetro de superficie de referencia; ~ **system** *n* GEN sistema *m* de referencia; ~ **telephonic power** *n* TELEP potencia *f* telefónica de referencia; ~ **terminal status interchange** *n* (*RTSI*) RADIO intercambio *m* de estado de terminal de referencia (*RTSI*); ~ **test method** *n* TRANS método *m* de prueba de referencia; ~ **unique word** *n* DATA palabra *f* única de referencia; ~**-usable field strength** *n* RADIO intensidad *f* de campo utilizable como referencia; ~**-usable power flux density** *n* RADIO densidad *f* de flujo de fuerza utilizable como referencia; ~ **value** *n* TEST valor *m* de referencia; ~ **variable** *n* CONTR variable *f* de referencia; ~ **voltage** *n* POWER tensión *f* de referencia; ~**-voltage generator** *n* POWER generador *m* de tensión de referencia

referencing *n* RADIO referencia *f*

referential: ~ **link** *n* COMP enlace *m* de referencia

refine *vt* COMP, DATA refinar

refinement *n* COMP, DATA refinación *f*

reflectance *n* ELECT, RADIO, TRANS, TV reflectancia *f*; ~ **diagram** TV diagrama *m* de reflectancia

reflected: ~ **binary code** *n* DATA, ELECT código *m* binario reflejado; ~ **current** *n* TRANS corriente *f* reflejada; ~ **pulse** *n* RADIO impulso *m* reflejado; ~ **voltage** *n* TRANS tensión *f* reflejada

reflecting: ~ **antenna** *n* RADIO antena *f* reflectante; ~ **satellite** *n* RADIO satélite *m* reflector

reflection *n* GEN reflexión *f*; ~ **coefficient** RADIO, TRANS coeficiente *m* de reflexión; ~ **factor** ELECT, RADIO, TRANS factor *m* de reflexión; ~ **gain** TRANS ganancia *f* por reflexión; ~ **loss** POWER, TRANS pérdida *f* de reflexión; ~ **mapping** COMP correlación *f* de reflexión; ~**-mode filter** RADIO filtro *m* de modo de reflexión; ~ **vector** COMP vector *m* de reflexión

reflectometer *n* ELECT, POWER, TEST, TRANS reflectómero *m*

reflector *n* GEN reflector *m*; ~ **antenna** RADIO antena *f* de reflector

reflex: ~ **amplifier** *n* ELECT amplificador *m* reflejo; ~ **klystron** *n* ELECT klistron *m* de reflexión

reflexive *adj* COMP, DATA reflexivo

reflow: ~ **soldering** *n* ELECT soldadura *f* por refusión de estaño
refracted: ~ **near-end method** *n* TRANS método *m* local refractado; ~ **near-field technique** *n* TRANS técnica *f* de campo próximo refractado; ~ **ray** *n* TRANS rayo *m* refractado; ~-**ray method** *n* TRANS método *m* de rayos refractados
refraction *n* RADIO, TRANS, TV refracción *f*; ~ **index** ELECT, RADIO, TRANS, TV índice *m* de refracción
refractive: ~ **index** *n* ELECT, RADIO, TRANS, TV índice *m* de refracción; ~-**index contrast** *n* TRANS contraste *m* del índice de refracción; ~-**index distribution** *n* TRANS distribución *f* del índice de refracción; ~-**index profile** *n* TRANS perfil *m* del índice de refracción; ~ **modulus** *n* RADIO módulo *m* de refracción
refractivity *n* ELECT, RADIO, TRANS, TV refractividad *f*
refractor *n* GEN medio *m* refringente
reframe *n* COMP recuperación *f*
reframing: ~ **time** *n* TELEP tiempo *m* de recuperación
refresh: ~ **rate** *n* COMP, DATA velocidad *f* de regeneración
refreshment: ~ **procedure** *n* COMP, DATA procedimiento *m* de regeneración
refrigerant *n* CLIM refrigerante *m*
refrigerate *vt* CLIM refrigerar
refrigeration *n* CLIM refrigeracion *f*; ~ **circuit** CLIM circuito *m* de refrigeración; ~ **engineering** CLIM tecnología *f* de refrigeración
refund *n* GEN reembolso *m*; ~ **chute** PER APP *of payphone* cajetín *m* de devolución de moneda
regenerate *vt* GEN *signal* regenerar
regeneration *n* GEN regeneración *f*; ~ **rate** COMP velocidad *f* de regeneración
regenerative: ~ **link** *n* TRANS enlace *m* regenerativo; ~ **pulse repeater** *n* TELEP repetidor *m* generador de impulsos; ~ **receiver** *n* RADIO receptor *m* regenerativo; ~ **repeater** *n* TRANS repetidor *m* reactivo; ~ **satellite** *n* RADIO satélite *m* regenerativo
regenerator *n* (*RG*) ELECT, SIG, SWG, TELEP recuperador *m* (*RG*); ~ **section** (*RS*) DATA sección *f* del regenerador; ~-**section overhead** DATA línea *f* aérea de la sección del regenerador; ~-**section termination** (*RST*) DATA terminación *f* de la sección del regenerador (*RST*); ~-**timing generator** (*RTG*) DATA generador *m* de temporización del regenerador *m* (*RTG*)
region *n* ELECT región *f*; ~ **of limited proportionality** RADIO región *f* de proporcionalidad limitada
regional: ~ **Bell operating company** *n* TELEP compañía *f* telefónica regional de sistema Bell; ~ **centre** *BrE n* TELEP centro *m* regional; ~ **processor** *n* (*RP*) SWG procesador *m* regional
register *n* DATA, RADIO, SWG, TELEP registrador *m*; ~ **addressing** SWG direccionamiento *m* de registro; ~-**clear signal** TELEP señal *f* de liberación del registrador; ~-**controlled system** SWG, TELEP sistema *m* de control por registradores; ~-**disconnect signal** TELEP señal *f* de desconexión del registro; ~ **end-of-selection signal** TELEP señal *f* de fin de selección del registrador; ~ **finder** SWG buscador *m* registrador; ~-**finder marker** SWG marcador *m* de buscador del registrador; ~ **function** TELEP, TRANS función *f* del registrador; ~ **length** DATA longitud *f* de registro; ~-**MS abstract-operation** DATA operación *f* abstracta de MS de registrador; ~ **organization** SWG organización *f* de registradores; ~-**preserve memory** SWG memoria *f* de conservación de registros; ~ **recall** TELEP, TRANS rellamada *f* del registrador; ~-**release signal** SIG, TELEP señal *f* de desconexión del registrador; ~-**reset signal** SIG, TELEP señal *f* de reposición a cero del registro; ~ **signal** SIG, TELEP señal *f* del registro; ~ **signalling** *BrE* SIG señalización *f* de registradores, TELEP señalización *f* de registradores, señalización *f* entre registradores; ~-**transfer level** SWG nivel *m* de transferencia del registro; ~ **translator** SWG registrador *m* traductor
registered *adj* GEN registrado, RADIO nominativo; ~ **access** *n* COMP acceso *m* registrado; ~ **earth station** *n* RADIO estación *f* terrena registrada; ~ **trademark** *n* GEN marca *f* registrada; ~ **user** *n* COMP usuario *m* registrado
registration *n* DATA, ELECT, RADIO, TV registro *m*; ~-**accepted signal** SIG señal *f* de registro aceptado; ~ **authority** DATA organismo *m* encargado del registro; ~-**completion signal** SIG señal *f* de registro concluido; ~-**hierarchical name tree** DATA árbol *m* de nombre de registro jerárquico; ~ **identifier** DATA identificador *m* de registro; ~ **of incoming calls** TELEP registro *m* de llamadas de entrada; ~ **request signal** SIG señal *f* de solicitud de registro
regression: ~ **equation** *n* GEN ecuación *f* de regresión
regroup *vt* COMP reagrupar
regular *adj* GEN *normal* uniforme; ~

replacement *n* TEST sustitución *f* regular; ~ **signalling** *BrE* **link** *n* SIG enlace de señalización regular; ~ **transmissions** *n pl* TRANS transmisiones *f pl* regulares

regularity: ~ **loss** *n* TELEP atenuación *f* de regularidad

regulate *vt* GEN regular

regulated: ~ **competition** *n* GEN competencia *f* regulada; ~ **line section** *n* TELEP sección *f* de regulación de línea

regulation *n* GEN regulación *f*; ~ **accuracy** GEN precisión *f* de regulación; ~ **relay** PARTS relé *m* de regulación

regulator *n* COMP, CONTR, ELECT regulador *m*

regulatory: ~ **administration** *n* GEN administración *f* reguladora

reignition *n* DATA, ELECT, RADIO reencendido *m*; ~ **voltage** ELECT tensión *f* de reencendido

reimburse *vt* GEN *person* reembolsar

reimbursement *n* GEN *of person* reembolso *m*

reimport *vt* DATA reimportar

reinforce *vt* GEN reforzar

reinforced: ~ **glass** *n* PARTS vidrio *m* reforzado; ~ **plastic** *n* PARTS plástico *m* armado

reinforcement *n* GEN reforzamiento *m*

reinitialization *n* COMP, DATA reiniciación *f*

reinitialize *vt* COMP, DATA reiniciar

reject[1] *n* COMP, TEST rechazo *m*; ~ **character** DATA carácter *m* de rechazo; ~ **frame** TELEP trama *f* de rechazo

reject[2] *vt* COMP, TEST rechazar

rejection *n* COMP, TEST rechazo *m*; ~ **number** TEST número *m* de rechazo; ~ **output** TELEP salida *f* de rechazo

rejective: ~ **condition** *n* TEST criterio *m* de rechazo

REL *abbr* (*release message*) TEST mensaje *m* de liberación

related: ~ **indication primitive** *n* DATA primitivo *m* de indicación relacionada

relation *n* TELEP relación *f*

relational: ~ **database** *n* (*RDB*) DATA base *f* de datos relacional

relationship *n* COMP parentesco *m*

relative: ~ **address** *n* COMP dirección *f* relativa; ~ **amplitude** *n* TELEP *of elementary echo* amplitud *f* relativa; ~ **build-up time** *n* RADIO *of telegraph signal* tiempo *m* de establecimiento relativo; ~ **density** *n* GEN densidad *f* relativa; ~ **distinguished name** *n* DATA nombre *m* caracterizado relativo; ~ **error** *n* TEST error *m* relativo; ~**-intensity noise** *n* (*RIN*) ELECT ruido *m* de intensidad relativa (*RIN*); ~ **level** *n* TELEP nivel *m* relativo; ~ **order** *n* COMP orden *f* relativa; ~ **power level** *n* TELEP nivel *m* relativo de potencia real o aparente; ~ **redundancy** *n* GEN redundancia *f* relativa; ~ **spectral sensitivity** *n* ELECT sensibilidad *f* espectral relativa; ~ **spectral sensitivity characteristic** *n* ELECT característica *f* de sensibilidad espectral relativa

relaxation: ~ **oscillator** *n* ELECT oscilador *m* de relajación

relay *n* PARTS, RADIO, TELEP, TRANS, TV relé *m*; ~**-armature stop** POWER parada *f* de armadura de relé; ~ **bar** PARTS barra *f* de relés; ~ **chain** PARTS cadena *f* de relés; ~ **coil** PARTS, POWER bobina *f* de relé; ~ **contact** PARTS contacto *m* de relé; ~ **contact unit** PARTS unidad *f* de contacto de relé; ~ **core** PARTS núcleo *m* de relé; ~ **driver** PARTS, POWER excitador *m* de relé; ~ **entity** DATA cuerpo *m* de relé; ~ **frame** TELEP marco *m* de relé; ~ **magnet** ELECT, PARTS electroimán *m* de relé; ~ **mounting bar** PARTS barra *f* de relés; ~ **network** RADIO red *f* de retransmisión; ~ **set** CONTR, PARTS, TELEP grupo *m* de relés; ~ **space** TELEP posición *f* de relés; ~ **spacing** TELEP separación *f* de relés; ~ **station** RADIO estación *f* repetidora; ~ **support** TELEP soporte *m* de relé; ~ **system** SWG, TELEP sistema *m* de retransmisión; ~ **test-point** PARTS punto *m* de prueba de relé; ~ **transmitter** RADIO transmisor *m* repetidor; ~ **tree** PARTS árbol *m* de relés; ~**-type switch** SWG, TELEP selector *m* de contactos de relé; ~ **winding** PARTS bobinado *m* de relé; ~ **yoke** ELECT, PARTS, TELEP culata *f* del relé

relaying: ~ **and routing function** *n* SWG función *f* de encaminamiento y de relevo

release[1]: ~ **condition** *n* PARTS posición *f* de reposo; ~ **current** *n* POWER corriente *f* de liberación; ~**-guard signal** *n* SIG, TELEP señal *f* de desbloqueo; ~ **indicator** *n* DATA, SIG indicador *m* de liberación; ~ **lever** *n* TELEP palanca *f* de desembrague; ~ **-link trunk** *n* (*RLT*) TELEP línea *f* troncal de enlaces de desconexión (*RLT*); ~ **mechanism** *n* ELECT mecanismo *m* de desconexión; ~ **message** *n* (*REL*) TEST mensaje de liberación *f*; ~ **rate** *n* COMP velocidad *f* de desconexión; ~ **signal** *n* DATA señal *f* de desconexión; ~ **time** *n* PARTS tiempo *m* de lanzamiento; ~ **trunk** *n* TELEP línea *f* troncal de desconexión; ~ **velocity** *n* COMP velocidad *f* de desconexión; ~**-velocity sensitivity** *n* COMP sensibilidad *f* de velocidad de desconexión; ~ **wire** *n* SWG hilo *m* activo

release[2] *vt* COMP, PARTS, TELEP *connection* desconectar (*DCN*); ◆ ~ **by shunting** PARTS desprender por shunt

released *adj* COMP, PARTS levante

relevance: ~ **feedback** *n* DATA realimentación *f* de aplicabilidad

relevant *adj* GEN importante; ~ **failure** *n* TEST fallo *m* significativo; ~ **time** *n* TEST tiempo *m* aplicable

reliability *n* GEN confiabilidad *f*; ~ **block diagram** TEST esquema *m* de bloques de fiabilidad; ~ **engineering** TEST ingeniería *f* de la fiabilidad; ~ **measure** TEST evaluación *f* de la fiabilidad; ~ **and quality measurement system** TEST sistema *m* de evaluación de la calidad y la fiabilidad

reliable *adj* GEN fiable; ~ **transfer** *n* DATA transferencia *f* fiable; ~**-transfer server** *n* COMP portador *m* de transferencia fiable; ~**-transfer service element** *n* DATA, TELEP elemento *m* de servicio de transferencia fiable

relief: ~ **telecommunications** *n pl* TELEP telecomunicaciones *f pl* de relevo

relieve *vt* COMP, TELEP aliviar

reload *vt* COMP recargar

reloading *n* COMP recarga *f*

relocatable: ~ **address** *n* COMP, TELEP dirección *f* reubicable; ~ **program** *n* COMP programa *m* reubicable

relocate *vt* COMP, DATA, RADIO reubicar

reluctance *n* (*R*) POWER reluctancia *f* (*R*); ~ **tuning** RADIO sintonización *f* de reluctancia

remanence *n* ELECT, POWER remanencia *f*

remanent: ~ **relay** *n* POWER relé *m* de remanencia

remeasure *vt* GEN reevaluar

reminder: ~ **alarm service** *n* APPL, TELEP servicio *m* de alarma recordatorio; ~ **call** *n* APPL, PER APP *of answering machine*, TELEP llamada *f* recordatoria; ~ **indication** *n* TELEP indicación *f* recordatoria; ~ **ring** *n* TELEP timbre *m* recordatorio; ~ **service** *n* TELEP servicio *m* recordatorio

remote: ~ **access** *n* DATA, TELEP acceso *m* remoto; ~**-access answering machine** *n* TELEP contestador *m* automático de acceso remoto; ~ **alarm** *n* RADIO, TELEP alarma *f* a distancia; ~**-alarm indication** *n* (*RAI*) TEST indicación *f* de alarma remota; ~ **announcement modification** *n* PER APP *of answering machine* modificación *f* de anuncio remoto; ~ **batch entry** *n* (*RBE*) DATA entrada *f* por lotes a distancia (*RBE*); ~ **batch processing** *n* COMP, DATA teleproceso *m* por lotes; ~ **blocking** *n* CONTR bloqueo *m* a distancia; ~ **boot** *n* ELECT arranque *m* a distancia; ~ **concentration unit** *n* TRANS unidad *f* de concentración remota; ~ **concentrator** *n* TRANS concentrador *m* remoto; ~ **console** *n* DATA pupitre *m* de proceso a distancia; ~ **consultation** *n* GEN consulta *f* a distancia; ~ **control** *n* GEN accionamiento *m* a distancia; ~**-controlled exchange** *n* TELEP centralita *f* telecomandada; ~ **controller** *n* PER APP *of answering machine* control *m* remoto; ~**-control monitoring scan system** *n* (*RCMSS*) CONTR sistema *m* de vigilancia de barrido por control remoto (*RCMSS*); ~**-control unit** *n* (*RCU*) CONTR unidad *f* de control remoto; ~**-control warning system** *n* CONTR sistema *m* de alarma por control remoto; ~ **CSE** *n* TELEP CSE *m* remoto; ~ **cutoff tube** *n* ELECT válvula *f* de corte remoto; ~ **database access** *n* (*RDA*) DATA acceso *m* remoto a base de datos (*RDA*); ~ **data processing** *n* COMP, DATA, RADIO teleprocesamiento *m* de datos; ~ **data unit** *n* (*RDU*) DATA unidad *f* de datos remota (*RDU*); ~ **detection** *n* TEST teledetección *f*; ~ **digital concentrator** *n* TELEP concentrador *m* digital remoto; ~ **document access** *n* DATA, TELEP acceso *m* remoto a documentos; ~ **document management** *n* DATA, TELEP gestión *f* remota de documentos; ~ **error** *n* DATA error *m* remoto; ~ **exchange concentrator** *n* TELEP concentrador *m* de centralita remoto; ~**-gate thyristor** *n* ELECT tiristor *m* de puerta remota; ~ **indication** *n* DATA teleafiche *m*; ~ **interrogation** *n* PER APP *of answering machine* interrogación *f* a distancia; ~**-interrogation by mf ph** *n* PER APP *of answering machine* interrogación *f* remota por mando telefónico; ~**-interrogation by telecommand** *n* PER APP *of answering machine* interrogación *f* remota por telecomando; ~**-job entry** *n* (*RJE*) DATA entrada *f* de teletrabajo; ~ **line concentrator** *n* TRANS concentrador *m* de líneas remoto; ~ **loading** *n* TEST carga *f* remota; ~ **log-in** *n* ELECT procedimiento *m* de entrada remoto; ~ **maintenance** *n* COMP, DATA, TEST, TV telemantenimiento *m*; ~ **management** *n* TEST teledirección *f*; ~ **measuring** *n* TEST telemedida *f*; ~ **metering** *n* DATA, TV telemedición *f*; ~ **meter reading** *n* DATA, TV lectura *f* remota de medidor; ~ **monitoring** *n* TEST monitorización *f* a distancia; ~ **monitoring system** *n* TEST sistema *m* de seguimiento remoto; ~ **multiplexer** *n* TRANS multiplexador *m* remoto; ~ **operating centre** *BrE n* RADIO centro *m* operativo remoto; ~ **operating terminal** *n* PER APP terminal *m* operativo remoto; ~ **operation** *n* (*RO*) DATA, TELEP, TEST operación *f* remota (*RO*); ~ **operation service element** *n* DATA, TELEP, TEST elemento *m* de servicio de manejo remoto; ~ **payment** *n* TV telepago *m*; ~ **peripheral equipment** *n*

(*RPE*) COMP equipo *m* periférico remoto (*RPE*); ~ **playback** *n* PER APP *of answering machine* reproducción *f* a distancia; ~ **printing** *n* COMP teleimpresión *f*; ~ **processing** *n* CONTR, DATA procesamiento *m* remoto; ~ **reading** *n* GEN lectura *f* a distancia; ~ **resource** *n* DATA fuente *f* remota; ~ **secure voice system** *n* (*REMSEVS*) RADIO sistema *m* remoto de voz protegida (*REMSEVS*); ~ **security** *n* TRANS, TV seguridad *f* a distancia; ~ **security network** *n* TRANS, TV red *f* de seguridad a distancia; ~ **sensing** *n* GEN detección *f* a distancia; ~ **sensing satellite** *n* RADIO satélite *m* de teledetección; ~ **single-layer embedded test method** *n* TEST método *m* incorporado de prueba a distancia de una capa; ~ **single-layer test method** *n* TEST método *m* de prueba a distancia de una sola capa; ~ **site** *n* GEN sede *f* remota; ~ **spectrum-monitoring** *n* (*RSM*) RADIO control *m* espectral a distancia (*RSM*); ~ **start** *n* CONTR telearranque *m*; ~ **station** *n* DATA estación *f* a distancia; ~ **subscriber stage** *n* TELEP selector *m* remoto de abonado; ~ **subscriber switch** *n* TELEP selector *m* remoto de abonado; ~ **supervision** *n* CONTR, TV supervisión *f* a distancia; ~ **surveillance** *n* CONTR, TV televigilancia *f*; ~ **survey** *n* CONTR, TV supervisión *f* a distancia; ~ **switching** *n* SWG interruptor *m* a distancia; ~ **switching point** *n* SWG punto *m* remoto de conmutación; ~ **switching stage** *n* SWG etapa *f* remota de conmutación; ~ **switching system** *n* SWG sistema *m* remoto de conmutación; ~ **switching unit** *n* (*RSU*) SWG unidad *f* remota de conmutación; ~ **terminal** *n* PER APP terminal *m* a distancia; ~ **terminal unit** *n* PER APP terminal *m* remoto; ~ **test method** *n* TEST método *m* de prueba a distancia; ~ **transmission** *n* TRANS teletransmisión *f*; ~ **unit** *n* (*RU*) SWG unidad *f* remota

remotely[1] *adv* GEN a distancia

remotely[2]: ~**-connected** *adj* TELEP conectado a distancia; ~**-controlled exchange** *n* SWG centralita *f* de control remoto; ~**-controlled station** *n* RADIO estación *f* telecontrolada; ~**-generated display** *n* DATA visualización *f* generada a distancia; ~**-interrogable** *adj* PER APP interrogable a distancia

removable *adj* GEN cambiable, extraíble; ~ **cartridge** *n* COMP cartucho extraíble; ~ **keyed shroud** *n* ELECT protección *f* en clave desactivable; ~ **magnetic disk** *n* COMP disco *m* magnético extraíble

removal *n* ELECT, POWER retirada *f*

remove *vt* ELECT, POWER desbarbar, retirar; ◆ ~ **from service** TELEP retirar de servicio; ~ **the sheath** GEN retirar la cubierta

remreed: ~ **crosspoint** *n* SWG punto *m* de cruce de lengüeta rem

REMSEVS *abbr* (*remote secure voice system*) RADIO REMSEVS (*sistema remoto de voz protegida*)

remuneration: ~ **for exclusive use of circuits** *n* TELEP pago *m* por el uso exclusivo de circuitos; ~ **for shared use of circuits** *n* TELEP pago *m* por el uso compartido de circuitos

rename *vt* COMP *file* renominar, cambiar de nombre de

renaming *n* COMP cambio *m* de nombre

render *vt* COMP rendir; ~ **inoperative** GEN dejar fuera de servicio

rendering: ~ **technique** *n* COMP técnica *f* de representación

renewal *n* GEN renovación *f*; ~ **function** TEST función *f* de renovación

renovate *vt* GEN renovar

renovation *n* GEN renovación *f*

rental *n* GEN renta *f*; ~ **charge** GEN cargo *m* por alquiler

renting *n* GEN locación *f*

reorder: ~ **buffer** *n* COMP memoria *f* de reaprovisionamiento; ~ **tone** *n* TELEP tono *m* de congestión

repaginate *vt* COMP repaginar

repair *n* GEN reparación *f*; ~ **rate** TEST tasa *f* de reparaciones; ~ **shop** TELEP taller *m* de reparaciones; ~ **time** TELEP tiempo *m* de reparación

repaired: ~ **item** *n* TEST artículo *m* reparado

repairing: ~ **cables** *n* TRANS cables *m pl* de reparación

repeat: ~ **action key** *n* COMP tecla *f* de servicio de repetición; ~ **call** *n* TELEP llamada *f* de repetición; ~ **coil** *n* TELEP bobina *f* de repetición

repeatability *n* GEN repetibilidad *f*

repeated: ~ **call attempt** *n* TELEP reintento *m* de llamada

repeater *n* DATA, ELECT, SIG, TELEP, TRANS, TV repetidor *m*; ~ **deck** SIG cubierta *f* de repetidor; ~ **fuse** POWER fusible *m* de repetidor; ~**-jammer** ELECT perturbador *m* reemisor; ~ **span** TELEP, TRANS amplitud *f* de repetidor; ~ **station** TELEP, TRANS estación *f* repetidora

repeaterless *adj* GEN sin repetidor *m*

repeating: ~ **coil** *n* TELEP bobina *f* repetidora

reperforator *n* TELEP reperforador *m*

repertoire *n* TELEP repertorio *m*

repetency *n* TRANS repetencia *f*

repetition: ~ **frequency** *n* TELEP cadencia *f*; ~ **rate** *n* TELEP frecuencia *f* de repetición
repetitive: ~ **metering** *n* TELEP cómputo *m* repetitivo; ~-**peak forward-current** *n* ELECT corriente *f* directa de cresta repetitiva, POWER tensión *f* directa de cresta repetitiva; ~-**peak offstage current** *n* PARTS corriente *f* de cresta repetitiva en estado de no conducción; ~-**peak onstage current** *n* ELECT corriente *f* de cresta repetitiva en estado de conducción; ~-**peak reverse-voltage** *n* PARTS, TELEP tensión *f* inversa de cresta repetitiva
replace *vt* COMP sustituir; ◆ ~ **the handset** TELEP colgar; ~ **the receiver** TELEP reemplazar el receptor; ~ **regularly** TEST reemplazar regularmente
replacement *n* COMP sustitución *f*; ~ **board** ELECT placa *f* de intercambio; ~ **free of charge** GEN reposición *f* sin cargo
reply: **on no** ~ *phr* RADIO sin respuesta
report *n* GEN informe *m*
representation *n* GEN representación *f*
representative: ~ **sample** *n* GEN muestra *f* representativa
reprint *n* PER APP reproducción *f*
reproduce *vt* GEN reproducir
reproducibility *n* GEN reproducibilidad *f*
reproducible *adj* GEN reproducible
reproductibility *n* GEN reproducción *f*
reproduction *n* GEN reproducción *f*; ~ **ratio** TELEP razón *f* de reproducción
repudiation *n* DATA rechazo *m*
req *abbr* (*request*) COMP, DATA, RADIO, TELEP pedido *m*
requantization *n* COMP recuantificación *f*
request *n* (*req*) COMP, DATA, RADIO, TELEP pedido *m*; ~ **for comments** DATA petición *f* de comentarios; ~ **data transfer** DATA, TELEP transferencia *f* de datos pedidos; ~ **indicator** TELEP indicador *m* de petición; ~ **output** TELEP resultado *m* de una petición; ~ **primitive** DATA petición *f* de primitiva; ~ **for proposal** (*RFP*) GEN solicitud *f* de propuesta (*RFP*); ~-**repeat system** TELEP sistema *m* de demanda de repetición; ~ **to send** DATA solicitud *f* para enviar; ~ **for service** GEN pedido *m* de servicio; ~ **transmission time** TRANS petición *f* de tiempo de transmisión
requestor *n* COMP, DATA, RADIO, TELEP llamador *m*, solicitante *m*
required: ~ **time** *n* TELEP, TEST tiempo *m* de funcionamiento requerido; ~ **unrequired time** *n* TELEP, TEST tiempo *m* no requerido requerido
requirement *n* COMP, TEST requisito *m*
RER *abbr* (*residual error rate*) COMP, DATA, TELEP RER (*proporción de errores residuales*)
reradiation *n* RADIO rerradiación *f*
rerecord *vt* COMP, DATA regrabar
rerecordable *adj* COMP, DATA regrabable
reroute *vt* DATA, SWG, TELEP *traffic* reencaminar, desviar
rerouting *n* DATA, SWG, TELEP *of traffic* reencaminamiento *m*
rerun[1] *n* COMP, TELEP reanudación *f*; ~ **time** COMP tiempo *m* de reanudación
rerun[2] *vt* COMP, TELEP volver a ejecutar
RES *abbr* (*reserved field*) ELECT campo *m* reservado
rescue: ~-**service radio system** *n* RADIO sistema *m* de radio para servicio de salvamento
research *n* GEN investigación *f*; ~ **centre** *BrE* GEN centro *m* de investigación; ~ **and development** (*RD*) TEST investigación *f* y desarrollo *m* (*ID*)
researcher *n* GEN investigador *m*
reseller *n* RADIO revendedor *m*
reserve: ~ **circuit** *n* TELEP circuito *m* de reserva; ~ **power** *n* POWER potencia *f* de reserva; ~ **signalling link** *BrE n* SIG enlace *m* de señal de reserva
reserved: ~ **field** *n* (*RES*) ELECT campo *m* reservado; ~ **switching path PE** *n* TELEP EP *m* de trayectoria de conmutación reservada; ~ **word** *n* COMP, DATA palabra *f* reservada
reservoir *n* ELECT embalse *m*
reset[1]: ~-**band acknowledgement signal** *n* SIG señal *f* de reconocimiento de banda de reposición; ~-**band signal** *n* SIG señal *f* de banda de reposición; ~ **button** *n* GEN botón *m* de restauración; ~-**circuit signal** *n* SIG señal *f* de circuito de reposición; ~ **mode** *n* COMP modo de restauración
reset[2] *vt* COMP, DATA, PARTS reponer
resettability *n* RADIO reajustabilidad *f*
resetting *n* GEN puesta *f* a cero; ~ **magnet** SWG imán *m* de reposición; ~ **percentage** PARTS porcentaje *m* de reposición; ~ **pulse** ELECT impulso *m* de reposición; ~ **ratio** PARTS relación *f* de reposición; ~ **relay** PARTS relé *m* de reposición; ~ **value** PARTS valor *m* de reposición; ~ **voltage** ELECT tensión *f* de reposición
reshaping *n* ELECT transformación *f*
resident *adj* COMP, DATA residente
residential: ~ **market** *n* GEN mercado *m* de usuarios privados; ~ **telephone** *n* PER APP, TELEP teléfono *m* de uso privado
residual: ~ **binary error rate** *n* RADIO coeficiente *m* de error binario residual; ~ **bit error ratio** *n*

(*RBER*) RADIO ratio *m* de error de binario residual (*RBER*); ~ **echo level** *n* TELEP nivel *m* de eco residual; ~ **error** *n* COMP, DATA, TELEP error *m* residual; ~ **error rate** *n* (*RER*) COMP, DATA, TELEP proporción *f* de errores residuales (*RER*); ~ **magnetism** *n* ELECT magnetismo *m* remanente; ~ **noise** *n* COMP ruido *m* residual; ~ **plate** *n* PARTS lámina *f* de entrehierro; ~ **reflectance** *n* TRANS reflectancia *f* remanente; ~ **sideband** *n* (*RSB*) RADIO, TRANS, TV banda *f* lateral residual (*RSB*); ~ **stud** *n* PARTS tope *m* de entrehierro; ~ **voltage** *n* POWER tensión *f* residual

residue *n* GEN residuo *m*

resilience: ~ **to interference** *n* TRANS elasticidad *f* a la interferencia

resin: ~**-core solder** *n* ELECT estaño *m* de soldar con núcleo de resina; ~ **flux** *n* PARTS resina *f* para soldar; ~ **smear** *n* ELECT corrimiento *m* de resina

resist: ~ **foil** *n* ELECT capa *f* protectora; ~ **lacquer** *n* PARTS laca *f* protectora

resistance *n* COMP, ELECT resistencia *f*; ~**-capacitance** (*RC*) ELECT resistencia-capacitancia *f* (*RC*); ~**-capacitance coupling** ELECT acoplamiento *m* por resistencia-capacitancia; ~ **coupling** ELECT acoplamiento *m* por resistencia; ~ **standard** ELECT, POWER patrón *m* de resistencia; ~ **thermometer** TEST termómetro *m* de resistencia; ~ **under illumination** ELECT resistencia *f* a la iluminación

resistive: ~ **attenuator** *n* ELECT atenuador *m* resistivo; ~ **circuit** *n* ELECT circuito *m* resistivo; ~ **cut-off frequency** *n* ELECT frecuencia *f* de corte resistiva; ~ **load** *n* ELECT, POWER carga *f* resistiva; ~ **reverse current** *n* ELECT corriente *f* inversa resistiva; ~**-temperature detector** *n* CLIM detector *m* de temperatura por resistencia

resistivity *n* ELECT, POWER resistividad *f*

resize *vt* COMP redimensionar

resizing *n* COMP redimensionamiento *m*

resolution *n* COMP, DATA, ELECT definición *f*; ~ **chart** TV carta *f* de resolución

resolving: ~ **power** *n* ELECT, RADIO poder *m* resolutivo

resonance *n* GEN resonancia *f*; ~ **absorption** ELECT absorción *f* de resonancia; ~ **frequency** ELECT, RADIO, TELEP frecuencia *f* resonante; ~ **method of measurement** TEST método *m* de medición por resonancia

resonant: ~ **cavity** *n* ELECT, TRANS cavidad *f* resonante; ~ **circuit** *n* ELECT, POWER circuito *m* resonante; ~ **frequency** *n* ELECT, RADIO, TELEP frecuencia *f* resonante; ~**-mode filter** *n* RADIO filtro *m* de modo resonante; ~**-reed relay** *n* POWER relé *m* de láminas resonantes

resonate *vi* GEN resonar

resonator *n* ELECT, POWER resonador *m*; ~ **block** GEN bloque *m* de resonador

re-sort *vt* COMP reclasificar

resource *n* GEN recurso *m*; ~ **sharing** COMP compartimiento *m* de recursos; ~**-unavailable class** DATA clase *f* de recurso no disponible

responding: ~ **N-address** *n* DATA dirección *f* N respondiendo

response *n* GEN respuesta *f*; ~ **for end of transmission** (*ERR*) DATA respuesta *f* para finalizar la transmisión (*ERR*); ~ **identifier** (*RI*) TELEP identificador *m* de respuesta; ~ **output** TELEP salida *f* de respuesta; ~ **primitive** DATA primitivo *m* de respuesta; ~ **time** GEN tiempo *m* de respuesta; ~ **to continue to correct** (*CTR*) DATA respuesta *f* a continuar para corregir (*CTR*)

responser *n* RADIO receptor *m* de identificación

responsivity *n* ELECT, TRANS responsividad *f*

responsor *n* RADIO respondedor *m*

rest: ~ **position** *n* ELECT, POWER, TELEP, TRANS posición *f* de equilibrio; ~ **state** *n* ELECT, POWER, TELEP, TRANS estado *m* de reposo

restart[1] *n* COMP, DATA reiniciación *f*

restart[2] *vt* COMP, DATA reiniciar

restitution *n* ELECT, TELEP restitución *f*; ~ **delay** ELECT, TELEP retraso *m* de restitución

restorable: ~ **service** *n* RADIO servicio *m* recuperable

restoration *n* GEN recuperación *f*, restablecimiento *m*; ~ **control point** (*RCP*) TELEP punto *m* de control de recuperación (*RCP*); ~ **of service** GEN restablecimiento *m* del servicio

restore *vt* GEN *power, connection* restablecer

restoring *n* SWG restablecimiento *m*; ~ **button** GEN botón *m* de reposición; ~ **pulse** ELECT impulso *m* de reposición; ~ **relay** PARTS relé *m* de reposición; ~ **time** TRANS *of echo suppresser* duración *f* del retorno

restrappable *adj* POWER repuenteable

restrapping *n* POWER modificación *f* de puenteado

restrict *vt* RADIO, TELEP restringir, limitar

restricted *adj* GEN restringido, limitado; ~ **access** *n* TELEP acceso *m* restringido; ~ **channel** *n* TV canal *m* restringido; ~ **information-transfer service** *n* APPL servicio *m* restringido de transferencia de información; ~ **service** *n* GEN, TELEP servicio *m* restringido

restriction: ~ **in the outgoing direction** *n* TELEP restricción *f* en la dirección salida

restrictor *n* ELECT estrangulador *m*
result *n* GEN resultado *m*; **~-comparing register** GEN registro *m* de comparación de resultados
resynchronization *n* ELECT resincronización *f*
resynchronize *vt* ELECT resincronizar
retail *adj* GEN al por menor
retained: **~-data block** *n* DATA bloque *m* de datos retenidos; **~ signal** *n* SIG señal *f* retenida
retarder *n* PARTS retardador *m* de fraguado
retention *n* ELECT retención *f*; **~ period** RADIO periodo *m* de retención
retentivity *n* ELECT retentividad *f*
retest *n* TELEP, TEST nuevo ensayo *m*
retime *vt* ELECT reajustar el ritmo de, TELEP repetir el reglaje de
retiming *n* ELECT reajuste *m* del ritmo, TELEP repetición *f* del reglaje
retouch *n* DATA *of scanner* opacado *m*
retrace *n* COMP, ELECT, TV retroceso *m*
retractable *adj* DATA *scanner* retráctil
retrain: **~ negative** *n* (*RTN*) DATA, TELEP negativo *m* de reinstrucción (*RTN*); **~ positive** *n* (*RTP*) DATA, TELEP positivo *m* de reinstrucción (*RTP*)
retransmission *n* RADIO, TELEP, TRANS, TV retransmisión *f*; **~ buffer** COMP, TRANS memoria *f* intermedia de retransmisión
retransmit *vt* RADIO, TELEP, TRANS, TV retransmitir
retransmitted: **~ condensed burst time plan** *n* (*RCTP*) DATA plan *m* temporal condensado de retransmisión por ráfagas (*RCTP*)
retrieval *n* COMP, DATA, RADIO recuperación *f*; **~ port** DATA puerta *f* de recuperación; **~ service** COMP servicio *m* de recuperación
retrieve *vt* COMP, DATA, RADIO recuperar
retrofit *n* ELECT, PARTS actualización *f* retroactiva
retroreflector: **~ array** *n* RADIO matriz *f* retrorreflectora
retry: **~ interval** *n* RADIO intervalo *m* de reintento
return[1]: **~-beam vidicon** *n* (*RBV*) TELEP vidicón *m* con haz de retorno (*RBV*); **~ block** *n* TELEP bloqueo *m* de retorno; **~ conductor** *n* POWER conductor *m* de retorno; **~ current** *n* TRANS corriente *f* de retorno; **~ flap** *n* PER APP *of payphone* cajetilla *f* de devolución de moneda; **~ loss** *n* TELEP, TRANS atenuación *f* por reflexión; **~ path** *n* TRANS, TV vía *f* de retorno; **~-to-reference recording** *n* COMP grabación *f* con retorno a un estado de referencia; **~ to service** *n* TEST retorno *m* al servicio; **~-service advice** *n* TELEP aviso *m* de servicio de retorno; **~ wire** *n* POWER hilo *m* de retorno; **~ to zero** *n* COMP, ELECT, TELEP retorno *m* a cero; **~-to-zero recording** *n* COMP grabación *f* por retorno a cero
return[2]: **~ sb's call** *vt* TELEP devolver a alguien la llamada; **~ to home position** *vt* SWG volver a la posición de reposo
returned: **~-content entry** *n* DATA entrada *f* de contenido reenviado
reusable *adj* COMP, RADIO reutilizable; **~ program** *n* COMP programa *m* reutilizable; **~ rubber** *n* ELECT goma *f* reutilizable
reuse: **~ of frequency channel** *n* RADIO reutilización *f* de canal de frecuencia; **~ pattern** *n* RADIO modelo *m* de reutilización
revenue *n* GEN entradas *f pl* brutas
reverb: **~ level** *n* COMP nivel *m* de reverbero; **~ time** *n* TELEP tiempo *m* de reverbero; **~ type** *n* COMP tipo *m* de reverbero
reverberant: **~ field** *n* TELEP campo *m* reverberante
reverberated: **~ field** *n* TELEP campo *m* repercutido
reverberation *n* COMP, ELECT, TELEP reverberación *f*; **~ level** COMP nivel *m* de reverberación; **~ time** TELEP tiempo *m* de reverberación; **~ type** COMP tipo *m* de reverberación
reversal *n* POWER, TELEP *of answer signal*, TRANS inversión *f*; **~ control channel** TELEP canal *m* de control por inversión
reverse[1]: **~ bias** *n* PARTS, RADIO polarización *f* inversa; **~ biasing** *n* PARTS, RADIO polarización *f* inversa; **~-blocking current** *n* ELECT corriente *f* inversa de bloqueo; **~-blocking diode thyristor** *n* ELECT, PARTS tiristor *m* diodo bloqueado en inversa; **~-blocking state** *n* ELECT, PARTS estado *m* de bloqueo inverso; **~-charge acceptance** *BrE n* (*cf collect-call acceptance AmE*) APPL, TELEP aceptación *f* de cobro en destino; **~-charge acceptance not subscribed signal** *BrE n* (*cf collect-call acceptance not subscribed signal AmE*) SIG, TELEP señal *f* no suscrita de aceptación de cobro en destino; **~-charge call** *BrE n* (*cf collect-call AmE*) APPL, TELEP llamada *f* pagadera en destino; **~ charging** *n* APPL cobro *m* en destino, TELEP cobro *m* revertido; **~-charging acceptance** *BrE n* (*cf collect-calling acceptance AmE*) APPL, TELEP aceptación de cobro en destino; **~-charging request indicator** *BrE n* (*cf collect-calling request indicator AmE*) APPL indicador *m* de solicitud de pago en destino, TELEP indicador *m* de solicitud de cobro en destino; **~-conducting diodethyristor** *n* ELECT, PARTS tiristor *m* diodo conductor eninversa; **~ current** *n* ELECT contra

corriente; **~-current circuit** *n* ELECT circuito *m* de corriente inversa; **~-current relay** *n* ELECT relé de corriente inversa; **~ DC resistance** *n* ELECT, POWER resistencia *f* inversa; **~ direction** *n* ELECT dirección *f* inversa; **~ display** *n* COMP visualización *f* inversa; **~ electrode current** *n* ELECT corriente *f* de electrodos inversa; **~ engineering** *n* COMP, ELECT ingeniería *f* invertida; **~ key** *n* COMP conmutador *m* de polaridad; **~ polarity** *n* POWER polaridad *f* invertida; **~ priority** *n* COMP prioridad *f* invertida; **~ recovery** *n* ELECT recuperación *f* invertida; **~-recovery current** *n* ELECT corriente *f* de recuperación en sentido inverso; **~ routing** *n* SWG encaminamiento *m* inverso; **~ routing address** *n* DATA, SWG dirección *f* de encaminamiento invertido; **~ video** *n* COMP vídeo *m* inverso; **~ voltage** *n* ELECT tensióninversa

reverse[2] *vt* COMP, ELECT invertir; ◆ **~ the charges** *BrE* (*cf call collect AmE*) APPL, TELEP cobrar en destino

reversed: **~ charging** *n* APPL, TELEP cobro *m* revertido

reversible: **~ blade** *n* ELECT conductor *m* plano reversible; **~ flexible disk** *n* COMP disco *m* flexible reversible; **~ process** *n* GEN proceso *m* reversible

reversing: **~ switch** *n* POWER conmutador *m* inversor

revert *vi* COMP revertir

revisable *adj* DATA revisable

revision: **~ state** *n* GEN estado *m* de revisión

revolver *n* DATA revólver *m*; **~ loop** DATA bucle *m* rotatorio

revolving *adj* GEN giratorio

rewritable: **~ compact disc** *n* COMP, DATA disco *m* compacto reescribible; **~ optical disk** *n* COMP, DATA disco *m* óptico reescribible

rewrite[1] *n* DATA reescritura *f*

rewrite[2] *vt* DATA reescribir

RF *abbr* (*radio frequency*) TV RF (*radiofrecuencia*); **~ amplifier** *n* TELEP amplificador *m* de RF; **~ channel** *n* RADIO canal *m* de RF; **~ head** *n* RADIO cabeza *f* de RF; **~ scanner** *n* RADIO digitalizador *m* de RF; **~ test** *n* RADIO, TEST prueba *f* RF

RFI *abbr* (*radio-frequency interference*) RADIO RFI (*interferencia de radiofrecuencia*); **~ caulking** *n* RADIO calafateo *m* de RFI

RFP *abbr* (*request for proposal*) GEN RFP (*solicitud de propuesta*)

RFS *abbr* (*receive-frame synchronization procedure*) TRANS RFS (*procedimiento de sincronización del marco de recepción*)

RG *abbr* (*regenerator*) ELECT, SIG, SWG, TELEP RG (*recuperador*)

RGB *abbr* (*red-green-blue*) TV RGB (*rojo-verde-azul*); **~ colour** *BrE* **model** *n* TV modelo *m* de color RGB

rheostat *n* PARTS, POWER reóstato *m*

rhodium: **~ plating** *n* GEN rodiodeposición *f*

rhombic: **~ antenna** *n* RADIO antena *f* rómbica

rhythm *n* COMP ritmo *m*; **~ composer** COMP compositor *m* de ritmo; **~ machine** COMP máquina *f* de ritmo; **~ pattern** COMP modelo *m* de ritmo; **~ programmer** COMP programador *m* de ritmo; **~ volume slider** COMP mando *m* deslizante de volumen del ritmo

rhythmic: **~ guide** *n* COMP guía *f* rítmica

RI *abbr* (*response identifier*) TELEP identificador *m* de respuesta

ribbon *n* GEN cinta *f*; **~ cable** TRANS cable *m* de tiras; **~ matrix** ELECT matriz *f* de cinta; **~ microphone** TELEP micrófono *m* de cinta

rich: **~ text format** *n* (*RTF*) COMP formato de texto muy elaborado (*RTF*)

Rician: **~-fading channel** *n* ELECT canal *m* desvanescente de Rician

ridge: **~ waveguide** *n* DATA, RADIO, TRANS guía *f* de ondas ondulado interiormente

right: **~-hand polarization** *n* RADIO polarización *f* a derecha; **~ parenthesis** *n* GEN paréntesis *m* de cierre; **~-sizing** *n* ELECT dimensionamiento *m* adecuado

rights *n pl* GEN derechos *m pl* de suscripción

rigid: **~ flex card** *n* ELECT tarjeta *f* flexible rígida; **~ plastic** *n* PARTS plástico *m* rígido

RIN *abbr* (*relative-intensity noise*) ELECT RIN (*ruido de intensidad relativa*)

ring[1] *n* DATA, ELECT, PARTS *of wire* anillo *m*; **~-access control** ELECT control *m* de acceso de señal de llamada; **~ adaptor** DATA adaptador *m* de anillo; **~ antenna** RADIO antena *f* en anillo; **~-back** TELEP retorno *m* de llamada; **~-back signal** TELEP señal *f* de llamada; **~-back tone** TELEP tono *m* de retorno de llamada; **~ configuration** TRANS configuración *f* en anillo; **~ current** TELEP corriente *f* circular; **~-current set** TELEP dispositivo *m* de corriente de llamada; **~-down signalling** *BrE* GEN señalización *f* manual; **~ filter** RADIO filtro *m* en anillo; **~-forward signal** SIG, TELEP señal *f* de llamada; **~ indicator** TELEP indicador *m* de timbre; **~ main** POWER canalización *f* circular; **~ modulator** ELECT modulador *m* doble en anillo; **~ network** DATA, TRANS red *f* en anillo; **~ oscillator** ELECT oscilador *m* en anillo; **~**

redundancy RADIO, TRANS redundancia *f* de llamada; ~ **start** TELEP arranque *m* de señal de llamada; ~ **switch** RADIO conmutador *m* de anillo; ~ **tone** TELEP tonalidad *f* de llamada; ~**-tone generator** TELEP generador *m* de tono de llamada; ~ **topology** DATA topología *f* de anillo; ~ **translator** TRANS traslador *m* de anillo; ~ **tripping** TELEP corte *m* de la señal de llamada; ~**-trip relay** TELEP relé *m* de corte de señal de llamada; ~ **wire** TELEP conductor *m* de la señal de llamada; ~ **wrap** ELECT enrollamiento *m* de llamada

ring[2] *vt* TELEP llamar por teléfono, llamar al teléfono, telefonear; ◆ ~ **again** (*RAG*) TELEP repetir la llamada (*RAG*)

ring back *vt* TELEP devuelva la llamada

ringer *n* TELEP timbre *m*; ~ **sensitivity** TELEP sensibilidad *f* del dispositivo de llamada

ringing *n* RADIO, TELEP llamada *f*; ~ **current** TELEP corriente *f* de llamada; ~**-current set** TELEP dispositivo *m* de corriente de llamada; ~**-current transformer** TELEP transformador *m* de corriente de llamada; ~ **duration** TELEP duración *f* de la señal de llamada; ~ **generator** TELEP generador *m* de señal de llamada; ~ **interruptor** TELEP interruptor *m* de señal de llamada; ~ **machine** TELEP máquina *f* de llamada; ~ **oscillator** RADIO oscilador *m* excitado por choque; ~ **period** TELEP periodo *m* de la llamada; ~**-period adjustment** PER APP, TELEP ajuste *m* del periodo de la señal de llamada; ~ **periodicity** TELEP periodicidad *f* de la señal de llamada; ~ **pilot indicator** TELEP indicador *m* de señal de llamada; ~ **position** TELEP posición *f* de llamada; ~ **set** TELEP indicador *m* acústico; ~ **signal** TELEP señal *f* de llamada; ~ **time** TELEP, TV hora *f* de llamada; ~ **tone** TELEP tono *m* de llamada; ~ **voltage** TELEP tensión *f* de llamada; ~**-voltage generator** TELEP generador *m* de la tensión de llamada

RIP *abbr* (*routing information protocol*) DATA RIP (*protocolo de información de encaminamiento*)

ripcord *n* ELECT disparador *m*

ripple *n* ELECT ondulación *f*; ~ **clocking arrangement** ELECT disposición *f* de sincronización de la ondulación; ~ **component** POWER componente *m* ondulatorio; ~**-through counter** TEST contador *m* de transporte ondulante; ~ **voltage** DATA tensión *f* de ondulación

RISC *abbr* (*reduced-instruction-set computer*) COMP, DATA ordenador *m* con programación reducida *Esp*, computadora *f* con programación reducida *AmL*, computador *m* con programación reducida *AmL*

rise: ~ **time** *n* GEN tiempo *m* de ascenso; ~**-time coefficient** *n* PARTS coeficiente *m* de tiempo de subida

riser: ~ **shaft** *n* ELECT eje *m* del elevador

risk *n* GEN riesgo *m*; ~ **analysis** TEST análisis *m* de riesgos; ~ **assessment** TEST estimación *f* de riesgos, evaluación *f* de riesgos; ~ **economy** TEST economía *f* de riesgo; ~ **evaluation** TEST evaluación *f* de riesgos; ~ **identification** TEST identificación *f* de riesgos; ~ **level** TEST nivel *m* de riesgo; ~ **management** GEN gestión *f* de riesgos; ~ **quantification** TEST cuantificación *f* de riesgos

Rivest: ~**, Shamir, Adleman algorithm** *n* (*RSA*) TELEP algoritmo *m* de Rivest, Shamir, Adleman (*RSA*)

rivet *n* TEST remache *m*

RJE *abbr* (*remote-job entry*) DATA entrada *f* de teletrabajo

RLE *abbr* (*run-length encoding*) COMP RLE (*codificación de longitud de tirada*)

RLRE *abbr* (*A-release-response APDU*) DATA RLRE (*APDU de liberación-respuesta-A*)

RLRQ *abbr* (*A-release-request APDU*) DATA RLRQ (*APDU de liberación-petición-A*)

RLT *abbr* (*release-link trunk*) TELEP RLT (*línea troncal de enlaces de desconexión*)

rms *abbr* (*root mean square*) GEN media *f* cuadrática; ~ **value** *n* GEN valor *m* rms

RN *abbr* (*receipt notification*) DATA, TELEP RN (*notificación de recepción*)

RNR: ~ **frame** *n* TELEP trama *f* RNR

RO *abbr* (*remote operation*) DATA, TELEP, TEST RO (*operación remota*)

roadside: ~ **cabinet** *n* TV unidad *f* móvil

roam *vi* RADIO, TELEP indicar la posición del móvil

roamer *n* RADIO, TELEP indicador *m* de la posición del móvil

roaming *n* RADIO, TELEP indicación *f* de la posición del móvil; ~ **capability** RADIO capacidad *f* de indicación de la posición del móvil; ~ **number** RADIO número *m* de indicación de la posición del móvil; ~ **station** RADIO estación *f* de indicación de la posición del móvil; ~ **subscriber** RADIO abonado *m* de vagabundeo

robbed: ~**-bit signalling** *BrE n* DATA, SIG señalización *f* de bitios robados

Rochelle: ~ **salt** *n* ELECT sal *f* de Rochelle

rocker: ~ **button** *n* POWER pulsador *m* de báscula

rocky: ~**-point effect** *n* ELECT efecto *m* Rocky Point

rod *n* ELECT varilla *f*; ~ **antenna** RADIO antena *f* de varilla; ~ **reflector** RADIO reflector *m* de varilla; ~**-in-tube technique** ELECT, TRANS técnica *f* de varilla en tubo
rodent: ~ **protection** *n* ELECT protección *f* contra roedores
roll: ~ **axis** *n* RADIO eje *m* de balanceo; ~**-off** *n* RADIO barrena *f*; ~**-out** *n* RADIO descarga *f* de la memoria principal
rolling: ~**-code band splitting** *n* DATA división *f* de banda por código circular; ~**-track ladder** *n* PARTS escalera *f* de ruedas
rolls: **in** ~ *adj* ELECT en archivos
ROLR *abbr* (*receive objective loudness rating*) TELEP ROLR (*valoración de sonoridad del objetivo de recepción*)
ROM *abbr* (*read-only memory*) COMP ROM (*memoria sólo de lectura*)
rooftop: ~ **aerial** *n* TV antena *f* exterior
room *n* COMP cuarto *m*; ~ **acoustics** *n pl* TELEP acústica *f* arquitectónica; ~ **noise** TELEP ruido *m* de ambiente; ~ **temperature** CLIM temperatura *f* ambiente
root *n* COMP base *f*; ~ **circle** GEN circunferencia *f* interna; ~ **directory** COMP directorio *m* raíz; ~ **mean square** (*rms*) GEN media *f* cuadrática; ~**-mean square value** (*rms value*) GEN raíz *f* cuadrada del valor medio cuadrático (*valor rms*); ~ **node** DATA nodo *m* de base; ~ **sum square** GEN raíz *f* cuadrada de la suma de los cuadrados; ~ **top node** COMP nodo *m* de la base de parte superior; ~ **tree** DATA árbol *m* base
rotary: ~ **attenuator** *n* RADIO atenuador *m* rotativo; ~ **control** *n* PARTS control *m* rotativo; ~ **direction** *n* TELEP dirección *f* rotativa; ~ **exchange** *n* SWG central *f* rotativa; ~ **ferrule** *n* ELECT manguito *m* de empalme rotativo; ~**-head digital audio tape** *n* (*R-DAT*) COMP, DATA cinta *f* de audio digital de cabeza rotatoria (*R-DAT*); ~ **phase changer** *n* RADIO desfasador *m* rotativo; ~ **phase shifter** *n* RADIO desfasador *m* rotativo; ~ **selector** *n* SWG selector *m* giratorio; ~ **switch** *n* GEN conjuntor *m* rotatorio; ~ **system** *n* SWG sistema *m* conmutador rotatorio
rotatable: ~ **aerial** *n* RADIO antena *f* giratoria; ~ **antenna** *n* RADIO antena *f* giratoria
rotate *vt* COMP girar
rotating: ~ **field** *n* POWER, RADIO, TELEP campo *m* giratorio; ~**-field alternator** *n* POWER, RADIO, TELEP alternador *m* de campo giratorio; ~ **joint** *n* PARTS junta *f* giratoria
rotational: ~ **delay** *n* COMP demora *f* rotacional
rotor *n* TELEP rodete *m*
roughness *n* GEN aspereza *f*
round: ~**-the-world echo** *n* RADIO eco *m* de circunvalación terrestre
route *n* ELECT, TELEP, TRANS ruta *f*; ~ **analysis** TELEP estudio *m* de rumbo; ~ **blocking** TELEP bloqueo *m* de ruta; ~**-choice data** *n pl* TELEP datos *m pl* de encaminamiento; ~ **congestion** SWG, TELEP congestión *f* de ruta; ~**-congestion counter** SWG, TELEP contador *m* de congestiones de ruta; ~ **determination** TELEP determinación *f* de ruta; ~**-idle marking** TELEP marcación *f* de ruta libre; ~ **information digits** *n pl* TELEP dígitos *m* de información de ruta; ~ **list** TELEP lista *f* de ruta; ~ **lockout supervision** TELEP supervisión *f* de cierre de ruta; ~ **marker** TELEP marcador *m* de ruta; ~ **numbering** TELEP numeración *f* de ruta; ~ **relay** TELEP relé *m* de ruta; ~ **seizure** TELEP ocupación *f* de ruta; ~ **supervision** TELEP supervisión *f* de ruta; ~**-supervision panel** TELEP panel *m* de supervisión de ruta; ~ **test** TELEP prueba *f* de ruta; ~**-test block** TELEP bloque *m* de prueba de ruta
router *n* DATA, ELECT, TELEP, TRANS encaminador *m*
routine *n* COMP, DATA rutina *f*; ~ **acquisition mode** DATA modo *m* de adquisición rutinario; ~ **maintenance** GEN conservación *f* ordinaria; ~ **monitoring** TEST supervisión *f* ordinaria; ~ **repetition** TELEP repetición *f* rutinaria; ~ **task** COMP tarea *f* rutinaria; ~ **test** TEST prueba *f* de rutina
routiner *n* TEST comprobador *m* rutinario
routing *n* COMP, DATA, SWG, TELEP, TRANS *of circuit* encaminamiento *m*; ~ **address** DATA, SWG dirección *f* de encaminamiento; ~ **code** TELEP código *m* de encaminamiento; ~ **control** DATA control *m* de encaminamiento; ~ **data** *n pl* TELEP datos *m pl* de encaminamiento; ~ **delay** DATA retardo *m* de encaminamiento; ~ **device** DATA dispositivo *m* de encaminamiento; ~ **digits** *n pl* SWG dígitos *m* de recorrido; ~ **diversity** TRANS diversidad *f* de encaminamiento; ~ **information protocol** (*RIP*) DATA protocolo *m* de información de encaminamiento (*RIP*); ~ **label** TELEP etiqueta *f* de encaminamiento; ~ **number** TELEP indicativo *m* de encaminamiento; ~**-number area** TELEP área *f* de indicador de encaminamiento; ~ **program** TELEP programa *m* de encaminamiento; ~ **slip** GEN hoja *f* de ruta; ~ **table maintenance protocol** (*RTMP*) COMP protocolo *m* de mantenimiento de tabla de encaminamiento (*RTMP*); ~ **wings** *n pl* ELECT alas *f pl* de encaminamiento

row *n* COMP, TELEP fila *f*; ~ **aisle** TELEP pasillo *m* entre filas; ~ **chute** TELEP canal *m* de fila; ~-**end cover** TELEP cubierta *f* de extremo de fila; ~ **length** TELEP longitud *f* de fila; ~ **mechanical structure** TELEP estructura *f* mecánica de fila; ~ **pitch** DATA, TELEP grado *m* de separación entre filas; ~ **spacing** TELEP distancia *f* entre las hileras

RP *abbr* (*regional processor*) SWG procesador *m* regional

RPE *abbr* (*remote peripheral equipment*) COMP RPE (*equipo periférico remoto*)

RPF *abbr* (*reference publication format*) DATA RPF (*formato de publicación de referencia*)

RPOA *abbr* (*recognized private operating agency*) TELEP RPOA (*agencia privada autorizada*); ~ **out-of-order signal** *n* TELEP señal *f* de avería RPOA; ~ **selection indicator** *n* TELEP indicador *m* de selección RPOA; ~ **transit-network identity** *n* TELEP identidad *f* de red de tránsito RPOA

RR: ~ **hunting** *n* TELEP búsqueda *f* por RR

R-reference: ~ **point** *n* DATA punto *m* de referencia R

RS *abbr* COMP (*Reed-Solomon*) RS (*Reed-Solomon*), DATA sección *f* del regenerador; ~ **code** *n* COMP código *m* RS

RSA *abbr* (*Rivest, Shamir, Adleman algorithm*) TELEP RSA (*algoritmo de Rivest, Shamir, Adleman*)

RSB *abbr* (*residual sideband*) RADIO, TRANS, TV RSB (*banda lateral residual, VSB*)

RSE: ~ **test method** *n* DATA método *m* de comprobación RSE

RSM *abbr* (*remote spectrum-monitoring*) RADIO RSM (*control espectral a distancia*)

RST *abbr* (*regenerator-section termination*) DATA RST (*terminación de la sección del regenerador*)

RSU *abbr* (*remote switching unit*) SWG unidad *f* remota de conmutación

RTF *abbr* (*rich text format*) COMP RTF (*formato de texto muy elaborado*)

RTG *abbr* (*regenerator-timing generator*) DATA RTG (*generador de temporización del regenerador*)

RTMP *abbr* (*routing table maintenance protocol*) COMP RTMP (*protocolo de mantenimiento de tabla de encaminamiento*)

RTN *abbr* (*retrain negative*) DATA, TELEP RTN (*negativo de reinstrucción*)

RTP *abbr* (*retrain positive*) DATA, TELEP RTP (*positivo de reinstrucción*)

RTSE: ~ **provider** *n* DATA, TELEP proveedor *m* RTSE; ~ **user** *n* DATA, TELEP usuario *m* de RTSE

RTSI *abbr* (*reference terminal status interchange*) RADIO RTSI (*intercambio de estado de terminal de referencia*)

RU *abbr* (*remote unit*) SWG unidad *f* remota

rub *vt* COMP descafilar

rub out *vt* COMP, RADIO borrar

rubber *n* PARTS caucho *m*; ~ **banding** COMP tira *f* de goma; ~ **gasket** COMPON guarnición *f* de caucho; ~ **shock-absorber** GEN amortiguador *m* de goma

ruggedize *vt* GEN reforzar

ruggedness *n* GEN robustez *f*

rule: ~-**based security policy** *n* DATA política *f* de seguridad basada en normas

ruler *n* COMP soberano *m*

run *n* COMP, TEST *process* ejecución *f*, *sequence* ciclo *m*; ~ **duration** COMP duración *f* de un ciclo de producción; ~-**length encoding** (*RLE*) COMP codificación *f* de longitud de tirada (*RLE*); ~ **stream** COMP corriente *f* de proceso; ~ **time** COMP periodo *m* de ejecución

running *n* COMP *of process, program* ejecución *f*; ~ **bar** COMP guía *f* de deslizamiento, DATA riel *m* de guía, PARTS, RADIO, TELEP, TRANS guía *f* de deslizamiento; ~ **costs** *n pl* GEN costes *m pl* (*AmL* costos) de funcionamiento; ~ **diagram** COMP diagrama *m* de marcha; ~ **list** GEN tabla *f* de colocación de hilos; ~ **text** GEN texto *m* corrido; ~ **time** COMP tiempo *m* de utilización de la máquina; ~ **time meter** TELEP cuentatiempo *m*

rupture *n* PARTS rotura *f*; ~ **strain** PARTS alargamiento *m* hasta rotura

rural: ~ **automatic exchange** *n* SWG, TELEP central *f* automática rural; ~ **exchange** *n* TELEP central *f* rural; ~ **network** *n* TELEP, TRANS red *f* rural; ~ **switch** *n* SWG conmutador *m* rural; ~ **telephone system** *n* TELEP sistema *m* telefónico para zonas rurales; ~ **telephony** *n* TELEP telefonía *f* rural

rust *n* PARTS roña *f*; ~-**preventing agent** GEN agente *m* anticorrosivo; ~ **preventive** GEN anticorrosivo *m*; ~-**protecting oil** GEN aceite *m* antioxidante

rustproof *adj* PARTS inoxidable

R-wire *n* TELEP cable *m* R

RX *abbr* (*receiver*) DATA, PER, RADIO, TELEP receptor *m*; ~ **sens** (*receiver sensibility*) RADIO sensibilidad *f* del receptor

S

S: ~ **stage** *n* SWG, TELEP etapa *f* S; ~ **switch** *n* SWG, TELEP conmutador *m* S

SA *abbr* (*subaddress*) APPL, DATA, TELEP SUB (*subdirección*)

SABM *abbr* (*set asynchronous balanced mode*) TELEP SABM (*modo equilibrado asincrónico de puesta a uno*)

SABME *abbr* (*set asynchronous balanced mode extended*) TELEP SABME (*modo equilibrado asincrónico de puesta a uno extendido*)

safeguard *n* COMP, DATA salvaguarda *f*

safety *n* COMP, DATA seguridad *f*; ~ **helmet** GEN casco *m* de protección; ~ **lock** TEST cerradura *f* de seguridad; ~ **shutdown system** CONTR sistema *m* de parada de seguridad

sales: ~ **engineer** *n* GEN técnico *m* de ventas; ~ **executive** *n* GEN ejecutivo *m* de ventas; ~ **force** *n* GEN equipo *m* de vendedores; ~ **outlet** *n* GEN punto *m* de venta; ~ **representative** *n* GEN representante *m* comercial

sample[1] *n* GEN muestra *f*; ~ **division** TRANS división *f* de muestra; ~ **and hold** CONTR muestreo *m* y retención; ~ **and hold action** CONTR acción *f* de muestreo y retención; ~ **number** TRANS número *m* de muestra; ~ **point** TEST punto *m* de espacio muestral; ~ **precision** TELEP precisión *f* de muestra; ~ **preparation** TEST preparación *f* de muestra; ~ **pulse** TEST impulso *m* de muestreo; ~ **size** TEST volumen *m* de muestra; ~ **space** TEST espacio *m* muestral

sample[2] *vt* GEN muestrear

sampled: ~ **signal** *n* SIG, TRANS señal *f* muestreada

sampler *n* GEN muestreador *m*; ~ **module** TEST módulo *m* de conmutador electrónico

samples: ~ **per second** *n pl* TRANS muestras *f pl* por segundo

sampling *n* GEN muestreo *m*; ~ **error** TRANS error *m* de muestreo; ~ **fraction** TRANS fracción *f* de muestreo; ~ **frequency** TRANS frecuencia *f* de muestreo; ~ **gate** TRANS puerta *f* de muestreo; ~ **inspection** TEST inspección *f* por muestreo; ~ **interval** TRANS intervalo *m* de muestreo; ~ **plan** TEST plan *m* de muestreo; ~ **procedure** TEST procedimiento *m* de muestreo; ~ **process** TRANS proceso *m* de muestreo; ~ **rate** TRANS frecuencia *f* de muestreo; ~ **scheme** TEST programa *m* de muestreo; ~ **system** TEST sistema *m* de muestreo; ~ **test** TEST ensayo *m* de muestreo; ~ **time** TRANS tiempo *m* de muestreo; ~ **unit** TEST unidad *f* de muestreo

sandwich: ~ **line** *n* TRANS línea *f* de construcción emparedada

SAP *abbr* (*service access point*) DATA punto *m* de acceso a servicio

SAPI *abbr* (*service access point identifier*) DATA identificador *m* de punto de acceso a servicio

SAR *abbr* (*segmentation and reassembly*) DATA SAR (*segmentación y rearmado*)

SAS *abbr* (*single-attachment station*) ELECT SAS (*estación de acoplamiento simple*)

satellite *n* RADIO, TV satélite *m*; ~ **aerial** RADIO, TV antena *f* de satélite artificial; ~ **antenna** RADIO, TV antena *f* de satélite artificial; ~ **attitude** RADIO, TV actitud *f* de satélite; ~ **band** RADIO banda *f* satélite; ~ **broadcast** RADIO emisión *f* por satélite, TV emisión *f* por vía satélite; ~ **broadcasting** RADIO, TV transmisión *f* por vía satélite; ~ **broadcasting service** RADIO, TV radiodifusión *f* por satélite; ~ **business system** (*SBS*) RADIO sistema *m* comercial de satélite (*SBS*); ~ **business telecommunications network** TELEP, TRANS red *f* de telecomunicaciones comerciales por satélite; ~ **channel** RADIO, TV canal *m* por satélite; ~ **closet** ELECT armario *m* satélite; ~ **communications** *n pl* RADIO, TELEP, TV comunicaciones *f pl* por satélite; ~ **conference** TV conferencia *f* por vía satélite; ~ **control centre** *BrE* (*SCC*) RADIO centro *m* de control de satélites; ~ **exchange** SWG central *f* satélite; ~ **link** RADIO, TV enlace *m* por satélite; ~**-link noise temperature** RADIO temperatura *f* de ruido del enlace por satélite; ~ **location** RADIO localización *f* vía satélite; ~ **loopback mode** RADIO modo *m* de retrobucle de satélite; ~ **master antenna television** (*SMATV*) TV televisión *f* de antena colectiva vía satélite (*SMATV*); ~ **master antenna television system** TV sistema *m* de televisión de antena colectiva vía satélite; ~ **multiservice system** RADIO, TV sistema *m* de multiservicio vía satélite; ~ **network** RADIO, TV red *f* por vía satélite; ~ **news gathering** RADIO, TV recopilación *f* de noticias por vía satélite; ~

outage RADIO, TV indisponibilidad *f* de satélite; ~ **paging system** RADIO sistema *m* de radiobúsqueda por vía satélite; ~ **programme** *BrE* TV programa *m* por vía satélite; ~ **range** RADIO, TV alcance *m* de satélite; ~ **receiver antenna** RADIO antena *f* receptora de satélite; ~ **relay system** RADIO sistema *m* de relé satélite; **~-to-~ link** RADIO, TV enlace *m* entre satélites (*ISL*); ~ **service** RADIO, TV servicio *m* de satélite; ~ **signalling** *BrE* **channel** SIG canal *m* de señal de satélite; ~ **spot beam** RADIO rayo *m* centrado de satélite; ~ **switch** RADIO, SWG conmutador *m* de satélite; ~ **switching** RADIO, SWG conmutación *f* de satélite; ~ **switching centre** *BrE* (*SSC*) RADIO, SWG centro *m* de conmutación de satélite (*SSC*); ~ **system** RADIO, TV sistema *m* de satélite; ~ **telecast** TV programa *m* de televisión por satélite; ~ **television broadcasting** TV transmisión *f* de televisión por vía satélite; ~ **television channel** RADIO, TV canal *m* de televisión por vía satélite; ~ **tracking** RADIO seguimiento *m* de satélites; ~ **transmission** RADIO, TRANS, TV transmisión *f* por satélite; ~ **transmission path** RADIO, TRANS, TV trayectoria *f* de transmisión por vía satélite; ~ **transmit antenna** RADIO antena *f* de transmisión por vía satélite; ~ **transmit antenna gain** RADIO ganancia *f* de antena de transmisión por vía satélite; ~ **transmit power** RADIO potencia *f* de transmisión de satélite; ~ **transponder** RADIO transpondedor *m* de satélite

SATS *abbr* (*selected abstract test suite*) DATA, TEST SATS (*batería de pruebas abstractas seleccionadas*)

saturated *adj* ELECT saturado

saturation *n* ELECT saturación *f*; ~ **gain** ELECT ganancia *f* de saturación; ~ **power** ELECT poder *m* de saturación; ~ **resistance** ELECT resistencia *f* de saturación; ~ **state** ELECT estado *m* de saturación; ~ **voltage** ELECT voltaje *m* de saturación

save[1]: ~ **operation** *n* COMP operación *f* de preservación; ~ **symbol** *n* COMP símbolo *m* de salvar

save[2] *vt* COMP salvar, TELEP *message* guardar; ◆ ~ **to cartridge** COMP salvar en cartucho; ~ **to cassette** COMP salvar en cinta; ~ **to diskette** COMP salvar en disquete

saved: ~ **number recall** *n* TELEP rellamada *f* a número memorizado; ~ **number redial** *n* TELEP remarcado *m* de número memorizado

saving: ~ **of messages** *n* PER APP *by answering machine*, TELEP conservación *f* de mensajes; ~ **in time** *n* GEN ahorro *m* de tiempo

SAW *abbr* (*surface acoustic wave*) RADIO, TELEP, TRANS OAS (*onda acústica de superficie*); ~ **device** *n* ELECT, PARTS, TRANS dispositivo *m* OAS

sawtooth: ~ **current** *n* ELECT corriente *f* en diente de sierra; ~ **generator** *n* ELECT generador *m* en diente de sierra; ~ **keyboard** *n* PER APP teclado *m* en diente de sierra; ~ **pulse** *n* ELECT impulso *m* en diente de sierra; ~ **voltage** *n* ELECT tensión *f* en diente de sierra; ~ **wave** *n* ELECT, TRANS oscilación *f* en diente de sierra

SB *abbr* (*sideband*) RADIO, TELEP, TRANS SB (*banda lateral*)

SB1 *abbr* (*scrambled binary ones*) DATA SB1 (*binarios cifrados*)

SBS *abbr* (*satellite business system*) RADIO SBS (*sistema comercial de satélite*)

SBTP *abbr* (*simulator burst time plan*) DATA SBTP (*plan simulador de tiempo de ráfaga*)

SBV *abbr* (*syntax-based videotext*) TELEP SBV (*videotexto basado en sintaxis*)

S/C *abbr* (*short circuit*) COMP, ELECT, POWER S/C (*cortocircuito*)

scalar *adj* PARTS escalado

scale *n* GEN escala *f*; ~ **of charges** APPL escala *f* de cargas; ~ **discrimination** TEST resolución *f* de escala; ~ **division** TEST división *f* de escala; ~ **drawing** GEN dibujo *m* a escala; ~ **numbering** TEST numeración *f* de escala; ~ **range** TEST límites *m pl* de escala

scaling *n* PARTS oxidación *f*; ~ **down** TEST reducción *f* de escala; ~ **factor** GEN factor *m* de desmultiplicación de impulsos

scan *n* COMP, ELECT, PER APP, RADIO exploración *f*; ~ **converter** RADIO convertidor *m* de exploración; ~ **line** COMP línea *f* de exploración

scanned: ~ **image** *n* COMP, PER APP imagen *f* explorada

scanner *n* COMP analizador *m*, escáner *m*, PER APP, RADIO antena *f* exploradora; ~ **distributor** ELECT distribuidor *m* de exploración; ~ **signal** SIG señal *f* de explorador; ~ **unit** SWG unidad *f* exploradora

scanning *n* COMP, ELECT, PER APP, RADIO exploración *f*; ~ **antenna** RADIO antena *f* exploradora; ~ **capability** PER APP *for character recognition* capacidad *f* de digitalización; **~-corrected antenna** RADIO antena *f* corregida por exploración; ~ **density** TELEP densidad *f* de exploración; ~ **electron microscope** (*SEM*) ELECT microscopio *m* electrónico de exploración (*SEM*); ~ **field**

TELEP campo *m* de exploración; ~ **interval** RADIO intervalo *m* de exploración; ~ **line** TV línea *f* de exploración; ~ **line frequency** TV frecuencia *f* de línea de exploración; ~ **line period** TV periodo *m* de línea de exploración; ~ **pitch** TELEP definición *f* de exploración; ~ **program** COMP programa *m* de exploración; ~ **rate** PER APP *for character recognition* régimen *m* de exploración; ~ **shift** TELEP turno *m* de exploración; ~ **speed** RADIO velocidad *f* de exploración; ~ **spot** COMP, TV punto *m* de exploración

SCART: ~ **socket television** *n* TV televisión *f* por cable SCART

scatter: ~ **diagram** *n* COMP diagrama *m* de dispersión; ~ **graph** *n* COMP gráfico *m* de dispersión

scattering: ~ **coefficient** *n* RADIO, TRANS coeficiente *m* de dispersión; ~ **loss** *n* RADIO, TRANS pérdida *f* por dispersión

scatterometer *n* TEST dispersiómetro *m*

SCC *abbr* (*satellite control centre*) RADIO centro *m* de control de satélites

SCCN *abbr* (*system coordination and control network*) TEST SCCN (*red de coordinación y control de sistema*)

SCCP *abbr* (*signalling connection control part*) SIG, TELEP SCCP (*dispositivo de control de la conexión de señalización*); ~ **method indicator** *n* SIG, TELEP indicador *m* del método SCCP

SCH *abbr* (*seizures per circuit per hour*) TELEP SCH (*ocupaciones por circuito por hora*)

scheduled: ~ **maintenance** *n* TEST mantenimiento *m* programado; ~ **operating time** *n* RADIO tiempo *m* de programación programado; ~ **reporting signal** *n* SIG señal *f* para informes programados; ~ **stop** *n* TEST parada *f* programada

schematic: ~ **representation** *n* GEN representación *f* esquemática

scheme: ~ **plan** *n* GEN plano *m* de disposición

Schmitt: ~ **limiter** *n* ELECT limitador *m* de Schmitt; ~ **trigger** *n* ELECT disparador *m* de Schmitt

Schottky: ~ **diode** *n* PARTS diodo *m* de Schottky; ~ **effect** *n* ELECT efecto *m* Schottky

scintillation *n* RADIO, TV titilación *f*

scissoring *n* COMP recortado *m*

scope *n* TEST *range* gama *f*, alcance *m*, TV, *oscilloscope* pantalla *f* de tubo de rayos catódicos

Scott: ~**-connected transformer** *n* PARTS, POWER transformador *m* con conexión Scott

scouring: ~ **tool** *n* PARTS herramienta *f* de limpieza

SCPC *abbr* (*single channel per carrier*) RADIO SCPC (*canal único por portadora*)

SCR *abbr* (*silicon-controlled rectifier*) ELECT, PARTS RCS (*rectificador controlado de silicio*); ~ **crosspoint** *n* ELECT, PARTS punto *m* de cruce RCS

scramble *vt* COMP, DATA, RADIO, TELEP, TV distorsionar

scrambled: ~ **binary ones** *n* (*SB1*) DATA binarios *m* cifrados (*SB1*); ~ **message** *n* TELEP transmisión *f* secreta

scrambler *n* COMP, DATA, RADIO, TELEP, TV distorsionador *m*, mezclador *m*

scrambling *n* COMP, DATA, RADIO, TELEP, TV distorsionado *m*; ~ **control** DATA control *m* de encriptación; ~ **sequence** GEN secuencia *f* de codificación

scrap *n* GEN fragmento *m*

scrapping *n* PARTS desguace *m*

scratching *n* PARTS, TEST rasponazos *m pl*; ~ **noise** TELEP crepitación *f*

screen *n* GEN pantalla *f*; ~ **background** COMP fondo *m* de pantalla; ~ **burn** ELECT debilitación de la luminosidad de la pantalla fluorescente; ~ **copy** COMP, DATA copia *f* de pantalla; ~ **display** COMP pantalla *f*; ~ **effect** GEN efecto *m* de pantalla; ~ **efficiency** ELECT rendimiento *m* de pantalla; ~ **grid** ELECT rejilla *f* de pantalla; ~ **image** COMP imagen *f* de pantalla; ~ **locking** COMP bloqueo *m* de pantalla; ~ **page** COMP, DATA, TV página *f* en pantalla; ~ **printing** COMP, PARTS estampación *f* con estarcido; ~ **terminal** COMP, TV terminal *m* de pantalla

screened: ~ **cable** *n* PARTS, TELEP cable *m* apantallado

screening: ~ **effect** *n* GEN efecto *m* de apantallamiento; ~ **indicator** *n* (*SI*) TV indicador *m* de apantallamiento; ~ **inspection** *n* TEST inspección *f* de selección; ~ **test** *n* TEST prueba *f* de selección

screensaver *n* COMP salvapantallas *m*

screw *n* PARTS tornillo *m*; ~ **base** PARTS casquillo *m* de rosca; ~ **bolt** PARTS tornillo *m*; ~ **cap** PARTS tapón *m* de tuerca; ~ **clamp** PARTS prensa *f* de tornillo; ~ **coupling** PARTS acoplamiento *m* roscado; ~ **cover** PARTS tapa *f* roscada; ~ **with hexagonal head** PARTS tornillo *m* de cabeza hexagonal; ~ **joint** PARTS junta *f* roscada; ~ **terminal** POWER terminal *m* de tornillo; ~ **terminal block** POWER bloque *m* de terminales de tornillo; ~ **top** PARTS terraja *f*

screwdriver *n* GEN destornillador *m*

screwed: ~ **connection** *n* POWER conexión *f* roscada

screwlock *n* PARTS cierre *m* de rosca

scribe *vt* PARTS trazar

scribed: ~ **line** *n* PARTS línea *f* trazada con gramil; ~ **mark** *n* PARTS marcado *m* de buril

scriber *n* PARTS punta *f* de trazar

scribing *n* PARTS trazado *m*; ~ **template** PARTS plantilla *f* de trazado

script *n* COMP, DATA guión *m*; ~-**based reasoning** COMP, DATA razonamiento con un guión como guía; ~ **system** COMP, DATA sistema *m* de rodaje

scriptable *adj* COMP, DATA emisible

SCS *abbr* (*system-conformance statement*) DATA, TEST SCS (*enunciado de conformidad de sistema*)

SCSI *abbr* (*small computer system interface*) COMP SCSI (*interfaz de pequeño sistema informático*)

SCTR *abbr* (*system conformance test report*) DATA, TEST SCTR (*informe de prueba de conformidad de sistema*)

SCU *abbr* (*system control signal unit*) SIG SCU (*unidad de señal de control de sistema*)

SD *abbr* (*signal degradation*) SIG degradación *f* de señal

SDA *abbr* (*signal distortion analyser*) TEST SDA (*analizador de distorsión de señal*)

SDH *abbr* (*synchronous digital hierarchy*) TELEP, TRANS jerarquía *f* digital síncrona; ~ **management network** *n* (*SMN*) TRANS red *f* de gestión de jerarquía digital síncrona; ~ **management subnetwork** *n* (*SMS*) TRANS subred *f* de gestión de jerarquía digital síncrona

S-distortion *n* ELECT distorsión *f* de línea

SDL *abbr* (*specification and description language*) COMP, DATA SDL (*lenguaje de descripción y especificación*)

SDLC *abbr* (*synchronous data-link control*) DATA SDLC (*procedimiento de control de transmisión síncrona*)

SDN *abbr* (*shared-data network*) DATA SDN (*red de datos compartidos*)

SDTNX: ~ **code** *n* (*selective do-not-transmit code*) DATA código *m* SDTNX (*código selectivo de no transmitir*)

SDU *abbr* (*service data unit*) DATA unidad *f* de datos de servicio

sea: ~ **clutter** *n* RADIO ecos *m* de mar; ~ **return** *n* RADIO retorno *m* de mar

seal *n* ELECT, PARTS sello *m*

sealant: ~ **applicator** *n* PARTS aplicador *m* de obturador

sealed *adj* ELECT, PARTS sellado, estanco

sealing *n* ELECT, PARTS sellado *m*; ~ **pliers** *n pl* GEN alicates *m pl* de precintar; ~ **ring** PARTS anillo *m* sellador

search[1] *n* COMP, RADIO búsqueda *f*; ~ **and rescue** (*SRR*) RADIO búsqueda *f* y rescate (*SRR*); ~-**and-rescue operation** RADIO operación *f* de búsqueda y rescate; ~ **coil** RADIO bobina *f* exploradora, bobina *f* buscadora; ~ **condition** RADIO situación *f* de búsqueda; ~ **engine** COMP, DATA dispositivo *m* de búsqueda; ~ **field** COMP, DATA campo *m* de búsqueda; ~ **key** COMP clave *f* de búsqueda; ~ **mark** COMP marca *f* de búsqueda; ~ **mode** (*SMA*) DATA modo *m* de búsqueda; ~ **mode acquisition** (*SMA*) DATA adquisición *f* de modo de búsqueda (*SMA*); ~ **mode acquisition procedure** DATA procedimiento *m* de adquisición en modo búsqueda; ~ **radar** RADIO radar *m* de exploración; ~ **subgroup** TELEP subgrupo *m* de búsqueda

search[2] *vt* COMP, DATA, RADIO búscar; ◆ ~ **and replace** COMP, DATA buscar y cambiar

second: ~ **adjacent channel** *n* RADIO canal *m* lateral secundario; ~ **approximation diode** *n* ELECT, PARTS diodo *m* de aproximación secundario; ~-**attempt routing** *n* SIG, TELEP encaminamiento *m* alternativo; ~ **choice** *n* TELEP segunda elección *f*; ~ **data multiplexer** *n* DATA, TRANS multiplexor *m* de datos secundario; ~ **dial tone** *n* TELEP tono *m* de marcar secundario; ~ **distribution link** *n* RADIO segundo enlace *m* de distribución; ~ **harmonic distortion** *n* ELECT distorsión *f* del segundo armónico; ~ **satellite hop** *n* RADIO segundo trayecto *m* de satélite; ~-**surface mirror** *n* RADIO espejo *m* de segunda superficie; ~-**tier trunk exchange area** *n* SWG segunda zona *f* telefónica de la central de línea privada; ~ **time around echo** *n* RADIO eco *m* de segunda vuelta

secondary: ~ **breakdown** *n* PARTS ruptura *f* secundaria; ~ **coating** *n* PARTS, TRANS cobertura *f* secundaria; ~ **dial tone** *n* TELEP tono *m* de marcar secundario; ~ **distribution link** *n* RADIO enlace *m* de distribución secundario; ~ **electron emission** *n* ELECT emisión *f* secundaria; ~ **electron emission current** *n* ELECT corriente *f* de emisión secundaria; ~ **electron emission factor** *n* ELECT factor *m* de emisión secundaria; ~ **failure** *n* TEST fallo *m* secundario; ~ **fleet** *n* RADIO flota *f* secundaria; ~ **flow direction** *n* GEN dirección *f* de flujo secundario, TRANS emisión *f* de flujo secundario; ~ **frequency standard** *n* RADIO patrón *m* secundario de frecuencia; ~ **function** *n* DATA función *f* secundaria; ~ **grade of reception quality** *n* RADIO grado *m* secundario de la calidad de recepción; ~ **ionic emission** *n* ELECT emisión *f* iónica secundaria; ~ **multiplex coupler** *n* TELEP acoplador *m* múltiplex secundario; ~ **network** *n* TELEP red *f* secundaria; ~ **processor unit** *n* COMP unidad *f* de procesador secundario; ~ **radar** *n* RADIO radar *m* secundario; ~ **radiation** *n* RADIO radiación secundaria; ~ **reference burst code** *n* DATA código *m* de ráfaga de referencia secundario; ~ **refrigerant** *n* CLIM refrigerante *m* secundario; ~ **relay** *n* PARTS relé *m* auxiliar; ~ **route** *n* TELEP, TRANS vía *f* auxiliar; ~ **transmitter** *n* RADIO transmisor *m* secundario; ~ **winding** *n* PARTS devanado *m* secundario

secrecy: ~ **button** *n* PER APP botón *m* de comunicación secreta

secret: ~ **code** *n* RADIO código *m* secreto

section *n* GEN sección *f*, tramo *m*; ~ **adaptation** DATA, TELEP adaptación *f* de sección; ~ **spacing** DATA, TELEP separación *f* de secciones; ~ **termination** DATA, TELEP terminación *f* de sección

sector *n* GEN sector *m*; ~ **attributes** *n pl* COMP atributos *m pl* de sector; ~ **cell** RADIO celda *f* sectorial; ~ **display** RADIO presentación *f* de sector; ~ **scanning** RADIO exploración *f* sectorial; **~-scanning beacon** RADIO radiofaro *m* de exploración sectorial; **~-shaped conductor** PARTS conductor *m* de sección sectorial

sectors: ~ **per track** *n pl* (*spt*) COMP sectores *m pl* por pista (*spt*)

secure[1] *adj* COMP, DATA, TELEP *line* seguro, fijo; ~ **access** *n* DATA acceso *m* protegido; ~ **storage** *n* DATA almacenamiento *m* protegido; ~ **telephone** *n* TELEP teléfono *m* secreto

secure[2] *vt* COMP, DATA, TELEP segurar, protegir, fijar

secured: ~ **telephone unit** *n* TELEP unidad *f* telefónica protegida

security *n* GEN seguridad *f*; ~ **arrangements** *n pl* TEST medidas *f pl* de seguridad; ~ **audit** DATA comprobación *f* de seguridad; ~ **audit trail** DATA trayectoria *f* de comprobación de seguridad; ~ **label** DATA etiqueta *f* de seguridad; ~ **management information base** (*SMIB*) DATA base *f* de información para la gestión de la seguridad; ~ **network** TRANS red *f* de seguridad; ~ **policy** GEN plan *m* de seguridad; ~ **service** GEN servicio *m* de seguridad

Seebeck: ~ **effect** *n* ELECT *thermoelectrics* efecto *m* Seebeck

S-effect *n* ELECT efecto *m* S

segment *n* COMP, DATA, TELEP segmento *m*; ~ **terminator** COMP, DATA terminador *m* de segmento; ~ **type** (*ST*) COMP, DATA tipo *m* de segmento

segmentation *n* COMP, DATA, TELEP segmentación *f*; ~ **permitted flag** (*SPF*) DATA bandera *f* reglamentaria de segmentación; ~ **and reassembly** (*SAR*) DATA segmentación *f* y rearmado (*SAR*); ~ **and reassembly sublayer** DATA subcapa *f* de segmentación y rearmado

segmented *adj* COMP, DATA, TELEP segmentado; ~ **coding law** *n* TELEP, TRANS ley *f* de codificación por segmentos; ~ **encoding law** *n* TELEP, TRANS ley *f* de codificación por segmentos; ~ **multiprocessor system** *n* DATA sistema *m* de multiprocesador segmentado

segmenting *n* COMP, DATA, TELEP segmentación *f*

seizing: ~ **acknowledgement signal** *n* SIG, TELEP señal *f* de reconocimiento de ocupación; ~ **signal** *n* SIG, TELEP señal *f* de ocupación

seizure *n* TELEP toma *f*, ocupación *f*; ~ **circuit** TELEP circuito *m* de ocupación; ~ **condition** TELEP estado *m* de ocupación; ~ **counter** TELEP contador *m* de ocupaciones; ~ **indicator** TELEP indicador *m* de ocupación; ~ **lamp** TELEP lámpara *f* de ocupación; ~ **relay** PARTS, TELEP relé *m* de ocupación; ~ **signal** SIG, TELEP señal *f* de toma de línea; ~ **supervision** TELEP supervisión *f* de ocupaciones; ~ **time** GEN tiempo *m* de ocupación; ~ **wire** PARTS, TELEP hilo *m* de ocupación

seizures: ~ **per circuit per hour** *n pl* (*SCH*) TELEP ocupaciones *f pl* por circuito por hora (*SCH*)

SELCAL *abbr* (*selective calling system*) RADIO, TELEP sistema *m* de llamada selectiva

select[1]: ~ **button** *n* COMP botón *m* de selección; **~-on-test value** *n* TEST valor *m* seleccionado en pruebas

select[2] *vt* COMP elegir, SWG, TELEP, TV seleccionar

selected: ~ **abstract test suite** *n* (*SATS*) DATA, TEST batería *f* de pruebas abstractas seleccionadas (*SATS*); ~ **executable test suite** *n* (*SETS*) DATA, TEST batería *f* de pruebas ejecutables seleccionadas (*SETS*); ~ **number redial** *n* TELEP repetición *f* automática de números externos; ~ **position** *n* TELEP posición *f* seleccionada

selecting: ~ **bar** *n* PARTS, SWG barra *f* selectora; **~-bar coil** *n* PARTS, SWG bobina *f* de barra selectora; **~-bar damper** *n* PARTS, SWG amortiguador *m* de horizontal; **~-bar position** *n* PARTS, SWG posición *f* de horizontal; **~-bar spring set** *n* PARTS, SWG grupo *m* de muelles de contacto horizontal; ~

finger *n* PARTS, SWG dedo *m* selector; **~-finger base** *n* PARTS, SWG fijación *f* de indicador; **~ finger support** *n* PARTS, SWG fijación *f* de indicador; **~ unit** *n* PARTS, SWG unidad *f* selectora

selection: **~ acknowledgement output** *n* TELEP salida *f* de reconocimiento de selección; **~ allotter** *n* TELEP cadena *f* de selección; **~ box** *n* COMP caja *f* de selección; **~ code** *n* TELEP código *m* de selección; **~ identity** *n* TELEP identidad *f* de selección; **~ input** *n* TELEP entrada *f* de selección; **~ signal** *n* SIG, TELEP señal *f* de selección; **~ stage** *n* SWG, TELEP etapa *f* de selección; **~ type** *n* TELEP tipo *m* de selección

selective: **~ accounting** *n* TELEP contabilización *f* selectiva; **~ back-alarm indication** *n* RADIO indicación *f* de alarma posterior selectiva; **~ calling** *n* RADIO, TELEP llamada *f* selectiva; **~ calling system** *n* (*SELCAL*) RADIO, TELEP sistema *m* de llamada selectiva; **~ cancellation** *n* RADIO cancelación *f* selectiva; **~ do-not-transmit code** *n* (*SDTNX code*) DATA código *m* selectivo de no transmitir (*código SDTNX*); **~ erasing** *n* DATA borrado *m* selectivo; **~ field-protection** *n* DATA, RADIO protección *f* selectiva de campo; **~ jamming** *n* ELECT, RADIO interferencia *f* selectiva; **~ reject** *n* TELEP rechazo *m* selectivo; **~-reject frame** *n* (*SREJ*) TELEP trama *f* de rechazo selectivo (*SREJ*); **~ ringing** *n* TELEP llamada *f* selectiva

selectivity *n* RADIO *of receiver*, TELEP selectividad *f*

selector *n* DATA, SWG, TELEP selector *m*; **~ mechanism** SWG, TELEP mecanismo *m* selector; **~ multiple** SWG, TELEP múltiple *m* de selectores; **~ noise** SWG, TELEP ruido *m* de selectores; **~ rack** SWG, TELEP armazón *m* de selectores; **~ stage** SWG, TELEP etapa *f* de conmutación

selenium: **~ rectifier** *n* PARTS, POWER rectificador *m* de selenio

self: **~-adapting** *adj* COMP autoadaptable; **~-adjusting** *adj* COMP autoajustable; **~-boot** *n* ELECT autoarranque *m*; **~-breaking contact** *n* PARTS contacto *m* autointerruptor; **~-carrying cable** *n* TRANS cable *m* autotransportado; **~-checking** *adj* COMP autocontrol; **~-cleaning contact** *n* PARTS contacto autolimpiante; **~-clocking signal** *n* SIG señal *f* de autotemporización; **~-contained** *adj* COMP, ELECT autónomo; **~-discharge current** *n* POWER corriente *f* de autodescarga; **~-excited transmitter** *n* RADIO emisor *m* autoexcitado; **~-healing** *adj* POWER autoregenerable; **~-heating** *adj* POWER de combustión espontánea; **~-identification** *n* RADIO autoidentificación *f*; **~-inductance** *n* POWER autoinductancia *f*; **~-locking** *adj* GEN de autoretención; **~-maintained discharge** *n* ELECT, POWER descarga *f* autónoma; **~-metering** *n* APPL automedición *f*; **~-neutralization frequency** *n* ELECT frecuencia *f* de autoneutralización; **~-operating springset** *n* PARTS, SWG grupo *m* de muelles de contacto autoaccionados; **~-optimization** *n* ELECT autooptimización *f*; **~-organizing program** *n* COMP programa *f* autoorganizante; **~-orthogonal convolutional coding** *n* DATA codificación *f* convolucional autooctogonal; **~-oscillating** *adj* ELECT autooscilante; **~-oscillation** *n* ELECT autooscilación *f*; **~-phase modulation** *n* TRANS modulación *f* de autofase; **~-quenched counter tube** *n* RADIO tubo *m* contador autoextintor; **~-quenching oscillator** *n* ELECT, PARTS oscilador de extinciones; **~-referential** *adj* COMP, DATA autorreferente; **~-registration** *n* DATA autorregistro *m*; **~-relative address** *n* COMP dirección *f* autorreferida; **~-resonant frequency** *n* RADIO frecuencia *f* autorresonante; **~-supporting antenna** *n* RADIO antena *f* autoestable; **~-supporting cable** *n* TRANS cable *m* autosoportado; **~-sustained oscillation** *n* GEN oscilación *f* autosostenida; **~-test** *n* TEST autoprueba *f*

self-reset *n* COMP autorreposición *f*

Sellmeir: **~'s equation** *n* TRANS ecuación *f* de Sellmeir

selsyn *adj* ELECT, POWER autosincrónico

SEM *abbr* (*scanning electron microscope*) ELECT SEM (*microscopio electrónico de exploración*)

semantic: **~ database** *n* DATA base *f* de datos semántica; **~ interpretation** *n* DATA interpretación *f* semántica; **~ network** *n* COMP, DATA red *f* semántica

semantically: **~ invalid event** *n* DATA, TEST evento *m* semánticamente inválido

semantics *n* GEN semántica *f*

semaphore *n* ELECT, TELEP semáforo *m*

semation *n* TELEP semación *f*

semator *n* TELEP semator *m*

semiadditive: **~ process** *n* ELECT proceso *m* semiaditivo

semiassembled: **~ representation** *n* GEN representación *f* en hileras

semiassociated: **~ network** *n* SWG red *f* semiasociada

semiautomatic *adj* SWG, TELEP semiautomático; **~ observation** *n* TEST observación *f* semiautomática; ~ **switching system** *n* SWG sistema *m* de conmutación semiautomática; ~ **system** *n* SWG, TELEP sistema *m* semiautomático

semicircular: ~ **error** *n* RADIO error *m* semicircular

semiconductive: ~ **floor** *n* TEST piso *m* semiconductor

semiconductor *n* ELECT, PARTS semiconductor *m*; ~ **amplifier** ELECT, PARTS amplificador *m* de semiconductores; ~ **contact** ELECT, PARTS contacto *m* semiconductor; ~ **crosspoint** ELECT, PARTS, SWG punto *m* de cruce de semiconductores; ~ **device** ELECT, PARTS dispositivo *m* semiconductor; ~ **diode** ELECT, PARTS diodo *m* semiconductor; ~ **laser** ELECT, PARTS, TRANS láser *m* semiconductor; ~ **laser diode amplifier** (*SLDA*) ELECT, PARTS, TRANS amplificador *m* de láser diódico semiconductor (*SLDA*); ~ **rectifier diode** ELECT, PARTS diodo *m* rectificador de semiconductor; ~ **rectifier stack** ELECT, PARTS conjunto *m* rectificador de semiconductores; ~ **storage** DATA, ELECT, PARTS memoria *f* de semiconductores

semiduplex *adj* TELEP semiduplex

semigraphic *adj* TEST semigráfico

semiloop: ~ **loss** *n* TELEP pérdida *f* de semibucle

semimajor: ~ **axis** *n* RADIO eje *m* semimayor

semipermanent: ~ **access** *n* DATA acceso *m* semipermanente; ~ **connection** *n* TELEP conexión *f* semipermanente; ~ **data** *n pl* TELEP datos *m pl* semipermanentes; ~ **link** *n* TELEP enlace *m* semipermanente

semirigid: ~ **plastic** *n* PARTS plástico *m* semirrígido

semitone *n* COMP semitono *m*

send[1]: ~ **button** *n* RADIO botón *m* de transmisión; ~ **event** *n* DATA evento *m* de transmisión; ~ **next digit signal** *n* SIG, TELEP señal *f* de transmisión de dígito siguiente; ~ **reference station** *n* RADIO, TELEP estación *f* de transmisión de referencia; ~ **special information tone signal** *n* SIG transmisión *f* de señal de tono de información especial

send[2] *vt* COMP, DATA, RADIO, TELEP, TRANS transmitir; ◆ ~ **by wire** TELEP enviarpor telegrama; ~ **special information tone** TELEP transmitir tono de información especial

sender *n* COMP, DATA, RADIO, TELEP, TRANS transmisor *m*; ~ **identification** DATA, TELEP identificación *f* de emisor; ~ **transfer time** TELEP tiempo *m* de transferencia de emisor

senderized: ~ **operation** *n* SIG, TELEP operación *f* emisorizada

sending *n* COMP, DATA, RADIO, TELEP, TRANS transmisión *f*; ~ **application entity** DATA entidad *f* de aplicación emisora; ~ **sensitivity** TELEP *of local telephone circuit* sensibilidad *f* de emisión

sense *vt* TEST detectar

sensing *n* TEST detección *f*; ~ **element** TEST elemento *m* detector; ~ **point** TEST punto *m* sensor; ~ **station** COMP, TEST estación *f* de detección; ~ **unit** TEST unidad *f* sensora

sensitive *adj* GEN sensible, confidencial; ~ **data** *n pl* COMP, DATA datos *m pl* confidenciales; ~ **volume** *n* RADIO volumen *m* sensible

sensitivity *n* GEN sensibilidad *f*; ~ **factor** RADIO factor *m* de sensibilidad; ~ **time control** (*STC*) CONTR atenuador *m* (*STC*)

sensitize *vt* GEN sensibilizar

sensor *n* GEN sensor *m*; ~ **system** GEN sistema *m* sensor; ~ **wire** PARTS hilo *m* sensor

sentence *n* COMP, DATA sentencia *f*

separate: ~ **charging system** *n* POWER sistema *m* con carga independiente; ~ **common-channel signalling** *BrE n* SIG señalización *f* de canal común separada; ~ **cooling** *n* GEN refrigeración *f* independiente

separated: ~ **representation** *n* GEN representación *f* independiente; ~ **video** *n* (*S-video*) COMP, TV vídeo *m* independiente

separation *n* GEN despegue *m*, separación *f*

separator *n* DATA, ELECT, POWER, RADIO, TELEP separador *m*

septate: ~ **waveguide** *n* DATA, RADIO, TRANS guíaondas *m* tabicado

septet *n* COMP septeto *m*

septum *n* RADIO *of waveguide* septum *m*

sequence *n* GEN secuencia *f*; ~ **adjustment** SWG ajuste *m* de secuencia; ~ **analysis** GEN análisis *m* de secuencia; ~ **behaviour** *BrE* COMP, TELEP comportamiento *m* de secuencia; ~ **chart** TELEP organigrama *m* de funcionamiento secuencial; ~**-connected rectifier** PARTS, POWER rectificador *m* conectado a una secuencia; ~ **data** *n pl* COMP, DATA datos *m pl* secuenciales; ~ **of events** TEST serie *f* de acontecimientos; ~ **function** SWG función *f* secuencial; ~ **header** COMP encabezador *m* de secuencia; ~ **number** (*SN*) COMP, DATA, SWG número *m* de orden; ~ **number protection** (*SNP*) COMP, DATA, SWG protección *f* de número de orden; ~ **recorder** COMP registrador *m* de secuencia; ~ **requirements** *n pl* SWG requisitos *m pl* de funcionamiento secuencial; ~ **spring set** PARTS, SWG grupo *m* de muelles de funcionamiento secuencial; ~ **switch** POWER, SWG conmutador *m* de secuencia; ~ **type** DATA tipo *m* de secuencia

sequencer *n* GEN secuenciador *m*; **~ program** DATA programa *m* secuenciador

sequencing *n* GEN control *m* secuencial; **~ system** DATA sistema *m* secuencial

sequential *adj* GEN secuencial; **~ access** *n* COMP, DATA acceso *m* secuencial; **~ access storage** *n* COMP, DATA memoria *f* de acceso secuencial; **~ analysis** *n* GEN análisis *f* secuencial; **~ assembling** *n* COMP montaje *m* secuencial; **~ character file** *n* COMP archivo *m* de caracteres secuenciales; **~ character file manager** *n* COMP gestor *m* de archivo de caracteres secuenciales; **~ circuit** *n* DATA, ELECT circuito secuencial; **~ coding** *n* COMP, TELEP codificación *f* secuencial; **~ control** *n* COMP, CONTR control *m* en secuencia; **~ decoding** *n* TELEP decodificación *f* secuencial; **~ hunting** *n* TELEP búsqueda *f* secuencial; **~ logic element** *n* ELECT elemento *m* lógico secuencial; **~ machine** *n* COMP, DATA máquina *f* secuencial; **~ monitoring** *n* RADIO control *m* secuencial; **~ operation** *n* COMP operación *f* secuencial; **~ program** *n* CONTR programa *m* secuencial; **~ reference burst distribution** *n* DATA distribución *f* secuencial de ráfaga de referencia; **~ sampling** *n* TEST muestreo *m* secuencial; **~ scanning** *n* COMP, TELEP exploración *f* por líneas sucesivas; **~ seizure** *n* TELEP ocupación *f* secuencial; **~ selection** *n* TELEP selección *f* secuencial; **~ sensing** *n* TEST detección *f* secuencial; **~ statement** *n* DATA sentencia *f* secuencial; **~ switching** *n* SWG conmutación *f* secuencial; **~ test** *n* TEST prueba *f* secuencial; **~ tone coding** *n* DATA codificación *f* por tonos secuenciales; **~ transmission** *n* RADIO transmisión *f* en secuencia, TRANS transmisión *f* secuencial

serial *adj* GEN en serie, serial; **~ access** *n* COMP acceso *m* en serie; **~ adder** *n* ELECT sumador *m* en serie; **~ call** *n* TELEP llamada *f* en serie; **~ call key** *n* TELEP botón *m* de llamadas en serie; **~ carry** *n* DATA arrastre *m* en serie; **~ interface** *n* DATA, ELECT interfaz *f* en serie; **~ jack** *n* PARTS jack *m* en serie; **~-line Internet protocol** *n* DATA protocolo *m* Internet de línea en serie; **~ number** *n* TELEP número *m* de serie; **~-parallel conversion** *n* DATA conversión *f* serie-paralelo; **~ port** *n* COMP puerta *f* en serie; **~ printer** *n* DATA impresora *f* en serie; **~ programmable interface** *n* COMP interfaz *f* serie programable; **~-to-parallel converter** *n* ELECT, TELEP convertidor *m* de serie aparalelo; **~ transfer** *n* DATA transferencia *f* en serie; **~ transmission** *n* TELEP, TRANS transmisión *f* en serie

serializer *n* ELECT, TELEP convertidor de paralelo a serie

series *n* ELECT, POWER, TRANS serie *f*; **~ circuit** ELECT, POWER circuito *m* en serie; **~ connection** ELECT, POWER acoplamiento *m* en serie; **~ impedance** ELECT, POWER impedancia *f* en serie; **~ inductance** ELECT, POWER inductancia *f* en serie; **~ pass transistor** ELECT, PARTS, POWER transistor *m* de conducción en serie; **~ port** COMP puerta *f* en serie; **~ resistance** ELECT, POWER resistencia *f* en serie; **~ resistor** ELECT, PARTS, POWER resistor *m* en serie; **~ resonance** ELECT, POWER resonancia *f* en serie; **~ resonant circuit** ELECT, POWER circuito *m* resonante en serie; **~ T** RADIO T en serie

serve *vt* COMP, DATA, RADIO, TELEP, TV calentar

serveability *n* APPL, TELEP capacidad *f* de servicio; **~ performance** APPL, TELEP servibilidad *f*

server *n* COMP, DATA, SWG, TELEP, TV procesador *m* de interconexión; **~ centre** *BrE* GEN centro *m* de servidor; **~ command** COMP, DATA orden *f* de servidor; **~ software** COMP, DATA software *m* de servidor

service *n* GEN servicio *m*; **~ access point** (*SAP*) DATA punto *m* de acceso a servicio; **~ access point identifier** (*SAPI*) DATA identificador *m* de punto de acceso a servicio; **~ access switch function** DATA, SWG, TELEP función *f* de conmutación de acceso al servicio; **~ advice** TELEP recomendación *f* de servicio; **~ alarm** TELEP alarma *f* de servicio; **~ arc** RADIO arco *m* de servicio; **~ area** GEN zona *f* de servicio; **~ availability performance** TELEP características *f pl* de disponibilidad del servicio; **~ bits** *n pl* COMP bitios *m pl* de servicio; **~ call** TELEP llamada *f* de servicio; **~ channel message** TEST mensaje *m* de canal de servicio; **~ code** SWG, TELEP código *m* de servicio; **~-code prefix** SWG, TELEP prefijo *m* de código de servicio; **~-command point** SWG punto *m* de orden de servicio; **~ company** GEN empresa *f* de asistencia técnica; **~ component** COMP componente *m* de servicio; **~ conditions** *n pl* POWER condiciones *f pl* de funcionamiento; **~ data unit** (*SDU*) DATA unidad *f* de datos de servicio; **~ digit** TELEP, TRANS dígito *m* de servicio; **~ element** DATA, TELEP, TRANS elemento *m* de servicio; **~ identification** APPL identificación *f* de servicio; **~ incident** TEST perturbación *f* de servicio; **~ indicator** TELEP indicador *m* de servicio; **~ information** DATA información *f* sobre el servicio; **~ integration PABX** APPL PABX *f* de integración de servicio;

~ **interruption** TELEP, TEST interrupción *f* de servicio; ~ **interworking** TELEP interfuncionamiento *m* de servicio; ~ **invocation component** TELEP componente *m* de invocación de servicio; ~ **irregularity** TEST irregularidad *f* de servicio; ~ **jack** PARTS, TELEP jack *m* de servicio; ~-**jack unit** PARTS, TELEP unidad *f* de jacks de servicio; ~ **layer** TRANS nivel *m* de servicio; ~ **line** TELEP línea *f* de servicio; ~ **node** TELEP nodo *m* de servicio; ~ **notification** DATA, TELEP notificación *f* de servicio; ~ **observation** TEST observación *f* de servicio; ~ **offering** GEN oferta *f* de servicio; ~ **operability performance** TEST característica *f* de operabilidad de servicio; ~-**or-option not-available class** DATA clase *f* no disponible de servicio u opción; ~-**or-option not-implemented class** DATA clase *f* no operativa de servicio u opción; ~ **plan** TV plan *m* de servicio; ~ **primitive** TELEP primitivo *m* de servicio; ~ **program** COMP programa *m* de servicio; ~-**provider link** (*SPL*) TRANS vínculo *m* con el prestador de servicios; ~ **rate** COMP precio *m* de servicio; ~ **relationship** DATA relación *f* de servicio; ~-**reliability performance** TEST característica *f* de fiabilidad de servicio; ~ **representative** GEN representante *m* de servicio; ~ **sector** GEN sector *m* de servicios; ~ **signal** SIG, TELEP señal *f* de servicio; ~-**specific coordination function** DATA función *f* de coordinación específica de servicio; ~ **string advice** DATA cadena *f* de aviso de servicio; ~ **supervision** TELEP supervisión *f* de servicio; ~-**supervision desk** TELEP oficina *f* de supervisión de servicio; ~-**support performance** TELEP característica *f* de apoyo de servicio; ~ **technician** GEN técnico *m* de servicio; ~ **telegram** PER APP, TELEP telegrama *m* de servicio; ~ **telex call** PER APP, TELEP comunicación *f* télex de servicio; ~ **time** APPL tiempo *m* de servicio; ~-**transfer node** TELEP nodo *m* de servicio/transferencia; ~-**user** DATA, TELEP usuario *m* de servicio

serviceability *n* APPL, TELEP aptitud *f* para funcionar; ~ **performance** APPL, TELEP servibilidad *f*

serviceable *adj* APPL, TELEP apto para funcionar

services: ~ **rendered** *n pl* TELEP servicios *m pl* prestados

serving *n* APPL *of subscribers,* PARTS *of cable* revestimiento *m*; ~ **exchange** SWG central *f* de servicio; ~ **port** COMP, ELECT puerta *f* de protector

servoamplifier *n* ELECT, POWER servoamplificador *m*

servomechanism *n* CONTR servomecanismo *m*

servosystem *n* CONTR servosistema *m*

SES *abbr* (*severely-errored second*) DATA, RADIO segundo *m* sumamente impreciso

session *n* GEN sesión *f*; ~ **connection** DATA, TRANS conexión *f* de sesión; ~ **layer** DATA, TRANS capa *f* de sesión; ~ **level** DATA, TRANS nivel *m* de sesión; ~ **number** COMP número *m* de sesión; ~ **protocol** DATA, TRANS protocolo *m* de sesión; ~-**protocol data unit** (*SPDU*) DATA, TRANS unidad *f* de datos de protocolo de sesión; ~ **service** DATA, TRANS servicio *m* de sesión; ~-**service access point** (*SSAP*) DATA, TRANS punto *m* de acceso al servicio de sesión (*SSAP*)

set[1] *adj* COMP predeterminado, TELEP puesta a uno; ~ **asynchronous balanced mode** *n* (*SABM*) TELEP modo *m* equilibrado asincrónico de puesta a uno (*SABM*); ~ **asynchronous balanced mode extended** *n* (*SABME*) TELEP modo *m* equilibrado asincrónico de puesta a uno extendido (*SABME*); ~ **mark** *n* COMP marca *f* establecida; ~ **point** *n* CONTR señal *f* de referencia; ~ **pulse** *n* CONTR impulso *m* de puesta a uno; ~ **signal** *n* CONTR, SIG señal *f* de puesta a uno; ~ **status** *n* COMP estatus *m* de puesta a uno; ~-**top box** *n* TV *descrambling, connecting equipment* caja *f* de la parte superior del aparato; ~-**top decoder** *n* TV *descrambling, connecting equipment* descodificador *m* para colocar encima del televisor; ~-**of type** *n* DATA tipo *m* resaltado; ~-**valued attribute** *n* DATA atributo *m* de valor predeterminado

set[2]: ~ **of alternatives** *n* DATA conjunto *m* de alternativas

set off *vt* RADIO compensar, TEST disparar

set to *vt* DATA poner a

set up *vt* COMP, DATA, GEN, RADIO, TELEP montar

SETS *abbr* (*selected executable test suite*) DATA, TEST SETS (*batería de pruebas ejecutables seleccionadas*)

settable *adj* GEN ajustable

setting *n* COMP, TELEP *on machine* ajuste *m*, *action* regulación *f*; ~ **accuracy** GEN precisión *f* de ajuste; ~ **agent** PARTS *for plastics* agente *m* endurecedor; ~ **range** GEN límite *m* de regulación; ~ **ratio** PARTS relación *f* de ajuste; ~ **time** CONTR tiempo *m* de ajuste; ~ **up** GEN ajuste *m*; ~-**up time** CONTR, TELEP tiempo *m* de establecimiento de la conexión; ~ **value** GEN valor *m* de ajuste; ~ **to zero** TEST ajuste *m* a cero

settings *n pl* COMP calibraciones *f pl*

settling: ~ **time** *n* POWER tiempo *m* de establecimiento
setup *n* COMP, RADIO, TELEP preparación *f*; ~ **channel** SIG, TRANS canal *m* de ajuste inicial; ~ **date** GEN fecha *f* de conexión; ~ **diagram** COMP diagrama *m* de estructuración; ~ **time** GEN tiempo *m* de establecimiento
SEU *abbr* (*single event upset*) TEST SEU (*redisposición de evento simple*)
seven: ~**-segment display** *n* CONTR indicador *m* de siete segmentos; ~**-slot rack-mounting assembly** *n* ELECT, PARTS montaje *m* de bastidor de siete ranuras
severely: ~**-errored second** *n* (*SES*) DATA, RADIO segundo *m* sumamente impreciso
SF *abbr* (*signal failure*) SIG falta *f* de señal
SGC *abbr* (*signalling grouping channel*) SIG canal *m* de agrupamiento de señalización
SGML *abbr* (*standard generalized markup language*) COMP SGML (*lenguaje estándar generalizado de referencia*); ~ **document** *n* COMP documento *m* SGML
shade *n* COMP, ELECT, TV sombra *f*
shading *n* COMP, ELECT, TV sombreado *m*; ~ **correction** TV corrección *f* de sombra
shadow *n* COMP, RADIO sombra *f*; ~ **factor** RADIO factor *m* de sombra; ~ **grid** ELECT cuadrícula *f* de sombra; ~ **offset** COMP desnivel *m* de sombra; ~ **region** RADIO, TRANS zona *f* de sombra; ~ **zone** RADIO, TRANS zona *f* de sombra
shadowing *n* RADIO sombra *f*
shaft *n* PARTS tiro *m*; ~ **coupling** PARTS acoplamiento *m* de eje
shape[1]: ~ **conductor** *n* PARTS conductor *m* perfilado
shape[2] *vt* COMP, RADIO modelar
shaped: ~**-beam antenna** *n* RADIO antena *f* de haz perfilado; ~**-beam offset parabolic antenna** *n* RADIO antena *f* parabólica desplazada de haz perfilado; ~ **conductor** *n* PARTS conductor *m* perfilado
shaping: ~ **filter** *n* DATA filtro *m* corrector
shared: ~**-control mode** *n* CONTR, DATA modo *m* de control compartido; ~**-data network** *n* (*SDN*) DATA red *f* de datos compartidos (*SDN*); ~ **disk** *n* COMP disco *m* compartido; ~ **frequency** *n* RADIO frecuencia *f* compartida; ~ **resource** *n* COMP, DATA recurso *m* compartido; ~**-resource device** *n* COMP, DATA dispositivo *m* de recurso compartido; ~**-service line** *n* TELEP línea *f* de servicio compartido
shareware *n* COMP, DATA equipos *m* compartidos; ~ **package** COMP, DATA paquete *m* de equipos compartidos
sharing *n* RADIO compartición *f*; ~ **option** DATA posibilidad *f* de compartir
sharpness *n* DATA *of scanner* definición *f* de imagen
shear: ~ **wave** *n* RADIO onda *f* rotacional
sheath *n* PARTS *of cable* forro *m*; ~ **current** TEST corriente *f* por la envoltura; ~ **cutter** PARTS cortador *m* de cubierta; ~**-retention clamp** ELECT grapa *f* de sujeción de envoltura; ~**-stripping tool** PARTS herramienta *f* para retirar la cubierta
sheathed *adj* PARTS revestido
sheathing *n* PARTS revestimiento *m*; ~ **layer** ELECT, PARTS capa *f* de revestimiento
sheet: ~ **feed** *n* PER APP *of printer* alimentación *f* hoja a hoja; ~ **feeder** *n* PER APP *of printer* alimentador *m* de hojas; ~ **metal** *n* PARTS chapa *f* metálica; ~ **metal duct** *n* PARTS conducto *m* de chapa metálica
shelf *n* TELEP estante *m*; ~ **chassis** TELEP armazón *m* de estante; ~ **fuse** PARTS, POWER fusible *m* de estante; ~ **fuse alarm** TELEP alarma *f* de fusible de estante; ~ **life** TEST vida *f* útil en almacenamiento
shell *n* COMPON cuerpo *m* cilíndrico
shellac *n* PARTS laca *f*
shelved: ~ **trolley** *n* PARTS carrito *m* de estantes
SHF *abbr* (*superhigh frequency*) RADIO FSA (*hiperfrecuencia*)
shield *n* POWER pantalla *f*; ~ **grid** POWER rejilla *f* de protección
shielded: ~ **cable** *n* PARTS, POWER cable *m* blindado; ~ **connector** *n* ELECT conector *m* blindado; ~ **cord** *n* PARTS cable *m* blindado; ~ **triple-faceplate outlet jack** *n* ELECT, PARTS jack *m* de salida blindado de chapa frontal triple; ~ **twisted pair** *n* (*STP*) ELECT, TRANS conductor *m* de doble torcido blindado; ~ **wire** *n* PARTS hilo *m* blindado
shielding *n* PARTS blindaje *m*; ~ **effect** TEC RAD, TV efecto *m* blindaje
shift *n* COMP *on keyboard* decalaje *m*, ELECT, PER APP, TELEP, TRANS desplazamiento *m*; ~ **clock** COMP reloj *m* de corrimiento; ~ **function** COMP función *f* de desplazamiento; ~**-in character** DATA carácter *m* cambiador de código; ~ **key** COMP *on keyboard*, PER APP tecla *f* de mayúsculas; ~ **lock** COMP, PER APP fijador *m* de mayúsculas; ~**-lock keyboard** COMP teclado *m* con seguro de cambio, PER APP teclado *m* con seguro de inversión; ~**-out character** DATA carácter *m* de desplazamiento hacia fuera
shifted: ~ **mode** *n* COMP, PER APP *of keyboard* caja *f* alta
ship: ~ **polling** *n* RADIO interrogación *f* de las

estaciones de barco; ~ **reporting system** *n* RADIO, SIG sistema *m* de información de estación de barco; **~-to-~ alerting** *n* RADIO alerta *f* entre barcos; **~-to-shore alerting** *n* RADIO alerta *f* barco-tierra; ~ **station** *n* RADIO estación *f* de barco; ~ **station identity** *n* DATA, RADIO, TELEP identidad *f* de estación de barco; ~ **station number** *n* DATA, RADIO, TELEP número *m* de estación de barco

shipboard: ~ **terminal** *n* RADIO terminal *m* de estación de barco

shock: **~-absorbing foam** *n* ELECT espuma *f* amortiguadora de impacto; **~-absorbing tray** *n* PARTS bandeja *f* de absorción de impacto; ~ **excitation** *n* ELECT, RADIO excitación *f* por choque; **~-excited oscillator** *n* ELECT, RADIO oscilador *m* excitado por choque; ~ **test** *n* TEST prueba *f* de choque; ~ **wave** *n* GEN onda *f* de choque

shore: **~-to-ship alerting** *n* RADIO alerta *f* tierra-barco; ~ **station** *n* RADIO estación *f* costera

short[1]: ~ **burst** *n* DATA ráfaga *f* corta; ~ **circuit** *n* (*S/C*) COMP, ELECT, POWER cortocircuito *m* (*S/C*); **~-circuit admittance** *n* ELECT, POWER admitancia *f* en cortocircuito; **~-circuit current** *n* ELECT, POWER corriente *f* de cortocircuito; **~-circuit impedance** *n* ELECT, POWER impedancia *f* de cortocircuito; **~-circuit test** *n* ELECT, POWER, TEST prueba *f* de cortocircuito; ~ **code** *n* TELEP *in telex* código *m* abreviado; ~ **code dialling** *BrE* *n* TELEP discado *m* de corto código; **~-distance backscattering** *n* POWER, RADIO retrodispersión *f* de corta distancia; **~-distance traffic** *n* TELEP tráfico *m* de corta distancia; **~-haul traffic** *n* TELEP tráfico *m* suburbano; **~-haul transmission** *n* TRANS transmisión *f* a corta distancia; **~-range transmitter** *n* RADIO transmisor *m* de corto alcance; **~-term duty** *n* APPL obligación *f* a corto plazo

short[2] *n* COMP, ELECT, POWER cortocircuito *m*

short[3] *vi* COMP, ELECT, POWER cortocircuitar

shortage *n* GEN escasez *f*

shortcut *n* COMP atajo *m*

shortened: ~ **address** *n* TELEP dirección *f* abreviada

shorthand: ~ **writing** *n* COMP estenografía *f*

shorting: ~ **bar** *n* ELECT barra *f* de cortocircuito; ~ **plunger** *n* ELECT, RADIO pistón *m* de cortocircuito

shortwave: ~ **infrared** *n* RADIO infrarrojo *m* de onda corta

shot: ~ **noise** *n* ELECT, TELEP ruido *m* granular

show *vt* GEN mostrar

shrink[1]: ~ **quad flatpack** *n* TRANS componente *m* plano de cuadrete de contracción; ~ **small-outline package** *n* DATA paquete *m* comprimido de configuración reducida; ~ **tube** *n* PARTS tubo *m* de contracción; ~ **tubing** *n* PARTS entubado *m* por contracción

shrink[2] *vt* PARTS encoger

shrink[3] *vi* PARTS contraerse

shrinkage *n* PARTS contracción *f*, encogimiento *m*

shrinking *n* PARTS contracción *f*

shroud *n* RADIO envoltura *f*

shuffle *vt* COMP mezclar

shunt *n* PARTS derivación *f*; ~ **capacitor** PARTS, POWER capacitor *m* en derivación; ~ **relay** PARTS relé *m* en derivación; ~ **resistance** PARTS, POWER resistencia *f* en derivación; ~ **resistor** PARTS, POWER resistor *m* de derivación; ~ **rheostat** PARTS reóstato en derivación; ~ **T** RADIO T en paralelo

shunting *n* POWER derivación *f*; **~-yard radio system** RADIO sistema *m* de radio por estación de derivación

shutdown *n* POWER parada *f*

shutter *n* TV obturador *m*

shuttle *n* COMP lanzadera *f*

SI *abbr* (*screening indicator*) TV indicador *m* de apantallamiento

sibling *n* COMP hermano *m*

side: ~ **aisle** *n* TELEP pasillo *m* lateral; ~ **chute** *n* TELEP canal *m* lateral; ~ **circuit** *n* TELEP circuito *m* real; ~ **cover** *n* TELEP cubierta *f* lateral; ~ **cutters** *n pl* GEN alicates *m pl* de corte lateral; **~-cutting pliers** *n pl* GEN alicates *m pl* de corte lateral; **~-determining** *n* SWG determinante *m* lateral; ~ **echo** *n* RADIO eco *m* lateral; ~ **frequency** *n* RADIO frecuencia *f* lateral; ~ **indicator** *n* SWG indicador *m* lateral; ~ **information** *n* COMP información *f* colateral; ~ **lobe** *n* RADIO lóbulo *m* lateral; **~-lobe performance** *n* RADIO funcionamiento *m* del lóbulo lateral; **~-lobe radiation** *n* RADIO radiación *f* del lóbulo lateral; **~-lobe suppression** *n* (*SLS*) RADIO supresión *f* del lóbulo lateral (*SLS*); **~-looking airborne radar** *n* (*SLAR*) RADIO radar *m* aéreo de exploración lateral (*SLAR*); ~ **mode** *n* TRANS modo *m* lateral; **~-mode suppression** *n* TRANS supresión *f* de modo lateral; **~-mode suppression ratio** *n* (*SMSR*) TRANS ratio *f* de supresión de modo lateral (*SMSR*); ~ **panel** *n* TV panel *m* lateral; ~ **plate** *n* TELEP placa *f* lateral; ~ **shifting** *n* SWG cambio *m* de lado; ~ **tone** *n* TELEP efecto *m* local; **~-tone objective loudness rating** *n* (*SOLR*) TELEP régimen *m* de sonoridad objetiva de efecto local (*SOLR*)

sideband *n* (*SB*) RADIO, TELEP, TRANS banda *f* lateral (*SB*); ~ **with carrier** RADIO, TELEP, TRANS banda *f* lateral con portadora; ~ **interference** RADIO, TELEP, TRANS interferencia *f* de banda lateral; ~ **splatter** RADIO, TELEP, TRANS emisiones *f pl* espurias de banda lateral

sidereal: ~ **period of revolution** *n* RADIO periodo *m* sidéreo de revolución; ~ **period of rotation** *n* RADIO periodo *m* sidéreo de rotación

sideways: ~**-looking airborne radar** *n* RADIO radar *m* aéreo de exploración lateral (*SLAR*)

sieve *n* ELECT tamiz *m*

sift *vt* ELECT tamizar

SIG *abbr* (*signature element*) SIG, TELEP SIG (*elemento de firma*)

sign *n* GEN signo *m*; ~**-binary character** COMP carácter *m* de signo binario; ~ **bit** COMP bit *m* de signo; ~ **character** COMP carácter *m* de signo; ~ **digit** COMP dígito *m* de signo

signal *n* COMP, RADIO, SIG, TELEP, TRANS señal *f*; ~ **amplitude** SIG amplitud *f* de señal; ~ **analyser** *BrE* COMP, SIG analizador *m* de señales; ~ **analysis** COMP, SIG análisis *m* de señales; ~ **attenuation** SIG atenuación *f* de señal; ~ **balance ratio** SIG *of sinusoidal signal generators* ratio *f* de equilibrio de señal; ~ **cable chute** TELEP canal *m* de cable de señales; ~ **characteristic** SIG característica *f* de señal; ~ **circuit** SIG, TELEP circuito *m* de señal; ~ **code** SIG código *m* de señal; ~ **comparator** RADIO comparador *m* de señales; ~ **component** SIG componente *m* de señal; ~ **compression** SIG compresión *f* de señales; ~ **conditioning** SIG condicionamiento *m* para señal; ~ **conversion** SIG, TELEP conversión *f* de señales; ~**-conversion equipment** SIG, TELEP equipo *m* convertidor de señales; ~ **converter** SIG, TELEP convertidor *m* de señales; ~**-to-crosstalk ratio** TRANS relación *f* señal/diafonía; ~ **current** ELECT corriente *f* de señal; ~ **degradation** (*SD*) SIG degradación *f* de señal; ~ **delay** SIG tiempo *m* de propagación de señal; ~ **detection** SIG detección *f* de señales; ~ **detector** SIG detector *m* de señales; ~ **device** SIG señalizador *m*; ~ **digitalization** SIG, TRANS digitalización *f* de señal; ~ **diode** ELECT diodo *m* de señal; ~ **distance** SIG distancia *f* de señal; ~ **distortion** GEN distorsión *f* de señal; ~ **distortion analyser** *BrE* (*SDA*) TEST analizador *m* de distorsión de señal (*SDA*); ~ **distribution** SIG distribución *f* de señales; ~ **duration** SIG duración *f* de señal; ~ **earth** *BrE* (*cf signal ground AmE*) SIG, TRANS entrada *f* de tierra de señalización; ~ **earth input** *BrE* (*cf signal ground input AmE*) SIG, TRANS entrada *f* de tierra de señalización; ~ **electrode** ELECT electrodo *m* de señal; ~ **element** SIG, TELEP elemento *m* de señal; ~**-element timing** SIG, TELEP temporización *f* de componente de señal; ~ **envelope** SIG envolvente *m* de señal; ~ **expansion** SIG expansión *f* de señal; ~ **extension** SIG extensión *f* de señal; ~ **extraction** SIG, TRANS extracción *f* de señal; ~ **failure** (*SF*) SIG falta *f* de señal; ~ **flow diagram** SIG diagrama *m* de circulación de señales; ~ **generator** SIG, TEST generador *m* de señales; ~ **ground** *AmE* (*cf signal earth BrE*) SIG, TRANS retorno *m* de tierra de circuito de señal; ~**-identification fingerprinting** SIG toma *f* de huellas de identificación de señal; ~ **indicator** SIG, TEST, TRANS indicador *m* de señal; ~**-indicator relay** SIG, TEST, TRANS relé *m* de indicador de señal; ~ **insertion** SIG, TRANS inserción *f* de señal; ~ **interface** SIG interfaz *f* de señalización; ~**-interface converter** SIG convertidor *m* de interfaz de señalización; ~**-interference ratio** RADIO relación *f* señal-interferencia; ~ **lamp** TELEP indicador *m* luminoso; ~ **level** SIG, TRANS nivel *m* de señal; ~ **library** SWG biblioteca *f* de señales; ~ **message** SIG mensaje *m* de señal; ~**-to-noise ratio** (*SNR*) ELECT, SIG, TEST, TRANS relación *f* señal/ruido; ~ **path** SIG trayecto *m* de señal, TRANS circuito *m* de señal; ~ **power** SIG potencia *f* de señal; ~ **processing** SIG procesamiento *m* de señales; ~ **processor** SIG procesador *m* de señales; ~ **processor unit** SIG unidad *f* de procesador de señales; ~ **quantization** SIG cuantificación *f* de una señal; ~ **receiver** RADIO, SIG receptor *m* de señales; ~ **regeneration** SIG regeneración *f* de señal; ~ **restoration** SIG restablecimiento *m* de una señal; ~ **sender** SIG, TRANS emisor *m* de señales; ~ **shaping** ELECT, SIG configuración *f* de señal; ~**-shaping filter** ELECT, SIG filtro *m* de formación de señal; ~**-shaping network** TELEP red *f* de formación de señal; ~ **state** GEN estado *m* de señal; ~ **and state recorder** SWG registrador *m* de señal y estado; ~ **strength** SIG intensidad *f* de señal; ~**-strength meter** SIG medidor *m* de intensidad de señal; ~ **surface** COMP superficie *f* de señal; ~ **synthesis** DATA, SIG síntesis *f* de señal; ~ **test** SIG, TEST prueba *f* de señal; ~**-test circuit** SIG, TEST circuito *m* de prueba de señal; ~ **threshold** SIG umbral *m* de señal; ~ **tracer** SIG rastreador *m* de señal; ~ **train** SIG tren *m* de señales; ~ **transfer** SIG, SWG, TELEP, TRANS transmisión *f* de señales; ~**-transfer point** SIG, SWG, TELEP, TRANS punto *m* de transmisión de señales; **~-transfer point switch** SIG, SWG, TELEP, TRANS conmutador *m* de punto de transmisión de señales; **~ transformation** ELECT, SIG transformación *f* de señales; **~-transit time** ELECT tiempo *m* de tránsito de señal; **~ unit** (*SU*) SIG unidad *f* de señal (*SU*); **~-unit alignment** SIG alineación *f* de unidad de señal; **~-unit error-rate monitoring** SIG, TELEP control *m* de la tasa de error por unidad de señal; **~ weakening** SIG debilitamiento *m* de señal; **~-to-weighted rms noise ratio** TRANS relación *f* de ruido señal/rms ponderado

signalling *BrE* (*AmE* **signaling**) *n* COMP, DATA, RADIO, SIG, TELEP señalización *f*; **~ adaptor group** SIG grupo *m* adaptador de señalización; **~ at 2600 Hz** SIG señalización *f* a 2600 Hz; **~ bit** SIG, TRANS bitio *m* de señalización; **~ cable** SIG cable *m* de señalización; **~ channel** SIG, TELEP canal *m* de señalización; **~ conditions** *n pl* SIG condiciones *f pl* de señalización; **~ connection control part** (*SCCP*) SIG, TELEP dispositivo *m* de control de conexión de señalización (*SCCP*); **~ control part** SIG dispositivo *m* de control de señalización; **~ converter** SIG convertidor *m* de señalización; **~ data link** SIG enlace *m* de datos de señalización; **~ description** SIG descripción *f* de señalización; **~ destination point** SIG punto *m* de destino de señalización; **~ detector** SIG detector *m* de señalización; **~ diagram** SIG diagrama *m* de señalización; **~ fault** SIG fallo *m* de señalización; **~ grouping channel** (*SGC*) SIG canal *m* de agrupamiento de señalización; **~ information** SIG información *f* de señalización; **~ information field** SIG campo *m* de información de señalización; **~ interworking** SIG interfuncionamiento *m* de señalización; **~ link** SIG, TELEP enlace *m* de señalización; **~-link blocking** SIG bloqueo *m* de enlace de señalización; **~-link code** SIG código *m* de enlace de señalización; **~-link error monitoring** SIG control *m* de error de enlace de señalización; **~-link failure** SIG falta *f* de enlace de señalización; **~-link function** SIG función *f* de enlace de señalización; **~-link group** SIG grupo *m* de enlace de señalización; **~-link management functions** *n pl* SIG funciones *f pl* de gestión de enlace de señalización; **~-link restoration** SIG restablecimiento *m* de vínculo de señalización, TELEP restablecimiento *m* de enlace de señalización; **~-link selection field** SIG campo *m* de selección de vínculo de señalización; **~-link set** SIG posición *f* de vínculo de señalización; **~-link unblocking** SIG desbloqueo *m* de vínculo de señalización; **~ message** SIG mensaje *m* de señalización; **~-message discrimination** SIG discriminación *f* de mensaje de señalización; **~-message distribution** SIG distribución *f* de mensaje de señalización; **~-message handling functions** *n pl* SIG funciones *f pl* de manipulación de mensaje de señalización; **~-message route** SIG ruta *f* de mensaje de señalización; **~-message routing** SIG encaminamiento *m* de mensaje de señalización; **~ network** SIG, SWG, TRANS red *f* de señalización; **~-network functions** *n pl* SIG funciones *f pl* de red de señalización; **~-network management functions** *n pl* SIG funciones *f pl* de gestión de red de señalización; **~-network management signals** *n pl* SIG señales *f pl* de gestión de red de señalización; **~ originating point** SIG punto *m* de origen de señalización; **~ point** SIG, TELEP punto *m* de señalización; **~-point code** SIG código *m* de punto de señalización; **~ protocol** SIG protocolo *m* de señalización; **~ receiver PE** SIG EP *m* receptor de señalización; **~ relation** SIG relación *f* de señalización; **~ route** SIG ruta *f* de señalización; **~-route management functions** *n pl* SIG funciones *f pl* de gestión de ruta de señalización; **~-route set** SIG posición *f* de ruta de señalización; **~-route-set test procedure** SIG procedimiento *m* de prueba de posición de ruta de señalización; **~ routing** SIG encaminamiento *m* de señalización; **~ sender PE** SIG EP *m* emisor de señalización; **~ sequence** SIG secuencia *f* de señalización; **~ state** SIG estado *m* de señalización; **~ system** SIG sistema *m* de señalización; **~ terminal equipment** (*STE*) TELEP equipo *m* terminal de señalización; **~ test** SIG prueba *f* de señalización; **~ time slot** SIG intervalo *m* de tiempo de señalización; **~-traffic flow control** DATA, SIG control *m* de flujo de tráfico de señalización; **~-traffic management functions** *n pl* SIG funciones *f pl* de gestión de tráfico de señalización; **~-transfer point** SIG, SWG, TELEP, TRANS punto *m* de transferencia de señalización; **~ virtual channel** SIG, TRANS canal *m* virtual de señalización; **~ voltage** POWER, SIG tensión *f* de señalización

signatory *n* TV signatario *m*

signature: **~-adapted processing** *n* RADIO procesamiento *m* adaptado a la firma; **~ element** *n* (*SIG*) SIG, TELEP elemento *m* de firma (*SIG*)

signed: ~ **response element** *n* TELEP elemento *m* de respuesta firmado
significance: ~ **level** *n* TEST nivel *m* de significación; ~ **test** *n* TEST prueba *f* de significación
significant *adj* GEN significativo; ~ **bit** *n* DATA bitio *m* significativo; ~ **condition** *n* SIG, TELEP, TEST condición *f* significativa; ~ **instant** *n* SIG, TELEP, TEST instante *m* significativo; ~ **interval** *n* SIG, TELEP, TEST intervalo *m* significativo
silence: ~ **elimination** *n* TELEP, TEST, TRANS eliminación *f* de silencio
silencer *BrE n* (*cf muffler AmE*) TELEP, TRANS silenciador *m*
silent *adj* GEN silencioso; ~ **alert vibrator** *n* RADIO vibrador *m* de alerta silenciosa; ~ **period** *n* TELEP período *m* de silencio; ~ **zone** *n* RADIO zona *f* de silencio
silica *n* PARTS sílice *f*; ~ **cladding** PARTS revestimiento *m* de sílice; ~ **coating** PARTS cobertura *f* de sílice; ~ **fibre** *BrE* PARTS, TRANS fibra *f* de sílice
silicium: ~ **nitride passivation** *n* PARTS pasivación *f* de nitruro de silicio
silicon *n* COMP, PARTS sílice *f*; ~**-controlled rectifier** (*SCR*) ELECT, PARTS rectificador *m* controlado de silicio (*RCS*); ~**-controlled rectifier crosspoint** ELECT, PARTS punto *m* de cruce de rectificador controlado por silicio; ~ **diode** PARTS diodo *m* de silicio; ~ **wafer** PARTS pastilla *f* de silicio
silk: ~**-screened lettering** *n* ELECT rotulación *f* serigrafiada
silumin *n* PARTS siluminio *m*
silver: ~**-electroplating** *n* PARTS plateado *m* por electrólisis; ~ **mica capacitor** *n* ELECT, PARTS capacitor *m* de mica plateada; ~**-plating** *n* GEN plateado *m*
SIM *abbr* (*subscriber identity module*) RADIO SIM (*módulo de identidad del abonado*)
simple: ~ **authentication** *n* DATA autentificación *f* simple; ~ **mail transfer protocol** *n* (*SMTP*) DATA protocolo *m* simple de transferencia de correo (*SMTP*); ~ **modulation** *n* RADIO modulación *f* simple; ~ **multipoint circuit** *n* TELEP circuito *m* simple de multipunto; ~ **network management protocol** *n* (*SNMP*) DATA protocolo *m* simple de gestión de red (*SNMP*); ~ **parameter argument** *n* TELEP argumento *m* de parámetro simple; ~ **protocol for ATM network signalling** *BrE n* TRANS señalización *f* de protocolo simple para red ATM; ~ **random sample** *n* TEST muestra *f* aleatoria simple; ~ **signal** *n* SIG, TELEP señal *f* sencilla; ~ **transmission** *n* TRANS transmisión *f* simple
simplex *adj* RADIO, TELEP, TRANS simplex; ~ **circuit** *n* TELEP circuito *m* simplex; ~ **operation** *n* TELEP funcionamiento *m* simplex; ~ **pair** *n* TELEP par *m* simplex; ~ **traffic** *n* RADIO tráfico *m* en simplex; ~ **transmission** *n* TRANS transmisión *f* en simplex; ~ **working** *n* TELEP funcionamiento *m* en simplex
simulate *vt* GEN simular
simulated: ~ **speech** *n* TELEP señal *f* vocal simulada
simulation *n* GEN simulación *f*
simulator *n* COMP, TELEP simulador *m*; ~ **burst time plan** (*SBTP*) DATA plan *m* simulador de tiempo de ráfaga (*SBTP*)
simulcast *n* RADIO, TV emisión *f* simultánea
simultaneous: ~ **call** *n* TELEP llamada *f* simultánea; ~**-call terminal** *n* TELEP terminal *m* de llamada simultánea; ~ **note capacity** *n* COMP *in sound reproduction* capacidad *f* de nota simultánea; ~ **note output** *n* COMP *in sound reproduction* salida *f* de nota simultánea; ~ **operation** *n* TELEP operación *f* simultánea
SINAD: ~ **ratio** *n* (*signal-to-noise and distortion ratio*) GEN relación *f* SINAD (*relación señal-ruido y distorsión*)
sine: ~**-cosine oscillator** *n* ELECT oscilador *m* de onda de seno o coseno; ~**-squared pulse** *n* ELECT, TELEP impulso *m* en seno cuadrado; ~ **vibration test** *n* TEST prueba *f* de onda sinusoidal; ~ **wave** *n* COMP, ELECT, TELEP onda *f* sinusoidal; ~**-wave pulse** *n* COMP, ELECT, TELEP impulso *m* de onda sinusoidal; ~**-wave signal** *n* COMP, ELECT, TELEP señal *f* de onda sinusoidal
singing *n* PARTS ondas *f pl* sinusoidales; ~ **point** TELEP amplificación *f* máxima
single: ~**-association control function** *n* DATA función *f* de control de asociación simple; ~**-association object** *n* DATA objeto *m* de asociación simple; ~**-attachment station** *n* (*SAS*) ELECT estación *f* de acoplamiento simple (*SAS*); ~**-billing system** *n* APPL sistema *m* de facturación simple; ~**-button calling** *n* TELEP marcación *f* por nombre; ~**-carrier algorithm** *n* COMP, RADIO, TRANS algoritmo *m* de portadora única; ~**-carrier operation** *n* COMP, RADIO, TRANS operación *f* de portadora única; ~**-cartridge adaptor** *n* COMP adaptador *m* de cartucho único; ~**-cavity mould** *BrE n* PARTS matriz *f* de cavidad única; ~**-channel carrier** *n* RADIO portadora *f* monocanal; ~ **channel per carrier** *n* (*SCPC*) RADIO canal *m* único por portadora (*SCPC*);

~-**channel recorder** *n* TEST grabador *m* de un solo canal; ~-**circuit system** *n* GEN sistema *m* de circuito único; ~-**compartment charger** *n* POWER cargador *m* de compartimento único; ~ **conductor** *n* ELECT, PARTS conductor *m* único; ~-**conductor cable** *n* ELECT, PARTS cable *m* monoconductor; ~-**constraint table** *n* DATA tabla *f* de restricción simple; ~ **contact** *n* PARTS contacto *m* simple; ~-**control bistable trigger circuit** *n* ELECT báscula *f* biestable con una entrada; ~-**cord switchboard** *n* SWG, TELEP conmutador *m* manual de monocordio; ~-**core cable** *n* PARTS cable *m* unipolar; ~-**crystal** *adj* ELECT monocristal; ~ **current** *n* POWER, TRANS corriente *f* simple; ~-**current transmission** *n* POWER transmisión *f* por corriente simple; ~-**device housing** *n* ELECT, PARTS recipiente *m* de un solo dispositivo; ~-**digit dialling** *BrE n* TELEP marcación *f* de un solo dígito; ~-**duct block** *n* PARTS, TRANS bloque *m* de un solo conducto; ~-**ended amplifier** *n* POWER amplificador *m* sin transformador de salida; ~-**ended control** *n* CONTR, TRANS control *m* asimétrico; ~ **error** *n* TEST error *m* simple; ~-**event upset** *n* (*SEU*) TEST redisposición *f* de evento simple (*SEU*); ~-**fee metering** *n* TELEP cómputo *m* simple; ~-**fibre** *BrE* **cable** *n* PARTS cable *m* de una sola fibra; ~-**fibre** *BrE* **jacket** *n* PARTS envoltura *f* de una sola fibra; ~-**frequency interference** *n* RADIO, SIG, TELEP interferencia *f* de frecuencia única; ~-**frequency laser** *n* ELECT, TRANS láser *m* de frecuencia única; ~-**frequency operation** *n* RADIO, SIG, TELEP operación *f* en una frecuencia; ~-**frequency return loss** *n* RADIO, SIG, TELEP pérdida *f* de retorno de frecuencia única; ~-**frequency signal** *n* RADIO, SIG, TELEP señal *f* de frecuencia única; ~-**gang faceplate** *n* ELECT, PARTS chapa *f* de salida simple; ~ **heterojunction** *n* TRANS heterounión *f* simple; ~ **industrial infrastructure** *n* TV infraestructura *f* industrial única; ~ **lashing** *n* GEN ligadura *f* simple; ~-**level continuous sampling** *n* TEST muestreo *m* continuo simple; ~-**line business** *n* GEN negocio *m* de línea única; ~ **in-line package** *n* (*SIP*) GEN encapsulado *m* con una sola línea de conexiones (*MBS*); ~-**line representation** *n* GEN representación *f* de línea única; ~-**link procedure** *n* (*SLP*) TELEP procedimiento *m* de enlace único (*SLP*); ~ **longitudinal mode** *n* TRANS modo *m* longitudinal simple; ~-**mode coupler** *n* TRANS acoplador *m* monomodal; ~-**mode fibre** *BrE n* DATA, PARTS, TRANS fibra *f* monomodo; ~-**mode laser** *n* DATA, ELECT, TRANS láser *m* monomodal; ~-**page feeding** *n* COMP, DATA *of printer, scanner* alimentación *f* hoja por hoja; ~-**pair cable box** *n* PARTS caja *f* de cable de un solo par; ~-**phase** *adj* POWER monofásico; ~-**phase alternating current** *n* POWER corriente *f* alterna monofásica; ~-**phase reactance** *n* POWER reactancia *f* monofásica; ~-**point failure** *n* TEST fallo *m* de punto único; -**precision** *adj* COMP de precisión simple; ~-**processor system** *n* SWG sistema *m* monoprocesador; ~-**progression sequential coding** *n* TELEP codificación *f* secuencial de progresión simple; ~ **sampling** *n* TEST muestreo *m* simple; ~-**segment message** *n* (*SSM*) DATA mensaje *m* de segmento único (*SSM*); ~-**server queue** *n* COMP, DATA cola *f* de un solo servidor; ~-**sheet feeding** *n* COMP, DATA *of printer, scanner* alimentación *f* hoja por hoja; ~-**shot element** *n* ELECT multivibrador *m* monoestable; ~-**shot flip-flop** *n* ELECT, PARTS báscula *f* monoestable; ~ **sideband** *n* RADIO, TRANS banda *f* lateral única; ~-**sideband emission** *n* RADIO, TRANS emisión *f* de banda lateral única; ~-**sideband transmission** *n* RADIO, TRANS transmisión *f* de banda lateral única; ~-**sided distribution frame** *n* SWG repartidor *m* de una sola cara; ~-**sided rack** *n* TELEP bastidor *m* unilateral; ~-**stroke control key** *n* COMP tecla *f* de control de un solo toque; ~ **sub-burst burst** *n* DATA ráfaga *f* de subráfaga *f* única; ~ **tone** *n* GEN tono *m* único; ~-**valued attribute** *n* DATA atributo *m* unívoco; ~-**voice speech recognition** *n* APPL, COMP, TELEP reconocimiento *m* de una sola voz; ~-**winding transformer** *n* POWER transformador *m* de bobinado sencillo; ~ **wire** *n* ELECT, PARTS hilo *m* único; ~-**wire circuit** *n* ELECT, PARTS circuito *m* monofilar; ~-**wire transmission line** *n* RADIO guíaondas *m* unifilar

single-precision *adj* COMP de precisión simple

sintering *n* PARTS sinterización *f*

sinusoidal: ~ **oscillator** *n* ELECT oscilador *m* sinusoidal; ~ **path** *n* ELECT trayectoria *f* sinusoidal; ~ **signal** *n* ELECT, SIG señal *f* sinusoidal

SIP *abbr* (*single in-line package*) GEN MBS (*encapsulado con una sola línea de conexiones*)

siren *n* TELEP sirena *f*

site *n* COMP, RADIO emplazamiento *m*; ~ **attenuation** RADIO atenuación *f* local; ~ **error** RADIO error *m* local; ~-**error susceptibility** RADIO susceptibilidad *f* de error local; ~ **investigation** RADIO estudio *m* de emplazamiento

siting *n* GEN *of antenna* emplazamiento *m*
situate *vt* RADIO situar
situated *adj* RADIO situado
situation *n* RADIO situación *f*
sixty: **~-four kbps restricted bearer service** *n* APPL servicio *m* portador restringido a 64 kbps; **~-four kbps restricted service** *n* APPL servicio *m* restringido a 64 kbps; **~-four kbps unrestricted bearer service** *n* APPL servicio *m* portador no restringido a 64 kbps
size *n* COMP, DATA *of file*, SIG, SWG tamaño *m*; **~ alteration** SWG alteración *f* de tamaño
skeletal: **~ code** *n* COMP código *m* esquemático
skew: **~ ray** *n* TRANS rayo *m* oblicuo
skewness *n* GEN *of distribution* oblicuidad *f*, asimetría *f*
skiatron *n* ELECT skiatrón *m*
skin: **~ effect** *n* POWER efecto *m* pelicular; **~ target** *n* RADIO blanco *m* pasivo
skinner *n* TELEP ramificación *f* de forma de cable
skinny: **~ wire set** *n* TELEP instalación *f* de cable desnudo
skip[1]: **~ area** *n* RADIO zona *f* saltada, TELEP zona *f* de silencio; **~ distance** *n* RADIO distancia *f* de salto, TELEP zona *f* de silencio; **~ lot sampling** *n* TEST muestreo *m* intermitente; **~ zone** *n* RADIO zona *f* de silencio, TELEP zona *f* de salto
skip[2] *vt* COMP, RADIO, TELEP omitir
skirt *n* RADIO faldón *m*
sky: **~-noise temperature** *n* RADIO temperatura *f* de ruido de origen celeste; **~ wave** *n* RADIO onda *f* espacial
slab: **~ interferometry** *n* RADIO, TRANS interferometría *f* plana
slant: **~ range** *n* RADIO distancia *f* oblicua
SLAR *abbr* (*side-looking airborne radar*) RADIO SLAR (*radar aéreo de exploración lateral*)
slave[1] *adj* ELECT esclavo, subordinado; **~ clock** *n* ELECT reloj *m* subordinado; **~ flip-flop** *n* PARTS báscula *f* subordinada; **~ station** *n* DATA estación *f* esclava; **~ transmitter** *n* RADIO emisor *m* esclavo
slave[2] *n* ELECT esclavo *m*
SLC *abbr* (*subscriber-line circuit*) SWG, TELEP circuito *m* de línea de abonado
SLD *abbr* (*superluminescent LED*) ELECT, PARTS, TRANS LED *m* superluminiscente
SLDA *abbr* (*semiconductor laser diode amplifier*) ELECT, PARTS, TRANS SLDA (*amplificador de láser diódico semiconductor*)
sleeve *n* ELECT, PARTS, POWER manguito *m*, camisa *f*; **~ contact** PARTS contacto *m* de manguito; **~ contact unit** POWER bloque *m* de contactos de manguito
sleeving *n* ELECT, PARTS, POWER cubierta *f* aislante
slew *n* COMP *of printer* salto *m* de papel
SLIC *abbr* (*subscriber line interface circuit*) SWG SLIC (*circuito de interconexión de línea de abonado*)
slicer *n* ELECT seccionador *m* de señal, PARTS rebanador *m*
slide: **~ calliper** *BrE n* TEST calibrador *m* de cursor, pie *m* de rey; **~ contact** *n* COMP cursor *m*; **~ effect** *n* COMP efecto *m* deslizante; **~ panel** *n* PARTS panel *m* deslizante; **~ scanner** *n* COMP, DATA explorador *m* de transparencias; **~ show** *n* COMP multivisión *f*; **~ switch** *n* PARTS conmutador *m* deslizante; **~-in unit** *n* PARTS grupo *m* conectable
slider *n* COMP cursor *m*; **~ control** COMP control *m* de cursor
sliding: **~ calliper** *BrE n* TEST calibre *m* de cursor, pie *m* de rey; **~ contact** *n* PARTS contacto *m* deslizante; **~ door** *n* PARTS puerta *f* corredera; **~-frequency generator** *n* POWER generador *m* de frecuencia inestable; **~ lid** *n* PARTS tapa *f* deslizante; **~ scale** *n* COMP escala *f* móvil
slim: **~ rack** *n* TELEP bastidor *m* estrecho
slip *n* DATA deslizamiento *m*, RADIO desviación *f*, TRANS deslizamiento *m*; **~-joint pliers** *n pl* GEN alicates *m pl* ajustables; **~ ring** POWER anillo *m* colector
slipped: **~ bank** *n* TELEP banco *m* de contactos salteados
SLM *abbr* (*spatial light modulator*) ELECT modulador *m* de luz espacial
slope *n* COMP, ELECT, RADIO pendiente *f*
sloping: **~-V antenna** *n* RADIO antena *f* en V inclinada
slot *n* GEN ranura *f*; **~ antenna** RADIO antena *f* ranurada; **~-fed dipole** RADIO dipolo *m* ranurado; **~ radiator** RADIO ranura *f* radiante; **~ waveguide** (*SWG*) DATA, RADIO, TRANS portaondas *m* de ranura (*SWG*)
slotted: **~ ALOHA system** *n* RADIO, TELEP sistema *m* ALOHA de línea ranurada; **~-core cable** *n* TRANS cable *m* ranurado; **~-core optical fibre** *BrE* **cable** *n* TRANS cable *m* de fibra óptica ranurado; **~ line** *n* RADIO línea *f* ranurada; **~ measuring line** *n* RADIO línea *f* ranurada de medida; **~ screwdriver** *n* PARTS destornillador *m* para tornillos ranurados; **~ system** *n* RADIO, TELEP sistema *m* de línea ranurada; **~ washer** *n* PARTS arandela *f* abierta; **~ waveguide** *n* DATA, RADIO, TRANS guíaondas *m* ranurado; **~-waveguide antenna** *n* RADIO antena *f* de guíaondas ranurado
slow: **~-access storage** *n* COMP

almacenamiento *m* de acceso lento; **~-attack sound** *n* COMP sonido *m* de tránsito lento; **~-scan television** *n* TV televisión *f* de exploración lenta; **~-scan terminal** *n* TELEP terminal *m* de digitalización lenta; **~-scan videoconferencing** *n* TV videoconferencia *f* de exploración lenta; **~-wave structure** *n* ELECT estructura *f* de onda lenta

SLP *abbr* (*single-link procedure*) TELEP SLP (*procedimiento de enlace único*)

SLS *abbr* (*side-lobe suppression*) RADIO SLS (*supresión del lóbulo lateral*)

slug *n* PARTS anillo *m* de retardo

SMA *abbr* (*search mode acquisition*) DATA SMA (*adquisición en modo de búsqueda*); **~ procedure** *n* DATA procedimiento *m* SMA

small: **very ~ aperture terminal** *n* (*VSAT*) DATA, TRANS, TV terminal *m* de apertura muy pequeña (*VSAT*); **~-cell network** *n* RADIO red *f* de pequeñas celdas; **~ computer system interface** *n* (*SCSI*) COMP interfaz *f* de pequeño sistema informático (*SCSI*); **~-outline integrated circuit** *n* (*SOIC*) ELECT circuito *m* integrado de contorno pequeño (*SOIC*); **~-outline package** *n* DATA paquete *m* de contorno pequeño; **~ red plastic clip** *n* ELECT sujetador *m* pequeño de plástico rojo; **~-scale integration** *n* (*SSI*) ELECT, PARTS, TELEP, TRANS integración *f* a pequeña escala (*IPE*); **~-signal amplifier** *n* ELECT, SIG amplificador *m* de pequeña señal; **~-signal capacitance** *n* ELECT capacitancia *f* para señal débil; **~-signal gain** *n* ELECT ganancia *f* para señal débil; **~-signal transistor** *n* ELECT, PARTS transistor *m* de pequeña señal; **~ telecommunications centre** *BrE* *n* ELECT pequeño centro *m* de telecomunicaciones

SMAP *abbr* (*systems management application protocol*) DATA SMAP (*protocolo de aplicación de gestión de sistemas*)

smart *adj* COMP inteligente; **~ chip** *n* ELECT microplaca *f* con memoria; **~ frequency selection** *n* RADIO selección *f* de frecuencia inteligente; **~ network** *n* APPL, TRANS red *f* inteligente

smartcard *n* APPL tarjeta *f* inteligente, COMP tarjeta *f* con memoria; **~ access** APPL acceso *m* con tarjeta inteligente; **~ reader** APPL lector *m* de tarjetas inteligentes

SMATV *abbr* (*satellite master antenna television*) TV SMATV (*televisión de antena colectiva vía satélite*); **~ system** *n* TV sistema *m* SMATV

SME *abbr* (*systems management equipment*) COMP SME (*equipo de gestión de sistemas*)

smearing *n* ELECT borrosidad *f*

smectic *adj* ELECT esméctico

S-meter *n* TEST medidor *m* en unidades S

SMF *abbr* (*submultiframe*) TELEP bastidor *m* auxiliar múltiple

SMI *abbr* (*structure of management information*) DATA SMI (*estructura de información de la dirección*)

SMIB *abbr* (*security management information base*) DATA base *f* de información para la gestión de la seguridad

SMN *abbr* (*SDH management network*) TRANS red *f* de gestión de jerarquía digital síncrona

SMO *abbr* (*stabilized master oscillator*) TEST SMO (*oscilador maestro estabilizado*)

smooth *adj* COMP, POWER, RADIO alisado

smoothed: **~ voltage** *n* POWER tensión *f* aplanada

smoothing *n* COMP, POWER, RADIO alisado *m*; **~ capacitor** PARTS, POWER capacitor *m* filtrador; **~ filter** PARTS, POWER filtro *m* de nivelación

SMPTE *abbr* (*Society of Motion Picture Technicians*) TRANS SMPTE (*Sociedad de Técnicos Cinematográficos*); **~ time code** *n* COMP, TV código *m* de tiempos de la SMPTE

SMS *abbr* (*SDH management subnetwork*) TRANS subred *f* de gestión de jerarquía digital síncrona

SMSR *abbr* (*side-mode suppression ratio*) TRANS SMSR (*ratio de supresión de modo lateral*)

SMTP *abbr* (*simple mail transfer protocol*) DATA SMTP (*protocolo simple de transferencia de correo*)

SMU *abbr* (*source-measure unit*) TEST SMU (*unidad de medición de fuente*)

SN *abbr* (*sequence number*) COMP, DATA, SWG número *m* de orden

snap: **~-action contact** *n* PARTS contacto *m* de acción rápida; **~ gauge** *BrE* *n* TEST calibre *m* de mordaza, calibre *m* exterior; **~ locking** *n* PARTS cierre *m* de presión

snap in *vt* PARTS encajar a presión

snap into *vt* PARTS encajar a presión

snapshot *n* COMP imagen *f* instantánea; **~ program** COMP, DATA programa *m* de análisis selectivo; **~ semantics** COMP, DATA semántica *f* selectiva; **~ trace program** COMP programa *m* de rastreo selectivo

SNDCF *abbr* (*subnetwork dependent convergence function*) DATA función *f* de convergencia dependiente de red secundaria

sneak: **~ current** *n* ELECT corriente *f* parásita

Snell: **~'s law** *n* TRANS ley *f* de Snell

snipe: **~-nose pliers** *n pl* GEN alicates *m pl* de punta larga

SNMP *abbr* (*simple network management*

protocol) DATA SNMP (*protocolo simple de gestión de red*)

snooze: ~ **function** *n* PER APP *of answering machine* función *f* despertador

snow *n* RADIO nieve *f*

SNP *abbr* (*sequence number protection*) COMP, DATA, SWG protección *f* de número de orden

SNPA *abbr* (*subnetwork point of attachment*) DATA SNPA (*punto de incorporación a subred*)

SNR *abbr* (*signal-to-noise ratio*) ELECT, SIG, TEST, TRANS relación *f* señal/ruido

Society: ~ **of Motion Picture Technicians** *abbr* (*SMPTE*) TRANS Sociedad *f* de Técnicos Cinematográficos (*SMPTE*)

socket *n* COMP, ELECT, TV tomacorriente *f*; ~**-connector block** PARTS bloque *m* conector de cubo; ~ **outlet** POWER enchufe *m*

soft: ~ **case** *n* GEN caja *f* blanda; ~ **copy** *n* COMP, TV copia *f* en pantalla; ~**-copy image** *n* COMP, TV imagen *f* de visualización transitoria; ~ **decision** *n* DATA decisión *f* de programa; ~**-decision decoding** *n* DATA decodificación *f* de decisión de programa; ~ **key** *n* COMP tecla *f* blanda; ~ **logic** *n* COMP lógica *f* débil; ~ **magnetic material** *n* ELECT, PARTS material *m* débilmente magnético; ~ **rubber** *n* PARTS caucho *m* blando; ~ **shorting** *n* POWER cortocircuitación *f* leve

softener *n* PARTS ablandador *m*

software *n* COMP, DATA soporte *m* lógico; ~ **application** APPL, COMP, DATA aplicación *f* de elementos de programación; ~ **company** COMP, DATA empresa *f* de programación; ~ **engineer** COMP, DATA ingeniero *m* de programación; ~ **engineering** COMP, DATA ingeniería *f* de soporte lógico; ~ **lockout** COMP, DATA bloqueo *m* de elementos de programación; ~ **machine** COMP, DATA máquina *f* de software; ~ **message distribution machine** COMP, DATA máquina *f* de distribución de mensaje de software; ~**-only video playback** COMP, TV reproducción *f* de vídeo sólo por software; ~ **package** COMP, DATA paquete *m* de software; ~ **piracy** COMP, DATA piratería *f* informática; ~ **product** COMP, DATA producto *m* de programación; ~ **resource** COMP, DATA recurso *m* de programación; ~ **tool** COMP, DATA dispositivo *m* de soporte lógico

SOIC *abbr* (*small-outline integrated circuit*) ELECT SOIC (*circuito integrado de contorno pequeño*)

solar: ~ **array** *n* POWER panel *m* de células solares; ~ **battery** *n* POWER batería *f* solar; ~ **interference** *n* RADIO interferencia *f* solar; ~ **panel** *n* POWER panel *m* solar; ~ **power** *n* POWER potencia *f* solar; ~ **quantization** *n* SIG cuantización *f* solar

solder[1] *n* ELECT, PARTS, TRANS soldadura *f*; ~ **bridge** PARTS puente *m* de soldadura; ~ **pad** PARTS superficie *f* de soldadura; ~**-plated quick clip** PARTS pinza *f* rápida estañada; ~ **resist** PARTS capa *f* protectora contra soldadura; ~**-resist lacquer** PARTS laca *f* protectora contra soldadura; ~ **spatter** PARTS salpicadura *f* de estaño; ~ **splash** PARTS salpicadura *f* de estaño

solder[2] *vt* ELECT, PARTS, TRANS soldar

solderability *n* ELECT, PARTS, TRANS soldabilidad *f*

solderable *adj* ELECT, PARTS, TRANS soldable

soldered: ~ **connection** *n* PARTS conexión *f* soldada; ~ **joint** *n* PARTS junta *f* soldada

soldering *n* ELECT, PARTS, TRANS soldadura *f*; ~ **height** PARTS altura *f* de soldadura; ~ **lacquer** PARTS laca *f* de soldadura; ~ **lug** PARTS espiga *f* de soldar, terminal *m* para soldar; ~ **point** PARTS punto *m* de soldadura; ~ **tag** PARTS espiga *f* de soldar; ~ **time** PARTS tiempo *m* de soldadura; ~ **tin** PARTS estaño *m* para soldar; ~ **touch-up** PARTS retoque *m* de soldadura; ~ **wire** PARTS hilo *m* de estaño para soldar

solenoid *n* POWER solenoide *m*; ~ **valve** PARTS válvula *f* de solenoide

solicited: ~**-information indicator** *n* DATA indicador *m* de información solicitada

solid *adj* GEN compacto; ~ **conductor** *n* PARTS conductor *m* macizo; ~ **copper-weld conductor** *n* PARTS conductor *m* macizo cobre soldado; ~ **electrolyte capacitor** *n* PARTS capacitor *m* de electrólito sólido; ~ **insulation** *n* GEN, PARTS aislamiento *m* sólido; ~**-state** PARTS de estado sólido; ~**-state amplifier** *n* ELECT amplificador *m* de estado sólido; ~**-state automatic terminal information system** *n* COMP sistema *m* de información de terminal automática de estado sólido; ~**-state device** *n* ELECT, PARTS dispositivo *m* de estado sólido; ~**-state disk** *n* ELECT disco *m* de estado sólido; ~**-state physics** *n* GEN física *f* de estado sólido; ~**-state power amplifier** *n* PARTS, POWER amplificador *m* de potencia de estado sólido; ~**-state rectifier** *n* PARTS, POWER rectificador *m* de estado sólido; ~**-state relay** *n* ELECT, PARTS, POWER relé *m* de estado sólido

solidity *n* GEN solidez *f*

solidus *n* GEN barra *f* oblicua

SOLR *abbr* (*side-tone objective loudness rating*) TELEP SOLR (*régimen de sonoridad objetiva de efecto local*)

son *n* COMP hijo *m*

sonar *n* RADIO sonar *m*; ~ **signal** RADIO señal *f* de sonar

S1 *abbr* (*unscrambled double dibit*) DATA S1 (*dibit doble descifrado*); ~-**field** *n* SWG campo *m* S1

song: ~ **memory** *n* COMP memoria *f* de canción

sonic *adj* TELEP sónico

sort *vt* POWER clasificar

sorter *n* COMP, POWER clasificador *m*

sound: ~ **absorber** *n* GEN amortiguador *m* de sonido; ~ **absorption** *n* GEN absorción *f* sonora; ~ **animation** *n* DATA *of scanner* animación *f* con sonido; ~-**attenuating cover** *n* PARTS, TELEP cubierta *f* de atenuación acústica; ~ **attenuation** *n* TELEP amortiguación *f* de sonido; ~ **attenuator** *n* TELEP amortiguador *m* de sonido; ~ **board** *n* COMP tablero *m* sonoro; ~ **broadcasting** *n* RADIO radiodifusión *f* sonora; ~-**broadcast transmitter** *n* RADIO transmisor *m* de radiodifusión de sonido; ~ **carrier** *n* TV portadora *f* de sonido; ~-**carrier frequency** *n* TV frecuencia *f* de portadora de sonido; ~ **channel** *n* TV canal *m* de sonido; ~ **density** *n* TELEP densidad *f* acústica; ~ **effect** *n* COMP, TELEP efecto *m* sonoro; ~ **engineer** *n* GEN ingeniero *m* de sonido; ~ **fade-out** *n* RADIO, TV desvanecimiento *m* de sonido; ~ **file** *n* DATA archivo *m* de sonido; ~ **insulation** *n* GEN insonorización *f*; ~ **intensity** *n* GEN intensidad *f* sonora; ~ **level** *n* GEN nivel *m* acústico; ~-**level control** *n* APPL control *m* de nivel acústico, PER APP control *m* de nivel de sonido; ~-**level meter** *n* TEST sonómetro *m*; ~ **module** *n* COMP módulo *m* de sonido; ~ **operator** *n* COMP operador *m* de sonido; ~ **power** *n* GEN potencia *f* sonora; ~-**powered telephone** *n* TELEP teléfono *m* autoalimentado; ~ **pressure** *n* TELEP presión *f* acústica; ~-**programme** *BrE* **circuit** *n* TV circuito *m* de programa de sonido; ~ **programme** *BrE* **circuit section** *n* TV sección *f* de circuito de programa sonoro; ~ **quality** *n* COMP, RADIO, TV calidad *f* de sonido; ~ **recording** *n* COMP, RADIO, TV registro *m* sonoro; ~ **reproduction** *n* COMP, RADIO, TV reproducción *f* de sonido; ~ **signal** *n* SIG señal *f* de sonido; ~ **source** *n* COMP, TELEP fuente *f* sonora; ~ **spectrogram** *n* RADIO, TELEP, TRANS espectrograma *m* de sonidos; ~ **spectrum** *n* RADIO, TELEP, TRANS espectro *m* sonoro; ~ **synthesis** *n* COMP síntesis *f* de sonido; ~ **synthesizer** *n* COMP sintetizador *m* de sonido; ~ **system** *n* COMP sistema *m* de sonido; ~ **and television programme** *BrE* **connection** *n* TV conexión *f* con programa de sonido e imagen; ~ **transmission** *n* APPL, TRANS transmisión *f* sonora; ~ **wave** *n* GEN onda *f* sonora

soundproof *vt* GEN insonorizar

soundproofing *n* GEN insonorización *f*

soundtrack *n* COMP, PER APP, TV pista *f* de sonido

source *n* GEN fuente *f*; ~ **block** ELECT bloque *m* emisor; ~ **document** COMP documento *m* fuente; ~ **of failure** TEST origen *m* de fallo; ~ **of fault** TEST origen *m* de fallo; ~ **identifier** TELEP identificador *m* de fuente; ~ **impedance** POWER impedancia *f* de fuente; ~ **of interference** POWER origen *m* de la interferencia; ~ **language** COMP lenguaje *m* fuente; ~ **link** COMP enlace *m* de origen; ~-**measure unit** (*SMU*) TEST unidad *f* de medición de fuente (*SMU*); ~ **node** COMP nodo *m* fuente; ~-**power efficiency** POWER, TRANS eficiencia *f* de la fuente de energía; ~ **program** COMP programa *m* fuente; ~ **region** PARTS *of field-effect transistor* región *f* de origen; ~ **rooting transparent** (*SRT*) ELECT transparente de sonido fundamental de la fuente (*SRT*); ~/**sink relationship** TELEP relación *f* fuente/dispositivo consumidor; ~ **stream** COMP cadena *f* fuente; ~ **system** SWG sistema *m* fuente

SP *abbr* (*space character*) DATA carácter *m* de espacio

space *n* GEN espacio *m*; ~ **allocation** GEN atribución *f* de espacio; ~ **bar** COMP, PER APP barra *f* espaciadora; ~ **bus** DATA bus *m* espaciador; ~ **character** (*SP*) DATA carácter *m* de espacio; ~ **charge** ELECT carga *f* espacial; ~-**charge-controlled tube** ELECT, PARTS tubo *m* controlado por carga espacial; ~-**charge debunching** ELECT desagrupamiento *m* de carga espacial; ~-**charge limited current** ELECT corriente *f* limitada por carga espacial; ~-**charge region** ELECT, PARTS región *f* de carga espacial; ~-**charge wave tube** ELECT, PARTS tubo *m* de onda de carga espacial; ~ **cloth** RADIO tejido *m* absorbente; ~ **communications** *n pl* GEN comunicaciones *f pl* espaciales; ~-**diversity reception** RADIO recepción *f* de diversidad espacial; ~ **division** SWG, TELEP división *f* en el espacio; ~-**division switching** SWG, TELEP conmutación *f* de división espacial; ~-**division system** SWG, TELEP sistema *m* de división espacial; ~ **period** DATA periodo *m* espacial; ~ **probe** RADIO, TV sonda *f* espacial; ~ **radiocommunication** RADIO, TV radiocomunicación *f* espacial; ~ **segment** RADIO segmento *m* espacial; ~ **shuttle** RADIO, TV lanzadera *f* espacial; ~-**time**-~ **network**

SWG, TRANS red *f* espacio-tiempo-espacio; ~ **stage** SWG, TELEP etapa *f* espacial; ~ **station** RADIO, TV estación *f* espacial; ~ **switching** SWG conmutación *f* espacial; ~ **system** RADIO sistema *m* espacial; ~**-time correlation** DATA, SWG correlación *f* espacio-temporal; ~ **vehicle** RADIO vehículo *m* espacial; ~ **wave** RADIO onda *f* espacial

spaced: ~ **carrier operation** *n* RADIO explotación *f* por portadoras distintas; ~**-division switching system** *n* SWG, TELEP sistema *m* de conmutación espacial; ~**-division system** *n* SWG, TELEP sistema *m* de división espacial; ~**-out letters** *n pl* PER APP letras *f pl* espaciadas

spacer: ~ **plate** *n* PARTS placa *f* separadora

spacing *n* COMP espaciamiento *m*, RADIO espaciado *m*, TV espaciamiento *m*; ~ **bar** PARTS barra *f* espaciadora; ~ **dimension** TELEP dimensión *f* de separación; ~ **error** RADIO error *m* de separación; ~ **ring** PARTS anillo *m* separador; ~ **washer** PARTS arandela *f* espaciadora

spacistor *n* ELECT, PARTS espacistor *m*

span *n* GEN vano *m*

Spanish: ~ **Institute for Normalization and Certification** *n* (*AENOR*) GEN Asociación *f* Española de Normalización y Certificación (*AENOR*)

spanner *BrE n* (*cf wrench AmE*) PARTS llave *f* de tuercas, TEST llave *f* fija

spanning: ~**-tree algorithm** *n* (*STA*) ELECT algoritmo *m* de árbol espaciador (*STA*)

spare: ~ **line** *n* TELEP línea *f* libre; ~ **number** *n* TELEP número *m* disponible; ~ **part** *n* PARTS pieza *f* de recambio; ~**-parts inventory** *n* PARTS existencias *f pl* de repuestos; ~**-parts kit** *n* PARTS juego *m* de piezas de repuesto; ~ **satellite** *n* RADIO satélite *m* de reserva

spark *n* POWER chispa *f*; ~ **gap** POWER distancia *f* explosiva; ~**-over** POWER *of protector* descarga *f* disruptiva; ~ **quench** POWER amortiguador *m* de chispas; ~ **quencher** POWER apagachispas *m*; ~**-quenching capacitor** PARTS, POWER capacitor *m* apagachispas; ~**-quenching resistor** PARTS, POWER resistor *m* apagachispas; ~ **transmitter** RADIO emisor *m* de chispa

sparking *n* POWER chispeo *m*

spatial: ~ **coherence** *n* TRANS coherencia *f* espacial; ~ **compression** *n* COMP compresión *f* espacial; ~**-light modulator** *n* (*SLM*) ELECT modulador *m* de luz espacial; ~**-multiplex switching** *n* SWG, TELEP conmutación *f* en múltiplex espacial; ~ **predictive concealment** *n* COMP ocultamiento *m* predictivo espacial; ~ **resolution** *n* COMP, TELEP resolución *f* espacial; ~ **switching** *n* SWG conmutación *f* espacial

SPDU *abbr* (*session protocol data unit*) DATA unidad *f* de datos de protocolo de sesión

speaker *n* COMP locutor *m*, TELEP, *device* parlante *m*; ~ **circuit** TELEP canal *m* de servicio; ~ **recognition** APPL, COMP, TELEP reconocimiento *m* de hablante

speakerphone *n* TELEP teléfono *m* manos libres

speaking: ~ **clock** *n* TELEP reloj *m* parlante; ~ **guide** *n* TELEP guía *f* de conversación; ~ **position** *n* TELEP posición *f* de conversación; ~ **rate** *n* TELEP velocidad *f* de conversación

special: ~ **character** *n* DATA carácter *m* especial; ~**-character combination** *n* DATA combinación *f* de carácter especial; ~ **connector** *n* PARTS conector *m* especial; ~ **dial tone** *n* TELEP tono *m* de marcar especial; ~**-effects generator** *n* COMP generador *m* de efectos especiales; ~ **information tone** *n* TELEP tono *m* de información especial; ~ **intercept** *n* TELEP interceptación *f* especial; ~**-purpose computer** *n* COMP ordenador *m* especializado *Esp*, computador *m* especializado *AmL*, computadora *f* especializada *AmL*; ~**-purpose language** *n* GEN lenguaje *m* especializado; ~ **ringing tone** *n* TELEP tono *m* de llamada especial; ~**-sensing microwave imager** *n* (*SSMI*) ELECT generador *m* de imágenes de microondas con sensor especial (*SSMI*); ~ **service** *n* TELEP servicio *m* especial; ~ **telephone set** *n* PER APP, TELEP aparato *m* telefónico especial

specialized: ~ **line** *n* TELEP línea *f* especializada; ~ **telephone** *n* PER APP, TELEP teléfono *m* especializado

specific *adj* ELECT, GEN específico; ~ **detectivity** *n* ELECT, TEST, TRANS detectividad *f* específica; ~ **heat** *n* ELECT calor *m* específico; ~ **material** *n* PARTS material *m* específico; ~ **mean repair-time** *n* TEST tiempo *m* medio de reparación específico; ~ **negative-recorded announcement without information** *n* TELEP anuncio *m* de registro informativo específico sin información; ~ **positive-recorded announcement without information** *n* TELEP anuncio *m* específico de registro positivo sin información; ~ **recorded announcement** *n* TELEP anuncio *m* específico registrado; ~ **resistance** *n* ELECT, POWER resistividad *f* específica; ~ **trunk-line selection** *n* SWG selección *f* de línea urbana específica; ~ **value** *n* DATA valor *m* específico

specification *n* TEST especificación *f*; ~ **and description language** (*SDL*) COMP, DATA lenguaje *m* de descripción y especificación

(*SDL*); ~ **language** GEN lenguaje *m* de especificación; ~ **of materials** PARTS especificación *f* de materiales

specified *adj* TEST especificado

specify *vt* TEST especificar

specimen *n* TEST muestra *f* de producto

speckle: ~ **noise** *n* TRANS ruido *m* de mácula

spectral: ~ **analysis** *n* TEST análisis *m* espectral; ~ **bandwidth** *n* RADIO, TRANS ancho *m* de banda espectral; ~ **characteristic** *n* ELECT, TRANS característica *f* espectral; ~ **component** *n* PARTS, TRANS componente *m* espectral; ~ **density** *n* DATA, TRANS densidad *f* espectral; ~ **distribution** *n* TRANS distribución *f* espectral; ~ **efficiency** *n* TRANS eficiencia *f* espectral; ~ **filter** *n* PARTS, TRANS filtro *m* espectral; ~ **irradiance** *n* GEN irradiancia *f* espectral; ~ **line** *n* TRANS raya *f* espectral; ~ **occupancy** *n* TRANS ocupancia *f* espectral; ~ **radiance** *n* ELECT, TRANS radiancia *f* espectral; ~ **radiant emittance** *n* ELECT, TRANS emitancia *f* radiante espectral; ~ **resources** *n pl* RADIO recursos *m pl* espectrales; ~ **responsivity** *n* TRANS responsividad *f* espectral; ~ **width** *n* TRANS anchura *f* espectral; ~ **window** *n* TRANS ventana *f* de prueba espectral

spectrogram *n* RADIO, TELEP, TRANS espectrograma *m*

spectrometer *n* RADIO, TELEP, TRANS espectrómetro *m*

spectrometry *n* RADIO, TELEP, TRANS espectrometría *f*

spectrum *n* COMP, RADIO, TRANS espectro *m*; ~ **allocation** RADIO adjudicación *f* de espectro; ~ **amplitude** RADIO amplitud *f* de espectro; ~ **analyzer** RADIO, TEST analizador *m* de espectro; ~-**analyzer monitoring** RADIO, TEST supervisión *f* de analizador de espectro; ~ **congestion** RADIO congestión *f* de espectro; ~ **efficiency** RADIO, TRANS eficiencia *f* de espectro; ~ **expansion** RADIO expansión *f* de espectro; ~ **of harmonics** COMP espectro *m* de armónicos; ~ **line** TRANS línea *f* espectral; ~ **plan** TV plano *m* espectral; ~ **resources** *n pl* RADIO recursos *m pl* espectrales; ~ **reuse** RADIO reutilización *f* de espectro; ~ **spreading** RADIO expansión *f* de espectro

specular: ~ **reflection** *n* TRANS reflexión *f* especular; ~-**reflection coefficient** *n* TRANS coeficiente *m* de reflexión especular

speech *n* RADIO habla *f*, TELEP voz *f*; ~-**activity factor** TELEP factor *m* de actividad vocal; ~ **analysis** TELEP análisis *m* de voz; ~ **band** TRANS banda *f* de frecuencias vocales; ~-**capture device** COMP mecanismo *m* de captación de voz; ~ **channel** TRANS canal *m* de voz; ~ **ciphering** COMP, TELEP cifrado *m* de voz; ~ **circuit** TELEP, TRANS circuito *m* de voz; ~ **code** TRANS código *m* vocal; ~ **coding** COMP, RADIO, TELEP codificación *f* de voz; ~ **compression** TELEP compresión *f* de voz; ~ **control** CONTR control *m* de conversación, PER APP control *m* por voz, TELEP control *m* de conversación; ~-**data network** DATA, TRANS red *f* de transmisión de voz y datos; ~ **detector** RADIO, TELEP detector *m* de palabra; ~-**digit signalling** *BrE* SIG señalización *f* por voz digitalizada; ~ **direction** TELEP dirección *f* de conversación; ~ **distortion** TELEP distorsión *f* de voz; ~ **encoder** COMP, RADIO, TELEP codificador *m* de voz; ~ **encoding** COMP, RADIO, TELEP codificación *f* de voz; ~ **generation** COMP, RADIO, TELEP generación *f* de la palabra; ~-**grade private wire** TRANS circuito *m* telefónico privado escalonado; ~ **intelligibility** COMP, RADIO, TELEP inteligibilidad *f* de voz; ~ **memory** TELEP memoria *f* de señales vocales; ~ **path** TELEP vía *f* de conversación; ~-**plus duplex** TELEP bivocal *m*; ~-**plus-duplex equipment** TELEP equipo *m* bivocal; ~-**plus simplex** TELEP univocal *m*; ~-**plus-simplex equipment** TELEP equipo *m* univocal; ~ **processing** APPL, TELEP procesamiento *m* de la palabra; ~-**processing card** APPL tarjeta *f* de procesamiento de voz; ~ **production** COMP, RADIO, TELEP producción *f* de voz; ~ **recognition** APPL, COMP, TELEP reconocimiento *m* de voz; ~ **recording** TELEP registro *m* de voz; ~ **sample** TRANS muestra *f* de voz; ~ **scrambler** RADIO, TELEP codificador *m* de la voz para comunicación secreta; ~ **service** APPL servicio *m* de transmisión de la palabra; ~ **signal** SIG, TELEP señal *f* vocal; ~ **sound** COMP fonema *m*; ~ **switchover** RADIO cambio *m* del sentido de la voz; ~ **synthesis** APPL, COMP, TELEP síntesis *f* de sonidos vocales; ~ **technology** APPL, COMP, RADIO, TELEP tecnología *f* de la voz; ~ **wire** PARTS hilo *m* de conversación

speed *n* GEN velocidad *f*; ~ **calling** TELEP marcación *f* abreviada; ~ **dialling** *BrE* TELEP marcación *f* abreviada; ~ **gate** RADIO ventana *f* de velocidad; ~ **indicator** RADIO indicador *m* de velocidad; ~ **regulator** TELEP regulador *m* de velocidad; ~ **search** RADIO exploración *f* en velocidad; ~ **of sound** GEN velocidad *f* del sonido; ~-**up capacitor** PARTS, POWER capacitor *m* acelerador

spell *vt* GEN deletrear

spellchecker *n* COMP corrector *m* ortográfico

spelling *n* GEN deletreo *m*; ~ **mistake** GEN error *m* de ortografía
SPF *abbr* (*segmentation permitted flag*) DATA bandera *f* reglamentaria de segmentación
spherical: ~ **aberration** *n* ELECT, TRANS aberración *f* esférica; ~ **antenna** *n* RADIO antena *f* esférica; ~ **mirror** *n* TRANS espejo *m* esférico
spigot *n* PARTS espiga *f*
spike: ~**-leakage energy** *n* ELECT energía *f* de fuga de punta
spin *vi* COMP girar
spiral *adj* GEN en espiral; ~ **antenna** *n* RADIO antena *f* espiral; ~**-ratchet screwdriver** *n* PARTS destornillador *m* helicoidal de trinquete; ~ **waveguide** *n* DATA, TRANS guíaondas *m* en espiral; ~ **wrap** *n* PARTS forro *m* en espiral
SPITE *abbr* (*switching process interworking telephony event*) TELEP sistema *m* de telefonía por interconexión de procesos
SPL *abbr* (*service provider link*) TRANS vínculo *m* con el prestador de servicios
splice[1] *n* ELECT, TRANS empalme *m*; ~ **chamber** ELECT cámara *f* de empalme; ~ **closure** ELECT caja *f* de empalme; ~ **connector** PARTS conector *m* de empalme; ~ **loss** TRANS pérdida *f* de empalme; ~**protection** TRANS protección *f* delempalme; ~**tray** PARTS cubeta *f* de empalme; ~ **tray method** PARTS método *m* de cubeta de empalme
splice[2] *vt* ELECT, TRANS empalmar
splicer *n* PARTS empalmador *m*; ~ **cable** PARTS cable *m* de soldador
splicing: ~ **facility** *n* COMP equipo *m* de empalmar; ~ **pigtail** *n* PARTS cola *f* de fibra de empalme
split *n* COMP hendidura *f*; ~ **bar** COMP barra *f* hendida; ~ **baud rate** DATA velocidad *f* de transmisión dividida; ~ **channel** SWG canal *m* dividido; ~ **connection** TELEP conexión *f* dividida; ~ **feed** POWER, TELEP alimentación *f* dividida; ~ **headphones** TELEP casco *m* de dos auriculares independientes; ~ **mould** *BrE* PARTS molde *m* seccionado; ~ **pin** PARTS pasador *m* de aletas; ~ **point** COMP punto *m* hendido
splitter *n* COMP, ELECT separador *m*
splitting *n* COMP, ELECT, TELEP hendidura *f*
SPN *abbr* (*subscriber premises network*) TRANS red *f* de instalación de abonado
spoken: ~ **message** *n* PER *of answer machine*, TELEP mensaje *m* hablado
spoking *n* COMP efecto *m* radial
sponsor *n* GEN patrocinado *m*
sponsoring *n* GEN patrocinio *m*; ~ **authority** GEN autoridad *f* patrocinadora
spontaneous: ~ **emission** *n* ELECT, TRANS emisión *f* espontánea; ~ **output** *n* TELEP salida *f* espontánea
spool *n* PARTS carrete *m*
sporadic: ~ **E layer** *n* RADIO capa *f* E esporádica; ~ **failure** *n* TEST fallo *m* esporádico; ~ **ionization** *n* RADIO ionización *f* esporádica
spot *n* COMP emplazamiento *m*, ELECT punto *m* terminal; ~ **beam** RADIO haz *m* localizado, TV haz *m* dirigido; ~**-beam coverage** RADIO cobertura *f* del haz del punto de exploración; ~ **check** TEST control *m* por muestreo; ~ **distortion** ELECT distorsión *f* del punto explorador; ~ **jamming** ELECT perturbación *f* de un canal, RADIO perturbación *f* de una frecuencia; ~**-noise factor** RADIO factor *m* de ruido puntual; ~**-noise figure** RADIO cifra *f* de ruido puntual; ~**-noise temperature** RADIO temperatura *f* de ruido puntual; ~ **sample** TEST muestra *f* puntual; ~ **size** TV tamaño *m* de punto; ~ **wobble** TV fluctuación *f* del punto explorador
spotlight: ~ **effect** *n* COMP efecto *m* de iluminación concentrada
spout *n* CLIM, PARTS canilla *f*
spray *vt* PARTS pulverizar ; ~**-coat** PARTS pintar con pistola; ~**-finish** PARTS barnizar con pistola; ~**-lacquer** PARTS laquear con pistola
spread *n* ELECT propagación *f*, RADIO alcance *m*; ~ **spectrum** (*SS*) RADIO espectro *m* extendido (*SS*); ~**-spectrum modulation** RADIO modulación *f* de espectro expandido; ~**-spectrum signal** RADIO señal *f* de espectro extendido; ~**-spectrum system** RADIO sistema *m* de espectro extendido
spreading: ~ **loss** *n* RADIO, TRANS pérdida *f* por dispersión; ~ **resistance** *n* ELECT resistencia *f* de dispersión
spreadsheet *n* COMP hoja *f* de cálculo
spring: ~ **adjuster** *n* PARTS, TEST doblador *m* de muelles de contacto; ~ **bender** *n* PARTS, TEST doblador *m* de muelles de contacto; ~ **contact** *n* PARTS contacto *m* de resorte; ~ **holder** *n* PARTS soporte *m* de muelles; ~**-loaded switch** *n* POWER, SWG conmutador *m* de resorte; ~**-set load** *n* PARTS carga *f* de juego de resortes; ~**-set mounting plate** *n* PARTS placa *f* de montaje con juego de resortes; ~ **steel** *n* PARTS acero *m* para muelles; ~ **washer** *n* PARTS arandela *f* elástica
sprocket *n* PARTS rueda *f* de cadena; ~ **hole** PARTS perforación *f* marginal; ~ **track** PARTS pista *f* de arrastre

spt *abbr* (*sectors per track*) COMP spt (*sectores por pista*)

SPTV *abbr* (*still-picture television*) TV SPTV (*televisión de imagen fija*)

spudger *n* ELECT, PARTS perforador *m* inicial

spurious: ~ **count** *n* RADIO impulso *m* parásito; ~ **emission** *n* RADIO emisión *f* espuria, TRANS emisión *f* falsa; ~ **frequency-conversion product** *n* RADIO producto *m* de conversión de frecuencia espuria; ~ **intermodulation product** *n* RADIO, TRANS producto *m* de intermodulación parásita; ~ **mixing product** *n* TRANS producto *m* de una mezcla espuria; ~ **modulation** *n* RADIO modulación *f* parásita, TRANS modulación *f* espuria; ~ **radiation** *n* RADIO radiación *f* parásita, TRANS radiación *f* espuria; ~ **response** *n* RADIO respuesta *f* espuria; **~-response rejection ratio** *n* RADIO ratio *f* de rechazo de respuesta espuria; ~ **satellite emission** *n* RADIO emisión *f* espuria de satélite; ~ **signal** *n* SIG señal *f* espuria; ~ **switch-off** *n* (*SSO*) TEST desconexión *f* espuria (*SSO*)

SQL *abbr* (*structured query language*) DATA SQL (*lenguaje de interrogación estructurado*)

square: ~ **bracket** *n* GEN corchete *m*; **~-law detection** *n* ELECT detección *f* cuadrática; **~-law detector** *n* ELECT detector *m* cuadrático; **~-pulse generator** *n* ELECT generador *m* de impulsos rectangulares; ~ **signal** *n* ELECT señal *f* cuadrada; ~ **wave** *n* ELECT, SIG onda *f* rectangular; ~ **waveform** *n* ELECT, SIG onda *f* cuadrada; **~-wave generator** *n* ELECT, SIG generador *m* de onda cuadrada; **~-wave response** *n* ELECT, SIG, TRANS respuesta *f* con onda rectangular; **~-wave response characteristic** *n* ELECT, SIG, TRANS característica *f* de respuesta con onda rectangular; **~-wave voltage** *n* ELECT, SIG voltaje *m* de onda cuadrada

squareax *n* TRANS supresión *f* de cuadratura

squealing *n* RADIO chillido *m*

squeeze *vt* COMP comprimir

squelch *n* RADIO supresión *f* de ruido; ~ **control** RADIO control *m* de silenciamiento

squint: ~ **angle** *n* RADIO ángulo *m* de deriva; ~ **antenna** *n* RADIO antena *f* de ángulo de conmutación de lóbulo

SRD *abbr* (*superradiant diode*) ELECT, PARTS, TRANS *fibre optics* diodo *m* superradiante

S-reference: ~ **point** *n* DATA punto *m* de referencia S

SREJ *abbr* (*selective-reject frame*) TELEP SREJ (*trama de rechazo selectivo*)

SRR *abbr* (*search and rescue*) RADIO SRR (*búsqueda y rescate*); ~ **operation** *n* RADIO operación *f* SRR

SRS *abbr* (*stimulated Raman scattering*) RADIO SRS (*dispersión de Raman estimulada*)

SRT *abbr* (*source rooting transparent*) ELECT SRT (*transparente de sonido fundamental de la fuente*)

SS *abbr* (*spread spectrum*) RADIO SS (*espectro extendido*)

SSA *abbr* (*sun sensor assembly*) RADIO SSA (*montaje de sensor solar*)

SSAP *abbr* (*session service access point*) DATA, TRANS SSAP (*punto de acceso al servicio de sesión*)

SSB *abbr* (*single sideband*) RADIO BLU (*banda lateral única*); ~ **emission** *n* RADIO, TRANS emisión *f* en BLU

SSC *abbr* (*satellite switching centre BrE* RADIO, SWG SSC (*centro de conmutación de satélite*)

SSI *abbr* (*small-scale integration*) ELECT, PARTS, TELEP, TRANS IPE (*integración a pequeña escala*)

SSM *abbr* (*single-segment message*) DATA SSM (*mensaje de segmento único*)

SSMI *abbr* (*special sensing microwave imager*) ELECT SSMI (*generador de imágenes de microondas con sensor especial*)

SSO *abbr* (*spurious switch-off*) TEST SSO (*desconexión espuria*)

SSR *abbr* (*steady-state reception*) RADIO SSR (*recepción de régimen estacionario*)

S/SYS *abbr* (*storage system*) DATA S/SYS (*sistema de almacenamiento*)

ST *abbr* (*segment type*) COMP, DATA ST (*tipo de segmento*); ~ **segment** *n* DATA segmento *m* ST; ~ **signal** *n* SIG señal *f* ST

STA *abbr* (*spanning-tree algorithm*) ELECT STA (*algoritmo de árbol espaciador*)

stability *n* GEN estabilidad *f*; ~ **loss** TELEP pérdida *f* de estabilidad

stabilize *vt* GEN estabilizar

stabilized: ~ **master oscillator** *n* (*SMO*) TEST oscilador *m* maestro estabilizado (*SMO*)

stable *adj* GEN estable; ~ **state** *n* ELECT estado *m* estable; ~ **testing state** *n* TEST estado *m* de comprobación estable

stack *n* COMP pila *f*; ~ **pointer** COMP indicador *m* de pila

stackable *adj* COMP, PARTS apilable

staff *n* GEN gancho *m* de pudelar; **~-locater system** RADIO sistema *m* de radiobúsqueda, TELEP sistema *m* buscapersonas

staffing *n* GEN dotación *f* de personal

stage *n* GEN etapa *f*

stagger: **~-tuned circuit** *n* ELECT circuito *m* de sintonía escalonada
staggering *n* GEN disposición *f* en zigzag
stain *vt* PARTS teñir
staining *n* PARTS tintura *f*
stamp *vt* PARTS estampar
stamper *n* PARTS estampador *m*
stand: **~-alone** *n* ELECT *computer* computador *m* solo *AmL*, computadora *f* sola *AmL*, ordenador *m* solo *Esp*; **~-alone bridge** *n* ELECT puente *m* independiente; **~-alone DSA** *n* DATA arquitectura *f* de sistema distribuido independiente; **~-alone exchange** *n* SWG central *f* autónoma; **~-alone rack** *n* TELEP bastidor *m* independiente; **~-alone satellite** *n* RADIO satélite *m* independiente; **~-alone unit** *n* ELECT unidad *f* independiente
standard *n* GEN norma *f*; **~ capacitor** POWER capacitor *m* patrón; **~ cell** ELECT pila *f* patrón; **~-cell circuit** ELECT circuito *m* de pila patrón; **~ component** ELECT, PARTS componente *m* estándar; **~ deviation** GEN desviación *f* característica; **~ error** GEN error *m* típico; **~-file dialog box** COMP, DATA cuadro *m* de diálogo de archivo estándar; **~ frequency** RADIO frecuencia *f* contrastada; **~-frequency emission** RADIO emisión *f* de frecuencia normal; **~-frequency satellite service** RADIO servicio *m* de satélite de frecuencia normal; **~-frequency and time-signal station** RADIO estación *f* de frecuencia normal y señal horaria; **~ generalized markup language** (*SGML*) COMP lenguaje *m* estándar generalizado de referencia (*SGML*); **~-grade performance channel** TRANS canal *m* de calidad funcional normal; **~ inductor** ELECT, PARTS patrón *m* de inductancia; **~ interface** GEN interfaz *f* normalizada; **~ keyboard pitch** COMP tono *m* normal de teclado; **~ loss** TRANS *of coaxial cable* pérdida *f* estándar; **~ modular unit** TELEP unidad *f* modular estándar; **~ NATO data message** RADIO mensaje *m* estándar de datos de la OTAN; **~ NATO data model** RADIO modelo *m* estándar de datos de la OTAN; **~ oscillator** ELECT, PARTS oscilador *m* estándar; **~ package** DATA paquete *m* estándar; **~ pitch** COMP altura *f* de tono normal; **~ population** TEST población *f* estándar; **~ propagation** RADIO propagación *f* normal; **~ radio atmosphere** RADIO atmósfera *f* radioeléctrica normal; **~ refraction** RADIO refracción *f* normal; **~ refractivity vertical gradient** RADIO gradiente *m* normal de vertical de refractividad; **~ resistor** PARTS, POWER resistor *m* normal; **~ sheet** GEN hoja *f* de normas; **~ telephone set** TELEP aparato *m* telefónico estándar; **~ time-signal emission** RADIO emisión *f* normal de señal horaria; **~ time-signal emission station** RADIO estación *f* de emisión de señal horaria normal; **~ value** GEN valor *m* indicativo; **~ voice jack** PARTS, TELEP jack *m* de voz normal; **~ wave error** RADIO, TRANS error *m* tipo de polarización
standardization *n* GEN normalización *f*
standardize *vt* GEN normalizar
standardized: **~ abstract test-suite** *n* DATA, TEST sucesión *f* de pruebas abstractas normalizadas; **~ option** *n* TELEP opción *f* normalizada
stand-by[1] *adj* COMP, RADIO a la escucha; **~ conditions** *n pl* GEN condiciones *f pl* de espera; **~ equipment** *n* TELEP equipo *m* de reserva; **~ mode** *n* TELEP, TEST modo *m* de espera; **~ phase** *n* TEST fase *f* de espera; **~ power** *n* POWER energía *f* de reserva; **~ power plant** *n* POWER grupo *m* electrógeno de reserva; **~ power supply** *n* POWER unidad *f* de energía de reserva; **~ processor** *n* SWG procesador *m* de reserva; **~/ready acknowledgement signal** *n* SIG señal *f* de reconocimiento tiempo de espera/tiempo de reserva, señal *f* de reconocimiento listo/reservado; **~/ready signal** *n* SIG señal *f* de espera/invitación a transmitir, señal *f* estado de espera/tiempo de reserva; **~ redundancy** *n* TELEP, TEST redundancia *f* pasiva; **~ set** *n* GEN grupo *m* de reserva; **~ state** *n* TEST estado *m* de reserva; **~ system** *n* GEN sistema *m* de reserva; **~ time** *n* TELEP tiempo *m* de espera, tiempo *m* en reserva; **~ unit** *n* GEN unidad *f* de reserva
stand-by[2] *n* TELEP recurso *m* seguro
standing: **~ current** *n* ELECT corriente *f* estacionaria; **~ wave** *n* RADIO, TRANS onda *f* estacionaria; **~-wave meter** *n* TEST indicador *m* de ondas estacionarias; **~-wave ratio** *n* (*SWR*) RADIO, TRANS coeficiente *m* de onda estacionaria (*ROE*)
stand-off: **~ ratio** *n* POWER ratio *f* de compensación
standstill *n* COMP parada *f*, parada *f* completa
star: **~ architecture** *n* TRANS, TV arquitectura *f* en estrella; **~ button** *n* TELEP botón *m*; **~ configuration** *n* TRANS configuración *f* en estrella; **~ connection** *n* POWER, TRANS conexión *f* en estrella; **~ coupler** *n* TRANS acoplador *m* en estrella; **~ distribution** *n* TRANS distribución *f* en estrella; **~ indicator** *n* TELEP indicador *m* visual; **~ network** *n* POWER, TRANS, TV red *f* en estrella; **~ service** *n* PER *on answering machine* servicio *m* en estrella; **~**

structure *n* TRANS, TV estructura *f* en estrella; ~ **switch** *n* SWG conmutador *m* en estrella; ~ **topology** *n* DATA topología *f* en estrella; **~-wired environment** *n* ELECT entorno *m* cableado en estrella

start *n* GEN comienzo *m*; **~-dialling** *BrE* **signal** SIG, TELEP señal *f* para transmitir; **~-of-block signal** SIG, TRANS señal *f* de comienzo de bloque; **~-of-heading character** DATA carácter *m* de comienzo de encabezamiento; ~ **signal** DATA, SIG señal *f* de arranque; **~-stop apparatus** TELEP aparato *m* arrítmico; **~-stop button** COMP botón *m* de arranque-parada; **~-stop distortion** TELEP distorsión *f* arrítmica; **~-stop format** DATA formato *m* arrítmico; **~-stop system** TELEP sistema *m* de arranque-parada, sistema *m* de marcha parada; **~-stop transmission** TELEP, TRANS transmisión *f* por arranque-parada; ~ **of text** COMP, DATA comienzo *m* de texto; **~-up** COMP arranque *m*; **~-up circuit** POWER circuito *m* de arranque; **~-up disk** COMP, DATA disco *m* de arranque

start up *vt* COMP, GEN arrancar

starter: ~ **gap** *n* ELECT intervalo *m* de encendido; ~ **transfer current** *n* ELECT corriente *f* de transferencia de cebado

starting: ~ **electrode** *n* ELECT electrodo *m* de cebado; ~ **level** *n* GEN nivel *m* inicial; ~ **position** *n* GEN posición *f* inicial

state *n* GEN estado *m*; ~ **of alert** TELEP estado *m* de alerta; ~ **symbol** TELEP símbolo *m* de estado; **~- and time-dependent routing** TRANS encaminamiento *m* dependiente del estado y del tiempo; ~ **of transition** TELEP estado *m* de transición; **~-transition diagram** TEST diagrama *m* de transición de estados

statement *n* COMP, DATA, TELEP declaración *f*

static *adj* TRANS estático; ~ **chaining** *n* DATA encadenamiento *m* estático; ~ **characteristic** *n* ELECT característica *f* estática; ~ **conformance requirement** *n* DATA requisito *m* de conformidad estática; ~ **conformance review** *n* DATA revisión *f* de conformidad estática; ~ **dump** *n* COMP vuelco *m* estático; ~ **field** *n* ELECT campo *m* estático; ~ **interference** *n* RADIO inteferencia *f* estática; ~ **memory** *n* COMP memoria *f* estática; ~ **RAM** *n* COMP RAM *f* estática; ~ **random-access memory** *n* COMP memoria *f* estática de acceso aleatorio; ~ **rectifier** *n* POWER rectificador *m* estático; ~ **relay** *n* ELECT, PARTS, POWER relé *m* estático; ~ **semantics** *n* COMP, DATA semántica *f* estática; ~ **storage** *n* DATA memoria *f* estática; ~ **switch** *n* POWER, SWG conmutador *m* estático

static *n* GEN ruido *m* atmosférico, POWER electricidad *f* estática

station *n* RADIO puesto *m*; ~ **conference** RADIO conferencia *f* de estación; ~ **keeping** RADIO control *m* de posición, mantenimiento *m* de órbita; **~-keeping lifetime** RADIO duración *f* de mantenimiento de órbita; **~-keeping satellite** RADIO satélite *m* de mantenimiento de órbita; ~ **management** TRANS administración *f* de estación

stationary *adj* GEN estacionario; ~ **high-altitude relay platform** *n* TRANS plataforma *f* relé estacionaria de gran altitud; ~ **information source** *n* GEN fuente *f* estacionaria de información; ~ **message source** *n* GEN fuente *f* estacionaria de mensajes; ~ **phase** *n* ELECT fase *f* estacionaria; ~ **satellite** *n* RADIO satélite *m* geoestacionario; ~ **wave** *n* ELECT onda *f* estacionaria

statistic *n* GEN estadístico *m*

statistical: ~ **coverage limit** *n* TEST límite *m* de cobertura estadística; ~ **delay of ignition** *n* ELECT retardo *m* estadístico de la ignición; ~ **frame traffic information** *n* ELECT información *f* estadística sobre tráfico de cuadros; ~ **method** *n* APPL método *m* estadístico; ~ **quality control** *n* TEST control *m* estadístico de calidad; ~ **test** *n* TEST prueba *f* estadística

statistics *n pl* GEN estadística *f*; ~ **counter** *n* TEST contador *m* de estadística

stator *n* ELECT, PARTS estator *m*

status *n* GEN categoría *f*; ~ **channel** TELEP canal *m* de estado; ~ **check** RADIO comprobación *f* de estado; ~ **and control code** CONTR, DATA código *m* de estado y control; ~ **data** *n pl* DATA, TEST datos *m pl* de estado; ~ **field** SWG *first party connected*, TELEP ventanilla *f* de estado; ~ **flag** COMP banderola *f* de estado; ~ **indication** COMP, TEST indicación *f* de estatus; ~ **lamp** TEST lámpara *f* de estado; ~ **message** RADIO mensaje *m* de estado; ~ **monitoring** CONTR, TEST supervisión *f* de estado

stay *n* PARTS travesaño *m*; ~ **bar** PARTS barra *f* tensora; ~ **iron** PARTS hierro *m* tensor de soporte; ~ **wire** PARTS cable *m* de retenida

STC *abbr* (*sensitivity time control*) CONTR STC (*atenuador*)

STE *abbr* (*signalling terminal equipment*) *BrE* TELEP equipo *m* terminal de señalización

steady: ~ **state** *n* GEN estado *m* estacionario; **~-state condition** *n* ELECT, TRANS condición *f* de régimen permanente; **~-state launching conditions** *n* ELECT condiciones *f pl* de establecimiento de estado estacionario; **~-state**

reception *n* (*SSR*) RADIO recepción *f* de régimen estacionario (*SSR*); **~-state value** *n* GEN valor *m* de régimen permanente

steel: **~ armour** *BrE n* PARTS blindaje *m* de acero; **~ tape** *n* PARTS fleje *m* de acero; **~ tape armour** *BrE n* PARTS blindaje *m* de fleje de acero

steer *vt* RADIO dirigir

steerability *n* RADIO gobernabilidad *f*

steerable *adj* RADIO gobernable, orientable; **~ antenna** *n* RADIO antena *f* orientable; **~-beam antenna** *n* RADIO antena *f* de haz orientable

steered: **~ flip-flop** *n* ELECT, PARTS basculador *m* orientado; **~ input** *n* ELECT entrada *f* orientada

steering *n* RADIO dirección *f*; **~ software** COMP, DATA *Internet* software *m* de orientación

step: **~ action** *n* CONTR acción *f* escalonada; **~-down transformation** *n* POWER transformación *f* reductora; **~-down transformer** *n* POWER transformador *m* reductor; **~ function** *n* TELEP función *f* por pasos; **~-function response** *n* TELEP respuesta *f* a la función escalón; **~-index fibre** *BrE n* DATA, PARTS, TRANS fibra *f* de índice escalonado; **~-index optical fibre** *BrE n* DATA, PARTS, TRANS fibra *f* óptica de índice escalonado; **~-index profile** *n* ELECT, TRANS perfil *m* de índice escalonado; **~ length** *n* COMP longitud *f* de paso; **~ operation** *n* CONTR operación *f* por pasos; **~ recording** *n* COMP registro *m* de pasos; **~ response** *n* CONTR respuesta *f* escalonada; **~-by-~ automatic system** *n* TELEP sistema *m* automático paso a paso; **~-by-~ connection** *n* POWER conexión *f* escalonada; **~-by-~ instruction** *n* COMP instrucciones *f pl* paso a paso; **~-by-~ office** *n* TELEP oficina *f* paso a paso; **~-by-~ operation** *n* COMP operación *f* paso a paso; **~-by-~ selector** *n* SWG selector *m* paso a paso; **~-by-~ system** *n* SWG sistema *m* paso a paso; **~-stress test** *n* TEST ensayo *m* por aplicación de esfuerzos escalonados; **~-strobe marker** *n* RADIO marcador *m* estroboscópico en escalón; **~ tracking** *n* RADIO seguimiento *m* gradual; **~-track system** *n* RADIO sistema *m* de vía gradual; **~-up transformation** *n* POWER transformación *f* elevadora; **~-up transformer** *n* POWER transformador *m* elevador de voltaje; **~ write** *n* COMP escritura *f* por pasos

step down *vt* POWER reducir

step up *vt* POWER elevar

stepper: **~ motor** *n* POWER motor *m* de avance por pasos

stepping *n* GEN progresión *f*; **~ magnet** TELEP imán *m* de avance; **~ relay** PARTS relé *m* de acción escalonada

stepwise *adv* GEN escalonadamente

stereo *adj* COMP, PER, TELEP, TV estéreo; **~ delay** *n* COMP retardo *m* estéreo; **~ input** *n* COMP, PER APP entrada *f* en estéreo; **~ output** *n* COMP, PER APP salida *f* en estéreo; **~ panning** *n* COMP toma *f* panorámica en estéreo; **~ telephony** *n* TELEP, TV telefonía *f* estereofónica; **~ vision** *n* COMP, TV visión *f* estereofónica

stereophonic *adj* COMP, PER, TELEP, TV estereofónico; **~ broadcast** *n* TV emisión *f* estereofónica; **~ circuit** *n* COMP circuito *m* estereofónico

stereophony *n* COMP, PER, TELEP, TV estereofonía *f*

stereoscopic *adj* COMP, TV estereoscópico

stick *vi* ELECT adherirse

stiffening: **~ bar** *n* PARTS barra *f* de refuerzo

still: **~ frame** *n* COMP, TV imagen *f* fija; **~-picture television** *n* (*SPTV*) TV televisión *f* de imagen fija (*SPTV*); **~-picture videophony** *n* RADIO, TELEP, TV videofonía *f* de imagen fija; **~-picture videotelephony** *n* RADIO, TELEP, TV videotelefonía *f* de imagen fija

stimulated: **~ emission** *n* ELECT, TRANS emisión *f* estimulada; **~ Raman scattering** *n* (*SRS*) RADIO dispersión *f* Raman estimulada (*SRS*)

stipulation *n* GEN estipulación *f*

STL *abbr* (*studio-transmitter link*) RADIO STL (*enlace entre estudio y transmisor*)

stochastic *adj* GEN estocástico; **~ process** *n* TRANS proceso *m* estocástico

stock: **~ diameter** *n* TRANS diámetro *m* normal

stool *n* GEN taburete *m*

stop[1]: **~ band** *n* TRANS banda *f* de supresión; **~ bit** *n* DATA bit *m* de parada; **~ button** *n* COMP botón *m* de parada; **~ code** *n* SIG código *m* de parada; **~ frame** *n* COMP, TV cuadro *m* de parada; **~ instruction** *n* COMP instrucción *f* de parada; **~-send signal** *n* SIG, TELEP señal *f* de parar la transmisión; **~ signal** *n* DATA, TELEP señal *f* de parada; **~ washer** *n* PARTS arandela *f* de tope

stop[2] *vt* COMP, TV desconectar

stopped *adj* COMP, TV desconectado

stopper: **~ circuit** *n* TELEP circuito *m* eliminador de parásitos

storability *n* TEST posibilidad *f* de memorización

storable: **~ message** *n* PER, RADIO, TELEP mensaje *m* almacenable

storage *n* COMP, TELEP memoria *f*, *within a network* almacenamiento *m*; **~ allocation** COMP, DATA asignación *f* de almacenamiento; **~-allocation system** COMP, DATA sistema *m* de asignación de memoria; **~-and-transfer system** (*ST/SYS*) DATA sistema *m* de almacenamiento y transferencia (*ST/SYS*); **~ battery** POWER acumulador *m*; **~ block** COMP bloque *m* de memoria; **~ camera tube** ELECT iconoscopio *m*; **~ capacity** DATA capacidad *f* de memoria; **~ cell** DATA celda *f* de memoria; **~ control** DATA control *m* de memoria; **~ dump** COMP vaciamiento *m* de memoria; **~ element** DATA elemento *m* de almacenamiento; **~ file** DATA, TELEP fichero *m* inactivo; **~ management** DATA gestión *f* de memoria; **~ medium** COMP soporte *m* de memoria; **~ protection** DATA protección *f* de almacenamiento; **~ register** TELEP registro *m* de almacenamiento; **~ surface** ELECT superficie *f* de almacenamiento; **~ system** (*S/SYS*) DATA sistema *m* de almacenamiento (*S/SYS*); **~ target** ELECT objetivo *m* de almacenamiento; **~ tube** ELECT tubo *m* de almacenamiento de memoria

store[1]: **~-all** *n* COMP memoria *f* para todo; **~-and-forward** *n* APPL, DATA almacenamiento *m* y retransmisión; **~-and-forward conversion facility** *n* APPL servicio *m* de conversión por almacenamiento y reexpedición; **~-and-forward facility** *n* APPL servicio *m* de almacenamiento y reexpedición; **~-and-forward mode** *n* DATA modo *m* de conmutación de mensajes; **~-and-forward switching network** *n* SIG, SWG red *f* de conmutación con escala; **~-and-forward transmission** *n* APPL, TRANS transmisión *f* por almacenamiento y reexpedición; **~-and-forward unit** *n* APPL, SWG unidad *f* de almacenamiento y reexpedición

store[2] *vt* COMP, DATA, PER *on answering machine*, RADIO almacenar

stored *adj* COMP, DATA memorizado, PER, *on answering machine* almacenado, RADIO memorizado; **~ data transmission** *n* DATA, TRANS transmisión *f* de datos almacenados; **~ message** *n* DATA, PER APP mensaje *m* memorizado, RADIO mensaje *m* almacenado, TELEP mensaje *m* memorizado; **~ number** *n* PER APP número *m* memorizado; **~-program computer** *n* COMP, CONTR ordenador *m* con programa grabado *Esp*, computador *m* con programa grabado *AmL*, computadora *f* con programa grabado *AmL*; **~-programme control** *BrE n* SWG, TELEP control *m* por programa registrado en memoria, CONTR, SWG, TELEP control *m* por programa grabado; **~-programme control exchange** *BrE n* CONTR, SWG centralita *f* controlada por programa grabado; **~-programme control PABX** *BrE n* CONTR, SWG, TELEP PABX *m* de control para programas almacenados; **~-programme switching system** *BrE n* CONTR, SWG, TELEP sistema *m* de conmutación por programa grabado, sistema *m* de conmutación de programa almacenado

storing *n* COMP, DATA, PARTS, RADIO almacenamiento *m*; **~ of digits** TELEP memorización *f* de dígitos; **~ of voice data** COMP memorización *f* de datos de voz

stoving: **~ lacquer** *n* PARTS, TV esmalte *m* de secado en horno

STP *abbr* (*shielded twisted pair*) ELECT, TRANS conductor *m* de doble torcido blindado

straight: **~ bank** *n* TELEP banco *m* de contactos alineados; **~ binary code** *n* DATA, ELECT código *m* binario rectilíneo; **~ joint** *n* ELECT, PARTS empalme *m* recto; **~ splice** *n* ELECT, PARTS unión *f* recta

straighten *vt* PARTS rectificar

straightforward: **~ migration path** *n* ELECT trayectoria *f* de migración directa

strain *n* GEN alargamiento *m*; **~ gauge** *BrE* PARTS medidor *m* de deformación; **~ relief** PARTS, POWER protección *f* contra tirones; **~-relief bracket** PARTS corchete *m* de anclaje; **~-relief mechanism** PARTS mecanismo *m* de protección contra tirones; **~-relief mounting bracket** PARTS corchete *m* de montaje de anclaje

strand *n* ELECT, PARTS torón *m*

stranded *adj* ELECT, PARTS trenzado; **~ conductor** *n* ELECT, PARTS conductor *m* trenzado; **~ multifibre building** *BrE n* ELECT elemento *m* de fibras múltiples trenzadas; **~ wire** *n* ELECT, PARTS hilo *m* cableado

stranding *n* ELECT, PARTS cableado *m*

strap *n* ELECT, PARTS, POWER, SWG tira *f*

strappable *adj* PARTS, POWER puenteable

strapped *adj* PARTS, POWER puenteado

strapping *n* PARTS conexión *f* de puente, POWER conexión *f* volante; **~ block** PARTS, POWER bloque *m* de puenteado; **~ diagram** POWER diagrama *m* de puenteado; **~ options** POWER opciones *f pl* de puenteado; **~ plug** PARTS, POWER clavija *f* de puenteado

strategic: **~ plan** *n* GEN plan *m* estratégico

strategy *n* GEN estrategia *f*

stratified: **~ language** *n* COMP lenguaje *m* estratificado; **~ sample** *n* TEST muestra *f* estratificada; **~ sampling** *n* TEST muestreo *m* estratificado

stray: **~ capacitance** *n* POWER capacitancia *f* parásita; **~ current** *n* POWER corriente *f* vagabunda; **~ emission** *n* ELECT emisión *f* parásita; ~ **pick-up** *n* RADIO captación *f* de componentes parásitos

streaking *n* COMP, TV falsa imagen *f*

stream *n* RADIO corriente *f*

streamer *n* COMP cinta *f* de serpentina

streetside: ~ **cabinet** *n* TV unidad *f* móvil

strength *n* GEN potencia *f*; ~ **test** TEST prueba *f* de resistencia

stress *n* PARTS esfuerzo *m*; ~ **analysis** TEST análisis *m* de esfuerzos; ~ **crack** PARTS grieta *f* por tensiones; ~**-relief loop** ELECT bucle *m* de atenuación de tensiones

stretch *vt* GEN estirar

stretched: ~ **text** *n* COMP texto *m* ampliado

striking *n* ELECT, POWER choque *m*; ~ **voltage** ELECT tensión *f* de encendido, tensión *f* de cebado

string *n* DATA cadena *f*

strip *n* PARTS, TELEP *of plug-sleeve indicators* cinta *f*; ~ **connector** ELECT, PARTS conector *m* de regleta; ~ **label** PARTS tira *f* de rotulación; ~**-mounted set** PARTS, POWER montaje *m* sobre platina

stripline *n* RADIO línea *f* plana

stripped: ~ **wire** *n* PARTS hilo *m* pelado

stripper *n* PARTS *knife* pelacables *m*

stripping *n* PARTS rascado *m*; ~ **force** PARTS fuerza *f* de pelado; ~ **knife** PARTS cuchilla *f* pelacables; ~ **machine** PARTS desmoldeadora *f*; ~ **pliers** ELECT alicates *m pl* pelacables; ~ **test** PARTS, TEST prueba *f* de pelado; ~ **tool** PARTS herramienta *f* de pelado

strobe[1] *n* GEN estroboscopio *m*; ~ **effect** COMP efecto *m* estroboscópico, TV efecto *m* de arco iris; ~ **marker** RADIO trazo *m* estroboscópico; ~ **pulse** ELECT impulso *m* de fijación

strobe[2] *vt* ELECT activar por un impulso

strobing *n* ELECT efecto *m* estroboscópico

stroboscope *n* ELECT, TEST estroboscopio *m*

stroke *n* COMP *keystrike*, PARTS carrera *f*; ~ **centre** *BrE* **line** COMP línea *f* central de trazo; ~**-cleaving tool** PARTS herramienta *f* separadora de trazos; ~ **edge** COMP borde *m* de trazo; ~ **width** COMP anchura *f* de trazo; ~ **writing** TV escritura *f* por teclado

strong: ~ **authentication** *n* DATA autentificación *f* convincente

Strowger: ~ **selector** *n* TELEP selector *m* Strowger

structural: ~ **method** *n* APPL método *m* estructural

structuration *n* GEN estructuración *f*; ~ **organization** COMP organización *f* de estructuración

structure *n* GEN estructura *f*; ~ **document** COMP documento *m* de estructura; ~ **flow chart** COMP organigrama *m* de la estructura; ~ **of management information** (*SMI*) DATA estructura *f* de información de dirección (*SMI*); ~ **specification** GEN especificación *f* de la estructura

structured: ~ **full-text query language** *n* DATA lenguaje *m* de interrogación de texto íntegro estructurado; ~ **query language** *n* (*SQL*) DATA lenguaje *m* de interrogación estructurado (*SQL*); ~ **type** *n* DATA tipo *m* estructurado

structuring *n* GEN estructuración *f*

STS *abbr* (*space-time-space*), TRANS STS (*espacio-tiempo-espacio*); ~ **network** *n* SWG, TRANS red *f* STS

ST/SYS *abbr* (*storage-and-transfer system*) DATA ST/SYS (*sistema de almacenamiento y transferencia*)

stub *n* RADIO, TELEP guarda *f*; ~ **cable** POWER cable *m* terminal

stuck *adj* ELECT adherido

stud *n* GEN prisionero *m*

studio: ~**-photographic experts group** *n* COMP, DATA grupo *m* de expertos en fotografía de estudio; ~**-transmitter link** *n* (*STL*) RADIO enlace *m* entre estudio y transmisor (*STL*)

study *n* GEN estudio *m*

stuffing *n* DATA relleno *m*; ~ **character** DATA carácter *m* de empaquetadura; ~ **device** DATA dispositivo *m* de relleno; ~ **digit** DATA dígito *m* de relleno; ~ **rate** DATA tasa *f* de relleno

S2: ~**-field** *n* SWG campo *m* S2

style *n* COMP estilo *m*

stylus *n* APPL estilete *m*, COMP aguja *f*, TELEP punzón *m*

SU *abbr* (*signal unit*) SIG SU (*unidad de señal*)

SUB *abbr* (*subaddressing*) APPL, DATA, TELEP SUB (*subdirección*)

subactivity *n* GEN subactividad *f*

subaddress *n* (*SA*) APPL, DATA, TELEP subdirección *f* (*SUB*)

subaddressing *n* (*SUB*) APPL, DATA, TELEP subdirección *f* (*SUB*)

subassembly *n* DATA subconjunto *m*

subaudio: ~ **frequency** *n* GEN frecuencia *f* infraacústica

sub-band *n* COMP, RADIO, SWG subbanda *f*; ~ **codec** COMP codec *m* de subbanda; ~ **coding** SIG codificación *f* de subbanda

subcarrier *n* GEN subportadora *f*; ~ **frequency** RADIO, TELEP, TRANS frecuencia *f* de subportadora; ~ **frequency modulation** RADIO, TELEP, TRANS modulación *f* de frecuencia de subportadora

sub-Clos: ~ **arrangement** *n* TELEP, TRANS

acuerdo *m* sub-Clos; ~ **network** *n* TELEP, TRANS red *f* sub-Clos
subclutter: ~ **visibility** *n* RADIO visibilidad *f* bajo los ecos parásitos
subcontracting *n* GEN subcontratación *f*
subcontractor *n* GEN subcontratista *m*
subcontrol: ~ **station** *n* TELEP estación *f* subdirectriz
subdirectory *n* COMP subdirectorio *m*
subdistribution: ~ **cabinet** *n* TRANS gabinete *m* de subdistribución; ~ **cupboard** *n* TRANS armario *m* de subdistribución; ~ **frame** *n* TRANS trama *f* de subdistribución
subfleet *n* RADIO flota *f* de apoyo
subfunction *n* GEN subfunción *f*
subgroup *n* TELEP, TRANS subgrupo *m*
subharmonic *adj* ELECT subarmónico
subheader *n* COMP subtítulo *m*
subheading *n* COMP subtítulo *m*
sub-hub *n* TRANS subcubo *m*
subject: ~ **to a charge** *adj* TELEP tasable
subjective: ~ **test** *n* TEST ensayo *m* subjetivo
subjob *n* COMP tarea *f*
submarine: ~ **cable** *n* TRANS cable *m* submarino
submenu *n* DATA *Internet* submenú *m*
submerged: ~ **optical repeater** *n* TRANS repetidor *m* óptico sumergido; ~ **repeater** *n* TRANS repetidor *m* sumergido
submicron: ~ **structure** *n* GEN estructura *f* submicrónica
subminiature *n* ELECT subminiatura *f*
submission: ~ **identifier** *n* DATA, TELEP identificador *m* de conformidad de adjudicación; ~ **time** *n* DATA *duration* tiempo *m* de confirmación
submit *vt* COMP someter
submultiframe *n* (*SMF*) TELEP bastidor *m* auxiliar múltiple
submultiplex *vt* TRANS submultiplexar
subnetwork *n* DATA, TRANS, TV subred *f*; ~ **address** DATA dirección *f* de subred; ~**-dependent convergence function** (*SNDCF*) DATA función *f* de convergencia dependiente de la red secundaria; ~ **functionality** DATA funcionalidad *f* de subred; ~ **point of attachment** (*SNPA*) DATA punto *m* de incorporación a subred (*SNPA*); ~ **point-of-attachment address** DATA dirección *f* de punto de incorporación a subred; ~ **service** DATA servicio *m* de subred
subnotebook *n* COMP subagenda *f*
suboperation *n* GEN suboperación *f*
subordinate: ~ **exchange** *n* TELEP centralita *f* subordinada
subpopulation *n* TEST subpoblación *f*
subprogram *n* COMP, DATA subprograma *m*
subrack *n* TELEP subbastidor *m*
subreflector *n* RADIO subreflector *m*
subrefraction *n* RADIO subrefracción *f*
subroute *n* TELEP subruta *f*
subroutine *n* COMP subrutina *f*
subsampling *n* COMP, TELEP submuestreo *m*
subsatellite: ~ **point** *n* RADIO punto *m* de subsatélite
subscribe *vi* COMP, DATA, RADIO, TELEP abonarse, TV suscribirse
subscriber *n* COMP, DATA, RADIO, TELEP, TV abonado *m*; ~ **access functional subsystem** TELEP subsistema *m* funcional de acceso de abonados; ~ **agreement holder** COMP, DATA, RADIO, TELEP, TV titular *m* del contrato de abono; ~ **and analysis database** TELEP base *f* de datos de abonados y análisis; ~ **base** COMP, DATA, RADIO, TELEP, TV número *m* total de abonados; ~ **box** TV casilla *f* de abonado; ~**-busy signal** SIG señal *f* de abonado ocupado; ~ **calling rate** TELEP tasa *f* de llamadas del abonado; ~ **channel** TELEP *multiplexed DTE/DCE interface* canal *m* de abonado; ~ **circuit** TELEP circuito *m* de abonado; ~ **connection system** PER, TELEP sistema *m* de conexión del abonado; ~ **connection unit** PER, TELEP unidad *f* de conexión del abonado; ~**-dialled** *BrE* **traffic** TELEP tráfico *m* originado por abonados; ~ **digital-access unit** TELEP unidad *f* de acceso digital del abonado; ~ **digital termination** TELEP equipo *m* terminal digital del abonado; ~ **equipment** PARTS, TELEP equipo *m* del abonado; ~ **equipment keeper** TELEP contacto *m* del equipo del abonado; ~ **identity module** (*SIM*) RADIO módulo *m* de identidad del abonado (*SIM*); ~ **installation** DATA, TELEP instalación *f* del abonado; ~ **line busy** SIG línea *f* de abonado ocupada;; ~ **line circuit** (*SLC*) SWG, TELEP circuito *m* de línea de abonado; ~ **line free, no charge** TELEP línea *f* de abonado gratuita, sin cargo; ~ **line interface circuit** (*SLIC*) SWG circuito *m* de interconexión de línea de abonado (*SLIC*); ~ **line junctor** TELEP juntor *m* de línea de abonado; ~ **line out of order** TELEP línea *f* de abonado averiada; ~ **line PE** TELEP elemento *m* procesador de la línea de abonado; ~ **loop** COMP, DATA, RADIO, TELEP línea *f* de abonado; ~ **metering** TELEP contador *m* de abonado; ~ **multiple test unit** TELEP probador *m* múltiple de abonados; ~ **number** DATA, TELEP número *m* de abonado; ~ **percentage** TELEP porcentaje *m* de abonados; ~ **premises network** (*SPN*) TRANS red *f* de

instalación del abonado; ~ **service** COMP, DATA, RADIO, TELEP, TV servicio *m* de abonado; ~ **stage** SWG paso *m* de abonado; **~-state signalling** *BrE* TELEP señalización *f* del estado del abonado; **~'s alphanumerical display** TELEP representación *f* alfanumérica del abonado; **~'s call meter** TELEP contador *m* del abonado; **~'s check meter** PER, TELEP contador *m* domiciliario del abonado; **~'s dialling** *BrE* **habits** TELEP comportamiento *m* del abonado; **~'s directory number** TELEP número *m* en guía del abonado; **~'s fax position** TELEP instalación *f* de fax del abonado; **~'s installation** DATA, TELEP instalación *f* del abonado; **~'s lead-in** TELEP conexión *f* del abonado; **~'s line** COMP, DATA, RADIO, TELEP, TRANS, TV línea *f* del abonado; **~'s line equipment** COMP, DATA, RADIO, TELEP, TRANS, TV equipo *m* de línea del abonado; **~'s main station** TELEP teléfono *m* principal del abonado; **~'s meter** PER, TELEP contador *m* del abonado; **~'s national telex number** TELEP número *m* de télex nacional del abonado; **~'s number** DATA, TELEP número *m* del abonado; **~'s private meter** PER, TELEP contador *m* privado del abonado; **~'s service centre** *BrE* *BrE* TELEP centro *m* de servicio de abonados; **~'s station** TELEP aparato *m* del abonado; **~'s store** SWG registro *m* del abonado; ~ **switch** SWG paso *m* de abonado; ~ **switching network** SWG, TELEP red *f* de conmutación de abonados; ~ **terminal** TELEP, TV terminal *f* del abonado; ~ **terminal installation** TELEP, TV instalación *f* terminal del abonado; ~ **test responder** TELEP respondedor *m* de prueba de abonado, TEST, *for subscriber lines and sets* respondedor *m* de prueba del abonado; ~ **traffic rate** TELEP intensidad *f* de tráfico de abonado; **~-transferred signal** SIG señal *f* transferida por el abonado; ~ **trunk dialling** *BrE* **access code** TELEP código *m* de acceso a la selección automática interurbana; ~ **unit** TV unidad *f* de abonado

subscript *n* COMP, DATA subíndice *m*

subscription *n* COMP, DATA, RADIO, TELEP, TV abono *m*, suscripción *f*; ~ **area** COMP, DATA, RADIO, TELEP, TV área *f* de abono; ~ **fee** COMP, DATA, RADIO, TELEP, TV cuota *f* de abono; ~ **price** COMP precio *m* de abono, DATA, RADIO, TELEP, TV precio *m* de suscripción

subsequent *adj* GEN subsiguiente; ~ **address message** *n* SIG, TELEP mensaje *m* subsiguiente de dirección; ~ **handover** *n* RADIO transferencia *f* subsiguiente; ~ **signal unit** *n* SIG unidad *f* de señal subsiguiente

subset *n* DATA subconjunto *m*

subsidiary *adj* GEN subsidiario

substatus *n* COMP estado *m* secundario

substitute: ~ **character** *n* DATA carácter *m* de sustitución; ~ **route** *n* TELEP ruta *f* de sustitución

substrate *n* ELECT, PARTS substrato *m*; ~ **current** ELECT corriente *f* de substrato

substring *n* DATA subcadena *f*

subsynchronous: ~ **satellite** *n* RADIO satélite *m* subsincrónico

subsystem *n* GEN subsistema *m*

subtelephone: ~ **frequency** *n* TELEP frecuencia *f* infratelefónica; ~ **telegraphy** *n* TELEP telegrafía *f* infratelefónica

subterranean: ~ **propagation** *n* RADIO propagación *f* subterránea

subtest *n* TEST prueba *f* parcial

subtitle[1] *n* TV subtítulo *m*

subtitle[2] *vt* TV subtitular

subtitling *n* TV subtitulado *m*

subtracter *n* ELECT sustractor *m*

subtractive: ~ **primary colour** *BrE* *n* COMP color *m* primario sustractivo; ~ **process** *n* ELECT proceso *m* sustractivo; ~ **synthesis** *n* COMP síntesis *f* sustractiva

subtree *n* DATA, TELEP *of MCS provider* subárbol *m*

subtype *n* DATA subtipo *m*

subunit *n* ELECT, TELEP subunidad *f*

subvariable *n* COMP subvariable *f*

subzone: ~ **centre** *BrE* *n* SWG central *f* de tránsito, central *f* de subzona

successful: ~ **call** *n* TELEP llamada *f* lograda; ~ **call attempt** *n* RADIO, TELEP intento *m* fructífero de llamada; ~ **injection** *n* RADIO *of satellite* inyección *f* lograda; ~ **traffic** *n* RADIO tráfico *m* conseguido

successfully: ~ **received** *adj* DATA bien recibido

successive: ~ **call phases** *n pl* TELEP fases *f pl* sucesivas de llamada

sudden: ~ **failure** *n* TEST fallo *m* repentino; ~ **total failure** *n* TEST fallo *m* total repentino

suite *n* COMP *of software*, PARTS, TELEP serie *f*; ~ **fuse** PARTS, POWER fusible *m* de fila de bastidores; **~-fuse alarm** PARTS, POWER alarma *f* de fusible de fila de bastidores; ~ **supervision equipment** TELEP equipo *m* de control de fila de bastidores; ~ **terminal equipment** TELEP equipo *m* terminal de bastidores

summary: ~ **abstract-operation** *n* DATA operación *f* abstracta sumaria

summation: ~ **check** *n* COMP comprobación *f* de suma

summer *n* ELECT viga *f* maestra
summing: ~ **integrator** *n* ELECT integrador *m* adicionador; ~**-scaling amplifier** *n* POWER amplificador *m* sumador gradual
sun: ~**-sensor assembly** *n* (*SSA*) RADIO montaje *m* de sensor solar (*SSA*); ~**-sensor head** *n* RADIO cabeza *f* de sensor solar
S-universal: ~ **access** *n* TELEP *ISDN* acceso *m* universal S; ~ **interface** *n* TELEP *ISDN* interconexión *f* universal S; ~ **interface card** *n* TELEP *ISDN* tarjeta *f* de interconexión universal S
sunshield *n* RADIO escudo *m* solar
super: ~**-audio frequency** *n* GEN frecuencia *f* ultraacústica; ~**-telephone frequency** *n* TELEP, TRANS frecuencia *f* ultratelefónica; ~**-telephone telegraphy** *n* TELEP, TRANS telegrafía *f* ultratelefónica
superconductor *n* ELECT, TRANS superconductor *m*; ~ **cable** ELECT, TRANS cable *m* superconductor; ~ **line** ELECT, TRANS línea *f* superconductora
supergroup *n* TELEP, TRANS supergrupo *m*; ~ **distribution frame** TELEP, TRANS repartidor *m* de grupo secundario; ~ **link** TELEP enlace *m* de grupos secundarios; ~ **section** TELEP sección *f* de grupo secundario; ~ **translating equipment** TELEP equipo *m* de transposición de grupo secundario
superheterodyne *adj* RADIO superheterodino; ~ **reception** *n* RADIO recepción *f* superheterodina
superhigh: ~ **frequency** *n* (*SHF*) RADIO hiperfrecuencia *f* (*FSA*)
superimpose *vt* COMP, TELEP sobreponer
superimposing *n* COMP, TELEP sobreposición *f*
superimposition *n* COMP, TELEP sobreposición *f*
superlattice *n* COMP, TELEP superred *f*
superluminescence *n* ELECT, TRANS *fibre optics* superluminiscencia *f*
superluminescent: ~ **LED** *n* (*SLD*) ELECT, PARTS, TRANS *fibre optics* LED *m* superluminiscente; ~ **light-emitting diode** *n* ELECT, PARTS, TRANS *fibre optics* diodo *m* emisor de luz superluminiscente
supermastergroup *n* TELEP grupo *m* cuaternario; ~ **link** TELEP enlace *m* de grupo cuaternario; ~ **section** TELEP sección *f* de grupo cuaternario
superphantom: ~ **balanced telegraph circuit** *n* TELEP circuito *m* telegráfico equilibrado superfantasma; ~ **circuit** *n* TRANS circuito *m* superfantasma
superpose *vt* COMP, TELEP superponer
superposed: ~ **circuit** *n* TELEP circuito *m* superpuesto
superposition *n* COMP, TELEP superposición *f*
superradiance *n* TRANS *fibre optics* superradiancia *f*
superradiant: ~ **diode** *n* (*SRD*) ELECT, PARTS, TRANS *fibre optics* diodo *m* superradiante
superrefraction *n* RADIO superrefracción *f*
superregenerative: ~ **reception** *n* RADIO recepción *f* superregenerativa
superscript *n* COMP, DATA superíndice *m*
supersynchronous: ~ **satellite** *n* RADIO satélite *m* supersincrónico
supertwist: ~ **nematic** *n* TRANS nemática *f* de superalabeo
supervise *vt* GEN supervisar
supervision *n* GEN supervisión *f*
supervisor *n* COMP, CONTR, TELEP, TEST *person* supervisor *m*
supervisory: ~ **aid** *n* CONTR, TEST ayuda *f* supervisora; ~ **announcement** *n* TELEP anuncio *m* supervisor; ~ **message** *n* TELEP mensaje *m* supervisor; ~ **panel** *n* CONTR panel *m* de supervisión; ~ **program** *n* COMP programa *m* supervisor; ~ **relay** *n* PARTS, TELEP relé *m* de supervisión; ~ **timer** *n* SWG temporizador *m* supervisor; ~ **tone** *n* CONTR, SWG tono *m* supervisor; ~ **unit** *n* CONTR, TELEP, TEST unidad *f* de supervisión
supplement *vt* GEN suplementar
supplementary: ~ **document** *n* GEN documento *m* suplementario; ~ **service** *n* APPL *ISDN*, RADIO, TELEP servicio *m* suplementario
supplier *n* GEN *seller* distribuidor *m*; ~ **evaluation** TEST evaluación *f* de proveedores; ~ **valuation** TEST valoración *f* de proveedor
supply[1]: ~ **current** *n* ELECT, POWER corriente *f* de alimentación; ~ **voltage** *n* ELECT, POWER tensión *f* de alimentación
supply[2] *vt* GEN suministrar
support *n* GEN *backup, assistance* apoyo *m*; ~ **activities** *n pl* COMP actividades *f pl* de apoyo; ~ **cable** RADIO cable *m* sustentador; ~ **system** TELEP sistema *m* de apoyo
supporting: ~ **bar** *n* PARTS barra *f* de soporte; ~ **comb** *n* PARTS, TELEP peine *m* de apoyo; ~ **plate** *n* PARTS placa *f* de soporte; ~ **rod** *n* PARTS varilla *f* de soporte
suppress *vt* ELECT, RADIO, TELEP suprimir
suppressed *adj* ELECT suprimido, RADIO, TELEP anulado, suprimido; ~ **aerial** *n* RADIO antena *f* empotrada; ~ **carrier** *n* RADIO onda *f* portadora suprimida; ~**-carrier emission** *n* RADIO emisión *f* de onda portadora suprimida; ~

demand *n* RADIO demanda *f* suprimida; ~ **frequency band** *n* TRANS banda *f* atenuada

suppression *n* ELECT, RADIO, TELEP supresión *f*; ~ **hangover time** TELEP tiempo *m* de persistencia de la supresión; ~ **loss** TELEP atenuación *f* de bloqueo; ~ **operate time** TELEP tiempo *m* de operación de la supresión

suppressor: ~ **grid** *n* ELECT, PARTS rejilla *f* supresora

supraconductor *n* ELECT, TRANS supraconductor *m*

supraphonic *adj* ELECT suprafónico

surf *vt* COMP *Internet* navegar, surfear

surface: ~ **acoustic wave** *n* (*SAW*) ELECT, TRANS onda *f* acústica de superficie (*OAS*); ~ **acoustic wave device** *n* ELECT, PARTS, TRANS dispositivo *m* de onda acústica de superficie; ~**-charge effect** *n* ELECT efecto *m* de las cargas superficiales; ~ **connection** *n* ELECT conexión *f* de superficie; ~ **current** *n* ELECT corriente *f* superficial; ~ **duct** *n* RADIO conducto *m* superficial; ~**-emitting LED** *n* PARTS, TRANS diodo *m* emisor de luz de emisión superficial, diodo *m* electroluminiscente de emisión superficial; ~**-emitting light-emitting diode** *n* PARTS, TELEP, TRANS diodo *m* emisor de luz de emisión superficial, diodo *m* electroluminiscente de emisión superficial; ~ **finish** *n* PARTS acabado *m* de una superficie; ~**-finish drawing** *n* PARTS dibujo *m* del acabado de una superficie; ~ **hardness** *n* PARTS dureza *f* de una superficie; ~ **layer** *n* PARTS capa *f* superficial; ~ **leakage** *n* POWER fuga *f* superficial; ~**-leakage path** *n* POWER recorrido *m* de la fuga superficial; ~**-mounted** *adj* ELECT, GEN de montaje exterior; ~**-mounted component** *n* ELECT, PARTS componente *m* de montaje exterior; ~**-mounted device** *n* ELECT, PARTS dispositivo *m* de montaje exterior; ~**-mounted technology** *n* PARTS tecnología *f* de montaje exterior; ~ **mounting** *n* ELECT, GEN montaje *m* de superficie; ~ **piezoelectric permittivity** *n* ELECT permitividad *f* piezoeléctrica superficial; ~ **resistance** *n* ELECT, POWER resistencia *f* superficial; ~ **resistivity** *n* ELECT, POWER resistividad *f* superficial; ~ **roughness** *n* GEN, PARTS rugosidad *f* de superficie; ~ **tension** *n* GEN tensión *f* superficial; ~ **wave** *n* ELECT, RADIO onda *f* de superficie; ~**-wave component** *n* ELECT componente *m* de onda de superficie; ~**-wave technology** *n* ELECT, RADIO tecnología *f* de las ondas de superficie

surfer *n* COMP *Internet* navegador *m*

surge *n* ELECT, POWER sobrevoltaje *m*; ~ **absorber** ELECT, PARTS, POWER absorbedor *m* de ondas; ~ **arrester** ELECT, PARTS *lightning protection*, POWER pararrayos *m*; ~ **current** ELECT, POWER sobrecorriente *f*; ~ **diverter** PARTS *lightning protection* pararrayos *m*; ~ **forward current** ELECT, POWER corriente *f* de cresta de sobrecarga accidental; ~ **suppressor** PARTS, POWER eliminador *m* de sobrevoltaje; ~**-voltage protector** PARTS, POWER protector *m* contra sobretensión transitoria

surplus: ~ **solder collector** *n* ELECT aspirador *m* de estaño

surveillance: ~ **radar** *n* RADIO radar *m* de vigilancia

survey *n* GEN *investigation* levantamiento *m*; ~ **diagram** GEN *overview* esquema *m* sinóptico

survival: ~ **craft station** *n* RADIO estación *f* de embarcación de salvamento

suspended: ~ **process** *n* TELEP proceso *m* suspendido

suspension: ~ **strand** *n* PARTS cable *m* de suspensión

sustain *n* COMP sostén *m*; ~ **footswitch** COMP pedal *m* de sostén; ~ **pedal** COMP pedal *m* de sostén; ~**-pedal assignment** COMP asignación *f* de pedal de sostén; ~ **switch** COMP interruptor *m* de sostén

sustaining: ~ **voltage** *n* ELECT tensión *f* de mantenimiento

SUT *abbr* (*system under test*) DATA SUT (*sistema en comprobación*); ~ **operator** *n* DATA operador *m* SUT

S-video *abbr* (*separated video*) COMP, TV vídeo *m* independiente

swab *n* ELECT mechón *m*

swan: ~ **feed** *n* RADIO alimentación *f* escalonada

swap: ~ **repair** *n* TELEP reparación *f* por intercambio

swapping *n* COMP intercambio *m*

sweep[1]: ~ **antenna** *n* RADIO antena *f* de barrido electrónico; ~ **frequency** *n* ELECT, RADIO frecuencia *f* de barrido; ~ **jamming** *n* ELECT, RADIO perturbación *f* con barrido de zona; ~ **microscope** *n* ELECT microscopio *m* de barrido; ~ **oscillator** *n* ELECT, PARTS oscilador *m* de barrido; ~ **signal** *n* COMP señal *f* de barrido; ~ **speed** *n* ELECT, RADIO velocidad *f* de barrido; ~ **time** *n* ELECT tiempo *m* de barrido, ELECT tiempo *m* de exploración

sweep[2] *vt* ELECT, RADIO barrer

swept: ~ **frequency** *n* RADIO frecuencia *f* con barrido

SWG *abbr* (*slot waveguide*) DATA, RADIO, TRANS SWG (*portaondas de ranura*)

swing *n* COMP amplitud *f*, oscilación *f*, RADIO

vibración *f*; **~-arm feature** ELECT configuración *f* de brazo oscilante; **~ error** RADIO error *m* de balanceo

swipe: **~ card** *n* APPL, TELEP tarjeta *f* de la palanca de arranque

switch *n* COMP, ELECT, PARTS, POWER, RADIO, SWG, TRANS interruptor *m*; **~ driver** RADIO oscilador *m* de mando; **~ hook** PER, RADIO gancho *m* conmutador, TELEP horquilla *f*; **~ indicator** GEN, SWG indicador *m* de aguja; **~ multiple** TELEP múltiple *m* de selectores; **~-off capability** GEN posibilidad *f* de desconexión; **~ rack** SWG bastidor *m* de selectores, TELEP bastidor *m* de conmutadores; **~ unit** SWG caja *f* de conmutación

switch off *vt* COMP, ELECT, PER, POWER, RADIO, SWG, TELEP desconectar

switch on *vt* COMP, ELECT, PER, POWER, RADIO, SWG, TELEP conectar

switch over *vt* ELECT conmutar, POWER, SWG cambiar

switchboard *n* SWG, TELEP panel *m* de conmutación, centralita *f*; **~ equipment** SWG, TELEP equipo *m* de cuadro conmutador; **~ floor plan** TELEP plano *m* de colocación de cuadros conmutadores; **~ operator** GEN operador *m* de cuadro conmutador; **~ PE** SWG EP *m* de cuadro de distribución; **~ row** SWG, TELEP fila *f* de cuadros conmutadores; **~ suite** SWG, TELEP fila *f* de cuadros conmutadores

switched: **~-beam direction finder** *n* RADIO detector *m* de dirección de haz conmutado; **~-capacitor filter** *n* ELECT, PARTS filtro *m* de capacitor conmutado; **~ circuit** *n* ELECT, TRANS circuito *m* conmutado; **~-loop console** *n* SWG consola *f* de bucle conmutado; **~ multimegabit data service** *n* DATA servicio *m* de líneas conmutadas de datos de multimegabits; **~ network** *n* SWG, TRANS, TV red *f* conmutada; **~-network layer** *n* SWG, TRANS, TV capa *f* de red conmutada; **~ power supply** *n* POWER suministro *m* de energía conmutada; **~ service** *n* SWG servicio *m* conmutado; **~-star network** *n* POWER, SWG, TRANS, TV red *f* conmutada en estrella; **~ telephone network** *n* SWG, TELEP red *f* telefónica de líneas conmutadas; **~-transit country** *n* TELEP país *m* de tránsito conmutado; **~ virtual circuit** *n* ELECT, SIG, SWG, TRANS circuito *m* virtual conmutado

switcher *n* SWG, TELEP, TV interruptor *m*

switchgear *n* PARTS, POWER, SWG, TELEP equipo *m* de conmutación y distribución

switching *n* COMP, ELECT, POWER, RADIO, SWG, TELEP conmutación *f*; **~-and-control unit** SWG unidad *f* de conmutación y control; **~ centre** *BrE* SWG, TELEP, TV centro *m* de conmutación; **~ complex** SWG, TELEP complejo *m* de conmutación; **~-complex control software machine** SWG ordenador *m* con programas para control del complejo de conmutación; **~ component** PARTS, POWER, SWG, TELEP componente *m* de conmutación; **~ delay** SWG retardo *m* de conmutación; **~ device** PARTS, POWER, SWG, TELEP dispositivo *m* de conmutación; **~ diode** ELECT, PARTS, SWG diodo *m* conmutador; **~ element** ELECT, SWG elemento *m* de conmutación; **~ equipment** SWG, TELEP equipo *m* de conmutación; **~-equipment congestion signal** SIG, SWG señal *f* de congestión de equipo de conmutación; **~ exchange** SWG, TELEP central *f* telefónica; **~ function** SWG, TELEP función *f* de conmutación; **~ matrix** SWG, TELEP matriz *f* de conmutación; **~ mechanism** SWG, TELEP mecanismo *m* de conmutación; **~ module PE** SWG EP *m* de módulo de conmutación; **~ multiplexer** SWG multiplexor *m* de conmutación; **~ network** SWG, TELEP red *f* de conmutación; **~-network complex** SWG complejo *m* de red de conmutación; **~ node** SIG, SWG nodo *m* de conmutación; **~ path** SWG trayecto *m* de conexión; **~ path PE** SWG EP *m* de trayecto de conexión; **~ pattern** SWG modelo *m* de conexión; **~ point** SWG, TELEP punto *m* de conmutación; **~ process** SWG, TELEP proceso *m* de conmutación; **~-process interworking-telephony event** (*SPITE*) TELEP sistema *m* de telefonía por interconexión de procesos; **~ processor** DATA, SWG procesador *m* de conmutación; **~ rate** SWG frecuencia *f* de basculación, TEST ritmo *m* de conmutación; **~ room** TELEP sala *f* de conmutación; **~ sequence** SWG proceso *m* de conmutación, TELEP proceso *m* de conexión; **~ stage** SWG, TELEP etapa *f* de conmutación; **~ statistical multiplexer** SWG multiplexor *m* estadístico de conmutación; **~ system** SWG, TELEP sistema *m* de conmutación; **~-system processor** SWG, TELEP procesador *m* de sistema de conmutación; **~ technique** SWG técnica *f* de conmutación; **~ technology** SWG técnica *f* de conmutación; **~ terminal** SWG terminal *m* de conmutación; **~ test** SWG, TEST prueba *f* de conmutación; **~ time** SWG, TELEP, TEST tiempo *m* de conmutación; **~ unit** SWG unidad *f* conmutadora; **~ variable** SWG variable *f* lógica

switchless *adj* ELECT, POWER, SWG, TELEP sin conmutador

switchover: ~ **diode** *n* ELECT, SWG diodo *m* de comunicación

swivel *n* COMP, PARTS placa *f* giratoria

SWR *abbr* (*standing-wave ratio*) RADIO, TRANS ROE (*coeficiente de onda estacionaria*)

syllabic: ~ **companding** *n* SIG, TRANS compansión *f* silábica

syllable *n* GEN sílaba *f*; ~ **articulation** TELEP articulación *f* silábica; ~-**articulation test** TELEP, TEST prueba *f* de articulación silábica

symbiosis *n* ELECT simbiosis *f*

symbol *n* GEN símbolo *m*; ~ **description** GEN descripción *f* de símbolo; ~ **location** DATA posición *f* de símbolo; ~ **name** GEN nombre *m* de símbolo; ~ **pattern** DATA configuración *f* de símbolos; ~ **rate** DATA, TELEP, TRANS velocidad *f* digital de línea; ~ **string** GEN serie *f* de símbolos; ~ **striping** GEN sombreado *m* de símbolo

symbolic: ~ **address** *n* COMP dirección *f* simbólica; ~ **addressing** *n* COMP direccionamiento *m* simbólico

symmetric *adj* GEN simétrico

symmetrical *adj* GEN simétrico; ~ **binary channel** *n* ELECT, TELEP canal *m* binario simétrico; ~ **binary code** *n* DATA, ELECT, TELEP, TRANS código *m* binario simétrico; ~ **channel** *n* ELECT, TELEP canal *m* simétrico; ~ **deflection** *n* ELECT deflexión *f* simétrica; ~-**pair cable** *n* TRANS cable *m* de pares simétricos; ~ **short-circuit current** *n* POWER corriente *f* de corto circuito simétrica; ~ **time matrix** *n* SWG matriz *f* de tiempo simétrica

symmetry *n* GEN simetría *f*

sync: ~ **clock** *n* COMP reloj *m* de sincronización; ~ **code** *n* COMP, DATA código *m* de sincronización; ~ **start** *n* COMP arranque *m* sincrónico; ~ **word** *n* DATA palabra *f* de sincronización

synch[1] *adj* GEN sincrónico; ~ **clock** *n* COMP reloj *m* de sincronización; ~ **start** *n* COMP arranque *m* sincrónico

synch[2] *vt* GEN, TELEP sincronizar

synchro: ~ **capacitor** *n* ELECT, PARTS sincrocapacitor *m*; ~ **transmitter** *n* RADIO sincrotransmisor *m*

synchronism *n* RADIO, TV sincronismo *m*; ~ **voltage** ELECT tensión *f* de sincronismo

synchronization *n* GEN sincronización *f*; ~ **bit** TELEP bitio *m* de sincronización; ~ **character** DATA carácter *m* de sincronización; ~ **clock** COMP reloj *m* de sincronización; ~ **code** COMP, DATA código *m* de sincronización; ~ **data** *n pl* COMP datos *m pl* de sincronización; ~ **error** TRANS error *m* de sincronización; ~ **information** COMP información *f* de sincronización; ~ **network** TRANS red *f* de sincronización; ~ **pulse** GEN pulso *m* de sincronización; ~ **sensitivity** GEN sensibilidad *f* de sincronización; ~ **signal** SIG señal *f* de sincronización; ~-**signal unit** SIG unidad *f* de señal de sincronización; ~ **and time-base station** TELEP estación *f* de sincronización y base de tiempo

synchronize *vt* GEN sincronizar

synchronized *adj* GEN sincronizado; ~ **multivibrator** *n* ELECT multivibrador *m* sincronizado; ~ **network** *n* TELEP red *f* sincronizada; ~-**oscillator system** *n* POWER sistema *m* de oscilador sincronizado; ~ **playback** *n* COMP reproducción *f* sincronizada; ~-**ringing signal** *n* TELEP señal *f* de llamada intermitente; ~ **satellite** *n* RADIO satélite *m* sincronizado; ~ **start** *n* COMP arranque *m* sincronizado

synchronizing *n* GEN sincronización *f*; ~ **pulse regenerator** TV regenerador *m* de impulso de sincronización; ~ **signal** COMP, SIG señal *f* de sincronización

synchronous *adj* GEN síncrono; ~ **capacitor** *n* ELECT, PARTS condensador *m* síncrono; ~ **circuit** *n* TRANS circuito *m* sincrónico; ~ **computer** *n* COMP ordenador *m* síncrono *Esp*, computador *m* síncrono *AmL*, computadora *f* síncrona *AmL*; ~ **condenser** *n* ELECT, PARTS condensador *m* síncrono; ~ **correction** *n* TELEP corrección *f* síncrona; ~ **data-link control** *n* (*SDLC*) DATA protocolo *m* de control de transmisión sincrona (*SDLC*); ~ **data network** *n* TELEP red *f* de datos sincrónica; ~ **data transmission** *n* DATA, TRANS transmisión *f* de datos síncrona; ~ **data unit** *n* DATA, ELECT unidad *f* de datos sincrónica; ~ **digital hierarchy** *n* (*SDH*) TELEP, TRANS jerarquía *f* digital síncrona; ~ **equipment management function** *n* APPL función *f* de gestión de equipo sincrónico; ~ **execution** *n* COMP ejecución *f* sincrónica; ~-**frequency encoding technique** *n* DATA técnica *f* de codificación de frecuencia sincrónica; ~ **idle character** *n* DATA carácter *m* de reserva de sincronización; ~ **link** *n* DATA enlace *m* sincrónico; ~ **motor** *n* TELEP motor *m* síncrono; ~ **multiplexer** *n* TRANS multiplexor *m* síncrono; ~ **network** *n* TELEP red *f* sincrónica; ~ **optical network** *n* COMP, TRANS red *f* óptica síncrona; ~ **port** *n* COMP, TRANS puerto *m* síncrono; ~ **satellite** *n* RADIO satélite *m* geosíncrono; ~ **transfer mode** *n* DATA, TRANS modo *m* de transferencia síncrona; ~ **transfer module** *n* DATA, TRANS módulo *m* de transferencia síncrono; ~ **transmission** *n* DATA, TRANS transmisión *f* síncrona; ~ **transport module** *n* DATA, TRANS módulo *m* de transporte síncrono; ~ **transport module-n** *n* DATA, TRANS módulo-n de transporte síncrono; ~ **voltage** *n* ELECT tensión *f* sincrónica

syncword *n* DATA palabra *f* de sincronización

synergetic *adj* GEN sinergético

synergic *adj* GEN sinérgico

synergy *n* GEN sinergía *f*; ~ **effect** GEN efecto *m* de sinergía

synonym *n* GEN sinónimo *m*

synonymous: ~ **name** *n* DATA sinónimo *m*

synopsis *n* COMP, DATA sinopsis *f*

syntactic *adj* COMP, DATA, TELEP sintáctico; ~ **analyser** *BrE n* TELEP analizador *m* sintáctico; ~ **analysis** *n* TELEP análisis *m* sintáctico; ~ **structure** *n* COMP estructura *f* sintáctica

syntactically: ~**-invalid test event** *n* DATA, TEST evento *m* de prueba sintácticamente inválida

syntax *n* GEN sintaxis *f*; ~**-based videotext** (*SBV*) TELEP videotexto *m* basado en sintaxis (*SBV*); ~ **diagram** TELEP diagrama *m* sintáctico; ~ **identifier** DATA, TELEP identificador *m* sintáctico; ~**-matching service** DATA servicio *m* de adaptación de la sintaxis; ~ **version** DATA, TELEP versión *f* sintáctica

synthesis *n* GEN síntesis *f*

synthesize *vt* COMP, ELECT, TELEP sintetizar

synthesizer *n* COMP, ELECT, TELEP sintetizador *m*; ~ **driver** COMP amplificador *m* previo de sintetizador; ~ **settling time** TELEP tiempo *m* de establecimiento de sintetizador; ~ **unit** COMP unidad *f* sintetizadora

synthetic: ~ **display** *n* PER APP presentación *f* sintética; ~ **image** *n* DATA, TV imagen *f* sintética; ~ **video signal** *n* RADIO, SIG, TV señal *f* de vídeo sintética; ~ **voice** *n* PER *on answer machine* voz *f* sintética

system *n* GEN sistema *m*; ~ **address** TELEP dirección *f* de sistema; ~**-availability information point** TELEP punto *m* de información sobre la disponibilidad del sistema; ~ **configuration** COMP configuración *f* de sistema; ~**-conformance statement** (*SCS*) DATA, TEST enunciado *m* de conformidad de sistema (*SCS*); ~**-conformance test report** (*SCTR*) DATA, TEST informe *m* de prueba de conformidad de sistema (*SCTR*); ~ **console** TELEP consola *f* de sistema; ~ **control** COMP control *m* de sistema; ~**-control signal unit** (*SCU*) SIG unidad *f* de señal de control de sistema (*SCU*); ~**-control station** TELEP estación *f* de control de sistema; ~ **coordination and control network** (*SCCN*) TEST red *f* de coordinación y control de sistema (*SCCN*); ~ **data** *n pl* COMP datos *m pl* sobre el sistema; ~ **deviation** CONTR desviación *f* de sistema; ~ **diagram** COMP diagrama *m* de sistema; ~ **dossier** DATA *Internet* expediente *m* de sistema; ~ **flow chart** COMP organigrama *m* de sistema; ~ **font** COMP juego *m* de caracteres de sistema; ~ **generation** COMP generación *f* de sistema; ~ **interconnection point** TELEP punto *m* de interconexión de sistema; ~ **load** GEN carga *f* de sistema; ~ **loss** GEN atenuación *f* de sistema; ~ **management** COMP gestión *f* de sistema; ~ **monitor** COMP supervisión *f* de sistema, CONTR supervisor *m* de sistema; ~**-noise temperature** TRANS temperatura *f* de ruido de sistema; ~ **production time** GEN tiempo *m* productivo de sistema; ~ **provider** APPL proveedor *m* de sistema; ~ **record** GEN registro *m* de sistema; ~ **recovery** DATA recuperación *f* de sistema; ~ **resilience** COMP elasticidad *f* de sistema; ~ **software** COMP, DATA software *m* de base; ~ **testing** COMP, TEST prueba *f* de sistema; ~ **title** DATA título *m* de sistema; ~ **under test** (*SUT*) DATA sistema *m* en comprobación (*SUT*)

systematic: ~ **effect** *n* TEST efecto *m* sistemático; ~ **error** *n* TEST error *m* sistemático; ~ **failure** *n* TEST fallo *m* sistemático; ~ **sampling** *n* TEST muestreo *m* sistemático; ~ **variation** *n* TEST variación *f* sistemática

systematics *n* TEST sistemática *f*

systematization *n* GEN sistematización *f*

systematize *vt* GEN sistematizar

systems: ~ **analysis** *n* COMP, TEST análisis *m* de sistemas; ~ **designer** *n* GEN analista *m* de sistemas; ~ **engineer** *n* GEN ingeniero *m* de sistemas; ~ **engineering** *n* GEN ingeniería *f* de sistemas; ~ **integration** *n* DATA integración *f* de sistemas; ~ **integrator** *n* DATA integrador *m* de sistemas; ~**-managed object** *n* DATA objeto *m* gestionado por sistemas; ~ **management** *n* GEN gestión *f* de sistemas; ~**-management application process** *n* COMP proceso *m* de aplicación de gestión de sistemas; ~**-management application protocol** *n* (*SMAP*) DATA protocolo *m* de aplicación de gestión de sistemas (*SMAP*); ~**-management application service element** *n* COMP elemento *m* del servicio de aplicación de gestión de sistemas; ~**-management equipment** *n* (*SME*) COMP equipo *m* de gestión de sistemas (*SME*); ~**-management function** *n* COMP función *f* de gestión de sistemas; ~**-management functional area** *n* COMP área *f* funcional de gestión de sistemas; **~-management functional unit package** *n* COMP paquete *m* de unidades funcionales de gestión de sistemas; **~-management operation** *n* COMP operación *f* de gestión de sistemas; **~-management service** *n* COMP servicio *m* de gestión de sistemas; **~ operator** *n* GEN operador *m* de sistemas; **~ planning** *n* COMP planificación *f* de sistemas; **~ and procedures** *n pl* COMP sistemas *m pl* y procedimientos; **~ programmer** *n* INFO programador *m* de sistemas; **~-support engineer** *n* GEN ingeniero *m* de apoyo de sistemas

systolic: **~ architecture** *n* APPL, TRANS arquitectura *f* sistólica

T

T: ~ **network** *n* POWER red *f* en T; ~ **reference point** *n* DATA punto *m* de referencia T; ~ **stage** *n* SWG paso *m* T, TELEP etapa *f* T; ~ **wire** *n* TELEP hiloconectado a un enchufe

TA *abbr* (*terminal adaptor*) DATA, PER, TELEP adaptador *m* de borne

tab *n* COMP, PARTS apéndice *m*

table: ~ **cassette** *n* RADIO *mobile* grabador *m* de mesa; ~ **of contents** *n* GEN *documentation* índice *m*; ~ **generation** *n* GEN generación *f* de tabla; ~ **instrument** *n* PER APP aparato *m* telefónico de mesa, TELEP teléfono *m* de mesa; ~ **look-up** *n* GEN consulta *f* de tablas; ~ **scanning** *n* GEN digitalización *f* de tabla; ~ **set** *n* PER APP aparato *m* telefónico de sobremesa, TELEP teléfono *m* de mesa; ~ **structure** *n* GEN estructura *f* de tabla

tablet *n* COMP tableta *f*

tabletop: ~ **device** *n* PER APP aparato *m* de sobremesa

tabulation *n* COMP tabulación *f*

tabulator *n* COMP tabulador *m*

tacan *n* RADIO tacán *m*

tachograph *n* GEN tacógrafo *m*

tactical: ~ **air navigation** *n* RADIO sistema *m* táctico de navegación aérea

tactile *adj* COMP táctil

tag *n* COMP *label*, ELECT, PARTS, TRANS etiqueta *f*; ~ **bit** DATA bitio *m* de banderola; ~ **block** DATA, ELECT bloque *m* de terminales; ~ **rash** ELECT serie *f* de etiquetas; ~ **strip** ELECT regleta *f* de conexiones

tag *vt* COMP, ELECT, PARTS, TRANS etiquetar, marcar

tagged: ~-**image file format** *n* (*TIFF*) COMP, DATA formato *m* de archivo de imagen con nombre simbólico (*TIFF*)

tagger *n* COMP etiquetador *m*

tagging *n* COMP *labels* adjudicación *f* de nombres

tailing *n* TELEP persistencia *f*

tailor *vt* GEN adaptar especialmente

tailored *adj* GEN adaptado especialmente

tailoring *n* GEN adaptación *f* especial

take apart *vt* GEN desensamblar

takeover *n* COMP absorción *f*

take-up: ~ **factor** *n* TELEP factor *m* de recepción

TA-LB *abbr* (*B-ISDN terminal adaptor*) TELEP adaptador terminal B de la red de transmisión digital de RDSI-BA

talk *n* DATA *Internet*, RADIO, TELEP charla *f*; ~ **battery** RADIO, TELEP batería *f* de conversación; ~ **set** ELECT aparato *m* de comunicación; ~ **spurt** RADIO chorro *m* de conversación, TELEP acceso *m* de conversación; ~ **time** RADIO, TELEP tiempo *m* de conversación

talker: ~ **echo** *n* TELEP eco *m* captado por hablante

talking: ~ **state** *n* RADIO, TELEP estado *m* de conversación

tally: ~ **reader** *n* DATA lectora *f* de lista de control

tamed: ~ **frequency modulation** *n* RADIO, TRANS modulación *f* de frecuencia regularizada

tamper with *vt* TEST violar

tamperproof *adj* TEST a prueba de manipulaciones imprudentes

T & C *abbr* (*tracking and command*) TEST T & C (*seguimiento y mando*)

tandem *adj* TELEP en tándem; ~ **completing trunk** *n* TRANS circuito *m* troncal complementario de central tándem; ~ **exchange** *n* DATA, SWG, TELEP central *f* tándem; ~ **PBX** *n* SWG, TELEP, TRANS centralita *f* privada en tándem; ~ **satellite** *n* RADIO satélite *m* en serie; ~ **selector** *n* SWG, TELEP selector *m* de tránsito

tangential: ~ **wave path** *n* RADIO camino *m* tangencial de onda

tangled: ~ **hierarchy** *n* COMP jerarquía *f* confusa; ~ **taxonomy** *n* COMP taxonomía *f* confusa

tangling *n* ELECT confusión *f*

tank: ~ **circuit** *n* ELECT circuito *m* tanque

tantalum: ~ **capacitor** *n* ELECT, PARTS condensador *m* de tantalio

tap[1] *n* ELECT, TELEP *on telephone line* sangría *f*

tap[2] *vt* ELECT, TELEP interceptar, escuchar clandestinamente

tape *n* APPL *audio, video*, COMP *audio*, ELECT, PARTS *insulating* cinta *f*; ~ **cartridge drive** COMP impulsor *m* de cartucho de cinta; ~ **cassette drive system** COMP sistema *m* impulsor de casete; ~ **drive** COMP mecanismo *m* impulsador de cinta; ~ **interface** COMP interfaz *f* de cinta; ~ **memory** COMP memoria *f* de cinta; ~ **printer** TELEP impresora *f* de cinta; ~ **recorder** TELEP grabador *m* de cinta; ~ **relay**

TELEP retransmisión *f* por cinta; **~-relay vehicle** (*TRV*) TV vehículo *m* de retransmisión por cinta (*TRV*); **~ reproducer** COMP reproductor *m* de cintas; **~ station** COMP estación *f* de cinta; **~ synchronization** COMP sincronización *f* de cinta; **~ transport unit** (*TTU*) APPL, COMP, TV unidad *f* de arrastre de cinta

taper *n* ELECT, RADIO ahusamiento *m*; **~ curve** ELECT curva *f* de reparto de resistencia

tapered *adj* PARTS cónico

tapping *n* TELEP escucha *f* clandestina; **~ equipment** TELEP equipo *m* de escucha

tar *n* DATA *file compression* brea *f*

target *n* ELECT, RADIO blanco *m*; **~ acquisition** RADIO adquisición *f* de blanco; **~ cell** RADIO elemento *m* objeto; **~ direction** RADIO dirección *f* de blanco; **~ echo** RADIO eco *m* de blanco; **~ identification** RADIO identificación *f* de blanco; **~ language** COMP lenguaje *m* objeto; **~ link** COMP enlace *m* de blanco; **~ lock-on** RADIO enganche *m* de blanco; **~ node** COMP nodo *m* de blanco; **~ programme** *BrE* TELEP programa *m* objeto; **~ signature** RADIO signatura *f* de blanco; **~ specification** TEST especificación *f* de objetivo; **~ tracking** RADIO seguimiento *m* de blanco; **~ transit delay** (*TTD*) DATA retardo *m* de tránsito de blanco (*TTD*); **~ zone** RADIO, TRANS zona *f* meta

tariff *n* APPL, TELEP tarifa *f*; **~ analysis** TELEP análisis *m* de tarifa; **~ area** TELEP zona *f* de tarifa; **~ determination** TELEP determinación *f* de tarifa; **~-determination point** TELEP punto *m* de determinación de tarifa; **~ generator** TELEP generador *m* de tarifa; **~-message receiver** TELEP receptor *m* de mensaje de tarifa; **~-message sender** TELEP emisor *m* de código de tarifa; **~ policy** APPL, TELEP política *f* arancelaria; **~ pulse meter** TELEP contador *m* de impulsos de cómputo; **~ reduction** APPL reducción *f* de tarifas; **~ setting** APPL, TELEP fijación *f* de tarifas; **~ structure** APPL, TELEP composición *f* de tarifas; **~ switching** TELEP conmutación *f* de tarifas; **~-switching class** TELEP clase *f* de conmutación de tarifas; **~ time-slot** TELEP intervalo *m* de tarifa; **~ zone** TELEP zona *f* de tarifa

task: **~-description data** *n pl* TELEP datos *m pl* de descripción de tarea; **~ specifications** *n pl* GEN *documentation* especificaciones *f pl* de tarea; **~ symbol** *n* TELEP símbolo *m* de tarea

taut: **~-band milliammeter** *n* TEST instrumento *m* de cintas tensas

taxi: **~ radio system** *n* RADIO sistema *m* de radiotaxi

TB: **~ cell** (*transmitter blocker cell*) RADIO, TRANS célula *f* TB

T-branching: **~ device** *n* ELECT elemento *m* de bifurcación en T, PARTS, *fibre optics* dispositivo *m* de bifurcación en T

TC *abbr* ELECT (*telecommunications closet*) gabinete *m* de telecomunicaciones, TELEP (*trunk code*) código *m* interurbano

TCE *abbr* (*transit connection element*) DATA, TRANS elemento *m* de comunicación en tránsito

TCI-DCS *abbr* (*transmitting subscriber identification-DCS*) DATA, TELEP TCI-DCS (*DCS de identificación del abonado emisor*)

TCL *abbr* (*transverse conversion loss*) TELEP TCL (*pérdida de conversión transversal*)

TCM *abbr* (*trellis-coded modulation*) TRANS TCM (*modulación codificada en enrejado*)

T-coupler *n* ELECT, PARTS acoplador *m* en T

TCR *abbr* (*telemetry command and ranging*) TEST TCR (*orden de telemetría y determinación de distancias*)

TCRF *abbr* (*transit connection-related function*) DATA, TRANS TCRF (*función de tránsito relacionada con la conexión*)

Tcu *abbr* (*cross-office transfer time*) SWG Tcu (*tiempo de transferencia en una centralita*)

TDC *abbr* (*telex destination code*) DATA, TELEP TDC (*código télex de destino*)

TDD *abbr* (*telecommunications device for the deaf*) GEN TDD (*aparato de telecomunicaciones para sordos*)

TDI *abbr* (*trade data interchange*) DATA intercambio *m* de datos comerciales

TDMA *abbr* (*time-division multiple access*) RADIO, SIG, TELEP, TRANS TDMA (*acceso múltiple por división de tiempo*); **~ burst time plan** *n* RADIO, SIG, TELEP, TRANS plan *m* de temporización de ráfagas con TDMA; **~ mode** *n* RADIO, SIG, TELEP, TRANS modo *m* TDMA

TE *abbr* (*terminal equipment*) COMP, DATA, PER, TELEP TE (*equipo terminal*); **~ mode** *n* TRANS modo *m* TE

teachware *n* COMP programas *m pl* de enseñanza

tear: **~-off menu** *n* COMP menú *m* de separación

tearing: **~ resistance** *n* PARTS resistencia *f* a desgarro

technical: **~ acknowledgement** *n* DATA reconocimiento *m* técnico; **~ breakdown** *n* TEST interrupción *f* en el servicio técnico; **~ delay** *n* TEST retraso *m* técnico; **~-delay time** *n* TEST tiempo *m* de retraso técnico; **~ expression** *n* GEN expresión *f* técnica; **~ fitter** *n* GEN montador *m* técnico; **~ language** *n* GEN lenguaje *m* técnico; **~ manager** *n* COMP director *m* técnico; ~ **manual** *n* COMP manual *m* técnico

technological *adj* GEN tecnológico

technology: ~ **transfer** *n* COMP transferencia *f* de tecnología

TEI *abbr* (*terminal end-point identifier*) DATA identificador *m* del punto final terminal

telecom *abbr* (*telecommunications*) GEN telecomunicaciones *f pl*

telecommand[1] *n* COMP, CONTR, PER, RADIO, TELEP telecomando *m*

telecommand[2] *vt* COMP, CONTR, PER, RADIO, TELEP mandar a distancia

telecommunicate *vi* GEN comunicarse a distancia

telecommunication *n* COMP, RADIO, TELEP telecomunicación *f*; ~ **circuit** RADIO, TELEP circuito *m* de telecomunicación

telecommunications *n pl* (*telecom*) GEN telecomunicaciones *f pl*; ~ **administration** *n* TELEP administración *f* de telecomunicaciones; ~ **board** *n* TELEP consejo *m* de telecomunicaciones; ~ **cable** *n* PARTS cable *m* de telecomunicaciones; ~ **closet** *n* (*TC*) ELECT gabinete *m* de telecomunicaciones; ~ **common carrier** *n* TELEP portadora *f* común de telecomunicaciones; ~ **device for the deaf** *n* (*TDD*) GEN aparato *m* de telecomunicaciones para sordos (*TDD*); ~ **and electronic equipment technician** *n* GEN técnico *m* de telecomunicaciones y equipos electrónicos; ~ **engineer** *n* GEN ingeniero *m* de telecomunicaciones; ~ **equipment** *n* GEN equipo *m* de telecomunicaciones; ~ **exchange** *n* TELEP central *f* de telecomunicaciones; ~ **management network** *n* (*TMN*) DATA, TELEP, TRANS red *f* de gestión de telecomunicaciones; ~ **network** *n* GEN red *f* de telecomunicaciones; ~ **operator** *n* GEN operador *m* de telecomunicaciones; ~ **satellite** *n* RADIO, TELEP, TV satélite *m* de telecomunicaciones; ~ **security** *n* GEN seguridad *f* de telecomunicaciones; ~ **service** *n* GEN servicio *m* de telecomunicaciones; ~ **system** *n* GEN sistema *m* de telecomunicaciones; ~ **terminal** *n* RADIO, TELEP terminal *m* de telecomunicaciones; ~ **traffic** *n* SWG, TELEP tráfico *m* de telecomunicaciones

telecommuting *n* GEN teleconmutación *f*

teleconference *n* APPL, RADIO, TRANS teleconferencia *f*

telecontrol *n* COMP, CONTR, RADIO, TELEP control *m* a distancia

telecopy *vt* COMP enviar por fax

telefax *n* COMP, TELEP telefax *m*

telegram *n* TELEP telegrama *m*

telegraph: ~ **alphabet** *n* TELEP alfabeto *m* telegráfico; ~ **channel** *n* TELEP canal *m* telegráfico; ~ **circuit** *n* TELEP circuito *m* telegráfico; ~ **code** *n* TELEP código *m* telegráfico; ~ **conversion** *n* TELEP conversión *f* telegráfica; ~ **converter** *n* TELEP convertidor *m* telegráfico; ~ **demodulator** *n* TELEP desmodulador *m* telegráfico; ~ **distortion** *n* TELEP distorsión *f* telegráfica; ~ **magnifier** *n* TELEP amplificador *m* telegráfico; ~ **modulation** *n* TELEP modulación *f* telegráfica; ~ **modulator** *n* TELEP modulador *m* telegráfico; ~ **network** *n* TELEP, TRANS red *f* telegráfica; ~ **noise** *n* TELEP ruido *m* telegráfico; ~ **office** *n* (*TO*) TELEP oficina *f* de telégrafos (*TO*); ~ **pole** *n* TELEP poste *m* de telégrafos; ~ **relay** *n* PARTS relé *m* telegráfico; ~ **repeater** *n* TELEP repetidor *m* telegráfico; ~ **service** *n* TELEP servicio *m* telegráfico; ~**-service correspondence** *n* TELEP correspondencia *f* de servicio telegráfico; ~ **signal** *n* TELEP señal *f* telegráfica; ~**-signal element** *n* SIG elemento *m* de señal telegráfica; ~ **switchboard** *n* SWG, TELEP conmutador *m* telegráfico manual; ~ **transfer** *n* TELEP transferencia *f* telegráfica; ~ **transmitter** *n* TELEP transmisor *m* telegráfico; ~ **word** *n* TELEP palabra *f* telegráfica

telegraphic: ~ **address** *n* TELEP dirección *f* telegráfica

telegraphy *n* RADIO, TELEP, TRANS telegrafía *f*

teleguidance *n* RADIO teleguía *f*

teleinformatics *n* COMP, DATA, RADIO teleinformática *f*

telemaintenance *n* COMP, DATA mantenimiento *m* a distancia

telemanagement *n* APPL telegestión *f*

telemarketing *n* GEN telecomercialización *f*

telematic: ~ **agent** *n* (*TLMA*) DATA, TELEP agente *m* telemático; ~ **file transfer** *n* TELEP transferencia *f* telemática de fichero; ~ **protocol architecture** *n* (*TPA*) TELEP arquitectura *f* de protocolo telemático (*TPA*); ~ **server** *n* DATA servidor *m* telemático; ~ **terminal** *n* DATA terminal *m* telemático

telematics *n* COMP, DATA, RADIO telemática *f*; ~ **service** COMP, DATA, RADIO servicio *m* de telemática

telemeasuring: ~ **device** *n* DATA dispositivo *m* telemedidor

telemetering *n* DATA, RADIO, TEST telemedición *f*

telemetry *n* DATA, RADIO, TEST telemetría *f*; ~ **command and ranging** (*TCR*) TEST orden *f* de telemetría y determinación de distancias (*TCR*)

telemonitoring *n* RADIO, TEST supervisión *f* a distancia

telephone[1] *n* TELEP teléfono *m*; ~ **answering machine** PER, TELEP contestador *m* automático; ~ **attachment** PER APP conexión *f* por teléfono; ~ **bill** TELEP factura *f* telefónica; ~ **book** TELEP agenda *f* telefónica; ~ **box** TELEP cabina *f* telefónica; ~ **carrier** TELEP portadora *f* telefónica; ~ **charge** TELEP tasación *f* telefónica; ~**-charge analyser** *BrE* TELEP analizador *m* de tasación de abonado; ~**-charge printer** TELEP impresora *f* de tasación de abonado; ~ **circuit** TELEP circuito *m* telefónico; ~**-circuit loss** TELEP atenuación *f* de circuito telefónico; ~ **company** TELEP compañía *f* telefónica; ~ **connection** TELEP enlace *m* telefónico; ~**-connection socket** PER, TELEP caja *f* de conexión telefónica; ~ **consumption** TELEP consumo *m* telefónico; ~ **conversation** TELEP conversación *f* telefónica; ~ **cord** PER, TELEP cable *m* telefónico; ~ **country code** TELEP código *m* telefónico de país; ~ **density** TELEP densidad *f* telefónica; ~ **engineering** TELEP técnica *f* telefónica; ~ **equipment** PER APP equipo *m* telefónico; ~**-extension cord** PER, TELEP cable *m* de extensión telefónica; ~**-extension line** TELEP línea *f* de extensión telefónica; ~ **frequency** TELEP frecuencia *f* vocal, TRANS frecuencia *f* telefónica; ~ **hybrid** TELEP híbrido *m* telefónico; ~ **installation** TELEP instalación *f* telefónica; ~ **instrument** PER APP instrumento *m* telefónico; ~**-instrument cord** PER, TELEP cable *m* de instrumento telefónico; ~**-interface equipment** TELEP equipo *m* de interfaz telefónico; ~**-interference factor** TELEP factor *m* de interferencia telefónica; ~ **jack** PARTS, PER, SWG, TELEP jack *m* de teléfono; ~ **keypad** PER APP teclado *m* de teléfono; ~ **link** TELEP enlace *m* telefónico; ~ **message** TELEP mensaje *m* telefónico; ~ **monitoring** TELEP, TEST supervisión *f* telefónica; ~ **number** TELEP número *m* de teléfono; ~ **relation** TELEP relación *f* telefónica; ~ **service** TELEP servicio *m* telefónico; ~ **set** PER, TELEP aparato *m* telefónico; ~ **set with memory** PER APP teléfono *m* con memoria, TELEP aparato *m* telefónico; ~ **signal** SIG señal *f* telefónica; ~ **socket** PARTS, PER, SWG, TELEP enchufe *m* de teléfono; ~ **station** RADIO, TELEP estación *f* telefónica; ~ **subscriber** TELEP abonado *m* telefónico; ~ **subscription** TELEP abono *m* al teléfono; ~ **tapping** TELEP intervención *f* de télefonos; ~ **traffic** TELEP tráfico *m* telefónico; ~**-traffic analyser** *BrE* TEST analizador *m* de tráfico telefónico; ~**-type channel** RADIO, TELEP canal *m* tipo telefónico; ~**-type circuit** RADIO, TELEP circuito *m* tipo telefónico; ~**-type connection** RADIO, TELEP conexión *f* tipo telefónico; ~**-user part** DATA, RADIO, TELEP parte *f* del usuario del servicio telefónico; ~**-service access unit** TELEP unidad *f* de acceso al servicio telefónico; ◆ **on the** ~ TELEP al teléfono; **over the** ~ TELEP por teléfono

telephone[2] *vt* TELEP llamar por teléfono, llamar al teléfono, telefonear

telephone[3] *vi* TELEP llamar por teléfono, llamar al teléfono, telefonear

telephonist *BrE n* SWG, TELEP operador *m*

telephony *n* TELEP telefonía *f*; ~**-baseband test** TEST prueba *f* de banda base de telefonía; ~ **head-end** TV sistema de alimentación por telefonía; ~ **performance** TELEP cualidades *f pl* técnicas de la telefonía; ~**-rated device** PER APP dispositivo *m* nominal telefónico

telephoto: ~ **service** *n* TELEP servicio *m* de telefoto

Telepoint *n* POWER Telepoint *m*

teleport *n* GEN telepuerta *f*

telepresence *n* COMP telepresencia *f*

teleprinter *BrE n* (*cf teletypewriter AmE*) APPL, PER, TELEP teleimpresor *m*; ~ **service** *BrE* (*cf teletypewriter service AmE*) APPL, PER, TELEP servicio *m* de télex

teleprinting *n* PER *newspapers, magazines* teleimpresión *f*

teleprocessing *n* GEN teleproceso *m*; ~ **network** COMP, TRANS red *f* de teleproceso

teleprompter *AmE n* (*cf autocue BrE*) TV apuntador *m*

telesales *n* APPL, TV ventas *f pl* por teléfono

telesecurity *n* APPL teleseguridad *f*

teleservice *n* APPL servicio *m* por teléfono

teleshopping *n* APPL, TV compra *f* a distancia

telesignalling: ~ **transmitter** *BrE n* APPL transmisor *m* de teleseñales

teletext *n* (*TTX*) APPL teletexto *m* (*TTX*); ~ **access unit** (*TTXAU*) APPL unidad *f* de acceso a teletexto (*TTXAU*); ~ **basic-control-function index** TELEP índice *m* de función de control básico de teletexto; ~ **basic-graphic-character index** TELEP índice *m* de caracteres gráficos básicos de teletexto; ~ **call** TELEP llamada *f* por teletexto; ~ **character index** TELEP índice *m* de caracteres de teletexto; ~ **control-function index** TELEP índice *m* de función de control de teletexto; ~ **document** TELEP documento *m* de teletexto; ~ **graphic-character index** TELEP índice *m* de caracteres

gráficos de teletexto; ~ **page** TELEP página *f* de teletexto; ~ **service** APPL, TV servicio *m* de teletexto; ~ **terminal** COMP, PER, RADIO terminal *m* de teletexto
teletrack *n* COMP televía *f*
teletraffic *n* SWG, TELEP teletráfico *m*; ~ **engineering** TELEP técnica *f* de teletráfico
teletuition *n* TV educación *f* a distancia
teletype *n* (*TTY*) COMP, PER, RADIO, TELEP teletipo *m* (*TTY*); ~ **machine** PER APP teletipo *m* (*TTY*); ~**-order channel** TRANS canal *m* de teletipo
teletypewriter *AmE n* (*cf teleprinter BrE*) APPL, PER, TELEP teleimpresor *m*; ~ **service** *AmE* (*cf teleprinter service BrE*) APPL, PER, TELEP servicio *m* de teletipo
television *n* (*TV*) TV televisión *f* (*TV*); ~ **aerial** TV antena *f* de televisión; ~ **broadcaster** TV transmisora *f* de televisión; ~ **broadcasting** TV radiodifusión *f* televisiva; ~ **channel** TV canal *m* de televisión; ~ **circuit** TV circuito *m* de televisión; ~**-circuit section** TV sección *f* de circuito de televisión; ~ **home** TV hogar con receptor de televisión; ~ **monitoring** TEST, TV supervisión *f* de la televisión; ~**-picture transmitter** TV transmisor *m* de imagen de televisión; ~ **programme** *BrE* TV programa *m* de televisión; ~ **receive-only antenna** TV antena *f* de televisión sólo para recepción; ~ **receiver** TV receptor *m* de televisión; ~ **set** TV receptor *m* de televisión; ~**-sound carrier** TV portadora *f* de sonido de televisión; ~**-sound transmitter** TV transmisor *m* de sonido de televisión; ~ **system** TV sistema *m* de televisión; ~**-system converter** TV convertidor *m* del sistema de televisión; ~ **transmitter** TV transmisor *m* de televisión; ~ **tube** PART, TV tubo *m* de rayos catódicos para televisión; ~ **tuner** TV sintonizador *m* de televisión; ~ **viewer** TV teleespectador *m*
teleworking *n* GEN teletrabajo *m*
telewriting *n* COMP, DATA, PER, RADIO, TV teleautógrafo *m*
telex *n* (*TLX*) DATA, RADIO, TELEP télex *m* (*TLX*); ~ **access unit** (*TLXAU*) DATA, RADIO, TELEP unidad *f* de acceso télex (*TLXAU*); ~ **destination code** (*TDC*) DATA, TELEP código *m* télex de destino (*TDC*); ~ **exchange** SWG, TELEP central *f* télex; ~ **message** TELEP mensaje *m* por télex; ~ **network identification code** DATA código *m* de identificación de la red télex; ~ **packet interworking function** (*TPIWF*) DATA función *f* de interoperación de paquetes de télex (*TPIWF*); ~ **position** TELEP posición *f* télex; ~ **relation** TELEP relación *f* de télex; ~ **switching centre** *BrE* SWG, TELEP centro *m* de conmutación de télex; ~ **terminal** DATA, RADIO, TELEP terminal *m* de télex; ~ **traffic** TELEP tráfico *m* de télex
telnet *n* DATA *Internet* telnet *f*
TEM *abbr* (*transverse electromagnetic*) TRANS TEM (*electromagnético transversal*); ~ **mode** *n* TRANS modo *m* TEM
temperature: ~ **coefficient** *n* CLIM coeficiente *m* de temperatura; ~ **control** *n* CLIM regulación *f* de temperatura; ~ **cycling** *n* CLIM, TEST ciclado *m* de temperatura; ~ **difference** *n* CLIM diferencia *f* de temperatura; ~ **error** *n* CLIM, TEST error *m* de temperatura; ~ **inversion** *n* CLIM, RADIO inversión *f* de temperatura; ~**-limited state** *n* CLIM, ELECT estado *m* de temperatura limitada; ~ **sensitivity** *n* CLIM sensibilidad *f* a la temperatura; ~ **sensor** *n* CLIM, CONTR termistor *m*
tempering *n* ELECT templado *m*
template *n* COMP *word processing*, PARTS, TRANS plantilla *f*
tempo: ~ **control** *n* COMP control *m* de tiempo; ~ **controller** *n* COMP controlador *m* de tiempo
temporal: ~ **coherence** *n* TRANS coherencia *f* temporal; ~ **compression** *n* COMP, DATA compresión *f* temporal; ~ **database** *n* DATA base *f* de datos temporal; ~ **logic** *n* DATA lógica *f* temporal; ~ **predictive concealment** *n* COMP ocultamiento *m* predictivo provisional; ~ **reasoning** *n* DATA razonamiento *m* temporal
temporary: ~**-buffer memory** *n* COMP memoria *f* intermedia temporal; ~ **call-forwarding** *n* PER *on answering machine* desvío *m* temporal de llamada; ~ **fault** *n* TELEP, TEST fallo *m* temporal; ~ **interruption** *n* TELEP, TEST interrupción *f* temporal; ~ **malfunction** *n* TELEP, TEST fallo *m* temporal; ~ **mobile-station identity** *n* DATA, RADIO, TELEP identidad *f* temporal de estación móvil; ~ **perturbation** *n* TELEP, TEST perturbación *f* temporal; ~ **transfer** *n* TELEP transferencia *f* temporal; ~ **window** *n* COMP ventana *f* temporal
ten: ~**-number memory** *n* PER APP memoria *f* de diez números
tens: ~ **counter** *n* ELECT contador *m* de decenas; ~ **digit** *n* COMP dígito *m* de decenas; ~ **relay** *n* PARTS, TELEP relé *m* de decena
tensile: ~ **force** *n* PARTS fuerza *f* a la tracción; ~ **strength** *n* GEN resistencia *f* a la tracción; ~ **stress** *n* PARTS esfuerzo *m* de tracción; ~ **test** *n* PARTS, TEST prueba *f* de tracción; ~**-yield limit** *n* PARTS límite *m* de elasticidad
tension: ~ **spring** *n* GEN resorte *m* tensor

term *n* GEN término *m*

terminal *n* COMP, DATA, ELECT, PARTS *semiconductors*, PER *contact*, POWER, RADIO, TELEP terminal *m*; **~ adaptor** (*TA*) DATA, PER, TELEP adaptador *m* de borne; **~ administration** TELEP administración *f* temporal; **~ board** POWER tablero *m* de terminales, TELEP cuadro *m* de terminales; **~ box** POWER caja *f* de bornes, TELEP caja *f* de conexión; **~ called** TELEP terminal *m* llamada; **~ call forwarding** APPL transmisión *f* de una llamada terminal; **~ category PE** TELEP EP *m* de categoría terminal; **~ circuit** POWER, TELEP circuito *m* terminal; **~ clip** POWER terminal *m* de pinza; **~ connector** ELECT conector *m* terminal; **~ country** TELEP país *m* terminal; **~ coupling loss** TELEP pérdida *f* de acoplamiento terminal; **~ diagram** GEN diagrama *m* terminal; **~ drawing** GEN dibujo *m* de conexión; **~ emulation** DATA *Internet* emulación *f* de terminal; **~ emulation software** COMP, DATA software *m* de emulación de terminal; **~ emulator** DATA emulador *m* de terminal; **~ end-point identifier** (*TEI*) DATA identificador *m* del punto final terminal; **~-engaged signal** SIG, TELEP señal *f* de terminal ocupada; **~ equipment** (*TE*) COMP, DATA, PER, TELEP equipo *m* terminal (*TE*); **~ equipment package** TELEP paquete *m* de equipo de terminal; **~ equipment PE** PER APP EP *m* de equipo terminal; **~ exchange** TELEP, TRANS central *f* terminal; **~ field** GEN campo *m* de terminales; **~ impedance** ELECT impedancia *f* terminal; **~ installation for data transmission** TELEP instalación *f* de terminal para transmisión de datos; **~ international centre** *BrE* TELEP centro *m* terminal internacional; **~ international exchange** TELEP central *f* terminal internacional; **~ national centre** *BrE* TELEP centro *m* terminal nacional; **~ national section** TELEP sección *f* terminal nacional; **~-node controller** (*TNC*) TRANS controlador *m* de nodo terminal (*TNC*); **~-number diagram** GEN diagrama *m* de número terminal; **~-number drawing** GEN dibujo *m* de número terminal; **~ pair** POWER par *m* terminal; **~ screw** POWER terminal *m* de tornillo; **~-server manager** (*TSM*) DATA gestor *m* de servidor terminal (*TSM*); **~ share** TELEP terminal *f* compartida; **~ strip** PARTS regleta *f* de conexiones; **~ subaddressing** APPL *ISDN* subdirección *f* terminal; **~ table** POWER, TELEP tabla *f* de terminales; **~ voltage** POWER tensión *f* en los bornes

terminate[1]: **~-and-stay resident** *n* DATA, TELEP residente *m* de terminar y permanecer; **~-and-stay resident program** *n* COMP programa *m* residente de terminar y permanecer

terminate[2] *vt* TELEP, TRANS terminar; ◆ **~ to a station** TELEP, TRANS conectar a una impedancia terminal

terminate[3] *vi* TELEP, TRANS terminarse

terminated: **~ level** *n* POWER, TEST nivel *m* de prueba adaptado; **~ reading** *n* POWER, TEST lectura *f* terminal

terminating: **~ equipment** *n* COMP, DATA, PER, TELEP aparato *m* terminal; **~ exchange** *n* SWG, TELEP central *f* de destino; **~ impedance** *n* ELECT impedancia *f* terminal; **~ junctor** *n* PARTS, SWG, TELEP conjuntor *m* terminal; **~ PBX** *n* TELEP PBX *m* terminal; **~ resistor** *n* ELECT, PARTS, TRANS resistor *m* terminal; **~ set** *n* ELECT, PARTS, TRANS terminador *m*; **~ stage** *n* SWG etapa *f* terminal; **~ traffic** *n* TELEP tráfico *m* terminal; **~ unit** *n* DATA terminador *m*

termination *n* DATA, ELECT, RADIO, TELEP, TRANS terminación *f*; **~ joint** PARTS empalme *m* terminal; **~ joint box** PARTS caja *f* de empalme terminal

terminator *n* DATA *of scanner* terminador *m*

terrain: **~ data** *n pl* RADIO datos *m pl* de terreno; **~ and surface database** *n* RADIO base *f* de datos de superficie y terreno

terrestrial: **~ hypothetical reference circuit** *n* RADIO circuito *m* de referencia hipotética terrestre; **~ interface** *n* RADIO, TRANS interfaz *f* terrestre; **~ interface module** *n* TRANS módulo *m* de interfaz terrestre; **~ radiocommunication** *n* RADIO radiocomunicación *f* terrestre; **~ standard clock** *n* TRANS reloj *m* de estándar terrestre; **~ station** *n* RADIO estación *f* terrestre

test[1]: **~ access** *n* TEST acceso *m* para pruebas y medidas; **~ access-point** *n* TEST punto *m* de acceso para pruebas y medidas; **~ adaptor** *n* TEST adaptador *m* para prueba; **~ alarm inhibition** *n* TEST inhibición *f* de alarma para prueba; **~ answer** *n* TEST respuesta *f* a la prueba; **~bar** *n* SWG, TEST barra *f* de prueba; **~ bed** *n* POWER, TEST banco *m* de pruebas; **~ bench** *n* POWER, TEST banco *m* de pruebas; **~ block** *n* TEST bloque *m* de prueba; **~ board** *n* ELECT panel *m* de pruebas y medidas, placa *f* para prueba; **~ body** *n* DATA, TEST cuerpo *m* de prueba; **~box** *n* TEST caja *f* de prueba; **~ call indicator** *n* TEST indicador *m* de llamada de prueba; **~ call of the subscriber-to-subscriber type** *n* TEST llamada *f* de prueba de tipo abonado a abonado; **~ call of type 1** *n* TEST llamada *f* de prueba de tipo 1; ~ **call of type 2** *n* TEST llamada *f* de prueba de tipo 2; ~ **call of type 3** *n* TEST llamada *f* de prueba de tipo 3; ~**card** *n* TV tarjeta *f* de prueba; ~**case** *n* TEST casilla *f* de prueba; ~ **case error** *n* DATA, TEST error *m* en la casilla de prueba; ~ **case identifier** *n* TEST identificador *m* de la casilla de prueba; ~ **case variable** *n* DATA variable *f* de la casilla de prueba; ~ **centre** *n* TEST centro *m* de pruebas; ~ **circuit** *n* TELEP, TEST circuito *m* de prueba; ~ **coordination procedure** *n* TEST procedimiento *m* de coordinación de pruebas; ~ **coupon** *n* TEST cupón *m* para pruebas; ~ **customer** *n* TEST abonado *m* de prueba; ~ **desk** *n* TEST mesa *f* de pruebas; ~ **duration** *n* TEST duración *f* de ensayo; ~ **engineering** *n* TEST técnica *f* de ensayos; ~ **equipment** *n* TEST equipo *m* de pruebas; ~ **event** *n* DATA, TEST evento *m* de prueba; ~ **group** *n* DATA, TEST grupo *m* de prueba; ~ **group objective** *n* DATA, TEST objetivo *m* de grupo de prueba; ~ **group reference** *n* DATA, TEST referencia *f* de grupo de prueba; ~ **handset** *n* TELEP, TEST microteléfono *m* de prueba; ~ **indicator** *n* DATA, TEST indicador *m* de prueba; ~ **instructions** *n pl* TEST instrucciones *f pl* de prueba; ~ **jack** *n* ELECT, TELEP jack *m* de prueba; ~ **kit** *n* TEST equipo *m* de prueba; ~ **laboratory** *n* DATA, TEST laboratorio *m* de ensayos; ~ **jackstrip** *n* TELEP, TEST regleta *f* de jacks de prueba; ~ **laboratory checklist** *n* DATA lista *f* de control del laboratorio de ensayos, TEST lista *f* de control del laboratorio de pruebas; ~ **laboratory manager** *n* DATA director *m* del laboratorio de ensayos, TEST director *m* del laboratorio de pruebas; ~**line** *n* TELEP línea *f* de pruebas; ~ **load** *n* TEST carga *f* de prueba; ~ **loop** *n* TEST bucle *m* de prueba; ~ **management PDU** *n* DATA, TEST PDU *m* de gestión de pruebas; ~ **management protocol** *n* DATA protocolo *m* de gestión de pruebas, TEST protocolo *m* de gestión de prueba; ~ **and measurement set** *n* TEST equipo *m* para puebas y medidas; ~ **menu** *n* TEST menú *m* de prueba; ~ **message** *n* DATA, TEST mensaje *m* de prueba; ~ **method** *n* TEST método *m* de control; ~ **mode** *n* TEST modo *m* de ensayo; ~ **and operation bus** *n* TELEP, TEST bus *m* de prueba y operación; ~ **and operation point** *n* TELEP, TEST punto *m* de prueba y operación; ~ **operator** *n* DATA operador *m* de ensayo, TEST operador *m* de pruebas; ~ **outcome** *n* TEST resultado *m* de la prueba; ~ **pattern** *n* TEST, TV carta *f* de ajuste; ~**piece** *n* TEST pieza *f* de ensayo; ~**point** *n* TEST punto *m* de prueba; ~ **position** *n* TEST posición *f* de pruebas; ~ **postamble** *n* TEST postámbulo *m* de prueba; ~ **preamble** *n* TEST preámbulo *m* de prueba; ~ **probe** *n* TEST sonda *f* de ensayo; ~ **procedure** *n* TEST método *m* de prueba; ~ **prod** *n* TEST punta *f* de prueba; ~ **purpose** *n* TEST finalidad *f* de la prueba; ~ **rack** *n* TEST mesa *f* de corte; ~ **realization** *n* TEST realización *f* de prueba; ~ **realizer** *n* TEST realizador *m* de prueba; ~ **record** *n* TEST planilla *f* de registro de pruebas; ~ **relay** *n* PARTS, TEST relé *m* verificador; ~ **result** *n* TEST resultado *m* del ensayo; ~ **setup** *n* TEST disposición *f* de prueba, montaje *m* para pruebas; ~ **stand** *n* TEST banco *m* de pruebas; ~ **step identifier** *n* TEST identificador *m* de operación de control; ~ **step library** *n* TEST biblioteca *f* de operacion de control; ~ **step group** *n* TEST grupo *m* de operación de control; ~ **step objective** *n* TEST objetivo *m* de la operación de control; ~ **step** *n* TEST operación *f* de control; ~ **step group reference** *n* DATA, TEST referencia *f* de grupo de operación de control; ~ **room** *n* TEST sala *f* de pruebas; ~ **signal** *n* SIG, TEST señal *f* de prueba; ~ **suite** *n* TEST sucesión *f* de pruebas; ~ **suite constant** *n* TEST constante *f* de sucesión de pruebas; ~ **suite parameter** *n* TEST parámetro *m* de sucesión de pruebas; ~ **suite structure** *n* TEST estructura *f* de la batería de pruebas; ~ **suite structure and ~ purposes** *n* TEST estructura *f* de la batería de pruebas y fines del ensayo; ~ **suite variable** *n* TEST variable *f* de sucesión de pruebas; ~ **system** *n* TEST sistema *m* de prueba; ~ **technology** *n* TEST tecnología *f* de prueba; ~ **traffic** *n* TELEP, TEST, TRANS tráfico *m* de prueba; ~ **transformer** *n* TEST transformador *m* de prueba; ~ **verdict** *n* TEST resolución *f* de la prueba; ~ **wire** *n* SWG, TELEP hilo *m* de prueba

test[2] *vt* GEN probar; ♦ **~-load** TEST cargar para prueba

test out *vt* COMP, TEST probar

testability *n* TEST facilidad *f* de prueba

tester *n* TEST comprobador *m*

testing: ~ **cycle** *n* TEST ciclo *m* de prueba; ~ **device** *n* TEST dispositivo *m* de prueba; ~ **equipment** *n* TELEP, TEST equipo *m* de comprobación; ~ **manager** *n* TEST jefe *m* de pruebas; ~ **point** *n* SIG, SWG, TEST, TRANS punto *m* de prueba; ~ **point line signalling** *BrE n* SIG, TEST señalización *f* de línea del punto de prueba; ~ **point switching and interregister signalling** *BrE n* SWG señal *f* de conmutación de punto de prueba e interregistro; ~ **point transmission** *n* TEST transmisión *f* del punto de prueba; ~ **position** *n* TELEP, TEST posición *f* de pruebas; ~ **programme** *BrE* *n* TEST programa *m* de pruebas; ~ **state** *n* TEST estado *m* de prueba; ~ **technique** *n* TEST técnica *f* de prueba

testware *n* COMP, DATA equipo *m* de pruebas

tether *n* ELECT, PARTS resistencia *f*

tetrode *n* ELECT, PARTS tetrodo *m*; ~ **field-effect transistor** ELECT, PARTS transistor *m* tetrodo de efecto de campo; ~ **transistor** ELECT, PARTS transistor *m* de tetrodo

texel *n* COMP texel *m*

text *n* COMP, DATA, TELEP texto *m*; ~ **block** TELEP bloque *m* de texto; ~ **box** COMP zona *f* de texto; ~ **editing** APPL, COMP edición *f* de textos; ~ **editor** COMP, DATA *person* editor *m* de textos; ~**-entry field** COMP campo *m* de entrada de texto; ~ **file** DATA archivo *m* de texto; ~ **mailbox** APPL buzón *m* para textos; ~ **memory** COMP memoria *f* de texto; ~ **message** RADIO mensaje *m* con texto; ~ **processing** COMP procesado *m* de texto; ~ **search** DATA búsqueda *f* textual; ~**-space entry** COMP espacio *m* para texto; ~ **string** TELEP cadena *f* de texto

texture: ~ **finish** *n* PARTS acabado *m* texturado; ~ **lacquer** *n* GEN pintura *f* con efecto de relieve; ~ **mapping** *n* COMP encuadramiento *m* de textura

TFEL *abbr* (*thin-film electroluminescent display*) ELECT, PARTS ELCD (*pantalla electroluminiscente de película delgada*)

TFT *abbr* (*thin-film transistor*) ELECT, PARTS TFT (*transistor de capa delgada*)

theme: ~ **channel** *n* TV canal *m* temático

theoretical: ~ **duration** *n* TELEP duración *f* teórica; ~ **pitch diameter** *n* PARTS diámetro *m* efectivo teórico

thermal *adj* GEN térmico; ~ **agitation noise** *n* ELECT ruido *m* de agitación térmica; ~ **breakdown** *n* ELECT, PARTS disgregación *f* térmica; ~ **capacitance** *n* ELECT, PARTS capacitancia *f* térmica; ~ **characteristic** *n* ELECT, PARTS característica *f* térmica; ~ **conductivity** *n* ELECT, PARTS conductividad *f* térmica; ~ **expansion** *n* ELECT, PARTS dilatación *f* térmica; ~ **expansion coefficient** *n* ELECT, PARTS coeficiente *m* de expansión térmica; ~ **instability** *n* ELECT, PARTS termoinestabilidad; ~ **noise** *n* ELECT, PARTS ruido *m* térmico; ~ **radiation** *n* GEN radiación *f* térmica; ~ **residual voltage** *n* ELECT, PARTS tensión *f* térmica residual; ~ **resistance** *n* ELECT, PARTS resistencia *f* térmica; ~ **runaway** *n* ELECT embalamiento *m* térmico; ~ **printer** *n* DATA, PER APP impresora *f* térmica

thermally: ~**-operated** *adj* GEN de accionamiento térmico

thermionic: ~ **emission** *n* ELECT emisión *f* termoiónica; ~ **emission efficiency** *n* ELECT eficiencia *f* de emisión termoiónica

thermistor *n* ELECT, PARTS termistor *m*; ~ **mount** PARTS, TEST montura *f* de termistor

thermo: ~**-shrinkable sheathing** *n* PARTS, TRANS envoltura *f* termoencogible

thermo couple *n* ELECT par *m* térmico, PARTS par *m* termoeléctrico; ~ **wattmeter** TEST medidor *m* de vataje del par termoeléctrico

thermodynamic *adj* GEN termodinámico

thermodynamics *n* GEN termodinámica *f*

thermoelectric *adj* GEN termoeléctrico; ~ **effect** *n* ELECT efecto *m* termoeléctrico

thermoelectricity *n* GEN termoelectricidad *f*

thermometer: ~ **probe** *n* CONTR sonda *f* termométrica

thermoplastic *n* GEN termoplástico *m*

thermosetting: ~ **plastic** *n* PARTS plástico *m* termoendurecible

thermostat *n* CLIM, CONTR, PARTS termostato *m*

Thévenin: ~ **equivalent circuit** *n* POWER circuito *m* equivalente de Thévenin; ~**'s theorem** *n* POWER teorema *m* de Thévenin

thick: ~ **coax** *n* PARTS, TRANS coaxial *m* grueso; ~ **coaxial cable** *n* PARTS, TRANS cable *m* coaxial grueso; ~**-film integrated circuit** *n* ELECT, PARTS circuito *m* integrado de película gruesa; ~**-film resistor** *n* ELECT, PARTS resistor *m* de capa gruesa; ~**-layer integrated circuit** *n* ELECT, PARTS circuito *m* integrado de capas gruesas

thickness *n* PARTS espesor *m*

thin: ~ **coax** *n* PARTS, TRANS coaxial *m* delgado; ~ **coaxial cable** *n* PARTS, TRANS cable *m* coaxial delgado; ~ **Ethernet** *n* ELECT Ethernet *m* delgado; ~**-film capacitor** *n* ELECT, PARTS capacitor *m* de capa delgada; ~**-film electroluminescent display** *n* (*TFEL*) ELECT, PARTS pantalla *f* electroluminiscente de película delgada (*ELCD*); ~**-film resistor** *n* ELECT, PARTS resistor *m* de capa delgada; ~**-film transistor** *n* (*TFT*) ELECT, PARTS transistor *m* de capa delgada (*TFT*); ~**-film waveguide** *n* DATA, TRANS guíaondas *m* de película delgada; ~**-layer capacitor** *n* ELECT, PARTS condensador *m* de película delgada; ~**-route network** *n* TRANS red *f* de ruta estrecha; ~**-route TDMA** *n* RADIO TDMA *f* de ruta estrecha; ~ **small-outline package** *n* DATA paquete *m* delgado de contorno reducido; ~**-wall counter tube** *n* RADIO tubo *m* contador de pared delgada; ~**-wire Ethernet multiport**

repeater *n* DATA, ELECT repetidor *m* multipuerta Ethernet de alambre fino
thinner *n* PARTS diluyente *m*
thin: **~-net tap** *n* ELECT, PARTS derivación *f* de red fina
third: **~-angle projection** *n* GEN proyección *f* a la americana; **~ harmonic distortion** *n* (*3HD*) ELECT distorsión *f* del tercer armónico (*3HD*); **~ party** *n* TELEP tercero *m*; **~-party billing** *n* TELEP facturación *f* a tercero; **~-party billing call** *n* TELEP comunicación *f* de facturación a tercero; **~-party warning tone** *n* TELEP tono *m* de alerta para terceros
Thomson: **~ effect** *n* GEN *physics* efecto *m* Thomson
thoriated: **~-tungsten cathode** *n* ELECT cátodo *m* de tungsteno toriado
thread: **~ gauge** *BrE n* TEST calibre *m* hembra de roscas, calibre *m* para rosca
threaded: **~ bushing** *n* ELECT casquillo *m* roscado, PARTS manguito *m* roscado; **~ coupler** *n* ELECT, PARTS acoplador *m* roscado
threading: **~ ring** *n* PARTS anillo *m* de roscar
three: **~-coil relay** *n* PARTS relé *m* de tres bobinas; **~-condition cable code** *n* TELEP, TRANS código *m* trivalente para cable; **~-condition telegraph code** *n* TELEP, TRANS código *m* telegráfico trivalente; **~-conductor plug** *n* PARTS clavija *f* de tres conductores; **~-dimensional guide** *n* TRANS guía *f* tridimensional; **~-party conference** *n* APPL, TELEP comunicación *f* tripartita; **~-party service** *n* APPL, TELEP servicio *m* tripartito; **~-phase circuit** *n* POWER circuito *m* trifásico; **~-phase rectifier** *n* PARTS, POWER rectificador *m* trifásico; **~-way call** *n* PER, RADIO, TELEP llamada *f* tridireccional; **~-way call chaining** *n* TELEP cadena *f* de llamada de tres direcciones; **~-way calling** *n* PER, RADIO, TELEP llamadas *f pl* de tres direcciones; **~-way conversation** *n* PER, RADIO, TELEP conversación *f* tridireccional; **~-way teleconference** *n* APPL, TELEP teleconferencia *f* de tres direcciones; **~-wire line** *n* SWG, TELEP línea *f* trifilar
threshold *n* ELECT, POWER, RADIO, TRANS umbral *m*; **~ of audibility** RADIO, TELEP umbral *m* de audibilidad; **~ current** ELECT, POWER, TRANS corriente *f* umbral; **~-current density** ELECT, POWER, TRANS densidad *f* de corriente umbral; **~ decoding** DATA descodificación *f* de umbral; **~ of discomfort** TELEP *hearing level* límite *m* superior de audibilidad; **~ element** ELECT elemento *m* de umbral; **~-extension demodulator** ELECT, RADIO desmodulador *m* de extensión umbral; **~ frequency** ELECT, TRANS frecuencia *f* crítica; **~ function** ELECT función *f* de umbral; **~ gate** ELECT circuito *m* umbral; **~ of hearing** TELEP umbral *m* de audibilidad; **~ operation** ELECT, RADIO operación *f* umbral; **~ of pain** TELEP *hearing level* umbral *m* de sensación dolorosa; **~ of profitability** GEN umbral *m* de rentabilidad; **~ value** GEN valor *m* umbral; **~ voltage** ELECT, POWER voltaje *m* de entrada; **~ wavelength** ELECT, TRANS longitud *f* de onda de entrada
thresholding *n* ELECT, POWER formación *f* de umbrales
throat: **~ microphone** *n* RADIO, TELEP, TRANS laringófono *m*
throttle: **~ valve** *n* CLIM válvula *f* reguladora
through[1]: **~ connection** *n* ELECT, POWER, TELEP conexión *f* transversal; **~-connection delay** *n* TELEP retardo *m* de conexión directa; **~-connection hole** *n* ELECT agujero *m* de pasante de conexión; **~-dialling number** *n* TELEP número *m* de marcado directo; **~-15-supergroup** *n* TELEP punto *m* de conexión de montaje de grupo secundario 15 directo; **~-group connection point** *n* TELEP punto *m* de conexión de grupo directo; **~ hypergroup** *n* TRANS hipergrupo *m* de conexión directa; **~ line** *n* TELEP línea *f* directa; **~ mastergroup** *n* TRANS grupo *m* terciario de conexión directa; **~-mastergroup connection point** *n* TELEP punto *m* de conexión de grupo terciario directo; **~ supergroup** *n* TRANS grupo *m* secundario de conexión directa; **~-supergroup connection point** *n* TELEP punto *m* de conexión de grupo secundario directo; **~-transmission loss** *n* TELEP, TRANS atenuación *f* de transmisión directa
through[2]: **~-connect** *vt* ELECT, POWER, TELEP conectar directamente
throughline *n* TRANS línea *f* directa
throughput *n* COMP, DATA, RADIO rendimiento *m* total; **~ class** DATA clase *f* de producción
Thruline: **~ wattmeter** *n* TEST vatímetro *m* de Thruline
thruput *AmE see throughput*
thumbnail *n* COMP *graphics* chincheta *f*
thumbscrew *n* ELECT, PARTS tornillo *m* de mariposa
thumbwheel *n* ELECT rueda *f* accionada por el pulgar; **~ switch** ELECT, SWG interruptor *m* de rueda moleteada
thyatron *n* ELECT tiratrón *m*
thyristor *n* ELECT, PARTS, POWER tiristor *m*; ~

rectifier unit POWER unidad *f* rectificadora del tiristor

tick: **~-mark label** *n* COMP rótulo *m* de marca de tictac; **~ tone** *n* TELEP tono *m* de tictac; **~-tone generator** *n* TELEP generador *m* de tono de tictac

ticker: **~ tone** *n* TELEP tono *m* de vibrador

tie *n* COMP, PARTS, TELEP ligadura *f*; **~ bar** PARTS, TELEP traviesa *f* de separación; **~ button** COMP botón *m* de enlace; **~-circuit interface** TRANS interfaz *f* de circuito de interconexión; **~ line** SWG, TELEP, TRANS línea *f* de enlace; **~-line traffic** SWG, TELEP, TRANS tráfico *m* de línea de interconexión; **~ trunk** TELEP línea *f* privada; **~ wrap** PARTS, POWER arrollamiento *m* de unión

TIFF *abbr* (*tagged-image file format*) COMP, DATA TIFF (*formato de archivo de imagen con nombre simbólico*)

tight: **~ buffer** *n* TRANS *fibre optics* compensador *m* estrecho; **~ jacket** *n* TRANS *fibre optics* cubierta *f* estrecha; **~-jacketed cable** *n* TRANS cable *m* revestido compacto

tighten *vt* CONTR *conditions* endurecer, PARTS tensar, TEST, *conditions* endurecer

tightened *adj* PARTS tensado

tightening *n* GEN apriete *m*

tilde *n* COMP tilde *f*

tilt *n* TV distorsión *f* de imagen

tilted: **~ cable television** *n* TV televisión *f* por cable SCART inclinado; **~ CATV** *n* TV CATV *m* inclinado; **~ front** *n* ELECT, PARTS frente *m* inclinado

timbre *n* COMP, TELEP timbre *m*; **~ envelope** COMP envolvente *m* de timbre; **~ modulation** COMP modulación *f* de timbre

time[1]: **~ and charges** *n* TELEP aviso *m* de importe; **~-and-distance metering** *n* TELEP cómputo *m* por tiempo y zona; **~-assigned speech interpolation** *n* SIG, TRANS interpolación *f* de palabra mediante tiempos asignados; **~-assignment speech** *n* TRANS conversación *f* por asignación de tiempos; **~ axis** *n* TEST, TRANS eje *m* de tiempo; **~ balance** *n* TEST equilibrio *m* de tiempos; **~ base** *n* COMP, ELECT base *f* de tiempo; **~-base generator** *n* ELECT generador *m* de base de tiempo; **~-base jitter** *n* ELECT fluctuación *f* de la base de tiempos; **~-based structuring language** *n* COMP lenguaje *m* estructurador basado en tiempos; **~ between failures** *n* COMP, TEST tiempo *m* entre fallos; **~ characteristic** *n* COMP, ELECT, TEST, TRANS característica *f* temporal; **~ charging** *n* TELEP tasación *f* de la duración; **~ code** *n* COMP, DATA, RADIO código *m* de tiempo; **~ coefficient** *n* TEST coeficiente *m* de tiempo; **~ comparison** *n* RADIO, TRANS comparación *f* de tiempos; **~ compression** *n* COMP, DATA compresión *f* del tiempo; **~ congestion** *n* RADIO, TELEP congestión *f* temporal; **~-consistent busy hour** *n* TELEP hora *f* cargada acorde al tiempo; **~ constant** *n* COMP, ELECT, TEST, TRANS constante *f* de tiempo; **~-coordinate system** *n* COMP sistema *m* coordinado por tiempos; **~ correlation** *n* DATA, SWG correlación *f* temporal; **~-critical printing application** *n* APPL aplicación *f* de impresión de tiempos críticos; **~ delay** *n* DATA, ELECT retardo; **~-delay circuit** *n* ELECT circuito *m* retardador; **~-delay relay** *n* ELECT, PARTS relé *m* de retardo; **~-derived channel** *n* TRANS canal *m* derivado en tiempo; **~ dispersion** *n* COMP, ELECT, TEST, TRANS dispersión *f* de tiempos; **~ displacement** *n* COMP, ELECT, TEST, TRANS desplazamiento *m* de tiempos; **~ diversity** *n* DATA, RADIO diversidad *f* temporal; **~-diversity reception** *n* DATA, RADIO recepción *f* de diversidad temporal; **~ division** *n* RADIO, SIG, SWG, TELEP, TRANS división *f* de tiempo; **~-division exchange** *n* SWG central *f* temporizadora; **~-division highway** *n* TELEP autopista *f* por división de tiempo; **~-division highway inswitching** *n* SWG, TELEP autopista *f* por división de tiempo en conmutación; **~-division multiple access** *n* (*TDMA*) RADIO, SIG, TELEP, TRANS acceso *m* múltiple por división de tiempo (*TDMA*); **~-division multiple access burst ~** *n* SIG tiempo *m* en ráfagas de acceso múltiple por división de tiempo; **~-division multiple access burst ~plan** *n* RADIO, TELEP, TRANS plan *m* temporizado en ráfagas de acceso múltiple por división de tiempo; **~-division multiple access mode** *n* RADIO, SIG, TELEP, TRANS modo *m* de acceso múltiple por división de tiempo; **~-division multiplex** *n* TELEP, TRANS múltiplex *m* por división de tiempo; **~-division multiplex transmitter** *n* COMP, TELEP, TRANS transmisor *m* múltiplex por división de tiempo; **~-division multiplexing** *n* TELEP, TRANS transmisión *f* múltiple con división de tiempo; **~-division network** *n* SWG, TRANS red *f* cronizadora; **~-division switching** *n* DATA, SWG, TELEP, TRANS conmutación *f* temporal; **~-division switching system** *n* DATA, SWG, TELEP, TRANS sistema *m* de conmutación temporal; **~-division system** *n* DATA, SWG, TELEP, TRANS sistema *m* de división de tiempo; **~ domain** *n* GEN dominio *m* temporal; **~-domain analysis** *n* DATA análisis *m* de dominio de tiempo; **~-domain**

characteristics *n pl* TEST características *f pl* de dominio de tiempo; **~-domain filter** *n* ELECT filtro *m* de dominio de tiempo; **~-domain reflectometer** *n* ELECT, POWER, TEST, TRANS reflectómetro *m* de dominio de tiempo; **~-domain reflectometry** *n* ELECT, POWER, TEST, TRANS reflectometría *f* de dominio de tiempo; **~ duration** *n* COMP, GEN duración *f* temporal; **~-insertion unit** *n* DATA unidad *f* de inserción temporal; **~ interval** *n* GEN intervalo *m* de tiempo; **~-interval error** *n* TELEP error *m* de intervalo de tiempo; **~-interval meter** *n* TEST medidor *m* de intervalo de tiempo; **~ lag** *n* GEN tiempo *m* de retardo; **~ marker** *n* RADIO cronometrador *m*; **~ match** *n* TEST comparación *f* de tiempos; **~ measurement** *n* TELEP, TEST medición *f* de tiempo; **~-measurement pulse** *n* TELEP impulso *m* temporizador; **~ meter** *n* APPL, TEST medidor *m* de tiempo; **~ multiplex switching** *n* SWG conmutación *f* múltiplex en el tiempo; **~ multiplexing** *n* COMP, TELEP, TRANS multiplexión *f* en el tiempo; **~ of persistence** *n* ELECT tiempo *m* de persistencia; **~ of start** *n* GEN momento *m* de arranque; **~-out** *n* DATA, ELECT, RADIO, TELEP retraso *m*; **~-out device** *n* PARTS desconexión *f* temporizada; **~-out disable facility** *n* ELECT instalación *f* con desconexión temporizada; **~-out event** *n* DATA evento *m* de desconexión temporizada; **~-out program** *n* COMP programa *m* de desconexión temporizada; **~-out supervision** *n* DATA, ELECT, RADIO, TELEP supervisión *f* del transcurso del intervalo de retardo; **~ program** *n* COMP programa *m* de temporización; **~-quantized control** *n* CONTR, TELEP control *m* por cuantificación del tiempo; **~ queue** *n* DATA, ELECT, RADIO, TELEP cola *f* de tiempo; **~ recording** *n* GEN cronografía *f*; **~ register** *n* GEN registro *m* de tiempo; **~-related charging** *n* TELEP cobro *m* por tiempo; **~-related tariffing** *n* TELEP tarificación *f* por tiempo; **~ relay** *n* PARTS relé *m* temporizado; **~ release** *n* DATA, ELECT, RADIO, TELEP desconexión *f* temporizada; **~ response** *n* CONTR respuesta *f* en función del tiempo; **~ sequence** *n* GEN secuencia *f* temporal; **~ sharing** *n* COMP, DATA distribución *f* de tiempo; **~-sharing program** *n* COMP, DATA programa *m* de tiempo compartido; **~-sharing system** *n* DATA sistema *m* de tiempo compartido; **~ shift** *n* GEN crono desplazamiento *m*; **~ signal** *n* RADIO, SIG señal *f* de tiempo; **~-signal control** *n* PER APP control *m* de señal horaria; **~-signal receiver** *n* RADIO, SIG receptor *m* de señal horaria; **~-signal satellite service** *n* RADIO, SIG servicio *m* de señales horarias por satélite; **~ signature** *n* COMP sintonía *f* de tiempo; **~ slicing** *n* COMP fraccionamiento *m* de tiempo; **~ slot** *n* RADIO, SIG, SWG, TRANS, TV segmento *m* de tiempo; **~-slot code** *n* TRANS código *m* de segmento de tiempo; **~-slot interchange** *n* SWG intercambio *m* de segmento de tiempo; **~-slot interchanger** *n* DATA, SWG intercambiador *m* de segmento de tiempo; **~-slot interface** *n* TRANS interfaz *m* de segmento de tiempo; **~-slot sequence integrity** *n* TRANS integridad *f* de secuencia de segmento de tiempo; **~-slot sequence integrity** *n* TELEP integridad *f* de secuencia del intervalo de tiempo; **~-slot structure** *n* RADIO estructura *f* del intervalo de tiempo; **~-space-~ network** *n* SWG red *f* de tiempo-espacio-tiempo, TRANS red *f* tiempo-espacio-tiempo; **~span** *n* GEN espacio *m* temporal; **~ stage** *n* SWG, TELEP etapa *f* temporizadora; **~ stamp** *n* COMP, SWG, TEST reloj *m* fechador; **~ standard** *n* RADIO señal *f* horaria; **~ step** *n* RADIO paso *m* de tiempo; **~ supervision** *n* TELEP supervisión *f* temporizada; **~-supervision flip-flop** *n* PARTS, TELEP flip-flop de temporización *m*; **~-supervision of a process PE** *n* TELEP supervisión *f* temporizada de un PE de proceso; **~switch** *n* DATA, SWG conmutador *m* de tiempo; **~switching** *n* DATA, SWG, TRANS conmutación *f* temporal; **~-tag command** *n* CONTR orden *f* de ordenación temporal; **~ to failure** *n* TEST tiempo *m* hasta el fallo; **~ to first failure** *n* COMP tiempo *m* hasta el primer fallo; **~ to half value** *n* TEST tiempo *m* a valor medio; **~ to saturation** *n* TEST tiempo *m* hasta saturación; **~ unit** *n* GEN unidad *f* de tiempo; **~ value** *n* GEN valor *m* tiempo; **~-variant amplifier** *n* COMP amplificador *m* de variación temporal; **~-variant filter** *n* COMP filtro *m* de variación temporal; **~ zone** *n* GEN huso *m* horario; **~-zone metering** *n* TELEP cómputo *m* por tiempo y zona

time^2 *vt* COMP, DATA, RADIO, TELEP, TEST contar el tiempo de; ◆ **~ the cycling** DATA temporizar el ciclo; **~-tag** DATA marcar el tiempo

timed: **~ call** *n* TELEP comunicación *f* con tasa de duración; **~ reminder** *n* PER, TELEP recordatorio *m* temporizado

timer *n* GEN cronómetro *m*; **~ relay** ELECT, PARTS relé *m* de temporizador; **~ system** GEN sistema *m* de temporizador

timescale *n* COMP, RADIO, TRANS trama *f* temporal; **~ difference** COMP, RADIO, TRANS diferencia *f* de trama temporal; **~ reading**

COMP, TRANS lectura *f* de trama temporal; ~ **in synchronization** RADIO trama *f* temporal en sincronización; ~ **unit** COMP, RADIO, TRANS unidad *f* de trama temporal

timetable: ~ **slot** *n* RADIO, TV intervalo *m* de horario

timing *n* GEN cronización *f*; ~ **chart** GEN diagrama *m* de tiempos; ~ **circuit** GEN circuito *m* temporizador; ~ **control** COMP sincronización *f*, CONTR temporización *f*; ~**-correction device** COMP dispositivo *m* corrector de distorsiones geométricas; ~ **diagram** GEN cronograma *m*; ~ **generator** TRANS generador *m* de medición de tiempo; ~ **information** DATA información *f* de ritmo, TRANS información *f* de temporización; ~ **jitter** TRANS fluctuación *f* de la temporización; ~ **pulse** ELECT, TELEP pulso *m* de sincronización; ~ **recovery** ELECT, TELEP, TRANS recuperación *f* del ritmo; ~**-recovery circuit** ELECT, TRANS circuito *m* de recuperación de la temporización; ~ **relationship** TRANS relación *f* de ritmo; ~ **relay** ELECT, PARTS relé *m* de temporización; ~ **signal** SIG señal *f* de temporización

tin *n* PARTS estaño *m*

tinned: ~ **conductor** *n* ELECT, PARTS conductor *m* estañado; ~ **copper** *n* ELECT, PARTS cobre *m* estañado; ~**-copper braid shield** *n* ELECT, PARTS blindaje *m* de trenza de cobre estañado; ~ **wire** *n* GEN alambre *m* estañado

tinning *n* PARTS estañado *m*

tinsel: ~ **conductor** *n* ELECT, PARTS conductor *m* de oropel

tip *n* ELECT, PARTS *wire*, TELEP hilo *m* de punta; ~ **node** COMP nodo *m* terminal

titanium: ~ **waveguide channel** *n* TRANS canal *m* guíaondas con difusión de titanio

title *n* COMP, DATA título *m*; ~ **bar** COMP barra *f* de título; ~ **page** COMP portada *f*; ~ **space** GEN espacio *m* para título

T-junction *n* RADIO cruce *m* de carreteras en T

TLMA *abbr* (*telematic agent*) DATA, TELEP agente *m* telemático

TLX *abbr* (*telex*) DATA, RADIO, TELEP TLX (*télex*)

TLXAU *abbr* (*telex access unit*) DATA, RADIO, TELEP TLXAU (*unidad de acceso télex*)

TMM *abbr* (*transverse magnetic mode*) RADIO, TRANS TMM (*modo magnético transversal*)

TMN *abbr* (*telecommunications management network*) DATA, TELEP, TRANS red *f* de gestión de telecomunicaciones

TMUX *abbr* (*transmultiplexer*) ELECT, TELEP, TRANS transmultiplexor *m*

TMUX-P *abbr* (*type-P transmultiplexer*) TELEP TMUX-P (*transmultiplexor tipo P*)

TMUX-S *abbr* (*type-S transmultiplexer*) TELEP TMUX-S (*transmultiplexor tipo S*)

TNC *abbr* (*terminal node controller*) TRANS TNC (*controlador de nodo terminal*)

TO *abbr* (*telegraph office*) TELEP TO (*oficina de telégrafos*)

TOA *abbr* (*type of address*) DATA, TELEP TOA (*tipo de dirección*)

toggle[1]: ~ **menu** *n* COMP, ELECT, POWER, SWG menú *m* de conmutación; ~ **switch** *n* COMP, ELECT, POWER, SWG interruptor *m* de codillo

toggle[2] *vt* COMP, ELECT, POWER, SWG bascular

toggling *n* COMP, ELECT, POWER, SWG conmutación *f*

token *n* GEN testigo *m*; ~ **passing** DATA relevo *m* de ficha, TRANS pase *m* de testigo; ~ **ring** COMP, DATA, TRANS red *f* en anillo; ~ **ring network** COMP, DATA, TRANS red *f* de ficha circulante

tolerance *n* TEST tolerancia *f*; ~ **limit** TEST límite *m* de tolerancia; ~ **parameter** TEST parámetro *m* de tolerancia; ~ **range** TEST intervalo *m* de tolerancia; ~ **zone** TEST zona *f* de tolerancia

toll: ~ **call** *n* TELEP llamada *f* interurbana; ~**-call barring** *n* TELEP bloqueo *m* de llamada interurbana; ~**-free call** *AmE n* (*cf freephone call BrE*) TELEP comunicación *f* interurbana no tasada; ~**-free number** *AmE n* (*cf freephone number BrE*) TELEP número *m* de llamada gratuita; ~**-free service** *AmE n* (*cf freephone service BrE*) TELEP servicio *m* sin cargo; ~ **switch** *n* SWG cuadro *m* interurbano; ~ **system** *n* APPL sistema *m* de tarificación; ~ **ticket** *n* TELEP tarjeta *f* de tarificación; ~ **ticketing** *n* TELEP tasación *f* automática con tarjetas

TON *abbr* (*type of number*) TELEP tipo *m* de número

tonal *adj* COMP, PER, TELEP tonal

T1: ~ **carrier** *n* (*T1C*) TRANS portadora *f* T1 (*T1C*); ~ **digital carrier** *n* TRANS portadora *f* digital T1

tone *n* GEN tono *m*; ~ **above band** RADIO tono *m* por encima de la banda; ~ **below band** RADIO tono *m* por debajo de la banda; ~ **call** RADIO, TELEP llamada *f* por tono; ~**-coded carrier** RADIO portadora *f* de tono; ~**-coded information** TELEP información *f* de tono; ~**-code sender** TELEP emisor *m* de código de tono; ~ **colour** *BrE* COMP timbre *m*; ~ **control** COMP, CONTR control *m* de tono; ~ **generation** COMP, ELECT generación *f* de tonos; ~ **generator** COMP, ELECT generador *m* de tono; ~ **in band** RADIO banda *f* de tono incorporado; ~ **message** TELEP información *f* de tono; ~**-off**

condition COMP, ELECT, RADIO, TELEP condición *f* tono ausente; **~-on condition** COMP, ELECT, RADIO, TELEP condición *f* tono presente; **~ on hold** TELEP tono *m* de retención de llamada; **~ pager** GEN localizador *m* de tono; **~-plus-voice paging** RADIO busca *f* de personas por tono más voz; **~ ringing** TELEP generación *f* de señal de llamada; **~-and-ringing set** SIG, TELEP dispositivo *m* de señales de llamada y de tonos; **~ signal** RADIO señal *f* acústica; **~ signalling** *BrE* SIG, TELEP señalización *f* de tonalidad; **~ squelch system** RADIO sistema *m* de sintonía silenciosa; **~ wedge** TELEP escala *f* de matices

tones: **~ and auxiliaries group** *n* TELEP *in CSN* grupo *m* de tonos y auxiliares

tool *n* GEN herramienta *f*; **~ cabinet** GEN armario *m* de herramientas; **~ palette** COMP, DATA bandeja *f* de herramientas; **~ steel** PARTS acero *m* para herramientas

toolbox *n* COMP, DATA *Internet* barra *f* de herramientas

tooling: **~ hole** *n* ELECT agujero *m* de posición; **~ pin** *n* ELECT, PARTS pivote *m* guía; **~ radius** *n* ELECT, PARTS radio *m* de estampación

toolkit *n* PARTS juego *m* de herramientas

top: **~ capacitive loading** *n* RADIO capacidad *f* terminal; **~ field** *n* TV *MPEG* campo *m* terminal; **~ MCS provider** *n* DATA, TELEP proveedor *m* de MCS terminal; **~ node** *n* COMP nodo *m* terminal; **~ panel** *n* COMP panel *m* superior

topic *n* COMP tópico *m*

topocentric: **~ angle** *n* RADIO ángulo *m* topocéntrico

topology *n* ELECT topología *f*

torn: **~-tape relay** *n* TELEP, TRANS escala *f* por cinta cortada

toroidal: **~ antenna** *n* RADIO antena *f* toroidal; **~ core** *n* ELECT núcleo *m* toroidal

torque: **~ differential receiver** *n* POWER receptor *m* por diferencial de torsión

torsion: **~-head wattmeter** *n* TEST vatímetro *m* de cabeza de torsión

Torx: **~ screwdriver** *n* PARTS destornillador *m* Torx

total: **~ anode-power input** *n* POWER, ELECT potencia *f* total anódica de entrada; **~ board thickness** *n* ELECT, PARTS espesor *m* total de la placa; **~ charge** *n* APPL, COMP, RADIO, TELEP cómputo *m* total; **~ configuration** *n* SWG configuración *f* total; **~ cost** *n* APPL, COMP, RADIO, TELEP coste *m* (*AmL* costo) total; **~ holding time** *n* TELEP tiempo *m* total de retención; **~ internal reflection** *n* ELECT, TRANS *fibre optics* reflexión *f* interna total; **~ leakage power** *n* ELECT, POWER potencia *f* de fuga total; **~ level** *n* COMP nivel *m* total; **~ loss** *n* RADIO *of link* pérdida *f* total; **~ plate-power input** *n* ELECT entrada *f* de alimentación anódica total, POWER potencia *f* total anódica de entrada; **~ quality management** *n* (*TQM*) TEST gestión *f* de calidad total (*TQM*); **~ reflection** *n* ELECT, TRANS *fibre optics* reflexión *f* total; **~ reflector current** *n* ELECT corriente *f* de reflexión total; **~ scanning-line length** *n* TELEP longitud *f* de línea de exploración total; **~ starting time** *n* ELECT tiempo *m* de inicio total; **~ transmission time** *n* TELEP, TRANS tiempo *m* de transmisión total

touch: **~ control** *n* COMP control *m* táctil; **~ response** *n* COMP respuesta *f* a la llamada por tonos; **~ screen** *n* COMP pantalla *f* táctil; **~-screen kiosk** *n* COMP subestación *f* de pantalla táctil; **~-tone telephone** *n* TELEP teléfono *m* de botonera; **~-up lacquer** *n* PARTS laca *f* de retoque; **~-up paint** *n* PARTS pintura *f* de retoque

tough *adj* PARTS duro

toughness *n* PARTS dureza *f*

tower: **~ body** *n* POWER fuste *m*

TP *abbr* (*transaction processing*) DATA procesado *m* de transacciones

TPA *abbr* (*telematic protocol architecture*) TELEP TPA (*arquitectura de protocolo telemático*)

TPASE *abbr* (*transaction processing application service element*) DATA TPASE (*elemento del servicio de aplicación de procesamiento de transacciones*)

TPDU *abbr* (*transport protocol data unit*) DATA unidad *f* de transferencia de datos del protocolo

TPE *abbr* (*transmission path endpoint*) DATA, TRANS punto *m* final de la vía de transmisión

TPI *abbr* (*tracks per inch*) COMP PPP (*pistas por pulgada*)

TPIWF *abbr* (*telex packet interworking function*) DATA TPIWF (*función de interfuncionamiento de paquetes de télex*)

TPPM *abbr* (*transaction processing protocol machine*) DATA TPPM (*máquina de protocolo de procesamiento de transacciones*)

TPSP *abbr* (*transaction processing service provider*) DATA TPSP (*proveedor del servicio de procesamiento de transacciones*)

TPSU *abbr* (*transaction processing service user*) DATA TPSU (*usuario del servicio de procesamiento de transacciones*); **~ invocation** *n* DATA invocación *f* de TPSU; **~ title** *n* DATA título *m* de TPSU

TPSUI *abbr* (*transaction processing service user invocation*) DATA TPSUI (*invocación del usuario del servicio de procesamiento de transacciones*)

TQM *abbr* (*total quality management*) TEST TQM (*gestión de calidad total*)

TR *abbr* (*tributary*) DATA, TELEP tributario *m*; ~ **cell** *n* RADIO célula *f* TR; ~ **switch** *n* RADIO, SWG, TRANS conmutador *m* TR; ~ **tube** *n* ELECT, RADIO tubo *m* ER

trace[1]: ~ **interrupt** *n* SWG interrupción *f* de rastreo; ~ **level** *n* SWG nivel *m* de rastreo; ~ **unit** *n* SWG unidad *m* de rastreo

trace[2] *vt* RADIO, TEST rastrear; ◆ ~ **a call** TELEP localizar una llamada; ~ **a fault** DATA, TELEP, TEST localizar una avería

traceability *n* ELECT rastreabilidad *f*, SWG trazabilidad *f*, TEST rastreabilidad *f*

traceable: ~ **tone** *n* ELECT tono *m* rastreable

tracer: ~ **bit** *n* SWG bit *m* de rastreo

tracing: ~ **function** *n* RADIO función *f* de seguimiento

track *n* COMP, DATA, RADIO, TV pista *f*; ~ **density** COMP densidad *f* de pista; ~**-descriptor sector** COMP sector *m* de descriptor de pista; ~ **mix** COMP combinación *f* de pista; ~ **number** COMP número *m* de pista; ~ **pitch** COMP distancia *f* entre pistas; ~ **production** RADIO producción *f* transversal; ~ **type** COMP tipo *m* de pista

trackball *n* COMP boleador *m*

tracker *n* ELECT indicador *m* de recorrido, RADIO rastreador *m*

tracking *n* GEN seguimiento *m*; ~ **antenna** RADIO antena *f* de rastreo; ~ **and command** (*TC*) TEST seguimiento *m* y mando (*TC*); ~ **error** RADIO error *m* de seguimiento; ~ **symbol** COMP símbolo *m* de seguimiento; ~ **system** RADIO sistema *m* de seguimiento; ~, **telemetry and command** (*TTC*) RADIO rastreo, *m* telemetría y órdenes (*TTC*)

tracks: ~ **per inch** *n pl* (*TPI*) COMP pistas *f pl* por pulgada (*PPP*)

trade: ~**-data interchange** *n* (*TDI*) DATA intercambio *m* de datos comerciales

trader: ~**'s turret** *n* TELEP consola *f* de carguero

traditional: ~ **telephone network** *n* TELEP, TRANS red *f* telefónica tradicional

traffic *n* GEN tráfico *m*; ~ **analyser** *BrE* DATA, TELEP, TEST analizador *m* de tráfico; ~ **analysis** DATA, TELEP, TEST análisis *m* de tráfico; ~ **area** RADIO zona *f* de tráfico; ~ **bureau** RADIO oficina *f* de coordinación de publicidad; ~ **capacity** TRANS capacidad *f* de tráfico; ~ **carried** TELEP tráfico *m* conducido; ~**-carrying burst** DATA ráfaga *f* portadora de tráfico; ~**-carrying device** DATA, TELEP dispositivo *m* de conducción de tráfico; ~ **channel** DATA, RADIO canal *m* de tráfico; ~ **circuit** TELEP circuito *m* de tráfico; ~ **control** DATA, TELEP control *m* de tráfico; ~ **controller** DATA, TELEP controlador *m* de tráfico; ~ **demand** RADIO demanda *f* de tráfico; ~ **density** TELEP densidad *f* de tráfico; ~ **distribution** SWG, TELEP distribución *f* de tráfico; ~**-distribution imbalance** SWG, TELEP desequilibrio *m* de distribución de tráfico; ~**-division system** SWG sistema *m* de división de tráfico; ~ **flow** DATA, TELEP volumen *m* de tráfico; ~**-flow control** DATA, TELEP control *m* de intensidad de tráfico; ~ **growth** TELEP aumento *m* de tráfico; ~ **handling** DATA tramitación *f* de tráfico, TELEP encaminamiento *m* de tráfico, encauzamiento *m* de tráfico; ~**-handling capacity** DATA, TELEP capacidad *f* de manipulación de tráfico; ~**-handling level** DATA, TELEP nivel *m* de manipulación de tráfico; ~**-handling program** DATA, TELEP programa *m* de manipulación de tráfico; ~ **imbalance** SWG, TELEP desequilibrio *m* de tráfico; ~ **intensity** TELEP intensidad *f* de tráfico; ~**-intensity load** TELEP carga *f* de intensidad de tráfico; ~ **level** TELEP nivel *m* de ocupación; ~**-level counter** TELEP contador *m* de nivel de ocupación; ~ **load** SWG, TELEP intensidad *f* de tráfico; ~**-load imbalance** SWG, TELEP desequilibrio *m* de intensidad de tráfico; ~ **loading** SWG, TELEP intensificación *f* de tráfico; ~ **lost** TELEP tráfico *m* desaprovechado; ~ **matrix** TELEP matriz *f* de tráfico; ~ **measurement** ELECT, TELEP, TEST medición *f* de tráfico; ~ **meter** ELECT, TELEP, TEST contador *m* de tráfico; ~ **metering** ELECT, TELEP, TEST medición *f* de tráfico; ~ **monitor** ELECT, TELEP, TEST supervisor *m* de tráfico; ~ **monitoring** ELECT, TELEP, TEST supervisión *f* de tráfico; ~ **observation** ELECT, TELEP, TEST observación *f* de tráfico; ~**-observation desk** TELEP puesto *m* de observación de tráfico; ~ **offered** TELEP tráfico *m* ofrecido; ~**-operator position** SWG, TELEP puesto *m* de operador de tráfico; ~ **padding** DATA, SWG relleno *m* de tráfico; ~ **path** RADIO canal *m* de tráfico; ~ **pattern** TELEP estructura *f* de tráfico; ~ **peak** TELEP cresta *f* de tráfico; ~ **performance** TELEP comportamiento *m* de tráfico; ~ **record** TELEP registro *m* de tráfico; ~ **recorder** TELEP registrador *m* de tráfico; ~ **recording** TELEP registro *m* de tráfico; ~ **relation** TELEP relación *f* de tráfico; ~ **rerouting** TELEP desviación *f* de tráfico; ~ **research** TELEP investigación *f* de tráfico; ~ **route** TELEP ruta *f* de tráfico; ~**-route tester** TELEP, TEST verificador *m* de vía de tráfico; **~-route testing** (*TRT*) TELEP, TEST verificación *f* de ruta de tráfico (*TRT*); **~ routing** TELEP encaminamiento *m* de tráfico; **~-routing strategy** TELEP estrategia *f* de encaminamiento de tráfico; **~ sharing** RADIO tráfico *m* compartido; **~ signalling** *BrE* **equipment** SIG equipo *m* de señales de tráfico; **~ signalling** *BrE* **installation** SIG instalación *f* de señales de tráfico; **~-signal system** GEN sistema *m* de señales de tráfico; **~ to special services** TELEP tráfico *m* a servicios especiales; **~ statistics** *n pl* DATA, RADIO, TELEP, TRANS estadística *f* de tráfico; **~ study** TELEP estudio *m* de tráfico; **~ superintendent** DATA, TELEP jefe *m* de tráfico; **~ supervision** ELECT, TELEP, TEST supervisión *f* de tráfico; **~-supervision desk** TELEP mesa *f* de supervisión de tráfico; **~ terminal** RADIO terminal *m* de tráfico; **~ tester** TELEP, TEST probador *m* de ocupaciones; **~ theory** TELEP teoría *f* de tráfico; **~-unit price procedure** TELEP procedimiento *m* de tasación de la unidad de tráfico; **~ volume** SWG, TELEP volumen *m* de tráfico

trafficability: **~ performance** *n* TELEP comportamiento *m* de la capacidad de tráfico

trail: **~ marker** *n* COMP marcador *m* de cola

trailer: **~ record** *n* DATA registro *m* de cola

trailing: **~ contact** *n* PARTS, POWER contacto *m* de acompañamiento; **~ edge** *n* GEN flanco *m* posterior; **~-edge position** *n* PARTS, TELEP posición *f* de flanco posterior; **~ noise** *n* RADIO ruido *m* retrasado

train *n* TRANS, TV tren *m*; **~ of events** GEN serie *f* de sucesos; **~ radio system** RADIO sistema *m* de radio de un tren

transaction *n* GEN transacción *f*; **~ branch** DATA sector *m* de transacción; **~-branch identifier** DATA identificador *m* de sector de transacción; **~ commitment** DATA decisión *f* de transacción; **~ identifier** DATA, TELEP identificador *m* de transacción; **~ logging** DATA registro *m* cronológico de transacciones; **~-management subsystem** DATA subsistema *m* de gestión de transacción; **~ processing** (*TP*) DATA procesado *m* de transacciones; **~ processing application service element** (*TPASE*) DATA elemento *m* del servicio de aplicación de procesamiento de transacciones (*TPASE*); **~-processing channel** DATA canal *m* de procesamiento de transacciones; **~-processing protocol machine** (*TPPM*) DATA máquina *f* de protocolo de procesamiento de transacciones (*TPPM*); **~-processing service provider** (*TPSP*) DATA proveedor *m* del servicio de procesamiento de transacciones (*TPSP*); **~-processing service user** (*TPSU*) DATA usuario *m* del servicio de procesamiento de transacciones (*TPSU*); **~-processing service user invocation** (*TPSUI*) DATA invocación *f* del usuario del servicio de procesamiento de transacciones (*TPSUI*); **~ recovery** DATA recuperación *f* de transacción; **~ roll-back** DATA vuelta *f* atrás de la transacción

transactional: **~ set header** *n* DATA título *m* del conjunto transaccional; **~ set header in ST segment** *n* DATA segmento *m* ST del título del conjunto transaccional; **~ system** *n* DATA sistema *m* transaccional

transadmittance *n* ELECT transadmitancia *f*

transceiver *n* DATA, RADIO transceptor *m*

transcode *vt* ELECT, TELEP, TEST, TRANS transcodificar

transcoder *n* ELECT, TELEP, TEST, TRANS transcodificador *m*

transcoding *n* ELECT, TELEP, TEST, TRANS transcodificación *f*; **~ gain** ELECT ganancia *f* de transcodificación

transconductance *n* ELECT transconductancia *f*

transcriber *n* TELEP registrador *m*

transducer *n* ELECT, PARTS, POWER, RADIO, TELEP, TEST transductor *m*; **~ gain** TELEP ganancia *f* de transductor; **~ loss** TELEP atenuación *f* transductiva

transductor *n* ELECT, PARTS, POWER, RADIO, TELEP, TEST transductor *m*; **~ amplifier** POWER amplificador *m* magnético

transfer[1] *n* DATA, TELEP transferencia *f*; **~-allowed acknowledgement signal** SIG, TELEP señal *f* de reconocimiento de transferencia permitida; **~-allowed signal** SIG, TELEP señal *f* de transferencia permitida; **~-allowed procedure** TELEP procedimiento *m* de transferencia permitida; **~ block** SWG bloque *m* de transferencia; **~ bus** SWG bus *m* de transferencia; **~ channel** SWG, TELEP canal *m* de transferencia; **~ characteristic** ELECT característica *f* de transferencia; **~-charge call** TELEP llamada *f* a cobro revertido; **~ control** SWG control *m* de transferencia; **~ of control** TELEP transferencia *f* de control; **~-control block** SWG bloque *m* de control de transferencia; **~ current** ELECT corriente *f* de transferencia; **~ function** APPL, CONTR, DATA, TRANS función *f* de transferencia; **~ impedance** ELECT impedancia *f* de transferencia; **~ link** TELEP enlace *m* de transferencia; **~ message** PER APP mensaje *m* de transferencia; **~ mode** COMP modo *m* de transferencia; ~ **multiplexer** SWG multiplexor *m* de transferencia; ~ **node** TELEP nodo *m* de transferencia; ~ **prohibited** TELEP transferencia *f* prohibida; ~-**prohibited procedure** TELEP procedimiento *m* de prohibición de transferencia; ~-**prohibited signal** SIG, TELEP señal *f* de prohibición de transferencia; ~ **rate** COMP, DATA velocidad *f* de traslado; ~ **subsystem** SWG subsistema *m* de transferencia; ~ **syntax** TELEP sintaxis *f* de transferencia; ~ **system** DATA sistema *m* de transferencia; ~ **time** ELECT, TELEP tiempo *m* de transferencia; ◆ ~ **allowed** TELEP transferencia permitida; ~ **controlled** TELEP transferencia controlada; ~ **restricted** TELEP transferencia restringida

transfer[2] *vt* DATA, TELEP transferir; ◆ ~ **the charges** TELEP poner una conferencia a cobro revertido

transferred: ~-**charge call** *n* TELEP llamada *f* a cobro revertido; ~ **charging** *n* TELEP cobro *m* revertido

transform *vt* POWER transformar

transformer *n* ELECT, PARTS, POWER, RADIO, TELEP, TEST, TV transformador *m*; ~ **coupling** POWER acoplamiento *m* por transformador

transforming: ~ **section** *n* RADIO sección *f* de adaptación

transhorizon: ~ **propagation** *n* RADIO propagación *f* transhorizonte; ~ **radio relay system** *n* RADIO sistema *m* de radioenlace transhorizonte

transient *adj* GEN transitorio; ~ **decay current** *n* ELECT corriente *f* residual; ~ **fault** *n* TEST avería *f* transitoria; ~ **phenomenon** *n* GEN fenómeno *m* transitorio; ~ **radiation effect on electronics** *n* ELECT efecto *m* de radiación transitoria sobre la electrónica; ~ **suppressor** *n* POWER supresor *m* de transitorios; ~ **testing state** *n* DATA, TEST estado *m* de pruebas en régimen transitorio

transients *n pl* TRANS transitorios *m pl*

transionospheric: ~ **propagation** *n* RADIO propagación *f* transionosférica

transistor *n* ELECT, PARTS, POWER transistor *m*; ~ **analyzer** TEST analizador *m* de transistores; ~ **base** ELECT, PARTS base *f* de transistor; ~ **collector** ELECT colector *m* de transistor; ~ **emitter** ELECT emisor *m* de transistor; ~ **flip-flop** ELECT, PARTS basculador *m* transistorizado; ~-**input resistance** ELECT resistencia *f* de entrada; ~-~ **logic** (*TTL*) ELECT circuito *m* lógico de transistor a transistor (*CLTT*)

transit: ~ **administration** *n* TELEP, TRANS administración *f* de tránsito; ~ **angle** *n* ELECT ángulo *m* de tránsito; ~ **connection element** *n* (*TCE*) DATA, TRANS elemento *m* de comunicación en tránsito; ~ **connection-related function** *n* (*TCRF*) DATA, TRANS función *f* de tránsito relacionada con la conexión (*TCRF*); ~ **country** *n* TELEP país *m* de tránsito; ~-**delay indication** *n* DATA indicación *f* de retardo de tránsito; ~ **exchange** *n* DATA, SWG, TELEP central *f* de tránsito; ~-**network identification** *n* TRANS identificación *f* de red de tránsito; ~ **proceed-to-send signal** *n* SIG, TELEP señal *f* de invitación a transmitir en tráfico de tránsito; ~ **routing** *n* TELEP ruta *f* de tránsito; ~-**seizure signal** *n* SIG, TELEP señal *f* de toma de tránsito; ~ **share** *n* TELEP tasa *f* de tránsito; ~ **stage** *n* TELEP etapa *f* de tránsito; ~ **switching centre** *BrE n* DATA, SWG, TELEP centro *m* de conmutación de tránsito; ~ **through-connect signal** *n* SIG, TELEP señal *f* de tránsito de conexión directa; ~ **time** *n* ELECT, POWER, TELEP, TEST tiempo *m* de tránsito; ~ **time jitter** *n* ELECT, POWER, TELEP, TEST fluctuación *f* de fase de tiempo de tránsito; ~ **time spread** *n* ELECT, POWER, TELEP, TEST fluctuación *f* de tiempo de tránsito; ~ **traffic** *n* TELEP tráfico *m* de tránsito

transition *n* TELEP *in SDL* transición *f*

transitional: ~ **state** *n* TELEP estado *m* transitorio

transitron *n* ELECT, PARTS transitrón *m*; ~ **oscillator** ELECT, PARTS oscilador *m* transitrón

translate *vt* COMP *program code*, DATA, ELECT, TELEP, TV traducir

translation *n* COMP *of language, image*, DATA *of program code*, ELECT, TELEP, TV traducción *f*; ~ **circuit** TELEP circuito *m* de traducción; ~ **error** DATA error *m* de traducción; ~ **register** TELEP registro *m* de traducción; ~ **relay** PARTS, TELEP relé *m* de traducción; ~ **store** SWG almacenamiento *m* de traslación

translator *n* COMP *software package* traductor *m*

transliterate *vt* COMP, DATA transliterar

transmission *n* COMP, DATA, RADIO, TELEP, TRANS transmisión *f*; ~ **bearer** TRANS, TV portador *m* de transmisión; ~ **bit rate** RADIO, TRANS velocidad *f* de tráfico binario de transmisión; ~ **bit slip** RADIO, TRANS deslizamiento *m* de tráfico binario de transmisión; ~ **breakdown** TEST, TRANS corte *m* de transmisión; ~ **bridge** RADIO, TRANS puente *m* de transmisión; ~ **buffer** TELEP, TRANS compensador *m* de transmisión; ~ **capability** DATA capacidad *f* de transmisión; ~ **channel** RADIO, TRANS, TV canal *m* de transmisión; ~ **characteristic** TRANS característica *f* de transmisión; ~ **code violation** TELEP, TRANS violación *f* de código de transmisión; ~ **coefficient** TRANS coeficiente *m* de transmisión; ~ **conditions** *n pl* COMP, TRANS condiciones *f pl* de transmisión; ~ **control protocol** COMP, DATA protocolo *m* de control de transmisión; ~ **convergence** TRANS convergencia *f* de transmisión; ~ **convergence sublayer** DATOS, TRANS subcapa *f* de convergencia de transmisión; ~ **definition** TELEP, TRANS definición *f* de transmisión; ~ **delay** DATA, TELEP *through a digital exchange*, TRANS retardo *m* de transmisión; ~ **equipment** TRANS equipo *m* de transmisión; ~ **error** TEST, TRANS error *m* de transmisión; ~ **highway** TRANS camino *m* de transmisión; ~ **interface** COMP, TRANS interfaz *f* de transmisión; ~ **layer** TRANS capa *f* de transmisión; ~ **line** RADIO, TRANS línea *f* de transmisión; ~ **link** TELEP, TRANS enlace *m* de transmisión; ~ **loss** TELEP, TRANS pérdida *f* de transmisión; ~ **loss of a-t-b** TELEP, TRANS atenuación *f* de a-t-b de transmisión; TELEP puntos *m pl* de mantenimiento de transmisión-línea internacional; ~ **measurement** TEST, TRANS medida *f* de transmisión; ~ **medium** TRANS, TV medio *m* de transmisión; ~ **network** TRANS red *f* de transmisión; ~ **node** TRANS nodo *m* de transmisión; ~ **overload** TRANS sobrecarga *f* de transmisión; ~ **path** RADIO, TRANS vía *f* de transmisión; ~ **path endpoint** (*TPE*) DATA, TRANS punto *m* final de la vía de transmisión; ~ **performance** TEST, TELEP, TRANS calidad *f* de transmisión; ~-**performance rating** TELEP, TEST, TRANS índice *m* de calidad de transmisión; ~ **plan** TRANS plan *m* de transmisión; ~-**priority code** DATA, TRANS código *m* de prioridad de transmisión; ~ **quality** TELEP, TEST, TRANS calidad *f* de transmisión; ~ **recipient** DATA, TRANS recipiente *m* de transmisión; ~-**reference point** (*TRP*) TELEP, TRANS punto *m* de referencia de transmisión (*TRP*); ~ **sender** DATA, TRANS emisor *m* de transmisión; ~ **sending** TELEP, TRANS envío *m* de transmisión; ~ **speed** TRANS velocidad *f* de transmisión; ~ **technique** TRANS técnica *f* de transmisión; ~ **test** TEST, TRANS prueba *f* de transmisión; ~ **test trunk** (*TTT*) TELEP, TEST línea *f* central de pruebas de transmisión (*TTT*), TRANS línea *f* troncal de pruebas de transmisión; ~ **time** TELEP, TRANS tiempo *m* de transmisión; ~ **window** TRANS ventana *f* de transmisión

transmissometer *n* TEST, TRANS transmisómetro *m*

transmit[1]: ~ **antenna** *n* RADIO antena *f* de transmisión; ~ **buffer** *n* COMP, TRANS memoria *f* intermedia de transmisión; ~ **channel** *n* TRANS canal *m* de transmisión; ~ **characteristic** *n* RADIO, TRANS característica *f* de transmisión; ~ **data** *n pl* DATA, TRANS datos *m pl* de transmisión; ~ **delay** *n* RADIO, TRANS retardo *m* de transmisión; ~ **direction** *n* TEST, TRANS dirección *f* de transmisión; ~ **fibre-optic terminal device** *n* TRANS dispositivo *m* terminal de fibra óptica de transmisión; ~ **flow control** *n* TELEP, TRANS control *m* del flujo de transmisión; ~-**frame acquisition and synchronization procedure** *n* DATA, TRANS procedimiento *m* de adquisición y sincronización de trama transmisión; ~ **gain** *n* TRANS ganancia *f* de transmisión; ~ **machine** *n* TELEP, TRANS máquina *f* de transmisión; ~ **multiframe timing** *n* TRANS temporización *f* de multitrama de la transmisión; ~ **objective loudness rating** *n* TELEP, TRANS índice *m* de sonoridad objetiva de la transmisión; ~-**receive cell** *n* RADIO célula *f* de transmisión-recepción; ~-**receive switch** *n* RADIO, SWG, TRANS conmutador *m* T-R

transmit[2] *vt* RADIO, TELEP, TRANS transmitir

transmittance *n* ELECT, TRANS transmisión *f*

transmitted: ~ **data** *n pl* DATA datos *m pl* transmitidos; ~-**phase advance** *n* TRANS avance *m* de fase transmitida

transmitter *n* (*XMTR*) COMP, DATA, ELECT, PARTS, PER APP, RADIO, TELEP, TRANS transmisor *m*; ~ **blocker cell** RADIO, TRANS cavidad *f* TB; ~ **cap** PER APP tapa *f* microfónica; ~ **distortion** TELEP, TRANS distorsión *f* en la emisión; ~ **error** RADIO, TRANS error *m* de transmisor; ~ **inset** PER APP cápsula *f* microfónica; ~-**multiplexer** TRANS multiplexor *m* de emisor; ~ **output power** TRANS potencia *f* de salida del emisor; ~ **power control** RADIO control *m* de potencia del emisor; ~ **relay** PARTS, TRANS relé *m* de emisor; ~ **turn-on signal** PER APP, SIG, TELEP, TRANS señal *f* de activación de transmisor; ~ **turn-on time** PER APP, TELEP, TRANS tiempo *m* de encendido de transmisor

transmitting: ~ **antenna** *n* RADIO antena *f* transmisora; ~ **channel** *n* TRANS canal *m* emisor; ~ **equipment** *n* COMP, TRANS equipo *m* emisor; ~ **subscriber identification-DCS** *n* (*TCI-DCS*) DATA, TELEP DCS *m* de identificación del abonado emisor (*TCI-DCS*)

transmultiplexer *n* (*TMUX*) ELECT, TELEP,

TRANS transmultiplexor *m*; ~ **channel** ELECT, TELEP, TRANS canal *m* transmultiplexor
transparency *n* DATA, TELEP, TRANS transparencia *f*
transparent: ~ **bearer service** *n* APPL, DATA servicio *m* transparente del portador; ~ **data-transfer phase** *n* DATA, TELEP, TRANS fase *f* de transferencia de datos transparentes; ~**-designation labelling** *BrE* **strip** *n* ELECT banda *f* de etiquetado de designación transparente, PARTS banda *f* de rotulación de designación transparente; ~ **mode** *n* DATA modalidad *f* transparente; ~ **repeater** *n* GEN repetidor *m* transparente
transponder *n* DATA, ELECT, RADIO respondedor *m*; ~ **allotment** RADIO atribución *f* de baliza respondedora; ~ **hopping** RADIO trayecto *m* de respondedor; ~ **reply efficiency** RADIO eficiencia *f* de respuesta de respondedor; ~ **system** DATA sistema *m* respondedor
transport: ~ **connection** *n* DATA, TELEP conexión *f* de transporte; ~ **connection end point identifier** *n* DATA, TELEP identificador *m* de punto final de conexión de transporte; ~ **entity** *n* DATA entidad *f* de transporte; ~ **handle** *n* GEN asa *f*; ~ **layer** *n* DATA, TRANS capa *f* de transmisión; ~ **locking device** *n* GEN dispositivo *m* de fijación para el transporte; ~ **network** *n* TELEP red *f* de transporte; ~ **protocol data unit** *n* (*TPDU*) DATA unidad *f* de transferencia de datos de protocolo; ~ **service** *n* DATA, TELEP, TRANS servicio *m* de transporte; ~**-service access point** *n* (*TSAP*) DATA, TELEP, TRANS punto *m* de acceso al servicio de transporte; ~**-service access-point address** *n* DATA dirección *f* de punto de acceso al servicio de transporte, TELEP, TRANS dirección *f* del punto de acceso al servicio de transporte; ~**-service data unit** *n* (*TSDU*) DATA, TELEP, TRANS unidad *f* de datos del servicio de transporte (*TSDU*); ~ **unit** *n* GEN unidad *f* de transporte
transportability *n* TEST transportabilidad *f*
transportable: ~ **telephone** *n* TELEP teléfono *m* portátil
transpose *vt* COMP, ELECT, TELEP, TRANS transponer
transposed: ~ **line** *n* TELEP línea *f* con transposiciones, TRANS línea *f* de transmisión; ~ **transmission line** *n* TELEP línea *f* de transmisión con transposiciones, TRANS línea *f* de transmisión
transposition *n* COMP, ELECT, TELEP, TRANS transposición *f*; ~ **method of measurement** TEST método *m* de transmisión por transposición
transputer *n* COMP, DATA transordenador *m*
transversal: ~ **filter** *n* TRANS filtro *m* transversal
transverse: ~ **conversion loss** *n* (*TCL*) TELEP pérdida *f* de conversión transversal (*TCL*); ~ **conversion ratio** *n* TELEP razón *f* de conversión transversal; ~ **electric wave** *n* RADIO, TRANS onda *f* eléctrica transversal; ~ **gyrofrequency** *n* RADIO girofrecuencia *f* transversal; ~ **interferometry** *n* TRANS interferometría *f* transversal; ~ **judder** *n* TELEP salto *m* transversal, TRANS trepidación *f* transversal; ~ **junction** *n* TELEP enlace *m* transversal; ~ **magnetic mode** *n* (*TMM*) RADIO, TRANS modo *m* magnético transversal (*TMM*); ~ **mode** *n* TRANS modo *m* transverso; ~ **offset loss** *n* TRANS pérdida *f* por desplazamiento transversal; ~ **redundancy check** *n* DATA verificación *f* transversal por redundancia; ~ **reflection factor** *n* TELEP factor *m* de reflexión transversal; ~ **return loss** *n* (*TRL*) TELEP atenuación *f* de retorno transversal (*TRL*); ~ **route** *n* TELEP ruta *f* transversal; ~ **section** *n* TRANS sección *f* transversal; ~ **slot** *n* TRANS ranura *f* transversal; ~ **test method** *n* TEST método *m* de comprobación transversal; ~ **traffic** *n* TELEP tráfico *m* transversal; ~ **voltage** *n* POWER, TELEP *of protector* tensión *f* transversal; ~ **wave** *n* RADIO, TRANS onda *f* transversal
trap *n* COMP, ELECT, PER APP, RADIO trampa *f*; ~ **circuit** ELECT circuito *m* eliminador
trapezoidal: ~ **distortion** *n* ELECT distorsión *f* trapezoidal
trapped: ~ **mode** *n* RADIO, TRANS modo *m* de propagación guiado
trash *n* COMP información *f* parásita
travel *vt* COMP recorrer
travelling *BrE* (*AmE* **traveling**) *adj* ELECT, TRANS progresivo; ~ **wave** *n* RADIO, TRANS onda *f* progresiva; ~**-wave aerial** *n* RADIO, TRANS antena *f* de onda progresiva; ~**-wave amplifier** *n* ELECT, RADIO, TRANS amplificador *m* de ondas progresivas; ~**-wave antenna** *n* RADIO, TRANS antena *f* de onda progresiva; ~ **waveguide** *n* ELECT, RADIO, TRANS guíaondas *m* progresiva; ~**-waveguide amplifier** *n* ELECT, RADIO, TRANS amplificador *m* de guíaondas progresivo; ~**-wave tube** *n* ELECT, RADIO, TRANS tubo *m* de ondas progresivas; ~**-wave tube amplifier** *n* ELECT, RADIO, TRANS amplificador *m* de tubo de ondas progresivas; ~**-wave valve** *n* ELECT, RADIO, TRANS válvula *f* de onda progresiva

tree *n* COMP, DATA, ELECT árbol *m*; ~ **attachment** COMP, DATA unión *f* en árbol; ~ **and branch architecture** COMP, DATA arquitectura *f* en árbol y en rama; ~ **and branch network** COMP, DATA, TRANS red *f* ramificada; ~ **and branch structure** COMP, DATA, ELECT estructura *f* en árbol y en rama; ~ **distribution** TRANS distribución *f* ramificada; ~ **expansion** COMP, DATA expansión *f* en árbol; ~ **header** COMP, DATA cabecera *f* de árbol; ~ **identifier** COMP, DATA identificador *m* de árbol; ~ **leaf** COMP, DATA hoja *f* de árbol; ~ **network** COMP, DATA, TRANS red *f* en árbol; ~ **node** COMP, DATA nodo *m* de árbol; ~ **notation** COMP, DATA notación *f* en árbol; ~ **structure** COMP, DATA, TRANS estructura *f* en árbol; **~-structured view** COMP, DATA representación *f* estructurada en árbol; ~ **and tabular combined notation** (*TTCN*) COMP, DATA notación *f* combinada en árbol y tabular (*TTCN*); ~ **topology** COMP, DATA topología *f* de árbol

treed *adj* COMP, DATA, ELECT arborizado

trellis: **~-coded modulation** *n* (*TCM*) DATA, TRANS modulación *f* codificada en enrejado (*TCM*); ~ **coding** *n* DATA, TRANS codificación *f* en enrejado

tremolo *n* COMP trémolo *m*; ~ **control** COMP control *m* de trémolo

trend *n* GEN tendencia *f*

TRF: ~ **reception** *n* (*tuned radio frequency reception*) RADIO recepción *f* de radiofrecuencia sintonizada

triac *n* ELECT triac *m*, PARTS tiristor *m* bidireccional

trial *n* RADIO, TEST prueba *f*

triangular: ~ **pulse** *n* ELECT impulso *m* triangular; ~ **random noise** *n* RADIO ruido *m* errático triangular

triaxial: ~ **cable** *n* PARTS cable *m* triaxial

tributary[1] *adj* (*TR*) DATA, TELEP tributario; ~ **bearing** *n* RADIO acimut *m* astronómico; **~-motion display** *n* RADIO visualización *f* de movimiento verdadero; ~ **rms voltmeter** *n* TEST voltímetro *m* de media cuadrática verdadera; ~ **root mean square voltmeter** *n* TEST voltímetro *m* de media cuadrática verdadera

tributary[2] *n* (*TR*) DATA, TELEP tributario *m*

trickable: ~ **switch** *n* POWER, SWG conmutador *m* compensable

trickle: ~ **charge** *n* POWER carga *f* lenta; ~ **charger** *n* POWER carga *f* lenta

trigatron *n* ELECT trigatrón *m*

trigger[1] *n* COMP, ELECT *device*, TEST disparo *m*; ~ **circuit** ELECT circuito *m* activador; ~ **electrode** ELECT, PARTS electrodo *m* de disparo; ~ **gap** ELECT espacio *m* de cebado; ~ **input** ELECT entrada *f* disparadora; ~ **level** COMP nivel *m* de disparo; ~ **mechanism** ELECT mecanismo *m* de desconexión; ~ **nozzle** ELECT boquilla *f* disparadora; ~ **play** COMP juego *m* de disparo; ~ **pulse** COMP, ELECT impulso *m* activador; ~ **tube** ELECT, PARTS tubo *m* de impulsos electrónicos

trigger[2] *vt* COMP, ELECT *device*, TEST disparar

triggered: ~ **flip-flop** *n* ELECT, PARTS basculador *m* gatillado; ~ **time-base** *n* ELECT base *f* de tiempos gatillada

triggering *n* COMP, ELECT, TEST disparo *m*; ~ **lead pulse** ELECT impulso *m* disparador de línea; ~ **waveform** RADIO forma *f* de onda activadora

trim *vt* COMP, ELECT, POWER recortar

trimmer: ~ **capacitor** *n* ELECT, PARTS compensador *m* de sintonía

trimming *n* COMP, ELECT, POWER ajuste *m* fino; ~ **potentiometer** ELECT, PARTS potenciómetro *m* de ajuste; ~ **resistor** ELECT, PARTS resistor *m* de ajuste

Trinitron *n* COMP Trinitrón *m*

triode *n* ELECT, PARTS triodo *m*; ~ **field-effect transistor** ELECT, PARTS transistor *m* triodo de efecto de campo

trip[1] *n* ELECT dispositivo *m* de disparo

trip[2] *vt* ELECT, POWER *circuit breakers* enganchar; ◆ ~ **the ringing** TELEP disparar la llamada

triple: **~-access line** *n* SWG, TELEP línea *f* de triple acceso; **~-pass technology** *n* DATA *of scanner* tecnología *f* de paso triple

tristate *adj* ELECT de triestado

TRL *abbr* (*transverse return loss*) TELEP TRL (*atenuación de retorno transversal*)

trombone *n* PARTS trombón *m*; ~ **connection** TELEP conexión *f* trombón; ~ **loop connection** TELEP conexión *f* en bucle de trombón

tropicalized: ~ **equipment** *n* CLIM, PARTS equipo *m* tropicalizado

tropopause *n* RADIO tropopausa *f*

troposphere *n* RADIO troposfera *f*

tropospheric: ~ **mode** *n* RADIO modo *m* troposférico; ~ **propagation** *n* RADIO propagación *f* troposférica; ~ **radioduct** *n* RADIO radioducto *m* troposférico; ~ **reflection** *n* RADIO reflexión *f* troposférica; ~ **scatter** *n* RADIO dispersión *f* troposférica; ~ **wave** *n* RADIO onda *f* troposférica

trouble: **~-free maintenance** *n* ELECT mantenimiento *m* libre de averías; ~ **report** *n* TELEP relación *f* de incidentes

troubleshooting *n* COMP, ELECT, TELEP localización y reparación de averías

trough *n* ELECT artesa *f*
troughing *n* ELECT atarjea *f*
TRP *abbr* (*transmission reference point*) TELEP, TRANS TRP (*punto de referencia de transmisión*)
TRT *abbr* (*traffic-route testing*) TELEP, TEST TRT (*verificación de ruta de tráfico*)
true *adj* COMP verdadero, ELECT, *logic function*, RADIO real, TEST puro; ~ **bearing** *n* RADIO acimut *m* astronómico; ~-**motion display** *n* RADIO visualización *f* de movimiento verdadero; ~ **rms voltmeter** *n* TEST voltímetro *m* de media cuadrática verdadera; ~ **root mean square voltmeter** *n* TEST voltímetro *m* de media cuadrática verdadera
trueness *n* TEST exactitud *f*
Truline: ~ **wattmeter** *n* TEST vatímetro *m* de Truline
truncate *vt* COMP, DATA truncar
truncation *n* COMP, DATA ruptura *f*, truncación *f*
trunk *n* RADIO línea *f*; ~-**access-barred extension** TELEP extensión *f* bloqueada a la salida del tráfico interurbano; ~-**access-restricted extension** TELEP extensión *f* bloqueada al tráfico interurbano saliente; ~ **barring** TELEP bloqueo *m* del tráfico interurbano; ~ **cable** ELECT, TRANS cable *m* de unión; ~ **call** TELEP comunicación *f* interurbana; ~ **channel** TRANS canal *m* principal; ~ **circuit** ELECT, SWG, TELEP, TRANS circuito *m* troncal; ~ **circuit block** ELECT, SWG, TELEP, TRANS bloque *m* de circuitos troncales; ~ **code** (*TC*) TELEP código *m* interurbano; ~ **connection** TELEP conexión *f* interurbana; ~ **control station** TELEP estación *f* de control interurbano; ~ **dialling** *BrE* TELEP tráfico *m* interurbano automático; ~ **discrimination** TELEP bloqueo *m* de tráfico interurbano; ~ **discrimination control** CONTR, TELEP control *m* de bloqueo del tráfico interurbano; ~ **distribution frame** TRANS repartidor *m* principal; ~ **equipment** TRANS equipo *m* de central interurbana; ~ **exchange** SWG, TELEP, TRANS central *f* interurbana; ~ **flow** SWG, TELEP flujo *m* central; ~ **group** TELEP, TRANS grupo *m* de líneas interurbanas; ~**line** TELEP línea *f* interurbana; ~-**line relay set** TELEP juego *m* de relés de línea interurbana; ~ **marking** TELEP marcación *f* interurbana; ~ **network** SWG, TELEP red *f* interurbana; ~ **offer** SWG, TELEP enlace *m* de oferta; ~ **offering** SWG ofrecimiento *m* interurbano, TELEP aviso *m* de llamada; ~ **ordering service** TELEP servicio *m* de anotaciones; ~ **PE** TELEP PE *m* interurbano; ~ **prefix** TELEP prefijo *m* interurbano; ~ **record position** TELEP posición *f* de anotación, TELEP puesto *m* de anotación; ~-**routed call** TELEP llamada *f* encaminada por línea interurbana; ~ **switching** SWG conexión *f*; ~ **switching centre** *BrE* SWG centro *m* de conexión; ~-**switching exchange area** SWG zona *f* telefónica de conexiones; ~-**switching stage** SWG, TELEP paso *m* de conmutación interurbana; ~ **system** RADIO, TELEP sistema *m* de enlaces; ~ **telephone circuit** TELEP, TRANS circuito *m* telefónico interurbano; ~ **traffic** TELEP tráfico *m* interurbano; ~ **transit exchange** SWG central *f* interurbana de tránsito; ~-**unrestricted extension** TELEP extensión *f* abierta al tráfico interurbano; ◆ ~ **busy** TELEP ocupado por una comunicación interurbana; ~ **free** TELEP línea desocupada; ~ **seized** TELEP línea ocupada
trunked: ~ **dispatch system** *n* RADIO sistema *m* de expedición de enlaces; ~ **radio** *n* POWER, RADIO radio *f* enlazada; ~ **radio network** *n* RADIO red *f* de radio enlazada; ~ **system** *n* RADIO sistema *m* enlazado
trunking *n* RADIO circuito *m* troncal; ~ **diagram** TELEP esquema *m* de vías troncales; ~ **system** RADIO sistema *m* de vías troncales
trusted: ~ **functionality** *n* DATA, TEST funcionalidad *f* fiable
truth *n* COMP verdad *f*, ELECT validez *f*, RADIO, TEST veracidad *f*; ~ **table** ELECT tabla *f* de verdad
TRV *abbr* (*tape-relay vehicle*) TV TRV (*vehículo de retransmisión por cinta*)
TS: ~ **network** *n* TELEP red *f* TS
TSAP *abbr* (*transport-service access point*) DATA, TELEP, TRANS punto *m* de acceso al servicio de transporte
TSDU *abbr* (*transport-service data unit*) DATA, TELEP, TRANS TSDU (*unidad de datos del servicio de transporte*)
TSM *abbr* (*terminal server manager*) DATA TSM (*gestor de servidor terminal*)
TT & C *abbr* (*tracking, telemetry and command*) RADIO TT & C (*rastreo, telemetría y órdenes*)
TTCN *abbr* (*tree and tabular combined notation*) COMP, DATA TTCN (*notación combinada en árbol y tabular*); ~ **statement** *n* COMP, DATA instrucción *f* TTCN
TTD *abbr* (*target transit delay*) DATA TTD (*retardo de tránsito del blanco*)
TTG: ~ **laser** *n* ELECT láser *m* TTG
TTL *abbr* (*transistor-transistor logic*) ELECT CLTT (*circuito lógico de transistor a transistor*)
TTT *abbr* (*transmission test trunk*) TELEP, TEST TTT (*línea central de pruebas de transmisión*),

TRANS línea *f* troncal de pruebas de transmisión

TTU *abbr* (*tape transport unit*) APPL unidad *f* de arrastre de la cinta, COMP TTU (*unidad de arrastre de la cinta*), TV unidad *f* de arrastre de la cinta

TTX *abbr* (*teletext*) APPL, TV TTX (*teletexto*)

TTXAU *abbr* (*teletex access unit*) APPL TTXAU (*unidad de acceso a teletexto*)

TTY *abbr* (*teletype*) COMP, PER APP, RADIO, TELEP TTY (*teletipo*)

tube *n* ELECT, TRANS tubo *m*; ~ **efficiency** ELECT, TEST eficiencia *f* en el tubo; ~ **of force** POWER tubo *m* de campo, tubo *m* de fuerza; ~ **fuse** ELECT, PARTS, POWER fusible *m* tubular; ~ **noise** ELECT ruido *m* de tubo; ~ **socket** PARTS portatubo *m*, portaválvulas *m*; ~ **starting time** ELECT hora *f* de arranque del tubo; ~ **voltage drop** ELECT caída *f* de tensión en un tubo; ~ **wrap** PARTS envoltura *f* del tubo

tubing *n* ELECT, PARTS, POWER sistema *m* de tuberías

tubular: ~ **capacitor** *n* PARTS, POWER capacitor *m* tubular

tumbler: ~ **switch** *n* POWER interruptor *m* basculante

tunable: ~ **twin-guide laser** *n* ELECT láser *m* de doble guía sintonizable; ~ **voltmeter** *n* TEST voltímetro *m* sintonizable

tune *vt* COMP, ELECT, RADIO sintonizar

tuned: ~ **circuit** *n* POWER circuito *m* sintonizado; ~ **radio frequency reception** *n* (*TRF reception*) RADIO recepción *f* de radiofrecuencia sintonizada

tuner *n* COMP, ELECT, RADIO sintonizador *m*; ~ **circuit** ELECT circuito *m* sintonizador

tuning *n* COMP, ELECT, RADIO sintonización *f*; ~ **capability** COMP, ELECT, RADIO capacidad *f* sintonizadora; ~ **circuit** ELECT circuito *m* sintonizador; ~ **fork** COMP diapasón *m*; ~-**fork oscillator drive** RADIO oscilador *m* de diapasón; ~ **function** COMP función *f* de sintonización; ~ **position meter** ELECT medidor *m* de posición sintonizadora; ~ **sensitivity** ELECT sensibilidad *f* sintonizadora

tunnel: ~ **action** *n* PARTS efecto *m* túnel; ~ **diode** *n* ELECT, PARTS diodo *m* túnel; ~ **effect** *n* PARTS efecto *m* túnel

tunnelling *BrE* (*AmE* **tunneling**) *n* ELECT efecto *m* túnel; ~ **mode** *BrE* TRANS modo *m* de efecto túnel; ~ **ray** *BrE* TRANS rayo *m* de efecto túnel

turn[1] *n* PARTS vuelta *f*; ~-**off delay** PARTS retardo *m* de desconexión; ~-**off program** COMP programa *m* de desconexión; ~-**off thyristor** PARTS tiristor *m* de desconexión; ~-**off time** PARTS tiempo *m* de desconexión; ~-**on time** PARTS momento *m* de conexión; ~ **ratio** PARTS relación *f* de rotación

turn[2] *vt* PARTS tornear

turn on *vt* COMP *device*, ELECT, PER APP *device, machine*, POWER *supply*, RADIO, SWG abrir

turn off *vt* COMP *device*, ELECT, PER APP *device, machine*, POWER *supply*, RADIO, SWG cerrar

turnaround: ~ **time** *n* COMP, DATA tiempo *m* de retorno

turnbuckle *n* PARTS tensor *m*

turning *n* PARTS viraje *m*; ~ **relay** PARTS relé *m* de retorno; ~ **tool** PARTS herramienta *f* torneadora

turnstile: ~ **whip** *n* RADIO antena *f* de látigo en torniquete

turret *n* ELECT torreta *f*; ~ **lathe** PARTS torno *m* revólver

tutorial *n* COMP, DATA programa *m* de instrucción

TV *abbr* (*television*) RADIO, TV TV (*televisión*)

TVRO: ~ **antenna** *n* RADIO, TV antena *f* TVRO; ~ **terminal** *n* TV terminal *m* TVRO

twenty: ~-**four-hour hotline** *n* TELEP línea *f* de atención permanente

twin: ~ **AX** *n* ELECT, PARTS AX *m* doble; ~ **cable** *n* ELECT, PARTS cable *m* bipolar; ~-**channel stereo telephony** *n* TELEP telefonía *f* estereofónica de dos canales; ~ **conductor** *n* ELECT, PARTS conductor *m* dúplex; ~-**conductor cable** *n* ELECT, PARTS cable *m* de conductor dúplex; ~ **contacts** *n pl* PARTS contactos *m pl* dobles; ~ **group** *n* GEN grupo *m* doble

twine *vt* PARTS retorcer

twinning *n* ELECT apareamiento *m*, TV emparejamiento *m*

twins *n pl* COMP gemelos *m pl*

twin-T: ~ **oscillator** *n* ELECT, PARTS oscilador *m* de doble T

twist *n* ELECT, PARTS alabeo *m*; ~ **drill** PARTS broca *f* helicoidal

twisted *adj* ELECT, PARTS retorcido; ~ **nematic** *n* ELECT nemática *f* torcida; ~ **pair** *n* ELECT, TRANS conductor *m* doble torcido; ~-**pair cable** *n* ELECT, PARTS, TRANS cable *m* de par trenzado; ~-**pair cabling** *n* ELECT, PARTS, TRANS cableado *m* por pares; ~-**pair hub** *n* ELECT, PARTS cubo *m* de par retorcido; ~ **waveguide** *n* DATA, RADIO, TRANS guía *f* de ondas revirada; ~ **wire** *n* ELECT, PARTS hilo *m* retorcido

twisting *n* ELECT conexión *f* por torsión, PARTS torsión *f*

two: ~-**bit byte** *n* COMP, DATA dupleto *m*; ~-**condition cable code** *n* TELEP, TRANS código *m* cablegráfico bivalente; ~-**condition telegraph code** *n* TELEP, TRANS código *m* telegráfico bivalente; **~-conductor cable** *n* ELECT, PARTS, TRANS cable *m* bifilar; **~-control-point method** *n* RADIO método *m* de los dos puntos de control; **~-frequency channelling** *BrE* **plan** *n* TRANS plan *m* de canalización con dos frecuencias; **~-frequency glide path system** *n* RADIO sistema *m* de trayectoria de planeo de doble frecuencia; **~-frequency localizer system** *n* RADIO sistema *m* localizador de doble frecuencia; **~-frequency operation** *n* RADIO, SIG explotación *f* en dos frecuencias; **~-frequency operation automatic repeater mode** *n* TELEP modo *m* repetidor automático de explotación en dos frecuencias; **~-frequency operation non-repeater mode** *n* TELEP modo *m* de no repetición de explotación en dos frecuencias; **~-frequency relay system** *n* TELEP sistema *m* de retransmisión en dos frecuencias; **~-frequency signalling** *BrE* *n* SIG, TELEP señalización *f* con dos frecuencias; **~-frequency simplex** *n* RADIO, TELEP simplex *m* con dos frecuencias; **~-gang capacitor** *n* PARTS, POWER capacitor *m* bipolar; **~-gang condenser** *n* PARTS, POWER condensador *m* bipolar; **~-motion selector** *n* TELEP selector *m* de dos movimientos; **~-out-of-five code** *n* DATA código *m* dos de cinco; **~-party line** *n* TELEP línea *f* compartida por dos abonados; **~-piece cover** *n* PARTS, TELEP cubierta *f* de dos piezas; **~-ray diversity** *n* RADIO diversidad *f* por dos haces; **~-stage contact spring set** *n* PARTS grupo *m* de muelles de contacto de dos pasos; **~-stage relay** *n* PARTS relé *m* de dos pasos; **~-state register** *n* TELEP registrador *m* de dos etapas; **~-step action** *n* CONTR acción *f* de dos posiciones; **~-step contact spring set** *n* PARTS, SWG grupo *m* de muelles de contacto de dos pasos; **~-step relay** *n* PARTS relé *m* de dos tiempos; **~-terminal network** *n* ELECT, POWER, TRANS red *f* de dos bornes; **~-terminal pair** *n* ELECT, POWER, TRANS cuadripolo *m*; **~-terminal-pair network** *n* ELECT, POWER, TRANS cuadripolo *m*; **~-tone detector** *n* TEST discriminador *m* telegráfico; **~-way** *adj* DATA, TELEP bilateral; **~-way alternate interaction** *n* DATA interacción *f* bilateral alternada; **~-way circuit** *n* ELECT circuito *m* bidireccional; **~-way communication** *n* RADIO, TELEP, TV comunicación *f* bidireccional; **~-way directionality** *n* COMP bidireccionalidad *f*; **~-way link** *n* COMP conexión *f* bidireccional; **~-way paging system** *n* RADIO sistema *m* de búsqueda bidireccional; **~-way repeater** *n* TRANS repetidor *m* bidireccional; **~-way simplex** *n* TELEP *duplex* simplex *m* por canales conjugados; **~-way simplex connection** *n* TELEP *duplex* conexión *f* simplex por canales conjugados; **~-way simplex system** *n* TELEP sistema *m* simplex por canales conjugados; **~-wire bus** *n* TV bus *m* bifilar; **~-wire crosspoint** *n* SWG punto *m* de cruce bifilar; **~-wire repeater** *n* TRANS repetidor *m* bifilar; **~-wire switch** *n* SWG conmutador *m* bifilar; **~-wire switching** *n* SWG conmutación *f* bifilar; **~-wire switching system** *n* SWG sistema *m* de conmutación bifilar; **~-wire system** *n* SWG sistema *m* de dos hilos

two-B-plus-D: **~ arrangement** *n* TRANS montaje *m* de dos B más D

TY: **~ wrap** *n* PARTS, POWER revestimiento *m* de TY

type: **~ of address** *n* (*TOA*) DATA, TELEP tipo *m* de dirección (*TOA*); **~ A display** *n* RADIO presentación *f* tipo A; **~ approval** *n* TEST homologación *f*; **~ bar** *n* TELEP barra *f* de tipos; **~-bar carriage** *n* TELEP carro *m* de barra de tipos; **~ exchange** *n* TELEP central *f* tipo; **~ font** *n* DATA fuente *f* de tipos; **~ mismatch** *n* COMP desadaptación *f* de tipo; **~ of number** *n* (*TON*) TELEP tipo *m* de número; **~ page** *n* PER APP página *f* tipo; **~ P transmultiplexer** *n* (*TMUX-P*) TELEP transmultiplexor *m* tipo P (*TMUX-P*); **~ of sound-programme** *BrE* **circuit** *n* TELEP tipo *m* de circuito de programa de sonido; **~ S transmultiplexer** *n* (*TMUX-S*) TELEP transmultiplexor *m* tipo S (*TMUX-S*); **~ test** *n* TEST ensayo *m* de tipo; **~ of transmission** *n* TELEP tipo *m* de transmisión; **~ wheel** *n* PER APP rueda *f* de impresión

typeface *n* PER APP estilo *m* de letra de imprenta

typical: **~ prediction** *n* TELEP predicción *f* típica

typing: **~ mistake** *n* COMP error *m* de mecanografía

typography *n* COMP, DATA tipografía *f*

U

U: ~ **interface** *n* TELEP interfaz *f* en U
UA *abbr* (*user agent*) DATA, TELEP agente *m* usuario
UART *abbr* (*universal asynchronous receiver-transmitter*) RADIO UART (*receptor-transmisor asíncrono universal*)
UAS *abbr* (*unavailable second*) DATA segundo *m* no disponible
UAT *abbr* (*unavailability time*) DATA tiempo *m* de indisponibilidad
UAX *abbr* (*unit automatic exchange*) SWG central *f* automática unitaria
UCM *abbr* (*user communications manager*) COMP UCM (*gestor de comunicaciones de usuario*)
UCN-CNE: ~ **interface** *n* TELEP interfaz *f* UCN-CNE
UDC *abbr* (*universal data connector*) ELECT UDC (*conector de datos universal*)
UE *abbr* (*user element*) DATA, TELEP UE (*elemento de usuario*)
UHF *abbr* (*ultrahigh frequency*) TV UHF (*frecuencia ultra-alta*); ~ **network** *n* TV red *f* UHF
UI *abbr* (*unnumbered information*) DATA información *f* no numerada
UIH: ~ **control field** *n* TELEP campo *m* de control UIH
UITS *abbr* (*unacknowledged information-transfer service*) APPL servicio *m* de transferencia de información no reconocido
UJT *abbr* (*unijunction transistor*) ELECT, PARTS UJT (*transistor de una sola unión*)
U-link *n* POWER estribo *m*
ULSI: ~ **circuit** *n* ELECT circuito *m* ULSI
Ultimedia *n* COMP Ultimedia *m*
ultrahigh: ~ **frequency** *n* (*UHF*) TV frecuencia *f* ultra-alta (*UHF*)
ultralarge: ~ **scale integration circuit** *n* ELECT circuito *m* de integración de escala ultragrande
ultrashort: ~ **wave** *n* RADIO onda *f* ultracorta
ultrasonic *adj* RADIO ultrasónico; ~ **frequency** *n* RADIO frecuencia *f* ultrasónica
ultrasound *n* RADIO ultrasonido *m*
ultraviolet *adj* (*UV*) ELECT, TRANS ultravioleta (*UV*); ~ **radiation** *n* TRANS radiación *f* ultravioleta
umbrella: ~ **cell** *n* RADIO celda *f* en paraguas
UMTS *abbr* (*universal mobile telecommunication system*) RADIO UMTS (*sistema universal de telecomunicaciones móviles*)
UNA: ~ **segment** *n* DATA, TELEP segmento *m* UNA
unabsorbed: ~ **field strength** *n* RADIO potencia *f* de campo en ausencia de absorción
unacceptable: ~ **quality** *n* TEST calidad *f* inaceptable
unacknowledged: ~ **information-transfer service** *n* (*UITS*) APPL servicio *m* de transferencia de información no reconocido
unaffected: ~ **level** *n* TELEP nivel *m* no afectado
unallocated: ~-**number signal** *n* SIG, TELEP señal *f* de número no asignado
unamplified *adj* ELECT, POWER, TV no amplificado
unanswered: ~ **call** *n* PER APP, TELEP llamada *f* no atendida; ~-**call indication** *n* PER APP, TELEP indicación *f* de llamada no atendida
unassigned *adj* COMP no asignado
unattended: ~ **exchange** *n* TELEP central *f* no atendida; ~ **mode marking** *n* TELEP marcación *f* de ausencia de operador; ~ **station** *n* TRANS estación *f* no atendida
unauthorized: ~ **visitor** *n* DATA visitante *m* no autorizado
unavailability *n* DATA, TEST indisponibilidad *f*; ~ **time** (*UAT*) DATA tiempo *m* de indisponibilidad
unavailable: ~ **second** *n* (*UAS*) DATA segundo *m* no disponible
UNB: ~ **segment** *n* DATA, TELEP segmento *m* UNB
unbalanced *adj* GEN desequilibrado, asimétrico; ~ **input** *n* COMP entrada *f* asimétrica; ~ **operation normal response mode class** *n* DATA clase *f* de operación asimétrica de modo de respuesta normal; ~ **output** *n* COMP salida *f* asimétrica; ~ **system** *n* TRANS sistema *m* no balanceado
unbarred: ~ **extension** *n* TELEP extensión *f* abierta
unbind *vt* DATA abrir
unblock *vt* ELECT, RADIO, TELEP desbloquear
unblocking *n* ELECT, RADIO, TELEP desbloqueo *m*; ~ **signal** RADIO, SIG, TELEP señal *f* de desbloqueo

unbound: ~ **mode** *n* TRANS modo *m* no ligado
uncertainty: ~ **of measurement** *n* TEST incertidumbre *f* de medición; ~ **symbol PE** *n* TELEP EP *m* de símbolo de incertidumbre
unchained: ~ **sequence** *n* DATA secuencia *f* desencadenada
unclocked *adj* ELECT descronometrado
uncoded *adj* TELEP no codificado
uncoloured *BrE* (*AmE* **uncolored**) *adj* GEN incoloro
unconditional: ~ **call-forwarding** *n* TELEP desvío *m* de llamada incondicional, envío *m* de llamada incondicional; ~ **class of service** *n* DATA clase *f* de servicio incondicional
uncontrolled: ~ **slip** *n* RADIO, TELEP deslizamiento *m* sin control
uncover *vt* GEN descubrir
undefined *adj* GEN indefinido
undercurrent *n* POWER corriente *f* submarina; ~ **relay** PARTS relé *m* de mínima
undercut *n* ELECT, PARTS socavadura *f*
undergo *vt* TEST *trials* someterse a
underground: ~ **cable** *n* ELECT, PARTS cable *m* subterráneo; ~ **chamber** *n* TRANS cámara *f* subterránea; ~ **installation** *n* APPL instalación *f* subterránea; ~ **line** *n* TRANS línea *f* subterránea
underlap *n* TELEP no yuxtaposición de las líneas
undermodulation *n* TRANS submodulación *f*
undervoltage: ~ **alarm** *n* POWER alarma *f* de subtensión; ~ **relay** *n* PARTS relé *m* de mínima
underwater: ~ **communication** *n* RADIO, TELEP comunicación *f* submarina; ~ **propagation** *n* RADIO propagación *f* submarina
undetected *adj* COMP, TEST no detectado
undulating *adj* PARTS *surface* ondulante
unearthed *BrE adj* (*cf ungrounded AmE*) POWER desconectado de tierra, sin conexión a tierra
unfiltered: ~ **voltage** *n* POWER tensión *f* sin aplanar
unforeseen: ~ **event** *n* TEST caso *m* imprevisto; ~ **outcome** *n* TEST resultado *m* imprevisto; ~ **test outcome** *n* TEST resultado *m* imprevisto de prueba
unfriendly *adj* COMP poco propicio
unfurl *vt* RADIO *antenna* desplegar
ungrounded *AmE adj* (*cf unearthed BrE*) POWER desconectado de tierra, sin conexión a tierra
ungroup *vt* COMP desagrupar
unguarded: ~ **interval** *n* TELEP intervalo *m* desprotegido
UNH: ~ **segment** *n* DATA, TELEP segmento *m* UNH
unhook *vt* PER APP, TELEP desenganchar
UNI *abbr* (*user-network interface*) DATA, TRANS interfaz *f* usuario-red
unidentified: ~ **event** *n* TEST caso *m* no identificado
unidirectional *adj* COMP, RADIO, TELEP, TV unidireccional; ~ **antenna** *n* RADIO antena *f* unidireccional; ~ **communication** *n* RADIO, TELEP, TV comunicación *f* unidireccional; ~ **connection** *n* SWG, TELEP, TV conexión *f* unidireccional; ~ **network** *n* TV red *f* unidireccional; ~ **pulse** *n* TRANS impulso *m* unidireccional
unification *n* ELECT unificación *f*
uniform: ~ **encoding** *n* TELEP, TRANS codificación *f* uniforme; ~ **field** *n* GEN campo *m* uniforme; ~ **numbering** *n* TELEP numeración *f* uniforme; ~ **quantizing** *n* TELEP, TRANS cuantificación *f* uniforme; ~ **resource locator** *n* (*URL*) DATA localizador *m* de recursos uniforme (*URL*); ~ **spectrum random noise** *n* GEN ruido *m* blanco; ~ **wiring** *n* PARTS, RADIO, TELEP cableado *m* uniforme
uniformity: ~ **of luminance** *n* ELECT uniformidad *f* de luminancia
unigauge: ~ **loop** *BrE n* TELEP bucle *m* de patrón único
unijunction: ~ **diode** *n* ELECT, PARTS diodo *m* de una sola unión; ~ **transistor** *n* (*UJT*) ELECT, PARTS transistor *m* de una sola unión (*UJT*)
unilateral *adj* GEN unilateral
uninitialized *adj* COMP no iniciado
uninsulated *adj* ELECT sin aislamiento, POWER desnudo
unintelligible: ~ **crosstalk** *n* TELEP diafonía *f* ininteligible; ~ **crosstalk components** *n pl* TELEP componentes *m* de diafonía ininteligible
uninterrupted: ~ **duty** *n* GEN servicio *m* ininterrumpido; ~ **power supply** *n* POWER alimentación *f* eléctrica ininterrumpida
uninterruptible: ~ **power supply** *n* (*UPS*) POWER alimentación *f* eléctrica ininterrumpible (*UPS*)
unipolar: ~ **pulse** *n* TELEP impulso *m* unipolar; ~ **sequence of pulses** *n* TELEP secuencia *f* unipolar de impulsos; ~ **transistor** *n* ELECT, PARTS transistor *m* unipolar
unipole: ~ **antenna** *n* RADIO antena *f* monopolo
unique: ~ **word** *n* (*UW*) DATA palabra *f* singular (*UW*); ~**-word detection** *n* DATA detección *f* de palabra singular; ~**-word loss** *n* DATA atenuación *f* de palabra singular; ~ **word-loss alarm** *n* DATA alarma *f* de atenuación de palabra singular; ~ **word-missed detection** *n* DATA detección *f* de palabra singular ausente;

~-**word window** *n* DATA ventana *f* de palabra singular
uniselector *n* SWG selector *m* unidireccional
unison *n* COMP armonía *f*
unit *n* GEN unidad *f*; ~ **automatic exchange** (*UAX*) SWG central *f* automática unitaria; ~ **charge** TELEP carga *f* unidad; ~-**charged marking** TELEP marcaje *m* por cómputo simple; ~ **charge metering** TELEP medición *f* de carga unidad; ~ **design** PER APP, SWG diseño *m* de unidades; ~ **element** GEN elemento *m* unitario; ~-**element error rate** TELEP tasa *f* de error por elemento unitario; ~ **gain** ELECT ganancia *f* unitaria; ~ **impulse response** CONTR respuesta *f* de impulso unitario; ~-**indicating alarm** TELEP alarma *f* indicadora de unidad; ~ **interval** GEN intervalo *m* unitario; ~ **of measurement** TEST unidad *f* de medición; ~ **section** TV sección *f* unitaria; ~ **step** TELEP función *f* escalón unitario; ~ **string** DATA serie *f* unitaria; ~ **of time** GEN unidad *f* de tiempo; ~ **tube** ELECT tubo *m* unidad
United: ~ **Nations trade-data interchange** *n* (*UNTDI*) TELEP intercambio *m* de datos de comercio de las Naciones Unidas (*UNTDI*)
units: ~ **counter** *n* ELECT contador *m* de unidades; ~ **digit** *n* GEN cifra *f* de unidad; ~ **input** *n* ELECT entrada *f* de unidades
unitunnel: ~ **diode** *n* ELECT, PARTS diodo *m* unitunel
unity: ~ **gain** *n* ELECT ganancia *f* unidad
universal: ~-**access number** *n* TELEP número *m* de acceso universal; ~-**access number service** *n* TELEP servicio *m* de número de acceso universal; ~ **asynchronous receiver-transmitter** *n* (*UART*) RADIO receptor-transmisor *m* asíncrono universal (*UART*); ~ **data connector** *n* (*UDC*) ELECT conector *m* de datos universal (*UDC*); ~ **element** *n* COMP elemento *m* universal; ~ **meter** *n* TEST metro *m* universal; ~ **mobile telecommunication system** *n* (*UMTS*) RADIO sistema *m* universal de telecomunicaciones móviles (*UMTS*); ~ **night-service** *n* TELEP servicio *m* nocturno general; ~ **number** *n* TELEP número *m* universal; ~ **personal telecommunication** *n* (*UPT*) APPL telecomunicación *f* personal universal (*UPT*); ~ **switch** *n* SWG conmutador *m* universal; ~ **time** *n* GEN hora *f* universal; ~ **time coordinated** *n* (*UTC*) GEN hora *f* universal coordinada (*HUC*); ~ **transport system** *n* (*UTS*) ELECT sistema *m* universal de transporte (*UTS*)
universe *n* TEST universo *m*
Unix: ~-**to**-~ **copy protocol** *n* (*UUCP*) DATA protocolo *m* de copia de Unix-a-Unix (*UUCP*)
unknown *adj* COMP, DATA desconocido
unlabelled *BrE* (*AmE* **unlabeled**) *adj* COMP sin etiqueta
unlisted *adj* TELEP sin listar
unload *vt* COMP, ELECT, POWER, TELEP *volume* descargar
unloaded *adj* COMP, ELECT, POWER, TELEP descargado; ~ **cable** *n* TRANS cable *m* descargado
unloading *n* COMP, ELECT, POWER, TELEP *of volume* descarga *f*
unlock: ~ **key** *n* COMP llave *f* de apertura
unmanned: ~ **exchange** *n* TELEP central *f* automática
unmasked *adj* COMP sin máscara
unnotched: ~ **signal** *n* SIG señal *f* no regulada
unnumbered: ~ **acknowledgement** *n* TELEP trama *f* de acuse de recepción, reconocimiento *m* no numerado; ~ **acknowledgement frame** *n* TELEP trama *f* de acuse de recepción no numerada; ~ **format** *n* TELEP formato *m* no numerado; ~ **information** *n* (*UI*) DATA información *f* no numerada
unoccupied: ~ **line** *n* TELEP línea *f* libre
unpack *vt* COMP, DATA expandir
unperforated *adj* GEN *tape* no perforado
unperturbed: ~ **orbit** *n* RADIO órbita *f* no perturbada
unplug *vt* COMP, ELECT, PER APP, TEST desenchufar, desconectar
unplugged *adj* COMP, ELECT, PER APP, TEST desconectado, no conectado, desenchufado
unprogrammable *adj* COMP no programable
unqualified: ~ **event** *n* DATA evento *m* no cualificado
unrelated: ~ **to distance** *adj* GEN sin relación con la distancia
unreliable *adj* COMP no fiable
unrelocatable *adj* COMP intrasladable
unrestricted *adj* GEN abierto, irrestricto; ~ **bearer service** *n* APPL, TRANS servicio *m* de portador sin restricciones; ~ **digital-data ratio** *n* DATA relación *f* abierta de datos digitales; ~ **extension** *n* TELEP supletorio *m* con toma directa de la red; ~ **information-transfer service** *n* APPL, TELEP servicio *m* abierto de transferencia de información; ~ **service** *n* APPL, TELEP servicio *m* abierto
unscheduled: ~ **maintenance** *n* TELEP, TEST mantenimiento *m* no programado; ~ **stop** *n* TEST parada *f* no programada
unscramble *vt* COMP, DATA descifrar
unscrambled: ~ **binary ones** *n* (*USBI*) DATA

binarios *m* descifrados (*USB1*); ~ **double dibit** *n* (*S1*) DATA dibit *m* doble descifrado (*S1*)
unsecured *adj* COMP sin garantía
unserviceable: ~ **station** *n* TRANS estación *f* fuera de servicio
unshielded: ~ **twisted pair** *n* (*UTP*) ELECT, TRANS par *m* torcido descubierto (*UTP*)
unshifted: ~ **mode** *n* COMP, PER APP *keyboard* modo *m* sin inversión de letra
unsmoothed: ~ **voltage** *n* POWER tensión *f* no filtrada
unsolder *vt* ELECT, PARTS desoldar
unstable: ~ **state** *n* ELECT estado *m* inestable
unsuccessful: ~ **attempt** *n* RADIO, TELEP intento *m* frustrado; ~ **call** *n* TELEP llamada *f* frustrada; ~ **traffic** *n* RADIO, TELEP tráfico *m* frustrado
unsymmetrical *adj* POWER, TELEP asimétrico; ~ **grading** *n* TELEP interconexión *f* progresiva asimétrica
UNTDI *abbr* (*United Nations trade-data interchange*) TELEP UNTDI (*intercambio de datos de comercio de las Naciones Unidas*)
untested *adj* COMP, TEST no probado
untimed: ~ **call** *n* TELEP comunicación *f* sin tasa de duración
untrafficable: ~ **path** *n* TRANS vía *f* no traficable
unusable *adj* GEN no utilizable
unvoiced *adj* COMP sin voz
unwanted: ~ **echo** *n* RADIO eco *m* parásito; ~ **emission** *n* RADIO emisión *f* no deseada
unweighted *adj* COMP no ponderado
unwelcome: ~ **caller** *n* DATA *Internet* solicitante *m* indeseado
unwired: ~ **rack** *n* PARTS, TELEP bastidor *m* inalámbrico
unwrap[1]: ~ **test** *n* GEN, TEST prueba *f* de desenrollamiento
unwrap[2] *vt* TEST desenrollar
unwrapping *n* TEST desenrollamiento *m*; ~ **test** TEST prueba *f* de descubrimiento; ~ **tool** PARTS herramienta *f* de desenrollamiento
unzip *vt* COMP, DATA *files* descomprimir
up: ~-**converter** *n* DATA, RADIO convertidor *m* ascendente; ~-**down time** *n* TELEP tiempo *m* de funcionamiento-avería; ~-**path** *n* RADIO trayecto *m* de subida; ~-**path power control** *n* RADIO control *m* de potencia del trayecto de subida; ~-**state** *n* TELEP, TEST estado *m* útil; ~-**time** *n* TELEP tiempo *m* útil
UPC *abbr* (*usage parameter control*) APPL UPC (*control de parámetro de uso*)
update *vt* GEN actualizar
updating *n* GEN actualización *f*; ~ **procedure** GEN procedimiento *m* de actualización
upgrade[1] *n* COMP mejoramiento *m*; ~ **board** COMP comisión *f* de mejoramiento
upgrade[2] *vt* COMP mejorar
upgrading *n* COMP mejoramiento *m*
uplink *n* RADIO, TELEP enlace *m* ascendente; ~ **block** RADIO bloque *m* de enlace ascendente; ~ **fade** RADIO desvanecimiento *m* de enlace ascendente; ~ **transmission** RADIO, TRANS transmisión *f* de enlace ascendente; ~ **transmission-phase** RADIO, TRANS fase *f* de transmisión de enlace ascendente
upload *vt* DATA cargar
uploading *n* DATA carga *f*
upper *adj* COMP superior; ~ **boundary** *n* TRANS cota *f* superior; ~ **case** *n* GEN caja *f* alta; ~ **ionosphere** *n* RADIO ionosfera *f* superior; ~ **keyboard** *n* COMP teclado *m* superior; ~ **limit** *n* GEN límite *m* superior; ~ **quartile** *n* GEN cuartil *m* superior; ~ **sideband** *n* (*USB*) RADIO, TRANS banda *f* lateral superior (*USB*); ~ **subfield** *n* RADIO, TRANS campo *m* auxiliar superior; ~ **tester** *n* (*UT*) DATA, TEST probador *m* superior (*UT*)
UPS *abbr* (*uninterruptible power supply*) POWER UPS (*alimentación eléctrica ininterrumpible*)
upstream *adj* ELECT, POWER, TELEP contracorriente; ~ **failure indication** *n* TELEP indicación *f* de fallo en un punto anterior
UPT *abbr* (*universal personal telecommunication*) APPL UPT (*telecomunicación personal universal*)
uptime *n* GEN tiempo *m* útil
upward: ~ **compatibility** *n* COMP compatibilidad *f* ascendente
urban: ~ **area** *n* TV área *f* urbana; ~ **network** *n* TELEP, TRANS red *f* urbana
urea: ~ **resin** *n* PARTS resina *f* de urea
urethane: ~ **plastic** *n* PARTS plástico *m* uretánico
urgency: ~ **signal** *n* RADIO, SIG señal *f* de urgencia
urgent: ~ **alarm** *n* TELEP alarma *f* urgente; ~ **call** *n* TELEP llamada *f* urgente
URL *abbr* (*uniform resource locator*) DATA *Internet* URL (*localizador de recursos uniforme*)
usable: ~ **field strength** *n* RADIO potencia *f* de campo utilizable; ~ **power flux-density** *n* RADIO densidad *f* de flujo de potencia utilizable; ~ **scanning line length** *n* TELEP longitud *f* de línea de exploración utilizable
usage *n* GEN consumo *m*; ~ **parameter control** (*UPC*) APPL control *m* de parámetro de uso (*UPC*)
USB *abbr* (*upper sideband*) RADIO, TRANS USB (*banda lateral superior*)

USB1 *abbr* (*unscrambled binary ones*) DATA USB1 (*binarios descifrados*)

use[1] *n* GEN empleo *m*, uso *m*; ~ **of network connection TPDU** DATA uso *m* de TPDU de conexión de red

use[2] *vt* DATA usar

useful *adj* GEN útil; ~ **area** *n* RADIO zona *f* útil; ~ **life** *n* TELEP, TEST vida *f* útil; ~ **output power** *n* ELECT potencia *f* de salida aprovechable, POWER potencia *f* de salida utilizable

Usenet *n* DATA Usenet *f*

user *n* GEN usuario *m*; ~ **access** GEN acceso *m* de usuario; ~ **address** TELEP dirección *f* de usuario; ~ **agent** (*UA*) DATA, TELEP agente *m* usuario; ~ **class indicator** TELEP indicador *m* de clase de usuario; **~-communications manager** (*UCM*) COMP gestor *m* de comunicaciones de usuario (*UCM*); **~-created link** COMP enlace *m* creado por usuario; ~ **data** *n pl* COMP, DATA datos *m pl* de usuario; **~-data field** DATA campo *m* de datos de usuario; **~-defined** *adj* COMP definido por el usuario; ~ **element** (*UE*) DATA, TELEP elemento *m* de usuario (*UE*); ~ **facility** TELEP servicio *m* de usuario; **~-friendly** *adj* ELECT, GEN fácil de manejar; **~-friendliness** GEN facilidad *f* de manejo; **~'s guide** GEN guía *f* de usuario; ~ **handling-time** TELEP tiempo *m* de manejo por el usuario; **~-information transfer rate** DATA velocidad *f* de transferencia de información de usuario; ~ **interface** GEN interfaz *f* de usuario; **~'s manual** GEN manual *m* de usuario; ~ **name** COMP nombre *m* de usuario; **~-network interface** (*UNI*) DATA, TRANS interfaz *f* usuario-red; ~ **option** TELEP opción *f* de usuario; ~ **part** TELEP usuario *m*; ~ **premises** *n pl* GEN locales *m pl* de usuario; **~-programmable** *adj* COMP programable por el usuario; **~'s quality cost** TEST coste *m* (*AmL* costo) de calidad de usuario; ~ **service** TELEP servicio *m* de usuario; **~-signalling** *BrE* **bearer service** DATA, SIG servicio *m* portador de señalización al usuario; ~ **terminal** TELEP, TRANS, TV terminal *m* de usuario; **~-to-~** *adj* GEN de usuario ausuario; **~-to-~ indicator** DATA indicador *m* de usuario a usuario; **~-to-~ information** (*UUI*) DATA información *f* de usuario a usuario; **~-to-~ information message** DATA mensaje *m* de información de usuario a usuario; **~-to-~ signalling** *BrE* (*UUS*) RADIO, SIG señalización *f* de usuario a usuario

user-ASE *n* DATA ASE *m* de usuario

usher in *vt* COMP anunciar

UT *abbr* (*upper tester*) DATA, TEST UT (*probador superior*)

UTC *abbr* (*universal time coordinated*) GEN HUC (*hora universal coordinada*)

utility *n* COMP, DATA utilidad *f*; ~ **memory** COMP memoria *f* de trabajo; ~ **program** COMP programa *m* de utilidades; ~ **routine** COMP rutina *f* de ayuda operacional

UTP *abbr* (*unshielded twisted pair*) ELECT, TRANS UTP (*par torcido descubierto*)

UTS *abbr* (*universal transport system*) ELECT UTS (*sistema universal de transporte*)

utterance *n* COMP emisión *f*

UUCP *abbr* (*Unix-to-Unix copy protocol*) DATA UUCP (*protocolo de copia de Unix-a-Unix*)

UUI *abbr* (*user-to-user information*) DATA información *f* de usuario a usuario

UUS *abbr* (*user-to-user signalling*) RADIO, SIG señalización *f* de usuario a usuario

UV *abbr* (*ultraviolet*) ELECT, TRANS UV (*ultravioleta*); ~ **radiation** *n* TRANS radiación *f* ultravioleta

UW *abbr* (*unique word*) DATA UW (*palabra singular*); ~ **window** *n* DATA ventana *f* UW

V

V *abbr* (*voltage*) POWER V (*voltaje*)

vacancy *n* PARTS *defect* laguna *f*, TELEP línea *f* vacante; **~ marking** TELEP marcación *f* de línea vacante

vacant: **~ number** *n* TELEP número *m* no asignado, número *m* vacante; **~ position** *n* TELEP *equipment* posición *f* libre

vaccine *n* COMP, DATA *virus protection* vacuna *f*

vacuum *n* GEN vacío *m*; **~ factor** ELECT factor *m* de vacío; **~ lightning arrester** POWER pararrayos *m* de gas rarificado; **~ metallization** *BrE* PARTS metalización *f* al vacío; **~ suction** ELECT succión *f* al vacío; **~ tube** ELECT tubo *m* de vacío; **~ vaporization** ELECT vaporización *f* al vacío

VAD *abbr* (*vapour axial deposition*) ELECT deposición *f* axial de vapor

VADIS *abbr* (*video-audio digital interactive system*) COMP VADIS (*sistema interactivo digital vídeo-audio*)

valence *n* GEN valencia *f*; **~ band** PARTS banda *f* de valencia

valency *n* GEN valencia *f*

valid: **~ cell** *n* COMP célula *f* válida; **~ event** *n* DATA, TEST evento *m* válido; **~ MCSPDU** *n* TELEP MCSPDU *m* válido

validate *vt* DATA validar

validation *n* DATA validación *f*

validity *n* DATA validez *f*; **~ check** CONTR, DATA prueba *f* de validez

valley: **~ point** *n* PARTS punto *m* de valle; **~-point voltage** *n* PARTS, POWER tensión *f* de punto de valle

valuate *vt* GEN valorar

valuation *n* GEN tasación *f*, valoración *f*

value *n* GEN valor *m*; **~-added data service** APPL, DATA servicio *m* de datos de valor añadido; **~-added network** (*VAN*) DATA, TELEP, TRANS, TV red *f* de valor añadido; **~-added network services** (*VANS*) DATA, TELEP, TRANS, TV servicios *m pl* de red de valor añadido (*SRVA*); **~-added paging** RADIO buscapersonas *m* de valor añadido; **~ assignment** COMP asignación *f* de valor

valve *n* ELECT, PARTS *tube* válvula *f*; **~ base** ELECT base *f* de válvula, PARTS culote *m* de lámpara; **~ socket** ELECT portaválvulas *m*, PARTS zócalo *m* de tubo; **~ spindle** ELECT pivote *m* de válvula, PARTS vástago *m* de válvula

VAM *abbr* (*virtual access method*) SWG VAM (*método de acceso virtual*)

vamp *n* COMP remiendo *m*

VAN *abbr* (*value-added network*) DATA, TELEP, TRANS, TV red *f* de valor añadido

vandal: **~-proofing** *n* GEN protección *f* contra vandalismo

vane: **~ attenuator** *n* RADIO atenuador *m* de lámina; **~ wattmeter** *n* TEST watímetro *m* de paleta

VANS *abbr* (*value-added network services*) DATA, TELEP, TRANS, TV VANS (*servicios de red de valor añadido, SRVA*)

vapour *BrE* (*AmE* **vapor**) *n* GEN vapor *m*; **~ axial deposition** (*VAD*) ELECT deposición *f* axial de vapor; **~ concentration** CLIM concentración *f* de vapor; **~-phase soldering** PARTS soldadura *f* por fase vapor

vapourware *BrE* (*AmE* **vaporware**) *n* COMP, DATA instrumental *m* de vapor

varactor *n* ELECT, PARTS varactor *m*; **~ tuning** ELECT, PARTS ajuste *m* del varactor

variable[1]: **~-air capacitor** *n* ELECT, PARTS capacitor *m* variable de aire; **~ antenna-coupling** *n* RADIO acoplamiento *m* variable de antena; **~ bit rate** *n* COMP, DATA velocidad *f* variable de tráfico binario; **~ capacitor** *n* ELECT, PARTS condensador *m* variable; **~ data filter** *n* GEN filtro *m* digital variable; **~ delay** *n* GEN retardo *m* variable; **~ digital amplifier** *n* COMP amplificador *m* digital variable; **~-frequency drive** *n* CONTR arrastre *m* de frecuencia variable; **~-frequency oscillator** *n* ELECT, PARTS oscilador *m* de frecuencia variable; **~-frequency telemetering** *n* TEST telemetría *f* de frecuencia variable; **~-length code** *n* COMP, DATA, TELEP código *m* de longitud variable; **~-length coding** *n* COMP, DATA, TELEP codificación *f* de longitud variable; **~-length numbering** *n* TELEP numeración *f* de longitud variable; **~-length numbering scheme** *n* TELEP plan *m* de numeración de longitud variable; **~-length OGM** *n* PER APP OGM *m* de longitud variable; **~ method** *n* TEST método *m* variable; **~ mu tube** *n* ELECT tubo *m* de mu variable; **~ quality** *n* GEN calidad *f* variable; **~**

queue *n* TELEP cola *f* variable; **~-packet recording** *n* COMP registro *m* de paquete variable; **~-resistor network** *n* POWER red *f* de resistor variable; **~ text** *n* TELEP texto *m* variable

variable[2] *n* ELECT variable *f*

variance *n* GEN varianza *f*

variant *n* GEN variante *f*

variation *n* GEN variación *f*

varioplex *n* ELECT varioplex *m*

varistor *n* ELECT, PARTS varistor *m*

VBI *abbr* (*vertical blanking interval*) DATA, TV VBI (*intervalo de borrado vertical*)

Vbox *n* COMP Vbox *m*

VC *abbr* APPL, DATA, TELEP (*virtual container*) recipiente *m* virtual, DATA, TV (*virtual channel*) canal *m* virtual

VCC *abbr* (*virtual-channel connection*) DATA, TV conexión *f* de canal virtual

VCCE *abbr* (*virtual-channel connection endpoint*) DATA, TV final *m* de conexión de canal virtual

VCCRF *abbr* (*virtual-channel connection-related function*) DATA, TV VCCRF (*función de canal virtual relacionada con la conexión*)

VCF *abbr* (*video command freeze-picture request*) TV petición *f* de imagen inmóvil de la videoseñal de mando

VCI *abbr* (*virtual-channel identifier*) DATA, TV identificador *m* de canal virtual

VC-n *abbr* (*virtual container-n*) APPL, DATA, TELEP recipiente-n *m* virtual

VCO *abbr* (*voltage-controlled oscillator*) ELECT oscilador *m* con voltaje controlado

VCP *abbr* (*videocassette player*) TV VCP (*reproductor de cintas de vídeo*)

VCR *abbr* (*videocassette recorder*) TV grabador *m* de videocasete, videograbadora *f*

VCS *abbr* (*virtual-circuit switch*) SWG conmutador *m* de circuito virtual

VCU *abbr* (*video command fast-update request*) TV VCU (*petición de actualización rápida de instrucción de vídeo*)

VCXO *abbr* (*voltage-controlled crystal oscillator*) ELECT VCXO (*oscilador de cristal de regulación por tensión*)

vector *n* GEN vector *m*; **~ document** DATA *Internet* documento *m* vectorial; **~ line** RADIO línea *f* vectorial; **~ quantization** DATA, SIG cuantización *f* vectorial; **~ scanning** COMP exploración *f* vectorial; **~ synthesis** COMP síntesis *f* vectorial

vehicle: **~ fleet** *n* RADIO flotilla *f* de vehículos; **~ intelligent communication system** *n* (*VICS*) RADIO sistema *m* de comunicación de vehículos inteligentes (*VICS*); **~ localization subsystem** *n* RADIO subsistema *m* de localización de vehículos; **~ localization system** *n* RADIO sistema *m* de localización de vehículos; **~-mounted** *adj* RADIO montado en un vehículo

velocity: **~ gate** *n* RADIO puerta *f* de velocidad; **~ indicator** *n* RADIO indicador *m* de velocidad; **~ modulation** *n* ELECT modulación *f* de velocidad; **~ resolution** *n* RADIO resolución *f* de velocidad; **~ search** *n* RADIO exploración *f* en velocidad; **~-sensitive keyboard** *n* COMP teclado *m* sensible a la velocidad; **~ of sound** *n* GEN velocidad *f* de sonido

vendor *n* TEST proveedor *m*; **~ appraisal** TEST evaluación *f* de proveedores; **~-independent messaging** (*VIM*) RADIO mensajería *f* independiente del proveedor (*VIM*)

veneer *n* PARTS chapa *f*

ventilation *n* CLIM ventilación *f*

verbal: **~ message** *n* TELEP mensaje *m* verbal; **~-message transmission service** *n* TELEP, TRANS servicio *m* de transmisión de mensaje verbal

verification *n* DATA *of command*, TELEP *of message*, TEST verificación *f*; **~ of overhead information** TEST, TRANS verificación *f* de información aérea

verify *vt* DATA *command*, TELEP *message*, TEST verificar

verifying *n* DATA, TELEP, TEST verificación *f*; **~ instrument** TEST instrumento *m* de verificación

vernier *n* TEST vernier *m*

versatile *adj* GEN versátil

version *n* COMP, DATA, TELEP versión *f*

vertex *n* GEN vértice *m*; **~ angle** RADIO ángulo *m* de vértice; **~ feed** RADIO alimentación *f* en el vértice; **~ plate** RADIO *of aerial reflector* placa *f* de vértice

vertical[1] *adj* GEN vertical; **~ arrow** *n* COMP flecha *f* vertical; **~ blanking interval** *n* (*VBI*) DATA, TV intervalo *m* de borrado vertical (*VBI*); **~ coverage** *n* RADIO cobertura *f* vertical; **~ directivity pattern** *n* RADIO modelo *m* de directividad vertical; **~ displacement** *n* GEN desplazamiento *m* vertical; **~ format** *n* DATA *Internet* formato *m* vertical; **~ grating logic** *n* TV lógica *f* de trama vertical; **~ hold** *n* TV estabilidad *f* vertical; **~ interval time code** *n* (*VITC*) COMP, DATA código *m* de tiempo de intervalo vertical (*VITC*); **~ iron** *n* PARTS hierro *m* vertical; **~ multiple cable** *n* SWG cable *m* múltiple vertical; **~ polarization** *n* TV polarización *f* vertical; **~ position** *n* GEN posición *f* vertical; **~ scan** *n* COMP barrido *m* vertical; **~ tabulation character** *n* (*VT*) DATA,

RADIO carácter *m* de tabulación vertical; ~ **unit base** *n* PARTS armazón *m* de vertical; ~ **unit multiple** *n* SWG múltiple *m* de verticales

vertical[2] *n* GEN *equipment* vertical *f*

vertically: ~ **adjustable** *adj* PARTS regulable en altura

vestigial: ~ **sideband** *n* (*VSB*) RADIO, TRANS banda *f* lateral residual (*RSB*, *VSB*); **~-sideband emission** *n* RADIO, TRANS emisión *f* en banda lateral residual; **~-sideband transmission** *n* RADIO, TRANS transmisión *f* en banda lateral residual

VF *abbr* (*voice frequency*) GEN frecuencia *f* acústica; ~ **dialling** *BrE n* PER, TELEP discado *m* por señales de frecuencia vocal

V/F *abbr* (*voltage/frequency*) ELECT, POWER V/F (*tensión/frecuencia*)

VFT *abbr* (*voice-frequency telegraphy*) TELEP VFT (*telegrafía de frecuencia vocal*)

VGA *abbr* (*video graphics array*) COMP VGA (*matriz de vídeo gráfico*)

VHF *abbr* (*very high frequency*) TRANS VHF (*frecuencia muy alta*); ~ **network** *n* TV red *f* VHF; ~ **omnidirectional radio range** *n* (*VOR*) RADIO radiofaro *m* omnidireccional de VHF (*VOR*)

via: ~ **hole** *n* ELECT agujero *m* de paso

vibrating: **~-reed frequency meter** *n* TEST medidor *m* de frecuencia de lámina vibrante

vibration: ~ **mode** *n* ELECT modo *m* de vibración; ~ **test** *n* TEST prueba *f* de vibración

vibrator *n* COMP, ELECT, RADIO vibrador *m*

VICS *abbr* (*vehicle intelligent communication system*) RADIO VICS (*sistema vehículo de comunicación inteligente*)

video *n* APPL, COMP, TV vídeo *m*; ~ **acquisition** COMP adquisición *f* de vídeo; ~ **attribute** TV atributo *m* de vídeo; **~-audio digital interactive system** (*VADIS*) COMP sistema *m* interactivo digital vídeo-audio (*VADIS*); ~ **board** COMP cuadro *m* de vídeo; ~ **camera** TV videocámara *f*; ~ **capture** COMP captación *f* de vídeo; ~ **card** COMP tarjeta *f* de vídeo; ~ **carrier** TV portadora *f* de video; ~ **CD** TV videodisco *m* compacto; ~ **codec** COMP codificador/descodificador *m* de vídeo; ~ **command fast-update request** (*VCU*) TV petición *f* de actualización rápida de instrucción de vídeo (*VCU*); ~ **command freeze-picture request** (*VCF*) TV petición *f* de imagen inmóvil de la videoseñal de mando; ~ **compression** DATA, TV videocompresión *f*; ~ **data** COMP visualización *f* de datos, DATA, TV información *f* de vídeo; ~ **data buffering** COMP, DATA, TV almacenamiento *m* intermedio de información de vídeo; ~ **digitizer** COMP, TV digitalizador *m* de imágenes; ~ **display terminal** *AmE* TV unidad *f* terminal de vídeo; ~ **equipment** TV equipo *m* de vídeo; **~-frequency protection ratio** RADIO relación *f* de protección de videofrecuencias; ~ **graphics array** (*VGA*) COMP matriz *f* de vídeo gráfico (*VGA*); ~ **input** DATA, TV potencia *f* de entrada de video; ~ **link** TV videoenlace *m*; ~ **loop** TV circuito *m* de bucle de vídeo; ~ **mail** COMP videocorreo *m*; ~ **map** RADIO, TV imagen *f* electrónica de mapa; ~ **mapping** RADIO, TV superposición *f* de vídeo; **~-on-demand** (*VOD*) COMP, TV vídeo *m* a petición (*VOD*); ~ **outlet** TV salida *f* de señal de vídeo; ~ **processor** COMP videoprocesador *m*; ~ **projector** TV proyector *m* de vídeo; ~ **RAM** (*VRAM*) COMP RAM *f* de vídeo (*VRAM*); ~ **recorder** COMP, TV videograbador *m*; ~ **recording** COMP, TV registro *m* de vídeo; ~ **server** COMP videoservidor *m*; ~ **services** *n pl* TV videoservicios *m pl*; ~ **signal** RADIO, SIG, TV señal *f* de vídeo; ~ **source** TV origen *m* de la señal de vídeo; ~ **switch** SWG conmutador *m* mezclador de vídeo; ~ **track** COMP, TV pista *f* de vídeo; ~ **transmission** TRANS, TV transmisión *f* de vídeo; ◆ ~ **indicate active** TV indicación de vídeo activo; ~ **indicate ready-to-activate** TV indicación de vídeo listo para activar; ~ **indicate suppressed** TV indicación de vídeo suprimido

videocassette *n* COMP, TV videocasete *f*; ~ **player** (*VCP*) COMP, TV reproductor *m* de cintas de vídeo (*VCP*); ~ **recorder** (*VCR*) COMP, TV grabador *m* de videocasete

videocommunication *n* TV videocomunicación *f*; ~ **equipment** TV equipo *m* de videocomunicación; ~ **network** TV red *f* de videocomunicación

videoconference *n* APPL, COMP, RADIO, TV videoconferencia *f*

videoconferencing *n* APPL, COMP, RADIO, TV servicio *m* de videoconferencias

videodisc *BrE* (*AmE* **videodisk**) *n* TV videodisco *m*

videofrequency *n* TRANS, TV videofrecuencia *f*; ~ **band** TRANS, TV banda *f* de videofrecuencia; ~ **protection ratio** TV relación *f* de protección de videofrecuencia; ~ **signal-to-interference ratio** RADIO, TV relación *f* señal de videofrecuencia/interferencia

videography *n* COMP, RADIO, TV videografía *f*

videomessaging *n* COMP, TV videomensajería *f*

videophone *n* APPL, RADIO, TELEP, TV videoteléfono *m*; ~ **service** APPL, RADIO,

TELEP, TV servicio *m* de videoteléfono; ~ **switching system** SWG sistema *m* de conmutación de videófono

videophony *n* RADIO, TELEP, TV videofonía *f*

videotape *n* COMP, TV cinta *f* de vídeo; ~ **player** (*VTP*) COMP, TV reproductor *m* de cintas de video (*VTP*); ~ **recorder** (*VTR*) COMP, TV grabadora *f* de cinta de video

videotelephone *n* RADIO, TELEP, TV videoteléfono *m*

videotelephony *n* RADIO, TELEP, TV videotelefonía *f*

videotext *n* COMP, RADIO, TV videotexto *m*; ~ **access point** TELEP, TV punto *m* de acceso a videotexto; ~ **closed user group** COMP, TV grupo *m* cerrado de usuarios de videotexto; ~ **document** COMP documento *m* de videotexto; ~ **gateway** TELEP, TV medio *m* de acceso al videotexto; ~ **host** COMP, TELEP, TV ordenador *m* central de videotexto; ~ **interworking** COMP, TV interfuncionamiento *m* de videotexto; ~ **PAD** TV PAD *m* de videotexto; ~ **server** COMP, TELEP, TV servidor *m* de videotexto; ~ **service** APPL, COMP, TELEP, TV servicio *m* de videotexto; ~ **service centre** *BrE* APPL, COMP, TELEP, TV centro *m* de servicio de videotexto; ~ **system** COMP, TELEP, TV sistema *m* de videotexto; ~ **telesoftware** COMP, TELEP, TV teleprogramas *m pl* de videotexto; ~ **terminal** COMP, TELEP, TV terminal *m* de videotexto

vidicon *n* ELECT vidicón *m*

view *vt* COMP ver

viewed: ~ **from below** *adj* GEN visto desde abajo; ~ **from component side** *adj* GEN visto por el lado de los componentes; ~ **from inside** *adj* GEN visto desde adentro; ~ **from rear** *adj* GEN visto desde atrás; ~ **from side** *adj* GEN visto de lado; ~ **from soldering side** *adj* GEN visto desde el lado de la soldadura

viewer *n* COMP *device*, TV *person* espectador *m*

viewing: ~ **angle** *n* GEN ángulo *m* de visión; ~ **distance** *n* GEN distancia *f* de visión; ~ **filter** *n* COMP filtro *m* de visión

viewphone *n* APPL, RADIO, TELEP, TV videoteléfono *m*

VIM *abbr* (*vendor-independent messaging*) RADIO VIM (*mensajería independiente del proveedor*)

vinyl *n* PARTS vinilo *m*; ~ **acetate plastic** PARTS plástico *m* de acetato de vinilo; ~ **chloride plastic** PARTS plástico *m* de cloruro de vinilo

VIO *abbr* (*virtual input/output*) COMP VIO (*entrada/salida virtual*)

violation: ~ **pulse** *n* TRANS impulso *m* de violación

virgin: ~ **medium** *n* DATA soporte *m* virgen

virtual: ~ **access method** *n* (*VAM*) SWG método *m* de acceso virtual (*VAM*); ~ **address** *n* COMP dirección *f* virtual; ~ **addressing** *n* COMP direccionamiento *m* virtual; ~ **analog switching point** *n* SWG punto *m* de conmutación analógica virtual; ~ **area** *n* COMP área *f* virtual; ~ **call** *n* DATA, TELEP llamada *f* virtual; ~-**call characteristic** *n* DATA, TELEP característica *f* de llamada virtual; ~-**call switched circuit** *n* SWG circuito *m* de llamada virtual; ~ **channel** *n* SIG, TRANS canal *m* virtual; ~-**channel connection** *n* (*VCC*) DATA, TRANS conexión *f* de canal virtual; ~-**channel connection endpoint** *n* (*VCCE*) DATA final *m* de conexión con canal virtual, TRANS final *m* de conexión de canal virtual; ~-**channel connection-related function** *n* (*VCCRF*) DATA, TRANS función *f* de canal virtual relacionada con la conexión (*VCCRF*); ~-**channel identifier** *n* (*VCI*) DATA, TRANS identificador *m* de canal virtual; ~ **circuit** *n* DATA, TRANS circuito *m* virtual; ~-**circuit bearer service** *n* DATA, TRANS servicio *m* portador de circuito virtual; ~-**circuit switch** *n* (*VCS*) SWG conmutador *m* de circuito virtual; ~-**circuit switching node** *n* SWG, TRANS nodo *m* de conmutación de circuito virtual; ~ **communication** *n* DATA, TELEP comunicación *f* virtual; ~ **connection** *n* TELEP conexión *f* virtual; ~ **container** *n* (*VC*) APPL, DATA, TELEP recipiente *m* virtual; ~ **container-n** *n* (*VC-n*) APPL, DATA, TELEP recipiente-n *m* virtual; ~ **decision value** *n* TELEP valor *m* de decisión virtual; ~ **device** *n* COMP dispositivo *m* virtual; ~ **DMA driver** *n* DATA accionador *m* DMA virtual; ~ **file store** *n* DATA memoria *f* de archivo virtual; ~ **height** *n* RADIO altura *f* virtual; ~ **image** *n* DATA, TV imagen *f* virtual; ~ **input/output** *n* (*VIO*) COMP entrada/salida *f* virtual (*VIO*); ~ **international connection point** *n* TELEP punto *m* virtual de conexión internacional; ~ **machine** *n* COMP computadora *f* virtual *AmL*, computador *m* virtual *AmL*, ordenador *m* virtual *Esp*; ~ **memory** *n* COMP, DATA memoria *f* virtual; ~ **mode** *n* COMP modalidad *f* virtual; ~ **path** *n* (*VP*) DATA, TRANS vía *f* de transmisión virtual; ~-**path connection** *n* (*VPC*) DATA conexión *f* con trayectoria virtual, TRANS conexión *f* de trayectoria virtual; ~-**path connection endpoint** *n* (*VPCE*) DATA final *m* de conexión con trayectoria virtual, TRANS final *m* de conexión de trayectoria virtual; ~-**path connection-related function** *n* (*VPCRF*) DATA,

TRANS función *f* de trayectoria virtual relacionada con la conexión (*VPCRF*); **~-path identifier** *n* (*VPI*) DATA, TRANS identificador *m* de trayectoria virtual; **~ private network** *n* COMP, TELEP red *f* privada virtual; **~ pushbutton** *n* COMP pulsador *m* virtual; **~ reality** *n* (*VR*) COMP, DATA realidad *f* virtual (*VR*); **~ SDA server** *n* TELEP servidor *m* SDA virtual; **~ storage** *n* COMP, DATA memoria *f* virtual; **~-studio system** *n* COMP sistema *m* de estudio virtual; **~ switching point** *n* SWG punto *m* de conmutación virtual; **~ terminal** *n* COMP, DATA, TELEP terminal *m* virtual; **~ unit** *n* COMP unidad *f* virtual; **~ zero time** *n* TELEP punto *m* cero virtual

virus *n* COMP, DATA virus *m*

visibility: **~ meter** *n* TEST medidor *m* de visibilidad

visible: **~ arc** *n* RADIO arco *m* visible; **~ light** *n* ELECT luz *f* visible; **~ radiation** *n* ELECT, TRANS *fibre optics* radiación *f* visible

visioconference *n* COMP, RADIO, TELEP, TV videoconferencia *f*

visiophone *n* APPL, RADIO, TELEP, TV videoteléfono *m*

visited: **~ location register** *n* RADIO registro *m* de posición visitada

visual: **~ alarm** *n* TEST alarma *f* visual; **~ call system** *n* RADIO, SIG sistema *m* de llamada visual; **~ carrier wave** *n* TV onda *f* portadora de imagen; **~ display unit** *BrE n* COMP, TV pantalla *f* visual; **~ indication** *n* COMP, TELEP, TEST indicación *f* visual; **~ indicator** *n* COMP, TELEP, TEST indicador *m* visual; **~ indicator strip** *n* COMP, TELEP, TEST regleta *f* de indicadores visuales; **~ inspection** *n* GEN inspección *f* visual; **~ read-out** *n* COMP lectura *f* visual; **~ signal** *n* COMP, SIG señal *f* visual; **~ telephone** *n* APPL, COMP, RADIO, TELEP, TV teléfono *m* visual; **~ telephone service** *n* APPL, RADIO, TELEP, TV servicio *m* de telefonía visual; **~ telephony** *n* COMP, RADIO, TELEP, TV telefonía *f* visual

VITC *abbr* (*vertical interval time code*) COMP, DATA VITC (*código de tiempo de intervalo vertical*)

Viterbi: **~ decoding** *n* TELEP decodificación *f* Viterbi

vitreous: **~ silica** *n* TRANS *fibre optics* sílice *f* vítrea

VLBI *abbr* (*very long baseline interferometry*) TEST VLBI (*interferometría con línea de base muy larga*)

VLSI *abbr* (*very large-scale integration*) ELECT, PARTS, TELEP, TRANS VLSI (*integración a escala muy grande*); **~ circuit** *n* ELECT, PARTS circuito *m* VLSI

VM *abbr* APPL (*voice mail*) VM (*correspondencia telefónica*), COMP (*virtual memory*), TELEP VM (*memoria virtual*), TELEP (*voice mail*) VM (*correspondencia telefónica*); **~ forwarding** *n* DATA envío *m* VM, TELEP transmisión *f* de memoria virtual

VMG *abbr* (*voice messaging*) APPL, DATA, TELEP MVG (*transmisión de mensajes hablados*)

VMGE *abbr* (*voice messaging environment*) APPL entorno *m* de mensajería vocal, DATA, TELEP VMGE (*entorno de mensajería por voz*)

VMGS-MS *abbr* (*voice messaging system message store*) APPL memoria *f* de mensaje de sistema de mensajería por voz, DATA VMGS-MS (*memoria de mensaje de sistema de mensajería por voz*), TELEP memoria *f* de mensaje de sistema de mensajería por voz

VMGS-UA *abbr* (*voice messaging system user agent*) APPL agente *m* de usuarios de sistema de mensajería por voz, DATA, TELEP VMGS-UA (*agente de usuario de sistema de mensajería por voz*)

VM-MS *abbr* (*voice messaging store*) APPL, DATA, TELEP VN-MS (*memoria de mensajería por voz*)

VN *abbr* (*voice notification*) TELEP NV (*notificación oral*)

vocal: **~ date-time stamping** *n* PER *on answer-machine message* impresión *f* sonora de día y fecha; **~ message** *n* COMP, TELEP mensaje *m* vocal; **~ track** *n* COMP pista *f* de voz; **~ user interface** *n* APPL interfaz *f* vocal de usuario

vocoder *n* COMP, RADIO, TELEP vocoder *m*

vocoding *n* COMP, RADIO, TELEP codificación *f* vocal

VOD *abbr* (*video-on-demand*) COMP, TV VOD (*vídeo a petición*)

voice *n* COMP, PER, TELEP voz *f*; **~-activated** *adj* CONTR, PER, TELEP activado por la voz; **~-activated phone** TELEP teléfono *m* activado por la voz; **~ activation** CONTR, PER, TELEP activación *f* por la voz; **~ amplifier** ELECT amplificador *m* de voz; **~ annotation** COMP, TELEP anotación *f* vocal; **~ band** DATA banda *f* de voz; **~-band data detector** DATA detector *m* de datos de banda de frecuencias vocales; **~-band data ratio** DATA relación *f* de datos de banda de voz; **~ body part** COMP, DATA parte *f* oral de un mensaje; **~ channel** TELEP, TRANS canal *m* de voz; **~ coder** COMP, RADIO, TELEP codificador *m* vocal; **~ command server** CONTR, PER, TELEP servidor *m* activado por la voz; ~ **command system** CONTR, PER, TELEP sistema *m* activado por la voz; ~ **control** CONTR, PER, TELEP modulación *f* por la palabra; ~**-controlled** *adj* CONTR, PER, TELEP activado por la voz; ~**-controlled operation** CONTR, SWG operación *f* controlada por la voz; ~ **and data integrated system** DATA, TELEP sistema *m* integrado de voz y datos; ~**-data integration** DATA integración *f* de datos de voz; ~**-data packet switch** SWG conmutador *m* de paquetes de datos de voz; ~ **dialler** *BrE* TELEP selector *m* por frecuencia vocal, teleselector *m* por frecuencia vocal; ~ **dialling** *BrE* TELEP teleselección *f* por frecuencia vocal; ~ **digitization** COMP, TELEP digitalización *f* de voz; ~ **frequency** (*VF*) GEN frecuencia *f* acústica; ~**-frequency band** TRANS banda *f* de frecuencia vocal; ~**-frequency code signalling** *BrE* SIG, TELEP señalización *f* de código de frecuencia vocal; ~**-frequency key sending** TELEP envío *m* de clave de frecuencia vocal; ~**-frequency signalling** *BrE* **receiver** SIG, TELEP receptor *m* de señales de frecuencia vocal; ~**-frequency telegraph** TELEP telégrafo *m* de frecuencia vocal; ~**-frequency telegraph circuit** TELEP, TRANS circuito *m* telegráfico de frecuencia vocal; ~**-frequency telegraphy** (*VFT*) TELEP telegrafía *f* de frecuencia vocal (*VFT*); ~ **grade** ELECT, TELEP calidad *f* telefónica; ~ **identification** APPL, CONTR identificación *f* de la voz; ~ **input device** COMP dispositivo *m* de entrada vocal; ~ **mail** (*VM*) APPL, TELEP correspondencia *f* telefónica (*VM*); ~ **mailbox** APPL, PER *of answer machine*, TELEP buzón *m* telefónico; ~**-mail system user** APPL, TELEP usuario *m* de sistema de correo por voz; ~ **message** COMP, DATA mensaje *m* por voz; ~**-message server** COMP, DATA servidor *m* de mensaje vocal; ~ **messaging** (*VMG*) APPL, DATA, TELEP transmisión *f* de mensajes hablados (*MVG*); ~**-messaging environment** (*VMGE*) APPL entorno *m* de mensajería vocal, DATA, TELEP entorno *m* de mensajería por voz (*VMGE*); ~**-messaging notification** APPL, DATA, TELEP notificación *f* de mensajería por voz; ~**-messaging service** APPL, DATA servicio *m* de mensajería por voz, RADIO servicio *m* de mensajes por voz, TELEP servicio *m* de mensajería por voz; ~**-messaging store** (*VM-MS*) APPL, DATA, TELEP memoria *f* de mensajería por voz (*VN-MS*); ~**-messaging system** APPL, TELEP sistema *m* de mensajería por voz; ~ **messaging system message store** (*VMGS-MS*) APPL memoria *f* de mensaje de sistema de mensajería por voz, DATA memoria *f* de mensaje de sistema de mensajería por voz (*VMGS-MS*), TELEP memoria *f* de mensaje de sistema de mensajería por voz; ~**-messaging system user** APPL, DATA, TELEP usuario *m* de sistema de mensajería por voz; ~**-messaging system user agent** (*VMGS-UA*) APPL agente *m* de usuarios de sistema de mensajería por voz, DATA, TELEP agente *m* de usuario de sistema de mensajería por voz (*VMGS-UA*); ~**-messaging user** APPL, DATA, TELEP usuario *m* de mensajería por voz; ~**-messaging user agent** TELEP agente *m* de usuario de mensajes por voz; ~ **network** TELEP, TRANS red *f* telefónica; ~ **notification** (*VN*) TELEP notificación *f* oral (*NV*); ~**-operated** *adj* CONTR, PER, TELEP activado por la voz; ~**-operated coder** COMP, RADIO, TELEP codificador *m* activado por la voz; ~**-operated relay** PARTS, SWG relé *m* accionado por la voz; ~**-operated switch** PARTS, SWG conmutador *m* activado por la voz; ~**-operated switching** CONTR, SWG conmutación *f* activada por la voz; ~**-order wire** TELEP, TRANS circuito *m* de órdenes de voz; ~**-order wire port** TELEP, TRANS puerta *f* de circuito de transferencia de voz; ~ **paging** RADIO búsqueda *f* de personas por altavoz; ~ **print** COMP, TELEP grabación *f* de voz; ~ **privacy** PER, TELEP privacidad *f* de la voz; ~ **prompt** PER *on answer machine* mensaje *m* guía oral; ~ **quality** ELECT, TELEP calidad *f* de transmisión de la voz; ~ **recognition** APPL, COMP, TELEP reconocimiento *m* de voz; ~ **recognizer** APPL, COMP, TELEP reconocedor *m* de voz; ~ **response** TELEP respuesta *f* telefónica; ~**-response system** TELEP sistema *m* de respuesta telefónica; ~**-response unit** APPL, TELEP unidad *f* de respuestas telefónicas; ~ **server** COMP servidor *m* local, DATA servidor *m* vocal; ~ **services** *n pl* GEN servicios *m pl* por voz; ~**-signal processing coupler** TELEP acoplador *m* de procesamiento de señal vocal; ~ **spurt** TELEP chorro *m* de voz; ~**-switching equipment** SWG equipo *m* de conmutación por la voz; ~ **synthesis** APPL, COMP, TELEP síntesis *f* de la voz; ~**-synthesis unit** APPL, COMP, TELEP unidad *f* de síntesis de la voz; ~ **transmission** TRANS transmisión *f* de la voz; ~**/cw** TELEP voz *f* /onda continua

voiced *adj* APPL sonorizado, COMP, TELEP vocalizado

void *n* GEN vacío *m*

volatile: ~ **fault** *n* TEST *dependability* fallo *m* transitorio; ~ **memory** *n* COMP, DATA memoria *f* volátil; ~ **storage** *n* COMP, DATA memoria *f* inestable

volatility *n* COMP volatilidad *f*, DATA inestabilidad *f*

volt *n* ELECT, POWER voltio *m*; ~**-ampere** ELECT, POWER voltamperio *m*

voltage *n* (*V*) POWER voltaje *m* (*V*), tensión *f*; ~ **adaptor switch** POWER conmutador *m* de adaptación de tensión; ~ **amplification factor** POWER factor *m* de amplificación de la tensión; ~ **amplifier** POWER amplificador *m* de voltaje; ~ **attenuation** POWER atenuación *f* de tensión; ~ **between lines** POWER tensión *f* entre conductores; ~**-controlled amplifier** POWER amplificador *m* regulado por tensión; ~**-controlled crystal oscillator** (*VCXO*) ELECT oscilador *m* de cristal de regulación por tensión (*VCXO*); ~**-controlled filter** COMP, POWER filtro *m* controlado por tensión; ~**-controlled oscillator** (*VCO*) ELECT oscilador *m* con voltaje controlado; ~**-controlled transistor** ELECT transistor *m* controlado por tensión; ~ **converter** POWER convertidor *m* de tensión; ~**-dependent resistor** PARTS, POWER resistor *m* dependiente de la tensión; ~ **discharge-current curve** TELEP *of protector* curva *f* de corriente de descarga de la tensión; ~ **divider** POWER divisor *m* de tensión; ~ **doubler** POWER doblador *m* de tensión; ~ **drop** ELECT, POWER caída *f* de tensión; ~ **feed** POWER alimentación *f* de tensión; ~ **fluctuation** POWER fluctuación *f* de tensión; ~ **gain** POWER ganancia *f* de tensión; ~ **jump** POWER salto *m* de tensión; ~ **level** POWER nivel *m* de tensión; ~ **limiter** POWER limitador *m* de tensión; ~ **loss** POWER pérdida *f* de tensión; ~ **monitor** CONTR, POWER monitor *m* de tensión; ~ **peak** POWER cresta *f* de tensión; ~ **rating** POWER tensión *f* de régimen; ~ **reference diode** ELECT, PARTS diodo *m* de referencia de voltaje; ~ **reference tube** ELECT, PARTS tubo *m* de tensión de referencia; ~ **regulation** ELECT regulación *f* de tensión, POWER regulación *f* de voltaje; ~ **regulator** ELECT, POWER regulador *m* de voltaje; ~**-regulator diode** PARTS diodo *m* regulador de voltaje; ~**-regulator tube** ELECT, POWER tubo *m* regulador de tensión; ~ **relay** PARTS, POWER relé *m* de tensión; ~**-saturation current** ELECT, POWER corriente *f* de saturación por tensión; ~ **selector** POWER selector *m* de tensión; ~ **simulator** POWER simulador *m* de tensión; ~ **spike** POWER punta *f* de tensión; ~ **stabilizer** ELECT, POWER estabilizador *m* de tensión; ~**-stabilizing circuit** ELECT circuito *m* estabilizador de tensión; ~**-stabilizing tube** ELECT, POWER tubo *m* estabilizador de tensión; ~ **standing wave** (*VSW*) RADIO, TRANS onda *f* estacionaria de tensión (*VSW*); ~ **standing wave ratio** (*VSWR*) RADIO, TRANS relación *f* de ondas estacionarias de tensión (*VSWR*); ~ **surge** POWER sobretensión *f*; ~ **test** POWER prueba *f* de tensión, TEST ensayo *m* de tensión; ~**-tuned magnetron** ELECT, RADIO magnetrón *m* sintonizado por tensión; ~ **variation** POWER, TEST variación *f* de tensión

voltage/frequency *n* (*V/F*) ELECT, POWER tensión/frecuencia *f* (*V/F*)

volume *n* GEN volumen *m*; ~ **change** COMP cambio *m* volumétrico; ~ **control** COMP, CONTR control *m* de volumen; ~ **envelope** COMP envolvente *m* de volumen; ~**-range control** CONTR control *m* de dinámica; ~ **resistance** POWER resistencia *f* volúmica; ~ **resistivity** ELECT, POWER resistividad *f* de volumen

volumetric: ~ **coverage** *n* RADIO cobertura *f* volumétrico

VOR *abbr* (*VHF omnidirectional radio range*) RADIO VOR (*radiofaro omnidireccional de VHF*)

vowel *n* GEN vocal *f*

voxel *n* COMP, TV voxel *m*

VP *abbr* (*virtual path*) DATA, TRANS vía *f* de transmisión virtual

VPC *abbr* (*virtual-path connection*) DATA conexión *f* con trayectoria virtual, TRANS conexión *f* de trayectoria virtual

VPCE *abbr* (*virtual-path connection endpoint*) DATA final *m* de conexión con trayectoria virtual, TRANS final *m* de conexión de trayectoria virtual

VPCRF *abbr* (*virtual-path connection-related function*) DATA, TRANS VPCRF (*función de trayectoria virtual relacionada con la conexión*)

VPI *abbr* (*virtual-path identifier*) DATA, TRANS identificador *m* de trayectoria virtual

VR *abbr* (*virtual reality*) COMP, DATA VR (*realidad virtual*)

VRAM *abbr* (*video RAM*) COMP VRAM (*RAM de vídeo*)

VSAT *abbr* (*very small aperture terminal*) DATA, TRANS, TV VSAT (*terminal de apertura muy pequeña*); ~ **network** *n* DATA, TRANS red *f* VSAT; ~ **terminal** *n* RADIO terminal *m* VSAT

VSB *abbr* (*vestigial sideband*) RADIO, TRANS RSB (*banda lateral residual*, *VSB*)

V-shaped: ~ **antenna** *n* RADIO antena *f* en V

VSW *abbr* (*voltage standing wave*) RADIO, TRANS VSW (*onda estacionaria de tensión*)

VSWR *abbr* (*voltage standing wave ratio*) RADIO, TRANS VSWR (*relación de ondas estacionarias de tensión*)

VT *abbr* (*vertical tabulation character*) DATA, RADIO carácter *m* de tabulación vertical

VTP *abbr* (*videotape player*) COMP, TV VTP (*reproductor de cintas de video*)

VTR *abbr* (*videotape recorder*) COMP, TV grabadora *f* de cinta de video

VU: **~-meter** *n* TEST volúmetro *m* VU

wafer *n* ELECT, PARTS oblea *f*

WAIS *abbr* (*wide-area information server*) DATA *Internet* WAIS (*servidor de información de área extendida*)

waiting: **~-allowed facility** *n* TELEP instalación *f* de espera permitida; **~ call** *n* TELEP llamada *f* en cola de espera; **~ queue** *n* TELEP cola *f* de espera; **~ state** *n* COMP estado *m* de espera; **~ time** *n* COMP tiempo *m* de espera

wake: **~-up service** *n* TELEP servicio *m* de despertador

walking: **~ strobe pulse** *n* RADIO impulso *m* estroboscópico automático

wall: **~ AC power socket** *n* COMP, ELECT, POWER enchufe *m* mural de alimentación de CA; **~ alternative current power socket** *n* COMP enchufe *m* mural de alimentación de corriente alterna; **~ bracket** *n* PER APP apoyo *m* empotrado; **~ cabinet** *n* TELEP caja *f* acústica de montaje mural; **~-mountable interconnect centre** *BrE n* (*WIC*) ELECT centro *m* de interconexión montable en pared (*WIC*); **~-mountable splice centre** *BrE n* (*WSC*) ELECT centro *m* de empalmes montable en pared (*WSC*); **~-mounted installation** *n* PER APP instalación *f* de montaje mural; **~-mounted telephone set** *n* PER, TELEP aparato *m* telefónico de pared; **~ mounting** *n* ELECT, PER, TELEP montaje *m* mural; **~ outlet** *n* COMP, ELECT, POWER, TELEP toma *f* corriente mural; **~ socket** *n* COMP, ELECT, POWER, TELEP enchufe *m* mural; **~ telephone** *n* TELEP teléfono *m* de pared; **~ telephone set** *n* TELEP teléfono *m* de pared; **~ terminal box** *n* COMP, ELECT, POWER, TELEP caja *f* terminal mural; **~ terminal strip** *n* POWER regleta *f* mural de terminales; **~ thickness** *n* PARTS espesor *m* de la pared

wallboard *n* PARTS fibra *f* prensada

wallplate *n* ELECT plancha *f* de fijación, PARTS placa *f* de asiento

WAN *abbr* (*wide-area network*) COMP, DATA, TELEP, TRANS WAN (*red de área extendida*); **~ terminal** *n* COMP terminal *m* WAN

wand: **~ reader** *n* COMP *bar codes* lápiz *m* de lectura

wanted: **~ emission** *n* ELECT, RADIO, SIG emisión *f* deseada; **~ signal** *n* ELECT, RADIO, SIG señal *f* rdeseada; **~-to-unwanted carrier power ratio** *n* ELECT, RADIO, SIG razón *f* de potencia de portadora deseada a portadora indeseada

warble *n* GEN ululación *f*

warning: **~ lamp** *n* CONTR luz *f* de aviso; **~ limit** *n* CONTR límite *m* de aviso; **~ tone** *n* TELEP tono *m* de aviso

warp *vt* COMP, ELECT, PARTS deformar

warping *n* COMP, ELECT, PARTS deformación *f*

washer *n* PARTS arandela *f*

waste *n* COMP despojos *m*; **~ traffic** TELEP tráfico *m* desaprovechado

watch: **~-keeping receiver** *n* RADIO receptor *m* de guardia; **~ receiver** *n* RADIO receptor *m* con reloj

watchdog: **~ timer** *n* ELECT temporizador *m* controlador de secuencia

watching *n* RADIO vigilancia *f*; **~ mode** RADIO modo *m* de vigilancia

water: **~-blocking compound** *n* PARTS compuesto *m* hidrófugo; **~ cooling** *n* GEN refrigeración *f* por agua; **~ load** *n* RADIO carga *f* líquida; **~ pressure** *n* GEN presión *f* del agua

waterproof *adj* PARTS impermeable

waterproofing *n* PARTS impermeabilización *f*

watertight *adj* PARTS estanco

WATS *abbr* (*wide-area telephone service*) TELEP servicio *m* telefónico de área extendida; **~ zone** *n* TELEP zona de servicio telefónico de coste reducido

wave *n* GEN onda *f*; **~ amplification** ELECT amplificación *f* de onda; **~ analyser** *BrE* TEST analizador *m* de onda; **~ coherence** ELECT coherencia *f* de onda; **~ coupling** ELECT acoplamiento *m* de onda; **~ diffraction** ELECT difracción *f* de onda; **~ dispersion** ELECT dispersión *f* de onda; **~ generation** ELECT generación *f* de ondas; **~ generator** ELECT generador *m* de ondas; **~ impedance** ELECT, RADIO impedancia *f* de onda; **~ interference** ELECT, RADIO interferencia *f*; **~ mechanics** GEN mecánica *f* ondulatoria; **~ number** TRANS *fibre optics* número *m* de onda; **~ optics** ELECT, TRANS *fibre optics* óptica *f* de onda; **~ polarization** ELECT, RADIO polarización *f* de onda; **~ propagation** ELECT, RADIO propagación *f* de ondas; **~ shadowing**

effect RADIO efecto *m* de ensombrecimiento de onda; ~ **shape** GEN forma *f* de onda; ~ **tilt** RADIO inclinación *f* de onda; ~ **train** ELECT, RADIO tren *m* de ondas; ~ **transmission** TRANS transmisión *f* de onda; ~ **trap** RADIO atrapaondas *m*, circuito *m* trampa de ondas, filtro *m* antiinterferencia, trampa *f* de ondas

waveform *n* GEN forma *f* de onda; ~ **corrector** ELECT corrector *m* de forma de onda; ~ **distortion** SIG distorsión *f* de armónico; ~ **monitor** CONTR monitor *m* de forma de onda

wavefront *n* GEN frente *m* de onda

waveguide *n* DATA, RADIO, TRANS guíaondas *m*; ~ **antenna** RADIO antena *f* de guíaondas; ~ **attenuator** RADIO, TRANS atenuador *m* de guíaondas; ~ **bolting** RADIO, TRANS tamizado *m* de guíaondas; ~**-to-coax adaptor** TRANS adaptador *m* guía-coaxil; ~ **connection** RADIO, TRANS conexión *f* de guíaondas; ~ **coupler** RADIO, TRANS acoplador *m* de guíaondas; ~ **dispersion** RADIO, TRANS dispersión *f* de guíaondas; ~**-fed aperture antenna** RADIO antena *f* de apertura alimentada por guíaondas; ~ **gasket** PARTS junta *f* de guíaondas; ~ **mode** RADIO, TRANS modo *m* de guíaondas; ~ **phase changer** RADIO desfasador *m* de guíaondas; ~ **phase shifter** RADIO desfasador *m* de guíaondas; ~ **shim** PARTS, RADIO, TRANS frisa *f* de guíaondas; ~ **switch** SWG conmutador *m* de guíaondas; ~ **wire** PARTS conductor *m* de guíaondas

wavelength *n* RADIO, TRANS longitud *f* de onda; ~ **division demultiplexer** TELEP, TRANS *fibre optics* desmultiplexor *m* por división en longitud de onda; ~ **division multiplex** TELEP, TRANS múltiplex *m* por división en longitud de onda; ~ **division multiplex coupler** TELEP, TRANS acoplador *m* de múltiplex por división en longitud de onda; ~ **division multiplexer** TELEP, TRANS multiplexor *m* por división en longitud de onda; ~ **division multiplexing** (*WDM*) TELEP, TRANS multiplexión *f* por división de longitud de onda (*WDM*); ~ **division switch** SWG conmutador *m* de división de longitud de onda; ~ **peak** GEN pico *m* de longitud de onda; ~ **switching** SWG conmutación *f* de longitud de onda

wavemeter *n* RADIO, TEST ondámetro *m*

WDM *abbr* (*wavelength division multiplexing*) TELEP, TRANS WDM (*multiplexión por división de longitud de onda*); ~ **coupler** *n* TRANS acoplador *m* WDM

weakly: ~**-guiding fibre** *BrE n* TRANS fibra *f* de guía débil

weakness: ~ **failure** *n* TEST fallo *m* por debilidad

wear: ~ **blade** *n* PARTS lámina *f* de desgaste; ~**-out failure** *n* TEST *dependability* fallo *m* por desgaste; ~**-out failure period** *n* TEST periodo *m* de fallos por desgaste; ~**-out life** *n* TEST vida *f* útil; ~ **and tear** *n* COMP uso *m* y desgaste

weather: ~ **line** *n* APPL, TELEP servicio *m* meteorológico; ~ **radar** *n* RADIO radar *m* meteorológico; ~**-sealed cable** *n* PARTS cable *m* con protección contra la intemperie

Web: **the** ~ *n* COMP el Web; ~ **document** *n* COMP, DATA *Internet* documento *m* de Web; ~ **file** *n* COMP, DATA fichero *m* de Web; ~**page** *n* COMP, DATA página *f* Web; ~ **server** *n* COMP, DATA servidor *m* Web

weight *n* GEN ponderación *f*

weighted: ~ **average** *n* GEN promedio *m* ponderado; ~ **keyboard** *n* COMP teclado *m* compensado; ~ **terminal coupling loss** *n* TELEP pérdida *f* de acoplamiento de terminal compensada

weighting *n* GEN ponderación *f*; ~ **filter** ELECT filtro *m* de compensación; ~ **network** TRANS red *f* compensadora

welcome: ~ **page** *n* COMP, DATA página *f* de bienvenida

welcoming: ~ **page** *n* COMP, DATA página *f* de bienvenida

weld *vt* PARTS soldar *f*

welding *n* PARTS soldadura *f*

wet: ~**-bulb thermometer** *n* ELECT termómetro *m* de depósito líquido; ~ **radome** *n* RADIO radomo *m* húmedo

wetting *n* ELECT *of contacts* humedecimiento *m*; ~ **current** ELECT, POWER corriente *f* humectante

what: ~ **you see is** ~ **you get** *phr* (*WYSIWYG*) COMP lo que se ve es lo que se obtiene (*WYSIWYG*)

Wheatstone: ~ **automatic system** *n* TELEP transmisor *m* automático Wheatstone; ~ **bridge** *n* POWER puente *m* de Wheatstone; ~ **system** *n* TELEP sistema *m* Wheatstone

wheel *n* COMP rueda *f*

whereabouts *n* RADIO paradero *m*

whip: ~ **antenna** *n* RADIO antena *f* de látigo

whistler: ~**-mode propagation** *n* RADIO propagación *f* en modo de silbidos atmosféricos

white: ~ **crushing** *n* COMP, TELEP, TRANS compresión *f* de blanco; ~ **lacquer** *n* ELECT laca *f* blanca; ~ **level** *n* COMP, TELEP, TRANS nivel *m* de blanco; ~ **noise** *n* RADIO, TELEP ruido *m* blanco

whitening: ~ **filter** *n* ELECT filtro *m* de blanqueo

Whois *n* DATA *Internet* Whois *m*

WIC *abbr* (*wall-mountable interconnect centre*) ELECT WIC (*centro de interconexión montable en pared*)

wide: **~-angle side lobe** *n* RADIO lóbulo *m* lateral de ángulo grande; **~ area** *n* RADIO área *f* extendida; **~ area information server protocol** *n* DATA protocolo *m* de servidor de información de área extendida; **~-area information server** *n* (*WAIS*) DATA *Internet* servidor *m* de información de área extendida (*WAIS*); **~-area network** *n* (*WAN*) COMP, DATA, TELEP, TRANS red *f* de área extendida (*WAN*); **~-area system** *n* DATA, TELEP, TRANS sistema *m* de área extendida; **~-area telephone service** *n* (*WATS*) TELEP servicio *m* telefónico de área extendida; **~-deviation angle modulation** *n* TRANS modulación *f* de ángulo de gran desviación; **~ flange** *n* ELECT brida *f* ancha, PARTS pestaña *f* ancha; **~ range** *n* GEN amplio *m* surtido

wideband: **~ amplifier** *n* ELECT amplificador *m* de banda ancha; **~ channel** *n* TRANS canal *m* de banda ancha; **~ circuit** *n* TRANS circuito *m* de banda ancha; **~ data transmission** *n* DATA, TRANS transmisión *f* de datos en banda ancha; **~ noise** *n* DATA ruido *m* de banda ancha; **~ radio relay system** *n* RADIO sistema *m* de relé de radio de banda ancha; **~ receiver** *n* RADIO receptor *m* de banda ancha; **~ signal** *n* ELECT, SIG, TRANS señal *f* de banda ancha; **~ switching network** *n* SWG, TRANS red *f* de conmutación de banda ancha; **~ telephony** *n* TELEP telefonía *f* de banda ancha; **~ transmission** *n* TRANS transmisión *f* de banda ancha; **~ video** *n* TV vídeo *m* de banda ancha

widespread *adj* GEN muy difundido

width *n* COMP, RADIO anchura *f*; **~ coding** RADIO codificación *f* de anchura; **~ of effective overall noise band** RADIO ancho *m* de banda de ruido general efectivo; **~ variation** COMP variación *f* de ancho

Wiegand: **~ effect** *n* DATA efecto *m* de Wiegand

wilco *phr* RADIO, TELEP recibido y atendido

wild: **~ connection** *n* DATA *Internet* conexión *f* desordenada

wildcard: **~ character** *n* COMP carácter *m* de sustitución

wind *vi* PARTS, POWER combarse

winding *n* PARTS, POWER alabeo *m*; **~ data** *n pl* POWER datos *m pl* de bobinado; **~ instructions** *n pl* POWER instrucciones *f pl* de bobinado; **~ layer** POWER capa *f* de bobinado; **~ resistance** POWER resistencia *f* de bobinado; **~ space** POWER espacio *m* para bobinado; **~ tapping** POWER toma *f* de bobinado; **~ wire** POWER hilo *m* para bobinado

window *n* COMP, DATA, RADIO, TELEP, TRANS, TV ventana *f*; **~ jamming** RADIO perturbación *f* con cintas reflectoras

windowpane *n* COMP cristal *m* de ventana

wink: **~ signal** *n* DATA, SIG señal *f* de unidad elemental de tiempo; **~ start signal** *n* DATA, SIG señal *f* de arranque de unidad elemental de tiempo

wipe *n* TV agrandamiento gradual de imagen

wipe out *vt* COMP *erase* borrar, SIG bloquear totalmente las señales

wiper *n* SWG escobilla *f* de contacto; **~ arm** SWG brazo *m* de contacto deslizante; **~ set** SWG grupo *m* de escobillas de contacto

wiping: **~ contact** *n* PARTS contacto *m* de frotamiento

wire[1]: **~ aerial** *n* RADIO antena *f* de hilo; **~ antenna** *n* RADIO antena *f* de hilo; **~ area** *n* POWER sección *f* transversal de hilo; **~ bonding** *n* PARTS puente *m* de continuidad; **~ end** *n* PARTS extremo *m* de hilo; **~ fastener** *n* PARTS fijador *m* de hilos; **~ frame model** *n* PARTS modelo *m* de trama de hilos; **~ fuse** *n* PARTS, POWER fusible *m* de alambre; **~ gage** *AmE n* PARTS calibre *m* de alambres; **~ gauge** *BrE n* PARTS calibrador *m* de alambres; **~ guide** *n* PARTS guíahilos *m*; **~ guide beam** *n* PARTS haz *m* de guíahilos; **~ guide strip** *n* PARTS regleta *f* de guíahilos; **~-minder** *n* ELECT vigilante *m* de cables; **~ netting** *n* PARTS enrejado *m* de alambre; **~-running diagram** *n* PARTS diagrama *m* de tendido de hilos; **~ spring** *n* SWG muelle *m* de hilo; **~-spring relay** *n* SWG relé *m* de muelle de hilo; **~ strippers** *n pl* PARTS pelacables *m*; **~ through connection** *n* ELECT, POWER conexión *f* alámbrica directa; **~ transpositions** *n pl* ELECT transposiciones *f pl* de hilos; **~ tweezers** *n pl* PARTS pinzas *f pl* para cables; **~-wound resistor** *n* PARTS resistor *m* devanado, POWER resistor *m* bobinado; **~-wrap connection** *n* PARTS conexión *f* enrollada; **~-wrap connection list** *n* PARTS lista *f* de conexiones enrolladas; **~-wrapping** *n* PARTS conexión *f* por enrollamiento; **~-wrap plane** *n* PARTS plano *m* de enrollamiento; **~-wrap terminal** *n* PARTS terminal *m* para conexión enrollada; **~-tapping** *n* TELEP espionaje *m* telefónico

wire[2] *vt* GEN cablear

wired: **~ handset** *n* RADIO, TELEP microteléfono *m* alámbrico; **~-logic system** *n* CONTR sistema *m* de lógica cableada; ~ **network** *n* RADIO, TELEP red *f* alámbrica; **~-programme** *BrE* **control system** *n* CONTR sistema *m* de control de programa cableado; ~ **rack** *n* TELEP bastidor *m* alámbrico; ~ **sequence controller** *n* COMP controlador *m* alámbrico de secuencia, CONTR controlador *m* alámbrico de frecuencia; ~ **telephone network** *n* TELEP red *f* telefónica alámbrica

wireless *adj* COMP, RADIO, TELEP inalámbrico, sin hilos; ~ **hearing-aid receiver** *n* RADIO receptor *m* de audífono inalámbrico; ~ **PBX** *n* SWG, TELEP PBX *m* inalámbrico; ~ **telegraphy** *n* RADIO, TELEP telegrafía *f* sin hilos; ~ **telephony** *n* RADIO, TELEP telefonía *f* sin hilos

wiretap *vt* TELEP *telephone* establecer una conexión espía

wiring *n* PARTS, POWER cableado *m*; ~ **board** COMP panel *m* de conexiones; ~ **change** PARTS cambio *m* de conexiones; ~ **check** PARTS, TEST prueba *f* de conexiones; ~ **closet** ELECT armario *m* de conexiones; ~ **concentrator** ELECT concentrador *m* de conexiones; ~ **diagram** PARTS diagrama *m* de conexiones; ~ **document** PARTS documento *m* de cableado; ~ **drawing** PARTS dibujo *m* de cableado; ~ **error** PARTS error *m* de cableado; ~ **fixture** PARTS dispositivo *m* de fijación para cableado; ~ **form** PARTS forma *f* de cableado; ~ **harness** ELECT cableado *m* preformado; ~ **jig** PARTS dispositivo *m* de fijación para cableado; ~ **table** PARTS tabla *f* de conexiones; ~ **test** PARTS prueba *f* de conexiones, TEST prueba *f* de cableado; ~ **tester** PARTS probador *m* de cableado, TEST probador *m* de conexiones; ~ **test set** PARTS probador *m* de cableado, TEST probador *m* de conexiones; ~ **unit** ELECT, PARTS unidad *f* de cableado

withdraw *vt* COMP, RADIO desconectar (*DCN*), retirar

withdrawal *n* COMP, RADIO retirada *f*, desconexión *f*

wooden: ~ **paddle** *n* PARTS paleta *f* de madera

woodscrew *n* PARTS tornillo *m* para madera

word *n* COMP, DATA palabra *f*; ~ **block** DATA bloque *m* de texto; ~ **length** DATA longitud *f* de palabra; ~ **processing** COMP, DATA tratamiento *m* de texto; **~-processing package** COMP, DATA paquete *m* de tratamiento de texto; ~ **processor** COMP, DATA procesador *m* de textos; ~ **rate** DATA frecuencia *f* de palabras; ~ **wrap** COMP escritura *f* continua; ~ **wrapping** COMP escritura *f* continua

words: ~ **per minute** *n pl* TELEP palabras *f pl* por minuto

work: ~ **area** *n* GEN área *f* de trabajo; ~ **file** *n* COMP archivo *m* de trabajo; ~ **flow** *n* GEN flujo *m* de trabajo; ~ **function** *n* GEN función *f* de trabajo; ~ **sheet** *n* COMP hoja *f* de cálculo; ~ **station** *n* COMP, DATA, PER APP estación *f* de trabajo

workability *n* DATA cualidad de trabajable

working: ~ **channel** *n* TRANS canal *m* activo; ~ **characteristics** *n pl* TELEP características *f pl* de funcionamiento; ~ **data** *n* DATA datos *m pl* de funcionamiento; ~ **diagram** *n* COMP diagrama *m* de trabajo; ~ **document** *n* GEN documento *m* de trabajo; ~ **frequency** *n* RADIO frecuencia *f* de comunicación; ~ **function** *n* GEN grupo *m* de trabajo, GEN marcha; ~ **point** *n* GEN secuencia *f* de trabajo; ~ **range** *n* TRANS alcance *m* útil; **~-station access method** *n* TELEP método *m* de acceso a estación de trabajo; ~ **standard** *n* TEST patrón *m* de trabajo; ~ **surface** *n* ELECT, PARTS área *m* de trabajo; ~ **temperature** *n* CLIM temperatura *f* de trabajo; ~ **voltage** *n* POWER tensión *f* de trabajo

workplace *n* GEN lugar *m* de trabajo

World: ~ **Wide Web** *n* (*WWW*) COMP, DATA telaraña *f* mundial (*WWW*)

worldwide: ~ **acceptance** *n* GEN aceptación *f* mundial; ~ **network** *n* TELEP red *f* mundial

worst: ~ **case** *n* TEST caso *m* más desfavorable; ~ **link** *n* TRANS enlace *m* más desfavorable

wow *n* ELECT fluctuación *f*

WPC: ~ **system** *n* (*wired programme control system*) CONTR sistema *m* de control de programa cableado

wrap *vt* PARTS *cable* encintar, *connection* revestir

wraparound *n* COMP curvatura *f* de cinta

wrapped: ~ **connection** *n* GEN conexión *f* arrollada

wrapping *n* GEN, PARTS *of cables* empaquetado *m*; ~ **fault** PARTS defecto *m* de enrollamiento; ~ **machine** PARTS máquina *f* para conexión arrollada; ~ **pin** PARTS espiga *f* de enrollado; ~ **programme** *BrE* PARTS programa *m* para conexiones enrolladas; ~ **tool** PARTS herramienta *f* de enrollado

wrench *AmE n* (*cf spanner BrE*) PARTS llave *f* de tuercas, TEST llave *f* fija

write[1]: ~ **action** *n* COMP acción *f* de escribir; **~-back** *n* COMP respuesta

write[2]: ~ **many read always** *phr* COMP escribir muchas veces, leer siempre; ~ **once read many** *phr* COMP escribir una vez, leer muchas

writing: ~ **bar** *n* PARTS, TELEP barra *f* de impresión; ~ **edge** *n* PARTS, TELEP barra *f* de impresión

wrong: ~ **connection** *n* TELEP comunicación *f* falsa

WSC *abbr* (*wall-mountable splice centre*) ELECT WSC (*centro de empalmes montable en pared*)

WWW *abbr* (*World Wide Web*) COMP, DATA WWW (*telaraña mundial*)

wye: ~ **connection** *n* POWER conexión *f* en estrella

WYSIWYG *abbr* (*what you see is what you get*) COMP WYSIWYG (*lo que se ve es lo que se obtiene*)

X

x: ~ **dB bandwidth** *n* RADIO ancho *m* de banda x dB

X *abbr* (*extension*) GEN X (*supletorio*)

X.410-1984: ~ **mode** *n* DATA modo *m* X.410-1984; ~ **mode presentation** *n* DATA presentación *f* en modo X.410-1984

XD *abbr* (*ex-directory*) GEN XD (*ex-directorio*)

xenon: ~ **lamp** *n* CLIM lámpara *f* de xenón

Xerox: ~ **Network Service** *n* (*XNS*) DATA Servicio *m* de Red Xerox (*XNS*)

XID *abbr* (*exchange of identification*) TELEP XID (*central de identificación*)

XMTR *abbr* (*transmitter*) COMP, DATA, ELECT, PARTS, PER, RADIO, TELEP, TRANS transmisor *m*

XNS *abbr* (*Xerox Network Service*) DATA XNS (*Servicio de Red Xerox*)

X-ray: ~ **laser** *n* ELECT láser *m* de rayos X; ~ **source** *n* ELECT generador *m* de rayos X

X-Y: ~ **plotter** *n* TEST trazador *m* de curvas X-Y; ~ **recorder** *n* TEST registrador *m* de curvas X-Y

Y

Y *abbr* (*yes*) DATA sí *m*

Yagi: ~ **aerial** *n* RADIO antena *f* Yagi; ~ **antenna** *n* RADIO antena *f* Yagi

Y-branching: ~ **device** *n* ELECT elemento *m* de bifurcación en Y, TRANS dispositivo *m* de bifurcación en Y

Y-connection *n* POWER conexión *f* en estrella

Y-coupler *n* ELECT, TRANS acoplador *m* en Y

yellow: ~ **book** *n* COMP libro *m* amarillo; ~ **pages** *n pl* TELEP páginas *f pl* amarillas

yes *n* (*Y*) DATA sí *m*; **~-or-no test** TELEP, TEST prueba *f* de sí o no

yield: ~ **limit** *n* PARTS límite *m* de rendimiento

yoke *n* ELECT, PARTS, TELEP yugo *m*; ~ **assembly** ELECT, PARTS montaje *m* de culata; ~ **coil** ELECT, PARTS bobina *f* de yugo

Young: **~'s modulus** *n* PARTS módulo *m* de Young

Z

Z: ~ **component** *n* RADIO componente *m* Z
zap *vt* COMP zapear
ZBR *abbr* (*zone bit recording*) COMP ZBR (*registro de bitio fuera de texto*)
Zener: ~ **diode** *n* PARTS diodo *m* Zener
zenith *n* RADIO cenit *m*
zero[1]: ~ **adjustment** *n* TEST ajuste *m* a cero; ~-**code suppression** *n* TRANS supresión *f* de código cero; ~ **crossing** *n* GEN energía *f* al cero absoluto, paso *m* porcero; ~ **error** *n* TEST error *m* de cero; ~-**filling** *n* COMP rellenado *m* con ceros; ~ **isochrone** *n* RADIO isócrono *m* de cero; ~ **justification** *n* TELEP justificación *f* de cero; ~ **loss circuit** *n* ELECT circuito *m* sin pérdidas; ~ **method** *n* TEST indicación *f* nula; ~ **level** *n* TEST nivel *m* cero; ~ **point** *n* TEST punto *m* cero; ~-**point entropy** *n* GEN entropía *f* al cero absoluto; ~ **position** *n* GEN posición *f* cero; ~ **potential** *n* POWER potencial *m* cero; ~ **restriction** *n* TRANS restricción *f* de ceros; ~ **setting** *n* TEST señal *f* de reposición; ~-**voltage relay** *n* PARTS, POWER relé *m* de tensión cero
zero[2] *vt* COMP, ELECT, TEST poner a cero; ◆ ~-**fill** COMP rellenar con ceros
zeroing *n* COMP reposición *f* a cero
zeroize *vt* COMP poner a cero
zigzag *n* COMP, POWER zigzag *m*; ~ **pattern** GEN diagrama *m* en zigzag; ~ **scanning order** COMP disposición *f* de digitalización en zigzag
zinc: ~ **stripping** *n* PARTS eliminación *f* del zincado
zip[1]: ~ **code** *AmE n* (*cf postcode BrE*) COMP código *m* postal
zip[2] *vt* COMP comprimir
zone *n* RADIO zona *f*; ~ **bit recording** (*ZBR*) COMP registro *m* de bitio fuera de texto (*ZBR*); ~ **boundaries** *n pl* RADIO límites *m pl* de zona; ~ **for transmission** RADIO, TRANS zona *f* para transmisión
zoned: ~ **lens** *n* RADIO lente *f* escalonada
zoom: ~ **box** *n* COMP cuadro *m* de diálogo del zoom; ~-**in** *n* COMP ampliación *f*; ~-**out** *n* COMP reducción *f*
zooming *n* COMP zoom *m*

Milton Keynes UK
Ingram Content Group UK Ltd.
UKHW042324061024
449327UK00004B/33